INFLAMMATION AND IMMUNITY IN DEPRESSION

INFLAMMATION AND IMMUNITY IN DEPRESSION

Basic Science and Clinical Applications

Edited by

BERNHARD T. BAUNE

ACADEMIC PRESS

An imprint of Elsevier

Academic Press is an imprint of Elsevier
125 London Wall, London EC2Y 5AS, United Kingdom
525 B Street, Suite 1800, San Diego, CA 92101-4495, United States
50 Hampshire Street, 5th Floor, Cambridge, MA 02139, United States
The Boulevard, Langford Lane, Kidlington, Oxford OX5 1GB, United Kingdom

Notices
Knowledge and best practice in this field are constantly changing. As new research and experience
broaden our understanding, changes in research methods, professional practices, or medical
treatment may become necessary.

Practitioners and researchers must always rely on their own experience and knowledge in evaluating
and using any information, methods, compounds, or experiments described herein. In using such
information or methods they should be mindful of their own safety and the safety of others, including
parties for whom they have a professional responsibility.

To the fullest extent of the law, neither the Publisher nor the authors, contributors, or editors, assume
any liability for any injury and/or damage to persons or property as a matter of products liability,
negligence or otherwise, or from any use or operation of any methods, products, instructions, or ideas
contained in the material herein.

Library of Congress Cataloging-in-Publication Data
A catalog record for this book is available from the Library of Congress

British Library Cataloguing-in-Publication Data
A catalogue record for this book is available from the British Library

ISBN 978-0-12-811073-7

For information on all Academic Press publications
visit our website at https://www.elsevier.com/books-and-journals

ELSEVIER   Book Aid International   Working together to grow libraries in developing countries
www.elsevier.com • www.bookaid.org

Publisher: Nikki Levy
Acquisition Editor: Joslyn Chaiprasert-Paguio
Editorial Project Manager: Timothy Bennett
Production Project Manager: Anusha Sambamoorthy
Cover Designer: Miles Hitchen

Typeset by SPi Global, India

Contents

12. Neuroimmunopharmacology at the Interface of Inflammation and Pharmacology Relevant to Depression

JOSHUA HOLMES, FRANCES CORRIGAN, MARK R. HUTCHINSON

13. The Gut-Brain-Microbe Interaction: Relevance in Inflammation and Depression

NATALIE PARLETTA

14. Childhood Trauma and Adulthood Immune Activation

MARIA A. NETTIS, VALERIA MONDELLI

15. Stress, Maltreatment, Inflammation, and Functional Brain Changes in Depression

KELLY DOOLIN, LEONARDO TOZZI, JOHANN STEINER, THOMAS FRODL

16. Structural Neuroimaging of Maltreatment and Inflammation in Depression

RONNY REDLICH, NILS OPEL, KATHARINA FÖRSTER, JENNIFER ENGELEN, UDO DANNLOWSKI

17. Biological Embedding of Childhood Maltreatment in Adult Depression

MAGDALENE C. JAWAHAR, BERNHARD T. BAUNE

18. Epigenetic Changes in the Immune Systems Following Early-Life Stress

CHRIS MURGATROYD

19. Mechanisms Linking Depression, Immune System and Epigenetics During Aging

STEVEN BRADBURN

20. Role of Inflammation in Neuropsychiatric Comorbidity of Obesity: Experimental and Clinical Evidence

CÉLIA FOURRIER, LUCILE CAPURON, NATHALIE CASTANON

21. Inflammation and Depression in Patients With Autoimmune Disease, Diabetes, and Obesity

JONATHAN M. GREGORY, MICHAEL MAK, ROGER S. MCINTYRE

22. Does Inflammation Link Clinical Depression and Coronary Artery Disease?

SILKE JÖRGENS, VOLKER AROLT

27. Inflammation as a Marker of Clinical Response to Treatment: A Focus on Treatment-Resistant Depression

REBECCA STRAWBRIDGE, ALLAN H. YOUNG, ANTHONY J. CLEARE

28. Clinical Trials of Anti-Inflammatory Treatments of Major Depression

NORBERT MÜLLER

33. Is There Still Hope for Treating Depression With Antiinflammatories?

BERNHARD T. BAUNE

34. Effects of Physical Exercise on Inflammation in Depression

HARRIS A. EYRE, KATARINA ARANDJELOVIC, DAVID A. MERRILL, AJEET B. SINGH, HELEN LAVRETSKY

35. Future Perspectives on Immune-Related Treatments

BERNHARD T. BAUNE

Contributors

Diana Ahmetspahic University of Münster, Münster, Germany

Judith Alferink University of Münster, Münster, Germany

George Anderson CRC Scotland & London, London, United Kingdom

Katarina Arandjelovic Deakin University, Geelong, VIC, Australia

Volker Arolt University Hospital Muenster, Münster, Germany

Bernhard T. Baune University of Adelaide, Adelaide, SA, Australia

Michael Eriksen Benros Mental Health Centre, Copenhagen University Hospital, Copenhagen, Denmark

Alessandra Borsini Institute of Psychiatry, Psychology and Neuroscience, London, United Kingdom

Steven Bradburn Manchester Metropolitan University, Manchester, United Kingdom

Dana Brinker University of Münster, Münster, Germany

Abigail R. Cannon Loyola University Chicago Health Sciences Division, Maywood, IL, United States

Lucile Capuron INRA; University of Bordeaux, Bordeaux, France

Joanne S. Carpenter University of Sydney, Camperdown, NSW, Australia

André F. Carvalho Faculty of Medicine, Department of Psychiatry, University of Toronto; Centre for Addiction & Mental Health (CAMH), Toronto, ON, Canada

Nathalie Castanon INRA; University of Bordeaux, Bordeaux, France

Mashkoor A. Choudhry Loyola University Chicago Health Sciences Division, Maywood, IL, United States

Liliana G. Ciobanu University of Adelaide, Adelaide, SA, Australia

Anthony J. Cleare Institute of Psychiatry, Psychology & Neuroscience, King's College London; South London and Maudsley NHS Foundation Trust, London, United Kingdom

Frances Corrigan University of Adelaide, Adelaide, SA, Australia

Udo Dannlowski University of Münster, Münster, Germany

Kelly Doolin Otto von Guericke University, Magdeburg, Germany; Institute of Neuroscience, Trinity College Dublin, Dublin, Ireland

Jennifer Engelen University of Marburg, Marburg, Germany

Harris A. Eyre University of Adelaide, Adelaide, SA, Australia; UCLA, Los Angeles, CA, United States; University of Melbourne, Melbourne; Deakin University, Geelong, VIC, Australia

Farheen Farzana Florey Institute of Neuroscience and Mental Health, University of Melbourne, Parkville, VIC, Australia

Sophie Flor-Henry University of Toronto, Toronto, ON, Canada

Katharina Förster University of Münster, Münster, Germany

Célia Fourrier INRA; University of Bordeaux, Bordeaux, France

Thomas Frodl Otto von Guericke University, Magdeburg, Germany; Institute of Neuroscience, Trinity College Dublin, Dublin, Ireland

Iria Grande University of Barcelona, Barcelona, Spain

Jonathan M. Gregory Western University, London, ON, Canada

Adam M. Hammer Loyola University Chicago Health Sciences Division, Maywood, IL, United States

Anthony J. Hannan Florey Institute of Neuroscience and Mental Health, University of Melbourne, Parkville, VIC, Australia

Ian B. Hickie University of Sydney, Camperdown, NSW, Australia

Joshua Holmes University of Adelaide, Adelaide, SA, Australia

Mark R. Hutchinson University of Adelaide, Adelaide, SA, Australia

Magdalene C. Jawahar University of Adelaide, Adelaide, SA, Australia

Silke Jörgens University Hospital Muenster, Münster, Germany

Ole Köhler-Forsberg Psychosis Research Unit, Aarhus University Hospital, Risskov; Department of Clinical Medicine, Aarhus University; Mental Health Centre, Copenhagen University Hospital, Copenhagen, Denmark

Femke Lamers VU University Medical Center & GGZ inGeest, Amsterdam Public Health Research Institute, Amsterdam, The Netherlands

Helen Lavretsky UCLA, Los Angeles, CA, United States

Julio Licinio South Australian Health and Medical Research Institute, Adelaide; Flinders University, Bedford Park, SA, Australia; South Ural State University Biomedical School, Chelyabinsk, Russian Federation

Christopher A. Lowry University of Colorado Boulder, Boulder; University of Colorado Anschutz Medical Campus, Aurora; Rocky Mountain Mental Illness Research Education and Clinical Center (MIRECC), Denver Veterans Affairs Medical Center (VAMC); Military and Veteran Microbiome Consortium for Research and Education (MVM-CoRE), Denver, CO, United States

Michael Maes Deakin University, Geelong, VIC, Australia

Michael Mak Western University, London, ON, Canada

Rocío Martin-Santos University of Barcelona, Barcelona, Spain

Roger S. McIntyre University of Toronto; University Health Network, Toronto, ON, Canada

David A. Merrill UCLA, Los Angeles, CA, United States

Yuri Milaneschi VU University Medical Center & GGZ inGeest, Amsterdam Public Health Research Institute, Amsterdam, The Netherlands

Valeria Mondelli Institute of Psychiatry, Psychology and Neuroscience, King's College London; National Institute for Health Research (NIHR) Mental Health Biomedical Research Centre at South London and Maudsley NHS Foundation Trust and King's College London, London, United Kingdom

Norbert Müller Ludwig-Maximilian University of Munich, Munich, Germany; Marion von Tessin Memory Center, Munich, Germany

Chris Murgatroyd Manchester Metropolitan University, Manchester, United Kingdom

Michael Musker South Australian Health and Medical Research Institute, Adelaide; Flinders University, Bedford Park, SA, Australia

Souhel Najjar Zucker School of Medicine at Hofstra/Northwell; Lenox Hill Hospital, New York, NY, United States

Maria A. Nettls Institute of Psychiatry, Psychology and Neuroscience, King's College London, London, United Kingdom

Naghmeh Nikkheslat King's College London, London, United Kingdom

Nils Opel University of Münster, Münster, Germany

Giovanni Oriolo University of Barcelona, Barcelona, Spain

Silky Pahlajani Lenox Hill Hospital, New York, NY, United States

Carmine M. Pariante King's College London; Institute of Psychiatry, Psychology and Neuroscience, London, United Kingdom

Natalie Parletta University of South Australia, Adelaide, SA, Australia

Brenda W.J.H. Penninx VU University Medical Center & GGZ inGeest, Amsterdam Public Health Research Institute, Amsterdam, The Netherlands

Charles L. Raison University of Wisconsin-Madison, Madison, WI, United States

Ronny Redlich University of Münster, Münster, Germany

Thibault Renoir Florey Institute of Neuroscience and Mental Health, University of Melbourne, Parkville, VIC, Australia

Graham A.W. Rook Centre for Clinical Microbiology, UCL (University College London), London, United Kingdom

Joshua D. Rosenblat University of Toronto, Toronto, ON, Canada

Kristi M. Sawyer Institute of Psychiatry, Psychology and Neuroscience, London, United Kingdom

Elizabeth M. Scott University of Sydney, Camperdown; Notre Dame Medical School, Sydney, NSW, Australia

Ajeet B. Singh Deakin University, Geelong, VIC, Australia

Gaurav Singhal University of Adelaide, Adelaide, SA, Australia

Johann Steiner Otto von Guericke University, Magdeburg, Germany

Rebecca Strawbridge Institute of Psychiatry, Psychology & Neuroscience, King's College London, London, United Kingdom

Catherine Toben University of Adelaide, Adelaide, SA, Australia

Leonardo Tozzi Otto von Guericke University, Magdeburg, Germany; Institute of Neuroscience, Trinity College Dublin, Dublin, Ireland

Eduard Vieta University of Barcelona, Barcelona, Spain

Ma-Li Wong South Australian Health and Medical Research Institute, Adelaide; Flinders University, Bedford Park, SA, Australia

Allan H. Young Institute of Psychiatry, Psychology & Neuroscience, King's College London; South London and Maudsley NHS Foundation Trust, London, United Kingdom

Patricia A. Zunszain King's College London; Institute of Psychiatry, Psychology and Neuroscience, London, United Kingdom

Preface

A pathophysiological role of the immune system in psychiatric disorders has long been discussed. A first significant number of publications appeared in PubMed in the second half of the 1960s followed by constantly high scientific interest over the past 2.5 half-decades in the relationship between depression and the immune system cumulating in over >9000 entries in PubMed since then. The *Macrophage Theory* by Smith published in 1991, which suggests an excessive secretion of macrophage monokines as the cause of depression, has stimulated a particular interest in inflammation-associated depressive symptoms. Despite constant scientific progress over two decades with seminal contributions by many researchers in the field (too many to be named here), a sharp increase in broader interest has occurred since 2010 with >420 PubMed entries in 2017 alone. The field has traditionally attracted researchers and clinicians from various contributing disciplines ranging from psychiatry, psychology, endocrinology, and immunology to neurosciences. What is this interest about? It is not only certainly the fascination about the details of the workings of the immune system itself but also how the classical understanding of the immune system as defense mechanism of the organism against microbes could contribute to abnormal behavior and complex psychological functions and how the immune and nervous systems might interact at cellular levels in particular. Moreover, clinician researchers in the psychological/psychiatric fields have shown a deep desire to help explain the pandemic presence of the heterogeneous phenomenology of depression and to find biological explanations for

psychiatric phenomenology more generally. Both the neurosciences and the increasing interest of researchers from the field of immunology itself in the cross talk between the immune and nervous systems have instigated further scientific endeavor and produced exciting new knowledge based on modern scientific technologies routed in both disciplinary depth and interdisciplinary inquiry.

The present book aims to capture key new developments in the field. By bringing together experts from a variety of disciplines and scientific and clinical backgrounds from around the world, this comprehensive collection of up-to-date knowledge on the involvement of the innate and adaptive immune system in depression is covered in three main areas: (A) basic science, (B) clinical science, and (C) treatment and future directions. The chapters in this book document an integrative perspective on the broad involvement of the immune system in depression extending the established theory of cytokine-induced depression (inflammation). While neuroinflammation is a widely accepted pathological feature of some forms of depression, knowledge on alterations of various types of T cells has extended our knowledge into the complex immunologic changes that can occur in depression. Moreover, the interrelationship between the immune and central nervous systems is shown by the increasing number of published neurobiological implications of neuroinflammation and immune alterations in brain development and brain function that have stimulated broad attention in recent years. Further relevance of this topic for clinical psychiatry and psychology has been derived from research that suggests that inflammation and immune markers not only contribute to the so-called sickness behavior but also may affect complex brain functions such as emotion processing, cognitive function, and behavior in depression. Additional excitement comes from research that relates immune function and inflammation in particular with functional and structural brain changes in depression as shown in neuroimaging studies. Finally, novel data on important clinical questions such as whether immune markers can serve as biomarkers for disease states and treatment response and the key question whether novel treatment approaches can be derived from this line of research have invigorated extended research programs and scientific debate.

The presented content of this book brings together knowledge on inflammation and immune function in depression including perspectives from neuroscience, immunology, psychology, psychiatry, and related fields that is written for the understanding of a broad audience in these fields and for interested students, researchers, and clinicians in related disciplines.

Bernhard T. Baune
Editor

1

Depression-Associated Cellular Components of the Innate and Adaptive Immune System

Diana Ahmetspahic, Dana Brinker, Judith Alferink
University of Münster, Münster, Germany

THE INNATE IMMUNE SYSTEM

The immune system, which defends the organism against potential pathogens and carcinogenesis, involves various tissues, cell types, and soluble components. It is generally divided into two major arms: innate immunity and adaptive immunity (Medzhitov et al., 2011). The innate immune system represents the first line of defense against invading microbial pathogens. In 1989, Charles Janeway Jr. postulated a revolutionary theory to explain what he called the "immunologist's dirty little secret" (Janeway, 2013). The classical self-nonself paradigm at that time did not explain why a foreign antigen alone was insufficient to elicit an adaptive immune response, and "dirty extracts" like mycobacteria, mineral oil, or aluminum hydroxide had to be coadministered for induction of an efficient T- and B-cell response. Janeway postulated two key features of innate immunity: the capacity to discriminate between "noninfectious self from infectious nonself" and to activate the

adaptive immune response. He suggested that germline-encoded receptors on innate immune cells may recognize evolutionarily conserved microbial patterns that would thus drive the adaptive response. It is meanwhile well established that within minutes of encountering pathogens, innate immune cells recognize distinct evolutionarily conserved molecular structures called pathogen-associated molecular patterns (PAMPs) (Medzhitov, 2009). In events of trauma or tissue injury, damage-associated molecular-pattern (DAMP) molecules released by dead, damaged, or stressed cells from the host itself can also trigger innate responses. DAMPs encompass a variety of molecule types, including heat-shock proteins, cytoplasmic complex proteins S100A8/S100A9, high-mobility group box 1 DNA-binding proteins in the nucleus, and ATP. These molecules bind various pattern-recognition receptors (PRRs) expressed within the cytosol or on the membranes of innate immune cells, such as toll-like receptors (TLRs), C-type lectin receptors (CLRs), leucine-rich

Inflammation and Immunity in Depression
https://doi.org/10.1016/B978-0-12-811073-7.00001-5

1

repeat (LRR)-containing (or NOD-like) receptors (NLRs), RIG-I-like receptors (RLRs), and AIM2-like receptors (ALRs). Subsequently, activated PRRs trigger signaling cascades, resulting in the release of cytokines and other factors by myeloid cells that orchestrate the recruitment of leukocytes to inflammatory sites and local effector functions (Iwasaki & Medzhitov, 2015). Myeloid cells that stem from common progenitors, descendants of hematopoietic stem cells in the bone marrow, are a highly heterogeneous population of innate immune cells, including granulocytes (e.g., neutrophils, basophils, eosinophils, and mast cells), monocytes/macrophages, and dendritic cells (DCs) (Akashi, Traver, Miyamoto, & Weissman, 2000) (Fig. 1). Current knowledge about the immune function of specific myeloid cell subsets, namely, neutrophils and monocytes, and their involvement in MDD is summarized in the following section.

ROLE OF NEUTROPHILS IN INNATE IMMUNITY AND MDD

Neutrophils (aka neutrophil polymorphonuclear granulocytes) account for 35%–80% of circulating leukocytes in mammals and are the ultimate first line of defense against acute infection. Peripherally circulating neutrophils are short-lived, terminally differentiated cells with a half-life of 8–12 h. It is currently discussed that distinct neutrophil subsets that exert inflammatory or antiinflammatory actions may exist under physiological and disease conditions (Kolaczkowska & Kubes, 2013). However, decisive markers for these functionally diverse neutrophil subsets have not been established. Activated neutrophils utilize various cytotoxic mechanisms to restrict pathogen replication, including engulfing invading pathogens by phagocytosis, degrading them by way of releasing granule-derived antimicrobial peptides, and

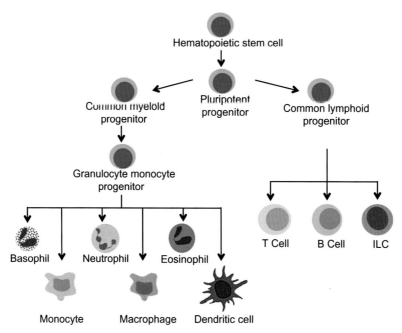

FIG. 1   Schematic overview of human hematopoiesis of myeloid and lymphoid lineages. *ILC*, innate lymphoid cell.

producing reactive oxygen species (ROS). Quite remarkably, activated neutrophils can also trap and kill pathogens by forming so-called neutrophil extracellular traps made up of extracellular weblike fibers of granule antimicrobial proteins and nuclear contents (Kolaczkowska & Kubes, 2013).

Originally, neutrophils were thought to act exclusively as effector cells in innate immunity. However, in recent years, they have been shown to also be capable of modulating macrophage and DC-dependent functions in adaptive immune responses via cell-cell interactions and soluble mediators (Mayadas, Cullere, & Lowell, 2014). Upon making contact with endothelial cells, neutrophils undergo phenotypic changes and upregulate adhesion molecules, chemokine receptors, and proteases, such as neutrophil elastase. When they populate inflammatory sites, they exhibit an extended half-life of up to 1–2 days, a greater capacity to migrate, cytotoxicity, and protease release, and they may also form neutrophil extracellular traps (Kolaczkowska & Kubes, 2013). Interestingly, neutrophils also leave sites of tissue injury and migrate back into the vasculature (de Oliveira, Rosowski, & Huttenlocher, 2016). The contribution of this "reverse migration" to resolution of inflammation has not yet been resolved.

Increased neutrophil counts and percentages have been observed in peripheral blood of patients with MDD (Darko, Rose, Gillin, Golshan, & Baird, 1988; Irwin, Smith, & Gillin, 1987; Kronfol & House, 1989; Maes, Lambrechts, Suy, Vandervorst, & Bosmans, 1994; McAdams & Leonard, 1993). A meta-analysis demonstrated a relative increase in circulating neutrophils that was paralleled by a decrease in lymphocytes in MDD patients (Zorrilla et al., 2001). Neutrophil-to-lymphocyte ratio (NLR) is an indicator of systemic inflammation whose prognostic value has been demonstrated in cancer, cardiovascular diseases, and inflammatory diseases (Isaac et al., 2016). Nonmedicated patients with MDD have been shown to express a high NLR, compared with healthy individuals, whereas medicated MDD patients (taking selective serotonin reuptake inhibitors (SSRIs)) had normalized physiological NLRs (Demircan, Gozel, Kilinc, Ulu, & Atmaca, 2016). A recent study evaluating NLRs and platelet-lymphocyte ratios (PLRs) of inpatients and outpatients with MDD did not reveal a relationship between severity of depression and NLR, but did demonstrate higher PLRs in patients suffering from MDD with psychotic features than in those suffering from other subtypes of depression (Kayhan, Gunduz, Ersoy, Kandeger, & Annagur, 2017).

Also neutrophil functions such as phagocytosis and ROS production have been investigated in MDD patients. Phagocytosis in immune cells can be assessed by *in vitro* uptake of fluorescently labeled microorganisms or particles coated with microbial cell-wall components (e.g., the yeast protein zymosan). In depressed individuals, phagocytosis of zymosan particles assessed with chemiluminescence was shown to be impaired in polymorphonuclear cells that include several types of granulocytes among those neutrophils. Reduced polymorphonuclear activity was normalized in response to effective therapy but not in nonresponders (O'Neill & Leonard, 1990). In a later study from the same group, phagocytosis was specifically investigated in monocytes and neutrophils from depressed patients. Neutrophil phagocytosis was reduced during acute phase of disease but returned to control values on recovery. Interestingly, monocyte phagocytosis was increased and returned to control values following remission (McAdams & Leonard, 1993). A subsequent meta-analysis also reported decreased cellular neutrophil function in MDD (Zorrilla et al., 2001). Maes et al. demonstrated that chemotaxis, superoxide release, and phagocytic capacity of circulating neutrophils were similar across diverse subtypes of depression and healthy volunteers (Maes et al., 1992). Also, no difference in neutrophil phagocytosis of *Escherichia coli* was detected by flow cytometry between hip fracture patients with versus without depressive symptoms (Duggal, Upton, Phillips, & Lord, 2016). Production of polymorphonuclear elastase, an index of neutrophil function, has also been assayed in MDD and found to be enhanced (Deger et al., 1996). Finally, patients with MDD, including elderly patients and patients with comorbidity in heart failure, have been reported to have increased plasma levels of neutrophil gelatinase-associated lipocalin (NGAL), an innate antibacterial factor expressed in activated neutrophils (Naude et al., 2014). Thus, NGAL has emerged as a potential biological link between depression and cardiovascular disease. The relationship between depression and cardiovascular disease will be reviewed in a following section.

MONOCYTES AND MDD

Monocytes are by far the most studied myeloid cell subset in relation to MDD. They are highly plastic regarding their phenotypes and population sizes. Monocytes can act as phagocytes via expression of various receptors, such as Fc, complement, and scavenger receptors (Ginhoux & Jung, 2014; Ziegler-Heitbrock, 2015). Generally, circulating monocytes represent 4%–10% of white blood cells in humans and mice. Monocyte numbers surge within minutes of stress exposure or physical exertion. This phenomenon has been explained by the existence of monocyte reservoirs in areas of slow blood velocity where monocytes adhere weakly to vascular endothelium and can thus be mobilized rapidly in response to catecholamines (Ziegler-Heitbrock, 2015).

Epitopes called cluster of differentiation (CD) molecules are commonly used to identify subsets of cell types, particularly immune system cell types. Based on their surface expression of the lipopolysaccharide (LPS) coreceptor CD14 and the FcγIII receptor CD16, circulating monocytes in humans have been grouped into three

monocyte classes: classical $CD14^+CD16^-$, intermediate $CD14^+CD16^+$, and nonclassical $CD14^{low}CD16^+$ monocytes. Classical monocytes (mouse analog, $Ly6C^{high}$) comprise some 80%–90% of all monocytes, while the other two $CD16^+$ monocyte classes (mouse analog, $Ly6C^{low}$) account for the remaining 10%–20% (Ziegler-Heitbrock, 2015). Classical monocytes are recruited to inflammatory sites, where they are highly phagocytic and modulate inflammation via release of ROS and cytokines, including the proinflammatory cytokines tumor necrosis factor (TNF) and interleukin (IL)-6 and the anti-inflammatory cytokine IL-10. Intermediate $CD14^+CD16^+$ monocytes are associated with active inflammatory conditions and can exert antitumoral actions. They are also phagocytic and produce various proinflammatory molecules, including not only ROS, TNF, IL-1β, IL-6, and IL-12 but also IL-10 (Ziegler-Heitbrock, 2015). Meanwhile, nonclassical monocytes (and their $Ly6C^{low}$ counterparts in mice), which have a low phagocytic capability, patrol blood vessels and are involved in tissue surveillance and homeostasis (Auffray et al., 2007; Cros et al., 2010).

Alterations in monocyte counts and functionality in patients with MDD have frequently been reported. For example, monocytosis and neutrophilia have been suggested as hallmarks of severe depression (Maes et al., 1992). However, both elevated and reduced and also unchanged monocyte counts and percentages in peripheral blood have been observed in depressed subjects (Darko et al., 1988; Lanquillon, Krieg, Bening-Abu-Shach, & Vedder, 2000; McAdams & Leonard, 1993; Schlatter, Ortuno, & Cervera-Enguix, 2004b; Seidel et al., 1996b). A reduced monocytic phagocytosis of fluorescently labeled particles or bacteria in the peripheral blood of patients with MDD along with lower surface expression of class II human leukocyte antigens (HLAs) and elevated respiratory burst activity was reported (Schlatter et al., 2004b). On the contrary, monocytes from MDD subjects have also been reported to have an increased capacity to engulf zymosan during active depression, but not during symptom remission (McAdams & Leonard, 1993).

Multiple studies, including meta-analyses have indicated that MDD is associated with elevated peripheral levels of cytokines, chemokines, and other inflammatory factors (Dowlati et al., 2010; Eyre, Stuart, & Baune, 2014; Goldsmith, Rapaport, & Miller, 2016; Haapakoski, Mathieu, Ebmeier, Alenius, & Kivimaki, 2015; Howren, Lamkin, & Suls, 2009; Kohler et al., 2017). Proinflammatory cytokines such as TNFα, IL-1β, IL-6, and chemokines CCL2 and CXCL8 (IL-8) are typically produced by activated monocytes (Ziegler-Heitbrock, 2015). Upregulated expression of proinflammatory mediators and chemokines (i.e., *IL-1*, *IL-6*, *TNF*, *CCL2*, and *CCL7*) has been demonstrated specifically in purified monocytes from nonmedicated MDD patients (Carvalho et al., 2014). In accordance, a meta-analysis has further demonstrated increased serum levels of CCL2 in depressed subjects (Eyre et al., 2014). Additionally, monocytes and other immune cells from MDD patients have been shown to exhibit an altered responsiveness to LPS stimulation (Humphreys, Schlesinger, Lopez, & Araya, 2006; Krause et al., 2012; Lisi et al., 2013). Stimulated whole blood from depressed patients has yielded high levels of IL-1β and IL-6, but not TNF, compared with control subjects (Schlatter et al., 2004a). In a large cohort study, proinflammatory cytokine production of LPS-stimulated whole blood cells was further associated with symptom severity of depression and anxiety (Vogelzangs, de Jonge, Smit, Bahn, & Penninx, 2016). Conversely, there is also evidence of reduced immune cell responsiveness in MDD, a phenomenon known as the depression-associated exhausted state of innate immunity (Krause et al., 2012). Krause et al. observed reduced interferon (IFN)-γ and IL-10 production in LPS-stimulated whole blood cultures with myeloid cells from MDD patients and found normalization of these differences in response to the antidepressant drug imipramine or the antiinflammatory drug celecoxib (Krause et al., 2012). Additionally, reduced expression of the proinflammatory factors prostaglandin E2 and cyclooxygenase-2 by LPS-stimulated monocytes has been observed in subjects with MDD (Lisi et al., 2013). These controversial findings regarding immune differences in patients with MDD could be due, at least in part, to variability across experimental designs, including the use of differing LPS concentrations. Notably, the ex vivo LPS-induced responsivity of immune cells may be LPS dose-dependent. Additionally, lifestyle and health factors (e.g., smoking, alcohol intake, body mass index, and chronic diseases) may greatly influence the basal cytokine levels and the production capacity of innate immune cells (Vogelzangs et al., 2016). Nonetheless, the majority of studies examining the issue point to depression-associated monocyte alterations.

INNATE LYMPHOID CELLS (ILCs) AND NATURAL KILLER (NK) CELLS

White blood cells derived from lymphoid tissues are known generally as lymphocytes. Lymphocytes include innate lymphocytes (ILCs), T cells, and B cells (Mjösberg & Spits, 2016). Until recently, the only known ILC was the NK (Vivier et al., 2011). ILCs defined as exhibiting lymphocyte morphology but lacking T-cell or B-cell lineage markers are modulated by cytokine signaling. They are present in mucosae and mucosal-associated lymphoid tissues where they are involved in the initiation and resolution of inflammatory responses and subsequent tissue repair. ILCs play a critical role in graft-versus-host disease, inflammatory bowel disease, asthma, atopic dermatitis, and multiple sclerosis. Based on their expression of transcription factors, cytokines, and cell surface markers, ILCs can be subdivided into three groups.

Group 1 ILCs act primarily against intracellular pathogens and produce IFN-γ; this group includes NK cells. Meanwhile, group 2 ILCs act mainly against parasites via production of IL-4 and IL-13. Group 3 is composed of lymphoid tissue inducer cells, which are important for secondary lymphoid tissue formation during embryogenesis, and cells that combat extracellular bacteria via release of IL-17 and IL-22 (Artis & Spits, 2015). Each ILC group has T-cell counterparts with some redundancy in functionality and transcription factor expression (Bando & Colonna, 2016). While the role of ILCs generally in MDD has not yet been studied, numerous studies have examined circulating NK cells in MDD.

NK cells, first described more than 40 years ago, were initially regarded as "nonspecific" immune cells with the capacity to kill tumor cells and virally infected cells without prior sensitization. However, the discovery of a complex system of activating (e.g., 2B4, NKG2D, NKp46, and NKp44) and inhibiting (e.g., killer-cell Ig-like receptors (KIRs)) NK cell receptors indicated that in fact, they are not truly nonspecific

(Vivier et al., 2011) (Fig. 2). In a simplified view, NK cell subsets in human peripheral blood can be distinguished based on the relative expression of CD16 (low-affinity Fc-receptor γ IIIA) and CD56 (neural cell adhesion molecule, NCAM). Some 90% of the circulating NK cells in humans are $CD56^{high}CD16^{low/high}$ NK cells, with the remaining ~10% being $CD56^{low}CD16^{high}$ NK cells that are localized primarily to secondary lymphoid organs (Lünemann, Lünemann, & Münz, 2009). $CD56^{high}$ NK cells are considered to be immune-regulatory cells that produce a wide variety of cytokines (e.g., GM-CSF, IFN-γ, IL-10 or IL-13, and TNF), whereas $CD56^{low}$ NK cells are the major killer population due to their ability to exert cytotoxic effector functions. In the presence of activating signals, NK cells exert cytotoxicity upon cells deemed to be invaders due to their failure to express self-identifying class I HLA molecules. NK cell cytotoxicity is mediated via three distinct pathways: exocytosis of granules with the release of perforin (a pore-forming protein) and granzymes, death receptor-

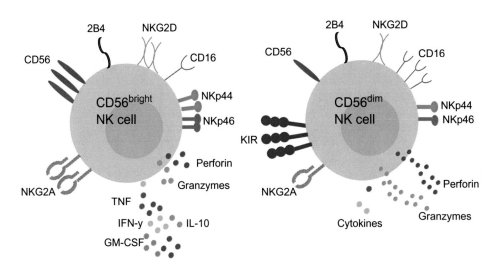

FIG. 2   Schematic overview of human NK cell subsets. Exemplary activation and inhibitory NK cell receptors on indicated NK cell subsets are shown. *KIR*, killer-cell Ig-like receptor; *NK*, natural killer cell.

induced apoptosis, and antibody-dependent cell-mediated cytotoxicity (Vivier et al., 2011).

NK CELLS AND DEPRESSION

Acute stressors, such as public speaking and stress tests, increase circulating NK cell counts in healthy individuals due, at least in part, to stress-induced catecholamines reducing β2-adrenergic receptor-mediated adhesion of NK cells to endothelial cells, resulting in a rise in circulating NK cells (Benschop, Schedlowski, Wienecke, Jacobs, & Schmidt, 1997; Segerstrom & Miller, 2004). Some research groups have reported evidence of increased circulating NK cells in MDD, while others have reported diminished NK cell numbers in MDD (Evans et al., 1992; Grosse et al., 2016; Ravindran, Griffiths, Merali, & Anisman, 1999; Rothermundt et al., 2001; Seidel et al., 1996a; Zorrilla et al., 2001). However, these incongruent studies differed methodologically in terms of cohorts, interventions, and evaluated outcomes.

A consistently reproduced finding has been reduced NK cell cytolytic activity (NKCA) in MDD, with lesser reduction being associated with a better treatment response (Irwin & Gillin, 1987; Zorrilla et al., 2001). Significant reductions of NKCA were observed in depressed men compared with nondepressed men, but not in women with MDD when compared with nondepressed women (Evans et al., 1992). Reduced NKCA appears to be particularly profound in patients with an early MDD onset age (Frank, Wieseler Frank, Hendricks, Burke, & Johnson, 2002). Variance in NKCA has partly been explained by depressed mood, somatic anxiety, and less diurnal variation (Maes et al., 1994). Sleep disturbance was found to correlate inversely with NKCA in both MDD patients (Cover & Irwin, 1994) and healthy individuals (Irwin, Smith, & Gillin, 1992). Electroconvulsive therapy, a treatment option for therapy-resistant MDD, has been shown to increase NKCA only transiently in medication-resistant MDD patients or MDD patients with psychotic features (Fluitman et al., 2011; Kronfol, Nair, Weinberg, Young, & Aziz, 2002).

Serotonergic mechanisms have been implicated in NK cell activation. Among depressed patients medicated with SSRIs, NKCA levels were augmented more in those patients whose symptoms were responsive to the treatment than in nonresponsive patients (Park, Lee, Jeong, Han, & Jeon, 2015). Moreover, SSRI therapy increased NKCA in depressed outpatients with a reduced NKCA at baseline, and the NKCA of mononuclear cells from MDD patients was increased upon *in vitro* incubation with SSRIs (Frank, Hendricks, Johnson, Wieseler, & Burke, 1999). Findings showing that enhanced intracellular ROS levels in monocytes correlate with reduced NKCA in MDD suggest that ROS may suppress NKCA (Frank et al., 2001). The reduction of NKCA may be dependent on serotonin, which was reported to clear ROS in the microenvironment consequently preserving NK cell survival and activity (Betten, Dahlgren, Hermodsson, & Hellstrand, 2001). Type 1A serotoninergic receptors on inhibitory monocytes are bound by ROS, thus protecting NK cells from oxidative damage at inflammatory sites (Hellstrand & Hermodsson, 1993). Additionally, reduced levels of the neuronal protein p11, which amplifies serotonin receptor-mediated signaling, have been associated with depression-like behavior (Svenningsson, Kim, Warner-Schmidt, Oh, & Greengard, 2013). Paradoxically, however, Svenningsson and colleagues found that reduction of p11 in NK cells and monocytes within 8 weeks of beginning citalopram antidepressant therapy predicted a positive antidepressant response in MDD patients (Svenningsson et al., 2014).

Various potential underlying mechanisms of reduced NKCA in MDD have been discussed. Clusters of genes involved in NK cell activation have been found to be downregulated selectively in patients with active, as opposed to remitted, MDD (Jansen et al., 2016). Meanwhile, rodent models of stress and depression have demonstrated an adrenergic influence on NK function and distribution (Engler et al., 2004; Tarr et al., 2012). For example, decreased NKCA in a rodent social confrontation model was associated with significantly increased NK-sensitive lung tumor metastasis in a manner that could be mitigated by a β-adrenergic antagonist pretreatment (Stefanski & Ben-Eliyahu, 1996). Additionally, augmented corticotropin-releasing factor (CRF) secretion in depression has been shown to affect NK cell function. Thus, treatment with synthetic CRF reduced NKCA in rats, while pretreatment of these animals by chemical sympathectomy completely abolished CRF-induced plasma catecholamine levels and reduction in splenic NK activity (Irwin, Hauger, Jones, Provencio, & Britton, 1990). Furthermore, stress and depression may impede NK cell migration. Restraint stress in rodents was reported to delay NK cell recruitment to the lungs during influenza infection, and this delay was associated with suppression particularly of chemokines (CCL2 and CCL3) (Hunzeker, Padgett, Sheridan, Dhabhar, & Sheridan, 2004).

THE ADAPTIVE IMMUNE SYSTEM AND ITS COMPONENTS

The adaptive immune response is mounted by T and B lymphocytes that recognize specific pathogens and enable immune memory to bolster subsequent encounters with recognized pathogens (Medzhitov et al., 2011). Broadly, adaptive immune responses consist of (i) the cellular immune response mounted by T cells and (ii) the humoral response mediated by B cells. B cells recognize native proteins via membrane-bound immunoglobulins, whereas T cells recognize antigenic peptides bound to HLA molecules expressed on antigen-presenting cells (APCs) via their antigen-specific T-cell receptors (TCRs) (Rajewsky, 1996; Zinkernagel & Doherty,

1979). APCs that comprise a heterogeneous group of immune cells, namely, DCs, macrophages, and B cells, exhibit the capacity to process and present antigens. When immature DCs that have been released from bone marrow encounter pathogens in peripheral tissues, they undergo a maturation program and migrate to the lymph nodes where they prime naive T cells (Steinman, 2012). In general, cellular and humoral responses may cooperate in immune defense. Adaptive immune responses are characterized by the capacity for immune memory that provides long-lasting protection and rapid responses to subsequent encounters with pathogens (Farber, Netea, Radbruch, Rajewsky, & Zinkernagel, 2016). Here, we will review T-helper cell subset functions that are potentially relevant for depression. The literature regarding altered cell-mediated adaptive immunity in depression will be reviewed in detail in the following section.

The human body contains about 10^{12} T cells (Miles, Douek, & Price, 2011). Committed lymphoid progenitors that arise in the bone marrow and do not yet express TCRs, CD4, or CD8 enter the thymus from the blood stream. The thymus is a multilobed retrosternal organ referred to as a "school for T cells" owing to its unique role in T-cell differentiation. A complex series of molecular and cellular events involving rearrangement of TCR genes and both negative and positive selection steps result in the production of a broad repertoire of T cells that are restricted by HLAs encoded by the major histocompatibility complex (MHC) gene region but lack self-reactivity. Only a minority of thymocytes that coexpress TCRs and CD4 or CD8 leave the thymus (Kurd & Robey, 2016). The thymus also governs the development of so-called regulatory T cells (T_{reg}) that control potentially autoreactive T cells that evaded negative selection (Sakaguchi, Vignali, Rudensky, Niec, & Waldmann, 2013). In addition to central tolerance mechanisms occurring in the thymus, various mechanisms of peripheral tolerance including deletion, functional inactivation (anergy) of potentially autoreactive T cells, and downregulation of potentially self-reactive TCRs in the periphery may prevent autoimmunity (Arnold, 2002). Most T cells express TCRs composed of a heterodimer of α and β chains associated with CD3 (αβ TCRs), an invariant complex that facilitates signal transduction (Wucherpfennig, Gagnon, Call, Huseby, & Call, 2010). Only a minority of T cells express γδ TCRs, and they, being first-line defense components, populate epithelial tissues commonly exposed to external agents (Chien, Meyer, & Bonneville, 2014).

Recognition of antigens by T cells expressing αβ TCRs depends on the presentation of antigenic peptides bound in the groove of HLA molecules, which are highly polymorphic cell surface molecules encoded by genetic loci in the MHC gene complex region on the short arm of chromosome 6 (Rammensee, Falk, & Rotzschke, 1993). Classical class I HLA molecules, namely, HLA-A, HLA-B, and HLA-C, are expressed by all human cell types except erythrocytes. Meanwhile, class II HLA molecules, namely, HLA-DR, HLA-DQ, and HLA-DP, are expressed constitutively on a subset of antigen-presenting cells, such as DCs, B cells, and macrophages, and can be induced in activated T cells in humans and some other cell types under inflammatory conditions. T cells expressing αβ TCRs consist of two major functionally distinct populations: CD8$^+$ and CD4$^+$ T cells that recognize peptides derived from proteins bound to class I and class II HLA molecules, respectively (Mak, 2007; Masopust, Vezys, Wherry, & Ahmed, 2007).

Historically, class I and class II HLA molecules have been associated with endogenous and exogenous peptide sources, respectively. The original assumption that class I HLAs present intracellular synthesized peptides exclusively was challenged by findings showing that exogenous antigens can also access the cytosolic MHC-I pathway, a phenomenon known as cross presentation (Cruz, Colbert, Merino, Kriegsman, & Rock, 2017). Moreover,

self-antigens can also be presented by class II HLAs and be subjected to autophagy (Roche & Furuta, 2015). Activation of T cells expressing αβ TCRs following antigen recognition in the context of a specific cytokine milieu induces a complex signaling cascade that may lead to cell proliferation, cytokine secretion, and differentiation into effector and/or memory cells of specialized phenotypes. Thus, CD8$^+$ and CD4$^+$ T-cell subsets are highly heterogeneous.

Activated CD8$^+$ T cells have the potential to develop into cytotoxic T lymphocytes (CTLs) that play key roles in limiting infection and tumor defense. Upon MHC-I-dependent recognition, CTLs kill target cells with lytic enzymes such as perforin and granzymes. Additionally, CTLs can eliminate infected cells by way of death-inducing ligand-receptor binding (Zhang & Bevan, 2011). Activated CD4$^+$ T cells differentiate into specific T-helper (Th) cell subtypes, principally Th1 and Th2, upon antigen recognition in the context of class II HLAs and an inductive cytokine milieu in the microenvironment (Fig. 3). The presence of IL-12 favors differentiation of CD4$^+$ T cells into Th1 cells, which produce among other cytokines IL-2 and IFN-γ. Meanwhile, the presence of IL-4 promotes the development of Th2 cells that secrete IL-4, IL-10, and IL-13. Th1 and Th2 differentiation depends on distinct transcription factors, most notably T-bet for Th1 cell differentiation and GATA-3 for Th2 development (Zhu & Paul, 2008; Ziegler, 2016). In recent years, considerable effort has been made to understand the diversity and functionality of additional CD4$^+$ Th cell subsets. Based on extensive multiparameter flow cytometry studies, several cell surface markers have been attributed to distinct T-cell functionality, such as CD45RA-mediated T-cell activation, CD27/28-mediated costimulation, and programmed cell death-1-mediated inhibition. Thus, the expression assays of CD45RA, the chemokine receptor CCR7, and the membrane-bound costimulatory molecules CD27 and CD28 have been utilized to

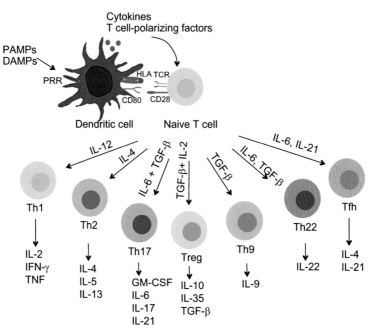

FIG. 3 Schematic overview of T-helper subset differentiation. Following antigen presentation by DCs, naive CD4[+] T cells differentiate into distinct Th cell subsets in the context of various cytokines. Exemplary differentiation cytokines are shown, and cytokines that are produced by T-cell subsets are shown. *PAMPs*, pathogen-associated molecular-pattern molecules; *DAMPs*, damage-associated molecular-pattern molecules; *PRR*, pattern-recognition receptor; *DC*, dendritic cell; *IL*, interleukin; *IFN*, interferon; *Th*, T-helper cell; T_{reg}, regulatory T cells; *Tfh*, follicular T-helper cell.

characterize naive and antigen-experienced T cells in humans (Mahnke, Brodie, Sallusto, Roederer, & Lugli, 2013). Functionally distinct CD4[+] T-cell subsets are also characterized by their chemokine receptor profiles. Accordingly, Th1 cells have been shown to express CCR5 and CXCR3, whereas Th2 cells have been found to express CCR3, CCR4, and CD294 (aka prostaglandin D2 receptor 2). Additional CD4[+] Th cell subsets include follicular helper (Tfh), Th9, Th17, and T_{reg} cells. Tfh cells express the chemokine receptor CXCR5 and can migrate to the follicle where they provide B cell help via expression of the CD40 ligand and production of IL-4 and IL-21 (Lefrancois & Marzo, 2006).

Numerous studies have investigated the Th1/Th2 dichotomy of CD4[+] Th cells in MDD, and some recent studies have focused on the potential role of Th17 cells and T_{reg} cells in

depression. Those bodies of work will be reviewed in the following section. In preparation for discussing those findings, key aspects of Th17 cell biology and T_{reg} cell biology are introduced here.

Th17 cells, discovered in 2005, were named for their capacity to produce the prototype cytokine IL-17. They also produce cytokines other than IL-17, such as granulocyte-macrophage colony-stimulating factor, IL-6, and IL-21. Th17 cells have a potent inflammatory influence and have been suggested to be implicated in the pathogenesis of autoimmune and inflammatory disorders, such as multiple sclerosis, rheumatoid arthritis, and allergic skin diseases (Korn, Bettelli, Oukka, & Kuchroo, 2009). Differentiation of Th17 cells is dependent on various transcription factors, such as the signal transducer and activator of transcription 3 (Stat3), retinoic acid receptor-related orphan receptor γ (RORγ), and nuclear factor kappa-light-chain-enhancer of activated B (NF-κB), and Th17 cells are promoted by transforming growth factor (TGF)-β, IL-6, and IL-23 (Crome, Wang, & Levings, 2010). The IL-17 cytokine family consists of six members, namely, IL-17A-F. IL17A, commonly referred to as simply IL-17, and IL-17F share high sequence homology and both mediate proinflammatory responses. While Th17 cells are the major sources of IL-17A and IL-17F in adaptive responses, T cells expressing γδ TCRs, NK cells, and NK-T cells are capable of producing IL-17 (Korn et al., 2009; Mahnke et al., 2013). Autoimmunity studies utilizing mice deficient in various IL17 cytokines or pharmacological blockade of IL17 have produced conflicting data on the role of Th17 cells and IL-17 in the pathogenesis of autoimmunity. While IL-17-producing lymphocytes have been shown to exhibit an enhanced capacity to migrate through the blood-brain barrier and to accumulate in CNS compartments of MS patients, the cytokine IL-17A appears to be largely dispensable for the development of central nervous system autoimmunity in mice (Becher & Segal, 2011).

Acting somewhat in opposition to Th17 cells, CD4$^+$ T$_{reg}$ cells play an important role in maintaining peripheral self-tolerance and controlling autoimmunity. T$_{reg}$ cells were first discovered nearly 50 years ago when they were identified as immune suppressor cells (Gershon & Kondo, 1970). For many years, they were a challenge to isolate due to a lack of appropriate markers. Finally, Sakaguchi et al. demonstrated in 1995 that IL-2 receptor α-chain (CD25) could be used as a surrogate marker for T$_{reg}$ cells (Sakaguchi, Sakaguchi, Asano, Itoh, & Toda, 1995). Subsequently, a variety of functionally and phenotypically diverse T$_{reg}$ cells, including CD25$^+$ and CD25$^-$ subsets, have been identified. Notably, two currently recognized major subsets of T$_{reg}$ cells are the (i) thymus-derived (aka naturally occurring) T$_{reg}$ cells that express FOXP3 (a forkhead/winged-helix transcription factor) and the (ii) induced (aka peripheral) T$_{reg}$ cells, commonly referred to as iT$_{reg}$ cells, that are induced in response to pathogens or inflammatory processes (Sakaguchi et al., 2013). T$_{reg}$ cells can exert suppressive effects via cell-cell contact or via soluble immunosuppressive factors, such IL-10 and TGF-β. T$_{reg}$ cells also employ IL-10 and TGF-β to suppress DC maturation and activation. Additionally, CD4$^+$CD25$^+$ T$_{reg}$ cells utilize cytotoxic T lymphocyte-associated antigen 4 and glucocorticoid-induced tumor necrosis factor receptor to suppress effector T-cell proliferation and IFN-γ production (Vignali, Collison, & Workman, 2008). The biological relevance of T$_{reg}$ cells in autoimmunity has been underscored by findings of a decreased suppressive function of CD4$^+$CD25$^+$ T$_{reg}$ cells in patients suffering from multiple sclerosis, type I diabetes, psoriasis, and myasthenia gravis, relative to that in healthy controls (Brusko, Putnam, & Bluestone, 2008). Such observations suggest that manipulations of T$_{reg}$ cell function could perhaps be used to treat human disease.

SUMMARY

In summary, this chapter highlighted depression-associated cellular components of the innate and adaptive immune system. Alterations of key features of myeloid cells, including chemotaxis, phagocytosis, production of nitric oxide, and expression of cytokines, and changes in essential T-cell effector functions could play important roles in the pathological processes underlying MDD. Future investigations that aim to resolve yet unknown immune mechanisms of MDD would benefit from approaches that dissect the temporal and spatial regulation of immune alterations and examine the interrelationship between the innate and acquired arms of the immune system in disease pathophysiology. Targeted research strategies that identify immune factors and cellular components at the interface of the innate and acquired immunity arms that are altered in various disease phases may help to identify predictive immune markers of affective disorders. Resolving the factors underlying MDD-associated alterations in innate and adaptive immunity may further lead to the identification of novel immune-based therapies for depression.

References

Akashi, K., Traver, D., Miyamoto, T., & Weissman, I. L. (2000). A clonogenic common myeloid progenitor that gives rise to all myeloid lineages [Research Support, Non-U.S. Gov't]. *Nature, 404*(6774), 193–197. https://doi.org/10.1038/35004599.

Arnold, B. (2002). Levels of peripheral T cell tolerance. *Transplant Immunology, 10*(2–3), 109–114.

Artis, D., & Spits, H. (2015). The biology of innate lymphoid cells. *Nature, 517*(7534), 293–301. https://doi.org/10.1038/nature14189.

Auffray, C., Fogg, D., Garfa, M., Elain, G., Join-Lambert, O., Kayal, S., et al. (2007). Monitoring of blood vessels and tissues by a population of monocytes with patrolling behavior. *Science, 317*(5838), 666–670. https://doi.org/10.1126/science.1142883.

Bando, J. K., & Colonna, M. (2016). Innate lymphoid cell function in the context of adaptive immunity. *Nature Immunology, 17*(7), 783–789. https://doi.org/10.1038/ni.3484.

Becher, B., & Segal, B. M. (2011). T(H)17 cytokines in autoimmune neuro-inflammation. *Current Opinion in Immunology, 23*(6), 707–712. https://doi.org/10.1016/j.coi.2011.08.005.

Benschop, R. J., Schedlowski, M., Wienecke, H., Jacobs, R., & Schmidt, R. E. (1997). Adrenergic control of natural killer cell circulation and adhesion. *Brain, Behavior, and Immunity, 11*(4), 321–332. https://doi.org/10.1006/brbi.1997.0499.

Betten, A., Dahlgren, C., Hermodsson, S., & Hellstrand, K. (2001). Serotonin protects NK cells against oxidatively induced functional inhibition and apoptosis [Research Support, Non-U.S. Gov't]. *Journal of Leukocyte Biology, 70*(1), 65–72.

Brusko, T. M., Putnam, A. L., & Bluestone, J. A. (2008). Human regulatory T cells: Role in autoimmune disease and therapeutic opportunities [Research Support, Non-U.S. Gov't Review]. *Immunological Reviews, 223*, 371–390. https://doi.org/10.1111/j.1600-065X.2008.00637.x.

Carvalho, L. A., Bergink, V., Sumaski, L., Wijkhuijs, J., Hoogendijk, W. J., Birkenhager, T. K., et al. (2014). Inflammatory activation is associated with a reduced glucocorticoid receptor alpha/beta expression ratio in monocytes of inpatients with melancholic major depressive disorder. *Translational Psychiatry, 4.* https://doi.org/10.1038/tp.2013.118.

Chien, Y. H., Meyer, C., & Bonneville, M. (2014). Gammadelta T cells: First line of defense and beyond [Research Support, N.I.H., Extramural Research Support, Non-U.S. Gov't Review]. *Annual Review of Immunology, 32*, 121–155. https://doi.org/10.1146/annurev-immunol-032713-120216.

Cover, H., & Irwin, M. (1994). Immunity and depression: Insomnia, retardation, and reduction of natural killer cell activity. *Journal of Behavioral Medicine, 17*(2), 217–223.

Crome, S. Q., Wang, A. Y., & Levings, M. K. (2010). Translational mini-review series on Th17 cells: Function and regulation of human T helper 17 cells in health and disease. *Clinical and Experimental Immunology, 159*(2), 109–119. https://doi.org/10.1111/j.1365-2249.2009.04037.x.

Cros, J., Cagnard, N., Woollard, K., Patey, N., Zhang, S. Y., Senechal, B., et al. (2010). Human CD14dim monocytes patrol and sense nucleic acids and viruses via TLR7 and TLR8 receptors. [Research Support, Non-U.S. Gov't]. *Immunity, 33*(3), 375–386. https://doi.org/10.1016/j.immuni.2010.08.012.

Cruz, F. M., Colbert, J. D., Merino, E., Kriegsman, B. A., & Rock, K. L. (2017). The biology and underlying mechanisms of cross-presentation of exogenous antigens on MHC-I molecules. *Annual Review of Immunology.* https://doi.org/10.1146/annurev-immunol-041015-055254.

Darko, D. F., Rose, J., Gillin, J. C., Golshan, S., & Baird, S. M. (1988). Neutrophilia and lymphopenia in major mood disorders. *Psychiatry Research, 25*(3), 243–251.

de Oliveira, S., Rosowski, E. E., & Huttenlocher, A. (2016). Neutrophil migration in infection and wound repair: Going forward in reverse. *Nature Reviews Immunology, 16*(6), 378–391. https://doi.org/10.1038/nri.2016.49.

Deger, O., Bekaroglu, M., Orem, A., Orem, S., Uluutku, N., & Soylu, C. (1996). Polymorphonuclear (PMN) elastase levels in depressive disorders. *Biological Psychiatry, 39*(5), 357–363. https://doi.org/10.1016/0006-3223(95)00176-X.

Demircan, F., Gozel, N., Kilinc, F., Ulu, R., & Atmaca, M. (2016). The impact of red blood cell distribution width and neutrophil/lymphocyte ratio on the diagnosis of major depressive disorder. *Neurology and Therapy, 5*(1), 27–33. https://doi.org/10.1007/s40120-015-0039-8.

Dowlati, Y., Herrmann, N., Swardfager, W., Liu, H., Sham, L., Reim, E. K., et al. (2010). A meta-analysis of cytokines in major depression [Meta-Analysis Research Support, Non-U.S. Gov't]. *Biological Psychiatry, 67*(5), 446–457. https://doi.org/10.1016/j.biopsych.2009.09.033.

Duggal, N. A., Upton, J., Phillips, A. C., & Lord, J. M. (2016). Development of depressive symptoms post hip fracture is associated with altered immunosuppressive phenotype in regulatory T and B lymphocytes. *Biogerontology*, *17*(1), 229–239. https://doi.org/10.1007/s10522-015-9587-7.

Engler, H., Dawils, L., Hoves, S., Kurth, S., Stevenson, J. R., Schauenstein, K., et al. (2004). Effects of social stress on blood leukocyte distribution: The role of alpha- and beta-adrenergic mechanisms [Comparative Study Research Support, Non-U.S. Gov't]. *Journal of Neuroimmunology*, *156*(1–2), 153–162. https://doi.org/10.1016/j.jneuroim.2004.08.005.

Evans, D. L., Folds, J. D., Petitto, J. M., Golden, R. N., Pedersen, C. A., Corrigan, M., et al. (1992). Circulating natural killer cell phenotypes in men and women with major depression. Relation to cytotoxic activity and severity of depression. *Archives of General Psychiatry*, *49*(5), 388–395.

Eyre, H. A., Stuart, M. J., & Baune, B. T. (2014). A phase-specific neuroimmune model of clinical depression. *Progress in Neuro-Psychopharmacology & Biological Psychiatry*, *54*, 265–274. https://doi.org/10.1016/j.pnpbp.2014.06.011.

Farber, D. L., Netea, M. G., Radbruch, A., Rajewsky, K., & Zinkernagel, R. M. (2016). Immunological memory: Lessons from the past and a look to the future [Historical Article Review]. *Nature Reviews Immunology*, *16*(2), 124–128. https://doi.org/10.1038/nri.2016.13.

Fluitman, S. B. A. H. A., Heijnen, C. J., Denys, D. A. J. P., Nolen, W. A., Balk, F. J., & Westenberg, H. G. M. (2011). Electroconvulsive therapy has acute immunological and neuroendocrine effects in patients with major depressive disorder. *Journal of Affective Disorders*, *131*(1–3), 388–392. https://doi.org/10.1016/j.jad.2010.11.035.

Frank, M. G., Hendricks, S. E., Bessette, D., Johnson, W., Wieseler Frank, J. L., & Burke, W. J. (2001). Levels of monocyte reactive oxygen species are associated with reduced natural killer cell activity in major depressive disorder. *Neuropsychobiology*, *44*(1), 1–6.

Frank, M. G., Hendricks, S. E., Johnson, D. R., Wieseler, J., & Burke, W. J. (1999). Antidepressants augment natural killer cell activity: In vivo and in vitro. *Neuropsychobiology*, *39*(1), 18–24. https://doi.org/10.1159/000026555.

Frank, M. G., Wieseler Frank, J. L., Hendricks, S. E., Burke, W. J., & Johnson, D. R. (2002). Age at onset of major depressive disorder predicts reductions in NK cell number and activity. *Journal of Affective Disorders*, *71*(1–3//1–3), 159–167. https://doi.org/10.1016/s0165-0327(01)00395-0.

Gershon, R. K., & Kondo, K. (1970). Cell interactions in the induction of tolerance: The role of thymic lymphocytes. *Immunology*, *18*(5), 723–737.

Ginhoux, F., & Jung, S. (2014). Monocytes and macrophages: Developmental pathways and tissue homeostasis. *Nature Reviews Immunology*, *14*(6), 392–404. https://doi.org/10.1038/nri3671.

Goldsmith, D. R., Rapaport, M. H., & Miller, B. J. (2016). A meta-analysis of blood cytokine network alterations in psychiatric patients: Comparisons between schizophrenia, bipolar disorder and depression. *Molecular Psychiatry*, *21*(12), 1696–1709. https://doi.org/10.1038/mp.2016.3.

Grosse, L., Carvalho, L. A., Birkenhager, T. K., Hoogendijk, W. J., Kushner, S. A., Drexhage, H. A., et al. (2016). Circulating cytotoxic T cells and natural killer cells as potential predictors for antidepressant response in melancholic depression. Restoration of T regulatory cell populations after antidepressant therapy. *Psychopharmacology*, *233*(9), 1679–1688. https://doi.org/10.1007/s00213-015-3943-9.

Haapakoski, R., Mathieu, J., Ebmeier, K. P., Alenius, H., & Kivimaki, M. (2015). Cumulative meta-analysis of interleukins 6 and 1beta, tumour necrosis factor alpha and C-reactive protein in patients with major depressive disorder. *Brain, Behavior, and Immunity*, *49*, 206–215. https://doi.org/10.1016/j.bbi.2015.06.001.

Hellstrand, K., & Hermodsson, S. (1993). Serotonergic 5-HT1A receptors regulate a cell contact-mediated interaction between natural killer cells and monocytes [Research Support, Non-U.S. Gov't]. *Scandinavian Journal of Immunology*, *37*(1), 7–18.

Howren, M. B., Lamkin, D. M., & Suls, J. (2009). Associations of depression with C-reactive protein, IL-1, and IL-6: A meta-analysis. [Meta-Analysis Review]. *Psychosomatic Medicine*, *71*(2), 171–186. https://doi.org/10.1097/PSY.0b013e3181907c1b.

Humphreys, D., Schlesinger, L., Lopez, M., & Araya, A. V. (2006). Interleukin-6 production and deregulation of the hypothalamic-pituitary-adrenal axis in patients with major depressive disorders. *Endocrine*, *30*(3), 371–376.

Hunzeker, J., Padgett, D. A., Sheridan, P. A., Dhabhar, F. S., & Sheridan, J. F. (2004). Modulation of natural killer cell activity by restraint stress during an influenza A/PR8 infection in mice [Research Support, U.S. Gov't, P.H.S.]. *Brain, Behavior, and Immunity*, *18*(6), 526–535. https://doi.org/10.1016/j.bbi.2003.12.010.

Irwin, M., & Gillin, J. C. (1987). Impaired natural killer cell activity among depressed patients. *Psychiatry Research*, *20*(2), 181–182. https://doi.org/10.1016/0165-1781(87)90010-2.

Irwin, M., Hauger, R. L., Jones, L., Provencio, M., & Britton, K. T. (1990). Sympathetic nervous system mediates central corticotropin-releasing factor induced suppression of natural killer cytotoxicity [Research Support, Non-U.S. Gov't Research Support, U.S. Gov't, Non-P.H.S. Research Support, U.S. Gov't, P.H.S.]. *The Journal of Pharmacology and Experimental Therapeutics*, *255*(1), 101–107.

Irwin, M., Smith, T. L., & Gillin, J. C. (1987). Low natural killer cytotoxicity in major depression [Research Support, U.S. Gov't, Non-P.H.S. Research Support, U.S. Gov't, P.H.S.]. *Life Sciences*, *41*(18), 2127–2133.

Irwin, M., Smith, T. L., & Gillin, J. C. (1992). Electroencephalographic sleep and natural killer activity in depressed patients and control subjects. *Psychosomatic Medicine, 54*(1), 10–21.

Isaac, V., Wu, C. Y., Huang, C. T., Baune, B. T., Tseng, C. L., & McLachlan, C. S. (2016). Elevated neutrophil to lymphocyte ratio predicts mortality in medical inpatients with multiple chronic conditions [Observational Study]. *Medicine (Baltimore). 95*(23). https://doi.org/10.1097/MD.0000000000003832.

Iwasaki, A., & Medzhitov, R. (2015). Control of adaptive immunity by the innate immune system [Research Support, N.I.H., Extramural Research Support, Non-U. S. Gov't Review]. *Nature Immunology, 16*(4), 343–353. https://doi.org/10.1038/ni.3123.

Janeway, C. A., Jr. (2013). Pillars article: Approaching the asymptote? Evolution and revolution in immunology. Cold spring harb symp quant biol. 1989. 54: 1-13 [Biography Classical Article Historical Article]. *Journal of Immunology, 191*(9), 4475–4487.

Jansen, R., Penninx, B. W., Madar, V., Xia, K., Milaneschi, Y., Hottenga, J. J., et al. (2016). Gene expression in major depressive disorder. *Molecular Psychiatry, 21*(3), 339–347. https://doi.org/10.1038/mp.2015.57.

Kayhan, F., Gunduz, S., Ersoy, S. A., Kandeger, A., & Annagur, B. B. (2017). Relationships of neutrophil-lymphocyte and platelet-lymphocyte ratios with the severity of major depression. *Psychiatry Research, 247*, 332–335. https://doi.org/10.1016/j.psychres.2016.11.016.

Kohler, C. A., Freitas, T. H., Maes, M., de Andrade, N. Q., Liu, C. S., Fernandes, B. S., et al. (2017). Peripheral cytokine and chemokine alterations in depression: A meta-analysis of 82 studies. *Acta Psychiatrica Scandinavica*. https://doi.org/10.1111/acps.12698.

Kolaczkowska, E., & Kubes, P. (2013). Neutrophil recruitment and function in health and inflammation [Research Support, Non-U.S. Gov't Review]. *Nature Reviews Immunology, 13*(3), 159–175. https://doi.org/10.1038/nri3399.

Korn, T., Bettelli, E., Oukka, M., & Kuchroo, V. K. (2009). IL-17 and Th17 cells [Research Support, Non-U.S. Gov't Review]. *Annual Review of Immunology, 27*, 485–517. https://doi.org/10.1146/annurev.immunol.021908.132710.

Krause, D. L., Riedel, M., Muller, N., Weidinger, E., Schwarz, M. J., & Myint, A. M. (2012). Effects of antidepressants and cyclooxygenase-2 inhibitor on cytokines and kynurenines in stimulated in vitro blood culture from depressed patients. *Inflammopharmacology, 20*(3), 169–176. https://doi.org/10.1007/s10787-011-0112-6.

Kronfol, Z., & House, J. D. (1989). Lymphocyte mitogenesis, immunoglobulin and complement levels in depressed patients and normal controls. *Acta Psychiatrica Scandinavica, 80*(2), 142–147.

Kronfol, Z., Nair, M. P., Weinberg, V., Young, E. A., & Aziz, M. (2002). Acute effects of electroconvulsive therapy on lymphocyte natural killer cell activity in patients with major depression. *Journal of Affective Disorders, 71*(1–3), 211–215.

Kurd, N., & Robey, E. A. (2016). T-cell selection in the thymus: A spatial and temporal perspective [Research Support, N.I.H., Extramural Review]. *Immunological Reviews, 271*(1), 114–126. https://doi.org/10.1111/imr.12398.

Lanquillon, S., Krieg, J. C., Bening-Abu-Shach, U., & Vedder, H. (2000). Cytokine production and treatment response in major depressive disorder. *Neuropsychopharmacology, 22*(4), 370–379. https://doi.org/10.1016/S0893-133X(99)00134-7.

Lefrancois, L., & Marzo, A. L. (2006). The descent of memory T-cell subsets [Research Support, N.I.H., Extramural Review]. *Nature Reviews Immunology, 6*(8), 618–623. https://doi.org/10.1038/nri1866.

Lisi, L., Camardese, G., Treglia, M., Tringali, G., Carrozza, C., Janiri, L., et al. (2013). Monocytes from depressed patients display an altered pattern of response to endotoxin challenge. *PLoS One, 8*(1). https://doi.org/10.1371/journal.pone.0052585.

Lünemann, A., Lünemann, J. D., & Münz, C. (2009). Regulatory NK-cell functions in inflammation and autoimmunity. *Molecular Medicine, 15*(9–10), 352–358. https://doi.org/10.2119/molmed.2009.00035.

Maes, M., Lambrechts, J., Suy, E., Vandervorst, C., & Bosmans, E. (1994). Absolute number and percentage of circulating natural killer, non-MHC-restricted T cytotoxic, and phagocytic cells in unipolar depression. *Neuropsychobiology, 29*(4), 157–163.

Maes, M., Stevens, W., DeClerck, L., Peeters, D., Bridts, C., Schotte, C., et al. (1992). Neutrophil chemotaxis, phagocytosis, and superoxide release in depressive illness. *Biological Psychiatry, 31*(12), 1220–1224.

Maes, M., Van der Planken, M., Stevens, W. J., Peeters, D., DeClerck, L. S., Bridts, C. H., et al. (1992). Leukocytosis, monocytosis and neutrophilia: Hallmarks of severe depression. *Journal of Psychiatric Research, 26*(2), 125–134.

Mahnke, Y. D., Brodie, T. M., Sallusto, F., Roederer, M., & Lugli, E. (2013). The who's who of T-cell differentiation: Human memory T-cell subsets [Research Support, N.I.H., Intramural Research Support, Non-U.S. Gov't Review]. *European Journal of Immunology, 43*(11), 2797–2809. https://doi.org/10.1002/eji.201343751.

Mak, T. W. (2007). The T cell antigen receptor: "The Hunting of the Snark" [Historical Article Review]. *European Journal of Immunology, 37*(Suppl. 1), S83–93. https://doi.org/10.1002/eji.200737443.

Masopust, D., Vezys, V., Wherry, E. J., & Ahmed, R. (2007). A brief history of CD8 T cells. [Historical Article Review]. *European Journal of Immunology, 37*(Suppl. 1), S103–110. https://doi.org/10.1002/eji.200737584.

Mayadas, T. N., Cullere, X., & Lowell, C. A. (2014). The multifaceted functions of neutrophils [Research Support, N.I.H., Extramural Research Support, Non-U.S.

Gov't Review]. *Annual Review of Pathology*, 9, 181–218. https://doi.org/10.1146/annurev-pathol-020712-164023.

McAdams, C., & Leonard, B. E. (1993). Neutrophil and monocyte phagocytosis in depressed patients. *Progress in Neuro-Psychopharmacology & Biological Psychiatry*, 17(6), 971–984.

Medzhitov, R. (2009). Approaching the asymptote: 20 years later [Research Support, Non-U.S. Gov't]. *Immunity*, 30(6), 766–775. https://doi.org/10.1016/j.immuni.2009.06.004.

Medzhitov, R., Shevach, E. M., Trinchieri, G., Mellor, A. L., Munn, D. H., Gordon, S., et al. (2011). Highlights of 10 years of immunology in Nature Reviews Immunology [Historical Article Review]. *Nature Reviews Immunology*, 11(10), 693–702. https://doi.org/10.1038/nri3063.

Miles, J. J., Douek, D. C., & Price, D. A. (2011). Bias in the alphabeta T-cell repertoire: Implications for disease pathogenesis and vaccination [Research Support, N.I.H., Extramural Review]. *Immunology and Cell Biology*, 89(3), 375–387. https://doi.org/10.1038/icb.2010.139.

Mjösberg, J., & Spits, H. (2016). Human innate lymphoid cells. *The Journal of Allergy and Clinical Immunology*. https://doi.org/10.1016/j.jaci.2016.09.009.

Naude, P. J., den Boer, J. A., Comijs, H. C., Bosker, F. J., Zuidersma, M., Groenewold, N. A., et al. (2014). Sex-specific associations between neutrophil gelatinase-associated lipocalin (NGAL) and cognitive domains in late-life depression [Clinical Trial Research Support, Non-U.S. Gov't]. *Psychoneuroendocrinology*, 48, 169–177. https://doi.org/10.1016/j.psyneuen.2014.06.016.

O'Neill, B., & Leonard, B. E. (1990). Abnormal zymosan-induced neutrophil chemiluminescence as a marker of depression. *Journal of Affective Disorders*, 19(4), 265–272.

Park, E.-J., Lee, J.-H., Jeong, D.-C., Han, S.-I., & Jeon, Y.-W. (2015). Natural killer cell activity in patients with major depressive disorder treated with escitalopram. *International Immunopharmacology*, 28(1), 409–413. https://doi.org/10.1016/j.intimp.2015.06.031.

Rajewsky, K. (1996). Clonal selection and learning in the antibody system [Review]. *Nature*, 381(6585), 751–758. https://doi.org/10.1038/381751a0.

Rammensee, H. G., Falk, K., & Rotzschke, O. (1993). MHC molecules as peptide receptors [Research Support, Non-U.S. Gov't Review]. *Current Opinion in Immunology*, 5(1), 35–44.

Ravindran, A. V., Griffiths, J., Merali, Z., & Anisman, H. (1999). Circulating lymphocyte subsets in obsessive compulsive disorder, major depression and normal controls. *Journal of Affective Disorders*, 52(1–3), 1–10.

Roche, P. A., & Furuta, K. (2015). The ins and outs of MHC class II-mediated antigen processing and presentation [Review]. *Nature Reviews Immunology*, 15(4), 203–216. https://doi.org/10.1038/nri3818.

Rothermundt, M., Arolt, V., Fenker, J., Gutbrodt, H., Peters, M., & Kirchner, H. (2001). Different immune patterns in melancholic and non-melancholic major

depression. *European Archives of Psychiatry and Clinical Neuroscience*, 251(2), 90–97.

Sakaguchi, S., Sakaguchi, N., Asano, M., Itoh, M., & Toda, M. (1995). Immunologic self-tolerance maintained by activated T cells expressing IL-2 receptor alpha-chains (CD25). Breakdown of a single mechanism of self-tolerance causes various autoimmune diseases [Research Support, Non-U.S. Gov't]. *Journal of Immunology*, 155(3), 1151–1164.

Sakaguchi, S., Vignali, D. A., Rudensky, A. Y., Niec, R. E., & Waldmann, H. (2013). The plasticity and stability of regulatory T cells [Research Support, N.I.H., Extramural Research Support, Non-U.S. Gov't Review]. *Nature Reviews Immunology*, 13(6), 461–467. https://doi.org/10.1038/nri3464.

Schlatter, J., Ortuno, F., & Cervera-Enguix, S. (2004a). Lymphocyte subsets and lymphokine production in patients with melancholic versus nonmelancholic depression. *Psychiatry Research*, 128(3), 259–265. https://doi.org/10.1016/j.psychres.2004.06.004.

Schlatter, J., Ortuno, F., & Cervera-Enguix, S. (2004b). Monocytic parameters in patients with dysthymia versus major depression. *Journal of Affective Disorders*, 78(3), 243–247. https://doi.org/10.1016/S0165-0327(02)00316-6.

Segerstrom, S. C., & Miller, G. E. (2004). Psychological stress and the human immune system: A meta-analytic study of 30 years of inquiry [Meta-Analysis Research Support, Non-U.S. Gov't Research Support, U.S. Gov't, P.H.S.]. *Psychological Bulletin*, 130(4), 601–630. https://doi.org/10.1037/0033-2909.130.4.601.

Seidel, A., Arolt, V., Hunstiger, M., Rink, L., Behnisch, A., & Kirchner, H. (1996a). Increased CD56 + natural killer cells and related cytokines in major depression. *Clinical Immunology and Immunopathology*, 78(1), 83–85.

Seidel, A., Arolt, V., Hunstiger, M., Rink, L., Behnisch, A., & Kirchner, H. (1996b). Major depressive disorder is associated with elevated monocyte counts. *Acta Psychiatrica Scandinavica*, 94(3), 198–204.

Stefanski, V., & Ben-Eliyahu, S. (1996). Social confrontation and tumor metastasis in rats: Defeat and beta-adrenergic mechanisms [Research Support, Non-U.S. Gov't]. *Physiology & Behavior*, 60(1), 277–282.

Steinman, R. M. (2012). Decisions about dendritic cells: Past, present, and future [Historical Article Review]. *Annual Review of Immunology*, 30, 1–22. https://doi.org/10.1146/annurev-immunol-100311-102839.

Svenningsson, P., Berg, L., Matthews, D., Ionescu, D. F., Richards, E. M., Niciu, M. J., et al. (2014). Preliminary evidence that early reduction in p11 levels in natural killer cells and monocytes predicts the likelihood of antidepressant response to chronic citalopram. *Molecular Psychiatry*, 19(9), 962–964. https://doi.org/10.1038/mp.2014.13.

Svenningsson, P., Kim, Y., Warner-Schmidt, J., Oh, Y. S., & Greengard, P. (2013). p11 and its role in depression and therapeutic responses to antidepressants [Research

Support, N.I.H., Extramural Research Support, Non-U.S. Gov't Review]. *Nature Reviews Neuroscience, 14*(10), 673–680. https://doi.org/10.1038/nrn3564.

Tarr, A. J., Powell, N. D., Reader, B. F., Bhave, N. S., Roloson, A. L., Carson, W. E., 3rd, et al. (2012). Beta-adrenergic receptor mediated increases in activation and function of natural killer cells following repeated social disruption [Research Support, N.I.H., Extramural]. *Brain, Behavior, and Immunity, 26*(8), 1226–1238. https://doi.org/10.1016/j.bbi.2012.07.002.

Vignali, D. A., Collison, L. W., & Workman, C. J. (2008). How regulatory T cells work [Research Support, N.I.H., Extramural Research Support, Non-U.S. Gov't Review]. *Nature Reviews Immunology, 8*(7), 523–532. https://doi.org/10.1038/nri2343.

Vivier, E., Raulet, D. H., Moretta, A., Caligiuri, M. A., Zitvogel, L., Lanier, L. L., et al. (2011). Innate or adaptive immunity? The example of natural killer cells. (Research Support, N.I.H., Extramural Research Support, Non-U.S. Gov't Review). *Science, 331*(6013), 44–49. https://doi.org/10.1126/science.1198687.

Vogelzangs, N., de Jonge, P., Smit, J. H., Bahn, S., & Penninx, B. W. (2016). Cytokine production capacity in depression and anxiety. *Translational Psychiatry, 6*(5). https://doi.org/10.1038/tp.2016.92.

Wucherpfennig, K. W., Gagnon, E., Call, M. J., Huseby, E. S., & Call, M. E. (2010). Structural biology of the T-cell receptor: Insights into receptor assembly, ligand recognition, and initiation of signaling. (Research Support, N.I.H.,

Extramural Research Support, Non-U.S. Gov't Review). *Cold Spring Harbor Perspectives in Biology, 2*(4). https://doi.org/10.1101/cshperspect.a005140.

Zhang, N., & Bevan, M. J. (2011). CD8(+) T cells: Foot soldiers of the immune system [Review]. *Immunity, 35*(2), 161–168. https://doi.org/10.1016/j.immuni.2011.07.010.

Zhu, J., & Paul, W. E. (2008). CD4 T cells: Fates, functions, and faults [Historical Article Research Support, N.I.H., Extramural Research Support, N.I.H., Intramural Review]. *Blood, 112*(5), 1557–1569. https://doi.org/10.1182/blood-2008-05-078154.

Ziegler, S. F. (2016). Division of labour by CD4(+) T helper cells. *Nature Reviews Immunology, 16*(7), 403. https://doi.org/10.1038/nri.2016.53.

Ziegler-Heitbrock, L. (2015). Blood monocytes and their subsets: Established features and open questions. *Frontiers in Immunology, 6*. https://doi.org/10.3389/fimmu.2015.00423.

Zinkernagel, R. M., & Doherty, P. C. (1979). MHC-restricted cytotoxic T cells: Studies on the biological role of polymorphic major transplantation antigens determining T-cell restriction-specificity, function, and responsiveness [Research Support, U.S. Gov't, P.H.S. Review]. *Advances in Immunology, 27*, 51–177.

Zorrilla, E. P., Luborsky, L., McKay, J. R., Rosenthal, R., Houldin, A., Tax, A., et al. (2001). The relationship of depression and stressors to immunological assays: A meta-analytic review. *Brain, Behavior, and Immunity, 15*(3), 199–226. https://doi.org/10.1006/brbi.2000.0597.

Childhood Microbial Experience, Immunoregulation, Inflammation, and Adult Susceptibility to Psychosocial Stressors and Depression

Graham A.W. Rook*, Charles L. Raison[†], Christopher A. Lowry[‡,§,¶,‖]

*Centre for Clinical Microbiology, UCL (University College London), London, United Kingdom
[†]University of Wisconsin-Madison, Madison, WI, United States
[‡]University of Colorado Boulder, Boulder, CO, United States
[§]University of Colorado Anschutz Medical Campus, Aurora, CO, United States
[¶]Rocky Mountain Mental Illness Research Education and Clinical Center (MIRECC), Denver Veterans Affairs Medical Center (VAMC), Denver, CO, United States
[‖]Military and Veteran Microbiome Consortium for Research and Education (MVM-CoRE), Denver, CO, United States

INTRODUCTION

This book deals with the role of inflammation and immunity in depression. Our exposures to microorganisms in early life impact on this theme in many ways, and Fig. 1 attempts to provide a simplified summary of the text that follows. Microbes modulate susceptibility to depression via effects on the immune system, local regulation of gut neuroendocrine systems, afferent neuronal signaling, and microbial metabolites that enter the circulation and act on distant targets including the brain. First, microbes drive the development of the immune system by providing the data that the immune system requires before it can function correctly. Second, microbes, not only mostly from mother and other family members but also to a lesser extent from the environment, constitute the symbiotic microbiotas that colonize our body surfaces, notably the gut, airways, skin, and mucosal surfaces of the urinary and reproductive tracts. Vertebrates, which evolved about 500 million years ago, rapidly developed a very

FIG. 1   Overview of the ways in which exposure to microorganisms and parasites can influence the brain via effects on the immune system and microbiota that modulate inflammation. Immunoregulatory microbial signals include metabolites and are illustrated in Fig. 2. Because the microbiota is critically involved in metabolism, obesity, and responses to psychosocial stressors, the role of these factors in depression is also modulated by microbial exposures. *Abbreviations*: CRP, C-reactive protein; IBD, inflammatory bowel disease.

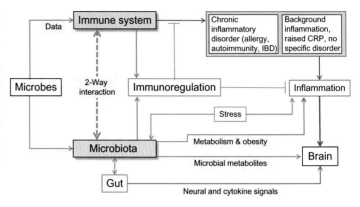

complex and diverse gut microbiota that took on crucial roles in the development and subsequent function of essentially all organ systems, including the gut, immune system, and brain (McFall-Ngai et al., 2013). Some of the effects on the brain are mediated by microbial metabolites that are only beginning to be explored. Most evolutionary biologists now think that the adaptive immune system (which invertebrates do not have) evolved to enable the vertebrate immune system to control and regulate this very complex and physiologically essential microbiota (we might almost say it evolved to help "farm" the microbiota) while simultaneously excluding pathogens, which we can define as organisms that cause damage or disrupt the human microbiota ecosystem (Pancer & Cooper, 2006). But communication between the microbiota and the immune system goes both ways, and in concert with the data from infections and environmental organisms, the microbiota plays a major role in setting up the regulatory mechanisms that limit and terminate inflammatory processes. Fig. 1 shows inflammation, highlighted in red, associated with chronic inflammatory disorders, such as allergies, autoimmune diseases, and inflammatory bowel diseases (IBDs), and also inflammation manifested as raised C-reactive protein or inflammatory cytokine levels in the absence of any of these diagnosable inflammatory disorders. These are all situations in which

failing immunoregulation contributes to inflammation, and all are associated with increased risk of psychiatric problems. Moreover, inflammation can also be caused by poorly regulated metabolism and obesity, in which the microbiota again plays a major role. Finally, Fig. 1 highlights inflammation due to stress. But the inflammatory response to stress and the subsequent behavioral changes are again modulated by the microbiota and attenuated by the regulatory arms of the immune system. In conclusion, childhood microbial exposures, by supplying and modifying the microbiota, modifying the regulation of the immune system, and modifying the regulation of the microbiota *by* the immune system, have major effects on our susceptibility to some psychiatric disorders.

In much of what follows, we are forced to discuss the increases in disorders of immunoregulation indicated in Fig. 1 as surrogates for depression, because these are easily diagnosed, common, and intensively studied, though data on depression are used when available, particularly toward the end of the chapter. But it should be remembered that, as indicated in Fig. 1, depression is frequently comorbid with the chronic inflammatory disorders and associated with raised levels of blood biomarkers of inflammation (Hodes, Menard, & Russo, 2016; Maes, 1999; Raison, Lowry, & Rook, 2010). For example, prior hospitalization due to an autoimmune disease has been associated with a 45% increased risk of subsequently developing mood disorder diagnosis (Benros et al., 2013). Migration, urbanization, modern medicine, and high-income lifestyle all lead to the loss of exposure to the organisms with which we coevolved ("old friends") and to increases in inflammatory and psychiatric disorders (reviewed and referenced in Rook, Raison, & Lowry, 2014). If we can understand these effects, we may be able to intervene and to counteract the trends toward higher incidences of inflammation and depression.

MICROBIAL EXPOSURES AND HUMAN EVOLUTION

Before analyzing these issues in greater depth, we need to identify the groups of organisms with which humans coevolved, and that might have become physiological necessities. The notion that modern life might deprive us of essential exposures initially emerged as "the hygiene hypothesis," following the observation that allergies are less frequent in children brought up with older siblings (Strachan, 1989). It was suggested that older siblings provided increased exposure to childhood infections that might somehow protect from allergic disorders. This was a valuable insight, but it implied a crucial role for the common infections of childhood and for hygiene, whereas neither implication is correct. The childhood infections are mostly "crowd" infections such as measles that either kill the host or induce solid sterilizing immunity, so they could not survive in isolated Paleolithic hunter-gatherer groups (Wolfe, Dunavan, & Diamond, 2007). They did not coevolve a necessary immunoregulatory role, and they do not protect from chronic inflammatory disorders and often actually trigger them (Bremner et al., 2008; Yoo, Tcheurekdjian, Lynch, Cabana, & Boushey, 2007). They are recent arrivals in human communities, endemic only since populations have increased

(Wertheim & Kosakovsky Pond, 2011). Meanwhile, implicating hygiene was a reasonable guess, but as will be demonstrated later, hygiene is a minor factor in the contemporary reduction in microbial exposures. Therefore, we prefer terms such as the "biodiversity" or "old friends" hypothesis (Rook, 2010; von Hertzen, Hanski, & Haahtela, 2011), which place emphasis on our evolutionary heritage and are leading to the identification of relevant organisms and mechanisms.

Microbiota

Humans evolved in small hunter-gatherer groups. As outlined in the introduction, humans, like all vertebrates, were colonized internally and externally by a vast range of symbiotic species including viruses, archaea, bacteria, fungi, protozoa, and even multicellular mites found in hair follicles and sebaceous glands. These diverse organisms constitute the microbiotas of epithelial linings, including the skin, genitourinary system, airways, oropharynx, and gut. At least 50%, perhaps more, of the cells that make up our bodies are microbial (Sender, Fuchs, & Milo, 2016), and they contribute far more genes, DNA, and metabolic pathways than are encoded in our human genomes (O'Hara & Shanahan, 2006). Studies of human metabolomics reveal that much of "our" metabolism is in fact microbial (Wikoff et al., 2009).

Spores

The issue of spores has been neglected. Spores are remarkably resistant and can remain viable for thousands, possibly millions of years (reviewed in Nicholson, 2002). They are relevant in two contexts. First, about 1/3 of the organisms in the gut microbiota are spore-forming, and spores are readily demonstrable in human feces (Hong et al., 2009a). Human feces average up to 10^4 spores/g, while soil contains approximately 10^6 spores/g (Hong et al., 2009b). Wherever humans have lived, the natural environment is inevitably seeded with human gut-adapted bacterial strains. A recent study revealed that the spore-forming strains within the human microbiota are more diverse than nonspore-forming bacteria and show a higher species turnover or a greater shift in relative abundance over the course of a year (Browne et al., 2016). Therefore, it is possible that when a gut organism becomes extinct as a result of dietary inadequacy or antibiotic misuse (Cox et al., 2014; Sonnenburg et al., 2016), it can be "reinstalled" via spores from the environment.

Other spore-forming organisms from the environment might also be important despite not being definite components of the human microbiota. Spores of *Bacillus subtilis* can germinate in the small bowels of mice and rabbits (Casula & Cutting, 2002; Tam et al., 2006) and also humans (Hong, To, et al., 2009b). Moreover, after germination, they replicate in the small bowel and then resporulate as they enter the colon. This might be very relevant to the "old friends" mechanism, particularly to the clear importance of exposure to animals, agricultural land, and green spaces. After germinating in the small bowel, these organisms will provide data to the immune system in the ileum where dendritic cells sample gut contents and where recently ingested organisms can constitute a significant proportion of the microbes present (Schulz & Pabst, 2013).

Environmental Microorganisms

In addition to spores of gut-adapted organisms discussed above (Browne et al., 2016; Mulder et al., 2009), our ancestors were also exposed to many other microorganisms from the natural environment, many of which would have had significant immunologic impact, even when not able to establish themselves within the microbiotas. Large epidemiological studies demonstrate that living close to the natural rural or coastal environment, often denoted "green space or "blue space," respectively,

reduces overall mortality, cardiovascular disease, and depressive symptoms and increases subjective feelings of well-being (Maas, Verheij, Groenewegen, de Vries, & Spreeuwenberg, 2006; Mitchell & Popham, 2008; Wheeler, White, Stahl-Timmins, & Depledge, 2012). The beneficial effects of exposure to green and blue space are particularly prominent in urban individuals of low socioeconomic status who tend to be most severely deprived of green space (Dadvand et al., 2012; Maas et al., 2006; Mitchell & Popham, 2008; Wheeler et al., 2012). It used to be assumed that these effects are explained by psychological mechanisms, but this view is untenable and supported only by experiments that lack relevant controls (Rook, 2013). While there undoubtedly are health benefits attributable to relaxation induced by exposure to the delights of nature and also benefits from accompanying exercise and sunlight, there is solid evidence for biological effects on the immune system mediated by exposures to environmental microorganisms.

Also, supporting the importance of natural environments in immunoregulation are studies demonstrating that exposure of pregnant mothers or infants to the farming environment protects the child against allergic disorders and juvenile forms of IBD (Radon et al., 2007; Riedler et al., 2001). This protection is attributable to airborne microbial biodiversity that can be assayed in children's bedrooms (Ege et al., 2011). Similarly, in a study performed in Finland, mere proximity to agricultural land rather than to urban agglomerations increased the biodiversity of skin microbiota; reduced atopic (allergic) sensitization; and increased release by blood cells of IL-10, an *anti*-inflammatory mediator (Hanski et al., 2012). It is important to note that in this study, hygiene was a constant, not a variable. The effect of the environment was seen in the presence of universally high levels of home hygiene. Most recently, studies of dust extracts obtained from Amish and Hutterite homes suggest that effects of sustained microbial exposures on innate immune function can explain the low incidence of asthma and allergic sensitization in Amish children (Stein et al., 2016).

Some of the relevant microbiota come from animals. Contact with cows and pigs protects against allergic disorders (Riedler et al., 2001; Sozanska, Blaszczyk, Pearce, & Cullinan, 2013). Contact with dogs, with which humans have coevolved for many millennia (Axelsson et al., 2013; Thalmann et al., 2013), also protects from allergic disorders (Aichbhaumik et al., 2008; Ownby, Johnson, & Peterson, 2002). Dogs greatly increase the microbial biodiversity of the home (Dunn, Fierer, Henley, Leff, & Menninger, 2013; Fujimura et al., 2010). In a developing country, the presence of animal feces in the home correlated with better ability to control background inflammation in adulthood (McDade et al., 2012), and in Russian Karelia (where the prevalence of childhood atopy is four times lower and type 1 diabetes is six times lower than in Finnish Karelia), house dust contained a sevenfold higher number of clones of animal-associated species than was present in Finnish Karelian house dust (Pakarinen et al., 2008). A recent study investigated 10,201 participants aged 26–54 years from 14 countries and generated a "biodiversity score" based on reported childhood exposures to farms, rural versus urban environment, cats, dogs, day care, bedroom sharing, and older siblings. It emerged that a high biodiversity score correlated with reduced allergic sensitization and improved lung health (Campbell et al., 2017). The routes and mechanisms involved in these effects on immunoregulation are discussed later.

Old Infections

Finally, there are certain "old" infections that established lifelong carrier states or subclinical infections and so were able to survive within small hunter-gatherer groups. This term "old infections" was used by Jared Diamond and his colleagues in their classic paper in 2007 (Wolfe et al., 2007). In order to persist in small hunter-gatherer groups, the old infections had to avoid inducing sterilizing immunity or killing the host. And in order to maintain the health of the host, they had to be tolerated. Thus, they drive regulatory anti-inflammatory responses. (These old infections must not be confused with "old friends," a term used to include *all* the categories of organism with which humans coevolved.) The old infections include ancestral strains of *Mycobacterium tuberculosis*, *Helicobacter pylori*, gut helminths and blood nematodes, and hepatitis A virus. Analysis of their phylogenetic trees and comparison with the human phylogenetic tree reveal how the old infections coevolved and spread over the globe with human populations (Galagan, 2014; Linz et al., 2007; Wolfe et al., 2007). There are many examples of the immunoregulatory roles of these old infections. Deworming in pregnancy increases the risk of eczema and wheeze in the resulting infant (Mpairwe et al., 2011). Tuberculin-positive children are less likely to have allergic rhinitis or positive allergen skin-prick tests (Obihara et al., 2005), and individuals carrying *H. pylori* are also somewhat protected from allergic disorders (Hussain et al., 2016).

These three categories of organism—microbiota, environmental organisms, and old infections—were constantly present and had to be tolerated and so coevolved roles in setting up immunoregulatory pathways.

INNATE AND ADAPTIVE IMMUNE SYSTEMS REGULATE MICROBIOTA

How does the immune system regulate the microbiota, and what is the role of host genetics? The microbiomes of monozygotic (MZ) twins are more similar than those of dizygotic (DZ) twins. Using a large cohort of MZ and DZ twins, it was possible to identify many microbial taxa of which the abundance was influenced by host genetics, including one (*Christensenella minuta*) that correlated with reduced obesity in the human subjects, and was shown to oppose obesity in an animal model (Goodrich et al., 2014). This important observation indicated that host genes can influence phenotype by controlling the organisms present in the microbiota (Goodrich et al., 2014, 2016). Much of this genetic effect is mediated via the immune system.

Innate Immune System and Microbiota

Some of this regulation of the microbiota is mediated via the innate immune system of which Toll-like receptors (TLR) and inflammasomes are essential components. If the gene encoding TLR5 is knocked out, mice develop the metabolic syndrome (hyperlipidemia, hypertension, insulin resistance, and adiposity) and altered microbiota. This altered microbiota will induce similar physiological changes when transferred to wild-type germ-free mice (Vijay-Kumar et al., 2010). Similar changes occur if components of inflammasomes are disabled. Inflammasomes are multiprotein oligomers of variable composition expressed in myeloid cells. They promote the maturation of the inflammatory cytokines IL-1β and IL-18 and play a major role in activating inflammatory responses. If various components of inflammasomes are knocked out, there are metabolic (Henao-Mejia et al., 2012) or inflammatory consequences (Elinav et al., 2011), and as in the TLR5 knockouts, these consequences are mediated by a changed microbiota. Several other genes have been shown to regulate the microbiota in a variety of animal models (Kostic, Howitt, & Garrett, 2013). So once again, it is evident that host genes within the immune system can influence phenotype by controlling the organisms present in the microbiota.

Adaptive Immune System and Microbiota

These experiments prove the role of the innate immune system, but other experiments make it clear that the adaptive immune system is just as crucial. For example, expression of major histocompatibility complex (MHC) class II on conventional dendritic cells (cDCs) is needed for effective control of the gut microbiota (Loschko et al., 2016). When MHC class II was not expressed on cDCs in a mouse model, there was chronic intestinal inflammation that could be reduced by antibiotic treatment. This inflammation did not occur in germ-free animals. Since the role of MHC class II is to present antigens to the T cells of the adaptive system, this is a formal proof of an essential role. Further evidence that the adaptive immune system is crucial comes from mice that lack both the transcription factor Tbet and the RAG2 gene. These animals cannot generate T cells or an adaptive receptor diversity. Such mice develop a severe colitis that can be treated with antibiotics or by infusion of regulatory T cells (Garrett & Glimcher, 2009). These observations confirm that both the innate and the adaptive immune systems are involved in maintaining and controlling the gut microbiota and so indirectly modulate the phenotype.

MECHANISMS OF IMMUNOREGULATION BY THE OLD FRIENDS

The previous section demonstrated that both arms of the immune system regulate the microbiota. The next question is how the microbial exposures that are the theme of this chapter, including the microbiota itself, influence the regulatory functions of the immune system. Conceptually, any input to the immune system might have direct immunoregulatory roles or might cause changes to the microbiota, leading secondarily to altered immunoregulation by the microbiota (Fig. 1). Some of the mechanisms of immunoregulation utilized by the three major categories of old friends are described here (Fig. 2).

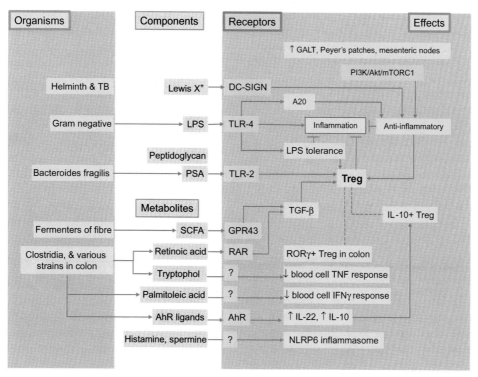

FIG. 2   Some mechanisms involved in microbial effects on immunoregulation and inflammation. Endotoxin (LPS) drives both inflammatory and anti-inflammatory pathways. The pathways shown are inevitably derived mostly from animal experiments, but human data are shown when available. References are provided in the main text. Pathways that *increase* inflammatory activity are not included on the figure. *Abbreviations*: GALT, gut-associated lymphoid tissue; PI3K, phosphatidylinositol-3 kinase; Akt, serine/threonine kinase Akt or protein kinase B (PKB); mTORC1, mammalian target of rapamycin complex 1; DC-SIGN, dendritic cell-specific intercellular adhesion molecule-3-grabbing nonintegrin; TLR-2 and TLR-4, Toll-like receptors 2 and 4; GPR43, G-protein coupled receptor 43; RAR, retinoic acid receptor; AhR, aryl hydrocarbon receptor; ILC1, ILC2, and ILC3, intestinal innate lymphoid cells types 1, 2, and 3; NLRP6, NOD-like receptor pyrin domain-containing protein 6; RORγ, RAR-related orphan receptor γ; SCFA, short-chain fatty acids; PSA, polysaccharide A of *B. fragilis*.

Immunoregulation by the Old Infections

The mechanisms used by helminths to achieve immunoregulation include modification of the bacterial microbiota (including increased *Lactobacillus* levels), modification of the phenotype of DC in the gut so that they tend to drive regulatory pathways, and release of molecules that directly drive the expansion of Treg populations (Grainger et al., 2010). The soluble egg antigen (SEA) from *Schistosoma mansoni* drives Treg development (Zaccone et al., 2009), possibly because it imitates the LewisX trisaccharide motif and thus mimics lacto-N-fucopentaose III, an immunomodulatory glycan found in human milk. Such fucosylated glycans bind DC-SIGN and induce expansion of Th2 and Treg responses as shown in Fig. 2 (Lowry et al., 2016). *H. pylori* is also an inducer of Treg, both locally, in the stomach and duodenum (Lundgren et al., 2005; Robinson et al., 2008), and systemically (Arnold et al., 2011; Lundgren et al., 2005), and some *H. pylori* strains also express DC-SIGN-binding Lewis antigen.

Although modern strains of *M. tuberculosis* are virulent, the ancestral strains with which humans coevolved were not. Human-infecting organisms resembling *M. canettii* probably evolved in Africa from environmental soil mycobacteria as much as 2.8 million years ago, in which case they might have infected human ancestors as far back as *Homo habilis* (Galagan, 2014). The *M. tuberculosis* complex evolved from these strains in Africa, at least 70,000 years ago, and accompanied the out-of-Africa human migrations. These organisms express a range of immunoregulatory molecules, some once again acting via DC-SIGN, reviewed elsewhere (Lowry et al., 2016).

Immunoregulation by Organisms From the Natural Environment

The immunoregulatory effects of exposure to the natural environment will operate both via the airways and the gut (Fig. 2). The airways contain a number of cellular sensor systems that can monitor the content of inhaled air. One of these involves the PI3K/Akt/mTORC1 signaling system that plays a role in inflammatory pathways via NF-κB. Many natural products from bacteria, algae, fungi, and higher plants can inhibit the activities of these protein kinases, and the overall effect is thought to be anti-inflammatory (Moore, 2015). Microbial metabolites also exert anti-inflammatory effects via the aryl hydrocarbon receptor (AhR). Tryptophan can be metabolized to produce AhR ligands that drive the production of IL-22 by activated DC, T cells, and innate lymphoid cells (ILC) that are abundant at mucosal surfaces (Zelante et al., 2014). But this pathway also activates host indoleamine-2,3-dioxygenase 1. This enzyme generates further tryptophan-derived AhR agonists that drive the production of TGF-β (Bessede et al., 2014) and Treg (Quintana et al., 2008).

A major microbial component in inhaled air is endotoxin (LPS), and the phenomenon known as "endotoxin tolerance" is important in the gut and the airways (Fig. 2). In a mouse model, frequent low doses of LPS increase the production of anti-inflammatory A20 (encoded by *Tnfaip3*). A20 is an ubiquitin-modifying enzyme that attenuates NF-κB activation and therefore reduces influx and activation of DC in the airways (Schuijs et al., 2015). Thus, exposure to LPS or to farm dust blocked a mouse model of allergic asthma induced by house dust mite, and this effect was dependent upon the expression of A20 in lung epithelial cells (Schuijs et al., 2015).

These observations are relevant to humans. A previously unrecognized genetic disorder has been described in families with an early onset systemic inflammatory disorder. The disease is caused by germ-line mutations in the gene that encodes A20 (Zhou et al., 2016). More evidence has come from a study of two culturally isolated farming communities in the United States. The Amish use traditional farming methods, while the Hutterites are industrialized. The Amish have much lower levels of asthma. Interestingly, peripheral blood cells from the Amish children express more *Tnfaip3* (Stein et al., 2016). Thus, it is suggested that chronic exposure of the airways to low-dose environmental microbiota sets up regulatory pathways within the airways. Endotoxin tolerance is likely to be an important part of this phenomenon.

Pulmonary neuroendocrine cells constitute another airway sensory system that is inevitably involved in the conditioning of the airway and its immune system. What is known is that stimulating the pulmonary neuroendocrine cells causes the release of neuropeptides that increase immune cell infiltrates (Branchfield et al., 2016).

Immunoregulation by the Microbiota

It used to be thought that the fetus was sterile before birth, but we now know that some transfer of maternal microbiota to the placenta and fetus starts in utero (Funkhouser & Bordenstein, 2013; Meropol & Edwards, 2015).

Animal experiments suggest that molecules derived from the maternal microbiota, some bound to maternal antibodies, cross the placenta and influence the development of the immune system (Gomez de Aguero et al., 2016). Some of these effects are mediated via the AhR and help to limit inflammatory responses to microbial molecules and translocation of intestinal microbes across the gut wall (Gomez de Aguero et al., 2016).

Germ-free animals show defects in the development of the immune system and of the gut itself (and also the brain, discussed later). This is particularly true of the gut-associated lymphoid tissue (GALT), Peyer's patches, and mesenteric lymph nodes (Round & Mazmanian, 2009). Bacterial strains that drive expansion of components of the immune system are beginning to be identified (Fig. 2). Segmented filamentous bacteria (SFB, provisionally known as *Candidatus savagella*) will expand Th17 cells, while in mouse models, certain members of the *Clostridia* (Atarashi et al., 2011) or *Bacteroides fragilis* (Round & Mazmanian, 2010) will drive Treg formation. There is a distinct subset of RORγ + Treg in the colon, and their formation is driven by a range of gut organisms from various different genera (Sefik et al., 2015). Interestingly, the RORγ transcription factor is also involved in driving Th17 cells in the small intestine, but this involves a different subset of gut organisms (Sefik et al., 2015). The microbiota also drives the development of intestinal innate lymphoid cells (ILC1, ILC2, and ILC3), and the transcriptomes of these cells are profoundly altered by antibiotic treatments (Gury-BenAri et al., 2016).

However, it is proving difficult to relate individual organisms to specific health problems, and the thinking is moving rapidly to the view that what really matters is the overall immunoregulatory potential and metabolome of the entire gut ecosystem and the resulting concentrations of certain critically important metabolites (Fig. 2). Progress is being made toward the identification of specific molecular signals from the microbiota to the immune system and brain (discussed later). Tryptophan metabolites such as indole-3-acetic acid have anti-inflammatory effects in the gut via AhR expressed on many cell types including DC (Lamas et al., 2016). Other microbial molecules also signal via AhR (Gomez de Aguero et al., 2016; Levy, Thaiss, & Elinav, 2016). AhR ligands influence the differentiation and function of Tregs by increasing the production of IL-10 and IL-22 (Goettel et al., 2016). The vitamin A metabolite retinoic acid enhances Treg and reduces Th17 via TGF-β and through the induction of histone acetylation at the FoxP3 promoter (Levy et al., 2016). Short-chain fatty acids (SCFA) help to drive Treg formation (Tan et al., 2016). Production of histamine from histidine and levels of spermine and taurine modulate the NLRP6 inflammasome (Levy et al., 2015). In the current context, it is of particular interest that the composition of the gut microbiota has strong influences on the release of cytokines by the donor's peripheral blood cells *in vitro* in the presence of bacterial and fungal stimuli (Schirmer et al., 2016). This shows that microbiota might influence background levels of inflammatory cytokines that are associated with depression. The strongest influences of the microbiota were on the production of IFNγ and TNFα, and the effects appeared to involve the tryptophan metabolite tryptophol, which has strong inhibitory effects on the TNF response, while the metabolism of palmitoleic acid was important for the IFNγ response (Schirmer et al., 2016).

Endotoxin Tolerance

Endotoxin tolerance, mentioned earlier in relation to the airways, is another mechanism that is important in the gut (Fig. 2). Animals can survive a potentially lethal dose of endotoxin if they have previously received one or more sublethal doses, and *in vitro* macrophages that have been exposed to endotoxin respond differently when challenged again, with less release of TNF and reduced NF-κB translocation (Biswas & Lopez-Collazo, 2009). Repeated low-dose endotoxin administration *in vivo* leads eventually to increased Treg activity (Caramalho et al., 2011; Wang et al., 2015). A recent study has suggested that this pathway is relevant to the fact that the prevalence of childhood atopy is fourfold higher in Finland than in a bordering area of Russia with a genetically similar population, while the prevalence of type 1 diabetes is sixfold higher (Kondrashova et al., 2005; Pakarinen et al., 2008). The endotoxin in the guts of Russian infants was mostly derived from *Escherichia coli*, which can drive endotoxin tolerance, whereas the endotoxin in the guts of Finnish infants was overwhelmingly derived from a *Bacteroides* species that releases an endotoxin that acts as an *inhibitor* of the agonist effects of *E. coli* endotoxin. This *Bacteroides* endotoxin might therefore fail to modulate the potential for cytokine induction, fail to evoke endotoxin tolerance, and fail to enhance Treg induction in Finnish infants (Vatanen et al., 2016).

Microbial Exosomes

Membrane vesicles (MV) of various types are generated by essentially all life forms, including gram-negative and gram-positive bacteria and mycobacteria, archaea, fungi, protozoa, helminths, and mammalian cells (Coakley, Maizels, & Buck, 2015; Pathirana & Kaparakis-Liaskos, 2016). It is now evident that these have a major role in the regulation of the immune system (Coakley et al., 2015; de Candia, De Rosa, Casiraghi, & Matarese, 2016). Moreover, they are also important for the development and function of the brain and can cross the blood-brain barrier (Kramer-Albers & Hill, 2016). MV from prokaryotic microbiota and parasites can enter the host from the gut (Coakley et al., 2015; de Candia et al., 2016). Furthermore, it has long been known that materials can enter the brain via the

nasal epithelium (Illum, 2004), and recent work indicates that in an animal model, synthetic MV can enter the brain by this route (Zhuang et al., 2011). It is too early to assess these findings, but if they are important, then some effects of changing exposures to microbes and parasites might operate in this way.

Other ways in which the immune system is regulated by the microbiota are considered below in relation to lifestyle changes that are causing our microbiota to differ from that with which humans evolved.

LIFESTYLE CHANGES THAT IMPAIR MICROBE-INDUCED IMMUNOREGULATION AND THAT MAY BE ASSOCIATED WITH DEPRESSION

A number of lifestyle factors, including a history of breast/formula feeding, vaginal/cesarean delivery, maternal obesity, and perinatal antibiotic exposure, modulate the development of the infant microbiota, yielding clear changes in microbiota composition during the first 1–3 years of life (Fig. 3). However, these effects on the infant microbiota do not necessarily cause detectable effects on the microbiota in adulthood (Falony et al., 2016). Nevertheless, persistent effects on the brain and on immune and metabolic systems are likely.

Perinatal and Early-Life Antibiotic Exposure

Antibiotics represent a particularly clear cause of dysbiosis, and work with antibiotics has also revealed why early dysbiosis, even when it does not persist into adulthood, is relevant to adult health.

Animal Experiments

In mice, periconceptual antibiotics also lead to weight gain later in life and to permanent changes in the immune system (Cox et al., 2014).

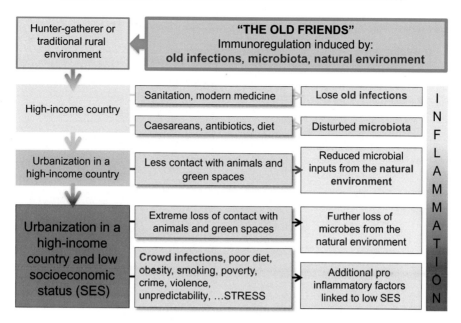

FIG. 3   The accumulation of factors that reduce microbial exposures as we progress from hunter-gatherer to urban life. The pro-inflammatory effects of reduced exposure to immunoregulation-inducing organisms are exacerbated by lifestyle problems in communities of low socioeconomic status, shown at the bottom of the figure. Abbreviations: SES, socioeconomic status. *Figure adapted, with permission from the publisher, from Rook, G. A., Raison, C.L., & Lowry, C. A. (2014). Microbial 'old friends', immunoregulation and socioeconomic status.* Clinical and Experimental Immunology, 177, 1–12, published by John Wiley & Sons Ltd on behalf of British Society for Immunology.

The brain is also affected. Exposing pregnant rats to succinylsulfathiazole, a nonabsorbable antibiotic, from 1 month before breeding until gestational day 15 leads to behavioral abnormalities in the offspring (Degroote, Hunting, Baccarelli, & Takser, 2016). Even the adult mouse brain can be modified by antibiotics. Administering a broad-spectrum antibiotic mixture to adult mice reduced hippocampal neurogenesis and memory retention (Möhle et al., 2016). These defects, which appeared to involve a monocyte subset in the brain, could be treated by reconstituting a normal microbiota, particularly when supplemented with a commercially available probiotic mixture (VSL#3) consisting of eight bacterial strains (Fig. 4).

In these models, dysbiosis can be taken to an experimental extreme by studying germ-free animals, which are informative even if of uncertain relevance to humans. Germ-free mice have abnormal stress responses that can be corrected by early restoration of the microbiota, but cannot be corrected by normalization of the microbiota in adulthood (Diaz Heijtz et al., 2011; Sudo et al., 2004). The germ-free state is also associated with abnormal upregulation of genes involved in myelination in the prefrontal cortex (Fig. 4) (Hoban et al., 2016).

Human Data

The relevance of perinatal antibiotic use to human health is most obvious in allergic disorders, which, being very common, have received most attention (Fujimura et al., 2016). Administration of antibiotics during the second or third trimesters of pregnancy or during the first months after birth can increase the risk of allergies in the infant (Korpela et al., 2016;

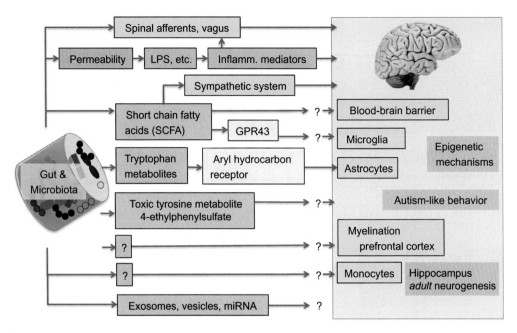

FIG. 4   Some mechanisms involved in the effects of the gut microbiota on the development and function of the brain. The pathways shown are inevitably derived from animal experiments. References and definitions are provided in the main text. Question marks imply that although the pathway has been demonstrated, the receptors or intermediary steps are unknown. References are provided in the main text. *Abbreviations*: GPR43, G-protein coupled receptor 43; LPS, lipopolysaccharide; SCFA, short-chain fatty acids.

Lapin et al., 2015; McKeever, Lewis, Smith, & Hubbard, 2002; Metsala et al., 2013). Similarly, just as in mice (Cox et al., 2014), epidemiological studies suggest that the perinatal antibiotics increase subsequent obesity in humans (Azad, Bridgman, Becker, & Kozyrskyj, 2014; Korpela et al., 2016; Trasande et al., 2013). As would be predicted by the overall adverse effects of antibiotics on the developing microbiota (as well as links between childhood antibiotic use and allergic/atopic disorders, obesity, and asthma—all associated with depression) (Kankaanranta, Kauppi, Tuomisto, & Ilmarinen, 2016; Maes, 1999; Raison et al., 2010), a large longitudinal cohort study in New Zealand found that children who received antibiotics in the first year of life had increased symptoms of depression at age 11, along with other behavioral disturbances, when compared with children not treated with

antibiotics early in life (Slykerman et al., 2017). These findings reinforce the likelihood that a "window of opportunity" exists in infancy when an appropriate microbiota must be in place, for healthy development of the brain, immune system, and metabolic regulation.

Caesarean Delivery

Cesarean delivery results in delayed transfer of maternal microbiota and delayed maturation of the infant microbiota (Dominguez-Bello et al., 2016). This can be partially corrected by exposing the neonate to maternal vaginal fluids at birth (Dominguez-Bello et al., 2016). There is evidence from human studies that cesarean delivery increases the risk of obesity in later life (Blustein et al., 2013; Yuan et al., 2016). This effect, like that due to perinatal antibiotics or the germ-free state, is probably due to early-life events, since a history of cesarean delivery did not cause detectable effects on the microbiota in adulthood (Falony et al., 2016). We know of no data demonstrating an increased risk for depression in individuals born by cesarean, but such an increase is a clear implication of the theoretical underpinnings informing this chapter.

Breast Feeding

Breast milk contains oligosaccharides that cannot be metabolized by the infant and that serve as nutrients for bifidobacteria (Garrido, Barile, & Mills, 2012; Zivkovic, German, Lebrilla, & Mills, 2011). Milk is also a source of *Bifidobacterium* and *Lactobacillus* species that seem to be transferred from the maternal gut to the breast (Jost, Lacroix, Braegger, Rochat, & Chassard, 2014; Melnik, John, Carrera-Bastos, & Schmitz, 2016). Finally, milk contains exosomes that carry microRNAs and TGF-β. These constituents are thought to drive FoxP3 expression and long-lasting Treg differentiation (Melnik et al., 2016; Saarinen, Vaarala, Klemetti, & Savilahti, 1999). Nutritional factors in breast milk include omega-3 fatty acids, which are important for the development of the brain. Moreover, whether via nutritional or immuno-regulatory pathways, via microbiota-induced effects, or via other mechanisms, breastfeeding is also important for the brain. Duration of breastfeeding is related to verbal and nonverbal intelligence later in life (Belfort et al., 2013), to better cognitive and motor development (Bernard et al., 2013), and to greater social mobility (Sacker, Kelly, Iacovou, Cable, & Bartley, 2013). Duration of breastfeeding is associated with reduced levels of circulating CRP in adulthood (Williams, Williams, & Poulton, 2006) and, as one would predict from this, has also been associated with a reduction in depression in adults who were breastfed as infants (Peus et al., 2012). Again, this is likely to be due to

an early-life effect, since a history of formula feeding did not cause detectable effects on the microbiota in adulthood (Falony et al., 2016).

Diet

The high-income urban diet, rich in fat and processed sugars, leads to a gut microbiome that differs markedly from that seen in communities living as traditional subsistence farmers or hunter-gatherers, and that also has a much reduced diversity (De Filippo et al., 2010; Gomez et al., 2016; Obregon-Tito et al., 2015; Rampelli et al., 2015; Schnorr et al., 2014). Diversity seems to be crucial, and there is now strong evidence that important dietary factors for the maintenance of diversity are fiber (polysaccharides that are fermented by gut microorganisms to yield short-chain fatty acids; SCFA) (Sonnenburg et al., 2016) and plant polyphenols (phenolic acids, flavonoids, stilbenoids, resveratrol, proanthocyanidins, curcuminoids, tannins, and lignans) (Vanamala, Knight, & Spector, 2015). Moreover, SCFA help to increase Treg (Tan et al., 2016). Thus, mothers failing to consume a diet rich in fiber and polyphenols might pass on an inappropriate microbiota to their infants. Excessive consumption of fat causes further problems. In a rat model, administering a high-fat diet during pregnancy caused dysbiosis in the offspring, with striking depletion of certain strains accompanied by behavioral social deficits. One of the depleted strains was *Lactobacillus reuteri*, and administering this organism could correct the behavioral deficits, as could cohousing with control mice (Buffington et al., 2016).

The nature of the fat consumed is also relevant. It is estimated that during human evolution, we consumed roughly equal quantities of omega-6 and omega-3 fatty acids, but thanks to recent dietary changes, we now consume 10–50 times more omega-6 than omega-3 (Blasbalg, Hibbeln, Ramsden, Majchrzak, & Rawlings, 2011; Kaliannan, Wang, Li, Kim, & Kang, 2015).

In a mouse model, feeding a diet high in omega-6 fatty acids resulted in high levels of metabolic endotoxemia (excessive absorption of LPS from the gut) and low-grade systemic inflammation that could be blocked by antibiotic treatment, implying that the effect was secondary to changed microbiota (Kaliannan et al., 2015). To our knowledge, it is not known whether maternal diet during pregnancy impacts the later development of depression in offspring, but an unhealthy diet during pregnancy independently increases the risk for offspring behavioral/emotional dysregulation at age 7 (Pina-Camacho, Jensen, Gaysina, & Barker, 2015). Overwhelming data now demonstrate that pro-inflammatory diets in adulthood are a risk factor for subsequent depression, and a recent randomized trial found that a dietary intervention designed to enhance healthy nutrition outperformed a social support protocol for reducing depressive symptoms in adults with MDD (Jacka et al., 2017).

Maternal Obesity

The fact that pro-inflammatory diets are associated with obesity provides a further link between diet and depression. It is extremely probable that the transfer of a dysfunctional microbiota from an obese mother to her offspring explains the fact that maternal obesity is a major risk factor for obesity in the child (Galley, Bailey, Kamp Dush, Schoppe-Sullivan, & Christian, 2014; Soderborg, Borengasser, Barbour, & Friedman, 2016). This, just like obesity due to perinatal antibiotic exposure discussed above, is likely to lead to increased levels of inflammatory mediators later in life. It is difficult to establish links between human maternal obesity and later depression in offspring, but the link with autism, another inflammation-associated disorder, is clear, and a recent review concluded that a link with depression was also probable (Edlow, 2017).

Immigration

Depression in immigrants also points to the importance of early-life events (Breslau, Borges, Hagar, Tancredi, & Gilman, 2009; Vega, Sribney, Aguilar-Gaxiola, & Kolody, 2004). Mexicans, Cubans, and African/Caribbean people have a two- to threefold increase in the prevalence of depression if immigration to the United States occurred when the individual was less than 13 years old or was born in the United States, compared with the prevalence in those who migrated after the age of 13 (Breslau et al., 2009; Vega et al., 2004). This implies that there is a protective effect of early childhood environmental influences, as has been shown for autoimmune disorders and IBD (Rook et al., 2014) and obesity (Trasande et al., 2013).

Stress

Perinatal stress causes changes in the regulation of the immune system and of the HPA axis and altered development of the brain, notably the hippocampus and amygdala. We reviewed these issues elsewhere (Rook, Lowry, & Raison, 2015). However, some aspects of these effects of stress are relevant to this chapter, because they are mediated via stress-induced changes in the microbiota. Stress alters the microbiota of experimental animals (Bailey et al., 2011; Kiliaan et al., 1998), and the same is true of the microbiota of severely stressed critically ill humans, where the changes are rapid and prolonged (Hayakawa et al., 2011). *Prenatal* stressors have been shown to alter the microbiome in rhesus monkeys by reducing the overall numbers of *Bifidobacteria* and *Lactobacilli* during adulthood (Bailey, Lubach, & Coe, 2004). In a rat model, the stress of maternal separation in the neonatal period had long-term effects on the diversity of the microbiota that was still apparent when the pups became adults (O'Mahony et al., 2009). Recently, it has been shown that when human mothers are stressed during pregnancy, the infants carry higher relative abundances of *Proteobacteria* some of which might be pathogens and lower relative abundances of lactic acid bacteria and *Bifidobacteria* (Zijlmans, Korpela, Riksen-Walraven, de Vos, & de Weerth, 2015). These findings, together with the effects on brain development and function of perinatal antibiotics and the germ-free state discussed earlier, make it very probable that part of the effect of perinatal stress on rates of depression is mediated via the microbiota.

DEPRESSION

A number of predictions can be made if a failure of immunoregulation, arising from deficient exposure to immunoregulatory microbial inputs during early life, is a risk factor for depression in adulthood. A first prediction is that lifestyle and environmental factors that promote immune dysregulation (and disorders of immune dysregulation) should also increase the risk for depression. As we've touched upon in the preceding sections, this appears to be widely true. Here, we also consider other predictions that stem from a link between immunodysregulation/disturbed microbial interactions and depression, including that (1) depression should be associated with chronic low-grade inflammation and exaggerated pro-inflammatory cytokine responses to psychosocial stressors during adulthood; (2) treatment with pro-inflammatory cytokines should induce depressive symptoms or clinical depression; (3) depressed patients should have a higher incidence and prevalence of autoimmune disorders, allergies, and other conditions associated with reduced FoxP3$^+$ Treg; (4) anti-inflammatory or immunoregulatory therapies should be effective in reducing symptoms of depression; and (5) depression should be associated with changes in the composition and function of the gut microbiota. Evidence supporting each of these predictions is described in the following paragraphs.

Chronic Stress-Induced Inflammation in Depression

A massive database, captured in several meta-analyses, confirms that depressed patients as a group have increased levels of circulating pro-inflammatory cytokines and downstream inflammatory markers (i.e., CRP), with the cytokine data being strongest for IL-6, CRP, and TNF. Depressed individuals have also been reported to present with a relative deficit in anti-inflammatory mediators and Treg (Chen et al., 2011; fully referenced in Raison et al., 2010). These peripheral changes are almost certainly reflected in immune functioning in the CNS, given data from noninvasive positron-emission tomography (PET) scanning that reveals the presence of inflammation in the brains of depressed individuals (Setiawan et al., 2015). In addition to these cross-sectional associations, studies now document that increased CRP or IL-6 *predicts* later depression in children (Khandaker, Pearson, Zammit, Lewis, & Jones, 2014) and adults in the United Kingdom (Gimeno et al., 2009) and also *predicts* later susceptibility to posttraumatic stress disorder (PTSD) in marines (Eraly et al., 2014). Depressed individuals (and those at high risk for depression via early-life adversity) have been shown in several studies to respond to laboratory-based psychosocial stressors with increased production of IL-6 and induction of the inflammatory signaling molecule NF-κB (Carpenter et al., 2010; Pace et al., 2006), and increases in inflammation in response to a laboratory stressor have been shown to predict increased depressive symptoms up to a year later (Aschbacher et al., 2012).

Proinflammatory Cytokines Induce Depression

Mercifully, for patients with hepatitis C, treatment with the cytokine interferon IFN-alpha has been supplanted by more effective and tolerable treatment options. But during the years of its clinical hegemony, IFN-alpha provided a unique model system for understanding behavioral and biological responses to chronic inflammation relevant to depression. Results from many studies have been quite consistent in demonstrating that IFN-alpha exposure produces a widespread increase in depressive and anxious symptoms, with a sizable minority of patients meeting full criteria for major depressive disorder (MDD) within a month of commencing treatment. As reviewed in Miller and Raison (2016), IFN-alpha treatment also produces all known biological changes associated with MDD more generally, including increased circulating pro-inflammatory cytokines, disruption of the diurnal cortisol rhythm and induction of glucocorticoid resistance, altered sleep physiology, and changes in monoamine metabolism, with many of these changes associating with increased depression during treatment (Miller & Raison, 2016). Supporting these findings are studies showing that even a single exposure to inflammatory stimuli (e.g., lipopolysaccharide or typhoid vaccine) induces depressive symptoms and depressive-style social cognitions in healthy volunteers, with these effects being strongest in women (Harrison et al., 2009; Moieni et al., 2015).

Depression is Associated With Autoimmune, Allergic and Inflammatory Bowel Disorders

A prospective population-based study in Denmark that included approximately 1.1 million people identified 145,217 individuals with depression and found that depression was associated with a higher incidence and prevalence of a wide range of autoimmune disorders, allergic disorders, IBDs, and enteropathies (Andersson et al., 2015). The association with bowel disorders is consistent in other studies of celiac disease and IBDs (Jackson, Eaton, Cascella, Fasano, & Kelly, 2012; Regueiro, Greer, & Szigethy, 2017).

Similarly, when a large birth cohort was followed prospectively for 31 years, it was found that in females, the presence of atopy, confirmed by skin-prick tests, increased the risk of depression up to 4.7-fold compared with nonatopic females from the same cohort (Timonen et al., 2003). The effect is large enough to explain why spring peaks in aeroallergens coincide with peaks in suicide (Postolache, Komarow, & Tonelli, 2008). A Finnish sample of 1337 monozygotic and 2506 dizygotic twin pairs suggested a shared genetic risk for atopy and depression (Wamboldt et al., 2000). Interestingly, depression is associated with reduced circulating FoxP3[+] Treg even in those patients who do not have manifest inflammatory disorders (Grosse et al., 2016).

Antidepressant Effects of Anti-Inflammatory and Immunoregulatory Treatments

Studies on the antidepressant effects of anti-inflammatory or immunoregulatory agents have done much to reinforce many years' evidence that only a subset of depressed patients evince increased inflammation and that making this distinction has very significant treatment implications. While meta-analyses suggest that nonspecific anti-inflammatory agents (i.e., those with biological effects relevant to depression other than inflammation, such as selective serotonin reuptake inhibitors) may produce small effect-size improvements in MDD generally, studies using specific cytokine antagonists (which have no off-target effects) paint a more precise and interesting picture. In particular, Raison et al. showed in a study of treatment-resistant MDD that treatment with the TNF antagonist infliximab had no overall effect on depressive symptoms when compared with placebo. On the other hand, infliximab performed as well as typical antidepressants in depressed subjects with baseline CRP concentrations >5 mg/L, suggesting that peripheral inflammatory processes were driving depressive symptomatology in these subjects (Raison et al., 2013). However, in depressed subjects with lower levels of CRP at baseline, placebo actually far outperformed infliximab—highlighting the dangers in making too easy assumptions about depression being "an inflammatory condition." Rather, it appears that a subset of depressed patients may come to their illness, at least in part, via chronic increased inflammation, while in other depressed patients, the immune system is either less relevant or actually abnormal in ways not subsumed under the moniker of "increased inflammation." Consistent with this possibility are recent findings that healthy patients with MDD and increased inflammation (as indexed by peripheral CRP) show differential glutamate activity in anterior striatum and different patterns of prefrontal-anterior striatum connectivity when compared with depressed patients with low levels of inflammation (Felger et al., 2016; Haroon et al., 2016). Further supporting the possibility that anti-inflammatory treatments may have antidepressant effects but only in those with elevated inflammation is a study by Rappaport et al. that found results identical to the infliximab study: depressed patients with elevated baseline CRP showed an antidepressant response to treatment with omega-3 fatty acids, whereas depressed subjects with low levels of inflammation obtained more benefit from placebo treatment (Rapaport et al., 2016).

Microbiota in Human Depression

So, is there any evidence of changes to the microbiota in human depression? There are few studies, and they are not conclusive. Patients suffering from MDD had reduced levels of *Faecalibacterium*, which showed a negative correlation with the severity of depressive symptoms (Jiang et al., 2015). This finding is of note as *F. prausnitzii*, the sole known species of the *Faecalibacterium* genus, is an abundant commensal that synthesizes butyrate and other SCFA implicated in driving immunoregulation

(Khan et al., 2012; Qiu, Zhang, Yang, Hong, & Yu, 2013; Tan et al., 2016). Consistent with the thesis of this paper, low relative abundances of *F. prausnitzii* have also been associated with Crohn's disease, obesity, and psoriasis (Newton et al., 2015; Qiu et al., 2013; Sokol et al., 2008). Patients also had increased levels of *Enterobacteriaceae* and *Alistipes* (Jiang et al., 2015). A second study also found some increases in members of the *Alistipes* group in the fecal microbiota of depressed individuals (Naseribafrouei et al., 2014). There is one report that transfer of fecal microbiota from depressed patients to germ-free mice results in depression-like behavioral changes in the latter (Zheng et al., 2016). Moreover, microbiota from patients with irritable bowel syndrome (IBS) will induce similar symptoms after transfer to rodents, including anxiety-like behavior when this symptom was present in the human donor (reviewed in Collins, 2016). The future may lie in seeking changes in levels of critical metabolites, rather than changes in the microbial composition.

However, there is good evidence that human behavior can be changed by modulation of the microbiota. A probiotic-rich fermented milk product or a matching placebo was given to women for 4 weeks. Both before and after this regimen, the women were exposed to emotive images of faces while undergoing functional magnetic resonance imaging (fMRI) of their brains. Consumption of the probiotic product was shown to have affected the activity of brain regions involved in central processing of emotion and sensation (Tillisch et al., 2013). Another randomized and blinded study used 20 healthy participants without current mood disorder who received a complex probiotic food supplement or placebo for 4 weeks. Assessment before and after the intervention indicated that the active preparation reduced negative thoughts associated with sad mood (Steenbergen, Sellaro, van Hemert, Bosch, & Colzato, 2015).

Emerging data suggest that microbe-based immunoregulatory interventions may also hold promise for the treatment of depression. A recent meta-analysis of five randomized trials of various probiotic formulations versus placebo found an overall effect size of 0.30 favoring probiotics in patients with MDD and an effect size for improving depressive symptoms of 0.25 in nonclinically depressed individuals, although effects were not significant for patients over 65 years of age for unknown reasons (Huang, Wang, & Hu, 2016). Less is known about the antidepressant effects of environmentally based microorganisms although both animal and human data suggest that a heat-killed preparation from the saprophytic microorganism, *M. vaccae*, may hold promise for the reduction of depressive and anxious symptoms (Lowry et al., 2007; O'Brien et al., 2004; Reber et al., 2016).

Animal Data

These human findings are supported by a mass of animal data, and two of these studies are of particular relevance. First, spleen cells from individual mice were tested *in vitro* for IL-6 output in the presence of endotoxin. Individual animals could then be classified as high or low IL-6 releasers. When the spleen cell donors were subsequently subjected to stress, only the high IL-6 releasers showed depression-like behavioral changes (Hodes et al., 2014). This is analogous to the human study mentioned in the previous paragraph (Pace et al., 2006). A still more relevant study exploited a model in which exposure to stress induces changes in the gut microbiota accompanied by an inflammatory colitis, exaggerated release of pro-inflammatory cytokines from mesenteric lymph node cells stimulated *ex vivo*, and anxiety-like behavioral changes (Reber et al., 2016). In this model, injections of a heat-killed environmental mycobacterium were able to block the colitis, the stress-induced exaggeration of inflammation, and the behavioral changes, and all these effects were shown

to be attributable to the induction of Treg (Reber et al., 2016). This simultaneous modulation of a peripheral inflammatory disorder (colitis) and a behavioral disorder by microbe-induced Treg is a striking finding (Fig. 5).

Microbial Metabolites and the Brain

Not all the effects of the microbiota are mediated indirectly via altered immunoregulation. Several neurotransmitters are derived from amino acids, and humans generate them using genes obtained by horizontal gene transfer from bacteria (Iyer, Aravind, Coon, Klein, & Koonin, 2004). It is therefore not surprising that organisms within the microbiota synthesize these mediators and also physiologically active variants (Fig. 4). Tyrosine (the precursor of dopamine, noradrenaline, and adrenaline) and tryptophan (the precursor of serotonin, neuroactive kynurenine metabolites, and melatonin) are two examples. When pregnant mice were exposed to poly(I:C) to mimic inflammation induced by virus infections, the offspring exhibited behavioral impairments (Hsiao et al., 2013). This was accompanied by dysbiosis and changes in blood levels of 4-ethylphenyl sulfate (a tyrosine metabolite) and indolepyruvate (a tryptophan metabolite). Administering 4-ethylphenyl sulfate to normal mice was able to elicit similar behavioral changes. Moreover, a probiotic preparation (*B. fragilis*) was able to normalize the blood levels of these metabolites.

Other tryptophan metabolites are agonists for the AhR, and several of these (indole, indoxyl-3-sulfate, indole-3-propionic acid, and indole-3-aldehyde) were found to exert anti-inflammatory effects in the CNS via AhR expressed on astrocytes (Rothhammer et al., 2016). Similarly, SCFA derived by the fermentation of fiber by organisms in the colon have been shown to regulate the development of microglia (Erny et al., 2015) and to be necessary for the development of an intact blood-brain barrier

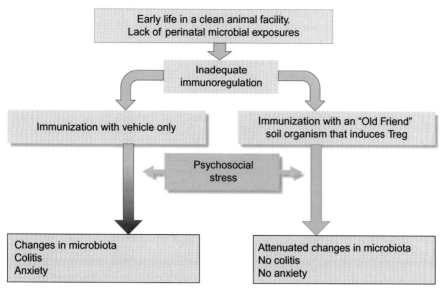

FIG. 5   An experimental model where stress leads to simultaneous development of altered microbiota, colitis, changes in brain chemistry, and behavioral changes resembling posttraumatic stress disorder (PTSD). The striking thing about this model is that the induction of regulatory T cells (Treg) by injecting an environmental organism blocks both the colitis and the PTSD-like state.

(Braniste et al., 2014). Thus, these microbial metabolites, already discussed above in relation to their indirect effects on the brain via the regulation of the immune system, are also relevant to brain function via direct effects on brain development (Fig. 4).

CONCLUSIONS

This chapter has followed the rather narrow brief of describing how susceptibility to depression may be modified by childhood exposures to microbes acting via effects on the regulation of the immune system that help to limit inflammation and via effects on the composition of the microbiota and gut-brain axis function. We hope to encourage greater awareness of the fact that environmental changes, particularly those associated with the 21st century urban lifestyle, can affect our brains via physical and biochemical pathways as well as via psychosocial ones.

Consequently, there are many dietary and life-style factors with the potential to increase the immunoregulatory influence of microbial exposure and reduce the risk of depression. These include (1) increased consumption of fresh plants and fermented foods, which may increase alpha diversity of the microbiome and its immunoregulatory potential (Selhub, Logan, & Bested, 2014; Sonnenburg et al., 2016; Tillisch et al., 2013); (2) increased consumption of prebiotics, which have been shown to selectively increase the abundance of bifidobacteria (Davis, Martinez, Walter, Goin, & Hutkins, 2011) and to attenuate negative outcomes of stress (Thompson et al., 2016); (3) minimized use of antibiotics, especially in early life (Slykerman et al., 2017); (4) increased exercise, which increases alpha diversity of the gut microbiome (O'Sullivan et al., 2015); (5) increased sleep (Benedict et al., 2016; Poroyko et al., 2016); and (6) increased exposure to the outdoor environment (Rook, 2013). It should be possible to modify indoor and urban environments so that their microbiotas resemble that of the natural environment with which we evolved (Hoisington, Brenner, Kinney, Postolache, & Lowry, 2015; Logan, 2015; Lowry et al., 2016; Stamper et al., 2016). This will require a much deeper understanding of the different roles of environmental organisms, followed by collaborations with town planners, architects, and designers of air-conditioning systems, building materials, and water supplies.

It is possible that not all the effects of microbial exposures on susceptibility to depression act via modulation of immunoregulation and inflammation. For example, Fig. 4 indicates the effects of the microbiota operating via microglia, astrocytes, myelination in the prefrontal cortex, neurogenesis in the hippocampus, and modification of the blood-brain barrier. It is not clear that all of these effects are dependent only on the balance of inflammation and immunoregulation, so it is entirely possible that we will learn to modulate the gut microbiota in ways that generate other signals that influence the development or function of the brain.

Anti-inflammatory treatments for depression are already showing efficacy (Kappelmann, Lewis, Dantzer, Jones, & Khandaker, 2018), and the time has come to test immunoregulation-inducing microorganisms. Ultimately, we anticipate the development of immunoregulatory immunizations and probiotics that will exert long-term therapeutic effects on those patients whose disorders are associated with inflammation or with inappropriate microbiota. Such immunizations or probiotics may also hold great promise when administered in early life during key developmental windows for protecting against the development of stress-related psychiatric disorders—including major depression—in adulthood.

References

Aichbhaumik, N., Zoratti, E. M., Strickler, R., Wegienka, G., Ownby, D. R., Havstad, S., et al. (2008). Prenatal exposure to household pets influences fetal immunoglobulin E production. *Clinical and Experimental Allergy, 38*(11), 1787–1794.

Andersson, N. W., Gustafsson, L. N., Okkels, N., Taha, F., Cole, S. W., Munk-Jorgensen, P., et al. (2015). Depression and the risk of autoimmune disease: A nationally representative, prospective longitudinal study. *Psychological Medicine, 45*(16), 3559–3569.

Arnold, I. C., Dehzad, N., Reuter, S., Martin, H., Becher, B., Taube, C., et al. (2011). *Helicobacter pylori* infection prevents allergic asthma in mouse models through the induction of regulatory T cells. *The Journal of Clinical Investigation, 121*(8), 3088–3093.

Aschbacher, K., Epel, E., Wolkowitz, O. M., Prather, A. A., Puterman, E., & Dhabhar, F. S. (2012). Maintenance of a positive outlook during acute stress protects against pro-inflammatory reactivity and future depressive symptoms. *Brain, Behavior, and Immunity, 26*(2), 346–352.

Atarashi, K., Tanoue, T., Shima, T., Imaoka, A., Kuwahara, T., Momose, Y., et al. (2011). Induction of colonic regulatory T cells by indigenous clostridium species. *Science, 331*, 337–341.

Axelsson, E., Ratnakumar, A., Arendt, M. L., Maqbool, K., Webster, M. T., Perloski, M., et al. (2013). The genomic signature of dog domestication reveals adaptation to a starch-rich diet. *Nature, 495*(7441), 360–364.

Azad, M. B., Bridgman, S. L., Becker, A. B., & Kozyrskyj, A. L. (2014). Infant antibiotic exposure and the development of childhood overweight and central adiposity. *International Journal of Obesity, 38*(10), 1290–1298.

Bailey, M. T., Dowd, S. E., Galley, J. D., Hufnagle, A. R., Allen, R. G., & Lyte, M. (2011). Exposure to a social stressor alters the structure of the intestinal microbiota: Implications for stressor-induced immunomodulation. *Brain, Behavior, and Immunity, 25*(3), 397–407.

Bailey, M. T., Lubach, G. R., & Coe, C. L. (2004). Prenatal stress alters bacterial colonization of the gut in infant monkeys. *Journal of Pediatric Gastroenterology and Nutrition, 38*(4), 414–421.

Belfort, M. B., Rifas-Shiman, S. L., Kleinman, K. P., Guthrie, L. B., Bellinger, D. C., Taveras, E. M., et al. (2013). Infant feeding and childhood cognition at ages 3 and 7 years: Effects of breastfeeding duration and exclusivity. *JAMA Pediatrics, 167*(9), 836–844.

Benedict, C., Vogel, H., Jonas, W., Woting, A., Blaut, M., Schurmann, A., et al. (2016). Gut microbiota and glucometabolic alterations in response to recurrent partial sleep deprivation in normal-weight young individuals. *Molecular Metabolism, 5*(12), 1175–1186.

Benros, M. E., Waltoft, B. L., Nordentoft, M., Ostergaard, S. D., Eaton, W. W., Krogh, J., et al. (2013). Autoimmune diseases and severe infections as risk factors for mood disorders: A nationwide study. *JAMA Psychiatry, 70*(8), 812–820.

Bernard, J. Y., De Agostini, M., Forhan, A., Alfaiate, T., Bonet, M., Champion, V., et al. (2013). Breastfeeding duration and cognitive development at 2 and 3 years of age in the EDEN mother-child cohort. *The Journal of Pediatrics, 163*(1), 36–42. e31.

Bessede, A., Gargaro, M., Pallotta, M. T., Matino, D., Servillo, G., Brunacci, C., et al. (2014). Aryl hydrocarbon receptor control of a disease tolerance defence pathway. *Nature, 511*(7508), 184–190.

Biswas, S. K., & Lopez-Collazo, E. (2009). Endotoxin tolerance: New mechanisms, molecules and clinical significance. *Trends in Immunology, 30*(10), 475–487.

Blasbalg, T. L., Hibbeln, J. R., Ramsden, C. E., Majchrzak, S. F., & Rawlings, R. R. (2011). Changes in consumption of omega-3 and omega-6 fatty acids in the United States during the 20th century. *The American Journal of Clinical Nutrition, 93*(5), 950–962.

Blustein, J., Attina, T., Liu, M., Ryan, A. M., Cox, L. M., Blaser, M. J., et al. (2013). Association of caesarean delivery with child adiposity from age 6 weeks to 15 years. *International Journal of Obesity, 37*(7), 900–906.

Branchfield, K., Nantie, L., Verheyden, J. M., Sui, P., Wienhold, M. D., & Sun, X. (2016). Pulmonary neuroendocrine cells function as airway sensors to control lung immune response. *Science, 351*(6274), 707–710.

Braniste, V., Al-Asmakh, M., Kowal, C., Anuar, F., Abbaspour, A., Toth, M., et al. (2014). The gut microbiota influences blood-brain barrier permeability in mice. *Science Translational Medicine, 6*(263).

Bremner, S. A., Carey, I. M., DeWilde, S., Richards, N., Maier, W. C., Hilton, S. R., et al. (2008). Infections presenting for clinical care in early life and later risk of hay fever in two UK birth cohorts. *Allergy, 63*(3), 274–283.

Breslau, J., Borges, G., Hagar, Y., Tancredi, D., & Gilman, S. (2009). Immigration to the USA and risk for mood and anxiety disorders: Variation by origin and age at immigration. *Psychological Medicine, 39*(7), 1117–1127.

Browne, H. P., Forster, S. C., Anonye, B. O., Kumar, N., Neville, B. A., Stares, M. D., et al. (2016). Culturing of 'unculturable' human microbiota reveals novel taxa and extensive sporulation. *Nature, 533*(7604), 543–546.

Buffington, S. A., Di Prisco, G. V., Auchtung, T. A., Ajami, N. J., Petrosino, J. F., & Costa-Mattioli, M. (2016). Microbial reconstitution reverses maternal diet-induced social and synaptic deficits in offspring. *Cell, 165*(7), 1762–1775.

Campbell, B., Raherison, C., Lodge, C. J., Lowe, A. J., Gislason, T., Heinrich, J., et al. (2017). The effects of growing up on a farm on adult lung function and allergic phenotypes: An international population-based study. *Thorax, 72*(3), 236–244.

Caramalho, I., Rodrigues-Duarte, L., Perez, A., Zelenay, S., Penha-Goncalves, C., & Demengeot, J. (2011). Regulatory T cells contribute to diabetes protection in lipopolysaccharide-treated non-obese diabetic mice. *Scandinavian Journal of Immunology, 74*(6), 585–595.

Carpenter, L. L., Gawuga, C. E., Tyrka, A. R., Lee, J. K., Anderson, G. M., & Price, L. H. (2010). Association between plasma IL-6 response to acute stress and early-life adversity in healthy adults. *Neuropsychopharmacology, 35*(13), 2617–2623.

Casula, G., & Cutting, S. M. (2002). *Bacillus* probiotics: Spore germination in the gastrointestinal tract. *Applied and Environmental Microbiology, 68*(5), 2344–2352.

Chen, Y., Jiang, T., Chen, P., Ouyang, J., Xu, G., Zeng, Z., et al. (2011). Emerging tendency towards autoimmune process in major depressive patients: A novel insight from Th17 cells. *Psychiatry Research, 188*(2), 224–230.

Coakley, G., Maizels, R. M., & Buck, A. H. (2015). Exosomes and other extracellular vesicles: The new communicators in parasite infections. *Trends in Parasitology, 31*(10), 477–489.

Collins, S. M. (2016). The intestinal microbiota in the irritable bowel syndrome. *International Review of Neurobiology, 131*, 247–261.

Cox, L. M., Yamanishi, S., Sohn, J., Alekseyenko, A. V., Leung, J. M., Cho, I., et al. (2014). Altering the intestinal microbiota during a critical developmental window has lasting metabolic consequences. *Cell, 158*(4), 705–721.

Dadvand, P., de Nazelle, A., Figueras, F., Basagana, X., Su, J., Amoly, E., et al. (2012). Green space, health inequality and pregnancy. *Environment International, 40*, 110–115.

Davis, L. M., Martinez, I., Walter, J., Goin, C., & Hutkins, R. W. (2011). Barcoded pyrosequencing reveals that consumption of galactooligosaccharides results in a highly specific bifidogenic response in humans. *PLoS One, 6*(9).

de Candia, P., De Rosa, V., Casiraghi, M., & Matarese, G. (2016). Extracellular RNAs: A secret arm of immune system regulation. *The Journal of Biological Chemistry, 291*(14), 7221–7228.

De Filippo, C., Cavalieri, D., Di Paola, M., Ramazzotti, M., Poullet, J. B., Massart, S., et al. (2010). Impact of diet in shaping gut microbiota revealed by a comparative study in children from Europe and rural Africa. *Proceedings of the National Academy of Sciences of the United States of America, 107*(33), 14691–14696.

Degroote, S., Hunting, D. J., Baccarelli, A. A., & Takser, L. (2016). Maternal gut and fetal brain connection: Increased anxiety and reduced social interactions in Wistar rat offspring following peri-conceptional antibiotic exposure. *Progress in Neuro-Psychopharmacology & Biological Psychiatry, 71*, 76–82.

Diaz Heijtz, R., Wang, S., Anuar, F., Qian, Y., Bjorkholm, B., Samuelsson, A., et al. (2011). Normal gut microbiota modulates brain development and behavior. *Proceedings of the National Academy of Sciences of the United States of America, 108*(7), 3047–3052.

Dominguez-Bello, M. G., De Jesus-Laboy, K. M., Shen, N., Cox, L. M., Amir, A., Gonzalez, A., et al. (2016). Partial restoration of the microbiota of cesarean-born infants via vaginal microbial transfer. *Nature Medicine, 22*(3), 250–253.

Dunn, R. R., Fierer, N., Henley, J. B., Leff, J. W., & Menninger, H. L. (2013). Home life: Factors structuring the bacterial diversity found within and between homes. *PLoS One, 8*(5).

Edlow, A. G. (2017). Maternal obesity and neurodevelopmental and psychiatric disorders in offspring. *Prenatal Diagnosis, 37*(1), 95–110.

Ege, M. J., Mayer, M., Normand, A. C., Genuneit, J., Cookson, W. O., Braun-Fahrlander, C., et al. (2011). Exposure to environmental microorganisms and childhood asthma. *The New England Journal of Medicine, 364*(8), 701–709.

Elinav, E., Strowig, T., Kau, A. L., Henao-Mejia, J., Thaiss, C. A., Booth, C. J., et al. (2011). NLRP6 inflammasome regulates colonic microbial ecology and risk for colitis. *Cell, 145*(5), 745–757.

Eraly, S. A., Nievergelt, C. M., Maihofer, A. X., Barkauskas, D. A., Biswas, N., Agorastos, A., et al. (2014). Assessment of plasma C-reactive protein as a biomarker of posttraumatic stress disorder risk. *JAMA Psychiatry, 71*(4), 423–431.

Erny, D., Hrabe de Angelis, A. L., Jaitin, D., Wieghofer, P., Staszewski, O., David, E., et al. (2015). Host microbiota constantly control maturation and function of microglia in the CNS. *Nature Neuroscience, 18*(7), 965–977.

Falony, G., Joossens, M., Vieira-Silva, S., Wang, J., Darzi, Y., Faust, K., et al. (2016). Population-level analysis of gut microbiome variation. *Science, 352*(6285), 560–564.

Felger, J. C., Li, Z., Haroon, E., Woolwine, B. J., Jung, M. Y., Hu, X., et al. (2016). Inflammation is associated with decreased functional connectivity within corticostriatal reward circuitry in depression. *Molecular Psychiatry, 21*(10), 1358–1365.

Fujimura, K. E., Johnson, C. C., Ownby, D. R., Cox, M. J., Brodie, E. L., Havstad, S. L., et al. (2010). Man's best friend? The effect of pet ownership on house dust microbial communities. *The Journal of Allergy and Clinical Immunology, 126*(2), 410–412. 412. e411-413.

Fujimura, K. E., Sitarik, A. R., Havstad, S., Lin, D. L., Levan, S., Fadrosh, D., et al. (2016). Neonatal gut microbiota associates with childhood multisensitized atopy and T cell differentiation. *Nature Medicine, 22*(10), 1187–1191.

Funkhouser, L. J., & Bordenstein, S. R. (2013). Mom knows best: The universality of maternal microbial transmission. *PLoS Biology, 11*(8).

Galagan, J. E. (2014). Genomic insights into tuberculosis. *Nature Reviews. Genetics, 15*(5), 307–320.

Galley, J. D., Bailey, M., Kamp Dush, C., Schoppe-Sullivan, S., & Christian, L. M. (2014). Maternal obesity is associated with alterations in the gut microbiome in toddlers. *PLoS One, 9*(11).

Garrett, W. S., & Glimcher, L. H. (2009). T-bet-/- RAG2-/- ulcerative colitis: The role of T-bet as a peacekeeper of host-commensal relationships. *Cytokine, 48*(1-2), 144–147.

Garrido, D., Barile, D., & Mills, D. A. (2012). A molecular basis for bifidobacterial enrichment in the infant gastrointestinal tract. *Advances in Nutrition, 3*(3), 415S–421S.

Gimeno, D., Kivimaki, M., Brunner, E. J., Elovainio, M., De Vogli, R., Steptoe, A., et al. (2009). Associations of C-reactive protein and interleukin-6 with cognitive symptoms of depression: 12-year follow-up of the Whitehall II study. *Psychological Medicine, 39*(3), 413–423.

Goettel, J. A., Gandhi, R., Kenison, J. E., Yeste, A., Murugaiyan, G., Sambanthamoorthy, S., et al. (2016). AHR activation is protective against colitis driven by T cells in humanized mice. *Cell Reports, 17*(5), 1318–1329.

Gomez, A., Petrzelkova, K. J., Burns, M. B., Yeoman, C. J., Amato, K. R., Vlckova, K., et al. (2016). Gut microbiome of coexisting BaAka pygmies and bantu reflects gradients of traditional subsistence patterns. *Cell Reports, 14*(9), 2142–2153.

Gomez de Aguero, M., Ganal-Vonarburg, S. C., Fuhrer, T., Rupp, S., Uchimura, Y., Li, H., et al. (2016). The maternal microbiota drives early postnatal innate immune development. *Science, 351*(6279), 1296–1302.

Goodrich, J. K., Davenport, E. R., Beaumont, M., Jackson, M. A., Knight, R., Ober, C., et al. (2016). Genetic determinants of the gut microbiome in UK twins. *Cell Host & Microbe, 19*(5), 731–743.

Goodrich, J. K., Waters, J. L., Poole, A. C., Sutter, J. L., Koren, O., Blekhman, R., et al. (2014). Human genetics shape the gut microbiome. *Cell, 159*(4), 789–799.

Grainger, J. R., Smith, K. A., Hewitson, J. P., McSorley, H. J., Harcus, Y., Filbey, K. J., et al. (2010). Helminth secretions induce de novo T cell Foxp3 expression and regulatory function through the TGF-beta pathway. *Journal of Experimental Medicine, 207*(11), 2331–2341.

Grosse, L., Hoogenboezem, T., Ambree, O., Bellingrath, S., Jorgens, S., de Wit, H. J., et al. (2016). Deficiencies of the T and natural killer cell system in major depressive disorder: T regulatory cell defects are associated with inflammatory monocyte activation. *Brain, Behavior, and Immunity, 54*, 38–44.

Gury-BenAri, M., Thaiss, C. A., Serafini, N., Winter, D. R., Giladi, A., Lara-Astiaso, D., et al. (2016). The spectrum and regulatory landscape of intestinal innate lymphoid cells are shaped by the microbiome. *Cell, 166*(5), 1231–1246. e1213.

Hanski, I., von Hertzen, L., Fyhrquist, N., Koskinen, K., Torppa, K., Laatikainen, T., et al. (2012). Environmental biodiversity, human microbiota, and allergy are interrelated. *Proceedings of the National Academy of Sciences of the United States of America, 109*(21), 8334–8339.

Haroon, E., Fleischer, C. C., Felger, J. C., Chen, X., Woolwine, B. J., Patel, T., et al. (2016). Conceptual convergence: Increased inflammation is associated with increased basal ganglia glutamate in patients with major depression. *Molecular Psychiatry, 21*(10), 1351–1357.

Harrison, N. A., Brydon, L., Walker, C., Gray, M. A., Steptoe, A., & Critchley, H. D. (2009). Inflammation causes mood changes through alterations in subgenual cingulate activity and mesolimbic connectivity. *Biological Psychiatry, 66*(5), 407–414.

Hayakawa, M., Asahara, T., Henzan, N., Murakami, H., Yamamoto, H., Mukai, N., et al. (2011). Dramatic changes of the gut flora immediately after severe and sudden insults. *Digestive Diseases and Sciences, 56*(8), 2361–2365.

Henao-Mejia, J., Elinav, E., Jin, C., Hao, L., Mehal, W. Z., Strowig, T., et al. (2012). Inflammasome-mediated dysbiosis regulates progression of NAFLD and obesity. *Nature, 482*(7384), 179–185.

Hoban, A. E., Stilling, R. M., Ryan, F. J., Shanahan, F., Dinan, T. G., Claesson, M. J., et al. (2016). Regulation of prefrontal cortex myelination by the microbiota. *Translational Psychiatry, 6*.

Hodes, G. E., Menard, C., & Russo, S. J. (2016). Integrating Interleukin-6 into depression diagnosis and treatment. *Neurobiology of Stress, 4*, 15–22.

Hodes, G. E., Pfau, M. L., Leboeuf, M., Golden, S. A., Christoffel, D. J., Bregman, D., et al. (2014). Individual differences in the peripheral immune system promote resilience versus susceptibility to social stress. *Proceedings of the National Academy of Sciences of the United States of America, 111*(45), 16136–16141.

Hoisington, A. J., Brenner, L. A., Kinney, K. A., Postolache, T. T., & Lowry, C. A. (2015). The microbiome of the built environment and mental health. *Microbiome, 3*, 60.

Hong, H. A., Khaneja, R., Tam, N. M., Cazzato, A., Tan, S., Urdaci, M., et al. (2009). *Bacillus subtilis* isolated from the human gastrointestinal tract. *Research in Microbiology, 160*(2), 134–143.

Hong, H. A., To, E., Fakhry, S., Baccigalupi, L., Ricca, E., & Cutting, S. M. (2009). Defining the natural habitat of *Bacillus* spore-formers. *Research in Microbiology, 160*(6), 375–379.

Hsiao, E. Y., McBride, S. W., Hsien, S., Sharon, G., Hyde, E. R., McCue, T., et al. (2013). Microbiota modulate behavioral and physiological abnormalities associated with neurodevelopmental disorders. *Cell, 155*(7), 1451–1463.

Huang, R., Wang, K., & Hu, J. (2016). Effect of probiotics on depression: A systematic review and meta-analysis of randomized controlled trials. *Nutrients, 8*(8), 483. https://doi.org/10.3390/nu8080483.

Hussain, K., Letley, D. P., Greenaway, A. B., Kenefeck, R., Winter, J. A., Tomlinson, W., et al. (2016). *Helicobacter pylori*-mediated protection from allergy is associated with IL-10-secreting peripheral blood regulatory T cells. *Frontiers in Immunology, 7*, 71.

Illum, L. (2004). Is nose-to-brain transport of drugs in man a reality? *The Journal of Pharmacy and Pharmacology, 56*(1), 3–17.

Iyer, L. M., Aravind, L., Coon, S. L., Klein, D. C., & Koonin, E. V. (2004). Evolution of cell-cell signaling in animals: Did late horizontal gene transfer from bacteria have a role? *Trends in Genetics, 20*(7), 292–299.

Jacka, F. N., O'Neil, A., Opie, R., Itsiopoulos, C., Cotton, S., Mohebbi, M., et al. (2017). A randomised controlled trial of dietary improvement for adults with major depression (the 'SMILES' trial). *BMC Medicine, 15*(1), 23.

Jackson, J. R., Eaton, W. W., Cascella, N. G., Fasano, A., & Kelly, D. L. (2012). Neurologic and psychiatric manifestations of celiac disease and gluten sensitivity. *The Psychiatric Quarterly, 83*(1), 91–102.

Jiang, H., Ling, Z., Zhang, Y., Mao, H., Ma, Z., Yin, Y., et al. (2015). Altered fecal microbiota composition in patients with major depressive disorder. *Brain, Behavior, and Immunity, 48*, 186–194.

Jost, T., Lacroix, C., Braegger, C. P., Rochat, F., & Chassard, C. (2014). Vertical mother-neonate transfer of maternal gut bacteria via breastfeeding. *Environmental Microbiology, 16*(9), 2891–2904.

Kaliannan, K., Wang, B., Li, X. Y., Kim, K. J., & Kang, J. X. (2015). A host-microbiome interaction mediates the opposing effects of omega-6 and omega-3 fatty acids on metabolic endotoxemia. *Scientific Reports, 5*.

Kankaanranta, H., Kauppi, P., Tuomisto, L. E., & Ilmarinen, P. (2016). Emerging comorbidities in adult asthma: Risks, clinical associations, and mechanisms. *Mediators of Inflammation*, 1–23. https://doi.org/10.1155/2016/3690628.

Kappelmann, N., Lewis, G., Dantzer, R., Jones, P. B., & Khandaker, G. M. (2018). Antidepressant activity of anti-cytokine treatment: A systematic review and meta-analysis of clinical trials of chronic inflammatory conditions. *Molecular Psychiatry, 23*(2), 335–343.

Khan, M. T., Duncan, S. H., Stams, A. J., van Dijl, J. M., Flint, H. J., & Harmsen, H. J. (2012). The gut anaerobe *Faecalibacterium prausnitzii* uses an extracellular electron shuttle to grow at oxic-anoxic interphases. *The ISME Journal, 6*(8), 1578–1585.

Khandaker, G. M., Pearson, R. M., Zammit, S., Lewis, G., & Jones, P. B. (2014). Association of serum interleukin 6 and C-reactive protein in childhood with depression and psychosis in young adult life: A population-based longitudinal study. *JAMA Psychiatry, 71*, 1121–1128.

Kiliaan, A. J., Saunders, P. R., Bijlsma, P. B., Berin, M. C., Taminiau, J. A., Groot, J. A., et al. (1998). Stress stimulates transepithelial macromolecular uptake in rat jejunum. *The American Journal of Physiology, 275*(5 Pt 1), G1037–1044.

Kondrashova, A., Reunanen, A., Romanov, A., Karvonen, A., Viskari, H., Vesikari, T., et al. (2005). A six-fold gradient in the incidence of type 1 diabetes at the eastern border of Finland. *Annals of Medicine, 37*(1), 67–72.

Korpela, K., Salonen, A., Virta, L. J., Kekkonen, R. A., Forslund, K., Bork, P., et al. (2016). Intestinal microbiome is related to lifetime antibiotic use in Finnish pre-school children. *Nature Communications, 7*.

Kostic, A. D., Howitt, M. R., & Garrett, W. S. (2013). Exploring host-microbiota interactions in animal models and humans. *Genes & Development, 27*(7), 701–718.

Kramer-Albers, E. M., & Hill, A. F. (2016). Extracellular vesicles: Interneural shuttles of complex messages. *Current Opinion in Neurobiology, 39*, 101–107.

Lamas, B., Richard, M. L., Leducq, V., Pham, H. P., Michel, M. L., Da Costa, G., et al. (2016). CARD9 impacts colitis by altering gut microbiota metabolism of tryptophan into aryl hydrocarbon receptor ligands. *Nature Medicine, 22*(6), 598–605.

Lapin, B., Piorkowski, J., Ownby, D., Freels, S., Chavez, N., Hernandez, E., et al. (2015). Relationship between prenatal antibiotic use and asthma in at-risk children. *Annals of Allergy, Asthma & Immunology, 114*(3), 203–207.

Levy, M., Thaiss, C. A., & Elinav, E. (2016). Metabolites: Messengers between the microbiota and the immune system. *Genes & Development, 30*(14), 1589–1597.

Levy, M., Thaiss, C. A., Zeevi, D., Dohnalova, L., Zilberman-Schapira, G., Mahdi, J. A., et al. (2015). Microbiota-modulated metabolites shape the intestinal microenvironment by regulating NLRP6 inflammasome signaling. *Cell, 163*(6), 1428–1443.

Linz, B., Balloux, F., Moodley, Y., Manica, A., Liu, H., Roumagnac, P., et al. (2007). An African origin for the intimate association between humans and *Helicobacter pylori*. *Nature, 445*(7130), 915–918.

Logan, A. C. (2015). Dysbiotic drift: Mental health, environmental grey space, and microbiota. *Journal of Physiological Anthropology, 34*, 23.

Loschko, J., Schreiber, H. A., Rieke, G. J., Esterhazy, D., Meredith, M. M., Pedicord, V. A., et al. (2016). Absence of MHC class II on cDCs results in microbial-dependent intestinal inflammation. *Journal of Experimental Medicine, 213*(4), 517–534.

Lowry, C. A., Hollis, J. H., de Vries, A., Pan, B., Brunet, L. R., Hunt, J. R., et al. (2007). Identification of an immune-responsive mesolimbocortical serotonergic system: Potential role in regulation of emotional behavior. *Neuroscience, 146*(2), 756–772.

Lowry, C. A., Smith, D. G., Siebler, P. H., Schmidt, D., Stamper, C. E., Hassell, J. E., Jr., et al. (2016). The microbiota, immunoregulation, and mental health: Implications for public health. *Current Environmental Health Reports, 3*(3), 270–286.

Lundgren, A., Stromberg, E., Sjoling, A., Lindholm, C., Enarsson, K., Edebo, A., et al. (2005). Mucosal FOXP3-expressing CD4+ CD25high regulatory T cells in *Helicobacter pylori*-infected patients. *Infection and Immunity, 73*(1), 523–531.

Maas, J., Verheij, R. A., Groenewegen, P. P., de Vries, S., & Spreeuwenberg, P. (2006). Green space, urbanity, and health: How strong is the relation? *Journal of Epidemiology and Community Health, 60*(7), 587–592.

Maes, M. (1999). Major depression and activation of the inflammatory response system. *Advances in Experimental Medicine and Biology, 461*, 25–46.

McDade, T. W., Tallman, P. S., Madimenos, F. C., Liebert, M. A., Cepon, T. J., Sugiyama, L. S., et al. (2012). Analysis of variability of high sensitivity C-reactive protein in lowland ecuador reveals no evidence of chronic low-grade inflammation. *American Journal of Human Biology, 24*(5), 675–681.

McFall-Ngai, M., Hadfield, M. G., Bosch, T. C., Carey, H. V., Domazet-Loso, T., Douglas, A. E., et al. (2013). Animals in a bacterial world, a new imperative for the life sciences. *Proceedings of the National Academy of Sciences of the United States of America, 110*(9), 3229–3236.

McKeever, T. M., Lewis, S. A., Smith, C., & Hubbard, R. (2002). The importance of prenatal exposures on the development of allergic disease: A birth cohort study using the West Midlands General Practice Database. *American Journal of Respiratory and Critical Care Medicine, 166*(6), 827–832.

Melnik, B. C., John, S. M., Carrera-Bastos, P., & Schmitz, G. (2016). Milk: A postnatal imprinting system stabilizing FoxP3 expression and regulatory T cell differentiation. *Clinical and Translational Allergy, 6*, 18.

Meropol, S. B., & Edwards, A. (2015). Development of the infant intestinal microbiome: A bird's eye view of a complex process. Birth Defects Research. Part C, Embryo Today, *105*(4), 228–239.

Metsala, J., Lundqvist, A., Virta, L. J., Kaila, M., Gissler, M., & Virtanen, S. M. (2013). Mother's and offspring's use of antibiotics and infant allergy to cow's milk. *Epidemiology, 24*(2), 303–309.

Miller, A. H., & Raison, C. L. (2016). The role of inflammation in depression: From evolutionary imperative to modern treatment target. *Nature Reviews Immunology, 16*(1), 22–34.

Mitchell, R., & Popham, F. (2008). Effect of exposure to natural environment on health inequalities: An observational population study. *Lancet, 372*(9650), 1655–1660.

Möhle, L., Mattei, D., Heimesaat, M. M., Bereswill, S., Fischer, A., Alutis, M., et al. (2016). Ly6C(hi) monocytes provide a link between antibiotic-induced changes in gut microbiota and adult hippocampal neurogenesis. *Cell Reports, 15*(9), 1945–1956.

Moieni, M., Irwin, M. R., Jevtic, I., Olmstead, R., Breen, E. C., & Eisenberger, N. I. (2015). Sex differences in depressive and socioemotional responses to an inflammatory challenge: Implications for sex differences in depression. *Neuropsychopharmacology, 40*(7), 1709–1716.

Moore, M. N. (2015). Do airborne biogenic chemicals interact with the PI3K/Akt/mTOR cell signalling pathway to benefit human health and wellbeing in rural and coastal environments? *Environmental Research, 140*, 65–75.

Mpairwe, H., Webb, E. L., Muhangi, L., Ndibazza, J., Akishule, D., Nampijja, M., et al. (2011). Anthelminthic treatment during pregnancy is associated with increased risk of infantile eczema: Randomised-controlled trial results. *Pediatric Allergy and Immunology, 22*(3), 305–312.

Mulder, I. E., Schmidt, B., Stokes, C. R., Lewis, M., Bailey, M., Aminov, R. I., et al. (2009). Environmentally-acquired bacteria influence microbial diversity and natural innate immune responses at gut surfaces. *BMC Biology, 7*, 79.

Naseribafrouei, A., Hestad, K., Avershina, E., Sekelja, M., Linlokken, A., Wilson, R., et al. (2014). Correlation between the human fecal microbiota and depression. *Neurogastroenterology and Motility, 26*(8), 1155–1162.

Newton, R. J., McLellan, S. L., Dila, D. K., Vineis, J. H., Morrison, H. G., Eren, A. M., et al. (2015). Sewage reflects the microbiomes of human populations. *MBio, 6*(2).

Nicholson, W. L. (2002). Roles of *Bacillus* endospores in the environment. *Cellular and Molecular Life Sciences, 59*(3), 410–416.

O'Brien, M. E., Anderson, H., Kaukel, E., O'Byrne, K., Pawlicki, M., Von Pawel, J., et al. (2004). SRL172 (killed *Mycobacterium vaccae*) in addition to standard chemotherapy improves quality of life without affecting survival, in patients with advanced non-small-cell lung cancer: Phase III results. *Annals of Oncology, 15*(6), 906–914.

O'Hara, A. M., & Shanahan, F. (2006). The gut flora as a forgotten organ. *EMBO Reports, 7*(7), 688–693.

O'Mahony, S. M., Marchesi, J. R., Scully, P., Codling, C., Ceolho, A. M., Quigley, E. M., et al. (2009). Early life stress alters behavior, immunity, and microbiota in rats: Implications for irritable bowel syndrome and psychiatric illnesses. *Biological Psychiatry, 65*(3), 263–267.

O'Sullivan, O., Cronin, O., Clarke, S. F., Murphy, E. F., Molloy, M. G., Shanahan, F., et al. (2015). Exercise and the microbiota. *Gut Microbes, 6*(2), 131–136.

Obihara, C. C., Beyers, N., Gie, R. P., Potter, P. C., Marais, B. J., Lombard, C. J., et al. (2005). Inverse association between *Mycobacterium tuberculosis* infection and atopic rhinitis in children. *Allergy, 60*(9), 1121–1125.

Obregon-Tito, A. J., Tito, R. Y., Metcalf, J., Sankaranarayanan, K., Clemente, J. C., Ursell, L. K., et al. (2015). Subsistence strategies in traditional societies distinguish gut microbiomes. *Nature Communications, 6*, 6505.

Ownby, D. R., Johnson, C. C., & Peterson, E. L. (2002). Exposure to dogs and cats in the first year of life and risk of allergic sensitization at 6 to 7 years of age. *JAMA, 288*(8), 963–972.

Pace, T. W., Mletzko, T. C., Alagbe, O., Musselman, D. L., Nemeroff, C. B., Miller, A. H., et al. (2006). Increased stress-induced inflammatory responses in male patients with major depression and increased early life stress. *The American Journal of Psychiatry, 163*(9), 1630–1633.

Pakarinen, J., Hyvarinen, A., Salkinoja-Salonen, M., Laitinen, S., Nevalainen, A., Makela, M. J., et al. (2008). Predominance of Gram-positive bacteria in house dust in the low-allergy risk Russian Karelia. *Environmental Microbiology, 10*(12), 3317–3325.

Pancer, Z., & Cooper, M. D. (2006). The evolution of adaptive immunity. *Annual Review of Immunology, 24*, 497–518.

Pathirana, R. D., & Kaparakis-Liaskos, M. (2016). Bacterial membrane vesicles: Biogenesis, immune regulation and pathogenesis. *Cellular Microbiology, 18*(11), 1518–1524.

Peus, V., Redelin, E., Scharnholz, B., Paul, T., Gass, P., Deuschle, P., et al. (2012). Breast-feeding in infancy and major depression in adulthood: A retrospective analysis. *Psychotherapy and Psychosomatics, 81*(3), 189–190.

Pina-Camacho, L., Jensen, S. K., Gaysina, D., & Barker, E. D. (2015). Maternal depression symptoms, unhealthy diet and child emotional-behavioural dysregulation. *Psychological Medicine, 45*(9), 1851–1860.

Poroyko, V. A., Carreras, A., Khalyfa, A., Khalyfa, A. A., Leone, V., Peris, E., et al. (2016). Chronic sleep disruption alters gut microbiota, induces systemic and adipose tissue inflammation and insulin resistance in mice. *Scientific Reports, 6*.

Postolache, T. T., Komarow, H., & Tonelli, L. H. (2008). Allergy: A risk factor for suicide? *Current Treatment Options in Neurology, 10*(5), 363–376.

Qiu, X., Zhang, M., Yang, X., Hong, N., & Yu, C. (2013). *Faecalibacterium prausnitzii* upregulates regulatory T cells and anti-inflammatory cytokines in treating TNBS-induced colitis. *Journal of Crohn's & Colitis, 7*(11), e558–568.

Quintana, F. J., Basso, A. S., Iglesias, A. H., Korn, T., Farez, M. F., Bettelli, E., et al. (2008). Control of T(reg) and T(H)17 cell differentiation by the aryl hydrocarbon receptor. *Nature, 453*(7191), 65–71.

Radon, K., Windstetter, D., Poluda, A. L., Mueller, B., von Mutius, E., & Koletzko, S. (2007). Contact with farm animals in early life and juvenile inflammatory bowel disease: A case-control study. *Pediatrics, 120*(2), 354–361.

Raison, C. L., Lowry, C. A., & Rook, G. A. W. (2010). Inflammation, sanitation and consternation: Loss of contact with co-evolved, tolerogenic micro-organisms and the pathophysiology and treatment of major depression. *Archives of General Psychiatry, 67*(12), 1211–1224.

Raison, C. L., Rutherford, R. E., Woolwine, B. J., Shuo, C., Schettler, P., Drake, D. F., et al. (2013). A randomized controlled trial of the tumor necrosis factor antagonist infliximab for treatment-resistant depression. *JAMA Psychiatry, 70*(1), 31–41.

Rampelli, S., Schnorr, S. L., Consolandi, C., Turroni, S., Severgnini, M., Peano, C., et al. (2015). Metagenome sequencing of the Hadza hunter-gatherer gut microbiota. *Current Biology, 25*(13), 1682–1693.

Rapaport, M. H., Nierenberg, A. A., Schettler, P. J., Kinkead, B., Cardoos, A., Walker, R., et al. (2016). Inflammation as a predictive biomarker for response to omega-3 fatty acids in major depressive disorder: A proof-of-concept study. *Molecular Psychiatry, 21*(1), 71–79.

Reber, S. O., Siebler, P. H., Donner, N. C., Morton, J. T., Smith, D. G., Kopelman, J. M., et al. (2016). Immunization with a heat-killed preparation of the environmental bacterium *Mycobacterium vaccae* promotes stress resilience in mice. *Proceedings of the National Academy of Sciences of the United States of America, 113*(22), E3130–3139.

Regueiro, M., Greer, J. B., & Szigethy, E. (2017). Etiology and treatment of pain and psychosocial issues in patients with inflammatory bowel diseases. *Gastroenterology, 152*(2), 430–439. e434.

Riedler, J., Braun-Fahrlander, C., Eder, W., Schreuer, M., Waser, M., Maisch, S., et al. (2001). Exposure to farming in early life and development of asthma and allergy: A cross-sectional survey. *Lancet, 358*(9288), 1129–1133.

Robinson, K., Kenefeck, R., Pidgeon, E. L., Shakib, S., Patel, S., Polson, R. J., et al. (2008). *Helicobacter pylori*-induced peptic ulcer disease is associated with inadequate regulatory T cell responses. *Gut, 57*(10), 1375–1385.

Rook, G. A. (2013). Regulation of the immune system by biodiversity from the natural environment: An ecosystem service essential to health. *Proceedings of the National Academy of Sciences of the United States of America, 110*(46), 18360–18367.

Rook, G. A., Lowry, C. A., & Raison, C. L. (2015). Hygiene and other early childhood influences on the subsequent function of the immune system. *Brain Research, 1617*, 47–62.

Rook, G. A., Raison, C. L., & Lowry, C. A. (2014). Microbiota, immunoregulatory old friends and psychiatric disorders. In: M. Lyte, & J. F. Cryan (Eds.), *Advances in experimental medicine and biology 817: Vol. 817. Microbial endocrinology: The microbiota-gut-brain axis in health and disease.* New York: Springer.

Rook, G. A. W. (2010). 99th Dahlem conference on infection, inflammation and chronic inflammatory disorders: Darwinian medicine and the 'hygiene' or 'old friends' hypothesis. *Clinical and Experimental Immunology, 160*(1), 70–79.

Rothhammer, V., Mascanfroni, I. D., Bunse, L., Takenaka, M. C., Kenison, J. E., Mayo, L., et al. (2016). Type I interferons and microbial metabolites of tryptophan modulate astrocyte activity and central nervous system inflammation via the aryl hydrocarbon receptor. *Nature Medicine, 22*(6), 586–597.

Round, J. L., & Mazmanian, S. K. (2009). The gut microbiota shapes intestinal immune responses during health and disease. *Nature Reviews Immunology, 9*(5), 313–323.

Round, J. L., & Mazmanian, S. K. (2010). Inducible Foxp3 + regulatory T-cell development by a commensal bacterium of the intestinal microbiota. *Proceedings of the National Academy of Sciences of the United States of America, 107*(27), 12204–12209.

Saarinen, K. M., Vaarala, O., Klemetti, P., & Savilahti, E. (1999). Transforming growth factor-beta1 in mothers' colostrum and immune responses to cows' milk proteins in infants with cows' milk allergy. *The Journal of Allergy and Clinical Immunology, 104*(5), 1093–1098.

Sacker, A., Kelly, Y., Iacovou, M., Cable, N., & Bartley, M. (2013). Breast feeding and intergenerational social mobility: What are the mechanisms? *Archives of Disease in Childhood, 98*(9), 666–671.

Schirmer, M., Smeekens, S. P., Vlamakis, H., Jaeger, M., Oosting, M., Franzosa, E. A., et al. (2016). Linking the human gut microbiome to inflammatory cytokine production capacity. *Cell, 167*(4), 1125–1136. e1128.

Schnorr, S. L., Candela, M., Rampelli, S., Centanni, M., Consolandi, C., Basaglia, G., et al. (2014). Gut microbiome of the Hadza hunter-gatherers. *Nature Communications, 5*, 3654.

Schuijs, M. J., Willart, M. A., Vergote, K., Gras, D., Deswarte, K., Ege, M. J., et al. (2015). Farm dust and endotoxin protect against allergy through A20 induction in lung epithelial cells. *Science, 349*(6252), 1106–1110.

Schulz, O., & Pabst, O. (2013). Antigen sampling in the small intestine. *Trends in Immunology, 34*(4), 155–161.

Sefik, E., Geva-Zatorsky, N., Oh, S., Konnikova, L., Zemmour, D., McGuire, A. M., et al. (2015). MUCOSAL IMMUNOLOGY. Individual intestinal symbionts induce a distinct population of RORgamma(+) regulatory T cells. *Science, 349*(6251), 993–997.

Selhub, E. M., Logan, A. C., & Bested, A. C. (2014). Fermented foods, microbiota, and mental health: Ancient practice meets nutritional psychiatry. *Journal of Physiological Anthropology, 33*, 2.

Sender, R., Fuchs, S., & Milo, R. (2016). Revised estimates for the number of human and bacteria cells in the body. *PLoS Biology, 14*(8).

Setiawan, E., Wilson, A. A., Mizrahi, R., Rusjan, P. M., Miler, L., Rajkowska, G., et al. (2015). Role of translocator protein density, a marker of neuroinflammation, in the brain during major depressive episodes. *JAMA Psychiatry, 72*(3), 268.

Slykerman, R. F., Thompson, J., Waldie, K. E., Murphy, R., Wall, C., & Mitchell, E. A. (2017). Antibiotics in the first year of life and subsequent neurocognitive outcomes. *Acta Paediatrica, 106*(1), 87–94.

Soderborg, T. K., Borengasser, S. J., Barbour, L. A., & Friedman, J. E. (2016). Microbial transmission from mothers with obesity or diabetes to infants: An innovative opportunity to interrupt a vicious cycle. *Diabetologia, 59*(5), 895–906.

Sokol, H., Pigneur, B., Watterlot, L., Lakhdari, O., Bermudez-Humaran, L. G., Gratadoux, J. J., et al. (2008). *Faecalibacterium prausnitzii* is an anti-inflammatory commensal bacterium identified by gut microbiota analysis of Crohn disease patients. *Proceedings of the National Academy of Sciences of the United States of America, 105*(43), 16731–16736.

Sonnenburg, E. D., Smits, S. A., Tikhonov, M., Higginbottom, S. K., Wingreen, N. S., & Sonnenburg, J. L. (2016). Diet-induced extinctions in the gut microbiota compound over generations. *Nature, 529*(7585), 212–215.

Sozanska, B., Blaszczyk, M., Pearce, N., & Cullinan, P. (2013). Atopy and allergic respiratory disease in rural Poland before and after accession to the European Union. *The Journal of Allergy and Clinical Immunology, 133*(5), 1347–1353.

Stamper, C. E., Hoisington, A. J., Gomez, O. M., Halweg-Edwards, A. L., Smith, D. G., Bates, K. L., et al. (2016). The microbiome of the built environment and human behavior: Implications for emotional health and well-being in postmodern Western societies. *International Review of Neurobiology, 131*, 289–323.

Steenbergen, L., Sellaro, R., van Hemert, S., Bosch, J. A., & Colzato, L. S. (2015). A randomized controlled trial to test the effect of multispecies probiotics on cognitive reactivity to sad mood. *Brain, Behavior, and Immunity, 48*, 258–264.

Stein, M. M., Hrusch, C. L., Gozdz, J., Igartua, C., Pivniouk, V., Murray, S. E., et al. (2016). Innate immunity and asthma risk in Amish and Hutterite farm children. *The New England Journal of Medicine, 375*(5), 411–421.

Strachan, D. P. (1989). Hay fever, hygiene, and household size. *British Medical Journal, 299*(6710), 1259–1260.

Sudo, N., Chida, Y., Aiba, Y., Sonoda, J., Oyama, N., Yu, X. N., et al. (2004). Postnatal microbial colonization programs the hypothalamic-pituitary-adrenal system for stress response in mice. *The Journal of Physiology, 558*(Pt 1), 263–275.

Tam, N. K., Uyen, N. Q., Hong, H. A., Duc le, H., Hoa, T. T., Serra, C. R., et al. (2006). The intestinal life cycle of *Bacillus subtilis* and close relatives. *Journal of Bacteriology, 188*(7), 2692–2700.

Tan, J., McKenzie, C., Vuillermin, P. J., Goverse, G., Vinuesa, C. G., Mebius, R. E., et al. (2016). Dietary fiber and bacterial SCFA enhance oral tolerance and protect against food allergy through diverse cellular pathways. *Cell Reports, 15*(12), 2809–2824.

Thalmann, O., Shapiro, B., Cui, P., Schuenemann, V. J., Sawyer, S. K., Greenfield, D. L., et al. (2013). Complete mitochondrial genomes of ancient canids suggest a European origin of domestic dogs. *Science, 342*(6160), 871–874.

Thompson, R. S., Roller, R., Mika, A., Greenwood, B. N., Knight, R., Chichlowski, M., et al. (2016). Dietary prebiotics and bioactive milk fractions improve NREM sleep, enhance REM sleep rebound and attenuate the stress-induced decrease in diurnal temperature and gut microbial alpha diversity. *Frontiers in Behavioral Neuroscience, 10*, 240.

Tillisch, K., Labus, J., Kilpatrick, L., Jiang, Z., Stains, J., Ebrat, B., et al. (2013). Consumption of fermented milk product with probiotic modulates brain activity. *Gastroenterology, 144*(7), 1394–1401. pp. 1401. e1391-1394.

Timonen, M., Jokelainen, J., Hakko, H., Silvennoinen-Kassinen, S., Meyer-Rochow, V. B., Herva, A., et al. (2003). Atopy and depression: Results from the Northern Finland 1966 Birth Cohort Study. *Molecular Psychiatry, 8*(8), 738–744.

Trasande, L., Blustein, J., Liu, M., Corwin, E., Cox, L. M., & Blaser, M. J. (2013). Infant antibiotic exposures and early-life body mass. *International Journal of Obesity, 37*(1), 16–23.

Vanamala, J. K., Knight, R., & Spector, T. D. (2015). Can your microbiome tell you what to eat? *Cell Metabolism, 22*(6), 960–961.

Vatanen, T., Kostic, A. D., d'Hennezel, E., Siljander, H., Franzosa, E. A., Yassour, M., et al. (2016). Variation in microbiome LPS immunogenicity contributes to autoimmunity in humans. *Cell, 165*(4), 842–853.

Vega, W. A., Sribney, W. M., Aguilar-Gaxiola, S., & Kolody, B. (2004). 12-Month prevalence of DSM-III-R psychiatric disorders among Mexican Americans: Nativity, social assimilation, and age determinants. *The Journal of Nervous and Mental Disease, 192*(8), 532–541.

Vijay-Kumar, M., Aitken, J. D., Carvalho, F. A., Cullender, T. C., Mwangi, S., Srinivasan, S., et al. (2010). Metabolic syndrome and altered gut microbiota in mice lacking Toll-like receptor 5. *Science, 328*(5975), 228–231.

von Hertzen, L., Hanski, I., & Haahtela, T. (2011). Natural immunity. Biodiversity loss and inflammatory diseases are two global megatrends that might be related. *EMBO Reports, 12*(11), 1089–1093.

Wamboldt, M. Z., Hewitt, J. K., Schmitz, S., Wamboldt, F. S., Rasanen, M., Koskenvuo, M., et al. (2000). Familial association between allergic disorders and depression in adult Finnish twins. *American Journal of Medical Genetics, 96*(2), 146–153.

Wang, J., Cao, H., Wang, H., Yin, G., Du, J., Xia, F., et al. (2015). Multiple mechanisms involved in diabetes protection by lipopolysaccharide in non-obese diabetic mice. *Toxicology and Applied Pharmacology, 285*(3), 149–158.

Wertheim, J. O., & Kosakovsky Pond, S. L. (2011). Purifying selection can obscure the ancient age of viral lineages. *Molecular Biology and Evolution, 28*(12), 3355–3365.

Wheeler, B. W., White, M., Stahl-Timmins, W., & Depledge, M. H. (2012). Does living by the coast improve health and wellbeing? *Health & Place, 18*(5), 1198–1201.

Wikoff, W. R., Anfora, A. T., Liu, J., Schultz, P. G., Lesley, S. A., Peters, E. C., et al. (2009). Metabolomics analysis reveals large effects of gut microflora on mammalian blood metabolites. *Proceedings of the National Academy of Sciences of the United States of America, 106*(10), 3698–3703.

Williams, M. J., Williams, S. M., & Poulton, R. (2006). Breast feeding is related to C reactive protein concentration in adult women. *Journal of Epidemiology and Community Health, 60*(2), 146–148.

Wolfe, N. D., Dunavan, C. P., & Diamond, J. (2007). Origins of major human infectious diseases. *Nature, 447*(7142), 279–283.

Yoo, J., Tcheurekdjian, H., Lynch, S. V., Cabana, M., & Boushey, H. A. (2007). Microbial manipulation of immune function for asthma prevention: Inferences from clinical trials. *Proceedings of the American Thoracic Society, 4* (3), 277–282.

Yuan, C., Gaskins, A. J., Blaine, A. I., Zhang, C., Gillman, M. W., Missmer, S. A., et al. (2016). Association between cesarean birth and risk of obesity in offspring in childhood, adolescence, and early adulthood. *JAMA Pediatrics, 170*(11), e162385.

Zaccone, P., Burton, O., Miller, N., Jones, F. M., Dunne, D. W., & Cooke, A. (2009). *Schistosoma mansoni* egg antigens induce Treg that participate in diabetes prevention in NOD mice. *European Journal of Immunology, 39*(4), 1098–1107.

Zelante, T., Iannitti, R. G., Fallarino, F., Gargaro, M., De Luca, A., Moretti, S., et al. (2014). Tryptophan feeding of the IDO1-AhR axis in host-microbial symbiosis. *Frontiers in Immunology, 5*, 640.

Zheng, P., Zeng, B., Zhou, C., Liu, M., Fang, Z., Xu, X., et al. (2016). Gut microbiome remodeling induces depressive-like behaviors through a pathway mediated by the host's metabolism. *Molecular Psychiatry, 21*(6), 786–796.

Zhou, Q., Wang, H., Schwartz, D. M., Stoffels, M., Park, Y. H., Zhang, Y., et al. (2016). Loss-of-function mutations in TNFAIP3 leading to A20 haploinsufficiency cause an early-onset autoinflammatory disease. *Nature Genetics, 48*(1), 67–73.

Zhuang, X., Xiang, X., Grizzle, W., Sun, D., Zhang, S., Axtell, R. C., et al. (2011). Treatment of brain inflammatory diseases by delivering exosome encapsulated anti-inflammatory drugs from the nasal region to the brain. *Molecular Therapy, 19*(10), 1769–1779.

Zijlmans, M. A., Korpela, K., Riksen-Walraven, J. M., de Vos, W. M., & de Weerth, C. (2015). Maternal prenatal stress is associated with the infant intestinal microbiota. *Psychoneuroendocrinology, 53*, 233–245.

Zivkovic, A. M., German, J. B., Lebrilla, C. B., & Mills, D. A. (2011). Human milk glycobiome and its impact on the infant gastrointestinal microbiota. *Proceedings of the National Academy of Sciences of the United States of America, 108*(Suppl. 1), 4653–4658.

Neuroendocrine Abnormalities in Major Depression: An Insight Into Glucocorticoids, Cytokines, and the Kynurenine Pathway

Naghmeh Nikkheslat, Carmine M. Pariante, Patricia A. Zunszain

King's College London, London, United Kingdom

INTRODUCTION

Major depressive disorder (MDD) is a debilitating condition and the most common mental illness affecting more than 300 million people worldwide (WHO, 2017). Despite the availability of numerous antidepressants, one-third of depressed patients do not respond to conventional treatments (Ferrari et al., 2013). Depression is heterogeneous in nature with multifactorial etiology resulting from complex interactions between genetic, psychological, and physiological factors.

Various theories have been developed to elucidate the pathophysiology of depression. As one of the oldest and major theories of depression, the monoamine-deficiency hypothesis claims depletion in monoamine neurotransmitters serotonin and norepinephrine in the brain as an etiologic feature of depression (Schildkraut, 1965). While the genetic approach

considers the heritability of depression (Levinson, 2006), interactions between environmental factors and major life stressors with genetic vulnerability have been suggested as stronger contributing factors in the development of this disorder (Firk & Markus, 2007). With growing advances in psychoneuroimmunology research, attention has been paid to abnormal brain-endocrine-immune interactions and alterations in inflammatory responses involved in the pathogenesis of depression. Considering endocrine aspects, the hypothalamic-pituitary-adrenal (HPA) axis hypothesis postulates dysregulation of the HPA axis and alteration in glucocorticoid response to stress being implicated in depression prior to monoamine abnormalities (Dinan, 1994). Focusing on immune aspects, the macrophage theory and cytokine hypothesis of depression postulate a critical role of excessive secretion of macrophage monokines (Smith, 1991) and elevated levels of proinflammatory cytokines and their action as neuromodulators (Schiepers, Wichers, & Maes, 2005) in depressed individuals. A more recent theory of depression considers the stress-induced reductions in neurogenesis as an important causal factor in precipitating episodes of depression (Jacobs, van Praag, & Gage, 2000). These latest theories have led to the inflammatory and neurodegenerative hypothesis, proposing that inflammatory processes lead to diminished neurogenesis and increased neurodegeneration in the brain of patients suffering from major depression (Maes et al., 2009).

Considering the hypotheses above, this chapter aims to explore the pathophysiological mechanisms associated with neuroendocrine abnormalities leading to inflammation and neurodegeneration in patients with depression by focusing on glucocorticoids and cytokine dysregulations as well as kynurenine pathway imbalances.

THE HPA AXIS AND GLUCOCORTICOID ABNORMALITIES IN MAJOR DEPRESSION

Alteration in HPA axis function has been one of the most consistent observations in MDD patients (Pariante & Lightman, 2008; Stetler & Miller, 2011). As a main physiological circuit connecting the brain with the endocrine system, the HPA axis is the central regulatory system in the control of stress responses. The HPA axis responds to a variety of stress stimuli including environmental, psychological, or physical challenges by secreting corticotropin-releasing hormone (CRH) and arginine vasopressin (AVP) from the hypothalamic paraventricular nucleus (PVN). The released neuropeptides stimulate the secretion of adrenocorticotropic hormone (ACTH) from the pituitary glands into the blood circulation, and this hormone then targets the adrenal cortex for the biosynthesis and release of glucocorticoids (Fig. 1).

Synthesized from cholesterol, the glucocorticoids cortisol (in humans) and corticosterone (in rodents) are the final products of the HPA axis. These corticosteroid hormones play a fundamental role in modulation and regulation of energy metabolism; stress-related homeostasis; cardiovascular function; and neuroendocrine, immune, and inflammatory responses. The widespread effects of the endogenous cortisol are governed by the interaction with their specific receptors known as glucocorticoid receptors (GR), which are expressed in almost all cells throughout the body including various regions of the brain and the HPA axis. Indeed, the intrinsic homeostasis of the HPA axis activity is maintained by negative feedback mechanisms mediated by GRs found at both hypothalamus and pituitary levels that, upon activation by increased concentrations of cortisol, inhibit the release of CRH, AVP, and ACTH (Fig. 1), ultimately leading to a decrease in cortisol production. This feedback mechanism can fail. Indeed, an inappropriate increase in cortisol levels has been found in a significant proportion of depressed patients, reflecting a dysregulated HPA axis function (Pariante & Nemeroff, 2012).

How scientists could explain alterations of HPA axis activity and overproduction of cortisol in depression will be addressed by introducing the phenomenon of "glucocorticoid resistance."

Glucocorticoid Resistance

In a subset of patients, depression is associated with hyperactivity of the HPA axis, with enlargement of pituitary and adrenal glands and hypercortisolism, which has been found to be correlated with cognitive deficits (Hinkelmann et al., 2009). As mentioned above and claimed by the glucocorticoid theory of depression, chronic stress, abnormal activation of the HPA axis, and excess cortisol largely contribute to the neuropathology of MDD. This theory has been linked to the hypotheses on monoamine dysfunction, diminished neurogenesis,

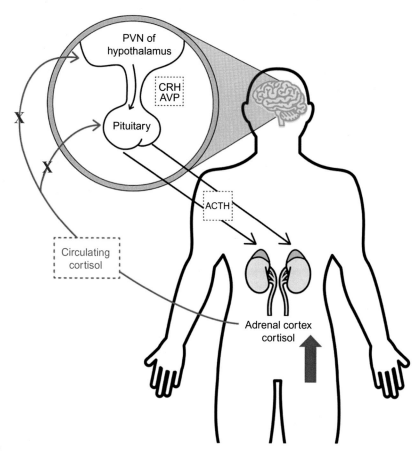

FIG. 1 HPA axis and glucocorticoid-mediated negative feedback regulation. *ACTH*, adrenocorticotropic hormone; *AVP*, arginine vasopressin; *CRH*, corticotropin-releasing hormone; *PVN*, hypothalamic paraventricular nucleus.

synaptic neuroplasticity, increased neurodegeneration, and regional brain alterations associated with MDD (Brown, McIntyre, Rosenblat, & Hardeland, 2017; Herbert et al., 2006; Holsboer, 2000).

Elevated cortisol concentrations have been found in cerebrospinal fluids (CSF), peripheral blood, saliva, and urine of patients with depression (Pariante & Nemeroff, 2012). Evidence suggests that failure of the HPA system in regulating cortisol production could be largely due to dysfunctionality of the GR and so impaired ability of cortisol to exert its feedback inhibition effect. This reduced sensitivity of the GR to

the effect of the hormone is known as "glucocorticoid resistance" and has been extensively reported in major depression (Pariante & Lightman, 2008; Raison & Miller, 2003; Stetler & Miller, 2011). The concept has been further explored by observations reporting the direct modulation of the GR in response to the effective antidepressants and normalization of the HPA axis disturbance associated with depression (Anacker, Zunszain, Carvalho, & Pariante, 2011; Carvalho et al., 2008; Carvalho & Pariante, 2008; Nikkheslat, Zunszain, Carvalho, Anacker, & Pariante, 2017; Pariante & Miller, 2001). While depression has been associated with GR dysfunction, cortisol primarily acts on the mineralocorticoid receptor (MR), with almost 10-fold higher affinity compared with the GR. The MR is expressed only in some parts of the body including the heart, kidneys, intestine, and limbic regions of the brain and is occupied by basal levels of endogenous corticosteroids. The MR is involved in the regulation of the HPA axis activity and cortisol circadian rhythm and is highly expressed and colocalized with the GR in the hippocampus, where both receptors operate in a complementary manner. While the GR shows lower affinity to cortisol, it becomes progressively occupied when the hormone concentrations are high, as observed after awakening when this receptor actively operates the modulation of cortisol peak morning responses. Similarly, during stressful conditions and following a significant increase in the production of glucocorticoids, the GR becomes critically responsible for the regulation of stress-related responses and plays a key role in stress copying and recovery processes (de Kloet, 2014; Pariante & Lightman, 2008). Therefore, studies on stress, endocrine activity, and associated psychiatric disorders have paid more attention into the GR due to its vital role in stress reactivity, and the glucocorticoid resistance in depression has been mostly studied in association with GR impairment, which will be further discussed in this chapter. However, this does not lessen the importance of the MR involvement in the control of stress responses, playing important roles in prevention and adaptation processes. Indeed, increasing evidence suggests that an imbalance in MR/GR could affect the initiation and management of the responses to stress that in turn could partly explain the enhanced vulnerability to stress-related conditions and the neurobiological disturbances associated with mental disorders (de Kloet, 2014; de Kloet, Derijk, & Meijer, 2007; ter Heegde, De Rijk, & Vinkers, 2015).

An appropriate and efficient cortisol response to stress via GR-dependent pathway is determined by not only GR affinity but also the expression, number, and functional properties of the receptor. The unbound GR resides predominantly in the cytoplasm of the cell where it is found in its inactive state stabilized and packaged within an assembly of chaperone proteins including heat-shock proteins and immunophilins. In response to a stressor and activation of the HPA axis, the concentration of cortisol increases, and after passive diffusion across the cell membrane, it binds to the cytosolic GR. As a ligand-dependent transcription factor, the GR gets activated upon binding to glucocorticoids. Following a series of events including the receptor conformational changes, dissociation from the multiprotein complex, and dimerization, GR translocation into the nucleolus is facilitated. In the nucleolus, the GR regulates gene expression through transactivation or transrepression activities. The GR homodimer performs its transactivation activity by positively altering the expression of target genes through direct binding to the highly conserved DNA glucocorticoid response elements. On the other hand, the GR translocation as a monomer enables its binding to other transcription factors that interact with their respective DNA response elements, which ultimately results in suppression of gene transcription. In fact, the transrepression activity of the GR is an indirect negative alteration of gene expression. The GR translocation and transcriptional activities may also be modulated by the phosphorylation state of the receptor required for glucocorticoid binding. Therefore, the GR ligand binding, transformation, phosphorylation, and translocation abilities as well as interaction with other transduction pathways are all crucial contributing factors for the efficacy of the receptor function in the regulation of gene expression (Anacker et al., 2011; Egeland, Zunszain, & Pariante, 2015; Pace & Miller, 2009).

The HPA axis abnormalities and the inappropriate dysregulated and inefficient cortisol elevation in MDD have prompted scientists to investigate molecular mechanisms associated with GR dysfunction in cellular and animal models of depression as well as human subjects. Alteration in glucocorticoid-mediated negative feedback inhibition has been observed due to acquired GR deficiency in the hippocampus (Boyle et al., 2005) and pituitary (Schmidt et al., 2009) in mice presenting HPA axis hyperactivity and depressive-like behavior. Earlier studies have shown a positive effect of antidepressant treatment in increasing GR expression in rat neuronal hippocampal cell cultures (Hery, Semont, Fache, Faudon, & Hery, 2000; Peiffer, Veilleux, & Barden, 1991). Similarly, an increase in hippocampal and hypothalamic GR transcript in association with reduced HPA axis reactivity has been observed in rats following escitalopram treatment (Flandreau et al., 2013). Human post-mortem studies have revealed decreased GR mRNA levels in the brain of depressed patients (Webster, Knable, O'Grady, Orthmann, & Weickert, 2002) and in suicide victims with a history of childhood abuse (McGowan et al., 2009). Interestingly, elevated GR expression levels were found in major depressed patients who were on antidepressant medications until death (Wang et al., 2014).

Evaluating the function of the GR rather than its simple quantification provides a better understanding of the GR-dependent glucocorticoid effects on HPA axis activity. Indirect examination of the GR function to evaluate the HPA axis response can be determined by *in vivo* dexamethasone (DEX) suppression test (DST), which utilizes the specificity of the synthetic glucocorticoid DEX to the GR. Following oral administration, DEX promotes inhibition of cortisol production through negative feedback mechanisms in healthy individuals (Hayes & Ettigi, 1983). A refined test with more specificity to depression is the combined DEX/CRH suppression test that measures GR-mediated feedback inhibition and HPA axis function by assessing suppression of ACTH to the CRH challenge following DEX pretreatment (Deuschle et al., 1998). Impaired cortisol suppression by DEX has been found in patients with MDD that reflects alteration in GR-mediated negative feedback (Juruena et al., 2006; Kinoshita, Kanazawa, Kikuyama, & Yoneda, 2016; Kunugi et al., 2006; Pariante & Miller, 2001).

The glucocorticoid resistance is not limited to the HPA axis, but to other tissues such as skin (Fitzgerald et al., 2006) and peripheral blood cells (Nikkheslat et al., 2015), which may also show impairment in GR function and a resultant glucocorticoid malfunction. DEX administration in healthy controls significantly suppressed the mitogen-induced lymphocyte proliferation and IL-1β production, but not in depressed patients (Maes et al., 1991). Similarly, following DEX administration, an increase in the number of neutrophils and a decrease in the number of lymphocytes were found in healthy individuals in contrast to the depressed nonsuppressors who showed resistance to the effect (Maes, Meltzer, Stevens, Cosyns, & Blockx, 1994). Indeed, the physiological effects of cortisol on the regulation of immune responses can be affected by alteration in glucocorticoid responsiveness. Cortisol is known as the most potent antiinflammatory hormone. Under normal conditions, when the genes associated with inflammatory cytokines are targeted, cortisol exerts its antiinflammatory response through the gene repressing property of the GR. However, in depression, despite increased levels of the stress hormone, some patients exhibit the elevation of inflammation similar to what is observed in chronic inflammatory diseases. The presence of these two opposite phenomena in MDD may be explained by the observations on reduced sensitivity of the GR presented in the peripheral blood cells, less able to respond effectively to the antiinflammatory effect of cortisol (Pariante, 2017). In the next section, we will discuss inflammation associated with depression in more detail.

INFLAMMATION AND CYTOKINE DYSREGULATION IN MAJOR DEPRESSION

Research over the past 20 years has consistently shown that depressed patients exhibit evidence of activated inflammatory responses and significantly high levels of inflammatory biomarkers (Dowlati et al., 2010; Felger & Lotrich, 2013; Howren, Lamkin, & Suls, 2009; Irwin & Miller, 2007; Pariante, 2017). Cytokines such as tumor necrosis factor-α (TNF-α), interleukin 1 (IL-1) and IL-6, and downstream acute-phase reactants more notably C-reactive protein (CRP) are among the most reported increased inflammatory biomarkers in peripheral blood in depression (Dowlati et al., 2010; Howren et al., 2009). Cerebrospinal cytokine elevation has been also observed in patients with major depression and suicide attempters and associated with clinical and symptoms severity (Levine et al., 1999; Lindqvist et al., 2009; Martinez, Garakani, Yehuda, & Gorman, 2012).

Recent investigations suggest that there is a bidirectional association between depression and inflammation and that not only depression is associated with activation of immune and inflammatory responses but also low-grade inflammation predisposes vulnerable individuals for the development of MDD. Indeed, chronic inflammation, a major characteristic and a common feature of inflammation-associated illnesses including cardiovascular disease, arthritis, diabetes, multiple sclerosis, and cancer, has been shown to be associated with a high prevalence of depression (Cowles, Pariante, & Nemeroff, 2009; Halaris, 2013; Moussavi et al., 2007; Nikkheslat et al., 2015; Polsky et al., 2005).

Various mechanisms have been studied to explain the role of cytokines in the pathophysiology of mood disorders and depression (Kim, Na, Myint, & Leonard, 2016). Cytokines influence the neuroendocrine function by having an impact on the HPA axis activity. The effect has been reported to be through stimulation of CRH, ACTH, and cortisol expression and release (Miller, Maletic, & Raison, 2009). Proinflammatory cytokines may generate glucocorticoid resistance by directly affecting the functional capacity of the GR. They may reduce GR ligand and DNA binding abilities, inhibit GR translocation to the nucleus, and influence GR protein-protein interactions. For example,

cytokines activate the mitogen-activated protein kinase-signaling pathway in the cytoplasm leading to phosphorylation of the receptor protein, thus diminishing GR transcriptional activity (Pace, Hu, & Miller, 2007; Raison & Miller, 2003). In addition, proinflammatory cytokines may increase the expression of the β-isoform of the GR, which is a nuclear localized homologous inert isoform of the receptor, as opposed to the cytoplasmic GRα isoform, the classic ligand-binding GR. Excess GRβ leads to the formation of inactive GRα/GRβ heterodimers, thus inhibiting transcriptional activity and decreasing sensitivity of the active GR (Carvalho et al., 2014; Oakley & Cidlowski, 2011).

Cytokines also stimulate acute-phase reactions. Activated acute-phase proteins such as CRP have been shown to be associated with alterations of the central nervous system (CNS) and the most commonly complained symptoms of depressive disorder, insomnia, and fatigue (Irwin & Miller, 2007). CRP has been found to be associated with increased BBB permeability leading to neuroinflammation (Hsuchou, Kastin, Mishra, & Pan, 2012). Indeed, the blood-brain barrier (BBB) hypothesis postulates the breakdown of the BBB and penetration of peripheral inflammatory molecules into the brain, a phenomenon that has been observed in some patients with psychiatric illnesses including depression (Shalev, Serlin, & Friedman, 2009). Activation of inflammatory responses, increased CRP, and disturbed blood-brain communication may explain to some extent the mechanisms underlying the neuropathophysiology of psychiatric conditions and MDD in relation to inflammation.

Neuroinflammation

Although it was once believed that due to the presence of the BBB the brain was protected from peripheral inflammatory activation, the concept has been challenged due to developments in psychoneuroimmunology research and evidence showing the presence of inflammatory responses within the CNS. In fact, cytokines may not passively cross the BBB due to their relatively large size and hydrophilic nature, at least under physiological conditions, but they are able to penetrate the brain via specific mechanisms or even be produced inside the brain, where they can adversely influence brain structure and function under pathological conditions. The communication between cytokines and the brain can be facilitated by active transport mechanisms, passive diffusion at circumventricular sites where the BBB is deficient, induction of specific adhesion molecules, and activation of peripheral afferent nerve terminals to release cytokines (Kronfol & Remick, 2000; Miller & Raison, 2016; Quan & Banks, 2007; Wichers & Maes, 2002).

Astrocytes and microglia as well as neurons can synthesize and secrete cytokines under stress-related pathophysiological conditions (Freidin, Bennett, & Kessler, 1992; Kronfol & Remick, 2000; Schobitz, De Kloet, & Holsboer, 1994). Alteration in their functional properties and abnormalities of these brain cells have been reported in suicide victims with depression (Nagy et al., 2015; Steiner et al., 2008) and in depressed patients, where neuroimaging studies revealed enhanced activation of microglial cells and astrocytes (Setiawan et al., 2015). When these brain cells are altered, they may further attract activated immune cells from the periphery, as shown, for example, in a mouse model of inflammatory liver injury, where microglia gets activated following TNF-α signaling, and the production of chemokines facilitates cerebral monocyte infiltration (D'Mello, Le, & Swain, 2009). Similarly, proinflammatory cytokines injected into mice hippocampus induced an exaggerated chemokine response by astrocytes, which markedly increased the infiltration of peripheral immune cells into the diseased brain (Hennessy, Griffin, & Cunningham, 2015). Similar cellular trafficking has been suggested by postmortem studies investigating the brain of depressed suicides who presented increased cerebral neuroinflammation associated with the perivascular macrophage recruitment (Torres-Platas, Cruceanu, Chen, Turecki, & Mechawar, 2014).

Elevated levels of glucocorticoids have been also linked to neuroinflammation and microglial activation in depression. In rats, chronic stress has been shown to alter the density and morphology and to activate microglia in stress-sensitive brain regions (Tynan et al., 2010). An increase in glucocorticoid production is perceived as a neuroendocrine warning signal and can induce a proinflammatory response by sensitizing the CNS innate immune effectors, including activating microglia (Frank, Watkins, & Maier, 2013; Nair & Bonneau, 2006). Therefore, glucocorticoids are not uniformly antiinflammatory and can act in a proinflammatory manner specifically in the CNS (Horowitz & Zunszain, 2015; Sorrells, Caso, Munhoz, & Sapolsky, 2009).

As mentioned before, the inflammatory hypothesis of depression proposes the association of inflammation with diminished neurogenesis and increased neurodegeneration in the brain of MDD patients (Maes et al., 2009). In fact, stress-induced activated inflammatory responses within the CNS may lead to oxidative stress and neuroprogressive processes in major depression (Moylan, Maes, Wray, & Berk, 2013). In animal models exposed to either acute or chronic stress, the excessive production of cytokines can cause diminished neurotrophic support and neurogenesis (Ben Menachem-Zidon et al., 2008; Koo & Duman, 2008), features that are enhanced by appropriate cytokine response within the physiological range (Borsini, Zunszain, Thuret, & Pariante, 2015; Goshen et al., 2007). Of note, neuroinflammation can enhance the oxidative status in the CNS, stimulating the production of nitric oxide, another characteristic observed in the pathophysiology of depression (Bakunina, Pariante, & Zunszain, 2015; Ida et al., 2008; Li et al., 2008).

Neuroinflammation and associated cellular and structural changes in the brain have been linked to behavioral alterations of depressive symptoms. As revealed by animal studies, cytokine-induced sickness behavior, resembling symptoms of depression in human, is observed following central or peripheral administration of proinflammatory cytokines including IL-1β, IL-6, and TNF-α (Dantzer, 2009; Kent, Bluthe, Kelley, & Dantzer, 1992). In humans, the use of interferon-α (IFN-α) in patients with cancer and hepatitis C as part of the therapeutic strategy for their condition is accompanied by neuropsychiatric side effects, with approximately half of the recipients developing clinically significant depressive symptoms (Capuron et al., 2002; Dieperink, Ho, Thuras, & Willenbring, 2003; Hauser et al., 2002; Udina et al., 2012) and some developing recurrent depression even years after their initial IFN-induced depressive episode (Chiu, Su, Su, & Chen, 2017). This suggests long-term adverse effects of cytokine-mediated inflammatory response in the periphery that may result in the neuroinflammation and future development of depression.

One of the postulated mechanisms by which neuroinflammation adversely affects the brain initiating a neurodegenerative process in depression is through the disturbance of the serotonergic system by proinflammatory cytokines (Dantzer, O'Connor, Lawson, & Kelley, 2011) and will be described in the next section.

Serotonin-Kynurenine Hypothesis and Neurodegeneration

Disruptions of the serotonergic and inflammatory systems are associated with alterations in the metabolism of tryptophan, the essential amino acid from which the neurotransmitter serotonin is produced (Fig. 2). The monoamine hypothesis considers the deficiency of brain

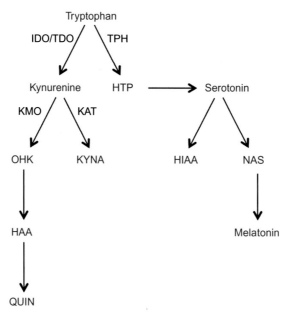

FIG. 2   Kynurenine and methoxyindole pathways of tryptophan metabolism. *IDO*, indoleamine-2,3-dioxygenase; *NAS*, N-acetyl serotonin; *QUIN*, quinolinic acid; *KAT*, kynurenine aminotransferase; *KMO*, kynurenine 3-monooxygenase or kynurenine 3-hydroxylase; *KYNA*, kynurenic acid; *TDO*, tryptophan-2,3-dioxygenase; *TPH*, tryptophan 5-hydroxylase; *HAA*, 3-hydroxyanthranilic acid; *OHK*, 3-hydroxykynurenin; *HIAA*, 5-hydroxyindoleacetic acid; *HTP*, 5-hydroxytryptophan.

serotonin as a key contributor in the pathogenesis of MDD, and targeting serotonin neurotransmission by medications is widely used as an effective treatment for mood and anxiety disorders and specifically depression (Belmaker & Agam, 2008; Nemeroff & Owens, 2002). Depressed patients have been shown to have up to 30% decreased levels of plasma tryptophan and reduced tryptophan availability in the brain (Badawy, 2010). The decrease in tryptophan concentration in MDD patients is postulated to be due to accelerations in the degradation of this amino acid through the kynurenine pathway of tryptophan metabolism (Badawy, 2013, 2017). Independent to a serotonin deficiency, other tryptophan metabolites may contribute to the pathogenesis of depression due to their neurotoxic actions and ability to generate oxidative radicals (Dantzer et al., 2011). An activated inflammatory response may alter the expression and activity of a key enzyme, indoleamine-2,3-dioxygenase (IDO), which activates the kynurenine pathway shunting it away from serotonin synthesis and shifting it toward the formation of neuroactive metabolites (Lapin & Oxenkrug, 1969; Oxenkrug, 2013; Wirleitner, Neurauter, Schrocksnadel, Frick, & Fuchs, 2003). In order to understand better the concept of serotonin-kynurenine theory of depression (Oxenkrug, 2013), a brief summary of two metabolic pathways associated with tryptophan in humans, the kynurenine pathway and the serotonin pathway, is reviewed in the following paragraphs and demonstrated in Fig. 2.

The serotonin or methoxyindole pathway of tryptophan metabolism leads to serotonin biosynthesis. This pathway is initiated by the catalytic activity of the enzyme tryptophan-5-hydroxylase (TPH) and the cofactor tetrahydrobiopterin, resulting in the hydroxylation of tryptophan and the formation of 5-hydroxytryptophan (HTP). A subsequent decarboxylation converts HTP into serotonin, which can be further converted to a stable metabolite known as 5-hydroxyindoleacetic acid catalyzed

by monoamine oxidase and aldehyde dehydrogenase. Alternatively, serotonin can act as a melatonin substrate when catalyzed by serotonin N-acetyltransferase; it is converted to N-acetyl serotonin, which in turn results in melatonin synthesis via O-methylation (Oxenkrug, 2010). Deficiencies of methoxyindole pathway metabolites have been linked to depressive symptoms due to the critical role of serotonin in the regulation of mood (Oxenkrug, 2007) and melatonin in the regulation of sleep (Brzezinski et al., 2005).

The kynurenine pathway of tryptophan metabolism is initiated by the activity of two enzymes: IDO, mentioned above, and tryptophan-2,3-dioxydase (TDO). Degradation of tryptophan by these rate-limiting enzymes induces the formation of kynurenine as an intermediate substrate, which can be converted to other catabolic products of the two distinct routes, either potentially neuroprotective or neurotoxic (Guillemin et al., 2007; Moffett & Namboodiri, 2003). The neuroprotective pathway is facilitated by kynurenine aminotransferase (KAT), which converts kynurenine into kynurenic acid. Kynurenic acid is an endogenous N-methyl-D-aspartate (NMDA) receptor antagonist, glutamate receptor blocker, and potentially neuroprotective metabolite. The neurotoxic pathway leads to the transformation of kynurenine into 3-hydroxykynurenine (OHK) catalyzed by kynurenine 3-monooxygenase (KMO) or kynurenine 3-hydroxylase. The subsequent step is facilitated by kynureninase activity to form 3-hydroxyanthranilic acid, the precursor of the quinolinic acid (QUIN). QUIN is an NMDA receptor agonist and induces oxidative stress, thus being a neurotoxic metabolite that can potentially lead to CNS excitotoxicity (Gabbay et al., 2010; Miller et al., 2009; Oxenkrug, 2010). In the competition for conversion of kynurenine into its metabolites, the activity of KMO directing the pathway through QUIN formation may overcome the KAT activity under pathophysiological conditions (Wichers et al., 2005). Elevation of the neurotoxins of the kynurenine pathway has been shown to be associated with hyperglutamatergic status in depression (Muller & Schwarz, 2008).

Depending on the cell type, kynurenine is produced or transported, and it is degraded into different metabolites. In the brain, the formation of neurotoxic OHK and QUIN occurs mainly in microglial cells, while the neuroprotective kynurenic acid is mainly produced by astrocytes, as these cells lack KMO (Dantzer et al., 2011). Peripheral kynurenine is carried through the BBB by protein transporters and reaches the CNS where it degrades into the neurotoxic metabolites if taken up by glial cells, potentially leading to neurodegeneration (Fukui, Schwarcz, Rapoport, Takada, & Smith, 1991). The switch of kynurenine neuroprotective pathway toward neurotoxic route has been proposed to be the result of inflammation-induced IDO overstimulation, a challenge that can support the neurodegeneration hypothesis of depression (Myint & Kim, 2003; Zunszain, Anacker, Cattaneo, Carvalho, & Pariante, 2011). The hypothesis has been suggested to be associated with depression when microglial kynurenine degradation is dominant over the astrocytic metabolism (Muller & Schwarz, 2008).

In normal physiological conditions, less than 5% of tryptophan is metabolized via the serotonin pathway, and more than 95% is shuttled through the kynurenine pathway in the liver. The fate of tryptophan depends on the comparative activities of the rate-limiting enzymes associated with each pathway (Gal & Sherman, 1980; Oxenkrug, 2007; Stone & Darlington, 2002). TDO is normally the acting enzyme, which is activated and metabolizes tryptophan into kynurenine in the liver; however, in stress-related pathophysiological situations such as infection or oxidative stress, IDO acts as the first rate-limiting enzyme. The activity of IDO enzyme may be directly modulated by the activation of inflammatory mediators. Immune activation and increased proinflammatory cytokines including IFN-γ, IL-1, IL-6, TNF-α, and CRP can enhance the activity of IDO enzyme (Zunszain et al., 2012), shifting tryptophan metabolism toward kynurenine formation rather than serotonin synthesis. This inflammatory-induced IDO overstimulation can be associated with depressive symptoms as it may lead to tryptophan depletion that in turn may result in serotonin deficiency (Halaris, 2013; Myint, Schwarz, & Muller, 2012; Wichers et al., 2005; Wichers & Maes, 2002). In animal models of depression, IDO has been shown as a critical molecular mediator of inflammation-induced depressive-like behavior (O'Connor et al., 2009). Findings from inflammation-associated disorders such as cardiovascular diseases support the association between IDO activity and activation of immune and inflammatory responses (Ozkan, Sukuroglu, Tulmac, Kisa, & Simsek, 2014). A decreased tryptophan level in the context of kynurenine pathway abnormalities has been observed in heart disease patients with comorbid depression and with elevated inflammation (Nikkheslat et al., 2015). Similarly, enhanced tryptophan degradation and increased kynurenine neurotoxic metabolites have been reported in IFN-α-treated patients who developed depressive symptoms (Baranyi et al., 2013; Raison et al., 2010).

The findings of the role of inflammation-related tryptophan degradation in the pathogenesis of MDD, together with treatment resistance complications observed in some depressed patients, have brought attention into looking at these mechanisms as a new therapeutic strategy in depression (Reus et al., 2015).

SUMMARY AND CONCLUSION

Alterations in HPA axis activity and ineffective glucocorticoid signaling can lead to inappropriate immune responses and dysregulated inflammatory processes in major depression. Abnormal communications between the periphery and the CNS can provoke neuroinflammation and may further alter the HPA axis and induce glucocorticoid resistance. Activated inflammatory mediators can induce depressive symptoms by directly affecting the brain, modulating the serotonergic system, and initiating neurodegenerative processes. Furthermore, an inflammation-mediated activation of the kynurenine pathway can result in reduced serotonin biosynthesis due to an increase of tryptophan breakdown, where a shift toward production of neurotoxic metabolites could lead to neurodegeneration.

Future Directions and Translational Significance

A comprehensive understanding of the causes and specific mechanisms involved in the pathogenesis of depression as a complex multifactorial disorder would be vital to achieve effective treatment strategies. Failure of regulatory responses in the control of stress and inflammation could be particularly important in the development of depression and may partly explain the complications associated with treatment resistance in some depressed individuals. Identifying inflammatory biomarkers to monitor vulnerability for depression or predict future adverse consequences in susceptible individuals could open the possibility for new promising therapeutic approaches. Furthermore, inflammation could be considered as a pharmacological target to develop new antidepressants, in addition to combination therapies of antiinflammatory and antidepressive medications. Mechanisms associated with neuroendocrine abnormalities and specifically glucocorticoids, cytokines, and the kynurenine pathway imbalances may be considered as potential targets for novel, more successful, and personalized treatment strategies in depression.

References

Anacker, C., Zunszain, P. A., Carvalho, L. A., & Pariante, C. M. (2011). The glucocorticoid receptor: Pivot of depression and of antidepressant treatment? *Psychoneuroendocrinology*, 36(3), 415–425. https://doi.org/10.1016/j.psyneuen.2010.03.007.

Badawy, A. A. (2010). Plasma free tryptophan revisited: What you need to know and do before measuring it. *Journal of Psychopharmacology*, 24(6), 809–815. https://doi.org/10.1177/0269881108098965.

Badawy, A. A. (2013). Tryptophan: The key to boosting brain serotonin synthesis in depressive illness. *Journal of Psychopharmacology*, 27(10), 878–893. https://doi.org/10.1177/0269881113499209.

Badawy, A. A. (2017). Kynurenine pathway of tryptophan metabolism: Regulatory and functional aspects. *International Journal of Tryptophan Research*, 10, https://doi.org/10.1177/1178646917691938.

Bakunina, N., Pariante, C. M., & Zunszain, P. A. (2015). Immune mechanisms linked to depression via oxidative stress and neuroprogression. *Immunology*, 144(3), 365–373. https://doi.org/10.1111/imm.12443.

Baranyi, A., Meinitzer, A., Stepan, A., Putz-Bankuti, C., Breitenecker, R. J., Stauber, R., et al. (2013). A biopsychosocial model of interferon-alpha-induced depression in patients with chronic hepatitis C infection. *Psychotherapy and Psychosomatics*, 82(5), 332–340. https://doi.org/10.1159/000348587.

Belmaker, R. H., & Agam, G. (2008). Major depressive disorder. *New England Journal of Medicine*, 358(1), 55–68. https://doi.org/10.1056/NEJMra073096.

Ben Menachem-Zidon, O., Goshen, I., Kreisel, T., Ben Menahem, Y., Reinhartz, E., Ben Hur, T., et al. (2008). Intrahippocampal transplantation of transgenic neural precursor cells overexpressing interleukin-1 receptor antagonist blocks chronic isolation-induced impairment in memory and neurogenesis. *Neuropsychopharmacology*, 33(9), 2251–2262. https://doi.org/10.1038/sj.npp.1301606.

Borsini, A., Zunszain, P. A., Thuret, S., & Pariante, C. M. (2015). The role of inflammatory cytokines as key modulators of neurogenesis. *Trends in Neurosciences*, 38(3), 145–157. https://doi.org/10.1016/j.tins.2014.12.006.

Boyle, M. P., Brewer, J. A., Funatsu, M., Wozniak, D. F., Tsien, J. Z., Izumi, Y., et al. (2005). Acquired deficit of forebrain glucocorticoid receptor produces depression-like changes in adrenal axis regulation and behavior. *Proceedings of the National Academy of Sciences of the United States of America*, 102(2), 473–478.

Brown, G. M., McIntyre, R. S., Rosenblat, J., & Hardeland, R. (2017). Depressive disorders: Processes leading to neurogeneration and potential novel treatments. *Progress in Neuro-Psychopharmacology and Biological Psychiatry*. https://doi.org/10.1016/j.pnpbp.2017.04.023.

Brzezinski, A., Vangel, M. G., Wurtman, R. J., Norrie, G., Zhdanova, I., Ben-Shushan, A., et al. (2005). Effects of exogenous melatonin on sleep: A meta-analysis. *Sleep Medicine Reviews*, 9(1), 41–50. https://doi.org/10.1016/j.smrv.2004.06.004.

Capuron, L., Gumnick, J. F., Musselman, D. L., Lawson, D. H., Reemsnyder, A., Nemeroff, C. B., et al. (2002). Neurobehavioral effects of interferon-alpha in cancer patients: Phenomenology and paroxetine responsiveness of symptom dimensions. *Neuropsychopharmacology*, 26(5), 643–652. https://doi.org/10.1016/s0893-133x(01)00407-9.

Carvalho, L. A., Bergink, V., Sumaski, L., Wijkhuijs, J., Hoogendijk, W. J., Birkenhager, T. K., et al. (2014). Inflammatory activation is associated with a reduced glucocorticoid receptor alpha/beta expression ratio in monocytes of inpatients with melancholic major depressive disorder. *Translational Psychiatry*, 4. https://doi.org/10.1038/tp.2013.118.

Carvalho, L. A., Juruena, M. F., Papadopoulos, A. S., Poon, L., Kerwin, R., Cleare, A. J., et al. (2008). Clomipramine in vitro reduces glucocorticoid receptor function in healthy subjects but not in patients with major depression. *Neuropsychopharmacology*, 33(13), 3182–3189. https://doi.org/10.1038/npp.2008.44.

Carvalho, L. A., & Pariante, C. M. (2008). In vitro modulation of the glucocorticoid receptor by antidepressants. *Stress*, 11(6), 411–424. https://doi.org/10.1080/10253890701850759.

Chiu, W. C., Su, Y. P., Su, K. P., & Chen, P. C. (2017). Recurrence of depressive disorders after interferon-induced depression. *Translational Psychiatry*, 7(2). https://doi.org/10.1038/tp.2016.274.

Cowles, M. K., Pariante, C. M., & Nemeroff, C. B. (2009). Depression in the medically ill. In C. M. Pariante, R. M. Nesse, D. Nutt, & L. Wolpert (Eds.), *Understanding depression*. New York, NY: Oxford University Press Inc.

Dantzer, R. (2009). Cytokine, sickness behavior, and depression. *Immunology and Allergy Clinics of North America*, 29(2), 247–264. https://doi.org/10.1016/j.iac.2009.02.002.

Dantzer, R., O'Connor, J. C., Lawson, M. A., & Kelley, K. W. (2011). Inflammation-associated depression: From serotonin to kynurenine. *Psychoneuroendocrinology*, 36(3), 426–436. https://doi.org/10.1016/j.psyneuen.2010.09.012.

de Kloet, E. R. (2014). From receptor balance to rational glucocorticoid therapy. *Endocrinology*, 155(8), 2754–2769. https://doi.org/10.1210/en.2014-1048.

de Kloet, E. R., Derijk, R. H., & Meijer, O. C. (2007). Therapy insight: Is there an imbalanced response of mineralocorticoid and glucocorticoid receptors in depression? Nature Clinical Practice. Endocrinology & Metabolism, 3(2), 168–179. https://doi.org/10.1038/ncpendmet0403.

Deuschle, M., Schweiger, U., Gotthardt, U., Weber, B., Korner, A., Schmider, J., et al. (1998). The combined dexamethasone/corticotropin-releasing hormone

stimulation test is more closely associated with features of diurnal activity of the hypothalamo-pituitary-adrenocortical system than the dexamethasone suppression test. *Biological Psychiatry*, 43(10), 762–766.

Dieperink, E., Ho, S. B., Thuras, P., & Willenbring, M. L. (2003). A prospective study of neuropsychiatric symptoms associated with interferon-alpha-2b and ribavirin therapy for patients with chronic hepatitis C. *Psychosomatics*, 44(2), 104–112.

Dinan, T. G. (1994). Glucocorticoids and the genesis of depressive illness. A psychobiological model. *British Journal of Psychiatry*, 164(3), 365–371.

D'Mello, C., Le, T., & Swain, M. G. (2009). Cerebral microglia recruit monocytes into the brain in response to tumor necrosis factor alpha signaling during peripheral organ inflammation. *Journal of Neuroscience*, 29(7), 2089–2102. https://doi.org/10.1523/jneurosci.3567-08.2009.

Dowlati, Y., Herrmann, N., Swardfager, W., Liu, H., Sham, L., Reim, E. K., et al. (2010). A meta-analysis of cytokines in major depression. *Biological Psychiatry*, 67(5), 446–457. https://doi.org/10.1016/j.biopsych.2009.09.033.

Egeland, M., Zunszain, P. A., & Pariante, C. M. (2015). Molecular mechanisms in the regulation of adult neurogenesis during stress. *Nature Reviews Neuroscience*, 16(4), 189.

Felger, J. C., & Lotrich, F. E. (2013). Inflammatory cytokines in depression: Neurobiological mechanisms and therapeutic implications. *Neuroscience*, 246, 199–229. https://doi.org/10.1016/j.neuroscience.2013.04.060.

Ferrari, A. J., Charlson, F. J., Norman, R. E., Patten, S. B., Freedman, G., Murray, C. J., et al. (2013). Burden of depressive disorders by country, sex, age, and year: Findings from the global burden of disease study 2010. *PLoS Medicine*, 10(11). https://doi.org/10.1371/journal.pmed.1001547.

Firk, C., & Markus, C. R. (2007). Review: Serotonin by stress interaction: A susceptibility factor for the development of depression? *Journal of Psychopharmacology*, 21(5), 538–544. https://doi.org/10.1177/0269881106075188.

Fitzgerald, P., O'Brien, S. M., Scully, P., Rijkers, K., Scott, L. V., & Dinan, T. G. (2006). Cutaneous glucocorticoid receptor sensitivity and pro-inflammatory cytokine levels in antidepressant-resistant depression. *Psychological Medicine*, 36(1), 37–43. https://doi.org/10.1017/s003329170500632x.

Flandreau, E. I., Bourke, C. H., Ressler, K. J., Vale, W. W., Nemeroff, C. B., & Owens, M. J. (2013). Escitalopram alters gene expression and HPA axis reactivity in rats following chronic overexpression of corticotropin-releasing factor from the central amygdala.

Psychoneuroendocrinology, 38(8), 1349–1361. https://doi.org/10.1016/j.psyneuen.2012.11.020.

Frank, M. G., Watkins, L. R., & Maier, S. F. (2013). Stress-induced glucocorticoids as a neuroendocrine alarm signal of danger. *Brain, Behavior, and Immunity, 33*, 1–6. https://doi.org/10.1016/j.bbi.2013.02.004.

Freidin, M., Bennett, M., & Kessler, J. (1992). Cultured sympathetic neurons synthesize and release the cytokine interleukin 1 beta. *Proceedings of the National Academy of Sciences of the United States of America, 89*(21), 10440–10443.

Fukui, S., Schwarcz, R., Rapoport, S. I., Takada, Y., & Smith, Q. R. (1991). Blood-brain barrier transport of kynurenines: Implications for brain synthesis and metabolism. *Journal of Neurochemistry, 56*(6), 2007–2017.

Gabbay, V., Klein, R. G., Katz, Y., Mendoza, S., Guttman, L. E., Alonso, C. M., et al. (2010). The possible role of the kynurenine pathway in adolescent depression with melancholic features. *Journal of Child Psychology and Psychiatry, and Allied Disciplines, 51*(8), 935–943. https://doi.org/10.1111/j.1469-7610.2010.02245.x.

Gal, E. M., & Sherman, A. D. (1980). L-Kynurenine: Its synthesis and possible regulatory function in brain. *Neurochemical Research, 5*(3), 223–239.

Goshen, I., Kreisel, T., Ounallah-Saad, H., Renbaum, P., Zalzstein, Y., Ben-Hur, T., et al. (2007). A dual role for interleukin-1 in hippocampal-dependent memory processes. *Psychoneuroendocrinology, 32*(8–10), 1106–1115. https://doi.org/10.1016/j.psyneuen.2007.09.004.

Guillemin, G. J., Cullen, K. M., Lim, C. K., Smythe, G. A., Garner, B., Kapoor, V., et al. (2007). Characterization of the kynurenine pathway in human neurons. *Journal of Neuroscience, 27*(47), 12884–12892. https://doi.org/10.1523/jneurosci.4101-07.2007.

Halaris, A. (2013). Inflammation, heart disease, and depression. *Current Psychiatry Reports, 15*(10), 400. https://doi.org/10.1007/s11920-013-0400-5.

Hauser, P., Khosla, J., Aurora, H., Laurin, J., Kling, M. A., Hill, J., et al. (2002). A prospective study of the incidence and open-label treatment of interferon-induced major depressive disorder in patients with hepatitis C. *Molecular Psychiatry, 7*(9), 942–947. https://doi.org/10.1038/sj.mp.4001119.

Hayes, P. E., & Ettigi, P. (1983). Dexamethasone suppression test in diagnosis of depressive illness. *Clinical Pharmacy, 2*(6), 538–545.

Hennessy, E., Griffin, E. W., & Cunningham, C. (2015). Astrocytes are primed by chronic neurodegeneration to produce exaggerated chemokine and cell infiltration responses to acute stimulation with the cytokines IL-1beta and TNF-alpha. *Journal of Neuroscience, 35*(22), 8411–8422. https://doi.org/10.1523/jneurosci.2745-14.2015.

Herbert, J., Goodyer, I. M., Grossman, A. B., Hastings, M. H., de Kloet, E. R., Lightman, S. L., et al. (2006). Do corticosteroids damage the brain? *Journal of Neuroendocrinology, 18*(6), 393–411. https://doi.org/10.1111/j.1365-2826.2006.01429.x.

Hery, M., Semont, A., Fache, M. P., Faudon, M., & Hery, F. (2000). The effects of serotonin on glucocorticoid receptor binding in rat raphe nuclei and hippocampal cells in culture. *Journal of Neurochemistry, 74*(1), 406–413.

Hinkelmann, K., Moritz, S., Botzenhardt, J., Riedesel, K., Wiedemann, K., Kellner, M., et al. (2009). Cognitive impairment in major depression: Association with salivary cortisol. *Biological Psychiatry, 66*(9), 879–885.

Holsboer, F. (2000). The corticosteroid receptor hypothesis of depression. *Neuropsychopharmacology, 23*(5), 477–501. https://doi.org/10.1016/s0893-133x(00)00159-7.

Horowitz, M. A., & Zunszain, P. A. (2015). Neuroimmune and neuroendocrine abnormalities in depression: Two sides of the same coin. *Annals of the New York Academy of Sciences, 1351*, 68–79. https://doi.org/10.1111/nyas.12781.

Howren, M. B., Lamkin, D. M., & Suls, J. (2009). Associations of depression with C-reactive protein, IL-1, and IL-6: A meta-analysis. *Psychosomatic Medicine, 71*(2), 171–186. https://doi.org/10.1097/PSY.0b013e3181907c1b.

Hsuchou, H., Kastin, A. J., Mishra, P. K., & Pan, W. (2012). C-reactive protein increases BBB permeability: Implications for obesity and neuroinflammation. *Cellular Physiology and Biochemistry, 30*(5), 1109–1119. https://doi.org/10.1159/000343302.

Ida, T., Hara, M., Nakamura, Y., Kozaki, S., Tsunoda, S., & Ihara, H. (2008). Cytokine-induced enhancement of calcium-dependent glutamate release from astrocytes mediated by nitric oxide. *Neuroscience Letters, 432*(3), 232–236. https://doi.org/10.1016/j.neulet.2007.12.047.

Irwin, M. R., & Miller, A. H. (2007). Depressive disorders and immunity: 20 years of progress and discovery. *Brain, Behavior, and Immunity, 21*(4), 374–383. https://doi.org/10.1016/j.bbi.2007.01.010.

Jacobs, B. L., van Praag, H., & Gage, F. H. (2000). Adult brain neurogenesis and psychiatry: A novel theory of depression. *Molecular Psychiatry, 5*(3), 262–269.

Juruena, M. F., Cleare, A. J., Papadopoulos, A. S., Poon, L., Lightman, S., & Pariante, C. M. (2006). Different responses to dexamethasone and prednisolone in the same depressed patients. *Psychopharmacology, 189*(2), 225–235. https://doi.org/10.1007/s00213-006-0555-4.

Kent, S., Bluthe, R. M., Kelley, K. W., & Dantzer, R. (1992). Sickness behavior as a new target for drug development. *Trends in Pharmacological Sciences, 13*(1), 24–28.

Kim, Y. K., Na, K. S., Myint, A. M., & Leonard, B. E. (2016). The role of pro-inflammatory cytokines in neuroinflammation, neurogenesis and the neuroendocrine system in major depression. *Progress in Neuro-Psychopharmacology and Biological Psychiatry*, *64*, 277–284. https://doi.org/10.1016/j.pnpbp.2015.06. 008.

Kinoshita, S., Kanazawa, T., Kikuyama, H., & Yoneda, H. (2016). Clinical application of DEX/CRH test and multi-channel NIRS in patients with depression. *Behavioral and Brain Functions*, *12*(1), 25.

Koo, J. W., & Duman, R. S. (2008). IL-1beta is an essential mediator of the antineurogenic and anhedonic effects of stress. *Proceedings of the National Academy of Sciences of the United States of America*, *105*(2), 751–756. https://doi.org/10.1073/pnas.0708092105.

Kronfol, Z., & Remick, D. G. (2000). Cytokines and the brain: Implications for clinical psychiatry. *American Journal of Psychiatry*, *157*(5), 683–694.

Kunugi, H., Ida, I., Owashi, T., Kimura, M., Inoue, Y., Nakagawa, S., et al. (2006). Assessment of the dexamethasone/CRH test as a state-dependent marker for hypothalamic-pituitary-adrenal (HPA) axis abnormalities in major depressive episode: A multicenter study. *Neuropsychopharmacology*, *31*(1), 212–220. https://doi.org/10.1038/sj.npp.1300868.

Lapin, I. P., & Oxenkrug, G. F. (1969). Intensification of the central serotoninergic processes as a possible determinant of the thymoleptic effect. *Lancet*, *1*(7586), 132–136.

Levine, J., Barak, Y., Chengappa, K. N., Rapoport, A., Rebey, M., & Barak, V. (1999). Cerebrospinal cytokine levels in patients with acute depression. *Neuropsychobiology*, *40*(4), 171–176. https://doi.org/10.1159/000026615.

Levinson, D. F. (2006). The genetics of depression: A review. *Biological Psychiatry*, *60*(2), 84–92. https://doi.org/10.1016/j.biopsych.2005.08.024.

Li, J., Ramenaden, E. R., Peng, J., Koito, H., Volpe, J. J., & Rosenberg, P. A. (2008). Tumor necrosis factor alpha mediates lipopolysaccharide-induced microglial toxicity to developing oligodendrocytes when astrocytes are present. *Journal of Neuroscience*, *28*(20), 5321–5330. https://doi.org/10.1523/jneurosci.3995-07.2008.

Lindqvist, D., Janelidze, S., Hagell, P., Erhardt, S., Samuelsson, M., Minthon, L., et al. (2009). Interleukin-6 is elevated in the cerebrospinal fluid of suicide attempters and related to symptom severity. *Biological Psychiatry*, *66*(3), 287–292. https://doi.org/10.1016/j.biopsych.2009.01.030.

Maes, M., Bosmans, E., Suy, E., Vandervorst, C., DeJonckheere, C., & Raus, J. (1991). Depression-related disturbances in mitogen-induced lymphocyte responses and interleukin-1 beta and soluble interleukin-2 receptor production. *Acta Psychiatrica Scandinavica*, *84*(4), 379–386.

Maes, M., Meltzer, H. Y., Stevens, W., Cosyns, P., & Blockx, P. (1994). Multiple reciprocal relationships between in vivo cellular immunity and hypothalamic-pituitary-adrenal axis in depression. *Psychological Medicine*, *24*(1), 167–177.

Maes, M., Yirmyia, R., Noraberg, J., Brene, S., Hibbeln, J., Perini, G., et al. (2009). The inflammatory & neurodegenerative (I&ND) hypothesis of depression: Leads for future research and new drug developments in depression. *Metabolic Brain Disease*, *24*(1), 27–53. https://doi.org/10.1007/s11011-008-9118-1.

Martinez, J. M., Garakani, A., Yehuda, R., & Gorman, J. M. (2012). Proinflammatory and "resiliency" proteins in the CSF of patients with major depression. *Depression and Anxiety*, *29*(1), 32–38. https://doi.org/10.1002/da.20876.

McGowan, P. O., Sasaki, A., D'Alessio, A. C., Dymov, S., Labonte, B., Szyf, M., et al. (2009). Epigenetic regulation of the glucocorticoid receptor in human brain associates with childhood abuse. *Nature Neuroscience*, *12*(3), 342–348. https://doi.org/10.1038/nn.2270.

Miller, A. H., Maletic, V., & Raison, C. L. (2009). Inflammation and its discontents: The role of cytokines in the pathophysiology of major depression. *Biological Psychiatry*, *65*(9), 732–741. https://doi.org/10.1016/j.biopsych.2008.11.029.

Miller, A. H., & Raison, C. L. (2016). The role of inflammation in depression: From evolutionary imperative to modern treatment target. *Nature Reviews Immunology*, *16*(1), 22–34. https://doi.org/10.1038/nri.2015.5.

Moffett, J. R., & Namboodiri, M. A. (2003). Tryptophan and the immune response. *Immunology and Cell Biology*, *81*(4), 247–265. https://doi.org/10.1046/j.1440-1711.2003.t01-1-01177.x.

Moussavi, S., Chatterji, S., Verdes, E., Tandon, A., Patel, V., & Ustun, B. (2007). Depression, chronic diseases, and decrements in health: Results from the World Health Surveys. *Lancet*, *370*(9590), 851–858. https://doi.org/10.1016/s0140-6736(07)61415-9.

Moylan, S., Maes, M., Wray, N. R., & Berk, M. (2013). The neuroprogressive nature of major depressive disorder: Pathways to disease evolution and resistance, and therapeutic implications. *Molecular Psychiatry*, *18*(5), 595–606. https://doi.org/10.1038/mp.2012.33.

Muller, N., & Schwarz, M. J. (2008). A psychoneuroimmunological perspective to Emil Kraepelins dichotomy: Schizophrenia and major depression as inflammatory CNS disorders. *European Archives of Psychiatry and Clinical Neuroscience*, *258*(Suppl 2), 97–106. https://doi.org/10.1007/s00406-008-2012-3.

Myint, A. M., & Kim, Y. K. (2003). Cytokine-serotonin interaction through IDO: A neurodegeneration hypothesis of depression. *Medical Hypotheses*, *61*(5–6), 519–525.

Myint, A. M., Schwarz, M. J., & Muller, N. (2012). The role of the kynurenine metabolism in major depression. *Journal of Neural Transmission*, 119(2), 245–251. https://doi.org/10.1007/s00702-011-0741-3.

Nagy, C., Suderman, M., Yang, J., Szyf, M., Mechawar, N., Ernst, C., et al. (2015). Astrocytic abnormalities and global DNA methylation patterns in depression and suicide. *Molecular Psychiatry*, 20(3), 320–328. https://doi.org/10.1038/mp.2014.21.

Nair, A., & Bonneau, R. H. (2006). Stress-induced elevation of glucocorticoids increases microglia proliferation through NMDA receptor activation. *Journal of Neuroimmunology*, 171(1–2), 72–85. https://doi.org/10.1016/j.jneuroim.2005.09.012.

Nemeroff, C. B., & Owens, M. J. (2002). Treatment of mood disorders. *Nature Neuroscience*, 5(Suppl.), 1068–1070. https://doi.org/10.1038/nn943.

Nikkheslat, N., Zunszain, P. A., Carvalho, L. A., Anacker, C., & Pariante, C. M. (2017). Chapter 27 - Antidepressant actions on glucocorticoid receptors. In G. Fink (Ed.), *Stress: Neuroendocrinology and neurobiology* (pp. 279–286). San Diego: Academic Press. https://doi.org/10.1016/B978-0-12-802175-0.00027-9.

Nikkheslat, N., Zunszain, P. A., Horowitz, M. A., Barbosa, I. G., Parker, J. A., Myint, A. M., et al. (2015). Insufficient glucocorticoid signaling and elevated inflammation in coronary heart disease patients with comorbid depression. *Brain, Behavior, and Immunity*, 48, 8–18. https://doi.org/10.1016/j.bbi.2015.02.002.

Oakley, R. H., & Cidlowski, J. A. (2011). Cellular processing of the glucocorticoid receptor gene and protein: New mechanisms for generating tissue-specific actions of glucocorticoids. *Journal of Biological Chemistry*, 286(5), 3177–3184. https://doi.org/10.1074/jbc.R110.179325.

O'Connor, J. C., Lawson, M. A., Andre, C., Moreau, M., Lestage, J., Castanon, N., et al. (2009). Lipopolysaccharide-induced depressive-like behavior is mediated by indoleamine 2,3-dioxygenase activation in mice. *Molecular Psychiatry*, 14(5), 511–522. https://doi.org/10.1038/sj.mp.4002148.

Oxenkrug, G. F. (2007). Genetic and hormonal regulation of tryptophan–kynurenine metabolism. *Annals of the New York Academy of Sciences*, 1122(1), 35–49.

Oxenkrug, G. F. (2010). Tryptophan kynurenine metabolism as a common mediator of genetic and environmental impacts in major depressive disorder: The serotonin hypothesis revisited 40 years later. *Israel Journal of Psychiatry and Related Sciences*, 47(1), 56–63.

Oxenkrug, G. (2013). Serotonin-kynurenine hypothesis of depression: Historical overview and recent developments. *Current Drug Targets*, 14(5), 514–521.

Ozkan, Y., Sukuroglu, M. K., Tulmac, M., Kisa, U., & Simsek, B. (2014). Relation of kynurenine/tryptophan with immune and inflammatory markers in coronary artery disease. *Clínica y Laboratorio*, 60(3), 391–396.

Pace, T. W., Hu, F., & Miller, A. H. (2007). Cytokine-effects on glucocorticoid receptor function: Relevance to glucocorticoid resistance and the pathophysiology and treatment of major depression. *Brain, Behavior, and Immunity*, 21(1), 9–19. https://doi.org/10.1016/j.bbi.2006.08.009.

Pace, T. W., & Miller, A. H. (2009). Cytokines and glucocorticoid receptor signaling. Relevance to major depression. *Annals of the New York Academy of Sciences*, 1179, 86–105. https://doi.org/10.1111/j.1749-6632.2009.04984.x.

Pariante, C. M. (2017). Why are depressed patients inflamed? A reflection on 20 years of research on depression, glucocorticoid resistance and inflammation. *European Neuropsychopharmacology*. https://doi.org/10.1016/j.euroneuro.2017.04.001.

Pariante, C. M., & Lightman, S. L. (2008). The HPA axis in major depression: Classical theories and new developments. *Trends in Neurosciences*, 31(9), 464–468. https://doi.org/10.1016/j.tins.2008.06.006.

Pariante, C. M., & Miller, A. H. (2001). Glucocorticoid receptors in major depression: Relevance to pathophysiology and treatment. *Biological Psychiatry*, 49(5), 391–404.

Pariante, C. M., & Nemeroff, C. B. (2012). Unipolar depression. *Handbook of Clinical Neurology*, 106, 239–249. https://doi.org/10.1016/b978-0-444-52002-9.00014-0.

Peiffer, A., Veilleux, S., & Barden, N. (1991). Antidepressant and other centrally acting drugs regulate glucocorticoid receptor messenger RNA levels in rat brain. *Psychoneuroendocrinology*, 16(6), 505–515.

Polsky, D., Doshi, J. A., Marcus, S., Oslin, D., Rothbard, A., Thomas, N., et al. (2005). Long-term risk for depressive symptoms after a medical diagnosis. *Archives of Internal Medicine*, 165(11), 1260–1266. https://doi.org/10.1001/archinte.165.11.1260.

Quan, N., & Banks, W. A. (2007). Brain-immune communication pathways. *Brain, Behavior, and Immunity*, 21(6), 727–735. https://doi.org/10.1016/j.bbi.2007.05.005.

Raison, C. L., Dantzer, R., Kelley, K. W., Lawson, M. A., Woolwine, B. J., Vogt, G., et al. (2010). CSF concentrations of brain tryptophan and kynurenines during immune stimulation with IFN-alpha: Relationship to CNS immune responses and depression. *Molecular Psychiatry*, 15(4), 393–403. https://doi.org/10.1038/mp.2009.116.

Raison, C. L., & Miller, A. H. (2003). When not enough is too much: The role of insufficient glucocorticoid signaling in the pathophysiology of stress-related disorders. *American Journal of Psychiatry*, 160(9), 1554–1565.

Reus, G. Z., Jansen, K., Titus, S., Carvalho, A. F., Gabbay, V., & Quevedo, J. (2015). Kynurenine pathway dysfunction in the pathophysiology and treatment of depression: Evidences from animal and human studies. *Journal of Psychiatric Research*, 68, 316–328. https://doi.org/10.1016/j.jpsychires.2015.05.007.

Schiepers, O. J., Wichers, M. C., & Maes, M. (2005). Cytokines and major depression. *Progress in Neuro-Psychopharmacology and Biological Psychiatry*, 29(2), 201–217. https://doi.org/10.1016/j.pnpbp.2004.11.003.

Schildkraut, J. J. (1965). The catecholamine hypothesis of affective disorders: A review of supporting evidence. *American Journal of Psychiatry*, 122(5), 509–522.

Schmidt, M. V., Sterlemann, V., Wagner, K., Niederleitner, B., Ganea, K., Liebl, C., et al. (2009). Postnatal glucocorticoid excess due to pituitary glucocorticoid receptor deficiency: Differential short- and long-term consequences. *Endocrinology*, 150(6), 2709–2716. https://doi.org/10.1210/en.2008-1211.

Schobitz, B., De Kloet, E. R., & Holsboer, F. (1994). Gene expression and function of interleukin 1, interleukin 6 and tumor necrosis factor in the brain. *Progress in Neurobiology*, 44(4), 397–432.

Setiawan, E., Wilson, A. A., Mizrahi, R., Rusjan, P. M., Miler, L., Rajkowska, G., et al. (2015). Role of translocator protein density, a marker of neuroinflammation, in the brain during major depressive episodes. *JAMA Psychiatry*, 72(3), 268–275. https://doi.org/10.1001/jamapsychiatry.2014.2427.

Shalev, H., Serlin, Y., & Friedman, A. (2009). Breaching the blood-brain barrier as a gate to psychiatric disorder. *Cardiovascular Psychiatry and Neurology*. 2009. https://doi.org/10.1155/2009/278531.

Smith, R. S. (1991). The macrophage theory of depression. *Medical Hypotheses*, 35(4), 298–306.

Sorrells, S. F., Caso, J. R., Munhoz, C. D., & Sapolsky, R. M. (2009). The stressed CNS: When glucocorticoids aggravate inflammation. *Neuron*, 64(1), 33–39. https://doi.org/10.1016/j.neuron.2009.09.032.

Steiner, J., Bielau, H., Brisch, R., Danos, P., Ullrich, O., Mawrin, C., et al. (2008). Immunological aspects in the neurobiology of suicide: Elevated microglial density in schizophrenia and depression is associated with suicide. *Journal of Psychiatric Research*, 42(2), 151–157. https://doi.org/10.1016/j.jpsychires.2006.10.013.

Stetler, C., & Miller, G. E. (2011). Depression and hypothalamic-pituitary-adrenal activation: A quantitative summary of four decades of research. *Psychosomatic Medicine*, 73(2), 114–126. https://doi.org/10.1097/PSY.0b013e31820ad12b.

Stone, T. W., & Darlington, L. G. (2002). Endogenous kynurenines as targets for drug discovery and development. *Nature Reviews Drug Discovery*, 1(8), 609–620.

ter Heegde, F., De Rijk, R. H., & Vinkers, C. H. (2015). The brain mineralocorticoid receptor and stress resilience. *Psychoneuroendocrinology*, 52, 92–110. https://doi.org/10.1016/j.psyneuen.2014.10.022.

Torres-Platas, S. G., Cruceanu, C., Chen, G. G., Turecki, G., & Mechawar, N. (2014). Evidence for increased microglial priming and macrophage recruitment in the dorsal anterior cingulate white matter of depressed suicides. *Brain, Behavior, and Immunity*, 42, 50–59. https://doi.org/10.1016/j.bbi.2014.05.007.

Tynan, R. J., Naicker, S., Hinwood, M., Nalivaiko, E., Buller, K. M., Pow, D. V., et al. (2010). Chronic stress alters the density and morphology of microglia in a subset of stress-responsive brain regions. *Brain, Behavior, and Immunity*, 24(7), 1058–1068. https://doi.org/10.1016/j.bbi.2010.02.001.

Udina, M., Castellvi, P., Moreno-Espana, J., Navines, R., Valdes, M., Forns, X., et al. (2012). Interferon-induced depression in chronic hepatitis C: A systematic review and meta-analysis. *Journal of Clinical Psychiatry*, 73(8), 1128–1138. https://doi.org/10.4088/JCP.12r07694.

Wang, Q., Verweij, E. W., Krugers, H. J., Joels, M., Swaab, D. F., & Lucassen, P. J. (2014). Distribution of the glucocorticoid receptor in the human amygdala; changes in mood disorder patients. *Brain Structure & Function*, 219(5), 1615–1626. https://doi.org/10.1007/s00429-013-0589-4.

Webster, M. J., Knable, M. B., O'Grady, J., Orthmann, J., & Weickert, C. S. (2002). Regional specificity of brain glucocorticoid receptor mRNA alterations in subjects with schizophrenia and mood disorders. *Molecular Psychiatry*, 7(9), 985–994. 924. https://doi.org/10.1038/sj.mp.4001139.

Wichers, M. C., Koek, G. H., Robaeys, G., Verkerk, R., Scharpe, S., & Maes, M. (2005). IDO and interferon-alpha-induced depressive symptoms: A shift in hypothesis from tryptophan depletion to neurotoxicity. *Molecular Psychiatry*, 10(6), 538–544. https://doi.org/10.1038/sj.mp.4001600.

Wichers, M., & Maes, M. (2002). The psychoneuroimmunopathophysiology of cytokine-induced depression in humans. *International Journal of Neuropsychopharmacology*, 5(4), 375–388. https://doi.org/10.1017/s1461145702003103.

Wirleitner, B., Neurauter, G., Schrocksnadel, K., Frick, B., & Fuchs, D. (2003). Interferon-gamma-induced conversion of tryptophan: Immunologic and neuropsychiatric aspects. *Current Medicinal Chemistry*, 10(16), 1581–1591.

World Health Organisation. (2017). Accessed from: http://www.who.int/mediacentre/factsheets/fs 369/en/.

Zunszain, P. A., Anacker, C., Cattaneo, A., Carvalho, L. A., & Pariante, C. M. (2011). Glucocorticoids, cytokines and brain abnormalities in depression. *Progress in Neuro-Psychopharmacology and Biological Psychiatry*, 35(3), 722–729. https://doi.org/10.1016/j.pnpbp.2010.04.011.

Zunszain, P. A., Anacker, C., Cattaneo, A., Choudhury, S., Musaelyan, K., Myint, A. M., et al. (2012). Interleukin-1β: A new regulator of the kynurenine pathway affecting human hippocampal neurogenesis. *Neuropsychopharmacology*, 37(4), 939–949. https://doi.org/10.1038/npp.2011.277.

4

Neurovascular Dysfunction With BBB Hyperpermeability Related to the Pathophysiology of Major Depressive Disorder

Silky Pahlajani[†], Souhel Najjar*,[†]

*Zucker School of Medicine at Hofstra/Northwell, New York, NY, United States
[†]Lenox Hill Hospital, New York, NY, United States

INTRODUCTION

According to the World Health Organization (2017) press release, depression is the leading cause of disability in the world (http://www.who.int/mediacentre/news/releases/2017/world-health-day/en/). The standard treatment for major depressive disorder (MDD) has been antidepressants that alter levels of serotonin and/or norepinephrine (NE), based on the 1950s monoamine hypothesis (Higgins & George, 2013). However, as shown by the Sequenced Treatment Alternatives to Relieve Depression (STAR*D) study (www.star-d.org), one-third of patients with MDD fail at least two antidepressant trials within the first year of treatment (Insel & Wang, 2009; Trivedi et al., 2006). Cumulative clinical and experimental evidence suggests a multifactorial etiology leading to important paradigm shifts in the neurobiology

and pharmacotherapy of depression (Nasrallah, 2015). These include heterogeneous and interrelated mechanisms that affect genetic, neurotransmitter, immune, oxidative, and inflammatory systems (Najjar, Pearlman, Devinsky, Najjar, & Zagzag, 2013) replacing the traditional concept implying that depression is solely due to a "chemical imbalance." Although the predictive validity of individually elevated biomarkers for symptoms of MDD is limited, it becomes particularly higher when multiple composite biomarkers are evaluated (Belmaker & Agam, 2008). This was demonstrated in a study of 36 MDD patients that assessed a composite biomarker test consisting of nine individual assays: apolipoprotein, $\alpha 1$ antitrypsin, brain-derived neurotropic factor, CIII, cortisol myeloperoxidase, epidermal growth factor, prolactin, resistin, and soluble tumor necrosis factor α (TNFα) receptor type II. The predictive validity of this composite biomarker analysis for MDD symptoms revealed a high sensitivity and specificity of 91.7% and 81.3%, respectively (Papakostas et al., 2013).

Neuroinflammation and oxidative stress are implicated in the neurobiology of MDD (Najjar, Pearlman, Alper, Najjar, & Devinsky, 2013; Ng, Berk, Dean, & Bush, 2008; Ozcan, Gulec, Ozerol, Polat, & Akyol, 2004; Steiner et al., 2008). Various studies have investigated neuropathologic, genetic, CSF, and serological biomarkers of oxidative stress in humans and animal models with depressive behaviors. They revealed numerous neuroinflammatory abnormalities such as microglial activation and proliferation, astroglial loss and activation, upregulation of T helper 1 (Th1) cells and proinflammatory cytokines, and reduction in the $CD4^+CD25^+FOXP3^+$ regulatory T (T_{Reg}) cell counts (Banasr & Duman, 2008; Haroon, Raison, & Miller, 2011; Hong et al., 2013; Najjar, Pearlman, Alper, et al., 2013; Najjar, Pearlman, Devinsky, et al., 2013; Steiner et al., 2008). Postmortem studies in MDD have documented decreased levels of antioxidants such as glutathione and increased levels of lipid peroxidation products such as 4-hydroxy-2-nonenal in brain tissue (Gawryluk, Wang, Andreazza, Shao, & Young, 2010; Maes, Galecki, Chang, & Berk, 2011; Ng et al., 2008; Scapagnini, Davinelli, Drago, Lorenzo, & Oriani, 2012). Neuroinflammation and oxidative stress may also contribute to decreased serotonergic and increased glutamatergic tones in MDD. Hyperglutamatergia, in turn may further perpetuate oxidative stress and neuroinflammation via a positive feedback loop (Najjar, Pearlman, Alper, et al., 2013; Najjar, Pearlman, Devinsky, et al., 2013). Additionally, experimental data suggest that reactive oxygen species (ROS)-mediated oxidative stress and neuroinflammation exhibit a bidirectional relationship (Maier, 2003; Najjar, Pearlman, Alper, et al., 2013). ROS can activate microglia and increase synthesis of pro-inflammatory cytokines by stimulating nuclear factor κB (NFκB), which can in turn perpetuate oxidative stress (Anderson et al., 2013; Block, Zecca, & Hong, 2007; Najjar, Pearlman, Devinsky, et al., 2013; Salim, Chugh, & Asghar, 2012).

It is well established that neurovascular unit dysfunction with blood-brain barrier (BBB) hyperpermeability can occur as the result of oxidative stress and neuroinflammation, causing neuronal dysfunction in many neurological disorders, such as multiple sclerosis, stroke, traumatic brain injury, epilepsy, and Alzheimer's disease (Carmeliet & De Strooper, 2012; Friedman & Kaufer, 2011; Khatri, Mckinney, Swenson, & Janardhan, 2012; Liu, Thom, et al., 2012; Lund et al., 2013; Shlosberg, Benifla, Kaufer, & Friedman, 2010; Skaper, 2007; Wang, Tang, & Yenari, 2007). It is also widely known that MDD is often comorbid with neurological disorders associated with BBB abnormalities and medical conditions associated with vascular endothelial dysfunction, such as cardiovascular disease, diabetes, and metabolic syndrome (Ford et al., 1998; Le Mellédo, Mahil, & Baker, 2004; Najjar, Pearlman, Devinsky, et al., 2013; Pan et al., 2012; Rugulies, 2002; Van der Kooy et al., 2007). Furthermore, large epidemiological studies confirmed a bidirectional relationship between MDD and medical conditions characterized by vascular endotheliopathy (Barth, Schumacher, & Herrmann-Lingen, 2004; Halaris, 2009, 2013; Pan et al., 2012; Saleptsis, Lambropoulos, Halaris, Angelopoulos, & Giannoukas, 2011). However, it is not well established whether primary MDD can be purely related to neurovascular dysfunction with BBB hyperpermeability, independent of neurological comorbidity. Evidence through 2009, reviewed by Shalev and colleagues, provides limited data linking BBB hyperpermeability to psychiatric disorders in general (Shalev, Serlin, & Friedman, 2009).

We review the recent clinical and experimental data implicating oxidative stress, eNOS uncoupling, reduced endothelial NO levels, and inflammation in the pathophysiology of peripheral vascular endothelial dysfunction associated with MDD. In the following sections, we also provide experimental and indirect evidence suggesting that neurovascular unit dysfunction with BBB hyperpermeability may occur in a subset of individuals with MDD. We also provide a theoretical integration of the human and animal data implicating oxidative stress and neuroinflammation in the pathophysiology of putative neurovascular endotheliopathy in MDD (Fig. 1).

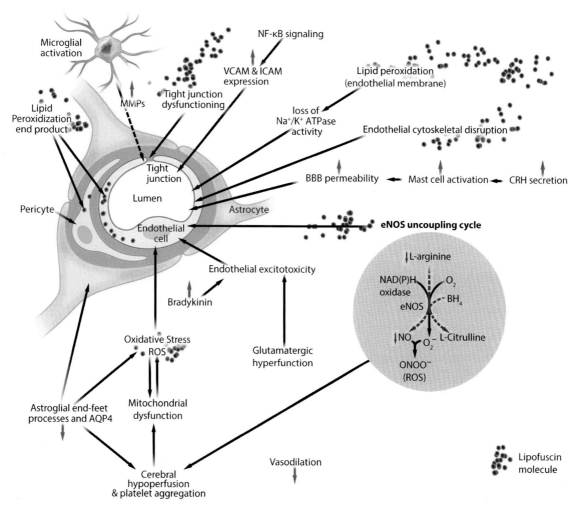

FIG. 1   Theoretical integration of human and animal data linking oxidative stress and neuroinflammation to neurovascular unit dysfunction and BBB hyperpermeability in major depressive disorder. *Abbreviations*: AQP4, aquaporin 4; BH2, dihydrobiopterin; BH4, tetrahydrobiopterin; eNOS, endothelial nitric oxide synthase; MMP, matrix metalloproteinase; NAD(P)H, nicotinamide adenosine dinucleotide phosphate; Na^+/K^+ ATPase, sodium-potassium adenosine triphosphatase; NFκB, nuclear factor κB; NO, nitric oxide; $ONOO^-$, peroxynitrite; O_2^-, superoxide; ROS, reactive oxygen species. *Modified from Abbott, N. J., Rönnbäck, L., & Hansson, E. (2006). Astrocyte-endothelial interactions at the blood-brain barrier.* Nature Reviews Neuroscience, 7(1), 41–53.

NEUROVASCULAR DYSFUNCTION

The neurovascular unit is made up of cerebral microvessels, glial cells, and neurons. Glial cells include astroglia, microglia, and oligodendroglia. It is the epicenter of several tightly controlled, dynamic, and complex cellular interactions between glia and neurons and the coupling of neuronal activity with endothelium-dependent cerebral blood flow (Abbott & Friedman, 2012). The evidence supports an association between MDD and neurovascular dysfunction is indirect and based on data primarily derived from studies documenting peripheral vascular endothelial dysfunction in MDD and epidemiological studies associating MDD with vascular disorders.

Dynamic nuclear imaging can evaluate endothelial dysfunction by measuring the relative uptake ratio (RUR) of blood flow in the brachial artery after hyperemic challenge. RUR is a measure of the vascular dilatory response in which a lower RUR suggests poorer vascular endothelial function. In a prospective cohort of 23 patients with MDD, 23 with minor depressive disorder, and 277 nondepressed controls, the mean RUR was significantly lower in participants with MDD (unadjusted mean $= 3.13$, SD $= 1.51$) or minor depressive disorder (unadjusted mean $= 3.38$, SD $= 1.00$) compared with nondepressed controls (unadjusted mean $= 4.22$, SD $= 1.74$) ($F = 6.68$, $P = .001$) (Lavoie, Pelletier, Arsenault, Dupuis, & Bacon, 2010). Even after adjusting for age, sex, socioeconomic factors, medical comorbidity, and medications ($F = 5.19$, $P = .006$), this effect remained statistically significant (Lavoie et al., 2010). One study measured the percentage of apoptotic nuclei in human umbilical vein endothelial cells, known as endothelial proapoptotic activity. The percentage of proapoptotic nuclei in participants with MDD was significantly higher compared with nondepressed controls (4.4% vs. 2.3%, $P \leq 0.001$), even after adjusting for age and cardiovascular comorbidity (Politi, Brondino, & Emanuele, 2008).

Epidemiological studies assessing vascular endothelial dysfunction in MDD revealed a strong and bidirectional association between MDD and medical conditions with vascular endothelial pathology (Najjar, Pearlman, Devinsky, et al., 2013; Serlin, Levy, & Shalev, 2011). A meta-analysis of 16,221 participants documented a significantly increased risk of MDD among individuals with major vascular diseases compared with those without vascular disease: diabetes (odds ratio (OR) 1.51, 95% confidence interval (CI) 1.30–1.76, $P < .0005$, 15 studies), cardiovascular disease (OR 1.76, 95% CI 1.08–1.80, $P < .0005$, 10 studies), and stroke (OR 2.11, 95% CI 1.61–2.77, $P < .0005$, 10 studies). In addition, MDD was found to be common among individuals with two or more classic risk factors for vascular disease compared with those with one or no risk factor (OR 1.49, 95% CI 1.27–1.7, $P < .0005$, 18 studies) (Valkanova & Ebmeier, 2013). After statistical adjustments for chronic illness and disability, these findings remained significant. Results from a meta-analysis assessing this association in the opposite direction indicate that MDD not only is an independent risk factor for cardiovascular disease (relative risk (RR) 2.69, 95% CI 1.63–4.43, $P < .001$, 11 studies) (Rugulies, 2002) but also is associated with a threefold increase in cardiovascular disease mortality rate (OR 2.61, 95% CI 1.53–4.47, $P = .0004$). Other studies also reported similar results (Barth et al., 2004; Carney, Freedland, Miller, & Jaffe, 2002; Najjar, Pearlman, Devinsky, et al., 2013; Van der Kooy et al., 2007).

BLOOD-BRAIN BARRIER UNIT HYPERPERMEABILITY

Components of the BBB include the neurovascular endothelium, extracellular matrix basal lamina, and astrocytic endfeet processes. The BBB restricts the entry of peripheral inflammatory mediators, such as cytokines and antibodies, thereby protecting neural transmission, limiting cross talk between brain innate immunity and peripheral adaptive immunity, and securing the brain's immune-privileged status (Friedman & Kaufer, 2011; Najjar, Pearlman, Devinsky, et al., 2013; Shalev et al., 2009). Neurovascular endothelial cells regulate influx of essential nutrients, efflux of toxic substances, and ionic homeostasis of brain interstitial fluid and prevent brain influx of peripheral neuroactive substances, neurotransmitters, and water-soluble molecules (Abbott, Rönnbäck, & Hansson, 2006; Najjar, Pearlman, Devinsky, et al., 2013). Findings from studies assessing cerebrospinal fluid (CSF)-to-serum ratios of various molecules and endothelial expression of P-glycoprotein indicate that BBB hyperpermeability can occur in a subset of individuals with MDD (Najjar, Pearlman, Devinsky, et al., 2013).

In some MDD patients, an elevated CSF-to-serum albumin ratio is suggestive of mild hyperpermeability of blood-brain and/or blood-CSF barriers (Bechter et al., 2010; Gudmundsson et al., 2007). In a cross-sectional study of elderly women without dementia (11 MDD, 3 dysthymia, and 70 nondepressed controls), an elevated mean CSF-to-serum albumin ratio was found among those with MDD or dysthymia relative to nondepressed controls (7.1×10^{-3} vs. 5.4×10^{-3}, age-adjusted $P < .015$) (Gudmundsson et al., 2007; Najjar, Pearlman, Devinsky, et al., 2013). In another study (24 affective disorders, 4100 age-matched controls), 37.5% of the affective disorder group (9 of 24) were found to have an increased mean CSF-to-serum albumin ratio; this value was 22%–89% above the upper limit of healthy age-matched controls (8.7×10^{-3} vs. 5.0×10^{-3}) (Bechter et al., 2010; Najjar, Pearlman, Devinsky, et al., 2013). A third study (99 MDD) found a positive association between increased CSF-to-serum ratios of albumin and urate to EEG slowing (a measure of cerebral dysfunction) and suicidality (Najjar, Pearlman, Devinsky, et al., 2013; Niklasson & Agren, 1984). Increased permeability of blood-brain and blood-CSF barriers may be related to elevated levels of S100B protein, a marker of glial activation (Falcone et al., 2010; Schroeter, Abdul-Khaliq, Krebs, Diefenbacher, & Blasig, 2008), and pro-inflammatory cytokines in serum, CSF, and neuropathologic specimen from individuals with MDD (Liu, Ho, & Mak, 2012; Miller, Buckley, Seabolt, Mellor, & Kirkpatrick, 2011). Elevated levels of these molecules may reflect increased synthesis and increased efflux from brain parenchyma into the blood (BBB hyperpermeability), and from blood into the CSF (blood-CSF hyperpermeability) (Najjar, Pearlman, Devinsky, et al., 2013; Shalev et al., 2009).

Reduced expression of BBB endothelial P-glycoprotein (a multidrug efflux transporter) was documented in some individuals with MDD. Reduced expression or function of P-glycoprotein may facilitate BBB permeability, permitting neurotoxic substances to enter (Hawkins, Sykes, & Miller, 2010). Furthermore, administering antidepressants after chronic stress exposure can enhance P-glycoprotein function demonstrated by positron emission tomography (PET) utilizing the [(11)C]-verapamil radioligand for P-glycoprotein in humans with MDD and in Wistar rats exhibiting depressive-like behavior (de Klerk et al., 2010). A human genetics study of 631 MDD individuals and 110 controls documented a significant association ($P = .034$) between alterations in the P-glycoprotein encoding gene ATP-binding cassette, subfamily B member 1 (ABCB1), and MDD (Fujii et al., 2012; Najjar, Pearlman, Devinsky, et al., 2013).

THEORETICAL INTEGRATION WITH OXIDATIVE AND NEUROINFLAMMATORY MECHANISMS

Oxidative Stress

"Oxidative stress" is often defined as an imbalance between excessive generation of ROS (free radicals) and the available antioxidant defenses and repair capacities (Halliwell, 2001). Typical examples of ROS include hydrogen peroxide ($H_2O_2^-$), hydroxyl radical (HO^-), superoxide (O_2^-), and peroxynitrite ($ONOO^-$). When nitric oxide (NO) combines with O_2^-, it produces the strongly oxidant compound $ONOO^-$ (Maes, Galecki, et al., 2011; Ng et al., 2008). Moderate levels of ROS play an important role in physiological cell signaling processes and even protect against oxidative stress, facilitating recovery and limiting injury. However, persistently elevated ROS levels can damage biological molecules like DNA, proteins, and lipids, resulting in oxidative injury (Dröge, 2002; Najjar, Pearlman, Devinsky, et al., 2013). Compared with other organs, the brain's oxygen demand is particularly high, and endogenous antioxidant mechanisms are limited. Furthermore, the brain has a high metabolic rate, high levels of peroxidizable polyunsaturated lipid content, and reduced forms of transition minerals that can induce lipid peroxidation and convert $H_2O_2^-$ to HO^-, increasing its susceptibility to oxidative damage (Maes, Mihaylova, Kubera, Leunis, & Geffard, 2011; Ng et al., 2008; Scapagnini et al., 2012). It is hypothesized that oxidative stress associated with MDD can cause neurovascular dysfunction through numerous mechanisms, including endothelial nitric oxide synthase (eNOS) dysfunction and uncoupling, leading to an imbalance between endothelial synthesis of beneficial NO and harmful O_2^- (Najjar, Pearlman, Devinsky, et al., 2013) (Fig. 1).

The cellular source of NO, endothelial versus nonendothelial, and concentration levels, determine whether NO will have a protective or injurious effect on vascular endothelia cell function (Förstermann, 2006). There are three conventional isoforms of NOS, eNOS, neuronal NOS (nNOS), and inducible NOS (iNOS). NOS synthesis in the brain is regulated by eNOS, constitutively expressed in endothelial cells and astrocytes, and nNOS expressed in neurons (Iwase, 2000; Lacza, Pankotai, & Busija, 2009; Wiencken & Casagrande, 1999). Vascular

smooth muscle tone is regulated by eNOS, and neurotransmission is modulated by nNOS. Some studies provide evidence of a fourth variant, mitochondrial NOS (mtNOS), which is expressed in the inner mitochondrial membrane or matrix (Lacza et al., 2009, 2001). The mechanisms regulating the activity of nNOS and iNOS differ; Ca^{2+} influx results in increased nNOS activity (Andrew & Mayer, 1999), whereas pro-inflammatory cytokines (Galea, Feinstein, & Reis, 1992) and NFκB signaling (Chuang, 2010) lead to increased iNOS activity. Pathological inflammatory states, such as those associated with brain injury, are associated with increased synthesis of NO by inflammatory cells such as activated microglia expressing iNOS (Najjar, Pearlman, Devinsky, et al., 2013; Pun, Lu, & Moochhala, 2009). When nonendothelium-derived NO combines with O_2^-, it forms $ONOO^-$, a highly reactive free radical that can damage neural tissue and disrupt BBB integrity leading to neurovascular endothelial dysfunction (Lehner et al., 2011; Najjar, Pearlman, Devinsky, et al., 2013; Stuehr, Santolini, Wang, Wei, & Adak, 2004).

Several factors influence the ability of eNOS to produce beneficial endothelium-derived NO. These are endothelial levels of arginine (eNOS substrate) (Liu & Huang, 2008), Ca^{2+}, and tetrahydrobiopterin (BH4) (eNOS cofactor) (Antoniades et al., 2006; Joynt, Whellan, & Oconnor, 2003; Najjar, Pearlman, Devinsky, et al., 2013; Szabó, Ischiropoulos, & Radi, 2007). eNOS mediates NO synthesis via oxidative conversion of L-arginine to L-citrulline (Najjar, Pearlman, Devinsky, et al., 2013). Endothelium-derived NO increases cellular levels of cyclic guanosine monophosphate, which triggers mechanisms involving endothelium-dependent vasodilation and platelet aggregation inhibition, thereby increasing cerebral blood flow (Joynt et al., 2003;Pun et al., 2009; Stuehr et al., 2004). Data from *in vitro* studies indicate that endothelium-derived NO can inhibit synthesis of 20-hydroxyeicosatetraenoic acid that promotes vasoconstriction, resulting in cerebrovascular dilation (Attwell et al., 2010; Sun, Falck, Okamoto, Harder, & Roman, 2000). Endothelium-derived NO can also scavenge free radicals and limit oxidative injury to endothelial vasculature (Pun et al., 2009; Stuehr et al., 2004). Downregulation of eNOS activity can lower endothelial NO levels, potentially contributing to several pathophysiological abnormalities associated with MDD such as (1) decreased cerebral blood flow; (2) increased platelet aggregation, which may contribute to the increased risk of cardiovascular disease in MDD; (3) increased oxidative stress; and (4) decreased vascular reactivity (Joynt et al., 2003; Najjar, Pearlman, Devinsky, et al., 2013; Pun et al., 2009; Stuehr et al., 2004).

In conditions associated with oxidative stress such as MDD (Maes, Galecki, et al., 2011; Maes, Mihaylova, et al., 2011; Najjar, Pearlman, Devinsky, et al., 2013), ROS may increase oxidative conversion of endothelial BH4 to dihydrobiopterin (BH2). Decreased and increased endothelial levels of BH4 and BH2, respectively, shift the eNOS substrate from L-arginine to molecular oxygen, resulting in "eNOS uncoupling" leading to decreased and increased endothelial synthesis of NO and O_2^- synthesis (Chrapko et al., 2004; Najjar, Pearlman, Devinsky, et al., 2013; Szabó et al., 2007). O_2^- then combines with residual NO to form $ONOO^-$, which in turn oxidizes BH4, and further decreases its levels via a positive feedback loop (Chen, Wang, et al., 2010; Najjar, Pearlman, Devinsky, et al., 2013; Szabó et al., 2007) (Fig. 1).

Upregulation of iNOS and nNOS and downregulation of eNOS expression can perpetuate neuronal injury, as evidenced by data from *in vitro* animal models with neurological deficits (Huang et al., 1994; Khan et al., 2011). In a study with murine models of ischemic stroke, knocking out iNOS and nNOS reduced the infarcted zone, whereas knocking out eNOS expanded the size of infract, compared with wild-type

mice (Lu et al., 2011; Najjar, Pearlman, Devinsky, et al., 2013; Samdani, Dawson, & Dawson, 1997). In animal models of traumatic brain injury, increased neural levels of $ONOO^-$ are associated with BBB disruption and neurobehavioral deficits, whereas treatment with an antioxidant such as S-nitrosoglutathione downregulates endothelial $ONOO^-$ synthesis and enhances neural repair mechanisms, thereby improving neurovascular unit function (Khan et al., 2011).

Cumulative evidence derived from numerous clinical and experimental studies suggests that eNOS uncoupling can contribute to vascular endothelial dysfunction in both cardiovascular diseases and MDD. In cardiovascular diseases, the mechanisms of eNOS uncoupling-induced vascular endothelial dysfunction are thought to involve (1) increased O_2^- synthesis (2), increased $ONOO^-$ production, and (3) decreased BH4 levels (Antoniades et al., 2006; Chen, Druhan, et al., 2010; Lavoie et al., 2010; Masano et al., 2008; Najjar, Pearlman, Devinsky, et al., 2013; Stuehr et al., 2004; Szabó et al., 2007). However, in MDD, the evidence of potential contribution of eNOS uncoupling to vascular endothelial dysfunction is primarily indirect and derived from some clinical and experimental studies. Individuals with MDD have significantly reduced eNOS activity in platelets and reduced serum NO levels (Luiking, Ten Have, Wolfe, & Deutz, 2012; Stuehr et al., 2004), which normalized after treatment with selective antidepressants (Ikenouchi-Sugita et al., 2011; Lopez-Vilchez et al., 2016; Matchkov, Kravtsova, Wiborg, Aalkjaer, & Bouzinova, 2015). One study showed decreased eNOS expression in cultured endothelial cells exposed to serum of 12 patients with untreated MDD compared with controls (P < 0.01); measurements after escitalopram treatment at 8 and 24 weeks showed restoration of eNOS activity similar to that in healthy control samples (Lopez-Vilchez et al., 2016). In the CREATE clinical trial, MDD patients were randomized to the citalopram treatment group ($n=36$) or placebo ($n=21$); those treated with citalopram for 12 weeks showed a statistically significant increase in serum NO levels compared with the placebo group ($P=.005$) (Najjar, Pearlman, Devinsky, et al., 2013; Van Zyl et al., 2008). Similar findings were documented following a 2-month treatment with paroxetine (Lara, Archer, Baker, & Le Mellédo, 2003). Another study utilizing a chronic stress mouse model found that treatment with fluoxetine restored previously insufficient levels of aortic endothelial NO (Isingrini, Belzung, Freslon, Machet, & Camus, 2011). Collectively, these findings suggest that eNOS uncoupling can occur in MDD and eNOS *re*coupling may actually be one of the mechanisms contributing to the therapeutic effects of antidepressants (Gibson, Korade, & Shelton, 2012; Najjar, Pearlman, Devinsky, et al., 2013; Ng et al., 2008; Scapagnini et al., 2012).

Another mechanism that suggests eNOS uncoupling plays a role in the neurobiology of MDD is the antidepressant effect of L-methylfolate. *in vitro* studies have shown that L-methylfolate can reverse eNOS uncoupling by upregulating BH4 synthesis (Antoniades et al., 2006; Najjar, Pearlman, Devinsky, et al., 2013). Randomized control trial showed that augmenting a selective serotonin reuptake inhibitor (SSRI) with 15 mg/day of L-methylfolate increases its antidepressant efficacy compared with SSRI plus placebo (Najjar, Pearlman, Devinsky, et al., 2013; Papakostas et al., 2012). The authors of this study suggested that the superior efficacy of the augmentation is attributed to increased BH4-mediated activation of the rate-limiting enzymes of monoamine synthesis, that is, norepinephrine, serotonin, and dopamine. However, we hypothesize that the superior antidepressant effects derived from augmentation with L-methylfolate may also be associated with the ability of BH4 to reverse eNOS uncoupling (Najjar, Pearlman, Devinsky, et al., 2013).

Literature assessing cerebral perfusion in MDD via single-photon emission tomography (SPECT) scans has shown reduced cerebral blood flow in selective regions, such as anterior cingulate, prefrontal, temporal, and occipital cortices, as well as thalamic nuclei. These abnormalities have been traditionally attributed to abnormal neural activity and fluctuations in emotional states (Chrapko et al., 2004; Diener et al., 2012; Nagafusa et al., 2012; Orosz et al., 2012; Smith & Cavanagh, 2005); however, they might also be related to eNOS uncoupling-mediated reduced endothelial vasodilator NO (Gardner et al., 2003; Moreno et al., 2013; Srivastava, Bath, & Bayraktutan, 2011; Tobe, 2013). Chronic cerebral hypoperfusion can in turn impair mitochondrial oxidative pathways, leading to increased endothelial ROS synthesis that can perpetuate eNOS uncoupling and further reduce endothelial NO levels and cerebral blood flow in a positive feedback loop (Aliev et al., 2013, 2009; Antoniades et al., 2006; Chen, Wang, et al., 2010; Lavoie et al., 2010; Liu & Zhang, 2012). Furthermore, SSRIs may restore cerebral hypoperfusion in MDD individuals via vasodilatory mechanisms involving upregulation of eNOS-mediated endothelial NO synthesis (Lopez-Vilchez et al., 2016; Ofek et al., 2012). We suggest that nonheritable factors, such as oxidative mechanisms, may be the primary cause of eNOS uncoupling associated with MDD (Najjar, Pearlman, Devinsky, et al., 2013). Several genetic studies have shown nonsignificant associations between eNOS gene polymorphisms and MDD (Ikenouchi-Sugita et al., 2011; Zeman et al., 2010). In an environment of high oxidative stress, BBB endothelial cells not only perpetuate eNOS uncoupling but also themselves can be the object of oxidative damage resulting in reduced BBB integrity (Friedman & Kaufer, 2011; Lehner et al., 2011; Najjar, Pearlman, Devinsky, et al., 2013). We recently reported a case of a patient with chronic and refractory MDD whose SPECT scan demonstrated moderate to severe bifrontal hypoperfusion associated with lipofuscin granule accumulation, a marker of oxidative stress in aging and neurodegenerative disorders (Gray & Woulfe, 2005; Keller et al., 2004; Liu, Ho, et al., 2012; Reeg & Grune, 2015), in the endothelium of the neurovascular unit documented by brain biopsy (Najjar et al., 2014). Experimental studies of neurological disorders known to be associated with BBB hyperpermeability suggest that there are several other mechanisms by which oxidative stress can disrupt neurovascular unit integrity. These mechanisms include (Najjar, Pearlman, Devinsky, et al., 2013): (a) activation of metalloproteinase (MMP)-2/-9 directly or indirectly through pro-inflammatory cytokines (Lehner et al., 2011); (b) downregulation of endothelial expression of E-cadherin (Pun et al., 2009); (c) alteration of the expression, distribution, and phosphorylation of BBB tight junction proteins (claudin, occluding, and ZO proteins) by molecules such as phosphatidylinositol-3-kinase γ (Haorah, Knipe, Leibhart, Ghorpade, & Persidsky, 2005; Jin et al., 2011; Lochhead et al., 2010); (d) alteration of endothelial cytoskeletal structure; (e) induction of endothelial NMDAR subunit expression, that is, NMDA receptor subunit 1 (NR1) subunit, leading to endothelial excitotoxicity (Betzen et al., 2009); and (f) impairment of vascular endothelial mitochondrial oxidative metabolism (Enciu, Gherghiceanu, & Popescu, 2013; Scapagnini et al., 2012). Future human studies are warranted to investigate the relevance of these mechanisms to the neurobiology of MDD (Najjar, Pearlman, Devinsky, et al., 2013).

Neuroinflammation

Previous data support that peripheral inflammatory mediators may cause neuroinflammation that can damage neurovascular unit function and increase BBB permeability leading to depression (Najjar, Pearlman, Devinsky, et al., 2013; Quan & Banks, 2007) (Fig. 1). Astroglia are an essential part of the neurovascular unit. These glial cells regulate BBB integrity, energy metabolism, blood flow, and neuronal signaling, among many other functions (Najjar, Pearlman, Devinsky, et al., 2013; Serlin et al., 2011). Numerous postmortem studies in subjects with MDD have documented reduced astroglial density in functionally relevant areas of the brain, such as the hippocampus, prefrontal and cingulate cortices, and amygdala (Cotter et al., 2002; Cotter, Hudson, & Landau, 2005; Rajkowska & Stockmeier, 2013; Shalev et al., 2009). Studies have also shown reduced coverage of blood vessels by astroglial processes in orbitofrontal gray matter of patients with MDD compared with controls; evidenced by reduced expression of aquaporin 4 (AQP4), a water channel located in astrocytic endfeet (Rajkowska, Hughes, Stockmeier, Miguel-Hidalgo, & Maciag, 2013). Recent data from an animal model of depression suggest that AQP4 not only is essential for brain water and ion homeostasis function but also contributes to the antidepressant effects of some SRRIs such as fluoxetine (Benedetto et al., 2016). In a gene expression study, hippocampal mRNA from 13 MDD subjects revealed a downregulation of genes involved in glial syncytial function, including AQP4, compared with 10 healthy controls (Medina et al., 2016). Furthermore, decreased AQP4 expression has also been associated with oxidative stress in mice with depressive-like behaviors (Liu, Lu, et al., 2012; Najjar, Pearlman, Devinsky, et al., 2013). Abnormalities on PET and SPECT scans of individuals with MDD may also be related to reduced AQP4 density (Najjar, Pearlman, Devinsky, et al., 2013; Serlin et al., 2011). Thus, reduced AQP4 density may impair essential glial-vascular regulatory pathways in the neurovascular unit leading to BBB hyperpermeability and possibly worsening of depressive symptoms (Benedetto et al., 2016; Najjar, Pearlman, Devinsky, et al., 2013).

Microglia serve critical functions in generating appropriate CNS immune responses and regulating synaptic pruning during brain development (Kabba et al., 2017; Kettenmann, Hanisch, Noda, & Verkhratsky, 2011). Microglial activation and proliferation (MAP) is initiated as a response to neuronal injury. Transient MAP can be neuroprotective and enhance recovery via the M2 phenotype. However, persistent MAP can be harmful and perpetuate neuronal injury by activating the M1 phenotype, producing excessive inflammatory mediators and ROS (Kabba et al., 2017). The harmful effects of MAP have been implicated in many psychiatric disorders (Berk et al., 2008; Mondelli, Vernon, Turkheimer, Dazzan, & Pariante, 2017), particularly in MDD in both humans and animal models (Frick, Williams, & Pittenger, 2013; Kreisel et al., 2013; Wachholz et al., 2016). Previous postmortem studies assessing MAP in the brains of subjects with MDD have yielded mixed results (Bayer, Buslei, Havas, & Falkai, 1999; Steiner et al., 2008; Steiner et al., 2011). However, more recent neuropathologic evidence suggests a strong mechanistic link between increased MAP and depression as well as suicide (Schnieder et al., 2014; Torres-Platas, Cruceanu, Chen, Turecki, & Mechawar, 2014). One study found a positive correlation between suicidality and both MAP density and increased microglial quinolinic acid expression (Najjar, Pearlman, Devinsky, et al., 2013). Persistent psychological stress in rats has been shown to induce MAP in the prefrontal cortex, hippocampus, and amygdala (Frick et al., 2013; Najjar, Pearlman, Devinsky, et al., 2013). One large meta-analysis confirmed that serum levels of pro-inflammatory cytokines, such as soluble interleukin-2 receptor (sIL-2R), interleukin 6 (IL-6), and TNFα, were consistently and significantly higher in individuals with MDD compared with healthy controls (Liu, Ho, et al., 2012; Miller, Maletic, & Raison, 2009). A recent brain PET imaging with radioligand [^{18}F]FEPPA in individuals with a major depressive episode found a significant increase in the expression of TSPO, a translocator protein considered a biomarker for neuroinflammation as in the brain it's primarily expressed by microglia (Setiawan et al., 2015). Experimental and human data showed that MAP and elevated pro-inflammatory cytokine levels can increase BBB permeability via several mechanisms, resulting in increased cross talk between innate and adaptive immunity, promoting a vicious cycle of MAP upregulation and brain cytokine production in a positive feedback loop (Betzen et al., 2009; Liu, Luiten, Eisel, Dejongste, & Schoemaker, 2013; Najjar, Pearlman, Devinsky, et al., 2013; Pun et al., 2009; Serlin et al., 2011). Activated microglia increase ROS synthesis (Dong, Zhang, & Qian, 2014; Hendriksen, Bergeijk, Oosting, & Redegeld, 2017), activate iNOS (Anderson et al., 2013; Salim et al., 2012), and promote COX2 expression within the neurovascular unit (Najjar, Pearlman, Devinsky, et al., 2013). These mechanisms are shown to increase BBB permeability *in vitro* (Pun et al., 2009; Stuehr et al., 2004). MAP and pro-inflammatory cytokines can activate matrix metalloproteinases (MMPs), which are proteolytic enzymes critical for tissue repair and extracellular matrix remodeling (Lehner et al., 2011; Shalev et al., 2009). Additionally, MMPs are shown *in vitro* to disrupt the BBB by damaging endothelial tight junction proteins leading to BBB leakage (Lehner et al., 2011). Data from 69 MDD patients showed significant correlations between serum levels of MMP-9 and depression severity, suggesting that MMP-9 activation may in part contribute to BBB disruption and overall pathophysiology of MDD (Yoshida et al., 2012). Several human and animal studies documented that pro-inflammatory cytokines, such as TNFα, interferon γ (IFNγ), and IL-1β, can induce expression of intercellular adhesion molecule 1 (ICAM-1) on the luminal surface of BBB endothelial cells (Dietrich, 2002; Haraldsen, Kvale, Lien, Farstad, & Brandtzaeg, 1996; Henninger et al., 1997; Isogai, Tanaka, & Asamura, 2004; Li et al., 2012; Pu et al., 2003; Zameer & Hoffman, 2003), resulting in a dose-dependent increase in BBB permeability (Najjar, Pearlman,

Devinsky, et al., 2013). A postmortem study found ICAM-1 expression to be significantly higher in the deep white matter of the dorsolateral prefrontal cortex in individuals with MDD compared with controls (Thomas et al., 2003). One study showed reduction in serum levels and vascular endothelial expression of both ICAM-1 and vascular cell adhesion molecule 1 (VCAM-1), in response to SSRIs (Lekakis et al., 2010). Although these findings suggest that increased expression of cellular adhesion molecules may be one of the mechanisms associated with BBB hyperpermeability in MDD (Miguel-Hidalgo et al., 2011; Najjar, Pearlman, Devinsky, et al., 2013; Schaefer et al., 2004), other studies did not support this assumption. Indeed, one neuropathologic study showed decreased expression of VCAM-1 and ICAM-1 in the orbitofrontal cortex of depressed subjects (Miguel-Hidalgo et al., 2011), and another study found SSRIs to have no effect on plasma levels of ICAM or VCAM (Dawood, Barton, Lambert, Eikelis, & Lambert, 2016). Further, *in vitro* studies showed that elevated pro-inflammatory cytokine such as TNFα may increase risk and severity of depression and increase BBB permeability (Liu et al., 2013) via other mechanisms related to excess generation of ROS by impairing vascular endothelial mitochondrial oxidative metabolism and reducing mitochondrial density (Gajewski, Rzodkiewicz, & Maśliński, 2017; Ott, Gogvadze, Orrenius, & Zhivotovsky, 2007; Scapagnini et al., 2012). Indeed, mounting human and animal data implicated mitochondrial dysfunction not only in the pathogenesis of MDD (Berk et al., 2013; Della et al., 2012; Moreno et al., 2013; Tobe, 2013) but also in the severity of symptoms (Karabatsiakis et al., 2014). One study showed a significant and negative correlation between mitochondrial respiration and severity of depressive symptoms, such as fatigue, lack of energy, and difficulty with concentration (Karabatsiakis et al., 2014). *In vitro*, vascular dysfunction caused by oxidative stress has been linked to mitochondrial abnormalities (Aliev et al., 2013). The potential contribution of

vascular endothelial mitochondrial abnormalities to neurovascular endothelial dysfunction and BBB hyperpermeability in MDD remains to be elucidated.

Another mechanism that may contribute to BBB hyperpermeability involves the polypeptide bradykinin, a pro-inflammatory mediator that induces vasodilation and capillary permeability (Golias, Charalabopoulos, Stagikas, Charalabopoulos, & Batistatou, 2007; Najjar, Pearlman, Devinsky, et al., 2013). Human data linking bradykinin abnormalities to MDD are primarily limited to the finding of functional single nucleotide polymorphisms of the bradykinin receptor B2 (BDKRB2) gene (Gratacòs et al., 2009; Najjar, Pearlman, Devinsky, et al., 2013). Upregulation of bradykinin activity and bradykinin B1 receptor expression was associated with depression-like behavior in mice, which improved following the administration of selective bradykinin B1 receptor antagonists (Viana et al., 2010). Experimental studies showed that increased expression of B1 and B2 receptors could increase BBB permeability, promote oxidative stress, and induce inflammation (Najjar, Pearlman, Devinsky, et al., 2013; Prat et al., 2000). Bradykinin activation can increase production of IL-6 via NFκB pathway in astroglia and promote opening of the BBB (Shalev et al., 2009). Increased endothelial Ca^{2+} influx via the B2 receptor stimulation can activate endothelial oxidant enzymes causing endothelial oxidative injury by increased ROS synthesis (Pun et al., 2009). Furthermore, accumulation of ROS within the endothelial cells can increase the BBB susceptibility to the toxic effects of bradykinin, exaggerating BBB hyperpermeability (Maes, Mihaylova, et al., 2011). *In vitro* human studies showed that inflammation-mediated kinin B1 receptor upregulation could increase BBB permeability (Prat et al., 2000). Similarly, inhibiting kinin B1 receptor is shown to limit BBB hyperpermeability and reduce inflammation in mice (Raslan et al., 2010). Activation of bradykinin can also stimulate phospholipase A2, which induces arachidonic acid metabolism and release, leading to increased malondialdehyde (Maes et al., 2013; Maes, Mihaylova, et al., 2011) and NO production (Kuhr, Lowry, Zhang, Brovkovych, & Skidgel, 2010; Pun et al., 2009) that may increase BBB permeability.

Glutamatergic hyperfunction may contribute to neurovascular unit dysfunction in MDD. Several neuroimaging, CSF, and neuropathologic studies have documented glutamatergic hyperfunction in individuals with MDD (Hashimoto, 2009; Hashimoto, Sawa, & Iyo, 2007; Jaso et al., 2016; Najjar, Pearlman, Devinsky, et al., 2013). Neuroinflammation may contribute to an imbalance between glutamate release and uptake via several mechanisms: (a) Microglial synthesis of quinolinic acid promotes microglial glutamate release via NMDA receptor activation (Miller et al., 2009) and inhibits astroglial glutamate uptake (Kim & Na, 2016); (b) astrocytic injury and activation promotes astroglial glutamate release, which in turn perpetuates microglial activation (Kim & Na, 2016); (c) reduced expression of astroglial excitatory amino acid transporters limits glutamate reuptake (Najjar, Pearlman, Devinsky, et al., 2013; Rao, Kellom, Reese, Rapoport, & Kim, 2012); and (d) upregulated microglial expression of Xc antiporter systems can enhance microglial glutamate release (Najjar, Pearlman, Devinsky, et al., 2013). Several postmortem investigations documented alteration in the expression of glutamatergic N-methyl-D-aspartate receptor (NMDAR) subunits, implicating glutamate neurotransmission dysregulation in subjects with MDD compared with healthy controls. These alterations are (a) hippocampus, increase or no change of NR1 subunit in the hippocampus and increase of NR2A and NR2B subunit (Oh, Kim, & Park, 2010; Oh, Park, Park, & Kim, 2012; Toro & Deakin, 2005); (b) prefrontal cortex, decrease or no change in NR1 subunit and decrease of NR2A and NR2B subunit (Beneyto & Meador-Woodruff, 2007; Feyissa, Chandran, Stockmeier, & Karolewicz, 2009);

and (c) lateral amygdalae, increase of NR2A subunit expression (Karolewicz et al., 2008). Excess glutamate can bind to its dysregulated BBB endothelial ionic NMDAR subunits and metabotropic glutamate receptors (mGluRs), resulting in increased endothelial Ca^{2+} influx, leading to oxidative stress-related endothelial injury and increased BBB permeability (Betzen et al., 2009; Najjar, Pearlman, Devinsky, et al., 2013; Pun et al., 2009; Scott, Bowman, Smith, Flower, & Bolton, 2007; Viana et al., 2010). ONOO-mediated oxidative injury of murine cerebrovascular endothelial cells is shown to be associated with upregulation of NMDAR expression, which in turn increases the susceptibility of vascular endothelium to glutamate toxicity (Betzen et al., 2009; Najjar, Pearlman, Devinsky, et al., 2013). In animal studies, oxidative stress is shown to specifically upregulate NR1 subunit expression, resulting in a positive feedback loop, which makes the BBB endothelium more vulnerable to both glutamate excitotoxicity and oxidative stress. The severity of NMDAR-induced oxidative damage is also shown to be attenuated by glutamate receptor antagonists (Betzen et al., 2009). Activation of NMDAR-induced oxidative injury can also adversely affect endothelial eNOS function and reduce endothelial synthesis of NO (Lemaistre et al., 2012), potentially contributing to cerebral hypoperfusion (Najjar, Pearlman, Devinsky, et al., 2013). BBB breakdown may disrupt the endothelium-bound glutamate efflux transporters, leading to elevated levels of glutamate in the CNS (Shlosberg et al., 2010); hyperglutamatergia may in turn increase BBB susceptibility to the harmful effects of bradykinin (Najjar, Pearlman, Devinsky, et al., 2013). Elevations in bradykinin-induced endothelial Ca^{2+} can be blocked by glutamate receptor antagonists (Najjar, Pearlman, Devinsky, et al., 2013; Pun et al., 2009). Neuronal dysfunction in MDD may be attributable to multiple mechanisms such as, increased endothelial NMDAR expression, BBB permeability, and elevated CNS glutamate levels (Najjar, Pearlman, Devinsky, et al., 2013).

Mast cells are tissue-bound granulated cells that play a critical role in the development of allergic reactions and inflammation (Theoharides et al., 2012). Similar to basophils, they contain high levels of histamine and heparin (Najjar, Pearlman, Devinsky, et al., 2013). Mast cells are abundant in the hypothalamic region, especially in the median eminence (Theoharides, Stewart, Panagiotidou, & Melamed, 2016). Inflammatory pathways involving mast cells activation have been linked to depression neurobiology (Georgin-Lavialle et al., 2016). Depressive symptoms are present in approximately 40%–70% of individuals with mastocytosis, a rare disorder characterized by mast cell accumulation and activation (Moura et al., 2012). Mast cells are located adjacent to corticotrophin-releasing factor (CRF)-positive neurons, and increased CRF secretion associated with MDD can lead to mast cell activation in MDD (Shalev et al., 2009; Theoharides et al., 2016). Evidence suggests various mechanisms through which mast cells may contribute to brain inflammation and BBB hyperpermeability: (a) release of inflammatory mediators such as IL-6 and TNFα, (b) stimulation of MAP, (c) facilitation of NMDAR-induced neuronal excitotoxicity, and (d) stimulation of release of VEGF that can increase BBB permeability and enhance transendothelial migration of inflammatory cells into the brain (Theoharides et al., 2016; Theoharides & Zhang, 2011).

CONCLUSIONS

Increasing evidence suggests that the neurobiology of depression has a multifactorial etiology. It is well established that MDD is often comorbid with neurological disorders associated with BBB abnormalities and medical conditions characterized by vascular endothelial dysfunction. We further reviewed recent clinical and experimental data implicating the potential role of inflammation, oxidative stress, eNOS uncoupling, and reduced endothelial NO synthesis in peripheral vascular endothelial dysfunction in a subset of individuals with primary MDD. Peripheral inflammation may cause neuroinflammation damaging astroglia, increasing microglial activation and proliferation with resultant neurovascular unit dysfunction and BBB hyperpermeability. Excessive generation of ROS and reduced antioxidant defenses lead to oxidative stress via mechanisms that affect endothelial NO synthesis, influenced by endothelial levels of arginine and BH4. Vascular endothelial dysfunction induced by eNOS uncoupling involves increased $ONOO^-$ production and O_2^- synthesis and decreased BH4 levels. Our theoretical integration of the human and animal data suggests that similar inflammatory and oxidative mechanisms may also contribute to the putative neurovascular endothelial dysfunction and BBB hyperpermeability in MDD. Future research should aim to develop methods that can quantify subtle increases in BBB permeability or extent of endothelial dysfunction. This can allow us to compare measurable differences in neurovascular unit dysfunction across various groups such as individuals with neurologic disorders and comorbid MDD and those with MDD alone. Clinical implications may be the development of novel treatments for MDD that target underlying inflammation, increase antioxidant defenses, or affect cell signaling mechanisms resulting in neurovascular unit dysfunction and increased BBB permeability.

References

Abbott, N. J., & Friedman, A. (2012). Overview and introduction: the blood-brain barrier in health and disease. *Epilepsia*, 53, 1–6. https://doi.org/10.1111/j.1528-1167.2012. 03696.x.

Abbott, N. J., Rönnbäck, L., & Hansson, E. (2006). Astrocyte-endothelial interactions at the blood-brain barrier. *Nature Reviews Neuroscience*, 7(1), 41–53. https://doi.org/10.1038/nrn1824.

Aliev, G., Obrenovich, M. E., Tabrez, S., Jabir, N. R., Reddy, V. P., Li, Y., et al. (2013). Link between cancer and Alzheimer disease via oxidative stress induced by nitric oxide-dependent mitochondrial DNA overproliferation and deletion. *Oxidative Medicine and Cellular Longevity*, *2013*, 1–19. https://doi.org/10.1155/2013/962984.

Aliev, G., Palacios, H. H., Walrafen, B., Lipsitt, A. E., Obrenovich, M. E., & Morales, L. (2009). Brain mitochondria as a primary target in the development of treatment strategies for Alzheimer disease. *The International Journal of Biochemistry & Cell Biology*, *41*(10), 1989–2004. https://doi.org/10.1016/j.biocel.2009.03.015.

Anderson, G., Berk, M., Dodd, S., Bechter, K., Altamura, A. C., Dell'osso, B., et al. (2013). Immuno-inflammatory, oxidative and nitrosative stress, and neuroprogressive pathways in the etiology, course and treatment of schizophrenia. *Progress in Neuro-Psychopharmacology and Biological Psychiatry*, *42*, 1–4. https://doi.org/10.1016/j.pnpbp.2012.10.008.

Andrew, P., & Mayer, B. (1999). Enzymatic function of nitric oxide synthases. *Cardiovascular Research*, *43*(3), 521–531. https://doi.org/10.1016/s0008-6363(99)00115-7.

Antoniades, C., Shirodaria, C., Warrick, N., Cai, S., de Bono, J., Lee, J., et al. (2006). 5-Methyltetrahydrofolate rapidly improves endothelial function and decreases superoxide production in human vessels: effects on vascular tetrahydrobiopterin availability and endothelial nitric oxide synthase coupling. *Circulation*, *114*(11), 1193–1201. https://doi.org/10.1161/circulationaha.106.612325.

Attwell, D., Buchan, A. M., Charpak, S., Lauritzen, M., Macvicar, B. A., & Newman, E. A. (2010). Glial and neuronal control of brain blood flow. *Nature*, *468*(7321), 232–243. https://doi.org/10.1038/nature09613.

Banasr, M., & Duman, R. S. (2008). Glial loss in the prefrontal cortex is sufficient to induce depressive-like behaviors. *Biological Psychiatry*, *64*(10), 863–870. https://doi.org/10.1016/j.biopsych.2008.06.008.

Barth, J., Schumacher, M., & Herrmann-Lingen, C. (2004). Depression as a risk factor for mortality in patients with coronary heart disease: a meta-analysis. *Psychosomatic Medicine*, *66*(6), 802–813. https://doi.org/10.1097/01.psy.0000146332.53619.b2.

Bayer, T. A., Buslei, R., Havas, L., & Falkai, P. (1999). Evidence for activation of microglia in patients with psychiatric illnesses. *Neuroscience Letters*, *271*(2), 126–128. https://doi.org/10.1016/s0304-3940(99)00545-5.

Bechter, K., Reiber, H., Herzog, S., Fuchs, D., Tumani, H., & Maxeiner, H. (2010). Cerebrospinal fluid analysis in affective and schizophrenic spectrum disorders: identification of subgroups with immune responses and blood-CSF barrier dysfunction. *Journal of Psychiatric Research*, *44*(5), 321–330. https://doi.org/10.1016/j.jpsychires.2009.08.008.

Belmaker, R., & Agam, G. (2008). Major depressive disorder. *New England Journal of Medicine*, *358*(1), 55–68. https://doi.org/10.1056/nejmra073096.

Benedetto, B. D., Malik, V. A., Begum, S., Jablonowski, L., Gómez-González, G. B., Neumann, I. D., et al. (2016). Fluoxetine requires the endfeet protein aquaporin-4 to enhance plasticity of astrocyte processes. *Frontiers in Cellular Neuroscience*. *10*, https://doi.org/10.3389/fncel.2016.00008.

Beneyto, M., & Meador-Woodruff, J. H. (2007). Lamina-specific abnormalities of NMDA receptor-associated postsynaptic protein transcripts in the prefrontal cortex in schizophrenia and bipolar disorder. *Neuropsychopharmacology*, *33*(9), 2175–2186. https://doi.org/10.1038/sj.npp.1301604.

Berk, M., Copolov, D. L., Dean, O., Lu, K., Jeavons, S., Schapkaitz, I., et al. (2008). N-Acetyl cysteine for depressive symptoms in bipolar disorder—a double-blind randomized placebo-controlled trial. *Biological Psychiatry*, *64*(6), 468–475. https://doi.org/10.1016/j.biopsych.2008.04.022.

Berk, M., Williams, L. J., Jacka, F. N., O'Neil, A., Pasco, J. A., Moylan, S., et al. (2013). So depression is an inflammatory disease, but where does the inflammation come from? *BMC Medicine*. *11*(1)https://doi.org/10.1186/1741-7015-11-200.

Betzen, C., White, R., Zehendner, C. M., Pietrowski, E., Bender, B., Luhmann, H. J., et al. (2009). Oxidative stress upregulates the NMDA receptor on cerebrovascular endothelium. *Free Radical Biology and Medicine*, *47*(8), 1212–1220. https://doi.org/10.1016/j.freeradbiomed.2009.07.034.

Block, M. L., Zecca, L., & Hong, J. (2007). Microglia-mediated neurotoxicity: uncovering the molecular mechanisms. *Nature Reviews Neuroscience*, *8*(1), 57–69. https://doi.org/10.1038/nrn2038.

Carmeliet, P., & De Strooper, B. (2012). Alzheimer's disease: a breach in the blood-brain barrier. *Nature*, *485*, 451–452. https://doi.org/10.1038/485451a.

Carney, R. M., Freedland, K. E., Miller, G. E., & Jaffe, A. S. (2002). Depression as a risk factor for cardiac mortality and morbidity. *Journal of Psychosomatic Research*, *53*(4), 897–902. https://doi.org/10.1016/s0022-3999(02)00311-2.

Chen, C., Wang, T., Varadharaj, S., Reyes, L. A., Hemann, C., Talukder, M. A., et al. (2010). S-glutathionylation uncouples eNOS and regulates its cellular and vascular function. *Nature*, *468*(7327), 1115–1118. https://doi.org/10.1038/nature09599.

Chen, W., Druhan, L. J., Chen, C., Hemann, C., Chen, Y., Berka, V., et al. (2010). Peroxynitrite induces destruction of the tetrahydrobiopterin and heme in endothelial nitric oxide synthase: transition from reversible to irreversible enzyme inhibition. *Biochemistry*, *49*(14), 3129–3137. https://doi.org/10.1021/bi9016632.

Chrapko, W. E., Jurasz, P., Radomski, M. W., Lara, N., Archer, S. L., & Mellédo, J. L. (2004). Decreased platelet nitric oxide synthase activity and plasma nitric oxide metabolites in major depressive disorder. *Biological Psychiatry*, *56*(2), 129–134. https://doi.org/10.1016/j.biopsych.2004.03.003.

Chuang, Y. C. (2010). Mitochondrial dysfunction and oxidative stress in seizure induced neuronal cell death. *Acta Neurologica Taiwanica*, *19*(1), 3–15.

Cotter, D., Hudson, L., & Landau, S. (2005). Evidence for orbitofrontal pathology in bipolar disorder and major depression, but not in schizophrenia. *Bipolar Disorders*, *7*(4), 358–369. https://doi.org/10.1111/j.1399-5618.2005.00230.x.

Cotter, D., Mackay, D., Chana, G., Beasley, C., Landau, S., & Everall, I. P. (2002). Reduced neuronal size and glial cell density in area 9 of the dorsolateral prefrontal cortex in subjects with major depressive disorder. *Cerebral Cortex*, *12*(4), 386–394. https://doi.org/10.1093/cercor/12.4.386.

Dawood, T., Barton, D. A., Lambert, E. A., Eikelis, N., & Lambert, G. W. (2016). Examining endothelial function and platelet reactivity in patients with depression before and after SSRI therapy. *Frontiers in Psychiatry*. *7*. https://doi.org/10.3389/fpsyt.2016.00018.

de Klerk, O. L., Bosker, F. J., Willemsen, A. T., Waarde, A. V., Visser, A. K., de Jager, T., et al. (2010). Chronic stress and antidepressant treatment have opposite effects on P-glycoprotein at the blood-brain barrier: an experimental PET study in rats. *Journal of Psychopharmacology*, *24*(8), 1237–1242. https://doi.org/10.1177/0269881109349840.

Della, F. P., Abelaira, H. M., Réus, G. Z., Antunes, A. R., Santos, M. A., Zappelinni, G., et al. (2012). Tianeptine exerts neuroprotective effects in the brain tissue of rats exposed to the chronic stress model. *Pharmacology Biochemistry and Behavior*, *103*(2), 395–402. https://doi.org/10.1016/j.pbb.2012.09.018.

Diener, C., Kuehner, C., Brusniak, W., Ubl, B., Wessa, M., & Flor, H. (2012). A meta-analysis of neurofunctional imaging studies of emotion and cognition in major depression. *NeuroImage*, *61*(3), 677–685. https://doi.org/10.1016/j.neuroimage.2012.04.005.

Dietrich, J. (2002). The adhesion molecule ICAM-1 and its regulation in relation with the blood-brain barrier. *Journal of Neuroimmunology*, *128*(1–2), 58–68. https://doi.org/10.1016/s0165-5728(02)00114-5.

Dong, H., Zhang, X., & Qian, Y. (2014). Mast cells and neuroinflammation. *Medical Science Monitor Basic Research*, *20*, 200–206. https://doi.org/10.12659/MSMBR.893093.

Dröge, W. (2002). Free radicals in the physiological control of cell function. *Physiological Reviews*, *82*(1), 47–95. https://doi.org/10.1152/physrev.00018.2001.

Enciu, A., Gherghiceanu, M., & Popescu, B. O. (2013). Triggers and effectors of oxidative stress at blood-brain barrier level: relevance for brain ageing and neurodegeneration. *Oxidative Medicine and Cellular Longevity*, *2013*, 1–12. https://doi.org/10.1155/2013/297512.

Falcone, T., Fazio, V., Lee, C., Simon, B., Franco, K., Marchi, N., et al. (2010). Serum S100B: a potential biomarker for suicidality in adolescents? *PLoS ONE*. *5*(6) https://doi.org/10.1371/journal.pone.0011089.

Feyissa, A. M., Chandran, A., Stockmeier, C. A., & Karolewicz, B. (2009). Reduced levels of NR2A and NR2B subunits of NMDA receptor and PSD-95 in the prefrontal cortex in major depression. *Progress in Neuro-Psychopharmacology and Biological Psychiatry*, *33*(1), 70–75. https://doi.org/10.1016/j.pnpbp.2008.10.005.

Ford, D. E., Mead, L. A., Chang, P. P., Cooper-Patrick, L., Wang, N. Y., & Klag, M. J. (1998). Depression is a risk factor for coronary artery disease in men: the precursors study. *Archives of Internal Medicine*, *158*(13), 1422–1426. https://doi.org/10.1001/archinte.158.13.1422.

Förstermann, U. (2006). Janus-faced role of endothelial NO synthase in vascular disease: uncoupling of oxygen reduction from NO synthesis and its pharmacological reversal. *Biological Chemistry*. *387*(12). https://doi.org/10.1515/bc.2006.190.

Frick, L. R., Williams, K., & Pittenger, C. (2013). Microglial dysregulation in psychiatric disease. *Clinical and Developmental Immunology*, *2013*, 1–10. https://doi.org/10.1155/2013/608654.

Friedman, A., & Kaufer, D. (2011). Blood-brain barrier breakdown and blood-brain communication in neurological and psychiatric diseases. *Cardiovascular Psychiatry and Neurology*, *2011*, 1–2. https://doi.org/10.1155/2011/431470.

Fujii, T., Ota, M., Hori, H., Sasayama, D., Hattori, K., Teraishi, T., et al. (2012). Association between the functional polymorphism (C3435T) of the gene encoding P-glycoprotein (ABCB1) and major depressive disorder in the Japanese population. *Journal of Psychiatric Research*, *46*(4), 555–559. https://doi.org/10.1016/j.jpsychires.2012.01.012.

Gajewski, M., Rzodkiewicz, P., & Maśliński, S. (2017). The human body as an energetic hybrid? New perspectives for chronic disease treatment? *Reumatologia*, *2*, 94–99. https://doi.org/10.5114/reum.2017.67605.

Galea, E., Feinstein, D. L., & Reis, D. J. (1992). Induction of calcium-independent nitric oxide synthase activity in primary rat glial cultures. *Proceedings of the National Academy of Sciences*, *89*(22), 10945–10949. https://doi.org/10.1073/pnas.89.22.10945.

Gardner, A., Pagani, M., Wibom, R., Nennesmo, I., Jacobsson, H., & Hallstrom, T. (2003). Alterations of rCBF and mitochondrial dysfunction in major depressive disorder: a case report. *Acta Psychiatrica Scandinavica*, 107(3), 233–239. https://doi.org/10.1034/j.1600-0447.2003.02188.x.

Gawryluk, J. W., Wang, J., Andreazza, A. C., Shao, L., & Young, L. T. (2010). Decreased levels of glutathione, the major brain antioxidant, in post-mortem prefrontal cortex from patients with psychiatric disorders. *The International Journal of Neuropsychopharmacology*, 14(01), 123–130. https://doi.org/10.1017/s1461145710000805.

Georgin-Lavialle, S., Moura, D. S., Salvador, A., Chauvet-Gelinier, J., Launay, J., Damaj, G., et al. (2016). Mast cells' involvement in inflammation pathways linked to depression: evidence in mastocytosis. *Molecular Psychiatry*, 21(11), 1511–1516. https://doi.org/10.1038/mp.2015.216.

Gibson, S. A., Korade, Ž., & Shelton, R. C. (2012). Oxidative stress and glutathione response in tissue cultures from persons with major depression. *Journal of Psychiatric Research*, 46(10), 1326–1332. https://doi.org/10.1016/j.jpsychires.2012.06.008.

Golias, C., Charalabopoulos, A., Stagikas, D., Charalabopoulos, K., & Batistatou, A. (2007). The kinin system-bradykinin: biological effects and clinical implications. Multiple role of the kinin system-bradykinin. *Hippokratia*, 11(3), 124–128.

Gratacòs, M., Costas, J., Cid, R. D., Bayés, M., González, J. R., Baca-Garcia, E., et al. (2009). Identification of new putative susceptibility genes for several psychiatric disorders by association analysis of regulatory and non-synonymous SNPs of 306 genes involved in neurotransmission and neurodevelopment. *American Journal of Medical Genetics Part B: Neuropsychiatric Genetics*, 150B(6), 808–816. https://doi.org/10.1002/ajmg.b.30902.

Gray, D. A., & Woulfe, J. (2005). Lipofuscin and aging: a matter of toxic waste. *Science of Aging Knowledge Environment*. 2005(5). https://doi.org/10.1126/sageke.2005.5.re1.

Gudmundsson, P., Skoog, I., Waern, M., Blennow, K., Pálsson, S., Rosengren, L., et al. (2007). The relationship between cerebrospinal fluid biomarkers and depression in elderly women. *The American Journal of Geriatric Psychiatry*, 15(10), 832–838. https://doi.org/10.1097/jgp.0b013e3180547091.

Halaris, A. (2009). Co-morbidity between depression and cardiovascular disease. *International Angiology: A Journal of the International Union of Angiology*, 28(2), 92–99. Retrieved from April 16, 2009. https://www.ncbi.nlm.nih.gov/pubmed/19367238.

Halaris, A. (2013). Inflammation, heart disease, and depression. *Current Psychiatry Reports*. 15(10). https://doi.org/10.1007/s11920-013-0400-5.

Halliwell, B. (2001). Role of free radicals in the neurodegenerative diseases. *Drugs & Aging, 18*(9), 685–716. https://doi.org/10.2165/00002512-200118090-00004.

Haorah, J., Knipe, B., Leibhart, J., Ghorpade, A., & Persidsky, Y. (2005). Alcohol-induced oxidative stress in brain endothelial cells causes blood-brain barrier dysfunction. *Journal of Leukocyte Biology*, 78(6), 1223–1232. https://doi.org/10.1189/jlb.0605340.

Haraldsen, G., Kvale, D., Lien, B., Farstad, I. N., & Brandtzaeg, P. (1996). Cytokine-regulated expression of E-selectin, intercellular adhesion molecule-1 (ICAM-1), and vascular cell adhesion molecule-1 (VCAM-1) in human microvascular endothelial cells. *Journal of Immunology*, 156(7), 2558–2565.

Haroon, E., Raison, C. L., & Miller, A. H. (2011). Psychoneuroimmunology meets neuropsychopharmacology: translational implications of the impact of inflammation on behavior. *Neuropsychopharmacology*, 37(1), 137–162. https://doi.org/10.1038/npp.2011.205.

Hashimoto, K. (2009). Emerging role of glutamate in the pathophysiology of major depressive disorder. *Brain Research Reviews*, 61(2), 105–123. https://doi.org/10.1016/j.brainresrev.2009.05.005.

Hashimoto, K., Sawa, A., & Iyo, M. (2007). Increased levels of glutamate in brains from patients with mood disorders. *Biological Psychiatry*, 62(11), 1310–1316. https://doi.org/10.1016/j.biopsych.2007.03.017.

Hawkins, B. T., Sykes, D. B., & Miller, D. S. (2010). Rapid, reversible modulation of blood-brain barrierP-glycoprotein transport activity by vascular endothelial growth factor. *Journal of Neuroscience*, 30(4), 1417–1425. https://doi.org/10.1523/jneurosci.5103-09.2010.

Hendriksen, E., Bergeijk, D. V., Oosting, R. S., & Redegeld, F. A. (2017). Mast cells in neuroinflammation and brain disorders. *Neuroscience & Biobehavioral Reviews*, 79, 119–133. https://doi.org/10.1016/j.neubiorev.2017.05.001.

Henninger, D. D., Panes, J., Eppihimer, M., Russell, J., Gerritsen, M., Anderson, D. C., et al. (1997). Cytokine-induced VCAM-1 and ICAM-1 expression in different organs of the mouse. *Journal of Immunology*, 158, 1825–1832.

Higgins, E. S., & George, M. S. (2013). *Neuroscience of clinical psychiatry: The pathophysiology of behavior and mental illness* (2nd ed.). Philadelphia: Lippincott Williams and Wilkins [chapter 21].

Hong, M., Zheng, J., Ding, Z., Chen, J., Yu, L., Niu, Y., et al. (2013). Imbalance between Th17 and Treg cells may play an important role in the development of chronic unpredictable mild stress-induced depression in mice. *Neuroimmunomodulation*, 20(1), 39–50. https://doi.org/10.1159/000343100.

Huang, Z., Huang, P., Panahian, N., Dalkara, T., Fishman, M., & Moskowitz, M. (1994). Effects of cerebral ischemia in mice deficient in neuronal nitric oxide synthase. *Science*, 265(5180), 1883–1885. https://doi.org/10.1126/science.7522345.

Ikenouchi-Sugita, A., Yoshimura, R., Kishi, T., Umene-Nakano, W., Hori, H., Hayashi, K., et al. (2011). Three polymorphisms of the eNOS gene and plasma levels of metabolites of nitric oxide in depressed Japanese patients: a preliminary report. *Human Psychopharmacology: Clinical and Experimental*, 26(7), 531–534. https://doi.org/10.1002/hup.1239.

Insel, T., & Wang, P. (2009). The STAR*D trial: revealing the need for better treatments. *Psychiatric Services*, 60(11). https://doi.org/10.1176/appi.ps.60.11.1466.

Isingrini, E., Belzung, C., Freslon, J., Machet, M., & Camus, V. (2011). Fluoxetine effect on aortic nitric oxide-dependent vasorelaxation in the unpredictable chronic mild stress model of depression in mice. *Psychosomatic Medicine*, 74(1), 63–72. https://doi.org/10.1097/psy.0b013e31823a43e0.

Isogai, N., Tanaka, H., & Asamura, S. (2004). Thrombosis and altered expression of intercellular adhesion molecule-1 (Icam-1) after avulsion injury in rat vessels. *Journal of Hand Surgery*, 29(3), 228–232. https://doi.org/10.1016/j.jhsb.2004.03.001.

Iwase, K. (2000). Induction of endothelial nitric-oxide synthase in rat brain astrocytes by systemic lipopolysaccharide treatment. *Journal of Biological Chemistry*, 275(16), 11929–11933. https://doi.org/10.1074/jbc.275.16.11929.

Jaso, B. A., Niciu, M. J., Iadarola, N. D., Lally, N., Richards, E. M., Park, M., et al. (2016). Therapeutic modulation of glutamate receptors in major depressive disorder. *Current Neuropharmacology*, 15(1), 57–70. https://doi.org/10.2174/1570159x14666160321123221.

Jin, R., Song, Z., Yu, S., Piazza, A., Nanda, A., Penninger, J. M., et al. (2011). Phosphatidylinositol-3-kinase gamma plays a central role in blood-brain barrier dysfunction in acute experimental stroke. *Stroke*, 42(7), 2033–2044. https://doi.org/10.1161/strokeaha.110.601369.

Joynt, K. E., Whellan, D. J., & Oconnor, C. M. (2003). Depression and cardiovascular disease: mechanisms of interaction. *Biological Psychiatry*, 54(3), 248–261. https://doi.org/10.1016/s0006-3223(03)00568-7.

Kabba, J. A., Xu, Y., Christian, H., Ruan, W., Chenai, K., Xiang, Y., et al. (2017). Microglia: housekeeper of the central nervous system. *Cellular and Molecular Neurobiology*. https://doi.org/10.1007/s10571-017-0504-2.

Karabatsiakis, A., Böck, C., Salinas-Manrique, J., Kolassa, S., Calzia, E., Dietrich, D. E., et al. (2014). Mitochondrial respiration in peripheral blood mononuclear cells correlates with depressive subsymptoms and severity of major depression. *Translational Psychiatry*. 4(6). https://doi.org/10.1038/tp.2014.44.

Karolewicz, B., Szebeni, K., Gilmore, T., Maciag, D., Stockmeier, C. A., & Ordway, G. A. (2008). Elevated levels of NR2A and PSD-95 in the lateral amygdala in depression. *The International Journal of Neuropsychopharmacology*, 12(02), 143–153. https://doi.org/10.1017/s1461145708008985.

Keller, J. N., Dimayuga, E., Chen, Q., Thorpe, J., Gee, J., & Ding, Q. (2004). Autophagy, proteasomes, lipofuscin, and oxidative stress in the aging brain. *The International Journal of Biochemistry & Cell Biology*, 36(12), 2376–2391. https://doi.org/10.1016/j.biocel.2004.05.003.

Kettenmann, H., Hanisch, U., Noda, M., & Verkhratsky, A. (2011). Physiology of microglia. *Physiological Reviews*, 91(2), 461–553. https://doi.org/10.1152/physrev.00011.2010.

Khan, M., Sakakima, H., Dhammu, T. S., Shunmugavel, A., Im, Y., Gilg, A. G., et al. (2011). S-Nitrosoglutathione reduces oxidative injury and promotes mechanisms of neurorepair following traumatic brain injury in rats. *Journal of Neuroinflammation*, 8(1), 78. https://doi.org/10.1186/1742-2094-8-78.

Khatri, R., Mckinney, A. M., Swenson, B., & Janardhan, V. (2012). Blood-brain barrier, reperfusion injury, and hemorrhagic transformation in acute ischemic stroke. *Neurology*. 79(13 Suppl 1). https://doi.org/10.1212/wnl.0b013e3182697e70.

Kim, Y., & Na, K. (2016). Role of glutamate receptors and glial cells in the pathophysiology of treatment-resistant depression. *Progress in Neuro-Psychopharmacology and Biological Psychiatry*, 70, 117–126. https://doi.org/10.1016/j.pnpbp.2016.03.009.

Kreisel, T., Frank, M. G., Licht, T., Reshef, R., Ben-Menachem-Zidon, O., Baratta, M. V., et al. (2013). Dynamic microglial alterations underlie stress-induced depressive-like behavior and suppressed neurogenesis. *Molecular Psychiatry*, 19(6), 699–709. https://doi.org/10.1038/mp.2013.155.

Kuhr, F., Lowry, J., Zhang, Y., Brovkovych, V., & Skidgel, R. (2010). Differential regulation of inducible and endothelial nitric oxide synthase by kinin B1 and B2 receptors. *Neuropeptides*, 44(2), 145–154. https://doi.org/10.1016/j.npep.2009.12.004.

Lacza, Z., Pankotai, E., & Busija, D. W. (2009). Mitochondrial nitric oxide synthase: current concepts and controversies. *Frontiers in Bioscience*, 14, 4436. https://doi.org/10.2741/3539.

Lacza, Z., Puskar, M., Figueroa, J. P., Zhang, J., Rajapakse, N., & Busija, D. W. (2001). Mitochondrial nitric oxide synthase is constitutively active and is functionally upregulated in hypoxia. *Free Radical Biology and Medicine*, 31 (12), 1609–1615. https://doi.org/10.1016/s0891-5849(01)00754-7.

Lara, N., Archer, S. L., Baker, G. B., & Le Mellédo, J. M. (2003). Paroxetine-induced increase in metabolic end products of nitric oxide. *Journal of Clinical Psychopharmacology*, 23(6), 641–645. https://doi.org/10.1097/01.jcp.0000085416.08426.1d.

Lavoie, K. L., Pelletier, R., Arsenault, A., Dupuis, J., & Bacon, S. L. (2010). Association between clinical depression and endothelial function measured by forearm hyperemic reactivity. *Psychosomatic Medicine*, 72(1), 20–26. https://doi.org/10.1097/psy.0b013e3181c2d6b8.

Le Mellédo, J.-M., Mahil, N., & Baker, G. B. (2004). Nitric oxide: a key player in the relation between cardiovascular disease and major depressive disorder? *Journal of Psychiatry and Neuroscience*, 29(6), 414–416.

Lehner, C., Gehwolf, R., Tempfer, H., Krizbai, I., Hennig, B., Bauer, H., et al. (2011). Oxidative stress and blood-brain barrier dysfunction under particular consideration of matrix metalloproteinases. *Antioxidants & Redox Signaling*, 15(5), 1305–1323. https://doi.org/10.1089/ars.2011.3923.

Lekakis, J., Ikonomidis, I., Papoutsi, Z., Moutsatsou, P., Nikolaou, M., Parissis, J., et al. (2010). Selective serotonin re-uptake inhibitors decrease the cytokine-induced endothelial adhesion molecule expression, the endothelial adhesiveness to monocytes and the circulating levels of vascular adhesion molecules. *International Journal of Cardiology*, 139(2), 150–158. https://doi.org/10.1016/j.ijcard.2008.10.010.

Lemaistre, J. L., Sanders, S. A., Stobart, M. J., Lu, L., Knox, J. D., Anderson, H. D., et al. (2012). Coactivation of NMDA receptors by glutamate and D-serine induces dilation of isolated middle cerebral arteries. *Journal of Cerebral Blood Flow & Metabolism*, 32(3), 537–547. https://doi.org/10.1038/jcbfm.2011.161.

Li, J., Ye, L., Wang, X., Liu, J., Wang, Y., Zhou, Y., et al. (2012). (−)-Epigallocatechin gallate inhibits endotoxin-induced expression of inflammatory cytokines in human cerebral microvascular endothelial cells. *Journal of Neuroinflammation*. 9(1). https://doi.org/10.1186/1742-2094-9-141.

Liu, H., Luiten, P. G., Eisel, U. L., Dejongste, M. J., & Schoemaker, R. G. (2013). Depression after myocardial infarction: TNF-α-induced alterations of the blood-brain barrier and its putative therapeutic implications. *Neuroscience & Biobehavioral Reviews*, 37(4), 561–572. https://doi.org/10.1016/j.neubiorev.2013.02.004.

Liu, H., & Zhang, J. (2012). Cerebral hypoperfusion and cognitive impairment: the pathogenic role of vascular oxidative stress. *International Journal of Neuroscience*, 122(9), 494–499. https://doi.org/10.3109/00207454.2012.686543.

Liu, J. Y., Thom, M., Catarino, C. B., Martinian, L., Figarella-Branger, D., Bartolomei, F., et al. (2012). Neuropathology of the blood-brain barrier and pharmaco-resistance in human epilepsy. *Brain*, 135(10), 3115–3133. https://doi.org/10.1093/brain/aws147.

Liu, L., Lu, Y., Kong, H., Li, L., Marshall, C., Xiao, M., et al. (2012). Aquaporin-4 deficiency exacerbates brain oxidative damage and memory deficits induced by long-term ovarian hormone deprivation and D-galactose injection. *The International Journal of Neuropsychopharmacology*, 15(01), 55–68. https://doi.org/10.1017/s1461145711000022.

Liu, V. W. T., & Huang, P. L. (2008). Cardiovascular roles of nitric oxide: a review of insights from nitric oxide synthase gene disrupted mice. *Cardiovascular Research*, 77(1), 19–29. https://doi.org/10.1016/j.cardiores.2007.06.024.

Liu, Y., Ho, R. C., & Mak, A. (2012). Interleukin (IL)-6, tumour necrosis factor alpha (TNF-α) and soluble interleukin-2 receptors (sIL-2R) are elevated in patients with major depressive disorder: a meta-analysis and meta-regression. *Journal of Affective Disorders*, 139(3), 230–239. https://doi.org/10.1016/j.jad.2011.08.003.

Lochhead, J. J., Mccaffrey, G., Quigley, C. E., Finch, J., Demarco, K. M., Nametz, N., et al. (2010). Oxidative stress increases blood-brain barrier permeability and induces alterations in occludin during hypoxia-reoxygenation. *Journal of Cerebral Blood Flow & Metabolism*, 30(9), 1625–1636. https://doi.org/10.1038/jcbfm.2010.29.

Lopez-Vilchez, I., Diaz-Ricart, M., Navarro, V., Torramade, S., Zamorano-Leon, J., Lopez-Farre, A., et al. (2016). Endothelial damage in major depression patients is modulated by SSRI treatment, as demonstrated by circulating biomarkers and an in vitro cell model. *Translational Psychiatry*. 6(9). https://doi.org/10.1038/tp.2016.156.

Lu, Q., Xia, N., Xu, H., Guo, L., Wenzel, P., Daiber, A., et al. (2011). Betulinic acid protects against cerebral ischemia-reperfusion injury in mice by reducing oxidative and nitrosative stress. *Nitric Oxide*, 24(3), 132–138. https://doi.org/10.1016/j.niox.2011.01.007.

Luiking, Y. C., Ten Have, G. A. M., Wolfe, R. R., & Deutz, N. E. P. (2012). Arginine de novo and nitric oxide production in disease states. *American Journal of Physiology Endocrinology and Metabolism*, 303(10), E1177–E1189. https://doi.org/10.1152/ajpendo.00284.2012.

Lund, H., Krakauer, M., Skimminge, A., Sellebjerg, F., Garde, E., Siebner, H. R., et al. (2013). Blood-brain barrier permeability of normal appearing white matter in relapsing-remitting multiple sclerosis. *PLoS ONE, 8*(2) https://doi.org/10.1371/journal.pone.0056375.

Maes, M., Galecki, P., Chang, Y. S., & Berk, M. (2011). A review on the oxidative and nitrosative stress (O&NS) pathways in major depression and their possible contribution to the (neuro)degenerative processes in that illness. *Progress in Neuro-Psychopharmacology and Biological Psychiatry, 35*(3), 676–692. https://doi.org/10.1016/j.pnpbp.2010.05.004.

Maes, M., Kubera, M., Mihaylova, I., Geffard, M., Galecki, P., Leunis, J., et al. (2013). Increased autoimmune responses against auto-epitopes modified by oxidative and nitrosative damage in depression: implications for the pathways to chronic depression and neuroprogression. *Journal of Affective Disorders, 149*(1–3), 23–29. https://doi.org/10.1016/j.jad.2012.06.039.

Maes, M., Mihaylova, I., Kubera, M., Leunis, J., & Geffard, M. (2011). IgM-mediated autoimmune responses directed against multiple neoepitopes in depression: new pathways that underpin the inflammatory and neuroprogressive pathophysiology. *Journal of Affective Disorders, 135*(1–3), 414–418. https://doi.org/10.1016/j.jad.2011.08.023.

Maier, S. F. (2003). Bi-directional immune-brain communication: implications for understanding stress, pain, and cognition. *Brain, Behavior, and Immunity, 17*(2), 69–85. https://doi.org/10.1016/s0889-1591(03)00032-1.

Masano, T., Kawashima, S., Toh, R., Satomi-Kobayashi, S., Shinohara, M., Takaya, T., et al. (2008). Beneficial effects of exogenous tetrahydrobiopterin on left ventricular remodeling after myocardial infarction in rats. *Circulation Journal, 72*(9), 1512–1519. https://doi.org/10.1253/circj.cj-08-0072.

Matchkov, V. V., Kravtsova, V. V., Wiborg, O., Aalkjaer, C., & Bouzinova, E. V. (2015). Chronic selective serotonin reuptake inhibition modulates endothelial dysfunction and oxidative state in rat chronic mild stress model of depression. *American Journal of Physiology—Regulatory, Integrative and Comparative Physiology. 309*(8). https://doi.org/10.1152/ajpregu.00337.2014.

Medina, A., Watson, S. J., Bunney, W., Myers, R. M., Schatzberg, A., Barchas, J., et al. (2016). Evidence for alterations of the glial syncytial function in major depressive disorder. *Journal of Psychiatric Research, 72*, 15–21. https://doi.org/10.1016/j.jpsychires.2015.10.010.

Miguel-Hidalgo, J. J., Overholser, J. C., Jurjus, G. J., Meltzer, H. Y., Dieter, L., Konick, L., et al. (2011). Vascular and extravascular immunoreactivity for intercellular adhesion molecule 1 in the orbitofrontal cortex of subjects with major depression: age-dependent changes. *Journal of*

Affective Disorders, 132(3), 422–431. https://doi.org/10.1016/j.jad.2011.03.052.

Miller, A. H., Maletic, V., & Raison, C. L. (2009). Inflammation and its discontents: the role of cytokines in the pathophysiology of major depression. *Biological Psychiatry, 65*(9), 732–741. https://doi.org/10.1016/j.biopsych.2008.11.029.

Miller, B. J., Buckley, P., Seabolt, W., Mellor, A., & Kirkpatrick, B. (2011). Meta-analysis of cytokine alterations in schizophrenia: clinical status and antipsychotic effects. *Biological Psychiatry, 70*(7), 663–671. https://doi.org/10.1016/j.biopsych.2011.04.013.

Mondelli, V., Vernon, A. C., Turkheimer, F., Dazzan, P., & Pariante, C. M. (2017). Brain microglia in psychiatric disorders. *The Lancet Psychiatry.* https://doi.org/10.1016/s2215-0366(17)30101-3.

Moreno, J., Gaspar, E., López-Bello, G., Juárez, E., Alcázar-Leyva, S., González-Trujano, E., et al. (2013). Increase in nitric oxide levels and mitochondrial membrane potential in platelets of untreated patients with major depression. *Psychiatry Research, 209*(3), 447–452. https://doi.org/10.1016/j.psychres.2012.12.024.

Moura, D. S., Sultan, S., Georgin-Lavialle, S., Barete, S., Lortholary, O., Gaillard, R., et al. (2012). Evidence for cognitive impairment in mastocytosis: prevalence, features and correlations to depression. *PLoS ONE, 7*(6). https://doi.org/10.1371/journal.pone.0039468.

Nagafusa, Y., Okamoto, N., Sakamoto, K., Yamashita, F., Kawaguchi, A., Higuchi, T., et al. (2012). Assessment of cerebral blood flow findings using 99mTc-ECD single-photon emission computed tomography in patients diagnosed with major depressive disorder. *Journal of Affective Disorders, 140*(3), 296–299. https://doi.org/10.1016/j.jad.2012.03.026.

Najjar, S., Pearlman, D. M., Alper, K., Najjar, A., & Devinsky, O. (2013). Neuroinflammation and psychiatric illness. *Journal of Neuroinflammation. 10*(1). https://doi.org/10.1186/1742-2094-10-43.

Najjar, S., Pearlman, D. M., Devinsky, O., Najjar, A., & Zagzag, D. (2013). Neurovascular unit dysfunction with blood-brain barrier hyperpermeability contributes to major depressive disorder: a review of clinical and experimental evidence. *Journal of Neuroinflammation. 10*(1). https://doi.org/10.1186/1742-2094-10-142.

Najjar, S., Pearlman, D. M., Hirsch, S., Friedman, K., Strange, J., Reidy, J., et al. (2014). Brain biopsy findings link major depressive disorder to neuroinflammation, oxidative stress, and neurovascular dysfunction: a case report. *Biological Psychiatry. 75*(12). https://doi.org/10.1016/j.biopsych.2013.07.041.

Nasrallah, H. A. (2015). 10 recent paradigm shifts in the neurobiology and treatment of depression. *Current Psychiatry*, *14*(2), 10–13. Retrieved from. http://go.galegroup. com/ps/i.do?p=HRCA&sw=w&u=nysl_me_lijm& v=2.1&it=r&id=GALE%7CA402477508&asid=beaea 719eeaffd996f551909bba119fc.

Ng, F., Berk, M., Dean, O., & Bush, A. I. (2008). Oxidative stress in psychiatric disorders: evidence base and therapeutic implications. *The International Journal of Neuropsychopharmacology*. *11*(06). https://doi.org/10.1017/ s1461145707008401.

Niklasson, F., & Agren, H. (1984). Brain energy metabolism and blood-brain barrier permeability in depressive patients: analyses of creatine, creatinine, urate, and albumin in CSF and blood. *Biological Psychiatry*, *19*(8), 1183–1206.

Ofek, K., Schoknecht, K., Melamed-Book, N., Heinemann, U., Friedman, A., & Soreq, H. (2012). Fluoxetine induces vasodilatation of cerebral arterioles by co-modulating NO/muscarinic signalling. *Journal of Cellular and Molecular Medicine*, *16*(11), 2736–2744. https://doi.org/10.1111/ j.1582-4934.2012.01596.x.

Oh, D., Kim, S. H., & Park, Y. C. (2010). P.2.b.005 the biological pathway underlying dysregulation of hippocampal 5HT1A-NR2B-GSK-3b function in major depression. *European Neuropsychopharmacology*, *20*, https://doi.org/ 10.1016/s0924-977x(10)70495-8.

Oh, D. H., Park, S. C., Park, Y. C., & Kim, S. H. (2012). Excessive activation of the loop between the NR2B subunit of the *N*-methyl-D-aspartate receptor and glycogen synthase kinase-3β in the hippocampi of patients with major depressive disorder. *Acta Neuropsychiatrica*, *24* (01), 26–33. https://doi.org/10.1111/j.1601-5215.2011. 00581x.

Orosz, A., Jann, K., Federspiel, A., Horn, H., Höfle, O., Dierks, T., et al. (2012). Reduced cerebral blood flow within the default-mode network and within total gray matter in major depression. *Brain Connectivity*, *2*(6), 303–310. https://doi.org/10.1089/brain.2012.0101.

Ott, M., Gogvadze, V., Orrenius, S., & Zhivotovsky, B. (2007). Mitochondria, oxidative stress and cell death. *Apoptosis*, *12*(5), 913–922. https://doi.org/10.1007/s10495-007- 0756-2.

Ozcan, M. E., Gulec, M., Ozerol, E., Polat, R., & Akyol, O. (2004). Antioxidant enzyme activities and oxidative stress in affective disorders. *International Clinical Psychopharmacology*, *19*(2), 89–95. https://doi.org/ 10.1097/00004850-200403000-00006.

Pan, A., Keum, N., Okereke, O. I., Sun, Q., Kivimaki, M., Rubin, R. R., et al. (2012). Bidirectional association between depression and metabolic syndrome: a systematic review and meta-analysis of epidemiological studies. *Diabetes Care*, *35*(5), 1171–1180. https://doi.org/10.2337/ dc11-2055.

Papakostas, G. I., Shelton, R. C., Kinrys, G., Henry, M. E., Bakow, B. R., Lipkin, S. H., et al. (2013). Assessment of a multi-assay, serum-based biological diagnostic test for major depressive disorder: a pilot and replication Study. *Molecular Psychiatry*, *18*(3), 332–339. https://doi. org/10.1038/mp.2011.166.

Papakostas, G. I., Shelton, R. C., Zajecka, J. M., Etemad, B., Rickels, K., Clain, A., et al. (2012). L-methylfolate as adjunctive therapy for SSRI-resistant major depression: results of two randomized, double-blind, parallel-sequential trials. *American Journal of Psychiatry*, *169*(12), 1267–1274. https://doi.org/10.1176/appi.ajp.2012. 11071114.

Politi, P., Brondino, N., & Emanuele, E. (2008). Increased proapoptotic serum activity in patients with chronic mood disorders. *Archives of Medical Research*, *39*(2), 242–245. https://doi.org/10.1016/j.arcmed.2007. 07.011.

Prat, A., Biernacki, K., Pouly, S., Nalbantoglu, J., Couture, R., & Antel, J. P. (2000). Kinin B1 receptor expression and function on human brain endothelial cells. *Journal of Neuropathology & Experimental Neurology*, *59*(10), 896–906. https://doi.org/10.1093/jnen/59.10.896.

Pu, H., Tian, J., Flora, G., Lee, Y. W., Nath, A., Hennig, B., et al. (2003). HIV-1 tat protein upregulates inflammatory mediators and induces monocyte invasion into the brain. *Molecular and Cellular Neuroscience*, *24*(1), 224–237. https://doi.org/10.1016/s1044-7431(03)00171-4.

Pun, P. B., Lu, J., & Moochhala, S. (2009). Involvement of ROS in BBB dysfunction. *Free Radical Research*, *43*(4), 348–364. https://doi.org/10.1080/10715760902751902.

Quan, N., & Banks, W. A. (2007). Brain-immune communication pathways. *Brain, Behavior, and Immunity*, *21*(6), 727–735. https://doi.org/10.1016/j.bbi.2007.05.005.

Rajkowska, G., Hughes, J., Stockmeier, C. A., Miguel-Hidalgo, J. J., & Maciag, D. (2013). Coverage of blood vessels by astrocytic endfeet is reduced in major depressive disorder. *Biological Psychiatry*, *73*(7), 613–621. https:// doi.org/10.1016/j.biopsych.2012.09.024.

Rajkowska, G., & Stockmeier, C. (2013). Astrocyte pathology in major depressive disorder: insights from human postmortem brain tissue. *Current Drug Targets*, *14*(11), 1225–1236. https://doi.org/10.2174/138945011131499 90156.

Rao, J. S., Kellom, M., Reese, E. A., Rapoport, S. I., & Kim, H. (2012). RETRACTED: dysregulated glutamate and dopamine transporters in postmortem frontal cortex from bipolar and schizophrenic patients. *Journal of Affective Disorders*, *136*(1–2), 63–71. https://doi.org/10.1016/j. jad.2011.08.017.

Raslan, F., Schwarz, T., Meuth, S. G., Austinat, M., Bader, M., Renné, T., et al. (2010). Inhibition of bradykinin receptor B1 protects mice from focal brain injury by reducing blood-brain barrier leakage and inflammation. *Journal of Cerebral Blood Flow & Metabolism*, *30*(8), 1477–1486. https://doi.org/10.1038/jcbfm.2010.28.

Reeg, S., & Grune, T. (2015). Protein oxidation in aging: does it play a role in aging progression? *Antioxidants & Redox Signaling*, 23(3), 239–255. https://doi.org/10.1089/ars.2014.6062.

Rugulies, R. (2002). Depression as a predictor for coronary heart disease. *American Journal of Preventive Medicine*, 23(1), 51–61. https://doi.org/10.1016/s0749-3797(02)00439-7.

Saleptsis, V. G., Lambropoulos, N., Halaris, A., Angelopoulos, N. V., & Giannoukas, A. D. (2011). Depression and atherosclerosis. *International Angiology: A Journal of the International Union of Angiology*, 30(2), 97–104.

Salim, S., Chugh, G., & Asghar, M. (2012). Inflammation in anxiety. *Advances in Protein Chemistry and Structural Biology*, 88, 1–25. https://doi.org/10.1016/b978-0-12-398314-5.00001-5.

Samdani, A. F., Dawson, T. M., & Dawson, V. L. (1997). Nitric oxide synthase in models of focal ischemia. *Stroke*, 28(6), 1283–1288. https://doi.org/10.1161/01.str.28.6. 1283.

Scapagnini, G., Davinelli, S., Drago, F., Lorenzo, A. D., & Oriani, G. (2012). Antioxidants as antidepressants. *CNS Drugs*, 26(6), 477–490. https://doi.org/10.2165/11633190-000000000-00000.

Schaefer, M., Horn, M., Schmidt, F., Schmid-Wendtner, M. H., Volkenandt, M., Ackenheil, M., et al. (2004). Correlation between sICAM-1 and depressive symptoms during adjuvant treatment of melanoma with interferon-α. *Brain, Behavior, and Immunity*, 18(6), 555–562. https://doi.org/10.1016/j.bbi.2004.02.002.

Schnieder, T. P., Trencevska, I., Rosoklija, G., Stankov, A., Mann, J. J., Smiley, J., et al. (2014). Microglia of prefrontal white matter in suicide. *Journal of Neuropathology & Experimental Neurology*, 73(9), 880–890. https://doi.org/10.1097/nen.0000000000000107.

Schroeter, M. L., Abdul-Khaliq, H., Krebs, M., Diefenbacher, A., & Blasig, I. E. (2008). Serum markers support disease-specific glial pathology in major depression. *Journal of Affective Disorders*, 111(2–3), 271–280. https://doi.org/10.1016/j.jad.2008.03.005.

Scott, G., Bowman, S., Smith, T., Flower, R., & Bolton, C. (2007). Glutamate-stimulated peroxynitrite production in a brain-derived endothelial cell line is dependent on N-methyl-ᴅ-aspartate (NMDA) receptor activation. *Biochemical Pharmacology*, 73(2), 228–236. https://doi.org/10.1016/j.bcp.2006.09.021.

Serlin, Y., Levy, J., & Shalev, H. (2011). Vascular pathology and blood-brain barrier disruption in cognitive and psychiatric complications of type 2 diabetes mellitus. *Cardiovascular Psychiatry and Neurology*, 2011, 1–10. https://doi.org/10.1155/2011/609202.

Setiawan, E., Wilson, A. A., Mizrahi, R., Rusjan, P. M., Miler, L., Rajkowska, G., et al. (2015). Role of translocator protein density, a marker of neuroinflammation, in the brain during major depressive episodes. *JAMA Psychiatry*, 72(3), 268. https://doi.org/10.1001/jamapsychiatry.2014.2427.

Shalev, H., Serlin, Y., & Friedman, A. (2009). Breaching the blood-brain barrier as a gate to psychiatric disorder. *Cardiovascular Psychiatry and Neurology*, 2009, 1–7. https://doi.org/10.1155/2009/278531.

Shlosberg, D., Benifla, M., Kaufer, D., & Friedman, A. (2010). Blood-brain barrier breakdown as a therapeutic target in traumatic brain injury. *Nature Reviews Neurology*, 6(7), 393–403. https://doi.org/10.1038/nrneurol.2010.74.

Skaper, S. D. (2007). The brain as a target for inflammatory processes and neuroprotective strategies. *Annals of the New York Academy of Sciences*, 1122(1), 23–34. https://doi.org/10.1196/annals.1403.002.

Smith, D. J., & Cavanagh, J. T. (2005). The use of single photon emission computed tomography in depressive disorders. *Nuclear Medicine Communications*, 26(3), 197–203. https://doi.org/10.1097/00006231-200503000-00004.

Srivastava, K., Bath, P. M., & Bayraktutan, U. (2011). Current therapeutic strategies to mitigate the eNOS dysfunction in ischaemic stroke. *Cellular and Molecular Neurobiology*, 32(3), 319–336. https://doi.org/10.1007/s10571-011-9777-z.

Steiner, J., Bielau, H., Brisch, R., Danos, P., Ullrich, O., Mawrin, C., et al. (2008). Immunological aspects in the neurobiology of suicide: elevated microglial density in schizophrenia and depression is associated with suicide. *Journal of Psychiatric Research*, 42(2), 151–157. https://doi.org/10.1016/j.jpsychires.2006.10.013.

Steiner, J., Walter, M., Gos, T., Guillemin, G. J., Bernstein, H., Sarnyai, Z., et al. (2011). Severe depression is associated with increased microglial quinolinic acid in subregions of the anterior cingulate gyrus: evidence for an immune-modulated glutamatergic neurotransmission? *Journal of Neuroinflammation*, 8(1), 94. https://doi.org/10.1186/1742-2094-8-94.

Stuehr, D. J., Santolini, J., Wang, Z., Wei, C., & Adak, S. (2004). Update on mechanism and catalytic regulation in the NO synthases. *Journal of Biological Chemistry*, 279(35), 36167–36170. https://doi.org/10.1074/jbc.r400017200.

Sun, C. W., Falck, J. R., Okamoto, H., Harder, D. R., & Roman, R. J. (2000). Role of cGMP versus 20-HETE in the vasodilator response to nitric oxide in rat cerebral arteries. *American Journal of Physiology Heart and Circulatory Physiology*, 279, 339–350.

Szabó, C., Ischiropoulos, H., & Radi, R. (2007). Peroxynitrite: biochemistry, pathophysiology and development of therapeutics. *Nature Reviews Drug Discovery*, 6(8), 662–680. https://doi.org/10.1038/nrd2222.

Theoharides, T. C., Alysandratos, K., Angelidou, A., Delivanis, D., Sismanopoulos, N., Zhang, B., et al. (2012).

Mast cells and inflammation. *Biochimica et Biophysica Acta (BBA) - Molecular Basis of Disease, 1822*(1), 21–33. https://doi.org/10.1016/j.bbadis.2010. 12.014.

Theoharides, T. C., Stewart, J. M., Panagiotidou, S., & Melamed, I. (2016). Mast cells, brain inflammation and autism. *European Journal of Pharmacology, 778*, 96–102. https://doi.org/10.1016/j.ejphar.2015.03.086.

Theoharides, T. C., & Zhang, B. (2011). Neuro-inflammation, blood-brain barrier, seizures and autism. *Journal of Neuroinflammation, 8*(1), 168. https://doi.org/10.1186/1742-2094-8-168.

Thomas, A. J., Perry, R., Kalaria, R. N., Oakley, A., Mcmeekin, W., & Obrien, J. T. (2003). Neuropathological evidence for ischemia in the white matter of the dorsolateral prefrontal cortex in late-life depression. *International Journal of Geriatric Psychiatry, 18*(1), 7–13. https://doi.org/10.1002/gps.720.

Tobe, E. (2013). Mitochondrial dysfunction, oxidative stress, and major depressive disorder. *Neuropsychiatric Disease and Treatment. 567*, https://doi.org/10.2147/ndt.s44282.

Toro, C., & Deakin, J. (2005). NMDA receptor subunit NRI and postsynaptic protein PSD-95 in hippocampus and orbitofrontal cortex in schizophrenia and mood disorder. *Schizophrenia Research, 80*(2–3), 323–330. https://doi.org/10.1016/j.schres.2005.07.003.

Torres-Platas, S. G., Cruceanu, C., Chen, G. G., Turecki, G., & Mechawar, N. (2014). Evidence for increased microglial priming and macrophage recruitment in the dorsal anterior cingulate white matter of depressed suicides. *Brain, Behavior, and Immunity, 42*, 50–59. https://doi.org/10.1016/j.bbi.2014.05.007.

Trivedi, M. H., Rush, A. J., Wisniewski, S. R., Nierenberg, A. A., Warden, D., Ritz, L., et al. (2006). Evaluation of outcomes with citalopram for depression using measurement-based care in STAR*D: implications for clinical practice. *American Journal of Psychiatry, 163*, 28–40. https://doi.org/10.1176/appi.ajp.163.1.28.

Valkanova, V., & Ebmeier, K. P. (2013). Vascular risk factors and depression in later life: a systematic review and meta-analysis. *Biological Psychiatry, 73*(5), 406–413. https://doi.org/10.1016/j.biopsych.2012.10.028.

Van der Kooy, K., Hout, H. V., Marwijk, H., Marten, H., Stehouwer, C., & Beekman, A. (2007). Depression and the risk for cardiovascular diseases: systematic review and meta analysis. *International Journal of Geriatric Psychiatry, 22*(7), 613–626. https://doi.org/10.1002/gps.1723.

Van Zyl, L. T., Lespérance, F., Frasure-Smith, N., Malinin, A. I., Atar, D., Laliberté, M., et al. (2008). Platelet and endothelial activity in comorbid major depression and coronary artery disease patients treated with citalopram: the Canadian Cardiac Randomized Evaluation of Antidepressant and Psychotherapy Efficacy Trial (CREATE) biomarker sub-study. *Journal of Thrombosis and Thrombolysis, 27*(1), 48–56. https://doi.org/10.1007/s11239-007-0189-3.

Viana, A. F., Maciel, I. S., Dornelles, F. N., Figueiredo, C. P., Siqueira, J. M., Campos, M. M., et al. (2010). Kinin B1 receptors mediate depression-like behavior response in stressed mice treated with systemic *E. coli* lipopolysaccharide. *Journal of Neuroinflammation, 7*(1), 98. https://doi.org/10.1186/1742-2094-7-98.

Wachholz, S., Eßlinger, M., Plümper, J., Manitz, M., Juckel, G., & Friebe, A. (2016). Microglia activation is associated with IFN-α induced depressive-like behavior. *Brain, Behavior, and Immunity, 55*, 105–113. https://doi.org/10.1016/j.bbi.2015.09.016.

Wang, Q., Tang, X., & Yenari, M. (2007). The inflammatory response in stroke. *Journal of Neuroimmunology, 184*(1–2), 53–68. https://doi.org/10.1016/j.jneuroim.2006. 11.014.

Wiencken, A., & Casagrande, V. (1999). Endothelial nitric oxide synthetase (eNOS) in astrocytes: another source of nitric oxide in neocortex. *Glia, 26*(4), 280–290. https://doi.org/10.1002/(sici)1098-1136(199906)26:4<280::aid-glia2>3.3.co;2-n.

World Health Organization (2017). *"Depression: Let's talk" says WHO, as depression tops list of causes of ill health. (2017).* www.who.int/mediacentre/news/releases/2017/world-health-day/en/.

Yoshida, T., Ishikawa, M., Niitsu, T., Nakazato, M., Watanabe, H., Shiraishi, T., et al. (2012). Decreased serum levels of mature brain-derived neurotrophic factor (BDNF), but not its precursor proBDNF, in patients with major depressive disorder. *PLoS ONE. 7*(8)https://doi.org/10.1371/journal.pone.0042676.

Zameer, A., & Hoffman, S. A. (2003). Increased ICAM-1 and VCAM-1 expression in the brains of autoimmune mice. *Journal of Neuroimmunology, 142*(1–2), 67–74. https://doi.org/10.1016/s0165-5728(03)00262-5.

Zeman, M., Jachymova, M., Jirak, R., Vecka, M., Tvrzicka, E., Stankova, B., et al. (2010). Polymorphisms of genes for brain-derived neurotrophic factor, methylenetetrahydrofolate reductase, tyrosine hydroxylase, and endothelial nitric oxide synthase in depression and metabolic syndrome. *Folia Biologica, 56*(1), 19–26.

Further Reading

Dantzer, R., Oconnor, J. C., Freund, G. G., Johnson, R. W., & Kelley, K. W. (2008). From inflammation to sickness and depression: when the immune system subjugates the brain. *Nature Reviews Neuroscience, 9*(1), 46–56. https://doi.org/10.1038/nrn2297.

Huang, Z., Huang, P. L., Ma, J., Meng, W., Ayata, C., Fishman, M. C., et al. (1996). Enlarged infarcts in endothelial nitric oxide synthase knockout mice are attenuated by nitro-L-arginine. *Journal of Cerebral Blood Flow & Metabolism,* 981–987. https://doi.org/10.1097/00004647-199609000-00023.

Iadecola, C., Zhang, F., Casey, R., Nagayama, M., & Ross, M. E. (1997). Delayed reduction of ischemic brain injury and neurological deficits in mice lacking the inducible nitric oxide synthase gene. *The Journal of Neuroscience: The Official Journal of the Society of Neuroscience, 17*, 9157–9164.

Ikenouchi-Sugita, A., Yoshimura, R., Hori, H., Umene-Nakano, W., Ueda, N., & Nakamura, J. (2009). Effects of antidepressants on plasma metabolites of nitric oxide in major depressive disorder: comparison between milnacipran and paroxetine. *Progress in Neuro-Psychopharmacology and Biological Psychiatry, 33*(8), 1451–1453. https://doi.org/10.1016/j.pnpbp.2009.07.028.

María, C. C. (2007). The biological significance of mtNOS modulation. *Frontiers in Bioscience, 12*(1), 1041. https://doi.org/10.2741/2124.

Miller, A. H., & Raison, C. L. (2015). The role of inflammation in depression: from evolutionary imperative to modern treatment target. *Nature Reviews Immunology, 16*(1), 22–34. https://doi.org/10.1038/nri.2015.5.

O'Connor, J. C., Lawson, M. A., André, C., Moreau, M., Lestage, J., Castanon, N., et al. (2008). Lipopolysaccharide-induced depressive-like behavior is mediated by indoleamine 2,3-dioxygenase activation in mice. *Molecular Psychiatry, 14*(5), 511–522. https://doi.org/10.1038/sj.mp.4002148.

Onore, C. E., Nordahl, C. W., Young, G. S., Water, J. A., Rogers, S. J., & Ashwood, P. (2012). Levels of soluble platelet endothelial cell adhesion molecule-1 and P-selectin are decreased in children with autism spectrum disorder. *Biological Psychiatry, 72*(12), 1020–1025. https://doi.org/10.1016/j.biopsych.2012.05.004.

Piche, T., Saint-Paul, M. C., Dainese, R., Marine-Barjoan, E., Iannelli, A., Montoya, M. L., et al. (2007). Mast cells and cellularity of the colonic mucosa correlated with fatigue and depression in irritable bowel syndrome. *Gut, 57*(4), 468–473. https://doi.org/10.1136/gut.2007.127068.

Qian, Y. (2014). Mast cells and neuroinflammation. *Medical Science Monitor Basic Research, 20*, 200–206. https://doi.org/10.12659/msmbr.893093.

Shimizu-Sasamata, M., Bosque-Hamilton, P., Huang, P. L., Moskowitz, M. A., & Lo, E. H. (1998). Attenuated neurotransmitter release and spreading depression-like depolarizations after focal ischemia in mutant mice with disrupted type I nitric oxide synthase gene. *The Journal of Neuroscience: The Official Journal of the Society of Neuroscience, 18*, 9564–9571.

Thanan, R., Oikawa, S., Hiraku, Y., Ohnishi, S., Ma, N., Pinlaor, S., et al. (2014). Oxidative stress and its significant roles in neurodegenerative diseases and cancer. *International Journal of Molecular Sciences, 16*(1), 193–217. https://doi.org/10.3390/ijms16010193.

Wightman, E. L., Haskell-Ramsay, C. F., Reay, J. L., Williamson, G., Dew, T., Zhang, W., et al. (2015). The effects of chronic trans-resveratrol supplementation on aspects of cognitive function, mood, sleep, health and cerebral blood flow in healthy, young humans. *British Journal of Nutrition, 114*(09), 1427–1437. https://doi.org/10.1017/s0007114515003037.

The Impact of Inflammation on Brain Function and Behavior in Rodent Models of Affective Disorders

Farheen Farzana, Thibault Renoir, Anthony J. Hannan

Florey Institute of Neuroscience and Mental Health, University of Melbourne, Parkville, VIC, Australia

INTRODUCTION

Major depressive disorder (MDD) is a common psychiatric illness and has been identified as the leading cause of disability worldwide, by the World Health Organization (WHO). This has been mainly attributed to the rise in the incidence of treatment-resistant depression (Al-Harbi, 2012) and partial efficacy and the longer time taken by the currently available antidepressants to manifest their effects (Trivedi et al., 2006). Furthermore, the fact that MDD can occur as a comorbid disorder in association with many other psychiatric and physical illnesses increases the complexity of its etiology, making it imperative to gain a better insight into the biological mechanisms that causally underlie the onset of this disorder.

In contrast to the stress-induced immunosuppressive role of depression proposed earlier (Irwin & Miller, 2007), recent years have witnessed a copious amount of research pointing toward the causative role of inflammatory mechanisms in the pathophysiology of MDD, thus paving the way for the field of psychoneuroimmunology (Glaser & Kiecolt-Glaser, 2005; Irwin, 2002; Leonard & Myint, 2009; Ziemssen & Kern, 2007). This has led to a dramatic paradigm shift beyond the "monoamine theory of depression" (Nutt, 2008) to the "macrophage theory of depression" (Smith, 1991) and the "cytokine hypothesis of depression" (Maes et al., 2009), with the aim to disentangle the complex etiology of MDD. Sufficient evidence has now accumulated supporting the concept that the brain recognizes the presence of proinflammatory cytokines as a molecular signal of sickness, which in turn generates an adaptive response by the body, termed "sickness behavior" (Dantzer, 2004, 2009; Kelley et al., 2003). Its symptoms include social avoidance, anhedonia, fatigue, depressed mood, and impaired concentration, all of which closely overlap with the symptoms of affective disorders (Dantzer, 2009). Evolutionarily, this bias has been proposed to prioritize shifting of energy sources solely for fighting an infection and increasing the host's security while the body recovers (Miller & Raison, 2015). However, chronic stress-induced systemic inflammation can damage tissues including the brain and contribute to the onset of major depressive episodes, reported in both clinical (Capuron et al., 2001; Raison, Capuron, & Miller, 2006) and preclinical studies (Dantzer, 2004).

Higher levels of several pro-inflammatory peripheral markers such as interleukin-1 (IL-1), interleukin-6 (IL-6), interleukin-1beta (IL-1β), C-reactive protein, tumor necrosis factor-alpha (TNF-α), and TNF-γ have been reported in depressed patients (Capuron & Dantzer, 2003; Maes, 1995). Notably, a higher incidence of depression has been reported in patients suffering from rheumatoid arthritis (Margaretten, Julian, Katz, & Yelin, 2011), chronic pain (Leo, 2005), psoriasis (Tyring et al., 2006), and cancer patients undergoing cytokine immunotherapy (Capuron et al., 2001), all of which have a common chronic inflammatory component, thus causally linking the disruption of the cross talk between the immune system and CNS with the neuropathologic mechanisms underlying MDD (Evans et al., 2005; Miller & Raison, 2015). Consistent with those findings, another controlled study in patients with MDD has shown baseline IL-6 levels to act as a possible criterion for dichotomizing patients into responders and nonresponders to the antidepressant amitriptyline (Lanquillon, Krieg, Bening-Abu-Shach, & Vedder, 2000).

Amid a number of studies and reviews published on the depressogenic effect of inflammatory activation, Raison and Miller elaborated on the differences in individual vulnerability as a major deciding factor, thus supporting the idea that inflammatory processes may contribute to only a subset of depressive presentations (Raison & Miller, 2011). In parallel to several studies that have demonstrated a higher incidence rate of depression in people experiencing childhood adversities (Caspi et al., 2003; Kessler & Magee, 1993), linking early-life adversities with increased vulnerability to depressive

disorders, mounting evidence also supports the role of early-life adversities as a major risk factor for developing a pro-inflammatory phenotype (Ehrlich, Ross, Chen, & Miller, 2016; Slopen et al., 2010), thus reinforcing the idea that inflammatory stimulation may be depressogenic only in vulnerable patients.

It is beyond the scope of this chapter to provide a comprehensive summary of the role and mode of action of individual cytokines that mediate depressive-like phenotype in animal models. However, there have been a number of recent reviews in the area focusing on the current standings on the role of cytokines as peripheral markers of depression (Lichtblau, Schmidt, Schumann, Kirkby, & Himmerich, 2013), brain-immune communication pathways along with the neuroimmune axis (Dantzer, O'Connor, Freund, Johnson, & Kelley, 2008), critical mechanisms underlying inflammation-induced depression using animal models, and etiologic role of neuroinflammation in depressive-like behaviors (Brites & Fernandes, 2015).

Over the past few decades, the role of gut microbiota has been increasingly recognized as a crucial mediator of the effects of stress on pro-inflammatory (Boulangé, Neves, Chilloux, Nicholson, & Dumas, 2016) and depressive-like state (Carabotti, Scirocco, Maselli, & Severi, 2015; Foster, Rinaman, & Cryan, 2017). In this review, we highlight the effect of stress-induced cytokines on affective behavioral impairments, targeting some of the mechanisms involved in mood regulation, such as altered serotonin signaling, activation of the kynurenine pathway leading to serotonin clearance and accumulation of kynurenine metabolites, and altered hippocampal neurogenesis (Fig. 1). We will also focus on recent findings on the role played by gut microbiome in mediating the inflammatory phenotype associated with behavioral alterations in rodent models. Finally, the need to conduct future studies on sexual dimorphism associated with inflammation-induced depression has

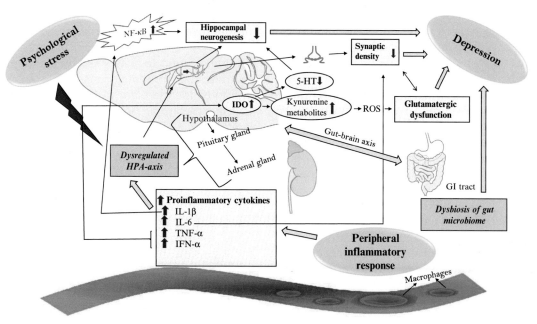

FIG. 1   Proposed mechanisms underlying the effects of stress-induced inflammation on molecular and cellular aspects of brain function associated with depression. While this diagram does not attempt to fully capture the extensive literature in this field nor all hypothesized pathways, it highlights key mechanistic insights derived from animal models of affective dysfunction and depression. *HPA axis*, hypothalamic-pituitary-adrenal axis; *NF-kB*, nuclear factor kappa-light-chain-enhancer of activated B cells; *5-HT*, 5-hydroxytryptamine or serotonin; *IDO*, indoleamine 2,3-dioxygenase; *ROS*, reactive oxygen species.

been emphasized, considering the crucial role played by sex/gender in governing the possible differences in mechanisms underlying the impact of inflammation on brain function and behavior.

STRESS, INFLAMMATION AND DEPRESSIVE-LIKE BEHAVIORS IN RODENT MODELS

Some of the most reproducible findings of biological psychiatry are the effects of major life stressors on hypothalamic-pituitary-adrenal axis (HPA axis) hyperactivity, glucocorticoid receptor resistance, elevated circulating levels of the glucocorticoid cortisol, and initiation of depressive symptoms (Silverman & Sternberg, 2012; Smith & Vale, 2006; Zunszain, Anacker,

Cattaneo, Carvalho, & Pariante, 2011). Even though glucocorticoids constitute the most effective anti-inflammatory therapy for asthma, rheumatoid arthritis, and other inflammatory disorders (Barnes, 2006; Coutinho & Chapman, 2011), there is a rich body of evidence demonstrating the role of stress-induced elevated levels of pro-inflammatory cytokines in disrupting the function and signaling of glucocorticoid receptors, leading to glucocorticoid receptor resistance (GCR) and the induction of HPA axis activity (Silverman & Sternberg, 2012; Slavich & Irwin, 2014; Turnbull & Rivier, 1999). Thus, although glucocorticoids (cortisol) have been shown to be elevated in patients with major depression, GCR fails to downregulate inflammatory responses by virtue of insufficient glucocorticoid signaling (Raison & Miller, 2003).

The availability of transgenic small animal models has served as an important pharmacological research tool for studying the mechanisms underlying the effect of stress on peripheral cytokines and the transmission of peripheral immune signals to the brain by merging ethologically valid behavioral assays with the latest technological advances in molecular biology (Remus & Dantzer, 2016). In this regard, a large body of preclinical evidence has been accumulated from studies establishing the role of certain cytokines in inducing depression-like and/or anxiety-like behaviors (Dunn & Swiergiel, 2005; Felger & Lotrich, 2013) and studies using several models of chronic stress to monitor the stress-induced changes in immune function (Miller, Maletic, & Raison, 2009).

Cytokine-Induced Modulation of HPA-Axis Activity in Rodent Models

The mechanisms underlying the HPA axis stimulating the effect of pro-inflammatory cytokines have been extensively studied in rodent models via the administration of human cytokines (Dunn, 2000). Both IL-1α and IL-1β have been shown to be far more potent at activating HPA axis, as compared with other cytokines like IL-6, TNF-α, and IFN-α (Dunn, 2000). Intraperitoneal injection of IL-1 has been reported to increase the circulating levels of corticosterone and activate HPA axis in mice, by activating the noradrenergic neurons in the hypothalamus, which regulate the secretion of corticotropin-releasing factor (CRF) (Dunn, & Swiergiel, 1998). In a separate study using a rat model, this effect of IL-1 has been shown to be blocked by acute treatment with indomethacin, a nonselective cyclooxygenase inhibitor, suggesting the role of prostaglandins in mediating the effect of IL-1 on the HPA axis (Rivier & Vale, 1991). Another study assessing the effect of IL-6 on the HPA axis via the intravenous injection of recombinant human IL-6 to conscious, freely moving rats reported a significant increase in plasma levels of adrenocorticotropic hormone (ACTH) in a dose-dependent manner. This effect was found to be blocked by the immuno-neutralization of endogenous CRF, suggesting CRF to be the mediator of the effect of IL-6 on adrenocorticotropic hormone secretion (Naitoh et al., 1988).

LPS-Mediated Behavioral Alterations in Rodent Models

Systemic administration of lipopolysaccharide (LPS), a cytokine inducer, has been extensively used to generate inflammation models of depression in rodents, to assess the subsequent changes in behavioral and molecular patterns (Remus & Dantzer, 2016). In addition to changes in peripheral pro-inflammatory markers, LPS has also been shown to directly influence CNS activity in rats, via the stimulation of central IL-1 production in all the brain regions, except the cerebellum (Quan, Sundar, & Weiss, 1994). A study investigating the role of LPS in activating the HPA axis in rats reported elevated plasma levels of ACTH, corticosterone, and IL-6 that were found to be normalized by the intravenous administration of an LPS antagonist, cationic antimicrobial protein 18 (Lenczowski, Van Dam, Poole, Larrick, & Tilders, 1997). In a separate study, activation of microglial cells in the brain has been reported to act as mediators of the LPS-induced depressive-like effects in mice, as attenuation of microglial activation with minocycline reversed the LPS-mediated depressive-like symptoms (Henry et al., 2008).

However, LPS-induced sickness can be a major confounding factor for studies evaluating the influence of inflammatory markers on behavioral alterations in animal models. This has been addressed by an interesting study investigating the differential effects of intraperitoneal injection of LPS on cytokine-induced sickness behavior (6h post-LPS) and cytokine-induced depressive-like behavior (24 h post-LPS) in mice, establishing the role of LPS in inducing depressive-like behavior independently of its sickness-inducing properties (Frenois et al., 2007). LPS-induced sickness evaluated at an early timepoint of 6 h solely impaired motor activity, while depressive-like phenotype observed at 24 h was confirmed by increased immobility time in the forced-swim test (FST), tail-suspension test (TST), reduced sucrose preference (with no motor impairment), and delayed cellular activity (as assessed by FosB/ΔFosB immunostaining) in specific brain areas, especially within the extended amygdala, hippocampus, and hypothalamus (Frenois et al., 2007). Moreover, LPS-mediated immobility in FST in mice was shown to be significantly reversed by the administration of desipramine (10 mg/kg), a tricycle antidepressant and fluoxetine (10 mg/kg), a serotonin reuptake inhibitor, and most notably by the cyclooxygenase inhibitors, pointing toward the role of prostaglandins in LPS-induced despair behavior (Jain, Kulkarni, & Singh, 2001).

In addition, the anxiogenic effects of LPS in mice, measured by decreased time spent in the illuminated portion of a light-dark box, reduced open-arm entries in a plus-maze test, and decreased contact with a novel stimulus object in an open-field test, have been well characterized (Lacosta, Merali, & Anisman, 1999). Specifically, a study investigating the effect of different doses of LPS in young male rats found an affective behavioral peak at an LPS dose of 200 μg/kg, emphasizing the dose dependency of LPS in eliciting anxiety-related behavioral deficits (Bassi et al., 2012).

Cytokine-Induced Behavioral Alterations in Rodent Models

In agreement with clinical studies showing elevated peripheral levels of IL-6 in male depressed patients with increased early-life stress (Pace et al., 2006), IL-6-deficient mice (IL-6$^{-/-}$) exhibited resistance to stress, assessed by reduced behavioral despair in FST and TST and resistance to helplessness in the learned helplessness paradigm (Chourbaji et al., 2006). Consistent with those findings, another study reported blockade of IL-6 receptor in a mouse model of social stress defeat, through intravenous injection of MR16-1, resulting in antidepressant-like effects in both TST and saccharin-preference test (Zhang et al., 2017).

In a similar manner, blockade of interleukin-1beta (IL-1β) receptor, either by using its inhibitor or IL1RI null (knockout) mice, resulted in attenuation of anhedonia induced by chronic unpredictable stress (CUS) (Koo & Duman, 2008). Mice exposed to CUS showed increased levels of IL-1β, concomitantly with behavioral alterations such as decreased sucrose preference, reduced social exploration, and elevated corticosterone levels, which were found to be blocked in IL-1 receptor knockout (IL-1rKO) mice (Goshen et al., 2008), thus affirming the antidepressant-like effect of IL-1 receptor antagonist (IL-1Ra) knockout (KO) mice (Wakabayashi, Kiyama, Kunugi, Manabe, & Iwakura, 2011). Also, ketamine-induced antidepressant effects have been shown to be mediated through a decrease in the level of pro-inflammatory cytokines, IL-1β and IL-6, in the prefrontal cortex and hippocampus of a rat model exposed to forced-swim stress (Yang et al., 2013).

Even though a study by Kaster, Gadotti, Calixto, Santos, and Rodrigues (2012) reported depressive-like behavior in mice following intracerebroventricular administration of tumor necrosis factor-α (TNF-α), a separate study using systemic administration (i.p. injection) of TNF-α in mice reported the induction of a strong sickness response characterized by reduced locomotor activity, accompanied by mild anhedonia signs (Biesmans et al., 2015). However, deletion of either TNF receptor 1 (TNFR1) or TNF receptor 2 (TNFR2) in mice has been associated with an antidepressant-like effect in FST and TST, with mice lacking TNFR2 showing a hedonic response in the sucrose preference test. Furthermore, no difference in anxiety levels in either of the null mutants was found, compared with their wild-type littermates (Simen, Duman, Simen, & Duman, 2006).

While the association between pro-inflammatory cytokines and psychiatric disorders has been the focus of much research, interleukin-10 (IL-10) has been shown to exert its antidepressant effect through its anti-inflammatory properties (Roque, Correia-Neves, Mesquita, Palha, & Sousa, 2009). Increased production of IL-10 resulting from a prolonged treatment of desipramine, a tricyclic antidepressant, has been shown to modulate depressive behavior in mice subjected to chronic mild stress (Kubera et al., 2001). Female mice lacking the expression of IL-10 (IL-10$^{-/-}$), an anti-inflammatory cytokine, showed longer immobilization time in the FST and decreased time spent in the center of the arena in the open-field test, suggestive of a depression-like and anxiety-like phenotype. Remarkably, these behavioral deficits were found to be reversed by an intraperitoneal injection of IL-10. Moreover, transgenic mice overexpressing IL-10 showed decreased depressive-like behavior in the FST in comparison with WT animals. Interestingly, no such differences were observed in male mice, hinting at a possible role of sex as a modulator of inflammation-induced depression (Mesquita et al., 2008).

MEDIATORS OF INFLAMMATION INDUCED DEPRESSIVE-LIKE BEHAVIORS

Given the findings that the activation of peripheral inflammatory responses has been causally linked with the onset of affective disorder, a rich body of evidence has been gathered on the targets of cytokines in the brain, using rodent studies, some of which have been highlighted below.

Cytokine-Induced Dysregulation of Serotonin (5-HT) Signaling

Serotonin transporter (5-HTT) has garnered much interest in neuropsychiatric disorders due to the accumulating evidence supporting the role played by 5-HTT polymorphisms (in particular the 5-HTTLPR long promoter region tandem repeat polymorphism) in increasing susceptibility to affective disorders (Caspi et al., 2003; Otte, McCaffery, Ali, & Whooley, 2007). Moreover, selective serotonin reuptake inhibitors (SSRIs), one of the most widely used class of antidepressant medications, act by blocking the serotonin transporter, thereby impairing its ability to mediate serotonin reuptake from the synaptic cleft and increasing serotonin availability (Schloss & Williams, 1998). Furthermore, 5-HTT animal models (with disrupted serotonin transporter signaling) have been reported to display depression- and anxiety-related behavior, thereby establishing the 5-HTT model as a gold standard for studying the etiology of affective disorders observed in the human population with a genetic predisposition to altered serotonin signaling (Holmes, Murphy, & Crawley, 2003).

In fact, serotonin transporter knockout (5-HTT KO) mice have been reported to display an increased immobility time in the forced-swim test (FST) and reduced time spent in light of light/dark box and open arms of the elevated plus maze (EPM), suggestive of a depressive- and anxiety-like phenotype (Murphy & Lesch, 2008). These findings were extended by our group to a Huntington's disease mouse model, which has been reported to display presymptomatic depressive-like behavior, prior to the onset of cognitive and motor deficits (Grote et al., 2005; Pang, Du, Zajac, Howard, & Hannan, 2009). While Huntingtin's disease (HD) is a genetic neurodegenerative disorder caused by an abnormal expansion of CAG trinucleotide repeats (Nance & Myers, 2001), the HD mutation has also been shown to add significantly to the genetic load for depression (Du, Pang, & Hannan, 2013). Reduced expression of specific serotonin receptors and 5-HTT in the hippocampus and cortex has been observed in the female R6/1 HD mice that displayed depressive-related behaviors on FST, TST, and novelty-suppressed feeding test (NSFT), suggesting altered serotonin signaling to be underlying mechanisms contributing to the development of depression in HD (Pang et al., 2009).

Dysregulation of serotonin signaling is, however, not limited to depressive-like behaviors, as alterations in 5-HT signaling have also been described in inflammatory disorders like inflammatory bowel disease (IBD), allergic airway inflammation, and rheumatoid arthritis (Shajib & Khan, 2015). Interestingly, basal levels of pro-inflammatory cytokines have been reported to be higher in 5-HTT mutant rats that were also reported to exhibit a heightened cytokine response to an immune challenge in the form of an intraperitoneal injection of lipopolysaccharide (LPS) (Macchi et al., 2013). Also, in agreement with clinical studies showing increased expression of 5-HTT and specific cytokines in depressed patients (Tsao, Lin, Chen, Bai, & Wu, 2006), rodent studies using LPS (a potent activator of inflammatory response) have also reported a strong association between increased cytokine levels, in particular TNF-α (Malynn, Campos-Torres, Moynagh, & Haase, 2013), and increased expression of 5-HTT in the frontal cortex (Schwamborn, Brown, & Haase, 2016), suggesting acceleration of serotonin clearance as a mediator of cytokine-induced 5-HTT-dependent behavioral deficits.

Cytokine-Induced Activation of Indoleamine 2,3 Dioxygenase (IDO)

In addition to serotonin clearance, reduced availability of the serotonin precursor, tryptophan (Trp), mainly through the activation of indoleamine 2,3-dioxygenase, an enzyme that breaks down tryptophan via the kynurenine pathway, has been strongly implicated in serotonin-deficiency-related depressive phenotype (Badawy, 2013; Capuron et al., 2002). The involvement of inflammatory response in speeding up the process of tryptophan degradation has been well supported by separate studies reporting interferon gamma (Takikawa, Tagawa, Iwakura, Yoshida, & Truscott, 1999), TNF-α (O'Connor et al., 2009), bacille Calmette-Guérin (BCG) infection (Moreau et al., 2005), and LPS-induced (Currier et al., 2000) activation of indoleamine 2,3-dioxygenase (IDO) in mice, thus emphasizing the role of IDO as an important mediator linking immunologic networks with the pathogenesis of depression. Furthermore, LPS-induced depressive-like behavior in mice was found to be attenuated by intraperitoneal injection of minocycline, a potent anti-inflammatory (Henry et al., 2008) and pretreatment with the competitive IDO inhibitor, 1-methyltryptophan (tryptophan analog). Most notably, IDO-deficient mice have been reported to show complete resistance to BCG-induced depressive-like behavior (O'Connor et al., 2009).

While enhanced tryptophan degradation and its reduced availability for the biosynthesis of neurotransmitter 5-hydroxytryptamine (5-HT, serotonin) have been strongly associated with an increased susceptibility to depression, especially in patients with a family history of depression (Riedel, Klaassen, & Schmitt, 2002), a separate study conducted in mice did not find a decrease in serotonin levels exhibiting BCG-induced depression-like behavior (O'Connor et al., 2009). This points toward the contribution of an alternate mechanism, possibly an elevation in the levels of kynurenine metabolites that result from the breakdown of tryptophan, as a mediator of IDO-induced depression-like behavior.

Cytokine-Induced Elevation of Kynurenine Metabolites and Oxidative Stress

Kynurenine, the breakdown product of tryptophan, has been shown to readily cross the blood-brain barrier (Fukui, Schwarcz, Rapoport, Takada, & Smith, 1991), where it is further metabolized to generate neuroactive glutamatergic compounds, namely, kynurenic acid or quinolinic acid, that act in an opposite manner. While kynurenic acid exerts its neuroprotective effect by acting as an antagonist of the glutamate receptor, N-methyl-D-aspartate (NMDA), quinolinic acid exerts its neurotoxic effect by acting as an agonist of the NMDA receptor (Stone, 2001). Consistent with that evidence, heightened glutamate receptor activity and induced excitotoxicity have been shown to cause significant neuronal damage in rats (Amori, Guidetti, Pellicciari, Kajii, & Schwarcz, 2009).

Clinical literature has also reported a significant reduction in the ratio of kynurenic acid to quinolinic acid (Meier et al., 2016) in depressed patients and elevated levels of quinolinic acid in the CSF of suicide attempters as compared with controls (Bay-Richter et al., 2015), confirming the dysregulation of kynurenine pathway (Laugeray et al., 2010), serotonergic deficiency, and glutamatergic overproduction to be critical biological components of depression (Müller & Schwarz, 2007; Remus & Dantzer, 2016). Using a mouse model, Walker et al. (2013) provided further supporting evidence, reporting that the blockade of quinolinic acid access to NMDA receptors via intraperitoneal injection of ketamine, just before the administration of LPS, could potentially reverse depressive-like behavior, measured by decreased immobility time in FST and increased sucrose preference.

Walker et al. (2013) also reported an elevation in yet another neurotoxic metabolite of kynurenine (i.e., 3-hydroxykynurenine or 3HK) in LPS-treated mice. 3HK has been shown to cause mitochondrial dysfunction through the generation of reactive oxygen species (Reyes-Ocampo et al., 2015) and hence may play an important role in oxidative stress-mediated depressive phenotype. In fact, our group has reaffirmed the role played by oxidative stress and glutamatergic excitotoxicity established by Anderson, Berk, Dean, Moylan, and Maes (2014) as a crucial mediator of depressive-like behavior in an HD mouse model. HD mice displayed depressive-like behavior, measured by an increased immobility time in FST that was shown to be reversed by the intraperitoneal injection of N-acetylcysteine (NAC), a regulator of glutamate homeostasis (Wright et al., 2016).

Altered Hippocampal Neurogenesis and Neural Plasticity

In the adult mammalian brain, the subgranular zone of the dentate gyrus (DG) is one of the brain regions where robust neurogenesis continues throughout life (Gould & Gross, 2002). However, this process is exquisitely sensitive to oxidative stress, and perturbations in the redox balance in the neurogenic microenvironment have been shown to lead to reduced neurogenesis (Huang, Zou, & Corniola, 2012). Moreover, chronic stress leads to a dysregulated HPA axis and elevated levels of cortisol (Stephens & Wand, 2012) that, in turn, has been associated with structural changes in the hippocampus (Campbell & MacQueen, 2006) and a selective loss of hippocampal volume, partly attributed to hippocampal neuronal death (Stockmeier et al., 2004) and decreased neurogenesis in animal models (Mirescu & Gould, 2006).

Under normal conditions, adult human neural stem cells express markers of inflammation and immunity in the zones of active neurogenesis that confer neuroprotection and positively regulate the remodeling of neural circuits. This includes NF-κB factor, a crucial mediator of neuronal plasticity and neuroprotection (Widera, Mikenberg, Kaltschmidt, & Kaltschmidt, 2006); IFN-γ, an enhancer of hippocampal neurogenesis shown to improve spatial learning and memory performance in a mouse model of Alzheimer's disease (Baron et al., 2008); and IL-6, an enhancer of progenitor cell survival in the dentate gyrus of adult mice (Bowen, Dempsey, & Vemuganti, 2011). However, under chronic stressful conditions, glia and other brain immune cells secrete excessively high levels of pro-inflammatory cytokines that exert their detrimental effects on cellular and neural plasticity and disrupt the delicate balance required for the neurophysiological actions of immune system on the brain (Belarbi & Rosi, 2013; Kohman & Rhodes, 2013; Kubera, Obuchowicz, Goehler, Brzeszcz, & Maes, 2011).

Depressive-like behavior and decreased hippocampal neurogenesis in chronic mild stress (CMS) models (Alonso et al., 2004; Kaster et al., 2012) have attributed a heightened immune response (Grippo, Francis, Beltz, Felder, & Johnson, 2005). This has been supported by a study, which showed the blockage of CMS-induced depressive-like behavior and reduced hippocampal neurogenesis in transgenic mice with a deletion of IL-1 receptor type 1 (IL-1RKO), compared with the CMS-exposed wild-type mice (Goshen et al., 2008). Further evidence of the anti-neurogenic role of inflammation comes from a study associating increased plasma levels of IL-6 and TNF-α and hippocampal levels of IL-6 and IL-1β and Iba1 (markers of activated microglia) with decreased hippocampal neurogenesis in a dextran sodium sulfate (DSS) mouse model of inflammatory bowel disease (IBD) (Zonis et al., 2015). The anti-neurogenic and anhedonic effects of stress have also been attributed to an increased IL-1β

signaling (Goshen et al., 2008; Koo & Duman, 2008) and higher levels of its downstream signaling factor, nuclear factor-κB (NF-κB), in the dentate gyrus of the hippocampus in adult rats exposed to chronic unpredictable stress (Koo, Russo, Ferguson, Nestler, & Duman, 2010). This was supported by attenuation of behavioral deficits, that is, increased latency to drink milk in novelty-induced hypophagia and decreased sucrose preference by the administration of inhibitors of NF-κB and IL-1β.

In addition to reduced neurogenesis, stress-induced immune response has also been implicated in the impairment of synaptic plasticity. A study investigating the role of IL-6 in modulating synaptic plasticity reported normalization in the alteration of dendritic spine density and the levels of a postsynaptic protein (PSD-95) in a social defeat stress model of mice by blocking IL-6, using MR16-1, an antimouse IL-6 receptor antibody. In addition, the antidepressant effects of MR16-1 were supported by a significant attenuation of increased immobility time in TST and decreased sucrose preference, but this was notably only observed via an intravenous and not an intracerebroventricular injection in mice (Zhang et al., 2017). Furthermore, in a social stress mouse model, a whole transcriptome analysis of the hemibrain, hippocampus, amygdala, medial prefrontal cortex, blood, and spleen revealed inhibition of biological pathways related to neurogenesis, synaptic plasticity, mTOR, Notch, and calcium signaling, accompanied by an activation of interferons, toll-like receptors, tumor necrosis factor-α (TNF-α), and interleukin signaling, with a simultaneous deactivation of signaling pathways implicated in anti-inflammatory responses and T-helper cell-type-2 immunity. This was further supported by epigenetic changes in DNA methylation patterns in the hemibrain, with stressed mice showing inhibited promoter methylation patterns of genes related to inflammation and innate immune responses and increased methylation patterns at promoter regions related to neurogenesis and synaptic plasticity (Deslauriers, Powell, & Risbrough, 2017).

Role of Gut-Microbiota-Brain Axis in Inflammation-Induced Depression

Accumulating evidence is strongly suggestive of dysbiosis of gut microbiome, an undesirable shift in the composition of gut microbiota, as a strong mediator of stress-related psychiatric and mood disorder (Hakansson & Molin, 2011). In addition, evidence supporting the role of intestinal bacteria in the etiology of inflammatory disorders comes from studies using animal models of inflammatory bowel disease (IBD), reporting a strong association between the disturbance of intestinal microflora and increased interstitial permeability to luminal constituents, termed "leaky gut syndrome" (Halfvarson et al., 2017; Hooper et al., 2001). On the other hand, oral probiotic administration in mice has been shown to confer protection from an acute dextran sodium sulfate (DSS)-induced increase in mucosal permeability, which was associated with colitis, thus confirming the role of gut microbiome in maintaining intestinal integrity (Ukena et al., 2007). This is in agreement with a clinical study reporting increased gastrointestinal permeability and high levels of serum IgM and IgA in response to LPS administration in MDD patients compared with healthy volunteers (Maes, Kubera, & Leunis, 2008).

Mounting preclinical evidence also suggests that alterations in gut microbiota can modulate CNS functions and behavior, by generating an increased immune response that manifests as chronic low-grade inflammation in affective disorders (Kelly et al., 2015). Gut microbiota produce several bioactive metabolic products, the majority of which include the short-chain fatty acids (SCFAs), namely, butyrate, acetate, and propionate, produced by the fermentation of nondigestible carbohydrates and dietary fiber in the cecum and large intestine (Russell, Hoyles, Flint, & Dumas, 2013). These SCFAs play a crucial role in maintaining gut integrity by preferentially fueling the enterocytes, inhibiting cell proliferation, and suppressing inflammation (Morrison & Preston, 2016). While *Butyricicoccus*, *Oscillospira*, and Firmicutes are the main butyrate producers; Bacteroidetes solely produces acetate and propionate. Hence, the ratio of Firmicutes to Bacteroidetes in the stool is considered a gauge of overall gut microbiota balance, with decreased ratios observed in antibiotic-associated diarrhea, Crohn's disease, and ulcerative colitis (Den Besten et al., 2013; Ott et al., 2004).

Interestingly, the antidepressant effect of blocking IL-6 has been shown to be mediated through the normalization of Firmicutes/Bacteroidetes ratio and a significant increase in the levels of *Butyricicoccus* and *Oscillospira* (analyzed using 16S ribosomal RNA gene sequencing) in stressed mice (Zhang et al., 2017). Mice infected with the noninvasive parasite *Trichuris muris* developed mild-to-moderate chronic gastrointestinal inflammation accompanied by increased peripheral levels of TNF-α and IFN-γ, reduced hippocampal brain-derived neurotrophic factor (BDNF) levels, and anxiety-like behavior that was assessed using the light/dark test (Bercik et al., 2010).

In another study exploring the link between IL-10, an anti-inflammatory cytokine, and colonization of gut bacteria, increased intestinal permeability accompanied by an increased mucosal secretion of TNF-α was observed in a homozygous IL-10 gene-deficient mice (IL-10$^{-/-}$). Reduced levels of colonic *Lactobacillus* sp. and an increase in colonic mucosal adherent and translocated aerobic bacteria have been attributed to the onset of colitis in IL-10$^{-/-}$ mice (Wang, Fang, & Hasselgren, 2001).

SEXUAL DIMORPHISM ASSOCIATED WITH INFLAMMATION-INDUCED DEPRESSION

The role played by sex differences in inflammation-induced depression has been investigated in a clinical study, which reported higher sensitivity to the behavioral effects (depressed mood and greater feelings of social disconnection) of inflammation in women following endotoxin exposure, despite a similar magnitude in cytokine (IL-6 and TNF-α) responses in both men and women (Moieni et al., 2015). A separate study investigating the incidence of MDD related to antiviral therapy for chronic hepatitis C also reported higher incidence rate of depression in women, in response to standardized doses of interferon-α (IFNα) (Udina et al., 2012). Women have also been reported to show a greater inflammatory reactivity compared with men (O'Connor, Motivala, Valladares, Olmstead, & Irwin, 2007), which is one possible explanation as to why women are twice as likely as men to experience stress-related psychiatric disorders like depression (Grigoriadis & Robinson, 2007; Seedat et al., 2009).

Preclinical studies aimed at understanding sex differences in affective disorders have also reported an increased immobile time in FST, accompanied by increased HPA axis activity in female rats as compared with male rats (Dalla, Pitychoutis, Kokras, & Papadopoulou-Daifoti, 2011; Drossopoulou et al., 2004). Consistent with those findings, we have reported alterations in depressive-like behavior and immune response in a mouse model of Huntington's disease (HD), in a sex-dependent manner (Renoir et al., 2011). In addition to an increased susceptibility to depressive-like behavior, female HD mice displayed an enhanced LPS-induced TNF-α gene expression in the hypothalamus and serum compared with male HD mice (Renoir, Pang, Shikano, Li, & Hannan, 2015). This differential immune response may contribute to the sexually dimorphic depressive-like behavior that was previously reported in HD mice (Renoir et al., 2011).

CONCLUSIONS

Despite the availability of a large variety of antidepressant medications and alternative psychotherapy-based approaches, depression continues to suffer a huge treatment gap worldwide, whereby a large number of individuals display treatment resistance or suffer from the "off-target" effects of the currently available antidepressant medications. Depression is a heterogeneous disorder characterized by a variable set of symptoms, with an etiology that is yet to be fully elucidated. The availability of animal models has established the role of prolonged inflammation and cytokines, in addition to the well-established role of a hyperactive HPA axis and glucocorticoid resistance, as crucial mediators of stress-induced depressive phenotypes. Also, recent years have witnessed an emerging role of the gut microbiome as a crucial node responsible for the regulation of stress-related responses by the gut-brain axis. Interestingly, mounting preclinical evidence suggests that the effect of the gut microbiome on the modulation of brain function and depressive-like behavior is mediated by the activation of pro-inflammatory cytokines, even though the detailed mechanisms mediating the interaction between gut microbiome, immune pathways, and behavioral alterations still remain to be elucidated. Hence, even though gut microbiome may appear to be a promising target for the development of novel treatment options in the form of probiotics for treating inflammation-induced depression, additional research in humans is sorely needed to establish a strong causal contribution of the gut microbiome in the modulation of inflammatory response and stress-related psychiatric disorders like depression.

Furthermore, most of the current preclinical studies investigating the underlying mechanisms of affective disorders involve male animals, despite a higher incidence of inflammation-induced depression reported in women. While this review highlights some of the preclinical and clinical studies reporting the crucial role played by sex differences in the etiology of inflammation-induced depression, there is an urgent need for future studies, in order to elucidate the sexually dimorphic mechanisms underlying the etiology of depression and associated psychiatric disorders.

Consistent with the emerging relationship between inflammation and depression, even though various classes of anti-inflammatory medications may appear to hold promise as 'novel antidepressants', it should be noted that these medications may only be effective in a subset of depressed patients, who have either experienced early-life adversities or display evidence of increased peripheral markers, as in the case of rheumatoid arthritis and cardiovascular disorders. Future studies exploring the genetic and epigenetic mechanisms may explain the contribution of early-life or adult trauma to an exaggerated or persistent response to inflammation that serves as a risk factor for the onset of depression and other stress-related psychiatric disorders.

In conclusion, while there have been numerous studies elaborating on the mechanisms that mediate the ability of inflammation to induce the symptoms of depression at the preclinical level, translation to clinical studies for the development of novel antidepressants relies heavily on the development of predictive biomarkers that can reliably inform us about who would respond to a given anti-inflammatory strategy. Nevertheless, gaining a better insight into how the immune system can be harnessed to improve the treatment of depression and other affective disorders provides hope for the development of novel antidepressant therapies.

References

Al-Harbi, K. S. (2012). Treatment-resistant depression: therapeutic trends, challenges, and future directions. *Patient Preference and Adherence, 6,* 369–388.

Alonso, R., Griebel, G., Pavone, G., Stemmelin, J., Le Fur, G., & Soubrié, P. (2004). Blockade of CRF(1) or V(1b) receptors reverses stress-induced suppression of neurogenesis in a mouse model of depression. *Molecular Psychiatry, 9*(3), 278–286 224.

Amori, L., Guidetti, P., Pellicciari, R., Kajii, Y., & Schwarcz, R. (2009). On the relationship between the two branches of the kynurenine pathway in the rat brain in vivo. *Journal of Neurochemistry, 109*(2), 316–325.

Anderson, G., Berk, M., Dean, O., Moylan, S., & Maes, M. (2014). Role of immune-inflammatory and oxidative and nitrosative stress pathways in the etiology of depression: therapeutic implications. *CNS Drugs, 28*(1), 1–10.

Badawy, A. A.-B. (2013). Tryptophan: the key to boosting brain serotonin synthesis in depressive illness. *Journal of Psychopharmacology, 27*(10), 878–893.

Barnes, P. J. (2006). Corticosteroids: the drugs to beat. *European Journal of Pharmacology, 533*(1–3), 2–14.

Baron, R., Nemirovsky, A., Harpaz, I., Cohen, H., Owens, T., & Monsonego, A. (2008). IFN-gamma enhances neurogenesis in wild-type mice and in a mouse model of Alzheimer's disease. *The FASEB Journal, 22*(8), 2843–2852.

Bassi, G. S., Kanashiro, A., Santin, F. M., de Souza, G. E. P., Nobre, M. J., & Coimbra, N. C. (2012). Lipopolysaccharide-induced sickness behaviour evaluated in different models of anxiety and innate fear in rats. *Basic & Clinical Pharmacology & Toxicology, 110*(4), 359–369.

Bay-Richter, C., Linderholm, K. R., Lim, C. K., Samuelsson, M., Träskman-Bendz, L., Guillemin, G. J., et al. (2015). A role for inflammatory metabolites as modulators of the glutamate N-methyl-D-aspartate receptor in depression and suicidality. *Brain, Behavior, and Immunity, 43,* 110–117.

Belarbi, K., & Rosi, S. (2013). Modulation of adult-born neurons in the inflamed hippocampus. *Frontiers in Cellular Neuroscience, 7,* 145.

Bercik, P., Verdu, E. F., Foster, J. A., Macri, J., Potter, M., Huang, X., et al. (2010). Chronic gastrointestinal inflammation induces anxiety-like behavior and alters central nervous system biochemistry in mice. *Gastroenterology, 139*(6), 2102–2112.e1.

Biesmans, S., Bouwknecht, J. A., Ver Donck, L., Langlois, X., Acton, P. D., De Haes, P., et al. (2015). Peripheral administration of tumor necrosis factor-alpha induces neuroinflammation and sickness but not depressive-like behavior in mice. *BioMed Research International, 2015,* 716920.

Boulangé, C. L., Neves, A. L., Chilloux, J., Nicholson, J. K., & Dumas, M. E. (2016). Impact of the gut microbiota on inflammation, obesity, and metabolic disease. *Genome Medicine, 8*(1), 42.

Bowen, K. K., Dempsey, R. J., & Vemuganti, R. (2011). Adult interleukin-6 knockout mice show compromised neurogenesis. *NeuroReport*, 22(3), 126–130.

Brites, D., & Fernandes, A. (2015). Neuroinflammation and depression: microglia activation, extracellular microvesicles and microRNA dysregulation. *Frontiers in Cellular Neuroscience*, 9, 476.

Campbell, S., & MacQueen, G. (2006). An update on regional brain volume differences associated with mood disorders. *Current Opinion in Psychiatry*, 19(1), 25–33.

Capuron, L., & Dantzer, R. (2003). Cytokines and depression: the need for a new paradigm. *Brain, Behavior, and Immunity*, 17(Suppl. 1), S119–124.

Capuron, L., Ravaud, A., Gualde, N., Bosmans, E., Dantzer, R., Maes, M., et al. (2001). Association between immune activation and early depressive symptoms in cancer patients treated with interleukin-2-based therapy. *Psychoneuroendocrinology*, 26(8), 797–808.

Capuron, L., Ravaud, A., Neveu, P. J., Miller, A. H., Maes, M., & Dantzer, R. (2002). Association between decreased serum tryptophan concentrations and depressive symptoms in cancer patients undergoing cytokine therapy. *Molecular Psychiatry*, 7(5), 468–473.

Carabotti, M., Scirocco, A., Maselli, M. A., & Severi, C. (2015). The gut-brain axis: interactions between enteric microbiota, central and enteric nervous systems. *Annals of Gastroenterology*, 28(2), 203–209.

Caspi, A., Sugden, K., Moffitt, T. E., Taylor, A., Craig, I. W., Harrington, H., et al. (2003). Influence of life stress on depression: moderation by a polymorphism in the 5-HTT gene. *Science (New York, N.Y.)*, 301(5631), 386–389.

Chourbaji, S., Urani, A., Inta, I., Sanchis-Segura, C., Brandwein, C., Zink, M., et al. (2006). IL-6 knockout mice exhibit resistance to stress-induced development of depression-like behaviors. *Neurobiology of Disease*, 23(3), 587–594.

Coutinho, A. E., & Chapman, K. E. (2011). The anti-inflammatory and immunosuppressive effects of glucocorticoids, recent developments and mechanistic insights. *Molecular and Cellular Endocrinology*, 335(1), 2–13.

Currier, A. R., Ziegler, M. H., Riley, M. M., Babcock, T. A., Telbis, V. P., & Carlin, J. M. (2000). Tumor necrosis factor-alpha and lipopolysaccharide enhance interferon-induced antichlamydial indoleamine dioxygenase activity independently. *Journal of Interferon & Cytokine Research*, 20(4), 369–376.

Dalla, C., Pitychoutis, P. M., Kokras, N., & Papadopoulou-Daifoti, Z. (2011). Sex differences in response to stress and expression of depressive-like behaviours in the rat. *Current Topics in Behavioral Neurosciences*, 8, 97–118.

Dantzer, R. (2004). Cytokine-induced sickness behaviour: a neuroimmune response to activation of innate immunity. *European Journal of Pharmacology*, 500(1), 399–411.

Dantzer, R. (2009). Cytokine, sickness behavior, and depression. *Immunology and Allergy Clinics of North America*, 29(2), 247–264.

Dantzer, R., O'Connor, J. C., Freund, G. G., Johnson, R. W., & Kelley, K. W. (2008). From inflammation to sickness and depression: when the immune system subjugates the brain. *Nature Reviews. Neuroscience*, 9(1), 46–56.

Den Besten, G., van Eunen, K., Groen, A. K., Venema, K., Reijngoud, D.-J., & Bakker, B. M. (2013). The role of short-chain fatty acids in the interplay between diet, gut microbiota, and host energy metabolism. *Journal of Lipid Research*, 54(9), 2325–2340.

Deslauriers, J., Powell, S. B., & Risbrough, V. B. (2017). Immune signaling mechanisms of PTSD risk and symptom development: insights from animal models. *Current Opinion in Behavioral Sciences*, 14, 123–132.

Drossopoulou, G., Antoniou, K., Kitraki, E., Papathanasiou, G., Papalexi, E., Dalla, C., & Papadopoulou-Daifoti, Z. (2004). Sex differences in behavioral, neurochemical and neuroendocrine effects induced by the forced swim test in rats. *Neuroscience*, 126(4), 849–857.

Du, X., Pang, T. Y. C., & Hannan, A. J. (2013). A tale of two maladies? Pathogenesis of depression with and without the Huntington's disease gene mutation. *Frontiers in Neurology*, 4.

Dunn, A. J. (2000). Cytokine activation of the HPA axis. *Annals of the New York Academy of Sciences*, 917, 608–617.

Dunn, A. J., & Swiergiel, A. H. (1998). The role of cytokines in infection-related behavior. *Annals of the New York Academy of Sciences*, 840, 577–585.

Dunn, A. J., & Swiergiel, A. H. (2005). Effects of interleukin-1 and endotoxin in the forced swim and tail suspension tests in mice. *Pharmacology, Biochemistry, and Behavior*, 81(3), 688–693.

Ehrlich, K. B., Ross, K. M., Chen, E., & Miller, G. E. (2016). Testing the biological embedding hypothesis: is early life adversity associated with a later pro-inflammatory phenotype? *Development and Psychopathology*, 28(4pt2), 1273–1283.

Evans, D. L., Charney, D. S., Lewis, L., Golden, R. N., Gorman, J. M., Krishnan, K. R. R., et al. (2005). Mood disorders in the medically ill: scientific review and recommendations. *Biological Psychiatry*, 58(3), 175–189.

Felger, J. C., & Lotrich, F. E. (2013). Inflammatory cytokines in depression: neurobiological mechanisms and therapeutic implications. *Neuroscience*, 246, 199–229.

Foster, J. A., Rinaman, L., & Cryan, J. F. (2017). Stress & the gut-brain axis: regulation by the microbiome. *Neurobiology of Stress*, 7, 124–136.

Frenois, F., Moreau, M., Connor, J. O., Lawson, M., Micon, C., Lestage, J., et al. (2007). Lipopolysaccharide induces delayed FosB/DeltaFosB immunostaining within the mouse extended amygdala, hippocampus and hypothalamus, that parallel the expression of depressive-like behavior. *Psychoneuroendocrinology*, 32(5), 516–531.

Fukui, S., Schwarcz, R., Rapoport, S. I., Takada, Y., & Smith, Q. R. (1991). Blood-brain barrier transport of kynurenines: implications for brain synthesis and metabolism. *Journal of Neurochemistry, 56*(6), 2007–2017.

Glaser, R., & Kiecolt-Glaser, J. K. (2005). Stress-induced immune dysfunction: implications for health. *Nature Reviews Immunology, 5*(3), 243–251.

Goshen, I., Kreisel, T., Ben-Menachem-Zidon, O., Licht, T., Weidenfeld, J., Ben-Hur, T., et al. (2008). Brain interleukin-1 mediates chronic stress-induced depression in mice via adrenocortical activation and hippocampal neurogenesis suppression. *Molecular Psychiatry, 13*(7), 717–728.

Gould, E., & Gross, C. G. (2002). Neurogenesis in adult mammals: some progress and problems. *The Journal of Neuroscience: The Official Journal of the Society for Neuroscience, 22*(3), 619–623.

Grigoriadis, S., & Robinson, G. E. (2007). Gender issues in depression. *Annals of Clinical Psychiatry, 19*(4), 247–255.

Grippo, A. J., Francis, J., Beltz, T. G., Felder, R. B., & Johnson, A. K. (2005). Neuroendocrine and cytokine profile of chronic mild stress-induced anhedonia. *Physiology & Behavior, 84*(5), 697–706.

Grote, H. E., Bull, N. D., Howard, M. L., van Dellen, A., Blakemore, C., Bartlett, P. F., et al. (2005). Cognitive disorders and neurogenesis deficits in Huntington's disease mice are rescued by fluoxetine. *European Journal of Neuroscience, 22*(8), 2081–2088.

Hakansson, A., & Molin, G. (2011). Gut microbiota and inflammation. *Nutrients, 3*(6), 637–682.

Halfvarson, J., Brislawn, C. J., Lamendella, R., Vázquez-Baeza, Y., Walters, W. A., Bramer, L. M., et al. (2017). Dynamics of the human gut microbiome in inflammatory bowel disease. *Nature Microbiology, 2*, 17004.

Henry, C. J., Huang, Y., Wynne, A., Hanke, M., Himler, J., Bailey, M. T., et al. (2008). Minocycline attenuates lipopolysaccharide (LPS)-induced neuroinflammation, sickness behavior, and anhedonia. *Journal of Neuroinflammation, 5*, 15.

Holmes, A., Murphy, D. L., & Crawley, J. N. (2003). Abnormal behavioral phenotypes of serotonin transporter knockout mice: parallels with human anxiety and depression. *Biological Psychiatry, 54*(10), 953–959.

Hooper, L. V., Wong, M. H., Thelin, A., Hansson, L., Falk, P. G., & Gordon, J. I. (2001). Molecular analysis of commensal host-microbial relationships in the intestine. *Science, 291*(5505), 881–884.

Huang, T.-T., Zou, Y., & Corniola, R. (2012). Oxidative stress and adult neurogenesis—effects of radiation and superoxide dismutase deficiency. *Seminars in Cell & Developmental Biology, 23*(7), 738–744.

Irwin, M. (2002). Psychoneuroimmunology of depression: clinical implications. *Brain, Behavior, and Immunity, 16*(1), 1–16.

Irwin, M. R., & Miller, A. H. (2007). Depressive disorders and immunity: 20 years of progress and discovery. *Brain, Behavior, and Immunity, 21*(4), 374–383.

Jain, N. K., Kulkarni, S. K., & Singh, A. (2001). Lipopolysaccharide-mediated immobility in mice: reversal by cyclooxygenase enzyme inhibitors. *Methods and Findings in Experimental and Clinical Pharmacology, 23*(8), 441–444.

Kaster, M. P., Gadotti, V. M., Calixto, J. B., Santos, A. R. S., & Rodrigues, A. L. S. (2012). Depressive-like behavior induced by tumor necrosis factor-α in mice. *Neuropharmacology, 62*(1), 419–426.

Kelley, K. W., Bluthé, R.-M., Dantzer, R., Zhou, J.-H., Shen, W.-H., Johnson, R. W., et al. (2003). Cytokine-induced sickness behavior. *Brain, Behavior, and Immunity, 17*(Suppl. 1), S112–118.

Kelly, J. R., Kennedy, P. J., Cryan, J. F., Dinan, T. G., Clarke, G., & Hyland, N. P. (2015). Breaking down the barriers: the gut microbiome, intestinal permeability and stress-related psychiatric disorders. *Frontiers in Cellular Neuroscience, 9*, 392.

Kessler, R. C., & Magee, W. J. (1993). Childhood adversities and adult depression: basic patterns of association in a US national survey. *Psychological Medicine, 23*(3), 679–690.

Kohman, R. A., & Rhodes, J. S. (2013). Neurogenesis, inflammation and behavior. *Brain, Behavior, and Immunity, 27C*, 22–32.

Koo, J. W., & Duman, R. S. (2008). IL-1beta is an essential mediator of the antineurogenic and anhedonic effects of stress. *Proceedings of the National Academy of Sciences of the United States of America, 105*(2), 751–756.

Koo, J. W., Russo, S. J., Ferguson, D., Nestler, E. J., & Duman, R. S. (2010). Nuclear factor-kappaB is a critical mediator of stress-impaired neurogenesis and depressive behavior. *Proceedings of the National Academy of Sciences of the United States of America, 107*(6), 2669–2674.

Kubera, M., Maes, M., Holan, V., Basta-Kaim, A., Roman, A., & Shani, J. (2001). Prolonged desipramine treatment increases the production of interleukin-10, an anti-inflammatory cytokine, in C57BL/6 mice subjected to the chronic mild stress model of depression. *Journal of Affective Disorders, 63*(1), 171–178.

Kubera, M., Obuchowicz, E., Goehler, L., Brzeszcz, J., & Maes, M. (2011). In animal models, psychosocial stress-induced (neuro)inflammation, apoptosis and reduced neurogenesis are associated to the onset of depression. *Progress in Neuro-Psychopharmacology & Biological Psychiatry, 35*(3), 744–759.

Lacosta, S., Merali, Z., & Anisman, H. (1999). Behavioral and neurochemical consequences of lipopolysaccharide in mice: anxiogenic-like effects. *Brain Research, 818*(2), 291–303.

Lanquillon, S., Krieg, J.-C., Bening-Abu-Shach, U., & Vedder, H. (2000). Cytokine production and treatment response in major depressive disorder. *Neuropsychopharmacology, 22*(4), 370–379.

Laugeray, A., Launay, J.-M., Callebert, J., Surget, A., Belzung, C., & Barone, P. R. (2010). Peripheral and cerebral metabolic abnormalities of the tryptophan–kynurenine pathway in a murine model of major depression. *Behavioural Brain Research, 210*(1), 84–91.

Lenczowski, M. J., Van Dam, A. M., Poole, S., Larrick, J. W., & Tilders, F. J. (1997). Role of circulating endotoxin and interleukin-6 in the ACTH and corticosterone response to intraperitoneal LPS. *The American Journal of Physiology, 273*(6 Pt 2), R1870–1877.

Leo, R. J. (2005). Chronic pain and comorbid depression. *Current Treatment Options in Neurology, 7*(5), 403–412.

Leonard, B. E., & Myint, A. (2009). The psychoneuroimmunology of depression. *Human Psychopharmacology, 24*(3), 165–175.

Lichtblau, N., Schmidt, F. M., Schumann, R., Kirkby, K. C., & Himmerich, H. (2013). Cytokines as biomarkers in depressive disorder: current standing and prospects. *International Review of Psychiatry, 25*(5), 592–603.

Macchi, F., Homberg, J. R., Calabrese, F., Zecchillo, C., Racagni, G., Riva, M. A., et al. (2013). Altered inflammatory responsiveness in serotonin transporter mutant rats. *Journal of Neuroinflammation, 10*, 116.

Maes, M. (1995). Evidence for an immune response in major depression: a review and hypothesis. *Progress in Neuro-Psychopharmacology and Biological Psychiatry, 19*(1), 11–38.

Maes, M., Kubera, M., & Leunis, J.-C. (2008). The gut-brain barrier in major depression: intestinal mucosal dysfunction with an increased translocation of LPS from gram negative enterobacteria (leaky gut) plays a role in the inflammatory pathophysiology of depression. *Neuro Endocrinology Letters, 29*(1), 117–124.

Maes, M., Yirmiya, R., Noraberg, J., Brene, S., Hibbeln, J., Perini, G., et al. (2009). The inflammatory & neurodegenerative (I&ND) hypothesis of depression: leads for future research and new drug developments in depression. *Metabolic Brain Disease, 24*(1), 27–53.

Malynn, S., Campos-Torres, A., Moynagh, P., & Haase, J. (2013). The pro-inflammatory cytokine TNF-α regulates the activity and expression of the serotonin transporter (SERT) in astrocytes. *Neurochemical Research, 38*(4), 694–704.

Margaretten, M., Julian, L., Katz, P., & Yelin, E. (2011). Depression in patients with rheumatoid arthritis: description, causes and mechanisms. *International Journal of Clinical Rheumatology, 6*(6), 617–623.

Meier, T. B., Drevets, W. C., Wurfel, B. E., Ford, B. N., Morris, H. M., Victor, T. A., et al. (2016). Relationship between neurotoxic kynurenine metabolites and reductions in right medial prefrontal cortical thickness in major depressive disorder. *Brain, Behavior, and Immunity, 53*, 39–48.

Mesquita, A. R., Correia-Neves, M., Roque, S., Castro, A. G., Vieira, P., Pedrosa, J., et al. (2008). IL-10 modulates depressive-like behavior. *Journal of Psychiatric Research, 43*(2), 89–97.

Miller, A. H., Maletic, V., & Raison, C. L. (2009). Inflammation and its discontents: the role of cytokines in the pathophysiology of major depression. *Biological Psychiatry, 65*(9), 732–741.

Miller, A. H., & Raison, C. L. (2015). The role of inflammation in depression: from evolutionary imperative to modern treatment target. *Nature Reviews Immunology, 16*(1), 22–34.

Mirescu, C., & Gould, E. (2006). Stress and adult neurogenesis. *Hippocampus, 16*(3), 233–238.

Moieni, M., Irwin, M. R., Jevtic, I., Olmstead, R., Breen, E. C., & Eisenberger, N. I. (2015). Sex differences in depressive and socioemotional responses to an inflammatory challenge: implications for sex differences in depression. *Neuropsychopharmacology, 40*(7), 1709–1716.

Moreau, M., Lestage, J., Verrier, D., Mormede, C., Kelley, K. W., Dantzer, R., et al. (2005). Bacille Calmette-Guérin inoculation induces chronic activation of peripheral and brain indoleamine 2,3-dioxygenase in mice. *The Journal of Infectious Diseases, 192*(3), 537–544.

Morrison, D. J., & Preston, T. (2016). Formation of short chain fatty acids by the gut microbiota and their impact on human metabolism. *Gut Microbes, 7*(3), 189–200.

Müller, N., & Schwarz, M. J. (2007). The immune-mediated alteration of serotonin and glutamate: towards an integrated view of depression. *Molecular Psychiatry, 12*(11), 988–1000.

Murphy, D. L., & Lesch, K.-P. (2008). Targeting the murine serotonin transporter: insights into human neurobiology. *Nature Reviews. Neuroscience, 9*(2), 85–96.

Naitoh, Y., Fukata, J., Tominaga, T., Nakai, Y., Tamai, S., Mori, K., et al. (1988). Interleukin-6 stimulates the secretion of adrenocorticotropic hormone in conscious, freely-moving rats. *Biochemical and Biophysical Research Communications, 155*(3), 1459–1463.

Nance, M. A., & Myers, R. H. (2001). Juvenile onset Huntington's disease—clinical and research perspectives. *Mental Retardation and Developmental Disabilities Research Reviews, 7*(3), 153–157.

Nutt, D. J. (2008). Relationship of neurotransmitters to the symptoms of major depressive disorder. *The Journal of Clinical Psychiatry, 69*(Suppl. E1), 4–7.

O'Connor, J. C., Lawson, M. A., André, C., Briley, E. M., Szegedi, S. S., Lestage, J., et al. (2009). Induction of IDO by Bacille Calmette-Guérin is responsible for development of murine depressive-like behavior. *Journal of Immunology, 182*(5), 3202–3212.

O'Connor, M.-F., Motivala, S. J., Valladares, E. M., Olmstead, R., & Irwin, M. R. (2007). Sex differences in monocyte expression of IL-6: role of autonomic mechanisms. *American Journal of Physiology. Regulative, Integrative and Comparative Physiology, 293*(1), R145–151.

Ott, S. J., Musfeldt, M., Wenderoth, D. F., Hampe, J., Brant, O., Fölsch, U. R., et al. (2004). Reduction in diversity of the colonic mucosa associated bacterial microflora in patients with active inflammatory bowel disease. *Gut*, *53*(5), 685–693.

Otte, C., McCaffery, J., Ali, S., & Whooley, M. A. (2007). Association of a serotonin transporter polymorphism (5-HTTLPR) with depression, perceived stress, and norepinephrine in patients with coronary disease: the heart and soul study. *The American Journal of Psychiatry*, *164*(9), 1379–1384.

Pace, T. W. W., Mletzko, T. C., Alagbe, O., Musselman, D. L., Nemeroff, C. B., Miller, A. H., et al. (2006). Increased stress-induced inflammatory responses in male patients with major depression and increased early life stress. *The American Journal of Psychiatry*, *163*(9), 1630–1633.

Pang, T. Y. C., Du, X., Zajac, M. S., Howard, M. L., & Hannan, A. J. (2009). Altered serotonin receptor expression is associated with depression-related behavior in the R6/1 transgenic mouse model of Huntington's disease. *Human Molecular Genetics*, *18*(4), 753–766.

Quan, N., Sundar, S. K., & Weiss, J. M. (1994). Induction of interleukin-1 in various brain regions after peripheral and central injections of lipopolysaccharide. *Journal of Neuroimmunology*, *49*(1–2), 125–134.

Raison, C. L., Capuron, L., & Miller, A. H. (2006). Cytokines sing the blues: inflammation and the pathogenesis of depression. *Trends in Immunology*, *27*(1), 24–31.

Raison, C. L., & Miller, A. H. (2003). When not enough is too much: the role of insufficient glucocorticoid signaling in the pathophysiology of stress-related disorders. *The American Journal of Psychiatry*, *160*(9), 1554–1565.

Raison, C. L., & Miller, A. H. (2011). Is depression an inflammatory disorder? *Current Psychiatry Reports*, *13*(6), 467–475.

Remus, J. L., & Dantzer, R. (2016). Inflammation models of depression in rodents: relevance to psychotropic drug discovery. *International Journal of Neuropsychopharmacology*, *19*(9), pyw028.

Renoir, T., Pang, T. Y., Shikano, Y., Li, S., & Hannan, A. J. (2015). Loss of the sexually dimorphic neuroinflammatory response in a transgenic mouse model of Huntington's disease. *Journal of Huntington's Disease*, *4*(4), 297–303.

Renoir, T., Zajac, M. S., Du, X., Pang, T. Y., Leang, L., Chevarin, C., et al. (2011). Sexually dimorphic serotonergic dysfunction in a mouse model of Huntington's disease and depression. *PLOS ONE*, *6*(7), e22133.

Reyes-Ocampo, J., Ramírez-Ortega, D., Cervantes, G. I. V., Pineda, B., Balderas, P. M. de O., González-Esquivel, D., et al. (2015). Mitochondrial dysfunction related to cell damage induced by 3-hydroxykynurenine and 3-hydroxyanthranilic acid: Non-dependent-effect of early reactive oxygen species production. *Neurotoxicology*, *50*, 81–91.

Riedel, W. J., Klaassen, T., & Schmitt, J. A. J. (2002). Tryptophan, mood, and cognitive function. *Brain, Behavior, and Immunity*, *16*(5), 581–589.

Rivier, C., & Vale, W. (1991). Stimulatory effect of interleukin-1 on adrenocorticotropin secretion in the rat: is it modulated by prostaglandins? *Endocrinology*, *129*(1), 384–388.

Roque, S., Correia-Neves, M., Mesquita, A. R., Palha, J. A., & Sousa, N. (2009). Interleukin-10: A Key Cytokine in Depression? *Cardiovascular Psychiatry and Neurology*, *2009*.

Russell, W. R., Hoyles, L., Flint, H. J., & Dumas, M.-E. (2013). Colonic bacterial metabolites and human health. *Current Opinion in Microbiology*, *16*(3), 246–254.

Schloss, P., & Williams, D. C. (1998). The serotonin transporter: a primary target for antidepressant drugs. *Journal of Psychopharmacology*, *12*(2), 115–121.

Schwamborn, R., Brown, E., & Haase, J. (2016). Elevation of cortical serotonin transporter activity upon peripheral immune challenge is regulated independently of p38 mitogen-activated protein kinase activation and transporter phosphorylation. *Journal of Neurochemistry*, *137*(3), 423–435.

Seedat, S., Scott, K. M., Angermeyer, M. C., Berglund, P., Bromet, E. J., & Brugha, T. S. (2009). Cross-national associations between gender and mental disorders in the World Health Organization world mental health surveys. *Archives of General Psychiatry*, *66*(7), 785–795.

Shajib, M. S., & Khan, W. I. (2015). The role of serotonin and its receptors in activation of immune responses and inflammation. *Acta Physiologica (Oxford, England)*, *213*(3), 561–574.

Silverman, M. N., & Sternberg, E. M. (2012). Glucocorticoid regulation of inflammation and its behavioral and metabolic correlates: from HPA axis to glucocorticoid receptor dysfunction. *Annals of the New York Academy of Sciences*, *1261*, 55–63.

Simen, B. B., Duman, C. H., Simen, A. A., & Duman, R. S. (2006). TNFα signaling in depression and anxiety: behavioral consequences of individual receptor targeting. *Biological Psychiatry*, *59*(9), 775–785.

Slavich, G. M., & Irwin, M. R. (2014). From stress to inflammation and major depressive disorder: a social signal transduction theory of depression. *Psychological Bulletin*, *140*(3), 774–815.

Slopen, N., Lewis, T. T., Gruenewald, T. L., Mujahid, M. S., Ryff, C. D., Albert, M. A., et al. (2010). Early life adversity and inflammation in African Americans and whites in the midlife in the United States survey. *Psychosomatic Medicine*, *72*(7), 694–701.

Smith, R. S. (1991). The macrophage theory of depression. *Medical Hypotheses*, *35*(4), 298–306.

Smith, S. M., & Vale, W. W. (2006). The role of the hypothalamic-pituitary-adrenal axis in neuroendocrine responses to stress. *Dialogues in Clinical Neuroscience*, *8*(4), 383–395.

Stephens, M. A. C., & Wand, G. (2012). Stress and the HPA axis. *Alcohol Research: Current Reviews*, *34*(4), 468–483.

Stockmeier, C. A., Mahajan, G. J., Konick, L. C., Overholser, J. C., Jurjus, G. J., Meltzer, H. Y., et al. (2004). Cellular changes in the postmortem hippocampus in major depression. *Biological Psychiatry, 56*(9), 640–650.

Stone, T. W. (2001). Endogenous neurotoxins from tryptophan. *Toxicon: Official Journal of the International Society on Toxinology, 39*(1), 61–73.

Takikawa, O., Tagawa, Y., Iwakura, Y., Yoshida, R., & Truscott, R. J. (1999). Interferon-gamma-dependent/independent expression of indoleamine 2,3-dioxygenase. Studies with interferon-gamma-knockout mice. *Advances in Experimental Medicine and Biology, 467*, 553–557.

Trivedi, M. H., Rush, A. J., Wisniewski, S. R., Nierenberg, A. A., Warden, D., Ritz, L., et al. (2006). Evaluation of outcomes with citalopram for depression using measurement-based care in STAR*D: implications for clinical practice. *The American Journal of Psychiatry, 163*(1), 28–40.

Tsao, C.-W., Lin, Y.-S., Chen, C.-C., Bai, C.-H., & Wu, S.-R. (2006). Cytokines and serotonin transporter in patients with major depression. *Progress in Neuro-Psychopharmacology & Biological Psychiatry, 30*(5), 899–905.

Turnbull, A. V., & Rivier, C. L. (1999). Regulation of the hypothalamic-pituitary-adrenal axis by cytokines: actions and mechanisms of action. *Physiological Reviews, 79*(1), 1–71.

Tyring, S., Gottlieb, A., Papp, K., Gordon, K., Leonardi, C., Wang, A., et al. (2006). Etanercept and clinical outcomes, fatigue, and depression in psoriasis: double-blind placebo-controlled randomised phase III trial. *Lancet, 367*(9504), 29–35.

Udina, M., Castellví, P., Moreno-España, J., Navinés, R., Valdés, M., Forns, X., et al. (2012). Interferon-induced depression in chronic hepatitis C: a systematic review and meta-analysis. *The Journal of Clinical Psychiatry, 73*(8), 1128–1138.

Ukena, S. N., Singh, A., Dringenberg, U., Engelhardt, R., Seidler, U., Hansen, W., et al. (2007). Probiotic Escherichia coli Nissle 1917 inhibits leaky gut by enhancing mucosal integrity. *PLoS One, 2*(12), e1308.

Wakabayashi, C., Kiyama, Y., Kunugi, H., Manabe, T., & Iwakura, Y. (2011). Age-dependent regulation of depression-like behaviors through modulation of adrenergic receptor α_{1a} subtype expression revealed by the analysis of interleukin-1 receptor antagonist knockout mice. *Neuroscience, 192*, 475–484.

Walker, A. K., Budac, D. P., Bisulco, S., Lee, A. W., Smith, R. A., Beenders, B., et al. (2013). NMDA receptor blockade by ketamine abrogates lipopolysaccharide-induced depressive-like behavior in C57BL/6J mice. *Neuropsychopharmacology, 38*(9), 1609–1616.

Wang, Q., Fang, C. H., & Hasselgren, P.-O. (2001). Intestinal permeability is reduced and IL-10 levels are increased in septic IL-6 knockout mice. *American Journal of Physiology.*

Regulatory, Integrative and Comparative Physiology, 281(3), R1013–R1023.

Widera, D., Mikenberg, I., Kaltschmidt, B., & Kaltschmidt, C. (2006). Potential role of NF-κB in adult neural stem cells: the underrated steersman? *International Journal of Developmental Neuroscience, 24*(2–3), 91–102.

Wright, D. J., Gray, L. J., Finkelstein, D. I., Crouch, P. J., Pow, D., Pang, T. Y., et al. (2016). N-acetylcysteine modulates glutamatergic dysfunction and depressive behavior in Huntington's disease. *Human Molecular Genetics, 25*(14), 2923–2933.

Yang, C., Hong, T., Shen, J., Ding, J., Dai, X.-W., Zhou, Z.-Q., et al. (2013). Ketamine exerts antidepressant effects and reduces IL-1β and IL-6 levels in rat prefrontal cortex and hippocampus. *Experimental and Therapeutic Medicine, 5*(4), 1093–1096.

Zhang, J., Yao, W., Dong, C., Yang, C., Ren, Q., Ma, M., et al. (2017). Blockade of interleukin-6 receptor in the periphery promotes rapid and sustained antidepressant actions: a possible role of gut–microbiota–brain axis. *Translational Psychiatry, 7*(5), e1138.

Ziemssen, T., & Kern, S. (2007). Psychoneuroimmunology—cross-talk between the immune and nervous systems. *Journal of Neurology, 254*(Suppl. 2), II8–11.

Zonis, S., Pechnick, R. N., Ljubimov, V. A., Mahgerefteh, M., Wawrowsky, K., Michelsen, K. S., et al. (2015). Chronic intestinal inflammation alters hippocampal neurogenesis. *Journal of Neuroinflammation, 12*, 65.

Zunszain, P. A., Anacker, C., Cattaneo, A., Carvalho, L. A., & Pariante, C. M. (2011). Glucocorticoids, cytokines and brain abnormalities in depression. *Progress in Neuro-Psychopharmacology & Biological Psychiatry, 35*(3), 722–729.

Further Reading

Abelaira, H. M., Réus, G. Z., & Quevedo, J. (2013). Animal models as tools to study the pathophysiology of depression. *Revista Brasileira de Psiquiatria, 35*(Suppl. 2), S112–120 [Sao Paulo, Brazil: 1999].

Anisman, H., & Merali, Z. (2002). Cytokines, stress, and depressive illness. *Brain, Behavior, and Immunity, 16*(5), 513–524.

De La Garza, R. (2005). Endotoxin- or proinflammatory cytokine-induced sickness behavior as an animal model of depression: focus on anhedonia. *Neuroscience and Biobehavioral Reviews, 29*(4–5), 761–770.

Dhabhar, F. S. (2009). Enhancing versus suppressive effects of stress on immune function: implications for immunoprotection and immunopathology. *Neuroimmunomodulation, 16*(5), 300–317.

Hooper, L. V., & Gordon, J. I. (2001). Commensal host-bacterial relationships in the gut. *Science (New York, N.Y.), 292*(5519), 1115–1118.

Hua, Y., Huang, X.-Y., Zhou, L., Zhou, Q.-G., Hu, Y., Luo, C.-X., et al. (2008). DETA/NONOate, a nitric oxide donor, produces antidepressant effects by promoting hippocampal neurogenesis. *Psychopharmacology*, *200*(2), 231–242.

Kumar, V., Bhat, Z. A., & Kumar, D. (2013). Animal models of anxiety: a comprehensive review. *Journal of Pharmacological and Toxicological Methods*, *68*(2), 175–183.

Madsen, K. L., Doyle, J. S., Jewell, L. D., Tavernini, M. M., & Fedorak, R. N. (1999). Lactobacillus species prevents colitis in interleukin 10 gene–deficient mice. *Gastroenterology*, *116*(5), 1107–1114.

Medzhitov, R. (2008). Origin and physiological roles of inflammation. *Nature*, *454*(7203), 428–435.

Nettle, D., Andrews, C., Reichert, S., Bedford, T., Kolenda, C., Parker, C., et al. (2017). Early-life adversity accelerates cellular ageing and affects adult inflammation: experimental evidence from the European starling. *Scientific Reports*, *7*. srep40794.

O'Connor, J. C., André, C., Wang, Y., Lawson, M. A., Szegedi, S. S., Lestage, J., et al. (2009). Interferon-γ and tumor necrosis factor-α mediate the upregulation of indoleamine 2,3-dioxygenase and the induction of depressive-like behavior in mice in response to bacillus Calmette-Guérin. *Journal of Neuroscience: The Official Journal of the Society for Neuroscience*, *29*(13), 4200–4209.

O'Connor, J. C., Lawson, M. A., André, C., Moreau, M., Lestage, J., Castanon, N., et al. (2009). Lipopolysaccharide-induced depressive-like behavior is mediated by indoleamine 2,3-dioxygenase activation in mice. *Molecular Psychiatry*, *14*(5), 511–522.

Okuda, S., Nishiyama, N., Saito, H., & Katsuki, H. (1996). Hydrogen peroxide-mediated neuronal cell death induced by an endogenous neurotoxin, 3-hydroxykynurenine. *Proceedings of the National Academy of Sciences of the United States of America*, *93*(22), 12553–12558.

Renoir, T., Pang, T. Y., Zajac, M. S., Chan, G., Du, X., Leang, L., et al. (2012). Treatment of depressive-like behaviour in Huntington's disease mice by chronic sertraline and exercise. *British Journal of Pharmacology*, *165*(5), 1375–1389.

Van Praag, H. M. (2005). Can stress cause depression? *The World Journal of Biological Psychiatry*, *6*(Suppl. 2), 5–22.

6

Neurogenesis, Inflammation, and Mental Health

Alessandra Borsini, Kristi M. Sawyer,
Patricia A. Zunszain, Carmine M. Pariante

Institute of Psychiatry, Psychology and Neuroscience, London, United Kingdom

INTRODUCTION

Recent research into the pathology of mental health disorders has highlighted their association with altered patterns of neurogenesis in the brain. Neurogenesis is the process by which new neurons are formed from neural stem cells or progenitor cells (Lee, Reif, & Schmitt, 2012). From here, it is possible for these newborn neurons to differentiate into specific cell subtypes. This process happens, for the most part, during fetal brain development; however, adult neurogenesis in humans also occurs in the hippocampus, where it is restricted to the subgranular zone (SGZ) of the dentate gyrus (DG) and in the subventricular zone (SVZ) of the lateral ventricles (Borsini, Zunszain, Thuret, & Pariante, 2015). Regulation of adult neurogenesis is thought to be involved in memory formation, cognition, and adult's ability to cope with stressors, which is of particular relevance to the pathogenesis of depression.

Adult neurogenesis gives rise to two different types of neural stem cells: type 1 cells are radial glia-like cells, and type 2 cells are nonradial neural progenitor cells (NPCs) (Lee et al., 2012). NPCs are under constant stimulation to proliferate, migrate, differentiate, and survive. However, pathological circumstances in the brain such as injury or infection can disrupt the usual pattern of proliferation, mainly through microglial activation and release of inflammatory mediators, such as cytokines, in the brain. Cytokines and providing vital immune protection in the brain and clearing dead or damaged neurons can also cause damage, ultimately leading to neuronal death (Borsini et al., 2015).

Cytokines are low-molecular-weight proteins or glycoproteins that are produced by a number of different cell types, notably white blood cells, in response to an inflammatory stimulus (Dinarello, 2000). There are various types of cytokines, including interleukins (ILs), so called because they are secreted by leukocytes and act on other similar cell types. Other types of cytokines include interferons (IFNs), which act by activating natural killer cells and macrophages, and tumor necrosis factors (TNFs), which are involved in cell death. In addition, peripheral

Inflammation and Immunity in Depression
https://doi.org/10.1016/B978-0-12-811073-7.00006-4

inflammation is associated with the impairment of hippocampal-dependent forms of synaptic plasticity (McAfoose & Baune, 2009), which may be relevant to cognitive impairment (Wilson, Finch, & Cohen, 2002).

Turning to neurogenesis, cytokines have recently been shown to regulate the proliferation, differentiation, and neurogenesis of NPCs, with research focusing mainly on this mechanism in the context of psychiatric disorders, such as depression (Makhija & Karunakaran, 2013). Indeed, recent research has shown that patients with depression show elevated levels of peripheral inflammatory markers (Zunszain, Hepgul, & Pariante, 2012). A metaanalysis specifically showed upregulation of IL-1β, IL-6, and TNF-α in both the serum and plasma of depressed patients (Dowlati et al., 2010). These cytokines are involved in the mechanisms underlying a range of cognitive processes, such as mood and learning. For example, IL-1, in combination with TNF-α, has been associated with the inhibition of the processes of memory consolidation and synaptic plasticity in the dentate gyrus of the hippocampus (Pickering & O'Connor, 2007).

To examine the effects of these cytokines further, this chapter, which comprises an update of the review published by Borsini et al. (2015), will now take some cytokines individually in order to illustrate their effects in the brain. We begin, first, by describing experiments focused on adult neurogenesis, followed by those investigating neurogenesis in fetal cells. In addition, where available, we provide information on downstream molecular mechanisms underlying the effects of cytokines in the brain.

IL-1α AND IL-1β

IL-1α and IL-1β are part of the interleukin-1 family, which is responsible for the control of pro-inflammatory responses, triggered by tissue injury associated with pathogenic infection or with release of danger-related molecules from damaged cells (Weber, Wasiliew, & Kracht, 2010).

Experiments *in vitro* have shown that IL-1α increases neurogenesis (Ling, Potter, Lipton, & Carvey, 1998), whereas in the majority of cases, IL-1β reduces the proliferation and neurogenesis of NPCs and enhances gliogenesis (Chen et al., 2013; Crampton, Collins, Toulouse, Nolan, & O'Keeffe, 2012; Zunszain, Anacker, et al., 2012).

The evidence for this mechanism is supported by experiments including cotreatment with an antagonist for the IL-1 receptor. One such study found that although IL-1β reduced proliferation and neurogenesis in rat adult NPCs from the DG, cotreatment with an IL-1 receptor antagonist alleviated this negative effect (Ryan, O'Keeffe, O'Connor, Keeshan, & Nolan, 2013). Another study also showed that IL-1β reduced the differentiation of rat neonatal NPCs into serotonergic neurons and that this effect was blocked by a cotreatment with an IL-1 receptor antagonist (Zhang, Xu, Cao, Li, & Huang, 2013). A similar abrogation of the inhibition by IL-1β can be achieved by coincubation with an inhibitor of NF-κB (Koo & Duman, 2008).

A similar pattern of effects has been observed in fetal NPCs. For example, treatment with recombinant IL-1β reduces neurogenesis but increased astrogliogenesis in both human fetal hippocampal NPCs (HNPCs) (Chen et al., 2013; Zunszain, Anacker, et al., 2012) and rat fetal HNPCs (Crampton et al., 2012). However, another study also found that treatment with IL-1α and IL-1β promotes differentiation of rat fetal mesencephalic NPCs into dopaminergic (DAergic) neurons (Ling et al., 1998). These findings may be explained by *in vivo* evidence, which indicates that treatment with both can promote reinnervation and differentiation of DAergic neurons in the striatum (STR) and mesencephalon (Hébert, Mingam, Arsaut, Dantzer, & Demotes-Mainard, 2005; Wang, Bankiewicz, Plunkett, & Oldfield, 1994) but inhibit hippocampal neurogenesis (Boehme et al., 2014).

Although IL-1β has been shown to reduce neurogenesis, some evidence collected by our own lab suggests that it promotes proliferation of human fetal HNPCs (Zunszain, Anacker, et al., 2012). However, these findings are conflicted by other studies (Crampton et al., 2012; Wang et al., 2007). One study showed that repeated, but not single, intrahippocampal injections of IL-1β enhance proliferation and that these effects did not occur if administered systemically (Seguin, Brennan, Mangano, & Hayley, 2009). A different study reported that the increase in proliferation caused by IL-1α treatment, which exerts its effects through the same IL-1 receptor, is only shown in young mice, but not old ones (McPherson, Aoyama, & Harry, 2011).

Regarding the mechanisms of these effects, IL-1β-induced suppressive effects on neurogenesis in human fetal HNPCs seem to be partially mediated by the activation of the kynurenine pathway and the accumulation of tryptophan metabolites, which have neurotoxic properties (Zunszain, Anacker, et al., 2012). Furthermore, as shown in rat fetal mesencephalic NPCs, where IL-1β treatment reduces proliferation, activation of phospho-38 mitogen-activated protein kinases (MAPK) can reverse this inhibition (Crampton et al., 2012). In addition, in rat fetal HNPCs, the reduction in proliferation seemed to occur via phosphorylation of the stress-activated protein kinase (SAPK) and c-Jun N-terminal kinase (JNK) system. In support of this finding, an inhibitor of this kinase abrogated IL-1β's inhibiting effect on proliferation (Wang et al., 2007). Indeed, another study using rat fetal HNPCs also found a mediating role of the glycogen synthase kinase 3β enzyme, as when inhibited, the effects of IL-1β were exaggerated. Activation of this enzyme is also associated with reduction in expression of the orphan nuclear receptor tailless homologue (TLX) protein. In fact, GSK-3β inhibition ameliorated the effects of IL-1β on TLX expression in both proliferating and differentiating cells (Green & Nolan, 2012).

In human fetal HNPCs, the inhibitory effect of IL-1β on neurogenesis seems to be mediated by activation of the STAT3 pathway (Chen et al., 2013).

IL-1β Induced by Lipopolysaccharide-Activated Human Monocyte-Derived Macrophages

To our knowledge, just one study to date has examined the role of IL-1β induced by lipopolysaccharide-monocyte-derived macrophage (LPS-MDM), in which they conduct conditioned media experiments. This study showed that IL-1β derived from LPS-MDMs stimulated the proliferation of human fetal cortical NPCs; however, this action was not affected by pretreatment with an IL-1 receptor antagonist, which may suggest that it was not occurring via an IL-1-dependent mechanism. In this treated culture, they also observed a reduction in neurogenesis but increased astrogliogenesis (Peng et al., 2008). It is possible that these effects are caused by the independent action of MDMs, which can also stimulate production of other molecules such as cyclin-dependent kinases, which are known to regulate the cell cycle in NPCs, and can thus affect cell proliferation (Ferguson, Callaghan, O'Hare, Park, & Slack, 2000).

IL-6

Interleukin-6 has recently received much research attention, due to its association in humans with depressive disorders (Bob et al., 2010; Howren, Lamkin, & Suls, 2009). The majority of preclinical experiments report that IL-6 either has no effect or reduces proliferation and gliogenesis while increasing neuronal differentiation (Islam, Gong, Rose-John, & Heese, 2009).

For example, one study found that IL-6 promoted neurogenesis in rat adult DG NPCs but did not cause any considerable modifications in astrogliogenesis (Oh et al., 2010). In contrast, another reported that IL-6 decreased neurogenesis and increased apoptosis in rat adult DG NPCs, without affecting proliferation and gliogenesis, and a blocking antibody to IL-6 restored neurogenesis (Monje, Toda, & Palmer, 2003).

Indeed, there is some evidence that IL-6 inhibits neurogenesis by reducing the expression of soluble molecules such as the sonic hedgehog (SHH) protein (Oh et al., 2010), a known promoter of neuronal differentiation. Furthermore, in humans, the administration of a neutralization antibody against the leukemia inhibitory factor (LIF), a member of the IL-6 family of cytokines, but not IL-6 itself, reduced astrogliogenesis in human fetal cortical NPCs (Lan et al., 2012). In rats, this also induced differentiation of fetal mesencephalic NPCs into dopaminergic neurons (Ling et al., 1998). Similarly, in humans, IL-6 promotes an increase in neurogenesis but has no effect on gliogenesis in both fetal NPCs and HNPCs (Johansson, Price, & Modo, 2008). In addition to this, in mice, a fusion of IL-6 and the soluble IL-6 receptor named "hyper IL-6" reduced cell proliferation and increases neurogenesis, although it had no effect on gliogenesis in fetal SVZ NSCs (Islam et al., 2009).

On a mechanistic level, evidence from mice has shown that IL-6 decreases neurogenesis in adult DG NPCs, through the activation of cyclin-dependent kinase inhibitor 1A (or p21) (Zonis et al., 2013). In contrast, hyper IL-6, as described above, increases neurogenesis in mouse fetal SVZ NSCs via activation or the MAPK and cAMP response element-binding (CREB) protein cascade (Islam et al., 2009).

Microglial-Derived IL-6-Family Cytokines

To date, we are aware of one study that investigated the role that IL-6 derived from microglia, as well as LIF and ciliary neurotrophic factor (CNTF), plays in cell proliferation and differentiation. This study used conditioned media (CM) experiments. This study found that microglia-derived IL-6 CM and LIF CM, but not CNTF CM, enhanced astrogliogenesis in rat fetal SVZ NPCs (Nakanishi et al., 2007). Evidence previously reported *in vivo* indicates that IL-6 and LIF exert their positive effects on gliogenesis via inhibition of neuronal differentiation. In fact, a receptor complex for IL-6 and LIF is involved in the activation of hairy enhancer of split (HES-1), a transcription factor, which negatively regulates neurogenesis and induces gliogenesis (Nakamura et al., 2000). However, as the above authors did not investigate the fate of NPCs, this possibility cannot be confirmed.

IL-4, IL-10, AND IL-11

Interleukins 4, 10, and 11 show antiinflammatory effects (Opal & DePalo, 2000). There is less evidence relating to the effect of these cytokines on neuronal proliferation, differentiation, and neurogenesis. However, those findings that have been reported are relatively consistent: IL-4 reduces neuronal proliferation but increases neuronal and glial differentiation, whereas IL-10 seems to show the opposite effects, where it increases or has no effect on proliferation and reduces or has no effect on differentiation of neurons or glia. IL-11 has been shown to increase neuronal differentiation (Butovsky et al., 2006; Kiyota et al., 2012; Perez-Asensio, Perpina, Planas, & Pozas, 2013). Differing results for IL-4 and IL-10 may be attributed to differences in modalities of cell treatment: experiments with IL-10 used cytokine-treated NPCs, whereas cytokine-treated microglia were used with both IL-4 and IL-10, which were then coincubated with NPCs (Kiyota et al., 2012). In addition, this study found that IL-4- and IL-10-treated microglia (TM) had different effects on both proliferation and differentiation, suggesting that along with the presence or absence of microglia, the type of microglia activator may also play a role in shaping the fate of the cell (Schwartz, 2003).

Another study reported that IL-10 does not affect proliferation or oligodendrogenesis. However, it did impair neurogenesis when used to treat murine adult SVZ NPCs. Indeed, this study found that the lack of IL-10 *in vivo* induced neuronal differentiation of SVZ NPCs and increased the incorporation of new neurons into the adult olfactory bulb (Perez-Asensio et al., 2013). The evidence regarding IL-11 is more sparse. However, similar to findings in adult cells, IL-11 has been shown to promote the differentiation of rat fetal mesencephalic NPCs into DAergic neurons (Ling et al., 1998); however, studies *in vivo* need to be completed to further our understanding of this mechanism (Guk & Kuprash, 2011).

IL-4- and IL-10-Treated Microglia

This review has identified two studies that have examined the role of IL-4-TM and IL-10-TM on cell generation. The first of these used murine adult SNZ NPCs and found that coculture with IL-4-TM, but not with LPS-TM, induced both neurogenesis and oligodendrogenesis (Butovsky et al., 2006). This finding indicates that specific stimuli are capable of producing different types of activated microglia, which will ultimately impact the fate of the cell. Indeed, quiescent microglia, which reside around the proliferating NPCs, may become activated by IL-4, but not by LPS, in order to support neurogenesis by the production of neurotrophic factors (Butovsky, Talpalar, Ben-Yaakov, & Schwartz, 2005; Chao, Molitor, & Hu, 1993).

The second of these studies (Kiyota et al., 2012) used murine fetal cortical NPCs, cultured with IL-4-TM and IL-10-TM. It found that IL-4-TM reduced proliferation of these cells, yet IL-10-TM increased proliferation. In addition to this, IL-4-TM increased neurogenesis and astrogliogenesis, while IL-10-TM showed the opposite pattern of effects. This finding provides further evidence that the type of inflammatory challenge is important in shaping cell fate.

Indeed, coculture with microglia alone in this experimental condition showed an antiapoptotic effect, which was strengthened by IL-4-TM.

IFN-α AND IFN-γ

Interferons are best known for their antiviral activity, but they are also involved in cell growth inhibition, immunosuppression, enhancement of natural killer cell function, and cell differentiation (Imanishi, 1994). In this section, we will focus on two types: interferons alpha (IFN-α) and gamma (IFN-γ). The literature regarding these two interferons is relatively consistent in terms of cell proliferation, differentiation, and neurogenesis. Broadly, it shows that both IFN-α and IFN-γ reduce cell proliferation (Mäkelä, Koivuniemi, Korhonen, Lindholm, & Sokka, 2010; Moriyama et al., 2011; Walter et al., 2011). In addition, there is some evidence that IFN-γ increases neuronal and glial differentiation (Johansson et al., 2008; Walter, Hartung, Dihné, Sanchez, & Brionne, 2012). Although relatively consistent, one study, which does report differences, identifies these differences in different cell types, in particular, human fetal STR NPCs versus human fetal HNPCs (Johansson et al., 2008).

In mice, IFN-α acts via the complement receptor 2 (CR2) and shows impaired proliferation of adult DG NPCs. In fact, in knockout mice for CR2 (CR2$^{-/-}$), treatment with IFN-α does not affect proliferation (Moriyama et al., 2011), indicating that this effect of IFN-α treatment is mediated by this receptor.

Also in mice, IFN-γ treatment combined with TNF-α does not affect neurogenesis in neonatal SVZ NPCs (Belmadani, Tran, Ren, & Miller, 2006), which may suggest that although IFN-γ has independently been shown to be proneurogenic (Whitney, Eidem, Peng, Huang, & Zheng, 2009), TNFα is capable of antagonizing this effect. Furthermore, the combination of these two molecules has been shown to enhance cell migration, apoptosis, and astrogliogenesis (Belmadani et al., 2006).

Turning to fetal cells, treatment with both rat and human IFN-γ enhances neurogenesis but has no effect on gliogenesis in human fetal STR NPCs. However, there is perhaps cell-type-specific influences on these effects, as in human fetal HNPCs, both human and rat IFN-γ increased gliogenesis, but interestingly had no effect on neurogenesis (Johansson et al., 2008). These differences could be attributed to higher levels of cytokine receptors expressed in the hippocampus, when compared with the STR (Robertson, Kong, Peng, Bentivoglio, & Kristensson, 2000). Indeed, in the above study, receptor expression for IFN-γ was threefold higher in the hippocampal cells than those from the STR (Johansson et al., 2008).

In rat fetal cells, IFN-γ treatment has been shown to reduce proliferation and enhance apoptosis of STR NPCs (Mäkelä et al., 2010). A similar effect has been observed in murine fetal hypothalamic NPCs, as well as an increase in neurogenesis and astrogliogenesis (Walter et al., 2011).

Interestingly, one of the abovementioned studies (Walter et al., 2011) reported the presence of cells that coexpressed beta-III-tubulin, a marker of mature neuron, and glial fibrillary acid protein (GFAP), a marker for astrocytes, but electrophysiological results showed that these cells were functionally distinct from mature neurons and astrocytes. This perhaps suggests that IFN-γ may cause a dysregulated phenotype of NPC-derived cells. These authors then replicated these findings and confirmed a phenotypic dysregulation, where cells coexpressed beta-III-tubulin and GFAP, which was induced by treatment with IFN-γ in rat fetal hypothalamic NPCs (Walter et al., 2012). On a mechanistic level, IFN-γ treatment has been associated with the upregulation of the STAT1 gene and SHH protein (Walter et al., 2011, 2012). This role for STAT1 and STAT2 signaling in the effects of IFN-γ has also been replicated *in vivo* (Schroder, Hertzog, Ravasi, & Hume, 2003).

IFN-γ-TM

We are aware of one study that has looked at the effect of IFN-γ-TM on cell differentiation. Coculture of IFN-γ-TM with murine adult SVZ NPCs causes increased differentiation of neurons and glia (Butovsky et al., 2006). This study used an IFN-γ-TM concentration of 20 ng/mL, which relates to the previous findings by the same group, showing that low concentrations (1–50 ng/mL) can have a neuroprotective effect on microglia (Butovsky et al., 2005).

TNF-α

In the majority of studies, TNF-α has a stimulatory effect on proliferation and gliogenesis but inhibitory effect on neurogenesis (Belmadani et al., 2006; Bernardino et al., 2008; Chen et al., 2013; Johansson et al., 2008; Lan et al., 2012; Peng et al., 2008; Widera, Mikenberg, Elvers, Kaltschmidt, & Kaltschmidt, 2006). There are few contradictions to these findings, although one finds no effect on gliogenesis (Widera et al., 2006). Where discrepancies do exist, it seems that increased neurogenesis can be seen in murine neonatal SVZ NPCs, so this may be as a result of different cell types being cultured (Bernardino et al., 2008; Zhang et al., 2012).

We have already discussed the antineurogenic effect of TNF-α above, in the context of its cotreatment with IFN-γ. In isolation, it has been shown to cause a reduction in neurogenesis in adult rat DG NPCs (Monje et al., 2003). As well as this, in SVZ NPCs, it increased proliferation and apoptosis, but had no effect on neurogenesis or astrogliogenesis (Widera et al., 2006).

In mice, TNF-α increases neuronal differentiation in neonatal SVZ NPCs (Zhang et al., 2012). Indeed, even a low dose of TNF-α (1 ng/m or mouse or human recombinant protein) has been shown to increase neuronal proliferation and neurogenesis in neonatal SVZ NPCs, however, doses from 10 to 100ng/mL induce apoptosis (Bernardino et al., 2008).

These conflicting data suggest that TNF-α exhibits species differences with regards to its effect on neurogenesis. In rats, it seems to show neuroprotective properties in the hippocampus (Cheng, Christakos, & Mattson, 1994) but is toxic to mouse-derived neurospheres, due to its interference with their formation (Neumann et al., 2002).

In human fetal STR NPCs and HNPCs, treatment with human or rat TNF-α inhibits neurogenesis but stimulates gliogenesis and apoptosis (Johansson et al., 2008). A similar pattern of impaired neurogenesis but enhanced astrogliogenesis was found in human fetal HNPCs (Chen et al., 2013) and human fetal cortical NPCs (Peng et al., 2008).

When trying to understand the underlying mechanism for these effects, it has been found that, in murine neonatal SVZ NPCs, activation of NF-kB signaling can increase the neuronal differentiation induced by TNF-α treatment. Indeed, cotreatment with a specific inhibitor of NF-kB blocks this effect (Zhang et al., 2012). On the other hand, in human fetal HNPCs, the reduction in neurogenesis caused by TNF-α seems to occur via the activation of the STAT-3 pathway (Chen et al., 2013).

TNF-α Induced by LPS-MDM

We are aware of one study (Peng et al., 2008) investigating the role of TNF-α induced by LPS-MDM, where CM experiments are used. In this case, cell proliferation and astrogliogenesis both increased, but a reduction in neurogenesis was shown by human fetal cortical NPCs. To support these findings, the authors also found that pretreatment with TNF-α receptors R1 and R2 partially reduced the effect of TNF-α on proliferation. The authors comment that this shows that although TNF-α is involved in the modulation of cell proliferation, MDM might have also contributed to the impaired neuronal differentiation.

CONCLUSIONS AND FUTURE DIRECTIONS

Taking these interleukins individually, it is clear that further experiments need to be conducted to reconcile the conflicting evidence that has emerged regarding their effects on neuronal proliferation, differentiation, and neurogenesis. Some of these discrepancies can be attributed to different cell types being used in experiments or cells derived from different species, in which it would be interesting to learn more about the distinct downstream molecular mechanisms in these cell types, which may underpin the differences observed. In addition, it is possible that stimulation with cytokines activate different mechanistic responses in distinct brain regions thus suggesting the need of a combination of additional *in vitro* and *in vivo* experiments in order to better understand this relationship.

Despite these conflicting results over the directionality of the effects, it is clear that changes in cytokine levels in the brain can have profound effects on brain functioning. These effects have been most associated with the pathogenesis of depression and some neurodegenerative diseases. There is evidence suggesting that depression itself is associated with impaired functioning of hippocampal neurogenesis mechanisms and that antidepressants and other antidepressant treatments, such as electroconvulsive therapy, can have a positive impact on the way in which and the rate at which neurogenesis occurs in the hippocampus (Hanson, Owens, & Nemeroff, 2011). Further, *in vivo* data are required to translate *in vitro* findings into clinical context, in order that treatments for depression can be developed with the neurogenesis theory in mind (Schoenfeld & Cameron, 2015).

Most interestingly, in clinical context, this field of research has added a new direction of focus for the pharmacotherapy of depression, particularly in treatment-resistant cases. Indeed, *in vivo* data from preclinical models have shown that antiinflammatory agents such as minocycline can influence neurogenesis (Lu et al., 2017; Zheng, Kaneko, & Sawamoto, 2015), which, in human patients, may contribute to improvements in depressive symptoms. Indeed, recent clinical studies have investigated the coprescription of drugs with antiinflammatory effects alongside traditional antidepressants, particularly for use in patients with treatment-resistant depression. One such class of antiinflammatory drugs that have been trialed are nonsteroidal antiinflammatory drugs (NSAIDs) such as acetylsalicylic acid (aspirin) (Köhler, Petersen, Mors, & Gasse, 2015) and celecoxib, a selective cyclooxygenase-2 inhibitor (Müller et al., 2006). Data tend to show that celecoxib in particular leads to therapeutic benefit and amelioration of psychiatric symptoms (Köhler et al., 2014; Na, Lee, Lee, Cho, & Jung, 2014), promoting antiinflammatory agents as new potential therapeutic antidepressant strategies for patients with depression and subchronic level of inflammation.

Therefore, future research should focus on immune products, such as cytokines, especially on those for which sufficient evidence has been accumulated on their molecular and cellular effects on mood and cognitive functions, ultimately allowing the possible translational link between *in vitro* data and clinical applications (McAfoose & Baune, 2009).

References

Belmadani, A., Tran, P. B., Ren, D., & Miller, R. J. (2006). Chemokines regulate the migration of neural progenitors to sites of neuroinflammation. *The Journal of Neuroscience: The Official Journal of the Society for Neuroscience, 26*(12), 3182–3191. https://doi.org/10.1523/JNEUROSCI.0156-06.2006.

Bernardino, L., Agasse, F., Silva, B., Ferreira, R., Grade, S., & Malva, J. O. (2008). Tumor necrosis factor-α modulates survival, proliferation, and neuronal differentiation in neonatal subventricular zone cell cultures. *Stem Cells, 26*(9), 2361–2371. https://doi.org/10.1634/stemcells.2007-0914.

Bob, P., Raboch, J., Maes, M., Susta, M., Pavlat, J., Jasova, D., et al. (2010). Depression, traumatic stress and interleukin-6. *Journal of Affective Disorders, 120*(1–3), 231–234. https://doi.org/10.1016/j.jad.2009.03.017.

Boehme, M., Guenther, M., Stahr, A., Liebmann, M., Jaenisch, N., Witte, O. W., et al. (2014). Impact of indomethacin on neuroinflammation and hippocampal neurogenesis in aged mice. *Neuroscience Letters, 572*, 7–12. https://doi.org/10.1016/j.neulet.2014.04.043.

Borsini, A., Zunszain, P. A., Thuret, S., & Pariante, C. M. (2015). The role of inflammatory cytokines as key modulators of neurogenesis. *Trends in Neurosciences, 38*(3), 145–157. https://doi.org/10.1016/j.tins.2014.12.006.

Butovsky, O., Talpalar, A. E., Ben-Yaakov, K., & Schwartz, M. (2005). Activation of microglia by aggregated β-amyloid or lipopolysaccharide impairs MHC-II expression and renders them cytotoxic whereas IFN-γ and IL-4 render them protective. *Molecular and Cellular Neuroscience, 29*(3), 381–393. https://doi.org/10.1016/j.mcn.2005.03.005.

Butovsky, O., Ziv, Y., Schwartz, A., Landa, G., Talpalar, A. E., Pluchino, S., et al. (2006). Microglia activated by IL-4 or IFN-γ differentially induce neurogenesis and oligodendrogenesis from adult stem/progenitor cells. *Molecular and Cellular Neuroscience, 31*(1), 149–160. https://doi.org/10.1016/j.mcn.2005.10.006.

Chao, C. C., Molitor, T. W., & Hu, S. (1993). Neuroprotective role of IL-4 against activated microglia. *Journal of Immunology, 151*(3), 1473–1481. Retrieved from http://www.ncbi.nlm.nih.gov/pubmed/8335941.

Chen, E., Xu, D., Lan, X., Jia, B., Sun, L., Zheng, J. C., et al. (2013). A novel role of the STAT3 pathway in brain inflammation-induced human neural progenitor cell differentiation. *Current Molecular Medicine, 13*(9), 1474–1484. Retrieved from http://www.ncbi.nlm.nih.gov/pubmed/23971732.

Cheng, B., Christakos, S., & Mattson, M. P. (1994). Tumor necrosis factors protect neurons against metabolic-excitotoxic insults and promote maintenance of calcium homeostasis. *Neuron, 12*(1), 139–153. Retrieved from http://www.ncbi.nlm.nih.gov/pubmed/7507336.

Crampton, S. J., Collins, L. M., Toulouse, A., Nolan, Y. M., & O'Keeffe, G. W. (2012). Exposure of foetal neural progenitor cells to IL-1β impairs their proliferation and alters their differentiation—a role for maternal inflammation? *Journal of Neurochemistry, 120*(6), 964–973. https://doi.org/10.1111/j.1471-4159.2011.07634.x.

Dinarello, C. A. (2000). Proinflammatory cytokines. *Chest, 118*(2), 503–508. Retrieved from http://www.ncbi.nlm.nih.gov/pubmed/10936147.

Dowlati, Y., Herrmann, N., Swardfager, W., Liu, H., Sham, L., Reim, E. K., et al. (2010). A meta-analysis of cytokines in major depression. *Biological Psychiatry, 67*(5), 446–457. https://doi.org/10.1016/j.biopsych.2009.09.033.

Ferguson, K. L., Callaghan, S. M., O'Hare, M. J., Park, D. S., & Slack, R. S. (2000). The Rb-CDK4/6 signaling pathway is critical in neural precursor cell cycle regulation. *The Journal of Biological Chemistry*, 275(43), 33593–33600. https://doi.org/10.1074/jbc.M004879200.

Green, H. F., & Nolan, Y. M. (2012). Unlocking mechanisms in interleukin-1β-induced changes in hippocampal neurogenesis—a role for GSK-3β and TLX. *Translational Psychiatry*, 2(11), e194. https://doi.org/10.1038/tp.2012.117.

Guk, K. D., & Kuprash, D. V. (2011). Interleukin-11, an IL-6 like cytokine. *Molekuliarnaia Biologiia*, 45(1), 44–55. Retrieved from http://www.ncbi.nlm.nih.gov/pubmed/21485496.

Hanson, N. D., Owens, M. J., & Nemeroff, C. B. (2011). Depression, antidepressants, and neurogenesis: a critical reappraisal. *Neuropsychopharmacology*, 36(13), 2589–2602. https://doi.org/10.1038/npp.2011.220.

Hébert, G., Mingam, R., Arsaut, J., Dantzer, R., & Demotes-Mainard, J. (2005). Cellular distribution of interleukin-1α-immunoreactivity after MPTP intoxication in mice. *Molecular Brain Research*, 138(2), 156–163. https://doi.org/10.1016/j.molbrainres.2005.04.019.

Howren, M. B., Lamkin, D. M., & Suls, J. (2009). Associations of depression with C-reactive protein, IL-1, and IL-6: a meta-analysis. *Psychosomatic Medicine*, 71(2), 171–186. https://doi.org/10.1097/PSY.0b013e3181907c1b.

Imanishi, J. (1994). Interferon-alpha, beta, gamma. *Gan to Kagaku Ryoho. Cancer & Chemotherapy*, 21(16), 2853–2858. Retrieved from http://www.ncbi.nlm.nih.gov/pubmed/7993128.

Islam, O., Gong, X., Rose-John, S., & Heese, K. (2009). Interleukin-6 and neural stem cells: more than gliogenesis. *Molecular Biology of the Cell*, 20(1), 188–199. https://doi.org/10.1091/mbc.E08-05-0463.

Johansson, S., Price, J., & Modo, M. (2008). Effect of inflammatory cytokines on major histocompatibility complex expression and differentiation of human neural stem/progenitor cells. *Stem Cells*, 26(9), 2444–2454. https://doi.org/10.1634/stemcells.2008-0116.

Kiyota, T., Ingraham, K. L., Swan, R. J., Jacobsen, M. T., Andrews, S. J., & Ikezu, T. (2012). AAV serotype 2/1-mediated gene delivery of anti-inflammatory interleukin-10 enhances neurogenesis and cognitive function in APP+PS1 mice. *Gene Therapy*, 19(7), 724–733. https://doi.org/10.1038/gt.2011.126.

Köhler, O., Benros, M. E., Nordentoft, M., Farkouh, M. E., Iyengar, R. L., Mors, O., et al. (2014). Effect of anti-inflammatory treatment on depression, depressive symptoms, and adverse effects. *JAMA Psychiatry*, 71(12), 1381. https://doi.org/10.1001/jamapsychiatry.2014.1611.

Köhler, O., Petersen, L., Mors, O., & Gasse, C. (2015). Inflammation and depression: combined use of selective serotonin reuptake inhibitors and NSAIDs or paracetamol and psychiatric outcomes. *Brain and Behavior: A Cognitive Neuroscience Perspective*, 5(8), e00338. https://doi.org/10.1002/brb3.338.

Koo, J. W., & Duman, R. S. (2008). IL-1 is an essential mediator of the antineurogenic and anhedonic effects of stress. *Proceedings of the National Academy of Sciences*, 105(2), 751–756. https://doi.org/10.1073/pnas.0708092105.

Lan, X., Chen, Q., Wang, Y., Jia, B., Sun, L., Zheng, J., et al. (2012). TNF-α affects human cortical neural progenitor cell differentiation through the Autocrine secretion of leukemia inhibitory factor. *PLoS One*, 7(12), e50783. https://doi.org/10.1371/journal.pone.0050783.

Lee, M. M., Reif, A., & Schmitt, A. G. (2012). Major depression: a role for hippocampal neurogenesis? *Current Topics in Behavioral Neurosciences*, 14, 153–179. https://doi.org/10.1007/7854_2012_226.

Ling, Z., Potter, E., Lipton, J., & Carvey, P. M. (1998). Differentiation of mesencephalic progenitor cells into dopaminergic neurons by cytokines. *Experimental Neurology*, 149, 411–423. Retrieved from https://www.semanticscholar.org/paper/Differentiation-of-mesencephalic-progenitor-cells-Ling-Potter/66c93cecc85b44c605c5f2b49a68107bf36376b2.

Lu, Y., Giri, P. K., Lei, S., Zheng, J., Li, W., Wang, N., et al. (2017). Pretreatment with minocycline restores neurogenesis in the subventricular zone and subgranular zone of the hippocampus after ketamine exposure in neonatal rats. *Neuroscience*, 352, 144–154. https://doi.org/10.1016/j.neuroscience.2017.03.057.

Mäkelä, J., Koivuniemi, R., Korhonen, L., Lindholm, D., & Sokka, A. (2010). Interferon-γ produced by microglia and the neuropeptide PACAP have opposite effects on the viability of neural progenitor cells. *PLoS One*, 5(6), e11091. https://doi.org/10.1371/journal.pone.0011091.

Makhija, K., & Karunakaran, S. (2013). The role of inflammatory cytokines on the aetiopathogenesis of depression. *The Australian and New Zealand Journal of Psychiatry*, 47(9), 828–839. https://doi.org/10.1177/0004867413488220.

McAfoose, J., & Baune, B. T. (2009). Evidence for a cytokine model of cognitive function. *Neuroscience & Biobehavioral Reviews*, 33(3), 355–366. https://doi.org/10.1016/j.neubiorev.2008.10.005.

McPherson, C. A., Aoyama, M., & Harry, G. J. (2011). Interleukin (IL)-1 and IL-6 regulation of neural progenitor cell proliferation with hippocampal injury: differential regulatory pathways in the subgranular zone (SGZ) of the adolescent and mature mouse brain. *Brain, Behavior, and Immunity*, 25(5), 850–862. https://doi.org/10.1016/j.bbi.2010.09.003.

Monje, M. L., Toda, H., & Palmer, T. D. (2003). Inflammatory blockade restores adult hippocampal neurogenesis. *Science*, 302(5651), 1760–1765. https://doi.org/10.1126/science.1088417.

Moriyama, M., Fukuhara, T., Britschgi, M., He, Y., Narasimhan, R., Villeda, S., et al. (2011). Complement receptor 2 is expressed in neural progenitor cells and regulates adult hippocampal neurogenesis. *Journal of Neuroscience: The Official Journal of the Society for*

Neuroscience, *31*(11), 3981–3989. https://doi.org/10.1523/JNEUROSCI.3617-10.2011.

Müller, N., Schwarz, M. J., Dehning, S., Douhe, A., Cerovecki, A., Goldstein-Müller, B., et al. (2006). The cyclooxygenase-2 inhibitor celecoxib has therapeutic effects in major depression: results of a double-blind, randomized, placebo controlled, add-on pilot study to reboxetine. *Molecular Psychiatry*, *11*(7), 680–684. https://doi.org/10.1038/sj.mp.4001805.

Na, K.-S., Lee, K. J., Lee, J. S., Cho, Y. S., & Jung, H.-Y. (2014). Efficacy of adjunctive celecoxib treatment for patients with major depressive disorder: a meta-analysis. *Progress in Neuro-Psychopharmacology and Biological Psychiatry*, *48*, 79–85. https://doi.org/10.1016/j.pnpbp.2013.09.006.

Nakamura, Y., Sakakibara, S. i., Miyata, T., Ogawa, M., Shimazaki, T., Weiss, S., et al. (2000). The bHLH gene hes1 as a repressor of the neuronal commitment of CNS stem cells. *Journal of Neuroscience: The Official Journal of the Society for Neuroscience*, *20*(1), 283–293. Retrieved from http://www.ncbi.nlm.nih.gov/pubmed/10627606.

Nakanishi, M., Niidome, T., Matsuda, S., Akaike, A., Kihara, T., & Sugimoto, H. (2007). Microglia-derived interleukin-6 and leukaemia inhibitory factor promote astrocytic differentiation of neural stem/progenitor cells. *European Journal of Neuroscience*, *25*(3), 649–658. https://doi.org/10.1111/j.1460-9568.2007.05309.x.

Neumann, H., Schweigreiter, R., Yamashita, T., Rosenkranz, K., Wekerle, H., & Barde, Y.-A. (2002). Tumor necrosis factor inhibits neurite outgrowth and branching of hippocampal neurons by a rho-dependent mechanism. *Journal of Neuroscience: The Official Journal of the Society for Neuroscience*, *22*(3), 854–862. Retrieved from http://www.ncbi.nlm.nih.gov/pubmed/11826115.

Oh, J., McCloskey, M. A., Blong, C. C., Bendickson, L., Nilsen-Hamilton, M., & Sakaguchi, D. S. (2010). Astrocyte-derived interleukin-6 promotes specific neuronal differentiation of neural progenitor cells from adult hippocampus. *Journal of Neuroscience Research*, *88*(13), 2798–2809. https://doi.org/10.1002/jnr.22447.

Opal, S. M., & DePalo, V. A. (2000). Anti-inflammatory cytokines. *Chest*, *117*(4), 1162–1172. Retrieved from http://www.ncbi.nlm.nih.gov/pubmed/10767254.

Peng, H., Whitney, N., Wu, Y., Tian, C., Dou, H., Zhou, Y., et al. (2008). HIV-1-infected and/or immune-activated macrophage-secreted TNF-α affects human fetal cortical neural progenitor cell proliferation and differentiation. *Glia*, *56*(8), 903–916. https://doi.org/10.1002/glia.20665.

Perez-Asensio, F. J., Perpina, U., Planas, A. M., & Pozas, E. (2013). Interleukin-10 regulates progenitor differentiation and modulates neurogenesis in adult brain. *Journal of Cell Science*, *126*(18), 4208–4219. https://doi.org/10.1242/jcs.127803.

Pickering, M., & O'Connor, J. J. (2007). Pro-inflammatory cytokines and their effects in the dentate gyrus. *Progress in Brain Research*, *163*, 339–354. https://doi.org/10.1016/S0079-6123(07)63020-9.

Robertson, B., Kong, G., Peng, Z., Bentivoglio, M., & Kristensson, K. (2000). Interferon-gamma-responsive neuronal sites in the normal rat brain: Receptor protein distribution and cell activation revealed by Fos induction. *Brain Research Bulletin*, *52*(1), 61–74. Retrieved from http://www.ncbi.nlm.nih.gov/pubmed/10779704.

Ryan, S. M., O'Keeffe, G. W., O'Connor, C., Keeshan, K., & Nolan, Y. M. (2013). Negative regulation of TLX by IL-1β correlates with an inhibition of adult hippocampal neural precursor cell proliferation. *Brain, Behavior, and Immunity*, *33*, 7–13. https://doi.org/10.1016/j.bbi.2013.03.005.

Schoenfeld, T. J., & Cameron, H. A. (2015). Adult neurogenesis and mental illness. *Neuropsychopharmacology*, *40*(1), 113–128. https://doi.org/10.1038/npp.2014.230.

Schroder, K., Hertzog, P. J., Ravasi, T., & Hume, D. A. (2003). Interferon: an overview of signals, mechanisms and functions. *Journal of Leukocyte Biology*, *75*(2), 163–189. https://doi.org/10.1189/jlb.0603252.

Schwartz, M. (2003). Macrophages and microglia in central nervous system injury: are they helpful or harmful? *Journal of Cerebral Blood Flow & Metabolism*, *23*(4), 385–394. https://doi.org/10.1097/01.WCB.0000061881.75234.5E.

Seguin, J. A., Brennan, J., Mangano, E., & Hayley, S. (2009). Proinflammatory cytokines differentially influence adult hippocampal cell proliferation depending upon the route and chronicity of administration. *Neuropsychiatric Disease and Treatment*, *5*, 5–14. Retrieved from http://www.ncbi.nlm.nih.gov/pubmed/19557094.

Walter, J., Hartung, H.-P., Dihné, M., Sanchez, P., & Brionne, T. (2012). Interferon gamma and sonic hedgehog signaling are required to dysregulate murine neural stem/precursor cells. *PLoS One*, *7*(8), e43338. https://doi.org/10.1371/journal.pone.0043338.

Walter, J., Honsek, S. D., Illes, S., Wellen, J. M., Hartung, H.-P., Rose, C. R., et al. (2011). A new role for interferon gamma in neural stem/precursor cell dysregulation. *Molecular Neurodegeneration*, *6*(1), 18. https://doi.org/10.1186/1750-1326-6-18.

Wang, J., Bankiewicz, K. S., Plunkett, R. J., & Oldfield, E. H. (1994). Intrastriatal implantation of interleukin-1. *Journal of Neurosurgery*, *80*(3), 484–490. https://doi.org/10.3171/jns.1994.80.3.0484.

Wang, X., Fu, S., Wang, Y., Yu, P., Hu, J., Gu, W., et al. (2007). Interleukin-1β mediates proliferation and differentiation of multipotent neural precursor cells through the activation of SAPK/JNK pathway. *Molecular and Cellular*

Neuroscience, 36(3), 343–354. https://doi.org/10.1016/j.mcn.2007.07.005.

Weber, A., Wasiliew, P., & Kracht, M. (2010). Interleukin-1 (IL-1) pathway. *Science Signaling, 3*(105), 1–5. Retrieved from http://stke.sciencemag.org/content/3/105/cm1.full.

Whitney, N. P., Eidem, T. M., Peng, H., Huang, Y., & Zheng, J. C. (2009). Inflammation mediates varying effects in neurogenesis: relevance to the pathogenesis of brain injury and neurodegenerative disorders. *Journal of Neurochemistry, 108*(6), 1343–1359. https://doi.org/10.1111/j.1471-4159.2009.05886.x.

Widera, D., Mikenberg, I., Elvers, M., Kaltschmidt, C., & Kaltschmidt, B. (2006). Tumor necrosis factor alpha triggers proliferation of adult neural stem cells via IKK/NF-kappaB signaling. *BMC Neuroscience, 7*(1), 64. https://doi.org/10.1186/1471-2202-7-64.

Wilson, C. J., Finch, C. E., & Cohen, H. J. (2002). Cytokines and cognition—the case for a head-to-toe inflammatory paradigm. *Journal of the American Geriatrics Society, 50*(12), 2041–2056. Retrieved from http://www.ncbi.nlm.nih.gov/pubmed/12473019.

Zhang, Y., Liu, J., Yao, S., Li, F., Xin, L., Lai, M., et al. (2012). Nuclear factor kappa B signaling initiates early differentiation of neural stem cells. *Stem Cells, 30*(3), 510–524. https://doi.org/10.1002/stem.1006.

Zhang, K., Xu, H., Cao, L., Li, K., & Huang, Q. (2013). Interleukin-1β inhibits the differentiation of hippocampal neural precursor cells into serotonergic neurons. *Brain Research, 1490*, 193–201. https://doi.org/10.1016/j.brainres.2012.10.025.

Zheng, L.-S., Kaneko, N., & Sawamoto, K. (2015). Minocycline treatment ameliorates interferon-alpha-induced neurogenic defects and depression-like behaviors in mice. *Frontiers in Cellular Neuroscience, 9*, 5. https://doi.org/10.3389/fncel.2015.00005.

Zonis, S., Ljubimov, V. A., Mahgerefteh, M., Pechnick, R. N., Wawrowsky, K., & Chesnokova, V. (2013). p21Cip restrains hippocampal neurogenesis and protects neuronal progenitors from apoptosis during acute systemic inflammation. *Hippocampus, 23*(12), 1383–1394. https://doi.org/10.1002/hipo.22192.

Zunszain, P. A., Anacker, C., Cattaneo, A., Choudhury, S., Musaelyan, K., Myint, A. M., et al. (2012). Interleukin-1β: a new regulator of the Kynurenine pathway affecting human hippocampal neurogenesis. *Neuropsychopharmacology, 37*(4), 939–949. https://doi.org/10.1038/npp.2011.277.

Zunszain, P. A., Hepgul, N., & Pariante, C. M. (2012). Inflammation and depression. *Current Topics in Behavioral Neurosciences, 14*, 135–151. https://doi.org/10.1007/7854_2012_211.

7

The Roles of T Cells in Clinical Depression

Catherine Toben, Bernhard T. Baune

University of Adelaide, Adelaide, SA, Australia

AN ACTIVE ROLE FOR T CELLS IN THE MODULATION OF INFLAMMATION ASSOCIATED DEPRESSION

Worldwide, the increasing prevalence of major depressive disorder (MDD) is cause for concern (Kessler & Bromet, 2013). According to the World Health Organization (WHO), this debilitating disorder is ranked as the fourth leading cause of disability and projected to be the second leading cause by 2020 (GBD 2015 Mortality and Causes of Death Collaborators, 2016). Taking into account the mental suffering and associated medical comorbidity, it is not surprising that MDD significantly increases the risk of suicide (Bostwick & Pankratz, 2000; Moller, 2003; Rihmer, 2007), decreases life expectancy in the general adult population (Zivin et al., 2015), and has a high socioeconomic burden on a struggling health-care system.

Over 30 years ago, reports on elevations within the peripheral blood of macrophage-released interleukin IL1 beta (IL1β) and tumor necrosis factor alpha (TNFα) cytokines, which affect hypothalamic function, led Smith to originally postulate the macrophage theory of depression (Smith, 1991). Elaboration of this hypothesis included the immune-modulating effects of T cells in the "macrophage-T-lymphocyte" model of depression (Maes, 1995; Maes, Smith, & Scharpe, 1995). This model takes into account the external and internal triggers of psychogenic stress-inducing depression via activation and consequent dysregulation of the hypothalamus-pituitary-adrenal (HPA) neuroimmune axis. Empirical findings have lent further support to these hypotheses with converging lines of evidence from clinical and experimental research pointing to an integral role for the immune system in the pathophysiology of depression (Fitzgerald et al., 2006; Harrison et al., 2009; Lanquillon, Krieg, Bening-Abu-Shach, & Vedder, 2000; Sluzewska, Sobieska, & Rybakowski, 1997). In particular, measurement of altered levels of T-cell-specific markers such as soluble IL2Rα (sIL2Rα) (Liu, Ho, & Mak, 2012) and CD3 + that defined lymphocytes within clinically depressed participants compared with healthy controls indicates the importance of defining the functions of T lymphocytes in neuroinflammatory mechanisms regulating MDD. However, most of the literature on

Inflammation and Immunity in Depression
https://doi.org/10.1016/B978-0-12-811073-7.00007-6

inflammation-associated MDD has focused on the activation of the initial and innate immune response including measurements of cytokines from innate and phagocytic cells such as monocytes/macrophages and natural killer cells (NKs). While cytokine profiling provides a basic line of evidence in terms of an immune response during MDD, it is a composite result of both innate and adaptive cell output and doesn't allow for a more meaningful interpretation of cell-specific contribution or the course of the T-cell response. Neurogenic stress-induced MDD can result in migration of peripherally derived lymphocytes across the blood-brain barrier (BBB) (Dhabhar, 2008), and meta-analyses (Herbert & Cohen, 1993) report a significant decrease in *in vitro* T-cell responses from both stressed and depressed individuals; however, the exact mechanisms of T lymphocytes between the brain and peripheral immune system in mediating risk and resilience to the development of clinical depression remain relatively unknown.

More recent research is providing evidence for central nervous systems (CNS) homeostasis and neuroinflammation being reliant on the intricate interplay between the innate and infiltrating adaptive immune T and B lymphocytes (Herkenham & Kigar, 2017). Whether in response to infectious or noninfectious/sterile disorders, T cells are being recognized as having important functions in this highly orchestrated response being either pathogenic or protective in nature. This includes continual immune surveillance, regulation of neurogenesis (Ziv & Schwartz, 2008), contribution to maintenance of cognition (Schwartz & Kipnis, 2011), and learning and emotional behavior as shown in T-cell-deficient mice (Kipnis, Cohen, Cardon, Ziv, & Schwartz, 2004). T-cell actions either through soluble mediators or via cell-contact-dependent mechanisms are highly flexible as their phenotype can interconvert between subsets. Cell surface markers distinguish between the main T-cell subsets of CD4+ (T helper) and CD8+ (T cytotoxic) T cells, while their specific

functionality is characterized by particular signatures of cell signaling mediators involved in humoral and cytolytic T-cell responses, respectively.

Arising from circulating bone marrow progenitors, T lymphocytes undergo expansion and T lineage commitment to either CD4+ or CD8+ single positive T cells via low-affinity selection for self-peptide-MHC within the thymus (Xing & Hogquist, 2012). The ability to distinguish between self and foreign Ag is paramount to minimizing CNS autoimmune responses and is regulated by central and peripheral tolerance mechanisms acting on maturing T cells as they move from the thymus into peripheral circulation. Centrally and within the thymus, clonal deletion, clonal diversion (selection of regulatory T cells) (Benoist & Mathis, 2012), and T-cell receptor (TCR) gene rearrangement or editing (Santori, Arsov, Lilic, & Vukmanovic, 2002) limit self-reactivity of the T-cell repertoire. However, not all self-antigens are intrathymically expressed, and controlling tolerance for CNS-autoreactive T cells in the periphery includes deletion or inactivation by regulatory T-cell (CD25+FoxP3+) anergy (induced state of unresponsiveness) mediated via inhibitory signals such as cytotoxic T-lymphocyte-associated antigen-4 (CTLA-4) (Walunas et al., 1994), Fas- and Bim-mediated peripheral deletion, and tolerogenic antigen-presenting cells (APCs) (Xing & Hogquist, 2012). Thymus-derived regulatory T cells (Tregs) develop with high avidity to self-Ag (Sakaguchi, Sakaguchi, Asano, Itoh, & Toda, 1995) and are suppressive in nature by attenuating CNS-autoreactive CD4+ T-cell hyperactivation, proliferation, and cytokine production within the CNS and its draining lymph nodes (Cervantes-Barragan et al., 2012; Lowther & Hafler, 2012).

Differentiation in the periphery of naive T cells into distinct T-cell subsets (T helper) and cytotoxic T cells (Tc) with specific functionality as characterized by signature effector molecules is mediated by effector, inducer, and stabilizer molecules (cytokines) and gene regulatory proteins (Allen, Turner, Bourges, Gleeson, & van Driel, 2011). Hereafter, T-cell activation occurs via recognition of antigen (Ag) in the context of APC-expressed major histocompatibility complex (MHC) either class II for CD4+ T cells or class I for CD8+ T cells. Within the CNS context, perivascular macrophages, dendritic cells (DCs) of either monocyte derivation or conventional CD8a+ spleen origin, or B lymphocytes all serve as APC whereby microglia are the main APC albeit with lower levels of MHC I and II molecules as compared with peripheral APCs. Microglia connect the innate with the adaptive immune response by Ag presentation and recruitment of T cells via secretion of innate immune-derived CKs such as IL1β, IL6, TNFα, monocyte chemoattractant protein-1 (MCP-1), and regulated on activation in normal T-cell expressed and secreted (RANTES). Activated microglia also increase their expression of MHC and T-cell costimulatory molecules such as CD28-inducing T-cell proliferation and secretion of effector CKs (Fig. 1).

Advances in neuroimaging have transformed the immune-privileged paradigm of the brain into one in which the brain is recognized as a "virtual secondary lymphoid organ" (Negi & Das, 2017). From this viewpoint, immune privilege within the brain is an active process by which immune surveillance and dynamic interaction with the peripheral immune system regulate immune cell activity within the CNS. In support of this, the BBB at the interface of the blood and brain tissue regulates lymphocytes in their migration not only by virtue of the tight junctions between cerebrovascular endothelial cells but also by inducing systemic immune deviation or tolerance to self-antigens. Further neuroimaging studies have identified a network of lymphatic vessels within the meninges draining into the deep cervical lymph nodes of the neck as being the direct link between the CNS and peripheral immune system (Aspelund et al., 2015; Louveau et al., 2015). Importantly during neuroinflammation, these cellular routes

enable the presentation of CNS peptide fragments by circulating APCs such as DCs to lymphocytes circulating between the brain and the periphery (Aspelund et al., 2015; Iliff, Goldman, & Nedergaard, 2015; Louveau et al., 2015).

Trafficking along these routes also enables continued and constitutive T-cell surveillance of the CNS for maintenance of neuronal function and homeostasis of CNS neurogenesis. Facilitating the CNS and peripheral immune bidirectional communication is the vast network of cytokines, chemokines, neuropeptides, and neurotransmitters shared between the two systems. Although the relative concentration of leukocytes circulating within the CNS is low compared with peripheral blood (Seabrook, Johnston, & Hay, 1998), circulating cerebrospinal fluid (CSF) contains leukocytes of which more than 90% are T cells and of which half are CD4+ T cells (Engelhardt & Ransohoff, 2012; Meeker, Williams, Killebrew, & Hudson, 2012; Ransohoff & Engelhardt, 2012). Other investigations into healthy human CSF have identified up to 150,000 CD4+ T central memory cells expressing CCR7, L-selectin, CD27, and the early activation marker CD69 (Kivisakk et al., 2003).

Within current paradigms, naive cells do not readily enter the CNS under steady-state conditions and are more dependent on microbial or other inflammatory stimuli for CNS Ag-induced activation. Instead, naive tolerogenic T cells require priming in peripheral lymph nodes for induction of a migratory program responsive to specific sets of adhesion molecules (selectins and integrins) and chemokine receptors facilitating CNS access (Beena, Hunter, & Harris, 2014). These unique "biological address codes" determine the combinational steps of T-cell extravasation and their positioning and surveillance activity in the CNS microenvironment.

Induction of MDD-associated chronic low-grade inflammation involves an adaptation of the innate immune response to the exposure of

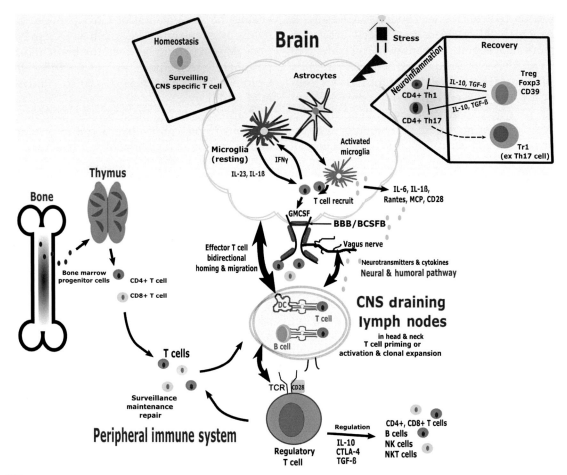

FIG. 1   Overview of continuous T cell immune surveillance pathways between the brain and periphery.

stressors not previously encountered in early evolution such as chronic psychosocial stressors and artificial environments. Microglia can sense both systemic and psychological stress via activation of either membrane-bound receptors (TLRs) by pathogen-associated molecular patterns (PAMPs) or cytosolic receptors (NOD-like receptors, NLRs) (Kigerl, de Rivero Vaccari, Dietrich, Popovich, & Keane, 2014) by damage-associated molecular patterns (DAMPs) (Martinon, Burns, & Tschopp, 2002). Subsequent activation of the inflammasome complex followed by release of pro-inflammatory cytokines including IL1β and IL18 enables the innate immune arm to initiate and regulate inflammation (Dinarello, 2000) by recruitment of the adaptive immune cells including T cells (Kaufmann et al., 2017; Singhal, Jaehne, Corrigan, Toben, & Baune, 2014). While immune-to-brain communication involves active participation in immune regulation by microglia, astrocytes, and neurons (Tian, Ma, Kaarela, & Li, 2012), brain-to-immune signaling is primarily mediated via noradrenergic innervation by the sympathetic nervous system (SNS) of lymphoid organs. Within this system, chronic stress leads to altered HPA-axis-mediated negative-feedback regulation of neuroinflammation by inducing glucocorticoid hyporesponsiveness of lymphocytes that leads to dysregulated T-cell function (Sternberg, 2006; Webster Marketon & Glaser, 2008).

T CELL SUBTYPES AND THEIR FUNCTIONAL PHENOTYPE IN MALADAPTIVE IMMUNE RESPONSES DURING CLINICAL DEPRESSION

The overall neuroinflammatory response within the CNS as regulated by immune-stress-specific pathways between the brain and immune system employs similar mediators of inflammation. The nature of the inflammatory response is either transient or long term and defined by either adaptive or maladaptive T-cell functionality. Furthermore, the magnitude of the psychogenic stress being either acute or chronic determines either the enhancement or suppression, respectively, of specific lymphocyte subset numbers (Dhabhar & McEwen, 1997; Pacheco-Lopez et al., 2009) and their functionality (Dhabhar, 2009, 2014; Segerstrom & Miller, 2004). Transient stress-induced CNS inflammation is beneficial in resolving the initial assault to the system and a return to homeostasis, while chronic stress that is most strongly associated with clinical depression (Phillips, Carroll, & Der, 2015) leads to maladaptive consequences including a dysregulated in vitro T-cell proliferative response and alterations in T-cell subtype ratios, numbers, and their functionality. These alterations affecting neuronal plasticity and integrity lead to altered behavior.

While some of the earlier studies (Maes, Meltzer, Stevens, Calabrese, & Cosyns, 1994; Maes et al., 1993) reported an increase in CD2 +HLA-DR+ T cells and CD7+CD25+ early-activated T cells in melancholic patients compared with controls and those with minor depression, this has not been substantiated with more recent studies. It is envisaged that as T-cell numbers increase, their activation status will also be altered during MDD-associated neuroinflammation. This would include potential upregulation of adhesion molecules enabling homing into specific CNS areas and interaction with different target cells. Although few studies report increased T-cell proliferation (Altshuler, Plaeger-Marshall, Richeimer, Daniels, & Baxter, 1989), further subtyping in others revealed the origin of the T-cell response to be accounted for by an increase in the percentage of CD4+ T cells and an increased CD4/CD8 ratio (Darko et al., 1989; Maes et al., 1992; Muller, Hofschuster, Ackenheil, Mempel, & Eckstein, 1993; Rothermundt et al., 2001; Seidel et al., 1996). These studies point to a function for CD4+ T cells in perpetuating neuroinflammation of MDD and can be regarded as being part of the maladaptive T-cell response.

The function of memory T cells in the pathophysiology and course of MDD in particular in relapsing-remitting depression remains to be elucidated. An early report by Maes et al. (1992) supported the presence of CD4 +CD45RA- memory T cells in single-episode depression although no difference between controls and MDD of CD4+CD45RO+ memory T cells was found in a separate longitudinal study over 43 days (Seidel et al., 1996). These early phenotypic studies have been followed by a plethora of animal studies from the Schwarz and Baruch laboratories in which in vitro programmed CNS-autoreactive effector memory T cells have been shown to be beneficial for CNS repair in various neurodegenerative conditions including spinal cord injury and Alzheimer's disease (Baruch & Schwartz, 2013; Schwartz & Shechter, 2010) and stress resilience (Cohen et al., 2006). The same remains to be determined for human counterparts. T cells generally acquire memory by antigen presentation via an APC; however, their presence during clinical depression in the absence of systemic inflammation suggests that other pathways of generation including psychogenic stress-induced signaling via lymphocyte adrenergic receptors could also affect T-cell number and activation status (Yu, Dimsdale, & Mills, 1999). This novel hypothesis has recently been supported by investigations using chronic stress mouse models in which antigen-independent immunologic memory was maintained by CD4+ T cells by varying repetitive stressors (Wang, Lavender, Watson, Arno, & Lehner, 2015). The idea of stress being able to induce resilience by the generation of effector memory T cells has been further substantiated by the conferral of antidepressant-like and anxiolytic effects from adoptively transferred lymphocytes from chronically stressed mice into stress naive (Brachman, Lehmann, Maric, & Herkenham, 2015) or chronically stressed Rag-2$^{-/-}$ mice (Scheinert, Haeri, Lehmann, & Herkenham, 2016). Although investigations are required in human cohorts, these findings suggest that T-cell memory, as generated under an antigen-free and within a chronic stress environment, is required to restore homeostasis and could be applied therapeutically.

Maladaptive T cells also appear to be regulated by telomere shortening that has been reported in depression cohorts (Simon et al., 2013) and more recently by Karabatsiakis, Kolassa, Kolassa, Rudolph, and Dietrich (2014) (Darrow et al., 2016). Although not controlling for covariates such as BMI or lifestyle factors, that is, tobacco or alcohol or hormonal variation within the small cohort of women, these studies found that CD8+ but not CD4+ T cells exhibit shortened telomeres in MDD when compared with controls. This implies that telomere length reduction as an index of cellular aging affects T-cell functioning within MDD, which is enhanced by increased severity and duration of symptoms (Verhoeven et al., 2014). These studies also indicate that any dysfunctional T-cell functions not only are simply a representation of the disease consequence but also may be attributable to their active participation in the immune response.

Under maladaptive conditions, dysregulation of T cells can also be attributed to altered enzymatic activity including dipeptidyl peptidase IV (DPPIV) activity involved in T-cell costimulation (Elgun, Keskinege, & Kumbasar, 1999) and found to be reduced during MDD. Purified lymphocytes including T cells have also been shown to have increased levels of antioxidase enzymes involved in oxidative stress responses and the cytoplasmic redox-sensitive transcription factor NF-κB (Lukic et al., 2014), which are both involved in regulating inflammatory conditions during MDD. Through as yet uncharacterized pathways, further dysregulation of T cells could be attributed to a decrease in the mRNA expression of neuroprotective growth factors (NGFs) in particular brain-derived neurotrophic factor (BDNF) and found in isolated lymphocytes from MDD participants (Pandey et al., 2010). Involved in facilitating synaptic plasticity and correlation with behavioral outcomes (Wolf et al., 2009), BDNF has long been associated with structural brain changes from postmortem studies. These more recent correlations within peripheral blood and MDD status provide new evidence of BDNF being a mediator of the CNS-immune bidirectional communication pathways regulating peripheral lymphocytes and associated neuroprotective mechanisms.

T CELL SUBTYPES AND THEIR FUNCTIONAL PHENOTYPE IN ADAPTIVE IMMUNE RESPONSES DURING CLINICAL DEPRESSION

Maladaptive T-cell responses occurring during current depression return to more adaptive responses during remission or recovery. To support this, a number of studies report unchanged *in vitro* measured nonspecific T-cell proliferation (Albrecht, Helderman, Schlesser, & Rush, 1985; Andreoli et al., 1992; Bauer et al., 1995; Hickie, Hickie, Lloyd, Silove, & Wakefield, 1993;

Schleifer, Keller, Bond, Cohen, & Stein, 1989; Wodarz et al., 1991), while other *in vitro* studies report a significant reduction in nonspecific T-cell proliferation in MDD compared with controls. While a number of studies investigating lymphocyte count in addition to T-cell proliferative responses show a decrease in lymphocyte numbers, the majority of studies find T-cell numbers to remain unchanged. Although the participant numbers are small in most studies, the results demonstrate a significant relationship between MDD and the measured T-cell response.

These cross-sectional studies also suggest that in most cases of inflammation-associated MDD, T-cell numbers may not be altered; however, it does not exclude their functionality from being affected. While most T-cell functional *in vitro* studies demonstrate altered mitogen-induced proliferation, whether this is representative of *in vivo* T-cell-mediated immune responses remains debatable. However, impaired *in vivo* delayed-type hypersensitivity (DTH) responses in depressed patients have been reported (Hickie et al., 1995). Conversely, other studies assessing wound repair involving a functioning immune response found that depressed patients exhibit longer wound healing time compared with nondepressed patients (Cole-King & Harding, 2001; Kiecolt-Glaser, Marucha, Malarkey, Mercado, & Glaser, 1995). Ambiguous *in vitro* T-cell proliferation results may exist across MDD studies due to the heterogeneous nature of clinical depression.

Further discrimination of T-cell subsets utilizing either immunofluorescent microscopy (Kronfol & House, 1989; Schleifer et al., 1989) or flow cytometry (Andreoli et al., 1992) has found no significant difference between CD4+ and CD8+ subpopulations within MDD compared with controls. So far, no studies have investigated direct T-cell subset proliferation or cell viability by incorporating carboxyfluorescein succinimidyl ester (CFSE) assessment or live/dead cell stains, respectively.

During the course of clinical depression, neuroinflammation is generally regarded as being characterized by reduced T-cell numbers with an activated phenotype. However, most of the literature reports an unchanged number of lymphocytes in MDD. In studies presenting with lymphopenia, this was found to be due to decreased NK cell counts but not T cells. Further in-depth investigations into the meta-analysis from 2001 (Zorrilla et al., 2001) showed a significant negative association for the number of lymphocytes and MDD but only for the fixed and not the random effect. While both measures were reported, the random-effect analyses would account for the psychiatric heterogeneity found in MDD.

Although our previous review (Toben & Baune, 2015) supported the latter findings, it also supported the previous meta-analyses (Zorrilla et al., 2001) in other findings of no significant association between CD4+ or CD8+ T cells and MDD and a significant association with MDD and higher CD4/CD8 ratio for both fixed and random effect. Similarly, in agreement with the meta-analysis, our review also showed a significant decreased cellular immune response (Zorrilla et al., 2001).

This hyporesponsiveness may be due to overproduction of pro-inflammatory CKs such as TNFα rendering T-cell refractory and unable to respond to Ag stimulation (Cope, Ettinger, & McDevitt, 1997), prostaglandin secretion, and the presence of sIL2Rs that may compete with IL2 for binding to cellular IL2Rs. Downregulation of immune function may also be due to HPA axis hyperactivity and lowered available L-tryptophan as found in MDD (Maes, Scharpe, et al., 1994).

As part of the adaptive T-cell function, programmed cell death or apoptosis is important in restoring homeostasis. Removal of functionally redundant T cells or those with autoreactive potential ensures tight control of T-cell numbers in line with cellular homeostasis. An MDD case control study (Ivanova et al., 2007) found lymphocytic apoptotic facilitating receptor CD95 activity in combination with nuclear fragmentation to be increased compared with controls. These observations in parallel with a decrease in CD4+ T-cell number suggest increased T-cell apoptosis to be an adaptive T-cell mechanism resulting in reduced T-cell number and responsiveness as seen during depression (Eilat, Mendlovic, Doron, Zakuth, & Spirer, 1999). Increased T-cell apoptosis in the context of increased immune activation within MDD may also be a consequence of tryptophan depletion (Mellor & Munn, 2003) or exposure to increased pro-inflammatory cytokines such as TNFα promoting apoptotic gene dysregulation (Hong et al., 2015) (Fig. 2).

REGULATION OF T CELL RESPONSES IN CLINICAL DEPRESSION

Growing interest lies with Tregs in playing a central function within low-grade inflammation observed in MDD. Expansion of thymus-derived natural Tregs (nTregs) and conversion of adaptive T cells or extrathymically induced CD4+ Tregs (iTregs) provide both active tolerance to self-antigens (Liu et al., 2006; Sakaguchi, Yamaguchi, Nomura, & Ono, 2008) and potently suppressing effector T cells (Adeegbe & Nishikawa, 2013), respectively, in addition to maintaining a Th1/Th2 cytokine balance (Xu et al., 2003). In line with the mono-amine theory of depression, MDD is characterized by low levels of 5-hydroxytryptamine (5HT). However, 5HT is not only a neurotransmitter but also an immune modulator influencing the function of NK, macrophages, pre-B cells, and T cells (Ahern, 2011). In particular, the 5HT receptor subtype $5HTR_{1a}R$ is associated with peripheral T-cell activities. Three studies reported nTregs, as defined by the panel CD4+ CD25+Foxp3+, to be decreased in MDD patients and therefore shown to be part of the maladaptive function of T cells. One study (Li et al., 2010) found diminished Tregs in parallel to an imbalanced Th1/Th2 cytokine ratio with lower levels of 5HT in the plasma and $5HTR_{1a}R$ mRNA in Tregs of MDD patients when compared with controls suggesting an important role for serotonin in maintaining Treg levels. Conversely, Himmerich et al. (2010) found 6 weeks of antidepressant treatment in a naturalistic type of study to increase Tregs and reduce serum IL1β levels when compared with prior antidepressant treatment levels. The precise mechanisms of serotonin-mediated T-cell modulation during depression remain to be investigated.

Other regulatory T-cell functions as mediated by Th17 cells (Grogan & Ouyang, 2012; Infante-Duarte, Horton, Byrne, & Kamradt, 2000) were observed by Chen et al. (2011) to minimize a coexisting autoimmunity within MDD. While their MDD patients exhibited a significant decrease in circulating Tregs, a significant increase in the autoimmune-inducing Th17 cells was found when compared with controls. Further investigation showed not only an increase in mRNA of the master transcription regulator of Th17 cells, RORt, but also an increase in serum IL17 from MDD patients compared with controls (Table 1). This study points to the importance in maintaining a balanced Th17/Treg ratio in reducing susceptibility to autoimmune-induced inflammation in MDD. A more recent and extensive immunophenotyping study in MDD participants found CD127low/CCR4+ Treg counts to be increased in the depressed population compared with healthy controls that was not related to sleep disturbance (Suzuki et al., 2017). Although this subset may contain recently activated T cells, this was reported to be controlled for by no difference found in the group analyses of monocyte subsets reactive to the activated T cells. This is in contrast to other studies that have measured decreases in Treg number and functional defects in those with depression compared with healthy

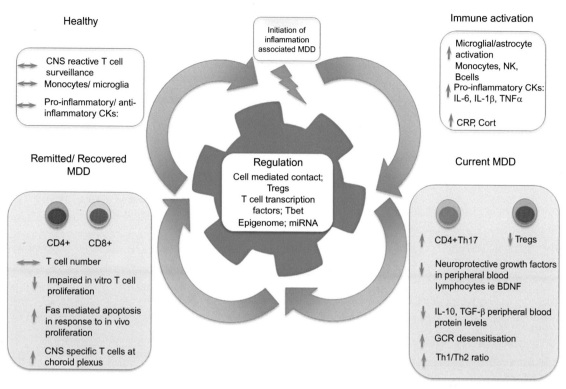

FIG. 2 Regulation of T-cell immune-modulating activity within MDD.

controls (Grosse et al., 2016; Li et al., 2010) and is more in line with those observations from elderly populations experiencing worse physical and mental health status (Ronaldson et al., 2016). This may be due to the aging process leading to decreased thymic T-cell output being compensated by an age-related increase in Treg cell levels (Vadasz, Haj, Kessel, & Toubi, 2013). However, it is important to differentiate between studies exploring either natural or inducible regulatory T cells as these subsets are different in terms of TCR repertoire, differentiation states, and their mechanisms of action (Bluestone & Abbas, 2003). It is also important to distinguish between effector memory T cells and Tregs either by staining with more specific cell markers or via assessment of group differences between reactive monocytes (Suzuki et al., 2017). A few studies point to a concurrent Treg

defect in the presence of a pro-inflammatory state in MDD, including those by Himmerich et al. (2010), Miller (2010), Ho, Yeh, Huang, and Liang (2015), and Sanna et al. (2014). Recently, Grosse et al. (2016) found an association between deficiencies in the proportion of natural Treg cells and an activated monocyte gene profile in older MDD participants. Unfortunately, the study design could not control for antidepressant or age-induced cellular dysfunction. However, this study does point to the reputed connection between a natural Treg deficiency and an increased age-related risk for autoimmunity within MDD (Euesden, Danese, Lewis, & Maughan, 2017). In contrast, natural Tregs are found to increase during healthy aging, while inducible Tregs decrease (Jagger, Shimojima, Goronzy, & Weyand, 2014). Unclear is whether this discrepancy is

TABLE 1   T-Cell Functions Between the Brain and Periphery and Associated With MDD

T-Cell Type	Effectors	Interaction		Association With MDD
CD4+ T cell		*Peripheral immune system*	*Central nervous system*	
Th1	IFNγ, lymphotoxin IL2, IL12, TNFα, IL1β, IL8, GM-CSF	Cell-mediated immune responses, delayed-type hypersensitivity, macrophage activation	Neurodegenerative	↑
Th2	IL4, IL5, IL13, IL9, IL25, IL10, IgE	Humoral and allergic immune responses	IL4 inhibits differentiation of meningeal myeloid cells, induces expression of BDNF by astrocytes and secretion of TGFβ, IGF1, and IL10 by microglia/macrophages	↓
Th9	IL9, IL10	Allergy, autoimmunity	Unknown	Unknown
Th17	IL17A, IL17F, IL21, IL22, GM-CSF	Both pathogenic and nonpathogenic	Structural alterations in the cortex	↑
Treg (CD25+ FoxP3+)	TGFβ, IL10, CTLA4 (IL4, IL13)	Regulation of CD8+ T cells, B cells, NKT cells, monocytes, macrophages, DCs by restricting proliferation, and CK production	Unknown	↓
CD8+ T cell				
Tc1	IL2, IFNγ	Cytotoxicity	Unknown	Unknown
Tc2	IL4, IL5, IL6, IL10	Cytotoxicity/regulation of CD4+ T-cell responses	Unknown	Unknown
Treg	IL4, IL6, IL10, TGFβ, IFNγ	Suppression/modulation of CD4+ T cells	Unknown	Unknown
NKT cell	IL4, IFNγ, TNFα, IL5, IL13, IL2, GM-CSF, IL6, RANTES, MIP1α,β, IL10, TGFβ, granzymes, perforin	Immune cell modulation	Unknown	Unknown

due to an MDD-specific effect. While these studies are setting the premise for an interrelationship between pathogenic T-cell effector cells and Tregs, the autoimmune processes within MDD remain to be determined but are likely to include host microbiome interactions and formation of neurotransmitter autoantibodies.

Being able to distinguish between adaptive immune responses leading to pathological autoimmunity or protective autoimmunity is important. The latter has been correlated with resistance to the development of autoimmune disease (Suzuki et al., 2017) in which autoreactive T cells can mitigate neuronal damage

(Kipnis et al., 2002, 2004; Schwartz & Kipnis, 2005). Further evidence for a role for CNS-reactive T cells in modulating behavior was found in a PTSD mouse model in which depletion of naturally occurring CD4+CD25+ T cells resulted in maladaptation to acute psychological stress (Cohen et al., 2006). These results further support the premise that Tregs maintain the tight regulation and balance between the risk of pathological autoimmunity and the requirement for CNS-protective-autoreactive T cells. Exact neuroprotective mechanisms remain to be determined but include homing and activation of CNS-specific T cells to the site of insult followed by modulation of the local immune response by resident microglia and astrocyte secretion of neurotrophic factors such as BDNF, vascular endothelial growth factor (VEGF), and insulin-like growth factor-1 (IGF-1) (Schwartz, 2003; Shaked, Porat, Gersner, Kipnis, & Schwartz, 2004) making individuals less vulnerable to stressors and resilient to MDD.

Regulating the balance between adaptive and maladaptive T-cell responses in MDD can be mediated by the cytotoxic T-lymphocyte-associated antigen-4 (CTLA-4) as it regulates T-cell activation and reactivity and thereby controls T-cell tolerance (Gribben et al., 1995). Single-nucleotide polymorphism (SNP) association studies have determined genetic heterogeneity at the CTLA-4 T-cell-associated locus to confer susceptibility to MDD in certain ethnic populations and as demonstrated between Korean and Caucasian (Jun, Pae, Chae, Bahk, & Kim, 2001) and separately in a Chinese Han cohort (Liu et al., 2011). Further studies within a Mexican-American cohort (Wong, Dong, Maestre-Mesa, & Licinio, 2008) have shown genes affecting T-cell function to be associated with MDD. Although a significant association between polymorphisms in T-bet, the transcription factor regulating T-cell differentiation and MDD, was identified, there was no associated or distinct polarization toward either a Th1 or a Th2 phenotype. These studies further

support the premise of a gene-environment interaction being key to T-cell dysregulation and mediating increased risk susceptibility to MDD pathogenesis.

Further regulation of T-cell differentiation is found in association between MDD and altered gene expression levels of epigenetic modifiers such as histone deacetylases (HDACs) (Hobara et al., 2010), which are involved in regulating T-bet-mediated T-cell differentiation (Helmstetter et al., 2015). Although cell surface signaling events may be similar between activation of naive and memory T cells, regulation of activation and differentiation is determined by differing global epigenetic patterns (Cuddapah, Barski, & Zhao, 2010).

CURRENT CHALLENGES FOR *IN VITRO* AND *IN VIVO* MEASUREMENT OF T CELLS AND THEIR FUNCTIONALITY

Two major challenges remain in acquiring the correct measurement of T-cell phenotype, number, and their functionality. Firstly, the nonstandardized collection of biospecimen or clinical data for MDD study participants impedes any general comparisons to be made across studies. Until recently, the majority of studies collected peripheral blood without regard for downstream analyses of T cells meaning most biorepositories contain blood unsuitable for downstream T-cell immunophenotyping. More recent studies include a blood collection protocol whereby freshly collected peripheral blood mononuclear cells (PBMCs) containing T cells are processed either within 24 h or within 8 h using a Ficoll-Paque-type density gradient centrifugation protocol. Controlled freezing and thawing at −80°C in specialized freezing media allow for future polychromatic flow cytometric analyses.

The second challenge lies in the inherently heterogenetic and unreliable nature of the clinically depressed cohort. A more standardized approach in biospecimen and clinical data collection between multicenter consortia will ensure more consistent characterizations of T-cell functionality across cohorts. Collected clinical data should contain information on variables affecting immune system responses including different forms of pharmacotherapy (Oertelt-Prigione, 2012) or nonpharmacotherapy interventions such as physical activity (Eyre & Baune, 2012).

Advancing the field in T-cell measurements includes the measurement of both gene and protein expression at the single-cell level. With technological advances in cellular measurements, isolated measurements of cell signaling mediators that act in networks of pleiotropism, redundancy, synergy, and antagonism will become obsolete. Paramount for future analyses will be to integrate measurements for cellular responses including transcriptomic and cellular platforms (Herkenham & Kigar, 2017) such as RNA-seq and fluorescent-activated cell sorting (FACS) (Vallejos, Richardson, & Marioni, 2016) to better define regulatory gene elements such as miRNA species within the functionally intricate cellular networks operating during different stages of MDD.

CONCLUDING REMARKS AND FUTURE OUTLOOK

T-cell functions in mediating a balance in inflammation-associated MDD are now well established and support a spectrum of functionality ranging between pathogenesis and homeostasis in their interactions between the CNS and periphery. Their function in supporting a healthy brain is currently better defined than their contributions to neuroimmune signaling leading to psychiatric dysfunction. In a proinflammatory MDD environment, peripheral T cells display maladaptive effects by increased percentages of CD4+ (Th17+) cells concurrent

with a decrease in Tregs. Clinical progression of MDD into remission finds T cells adapting as numbers remain stable as mediated via apoptosis and *in vitro* proliferative capacity diminishes. Further adaptive T-cell mechanisms lead to neurorepair and protection. Therefore, T-cell phenotypes and responses are part of a continuum in which T-cell plasticity is paramount in maintaining CNS-immune homeostasis.

Dynamically, altered levels of different T-cell subtypes and functionality are most likely associated with altered states or subgroups of clinical depression and in response to altered molecular signaling gradients (Dahl et al., 2014; Schmidt et al., 2016) for which cross-sectional studies cannot account for. Longitudinal studies would better serve the investigation of the T-cell functions within the heterogeneous spectrum of MDD and assist in stratification of MDD subtypes not accounted for by ICD-10 or DSMV criteria. These would include clinical stratification based on underlying physiology such as in stress-induced inflammation-associated MDD, whereby classification could be made using the HPA-axis-related biomarker C-reactive protein (CRP) (Kunugi, Hori, & Ogawa, 2015; Lichtenberg & Belmaker, 2010). Symptom severity has, for example, been found to be predictive of T-cell activation in women suffering PTSD postchildhood maltreatment (Lemieux, Coe, & Carnes, 2008). In MDD, it remains to be determined whether severity of symptoms could predict the type of T-cell activation status and the degree of polarization and differentiation.

Understanding the intricate function between T cells and their innate and adaptive immune cell counterparts is fundamental to developing innovative therapeutics in inflammation-associated MDD. Further multidimensional phenotyping of T-cell subsets integrating both genetic and cellular markers will enable peripheral blood T-cell population shifts to be better defined for diagnostic potential and ultimately treatment response. Obviously, therapeutic options cannot include broad inhibition of neuroimmune systems as physiological levels of immune mediators such as cytokines are required for synaptic homeostasis and plasticity.

Although CNS-autoreactive T cells are involved in "protective autoimmunity" (M. Schwartz & Baruch, 2014), their specificity in MDD patients remains to be determined for which further high-throughput sequencing of the TCR repertoire would help in identifying CNS specificity (Baruch & Schwartz, 2013). Future therapeutic treatment considerations for treatment-resistant depressive disorders could utilize this information in combination with the unique T-cell feature of acquired CNS memory in nonpathogenic conditions such as stress and program T cells via vaccination with CNS-specific antigens to ameliorate T-cell dysregulation. Conversely, alternative immunotherapy approaches could utilize T-cell "adoptive cell therapy," which is already being successfully applied within the cancer realm (Restifo, Dudley, & Rosenberg, 2012; Rosenberg & Restifo, 2015). In this scenario, lymphocytes would be programmed ex vivo toward Th2, Th1, or Treg polarities and reinfused into participants having undergone lymphodepletion. Alternatively, "stress-educated" T cells could be transferred as these have been shown to confer antidepressant effects (Brachman et al., 2015; Scheinert et al., 2016).

Nonpharmacological interventions such as exercise (Eyre & Baune, 2012), environmental enrichment, and social interaction (Ota & Duman, 2013) show promising results in promoting resolution of immune dysregulation via the production of antiinflammatory cytokines (Vukovic, Colditz, Blackmore, Ruitenberg, & Bartlett, 2012). Studies employing mindfulness-based therapies including yoga already report decreases in pro-inflammatory cytokines, for example, IL6 (Kiecolt-Glaser et al., 2010), and induction of antiinflammatory effects via the parasympathetic nervous system as mediated by T cells (O'Mahony, van der Kleij, Bienenstock, Shanahan, & O'Mahony, 2009) and influences on gene expression in lymphocytes (Qu, Olafsrud, Meza-Zepeda, & Saatcioglu, 2013).

Combining advanced bioinformatics with -omics technologies will enable T-cell-specific functional molecular and cellular signatures to be further defined. These T-cell biomarker signatures will be important in prediction, diagnosis, and prognosis including outcome of treatment response leading to the design of innovative and personalized therapeutic strategies. Future therapies will be aimed at modulating specific effector T cells rather than global manipulation of the immune response. Ideally, harnessing inherent inflammation resolving qualities of T cells could promote neuroregenerative and protective pathways and thereby remission of clinical depression.

ACKNOWLEDGMENT

We would like to acknowledge the artwork in Fig. 1 provided by Gaurav Singhal.

CONFLICT OF INTEREST

The authors have no conflicts of interest to declare.

References

Adeegbe, D. O., & Nishikawa, H. (2013). Natural and induced T regulatory cells in cancer. *Frontiers in Immunology, 4.* https://doi.org/10.3389/fimmu.2013.00190.

Ahern, G. P. (2011). 5-HT and the immune system. *Current Opinion in Pharmacology, 11*(1), 29–33. https://doi.org/10.1016/j.coph.2011.02.004.

Albrecht, J., Helderman, J. H., Schlesser, M. A., & Rush, A. J. (1985). A controlled study of cellular immune function in affective disorders before and during somatic therapy. *Psychiatry Research, 15*(3), 185–193.

Allen, S., Turner, S. J., Bourges, D., Gleeson, P. A., & van Driel, I. R. (2011). Shaping the T-cell repertoire in the periphery. *Immunology and Cell Biology, 89*(1), 60–69. https://doi.org/10.1038/icb.2010.133.

Altshuler, L. L., Plaeger-Marshall, S., Richeimer, S., Daniels, M., & Baxter, L. R., Jr. (1989). Lymphocyte function in major depression. *Acta Psychiatrica Scandinavica, 80*(2), 132–136.

Andreoli, A., Keller, S. E., Rabaeus, M., Zaugg, L., Garrone, G., & Taban, C. (1992). Immunity, major depression, and panic disorder comorbidity. *Biological Psychiatry*, *31*(9), 896–908.

Aspelund, A., Antila, S., Proulx, S. T., Karlsen, T. V., Karaman, S., Detmar, M., et al. (2015). A dural lymphatic vascular system that drains brain interstitial fluid and macromolecules. *The Journal of Experimental Medicine*, *212*(7), 991–999. https://doi.org/10.1084/jem.20142290.

Baruch, K., & Schwartz, M. (2013). CNS-specific T cells shape brain function via the choroid plexus. *Brain, Behavior, and Immunity*, *34*, 11–16. https://doi.org/10.1016/j.bbi.2013.04.002.

Bauer, M. E., Gauer, G. J., Luz, C., Silveira, R. O., Nardi, N. B., & von Muhlen, C. A. (1995). Evaluation of immune parameters in depressed patients. *Life Sciences*, *57*(7), 665–674.

Beena, J., Hunter, C. A., & Harris, T. H. (2014). Immune cell trafficking in the central nervous system. In P. K. Peterson, & M. Toborek (Eds.), *Neuroinflammation and neurodegeneration* (pp. 29–45): New York: Springer.

Benoist, C., & Mathis, D. (2012). Treg cells, life history, and diversity. *Cold Spring Harbor Perspectives in Biology*, *4*(9). https://doi.org/10.1101/cshperspect.a007021.

Bluestone, J. A., & Abbas, A. K. (2003). Natural versus adaptive regulatory T cells. *Nature Reviews. Immunology*, *3*(3), 253–257. https://doi.org/10.1038/nri1032.

Bostwick, J. M., & Pankratz, V. S. (2000). Affective disorders and suicide risk: a reexamination. *The American Journal of Psychiatry*, *157*(12), 1925–1932. https://doi.org/10.1176/appi.ajp.157.12.1925.

Brachman, R. A., Lehmann, M. L., Maric, D., & Herkenham, M. (2015). Lymphocytes from chronically stressed mice confer antidepressant-like effects to naive mice. *The Journal of Neuroscience*, *35*(4), 1530–1538. https://doi.org/10.1523/JNEUROSCI.2278-14.2015.

Cervantes-Barragan, L., Firner, S., Bechmann, I., Waisman, A., Lahl, K., Sparwasser, T., et al. (2012). Regulatory T cells selectively preserve immune privilege of self-antigens during viral central nervous system infection. *Journal of Immunology*, *188*(8), 3678–3685. https://doi.org/10.4049/jimmunol.1102422.

Chen, Y., Jiang, T., Chen, P., Ouyang, J., Xu, G., Zeng, Z., et al. (2011). Emerging tendency towards autoimmune process in major depressive patients: a novel insight from Th17 cells. *Psychiatry Research*, *188*(2), 224–230. https://doi.org/10.1016/j.psychres.2010.10.029.

Cohen, H., Ziv, Y., Cardon, M., Kaplan, Z., Matar, M. A., Gidron, Y., et al. (2006). Maladaptation to mental stress mitigated by the adaptive immune system via depletion of naturally occurring regulatory CD4+CD25+cells. *Journal of Neurobiology*, *66*(6), 552–563. https://doi.org/10.1002/neu.20249.

Cole-King, A., & Harding, K. G. (2001). Psychological factors and delayed healing in chronic wounds. *Psychosomatic Medicine*, *63*(2), 216–220.

Cope, A., Ettinger, R., & McDevitt, H. (1997). The role of TNF alpha and related cytokines in the development and function of the autoreactive T-cell repertoire. *Research in Immunology*, *148*(5), 307–312.

Cuddapah, S., Barski, A., & Zhao, K. (2010). Epigenomics of T cell activation, differentiation, and memory. *Current Opinion in Immunology*, *22*(3), 341–347. https://doi.org/10.1016/j.coi.2010.02.007.

Dahl, J., Ormstad, H., Aass, H. C., Malt, U. F., Bendz, L. T., Sandvik, L., et al. (2014). The plasma levels of various cytokines are increased during ongoing depression and are reduced to normal levels after recovery. *Psychoneuroendocrinology*, *45*, 77–86. https://doi.org/10.1016/j.psyneuen.2014.03.019.

Darko, D. F., Gillin, J. C., Risch, S. C., Bulloch, K., Golshan, S., Tasevska, Z., et al. (1989). Mitogen-stimulated lymphocyte proliferation and pituitary hormones in major depression. *Biological Psychiatry*, *26*(2), 145–155.

Darrow, S. M., Verhoeven, J. E., Revesz, D., Lindqvist, D., Penninx, B. W., Delucchi, K. L., et al. (2016). The association between psychiatric disorders and telomere length: a meta-analysis involving 14,827 persons. *Psychosomatic Medicine*, *78*(7), 776–787. https://doi.org/10.1097/PSY.0000000000000356.

Dhabhar, F. S. (2008). Enhancing versus suppressive effects of stress on immune function: implications for immunoprotection versus immunopathology. *Allergy, Asthma and Clinical Immunology*, *4*(1), 2–11. https://doi.org/10.1186/1710-1492-4-1-2.

Dhabhar, F. S. (2009). Enhancing versus suppressive effects of stress on immune function: implications for immunoprotection and immunopathology. *Neuroimmunomodulation*, *16*(5), 300–317. https://doi.org/10.1159/000216188.

Dhabhar, F. S. (2014). Effects of stress on immune function: the good, the bad, and the beautiful. *Immunologic Research*, *58*(2-3), 193–210. https://doi.org/10.1007/s12026-014-8517-0.

Dhabhar, F. S., & McEwen, B. S. (1997). Acute stress enhances while chronic stress suppresses cell-mediated immunity in vivo: a potential role for leukocyte trafficking. *Brain, Behavior, and Immunity*, *11*(4), 286–306. https://doi.org/10.1006/brbi.1997.0508.

Dinarello, C. A. (2000). Proinflammatory cytokines. *Chest*, *118*(2), 503–508.

Eilat, E., Mendlovic, S., Doron, A., Zakuth, V., & Spirer, Z. (1999). Increased apoptosis in patients with major depression: a preliminary study. *Journal of Immunology*, *163*(1), 533–534.

Elgun, S., Keskinege, A., & Kumbasar, H. (1999). Dipeptidyl peptidase IV and adenosine deaminase activity. Decrease in depression. *Psychoneuroendocrinology*, *24*(8), 823–832.

Engelhardt, B., & Ransohoff, R. M. (2012). Capture, crawl, cross: the T cell code to breach the blood-brain barriers. *Trends in Immunology*, 33(12), 579–589. https://doi.org/10.1016/j.it.2012.07.004.

Euesden, J., Danese, A., Lewis, C. M., & Maughan, B. (2017). A bidirectional relationship between depression and the autoimmune disorders—new perspectives from the National Child Development Study. *PLoS One*, 12(3). https://doi.org/10.1371/journal.pone.0173015.

Eyre, H., & Baune, B. T. (2012). Neuroimmunological effects of physical exercise in depression. *Brain, Behavior, and Immunity*, 26(2), 251–266. https://doi.org/10.1016/j.bbi.2011.09.015.

Fitzgerald, P., O'Brien, S. M., Scully, P., Rijkers, K., Scott, L. V., & Dinan, T. G. (2006). Cutaneous glucocorticoid receptor sensitivity and pro-inflammatory cytokine levels in antidepressant-resistant depression. *Psychological Medicine*, 36(1), 37–43. https://doi.org/10.1017/S003329170500632X.

GBD 2015 Mortality and Causes of Death Collaborators. (2016). Global, regional, and national life expectancy, all-cause mortality, and cause-specific mortality for 249 causes of death, 1980-2015: a systematic analysis for the Global Burden of Disease Study 2015. *Lancet*, 388(10053), 1459–1544. https://doi.org/10.1016/S0140-6736(16)31012-1.

Gribben, J. G., Freeman, G. J., Boussiotis, V. A., Rennert, P., Jellis, C. L., Greenfield, E., et al. (1995). CTLA4 mediates antigen-specific apoptosis of human T cells. *Proceedings of the National Academy of Sciences of the United States of America*, 92(3), 811–815.

Grogan, J. L., & Ouyang, W. (2012). A role for Th17 cells in the regulation of tertiary lymphoid follicles. *European Journal of Immunology*, 42(9), 2255–2262. https://doi.org/10.1002/eji.201242656.

Grosse, L., Carvalho, L. A., Birkenhager, T. K., Hoogendijk, W. J., Kushner, S. A., Drexhage, H. A., et al. (2016). Circulating cytotoxic T cells and natural killer cells as potential predictors for antidepressant response in melancholic depression. Restoration of T regulatory cell populations after antidepressant therapy. *Psychopharmacology*, 233(9), 1679–1688. https://doi.org/10.1007/s00213-015-3943-9.

Harrison, N. A., Brydon, L., Walker, C., Gray, M. A., Steptoe, A., & Critchley, H. D. (2009). Inflammation causes mood changes through alterations in subgenual cingulate activity and mesolimbic connectivity. *Biological Psychiatry*, 66(5), 407–414. https://doi.org/10.1016/j.biopsych.2009.03.015.

Helmstetter, C., Flossdorf, M., Peine, M., Kupz, A., Zhu, J., Hegazy, A. N., et al. (2015). Individual T helper cells have a quantitative cytokine memory. *Immunity*, 42(1), 108–122. https://doi.org/10.1016/j.immuni.2014.12.018.

Herbert, T. B., & Cohen, S. (1993). Depression and immunity: a meta-analytic review. *Psychological Bulletin*, 113(3), 472–486.

Herkenham, M., & Kigar, S. L. (2017). Contributions of the adaptive immune system to mood regulation: mechanisms and pathways of neuroimmune interactions. *Progress in Neuro-Psychopharmacology & Biological Psychiatry*, 79(Pt A), 49–57. https://doi.org/10.1016/j.pnpbp.2016.09.003.

Hickie, I., Hickie, C., Bennett, B., Wakefield, D., Silove, D., Mitchell, P., et al. (1995). Biochemical correlates of in vivo cell-mediated immune dysfunction in patients with depression: a preliminary report. *International Journal of Immunopharmacology*, 17(8), 685–690.

Hickie, I., Hickie, C., Lloyd, A., Silove, D., & Wakefield, D. (1993). Impaired in vivo immune responses in patients with melancholia. *The British Journal of Psychiatry*, 162, 651–657.

Himmerich, H., Milenovic, S., Fulda, S., Plumakers, B., Sheldrick, A. J., Michel, T. M., et al. (2010). Regulatory T cells increased while IL-1beta decreased during antidepressant therapy. *Journal of Psychiatric Research*, 44(15), 1052–1057. https://doi.org/10.1016/j.jpsychires.2010.03.005.

Ho, P. S., Yeh, Y. W., Huang, S. Y., & Liang, C. S. (2015). A shift toward T helper 2 responses and an increase in modulators of innate immunity in depressed patients treated with escitalopram. *Psychoneuroendocrinology*, 53, 246–255. https://doi.org/10.1016/j.psyneuen.2015.01.008.

Hobara, T., Uchida, S., Otsuki, K., Matsubara, T., Funato, H., Matsuo, K., et al. (2010). Altered gene expression of histone deacetylases in mood disorder patients. *Journal of Psychiatric Research*, 44(5), 263–270. https://doi.org/10.1016/j.jpsychires.2009.08.015.

Hong, S., Kim, E. J., Lee, E. J., San Koo, B., Min Ahn, S., Bae, S. H., et al. (2015). TNF-alpha confers resistance to Fas-mediated apoptosis in rheumatoid arthritis through the induction of soluble Fas. *Life Sciences*, 122, 37–41. https://doi.org/10.1016/j.lfs.2014.12.008.

Iliff, J. J., Goldman, S. A., & Nedergaard, M. (2015). Implications of the discovery of brain lymphatic pathways. *Lancet Neurology*, 14(10), 977–979. https://doi.org/10.1016/S1474-4422(15)00221-5.

Infante-Duarte, C., Horton, H. F., Byrne, M. C., & Kamradt, T. (2000). Microbial lipopeptides induce the production of IL-17 in Th cells. *Journal of Immunology*, 165(11), 6107–6115.

Ivanova, S., Semke, V., Vetlugina, T., Rakitina, N., Kudyakova, T., & Simutkin, G. (2007). Signs of apoptosis of immunocompetent cells in patients with depression. *Neuroscience and Behavioral Physiology*, 37(5), 527–530.

Jagger, A., Shimojima, Y., Goronzy, J. J., & Weyand, C. M. (2014). Regulatory T cells and the immune aging process: a mini-review. *Gerontology*, 60(2), 130–137. https://doi.org/10.1159/000355303.

Jun, T. Y., Pae, C. U., Chae, J. H., Bahk, W. M., & Kim, K. S. (2001). Polymorphism of CTLA-4 gene for major depression in the Korean population. *Psychiatry and Clinical Neurosciences, 55*(5), 533–537. https://doi.org/10.1046/j.1440-1819.2001.00901.x.

Karabatsiakis, A., Kolassa, I. T., Kolassa, S., Rudolph, K. L., & Dietrich, D. E. (2014). Telomere shortening in leukocyte subpopulations in depression. *BMC Psychiatry, 14.* https://doi.org/10.1186/1471-244x-14-192.

Kaufmann, F. N., Costa, A. P., Ghisleni, G., Diaz, A. P., Rodrigues, A. L. S., Peluffo, H., et al. (2017). NLRP3 inflammasome-driven pathways in depression: clinical and preclinical findings. *Brain, Behavior, and Immunity, 64,* 367–383. https://doi.org/10.1016/j.bbi.2017.03.002.

Kessler, R. C., & Bromet, E. J. (2013). The epidemiology of depression across cultures. *Annual Review of Public Health, 34,* 119–138. https://doi.org/10.1146/annurev-publhealth-031912-114409.

Kiecolt-Glaser, J. K., Christian, L., Preston, H., Houts, C. R., Malarkey, W. B., Emery, C. F., et al. (2010). Stress, inflammation, and yoga practice. *Psychosomatic Medicine, 72*(2), 113–121. https://doi.org/10.1097/PSY.0b013e3181cb9377.

Kiecolt-Glaser, J. K., Marucha, P. T., Malarkey, W. B., Mercado, A. M., & Glaser, R. (1995). Slowing of wound healing by psychological stress. *Lancet, 346*(8984), 1194–1196.

Kigerl, K. A., de Rivero Vaccari, J. P., Dietrich, W. D., Popovich, P. G., & Keane, R. W. (2014). Pattern recognition receptors and central nervous system repair. *Experimental Neurology, 258,* 5–16. https://doi.org/10.1016/j.expneurol.2014.01.001.

Kipnis, J., Cohen, H., Cardon, M., Ziv, Y., & Schwartz, M. (2004). T cell deficiency leads to cognitive dysfunction: implications for therapeutic vaccination for schizophrenia and other psychiatric conditions. *Proceedings of the National Academy of Sciences of the United States of America, 101*(21), 8180–8185. https://doi.org/10.1073/pnas.0402268101.

Kipnis, J., Mizrahi, T., Hauben, E., Shaked, I., Shevach, E., & Schwartz, M. (2002). Neuroprotective autoimmunity: naturally occurring CD4+CD25+ regulatory T cells suppress the ability to withstand injury to the central nervous system. *Proceedings of the National Academy of Sciences of the United States of America, 99*(24), 15620–15625. https://doi.org/10.1073/pnas.232565399.

Kivisakk, P., Mahad, D. J., Callahan, M. K., Trebst, C., Tucky, B., Wei, T., et al. (2003). Human cerebrospinal fluid central memory CD4+ T cells: evidence for trafficking through choroid plexus and meninges via P-selectin. *Proceedings of the National Academy of Sciences of the United States of America, 100*(14), 8389–8394. https://doi.org/10.1073/pnas.1433000100.

Kronfol, Z., & House, J. (1989). Lymphocyte, mitogenesis, immunoglobulin and complement levels in depressed patients and normal controls. *Acta Psychiatrica Scandinavica, 80*(2), 142–147.

Kunugi, H., Hori, H., & Ogawa, S. (2015). Biochemical markers subtyping major depressive disorder. *Psychiatry and Clinical Neurosciences, 69*(10), 597–608. https://doi.org/10.1111/pcn.12299.

Lanquillon, S., Krieg, J. C., Bening-Abu-Shach, U., & Vedder, H. (2000). Cytokine production and treatment response in major depressive disorder. *Neuropsychopharmacology, 22*(4), 370–379. https://doi.org/10.1016/S0893-133X(99)00134-7.

Lemieux, A., Coe, C. L., & Carnes, M. (2008). Symptom severity predicts degree of T cell activation in adult women following childhood maltreatment. *Brain, Behavior, and Immunity, 22*(6), 994–1003. https://doi.org/10.1016/j.bbi.2008.02.005.

Li, Y., Xiao, B., Qiu, W., Yang, L., Hu, B., Tian, X., et al. (2010). Altered expression of CD4(+)CD25(+) regulatory T cells and its 5-HT(1a) receptor in patients with major depression disorder. *Journal of Affective Disorders, 124*(1-2), 68–75. https://doi.org/10.1016/j.jad.2009.10.018.

Lichtenberg, P., & Belmaker, R. H. (2010). Subtyping major depressive disorder. *Psychotherapy and Psychosomatics, 79*(3), 131–135. https://doi.org/10.1159/000286957.

Liu, J., Li, J., Li, T., Wang, T., Li, Y., Zeng, Z., et al. (2011). CTLA-4 confers a risk of recurrent schizophrenia, major depressive disorder and bipolar disorder in the Chinese Han Population. *Brain, Behavior, and Immunity, 25*(3), 429–433.

Liu, W., Putnam, A. L., Xu-Yu, Z., Szot, G. L., Lee, M. R., Zhu, S., et al. (2006). CD127 expression inversely correlates with Fox P 3 and suppressive function of human CD4+ T reg cells. *The Journal of Experimental Medicine, 203*(7), 1701–1711. https://doi.org/10.1084/jem.20060772.

Liu, Y., Ho, R. C., & Mak, A. (2012). Interleukin (IL)-6, tumour necrosis factor alpha (TNF-alpha) and soluble interleukin-2 receptors (sIL-2R) are elevated in patients with major depressive disorder: a meta-analysis and meta-regression. *Journal of Affective Disorders, 139*(3), 230–239. https://doi.org/10.1016/j.jad.2011.08.003.

Louveau, A., Smirnov, I., Keyes, T. J., Eccles, J. D., Rouhani, S. J., Peske, J. D., et al. (2015). Structural and functional features of central nervous system lymphatic vessels. *Nature, 523*(7560), 337–341. https://doi.org/10.1038/nature14432.

Lowther, D. E., & Hafler, D. A. (2012). Regulatory T cells in the central nervous system. *Immunological Reviews, 248*(1), 156–169. https://doi.org/10.1111/j.1600-065X.2012.01130.x.

Lukic, I., Mitic, M., Djordjevic, J., Tatalovic, N., Bozovic, N., Soldatovic, I., et al. (2014). Lymphocyte levels of redox-sensitive transcription factors and antioxidative enzymes as indicators of pro-oxidative state in depressive patients. *Neuropsychobiology, 70*(1), 1–9.

Maes, M. (1995). Evidence for an immune response in major depression: a review and hypothesis. *Progress in Neuro-Psychopharmacology & Biological Psychiatry, 19*(1), 11–38.

Maes, M., Meltzer, H. Y., Stevens, W., Calabrese, J., & Cosyns, P. (1994). Natural killer cell activity in major depression: relation to circulating natural killer cells, cellular indices of the immune response, and depressive phenomenology. *Progress in Neuro-Psychopharmacology & Biological Psychiatry, 18*(4), 717–730.

Maes, M., Scharpe, S., Meltzer, H. Y., Okayli, G., Bosmans, E., D'Hondt, P., et al. (1994). Increased neopterin and interferon-gamma secretion and lower availability of L-tryptophan in major depression: further evidence for an immune response. *Psychiatry Research, 54*(2), 143–160.

Maes, M., Smith, R., & Scharpe, S. (1995). The monocyte-T-lymphocyte hypothesis of major depression. *Psychoneuroendocrinology, 20*(2), 111–116.

Maes, M., Stevens, W., DeClerck, L., Bridts, C., Peeters, D., Schotte, C., et al. (1992). Immune disorders in depression: higher T helper/T suppressor-cytotoxic cell ratio. *Acta Psychiatrica Scandinavica, 86*(6), 423–431.

Maes, M., Stevens, W. J., Declerck, L. S., Bridts, C. H., Peters, D., Schotte, C., et al. (1993). Significantly increased expression of T-cell activation markers (interleukin-2 and HLA-DR) in depression: further evidence for an inflammatory process during that illness. *Progress in Neuro-Psychopharmacology & Biological Psychiatry, 17*(2), 241–255.

Martinon, F., Burns, K., & Tschopp, J. (2002). The inflammasome: a molecular platform triggering activation of inflammatory caspases and processing of proIL-beta. *Molecular Cell, 10*(2), 417–426.

Meeker, R. B., Williams, K., Killebrew, D. A., & Hudson, L. C. (2012). Cell trafficking through the choroid plexus. *Cell Adhesion & Migration, 6*(5), 390–396. https://doi.org/10.4161/cam.21054.

Mellor, A. L., & Munn, D. H. (2003). Tryptophan catabolism and regulation of adaptive immunity. *Journal of Immunology, 170*(12), 5809–5813.

Miller, A. H. (2010). Depression and immunity: a role for T cells? *Brain, Behavior, and Immunity, 24*(1), 1–8. https://doi.org/10.1016/j.bbi.2009.09.009.

Moller, H. J. (2003). Suicide, suicidality and suicide prevention in affective disorders. *Acta Psychiatrica Scandinavica,* (suppl. 418), 73–80.

Muller, N., Hofschuster, E., Ackenheil, M., Mempel, W., & Eckstein, R. (1993). Investigations of the cellular immunity during depression and the free interval: evidence

for an immune activation in affective psychosis. *Progress in Neuro-Psychopharmacology & Biological Psychiatry, 17*(5), 713–730.

Negi, N., & Das, B. K. (2017). CNS: not an immunoprivilaged site anymore but a virtual secondary lymphoid organ. *International Reviews of Immunology, 37*(1), 57–68. https://doi.org/10.1080/08830185.2017.1357719.

Oertelt-Prigione, S. (2012). The influence of sex and gender on the immune response. *Autoimmunity Reviews, 11*(6-7), A479–485. https://doi.org/10.1016/j.autrev.2011.11.022.

O'Mahony, C., van der Kleij, H., Bienenstock, J., Shanahan, F., & O'Mahony, L. (2009). Loss of vagal anti-inflammatory effect: in vivo visualization and adoptive transfer. *American Journal of Physiology. Regulatory, Integrative and Comparative Physiology, 297*(4), R1118–1126. https://doi.org/10.1152/ajpregu.90904.2008.

Ota, K. T., & Duman, R. S. (2013). Environmental and pharmacological modulations of cellular plasticity: role in the pathophysiology and treatment of depression. *Neurobiology of Disease, 57*, 28–37. https://doi.org/10.1016/j.nbd.2012.05.022.

Pacheco-Lopez, G., Riether, C., Doenlen, R., Engler, H., Niemi, M. B., Engler, A., et al. (2009). Calcineurin inhibition in splenocytes induced by pavlovian conditioning. *The FASEB Journal, 23*(4), 1161–1167. https://doi.org/10.1096/fj.08-115683.

Pandey, G. N., Dwivedi, Y., Rizavi, H. S., Ren, X., Zhang, H., & Pavuluri, M. N. (2010). Brain-derived neurotrophic factor gene and protein expression in pediatric and adult depressed subjects. *Progress in Neuro-Psychopharmacology & Biological Psychiatry, 34*(4), 645–651.

Phillips, A. C., Carroll, D., & Der, G. (2015). Negative life events and symptoms of depression and anxiety: stress causation and/or stress generation. *Anxiety, Stress, and Coping, 28*(4), 357–371. https://doi.org/10.1080/10615806.2015.1005078.

Qu, S., Olafsrud, S. M., Meza-Zepeda, L. A., & Saatcioglu, F. (2013). Rapid gene expression changes in peripheral blood lymphocytes upon practice of a comprehensive-yoga program. *PLoS One, 8*(4). https://doi.org/10.1371/journal.pone.0061910.

Ransohoff, R. M., & Engelhardt, B. (2012). The anatomical and cellular basis of immune surveillance in the central nervous system. *Nature Reviews. Immunology, 12*(9), 623–635. https://doi.org/10.1038/nri3265.

Restifo, N. P., Dudley, M. E., & Rosenberg, S. A. (2012). Adoptive immunotherapy for cancer: harnessing the T cell response. *Nature Reviews. Immunology, 12*(4), 269–281. https://doi.org/10.1038/nri3191.

Rihmer, Z. (2007). Suicide risk in mood disorders. *Current Opinion in Psychiatry, 20*(1), 17–22. https://doi.org/10.1097/YCO.0b013e3280106868.

Ronaldson, A., Gazali, A. M., Zalli, A., Kaiser, F., Thompson, S. J., Henderson, B., et al. (2016). Increased percentages of regulatory T cells are associated with inflammatory and neuroendocrine responses to acute psychological stress and poorer health status in older men and women. *Psychopharmacology, 233*(9), 1661–1668. https://doi.org/10.1007/s00213-015-3876-3.

Rosenberg, S. A., & Restifo, N. P. (2015). Adoptive cell transfer as personalized immunotherapy for human cancer. *Science, 348*(6230), 62–68. https://doi.org/10.1126/science.aaa4967.

Rothermundt, M., Arolt, V., Fenker, J., Gutbrodt, H., Peters, M., & Kirchner, H. (2001). Different immune patterns in melancholic and non-melancholic major depression. *European Archives of Psychiatry and Clinical Neuroscience, 251*(2), 90–97.

Sakaguchi, S., Sakaguchi, N., Asano, M., Itoh, M., & Toda, M. (1995). Immunologic self-tolerance maintained by activated T cells expressing IL-2 receptor alpha-chains (CD25). Breakdown of a single mechanism of self-tolerance causes various autoimmune diseases. *Journal of Immunology, 155*(3), 1151–1164.

Sakaguchi, S., Yamaguchi, T., Nomura, T., & Ono, M. (2008). Regulatory T cells and immune tolerance. *Cell, 133*(5), 775–787. https://doi.org/10.1016/j.cell.2008.05.009.

Sanna, L., Stuart, A. L., Pasco, J. A., Jacka, F. N., Berk, M., Maes, M., et al. (2014). Atopic disorders and depression: findings from a large, population-based study. *Journal of Affective Disorders, 155*, 261–265. https://doi.org/10.1016/j.jad.2013.11.009.

Santori, F. R., Arsov, I., Lilic, M., & Vukmanovic, S. (2002). Editing autoreactive TCR enables efficient positive selection. *Journal of Immunology, 169*(4), 1729–1734.

Scheinert, R. B., Haeri, M. H., Lehmann, M. L., & Herkenham, M. (2016). Therapeutic effects of stress-programmed lymphocytes transferred to chronically stressed mice. *Progress in Neuro-Psychopharmacology & Biological Psychiatry, 70*, 1–7. https://doi.org/10.1016/j.pnpbp.2016.04.010.

Schleifer, S. J., Keller, S. E., Bond, R. N., Cohen, J., & Stein, M. (1989). Major depressive disorder and immunity. Role of age, sex, severity, and hospitalization. *Archives of General Psychiatry, 46*(1), 81–87.

Schmidt, F. M., Schroder, T., Kirkby, K. C., Sander, C., Suslow, T., Holdt, L. M., et al. (2016). Pro- and anti-inflammatory cytokines, but not CRP, are inversely correlated with severity and symptoms of major depression. *Psychiatry Research, 239*, 85–91. https://doi.org/10.1016/j.psychres.2016.02.052.

Schwartz, M. (2003). Macrophages and microglia in central nervous system injury: are they helpful or harmful? *Journal of Cerebral Blood Flow and Metabolism, 23*(4), 385–394.

Schwartz, M., & Baruch, K. (2014). Breaking peripheral immune tolerance to CNS antigens in neurodegenerative diseases: boosting autoimmunity to fight-off chronic neuroinflammation. *Journal of Autoimmunity, 54*, 8–14. https://doi.org/10.1016/j.jaut.2014.08.002.

Schwartz, M., & Kipnis, J. (2005). Protective autoimmunity and neuroprotection in inflammatory and noninflammatory neurodegenerative diseases. *Journal of the Neurological Sciences, 233*(1-2), 163–166. https://doi.org/10.1016/j.jns.2005.03.014.

Schwartz, M., & Kipnis, J. (2011). A conceptual revolution in the relationships between the brain and immunity. *Brain, Behavior, and Immunity, 25*(5), 817–819. https://doi.org/10.1016/j.bbi.2010.12.015.

Schwartz, M., & Shechter, R. (2010). Protective autoimmunity functions by intracranial immunosurveillance to support the mind: the missing link between health and disease. *Molecular Psychiatry, 15*(4), 342–354. https://doi.org/10.1038/mp.2010.31.

Seabrook, T. J., Johnston, M., & Hay, J. B. (1998). Cerebral spinal fluid lymphocytes are part of the normal recirculating lymphocyte pool. *Journal of Neuroimmunology, 91*(1-2), 100–107.

Segerstrom, S. C., & Miller, G. E. (2004). Psychological stress and the human immune system: a meta-analytic study of 30 years of inquiry. *Psychological Bulletin, 130*(4), 601–630. https://doi.org/10.1037/0033-2909.130.4.601.

Seidel, A., Arolt, V., Hunstiger, M., Rink, L., Behnisch, A., & Kirchner, H. (1996). Major depressive disorder is associated with elevated monocyte counts. *Acta Psychiatrica Scandinavica, 94*(3), 198–204.

Shaked, I., Porat, Z., Gersner, R., Kipnis, J., & Schwartz, M. (2004). Early activation of microglia as antigen-presenting cells correlates with T cell-mediated protection and repair of the injured central nervous system. *Journal of Neuroimmunology, 146*(1-2), 84–93.

Simon, N. M., Walton, Z., Prescott, J., Hoge, E., Keshaviah, A., Bui, T. H. E., et al. (2013). A cross-sectional examination of telomere length and telomerase in a well-characterized sample of individuals with major depressive disorder compared to controls. *Neuropsychopharmacology, 38*, S322–S323.

Singhal, G., Jaehne, E. J., Corrigan, F., Toben, C., & Baune, B. T. (2014). Inflammasomes in neuroinflammation and changes in brain function: a focused review. *Frontiers in Neuroscience, 8*. https://doi.org/10.3389/fnins.2014.00315.

Sluzewska, A., Sobieska, M., & Rybakowski, J. K. (1997). Changes in acute-phase proteins during lithium potentiation of antidepressants in refractory depression. *Neuropsychobiology, 35*(3), 123–127.

Smith, R. S. (1991). The macrophage theory of depression. *Medical Hypotheses, 35*(4), 298–306.

Sternberg, E. M. (2006). Neural regulation of innate immunity: a coordinated nonspecific host response to pathogens. *Nature Reviews. Immunology*, 6(4), 318–328. https://doi.org/10.1038/nri1810.

Suzuki, H., Savitz, J., Kent Teague, T., Gandhapudi, S. K., Tan, C., Misaki, M., et al. (2017). Altered populations of natural killer cells, cytotoxic T lymphocytes, and regulatory T cells in major depressive disorder: association with sleep disturbance. *Brain, Behavior, and Immunity*, 66, 193–200. https://doi.org/10.1016/j.bbi.2017.06.011.

Tian, L., Ma, L., Kaarela, T., & Li, Z. (2012). Neuroimmune crosstalk in the central nervous system and its significance for neurological diseases. *Journal of Neuroinflammation*, 9. https://doi.org/10.1186/1742-2094-9-155.

Toben, C., & Baune, B. T. (2015). An act of balance between adaptive and maladaptive immunity in depression: a role for T lymphocytes. *Journal of Neuroimmune Pharmacology*, 10(4), 595–609. https://doi.org/10.1007/s11481-015-9620-2.

Vadasz, Z., Haj, T., Kessel, A., & Toubi, E. (2013). Age-related autoimmunity. *BMC Medicine*, 11. https://doi.org/10.1186/1741-7015-11-94.

Vallejos, C. A., Richardson, S., & Marioni, J. C. (2016). Beyond comparisons of means: understanding changes in gene expression at the single-cell level. *Genome Biology*, 17. https://doi.org/10.1186/s13059-016-0930-3.

Verhoeven, J. E., Revesz, D., Epel, E. S., Lin, J., Wolkowitz, O. M., & Penninx, B. W. (2014). Major depressive disorder and accelerated cellular aging: results from a large psychiatric cohort study. *Molecular Psychiatry*, 19(8), 895–901. https://doi.org/10.1038/mp.2013.151.

Vukovic, J., Colditz, M. J., Blackmore, D. G., Ruitenberg, M. J., & Bartlett, P. F. (2012). Microglia modulate hippocampal neural precursor activity in response to exercise and aging. *The Journal of Neuroscience*, 32(19), 6435–6443. https://doi.org/10.1523/JNEUROSCI.5925-11.2012.

Walunas, T. L., Lenschow, D. J., Bakker, C. Y., Linsley, P. S., Freeman, G. J., Green, J. M., et al. (1994). CTLA-4 can function as a negative regulator of T cell activation. *Immunity*, 1(5), 405–413.

Wang, Y., Lavender, P., Watson, J., Arno, M., & Lehner, T. (2015). Stress-activated dendritic cells (DC) induce dual interleukin (IL)-15- and IL1beta-mediated pathways, which may elicit CD4+ memory T cells and interferon (IFN)-stimulated genes. *The Journal of Biological Chemistry*, 290(25), 15595–15609. https://doi.org/10.1074/jbc.M115. 645754.

Webster Marketon, J. I., & Glaser, R. (2008). Stress hormones and immune function. *Cellular Immunology*, 252(1-2), 16–26. https://doi.org/10.1016/j.cellimm.2007.09.006.

Wodarz, N., Rupprecht, R., Kornhuber, J., Schmitz, B., Wild, K., Braner, H., et al. (1991). Normal lymphocyte responsiveness to lectins but impaired sensitivity to in vitro glucocorticoids in major depression. *Journal of Affective Disorders*, 22(4), 241–248.

Wolf, S. A., Steiner, B., Akpinarli, A., Kammertoens, T., Nassenstein, C., Braun, A., et al. (2009). CD4-positive T lymphocytes provide a neuroimmunological link in the control of adult hippocampal neurogenesis. *Journal of Immunology*, 182(7), 3979–3984. https://doi.org/10.4049/jimmunol.0801218.

Wong, M., Dong, C., Maestre-Mesa, J., & Licinio, J. (2008). Polymorphisms in inflammation-related genes are associated with susceptibility to major depression and antidepressant response. *Molecular Psychiatry*, 13(8), 800–812.

Xing, Y., & Hogquist, K. A. (2012). T-cell tolerance: central and peripheral. *Cold Spring Harbor Perspectives in Biology*, 4(6), 2–15. https://doi.org/10.1101/cshperspect.a006957.

Xu, D., Liu, H., Komai-Koma, M., Campbell, C., McSharry, C., Alexander, J., et al. (2003). CD4+CD25+ regulatory T cells suppress differentiation and functions of Th1 and Th2 cells, Leishmania major infection, and colitis in mice. *Journal of Immunology*, 170(1), 394–399.

Yu, B. H., Dimsdale, J. E., & Mills, P. J. (1999). Psychological states and lymphocyte beta-adrenergic receptor responsiveness. *Neuropsychopharmacology*, 21(1), 147–152. https://doi.org/10.1016/S0893-133X(98)00133-X.

Ziv, Y., & Schwartz, M. (2008). Immune-based regulation of adult neurogenesis: implications for learning and memory. *Brain, Behavior, and Immunity*, 22(2), 167–176. https://doi.org/10.1016/j.bbi.2007.08.006.

Zivin, K., Yosef, M., Miller, E. M., Valenstein, M., Duffy, S., Kales, H. C., et al.Kim, H. M., (2015). Associations between depression and all-cause and cause-specific risk of death: a retrospective cohort study in the Veterans Health Administration. *Journal of Psychosomatic Research*, 78(4), 324–331. https://doi.org/10.1016/j.jpsychores.2015.01.014.

Zorrilla, E. P., Luborsky, L., McKay, J. R., Rosenthal, R., Houldin, A., Tax, A., et al. (2001). The relationship of depression and stressors to immunological assays: a meta-analytic review. *Brain, Behavior, and Immunity*, 15(3), 199–226. https://doi.org/10.1006/brbi.2000.0597.

Do Chemokines Have a Role in the Pathophysiology of Depression?

Gaurav Singhal, Bernhard T. Baune

University of Adelaide, Adelaide, SA, Australia

INTRODUCTION

Depression is a psychiatric syndrome characterized by a state of low mood, losing the sense of self, sadness, irritability, and loss of interest in all activities and events (Belmaker & Agam, 2008). Patients with MDD are vulnerable to develop life-threatening metabolic disorders such as type II diabetes (Holt, De Groot, & Golden, 2014; Katon et al., 2005), cancer (Krebber et al., 2014), cardiovascular disease (Air, Tully, Sweeney, & Beltrame, 2016; Glassman, 2008; Knol et al., 2006), and chronic obstructive pulmonary disease (Smith & Wrobel, 2014), as well as central nervous system (CNS) disorders, such as Alzheimer's disease (AD) (Chi et al., 2015) and multiple sclerosis (MS) (Théaudin & Feinstein, 2015). The underlying external causes of depression are many, ranging from psychosocioeconomic (Lorant et al., 2003) to substance abuse (Renner & Ciraulo, 1994). Additionally, age and gender differences can also determine the vulnerability to depression. For example, men are more prone to acquiring depressive disorder at an old age and have low socioeconomic support than women (Sonnenberg et al., 2013; Van Deurzen, Van Ingen, & Van Oorschot, 2015).

However, several intertwined mechanistic molecular pathways have also been proposed for the onset of depression, especially in old age. Chemokines, which are small chemotactic immune proteins that attract and draw in various immune cells, such as monocytes, natural killer (NK) cells, and T and B lymphocytes to the site of inflammation, have been implicated in altering these pathways (Fig. 1). These pathways include neuroinflammation with resultant neurodegeneration, impaired neurogenesis, alteration in the level of neurotransmitters, dysregulation in the hypothalamic-pituitary-adrenal (HPA) axis functioning, and irregularity in endocrine response (Di Prisco et al., 2013; Guyon, 2014; Heinisch & Kirby, 2010; Iwata, Ota, & Duman, 2013; Réaux-Le Goazigo, Van Steenwinckel, Rostène, & Parsadaniantz, 2013; Rostène, Kitabgi, & Parsadaniantz, 2007; Skrzydelski et al., 2007; Verburg-van Kemenade, Van der Aa, & Chadzinska, 2013).

Studies have shown that neuroinflammation in response to psychological and physical stressors can result in depression (Brites & Fernandes, 2015; Kim, Na, Myint, & Leonard, 2016). Furthermore, the peripheral infection can stimulate the innate immune system

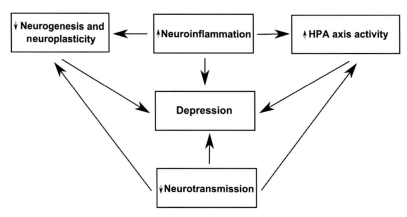

FIG. 1   The above illustration shows various intertwined mechanistic molecular pathways proposed for the onset of depression. HPA, hypothalamic-pituitary-adrenal.

exacerbating neuroinflammatory behavior and depressive-like behavior (Godbout et al., 2008). Neuroinflammation when chronic can result in impaired hippocampal neurogenesis, and this has been found to be correlated with neuroinflammation-induced depression (Tang, Lin, Pan, Guan, & Li, 2016). Several chemokines, such as CCL2, CCL7, CCL8, and CXCL1–11, have been shown to elicit dysregulated functioning with their level increasing in serum and cerebrospinal fluid during neuroinflammation and associated depression (Eyre et al., 2016; Murphy et al., 2000; Ramesh, MacLean, & Philipp, 2013; Stuart & Baune, 2014; Yamagami et al., 1999).

Impaired hippocampal neurogenesis has been long held responsible for the development of depressive-like symptoms, especially in old age (Eyre & Baune, 2012; Kim et al., 2016; Kubera, Obuchowicz, Goehler, Brzeszcz, & Maes, 2011; Tang et al., 2016). Downregulation of the BDNF in the hippocampus (Duman, 2004) further impairs neurogenesis (Kubera et al., 2011). Reduction in the hippocampal volume has a direct correlation with depression (Campbell, Marriott, Nahmias, & MacQueen, 2004; Sheline, Gado, & Kraemer, 2003; Videbech & Ravnkilde, 2004), with the course of illness determining the magnitude of loss in

hippocampal volume (MacQueen et al., 2003; Sheline, Sanghavi, Mintun, & Gado, 1999). The reduced hippocampal volume may dysregulate HPA axis affecting stress response and add to the vulnerability of acquiring depressive disorder (Frodl & O'Keane, 2013). Chemokines such as CXCL12, CXCL14, and CX3CL are important during the early developmental processes, migration of the neural stem/progenitor cells, and maintenance of the hippocampal integrity (Banisadr, Bhattacharyya, et al., 2011; Schönemeier et al., 2008; Sheridan et al., 2014; Tran, Banisadr, Ren, Chenn, & Miller, 2007). On the contrary, CCL11 negatively regulates hippocampal integrity and hence could be responsible for the aging-associated development of depression (Villeda et al., 2011).

Dysregulated neurotransmission pathways, such as those of norepinephrine, serotonin, glutamate, gamma-aminobutyric acid (GABA), and dopamine, in the brain have been shown to result in depression (Choudary et al., 2005; Coppen, 1967; Lambert, Johansson, Ågren, & Friberg, 2000; Santiago et al., 2014). Patients with major depressive disorder demonstrated significant downregulation of the transporter proteins SLC1A2 and SLC1A3 and decreased expression of L-glutamate-ammonia ligase, the enzyme that converts glutamate to nontoxic glutamine.

This resulted in elevated levels of extracellular glutamate affecting glutamate signaling (Choudary et al., 2005). Similarly, alterations of GABA interneurons and inhibitory neurotransmission might have a role to play in the development of stress-related major depression (Ghosal, Hare, & Duman, 2017). The deficiency of norepinephrine and dopamine has also been observed in the brain of depressed patients (Espay, LeWitt, & Kaufmann, 2014; Lambert et al., 2000). Chemokines such as CCL2, CCL5, CCL21, and CXCL12 have been shown to modulate serotonin, dopamine, acetylcholine, glutamate, and GABA neurotransmissions (Guyon, 2014; Heinisch & Kirby, 2010; Melik-Parsadaniantz & Rostene, 2008; Rostene, Dansereau, et al., 2011; Rostène et al., 2007), plausibly leading to depression through altering these pathways.

There is accumulating evidence suggesting that developmental changes in HPA axis functioning during childhood and adolescence could subsequently result in the onset of MDD (Apter-Levi et al., 2016; Colich, Kircanski, Foland-Ross, & Gotlib, 2015; Lopez-Duran et al., 2015). In addition, dysregulation of neurogenesis and reduction in the hippocampal volume after chronic stress can adversely affect the HPA axis functioning (Frodl & O'Keane, 2013), leading to excess production of cortisol (Schloesser, Manji, & Martinowich, 2009). Hyperactivity of HPA axis and hypercortisolemia are established etiologies of depression (Belanoff, Kalehzan, Sund, Fleming Ficek, & Schatzberg, 2001; Colla et al., 2007; Jokinen & Nordström, 2009). Furthermore, glucocorticoid-mediated reduction in tryptophan hydroxylase isoform (TPH2) has shown to impair neuroimmune regulation and hence could be relevant as an etiology for depression (Clark, Pai, Flick, & Rohrer, 2005). Recent literature has started pointing to the role of chemokines, such as CCL2, CCL3, CCL5, CXCL8, and CXCL12, in influencing the HPA axis activity and neuroendocrine functions (Besedovsky, Del Rey, Sorkin, & Dinarello, 1986; Rostene, Guyon, et al., 2011; Verburg-van Kemenade et al., 2013), suggestive of the possible role of chemokines in causing depression through alteration of HPA axis pathway.

In this chapter, we will first discuss the established biological roles of various chemokines in the modulation of CNS functions. In the end, we will discuss the proved and plausible pathways of chemokine action in the pathophysiology of depression.

WHAT ARE CHEMOKINES?

Chemokines are low-molecular-weight proteins that function as both chemotactic and adhesion molecules. They are highly specific chemoattractant, interact with their explicit receptors, and draw in well-defined leukocyte subsets at the sites of injury and inflammation. The term "chemokine" was first coined in 1992 for a bunch of small proteins with chemotactic functions (Murphy et al., 2000). Since then, numerous chemokines have been identified, structurally and functionally analyzed, and characterized by various researchers. All, except CX3CL1, which is a membrane-bound protein, are present in a secreted soluble form. Chemokines are activated and aid in the chemotaxis of the leukocyte subset under both physiological and pathological conditions after binding with their receptors present on leukocytes and have multiple ligands with a variable affinity (Cyster, 1999; Murphy et al., 2000). Chemokines act through common intracellular signaling mechanisms to increase intracellular calcium (Nelson & Gruol, 2004), and their direct signaling is through G-protein-coupled receptors (Baggiolini, Dewald, & Moser, 1997). Chemokines have been shown to induce the release of pro-inflammatory mediators and control of T-helper (T_h)-1/T_h-2 phenotypic polarization (Cyster, 1999).

In constitutive levels, chemokines are always expressed in certain organs and tissues and are required for basal immune cell migration to the local draining lymph nodes, where they activate more antigen-specific T cells. However, when under the influence of pro-inflammatory factors, such as LPS, TNF-α, and IL-1β, these proteins upregulate the inflammatory response by attracting immune cells (e.g., macrophages, fibroblasts, T cells, and NK cells) to the site of injury and inflammation (Groves & Jiang, 1995; Renner, Ivey, Redmann, Lackner, & MacLean, 2011; Rossi & Zlotnik, 2000; Zlotnik & Yoshie, 2000).

NOMENCLATURE OF CHEMOKINES

As the acronym suggests, CC chemokine subfamily comprises cysteine amino acids adjacent to each other and next to the amino termini. There are 27 known CC chemokine ligands (CCL) in mammals, numbered from 1 to 28 and where CCL9 and CCL10 are the same. Members of the subfamily may contain four or six cysteines between the amino termini and hence termed as C4-CC chemokines or C6-CC chemokines, respectively. While CCLs 1, 15, 21, and 23 are C6-CC, others are C4-CC.

CXC chemokines have an amino acid between the two cysteine residues adjoining amino termini. There are 17 known CXC chemokine ligands (CXCL) in mammals, numbered from 1 to 17. These CXCLs are categorized into ELR-positive or ELR-negative (ELR represents glutamic acid, leucine, and arginine amino acids in biochemistry). CXC chemokines with ELR motif before the first cysteine are ELR-positive, and they specifically induce migration of neutrophils. Unlike ELR-positive CXC chemokines, ELR-negative CXC chemokines have no ELR motif, and they specifically induce migration of lymphocytes.

CX3C chemokine, as the name suggests, contain three amino acids between the cysteine residues adjoining N- and C-termini, and there is only one CX3C chemokine known so far, CX3CL1. It is the only chemokine that is both membrane-bound and secreted in soluble form, thus acting as chemoattractant and adhesion molecules (Table 1; Jones, Beamer, & Ahmed, 2010; Stuart, Singhal, & Baune, 2015).

While some chemokines, such as CCL2, CCL3, CCL19, CCL21, CXCL8, CXCL12, and CX3CL1, are expressed in CNS in constitutive levels under physiological conditions, others are upregulated during pathological conditions, for example, at times of injury and inflammation (Jaerve & Müller, 2012).

STRUCTURE OF CHEMOKINES AND CHEMOKINES RECEPTORS, AND LIGAND EXPRESSION IN CNS

All chemokines have a tertiary structure, composed of a variable N-terminal signaling domain, followed by a regulated core domain consisting of "N Loop," a three-stranded β-sheet and a C-terminal helix. Most CXC chemokines form dimers where two three-stranded β-sheets interact with each other forming a six-stranded β-sheet platform with two C-terminal helixes (Kufareva, Salanga, & Handel, 2015). Conversely, CC chemokines have an elongated dimeric structure with two or three dicysteine sequences, where two cysteine residues adjoining amino ends interact with each other (Kufareva et al., 2015). The only CX3C chemokine, CX3CL1, or fractalkine can either exist in monomeric form (Mizoue, Bazan, Johnson, & Handel, 1999) or form dimers through an intermolecular β-sheet forming compact quaternary structure characteristic of CX3C chemokine (Hoover, Mizoue, Handel, & Lubkowski, 2000).

The receptors of chemokines are structurally similar. They are comparable in size with around 350 amino acids each, membrane-bound molecules, composed of 7-transmembrane

TABLE 1 Distribution of Chemokine Receptors Within and Outside the CNS

Chemokine	Cellular Distribution Within CNS	Cellular Distribution Outside CNS	References
CCR1	Mg, As, O, Ne, NSC/NPC	N, M, T, NK, B, Ms	(Eltayeb et al., 2007; Kan et al., 2012; Meucci et al., 1998; Murdoch & Finn, 2000; Nguyen et al., 2003; Stuart et al., 2015; Tran et al., 2007)
CCR2	Mg, As, Ne, NSC/NPC	M, T, B, Bs	(Banisadr, Gosselin, Mechighel, Rostène, et al., 2005; Banisadr, Queraud-Lesaux, et al., 2002; Murdoch & Finn, 2000; Stuart et al., 2015; Tran et al., 2007)
CCR3	Mg, As, O, Ne, NSC/NPC	Eo, Bs, T	(Flynn, Maru, Loughlin, Romero, & Male, 2003; Krathwohl & Kaiser, 2004; Murdoch & Finn, 2000; Stuart et al., 2015; van der Meer, Ulrich, Gonźalez-Scarano, & Lavi, 2000)
CCR4	Mg, As, Ne	T, P	(Flynn et al., 2003; Meucci et al., 1998; Murdoch & Finn, 2000; Stuart et al., 2015)
CCR5	Mg, As, NSC/NPC	T, M, DC	(Eltayeb et al., 2007; Ji, He, Dheen, & Tay, 2004; Kan et al., 2012; Murdoch & Finn, 2000; Nguyen et al., 2003; Spleiss, Appel, Boddeke, Berger, & Gebicke-Haerter, 1998; Stuart et al., 2015; Tran et al., 2007)
CCR6	Mg, As	T, B, DC	(Coughlan et al., 2000; Flynn et al., 2003; Murdoch & Finn, 2000; Stuart et al., 2015)
CCR7	Mg, As, Ne	T, B, DC	(Dijkstra, de Haas, Brouwer, Boddeke, & Biber, 2006; Gomez-Nicola, Pallas-Bazarra, Valle-Argos, & Nieto-Sampedro, 2010; Liu, Cao, Tang, Liu, & Tang, 2007; Murdoch & Finn, 2000; Stuart et al., 2015)
CCR8	Mg, As, Ne	M, thymus	(Liu et al., 2007; Murdoch & Finn, 2000; Stuart et al., 2015; Trebst et al., 2003)
CCR9	Mg, As, Ne	T, thymus	(de Haas, Boddeke, & Biber, 2008; Liu et al., 2007; Murdoch & Finn, 2000; Stuart et al., 2015)
CCR10	As, Ne	Placenta, liver	(Flynn et al., 2003; Liu et al., 2007; Murdoch & Finn, 2000; Stuart et al., 2015)
CXCR1	Mg, As, O, Ne, NSC/NPC	N, M, T, NK, Bs, Ms., En	(Flynn et al., 2003; Murdoch & Finn, 2000; Omari, John, Sealfon, & Raine, 2005; Puma, Danik, Quirion, Ramon, & Williams, 2001; Stuart et al., 2015; Weiss et al., 2010)
CXCR2	Mg, As, O, Ne	N, M, T, NK, Ms., En	(Flynn et al., 2003; Giovannelli et al., 1998; Murdoch & Finn, 2000; Omari et al., 2005; Stuart et al., 2015; Weiss et al., 2010)
CXCR3	Mg, As, O, Ne, NSC/NPC	Activated T	(Coughlan et al., 2000; Flynn et al., 2003; Murdoch & Finn, 2000; Omari et al., 2005; Stuart et al., 2015; Tran et al., 2007)
CXCR4	Mg, As, O, Ne, NSC/NPC	Myeloid, T, B, Ep, En, DC	(Banisadr, Fontanges, et al., 2002; Banisadr, Frederick, et al., 2011; Gottle et al., 2010; Murdoch & Finn, 2000; Stuart et al., 2015; Tran et al., 2007)
CXCR5	Mg, As, Ne, NSC/NPC	B	(Bagaeva, Rao, Powers, & Segal, 2006; Flynn et al., 2003; Murdoch & Finn, 2000; Petito, Roberts, Cantando, Rabinstein, & Duncan, 2001; Stuart et al., 2015; Weiss et al., 2010)

Continued

TABLE 1 Distribution of Chemokine Receptors Within and Outside the CNS—cont'd

Chemokine	Cellular Distribution Within CNS	Cellular Distribution Outside CNS	References
CXCR6	–	Activated T	(Stuart et al., 2015; Wilbanks et al., 2001)
CXCR7	As, O, Ne, NSC/NPC	–	(Gottle et al., 2010; Murdoch & Finn, 2000; Schönemeier et al., 2008; Stuart et al., 2015)
CXCL14 (unidentified)	NSC/NPC	–	(Banisadr, Bhattacharyya, et al., 2011; Stuart et al., 2015)
CX3CR1	Mg, As, Ne, NSC/NPC	Fractalkine, M, T	(Ji et al., 2004; Meucci et al., 1998; Murdoch & Finn, 2000; Stuart et al., 2015; Sunnemark et al., 2005)

N, neutrophil; M, monocyte/macrophage; T, T lymphocyte; B, B lymphocyte; NK, natural killer cell; Eo, eosinophil; Bs, basophil; Ms., mast cell; As, astrocyte; Nn, neurone; P, platelet; En, endothelial cell; Ep, epithelial cell; Hp, hepatocyte; DC, dendritic cell; Mg, microglia; As, astrocyte; O, oligodendrocyte; Ne, neuron; NSC/NPC, neural stem/progenitor cells.

domains, and coupled to G-proteins. They have a short extracellular acidic N-terminus, seven α-helical transmembrane domains with three intracellular and three extracellular hydrophilic loops, and the intracellular C-terminus. The N-terminus contains N-linked glycosylation sites that determine ligand-binding specificity. The C-terminus contains serine and threonine residues that act as phosphorylation sites for receptor regulation after activation. Extracellular loops 1 and 2 form a disulfide bridge through highly conserved cysteine residues. G-proteins are coupled at the C-terminus segment and possibly through the third intracellular loop (Murdoch & Finn, 2000). There are 10 known CC receptors (CCR1–10), 8 known CXC receptors (CXCR1–7 and 14), and 1 known CX3C receptor (CX3CR1) (Table 1; Murdoch & Finn, 2000; Stuart et al., 2015).

Both the developing and adult CNS have a wide range of chemokine receptors and their ligands (Bajetto, Bonavia, Barbero, Florio, & Schettini, 2001; Jaerve & Müller, 2012; Miller et al., 2008; Rostene, Dansereau, et al., 2011). Although direct signaling of chemokines is through G-protein-coupled receptors, these may be linked to ionotropic receptors and other intracellular signaling pathways that converge to increase intracellular calcium—indicating a potential role in contributing to synaptic plasticity or indeed to excitotoxicity. However, these functional roles remain speculative (Rostene, Dansereau, et al., 2011; Rostène et al., 2007).

BIOLOGICAL ROLE OF VARIOUS CHEMOKINES IN CNS MODULATION

CC Chemokines in CNS

CC chemokines attract various cell sorts, such as monocytes, NK cells, T cells, eosinophils, basophils, and DC, aiding in their movement to the locales of injury and inflammation (Table 2). CCL1, also known as I-309 in humans and TCA-3 in mice, is activated after its interaction with cell-surface chemokine receptor CCR8. The receptor CCR8 has been shown to be associated with phagocytic macrophages and activated microglia in MS lesions and directly correlated with demyelinating activity (Trebst et al., 2003). Expressions of CCL1 and CCR8 mRNA in the CNS of mice with experimental autoimmune encephalomyelitis (EAE) (Fischer et al., 2000; Godiska, Chantry, Dietsch, & Gray, 1995)

TABLE 2 Biological Characteristics of the CC Chemokines in CNS Functions

CC Chemokine and Synonyms	Receptor (s)	CNS Functions	Classical Peripheral Functions	References
CCL1 (Cytokine I-309, TCA-3)	CCR8	Microglial chemotaxis	mø, B Lø, NK, DC chemotaxis, transient increase in Ca levels in mø, and macrophages	(Devi, Laning, Luo, & Dorf, 1995; Le, Zhou, Iribarren, & Wang, 2004; Miller & Krangel, 1992; Murdoch & Finn, 2000; Ono et al., 2003; Roos et al., 1997; Stuart et al., 2015; Trebst et al., 2003)
CCL2 (MCP-1)	CCR2	NSC/NPC chemotaxis and differentiation, microglial phenotype modulation, HPA axis modulation	mø, T Lø, DC chemotaxis and activation	(Cazareth, Guyon, Heurteaux, Chabry, & Petit-Paitel, 2014; Le et al., 2004; Murdoch & Finn, 2000; Ono et al., 2003; Stuart et al., 2015)
CCL3 (MIP-1α)	CCR1	NSC/NPC, microglial chemotaxis	nø chemotaxis and activation	(Le et al., 2004; Marciniak et al., 2015; Murdoch & Finn, 2000; Ono et al., 2003; Stuart et al., 2015)
CCL4 (MIP-1β)	CCR1, CCR5	Microglial and CD4+ T cell subset chemotaxis	mø, T Lø, NK chemotaxis	(Le et al., 2004; Murdoch & Finn, 2000; Ono et al., 2003; Quandt & Dorovini-Zis, 2004; Stuart et al., 2015)
CCL5 (RANTES)	CCR5	Microglial chemotaxis, HPA axis modulation	T Lø, bø, eø chemotaxis and activation	(Le et al., 2004; Murdoch & Finn, 2000; Ono et al., 2003; Stuart et al., 2015)
CCL6 (C10, MRP-2)	CCR1	Astrocytes migration	mø, nø chemotaxis	(Le et al., 2004; Li et al., 2013; Murdoch & Finn, 2000; Ono et al., 2003; Stuart et al., 2015)
CCL7 (MARC, MCP-3)	CCR2	mø migration across BBB	mø, T Lø chemotaxis	(Le et al., 2004; Murdoch & Finn, 2000; Ono et al., 2003; Renner et al., 2011; Stuart et al., 2015; Torres et al., 2013)
CCL8 (MCP-2)	CCR1, CCR5, CCR2B	Unknown	Mast cells, bø, eø, mø, T Lø, NK chemotaxis and activation	(Le et al., 2004; Murdoch & Finn, 2000; Ono et al., 2003; Stuart et al., 2015)
CCL9/CCL10 (MRP-2, CCF18, MP-1)	CCR1	Unknown	DC chemotaxis	(Le et al., 2004; Murdoch & Finn, 2000; Ono et al., 2003; Stuart et al., 2015)
CCL11 (eotaxin)	CCR2, CCR3, CCR5	NSC/NPC chemotaxis impairs neurogenesis	eø, bø chemotaxis	(Le et al., 2004; Murdoch & Finn, 2000; Ono et al., 2003; Stuart et al., 2015; Wang et al., 2017)
CCL12 (MCP-5)	CCR2	Unknown	eø, mø, T Lø chemotaxis	(Le et al., 2004; Murdoch & Finn, 2000; Ono et al., 2003; Stuart et al., 2015)
CCL13 (MCP-4, NCC-1)	CCR2, CCR3, CCR5	Unknown	mø, T Lø, bø, eø chemotaxis	(Le et al., 2004; Murdoch & Finn, 2000; Ono et al., 2003; Stuart et al., 2015)
CCL14 (HCC-1, MCIF, NCC-2)	CCR1	Unknown	Activates mø	(Le et al., 2004; Murdoch & Finn, 2000; Ono et al., 2003; Stuart et al., 2015)

Continued

TABLE 2   Biological Characteristics of the CC Chemokines in CNS Functions—cont'd

CC Chemokine and Synonyms	Receptor (s)	CNS Functions	Classical Peripheral Functions	References
CCL15 (leukotactin-1, MIP-5, HCC-2, NCC-3)	CCR1, CCR3	Unknown	mø, T Lø, nø chemotaxis	(Le et al., 2004; Murdoch & Finn, 2000; Ono et al., 2003; Stuart et al., 2015)
CCL16 (LEC, NCC-4, LMC)	CCR1, CCR2, CCR5, CCR8	Unknown	mø, T Lø chemotaxis	(Le et al., 2004; Murdoch & Finn, 2000; Ono et al., 2003; Stuart et al., 2015)
CCL17 (TARC, dendrokine, ABCD-2)	CCR4	Unknown	T Lø chemotaxis	(Le et al., 2004; Murdoch & Finn, 2000; Ono et al., 2003; Stuart et al., 2015)
CCL18 (MIP-4, PARC, DC-CK1, AMAC-1)	GPR30	Unknown	T Lø chemotaxis	(Le et al., 2004; Murdoch & Finn, 2000; Ono et al., 2003; Stuart et al., 2015)
CCL19 (ELC, Exodus-3, Ckβ11)	CCR7	Unknown	DC, T, and B Lø	(Le et al., 2004; Murdoch & Finn, 2000; Ono et al., 2003; Stuart et al., 2015)
CCL20 (LARC, Exodus-1, Ckβ4)	CCR6	T Lø chemotaxis	T Lø, nø, DC chemotaxis	(Arima et al., 2012; Le et al., 2004; Murdoch & Finn, 2000; Ono et al., 2003; Stuart et al., 2015)
CCL21 (SLC, Exodus-2, Ckβ9, TCA-4, 6Ckine)	CCR7	Unknown	T Lø chemotaxis	(Le et al., 2004; Murdoch & Finn, 2000; Nagira et al., 1997; Ono et al., 2003; Stuart et al., 2015)
CCL22 (MDC)	CCR4	Unknown	mø, DC, and NK chemotaxis	(Godiska et al., 1997; Le et al., 2004; Murdoch & Finn, 2000; Ono et al., 2003; Stuart et al., 2015)
CCL23 (MPIF-1, Ckβ8, MIP-3)	CCR1	Unknown	T Lø, mø, nø chemotaxis	(Le et al., 2004; Murdoch & Finn, 2000; Ono et al., 2003; Stuart et al., 2015)
CCL24 (eotaxin-2, MPIF-2, Ckβ6)	CCR3	Unknown	T Lø, eø, bø, nø chemotaxis	(Le et al., 2004; Murdoch & Finn, 2000; Ono et al., 2003; Stuart et al., 2015)
CCL25 (TECK, Ckβ15)	CCR9	Unknown	mø, T Lø, DC, thymocyte chemotaxis	(Le et al., 2004; Murdoch & Finn, 2000; Ono et al., 2003; Stuart et al., 2015)
CCL26 (eotaxin-3, MIP-4α, IMAC, TSC-1)	CCR3	Unknown	eø, bø chemotaxis and activation	(Le et al., 2004; Murdoch & Finn, 2000; Ono et al., 2003; Stuart et al., 2015)
CCL27 (CTACK, ILC, PESKY, eskine, skinkine)	CCR10	Unknown	T Lø chemotaxis	(Le et al., 2004; Murdoch & Finn, 2000; Ono et al., 2003; Stuart et al., 2015)
CCL28 (MEC)	CCR3, CCR10	Unknown	T and B Lø, eø, IgA expressing cells	(Le et al., 2004; Murdoch & Finn, 2000; Ono et al., 2003; Stuart et al., 2015)

mø, monocyte/macrophage; Lø, lymphocyte; nø, neutrophil; bø, basophil; eø, eosinophil; DC, dendritic cell; NK, natural killer cell; NSC/NPC, neural stem/progenitor cell; PIC, peripheral immune cell; HPA axis, hypothalamus-pituitary-adrenal axis.

additionally propose that CCL1 has a part to play in various CNS functions and disorders and could indeed be associated with neuroinflammatory disorders. Alongside CCL2, CCL3, and CCL4, CCL1's presence has been noted in evolving brain abscesses, most likely leading to the influx of lymphocytes and monocytes and hence in the development of adaptive immune response (Kielian, Barry, & Hickey, 2001).

CCL2 is the best-known CC chemokine, otherwise called as monocyte chemoattractant protein-1 (MCP-1), which attracts monocytes expressing CCR2 receptor in the circulatory system helping them to enter the encompassing injured and inflamed tissues where they transform into tissue macrophages. Neurons, astrocytes, and microglia extensively express CCL2, particularly in the cerebral cortex, globus pallidus, hippocampus, paraventricular and supraoptic hypothalamic nuclei, lateral hypothalamus, substantia nigra, facial nuclei, motor and spinal trigeminal nuclei, gigantocellular reticular nucleus, and Purkinje cells of the cerebellum (Banisadr, Gosselin, Mechighel, Kitabgi, et al., 2005). MCP-1/CCL2 is constitutively expressed in cholinergic neurons, particularly in the magnocellular preoptic and oculomotor nuclei, and in dopaminergic neurons of the substantia nigra pars compacta (Banisadr, Gosselin, Mechighel, Kitabgi, et al., 2005). Also, MCP-1/CCL2 colocalized with melanin-concentrating hormone-expressing neurons in the lateral hypothalamic area, vasopressin in magnocellular neuronal cell bodies and processes in the supraoptic and paraventricular hypothalamic nuclei, as well as in processes in the internal layer of the median eminence and in the posterior pituitary (Banisadr, Gosselin, Mechighel, Kitabgi, et al., 2005). Such a vast presence of MCP-2 in the brain clearly indicates its part in the neuroinflammatory activity and neuroendocrine functions. CCL2 has been shown to activate microglia triggering the release of pro-inflammatory cytokines TNFα and IL-1β, and it induces a hyperpolarization

and a decrease in firing serotonergic neurons of the median and dorsal raphe nucleus (Cazareth et al., 2014), both concurrent with inflammatory conditions of the brain. CCL2 indeed has been found to be associated with several proved neuroinflammatory and neurodegenerative disorders, such as depression (Eyre et al., 2016), epilepsy (Fabene, Bramanti, & Constantin, 2010; Foresti et al., 2009), cerebral ischemia (Kim et al., 1995), AD (Hickman & Khoury, 2010), EAE (Ransohoff et al., 1993), traumatic brain injury (Semple, Bye, Rancan, Ziebell, & Morganti-Kossmann, 2010), and thiamine-deficiency-induced neuronal death (Ramesh et al., 2013). The increase in the expression of CCL2 mRNA was observed after systemic LPS injection causing upregulation of CCR2 expressing microglia and inflammatory monocytes, further suggesting that CCL2 attracts peripheral immune cell into the brain and promotes neuroinflammation (Cazareth et al., 2014). Contrary to the above, CCR2 deficiency in microglia aggravated amyloid deposition, possibly due to the decreased migration and recruitment of inflammatory monocytes to the site of amyloid deposition in a transgenic mouse model of AD (Naert & Rivest, 2011).

CCL3 has been shown to modulate hippocampal functions such as learning and memory. It is extensively expressed by microglia, astrocytes, and neurons in the CNS and found to be upregulated during neurodegenerative disorders, such as AD (Marciniak et al., 2015). While CCL3 significantly reduced basal synaptic transmission at the Schaffer collateral-CA1 synapse (Marciniak et al., 2015), the exact mechanism of CCL3 action regulating the synaptic plasticity involved in learning and memory functions remains to be elucidated. A clinical trial showed an increase in the expression of CCL3 (and CCL5) chemokine in the serum of patients with migraine and tension-type headaches (Domingues et al., 2016). CCL5 has also been shown to increase in the ischemic brain of humans (Fan et al., 2016) and the brain of mice with EAE (Di Prisco et al., 2014), strengthening the view that CCL5 chemokine toxicity could result in neuroinflammation.

CCL4, CCL12, and CCL18 are not defined extensively for their role in the CNS. However, CCL4 has been shown to improve adhesion of CD4+ T cell subsets to human brain microvessel endothelial cells (Quandt & Dorovini-Zis, 2004), suggesting it has a role to play in immune pathways in the brain. Similarly, CCL12 has been found to express highly in regions of the choroid plexus and the ependymal cell layer (Tabor-Godwin et al., 2010), and CCL18 presence has been noted in the CSF of the patients suffering from infectious meningitis and traumatic brain injury but not in AD lesions (Lautner et al., 2011).

CCL6 expression when reduced can adversely affect astrocyte migration to the site to inflammation and thereby reduce the repairability of damaged brain regions (Li et al., 2013). This could affect CCL7 production since CCL7 is predominantly produced by the astrocytes and its production rapidly increases upon TNF-α stimulation (Renner et al., 2011). The expression of CCL7 in the brain has been linked to the migration of effector macrophages across blood-brain barrier and the recruitment of IFN-γ producing CD8+ T cells in the brain (Renner et al., 2011; Torres et al., 2013), thus promoting neuroinflammation.

Enhanced expression of CCL8, CCL20, and CCL27 has been observed in MS brains (Arima et al., 2012; Banisor, Leist, & Kalman, 2005; Khaiboullina et al., 2015) and mouse brain samples with rising eosinophil count in blood after infection (Yu et al., 2015). While the mechanisms of action of CCL8 in CNS is not clear yet, CCL27 attracts memory T cells to the site of neuroinflammation (Khaiboullina et al., 2015). CCL20 has been shown to act as a catalyst by aiding with the infiltration of CD4+ T cells into the CNS (Arima et al., 2012). CCL20 is present in constitutive levels in the epithelium of choroid plexus; however, its concentration in choroid

plexus, GFAP-positive astrocytes, and inflamed tissues increases during MS (Arima et al., 2012).

It has been shown that higher levels of CCL11 in the blood stream can reduce hippocampal neurogenesis and accelerate aging (Villeda et al., 2011). CCL11 is released by activated astrocytes that trigger oxidative stress via microglial NOX1 activation and potentiate glutamate-mediated toxicity (Parajuli, Horiuchi, Mizuno, Takeuchi, & Suzumura, 2015). Indeed, the concentration of CCL11 has been shown to increase in the sera and cerebrospinal fluid of patients with psychiatric disorders (Parajuli et al., 2015). CCL11 chemokine has been shown to express highly at brain injury sites, and since it promotes migration and proliferation of neuronal progenitor cells, it can be hypothesized that CCL11 may play a vital role in the neonatal CNS (Wang et al., 2017).

CCL13, being a part of MCP family, promote chemotaxis of monocyte-derived macrophages and other inflammatory leukocytes to the site of injury and inflammation in the CNS (Stuart et al., 2015). On the other hand, CCL19, CCL21, and CCL22 chemokines have been shown to be expressed by infiltrating leukocytes and some astrocytes and microglia in the CNS of mice with EAE (Columba-Cabezas, Elena, & Aloisi, 2003; Dogan et al., 2011).

CXC Chemokines in CNS

Unlike CC chemokines that primarily attract monocytes and T lymphocytes, CXC chemokines, particularly CXCL1–CXCL8, ligands for CXCR2, have been shown to primarily attract neutrophils to regulate CNS inflammatory states (Murphy et al., 2000). However, whether neutrophil chemotaxis by CXC chemokines aids or deters neuronal survival and repairs under inflammatory conditions remains unclear (Jaerve & Müller, 2012; Stirling, Liu, Kubes, & Yong, 2009). For example, both neuroprotective and neurodegenerative effects of CXCR2 and its ligand CXCL1 have been reported in the mouse model of EAE (Kerstetter, Padovani-Claudio, Bai, & Miller, 2009; Omari, Lutz, Santambrogio, Lira, & Raine, 2009). Similarly, CXCL8 produced in the injured brain is essential for the migration and activation of leukocytes, in particular poly-morphonuclear neutrophils toward the lesioned brain region (Dirnagl, Iadecola, & Moskowitz, 1999). Activated neutrophils, in turn, produce a variety of toxic mediators, such as cytokines, reactive oxygen and nitrogen species, and lipid mediators that may lead to transient brain ische-mia and other neuroinflammatory disorders (Garau et al., 2005; Villa et al., 2007; Witko-Sarsat, Rieu, Descamps-Latscha, Lesavre, & Halbwachs-Mecarelli, 2000). However, CXCL8 has also been shown to be neuroprotective in a model of β-amyloid toxicity *in vitro* (Watson & Fan, 2005).

CXCL9–CXCL11 chemokines are evidently detrimental and exert pro-inflammatory effects through CXCR3-mediated chemotaxis of NK cells, T_h1 cells, and associated classically acti-vated (M1) pro-inflammatory monocyte-derived macrophages (Murphy et al., 2000). For example, CXCL10 that can cross the vascular endothelial cells from the CNS (Mordelet, Davies, Hillyer, Romero, & Male, 2007) shows reduction in migration when its CXCR3 receptor is blocked. This, in turn, has shown to reduce tissue damage and functional deficit in the mouse and rat models of EAE (Jenh et al., 2012).

CXCL12 has received a great deal of attention recently and found to be present in neurons with its receptors CXCR4 and CXCR7 expressed in diverse brain regions (Guyon, 2014). Its presence has also been reported in glial cells in the hippocampus, cerebral cortex substantia nigra, striatum, hypothalamus, and globus pallidus (Heinisch & Kirby, 2010). It has been shown to crosstalk with neurotransmitter systems, in par-ticular with GABAergic systems, glutamatergic systems, and serotonergic systems (Guyon, 2014; Heinisch & Kirby, 2010), and is important for neurogenesis and neuronal migration during development (Heinisch & Kirby, 2010).

Likewise, CXCL13 with its receptor CXCR5 modulates the maturation and proliferation of subgranular cells in the hippocampal dentate gyrus (Stuart, Corrigan, & Baune, 2014) (Table 3).

CX3C Chemokine in CNS

The only CX3C chemokine, that is, CX3CL1 (also known as fractalkine in humans and neuro-tactin in mice), is primarily found in neurons with its receptor CXCR1 expressed on microglia (Stuart et al., 2015) and can be neuroinflamma-tory or neuroprotective (Ferretti, Pistoia, & Corcione, 2014). It is extensively distributed in the hippocampal neurons and glial cells, as well as in the cerebral cortex, medulla, occipital pole, frontal lobe, temporal lobe, putamen, and spinal cord, where it primarily attracts monocytes and T lymphocytes and activates NK cells (Jiang et al., 1998; Le et al., 2004; Meucci et al., 1998; Murdoch & Finn, 2000; Nishiyori et al., 1998; Ono et al., 2003; Raport, Schweickart, Eddy, Shows, & Gray, 1995; Stuart et al., 2015).

CX3CL chemokine has been reported to be upregulated in the CA1, CA3, and dentate gyrus of the rat hippocampus following spatial learning, apparently to regulate glutamate-mediated neurotransmission tone; hence, CX3CL1 may have a role in the synaptic scaling (Sheridan et al., 2014). In addition, CX3CL1 has been shown to promote microglial and astro-cytic activation, pro-inflammatory cytokine secretion, expression of intracellular adhesion molecule (ICAM-1), and recruitment of CD4+ T cells into the CNS during neuroinflammatory diseases, such as MS and AD (Blauth, Zhang, Chopra, Rogan, & Markovic-Plese, 2015; Sheridan & Murphy, 2013). Indeed, a positive correlation has been observed in the plasma levels of soluble CX3CL1 and progression of AD (Kim et al., 2008). In mice models of EAE, CX3CL1 triggered the migration of lympho-cytes into the CNS (Mills, Alabanza, Mahamed, & Bynoe, 2012). However, it is the membrane-bound and not the soluble form of CX3CL1 that

TABLE 3 Biological Characteristics of CXC Chemokines in CNS Functions

CXC Chemokines and Synonyms	Receptor(s)	CNS Functions	Classical Peripheral Functions	References
CXCL1 (Gro-α, GRO1, NAP-3, KC)	CXCR2	NSC/NPC chemotaxis and differentiation	nø chemotaxis	(Le et al., 2004; Murdoch & Finn, 2000; Ono et al., 2003; Stuart et al., 2015; Tran, Ren, Veldhouse, & Miller, 2004)
CXCL2 (Gro-β, GRO2, MIP-2α)	CXCR2	Unknown	nø chemotaxis	(Le et al., 2004; Murdoch & Finn, 2000; Ono et al., 2003; Stuart et al., 2015; Wolpe et al., 1989)
CXCL3 (GRO3, MIP-2β)	CXCR2	Unknown	mø, nø chemotaxis	(Le et al., 2004; Murdoch & Finn, 2000; Ono et al., 2003; Stuart et al., 2015)
CXCL4 (platelet factor-4)	CXCR3B	Unknown	mø, nø, fibroblasts chemotaxis	(Le et al., 2004; Murdoch & Finn, 2000; Ono et al., 2003; Stuart et al., 2015)
CXCL5 (ENA-78)	CXCR2	Unknown	nø chemotaxis	(Le et al., 2004; Murdoch & Finn, 2000; Ono et al., 2003; Stuart et al., 2015)
CXCL6 (GCP-2)	CXCR1, CXCR2	Unknown	nø chemotaxis	(Le et al., 2004; Murdoch & Finn, 2000; Ono et al., 2003; Stuart et al., 2015)
CXCL7 (NAP-2, CTAPIII, β-Ta, PEP)	CXCR2	Unknown	nø chemotaxis	(Le et al., 2004; Murdoch & Finn, 2000; Ono et al., 2003; Stuart et al., 2015)
CXCL8 (IL-8, NAP-1, MDNCF, GCP-1)	CXCR1, CXCR2	NSC/NPC chemotaxis, HPA axis modulation	nø, eø, bø, T Lø, B Lø, NK, DC chemotaxis, nø, mø, bø activation	(Le et al., 2004; Murdoch & Finn, 2000; Ono et al., 2003; Stuart et al., 2015)
CXCL9 (MIG, CRG-10)	CXCR3	NSC/NPC differentiation, PIC infiltration	T Lø chemotaxis	(Le et al., 2004; Murdoch & Finn, 2000; Murphy et al., 2000; Ono et al., 2003; Stuart et al., 2015)
CXCL10 (IP-10, CRG-2)	CXCR3	PIC infiltration	mø, T Lø, NK, DC chemotaxis	(Le et al., 2004; Murdoch & Finn, 2000; Murphy et al., 2000; Ono et al., 2003; Stuart et al., 2015)
CXCL11 (IP-9, I-TAC, β-R1)	CXCR3, CXCR7	PIC infiltration	T Lø chemotaxis	(Le et al., 2004; Murdoch & Finn, 2000; Murphy et al., 2000; Ono et al., 2003; Stuart et al., 2015)
CXCL12 (SDF-1, PBSF)	CXCR4, CXCR7	NSC/NPC chemotaxis, enhances neurogenesis, modulates glutamate and GABA neurotransmission, HPA axis modulation	T Lø, mø chemotaxis, inhibits hematopoietic stem cell proliferation and differentiation, promotes angiogenesis	(Le et al., 2004; Murdoch & Finn, 2000; Ono et al., 2003; Stuart et al., 2015)

TABLE 3 Biological Characteristics of CXC Chemokines in CNS Functions—cont'd

CXC Chemokines and Synonyms	Receptor(s)	CNS Functions	Classical Peripheral Functions	References
CXCL13 (BCA-1, BLC)		Unknown	B Lø chemotaxis	(Le et al., 2004; Murdoch & Finn, 2000; Ono et al., 2003; Stuart et al., 2015)
CXCL14 (BRAK, bolekine)		Unknown	mø, NK, DC chemotaxis and activation, inhibits angiogenesis	(Le et al., 2004; Murdoch & Finn, 2000; Ono et al., 2003; Stuart et al., 2015)
CXCL15 (lungkine, WECHE)		Unknown	nø chemotaxis	(Le et al., 2004; Murdoch & Finn, 2000; Ono et al., 2003; Stuart et al., 2015)
CXCL16 (SRPSOX)	CXCR6	Unknown	T Lø, NK chemotaxis	(Le et al., 2004; Murdoch & Finn, 2000; Ono et al., 2003; Stuart et al., 2015)
CXCL17 (DMC, VCC-1)		Unknown	mø, DC chemotaxis	(Le et al., 2004; Murdoch & Finn, 2000; Ono et al., 2003; Stuart et al., 2015)

mø, monocyte/macrophage; Lø, lymphocyte; nø, neutrophil; bø, basophil; eø, eosinophil; DC, dendritic cell; NK, natural killer cell; NSC/NPC, neural stem/progenitor cell; PIC, peripheral immune cell; HPA axis, hypothalamus-pituitary–adrenal axis.

regulates microglial phagocytosis of Aβ and neuronal microtubule-associated protein tau (MAPT) phosphorylation (Lee et al., 2014). Conversely, when accumulated, this may result in instability of microtubules, the consequent loss of effective transport of molecules and organelles, and ultimately neuronal death (KoSIK, Joachim, & Selkoe, 1986).

Likewise, when CX3CR1 is deficit, this may affect downstream molecular cascades, such as that of microglia and subsequent release of pro-inflammatory cytokines (Maten, Henck, Wieloch, & Ruscher, 2017). For example, CX3CR1 deficiency has been shown to result in microglial hyperactivity in lipopolysaccharide-induced neuroinflammation (De Haas, Van Weering, De Jong, Boddeke, & Biber, 2007). On the contrary, few other studies reported that CX3CR1 deficiency in microglia enhances beneficial microglial activity, increases amyloid clearance, and prevents neuron loss in mice models of AD (Fuhrmann et al., 2010; Harrison et al., 1998; Liu, Condello, Schain, Harb, & Grutzendler, 2010). Other disparate findings observed in mice models of CX3CR1 deficiency include increased neurotoxicity following peripheral lipopolysaccharide injections in the CX3CR1 KO mice (Cardona et al., 2006) and decreased neurotoxicity with no harmful effects on microglia in mice models with focal cerebral ischemia (Dénes, Ferenczi, Halász, Környei, & Kovács, 2008) and no neurotoxic effects at all in neuroinflammatory conditions other than AD in mice (Jung et al., 2000). This evidence suggests that the mechanism of action of CX3CL1 and its receptor CX3CR1 is complex, and a clear understanding on their role in the CNS still needs to be developed.

MECHANISTIC LINKS BETWEEN CHEMOKINES ACTION AND DEVELOPMENT OF DEPRESSION

Neuroinflammatory Action of Chemokines

Chemokines help in the sustenance of inflammation that could result in the development of depression. Evidently, the level of many chemokines increases in serum and cerebrospinal fluid during neuroinflammation and subsequent psychiatric disorders (Stuart & Baune, 2014). Several pro-inflammatory factors, for example, IL33 (a pro-inflammatory product of inflammasome activation), interferon-γ, and TNF-α, induce the expression of several chemokine ligands and receptors such as CCL2, CXCL9, CXCL10, CXCL11, and CCR8 in the CNS (Gadani, Walsh, Smirnov, Zheng, & Kipnis, 2015; Murphy, Hoek, Wiekowski, Lira, & Sedgwick, 2002; Shurin et al., 2007). These chemokines attract monocytes, neutrophils, NK cells, T and B lymphocytes, and dendritic cells to the inflamed region, aggravating inflammation initially, however later helping with the removal of microbial and tissue debris, thereby reducing inflammation.

Some chemokines, for example, CXCL12, are expressed in constitutive levels and important for maintaining homeostasis within the brain (Banisadr, Skrzydelski, Kitabgi, Rostène, & Parsadaniantz, 2003); however, many others express and upregulate only during infection. The upregulation of chemokines induces leukocyte infiltration, some of which such as monocytes, T cells, and NK cells are known to play a prominent role under neuroinflammatory conditions. CCL2 has been found to attract monocytes to the brain, contributing to neuroinflammation with resultant depression (Eyre et al., 2016). While monocytes are the primary responders, CC chemokines such as CCL3, CCL4, and CCL5 recruit other leukocyte subsets such as neutrophils, eosinophils, basophils, and

microglia to the inflamed site as mentioned before. Classical pro-inflammatory chemokines, that is, CXCL1–8, activate leukocytes, especially neutrophils, and glia to a pro-inflammatory state and attract them to the target site (Murphy et al., 2000). Likewise, pro-inflammatory chemokines with a primarily chemotactic activity such as CCL2, CCL7, CCL8, CCL12, CCL13, and CXCL9–11 selectively attract pro-inflammatory cells to the CNS or choroid plexus (Yamagami et al., 1999), which in turn help in the sustenance of primary inflammatory response. When inflammation is chronic, it could lead to the development of depression. Interestingly, few chemokines elicit both neuroprotective and neuroinflammatory functions, for example, CXCL1 (Kerstetter et al., 2009; Omari et al., 2009), CXCL2 (Watson & Fan, 2005), CXCL8 (Dirnagl et al., 1999; Garau et al., 2005; Villa et al., 2007; Watson & Fan, 2005; Witko-Sarsat et al., 2000), CXCL12 (Jaerve & Müller, 2012), and CX3CL1 (Sheridan et al., 2014), making the process of deciphering chemokine pathways in psychiatric disorders disparate and complex. Nonetheless, an increase in the level of soluble interleukin-2 receptor in patients with MDD (Howren, Lamkin, & Suls, 2009; Liu, Ho, & Mak, 2012) is evidence of the mechanistic link between the chemokine-induced inflammatory pathways and the pathophysiology of depression. And as mentioned earlier too, chemokines, such as CCL2–8 and CXCL9–11, and their receptors are important mediators of the inflammatory processes in the brain (Thuc, Blondeau, Nahon, & Rovère, 2015).

Role of Chemokines During Brain Development and in Neuronal Plasticity

In addition to the alteration of the neuroinflammatory pathways in the brain, chemokines have been shown to modulate brain physiology by altering the neuronal plasticity. For example, CCL11 has been shown to negatively regulate hippocampal neurogenesis and hence could be responsible for the aging-associated depression (Villeda et al., 2011). Indeed, chemokines, such as CXCL12 and its receptors CXCR4 and CXCR7, may have a role in the early neurodevelopmental process (Schönemeier et al., 2008; Tran et al., 2007) since the presence of CXCL12 and its receptors have been noted in diverse brain regions (Guyon, 2014). It has also been reported in the glial cells of the hippocampus, cerebral cortex substantia nigra, striatum, hypothalamus, and globus pallidus (Heinisch & Kirby, 2010) and is important for neurogenesis and neuronal migration during development (Heinisch & Kirby, 2010). CXCL12 also interacts with various neurotransmitter systems, for example, GABAergic systems, glutamatergic systems, and serotonergic systems (Guyon, 2014; Heinisch & Kirby, 2010), and deficiency of CXCL12 could affect neurotransmission and neurodevelopmental processes. Similarly, CXCL13 and its receptor CXCR5 (Stuart et al., 2014) and CXCL14 may regulate the migration of the neural stem/progenitor cells and maintain hippocampal integrity (Banisadr, Bhattacharyya, et al., 2011). Fractalkine or CX3CL is found to be upregulated in the CA1, CA3, and dentate gyrus of the rat hippocampus following spatial learning; hence, it may have a role in the synaptic scaling (Sheridan et al., 2014). CX3CL is both neuroprotective and neurodegenerative (Ferretti et al., 2014) and has been shown to regulate the development and plasticity of neuronal circuits (Sheridan et al., 2014; Xiao, Xu, & Jiang, 2015). Mice deficit in CX3CL-CX3CR1 possessed more synapses but reduced microglial cells in the hippocampus during postnatal development (Paolicelli et al., 2011). Overall, depending on the neuroprotective or neurodegenerative effects on hippocampal plasticity, the chemokines may play a vital role in the pathophysiology of depression. More research in this direction is warranted to elucidate the link between neuroplasticity, chemokines, and depression.

Modulation of Neurotransmission by Chemokines

Chemokines, in particular CXCL12, has been shown to modulate serotonin, dopamine, GABA, and glutamate neurotransmission (Guyon, 2014; Heinisch & Kirby, 2010), thereby affecting neurogenesis and neuronal migration during development (Heinisch & Kirby, 2010). Since CXCL12 is present in diverse brain regions, like the hippocampus, cerebral cortex substantia nigra, striatum, hypothalamus, and globus pallidus (Guyon, 2014), the cross talk between CXCL12 and neurotransmitters is of great relevance to the pathophysiology of depression. For example, CXCL12 modulates glutamatergic and GABAergic neurotransmission in the rat substantia nigra at presynapses and hence regulates downstream dopaminergic neurons (Guyon et al., 2006); the deficiency of the latter has been shown to result in depression (Santiago et al., 2014). Indeed, CXCL12-CXCR4 exerts direct neurotransmitter-like postsynaptic effects on these dopaminergic neurons (Skrzydelski et al., 2007). Likewise, CXCL12-CX3CL1 regulates glutamatergic and GABAergic signaling of serotonergic neurons in the dorsal raphe nucleus of rats (Heinisch & Kirby, 2009, 2010). In addition, other chemokines, such as CCL2, CCL5, and CCL21, have also been implicated in modulating the release of neurotransmitters from neurons (see for reviews Melik-Parsadaniantz & Rostene, 2008; Rostène et al., 2007; Rostene, Dansereau, et al., 2011). For example, CCL2 is found along with a classical neurotransmitter acetylcholine in the substantial innominate and the oculomotor nucleus and with dopamine in the substantia nigra (Banisadr, Gosselin, Mechighel, Rostène, et al., 2005). Similarly, CCL5 has been shown to regulate the release of glutamate from the cortex and spinal cord in mice (Di Prisco, Summa, Chellakudam, Rossi, & Pittaluga, 2012). Altogether, it is apparent that this chemokine regulation of neurotransmitter pathways may have significant relevance to the etiology of mood disorders, such as depression; however, more findings are needed to substantiate our assertion further.

Regulation of HPA Axis and Neuroendocrine Functions by Chemokines

Recent literature has started pointing to the role of various chemokines, such as CCL2, CCL3, CCL5, CXCL8, and CXCL12, in the regulation of the HPA axis and associated functions such as stress, metabolism, feeding behaviors, reproduction, fluid/electrolyte balance, and neuroendocrine hormone secretions (Besedovsky et al., 1986; Rostene, Guyon, et al., 2011; Verburg-van Kemenade et al., 2013). Significant levels of CXCL8 have been observed in the paraventricular nucleus, where the corticotropin-releasing hormone is produced and responsible for the negative feedback on the HPA axis (Licinio, Wong, & Gold, 1992). Enhanced expression of CXCL1 (Sakamoto et al., 1996) has been noted in the paraventricular nucleus in response to LPS administration (Reyes, Walker, DeCino, Hogenesch, & Sawchenko, 2003; Sakamoto et al., 1996). It was observed that CXCL1 stimulates the release of adrenocorticotropic hormone from cultured pituitary neurons (Sawada et al., 1994), thus further supporting the role of chemokines in endocrine regulation. Overall, it can be asserted that available evidence indicates chemokines elevate the level of cortisol in brain regions, particularly the hippocampus, resulting in the development of the depressive disorder.

DISCUSSION

Chemokines have caught the attention of researchers in recent times because of their ability to modulate not just one but multiple pathways that could lead to depression. They have been shown to produce neuroinflammatory effects (Eyre et al., 2016; Thuc et al., 2015), modulate neuroplasticity (Banisadr, Bhattacharyya, et al., 2011; Heinisch & Kirby, 2010; Sheridan et al., 2014; Xiao et al., 2015), dysregulate neurotransmission (Guyon, 2014; Heinisch & Kirby, 2010), and alter HPA axis (Besedovsky et al., 1986; Rostene, Guyon, et al., 2011; Verburg-van Kemenade et al., 2013). All these pathways are intertwined, and some chemokines, such as CCL2, CCL5, and CXCL12, take part in modulating more than one pathway (Fig. 2).

Few chemokines, for example, CXCL1 (Kerstetter et al., 2009; Omari et al., 2009), CXCL2 (Watson & Fan, 2005), CXCL8 (Dirnagl et al., 1999; Garau et al., 2005; Villa et al., 2007; Watson & Fan, 2005; Witko-Sarsat et al., 2000), CXCL12 (Jaerve & Müller, 2012), and CX3CL1 (Sheridan et al., 2014), have shown beneficial, harmful, or neutral effects under different circumstances. Furthermore, while chemokines attract immune cells to the sites of inflammation and elicit neuroinflammatory effects as discussed in the previous sections, various other classical immune mediators, such as interferon-γ (Shurin et al., 2007), TNF (Murphy et al., 2002), and IL33 (Gadani et al., 2015), as well as neurotrophins, such as BDNF, NGF, and neurotrophin-3 (Ahmed, Tessarollo, Thiele, & Mocchetti, 2008; Avdoshina et al., 2011), have been shown to regulate the expression of chemokines and their receptors in the CNS. Neuroinflammation when chronic modulates the HPA axis to a hyperactive state. Chemokines, such as CCL2, CCL5, CXCL8, and CXCL12, also upregulate the HPA axis directly, in turn modifying HPA axis-associated functions such as stress, metabolisms, feeding behaviors, reproduction, fluid/electrolyte balance, and neuroendocrine hormone secretions (Rostene, Guyon, et al., 2011; Verburg-van Kemenade et al., 2013), which are generally affected during depression as well. Likewise, several chemokines, such as CXCL12, CCL2, CCL5, and CCL21, modulate the release of neurotransmitters from neurons (Melik-Parsadaniantz & Rostene, 2008;

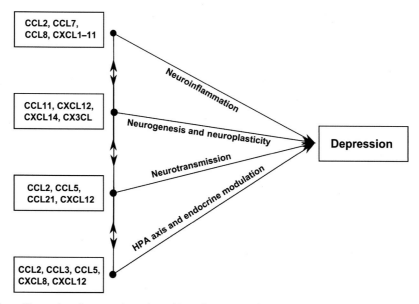

FIG. 2 The above illustration shows various chemokines that are involved in the modulation of molecular pathways leading to depression. HPA, hypothalamic-pituitary-adrenal.

Rostene, Dansereau, et al., 2011; Rostène et al., 2007), in turn upregulating or downregulating HPA axis functions and neuroplasticity. These multifunctional roles of chemokines and intertwined pathways underline the complexity of chemokine actions leading to depression.

inflammation leading to depression. As such, further research and clinical trials into the role of chemokines in depression are advisable before chemokine-targeted therapies could be developed and used for treating depression.

CONCLUDING REMARKS

In conclusion, chemokines provide a vital tool for regulating various innate immune pathways, in particular, those associated with neuroinflammation leading to depression. However, despite more than two decades of research in chemokines, their role in depression is not fully elucidated, and more research is needed to reveal the hidden intricacies of the recognized pathways. Although chemokine activation and chemotaxis pathways are complex and involve other immune factors such as cytokines, glial cells, neurotrophins, and leukocyte subsets, chemokines have been shown to play a vital role in their activation and the sustenance of chronic

CONFLICT OF INTEREST STATEMENT

The presented work is supported by the National Health and Medical Research Council, Australia (APP 1043771 to BTB). The funders had no role in study design, data collection and analysis, decision to publish, or preparation of the manuscript.

References

Ahmed, F., Tessarollo, L., Thiele, C., & Mocchetti, I. (2008). Brain-derived neurotrophic factor modulates expression of chemokine receptors in the brain. *Brain Research, 1227*, 1–11.

Air, T., Tully, P. J., Sweeney, S., & Beltrame, J. (2016). Epidemiology of cardiovascular disease and depression. In *Cardiovascular diseases and depression* (pp. 5–21): Springer.

Apter-Levi, Y., Pratt, M., Vakart, A., Feldman, M., Zagoory-Sharon, O., & Feldman, R. (2016). Maternal depression across the first years of life compromises child psychosocial adjustment; relations to child HPA-axis functioning. *Psychoneuroendocrinology, 64*, 47–56.

Arima, Y., Harada, M., Kamimura, D., Park, J.-H., Kawano, F., Yull, F. E., et al. (2012). Regional neural activation defines a gateway for autoreactive T cells to cross the blood-brain barrier. *Cell, 148*(3), 447–457.

Avdoshina, V., Becker, J., Campbell, L. A., Parsadanian, M., Mhyre, T., Tessarollo, L., et al. (2011). Neurotrophins modulate the expression of chemokine receptors in the brain. *Journal of Neurovirology, 17*(1), 58–62.

Bagaeva, L. V., Rao, P., Powers, J. M., & Segal, B. M. (2006). CXC chemokine ligand 13 plays a role in experimental autoimmune encephalomyelitis. *Journal of Immunology, 176*(12), 7676–7685.

Baggiolini, M., Dewald, B., & Moser, B. (1997). Human chemokines: an update. *Annual Review of Immunology, 15*(1), 675–705.

Bajetto, A., Bonavia, R., Barbero, S., Florio, T., & Schettini, G. (2001). Chemokines and their receptors in the central nervous system. *Frontiers in Neuroendocrinology, 22*(3), 147–184.

Banisadr, G., Queraud-Lesaux, F., Boutterin, M. C., Pelaprat, D., Zalc, B., Rostene, W., et al. (2002). Distribution, cellular localization and functional role of CCR2 chemokine receptors in adult rat brain. *Journal of Neurochemistry, 81*(2), 257–269.

Banisadr, G., Fontanges, P., Haour, F., Kitabgi, P., Rostene, W., & Melik Parsadaniantz, S. (2002). Neuroanatomical distribution of CXCR4 in adult rat brain and its localization in cholinergic and dopaminergic neurons. *The European Journal of Neuroscience, 16*(9), 1661–1671.

Banisadr, G., Skrzydelski, D., Kitabgi, P., Rostène, W., & Parsadaniantz, S. M. (2003). Highly regionalized distribution of stromal cell-derived factor-1/CXCL12 in adult rat brain: constitutive expression in cholinergic, dopaminergic and vasopressinergic neurons. *European Journal of Neuroscience, 18*(6), 1593–1606.

Banisadr, G., Gosselin, R. D., Mechighel, P., Rostène, W., Kitabgi, P., & Mélik Parsadaniantz, S. (2005). Constitutive neuronal expression of CCR2 chemokine receptor and its colocalization with neurotransmitters in normal rat brain: functional effect of MCP-1/CCL2 on calcium mobilization in primary cultured neurons. *Journal of Comparative Neurology, 492*(2), 178–192.

Banisadr, G., Gosselin, R. D., Mechighel, P., Kitabgi, P., Rostène, W., & Parsadaniantz, S. M. (2005). Highly regionalized neuronal expression of monocyte chemoattractant protein-1 (MCP-1/CCL2) in rat brain: evidence for its colocalization with neurotransmitters and neuropeptides. *Journal of Comparative Neurology, 489*(3), 275–292.

Banisadr, G., Bhattacharyya, B. J., Belmadani, A., Izen, S. C., Ren, D., Tran, P. B., et al. (2011). The chemokine BRAK/CXCL14 regulates synaptic transmission in the adult mouse dentate gyrus stem cell niche. *Journal of Neurochemistry, 119*(6), 1173–1182.

Banisadr, G., Frederick, T. J., Freitag, C., Ren, D., Jung, H., Miller, S. D., et al. (2011). The role of CXCR4 signaling in the migration of transplanted oligodendrocyte progenitors into the cerebral white matter. *Neurobiology of Disease, 44*(1), 19–27.

Banisor, I., Leist, T. P., & Kalman, B. (2005). Involvement of β-chemokines in the development of inflammatory demyelination. *Journal of Neuroinflammation, 2*(1), 7.

Belanoff, J. K., Kalehzan, M., Sund, B., Fleming Ficek, S. K., & Schatzberg, A. F. (2001). Cortisol activity and cognitive changes in psychotic major depression. *American Journal of Psychiatry, 158*(10), 1612–1616.

Belmaker, R., & Agam, G. (2008). Major depressive disorder. *New England Journal of Medicine, 358*(1), 55–68.

Besedovsky, H., Del Rey, A., Sorkin, E., & Dinarello, C. A. (1986). Immunoregulatory feedback between interleukin-I and glucocorticoid hormones. *Science, 233*, 652–655.

Blauth, K., Zhang, X., Chopra, M., Rogan, S., & Markovic-Plese, S. (2015). The role of fractalkine (CX3CL1) in regulation of CD4+ cell migration to the central nervous system in patients with relapsing-remitting multiple sclerosis. *Clinical Immunology, 157*(2), 121–132.

Brites, D., & Fernandes, A. (2015). Neuroinflammation and depression: microglia activation, extracellular microvesicles and microRNA dysregulation. *Frontiers in Cellular Neuroscience, 9*, 476.

Campbell, S., Marriott, M., Nahmias, C., & MacQueen, G. M. (2004). Lower hippocampal volume in patients suffering from depression: a meta-analysis. *American Journal of Psychiatry, 161*(4), 598–607.

Cardona, A. E., Pioro, E. P., Sasse, M. E., Kostenko, V., Cardona, S. M., Dijkstra, I. M., et al. (2006). Control of microglial neurotoxicity by the fractalkine receptor. *Nature Neuroscience, 9*(7), 917–924.

Cazareth, J., Guyon, A., Heurteaux, C., Chabry, J., & Petit-Paitel, A. (2014). Molecular and cellular neuroinflammatory status of mouse brain after systemic lipopolysaccharide challenge: importance of CCR2/CCL2 signaling. *Journal of Neuroinflammation, 11*(1), 132.

Chi, S., Wang, C., Jiang, T., Zhu, X.-C., Yu, J.-T., & Tan, L. (2015). The prevalence of depression in Alzheimer's disease: a systematic review and meta-analysis. *Current Alzheimer Research, 12*(2), 189–198.

Choudary, P. V., Molnar, M., Evans, S., Tomita, H., Li, J., Vawter, M., et al. (2005). Altered cortical glutamatergic and GABAergic signal transmission with glial involvement in depression. *Proceedings of the National Academy of Sciences, 102*(43), 15653–15658.

Clark, J. A., Pai, L.-Y., Flick, R. B., & Rohrer, S. P. (2005). Differential hormonal regulation of tryptophan hydroxylase-2 mRNA in the murine dorsal raphe nucleus. *Biological Psychiatry, 57*(8), 943–946.

Colich, N. L., Kircanski, K., Foland-Ross, L. C., & Gotlib, I. H. (2015). HPA-axis reactivity interacts with stage of pubertal development to predict the onset of depression. *Psychoneuroendocrinology, 55*, 94–101.

Colla, M., Kronenberg, G., Deuschle, M., Meichel, K., Hagen, T., Bohrer, M., et al. (2007). Hippocampal volume reduction and HPA-system activity in major depression. *Journal of Psychiatric Research, 41*(7), 553–560.

Columba-Cabezas, S., Elena, B. S., & Aloisi, A. (2003). Lymphoid chemokines CCL19 and CCL21 are expressed in the central nervous system during experimental autoimmune encephalomyelitis: implications for the maintenance of chronic neuroinflammation. *Brain Pathology, 13*(1), 38–51.

Coppen, A. (1967). The biochemistry of affective disorders. *The British Journal of Psychiatry, 113*(504), 1237–1264.

Coughlan, C. M., McManus, C. M., Sharron, M., Gao, Z., Murphy, D., Jaffer, S., et al. (2000). Expression of multiple functional chemokine receptors and monocyte chemoattractant protein-1 in human neurons. *Neuroscience, 97*(3), 591–600.

Cyster, J. G. (1999). Chemokines and cell migration in secondary lymphoid organs. *Science, 286*(5447), 2098–2102.

De Haas, A., Van Weering, H., De Jong, E., Boddeke, H., & Biber, K. (2007). Neuronal chemokines: versatile messengers in central nervous system cell interaction. *Molecular Neurobiology, 36*(2), 137–151.

de Haas, A. H., Boddeke, H. W., & Biber, K. (2008). Region-specific expression of immunoregulatory proteins on microglia in the healthy CNS. *Glia, 56*(8), 888–894.

Dénes, Á., Ferenczi, S., Halász, J., Környei, Z., & Kovács, K. J. (2008). Role of CX3CR1 (fractalkine receptor) in brain damage and inflammation induced by focal cerebral ischemia in mouse. *Journal of Cerebral Blood Flow & Metabolism, 28*(10), 1707–1721.

Devi, S., Laning, J., Luo, Y., & Dorf, M. E. (1995). Biologic activities of the beta-chemokine TCA3 on neutrophils and macrophages. *Journal of Immunology, 154*(10), 5376–5383.

Di Prisco, S., Summa, M., Chellakudam, V., Rossi, P. I., & Pittaluga, A. (2012). RANTES-mediated control of excitatory amino acid release in mouse spinal cord. *Journal of Neurochemistry, 121*(3), 428–437.

Di Prisco, S., Merega, E., Milanese, M., Summa, M., Casazza, S., Raffaghello, L., et al. (2013). CCL5-glutamate interaction in central nervous system: early and acute presynaptic defects in EAE mice. *Neuropharmacology, 75*, 337–346.

Di Prisco, S., Merega, E., Lanfranco, M., Casazza, S., Uccelli, A., & Pittaluga, A. (2014). Acute desipramine restores presynaptic cortical defects in murine experimental autoimmune encephalomyelitis by suppressing central CCL5 overproduction. *British Journal of Pharmacology, 171*(9), 2457–2467.

Dijkstra, I. M., de Haas, A. H., Brouwer, N., Boddeke, H. W., & Biber, K. (2006). Challenge with innate and protein antigens induces CCR7 expression by microglia in vitro and in vivo. *Glia, 54*(8), 861–872.

Dirnagl, U., Iadecola, C., & Moskowitz, M. A. (1999). Pathobiology of ischaemic stroke: an integrated view. *Trends in Neurosciences, 22*(9), 391–397.

Dogan, R.-N. E., Long, N., Forde, E., Dennis, K., Kohm, A. P., Miller, S. D., et al. (2011). CCL22 regulates experimental autoimmune encephalomyelitis by controlling inflammatory macrophage accumulation and effector function. *Journal of Leukocyte Biology, 89*(1), 93–104.

Domingues, R. B., Duarte, H., Senne, C., Bruniera, G., Brunale, F., Rocha, N. P., et al. (2016). Serum levels of adiponectin, CCL3/MIP-1α, and CCL5/RANTES discriminate migraine from tension-type headache patients. *Arquivos de Neuro-Psiquiatria, 74*(8), 626–631.

Duman, R. S. (2004). Role of neurotrophic factors in the etiology and treatment of mood disorders. *Neuromolecular Medicine, 5*(1), 11–25.

Eltayeb, S., Berg, A.-L., Lassmann, H., Wallström, E., Nilsson, M., Olsson, T., et al. (2007). Temporal expression and cellular origin of CC chemokine receptors CCR1, CCR2 and CCR5 in the central nervous system: insight into mechanisms of MOG-induced EAE. *Journal of Neuroinflammation, 4*(1), 14.

Espay, A. J., LeWitt, P. A., & Kaufmann, H. (2014). Norepinephrine deficiency in Parkinson's disease: the case for noradrenergic enhancement. *Movement Disorders, 29*(14), 1710–1719.

Eyre, H., & Baune, B. T. (2012). Neuroplastic changes in depression: a role for the immune system. *Psychoneuroendocrinology, 37*(9), 1397–1416.

Eyre, H. A., Air, T., Pradhan, A., Johnston, J., Lavretsky, H., Stuart, M. J., et al. (2016). A meta-analysis of chemokines in major depression. *Progress in Neuro-Psychopharmacology and Biological Psychiatry, 68*, 1–8.

Fabene, P. F., Bramanti, P., & Constantin, G. (2010). The emerging role for chemokines in epilepsy. *Journal of Neuroimmunology, 224*(1), 22–27.

Fan, Y., Xiong, X., Zhang, Y., Yan, D., Jian, Z., Xu, B., et al. (2016). MKEY, a peptide inhibitor of CXCL4-CCL5 heterodimer formation, protects against stroke in mice. *Journal of the American Heart Association, 5*(9) e003615.

Ferretti, E., Pistoia, V., & Corcione, A. (2014). Role of fractalkine/CX3CL1 and its receptor in the pathogenesis of inflammatory and malignant diseases with emphasis on B cell malignancies. *Mediators of Inflammation, 2014*.

Fischer, F. R., Santambrogio, L., Luo, Y., Berman, M. A., Hancock, W. W., & Dorf, M. E. (2000). Modulation of experimental autoimmune encephalomyelitis: effect of altered peptide ligand on chemokine and chemokine receptor expression. *Journal of Neuroimmunology, 110*(1), 195–208.

Flynn, G., Maru, S., Loughlin, J., Romero, I. A., & Male, D. (2003). Regulation of chemokine receptor expression in human microglia and astrocytes. *Journal of Neuroimmunology, 136*(1), 84–93.

Foresti, M. L., Arisi, G. M., Katki, K., Montañez, A., Sanchez, R. M., & Shapiro, L. A. (2009). Chemokine CCL2 and its receptor CCR2 are increased in the hippocampus following pilocarpine-induced status epilepticus. *Journal of Neuroinflammation, 6*(1), 40.

Frodl, T., & O'Keane, V. (2013). How does the brain deal with cumulative stress? A review with focus on developmental stress, HPA axis function and hippocampal structure in humans. *Neurobiology of Disease, 52*, 24–37.

Fuhrmann, M., Bittner, T., Jung, C. K., Burgold, S., Page, R. M., Mitteregger, G., et al. (2010). Microglial Cx3cr1 knockout prevents neuron loss in a mouse model of Alzheimer's disease. *Nature Neuroscience, 13*(4), 411–413.

Gadani, S. P., Walsh, J. T., Smirnov, I., Zheng, J., & Kipnis, J. (2015). The glia-derived alarmin IL-33 orchestrates the immune response and promotes recovery following CNS injury. *Neuron, 85*(4), 703–709.

Garau, A., Bertini, R., Colotta, F., Casilli, F., Bigini, P., Cagnotto, A., et al. (2005). Neuroprotection with the CXCL8 inhibitor repertaxin in transient brain ischemia. *Cytokine, 30*(3), 125–131.

Ghosal, S., Hare, B. D., & Duman, R. S. (2017). Prefrontal cortex GABAergic deficits and circuit dysfunction in the pathophysiology and treatment of chronic stress and depression. *Current Opinion in Behavioral Sciences, 14*, 1–8.

Giovannelli, A., Limatola, C., Ragozzino, D., Mileo, A. M., Ruggieri, A., Ciotti, M. T., et al. (1998). CXC chemokines interleukin-8 (IL-8) and growth-related gene product alpha (GROalpha) modulate Purkinje neuron activity in mouse cerebellum. *Journal of Neuroimmunology, 92*(1–2), 122–132.

Glassman, A. (2008). Depression and cardiovascular disease. *Pharmacopsychiatry, 41*(06), 221–225.

Godbout, J. P., Moreau, M., Lestage, J., Chen, J., Sparkman, N. L., O'Connor, J., et al. (2008). Aging exacerbates depressive-like behavior in mice in response to activation of the peripheral innate immune system. *Neuropsychopharmacology, 33*(10), 2341–2351.

Godiska, R., Chantry, D., Dietsch, G. N., & Gray, P. W. (1995). Chemokine expression in murine experimental allergic encephalomyelitis. *Journal of Neuroimmunology, 58*(2), 167–176.

Godiska, R., Chantry, D., Raport, C. J., Sozzani, S., Allavena, P., Leviten, D., et al. (1997). Human macrophage-derived chemokine (MDC), a novel chemoattractant for monocytes, monocyte-derived dendritic cells, and natural killer cells. *Journal of Experimental Medicine, 185*(9), 1595–1604.

Gomez-Nicola, D., Pallas-Bazarra, N., Valle-Argos, B., & Nieto-Sampedro, M. (2010). CCR7 is expressed in astrocytes and upregulated after an inflammatory injury. *Journal of Neuroimmunology, 227*(1–2), 87–92.

Gottle, P., Kremer, D., Jander, S., Odemis, V., Engele, J., Hartung, H. P., et al. (2010). Activation of CXCR7 receptor promotes oligodendroglial cell maturation. *Annals of Neurology, 68*(6), 915–924.

Groves, D., & Jiang, Y. (1995). Chemokines, a family of chemotactic cytokines. *Critical Reviews in Oral Biology & Medicine, 6*(2), 109–118.

Guyon, A. (2014). CXCL12 chemokine and its receptors as major players in the interactions between immune and nervous systems. *Frontiers in Cellular Neuroscience, 8*, 65.

Guyon, A., Skrzydelsi, D., Rovere, C., Rostene, W., Parsadaniantz, S. M., & Nahon, J. L. (2006). Stromal cell-derived factor-1alpha modulation of the excitability of rat substantia nigra dopaminergic neurones: presynaptic mechanisms. *Journal of Neurochemistry, 96*(6), 1540–1550.

Harrison, J. K., Jiang, Y., Chen, S., Xia, Y., Maciejewski, D., McNamara, R. K., et al. (1998). Role for neuronally derived fractalkine in mediating interactions between neurons and CX3CR1-expressing microglia. *Proceedings of the National Academy of Sciences, 95*(18), 10896–10901.

Heinisch, S., & Kirby, L. G. (2009). Fractalkine/CX3CL1 enhances GABA synaptic activity at serotonin neurons in the rat dorsal raphe nucleus. *Neuroscience, 164*(3), 1210–1223.

Heinisch, S., & Kirby, L. G. (2010). SDF-1α/CXCL12 enhances GABA and glutamate synaptic activity at serotonin neurons in the rat dorsal raphe nucleus. *Neuropharmacology, 58*(2), 501–514.

Hickman, S. E., & Khoury, J. E. (2010). Mechanisms of mononuclear phagocyte recruitment in Alzheimer's disease. *CNS & Neurological Disorders-Drug Targets (Formerly Current Drug Targets-CNS & Neurological Disorders), 9*(2), 168–173.

Holt, R. I., De Groot, M., & Golden, S. H. (2014). Diabetes and depression. *Current Diabetes Reports, 14*(6), 1–9.

Hoover, D. M., Mizoue, L. S., Handel, T. M., & Lubkowski, J. (2000). The crystal structure of the chemokine domain of fractalkine shows a novel quaternary arrangement. *Journal of Biological Chemistry, 275*(30), 23187–23193.

Howren, M. B., Lamkin, D. M., & Suls, J. (2009). Associations of depression with C-reactive protein, IL-1, and IL-6: a meta-analysis. *Psychosomatic Medicine, 71*(2), 171–186.

Iwata, M., Ota, K. T., & Duman, R. S. (2013). The inflammasome: pathways linking psychological stress, depression, and systemic illnesses. *Brain, Behavior, and Immunity, 31*, 105–114.

Jaerve, A., & Müller, H. W. (2012). Chemokines in CNS injury and repair. *Cell and Tissue Research, 349*(1), 229–248.

Jenh, C. H., Cox, M. A., Cui, L., Reich, E. P., Sullivan, L., Chen, S. C., et al. (2012). A selective and potent CXCR3 antagonist SCH 546738 attenuates the development of autoimmune diseases and delays graft rejection. *BMC Immunology, 13*(1), 2.

Ji, J. F., He, B. P., Dheen, S. T., & Tay, S. S. W. (2004). Expression of chemokine receptors CXCR4, CCR2, CCR5 and CX 3 CR1 in neural progenitor cells isolated from the subventricular zone of the adult rat brain. *Neuroscience Letters, 355*(3), 236–240.

Jiang, Y., Salafranca, M. N., Adhikari, S., Xia, Y., Feng, L., Sonntag, M. K., et al. (1998). Chemokine receptor expression in cultured glia and rat experimental allergic encephalomyelitis. *Journal of Neuroimmunology, 86*(1), 1–12.

Jokinen, J., & Nordström, P. (2009). HPA axis hyperactivity and attempted suicide in young adult mood disorder inpatients. *Journal of Affective Disorders, 116*(1), 117–120.

Jones, B. A., Beamer, M., & Ahmed, S. (2010). Fractalkine/ CX3CL1: a potential new target for inflammatory diseases. *Molecular Interventions, 10*(5), 263.

Jung, S., Aliberti, J., Graemmel, P., Sunshine, M. J., Kreutzberg, G. W., Sher, A., et al. (2000). Analysis of fractalkine receptor CX3CR1 function by targeted deletion and green fluorescent protein reporter gene insertion. *Molecular and Cellular Biology, 20*(11), 4106–4114.

Kan, A. A., van der Hel, W. S., Kolk, S. M., Bos, I. W., Verlinde, S. A., van Nieuwenhuizen, O., et al. (2012). Prolonged increase in rat hippocampal chemokine signalling after status epilepticus. *Journal of Neuroimmunology, 245* (1–2), 15–22.

Katon, W. J., Rutter, C., Simon, G., Lin, E. H., Ludman, E., Ciechanowski, P., et al. (2005). The association of comorbid depression with mortality in patients with type 2 diabetes. *Diabetes Care, 28*(11), 2668–2672.

Kerstetter, A. E., Padovani-Claudio, D. A., Bai, L., & Miller, R. H. (2009). Inhibition of CXCR2 signaling promotes recovery in models of multiple sclerosis. *Experimental Neurology, 220*(1), 44–56.

Khaiboullina, S. F., Gumerova, A. R., Khafizova, I. F., Martynova, E. V., Lombardi, V. C., Bellusci, S., et al. (2015). CCL27: novel cytokine with potential role in pathogenesis of multiple sclerosis. *BioMed Research International, 2015*.

Kielian, T., Barry, B., & Hickey, W. F. (2001). CXC chemokine receptor-2 ligands are required for neutrophil-mediated host defense in experimental brain abscesses1. *The Journal of Immunology, 166*(7), 4634–4643.

Kim, J. S., Gautam, S. C., Chopp, M., Zaloga, C., Jones, M. L., Ward, P. A., et al. (1995). Expression of monocyte chemoattractant protein-1 and macrophage inflammatory protein-1 after focal cerebral ischemia in the rat. *Journal of Neuroimmunology, 56*(2), 127–134.

Kim, T.-S., Lim, H.-K., Lee, J. Y., Kim, D.-J., Park, S., Lee, C., et al. (2008). Changes in the levels of plasma soluble fractalkine in patients with mild cognitive impairment and Alzheimer's disease. *Neuroscience Letters, 436*(2), 196–200.

Kim, Y.-K., Na, K.-S., Myint, A.-M., & Leonard, B. E. (2016). The role of pro-inflammatory cytokines in neuroinflammation, neurogenesis and the neuroendocrine system in major depression. *Progress in Neuro- Psychopharmacology and Biological Psychiatry, 64*, 277–284.

Knol, M., Twisk, J., Beekman, A., Heine, R., Snoek, F., & Pouwer, F. (2006). Depression as a risk factor for the onset of type 2 diabetes mellitus. *A meta-analysis, Diabetologia, 49*(5), 837–845.

KoSIK, K. S., Joachim, C. L., & Selkoe, D. J. (1986). Microtubule-associated protein tau (tau) is a major antigenic component of paired helical filaments in Alzheimer disease. *Proceedings of the National Academy of Sciences, 83*(11), 4044–4048.

Krathwohl, M. D., & Kaiser, J. L. (2004). Chemokines promote quiescence and survival of human neural progenitor cells. *Stem Cells, 22*(1), 109–118.

Krebber, A., Buffart, L., Kleijn, G., Riepma, I., Bree, R., Leemans, C., et al. (2014). Prevalence of depression in cancer patients: a meta-analysis of diagnostic interviews and self-report instruments. *Psycho-Oncology, 23*(2), 121–130.

Kubera, M., Obuchowicz, E., Goehler, L., Brzeszcz, J., & Maes, M. (2011). In animal models, psychosocial stress-induced (neuro) inflammation, apoptosis and reduced neurogenesis are associated to the onset of depression. *Progress in Neuro-Psychopharmacology and Biological Psychiatry, 35*(3), 744–759.

Kufareva, I., Salanga, C. L., & Handel, T. M. (2015). Chemokine and chemokine receptor structure and interactions: implications for therapeutic strategies. *Immunology and Cell Biology, 93*(4), 372–383.

Lambert, G., Johansson, M., Ågren, H., & Friberg, P. (2000). Reduced brain norepinephrine and dopamine release in treatment-refractory depressive illness: evidence in support of the catecholamine hypothesis of mood disorders. *Archives of General Psychiatry, 57*(8), 787–793.

Lautner, R., Mattsson, N., Schöll, M., Augutis, K., Blennow, K., Olsson, B., et al. (2011). Biomarkers for microglial activation in Alzheimer's disease. *International Journal of Alzheimer's Disease, 2011*.

Le, Y., Zhou, Y., Iribarren, P., & Wang, J. (2004). Chemokines and chemokine receptors: their manifold roles in homeostasis and disease. *Cellular & Molecular Immunology, 1*(2), 95–104.

Lee, S., Xu, G., Jay, T. R., Bhatta, S., Kim, K.-W., Jung, S., et al. (2014). Opposing effects of membrane-anchored CX3CL1 on amyloid and tau pathologies via the p38 MAPK pathway. *The Journal of Neuroscience, 34*(37), 12538–12546.

Li, M. D., Cao, J., Wang, S., Wang, J., Sarkar, S., Vigorito, M., et al. (2013). Transcriptome sequencing of gene expression in the brain of the HIV-1 transgenic rat. *PLoS One, 8*(3) e59582.

Licinio, J., Wong, M. L., & Gold, P. W. (1992). Neutrophil-activating peptide-1/interleukin-8 mRNA is localized in rat hypothalamus and hippocampus. *Neuroreport, 3*(9), 753–756.

Liu, J. X., Cao, X., Tang, Y. C., Liu, Y., & Tang, F. R. (2007). CCR7, CCR8, CCR9 and CCR10 in the mouse hippocampal CA1 area and the dentate gyrus during and after pilocarpine-induced status epilepticus. *Journal of Neurochemistry, 100*(4), 1072–1088.

Liu, Y., Ho, R. C.-M., & Mak, A. (2012). Interleukin (IL)-6, tumour necrosis factor alpha (TNF-α) and soluble interleukin-2 receptors (sIL-2R) are elevated in patients with major depressive disorder: a meta-analysis and meta-regression. *Journal of Affective Disorders, 139*(3), 230–239.

Liu, Z., Condello, C., Schain, A., Harb, R., & Grutzendler, J. (2010). CX3CR1 in microglia regulates brain amyloid deposition through selective protofibrillar amyloid-β phagocytosis. *The Journal of Neuroscience, 30*(50), 17091–17101.

Lopez-Duran, N. L., McGinnis, E., Kuhlman, K., Geiss, E., Vargas, I., & Mayer, S. (2015). HPA-axis stress reactivity in youth depression: evidence of impaired regulatory processes in depressed boys. *Stress, 18*(5), 545–553.

Lorant, V., Deliège, D., Eaton, W., Robert, A., Philippot, P., & Ansseau, M. (2003). *Socioeconomic inequalities in depression: A meta-analysis.* Oxford Univ Press.

MacQueen, G. M., Campbell, S., McEwen, B. S., Macdonald, K., Amano, S., Joffe, R. T., et al. (2003). Course of illness, hippocampal function, and hippocampal volume in major depression. *Proceedings of the National Academy of Sciences, 100*(3), 1387–1392.

Marciniak, E., Faivre, E., Dutar, P., Pires, C. A., Demeyer, D., Caillierez, R., et al. (2015). The chemokine MIP-1α/CCL3 impairs mouse hippocampal synaptic transmission, plasticity and memory. *Scientific Reports, 5.*

Maten, G., Henck, V., Wieloch, T., & Ruscher, K. (2017). CX 3 C chemokine receptor 1 deficiency modulates microglia morphology but does not affect lesion size and short-term deficits after experimental stroke. *BMC Neuroscience, 18*(1), 11.

Melik-Parsadaniantz, S., & Rostene, W. (2008). Chemokines and neuromodulation. *Journal of Neuroimmunology, 198*(1–2), 62–68.

Meucci, O., Fatatis, A., Simen, A. A., Bushell, T. J., Gray, P. W., & Miller, R. J. (1998). Chemokines regulate hippocampal neuronal signaling and gp120 neurotoxicity. *Proceedings of the National Academy of Sciences of the United States of America, 95*(24), 14500–14505.

Miller, M. D., & Krangel, M. S. (1992). The human cytokine I-309 is a monocyte chemoattractant. *Proceedings of the National Academy of Sciences of the United States of America, 89*(7), 2950–2954.

Miller, R. J., Rostene, W., Apartis, E., Banisadr, G., Biber, K., Milligan, E. D., et al. (2008). Chemokine action in the nervous system. *The Journal of Neuroscience : The Official Journal of the Society for Neuroscience, 28*(46), 11792–11795.

Mills, J. H., Alabanza, L. M., Mahamed, D. A., & Bynoe, M. S. (2012). Extracellular adenosine signaling induces CX3CL1 expression in the brain to promote experimental autoimmune encephalomyelitis. *Journal of Neuroinflammation, 9*(193), 2094–2099.

Mizoue, L. S., Bazan, J. F., Johnson, E. C., & Handel, T. M. (1999). Solution structure and dynamics of the CX3C chemokine domain of fractalkine and its interaction with an N-terminal fragment of CX3CR1†. *Biochemistry, 38*(5), 1402–1414.

Mordelet, E., Davies, H. A., Hillyer, P., Romero, I. A., & Male, D. (2007). Chemokine transport across human vascular endothelial cells. *Endothelium, 14*(1), 7–15.

Murdoch, C., & Finn, A. (2000). Chemokine receptors and their role in inflammation and infectious diseases. *Blood, 95*(10), 3032–3043.

Murphy, C. A., Hoek, R. M., Wiekowski, M. T., Lira, S. A., & Sedgwick, J. D. (2002). Interactions between hemopoietically derived TNF and central nervous system-resident glial chemokines underlie initiation of autoimmune inflammation in the brain. *The Journal of Immunology, 169*(12), 7054–7062.

Murphy, P. M., Baggiolini, M., Charo, I. F., Hebert, C. A., Horuk, R., Matsushima, K., et al. (2000). International union of pharmacology. XXII. Nomenclature for chemokine receptors. *Pharmacological Reviews, 52*(1), 145–176.

Naert, G., & Rivest, S. (2011). CC chemokine receptor 2 deficiency aggravates cognitive impairments and amyloid pathology in a transgenic mouse model of Alzheimer's disease. *The Journal of Neuroscience, 31*(16), 6208–6220.

Nagira, M., Imai, T., Hieshima, K., Kusuda, J., Ridanpää, M., Takagi, S., et al. (1997). Molecular cloning of a novel human CC chemokine secondary lymphoid-tissue chemokine that is a potent chemoattractant for lymphocytes and mapped to chromosome 9p13. *Journal of Biological Chemistry, 272*(31), 19518–19524.

Nelson, T. E., & Gruol, D. L. (2004). The chemokine CXCL10 modulates excitatory activity and intracellular calcium signaling in cultured hippocampal neurons. *Journal of Neuroimmunology, 156*(1), 74–87.

Nguyen, D., Höpfner, M., Zobel, F., Henke, U., Scherübl, H., & Stangel, M. (2003). Rat oligodendroglial cell lines express a functional receptor for the chemokine CCL3 (macrophage inflammatory protein-1alpha). *Neuroscience Letters, 351*(2), 71–74.

Nishiyori, A., Minami, M., Ohtani, Y., Takami, S., Yamamoto, J., Kawaguchi, N., et al. (1998). Localization of fractalkine and CX3CR1 mRNAs in rat brain: does fractalkine play a role in signaling from neuron to microglia? *FEBS Letters, 429*(2), 167–172.

Omari, K. M., John, G. R., Sealfon, S. C., & Raine, C. S. (2005). CXC chemokine receptors on human oligodendrocytes: implications for multiple sclerosis. *Brain, 128*(Pt 5), 1003–1015.

Omari, K. M., Lutz, S. E., Santambrogio, L., Lira, S. A., & Raine, C. S. (2009). Neuroprotection and remyelination after autoimmune demyelination in mice that inducibly overexpress CXCL1. *The American Journal of Pathology, 174*(1), 164–176.

Ono, S. J., Nakamura, T., Miyazaki, D., Ohbayashi, M., Dawson, M., & Toda, M. (2003). Chemokines: roles in leukocyte development, trafficking, and effector function. *Journal of Allergy and Clinical Immunology, 111*(6), 1185–1199 quiz 1200.

Paolicelli, R. C., Bolasco, G., Pagani, F., Maggi, L., Scianni, M., Panzanelli, P., et al. (2011). Synaptic pruning by microglia is necessary for normal brain development. *Science, 333* (6048), 1456–1458.

Parajuli, B., Horiuchi, H., Mizuno, T., Takeuchi, H., & Suzumura, A. (2015). CCL11 enhances excitotoxic neuronal death by producing reactive oxygen species in microglia. *Glia, 63*(12), 2274–2284.

Petito, C. K., Roberts, B., Cantando, J. D., Rabinstein, A., & Duncan, R. (2001). Hippocampal injury and alterations in neuronal chemokine co-receptor expression in patients with AIDS. *Journal of Neuropathology and Experimental Neurology, 60*(4), 377–385.

Puma, C., Danik, M., Quirion, R., Ramon, F., & Williams, S. (2001). The chemokine interleukin-8 acutely reduces Ca(2+) currents in identified cholinergic septal neurons expressing CXCR1 and CXCR2 receptor mRNAs. *Journal of Neurochemistry, 78*(5), 960–971.

Quandt, J., & Dorovini-Zis, K. (2004). The beta chemokines CCL4 and CCL5 enhance adhesion of specific CD4+ T cell subsets to human brain endothelial cells. *Journal of Neuropathology & Experimental Neurology, 63*(4), 350–362.

Ramesh, G., MacLean, A. G., & Philipp, M. T. (2013). Cytokines and chemokines at the crossroads of neuroinflammation, neurodegeneration, and neuropathic pain. *Mediators of Inflammation, 2013*.

Ransohoff, R. M., Hamilton, T. A., Tani, M., Stoler, M. H., Shick, H. E., Major, J. A., et al. (1993). Astrocyte expression of mRNA encoding cytokines IP-10 and JE/MCP-1 in experimental autoimmune encephalomyelitis. *The FASEB Journal, 7*(6), 592–600.

Raport, C. J., Schweickart, V. L., Eddy, R. L., Shows, T. B., & Gray, P. W. (1995). The orphan G-protein-coupled receptor-encoding gene V28 is closely related to genes for chemokine receptors and is expressed in lymphoid and neural tissues. *Gene, 163*(2), 295–299.

Réaux-Le Goazigo, A., Van Steenwinckel, J., Rostène, W., & Parsadaniantz, S. M. (2013). Current status of chemokines in the adult CNS. *Progress in Neurobiology, 104*, 67–92.

Renner, J. A., & Ciraulo, D. A. (1994). Substance abuse and depression. *Psychiatric Annals, 24*(10), 532–539.

Renner, N. A., Ivey, N. S., Redmann, R. K., Lackner, A. A., & MacLean, A. G. (2011). MCP-3/CCL7 production by astrocytes: implications for SIV neuroinvasion and AIDS encephalitis. *Journal of Neurovirology, 17*(2), 146–152.

Reyes, T. M., Walker, J. R., DeCino, C., Hogenesch, J. B., & Sawchenko, P. E. (2003). Categorically distinct acute stressors elicit dissimilar transcriptional profiles in the paraventricular nucleus of the hypothalamus. *Journal of Neuroscience, 23*(13), 5607–5616.

Roos, R. S., Loetscher, M., Legler, D. F., Clark-Lewis, I., Baggiolini, M., & Moser, B. (1997). Identification of CCR8, the receptor for the human CC chemokine I-309. *Journal of Biological Chemistry, 272*(28), 17251–17254.

Rossi, D., & Zlotnik, A. (2000). The biology of chemokines and their receptors. *Annual Review of Immunology, 18*(1), 217–242.

Rostène, W., Kitabgi, P., & Parsadaniantz, S. M. (2007). Chemokines: a new class of neuromodulator? *Nature Reviews Neuroscience, 8*(11), 895–903.

Rostene, W., Dansereau, M. A., Godefroy, D., Van Steenwinckel, J., Reaux-Le Goazigo, A., Melik-Parsadaniantz, S., et al. (2011). Neurochemokines: a menage a trois providing new insights on the functions of chemokines in the central nervous system. *Journal of Neurochemistry, 118*(5), 680–694.

Rostene, W., Guyon, A., Kular, L., Godefroy, D., Barbieri, F., Bajetto, A., et al. (2011). Chemokines and chemokine receptors: new actors in neuroendocrine regulations. *Frontiers in Neuroendocrinology, 32*(1), 10–24.

Sakamoto, Y., Koike, K., Kiyama, H., Konishi, K., Watanabe, K., Osako, Y., et al. (1996). Endotoxin activates a chemokinergic neuronal pathway in the hypothalamo-pituitary system. *Endocrinology, 137*(10), 4503–4506.

Santiago, R. M., Barbiero, J., Gradowski, R. W., Bochen, S., Lima, M. M., Da Cunha, C., et al. (2014). Induction of depressive-like behavior by intranigral 6-OHDA is directly correlated with deficits in striatal dopamine and hippocampal serotonin. *Behavioural Brain Research, 259*, 70–77.

Sawada, T., Koike, K., Kanda, Y., Sakamoto, Y., Nohara, A., Ohmichi, M., et al. (1994). In vitro effects of CINC/gro, a member of the interleukin-8 family, on hormone secretion by rat anterior pituitary cells. *Biochemical and Biophysical Research Communications, 202*(1), 155–160.

Schloesser, R. J., Manji, H. K., & Martinowich, K. (2009). Suppression of adult neurogenesis leads to an increased HPA axis response. *Neuroreport, 20*(6), 553.

Schönemeier, B., Kolodziej, A., Schulz, S., Jacobs, S., Hoellt, V., & Stumm, R. (2008). Regional and cellular localization of the CXCl12/SDF-1 chemokine receptor CXCR7 in the developing and adult rat brain. *Journal of Comparative Neurology*, 510(2), 207–220.

Semple, B. D., Bye, N., Rancan, M., Ziebell, J. M., & Morganti-Kossmann, M. C. (2010). Role of CCL2 (MCP-1) in traumatic brain injury (TBI): evidence from severe TBI patients and CCL2−/− mice. *Journal of Cerebral Blood Flow & Metabolism*, 30(4), 769–782.

Sheline, Y. I., Gado, M. H., & Kraemer, H. C. (2003). Untreated depression and hippocampal volume loss. *American Journal of Psychiatry*, 160(8), 1516–1518.

Sheline, Y. I., Sanghavi, M., Mintun, M. A., & Gado, M. H. (1999). Depression duration but not age predicts hippocampal volume loss in medically healthy women with recurrent major depression. *Journal of Neuroscience*, 19(12), 5034–5043.

Sheridan, G. K., & Murphy, K. J. (2013). Neuron-glia crosstalk in health and disease: fractalkine and CX3CR1 take centre stage. *Open Biology*, 3(12).

Sheridan, G. K., Wdowicz, A., Pickering, M., Watters, O., Halley, P., O'Sullivan, N. C., et al. (2014). CX3CL1 is up-regulated in the rat hippocampus during memory-associated synaptic plasticity. *Frontiers in Cellular Neuroscience*, 8.

Shurin, G. V., Yurkovetsky, Z. R., Chatta, G. S., Tourkova, I. L., Shurin, M. R., & Lokshin, A. E. (2007). Dynamic alteration of soluble serum biomarkers in healthy aging. *Cytokine*, 39(2), 123–129.

Skrzydelski, D., Guyon, A., Dauge, V., Rovere, C., Apartis, E., Kitabgi, P., et al. (2007). The chemokine stromal cell-derived factor-1/CXCL12 activates the nigrostriatal dopamine system. *Journal of Neurochemistry*, 102(4), 1175–1183.

Smith, M. C., & Wrobel, J. P. (2014). Epidemiology and clinical impact of major comorbidities in patients with COPD. *International Journal of Chronic Obstructive Pulmonary Disease*, 9, 871–888.

Sonnenberg, C., Deeg, D., Van Tilburg, T., Vink, D., Stek, M., & Beekman, A. (2013). Gender differences in the relation between depression and social support in later life. *International Psychogeriatrics*, 25(01), 61–70.

Spleiss, O., Appel, K., Boddeke, H. W., Berger, M., & Gebicke-Haerter, P. J. (1998). Molecular biology of microglia cytokine and chemokine receptors and microglial activation. *Life Sciences*, 62(17–18), 1707–1710.

Stirling, D. P., Liu, S., Kubes, P., & Yong, V. W. (2009). Depletion of Ly6G/Gr-1 leukocytes after spinal cord injury in mice alters wound healing and worsens neurological outcome. *The Journal of Neuroscience : The Official Journal of the Society for Neuroscience*, 29(3), 753–764.

Stuart, M., & Baune, B. (2014). Chemokines and chemokine receptors in mood disorders, schizophrenia, and cognitive impairment: a systematic review of biomarker studies. *Neuroscience & Biobehavioral Reviews*, 42, 93–115.

Stuart, M. J., Corrigan, F., & Baune, B. T. (2014). Knockout of CXCR5 increases the population of immature neural cells and decreases proliferation in the hippocampal dentate gyrus. *Journal of Neuroinflammation*, 11(1), 31.

Stuart, M. J., Singhal, G., & Baune, B. T. (2015). Systematic review of the neurobiological relevance of chemokines to psychiatric disorders. *Frontiers in Cellular Neuroscience*, 9, 357.

Sunnemark, D., Eltayeb, S., Nilsson, M., Wallstrom, E., Lassmann, H., Olsson, T., et al. (2005). CX3CL1 (fractalkine) and CX3CR1 expression in myelin oligodendrocyte glycoprotein-induced experimental autoimmune encephalomyelitis: kinetics and cellular origin. *Journal of Neuroinflammation*, 2, 17.

Tabor-Godwin, J. M., Ruller, C. M., Bagalso, N., An, N., Pagarigan, R. R., Harkins, S., et al. (2010). A novel population of myeloid cells responding to coxsackievirus infection assists in the dissemination of virus within the neonatal CNS. *Journal of Neuroscience*, 30(25), 8676–8691.

Tang, M.-m., Lin, W.-j., Pan, Y.-q., Guan, X.-t., & Li, Y.-c. (2016). Hippocampal neurogenesis dysfunction linked to depressive-like behaviors in a neuroinflammation induced model of depression. *Physiology & Behavior*, 161, 166–173.

Théaudin, M., & Feinstein, A. (2015). Depression and multiple sclerosis: clinical aspects, epidemiology, and management. In *Neuropsychiatric symptoms of inflammatory demyelinating diseases* (pp. 17–25). Springer.

Thuc, O., Blondeau, N., Nahon, J. L., & Rovère, C. (2015). The complex contribution of chemokines to neuroinflammation: switching from beneficial to detrimental effects. *Annals of the New York Academy of Sciences*, 1351(1), 127–140.

Torres, M., Guiton, R., Lacroix-Lamandé, S., Ryffel, B., Leman, S., & Dimier-Poisson, I. (2013). MyD88 is crucial for the development of a protective CNS immune response to toxoplasma gondii infection. *Journal of Neuroinflammation*, 10(1), 19.

Tran, P. B., Ren, D., Veldhouse, T. J., & Miller, R. J. (2004). Chemokine receptors are expressed widely by embryonic and adult neural progenitor cells. *Journal of Neuroscience Research*, 76(1), 20–34.

Tran, P. B., Banisadr, G., Ren, D., Chenn, A., & Miller, R. J. (2007). Chemokine receptor expression by neural progenitor cells in neurogenic regions of mouse brain. *The Journal of Comparative Neurology*, 500(6), 1007–1033.

Trebst, C., Staugaitis, S. M., Kivisäkk, P., Mahad, D., Cathcart, M. K., Tucky, B., et al. (2003). CC chemokine receptor 8 in the central nervous system is associated with phagocytic macrophages. *The American Journal of Pathology*, 162(2), 427–438.

van der Meer, P., Ulrich, A. M., Gonźalez-Scarano, F., & Lavi, E. (2000). Immunohistochemical analysis of CCR2, CCR3, CCR5, and CXCR4 in the human brain: potential mechanisms for HIV dementia. *Experimental and Molecular Pathology*, 69(3), 192–201.

Van Deurzen, I., Van Ingen, E., & Van Oorschot, W. J. (2015). Income inequality and depression: the role of social comparisons and coping resources. *European Sociological Review*, jcv007.

Verburg-van Kemenade, B., Van der Aa, L., & Chadzinska, M. (2013). Neuroendocrine-immune interaction: regulation of inflammation via G-protein coupled receptors. *General and Comparative Endocrinology*, 188, 94–101.

Videbech, P., & Ravnkilde, B. (2004). Hippocampal volume and depression: a meta-analysis of MRI studies. *American Journal of Psychiatry*, 161(11), 1957–1966.

Villa, P., Triulzi, S., Cavalieri, B., Di Bitondo, R., Bertini, R., Barbera, S., et al. (2007). The interleukin-8 (IL-8/CXCL8) receptor inhibitor reparixin improves neurological deficits and reduces long-term inflammation in permanent and transient cerebral ischemia in rats. *Molecular Medicine-Cambridge MA Then New York*, 13(3/4), 125.

Villeda, S. A., Luo, J., Mosher, K. I., Zou, B., Britschgi, M., Bieri, G., et al. (2011). The ageing systemic milieu negatively regulates neurogenesis and cognitive function. *Nature*, 477(7362), 90–94.

Wang, F., Baba, N., Shen, Y., Yamashita, T., Tsuru, E., Tsuda, M., et al. (2017). CCL11 promotes migration and proliferation of mouse neural progenitor cells. *Stem Cell Research & Therapy*, 8(1), 26.

Watson, K., & Fan, G.-H. (2005). Macrophage inflammatory protein 2 inhibits β-amyloid peptide (1–42)-mediated hippocampal neuronal apoptosis through activation of mitogen-activated protein kinase and phosphatidylinositol 3-kinase signaling pathways. *Molecular Pharmacology*, 67(3), 757–765.

Weiss, N., Deboux, C., Chaverot, N., Miller, F., Baron-Van Evercooren, A., Couraud, P. O., et al. (2010). IL8 and CXCL13 are potent chemokines for the recruitment of human neural precursor cells across brain endothelial cells. *Journal of Neuroimmunology*, 223(1–2), 131–134.

Wilbanks, A., Zondlo, S. C., Murphy, K., Mak, S., Soler, D., Langdon, P., et al. (2001). Expression cloning of the STRL33/BONZO/TYMSTR ligand reveals elements of CC, CXC, and CX3C chemokines. *The Journal of Immunology*, 166(8), 5145–5154.

Witko-Sarsat, V., Rieu, P., Descamps-Latscha, B., Lesavre, P., & Halbwachs-Mecarelli, L. (2000). Neutrophils: molecules, functions and pathophysiological aspects. *Laboratory Investigation*, 80(5), 617.

Wolpe, S. D., Sherry, B., Juers, D., Davatelis, G., Yurt, R. W., & Cerami, A. (1989). Identification and characterization of macrophage inflammatory protein 2. *Proceedings of the National Academy of Sciences*, 86(2), 612–616.

Xiao, F., Xu, J.-m., & Jiang, X.-h. (2015). CX3 chemokine receptor 1 deficiency leads to reduced dendritic complexity and delayed maturation of newborn neurons in the adult mouse hippocampus. *Neural Regeneration Research*, 10(5), 772–777.

Yamagami, S., Tamura, M., Hayashi, M., Endo, N., Tanabe, H., Katsuura, Y., et al. (1999). Differential production of MCP-1 and cytokine-induced neutrophil chemoattractant in the ischemic brain after transient focal ischemia in rats. *Journal of Leukocyte Biology*, 65(6), 744–749.

Yu, L., Wu, X., Wei, J., Liao, Q., Xu, L., Luo, S., et al. (2015). Preliminary expression profile of cytokines in brain tissue of BALB/c mice with *Angiostrongylus cantonensis* infection. *Parasites & Vectors*, 8(1), 328.

Zlotnik, A., & Yoshie, O. (2000). Chemokines: a new classification system and their role in immunity. *Immunity*, 12 (2), 121–127.

Inflammasomes Action as an Important Mechanism in Experimental and Clinical Depression

Gaurav Singhal, Bernhard T. Baune

University of Adelaide, Adelaide, SA, Australia

INTRODUCTION

Depression is a psychiatric syndrome that is usually comorbid with other systemic and central nervous system (CNS) diseases such as cancer (Krebber et al., 2014), Alzheimer's disease (AD) (Chi et al., 2015), cardiovascular disease (CVD) (Air, Tully, Sweeney, & Beltrame, 2016; Carney & Freedland, 2017), multiple sclerosis (MS) (Théaudin & Feinstein, 2015), chronic obstructive pulmonary disease (COPD) (Smith & Wrobel, 2014), diabetes (Holt, De Groot, & Golden, 2014; Katon et al., 2005), and osteoarthritis (Blixen & Kippes, 1999; Hawker et al., 2011). However, a common etiologic link between depression and comorbid disorders is chronic inflammation, either systemic or CNS or both (Iwata, Ota, & Duman, 2013). Studies have shown that chronic neuroinflammation in response to psychological and physical stressors impairs hippocampal neurogenesis resulting in depression (Brites & Fernandes, 2015; Kim, Na, Myint, & Leonard, 2016; Tang et al., 2016).

Several immune proteins and cells such as cytokines, chemokines, and T and B cells are found to play an active part in the sustenance of inflammation over a long period in response to extrinsic pathogen-associated molecular patterns (PAMPs) and intrinsic by-products of psychological stress and tissue injury, also called as damage-associated molecular patterns (DAMPs). These PAMPs and DAMPs are recognized by receptors present in the membranes (toll-like receptors, TLRs) or cytoplasm (NOD-like receptors, NLRs) of the immune cells, for example, T cells and glial cells, and nonimmune cells, for example, neurons (Singhal et al., 2014). Once sensitized, these cells secrete a number of pro-inflammatory cytokines, for example, IL-1β and TNF-α, that causes neuroinflammation, in turn leading to the onset of depression. This is supported by the clinical evidence of increase in the levels of pro-inflammatory cytokines in serum and cerebrospinal fluid during depression (Lehto et al., 2010; Levine et al., 1999; Liu, Ho, & Mak, 2012; Maes et al., 1997; Owen, Eccleston, Ferrier, & Young, 2001). The overexpression of the pro-inflammatory cytokines can also dysregulate the HPA axis pathway leading to further aggravation of depressive-like behavior (Goshen & Yirmiya,

2009). These pro-inflammatory cytokines also elicit a negative feedback and stimulate the production of anti-inflammatory cytokines (Singhal et al., 2014). However, the anti-inflammatory effect may be insufficient to reduce inflammation, if there is constant sensitization, in particular by DAMPs, as seen during old age, resulting in chronic neuro-inflammation and thereby depression.

Important target molecules in the pathophysiology of inflammation associated depression are inflammasomes. These are the intracellular protein complexes, assembled and activated within the glial cells (astrocytes and microglia) and neurons in the CNS after recognition of PAMPs, DAMPs, and stress by the TLRs and NLRs (Franchi, Warner, Viani, & Nuñez, 2009; Inohara, Chamaillard, McDonald, & Nunez, 2005; Kanneganti, Lamkanfi, & Núñez, 2007; Singhal et al., 2014; Velasquez & Rappaport, 2016). Once triggered, inflammasomes, in particular NLRP3 inflammasome, activate caspase-I enzyme that functions to cleave the precursors of IL-1β, IL-18, and IL-33 cytokines to their pro-inflammatory states (Singhal et al., 2014). These members of the IL-1 family of cytokines have also been shown to dysregulate the HPA axis pathway that has been implicated in the pathophysiology of depression (Goshen & Yirmiya, 2009). Indeed, the overexpression of IL-1β in the brain regions associated with the development of depressive disorder, that is, the prefrontal cortex and hippocampus, has been reported in depressed patients (Pandey et al., 2012). Glial cells also release other pro-inflammatory cytokines, the most important of them is TNF-α, that in association with inflammasomes induced activated IL-1 family cytokines adds to the inflammatory state within the brain (Zhang & An, 2007). Hence, under the influence of chronic stimulus, these pro-inflammatory cytokines can contribute toward aggravating the systemic inflammation and neuroinflammation, resulting in various systemic and CNS diseases and their comorbidity with depression.

Psychological stress is another contributing factor to depression. Exposure to prenatal stress has shown an increase in the vulnerability of neuroinflammation-induced depression by altering the morphology and activity of glial cells and increasing the expression of pro-inflammatory cytokines (Diz-Chaves, Astiz, Bellini, & Garcia-Segura, 2013; Goshen & Yirmiya, 2009; Kubera et al., 2011; Ślusarczyk et al., 2015). Psychological stress stimulates the assembly and activation of inflammasomes in the glial cells resulting in the overexpression of IL-1 family pro-inflammatory cytokines (Alcocer-Gómez et al., 2016; Iwata et al., 2013, 2016; Salminen, Ojala, Kaarniranta, & Kauppinen, 2012; Zhang et al., 2014). Moreover, neuroinflammation associated with aging further sensitizes the brain to the effects of stress (Sparkman & Johnson, 2008). Overall, it starts a vicious cycle of ever-increasing levels of cytokines in the brain, leading to chronic neuroinflammation and associated depression.

In this chapter, we will discuss the biological role that inflammasomes and the associated pro-inflammatory IL-1 family cytokines play in the hypersensitization of the innate immune system leading to chronic neuroinflammation and the subsequent development of depression.

WHAT ARE INFLAMMASOMES?

Inflammasomes are intracellular cytosolic protein complexes that are assembled in response to PAMPs and DAMPs and function to cleave the inactive precursor forms of an IL-1β, IL-18, and IL-33 (members of IL-1 family of cytokines) into their pro-inflammatory active forms. These pro-inflammatory cytokines have been shown to initiate, regulate, and maintain inflammation (Dinarello, 2000). When in the brain, these IL-1 family cytokines have been reported to cause neuroinflammation (Arend, Palmer, & Gabay, 2008; Felderhoff-Mueser et al., 2005; Liew, Pitman, & McInnes, 2010), in turn leading to a number of neuropsychiatric disorders, such as depression, dementia, and AD (Cacquevel, Lebeurrier, Cheenne, & Vivien, 2004; Licastro et al., 2000;

McAfoose & Baune, 2009). It has been suggested that inflammasome activity increases with age, indicating that inflammasomes are associated with neuroinflammation during neuronal aging (Chakraborty, Kaushik, Gupta, & Basu, 2010; Liu & Chan, 2014; Mawhinney et al., 2011; Simi, Lerouet, Pinteaux, & Brough, 2007) that results in psychiatric disorders like depression (Zhang et al., 2014). While neurons also secrete IL-1 family cytokines, activated microglia and astrocytes are the main source of these cytokines in the brain (Hanisch, 2002; Rothwell, Luheshi, & Toulmond, 1996).

Although a number of inflammasomes have been recognized based on molecular structure, such as NLRP1, NLRP2, NLRP3, NLRC4, and AIM2, it is the NLRP3 inflammasome that has been extensively implicated in neuroinflammatory disorders, in particular in depression.

MOLECULAR STRUCTURE OF INFLAMMASOMES

NLRs are intracellular pattern recognition receptors (PRRs) that function similar to TLRs and identify PAMPs and DAMPs that attack the cell. This recognition of PAMPs and DAMPs by NLRs in turn activates the innate and acquired immune responses (Franchi et al., 2009; Inohara et al., 2005; Kanneganti et al., 2007; Velasquez & Rappaport, 2016), which includes the assembly of inflammasome complexes of NLR family in the cytoplasm of the attacked cell. These include NLRP1, NLRP2, NLRP3, NLRP6, NLRP7, NLRP10, NLRP12, and NLRC4 (Pedra, Cassel, & Sutterwala, 2009).

NLRP inflammasomes, as the name suggests, consist of NACHT, LRR, and PYD domain-containing proteins 1, 2, 3, 6, 7, 10, or 12 (hence the nomenclature, see Fig. 1). In addition, they have the adapter protein ASC (apoptosis-associated speck-like protein containing caspase recruitment domain (CARD)), also known as PYCARD, and enzyme procaspases 1 and 5. The amino (N)-terminal of the NLRP inflammasome has pyrin (PYD) domain, followed by NACHT domain, NACHT-associated domain (NAD), and several leucine-rich repeats (LRRs). The LRRs are then bonded either to FIIND domain followed by CARD domain at the carboxy (C)-terminal (as in the case of NLRP1 inflammasome) or to a cardinal protein that consists of FIIND domain on N-terminal

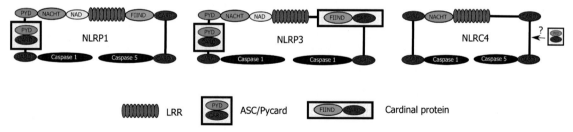

FIG. 1 Molecular structure of the NLRP1, NLRP3, and NLRC4 inflammasomes. All NLR inflammasomes consist of NACHT and LRR; however, the other components vary between different inflammasomes. NLRP1 inflammasome has pyrin (PYD) domain on the amino (N)-terminal bonded to a NACHT domain followed by NACHT-associated domain (NAD), several leucine-rich repeats (LRR), FIIND domain, and CARD at the carboxy (C)-terminal. The molecular structure of NLRP3 inflammasomes is similar as above, except that LRRs are linked to a cardinal protein that consists of FIIND domain on N-terminal and CARD domain on the C-terminal. NLRC4 inflammasomes have no PYD domain and instead have CARD domain on the N-terminal, followed by NACHT and LRR. NLRP inflammasomes interact with the adapter protein ASC (apoptosis-associated speck-like protein containing a CARD), also known as PYCARD, and procaspase enzymes 1 and 5, in turn converting proforms of IL-1β, IL-18, and IL-33 cytokines into their active pro-inflammatory forms within the cytoplasm of glial cells and neurons. Notably, enzyme caspase 5 may not be needed for the activation of enzyme caspase 1 in NLRP3 inflammasomes. Role of ASC in NLRC4 inflammasomes is still unclear, and it is yet to be elucidated how NLRC4 inflammasome interacts with caspase 5 enzyme at the C-terminal.

and CARD domain on the C-terminal (as in the case of NLRP3 inflammasome). When activated, NLRPs recruit ASC that has a CARD domain. ASC interacts with the CARD of procaspase 1 converting it into active caspase 1, which then cleave the precursors of IL-1β, IL-18, and IL-33 into their active forms, instigating an inflammatory response (Martinon, Burns, & Tschopp, 2002; Petrilli, Papin, & Tschopp, 2005). NLRC4 inflammasome, on the other hand, does not have PYD domain and instead has CARD domain on the N-terminal followed by NACHT domain and several LRRs on the C-terminal (see Fig. 1) that directly interacts with procaspases 1 and 5.

The NLR family, pyrin domain-containing 3 (NLRP3) is the most studied, largest, and best-characterized inflammasome during inflammation (Stutz, Golenbock, & Latz, 2009) and during depression (Alcocer-Gómez & Cordero, 2014; Zhang et al., 2015). As such, we will keep our discussion targeted at the role of the NLRP3 inflammasome in depression in this chapter.

NLRP3 INFLAMMASOME-MEDIATED INFLAMMATORY PATHWAYS IN THE BRAIN

Research suggests that glial pro-inflammatory cytokines that cause inflammation of the brain parenchyma are closely associated with the age-related depression and aggravate cognitive and memory deficit (Godbout et al., 2008; Huang et al., 2008; Mrak & Griffin, 2005). Increased expression of pro-inflammatory cytokines, TNF-α, IL-1β, and IL-6, in the brain of MDD patients has been reported in a number of experimental and meta-analytic studies (Dowlati et al., 2010; Hannestad, DellaGioia, & Bloch, 2011; Howren, Lamkin, & Suls, 2009; Maes et al., 1997). Chronic neuroinflammation has also been shown to be the cause of several depression comorbid neurodegenerative diseases, such as AD and PD (Heneka, O'Banion, Terwel, & Kummer, 2010; Hirsch, Vyas, & Hunot, 2012), and may result in cognitive

impairment and dementia (Chakraborty et al., 2010; Liu & Chan, 2014; Mawhinney et al., 2011; Simi et al., 2007). NLRP3 inflammasomes play a vital role in the pathophysiology of neuroinflammation by producing an IL-1 family of pro-inflammatory cytokines as mentioned before. In addition to PAMPs and DAMPs, reactive oxygen species (ROS) that are produced in dysfunctional mitochondria and enhanced NF-κB signaling in the aging brain could further trigger the assembly and activation of NLRP3 inflammasomes leading to greater inflammatory response (Salminen et al., 2012). Indeed, under cellular stress, mitochondria may undergo apoptosis and the released oxidized DNA acts as DAMP that initiates the assembly and activation of NLRP3 inflammasomes in the cytosol (Bakunina, Pariante, & Zunszain, 2015; Latz, Xiao, & Stutz, 2013). Likewise, mutant α-synuclein and Aβ fibrils, seen during PD and AD respectively, can also prime NLRP3 inflammasomes within the glial cells (Cedillos, 2013; Salminen et al., 2009; Tschopp & Schroder, 2010). Studies have found that several genes that signal inflammasome assembly and activation, for example, thioredoxin-interacting protein, P2X7, and pannexins, and those that signal TLRs activity, for example, CD14, TLR2, TLR4, TLR7, TOLLIP, and MYD88, upregulate in the hippocampus, postcentral gyrus, and superior frontal gyrus regions of the aging brain (Cribbs et al., 2012; Singhal et al., 2014).

MECHANISTIC LINK BETWEEN INFLAMMASOMES ACTION AND DEVELOPMENT OF DEPRESSIVE-LIKE BEHAVIOR

Dementia, cognitive impairment, and decline in spatial memory are often the result of neuroinflammatory changes within the brain and closely linked to depressive-like behavior. A rise in DAMPs with aging could prime NLRP3 inflammasomes that in turn increases the levels of pro-inflammatory cytokines, in particular

IL-1β, in the brain leading to neuroinflammation and associated age-related depression (Capuron & Miller, 2011; Godbout et al., 2008; Sparkman & Johnson, 2008; Wager-Smith & Markou, 2011). Several other studies have also found NLRP3 inflammasomes to have a role in the etiology of depression, especially during chronic mild stress (Alcocer-Gómez et al., 2016; Farooq et al., 2012; Velasquez & Rappaport, 2016; Zhang et al., 2014). The increase in the levels of IL-1 cytokines and neuroinflammation in the brain of depressed patients are reliable indicators for inflammasomes playing a role in the pathophysiology of MDD (Schwarz, Chiang, Müller, & Ackenheil, 2001). In studies on mice, lipopolysaccharide (LPS) injection stimulated activation of NLRP3 inflammasome leading to neuroinflammation and subsequent depressive-like behavior (Zhang et al., 2014; Zhu et al., 2017). On the contrary, mice that lacked caspase 1 were resistant to LPS-induced depressive-like behavior (Moon et al., 2009), further supporting inflammasome hypothesis of neuroinflammation and subsequent depressive-like behavior. Likewise, in humans, NLRP3 inflammasomes were detected in the blood mononuclear cells from depressive patients (Alcocer-Gómez et al., 2013), supporting the findings suggesting that inflammasome-related inflammation could be an ongoing process (Hohmann, Bechter, & Schneider, 2014). A recent review on the role of inflammasomes in MDD and its comorbidity with systemic illnesses discusses the central mediator role that inflammasomes play in the contribution of psychological and physical stressors to the development of depression and its association with systemic illnesses (Iwata et al., 2013). Several other findings have also supported the role of NLRP3 inflammasomes in the etiology of neuroinflammation and subsequent depression (Dowlati et al., 2010; Hannestad et al., 2011; Howren et al., 2009; Maes et al., 1997).

A common link between the chronic inflammatory diseases such as cancer (Il'yasova et al., 2005), diabetes (De Rekeneire et al., 2006), osteoarthritis (Stannus et al., 2013), and CVD (Volpato et al., 2001) and age-related depression (Godbout et al., 2008) is an increase in the level of pro-inflammatory cytokines such as TNF-α, IL-1β, and IL-6, both in the systemic circulation and in the brain. Pro-inflammatory cytokines therefore migrate from systemic circulation to the brain and vice versa, using pathways as described in a review by Capuron and Miller (Capuron & Miller, 2011). For example, increased levels of IL-1β in the brain of rodents were observed after LPS-induced systemic inflammation (Cunningham et al., 2005), and this has been shown to cause changes in behavior similar to depression (Pollak & Yirmiya, 2002). Inflammasomes may therefore be responsible for the comorbidity of systemic diseases such as type II diabetes (Grant & Dixit, 2013; Lee et al., 2013), obesity (Stienstra et al., 2011), CVD (Carney & Freedland, 2017; Connat, 2011), and cancer (Fallowfield, Ratcliffe, Jenkins, & Saul, 2001) with depression (see Fig. 2). It is important to note that significant changes in the number, density, and size of glial cells were also documented during MDD (Ye, Wang, Wang, & Wang, 2011), suggestive of the role of glial cells in the pathogenesis of the depression.

DISCUSSION

The 11 secreted factors of the IL-1 family of cytokines, including IL-1α, IL-18, and IL-33, are known to be involved in the host defense and in the immune regulation of inflammatory diseases (Arend et al., 2008; Barksby, Lea, Preshaw, & Taylor, 2007; Dinarello, 2000, 2009; Sims & Smith, 2010). These pro-inflammatory cytokines when overexpressed in the CNS have been shown to result in neuroinflammatory and neurodegenerative diseases, such as MDD, AD, and dementia (Arend et al., 2008; Cacquevel et al., 2004; Felderhoff-Mueser et al., 2005; Licastro et al., 2000; Liew et al., 2010; McAfoose & Baune, 2009; Tarkowski et al., 2003). In particular, during aging, there is an

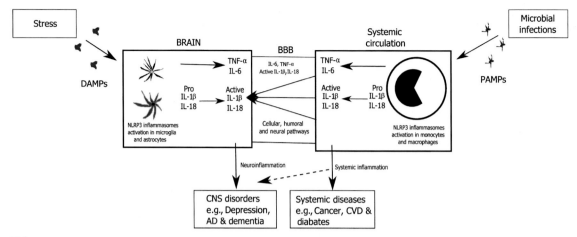

FIG. 2 The above illustration shows a link between systemic inflammatory conditions and CNS disorders via activation of NLRP3 inflammasomes leading to depression. PAMPs and DAMPs from trauma, infection, and metabolic waste prime NLRP3 inflammasomes in monocytes and macrophages in the systemic circulation and microglia and astrocytes in the CNS, leading to the release of pro-inflammatory cytokines TNF-α, IL-1β, IL-18, IL-33, and IL-6. These pro-inflammatory cytokines can migrate from systemic circulation to the brain via three pathways, that is, cellular, humoral, and neural as described by Capuron and Miller (Capuron & Miller, 2011). These cytokines activate granulocytes, monocytes/macrophages, and natural killer and T cells, hence contributing to the pathophysiology of neuroinflammation and associated psychiatric disorders, including depression. PAMPs, pathogen-associated molecular patterns; DAMPs, damage-associated molecular patterns; AD, Alzheimer's disease; CVD, cardiovascular diseases; BBB, blood-brain barrier.

increase in the levels of pro-inflammatory cytokines in the brain, which has been shown to be associated with dementia, decline in cognitive abilities, impairment of spatial memory, and age-related depression (Godbout et al., 2008). Evidence suggests that these pro-inflammatory cytokines are primarily overexpressed by the astrocytes and microglia in the brain resulting in neuroinflammation followed by neurodegeneration, a known etiology for cognitive and memory deficit and exacerbated sickness and depressive disorder during old age (Dong & Benveniste, 2001; Huang et al., 2008; Mrak & Griffin, 2005; Streit, 2005).

Transmembrane TLRs and cytosolic NLRs in the glial cells and neurons initiate an inflammasome cascade in response to PAMPs and DAMPs, leading to the release of caspase 1 enzyme that in turn cleaves the proforms of the IL-1 family of cytokines to their active forms (Franchi et al., 2009;

Inohara et al., 2005; Kanneganti et al., 2007). In addition to inducing acute neuroinflammatory response, this may also cause oxidative and nitrosative stress (Frank-Cannon, Alto, McAlpine, & Tansey, 2009). When the stimuli are continuous, potent, and self-replicating, it could result in chronic neuroinflammation followed by neurodegenerative changes and in short-term cognitive impairment and exacerbated sickness and depressive behavior. Indeed, chronic neuroinflammation that is generally seen during old age (Sparkman & Johnson, 2008) has been shown to be a cause of depression (Wager-Smith & Markou, 2011). Several meta-analytic studies have also confirmed neuroinflammation as an etiology of depression (Dowlati et al., 2010; Hannestad et al., 2011; Howren et al., 2009; Maes et al., 1997).

Recent findings point to the role of inflammasomes in the pathophysiology of neuroinflammation during neuronal aging and associated neurodegenerative diseases, cognitive impairment, dementia, and depression (Chakraborty et al., 2010; Liu & Chan, 2014; Mawhinney et al., 2011; Simi et al., 2007; Zhang et al., 2015). For example, several genes that signal inflammasome assembly and activation of caspase 1, such as thioredoxin-interacting protein, P2X7, and pannexins, and signaling of TLRs, such as CD14, TLR2, TLR4, TLR7, TOL-LIP, and MYD88, upregulate in the hippocampus, postcentral gyrus, and superior frontal gyrus during aging (Cribbs et al., 2012). The continuous sensitization of the TLRs and NLRs by PAMPs and DAMPs in the aging brain can therefore result in a sustained activation of the NLRP3 inflammasomes in the cytoplasm of glial cells causing neuroinflammation and associated neurodegenerative diseases. In addition, cellular stress can result in mitochondrial apoptosis, releasing ROS, and oxidized mitochondrial DNA, which in addition to enhanced NF-κB signaling with aging could further prime the NLRP3 inflammasomes in the brain resulting in an aggravated inflammatory response (Bakunina et al., 2015; Salminen et al., 2012). Furthermore, chronic mild stress has also been shown to activate NLRP3 inflammasomes resulting in neuroinflammation with subsequent occurrence of depression (Farooq et al., 2012; Zhang et al., 2015).

Research has shown that inflammasome-induced neuroinflammation occurs constantly in psychiatric patients (Hohmann et al., 2014). This is supported by both laboratory research and clinical findings. For example, NLRP3 inflammasome has been shown to play a role in the onset of LPS-induced mouse depressive-like behavior (Zhang et al., 2014). Likewise, blood mononuclear cells in depressive patients have bene shown to contain activated NLRP3 inflammasomes (Alcocer-Gómez et al., 2013). In another random controlled trial, caspase 1 knockout mice, when injected with LPS, showed no onset of depressive-like behavior (Moon et al., 2009).

Interestingly, inflammasomes have also been shown to be activated through TNF-α-dependent pathways (Alvarez & Munoz-Fernandez, 2013). Both TNF-α and IL-1β stimulate each other's secretion and elicit similar pro-inflammatory effects (Akira, Hirano, Taga, & Kishimoto, 1990; Ikejima et al., 1990; Knofler et al., 1997). In addition, TNF-α has been shown to induce secretion of IL-33 in keratinocytes (Taniguchi et al., 2013) and regulate expression of IL-18 in dendritic precursor-like cell line KG-1 and cardiomyocytes (Chandrasekar et al., 2003; Koutoulaki et al., 2010). Overall, this suggests that there could be a possible bidirectional cause-effect relationship between inflammasomes and TNF-α (Singhal et al., 2014).

Taken together, the above discussion suggests that NLRP3 inflammasomes act as the central mediator between psychological and physical stressors and the development of depression, as well as in the association between depression and systemic illnesses. As such, NLRP3 inflammasomes could potentially be looked upon to develop inflammasome-targeted therapies for depression in the near future.

CONCLUDING REMARKS

In conclusion, inflammasomes provide a vital tool for regulating various innate immune pathways, in particular, those associated with neuroinflammation leading to depression. However, despite recent understanding in inflammasomes-associated mechanistic links, their role in causing depression is not fully elucidated. However, it is clear that inflammasomes play a vital role in the onset and the sustenance of chronic inflammation (systemic and neuro) leading to depression. Further research and clinical trials into the role of inflammasomes in depression are advisable before inflammasome-targeted therapies could be developed for the treatment of patients with depression.

CONFLICT OF INTEREST STATEMENT

The presented work is supported by the National Health and Medical Research Council, Australia (APP 1043771 to BTB). The funders had no role in study design, data collection and analysis, decision to publish, or preparation of the manuscript.

References

Air, T., Tully, P. J., Sweeney, S., & Beltrame, J. (2016). Epidemiology of cardiovascular disease and depression. In *Cardiovascular diseases and depression* (pp. 5–21). Springer.

Akira, S., Hirano, T., Taga, T., & Kishimoto, T. (1990). Biology of multifunctional cytokines: IL 6 and related molecules (IL 1 and TNF). *The FASEB Journal*, 4(11), 2860–2867.

Alcocer-Gómez, E., & Cordero, M. D. (2014). NLRP3 inflammasome: a new target in major depressive disorder. *CNS Neuroscience & Therapeutics*, 20(3), 294–295.

Alcocer-Gómez, E., de Miguel, M., Casas-Barquero, N., Núñez-Vasco, J., et al. (2013). NLRP3 inflammasome is activated in mononuclear. *Blood Cells*.

Alcocer-Gómez, E., Ulecia-Morón, C., Marín-Aguilar, F., Rybkina, T., et al. (2016). Stress-induced depressive behaviors require a functional NLRP3 inflammasome. *Molecular Neurobiology*, 53(7), 4874–4882.

Alvarez, S., & Munoz-Fernandez, M. A. (2013). TNF-alpha may mediate inflammasome activation in the absence of bacterial infection in more than one way. *PLoS One*, 8(8).

Arend, W. P., Palmer, G., & Gabay, C. (2008). IL-1, IL-18, and IL-33 families of cytokines. *Immunological Reviews*, 223(1), 20–38.

Bakunina, N., Pariante, C. M., & Zunszain, P. A. (2015). Immune mechanisms linked to depression via oxidative stress and neuroprogression. *Immunology*, 144(3), 365–373.

Barksby, H., Lea, S., Preshaw, P., & Taylor, J. (2007). The expanding family of interleukin-1 cytokines and their role in destructive inflammatory disorders. *Clinical & Experimental Immunology*, 149(2), 217–225.

Blixen, C. E., & Kippes, C. (1999). Depression, social support, and quality of life in older adults with osteoarthritis. *Journal of Nursing Scholarship*, 31(3), 221–226.

Brites, D., & Fernandes, A. (2015). Neuroinflammation and depression: microglia activation, extracellular microvesicles and microRNA dysregulation. *Frontiers in Cellular Neuroscience*, 9, 476.

Cacquevel, M., Lebeurrier, N., Cheenne, S., & Vivien, D. (2004). Cytokines in neuroinflammation and Alzheimer's disease. *Current Drug Targets*, 5(6), 529–534.

Capuron, L., & Miller, A. H. (2011). Immune system to brain signaling: neuropsychopharmacological implications. *Pharmacology & Therapeutics*, 130(2), 226–238.

Carney, R. M., & Freedland, K. E. (2017). Reply: NLRP3 inflammasome as a mechanism linking depression and cardiovascular diseases. *Nature Reviews Cardiology*.

Cedillos, R. O. (2013). *Alpha-synuclein aggregates activate the Nlrp3 Inflammasome following vesicle rupture.* Master thesis Loyola University Chicago.

Chakraborty, S., Kaushik, D. K., Gupta, M., & Basu, A. (2010). Inflammasome signaling at the heart of central nervous system pathology. *Journal of Neuroscience Research*, 88(8), 1615–1631.

Chandrasekar, B., Colston, J. T., de la Rosa, S. D., Rao, P. P., et al. (2003). TNF-alpha and H_2O_2 induce IL-18 and IL-18R beta expression in cardiomyocytes via NF-kappa B activation. *Biochemical and Biophysical Research Communications*, 303(4), 1152–1158.

Chi, S., Wang, C., Jiang, T., Zhu, X.-C., et al. (2015). The prevalence of depression in Alzheimer's disease: a systematic review and meta-analysis. *Current Alzheimer Research*, 12(2), 189–198.

Connat, J. L. (2011). Inflammasome and cardiovascular diseases. *Annales de cardiologie et d'angéiologie*, 60(1), 48–54.

Cribbs, D. H., Berchtold, N. C., Perreau, V., Coleman, P. D., et al. (2012). Extensive innate immune gene activation accompanies brain aging, increasing vulnerability to cognitive decline and neurodegeneration: a microarray study. *Journal of Neuroinflammation*, 9, 179.

Cunningham, C., Wilcockson, D. C., Campion, S., Lunnon, K., et al. (2005). Central and systemic endotoxin challenges exacerbate the local inflammatory response and increase neuronal death during chronic neurodegeneration. *The Journal of Neuroscience*, 25(40), 9275–9284.

De Rekeneire, N., Peila, R., Ding, J., Colbert, L. H., et al. (2006). Diabetes, hyperglycemia, and inflammation in older individuals the health, aging and body composition study. *Diabetes Care*, 29(8), 1902–1908.

Dinarello, C. A. (2000). Proinflammatory cytokines. *Chest Journal*, 118(2), 503–508.

Dinarello, C. A. (2009). Immunological and inflammatory functions of the interleukin-1 family. *Annual Review of Immunology*, 27, 519–550.

Diz-Chaves, Y., Astiz, M., Bellini, M. J., & Garcia-Segura, L. M. (2013). Prenatal stress increases the expression of proinflammatory cytokines and exacerbates the inflammatory response to LPS in the hippocampal formation of adult male mice. *Brain, Behavior, and Immunity*, 28, 196–206.

Dong, Y., & Benveniste, E. N. (2001). Immune function of astrocytes. *Glia*, 36(2), 180–190.

Dowlati, Y., Herrmann, N., Swardfager, W., Liu, H., et al. (2010). A meta-analysis of cytokines in major depression. *Biological Psychiatry, 67*(5), 446–457.

Fallowfield, L., Ratcliffe, D., Jenkins, V., & Saul, J. (2001). Psychiatric morbidity and its recognition by doctors in patients with cancer. *British Journal of Cancer, 84*(8), 1011.

Farooq, R. K., Isingrini, E., Tanti, A., Le Guisquet, A.-M., et al. (2012). Is unpredictable chronic mild stress (UCMS) a reliable model to study depression-induced neuro-inflammation? *Behavioural Brain Research, 231*(1), 130–137.

Felderhoff-Mueser, U., Schmidt, O. I., Oberholzer, A., Bührer, C., et al. (2005). IL-18: a key player in neuroinflammation and neurodegeneration? *Trends in Neurosciences, 28*(9), 487–493.

Franchi, L., Warner, N., Viani, K., & Nuñez, G. (2009). Function of NOD-like receptors in microbial recognition and host defense. *Immunological Reviews, 227*(1), 106–128.

Frank-Cannon, T. C., Alto, L. T., McAlpine, F. E., & Tansey, M. G. (2009). Does neuroinflammation fan the flame in neurodegenerative diseases. *Molecular Neurodegeneration, 4*(47), 1–13.

Godbout, J. P., Moreau, M., Lestage, J., Chen, J., et al. (2008). Aging exacerbates depressive-like behavior in mice in response to activation of the peripheral innate immune system. *Neuropsychopharmacology, 33*(10), 2341–2351.

Goshen, I., & Yirmiya, R. (2009). Interleukin-1 (IL-1): a central regulator of stress responses. *Frontiers in Neuroendocrinology, 30*(1), 30–45.

Grant, R. W., & Dixit, V. D. (2013). Mechanisms of disease: inflammasome activation and the development of type 2 diabetes. *Frontiers in Immunology, 4*, 50.

Hanisch, U. K. (2002). Microglia as a source and target of cytokines. *Glia, 40*(2), 140–155.

Hannestad, J., DellaGioia, N., & Bloch, M. (2011). The effect of antidepressant medication treatment on serum levels of inflammatory cytokines: a meta-analysis. *Neuropsychopharmacology, 36*(12), 2452–2459.

Hawker, G. A., Gignac, M. A., Badley, E., Davis, A. M., et al. (2011). A longitudinal study to explain the pain-depression link in older adults with osteoarthritis. *Arthritis Care & Research, 63*(10), 1382–1390.

Heneka, M. T., O'Banion, M. K., Terwel, D., & Kummer, M. P. (2010). Neuroinflammatory processes in Alzheimer's disease. *Journal of Neural Transmission, 117*(8), 919–947.

Hirsch, E. C., Vyas, S., & Hunot, S. (2012). Neuroinflammation in Parkinson's disease. *Parkinsonism & Related Disorders, 18*, S210–S212.

Hohmann, H., Bechter, K., & Schneider, E. (2014). A small and validated cytokine panel supports inflammasome activation in cerebrospinal fluid of patients with major depression and schizophrenia. *Neurology, Psychiatry and Brain Research, 20*(1), 13–14.

Holt, R. I., De Groot, M., & Golden, S. H. (2014). Diabetes and depression. *Current Diabetes Reports, 14*(6), 1–9.

Howren, M. B., Lamkin, D. M., & Suls, J. (2009). Associations of depression with C-reactive protein, IL-1, and IL-6: a meta-analysis. *Psychosomatic Medicine, 71*(2), 171–186.

Huang, Y., Henry, C., Dantzer, R., Johnson, R., et al. (2008). Exaggerated sickness behavior and brain proinflammatory cytokine expression in aged mice in response to intracerebroventricular lipopolysaccharide. *Neurobiology of Aging, 29*(11), 1744–1753.

Ikejima, T., Okusawa, S., Ghezzi, P., Van Der Meer, J. W., et al. (1990). Interleukin-l induces tumor necrosis factor (TNF) in human peripheral blood mononuclear cells in vitro and a circulating TNF-like activity in rabbits. *Journal of Infectious Diseases, 162*(1), 215–223.

Il'yasova, D., Colbert, L. H., Harris, T. B., Newman, A. B., et al. (2005). Circulating levels of inflammatory markers and cancer risk in the health aging and body composition cohort. *Cancer Epidemiology Biomarkers & Prevention, 14*(10), 2413–2418.

Inohara, N., Chamaillard, M., McDonald, C., & Nunez, G. (2005). NOD-LRR proteins: role in host-microbial interactions and inflammatory disease. *Annual Review of Biochemistry, 74*, 355–383.

Iwata, M., Ota, K. T., & Duman, R. S. (2013). The inflammasome: pathways linking psychological stress, depression, and systemic illnesses. *Brain, Behavior, and Immunity, 31*, 105–114.

Iwata, M., Ota, K. T., Li, X.-Y., Sakaue, F., et al. (2016). Psychological stress activates the inflammasome via release of adenosine triphosphate and stimulation of the purinergic type 2X7 receptor. *Biological Psychiatry, 80*(1), 12–22.

Kanneganti, T.-D., Lamkanfi, M., & Núñez, G. (2007). Intracellular NOD-like receptors in host defense and disease. *Immunity, 27*(4), 549–559.

Katon, W. J., Rutter, C., Simon, G., Lin, E. H., et al. (2005). The association of comorbid depression with mortality in patients with type 2 diabetes. *Diabetes Care, 28*(11), 2668–2672.

Kim, Y.-K., Na, K.-S., Myint, A.-M., & Leonard, B. E. (2016). The role of pro-inflammatory cytokines in neuroinflammation, neurogenesis and the neuroendocrine system in major depression. *Progress in Neuro-Psychopharmacology and Biological Psychiatry, 64*, 277–284.

Knofler, M., Kiss, H., Mosl, B., Egarter, C., et al. (1997). Interleukin-1 stimulates tumor necrosis factor-alpha (TNF-alpha) release from cytotrophoblastic BeWo cells independently of induction of the TNF-alpha mRNA. *FEBS Letters, 405*(2), 213–218.

Koutoulaki, A., Langley, M., Sloan, A. J., Aeschlimann, D., et al. (2010). TNFα and TGF-β1 influence IL-18-induced IFNγ production through regulation of IL-18 receptor and T-bet expression. *Cytokine, 49*(2), 177–184.

Krebber, A., Buffart, L., Kleijn, G., Riepma, I., et al. (2014). Prevalence of depression in cancer patients: a meta-analysis of diagnostic interviews and self-report instruments. *Psycho-Oncology, 23*(2), 121–130.

Kubera, M., Obuchowicz, E., Goehler, L., Brzeszcz, J., et al. (2011). In animal models, psychosocial stress-induced (neuro) inflammation, apoptosis and reduced neurogenesis are associated to the onset of depression. *Progress in Neuro-Psychopharmacology and Biological Psychiatry, 35*(3), 744–759.

Latz, E., Xiao, T. S., & Stutz, A. (2013). Activation and regulation of the inflammasomes. *Nature Reviews Immunology, 13*(6), 397–411.

Lee, H. M., Kim, J. J., Kim, H. J., Shong, M., et al. (2013). Upregulated NLRP3 inflammasome activation in patients with type 2 diabetes. *Diabetes, 62*(1), 194–204.

Lehto, S. M., Niskanen, L., Herzig, K.-H., Tolmunen, T., et al. (2010). Serum chemokine levels in major depressive disorder. *Psychoneuroendocrinology, 35*(2), 226–232.

Levine, J., Barak, Y., Chengappa, K., & Rapoport, A. (1999). Cerebrospinal cytokine levels in patients with acute depression. *Neuropsychobiology, 40*(4), 171–176.

Licastro, F., Pedrini, S., Caputo, L., Annoni, G., et al. (2000). Increased plasma levels of interleukin-1, interleukin-6 and α-1-antichymotrypsin in patients with Alzheimer's disease: peripheral inflammation or signals from the brain? *Journal of Neuroimmunology, 103*(1), 97–102.

Liew, F. Y., Pitman, N. I., & McInnes, I. B. (2010). Disease-associated functions of IL-33: the new kid in the IL-1 family. *Nature Reviews Immunology, 10*(2), 103–110.

Liu, L., & Chan, C. (2014). The role of inflammasome in Alzheimer's disease. *Ageing Research Reviews*.

Liu, Y., Ho, R. C.-M., & Mak, A. (2012). Interleukin (IL)-6, tumour necrosis factor alpha (TNF-α) and soluble interleukin-2 receptors (sIL-2R) are elevated in patients with major depressive disorder: a meta-analysis and meta-regression. *Journal of Affective Disorders, 139*(3), 230–239.

Maes, M., Bosmans, E., De Jongh, R., Kenis, G., et al. (1997). Increased serum IL-6 and IL-1 receptor antagonist concentrations in major depression and treatment resistant depression. *Cytokine, 9*(11), 853–858.

Martinon, F., Burns, K., & Tschopp, J. (2002). The inflammasome: a molecular platform triggering activation of inflammatory caspases and processing of proIL-beta. *Molecular Cell, 10*(2), 417–426.

Mawhinney, L. J., de Rivero Vaccari, J. P., Dale, G. A., Keane, R. W., et al. (2011). Heightened inflammasome activation is linked to age-related cognitive impairment in Fischer 344 rats. *BMC Neuroscience, 12*(1), 123.

McAfoose, J., & Baune, B. T. (2009). Evidence for a cytokine model of cognitive function. *Neuroscience and Biobehavioral Reviews, 33*(3), 355–366.

Moon, M., McCusker, R., Lawson, M., Dantzer, R., et al. (2009). Mice lacking the inflammasome component caspase-1 are resistant to central lipopolysaccharide-induced depressive-like behavior. *Brain, Behavior, and Immunity, 23*, S50.

Mrak, R. E., & Griffin, W. S. T. (2005). Glia and their cytokines in progression of neurodegeneration. *Neurobiology of Aging, 26*(3), 349–354.

Owen, B., Eccleston, D., Ferrier, I., & Young, H. (2001). Raised levels of plasma interleukin-1β in major and postviral depression. *Acta Psychiatrica Scandinavica, 103*(3), 226–228.

Pandey, G. N., Rizavi, H. S., Ren, X., Fareed, J., et al. (2012). Proinflammatory cytokines in the prefrontal cortex of teenage suicide victims. *Journal of Psychiatric Research, 46*(1), 57–63.

Pedra, J. H., Cassel, S. L., & Sutterwala, F. S. (2009). Sensing pathogens and danger signals by the inflammasome. *Current Opinion in Immunology, 21*(1), 10–16.

Petrilli, V., Papin, S., & Tschopp, J. (2005). The inflammasome. *Current Biology, 15*(15), R581.

Pollak, Y., & Yirmiya, R. (2002). Cytokine-induced changes in mood and behaviour: implications for 'depression due to a general medical condition', immunotherapy and antidepressive treatment. *The International Journal of Neuropsychopharmacology, 5*(04), 389–399.

Rothwell, N. J., Luheshi, G., & Toulmond, S. (1996). Cytokines and their receptors in the central nervous system: physiology, pharmacology, and pathology. *Pharmacology & Therapeutics, 69*(2), 85–95.

Salminen, A., Ojala, J., Kaarniranta, K., & Kauppinen, A. (2012). Mitochondrial dysfunction and oxidative stress activate inflammasomes: impact on the aging process and age-related diseases. *Cellular and Molecular Life Sciences, 69*(18), 2999–3013.

Salminen, A., Ojala, J., Kauppinen, A., Kaarniranta, K., et al. (2009). Inflammation in Alzheimer's disease: amyloid-β oligomers trigger innate immunity defence via pattern recognition receptors. *Progress in Neurobiology, 87*(3), 181–194.

Schwarz, M. J., Chiang, S., Müller, N., & Ackenheil, M. (2001). T-helper-1 and T-helper-2 responses in psychiatric disorders. *Brain, Behavior, and Immunity, 15*(4), 340–370.

Simi, A., Lerouet, D., Pinteaux, E., & Brough, D. (2007). Mechanisms of regulation for interleukin-1β in neurodegenerative disease. *Neuropharmacology, 52*(8), 1563–1569.

Sims, J. E., & Smith, D. E. (2010). The IL-1 family: regulators of immunity. *Nature Reviews Immunology, 10*(2), 89–102.

Singhal, G., Jaehne, E. J., Corrigan, F., Toben, C., et al. (2014). Inflammasomes in neuroinflammation and changes in brain function: a focused review. *Frontiers in Neuroscience, 8*, 315.

Ślusarczyk, J., Trojan, E., Głombik, K., Budziszewska, B., et al. (2015). Prenatal stress is a vulnerability factor for altered morphology and biological activity of microglia cells. *Frontiers in Cellular Neuroscience, 9*, 82.

Smith, M. C., & Wrobel, J. P. (2014). Epidemiology and clinical impact of major comorbidities in patients with

COPD. *International Journal of Chronic Obstructive Pulmonary Disease, 9*, 871–888.

Sparkman, N. L., & Johnson, R. W. (2008). Neuroinflammation associated with aging sensitizes the brain to the effects of infection or stress. *Neuroimmunomodulation, 15*(4–6), 323–330.

Stannus, O. P., Jones, G., Blizzard, L., Cicuttini, F. M., et al. (2013). Associations between serum levels of inflammatory markers and change in knee pain over 5 years in older adults: a prospective cohort study. *Annals of the Rheumatic Diseases, 72*(4), 535–540.

Stienstra, R., van Diepen, J. A., Tack, C. J., Zaki, M. H., et al. (2011). Inflammasome is a central player in the induction of obesity and insulin resistance. *Proceedings of the National Academy of Sciences of the United States of America, 108*(37), 15324–15329.

Streit, W. J. (2005). Microglia and neuroprotection: implications for Alzheimer's disease. *Brain Research Reviews, 48*(2), 234–239.

Stutz, A., Golenbock, D. T., & Latz, E. (2009). Inflammasomes: too big to miss. *The Journal of Clinical Investigation, 119*(12), 3502.

Tang, M.-M., Lin, W.-J., Pan, Y.-Q., Guan, X.-T., et al. (2016). Hippocampal neurogenesis dysfunction linked to depressive-like behaviors in a neuroinflammation induced model of depression. *Physiology & Behavior, 161*, 166–173.

Taniguchi, K., Yamamoto, S., Hitomi, E., Inada, Y., et al. (2013). Interleukin-33 is induced by tumor necrosis factor-alpha and interferon-gamma in keratinocytes, and contributes to allergic contact dermatitis. *Journal of Investigational Allergology & Clinical Immunology, 23*(6), 428–434.

Tarkowski, E., Liljeroth, A. M., Minthon, L., Tarkowski, A., et al. (2003). Cerebral pattern of pro- and anti-inflammatory cytokines in dementias. *Brain Research Bulletin, 61*(3), 255–260.

Théaudin, M., & Feinstein, A. (2015). Depression and multiple sclerosis: clinical aspects, epidemiology, and management. In *Neuropsychiatric symptoms of inflammatory demyelinating diseases* (pp. 17–25). Springer.

Tschopp, J., & Schroder, K. (2010). NLRP3 inflammasome activation: the convergence of multiple signalling pathways on ROS production? *Nature Reviews Immunology, 10*(3), 210–215.

Velasquez, S., & Rappaport, J. (2016). Inflammasome activation in major depressive disorder: a pivotal linkage between psychological stress, purinergic signaling, and the kynurenine pathway. *Biological Psychiatry, 80*(1), 4–5.

Volpato, S., Guralnik, J. M., Ferrucci, L., Balfour, J., et al. (2001). Cardiovascular disease, interleukin-6, and risk of mortality in older women the women's health and aging study. *Circulation, 103*(7), 947–953.

Wager-Smith, K., & Markou, A. (2011). Depression: a repair response to stress-induced neuronal microdamage that can grade into a chronic neuroinflammatory condition? *Neuroscience & Biobehavioral Reviews, 35*(3), 742–764.

Ye, Y., Wang, G., Wang, H., & Wang, X. (2011). Brain-derived neurotrophic factor (BDNF) infusion restored astrocytic plasticity in the hippocampus of a rat model of depression. *Neuroscience Letters, 503*(1), 15–19.

Zhang, J.-M., & An, J. (2007). Cytokines, inflammation and pain. *International Anesthesiology Clinics, 45*(2), 27.

Zhang, Y., Liu, L., Peng, Y. L., Liu, Y. Z., et al. (2014). Involvement of inflammasome activation in lipopolysaccharide-induced mice depressive-like behaviors. *CNS Neuroscience & Therapeutics, 20*(2), 119–124.

Zhang, Y., Liu, L., Liu, Y.-Z., Shen, X.-L., et al. (2015). NLRP3 inflammasome mediates chronic mild stress-induced depression in mice via neuroinflammation. *International Journal of Neuropsychopharmacology, 18*(8).

Zhu, W., Cao, F.-S., Feng, J., Chen, H.-W., et al. (2017). NLRP3 inflammasome activation contributes to long-term behavioral alterations in mice injected with lipopolysaccharide. *Neuroscience, 343*, 77–84.

10

Pathways Driving Neuroprogression in Depression: The Role of Immune Activation

Giovanni Oriolo, Iria Grande*, Rocío Martin-Santos*,
Eduard Vieta*, André F. Carvalho[†,‡]*

*University of Barcelona, Barcelona, Spain
[†]Faculty of Medicine, Department of Psychiatry, University of Toronto, Toronto, ON, Canada
[‡]Centre for Addiction & Mental Health (CAMH), Toronto, ON, Canada

INTRODUCTION

Depression is a severe and common disorder, with an estimated 350 millions of people affected across the globe, and is considered the second leading cause of disability worldwide (Global Burden of Disease Study 2013 Collaborators, 2015). To date, available first-line antidepressants are classically thought to act inhibiting the uptake of monoaminergic neurotransmitters or otherwise modulating monoamine neurotransmission (Li, Frye, & Shelton, 2012). Taking into account that only approximately one-third of patients with major depressive disorder (MDD) achieve remission after an adequate trial with a first-line anti-depressant (Carvalho, Berk, Hyphantis, & McIntyre, 2014; Rush et al., 2006; Trivedi et al., 2006), a better understanding of its underlying pathophysiology is a crucial step to develop

mechanistically novel therapeutic trategies (Rosenblat, McIntyre, Alves, Fountoulakis, & Carvalho, 2015). Since the original hypothesis put forth by Smith (1991), accumulating evidence indicates that peripheral immune activation, which may concur to (neuro)inflammation, may contribute to the onset of a subset of cases of MDD (Gallagher, Kiss, Lanctot, & Herrmann, 2016; Miller & Raison, 2016; Valkanova, Ebmeier, & Allan, 2013).

A recent large-scale metaanalysis of 82 studies found that individuals with MDD have significantly elevated peripheral levels of interleukin-6 (IL-6), tumor necrosis factor (TNF)-α, IL-10, soluble IL-2 receptor (sIL-2R), C-C chemokine ligand 2 (CCL-2), IL-13, IL-18, IL-12, IL-1 receptor antagonist, and soluble TNF receptor 2 (sTNFR2) compared with healthy controls (Kohler et al., 2016). The role of immune activation in the pathoetiology of

Inflammation and Immunity in Depression
https://doi.org/10.1016/B978-0-12-811073-7.00010-6

MDD is further supported by the fact that approximately 28% of individuals with chronic hepatitis C may develop a depressive episode after interferon treatment (Machado et al., 2016; Udina et al., 2012). In addition, emerging evidence suggests that antiinflammatory agents may offer some therapeutic benefits for a subgroup of patients with MDD (Rosenblat et al., 2015). Moreover, it has been proposed that for a subset of patients, the course of MDD may be neuroprogressive and characterized by progressive cognitive impairment and treatment resistance (Moylan, Maes, Wray, & Berk, 2013). In general, neuroprogression refers to a progressive stage-related process of structural brain aberrations, which could result from reduced neurogenesis, dysfunction in neural plasticity, and increased apoptosis (Bakunina, Pariante, & Zunszain, 2015). This multifactorial phenomenon can be reflected by clinical features such as a progressive worsening of symptoms; psychosocial dysfunction; cognitive impairment; treatment refractoriness; and increased length, frequency, and severity of episodes over

the course of the illness (Bortolato, Carvalho, & McIntyre, 2014; Kessing & Andersen, 2017; Moylan et al., 2013).

An increasing amount of literature is focusing on the neuroprogressive nature of MDD (Kessing & Andersen, 2017; Moylan et al., 2013) (see Fig. 1). A recent metaanalysis provides support to the notion that depression may not be considered a prodromal manifestation of either all-cause dementia or Alzheimer's disease (AD), but is an independent risk factor for those neurodegenerative disorders (Gao et al., 2013). Furthermore, a large prospective study of individuals with depression has found that these individuals may display accelerated cellular aging as indicated by a progressive shortening of telomeres, and those with the most severe and chronic forms of MDD had the shortest telomere lengths pointing to a dose-response relationship (Verhoeven et al., 2014). A recent metaanalysis supported those findings showing that individuals with MDD may have shortened telomere lengths (Lin, Huang, & Hung, 2016). However, those peripheral findings do not

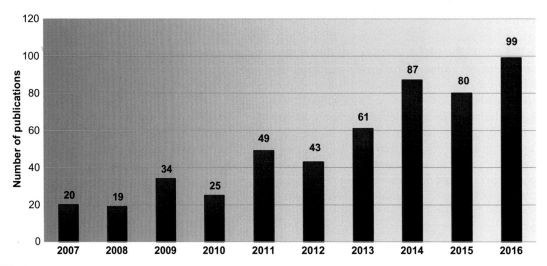

FIG. 1   Number of publications from 2007 to 2016 (31 December) concerning neuroprogression in depression. The search was done in PubMed, with the following string: "MDD" or "major depressive disorder" or "depression" or "depressive symptoms" and "neuroprogression" or "neurodegeneration."

provide a definitive proof that MDD is associated with progressive brain aging.

In this chapter, we critically review clinical evidence that seems to support a neuroprogressive course of MDD in at least a subset of cases. Then, we provide a critical overview of neurobiological pathways that could drive neuroprogression in MDD, with a clear focus on immune-inflammatory mechanisms. Finally, we illustrate possible integrative models of the role of the immune system in MDD-related neuroprogression.

DEPRESSION AS A NEUROPROGRESSIVE DISORDER

The concept of neuroprogression in MDD is sustained by clinical observations and data on anatomical structural changes in the central nervous system (CNS). Emil Kraepelin, in his first descriptions of manic-depressive disorders, suggested that faster cycling and shortening of wellness intervals were common features of such disease (Kraepelin, 1921). Nevertheless, the notion that unipolar depression is progressive in nature has been questioned (Kessing & Andersen, 2017; Oepen,

Baldessarini, & Salvatore, 2004), with several studies pointing to the overwhelming phenotypic heterogeneity of MDD (Kessing, Hansen, Andersen, & Angst, 2004; Kvist, Andersen, Angst, & Kessing, 2010). Actually, the course of MDD can be characterized by recurrences and relapses in a substantial proportion of patients (Beshai, Dobson, Bockting, & Quigley, 2011), but not all patients exhibit cognitive and structural neuroimaging alterations consistent with neuroprogression. Neuroprogression can be clinically defined by an increase of number, length, and severity of major depressive episodes (MDE); higher recurrence rates; progressive cognitive deterioration; and lower threshold for developing new episodes (Kessing, 2015; see Fig. 2). Furthermore, earlier onset MDD has been associated to worse prognosis, elevated recurrences, higher prevalence of comorbidities, and worse functional impairment (Kovacs, Obrosky, & George, 2016).

The majority of individuals do not achieve premorbid levels of psychosocial functioning after treatment with a first-line antidepressant agent (Stotland, 2012; Trivedi et al., 2006), which increases the risk of relapses and recurrences, which may lead to chronicity (Miller et al., 1998). Cognitive symptoms appear to be the

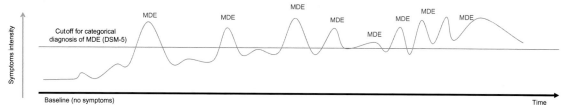

FIG. 2 Depression as neuroprogressive disease. The notion of neuroprogression is referred to a progressive stage-related process of neurodegeneration, which is the result of reduced neurogenesis, dysfunction in neural plasticity, and increased apoptosis. Major depressive disorder (MDD) may be associated with progressive changes in the central nervous system (CNS) in a subset of patients, resulting from toxic or adaptive neurobiological changes produced during depressive episode. Neuroprogression in MDD can be clinically defined by an increase of number, length, and severity of major depressive episodes (MDE); psychosocial dysfunction; cognitive impairment; treatment refractoriness; an augmentation of the risk of recurrence; and a reduction in resilience, that is, a minor threshold for developing new episodes. Thus, the development of the disease can be framed as a continuum that goes from a totally healthy state to a severe depressive state, without the necessity of categorical cutoff. Abbreviations: DSM, diagnostic and statistical manual of mental disorder; MDE, major depressive episode.

critical mediator of functional impairment (Carvalho, Miskowiak, et al., 2014). Hence, a growing body of evidence suggests that cognitive impairment often persists after the full resolution of affective symptoms (Bortolato et al., 2014). Persistent dysfunctions in the domain of attention, working memory, inhibitory control, and executive functions have been observed after symptomatic improvement of depressive symptoms (Bora, Harrison, Yücel, & Pantelis, 2013; Hasselbalch, Knorr, & Kessing, 2011). Interestingly, several studies indicate that early cognitive impairments may influence the long-term course of the disorder, and these abnormalities can be considered a premorbid "trait marker" of MDD in a subgroup of patients (Bortolato et al., 2014). In fact, in a cross-sectional study involving twin discordant for unipolar depression, it has been observed that worse cognitive performance may be evident before the onset of an MDE and was found to be associated with a more elevated genetic liability to the disorder (Christensen, Kyvik, & Kessing, 2006). Taken together, these results support the hypothesis that cognitive impairment may be an endophenotype for MDD and may also be related to a more pernicious course of this illness. The routine neuropsychological assessment of MDD patients should be incorporated on measurement-based care of those patients (McIntyre, 2013).

Another important clinical observation of neuroprogression in MDD is that the number of previous MDE may increase the risk of recurrences in a subset of patients (Kessing, Hansen, & Andersen, 2004; Solomon et al., 2000), and this risk may persist across the life span (Herrera-Guzmán et al., 2008). Additionally, Beshai et al. (2011) pointed out that the maintenance treatment of MDD, even when effective, seems not to fully prevent some individuals from experiencing recurrences (Beshai et al., 2011). With the advent of antidepressant treatment, the length of episodes has been generally reduced (Kessing & Andersen, 2017). However, several studies indicated that

subsequent episodes are usually longer and more resistant to antidepressant therapy in a subset of patients, suggesting an accumulation of underpinning pathophysiological aberrations (Beshai et al., 2011). Thus, it can be assumed that the duration of MDE increases with illness progression and that available treatments may limit the length of episodes (Kessing, 2015).

Lastly, the prevalence of life stressors in recurrent depression seems to be increased in earlier than later episodes, indicating that resilience and threshold for developing an MDE decreases during the progression of the disease (Goodwin, Jamison, & Ghaemi, 2007). Systematic observations by Kendler, Thornton, and Gardner (2000) and Slavich, Monroe, and Gotlib (2011) suggest that the first MDEs are linked to stressors closer than the later episodes, in which there is less need for stressful precipitants (Kendler et al., 2000; Slavich et al., 2011). Nevertheless, methodological issues have been highlighted, being such studies prone to recall bias for successive episodes, and increased dropout was observed in patients suffering stressful life events (Kessing & Andersen, 2017).

PERIPHERAL IMMUNE ACTIVATION IN DEPRESSION

Evidence

Pro-inflammatory cytokines like IL-1, IL-6, and TNF-α are soluble mediators usually produced by monocytes/macrophages activated by specific pathogen-associated molecular patterns (PAMPs) and/or damage-associated molecular patterns (DAMPs) and then drive local and systemic inflammatory adaptive responses against microbial pathogens. Accumulating preclinical data have demonstrated the direct implication of these cyto-kines in the promotion of a behavioral condition referred to as "sickness behavior" (Aubert, Goodall, Dantzer, & Gheusi, 1997; Avitsur, Cohen, & Yirmiya, 1997; Kent, Bluthé,

Kelley, & Dantzer, 1992; Kent, Bret-Dibat, Kelley, & Dantzer, 1996). This condition is characterized by a set of manifestations such as fatigue, pyrexia, malaise, anorexia, psychomotor retardation, and increased sensitivity to pain, as well as irritability, anhedonia, and social responsiveness (Dantzer, 2001; Maes, Scharpe, et al., 1992; Watkins & Maier, 2000). Due to phenomenological similarities between sickness behavior and MDD induced, some authors consider clinical depression as a consequence of sickness behavior (Dantzer, O'Connor, Freund, Johnson, & Kelley, 2008). Nevertheless, clinical differences between depression and sickness behavior have been extensively reviewed. For example, pyrexia is common in sickness behavior, but no evidence of pyrexia in MDD patients has been observed (Maes, Berk, et al., 2012). Furthermore, only a subset of patients with MDD would suffer from malaise and other somatic symptoms (Maes, 2009). Lastly, suicidal ideation, feelings of guilty, and unworthiness may be distinctive of depression (Charlton, 2000). Differences in the onset and course between sickness behavior and MDD have been further described. Sickness behavior would be characterized by an acute onset and an acute course, whereas MDD would have features of a progressive illness, with a more insidious onset and, at least in a subset of patients, a relapsing-remitting chronic course (Maes, Berk, et al., 2012).

As mentioned before (see "Introduction" section), results from a recent and comprehensive metaanalysis by Kohler et al. (2016) showed significantly higher plasma concentrations of TNF-α, IL-6, IL-10, IL-12, IL-13, IL-18, sIL-2R, sTNFR2, and CCL-2 in MDD patients compared with control subjects (Kohler et al., 2016), whereas higher mean levels of C-reactive protein (CRP) in MDD patients compared with healthy controls were observed in other metaanalysis (Haapakoski, Mathieu, Ebmeier, Alenius, & Kivimäki, 2015). The elevation of cytokines and chemokines involved in innate immunity supports the hypothesis of a monocyte-macrophage peripheral activation in MDD (Mills, 2015; Wohleb, Franklin, Iwata, & Duman, 2016). Depending on prime activity, macrophages can be divided in two different phenotypes, M1, which is involved in acute-phase pathogen defense, and M2, which primarily subserves immunoregulation and tissue remodeling (Kalkman & Feuerbach, 2016). The overactivation of M1 cells has been related to the pathophysiology of MDD (Miller & Raison, 2016; Yirmiya, Rimmerman, & Reshef, 2015).

Cell-mediated immunity (CMI) activation has also been associated with depression, as indicated by elevated levels of sIL-2R in patients with MDD (Kohler et al., 2016). IL-2 is predominantly secreted by the T-helper (Th) cell type 1 together with interferon-γ (IFN-γ), and both induce the suppression of Th2 responses, which have predominant immunoregulatory and antiinflammatory roles (Lichtblau, Schmidt, Schumann, Kirkby, & Himmerich, 2013). Moreover, IL-2 and IFN-γ drive the activation of monocytes and macrophages, which may further support and propagate immune responses (Haapakoski, Ebmeier, Alenius, & Kivimäki, 2016). This is consistent with increased levels of neopterin (a serum marker of CMI activation) in MDD patients compared with healthy controls, further pointing to a prominent role of IFN-γ-induced macrophage activation in the pathophysiology of MDD (Maes, 2011). For instance, it has been hypothesized that an imbalance between Th1/Th2 ratio could contribute to the excess of proinflammatory cytokines and thus to a chronic low-grade inflammatory state in MDD (Dantzer, O'Connor, Lawson, & Kelley, 2011; Lichtblau et al., 2013). In addition, increased levels of IL-10 in MDD patients suggest an imbalance in inflammatory response regulation. Actually, IL-10 is mainly secreted by regulatory T cells (Treg), which are involved in immune tolerance (Sakaguchi, Yamaguchi, Nomura, & Ono, 2008).

Taken together, these data point to a strong association between a putative pro-inflammatory state and CMI activation in MDD.

Possible Sources of Peripheral Immune Activation in Depression

Identifying possible modifiable sources of peripheral immune activation in depression may be crucial to develop suitable preventative and therapeutic interventions (Berk et al., 2013). Hence, several environmental triggers can lead to a low-grade inflammation and an activation of oxidative and nitrosative stress (O&NS) pathways.

There are reciprocal interactions between MDD and several somatic conditions. Taking into account that diabetes (Nunemaker, 2016), rheumatoid arthritis (Derdemezis, Voulgari, Drosos, & Kiortsis, 2012), coronary artery disease (Montecucco et al., 2017), and several types of cancer (Carvalho, Hyphantis, et al., 2014; Sethi, Shanmugam, Ramachandran, Kumar, & Tergaonkar, 2012) are associated with immune-inflammatory activation, it has been proposed that shared immune pathways could contribute to a higher prevalence of MDD in these patients compared with the general population (Rosenblat, Cha, Mansur, & McIntyre, 2014).

Well-known sources of peripheral inflammation are chronic psychosocial stressors and acute trauma (Kiecolt-Glaser et al., 2011); robust evidence from animal models and humans sustained increased pro-inflammatory cytokines such as TNF-α, IL-6, or IL-1, as a consequence of different stressors (Dobbin, Harth, McCain, Martin, & Cousin, 1991; Möller et al., 2013). Notwithstanding women who suffered early-childhood trauma showed higher blood levels of CRP and IL-6 and a greater likelihood of depression, compared with women without a history of childhood adversity, suggesting an enduring effect of stressors on the immune system, which may confer a heightened vulnerability to psychiatric disease (Miller & Cole, 2012).

A diet rich in refined sugar and saturated fats may further increase the susceptibility to MDD through the modulation of systemic inflammation (Lopez-Garcia et al., 2004). Furthermore, a sedentary lifestyle has been related to adiposity-driven inflammation and increased cardiovascular risk (Allison, Jensky, Marshall, Bertoni, & Cushman, 2012). An unhealthy diet may thus contribute to the bidirectional associations between obesity and clinical depression by means of immune-inflammatory activation (Luppino et al., 2010). Adipocytes, together with macrophages and lymphocytes accumulating in white adipose tissue, may further contribute to the production of inflammatory mediators (Capuron, Lasselin, & Castanon, 2016).

Diet and stress may also contribute to intestinal microbiota alterations (Berk et al., 2013; Macedo et al., 2016). It is increasingly recognized that there is a bidirectional communication between the gut and the brain and that this may contribute to core symptoms of neuropsychiatric disorder, such as MDD, and comorbidities with noncommunicable diseases (Kelly et al., 2015; Slyepchenko et al., 2017). The main pathways involved in the so-called brain-gut-microbiota axis include the vagus nerve, the synthesis of neurotransmitters by gut microbes, and the production of short-chain fatty acids (SCFAs), which may have distal influences upon the HPA axis and may also lead to immune activation (Dinan & Cryan, 2017). Recently, SCFAs like butyrate, acetate, or propionate have been identified as neurohormonal active metabolites produced by some microbiome phyla, which may be crucial for the maintenance of intestinal barrier (Peng, He, Chen, Holzman, & Lin, 2007) and in the control of bacteria translocation (Lewis et al., 2010). Both gut dysbiosis and an increase in gut permeability referred to as "the leaky gut" may disrupt neuroendocrine or neuroimmune signaling pathways, increasing the vulnerability to stress-related disorders, including MDD (Dinan, Borre, & Cryan, 2014; Macedo et al., 2016; Maes, Kubera, Mihaylova, et al., 2013). The gut gram-negative commensal bacteria can translocate when tight junction barrier is disrupted, leading to endotoxemia and immune activation. Notwithstanding, results from a recent study by Maes et al. (2012) showed that in 112 patients with MDD, increased levels of immunoglobulin (Ig) M and IgA against LPS are derived from commensal bacteria compared with 28 healthy controls, suggesting a direct implication of this mechanism in the pathophysiology of depression (Maes, Kubera, Leunis, & Berk, 2012).

Finally, other environmental triggers such as smoking, sleep disturbances, periodontal diseases, or the lack of vitamin D could also contribute to immune activation in MDD (Berk et al., 2013).

FROM PERIPHERAL IMMUNE ACTIVATION TO NEUROINFLAMMATION

Several mechanisms may drive neuroinflammation as a result of peripheral immune activation (Miller, Haroon, Raison, & Felger, 2013). The CNS parenchyma contains a limited immune repertoire, composed mostly of microglia and perivascular macrophages (Marin & Kipnis, 2016). Nevertheless, these cells are not the sole responsible for inflammatory activity in the brain. Actually, it has been observed that peripheral inflammatory activation may drive inflammation in CNS, interacting with neurons, astrocytes, and microglia, even when the immune condition is elicited by innocuous stimuli (Miller et al., 2013). Cytokines provide afferent information to the brain about the activity of the immune system and therefore elicit neuroendocrine responses (Besedovsky & Del Rey, 2011). Works on animal models have illustrated at least three different pathways by which cytokines or active immune cells can cross the blood-brain barrier (BBB) (Quan & Banks,

2007). In the so-called humoral route, circulating cytokines pass through leaky regions of BBB or through cytokine-specific active transporters. In the "neural pathway," cytokines bind receptors in afferent nerve fibers (e.g., IL-1β) through the vagus nerve (Erickson, Dohi, & Banks, 2012), which in turn relay the signal to different brain regions. Finally, the "cellular route" consists in the activation of microglia or endothelial cells in CNS vasculature that can induce chemokine production and attract immune cells to the brain (Miller & Raison, 2016).

Since it remains unclear whether the activation of inflammatory pathways in the CNS during depression primarily originates from the periphery or from a direct activation of inflammatory responses within the brain (Miller, Maletic, & Raison, 2009), it has been assumed that bidirectional pathways enable immune mediators to influence neuronal activity and, otherwise, also enable the brain to influence peripheral immune responses (Wohleb et al., 2016).

IMMUNE-INFLAMMATORY MECHANISMS AND NEUROPROGRESSION IN DEPRESSION

Direct Mechanisms

Evidence supports a role of cytokines in MDD-related neuroprogression (Maes, Mihaylova, Kubera, & Ringel, 2012). Actually, pro-inflammatory cytokines and CMI activation can contribute to neuroprogression in MDD interacting with the microglia contributing to neuroinflammation or directly affecting neurons (Haapakoski et al., 2016).

Microglia displays different essential functions in CNS, dynamically sustaining neuronal tasks during homeostasis and stress conditions (Wohleb et al., 2016). It has been highlighted how microglia are primed by genetic predisposition in multiple chronic disease states, leading to a stronger response to inflammatory stimulation, thus transforming an adaptive CNS inflammatory activation to persistent inflammation (Cunningham, 2013). When activated by several stimuli, such as cytokines, hormones, or pattern recognition receptors, microglia may contribute to amplify inflammatory signals through the release of pro-inflammatory mediators (i.e., IL-1β and TNF-α). At the same time, microglia can also activate inducible nitric oxide synthase (iNOS), release reactive oxygen species (ROS), and increase the synthesis of tryptophan catabolites (TRYCATs) (Yirmiya et al., 2015), which are pathways that can have a crucial role in neuroprogression (see "Indirect Mechanisms" section). Hence, such activation can elicit molecular changes in neurons, leading to neurotoxicity and cellular death, disruption of neural plasticity, or impaired functional brain connectivity and social behavior (Zhan et al., 2014). Importantly, suppression of hippocampal neurogenesis, which is considered an important mechanism underpinning MDD, has been related to LPS-induced microglial activation in rodents (Ekdahl, Claasen, Bonde, Kokaia, & Lindvall, 2003). Moreover, gut microbiota alterations may lead to neural and hormonal dysfunction, through alteration of microglia maturation and function (Cryan & Dinan, 2012; Erny et al., 2015). In humans, abnormal microglial activation and increased microglial cell numbers have been observed in depression and anxiety disorders (Serafini, Amore, & Rihmer, 2015). Setiawan et al. (2015) provide evidence of neuroimmune activation in MDD patients using a neuroimaging case-control study with positron-emission tomography (PET) and radio-labeled tracer for the translocator protein (TSPO). Increased TSPO density was found in the prefrontal cortex, insula, and ACC of depressed patients compared with healthy controls, indicating microglial activation. Interestingly, higher TSPO signals are correlated with depression severity (Setiawan, 2015). Taken together, these findings suggest

that microglial function may play a significant role in MDD-related neuroprogressive abnormalities.

In addition, pro-inflammatory cytokines may have direct CNS effects. For example, it has been shown that TNF-α can influence cellular viability, ionic homeostasis, and synaptic plasticity (Park & Bowers, 2010). Hence, TNF-α elicits apoptosis and neuronal damage through the activation of caspase-dependent mechanisms, thereby potentiating glutamate neurotoxicity (Zou & Crews, 2005). This mechanism can inhibit long-term potentiation (Carvalho et al., 2014; Pickering, Cumiskey, & O'Connor, 2005). Furthermore, in animal models, TNF-α overexpression has been shown to decrease the secretion of nerve growth factor (NGF) in the hippocampus, thus suppressing hippocampal plasticity (Fiore et al., 2000).

Subchronic elevations of IL-1 have been associated with a reduction in the neurotrophins BDNF and NGF (Song & Wang, 2011) in animal models, whereas chronic IL-1 exposure has been shown to disrupt hippocampal neurogenesis (Goshen et al., 2008). IL-1β, a pro-inflammatory cytokines activated by the stress-induced inflammasome complex NOD-like receptor pyrin-domain containing 3 (NLRP3) (Walsh, Muruve, & Power, 2014), can alter neuroplasticity via an activation of NF-κB pathway (Koo, Russo, Ferguson, Nestler, & Duman, 2010). Moreover, in animal models, this cytokine has been shown to exacerbate neuronal death (Patel et al., 2006). This was in line with the neurotrophic theory of depression developed by Duman, Heninger, and Nestler (1997), based on the observation of decreased neurogenesis and loss of glia cells in the hippocampus and other brain regions in patients with MDD. This abnormality may at least in part be due to a decrease in neurotrophic support and is reversible by antidepressant drug treatment (Duman, 2012; Duman et al., 1997; Duman & Monteggia, 2006).

There is a controversy surrounding putative effects of IL-6 on the CNS, with studies reporting both beneficial and detrimental effects of this cytokine in brain function, with relevance to both physiological and pathological conditions (Spooren et al., 2011). IL-6 has been related to microglial activation (Krady et al., 2008) and calcium (Ca^{2+}) excitotoxicity mediated by N-methyl-D-aspartate (NMDA) receptor activation, which may promote neuronal death. Nevertheless, whereas some authors have found that IL-6 enhances NMDA-induced excitotoxicity in cerebellar granule neurons (Conroy et al., 2004), several studies on animal models pointed to neuroprotector effects of IL-6 against excitotoxicity (Wang, Peng, Lu, Cao, & Qiu, 2009). Another mechanism by which IL-6 can influence NMDA neurotransmission is through TRYCAT pathway activation (Connor, Starr, O'Sullivan, & Harkin, 2008) (for more details on TRYCAT pathways and its role with neuroprogression, see "The Role of Indoleamine 2,3 Dioxygenase (IDO) and the Tryptophan Catabolism (TRYCAT) Pathway" section).

Similarly to IL-6, IL-2 may activate microglia and directly interact with NMDA receptors, thus altering synaptic plasticity (Shen, Zhu, Liu, Zhou, & Luo, 2006). Furthermore, IL-2 has been related to chronic myelin damage in patients with multiple sclerosis (MS) (Mott, Zandian, Allen, & Ghiasi, 2013).

Regarding IFN-γ, this cytokine can directly contribute to neurotoxicity and neuronal loss in neurodegenerative, neuroprogressive, and neuroimmune diseases (Lambertsen et al., 2004).

The imbalance in Th1/Th2 ratio due to an excess of CD4[+] Th1 cells has been hypothesized as the basis of the excess of pro-inflammatory cytokines observed in MDD (Slyepchenko et al., 2016). Th1 activation is related to the increase in pro-inflammatory activity, leading to the activation of macrophages that in turn sustain Th1 propagation (Maes, 2011). CD8[+] Th1 cells are another subtype of lymphocytes that drive immune cell proliferation through the release of IL-2, a pro-inflammatory cytokine whose soluble receptor levels are increased in MDD (Maes et al., 1990). Activated B cells, stimulated by IL-2, have also been observed in MDD (Maes, Scharpe, et al., 1992). These observations suggest a CMI activation biosignature in MDD.

Recently, potential interactions of Th17 lymphocytes with mechanisms associated with MDD and neuroprogression have been reviewed (Slyepchenko et al., 2016). Cytokine activity, mostly driven by IL-6 and transforming growth factor-β (TGF-β), may induce the differentiation of Th naive cells into Th17 cells, which in turn produce several types of cytokines, grouped in the IL-17 family and the chemokine ligand 20 (CCL-20). Th17 is an independent lineage of Th lymphocytes (Dinan & Cryan, 2013), and it has been shown that Th17 can migrate to CNS and is important in the clearance of bacterial infections and in the regulation neuroinflammatory processes (Holley & Kielian, 2012). Th17 cells regulate microglia activation and immune response within the CNS, through the release of IL-17 and IFN-γ. Such cytokines induce microglial activation and IL-1β production (Murphy, Lalor, Lynch, & Mills, 2010). Furthermore, IL-17 and TNF-α together contribute to the loss of oligodendrocytes in MS (Paintlia, Paintlia, Singh, & Singh, 2011), whereas in microglial cells, IL-17A production induces the release of IL-6 and the activation of iNOS, leading to O&NS (Meares, Ma, Qin, & Benveniste, 2012).

In keeping with this view, these findings shed a light on the important role of pro-inflammatory cytokines and CMI activation along with CNS microglial aberrations, in neuroprogressive mechanisms in a subset of patients with MDD.

Indirect Mechanisms

Cytokines and CMI can interact directly or through the modulation of microglial cells, with virtually every pathophysiological domain relevant to depression, including neurotransmitter metabolism, neuroendocrine function, and neural plasticity (Miller et al., 2009; Raison, Capuron, & Miller, 2006). Fig. 3 provided a wide-angle lens

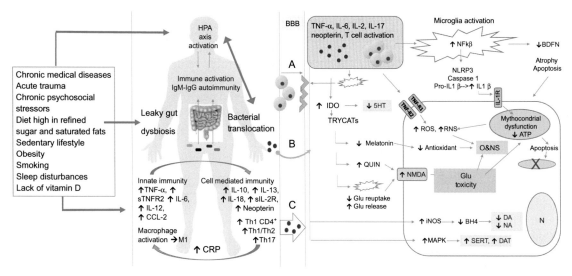

FIG. 3 Peripheral activation of innate and cell-mediated immune system has been observed in patients with major depressive disorder (MDD). Several environmental factor triggers can activate low-grade inflammation and HPA axis and microbiota imbalance. Pro-inflammatory cytokines and cell-mediated immunity (CMI) activation may contribute to neuroprogression in such patients communicating with the central nervous system (CNS) crossing the blood-brain barrier through the "cellular route" (A), the "neural route" (B), and the "humoral route" (C). Inflammatory molecules and cells can interact with the microglia contributing to neuroinflammation or directly affect neurons. Moreover, cytokines and CMI can interact with virtually every pathophysiological domain relevant to depression, including neurotransmitter metabolism, neuroendocrine function, and neural plasticity (see text for details). Abbreviations: *5HT*, serotonin; *ATP*, adenosine triphosphate; *BBB*, blood-brain barrier; *BDNF*, brain-derived neurotrophic factor; *BH4*, tetrahydrobiopterin; *CCL*, chemokine ligand; *CRP*, C-reactive protein; *DA*, dopamine; *DAT*, dopamine transporter; *Glu*, glutamate; *HPA*, hypothalamus-pituitary axis; *IDO*, indoleamine-2,3-dioxygenase; *Ig*, immunoglobulin; *IL*, interleukin; *iNOS*, inducible nitric oxide synthase; *N*, nucleus; *NA*, noradrenaline; *NLRP3*, NOD-like receptor pyrin-domain containing 3; *O&NS*, oxidative and nitrosative stress; *QUIN*, quinolinic acid; *RNS*, reactive nitrogen species; *ROS*, reactive oxygen species; *SERT*, serotonin transporter; *sIL-2R*, interleukin-2 soluble receptor; *Th*, T-helper lymphocytes; *TNF-α*, tumor necrosis factor-α; *TNF-R*, tumor necrosis factor receptor; *TRYCATs*, tryptophan catabolites.

view of the putative role of these processes to MDD-related neuroprogression.

The Role of Indoleamine 2,3 Dioxygenase (IDO) and the Tryptophan Catabolism (TRYCAT) Pathway

The higher peripheral levels of inflammatory mediators observed in depression may induce the IDO enzyme, with a more prominent role for IL-2, IL-6, and IFN-α (Anderson et al., 2013; Capuron et al., 2001, 2003). IDO activation drives tryptophan (TRP) catabolism into kynurenine (KYN), activating the so-called TRYCAT pathway, potentially depleting the availability of

serotonin (5-HT) in the brain (Miller et al., 2013). TRP is an essential amino acid and a precursor of 5-HT synthesis. Activation of the TRYCAT pathway may have a role in the emergence of depressive symptoms and neuroprogression (Morris, Carvalho, Anderson, Galecki, & Maes, 2015). Hence, KYN can be converted into kynurenic acid (KYNA) or 3-hydroxykynurenine (3-HK) and quinolinic acid (QUIN) in astrocytes and microglial cells, which modulate the production and release of several neurotransmitters (e.g., 5-HT, dopamine, glutamate, and melatonin) (Reus et al., 2015). TRYCATs have been related to impaired mitochondrial energy metabolism through the inhibition of enzymes involved in adenosine triphosphate (ATP) synthesis (Naoi, Ishiki, Nomura, Hasegawa, & Nagatsu, 1987). Furthermore, disruption in mitochondrial functions and integrity can promote O&NS (Maes, Galecki, Chang, & Berk, 2011). Recently, it was observed that TRYCAT pathway overactivation may disrupt the melatoninergic pathway, which may be involved in the pathophysiology of several medical conditions (Morris et al., 2015). Melatonin is an essential antiinflammatory and antioxidant molecule that can modulate mitochondrial function (Martín et al., 2002). Interestingly, melatonin is also produced by gut cells, thus reducing gut permeability and avoiding bacteria translocation (Trivedi & Jena, 2013) and gut dysbiosis, which has been observed in many medical conditions associated with MDD (Slyepchenko et al., 2017). Furthermore, 3-HK generates ROS contributing to O&NS, whereas QUIN is a known agonist of the NMDA receptor, contributing to excessive glutamatergic signaling, a mechanism related to MDD (Müller & Schwarz, 2007), and may cause the destruction of synaptic elements or apoptosis of nerve cells (Tavares et al., 2002).

Oxidative and Nitrosative Stress (O&NS)

A balance between the endogenous production of ROS or reactive nitrogen species (RNS) and the activity of antioxidant defense systems determines the magnitude of O&NS. When the generation of ROS or RNS exceeds the elimination capability of antioxidant mechanisms, either due to an increase in their output or due to a relative deficiency of antioxidant defense mechanisms, then O&NS ensues. MDD has been repeatedly associated with O&NS (Palta, Samuel, Miller, & Szanton, 2014), and the immune system activation may contribute to O&NS in depression (Moylan et al., 2014).

Inflammation and mitochondrial processes can increase the ROS and RNS production (Moylan et al., 2013). This can result in a widespread damage to different biomolecules, such as lipids, proteins, sugars, and DNA, causing

impairment in cellular function and rendering them potentially immunogenic (Maes, Mihaylova, Kubera, Leunis, & Geffard, 2011). Actually, the creation of modified neoepitopes due to the O&NS can induce a secondary autoimmune response and a persistent inflammatory activation, contributing to neuroprogression in MDD (Iseme et al., 2014). For example, protein nitrosylation secondary to nitrosative stress may induce IgM responses in depression. IgM-mediated responses can amplify inflammatory reactions and damage biomolecules, such as anchorage molecules, oleic acid, or complexes involved in cell signaling pathways, leading to aberrations in brain cellular function (Maes, Kubera, Leunis, et al., 2013). Indeed, autoimmune activity against 5-HT in MDD has been observed in patients with melancholic depression and was found to be associated with the number of depressive episode (Maes, Ringel, Kubera, Berk, & Rybakowski, 2012).

Oxidative stress is thought to be one of the main drivers of disease progression across a number of neurodegenerative diseases and may reduce neuronal viability, induce apoptosis, and inhibit mitochondrial respiratory processes (Tramutola, Lanzillotta, Perluigi, & Butterfield, 2017; Zhao & Zhao, 2013). In depression, heightened lipid peroxidation has been observed and was found to correlate with the severity of depression (Dimopoulos, Piperi, Psarra, Lea, & Kalofoutis, 2008), whereas DNA oxidative damage in MDD was related to the number of previous depressive episodes (Forlenza & Miller, 2006).

The inducible isoform of NOS (iNOS) is induced by cytokines and is primarily responsible for the inflammatory effects of NO (Maes et al., 2011). NO has regulatory roles in neuroendocrine and immune functions and influences 5-HT, DA, and glutamatergic neurotransmission (Moylan et al., 2013). Increased levels of NO and increased iNOS expression have been found in MDD patients with recurring depression (Akpinar, Yaman, Demirdas, & Onal, 2013; Gałecki et al., 2012).

Mitochondrial Dysfunction

Mitochondria play a crucial role in CNS physiology, and alterations in energy production have been related to several neurodegenerative disorders, such as Alzheimer's disease (Salminen et al., 2015), MS (Gounopoulos, Merki, Hansen, Choi, & Tsimikas, 2007), and Parkinson's disease (PD) (Trudler, Nash, & Frenkel, 2015). Furthermore, mitochondrial dysfunction may impair neurogenesis and cell survival (Voloboueva & Giffard, 2011) and is a mediator of O&NS. Importantly, changes in the size and distribution of mitochondria have been observed in MDD (Gardner & Boles, 2011), and a reduction in mitochondrial ATP generation was found in MDD patients compared with controls (Gardner et al., 2003). On the other hand, antidepressant treatment can reverse mitochondrial impairment observed in a genetic model of depression (Chen, Wegener, Madsen, & Nyengaard, 2013). Pro-inflammatory cytokines, such IL-6 or TNF-α, can amplify mitochondrial dysfunction and disrupt neural integrity (Voloboueva & Giffard, 2011), providing evidence for a role of mitochondria in neuroprogression associated with MDD.

Neural Plasticity and the Role of Brain Derived Neurotrophic Factor

Immune system activation has been implicated in impaired neuroplasticity. The pro-inflammatory cytokines IL-1, IL-6, and TNF-α, in normal condition, provide trophic support to neurons and enhance neurogenesis (Bernardino et al., 2008). Hence, evidence from animal studies suggests that peripheral inflammation and stress can decrease the expression of neurotrophic factors, such as BDNF, potentially resulting in reduced hippocampal plasticity and increased neuronal atrophy (Sahay & Hen, 2007; Wu et al., 2007). BDNF and other neurotrophins, such as vascular endothelial growth factor (Carvalho et al., 2015) or insulin growth factor (Sievers et al., 2014), may regulate hippocampal plasticity. In human studies, lower levels of BDNF have been related to severity and recurrent MDD (Lee, Kim, Park, & Kim, 2007), thought this result has not been confirmed in a recent metaanalysis (Molendijk et al., 2014). In addition, BDNF levels appear to be higher in treated depressive patients, compared with untreated ones (Shimizu et al., 2003).

Intriguing results highlighting a role for BDNF in neuroprogression associated with MDD were derived from studies on neonatal stress and early trauma. It is widely recognized that exposure to early-life stress (e.g., childhood abuse or neglect) may predispose to a later onset of MDD or anxiety (Roth & Sweatt, 2011; Vogt, Waeldin, Hellhammer, & Meinlschmidt, 2016). Altered gene expression and epigenetic mechanisms seem to contribute to sensitization, with epigenetic evidence suggesting a role for epigenetic alterations in the BDNF gene (Post, 2016). For example, intense stress experience in the first week of life in rodents can induce methylation of the promoter region of the BDNF gene and thus decrease BDNF production throughout the animal's life span (Roth, Lubin, Funk, & Sweatt, 2009), which may contribute to depressive-like behavior.

Other Indirect Pathways

HPA axis hyperactivity has been repeatedly associated with clinical depression and is considered one of the most consistently replicated findings in biological psychiatry (Pariante & Lightman, 2008). Cytokines involved in MDD and neuroprogression, such as IL-1, IL-6, TNF-α, and IFN-γ, stimulate the release of corticotropin-releasing hormone (CRH), adrenocorticotropic hormone (ACTH), and cortisol. A large metaanalysis of 354 studies found that 73% of depressed individuals may have elevated cortisol values compared with nondepressed patients (Stetler & Miller, 2011).

Other biological mechanisms that can link cytokines with depression are the disruption of BH4, an enzyme cofactor essential for the synthesis of 5-HT, dopamine (DA) and norepinephrine (NE) neurotransmitters (Haroon, Raison, & Miller, 2012), and the mitogen-activated protein kinase (MAPK) pathways, as p38, which have been found to increase the expression and function of reuptake transporters for 5-HT, DA, and NE (Miller et al., 2009).

Moreover, it has been observed that the endo-cannabinoid system (eCBs) plays an important regulatory roles in immunity, stress response, and neurogenesis, and this system appears to be dysregulated in depression (Bhattacharyya, Crippa, Martin-Santos, Winton-Brown, & Fusar-Poli, 2009; Campos, Moreira, Gomes, Del Bel, & Guimarães, 2012; Martin-Santos et al., 2012; Mechoulam & Parker, 2013). eCBs are derived from arachidonic acid (AA), a polyunsaturated fatty acid whose oxidation leads to the synthesis of the eicosanoid prostaglandins and leukotrienes (Alhouayek, Masquelier, & Muccioli, 2014). Eicosanoids are molecules involved in inflammatory modulation that can interact with eCBs signaling mechanisms, with consequences to the regulation of the immune system that could influence the onset and maintenance of depression (Boorman, Zajkowska, Ahmed, Pariante, & Zunszain, 2016).

INFLAMMATION AND THE PHENOTYPIC HETEROGENEITY OF DEPRESSION

Besides the increasing evidence that inflammation increases the risk of depression, it is not possible to claim that MDD is an inflammatory disease (Lotrich, 2015). Hence, not all patients with increased inflammation, such as chronically ill patients, develop MDD. Furthermore, it is also true that not all people with MDD show prominent immune activation (Steptoe, Kunz-Ebrecht, & Owen, 2003); it has become increasingly clearer that inflammatory markers are increased only in a subset of patients (Rosenblat et al., 2015).

Evidence supporting the heterogeneity of the depression phenotype comes from several sources. Hence, different types of symptoms usually described in MDD patients may be more frequently observed in patients who exhibit inflammatory activation. For example, fatigue, anhedonia, psychomotor retardation, reduced interest in the environment, and anorexia are manifestations that are observed in animal models of "sickness behavior" (Frenois, 2007; O'Connor et al., 2009). When the use of immunotherapy became a quasiexperimental model to investigate the physiopathology of cytokine-induced depression (Dantzer et al., 2008), many studies based on the dimensional evaluation of depressive symptoms provided indications that somatic complaints, psychomotor retardation, and neurovegetative alterations are more closely associated with immune activation in depression (Capuron et al., 2002; Mahajan, Avasthi, Grover, & Chawla, 2014; Whale et al., 2015). In the same line, Loftis et al. (2013) highlighted that somatic symptoms of depression can be significantly exacerbated by IFN-α treatment in HCV patients and may be predicted by higher TNF-α and lower serotonin levels at baseline (Loftis et al., 2013). Recently, it was observed in a group of MDD patients that increased peripheral levels of IL-6 were correlated with decreased psychomotor speed (Goldsmith et al., 2016). Chronic pain is also frequently associated with depression, in a bidirectional manner (Kroenke et al., 2011), but this manifestation is evident in only a subset of MDD patients. Actually, between 30% and 60% of patients with chronic pain also have comorbid MDD (Bair, Robinson, Katon, & Kroenke, 2003), and around 50% of MDD patients experience physical pain (Katona et al., 2005). Interestingly, increasing levels of IL-1β and TNF-α can promote directly or indirectly sensitization to pain, which indicate the presence of common pathways underpinning pain and depression symptoms (Walker, Kavelaars, Heijnen, & Dantzer, 2014).

Further evidence that chronic forms of depressive subtypes may differ not only in their symptom presentation but also in their biological correlates derives from a study by Lamers et al. (2013) in which differences in HPA axis functioning and inflammatory profile were observed in different types of depression. Patients with atypical depression showed higher inflammatory markers, triglycerides, and increased body mass index (BMI) than patients with melancholic depression, whereas the latter presented HPA axis hyperactivity compared with the former (Lamers et al., 2013).

Importantly, research has established a correlation between neurodegenerative diseases, such as Alzheimer's disease, multiple sclerosis, Parkinson's disease, and MDD. Actually, MDD is being considered more as a part of those biological processes involved in etiology and course of neurodegenerative illnesses, sharing elements of common pathways, than a mere comorbidity (Anderson & Maes, 2014; Leonard & Maes, 2012; Post, 2010). It has been highlighted that neuroinflammation is present on a molecular level in both conditions (Réus et al., 2016), thus providing a possible pathophysiological link. Nevertheless, the role of neuroinflammation as main driving force of CNS diseases is still on debate, as some consider it as a consequence rather than a cause of neurodegeneration (Wyss-Coray & Mucke, 2002).

In a recent observational longitudinal study of 811 outpatients with depression, it was observed that higher inflammation (defined as CRP levels higher than 3 mg/L) was associated with a longer duration of depressive symptoms. Furthermore, the symptom of a motivation was more common in "inflammatory" depression, whereas sadness was less frequently associated. Altogether, these observations provide initial yet provocative evidence that inflammation may contribute to the expression of phenotypic heterogeneity in MDD (Gallagher et al., 2016).

Inflammation can be involved in mood and cognitive symptoms across several psychiatric, neurological, and medical diseases (Swardfager, Rosenblat, Benlamri, & McIntyre, 2016). Thus, inflammation would not represent a specific trait, suggesting the need for a transdiagnostic dimensional approach. The framework proposed by the Research Domain Criteria (RDoC) matrix could thus aid in the investigation of the putative role of inflammation in neuropsychiatric diseases. This framework is based on the notion that data from genetics and clinical neuroscience may provide more reliable biosignatures and more specific treatments for behavioral and psychopathological manifestations that usually cross conventional diagnostic categories (Insel et al., 2010). In other words, it may be more precise to portray behavioral and affective alterations that may due to inflammation if they are associated with specific symptom dimensions and not to an MDD categorical diagnosis, as suggested by Miller and Raison in a recent review (Miller & Raison, 2016). One example is related to anhedonia, a core symptom of MDD, which can be considered as a transdiagnostic psychopathological domain, being present in several medical and neurological conditions. Inflammation has been related to anhedonia in animal models and humans, and downstream pathophysiological processes (TRYCAT pathway activation) have been identified as possible mediators of neurocircuitry dysfunction (Swardfager et al., 2016). More specifically, it has been associated with poorer cognitive performance in depression, with possible shared neurobiological substrates (McIntyre et al., 2015).

THERAPEUTIC PERSPECTIVES

In keeping with this view, a dimensional approach for MDD may contribute to the development of novel therapeutic strategies. A large metaanalysis of 14 RCTs that investigated the efficacy of antiinflammatory treatments for depression (10 considering nonsteroidal antiinflammatory drugs (NSAIDs) and 4 considering anticytokine agents, $N = 6262$) showed significant reductions in depressive symptoms in patients treated with NSAIDs versus placebo (Köhler et al., 2014). In particular, celecoxib, a selective inhibitor of cyclooxygenase-2 (COX-2), showed the strongest effect, in line with a recent metaanalysis that provided evidence for the efficacy of celecoxib on depressive symptoms, when used either alone or in combination with antidepressant drugs (Iyengar et al., 2013). Nevertheless, many trials were based on patients with chronic disease comorbidity, making it difficult to distinguish whether the improvement on depressive symptomatology was related to the amelioration of co-occurring somatic conditions. Anticytokine treatment is another suitable target used to control and decrease inflammation, with a more specific spectrum than NSAIDs (Hu, Wang, Pace, Wu, & Miller, 2005). An important RCT comparing the TNF inhibitor infliximab with placebo in MDD patients showed antidepressant response only in those patients that exhibit CRP levels more than 5 mg/L (Raison et al., 2013). A recent metaanalysis considering antidepressant activity of cytokine inhibitor treatment in chronic inflammatory conditions found robust improvement in depressive symptoms of anti-TNF-α drugs, namely, infliximab; etanercept; adalimumab; and anti-IL-6R, namely, tocilizumab (Kappelmann, Lewis, Dantzer, Jones, & Khandaker, 2016).

Taken together, these findings provide evidence on the key role of cytokines in pathogenesis of depression and sustain that drugs with antiinflammatory activity may be useful in that subset of patients with depression characterized by increased inflammation. In addition, considering the role of inflammation in neuroprogression, these drugs could offer neuroprotection and prevent the progressive course of depression, though such hypothesis needs to be tested.

Intriguingly, antiinflammatory agents could provide domain-specific treatment for specific symptoms, such as cognition or anhedonia. It may be the case of TNF-α, which has been associated with cognitive dysfunction in MDD (Bortolato, Carvalho, Soczynska, Perini, & McIntyre, 2015). Increase in TNF- α signaling has been related to memory loss and general cognitive impairment in models of neuropsychiatric disorders (McAfoose & Baune, 2009). In preclinical models of LPS-induced inflammation, a procognitive effect of TNF-α antagonists was observed (Raftery et al., 2012). As mentioned before, treatment with anti-TNF-α can improve depressive symptoms in those depressed patients with increased inflammation (Raison et al., 2013). In addition, those patients receiving anti-TNF-α drugs for no psychiatric conditions showed less prevalence of MDD comorbidity, compared with patients who do not receive such medication (Uguz, Akman, Kucuksarac, & Tufekci, 2009). Nevertheless, no specific data on the improvement of cognitive domain in MDD patients have been reported.

Further research should be directed to clarify the role of the immune system in MDD in order to identify novel mechanism and psychopharmacological targets that could span the classical boundaries of discrete mental disorders for refined diagnostics and improved treatment outcomes.

CONCLUSIONS

Major depressive disorder can be considered a neuroprogressive disease in a subset of patients associated with progressive changes in CNS mediated by inflammatory cytokines or CMI activation per se, through microglia activation or through a bidirectional influence on several pathways. Disruption in monoaminergic function through enhanced TRYCAT pathway or MAPK activity, increased O&NS and mitochondrial dysfunction, and activation of the HPA axis or epigenetic mechanisms involving the synthesis of BDNF are some of such pathways involved in altering neural physiology, synaptic plasticity, and neurogenesis. Importantly, due to observations of heterogeneity in MDD patients, the field is increasingly recognizing that depression is inflammatory in only a subset of patients, which may provide a pathway toward precision psychiatry. Finally, the observations that inflammation can be involved in mood and cognitive symptoms across several psychiatric, neurological, and medical disease (Swardfager et al., 2016) warrant the need for a transdiagnostic dimensional approach, which is partly provided by the RDoC framework. Drugs with antiinflammatory activity may be useful in depressed patients with increased inflammation and could provide domain-specific treatment for specific symptoms, such as cognition or anhedonia. Further translational research on this field is needed to identify new target drug and to promote individualized and more precise treatments for depression.

ACKNOWLEDGMENTS

This work has been done in part with the support of the Comissionat per a Universitats i Recerca del DIUE, Generalitat de Catalunya, Spain (grant number 2014SGR1135) (IP. R. Martín-Santos).

CONFLICTS OF INTEREST

Dr. I. Grande has received a Juan Rodés Contract (JR15/00012), Instituto de Salud Carlos III, Spanish Ministry of Economy and Competiveness, Barcelona, Spain, and has served as a consultant for Ferrer and as a speaker for AstraZeneca, Ferrer, and Janssen-Cilag.

Prof. E. Vieta has received research grants, CME-related honoraria, or consulting fees from AB-Biotics, Allergan, AstraZeneca, Ferrer, Forest Research Institute, Gedeon Richter, GlaxoSmithKline, Janssen, Lundbeck, Otsuka, Pfizer, Sanofi-Aventis, Sunovion, and Takeda and from Centro para la Investigación Biomédica en Red de Salud Mental (CIBERSAM), GrupsConsolidats de Recerca 2014 (SGR 398), Seventh European Framework Programme (ENBREC), and Stanley Medical Research Institute. The other authors declare no conflict of interests.

References

Akpinar, A., Yaman, G. B., Demirdas, A., & Onal, S. (2013). Possible role of adrenomedullin and nitric oxide in major depression. *Progress in Neuro-Psychopharmacology and Biological Psychiatry*, 46, 120–125. https://doi.org/10.1016/j.pnpbp.2013.07.003.

Alhouayek, M., Masquelier, J., & Muccioli, G. G. (2014). Controlling 2-arachidonoylglycerol metabolism as an antiinflammatory strategy. *Drug Discovery Today*, 19(3), 295–304. https://doi.org/10.1016/j.drudis.2013.07.009.

Allison, M. A., Jensky, N. E., Marshall, S. J., Bertoni, A. G., & Cushman, M. (2012). Sedentary behavior and adiposity-associated inflammation: the multi-ethnic study of atherosclerosis. *American Journal of Preventive Medicine*, 42(1), 8–13. https://doi.org/10.1016/j.amepre.2011.09.023.

Anderson, G., Kubera, M., Duda, W., Lasoń, W., Berk, M., & Maes, M. (2013). Increased IL-6 trans-signaling in depression: focus on the tryptophan catabolite pathway, melatonin and neuroprogression. *Pharmacological Reports*, 65(6), 1647–1654. https://doi.org/10.1016/S1734-1140(13)71526-3.

Anderson, G., & Maes, M. (2014). Oxidative/nitrosative stress and immuno-inflammatory pathways in depression: treatment implications. *Current Pharmaceutical Design*, 20(23), 3812–3847. Retrieved from http://www.ncbi.nlm.nih.gov/pubmed/24180395.

Aubert, A., Goodall, G., Dantzer, R., & Gheusi, G. (1997). Differential effects of lipopolysaccharide on pup retrieving and nest building in lactating mice. *Brain, Behavior, and Immunity*, 11(2), 107–118. https://doi.org/10.1006/brbi.1997.0485.

Avitsur, R., Cohen, E., & Yirmiya, R. (1997). Effects of interleukin-1 on sexual attractivity in a model of sickness behavior. *Physiology and Behavior*, 63(1), 25–30. https://doi.org/10.1016/S0031-9384(97)00381-8.

Bair, M. J., Robinson, R. L., Katon, W., & Kroenke, K. (2003). Depression and pain comorbidity. *Archives of Internal Medicine*, 163(20), 2433. https://doi.org/10.1001/archinte.163.20.2433.

Bakunina, N., Pariante, C. M., & Zunszain, P. A. (2015). Immune mechanisms linked to depression via oxidative stress and neuroprogression. *Immunology*, 144, 365–373. https://doi.org/10.1111/imm.12443.

Berk, M., Williams, L. J., Jacka, F. N., O'Neil, A., Pasco, J. A., Moylan, S., et al. (2013). So depression is an inflammatory disease, but where does the inflammation come from? *BMC Medicine*, 11(1), 200. https://doi.org/10.1186/1741-7015-11-200.

Bernardino, L., Agasse, F., Silva, B., Ferreira, R., Grade, S., & Malva, J. O. (2008). Tumor necrosis factor-alpha modulates survival, proliferation, and neuronal differentiation in neonatal subventricular zone cell cultures. *Stem Cells*, 26(9), 2361–2371. https://doi.org/10.1634/stemcells.2007-0914.

Besedovsky, H. O., & Del Rey, A. (2011). Central and peripheral cytokines mediate immune-brain connectivity. *Neurochemical Research*, 36(1), 1–6. https://doi.org/10.1007/s11064-010-0252-x.

Beshai, S., Dobson, K. S., Bockting, C. L. H., & Quigley, L. (2011). Relapse and recurrence prevention in depression: current research and future prospects. *Clinical Psychology Review*, *31*(8), 1349–1360. https://doi.org/10.1016/j.cpr.2011.09.003.

Bhattacharyya, S., Crippa, J. A., Martin-Santos, R., Winton-Brown, T., & Fusar-Poli, P. (2009). Imaging the neural effects of cannabinoids: current status and future opportunities for psychopharmacology. *Current Pharmaceutical Design*, *15*(22), 2603–2614. Retrieved from http://www.ncbi.nlm.nih.gov/pubmed/19689331.

Boorman, E., Zajkowska, Z., Ahmed, R., Pariante, C. M., & Zunszain, P. A. (2016). Crosstalk between endocannabinoid and immune systems: a potential dysregulation in depression? *Psychopharmacology*, *233*(9), 1591–1604. https://doi.org/10.1007/s00213-015-4105-9.

Bora, E., Harrison, B. J., Yücel, M., & Pantelis, C. (2013). Cognitive impairment in euthymic major depressive disorder: a meta-analysis. *Psychological Medicine*, *43*(10), 2017–2026. https://doi.org/10.1017/S0033291712002085.

Bortolato, B., Carvalho, A. F., & McIntyre, R. S. (2014). Cognitive dysfunction in major depressive disorder: a state-of-the-art clinical review. *CNS & Neurological Disorders: Drug Targets*, *13*(10), 1804–1818. https://doi.org/10.2174/1871527313666141130203823.

Bortolato, B., Carvalho, A. F., Soczynska, J. K., Perini, G. I., & McIntyre, R. S. (2015). The involvement of TNF-α in cognitive dysfunction associated with major depressive disorder: an opportunity for domain specific treatments. *Current Neuropharmacology*, *13*(5), 558–576. https://doi.org/10.2174/1570159X13666150630171433.

Campos, A. C., Moreira, F. A., Gomes, F. V., Del Bel, E. A., & Guimarães, F. S. (2012). Multiple mechanisms involved in the large-spectrum therapeutic potential of cannabidiol in psychiatric disorders. *Philosophical Transactions of the Royal Society of London. Series B, Biological Sciences*, *367*(1607), 3364–3378. https://doi.org/10.1098/rstb.2011.0389.

Capuron, L., Gumnick, J. F., Musselman, D. L., Lawson, D. H., Reemsnyder, A., Nemeroff, C. B., et al. (2002). Neurobehavioral effects of interferon-alpha in cancer patients: phenomenology and paroxetine responsiveness of symptom dimensions. *Neuropsychopharmacology: Official Publication of the American College of Neuropsychopharmacology*, *26*(1), 643–652. https://doi.org/10.1016/S0893-133X(01)00407-9.

Capuron, L., Lasselin, J., & Castanon, N. (2016). Role of adiposity-driven inflammation in depressive morbidity. *Neuropsychopharmacology*, 1–14. https://doi.org/10.1038/npp.2016.123.

Capuron, L., Neurauter, G., Musselman, D. L., Lawson, D. H., Nemeroff, C. B., Fuchs, D., et al. (2003). Interferon-alpha-induced changes in tryptophan metabolism: relationship to depression and paroxetine treatment. *Biological Psychiatry*, *54*(9), 906–914. https://doi.org/10.1016/S0006-3223(03)00173-2.

Capuron, L., Ravaud, A., Gualde, N., Bosmans, E., Dantzer, R., Maes, M., et al. (2001). Association between immune activation and early depressive symptoms in cancer patients treated with interleukin-2-based therapy. *Psychoneuroendocrinology*, *26*(8), 797–808. https://doi.org/10.1016/S0306-4530(01)00030-0.

Carvalho, A. F., Berk, M., Hyphantis, T. N., & McIntyre, R. S. (2014). The integrative management of treatment-resistant depression: a comprehensive review and perspectives. *Psychotherapy and Psychosomatics*, *83*(2), 70–88. https://doi.org/10.1159/000357500.

Carvalho, A. F., Hyphantis, T., Sales, P. M. G., Soeiro-de-Souza, M. G., Macêdo, D. S., Cha, D. S., et al. (2014). Major depressive disorder in breast cancer: a critical systematic review of pharmacological and psychotherapeutic clinical trials. *Cancer Treatment Reviews*, *40*, 349–355. https://doi.org/10.1016/j.ctrv.2013.09.009.

Carvalho, A. F., Kohler, C. A., McIntyre, R. S., Knochel, C., Brunoni, A. R., Thase, M. E., et al. (2015). Peripheral vascular endothelial growth factor as a novel depression biomarker: a meta-analysis. *Psychoneuroendocrinology*, *62*, 18–26. https://doi.org/10.1016/j.psyneuen.2015.07.002.

Carvalho, A. F., Miskowiak, K. K., Hyphantis, T. N., Köhler, C. A., Alves, G. S., Bortolato, B., et al. (2014). Cognitive dysfunction in depression—pathophysiology and novel targets. *CNS & Neurological Disorders Drug Targets*, *13* (10), 1819–1835.

Charlton, B. G. (2000). The malaise theory of depression: major depressive disorder is sickness behavior and antidepressants are analgesic. *Medical Hypotheses*, *54*(1), 126–130. https://doi.org/10.1054/mehy.1999.0986.

Chen, F., Wegener, G., Madsen, T. M., & Nyengaard, J. R. (2013). Mitochondrial plasticity of the hippocampus in a genetic rat model of depression after antidepressant treatment. *Synapse*, *67*(3), 127–134. https://doi.org/10.1002/syn.21622.

Christensen, M. V., Kyvik, K. O., & Kessing, L. V. (2006). Cognitive function in unaffected twins discordant for affective disorder. *Psychological Medicine*, *36*(8), 1119–1129. https://doi.org/10.1017/S0033291706007896.

Connor, T. J., Starr, N., O'Sullivan, J. B., & Harkin, A. (2008). Induction of indolamine 2,3-dioxygenase and kynurenine 3-monooxygenase in rat brain following a systemic inflammatory challenge: a role for IFN-gamma? *Neuroscience Letters*, *441*(1), 29–34. https://doi.org/10.1016/j.neulet.2008.06.007.

Conroy, S. M., Nguyen, V., Quina, L. A., Blakely-Gonzales, P., Ur, C., Netzeband, J. G., et al. (2004). Interleukin-6 produces neuronal loss in developing cerebellar granule neuron cultures. *Journal of Neuroimmunology, 155*(1–2), 43–54. https://doi.org/10.1016/j.jneuroim.2004.06.014.

Cryan, J. F., & Dinan, T. G. (2012). Mind-altering microorganisms: the impact of the gut microbiota on brain and behaviour. *Nature Reviews Neuroscience, 13*(10), 701–712. https://doi.org/10.1038/nrn3346.

Cunningham, C. (2013). Microglia and neurodegeneration: the role of systemic inflammation. *Glia, 61*(1), 71–90. https://doi.org/10.1002/glia.22350.

Dantzer, R. (2001). Cytokine-induced sickness behavior: mechanisms and implications. *The Annals of the New York Academy of Sciences, 933*, 222–234.

Dantzer, R., O'Connor, J. C., Freund, G. G., Johnson, R. W., & Kelley, K. W. (2008). From inflammation to sickness and depression: when the immune system subjugates the brain. *Nature Reviews Neuroscience, 9*(1), 46–56. https://doi.org/10.1038/nrn2297.

Dantzer, R., O'Connor, J. C., Lawson, M. A., & Kelley, K. W. (2011). Inflammation-associated depression: from serotonin to kynurenine. *Psychoneuroendocrinology, 36*(3), 426–436. https://doi.org/10.1016/j.psyneuen.2010.09.012.

Derdemezis, C. S., Voulgari, P. V., Drosos, A. A., & Kiortsis, D. N. (2012). Obesity, adipose tissue and rheumatoid arthritis: coincidence or more complex relationship? *Clinical and Experimental Rheumatology, 29*(4), 712–727. Retrieved from http://www.ncbi.nlm.nih.gov/pubmed/21640051.

Dimopoulos, N., Piperi, C., Psarra, V., Lea, R. W., & Kalofoutis, A. (2008). Increased plasma levels of 8-iso-PGF2α and IL-6 in an elderly population with depression. *Psychiatry Research, 161*(1), 59–66. https://doi.org/10.1016/j.psychres.2007.07.019.

Dinan, T., Borre, Y., & Cryan, J. (2014). Genomics of schizophrenia: time to consider the gut microbiome? *Molecular Psychiatry, 19*(12), 1252–1257. https://doi.org/10.1038/mp.2014.93.

Dinan, T. G., & Cryan, J. F. (2013). Melancholic microbes: a link between gut microbiota and depression? *Neurogastroenterology and Motility: The Official Journal of the European Gastrointestinal Motility Society, 25*(9), 713–719. https://doi.org/10.1111/nmo.12198.

Dinan, T. G., & Cryan, J. F. (2017). Gut instincts: microbiota as a key regulator of brain development, ageing and neurodegeneration. *The Journal of Physiology, 595*(2), 489–503. https://doi.org/10.1113/JP273106.

Dobbin, J. P., Harth, M., McCain, G. A., Martin, R. A., & Cousin, K. (1991). Cytokine production and lymphocyte transformation during stress. *Brain, Behavior, and Immunity, 5*(4), 339–348. Retrieved from http://www.ncbi.nlm.nih.gov/pubmed/1777728.

Duman, R. S. (2012). A neurotrophic hypothesis of depression: role of synaptogenesis in the actions of NMDA receptor antagonists. *Philosophical Transactions of the Royal Society of London. Series B, Biological Sciences, 367*(1601), 2475–2484. https://doi.org/10.1098/rstb.2011.0357.

Duman, R. S., Heninger, G. R., & Nestler, E. J. (1997). A molecular and cellular theory of depression. *Archives of General Psychiatry, 54*(7), 597. https://doi.org/10.1001/archpsyc.1997.01830190015002.

Duman, R. S., & Monteggia, L. M. (2006). A neurotrophic model for stress-related mood disorders. *Biological Psychiatry, 59*(12), 1116–1127. https://doi.org/10.1016/j.biopsych.2006.02.013.

Ekdahl, C. T., Claasen, J. H., Bonde, S., Kokaia, Z., & Lindvall, O. (2003). Inflammation is detrimental for neurogenesis in adult brain. *Proceedings of the National Academy of Sciences of the United States of America, 100*(23), 13632–13637. https://doi.org/10.1073/pnas.2234031100.

Erickson, M. A., Dohi, K., & Banks, W. A. (2012). Neuroinflammation: a common pathway in CNS diseases as mediated at the blood-brain barrier. *NeuroImmunoModulation, 19*(2), 121–130. https://doi.org/10.1159/000330247.

Erny, D., Hrabě de Angelis, A. L., Jaitin, D., Wieghofer, P., Staszewski, O., David, E., et al. (2015). Host microbiota constantly control maturation and function of microglia in the CNS. *Nature Neuroscience, 18*(7), 965–977. https://doi.org/10.1038/nn.4030.

Fiore, M., Angelucci, F., Alleva, E., Branchi, I., Probert, L., & Aloe, L. (2000). Learning performances, brain NGF distribution and NPY levels in transgenic mice expressing TNF-alpha. *Behavioural Brain Research, 112*(1–2), 165–175. https://doi.org/10.1016/S0166-4328(00)00180-7.

Forlenza, M. J., & Miller, G. E. (2006). Increased serum levels of 8-hydroxy-2′-deoxyguanosine in clinical depression. *Psychosomatic Medicine, 7*(68), 1–7. https://doi.org/10.1097/01.psy.0000195780.37277.2a.

Frenois, F. (2007). Lipopolysaccharide induces. *Psychoneuroendocrinology, 32*(5), 516–531. https://doi.org/10.1038/nmeth.2250.Digestion.

Gałecki, P., Gałecka, E., Maes, M., Chamielec, M., Orzechowska, A., Bobińska, K., et al. (2012). The expression of genes encoding for COX-2, MPO, iNOS, and sPLA2-IIA in patients with recurrent depressive disorder. *Journal of Affective Disorders, 138*(3), 360–366. https://doi.org/10.1016/j.jad.2012.01.016.

Gallagher, D., Kiss, A., Lanctot, K., & Herrmann, N. (2016). Depressive symptoms and cognitive decline: a longitudinal analysis of potentially modifiable risk factors in community dwelling older adults. *Journal of Affective Disorders, 190*, 235–240. https://doi.org/10.1016/j.jad.2015.09.046.

Gao, Y., Huang, C., Zhao, K., Ma, L., Qiu, X., Zhang, L., et al. (2013). Depression as a risk factor for dementia and mild cognitive impairment: a meta-analysis of longitudinal studies. *International Journal of Geriatric Psychiatry, 28*(5), 441–449. https://doi.org/10.1002/gps.3845.

Gardner, A., & Boles, R. G. (2011). Beyond the serotonin hypothesis: mitochondria, inflammation and neurodegeneration in major depression and affective spectrum disorders. *Progress in Neuro-Psychopharmacology and Biological Psychiatry*, 35(3), 730–743. https://doi.org/10.1016/j.pnpbp.2010.07.030.

Gardner, A., Johansson, A., Wibom, R., Nennesmo, I., Von Döbeln, U., Hagenfeldt, L., et al. (2003). Alterations of mitochondrial function and correlations with personality traits in selected major depressive disorder patients. *Journal of Affective Disorders*, 76(1–3), 55–68. https://doi.org/10.1016/S0165-0327(02)00067-8.

Global Burden of Disease Study 2013 Collaborators. (2015). Global, regional, and national incidence, prevalence, and years lived with disability for 301 acute and chronic diseases and injuries in 188 countries, 1990–2013: a systematic analysis for the Global Burden of Disease Study 2013. *Lancet*, 386(9995), 743–800. https://doi.org/10.1016/S0140-6736(15)60692-4.Global.

Goldsmith, D. R., Haroon, E., Woolwine, B. J., Jung, M. Y., Wommack, E. C., Harvey, P. D., et al. (2016). Inflammatory markers are associated with decreased psychomotor speed in patients with major depressive disorder. *Brain, Behavior, and Immunity*, 56, 281–288. https://doi.org/10.1016/j.bbi.2016.03.025.

Goodwin, F. K., Jamison, K. R., & Ghaemi, S. N. (2007). *Manic-depressive illness: Bipolar disorders and recurrent depression*. Oxford: Oxford University Press. Retrieved from https://global.oup.com/academic/product/manic-depressive-illness-9780195135794?cc=fr&lang=en&.

Goshen, I., Kreisel, T., Ben-Menachem-Zidon, O., Licht, T., Weidenfeld, J., Ben-Hur, T., et al. (2008). Brain interleukin-1 mediates chronic stress-induced depression in mice via adrenocortical activation and hippocampal neurogenesis suppression. *Molecular Psychiatry*, 13(7), 717–728. https://doi.org/10.1038/sj.mp.4002055.

Gounopoulos, P., Merki, E., Hansen, L. F., Choi, S.-H., & Tsimikas, S. (2007). Antibodies to oxidized low density lipoprotein: epidemiological studies and potential clinical applications in cardiovascular disease. *Minerva Cardioangiologica*, 55(6), 821–837. Retrieved from http://www.ncbi.nlm.nih.gov/pubmed/18091649.

Haapakoski, R., Ebmeier, K. P., Alenius, H., & Kivimäki, M. (2016). Innate and adaptive immunity in the development of depression: an update on current knowledge and technological advances. *Progress in Neuro-Psychopharmacology and Biological Psychiatry*, 66, 63–72. https://doi.org/10.1016/j.pnpbp.2015.11.012.

Haapakoski, R., Mathieu, J., Ebmeier, K. P., Alenius, H., & Kivimäki, M. (2015). Cumulative meta-analysis of interleukins 6 and 1β, tumour necrosis factor α and C-reactive protein in patients with major depressive disorder. *Brain, Behavior, and Immunity*, 49(October), 206–215. https://doi.org/10.1016/j.bbi.2015.06.001.

Haroon, E., Raison, C. L., & Miller, A. H. (2012). Psychoneuroimmunology meets neuropsychopharmacology: translational implications of the impact of inflammation on behavior. *Neuropsychopharmacology: Official Publication of the American College of Neuropsychopharmacology*, 37(1), 137–162. https://doi.org/10.1038/npp.2011.205.

Hasselbalch, B. J., Knorr, U., & Kessing, L. V. (2011). Cognitive impairment in the remitted state of unipolar depressive disorder: a systematic review. *Journal of Affective Disorders*, 134(1–3), 20–31. https://doi.org/10.1016/j.jad.2010.11.011.

Herrera-Guzmán, I., Gudayol-Ferré, E., Lira-Mandujano, J., Herrera-Abarca, J., Herrera-Guzmán, D., Montoya-Pérez, K., et al. (2008). Cognitive predictors of treatment response to bupropion and cognitive effects of bupropion in patients with major depressive disorder. *Psychiatry Research*, 160(1), 72–82. https://doi.org/10.1016/j.psychres.2007.04.012.

Holley, M. M., & Kielian, T. (2012). Th1 and Th17 cells regulate innate immune responses and bacterial clearance during central nervous system infection. *Journal of Immunology (Baltimore, Md: 1950)*, 188(3), 1360–1370. https://doi.org/10.4049/jimmunol.1101660.

Hu, F., Wang, X., Pace, T. W. W., Wu, H., & Miller, A. H. (2005). Inhibition of COX-2 by celecoxib enhances glucocorticoid receptor function. *Molecular Psychiatry*, 10(5), 426–428. https://doi.org/10.1038/sj.mp.4001644.

Insel, T., Cuthbert, B., Garvey, M., Heinssen, R., Pine, D. S., Quinn, K., et al. (2010). Research domain criteria (RDoC): toward a new classification framework for research on mental disorders. *American Journal of Psychiatry*, 167(7), 748–751. https://doi.org/10.1176/appi.ajp.2010.09091379.

Iseme, R. A., McEvoy, M., Kelly, B., Agnew, L., Attia, J., & Walker, F. R. (2014). Autoantibodies and depression. Evidence for a causal link? *Neuroscience and Biobehavioral Reviews*, 40, 62–79. https://doi.org/10.1016/j.neubiorev.2014.01.008.

Iyengar, R. L., Gandhi, S., Aneja, A., Thorpe, K., Razzouk, L., Greenberg, J., et al. (2013). NSAIDs are associated with lower depression scores in patients with osteoarthritis. *The American Journal of Medicine*, 126(11), 1017.e11–1017.e18. https://doi.org/10.1016/j.amjmed.2013.02.037.

Kalkman, H. O., & Feuerbach, D. (2016). Antidepressant therapies inhibit inflammation and microglial M1-polarization. *Pharmacology & Therapeutics*, 163, 82–93. https://doi.org/10.1016/j.pharmthera.2016.04.001.

Kappelmann, N., Lewis, G., Dantzer, R., Jones, P., & Khandaker, G. (2016). Antidepressant activity of anticytokine treatment: a systematic review and meta-analysis of clinical trials of chronic inflammatory conditions. *Nature Publishing Group*, 1–9. https://doi.org/10.1038/mp.2016.167.

Katona, C., Peveler, R., Dowrick, C., Wessely, S., Feinmann, C., Gask, L., et al. (2005). Pain symptoms in depression: definition and clinical significance. *Clinical Medicine (London, England)*, 5(4), 390–395. Retrieved from http://www.ncbi.nlm.nih.gov/pubmed/16138496.

Kelly, J. R., Kennedy, P. J., Cryan, J. F., Dinan, T. G., Clarke, G., & Hyland, N. P. (2015). Breaking down the barriers: the gut microbiome, intestinal permeability and stress-related psychiatric disorders. *Frontiers in Cellular Neuroscience*, 9, 392. https://doi.org/10.3389/fncel.2015.00392.

Kendler, K. S., Thornton, L. M., & Gardner, C. O. (2000). Stressful life events and previous episodes in the etiology of major depression in women: an evaluation of the "kindling" hypothesis. *American Journal of Psychiatry*, 157(8), 1243–1251. https://doi.org/10.1176/appi.ajp.157.8.1243.

Kent, S., Bluthé, R.-M., Kelley, K. W., & Dantzer, R. (1992). Sickness behavior as a new target for drug development. *Trends in Pharmacological Sciences*, 13(1), 24–28. https://doi.org/10.1016/0165-6147(92)90012-U.

Kent, S., Bret-Dibat, J. L., Kelley, K. W., & Dantzer, R. (1996). Mechanisms of sickness-induced decreases in food-motivated behavior. *Neuroscience and Biobehavioral Reviews*, 20(1), 171–175. https://doi.org/10.1016/0149-7634(95)00037-F.

Kessing, L. V. (2015). Course and cognitive outcome in major affective disorder. *Danish Medical Journal*, 62(11), B5160. Retrieved from http://www.ncbi.nlm.nih.gov/pubmed/26522485.

Kessing, L. V., & Andersen, P. K. (2017). Evidence for clinical progression of unipolar and bipolar disorders. *Acta Psychiatrica Scandinavica*, 135(1), 51–64. https://doi.org/10.1111/acps.12667.

Kessing, L. V., Hansen, M. G., & Andersen, P. K. (2004). Course of illness in depressive and bipolar disorders. Naturalistic study, 1994–1999. *British Journal of Psychiatry*, 185, 372–377.

Kessing, L. V., Hansen, M. G., Andersen, P. K., & Angst, J. (2004). The predictive effect of episodes on the risk of recurrence in depressive and bipolar disorders – a lifelong perspective. *Acta Psychiatrica Scandinavica*, 7(5), 339–344.

Kiecolt-Glaser, J. K., Gouin, J.-P., Weng, N.-P., Malarkey, W. B., Beversdorf, D. Q., & Glaser, R. (2011). Childhood adversity heightens the impact of later-life caregiving stress on telomere length and inflammation. *Psychosomatic Medicine*, 73(1), 16–22. https://doi.org/10.1097/PSY.0b013e31820573b6.

Köhler, O., Benros, M. E., Nordentoft, M., Farkouh, M. E., Iyengar, R. L., Mors, O., et al. (2014). Effect of anti-inflammatory treatment on depression, depressive symptoms, and adverse effects a systematic review and meta-analysis of randomized clinical trials. *JAMA*

Psychiatry, 71(12), 1381. https://doi.org/10.1001/jamapsychiatry.2014.1611.

Kohler, C., Freitas, T., Maes, M., de Andrade, N., Liu, C., Fernandes, B., et al. (2016). Peripheral cytokine and chemokine alterations in depression: a meta-analysis of 82 studies. *Acta Psychiatrica Scandinavica*, 135, 1–15. https://doi.org/10.1111/acps.12698.

Koo, J. W., Russo, S. J., Ferguson, D., Nestler, E. J., & Duman, R. S. (2010). Nuclear factor-kB is a critical mediator of stress-impaired neurogenesis and depressive behavior. *Proceedings of the National Academy of Sciences*, 107(6), 2669–2674. https://doi.org/10.1073/pnas.0910658107.

Kovacs, M., Obrosky, S., & George, C. (2016). The course of major depressive disorder from childhood to young adulthood: recovery and recurrence in a longitudinal observational study. *Journal of Affective Disorders*, 203, 374–381. https://doi.org/10.1016/j.jad.2016.05.042.

Krady, J. K., Lin, H. W., Liberto, C. M., Basu, A., Kremlev, S. G., & Levison, S. W. (2008). Ciliary neurotrophic factor and interleukin-6 differentially activate microglia. *Journal of Neuroscience Research*, 86(7), 1538–1547. https://doi.org/10.1002/jnr.21620.

Kraepelin, E. (1921). *Manic-depressive Insanity and Paranoia (Translated by M.Barclay)*. Edinburgh: Churchill Livingstone.

Kroenke, K., Wu, J., Bair, M. J., Krebs, E. E., Damush, T. M., & Tu, W. (2011). Reciprocal relationship between pain and depression: a 12-month longitudinal analysis in primary care. *The Journal of Pain: Official Journal of the American Pain Society*, 12(9), 964–973. https://doi.org/10.1016/j.jpain.2011.03.003.

Kvist, K., Andersen, P. K., Angst, J., & Kessing, L. V. (2010). Event dependent sampling of recurrent events. *Lifetime Data Analysis*, 16(4), 580–598. https://doi.org/10.1007/s10985-010-9172-y.

Lambertsen, K. L., Gregersen, R., Meldgaard, M., Clausen, B. H., Heibøl, E. K., Ladeby, R., et al. (2004). A role for interferon-gamma in focal cerebral ischemia in mice. *Journal of Neuropathology and Experimental Neurology*, 63(9), 942–955.

Lamers, F., Vogelzangs, N., Merikangas, K., De Jonge, P., Beekman, A., & Penninx, B. (2013). Evidence for a differential role of HPA-axis function, inflammation and metabolic syndrome in melancholic versus atypical depression. *Molecular Psychiatry*, 18(6), 692–699. https://doi.org/10.1038/mp.2012.144.

Lee, B. H., Kim, H., Park, S. H., & Kim, Y. K. (2007). Decreased plasma BDNF level in depressive patients. *Journal of Affective Disorders*, 101(1–3), 239–244. https://doi.org/10.1016/j.jad.2006.11.005.

Leonard, B., & Maes, M. (2012). Mechanistic explanations how cell-mediated immune activation, inflammation and oxidative and nitrosative stress pathways and their sequels and concomitants play a role in the pathophysiology of unipolar depression. *Neuroscience and Biobehavioral Reviews, 36*(2), 764–785. https://doi.org/10.1016/j.neubiorev.2011.12.005.

Lewis, K., Lutgendorff, F., Phan, V., Söderholm, J. D., Sherman, P. M., & McKay, D. M. (2010). Enhanced translocation of bacteria across metabolically stressed epithelia is reduced by butyrate. *Inflammatory Bowel Diseases, 16*(7), 1138–1148. https://doi.org/10.1002/ibd.21177.

Li, X., Frye, M. A., & Shelton, R. C. (2012). Review of pharmacological treatment in mood disorders and future directions for drug development. *Neuropsychopharmacology, 37*(1), 77–101. https://doi.org/10.1038/npp.2011.198.

Lichtblau, N., Schmidt, F. M., Schumann, R., Kirkby, K. C., & Himmerich, H. (2013). Cytokines as biomarkers in depressive disorder: current standing and prospects. *International Review of Psychiatry (Abingdon, England), 25*(5), 592–603. https://doi.org/10.3109/09540261.2013.813442.

Lin, P. Y., Huang, Y. C., & Hung, C. F. (2016). Shortened telomere length in patients with depression: a meta-analytic study. *Journal of Psychiatric Research, 76*, 84–93. https://doi.org/10.1016/j.jpsychires.2016.01.015.

Loftis, J. M., Patterson, A. L., Wilhelm, C. J., McNett, H., Morasco, B. J., Huckans, M., et al. (2013). Vulnerability to somatic symptoms of depression during interferon-alpha therapy for hepatitis C: a 16-week prospective study. *Journal of Psychosomatic Research, 74*(1), 57–63. https://doi.org/10.1016/j.jpsychores.2012.10.012.

Lopez-Garcia, E., Schulze, M. B., Fung, T. T., Meigs, J. B., Rifai, N., Manson, J. E., et al. (2004). Major dietary patterns are related to plasma concentrations of markers of inflammation and endothelial dysfunction. *The American Journal of Clinical Nutrition, 80*(4), 1029–1035. Retrieved from http://www.ncbi.nlm.nih.gov/pubmed/15447916.

Lotrich, F. E. (2015). Inflammatory cytokine-associated depression. *Brain Research, 1617*, 113–125. https://doi.org/10.1016/j.brainres.2014.06.032.

Luppino, F. S., de Wit, L. M., Bouvy, P. F., Stijnen, T., Cuijpers, P., Penninx, B. W. J. H., et al. (2010). Overweight, obesity, and depression: a systematic review and meta-analysis of longitudinal studies. *Archives of General Psychiatry, 67*(3), 220–229. https://doi.org/10.1001/archgenpsychiatry.2010.2.

Macedo, D., Filho, A. J. M. C., Soares de Sousa, C. N., Quevedo, J., Barichello, T., Júnior, H. V. N., et al. (2016). Antidepressants, antimicrobials or both? Gut microbiota dysbiosis in depression and possible implications of the antimicrobial effects of antidepressant drugs for antidepressant effectiveness. *Journal of Affective Disorders, 208*, 22–32. https://doi.org/10.1016/j.jad.2016.09.012.

Machado, M. O., Oriolo, G., Bortolato, B., Köhler, C. A., Maes, M., Solmi, M., et al. (2016). Biological mechanisms of depression following treatment with interferon for chronic hepatitis C: a critical systematic review. *Journal of Affective Disorders, 209*, 235–245. https://doi.org/10.1016/j.jad.2016.11.039.

Maes, M. (2009). "Functional" or "psychosomatic" symptoms, e.g. a flu-like malaise, aches and pain and fatigue, are major features of major and in particular of melancholic depression. *Neuro Endocrinology Letters, 30*(5), 564–573. Retrieved from http://www.ncbi.nlm.nih.gov/pubmed/20035251.

Maes, M. (2011). Depression is an inflammatory disease, but cell-mediated immune activation is the key component of depression. *Progress in Neuro-Psychopharmacology and Biological Psychiatry, 35*(3), 664–675. https://doi.org/10.1016/j.pnpbp.2010.06.014.

Maes, M., Berk, M., Goehler, L., Song, C., Anderson, G., Galecki, P., et al. (2012). Depression and sickness behavior are Janus-faced responses to shared inflammatory pathways. *BMC Medicine, 10*, 66. https://doi.org/10.1186/1741-7015-10-66.

Maes, M., Bosmans, E., Suy, E., Vandervorst, C., De Jonckheere, C., & Raus, J. (1990). Immune disturbances during major depression: upregulated expression of interleukin-2 receptors. *Neuropsychobiology, 24*(3), 115–120. Retrieved from http://www.ncbi.nlm.nih.gov/pubmed/2135065.

Maes, M., Galecki, P., Chang, Y. S., & Berk, M. (2011). A review on the oxidative and nitrosative stress (O&NS) pathways in major depression and their possible contribution to the (neuro)degenerative processes in that illness. *Progress in Neuro-Psychopharmacology and Biological Psychiatry, 35*(3), 676–692. https://doi.org/10.1016/j.pnpbp.2010.05.004.

Maes, M., Kubera, M., Leunis, J. C., Berk, M., Geffard, M., & Bosmans, E. (2013). In depression, bacterial translocation may drive inflammatory responses, oxidative and nitrosative stress (O&NS), and autoimmune responses directed against O&NS-damaged neoepitopes. *Acta Psychiatrica Scandinavica, 127*(5), 344–354. https://doi.org/10.1111/j.1600-0447.2012.01908.x.

Maes, M., Kubera, M., Mihaylova, I., Geffard, M., Galecki, P., Leunis, J. C., & Berk, M. (2013). Increased autoimmune responses against auto-epitopes modified by oxidative and nitrosative damage in depression: implications for the pathways to chronic depression and neuroprogression. *Journal of Affective Disorders, 149*(1–3), 23–29. https://doi.org/10.1016/j.jad.2012.06.039.

Maes, M., Kubera, M., Leunis, J. C., & Berk, M. (2012). Increased IgA and IgM responses against gut commensals in chronic depression: further evidence for increased bacterial translocation or leaky gut. *Journal of Affective Disorders*, 141(1), 55–62. https://doi.org/10.1016/j.jad.2012.02.023.

Maes, M., Mihaylova, I., Kubera, M., & Ringel, K. (2012). Activation of cell-mediated immunity in depression: association with inflammation, melancholia, clinical staging and the fatigue and somatic symptom cluster of depression. *Progress in Neuro-Psychopharmacology and Biological Psychiatry*, 36(1), 169–175. https://doi.org/10.1016/j.pnpbp.2011.09.006.

Maes, M., Mihaylova, I., Kubera, M., Leunis, J. C., & Geffard, M. (2011). IgM-mediated autoimmune responses directed against multiple neoepitopes in depression: new pathways that underpin the inflammatory and neuroprogressive pathophysiology. *Journal of Affective Disorders*, 135(1–3), 414–418. https://doi.org/10.1016/j.jad.2011.08.023.

Maes, M., Ringel, K., Kubera, M., Berk, M., & Rybakowski, J. (2012). Increased autoimmune activity against 5-HT: a key component of depression that is associated with inflammation and activation of cell-mediated immunity, and with severity and staging of depression. *Journal of Affective Disorders*, 136(3), 386–392. https://doi.org/10.1016/j.jad.2011.11.016.

Maes, M., Scharpe, S., Bosmans, E., Vandewoude, M., Suy, E., Uyttenbroeck, W., et al. (1992). Disturbances in acute phase plasma proteins during melancholia: additional evidence for the presence of an inflammatory process during that illness. *Progress in Neuro-Psychopharmacology & Biological Psychiatry*, 16(4), 501–515. Retrieved from http://www.ncbi.nlm.nih.gov/pubmed/1379370.

Mahajan, S., Avasthi, A., Grover, S., & Chawla, Y. K. (2014). Role of baseline depressive symptoms in the development of depressive episode in patients receiving antiviral therapy for Hepatitis C infection. *Journal of Psychosomatic Research*, 77(2), 109–115. https://doi.org/10.1016/j.jpsychores.2014.05.008.

Marin, I. A., & Kipnis, J. (2016). Central nervous system: (immunological) ivory tower or not? *Neuropsychopharmacology*, 1–8. https://doi.org/10.1038/npp.2016.122.

Martín, M., Macías, M., León, J., Escames, G., Khaldy, H., & Acuña-Castroviejo, D. (2002). Melatonin increases the activity of the oxidative phosphorylation enzymes and the production of ATP in rat brain and liver mitochondria. *The International Journal of Biochemistry & Cell Biology*, 34(4), 348–357. Retrieved from http://www.ncbi.nlm.nih.gov/pubmed/11854034.

Martin-Santos, R., Crippa, J. A., Batalla, A., Bhattacharyya, S., Atakan, Z., Borgwardt, S., et al. (2012). Acute effects of a single, oral dose of d9-tetrahydrocannabinol (THC) and cannabidiol (CBD) administration in healthy volunteers. *Current Pharmaceutical Design*, 18(32), 4966–4979. Retrieved from http://www.ncbi.nlm.nih.gov/pubmed/22716148.

McAfoose, J., & Baune, B. T. (2009). Evidence for a cytokine model of cognitive function. *Neuroscience & Biobehavioral Reviews*, 33(3), 355–366. https://doi.org/10.1016/j.neubiorev.2008.10.005.

McIntyre, R. S. (2013). Using measurement strategies to identify and monitor residual symptoms. *Journal of Clinical Psychiatry*, 74(Suppl. 2), 14–18. https://doi.org/10.4088/JCP.12084su1c.03.

McIntyre, R. S., Woldeyohannes, H. O., Soczynska, J. K., Maruschak, N. A., Wium-Andersen, I. K., Vinberg, M., et al. (2015). Anhedonia and cognitive function in adults with MDD: results from the International Mood Disorders Collaborative Project. *CNS Spectrums*, 21, 1–5. https://doi.org/10.1017/S1092852915000747.

Meares, G. P., Ma, X., Qin, H., & Benveniste, E. N. (2012). Regulation of CCL20 expression in astrocytes by IL-6 and IL-17. *Glia*, 60(5), 771–781. https://doi.org/10.1002/glia.22307.

Mechoulam, R., & Parker, L. A. (2013). The endocannabinoid system and the brain. *Annual Review of Psychology*, 64(1), 21–47. https://doi.org/10.1146/annurev-psych-113011-143739.

Miller, G. E., & Cole, S. W. (2012). Clustering of depression and inflammation in adolescents previously exposed to childhood adversity. *Biological Psychiatry*, 72(1), 34–40. https://doi.org/10.1016/j.biopsych.2012.02.034.

Miller, A. H., Haroon, E., Raison, C. L., & Felger, J. C. (2013). Cytokine targets in the brain: impact on neurotransmitters and neurocircuits. *Depression and Anxiety*, 30(4), 297–306. https://doi.org/10.1016/j.biotechadv.2011.08.021. Secreted.

Miller, I. W., Keitner, G. I., Schatzberg, A. F., Klein, D. N., Thase, M. E., Rush, A. J., et al. (1998). The treatment of chronic depression, Part 3: psychosocial functioning before and after treatment with sertraline or imipramine. *Journal of Clinical Psychiatry*, 59(11), 608–619. https://doi.org/10.4088/JCP.v59n1108.

Miller, A. H., Maletic, V., & Raison, C. L. (2009). Inflammation and its discontents: the role of cytokines in the pathophysiology of major depression. *Psychiatry: Interpersonal and Biological Processes*, 65(9), 732–741. https://doi.org/10.1016/j.biopsych.2008.11.029.Inflammation.

Miller, A. H., & Raison, C. L. (2016). The role of inflammation in depression: from evolutionary imperative to modern treatment target. *Nature Reviews Immunology*, 16(1), 22–34. https://doi.org/10.1038/nri.2015.5.

Mills, C. D. (2015). Anatomy of a discovery: m1 and m2 macrophages. *Frontiers in Immunology*, 6, 1–12. https://doi.org/10.3389/fimmu.2015.00212.

Molendijk, M. L., Spinhoven, P., Polak, M., Bus, B. A. A., Penninx, B. W. J. H., & Elzinga, B. M. (2014). Serum BDNF concentrations as peripheral manifestations of depression: evidence from a systematic review and meta-analyses on 179 associations (N = 9484). *Molecular Psychiatry*, *19*(7), 791–800. https://doi.org/10.1038/mp.2013.105.

Möller, M., Du Preez, J. L., Viljoen, F. P., Berk, M., Emsley, R., & Harvey, B. H. (2013). Social isolation rearing induces mitochondrial, immunological, neurochemical and behavioural deficits in rats, and is reversed by clozapine or N-acetyl cysteine. *Brain, Behavior, and Immunity*, *30*, 156–167. https://doi.org/10.1016/j.bbi.2012.12.011.

Montecucco, F., Liberale, L., Bonaventura, A., Vecchiè, A., Dallegri, F., & Carbone, F. (2017). The role of inflammation in cardiovascular outcome. *Current Atherosclerosis Reports*, *19*(3), 11. https://doi.org/10.1007/s11883-017-0646-1.

Morris, G., Carvalho, A., Anderson, G., Galecki, P., & Maes, M. (2015). The many neuroprogressive actions of tryptophan catabolites (TRYCATs) that may be associated with the pathophysiology of neuro-immune disorders. *Current Pharmaceutical Design*, *22*(8), 963–977. https://doi.org/10.2174/1381612822666151215102420.

Mott, K. R., Zandian, M., Allen, S. J., & Ghiasi, H. (2013). Role of interleukin-2 and herpes simplex virus 1 in central nervous system demyelination in mice. *Journal of Virology*, *87*(22), 12102–12109. https://doi.org/10.1128/JVI.02241-13.

Moylan, S., Berk, M., Dean, O. M., Samuni, Y., Williams, L. J., O'Neil, A., et al. (2014). Oxidative & nitrosative stress in depression: why so much stress? *Neuroscience and Biobehavioral Reviews*, *45*, 46–62. https://doi.org/10.1016/j.neubiorev.2014.05.007.

Moylan, S., Maes, M., Wray, N. R., & Berk, M. (2013). The neuroprogressive nature of major depressive disorder: pathways to disease evolution and resistance, and therapeutic implications. *Molecular Psychiatry*, *18*(5), 595–606. https://doi.org/10.1038/mp.2012.33.

Müller, N., & Schwarz, M. J. (2007). The immune-mediated alteration of serotonin and glutamate: towards an integrated view of depression. *Molecular Psychiatry*, *12*(11), 988–1000. https://doi.org/10.1038/sj.mp.4002006.

Murphy, Á. C., Lalor, S. J., Lynch, M. A., & Mills, K. H. G. (2010). Infiltration of Th1 and Th17 cells and activation of microglia in the CNS during the course of experimental autoimmune encephalomyelitis. *Brain, Behavior, and Immunity*, *24*(4), 641–651. https://doi.org/10.1016/j.bbi.2010.01.014.

Naoi, M., Ishiki, R., Nomura, Y., Hasegawa, S., & Nagatsu, T. (1987). Quinolinic acid: an endogenous inhibitor specific for type B monoamine oxidase in human brain synaptosomes. *Neuroscience Letters*, *74*(2), 232–236. https://doi.org/10.1016/0304-3940(87)90155-8.

Nunemaker, C. G. (2016). Considerations for defining cytokine dose, duration, and milieu that are appropriate for modeling chronic low-grade inflammation in type 2 diabetes. *Journal of Diabetes Research*, *2016*. 2846570. https://doi.org/10.1155/2016/2846570.

O'Connor, J. C., Lawson, M. A., André, C., Moreau, M., Lestage, J., Castanon, N., et al. (2009). Lipopolysaccharide-induced depressive-like behavior is mediated by indoleamine 2,3-dioxygenase activation in mice. *Molecular Psychiatry*, *14*(5), 511–522. https://doi.org/10.1038/sj.mp.4002148.

Oepen, G., Baldessarini, R. J., & Salvatore, P. (2004). On the periodicity of manic-depressive insanity, by Eliot Slater (1938): translated excerpts and commentary. *Journal of Affective Disorders*, *78*(1), 1–9. https://doi.org/10.1016/S0165-0327(02)00359-2.

Paintlia, M. K., Paintlia, A. S., Singh, A. K., & Singh, I. (2011). Synergistic activity of interleukin-17 and tumor necrosis factor-α enhances oxidative stress-mediated oligodendrocyte apoptosis. *Journal of Neurochemistry*, *116*(4), 508–521. https://doi.org/10.1111/j.1471-4159. 2010. 07136.x.

Palta, P., Samuel, L. J., Miller, E. R., & Szanton, S. L. (2014). Depression and oxidative stress: results from a meta-analysis of observational studies. *Psychosomatic Medicine*, *76*(1), 12–19. https://doi.org/10.1097/PSY.00000000000000009.

Pariante, C. M., & Lightman, S. L. (2008). The HPA axis in major depression: classical theories and new developments. *Trends in Neurosciences*, *31*(9), 464–468. https://doi.org/10.1016/j.tins.2008.06.006.

Park, K. M., & Bowers, W. J. (2010). Tumor necrosis factor-alpha mediated signaling in neuronal homeostasis and dysfunction. *Cellular Signalling*, *22*(7), 977–983. https://doi.org/10.1016/j.cellsig.2010.01.010.

Patel, H., Ross, F., Heenan, L., Davies, R., Rothwell, N., & Allan, S. (2006). Neurodegenerative actions of interleukin-1 in the rat brain are mediated through increases in seizure activity. *Journal of Neuroscience Research*, *83*, 385–391.

Peng, L., He, Z., Chen, W., Holzman, I. R., & Lin, J. (2007). Effects of butyrate on intestinal barrier function in a Caco-2 cell monolayer model of intestinal barrier. *Pediatric Research*, *61*(1), 37–41. https://doi.org/10.1203/01.pdr.0000250014.92242.f3.

Pickering, M., Cumiskey, D., & O'Connor, J. J. (2005). Actions of TNF-alpha on glutamatergic synaptic transmission in the central nervous system. *Experimental Physiology*, *90*(5), 663–670. https://doi.org/10.1113/expphysiol.2005.030734.

Post, R. M. (2010). Mechanisms of illness progression in the recurrent affective disorders. *Neurotoxicity Research*, *18*(3–4), 256–271. https://doi.org/10.1007/s12640-010-9182-2.

Post, R. M. (2016). Epigenetic basis of sensitization to stress, affective episodes, and stimulants: implications for illness progression and prevention. *Bipolar Disorders*, 18(4), 315–324. https://doi.org/10.1111/bdi.12401.

Quan, N., & Banks, W. A. (2007). Brain-immune communication pathways. *Brain, Behavior, and Immunity*, 21(6), 727–735. https://doi.org/10.1016/j.bbi.2007.05.005.

Raftery, G., He, J., Pearce, R., Birchall, D., Newton, J. L., Blamire, A. M., et al. (2012). Disease activity and cognition in rheumatoid arthritis: an open label pilot study. *Arthritis Research & Therapy*, 14(6), R263. https://doi.org/10.1186/ar4108.

Raison, C. L., Capuron, L., & Miller, A. H. (2006). Cytokines sing the blues: inflammation and pathogenesis of depression. *Trends in Immunology*, 27(1), 24–31. https://doi.org/10.1016/j.it.2005.11.006.Cytokines.

Raison, C. L., Rutherford, R. E., Woolwine, B. J., Shuo, C., Schettler, P., Drake, D. F., et al. (2013). A randomized controlled trial of the tumor necrosis factor-alpha antagonist infliximab in treatment resistant depression: role of baseline inflammatory biomarkers. *JAMA Psychiatry*, 70 (1), 31–41. https://doi.org/10.1001/2013.jamapsychiatry.4.A.

Reus, G. Z., Jansen, K., Titus, S., Carvalho, A. F., Gabbay, V., & Quevedo, J. (2015). Kynurenine pathway dysfunction in the pathophysiology and treatment of depression: evidences from animal and human studies. *Journal of Psychiatric Research*, 68, 316–328. https://doi.org/10.1016/j.jpsychires.2015.05.007.

Réus, G. Z., Titus, S. E., Abelaira, H. M., Freitas, S. M., Tuon, T., Quevedo, J., et al. (2016). Neurochemical correlation between major depressive disorder and neurodegenerative diseases. *Life Sciences*, 158, 121–129. https://doi.org/10.1016/j.lfs.2016.06.027.

Rosenblat, J. D., Cha, D. S., Mansur, R. B., & McIntyre, R. S. (2014). Inflamed moods: a review of the interactions between inflammation and mood disorders. *Progress in Neuro-Psychopharmacology and Biological Psychiatry*, 53, 23–34. https://doi.org/10.1016/j.pnpbp.2014.01.013.

Rosenblat, J. D., McIntyre, R. S., Alves, G. S., Fountoulakis, K. N., & Carvalho, A. F. (2015). Beyond monoamines-novel targets for treatment-resistant depression: a comprehensive review. *Current Neuropharmacology*, 13(5), 636–655. https://doi.org/10.2174/1570159X13666150630175044.

Roth, T. L., Lubin, F. D., Funk, A. J., & Sweatt, J. D. (2009). Lasting epigenetic influence of early-life adversity on the BDNF gene. *Biological Psychiatry*, 65(9), 760–769. https://doi.org/10.1016/j.biopsych.2008.11.028.

Roth, T. L., & Sweatt, J. D. (2011). Epigenetic marking of the BDNF gene by early-life adverse experiences. *Hormones and Behavior*, 59(3), 315–320. https://doi.org/10.1016/j.yhbeh.2010.05.005.

Rush, A. J., Trivedi, M. H., Wisniewski, S. R., Nierenberg, A. A., Stewart, J. W., Warden, D., et al. (2006). Acute and longer-term outcomes in depressed outpatients requiring one or several treatment steps: a STAR*D report. *American Journal of Psychiatry*, 163(11), 1905–1917. https://doi.org/10.1176/appi.ajp.163.11.1905.

Sahay, A., & Hen, R. (2007). Adult hippocampal neurogenesis in depression. *Nature Neuroscience*, 10(9), 1110–1115. https://doi.org/10.1038/nn1969.

Sakaguchi, S., Yamaguchi, T., Nomura, T., & Ono, M. (2008). Regulatory T cells and immune tolerance. *Cell*, 133(5), 775–787. https://doi.org/10.1016/j.cell.2008.05.009.

Salminen, A., Haapasalo, A., Kauppinen, A., Kaarniranta, K., Soininen, H., & Hiltunen, M. (2015). Impaired mitochondrial energy metabolism in Alzheimer's disease: impact on pathogenesis via disturbed epigenetic regulation of chromatin landscape. *Progress in Neurobiology*, 131, 1–20. https://doi.org/10.1016/j.pneurobio.2015.05.001.

Serafini, G., Amore, M., & Rihmer, Z. (2015). The role of glutamate excitotoxicity and neuroinflammation in depression and suicidal behavior: focus on microglia cells. *Neuroimmunology and Neuroinflammation*, 2(3), 127. https://doi.org/10.4103/2347-8659.157955.

Sethi, G., Shanmugam, M. K., Ramachandran, L., Kumar, A. P., & Tergaonkar, V. (2012). Multifaceted link between cancer and inflammation. *Bioscience Reports*, 32(1), 1–15. https://doi.org/10.1042/BSR201spi6;00136.

Setiawan, E. (2015). Increased translocator protein distribution volume, a marker of neuroinflammation, in the brain during major depressive episodes. *JAMA Psychiatry*, 72 (3), 37–54. https://doi.org/10.1016/bs.mcb.2015.01.016. Observing.

Setiawan, E., Wilson, A. A., Mizrahi, R., Rusjan, P. M., Miler, L., Rajkowska, G., et al. (2015). Role of translocator protein density, a marker of neuroinflammation, in the brain during major depressive episodes. *JAMA Psychiatry*, 72(3), 268–275. https://doi.org/10.1001/jamapsychiatry.2014.2427.

Shen, Y., Zhu, L. J., Liu, S. S., Zhou, S. Y., & Luo, J. H. (2006). Interleukin-2 inhibits NMDA receptor-mediated currents directly and may differentially affect subtypes. *Biochemical and Biophysical Research Communications*, 351(2), 449–454. https://doi.org/10.1016/j.bbrc.2006.10.047.

Shimizu, E., Hashimoto, K., Okamura, N., Koike, K., Komatsu, N., Kumakiri, C., et al. (2003). Alterations of serum levels of brain-derived neurotrophic factor (BDNF) in depressed patients with or without antidepressants. *Biological Psychiatry*, 54(1), 70–75. https://doi.org/10.1016/S0006-3223(03)00181-1.

Sievers, C., Auer, M. K., Klotsche, J., Athanasoulia, A. P., Schneider, H. J., Nauck, M., et al. (2014). IGF-I levels and depressive disorders: results from the Study of Health in Pomerania (SHIP). *European Neuropsychopharmacology*, 24(6), 890–896. https://doi.org/10.1016/j.euroneuro.2014.01.008.

Slavich, G. M., Monroe, S. M., & Gotlib, I. H. (2011). Early parental loss and depression history: associations with recent life stress in major depressive disorder. *Journal of Psychiatric Research*, 45(9), 1146–1152. https://doi.org/10.1016/j.jpsychires.2011.03.004.

Slyepchenko, A., Maes, M., Jacka, F. N., Köhler, C. A., Barichello, T., Mcintyre, R. S., et al. (2017). Gut microbiota, bacterial translocation, and interactions with diet: pathophysiological links between major depressive disorder and non-communicable medical comorbidities. *Psychotherapy and Psychosomatics*, 8686, 31–4631. https://doi.org/10.1159/000448957.

Slyepchenko, A., Maes, M., Kohler, C. A., Anderson, G., Quevedo, J., Alves, G. S., et al. (2016). T helper 17 cells may drive neuroprogression in major depressive disorder: proposal of an integrative model. *Neuroscience and Biobehavioral Reviews*, 64, 83–100. https://doi.org/10.1016/j.neubiorev.2016.02.002.

Smith, R. S. (1991). The macrophage theory of depression. *Medical Hypotheses*, 35(4), 298–306. https://doi.org/10.1016/0306-9877(91)90272-Z.

Solomon, D. A., Keller, M. B., Leon, A. C., Mueller, T. I., Lavori, P. W., Shea, M. T., et al. (2000). Multiple recurrences of major depressive disorder. *American Journal of Psychiatry*, 157(2), 229–233. https://doi.org/10.1176/appi.ajp.157.2.229.

Song, C., & Wang, H. (2011). Cytokines mediated inflammation and decreased neurogenesis in animal models of depression. *Progress in Neuro-Psychopharmacology and Biological Psychiatry*, 35(3), 760–768. https://doi.org/10.1016/j.pnpbp.2010.06.020.

Spooren, A., Kolmus, K., Laureys, G., Clinckers, R., De Keyser, J., Haegeman, G., et al. (2011). Interleukin-6, a mental cytokine. *Brain Research Reviews*, 67(1–2), 157–183. https://doi.org/10.1016/j.brainresrev.2011.01.002.

Steptoe, A., Kunz-Ebrecht, S., & Owen, N. (2003). Lack of association between depressive symptoms and markers of immune and vascular inflammation in middle-aged men and women. *Psychological Medicine*, 33(4), 667–674.

Stetler, C., & Miller, G. E. (2011). Depression and hypothalamic-pituitary-adrenal activation: a quantitative summary of four decades of research. *Psychosomatic Medicine*, 73(2), 114–126. https://doi.org/10.1097/PSY.0b013e31820ad12b.

Stotland, N. L. (2012). Recovery from depression. *The Psychiatric Clinics of North America*, 35(1), 37–49. https://doi.org/10.1016/j.psc.2011.11.007.

Swardfager, W., Rosenblat, J. D., Benlamri, M., & McIntyre, R. S. (2016). Mapping inflammation onto mood: inflammatory mediators of anhedonia. *Neuroscience and Biobehavioral Reviews*, 64, 148–166. https://doi.org/10.1016/j.neubiorev.2016.02.017.

Tavares, R. G., Tasca, C. I., Santos, C. E. S., Alves, L. B., Porciúncula, L. O., Emanuelli, T., et al. (2002). Quinolinic acid stimulates synaptosomal glutamate release and inhibits glutamate uptake into astrocytes. *Neurochemistry International*, 40(7), 621–627. https://doi.org/10.1016/S0197-0186(01)00133-4.

Tramutola, A., Lanzillotta, C., Perluigi, M., & Butterfield, D. A. (2017). Oxidative stress, protein modification and Alzheimer disease. *Brain Research Bulletin*, 133, 88–96. https://doi.org/10.1016/j.brainresbull.2016.06.005.

Trivedi, P. P., & Jena, G. B. (2013). Melatonin reduces ulcerative colitis-associated local and systemic damage in mice: investigation on possible mechanisms. *Digestive Diseases and Sciences*, 58(12), 3460–3474. https://doi.org/10.1007/s10620-013-2831-6.

Trivedi, M. H., Rush, A. J., Wisniewski, S. R., Nierenberg, A. A., Warden, D., Ritz, L., et al. (2006). Evaluation of outcomes with citalopram for depression using measurement-based care in STAR*D: implications for clinical practice. *American Journal of Psychiatry*, 163(1), 28–40. https://doi.org/10.1176/appi.ajp.163.1.28.

Trudler, D., Nash, Y., & Frenkel, D. (2015). New insights on Parkinson's disease genes: the link between mitochondria impairment and neuroinflammation. *Journal of Neural Transmission*, 122(10), 1409–1419. https://doi.org/10.1007/s00702-015-1399-z.

Udina, M., Castellví, P., Moreno-España, J., Navinés, R., Valdés, M., Forns, X., et al. (2012). Interferon-induced depression in chronic hepatitis C. *The Journal of Clinical Psychiatry*, 73(8), 1128–1138. https://doi.org/10.4088/JCP.12r07694.

Uguz, F., Akman, C., Kucuksarac, S., & Tufekci, O. (2009). Anti-tumor necrosis factor-alpha therapy is associated with less frequent mood and anxiety disorders in patients with rheumatoid arthritis. *Psychiatry and Clinical Neurosciences*, 63(1), 50–55.

Valkanova, V., Ebmeier, K. P., & Allan, C. L. (2013). CRP, IL-6 and depression: a systematic review and meta-analysis of longitudinal studies. *Journal of Affective Disorders*, 150(3), 736–744. https://doi.org/10.1016/j.jad.2013.06.004.

Verhoeven, J. E., Révész, D., Epel, E. S., Lin, J., Wolkowitz, O. M., Penninx, B. W. J. H., et al. (2014). Major depressive disorder and accelerated cellular aging: results from a large psychiatric cohort study. *Molecular Psychiatry*, 19(8), 895–901. https://doi.org/10.1038/mp.2013.151.

Vogt, D., Waeldin, S., Hellhammer, D., & Meinlschmidt, G. (2016). The role of early adversity and recent life stress in depression severity in an outpatient sample. *Journal of Psychiatric Research*, 83, 61–70. https://doi.org/10.1016/j.jpsychires.2016.08.007.

Voloboueva, L. A., & Giffard, R. G. (2011). Inflammation, mitochondria, and the inhibition of adult neurogenesis. *Journal of Neuroscience Research*, 89(12), 1989–1996. https://doi.org/10.1002/jnr.22768.

Walker, A. K., Kavelaars, A., Heijnen, C. J., & Dantzer, R. (2014). Neuroinflammation and comorbidity of pain and depression. *Pharmacological Reviews*, 66(1), 80–101. https://doi.org/10.1124/pr.113.008144.

Walsh, J. G., Muruve, D. A., & Power, C. (2014). Inflammasomes in the CNS. *Nature Publishing Group*, 15, 1–14. https://doi.org/10.1038/nrn3638.

Wang, X. Q., Peng, Y. P., Lu, J. H., Cao, B. B., & Qiu, Y. H. (2009). Neuroprotection of interleukin-6 against NMDA attack and its signal transduction by JAK and MAPK. *Neuroscience Letters*, 450(2), 122–126.

Watkins, L. R., & Maier, S. F. (2000). The pain of being sick: implications of immune-to-brain communication for understanding pain. *Annual Review of Psychology*, 51, 29–57. https://doi.org/10.1177/0969733007088355.

Whale, R., Fialho, R., Rolt, M., Eccles, J., Pereira, M., Keller, M., et al. (2015). Psychomotor retardation and vulnerability to interferon alpha induced major depressive disorder: prospective study of a chronic hepatitis C cohort. *Journal of Psychosomatic Research*, 79(6), 640–645. https://doi.org/10.1016/j.jpsychores.2015.06.003.

Wohleb, E. S., Franklin, T., Iwata, M., & Duman, R. S. (2016). Integrating neuroimmune systems in the neurobiology of depression. *Nature Reviews Neuroscience*, 17(8), 497–511. https://doi.org/10.1038/nrn.2016.69.

Wu, C. W., Chen, Y. C., Yu, L., Chen, H. I., Jen, C. J., Huang, A. M., et al. (2007). Treadmill exercise counteracts the suppressive effects of peripheral lipopolysaccharide on hippocampal neurogenesis and learning and memory. *Journal of Neurochemistry*, 103(6), 2471–2481. https://doi.org/10.1111/j.1471-4159.2007.04987.x.

Wyss-Coray, T., & Mucke, L. (2002). Inflammation in neurodegenerative disease—a double-edged sword. *Neuron*, 35(3), 419–432. https://doi.org/10.1016/S0896-6273(02)00794-8.

Yirmiya, R., Rimmerman, N., & Reshef, R. (2015). Depression as a microglial disease. *Trends in Neurosciences*, 38(10), 637–658. https://doi.org/10.1016/j.tins.2015.08.001.

Zhan, Y., Paolicelli, R. C., Sforazzini, F., Weinhard, L., Bolasco, G., Pagani, F., et al. (2014). Deficient neuron-microglia signaling results in impaired functional brain connectivity and social behavior. *Nature Neuroscience*, 17(3), 400–406. https://doi.org/10.1038/nn.3641.

Zhao, Y., & Zhao, B. (2013). Oxidative stress and the pathogenesis of Alzheimer's disease. *Oxidative Medicine and Cellular Longevity*, 2013, 316523.

Zou, J. Y., & Crews, F. T. (2005). TNF-alpha potentiates glutamate neurotoxicity by inhibiting glutamate uptake in organotypic brain slice cultures: neuroprotection by NFkB inhibition. *Brain Research*, 1034(1–2), 11–24. https://doi.org/10.1016/j.brainres.2004.11.014.

Further Reading

Maes, M., Leonard, B. E., Myint, A. M., Kubera, M., & Verkerk, R. (2011). The new "5-HT" hypothesis of depression: cell-mediated immune activation induces indoleamine 2,3-dioxygenase, which leads to lower plasma tryptophan and an increased synthesis of detrimental tryptophan catabolites (TRYCATs), both of which contribute to the onset of depression. *Progress in Neuro-Psychopharmacology and Biological Psychiatry*, 35(3), 702–721. https://doi.org/10.1016/j.pnpbp.2010.12.017.

11

Gene Expression of Inflammation Markers in Depression

Liliana G. Ciobanu, Bernhard T. Baune

University of Adelaide, Adelaide, SA, Australia

Compelling evidence suggests that depression is a multifaceted disorder with both genetic and environmental factors contributing to the onset and progression of the disease. Despite substantial heritability of depression estimated at 31%–42% (Sullivan, Neale, & Kendler, 2000), identification of the genetic underpinning of depression has been challenging. An intensive search for genetic factors of depression using a candidate gene approach pointed toward more than 200 genetic loci, mainly genes involved in neurotransmission and the hypothalamic-pituitary-adrenal axis (HPA); however, only a few of these findings have been successfully replicated (Rivera & McGuffin, 2015). Genome-wide association studies (GWAS), after several unsuccessful attempts (Flint & Kendler, 2014), recently revealed 18 novel loci associated with depression at genome-wide level: Two loci were found associated with severe depression in Han Chinese women (CONVERGE, 2015), and 15 loci were identified through 23andMe using self-report data of severely depressed individuals (Hyde et al., 2016), and one locus, recently identified, was found to be associated with late-onset depression (Power et al., 2017). Although each of the three GWAS validated their findings during replication studies, there was no overlap in genetic variants across these studies. The discrepancies between the findings may reflect a previously proposed highly heterogeneous etiological, diagnostic, and genetic architecture of depression (Levinson et al., 2014).

Gene expression of inflammation in depression presents as a relatively novel and promising approach to uncover the pathophysiology of depression and to possibly provide useful clinical information for predicting treatment response and for decision-making processes in depression treatment. Quantifying the abundance of mRNA molecules in a single cell or in a population of cells provides essential information on the biological activity and functions of genes. Studying gene expression in depression can be viewed as complementary to a gene discovery approach aiming at understanding the dynamic molecular changes in depression. Given that the level and patterns of gene expression are influenced by both genetic and environmental factors (Wright & Sullivan, 2014), such as age (van den Akker et al., 2014), sex (Jansen et al., 2014), smoking status (Charlesworth et al., 2010), and well-being (Fredrickson et al., 2013), association between gene expression and depression may reflect combined genetic and nongenetic effects. For a clinical research context, the identification of altered gene expression patterns in depression is of critical importance for (1) a better understanding of molecular underpinnings of depression, (2) establishing biology-based clinical markers of depression, (3) providing evidence-based grounds for the development of novel antidepressant treatments, and (4) developing biomarkers for predicting treatment outcome, all of which are urgently needed for a better diagnosis and for tailored treatments of affected individuals (Ferrari et al., 2013).

Rapidly advancing technologies, such as microarrays and RNA-sequencing that allow for genome-wide coverage, have become powerful tools to quantify levels of gene expression in various tissues relevant for the pathophysiology of depression. Although a substantial amount of high-throughput experiments explored gene expression patterns in different brain areas/cell types and in peripheral blood of depressed patients, the results are inconsistent (Mehta, Menke, & Binder, 2010). Many biological and technical factors potentially contribute to these inconsistencies ranging from dynamic nature of both progression of depression and levels of gene expression to difficulties associated with mRNA stability and statistical analysis of gene expression data. Although there is a limited success on replicating the findings at the individual gene level, a leitmotif of immune dysregulation in depression frequently emerges at a functional level. As reviewed in this chapter, studies on antidepressant treatment response also outline the role for immune-related genes in predicting the efficacy of drug interventions (see Table 4). While there is converging evidence suggesting that inflammation plays a role in depression, it is well known that inflammation is also involved in many somatic diseases, like cardiovascular

disorders (Montecucco et al., 2017), arthritis (Firestein & McInnes, 2017), diabetes (Stuart & Baune, 2012), asthma (Robinson et al., 2017), and cancer (Crusz & Balkwill, 2015). These disorders commonly coexist with depressive symptoms (Kang et al., 2015; Penninx, Milaneschi, Lamers, & Vogelzangs, 2013), suggesting that inflammation in a comorbid phenotype represents another layer of clinical and molecular complexity of depression that requires investigation. Moreover, the development of depression is characterized by different clinical phases, which seem to be accompanied by dynamic immunomodulatory processes (Eyre, Stuart, & Baune, 2014).

In this chapter, the current knowledge of gene expression of inflammation markers in depression in the brain and in peripheral tissue will be presented in the context of depression with and without comorbid medical and/or psychiatric conditions. The chapter will also discuss the molecular overlap between psychiatric disorders from a gene expression perspective. Additionally, for a clinical context, gene expression of inflammation candidates will be evaluated on the topics of response to antidepressant treatment and anti-inflammatory interventions.

GENE EXPRESSION PATTERNS OF IMMUNE DYSREGULATIONS IN THE BRAIN

Studying the gene expression patterns in postmortem brain tissues of individuals who suffered from depression provides us with valuable information about molecular changes occurring in depressed brains compared with healthy controls. Such studies substantially advanced our understanding of the pathophysiological mechanisms of depression. Gene expression signatures derived from various brain regions collectively point toward molecular processes involving inflammatory pathways, cell survival, apoptosis, and oxidative stress (Bakunina, Pariante, & Zunszain, 2015).

Structural and functional neuroimaging studies in humans suggest that the limbic system (predominantly, amygdala and hippocampus) and the prefrontal cortex (PFC) serve as primary brain areas responsible for disturbances in emotion processing and mood regulation in depression (Wise, Cleare, Herane, Young, & Arnone, 2014). Gene expression studies on human brain tissue, utilizing both candidate genes and genome-wide approaches, provide some support for a dysregulated immune signaling within the brain; however, the results lack consistency across the studies, which makes it challenging to specify how altered markers of inflammation found in different brain areas contribute to depression. For instance, the most commonly reported circulating markers of inflammation pro-inflammatory cytokines *IL1B*, *IL6*, *TNF*, or *INF* were upregulated within various areas of the PFC, such as the dorsolateral prefrontal cortex (DLPFC) (BA9) (Kang et al., 2007) and the anterior PFC (BA10) (Malki et al., 2015; Shelton et al., 2011) of depressed individuals, in the primary ventral regions of the PFC (BA 44, 45, 46, and 47) of depressed suicide victims (Klempan et al., 2009), in the orbitofrontal area (BA11) of adult suicide victims (Tonelli et al., 2008), and in BA8 (part of the frontal cortex involved in the management of uncertainty) and the anterior PFC (BA10) of teenage suicide victims (Pandey et al., 2012). However, none of these genes were replicated within the same brain area. Immune and apoptosis signaling along with synaptic and glutamatergic signaling was also found disrupted in the hippocampal subfields dentate gyrus (DG) and CA1 of middle-aged subjects diagnosed with major depressive disorder (MDD) (Duric et al., 2013). An interesting study proposing synchronized dysregulation of expression in depression across different brain areas found a shift in coordinated gene expression levels between the amygdala and cingulate cortex for 100–250

individual genes, including *IL1* and *CREB1* in male MDD patients (Gaiteri, Guilloux, Lewis, & Sibille, 2010). Furthermore, several transcription factors known to be involved in immune response were found dysregulated in the depressed brain. For instance, alterations in expression levels of *CREB1*, a transcriptional factor known to be involved in a wide variety of biological processes including immune response, is one of the most consistently replicated findings; however, the directionality of dysregulation is not consistent across different brain areas. Thus, Sibille et al. (2004) found that *CREB1* is downregulated in the DLPFC (BA9 and BA47) of depressed suicide subjects (caution: it did not survive correction for multiple testing), while Tochigi et al. (2008) observed upregulation of *CREB1* in the anterior PFC (BA10) of nonsuicide depressed subjects. This discrepancy may be explained by the presence of a suicide component in one study and its absence in another, pointing toward a differential role of *CREB1* in depression and suicide. Alternatively, it may indicate that *CREB1* is downregulated in the DLPFC and upregulated in the anterior PFC in depression. Another transcription factor *FOXD3*, which functions as a transcriptional repressor, was found upregulated in the DLPFC together with *TNFRSF11B*, *INFA6*, and *INFR1* (Kang et al., 2007). Moreover, it seems that posttranscriptional regulation by short noncoding microRNAs, affecting both the stability and translation of mRNAs, is also involved in depression. Thus, utilizing RNA-sequencing data derived from DG granule cells, it has been found that posttranscriptional regulation by *miR-182*, which is involved in a broad range of biological processes including the regulation of immune response, significantly contributed to disrupted signaling in the hippocampus (DG) in depression (Kohen, Dobra, Tracy, & Haugen, 2014; Table 1).

Although all these findings suggest that brain dysregulation of immune genes, involving *IL1B*, *IL6*, *TNF*, *INF*, *CREB1*, *FOXD3*, and *miR-182*, might play a role in depression, they do not

TABLE 1   Studies on the Dysregulation of Immune and Oxidative Stress Factors in *Brain Tissue*

Citation	Brain Area/Cell Type	Platform/ Genes Studied	Diagnostic Criteria	Sample Size	Medications	Main Findings
Chu et al. (2009)	Neurons from thalamus by LCM	Affimetrix Human Genome U133 Plus 2.0	DSM-IV	15 MDD cases (age range 30–65, M:F) 15 Schizophrenia cases 15 Bipolar disorder cases 15 Controls	Majority of subjects in patients group had histories of medications that alter neuronal function	Immune-related dysregulation found to be unique to SZ; not found in depression
Duric et al. (2013)	Hippocampus (DG granule, CA1 pyramidal cell layers)	HEEBO 48K human	DSM-IV	21 MDD cases (age range 30–87, 13M:8F) 18 Controls	Antidepressant use included as covariate in analysis	Dysregulation of immune response signaling and apoptosis
Gaiteri et al. (2010)	Amygdala and cingulate cortex	Affimetrix	DSM-IV	14 MDD cases (familial) (mean age 52.2 ± 12.2, M only) 14 Controls	Toxicology results were incorporated into analysis	A shift in coordinated gene expression levels between amygdala and cingulate cortex for between 100 and 250 individual genes, including *IL-1* and *CREB1*
Iwamoto et al. (2004)	PFC: BA10	Affimetrix HU95A	DSM-IV	15 MDD cases (mean age 46.5 ± 9.3, 9M:6F) 15 Schizophrenia cases 15 Bipolar disorder cases 15 Controls	13 Subjects used antidepressants, included as covariate in analysis	Distinct gene expression patterns across disorders with shared upregulation of transcription and translation genes
Kang et al. (2007)	DLPFC: BA9	Agilent Human 1A	DSM-IV	15 MDD cases (mean age 51 ± 14.99, 10M:4F) 15 Controls	An antidepressant medication was present in the blood of four depressed subjects	*FOXD3, TNFRSF11B, INFA6, INFR1* found upregulated in DLPFC of depressed subjects
Kim et al. (2016)	Hippocampus	RNA-seq	DSM-IV	15 MDD cases (mean age 46.5 ± 9.3, 9M:6F) 15 Schizophrenia cases 15 Bipolar disorder cases 15 Controls	NA	Activation of immune response is present in all disorders; however, it appears to be different across disorders; for MDD, there was upregulation of C1q and *IL1B*

TABLE 1 Studies on the Dysregulation of Immune and Oxidative Stress Factors in *Brain Tissue*—cont'd

Citation	Brain Area/Cell Type	Platform/ Genes Studied	Diagnostic Criteria	Sample Size	Medications	Main Findings
Klempan et al. (2009)	PFC: BA44, 45, 46, 47	Affimetrix HG-U133	DSM-IV	16 Depressed suicide and 8 nondepressed suicide cases 13 Controls	One subject in depressed suicide group known to take antidepressants	"Cytokinesis" and "immune cell activation" are the central GO terms for distinguishing between depressed suicides and controls
Kohen et al. (2014)	Hippocampus DG granule cells by LCM	RNA-seq	DSM-IV	Cases (age range 25–91): 17 MDD 17 Schizophrenia 16 Bipolar disorder 29 Controls	NA	Disrupted signaling by miR-182 (loss of DG miR-182 signaling) in SZ and MDD miR182 is involved in a broad range of biological processes, including immune response
Malki et al. (2015)	PFC: BA10	Affimetrix HG-U95A	DSM-IV	11 MDD cases (age <65, M:F) 15 Controls	Antidepressant use included as covariate in analysis	80% of dysregulated genes were functionally associated with of a key stress response signaling cascades, including *NF-kB*, *AP-1*, and *ERK/MAPK*
Shelton et al. (2011)	PFC: BA10	Affimetrix Human Exon 1.0 ST	DSM-IV 11 out of 14— melancholic subtype	14 MDD cases (age mean 47.2±14.0, 11M:3F) 14 Controls	Psychotropic drug-free	Increased inflammatory and apoptotic stress, including elevated cytokines
Sibille et al. (2004)	PFC: BA9, BA47	Affimetrix U133A	DSM-III-R SCID-1	19 Depressed suicide cases (mean age 44.6±21.2 75%M) 19 Controls	Psychotropic medication-free	*CREB1* was downregulated in suicide victims, however, did not reach the level of significance

Continued

TABLE 1   Studies on the Dysregulation of Immune and Oxidative Stress Factors in *Brain Tissue*—cont'd

Citation	Brain Area/Cell Type	Platform/ Genes Studied	Diagnostic Criteria	Sample Size	Medications	Main Findings
Szebeni et al. (2014)	Astrocytes and oligodendrocytes from temporal lobe (UF) or right BA10 by LCM	*SOD1, SOD2, GPX, CAT*	DSM-IV	12 MDD cases (mean age 51, M:F) 12 Controls	Information about toxicology is available, included as covariate in analysis	Oxidative defense enzymes *SOD1* and *SOD2*, catalase (*CAT*) and glutathione peroxidase (*GPX1*) lower in oligodendrocytes from MDD
Tonelli et al. (2008)	Orbitofrontal cortex: BA11	*TNF, IL1B, IL4, IL5, IL6, IL16*	Limited information about diagnosis	34 Suicide cases (9 cases of MDD) (mean age 52.3 ± 15, 20M:14F) 17 Controls	Limited information about toxicology	Elevated level of cytokines in suicide victims (*IL4* in females, *IL13* in males)
Tochigi et al. (2008)	PFC: BA10	Affimetrix HU95Av2	DSM-IV	11 MDD cases (nonsuicide) (mean age 46 ± 10, 6M:5F) 11 Bipolar disorder cases 13 Schizophrenia cases 15 Controls	Information about antipsychotic drugs is available	*CREB1* was upregulated in BA10 of nonsuicide MDD subjects

HEEBO, human exonic evidence-based oligonucleotide; LCM, laser capture microdissection; PFC, prefrontal cortex; DLPFC, dorsolateral prefrontal cortex; PMI, port-mortem interval; BA, Brodmann's area; UF, uncinate fasciculus; DG, dentate gyrus; DSM-IV, Diagnostic and Statistical Manual of Mental Disorders, version IV; SCID-1, Structural Clinical Interview for DSM.

provide solid evidence, as the majority of those individual players of immune signaling have not been replicated. To further explore the level of replication of the findings in gene expression studies, we reanalyzed data from 15 brain transcriptomic studies in depression and identified only seven genes of the immune response showing minimal replication (i.e., they were found dysregulated within the depressed brain but in different brain areas): *FGFR3, ENPP2, PTP4A2* (innate immune response), *CREB1, MOG, JUN* (toll-like receptor signaling pathways), and *LEPR*, which belongs to the gp130 family of cytokine receptors (cytokine-mediated signaling) (Ciobanu et al., 2016). These findings point to a possible involvement of immune factors going beyond the typically reported pro-inflammatory cytokines.

It has been suggested that inflammation is tightly linked with oxidative stress in depression, which may lead to cell death and further inflammation, creating a vicious circle, a mechanism that is not well understood (Bakunina et al., 2015). A recent integrative brain analysis of rat and human PFC transcriptomes revealed that 80% of dysregulated genes were functionally associated with a key stress response signaling cascade, involving *NF-kB*, *AP-1* (activator protein 1), and extracellular signal-regulated kinase (*ERK*)/*MAPK*, suggesting inflammation-mediated oxidative stress and further dysregulation of neuroplasticity and neurogenesis in the PFC in MDD (Malki et al., 2015). Furthermore, oxidative stress, measured by expression levels of four oxidative defense enzymes, (A + B) superoxide dismutases (*SOD1* and *SOD2*), (C) catalase (*CAT*), and (D) glutathione peroxidase (*GPX1*), which were significantly lower in depressed individuals, is found to contribute to telomere shortening in oligodendrocytes of the PFC (BA10) (Szebeni et al., 2014). These findings provide evidence for a possible etiologic link between inflammation, oxidative stress, telomere shortening, and white matter abnormalities previously observed in depression.

Although many individual players of the immune and the oxidative stress pathways have been found altered in brain tissue of patients diagnosed with depression, an inconsistency on the direction and brain areas of dysregulated genes precludes from firm conclusions on specific immune-related pathways in the brain in depression. The current state of knowledge suggests that multiple brain areas are possibly involved in depression-related immune dysregulation in a complex manner.

Overlap of Gene Expression Markers of Inflammation Across Psychiatric Disorders

Given that a cross disorder assessment may reveal joint biological features between depression, schizophrenia, and bipolar disorder, the comparative assessment of gene expression levels can shed some light on shared and distinct pathophysiological mechanisms of these disorders, which potentially can provide a molecular basis for developing diagnostic guidelines. Molecular comparison of different brain areas in schizophrenia, bipolar disorder, and MDD at a transcriptome level suggested that (1) the PFC in all three disorders have distinct gene expression signatures with shared upregulation of genes encoding proteins for transcription and translation (Iwamoto, Kakiuchi, Bundo, Ikeda, & Kato, 2004) and that (2) the activation of immune/inflammatory response in the hippocampus is present in all disorders. However, despite these similarities across disorders, several differences in the specifics of the dysregulated transcriptomes were found. For MDD, abnormal activation of the first component of the complement cascade C1q (hub genes in coexpression analysis *C1QA*, *C1QB*, and *C1QC*) and *IL1B* (Kim, Hwang, Webster, & Lee, 2016) was observed in the hippocampus, while a dysregulation of the immune-related response in the thalamus, including B-cell receptor signaling, was specific to schizophrenia, as it was not found in depression (Chu, Liu, & Kemether, 2009; Table 1).

Taken together, there is a plethora of findings pointing to inflammation and oxidative stress-related events in the depressed brains, including upregulation of pro-inflammatory cytokines and transcriptional and posttranscriptional regulators of the immune signaling in the PFC and hippocampus. Emerging evidence also suggests a possible causal link between inflammation, oxidative stress, and structural changes in the brain in depression. However, it is premature to propose a distinct inflammatory/oxidative stress transcriptomic signature of depression as of yet, as replication of these findings is minimal at present. Further exploration of transcriptomes across different brain areas at a single-cell-type level and peripheral blood has a great potential to discover how inflammation-induced molecular changes lead to structural abnormalities and impair neural circuits involved in emotional and cognitive processing in depression.

GENE EXPRESSION PATTERN OF IMMUNE DYSREGULATION IN THE PERIPHERY

Studying gene expression markers of depression in peripheral tissues is a promising approach to identify biomarkers that are potentially translatable into clinical practice for diagnostic and prognostic purposes. Dysregulated transcripts identified at candidate gene and genome-wide levels provide us with new insights into biological mechanisms of depression. However, similar to gene expression studies in postmortem brain, the findings in peripheral tissue lack consistency. In a recent systematic review, we reanalyzed the results obtained from 10 transcriptomic studies in depression and showed that only 2.8% of genes (21 out of 752) identified as statistically significantly differentially expressed in the periphery between depressed patients and healthy control subjects were replicated. Although a low level of replication at an individual gene level is discouraging, we made the observation that various factors of the immune response were consistently dysregulated. Among these 21 replicated genes, three were involved in the immune system: *IFIT3*, which is involved in the type 1 interferon signaling pathway, was found upregulated in MDD and downregulated in postpartum depression; *STAT3*, a transcription factor that mediates cellular responses to interleukins, and *SEMA3C*, which is also known to be involved in the immune response, were found upregulated in MDD (Ciobanu et al., 2016). This suggests that further investigation of how peripheral inflammation is linked to depression is a promising avenue.

It can be expected that peripheral gene expression markers cannot reliably predict expression patterns in the brain (and vice versa).

Despite a substantial overlap between brain and peripheral blood transcriptomes (Liew, Ma, Tang, Zheng, & Dempsey, 2006), the extent at which peripheral inflammation is a reflection of central nervous system inflammation in depressed individuals is not fully understood. One of the challenges in biomarker research for psychiatric disorders is to agree on the "best" peripheral tissue source. While separated cell types and stimulated blood provide better resolution compared with unstimulated blood, dysregulation of inflammatory and other immune-related genes seems to be detectable even in a "cocktail" of many unstimulated cell types, such as whole blood. For example, elements of a disrupted immune signaling were found in unchallenged monocytes (Carvalho et al., 2014), peripheral blood mononuclear cells (PBMCs) (Belzeaux et al., 2010; Savitz et al., 2013; Segman et al., 2010), dermal fibroblasts (Garbett et al., 2015), whole blood after lipopolysaccharide (LPS) stimulation (Spijker et al., 2010), and unstimulated whole blood (Jansen et al., 2016) of MDD patients. Although the aforementioned studies differ not only in cell type but also in the type of depression (postpartum and melancholic, induced by INF-α treatment), in the type of study design (targeted vs genome-wide), and among cohort characteristics (age groups and medications), there was some agreement in findings on gene expression markers of inflammation between the studies (Table 2). For instance, by studying a panel of 47 inflammatory-related genes, Carvalho et al. (2014) found that one of two identified clusters, consisting primarily of pro-inflammatory mediators (*IL1A*, *IL1B*, *IL6*, *PTX3*, *PDE4B*, *PTGS2*, and *TNF*), were upregulated in monocytes of 47 patients with melancholic depression. This was somewhat consistent with genome-wide findings of a dysregulated functional network centered on differentially expressed *TNF* in PBMCs from 21 current and recurrent moderately to severely affected MDD cases (Savitz et al., 2013). However, the latter study underscored that differentially

TABLE 2 Studies on the Dysregulation of Immune Factors in *Peripheral Tissue*

Citation	Tissue/ Cell Type	Platform/Genes Studied	Diagnostic Criteria	Sample Characteristics	Medications	Main Findings
Carvalho et al. (2014)	Monocytes	47 Inflammatory-related genes RT-qPCR	DSM-IV-TR HAMD 17 SCID-1	47 Melancholic MDD cases (age range 32–82, 20% males) 42 Controls	Medication-free for at least 1 week	34 Genes upregulated, 2 downregulated Cluster analysis: First cluster—upregulated pro-inflammatory mediators (*IL1A, IL1B, IL6, PTX3, TNF, PDE4B,* and *PTGS2*)
Garbett et al. (2015)	Dermal fibroblasts	Affimetrix HT HG-U133+	DSM-IV-TR	16 Current MDE cases Mean age 34.9, 4M:12F 16 Controls	Not relevant	13 "immune" genes were dysregulated: *CD74, HLA-DRA, HLA-DQB1, IL11, HLA-DPA1, S100B, HBEGF, HLA-DPB1, HLA-DQA1, MET, PCDH10, TNF19, GSTT1*
Felger et al. (2012)	PBMCs	Illumina HT-12	DSM-IV SCID MMSE MADRS	21 HCV patients 12M:9F	No antidepressants for at least 4 months prior 12 weeks of INF-α/ribavirin treatment	IFN-α treated patients with high depression scores showed upregulation in INF-α and *AP1* signaling and reduced prevalence of *CREB/ATF* motifs
Hepgul et al. (2016)	Whole blood	Affimetrix HG 1.1 ST	DSM-IV IDS	58 HCV patients 20 Patients developed IFN-α induced major depressive episode	24 weeks of INF-α treatment	More genes (506) were modulated in patients who developed depression with enrichment in inflammation- and oxidative stress-related pathways
Jansen et al. (2016)	Whole blood	Affimetrix U219	DSM-IV	882 Current MDD cases 331 Controls	Antidepressants used as covariates	Upregulation of IL-6 signaling and downregulation of NK-cell cytotoxicity pathways
Mostafavi et al. (2014)	Whole blood	Illumina HiSeq 2000	DSM-IV	463 Recurrent MDD cases 459 Controls	Antidepressants used as covariates	No significant single-gene association after multiple-testing correction; however, there was increased expression of interferon α/β pathway

Continued

TABLE 2   Studies on the Dysregulation of Immune Factors in *Peripheral Tissue*—cont'd

Citation	Tissue/ Cell Type	Platform/Genes Studied	Diagnostic Criteria	Sample Characteristics	Medications	Main Findings
Savitz et al. (2013)	PBMCs	Illumina Human HT-12 v4 fMRI scanning	DSM-IV-TR HDRS	21 MDD (current or recurrent moderate to severe) cases And eight adults with BP in a current MDE; age 35±10, 32% males 24 Controls	No psychotropic medications for at least 3 weeks (eight for fluoxetine)	12 Protein-coding *genes* (*ADM, APBB3, CD160, CFD, CITED2, CTSZ, IER5, NFKBIZ, NR4A2, NUCKS1, SERTAD1,* and *TFN*) were dysregulated. One functional network is centered around *TNF*; inflammatory genes correlate with gray matter volume of the hippocampus and caudate and thickness of subgenual ACC
Segman et al. (2010)	PBMCs	Affimetrix Human Gene Chip Exon ST 1.0	Edinburgh Postnatal Depression Scale	9 PD 10 Controls	Antidepressant naive	73 Differentially expressed genes, 71 downregulated, 12 of them involved in immune response (*HELLS, HIST2H2B, GBPI, IFIT3, IGJ, SERPING, IFIT1, IFIT2, LOC44203, CXCL10, TNFRSF1,* and *EREG*)
Spijker et al. (2010)	Whole blood ex vivo LPS- stimulated	Agilent	DSM-IV IDS-SR30	21 Single or recurrent MDD episode cases Mean age 42.6±11.5, M:F 21 Controls Validation: 13 MDD cases 14 Controls	No current antidepressants or benzodiazepines	Seven genes were proposed as diagnostic signature of depression, six of which are related to immune system and deal with cellular proliferation (*CAPRIN1, PROK2,* and *ZBTB16*) and differentiation (*CLEC4A, KRT23,* and *PLSCR1*)

LPS, lipopolysaccharide; *CRS*, chronic restraint stress; *HCV*, chronic hepatitis C virus; *HDRS*, Hamilton Depression Rating Scale; *MDE*, Major Depressive Episode, postpartum depression; *IDS-SR30*, The Inventory of Depressive Symptomology; *MINI*, Mini-International Neuropsychiatric Interview; *PHQ-9*, Patient Health Questionnaire-9; *MMSE*, mini-mental state examination; *MADRS*, Montgomery-Asberg Depression Rating Scale; *IDS*, Inventory of Depression Symptomatology.

expressed immune players were functionally linked with nondifferentially expressed *NF-kβ*, *TGFβ*, and *ERK*, indicating that differential expression analysis might be a suboptimal option for detecting complex gene-gene interactions in depression. In another study, one of the TNF receptors, *TNFRSF1*, together with interferon-induced proteins *IFIT1*, *IFIT2*, and *IFIT3*, and eight other genes involved in immune response (*HELLS*, *HIST2H2B*, *GBPI*, *IGJ*, *SERPING*, *LOC44203*, *CXCL10*, and *EREG*) were found differentially expressed in PBMCs of nine patients with postpartum depression (Segman et al., 2010). In contrast, among 40 studied candidate genes including *TNF*, *IL1B*, *IL2*, *IL4*, *IL6*, *IL8*, and *IL10*, only the anti-inflammatory cytokine *IL10* was statistically significantly elevated in PBMCs of 11 individuals suffering from a severe melancholic depressive episode (Belzeaux et al., 2010) To overcome the common noisiness of expression in basal blood, Spijker et al. (2010) challenged whole-blood cells of 21 individuals with a single MDD episode. They observed massive LPS-induced gene expression, among which were several cytokines, such as *TNF*, *NF-kB*, *IL1*, *IL6*, and *IL10*. Although none of these genes displayed a differential expression level between MDD patients and control subjects, six out seven genes in a proposed diagnostic signature of depression are related to the immune system and deal with cellular proliferation (*CAPRIN1*, *PROK2*, and *ZBTB16*) and differentiation (*CLEC4A*, *KRT23*, and *PLSCR1*). Environmental influences, like lifestyle and medication use, can confound gene expression findings. Dermal fibroblasts were proposed as an alternative experimental model to study depression-specific gene expression alterations less independent of environmental influences (Garbett et al., 2015). The authors argue that after several rounds of fibroblasts' division in the cell culture, many epigenetic changes disappear over time, leaving a more "pure" genetic model at hand. Findings in fibroblasts were consistent with previously described observations in other cell types, pointing to disrupted

molecular pathways related to cell-to-cell communication that are known to play a role in the adaptive and innate immune system. A set of 13 PR-qPCR validated immune-related genes were suggested to be associated with lifestyle-independent and medication-free status in depression (*CD74*, *HLA-DRA*, *HLA-DQB1*, *IL11*, *HLA-DPA1*, *S100B*, *HBEGF*, *HLA-DPB1*, *HLA-DQA1*, *MET*, *PCDH10*, *TNF19*, and *GSTT1*).

While we report on some agreement of gene expression markers of inflammation in the periphery across studies, the majority of immune-related genes have not been replicated. One of the major factors that may have led to the disagreement between the studies is low statistical power. Small sample sizes, ranging from only 9 to 47 depression cases (Table 2), are one of the main limiting factors affecting statistical power to detect and replicate dysregulated transcripts. To overcome this limitation, Jansen et al. (2016) performed the largest to date study on 882 patients with current MDD. Using both differential expression and coexpression clustering (weighted gene coexpression network analysis, WGCNA) (Zhang & Horvath, 2005) methods and accounting for 16 demographic and technical covariates, they identified that MDD is characterized by upregulated IL-6 signaling (*IL6R*, *STAT3*, *MAPK14*, and *RXRA*) and downregulated NK-cell activation (*GZMB*, *KLRK1*, *PRF1*, *SH2D1B*, *KLDR1*, *NFATC2*, *IL2RB*, *CALM1*, and *NCALD*). Longitudinal analysis at 2-year follow-up showed that the levels of expression for 15% of genes out of the 129 genes identified in cross-sectional analyses were reversed in those who remitted after a previous depression episode. This indicates that transient gene expression patterns are detectable in peripheral blood, and the results provide support for the potential success in the development of whole-blood gene expression-based biomarkers of depression. These results were also meta-analyzed with a recent RNA-seq study, in which alone no significant association with depression on 463 self-reported MDD cases accounting for 39 covariates were found. Binding the two largest datasets derived from microarrays and RNA-sequencing technologies together resulted in 12 differentially expressed genes at FDR (false discovery rate) <0.1 between MDD cases and healthy controls, 7 of which are known to be involved in activated immune signaling and oxidative stress (*CALM1*, *FCRL6*, *APOBEC3G*, *RAP2B*, *PIPOX*, *PRR5L*, and *KLRD1*), providing further support for the often reported peripheral inflammation in depression.

Although an intensive search for peripheral gene expression biomarkers of inflammation in depression at both the candidate gene and transcriptome levels identified some promising candidates, including *IL1B*, *TNF*, *IL6*, *INF*, *CREB1*, and *NF-kB*, the field is far away from claiming an immune signature of depression. Further understanding of transcriptomic alterations in depression with a particular focus on complex gene-gene interactions involved in the peripheral immune response will help us to delineate the molecular pathways involved in depression, which can be used not only as biomarkers but also possibly as therapeutic targets.

Peripheral Inflammation in Depression With Medical Comorbidity

Depressive symptoms often accompany somatic disorders, giving rise to a complex phenotype of depression with medical comorbidity. Several studies suggest an inflammatory link between somatic diseases and depressive symptoms informed by gene expression analyses (Table 3).

Depression and Hepatitis C

The observation that about one in four patients with chronic hepatitis C develops clinically significant depression during INF-α therapy (Udina et al., 2012) suggested that such patients can serve as a model to study molecular mechanisms relevant to depression, especially in the context of enhanced inflammation. By assessing transcriptomes of 21 chronic hepatitis

C virus patients after 12 weeks of INF-α/ribavirin treatment, Felger et al. (2012) found that patients with a higher depression score exhibited upregulation of genes involved in INF-α and AP1 signaling, while expression of CREB/ATE was downregulated. In recent work, Hepgul et al. (2016) showed that dysregulated gene expression before INF-α treatment in inflammation- (IL-1 signaling), oxidative stress- (NRF2-mediated oxidative stress response), and neuroplasticity-related pathways (axonal guidance signaling) differentiated patients who developed IFN-α-induced depression from those who did not. Moreover, inflammation (IL-1, IL-6, IL-8, and NF-kB signaling) and oxidative stress (NRF2-mediated oxidative stress response, p53 signaling, and production of nitric oxide and reactive oxygen species (ROS) in macrophages) were modulated in those who developed depression ($N=20$) compared with those who did not ($N=38$) at week 4 of the INF-α treatment. This work indicates that patients with hepatitis C and increased inflammation before the INF-α therapy are more likely to develop a depressive episode, and this inflammation can be modulated even after 4 weeks of INF-α treatment. It supports the current view that peripheral inflammation is involved in depression and its modulation can potentially be beneficial for depressed individuals (Baune, 2017).

Depression and Cancer

It has been long observed that the prevalence of depression among cancer patients increases with disease severity, suggesting a bidirectional relationship between cancer and depression (Hinz et al., 2010). However, the mechanisms of this relationship are not well understood (Spiegel & Giese-Davis, 2003). Growing amount of research suggests that altered immune response driven by pro-inflammatory cytokines is related to both conditions (Irwin, Olmstead, Ganz, & Haque, 2013; Seruga, Zhang, Bernstein, & Tannock, 2008). Studying gene

TABLE 3 Studies on Dysregulation of Immune and Oxidative Stress Markers in Depression Comorbid With Medical Diseases

Citation	Tissue/Cell Type	Platform/Genes Studied	Depression Diagnosis	Sample Characteristics	Comorbid illness	Main Findings
Cohen et al. (2012)	PBMCs	Illumina Human Ref-8 BeadArrays	CES-D	$N=47$ (mean age 59 ± 10, 77% males)	Metastatic RCC	RCC patients with elevated depressive symptoms showed upregulation of pro-inflammatory genes (*IL1B*, *TNF*, *IL6*, and *PTGS2*) and associated transcriptional regulators (*NF-kB* and *STAT*)
Felger et al. (2012)	PBMCs	Illumina Human HT-12 Expression BeadChips	MADRS	$N=21$ (mean age 47.8 ± 3.7, 50% males)	HCV	Patients with a higher depression score exhibited upregulation of genes involved in INF-α and AP1 signaling, while expression of CREB/ATE was downregulated
Hepgul et al. (2016)	Whole blood	PreAnalytiX, Hombrechtikon, CHE	MINI	$N=58$; 20 depressed cases, 38 nondepressed cases (mean age 44.7 ± 1.6, 78% males)	HCV	Dysregulated gene expression before INF-α treatment in inflammation (IL-1 signaling), oxidative stress (NRF2-mediated oxidative stress response), and neuroplasticity-related pathways differentiated patients who developed IFN-α induced depression from those who did not
Lutgendorf et al. (2008)	Microphages (CD68⁺) stimulated by norepinephrine and cortisol	*MMP9*	CES-D	56 Cancer cases (mean age 63.51 ± 11.01) 58% Within the range of clinical depression	Epithelial ovarian cancer	Increased level of *MMP9* was associated with increased depressive symptoms, chronic stress, and low social support

Continued

TABLE 3 Studies on Dysregulation of Immune and Oxidative Stress Markers in Depression Comorbid With Medical Diseases—cont'd

Citation	Tissue/Cell Type	Platform/Genes Studied	Depression Diagnosis	Sample Characteristics	Comorbid illness	Main Findings
Nikkheslat et al. (2015)	Dexamethasone inhibited of LPS-stimulated PBMCs	*IL6*	ICD-10	28 CHD with depression (mean age 69 ± 2, 54% males) 55 CHD without depression	CHD	Elevated level of *IL6* in CHD patients with depression
Wang et al. (2015)	CD4$^+$ T cells	Agilent SurePrint G3 Human GE	HDRS	6 Asthma + depression (M:F, mean age 38.14 ± 9.41) 6 Controls	Asthma	156 Transcripts were dysregulated in depressive asthma versus controls; 20 biological pathways were identified including "inflammation" and "immunity"
Zhou et al. (2007)	PBMCs	*hOGG1*	SDS	44 Depressed cases 48 Nondepressed cases 20 Controls	Acute leukemia	Antioxidant system (elevated level of *hOGG1*) is impaired in leukemic patients with depression

CHD, coronary heart disease; *ICD-10*, International Classification of Disease-10; *RCC*, renal cell carcinoma; *MADRS*, Montgomery-Asberg Depression Rating Scale; *HCV*, hepatitis C virus; *MINI*, Mini-International Neuropsychiatric Interview; *CES-D*, The Center for Epidemiological Studies-Depression; *HDRS*, Hamilton Depression Rating Scale; *SDS*, self-rating depression scale.

expression patterns in depressed cancer patients may shed some light on molecular underpinnings of inflammation in both disorders. Despite cumulative research on elevated levels of circulating pro-inflammatory cytokines in cancer patients with comorbid depression (Illman et al., 2005; Maccio & Madeddu, 2012; Musselman et al., 2001), the literature on gene expression markers of inflammation for these conditions is limited. We were able to find only three studies explicitly investigating the expression of immune- and oxidative stress-related genes in this comorbid condition. Thus, gene expression profiling of PBMCs from renal cell carcinoma patients with elevated depressive symptoms found upregulation of pro-inflammatory genes (*IL1B*, *TNF*, *IL6*, and *PTGS2*) and associated transcriptional regulators (*NF-kB* and *STAT* family transcription factors) that have previously been linked to cancer progression (Cohen et al., 2012). An upregulation of *hOGG1*, which indicates excessive accumulation of ROS and activation of the DNA base excision repair pathway, has been reported in the group of 92 MDD patients with acute leukemia (Zhou et al., 2007). Several lines of evidence suggest that matrix metalloproteinases (MMPs), which act on pro-inflammatory cytokines, chemokines, and other proteins (Nissinen & Kahari, 2014), also play a role in the regulation of various aspects of depression-related inflammation in both depression and cancer. It has been found that an increased level of *MMP9* in tumor-associated macrophages was associated with increased depressive symptoms, chronic stress, and low social support in ovarian cancer patients (Lutgendorf et al., 2008). In addition, elevated level of *MMP28* was associated with better response to nortriptyline in "pure" depression (Hodgson et al., 2016), providing further support for the role of MMPs in depression.

Depression and Asthma

Depression is often present in patients with asthma. It has been suggested that both conditions may be mediated via inflammatory pathways (Kankaanranta, Kauppi, Tuomisto, & Ilmarinen, 2016). While some meta-analyzed evidence suggests that the levels of cytokines IL-1, IL-4, IL-6, and TNF-α are elevated in asthmatic patients with depression (Jiang, Qin, & Yang, 2014), literature on gene expression markers is very limited. In fact, we found only one study investigating the transcriptome of asthmatic patients with depression compared with those without depression. In this study, it has been found that *PIC3R1*, which is involved in the acute immune response signaling via the transcription factor *STAT3* (Abell et al., 2005), was differentially expressed (upregulated) in CD4+ cells between asthmatic patients with and without depression (Wang et al., 2015). In addition, *PIC3R1* was also found upregulated in dermal fibroblasts of MDD patients (Garbett et al., 2015).

Depression and Cardiometabolic Disorders

An epidemiological and clinical relationship between cardiometabolic disorders and depression has long been suspected (Musselman, Betan, Larsen, & Phillips, 2003). However, the biological mechanisms underlying this relationship are not well understood. It has been suggested that cardiovascular disease (CVD) and depression are sharing a complex pattern of systemic immune activation and HPA axis hyperactivity (Baune et al., 2012). Many studies propose that elevated inflammatory markers, particularly C-reactive protein, IL-6, IL-1β, and TNF-α, may be a key biological molecule in the comorbidity of depression and CVD (Grippo & Johnson, 2002). However, the evidence at the gene expression level is limited. To the best of our knowledge, only one study reported hints for an overlapping biology between coronary heart disease (CHD) and depression stemming from gene expression analyses. In this study, the pro-inflammatory cytokine *IL6* was found to be differentially overexpressed in conjunction with insufficient glucocorticoid signaling in LPS-stimulated PBMCs of 28 CHD patients with comorbid depression compared with 55 CHD patients without depression (Nikkheslat et al., 2015). Although the comorbidity between type 2 diabetes and depression was observed in epidemiological studies and inflammation was suggested as a key biological mechanism (Stuart & Baune, 2012), we are not aware of any studies investigating shared biological pathways between type 2 diabetes and depression using gene expression analyses.

In order to enhance the understanding of the shared biology between depression and medical comorbid disorders as discussed earlier, further studies on gene expression patterns in medical disorders and depression are needed. A better understanding of common and distinct molecular features across these disease entities is fundamental for the development of targeted treatments for individuals suffering from both medical disease and depression.

INFLAMMATION AS A PREDICTOR OF A TREATMENT RESPONSE

Molecular mechanisms by which antidepressants exert their effects are not well understood. However, emerging evidence suggests that changes in gene expression may be involved in the pharmacological effects of antidepressants. Studies investigating gene expression levels in responders versus nonresponders to antidepressants provide valuable insights into the molecular basis of antidepressant therapy and may offer important information about prediction of different treatment regiments. To date, a limited number of studies have explored treatment-related changes of gene expression levels. The results are mixed and require replication; however, it is worth mentioning that various immune markers, including pro-inflammatory cytokines and their transcription factors were found to be involved in antidepressant response (Table 4). For instance, elevated mRNA levels of the pro-inflammatory cytokines TNF and IL1B genes in combination with PPT1 and HIST1H1E in PBMCs of 16 patients suffering from a severe major depression episode were able to predict response to 8 weeks antidepressant treatment on a genome-wide scale, suggesting that a pro-inflammatory response in the acute clinical phase may be associated with a better prognosis (Belzeaux et al., 2012). However, another study found an opposite effect. Using a candidate gene approach, Cattaneo et al. (2013) investigated the expression levels of genes belonging to glucocorticoid receptor function, neuroplasticity, and inflammation (*IL1A*, *IL1B*, *IL4*, *IL6*, *IL7*, *IL8*, *IL10*, *MIF*, and *TNF*) in 74 moderately severe

depressed patients aged 17–72 before and after 8 weeks of treatment with escitalopram or nortriptyline. They identified that elevated mRNA levels of *MIF*, *IL1B*, and *TNF* at baseline were strongly and negatively correlated with the treatment outcome, predicting a poor response to antidepressants. Although in this study, antidepressants reduced levels of *IL1B* and *MIF*, these changes were not associated with treatment response. In contrast, a reduction in *IL6* gene expression was associated with successful antidepressant response. This study outlined that changes in inflammation associated with antidepressant treatment response are not reflected by all cytokines at the same time.

We observed a similar pattern of disagreement across studies for the transcription factor *CREB1*. One study, investigating the behavior of 40 PBMC genes in response to 8 weeks of benzodiazepines, found that *CREB1* was upregulated and associated with treatment efficacy and clinical improvement (Belzeaux et al., 2010). However, the other two candidate gene studies, despite reporting a significant reduction of *CREB1* levels after 8 weeks of paroxetine (Iga et al., 2007) and unspecified (Lai, Hong, & Tsai, 2003) antidepressant treatment, failed to find an association between changes in *CREB1* levels and therapeutic response.

In addition to pro-inflammatory cytokines and transcription factor *CREB1*, several other immune-related genes were found to be involved in response to antidepressant treatment. By measuring genome-wide gene expression levels in whole blood of 63 depressed patients, it has been found that *IRF7*, a transcription factor regulating INF-α, was the most significantly differentially expressed gene, and its expression was upregulated by 8 weeks of citalopram treatment in depressed individuals who responded to treatment. These authors also found that the expression of *IRF7* was

TABLE 4   Studies on Immune Markers Predicting Antidepressant Treatment Response

Citation	Tissue/Cell Type	Platform/ Genes Studied	Depression Diagnostic Criteria	Sample Characteristics	Treatment	Main Findings
Belzeaux et al. (2010)	PBMCs	40 Candidate genes RT-qPCR	DSM-IV HDRS 17 SCOD-P	11 Severe melancholic MDE (current or recurrent) cases, age range 49–60, 4M:7F 11 Controls	8 Weeks of benzodiazepines	Baseline: *IL-10* was upregulated After treatment: Upregulation of *CREB1* was associated with treatment efficacy and clinical improvement
Belzeaux et al. (2012)	PBMCs	Agilent SurePrint G3	DSM-IV-TR HRSD-17	13 Severe MDE cases (age range 35–70) 13 Controls matched for age and sex	8 Weeks of antidepressant treatment (not specified)	Upregulated *TNF* and *IL1B* in combination with *PPT1* and *HIST1H1E* in PBMCs could predict treatment response
Cattaneo et al. (2013)	Leukocytes	15 Candidate genes, including *IL1A, IL1B, IL4, IL6, IL7, IL8, IL10, TNF, MIF*	DSM-IV MADRS HRSD-17	74 Moderately severe MDD cases Mean age 38.3 ± 10.9 31M:43F	8 Weeks of escitalopram or nortriptyline	Elevated baseline expression levels of *IL1B, TNF*, and *MIF* were strongly and negatively correlated with the treatment outcome, predicting lack of response
Eyre et al. (2016)	Leukocytes	Illumina HumanHT-12 v4	DSM-IV-TR HDRS-24 MMSE MADRS	35 LLD cases: 24 Remitters (mean age 67.2 ± 5.7) 11 Nonremitters (mean age 73.5 ± 9.3)	16 Weeks of methylphenidate, citalopram and placebo in combinations	Increased level *HLA-DRB5* can predict the antidepressant response in the elderly
Hodgson et al. (2016)	Whole blood	Illumina HumanHT-12 v4	ICD-10 DSM-IV MADRS	136 Moderate to severe depression cases (mean age 41.76 ± 12.35, 67.6% F)	8 Weeks of nortriptyline ($N=44$) or escitalopram ($N=92$)	MMP28 was associated with better response to nortriptyline Network analysis suggests that changes in gene expression are not drug-specific

Continued

TABLE 4 Studies on Immune Markers Predicting Antidepressant Treatment Response—cont'd

Citation	Tissue/Cell Type	Platform/ Genes Studied	Depression Diagnostic Criteria	Sample Characteristics	Treatment	Main Findings
Iga et al. (2007)	Leukocytes	CREB1 HDAC5	DSM-IV SIGH-D 17	25 MDD cases (mean age 41.1 ± 13.1) 25 Controls	8 Weeks of paroxetine	CREB1 was upregulated prior to treatment, with a significant reduction following 8 weeks of paroxetine treatment
Lai et al. (2003)	Lymphocytes	CREB1	DSM-IV	21 MDD cases 21 Controls	8 Weeks of antidepressant treatment (not specified)	After 8 weeks of antidepressant treatment, CREB was downregulated in MDD patients. No associated with therapeutic response
Mamdani et al. (2011)	Whole blood	Affimetrix, HG-U133 Plus2	DSM-IV SCID HAMD-21	63 MDD cases	8 Weeks of citalopram	IRF7 was upregulated in response to citalopram in individuals who responded to citalopram treatment

LLD, late-life depression; *MDE*, major depressive episode; *CBT*, cognitive behavioral therapy; *SCID-CV*, Structural Clinical Interview for DSM-IV, Clinical Version; *HAMD*, Hamilton Rating Scale for Depression; *MINI*, Mini-International Neuropsychiatric Interview; *PHQ-9*, Patient Health Questionnaire-9; *MADRS*, Montgomery-Asberg Depression Rating Scale; *HRSD-17*, 17-item Hamilton Rating Scale for Depression; *MMSE*, mini-mental state examination; *SIGH-D 17*, Structured Interview Guide for the 17-item Hamilton Depression Rating Scale.

downregulated in the prefrontal cortices of subjects who died during a current depressive episode, which was concordant with the blood results (Mamdani et al., 2011). However, these findings were not supported by another whole-blood transcriptome study on 136 moderate to severe depression cases. Instead, it was found that elevated levels of *MMP28*, which influences cell function and plays a central role in various biological processes including inflammation, were associated with better response to nortriptyline in depression (Hodgson et al., 2016). In addition, an increased level of *HLA-DRB5* (major histocompatibility complex, class II, DR beta 5) at baseline was identified to predict the antidepressant treatment response in late-life depression; however, no other studies reported this gene being involved in response to antidepressants yet (Eyre et al., 2016).

Given the clinical significance and molecular complexity of antidepressant response, future investigations of the behavior of immune-related genes and the exploration of gene expression patterns at a genome-wide level in depressed patients before and after antidepressant treatment are warranted.

ANTI-INFLAMMATORY THERAPY FROM A GENE EXPRESSION PERSPECTIVE

Based on the assumption that inflammation is linked to depressive symptoms, potential therapeutic effects of anti-inflammatory interventions in depressed patients were explored. While the clinical effects of broad-spectrum nonsteroidal anti-inflammatory drugs (Kohler et al., 2014) or more targeted anticytokine treatments (Kappelmann, Lewis, Dantzer, Jones, & Khandaker, 2016) have been investigated in several clinical trials, a consensus on the effectiveness and safety of these various anti-inflammatory approaches has not been reached yet. Studying gene expression patterns before and after drug administration can provide valuable insights into biological effects of anti-inflammatory therapy in depression. However, to date, there is limited literature on the effect of anti-inflammatory drugs on depression stemming from gene expression analyses. The only study of this kind was performed in a LPS-induced animal model of depression that showed intriguing results. Using a candidate gene approach, Teeling, Cunningham, Newman, and Perry (2010) found differential effects of anti-inflammatory agents on LPS-induced behavioral changes and on mRNA levels of pro-inflammatory cytokines in mice. Thus, indomethacin and ibuprofen reversed the effect of LPS on behavior without changing *IL6*, *IL1B*, and *TNF* mRNA levels neither in the periphery nor in the brain. In contrast, dexamethasone did not alter LPS-induced behavioral changes; however, it completely inhibited the production of cytokines. A selective COX-1 inhibitor, piroxicam, but not the selective COX-2 inhibitor, nimesulide, reversed the LPS-induced behavioral changes without affecting *IL6*, *IL1B*, and *TNF* levels of expression in the hippocampus. These findings may have implications for developing novel therapeutic strategies for depression. However, to date, no human studies investigating the effect of anti-inflammatory therapy on depression by measuring levels of gene expression in peripheral blood of depressed patients were performed.

SUMMARY AND CONCLUSIONS

In this chapter, we have covered important aspects of gene expression markers of inflammation in depression, including markers of inflammation in both the brain and periphery, and inflammatory markers of an antidepressant treatment response. We also reported on the inflammatory overlap between somatic disorders and depression (Fig. 1).

In summary, the current state of knowledge on gene expression of inflammation markers in depression is limited and needs expansion:

(1) A dysregulation of immune signaling at the gene expression level is apparent in the brain and is detectable at the periphery; however, the interplay between the brain and peripheral tissue needs further investigation.

(2) An upregulation of pro-inflammatory cytokines *IL1B*, *IL6*, *TNF*, and *INF* and the transcription factors *NF-kB* and *CREB1* has been observed in the PFC and hippocampus, as well as in peripheral blood cells of depressed patients, providing support for the cytokine hypothesis of depression. However, replication of these findings is limited. Furthermore, exploration of gene expression pattern at the genome-wide level suggests that immune dysregulation in depression goes beyond typically studied cytokines.

(3) Although peripheral inflammation is observed in depression with medical comorbidity and in a "pure" depression, the evidence is insufficient to propose common expression patterns of immune-related genes across somatic disorders and depression.

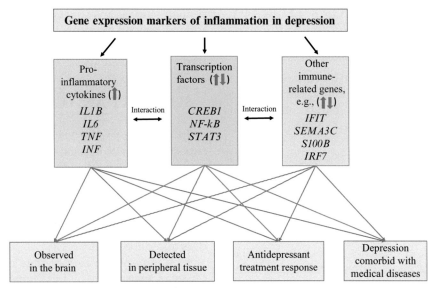

FIG. 1   Gene expression markers of inflammation in depression.

(4) A dysregulation of an immune/ inflammatory response in the brain was found in three mental disorders— schizophrenia, bipolar disorder, and depression; however, it appears that the molecular pattern of inflammation and immune dysregulation varies across these disorders.

(5) Blood-based gene expression levels of immune genes measured during treatment trials can potentially be used for tailoring treatment options for depressed patients. Although immune-related genes, like *IRF7*, *IL1B*, *TNF*, and *CREB1*, have been observed in response to antidepressant treatment, the results are mixed. This suggests that more work with a particular focus on the type of treatment, phase of disease, and participant's age group is needed.

(6) The potential efficacy of anti-inflammatory agents for treating depression is an area of an intense debate that requires solid evidence. To evaluate the effectiveness of anti-inflammatory therapies at a gene expression level, clinical trials measuring expression of genes before and after administration of anti-inflammatory agents in humans need to be conducted.

As a concluding remark, we suggest that an investment in exploring gene expression patterns in depression can be well paid off by producing long-awaited molecular insights into the biology of depression. Rapidly improving technologies and sophisticated bioinformatics methods for data analysis have the potential to detect altered pathways with a better precision for genes expressed even at low levels but relevant in depression, which appears to be a dynamic disorder of multiple players of small effects. In this chapter, we reported that inflammation, featuring upregulation of *IL1B*, *IL6*, *TNF*, *INF*, *NF-kB*, and *CREB1*-signaling in both the brain and periphery, is an important contributor to the pathophysiological mechanisms of depression. However, given that the majority of these findings were derived from candidate gene studies, it is natural to imply that depression-related inflammation may not be limited to these markers. Emerging high-throughput studies, providing genome-wide coverage, enable the discovery of dysregulated immune genes in depression beyond the commonly studied cytokines and their transcription factors. Furthermore, using network analyses enables the study of complex interactions between immune genes and the interactions of these genes with other biological candidates in depression at the transcriptome level. It is also worth noting that many gene expression analyses have not investigated inflammation (Hennings et al., 2015; Sequeira et al., 2009). While inflammation is currently a dominant speculation in the biology of depression, we must not ignore the findings that suggest other biological pathways. An improved understanding of multiple facets of gene expression alterations in depression can contribute to the likely complex molecular signature of the disease.

It is well established that depression is a multifactorial and dynamic disorder; therefore, there is a growing need for integrative analysis at multiple levels ranging from clinical phenotype to genotype, gene expression, epigenetic markers to structural changes in the brain, and circulating proteins associated with depression at different stages of disease progression across various age groups. The integration of multiple layers of data—although highly challenging—may finally have better predictive value than the analysis of single markers alone. This type of analysis is not widely occurring in depression research yet (as opposed to cancer research (Liu, Devescovi, Chen, & Nardini, 2013)), but the methodology of an integration of omics data is pursued (Suravajhala, Kogelman, & Kadarmideen, 2016). It is still speculation; however, in the future, such an integrated omics approach may equip clinicians with instruments necessary to provide better care for the patients suffering from depression.

References

Abell, K., Bilancio, A., Clarkson, R. W. E., Tiffen, P. G., Altaparmakov, A. I., Burdon, T. G., et al. (2005). Stat3-induced apoptosis requires a molecular switch in PI(3)K subunit composition. *Nature Cell Biology*, 7(4), 392–398. https://doi.org/10.1038/ncb1242.

Bakunina, N., Pariante, C. M., & Zunszain, P. A. (2015). Immune mechanisms linked to depression via oxidative stress and neuroprogression. *Immunology*, 144(3), 365–373.

Baune, B. T. (2017). Are non-steroidal anti-inflammatory drugs clinically suitable for the treatment of symptoms in depression-associated inflammation? *Current Topics in Behavioral Neurosciences*, 31, 303–319.

Baune, B. T., Stuart, M., Gilmour, A., Wersching, H., Heindel, W., Arolt, V., et al. (2012). The relationship between subtypes of depression and cardiovascular disease: a systematic review of biological models. *Translational Psychiatry*, 2(3).

Belzeaux, R., Bergon, A., Jeanjean, V., Loriod, B., Formisano-Tréziny, C., Verrier, L., et al. (2012). Responder and nonresponder patients exhibit different peripheral transcriptional signatures during major depressive episode. *Translational Psychiatry*, 2, e185.

Belzeaux, R., Formisano-Tréziny, C., Loundou, A., Boyer, L., Gabert, J., Samuelian, J. C., et al. (2010). Clinical variations modulate patterns of gene expression and define blood biomarkers in major depression. *Journal of Psychiatric Research*, 44(16), 1205–1213.

Carvalho, L. A., Bergink, V., Sumaski, L., Wijkhuijs, J., Hoogendijk, W. J., Birkenhager, T. K., et al. (2014). Inflammatory activation is associated with a reduced glucocorticoid receptor alpha/beta expression ratio in monocytes of inpatients with melancholic major depressive disorder. *Translational Psychiatry*, 4.

Cattaneo, A., Gennarelli, M., Uher, R., Breen, G., Farmer, A., Aitchison, K. J., et al. (2013). Candidate genes expression profile associated with antidepressants response in the GENDEP study: differentiating between baseline 'predictors' and longitudinal 'targets'. *Neuropsychopharmacology*, 38(3), 377–385.

Charlesworth, J. C., Curran, J. E., Johnson, M. P., Goring, H. H., Dyer, T. D., Diego, V. P., et al. (2010). Transcriptomic epidemiology of smoking: the effect of smoking on gene expression in lymphocytes. *BMC Medical Genomics*, 3, 29.

Chu, T. T., Liu, Y., & Kemether, E. (2009). Thalamic transcriptome screening in three psychiatric states. *Journal of Human Genetics*, 54(11), 665–675.

Ciobanu, L. G., Sachdev, P. S., Trollor, J. N., Reppermund, S., Thalamuthu, A., Mather, K. A., et al. (2016). Differential gene expression in brain and peripheral tissues in depression across the life span: a review of replicated findings. *Neuroscience and Biobehavioral Reviews*, 71, 281–293.

Cohen, L., Cole, S. W., Sood, A. K., Prinsloo, S., Kirschbaum, C., Arevalo, J. M., et al. (2012). Depressive symptoms and cortisol rhythmicity predict survival in patients with renal cell carcinoma: role of inflammatory signaling. *PLoS One, 7*(8).

CONVERGE. (2015). Sparse whole-genome sequencing identifies two loci for major depressive disorder. *Nature, 523*(7562), 588–591.

Crusz, S. M., & Balkwill, F. R. (2015). Inflammation and cancer: advances and new agents. *Nature Reviews. Clinical Oncology, 12*(10), 584–596.

Duric, V., Banasr, M., Stockmeier, C. A., Simen, A. A., Newton, S. S., Overholser, J. C., et al. (2013). Altered expression of synapse and glutamate related genes in post-mortem hippocampus of depressed subjects. *International Journal of Neuropsychopharmacology, 16*(1), 69–82.

Eyre, H. A., Eskin, A., Nelson, S. F., St Cyr, N. M., Siddarth, P., Baune, B. T., et al. (2016). Genomic predictors of remission to antidepressant treatment in geriatric depression using genome-wide expression analyses: a pilot study. *International Journal of Geriatric Psychiatry, 31*(5), 510–517.

Eyre, H. A., Stuart, M. J., & Baune, B. T. (2014). A phase-specific neuroimmune model of clinical depression. *Progress in Neuro-Psychopharmacology and Biological Psychiatry, 54*, 265–274.

Felger, J. C., Cole, S. W., Pace, T. W., Hu, F., Woolwine, B. J., Doho, G. H., et al. (2012). Molecular signatures of peripheral blood mononuclear cells during chronic interferon-alpha treatment: relationship with depression and fatigue. *Psychological Medicine, 42*(8), 1591–1603.

Ferrari, A. J., Charlson, F. J., Norman, R. E., Patten, S. B., Freedman, G., Murray, C. J. L., et al. (2013). Burden of depressive disorders by country, sex, age, and year: findings from the global burden of disease study 2010. *PLoS Medicine, 10*(11).

Firestein, G. S., & McInnes, I. B. (2017). Immunopathogenesis of rheumatoid arthritis. *Immunity, 46*(2), 183–196.

Flint, J., & Kendler, K. S. (2014). The genetics of major depression. *Neuron, 81*(5), 1214.

Fredrickson, B. L., Grewen, K. M., Coffey, K. A., Algoe, S. B., Firestine, A. M., Arevalo, J. M., et al. (2013). A functional genomic perspective on human well-being. *Proceedings of the National Academy of Sciences of the United States of America, 110*(33), 13684–13689.

Gaiteri, C., Guilloux, J. P., Lewis, D. A., & Sibille, E. (2010). Altered gene synchrony suggests a combined hormone-mediated dysregulated state in major depression. *PLoS One, 5*(4).

Garbett, K. A., Vereczkei, A., Kalman, S., Brown, J. A., Taylor, W. D., Faludi, G., et al. (2015). Coordinated messenger RNA/microRNA changes in fibroblasts of patients with major depression. *Biological Psychiatry, 77*(3), 256–265.

Grippo, A. J., & Johnson, A. K. (2002). Biological mechanisms in the relationship between depression and heart disease. *Neuroscience and Biobehavioral Reviews, 26*(8), 941–962.

Hennings, J. M., Uhr, M., Klengel, T., Weber, P., Pütz, B., Touma, C., et al. (2015). RNA expression profiling in depressed patients suggests retinoid-related orphan receptor alpha as a biomarker for antidepressant response. *Translational Psychiatry, 5*(3).

Hepgul, N., Cattaneo, A., Agarwal, K., Baraldi, S., Borsini, A., Bufalino, C., et al. (2016). Transcriptomics in interferon-α-treated patients identifies inflammation-, neuroplasticity- and oxidative stress-related signatures as predictors and correlates of depression. *Neuropsychopharmacology, 41*(10), 2502–2511.

Hinz, A., Krauss, O., Hauss, J. P., Höckel, M., Kortmann, R. D., Stolzenburg, J. U., et al. (2010). Anxiety and depression in cancer patients compared with the general population. *European Journal of Cancer Care, 19*(4), 522–529.

Hodgson, K., Tansey, K. E., Powell, T. R., Coppola, G., Uher, R., Zvezdana Dernovšek, M., et al. (2016). Transcriptomics and the mechanisms of antidepressant efficacy. *European Neuropsychopharmacology, 26*(1), 105–112.

Hyde, C. L., Nagle, M. W., Tian, C., Chen, X., Paciga, S. A., Wendland, J. R., et al. (2016). Identification of 15 genetic loci associated with risk of major depression in individuals of European descent. *Nature Genetics, 48*(9), 1031–1036.

Iga, J., Ueno, S., Yamauchi, K., Numata, S., Kinouchi, S., Tayoshi-Shibuya, S., et al. (2007). Altered HDAC5 and CREB mRNA expressions in the peripheral leukocytes of major depression. *Progress in Neuro-Psychopharmacology & Biological Psychiatry, 31*(3), 628–632.

Illman, J., Corringham, R., Robinson, D., Jr., Davis, H. M., Rossi, J. F., Cella, D., et al. (2005). Are inflammatory cytokines the common link between cancer-associated cachexia and depression? *Journal of Supportive Oncology, 3*(1), 37–50.

Irwin, M. R., Olmstead, R. E., Ganz, P. A., & Haque, R. (2013). Sleep disturbance, inflammation and depression risk in cancer survivors. *Brain, Behavior, and Immunity, 30*(Suppl), S58–S67.

Iwamoto, K., Kakiuchi, C., Bundo, M., Ikeda, K., & Kato, T. (2004). Molecular characterization of bipolar disorder by comparing gene expression profiles of postmortem brains of major mental disorders. *Molecular Psychiatry, 9*(4), 406–416.

Jansen, R., Batista, S., Brooks, A. I., Tischfield, J. A., Willemsen, G., van Grootheest, G., et al. (2014). Sex differences in the human peripheral blood transcriptome. *BMC Genomics, 15*(1), 33.

Jansen, R., Penninx, B. W. J. H., Madar, V., Xia, K., Milaneschi, Y., Hottenga, J. J., et al. (2016). Gene expression in major depressive disorder. *Molecular Psychiatry, 21*(3), 339–347.

Jiang, M., Qin, P., & Yang, X. (2014). Comorbidity between depression and asthma via immune-inflammatory pathways: a meta-analysis. *Journal of Affective Disorders, 166*, 22–29.

Kang, H. J., Adams, D. H., Simen, A., Simen, B. B., Rajkowska, G., Stockmeier, C. A., et al. (2007). Gene expression profiling in postmortem prefrontal cortex of major depressive disorder. *Journal of Neuroscience*, 27(48), 13329–13340.

Kang, H.-J., Kim, S.-Y., Bae, K.-Y., Kim, S.-W., Shin, I.-S., Yoon, J.-S., et al. (2015). Comorbidity of depression with physical disorders: research and clinical implications. *Chonnam Medical Journal*, 51(1), 8–18.

Kankaanranta, H., Kauppi, P., Tuomisto, L. E., & Ilmarinen, P. (2016). Emerging comorbidities in adult asthma: risks, clinical associations, and mechanisms. *Mediators of Inflammation*, 2016, 3690628.

Kappelmann, N., Lewis, G., Dantzer, R., Jones, P. B., & Khandaker, G. M. (2016). Antidepressant activity of anti-cytokine treatment: a systematic review and meta-analysis of clinical trials of chronic inflammatory conditions. *Molecular Psychiatry*, 23, 335–343.

Kim, S., Hwang, Y., Webster, M. J., & Lee, D. (2016). Differential activation of immune/inflammatory response-related co-expression modules in the hippocampus across the major psychiatric disorders. *Molecular Psychiatry*, 21(3), 376–385.

Klempan, T. A., Sequeira, A., Canetti, L., Lalovic, A., Ernst, C., ffrench-Mullen, J., et al. (2009). Altered expression of genes involved in ATP biosynthesis and GABAergic neurotransmission in the ventral prefrontal cortex of suicides with and without major depression. *Molecular Psychiatry*, 14(2), 175–189.

Kohen, R., Dobra, A., Tracy, J. H., & Haugen, E. (2014). Transcriptome profiling of human hippocampus dentate gyrus granule cells in mental illness. *Translational Psychiatry*, 4(3).

Kohler, O., Benros, M. E., Nordentoft, M., Farkouh, M. E., Iyengar, R. L., Mors, O., et al. (2014). Effect of anti-inflammatory treatment on depression, depressive symptoms, and adverse effects: a systematic review and meta-analysis of randomized clinical trials. *JAMA Psychiatry*, 71(12), 1381–1391.

Lai, I. C., Hong, C. J., & Tsai, S. J. (2003). Expression of cAMP response element-binding protein in major depression before and after antidepressant treatment. *Neuropsychobiology*, 48(4), 182–185.

Levinson, D. F., Mostafavi, S., Milaneschi, Y., Rivera, M., Ripke, S., Wray, N. R., et al. (2014). Genetic studies of major depressive disorder: why are there no GWAS findings, and what can we do about it? *Biological Psychiatry*, 76(7), 510–512.

Liew, C. C., Ma, J., Tang, H. C., Zheng, R., & Dempsey, A. A. (2006). The peripheral blood transcriptome dynamically reflects system wide biology: a potential diagnostic tool. *Journal of Laboratory and Clinical Medicine*, 147(3), 126–132.

Liu, Y., Devescovi, V., Chen, S., & Nardini, C. (2013). Multi-level omic data integration in cancer cell lines: advanced annotation and emergent properties. *BMC Systems Biology*, 7, 14.

Lutgendorf, S. K., Lamkin, D. M., Jennings, N. B., Arevalo, J. M., Penedo, F., DeGeest, K., et al. (2008). Biobehavioral influences on matrix metalloproteinase expression in ovarian carcinoma. *Clinical Cancer Research*, 14(21), 6839–6846.

Maccio, A., & Madeddu, C. (2012). Inflammation and ovarian cancer. *Cytokine*, 58(2), 133–147.

Malki, K., Pain, O., Tosto, M. G., Du Rietz, E., Carboni, L., & Schalkwyk, L. C. (2015). Identification of genes and gene pathways associated with major depressive disorder by integrative brain analysis of rat and human prefrontal cortex transcriptomes. *Translational Psychiatry*, 5.

Mamdani, F., Berlim, M. T., Beaulieu, M. M., Labbe, A., Merette, C., & Turecki, G. (2011). Gene expression biomarkers of response to citalopram treatment in major depressive disorder. *Translational Psychiatry*, 1(6).

Mehta, D., Menke, A., & Binder, E. B. (2010). Gene expression studies in major depression. *Current Psychiatry Reports*, 12(2), 135–144.

Montecucco, F., Liberale, L., Bonaventura, A., Vecchie, A., Dallegri, F., & Carbone, F. (2017). The role of inflammation in cardiovascular outcome. *Current Atherosclerosis Reports*, 19(3), 11.

Mostafavi, S., Battle, A., Zhu, X., Potash, J. B., Weissman, M. M., Shi, J., et al. (2014). Type I interferon signaling genes in recurrent major depression: increased expression detected by whole-blood RNA sequencing. *Molecular Psychiatry*, 19(12), 1267–1274.

Musselman, D. L., Betan, E., Larsen, H., & Phillips, L. S. (2003). Relationship of depression to diabetes types 1 and 2: epidemiology, biology, and treatment. *Biological Psychiatry*, 54(3), 317–329.

Musselman, D. L., Miller, A. H., Porter, M. R., Manatunga, A., Gao, F., Penna, S., et al. (2001). Higher than normal plasma interleukin-6 concentrations in cancer patients with depression: preliminary findings. *American Journal of Psychiatry*, 158(8), 1252–1257.

Nikkheslat, N., Zunszain, P. A., Horowitz, M. A., Barbosa, I. G., Parker, J. A., Myint, A. M., et al. (2015). Insufficient glucocorticoid signaling and elevated inflammation in coronary heart disease patients with comorbid depression. *Brain, Behavior, and Immunity*, 48, 8–18.

Nissinen, L., & Kahari, V. M. (2014). Matrix metalloproteinases in inflammation. *Biochimica et Biophysica Acta*, 1840(8), 2571–2580.

Pandey, G. N., Rizavi, H. S., Ren, X., Fareed, J., Hoppensteadt, D. A., Roberts, R. C., et al. (2012). Proinflammatory cytokines in the prefrontal cortex of teenage suicide victims. *Journal of Psychiatric Research*, 46(1), 57–63.

Penninx, B. W., Milaneschi, Y., Lamers, F., & Vogelzangs, N. (2013). Understanding the somatic consequences of depression: biological mechanisms and the role of depression symptom profile. *BMC Medicine*, 11(1), 129.

Power, R. A., Tansey, K. E., Buttenschøn, H. N., Cohen-Woods, S., Bigdeli, T., Hall, L. S., et al. (2017). Genome-wide association for major depression through age at onset stratification: major depressive disorder working group of the psychiatric genomics consortium. *Biological Psychiatry*, *81*(4), 325–335.

Rivera, M., & McGuffin, P. (2015). The successful search for genetic loci associated with depression. *Genome Medicine*, *7*(1), 92.

Robinson, D., Humbert, M., Buhl, R., Cruz, A. A., Inoue, H., Korom, S., et al. (2017). Revisiting type 2-high and type 2-low airway inflammation in asthma: current knowledge and therapeutic implications. *47*(2), 161–175.

Savitz, J., Frank, M. B., Victor, T., Bebak, M., Marino, J. H., Bellgowan, P. S., et al. (2013). Inflammation and neurological disease-related genes are differentially expressed in depressed patients with mood disorders and correlate with morphometric and functional imaging abnormalities. *Brain, Behavior, and Immunity*, *31*, 161–171.

Segman, R. H., Goltser-Dubner, T., Weiner, I., Canetti, L., Galili-Weisstub, E., Milwidsky, A., et al. (2010). Blood mononuclear cell gene expression signature of postpartum depression. *Molecular Psychiatry*, *15*(1). 93–100, 102.

Sequeira, A., Mamdani, F., Ernst, C., Vawter, M. P., Bunney, W. E., Lebel, V., et al. (2009). Global brain gene expression analysis links glutamatergic and GABAergic alterations to suicide and major depression. *PLoS One*, *4*(8).

Seruga, B., Zhang, H., Bernstein, L. J., & Tannock, I. F. (2008). Cytokines and their relationship to the symptoms and outcome of cancer. *Nature Reviews. Cancer*, *8*(11), 887–899.

Shelton, R. C., Claiborne, J., Sidoryk-Wegrzynowicz, M., Reddy, R., Aschner, M., Lewis, D. A., et al. (2011). Altered expression of genes involved in inflammation and apoptosis in frontal cortex in major depression. *Molecular Psychiatry*, *16*(7), 751–762.

Sibille, E., Arango, V., Galfalvy, H. C., Pavlidis, P., Erraji-Benchekroun, L., Ellis, S. P., et al. (2004). Gene expression profiling of depression and suicide in human prefrontal cortex. *Neuropsychopharmacology*, *29*(2), 351–361.

Spiegel, D., & Giese-Davis, J. (2003). Depression and cancer: mechanisms and disease progression. *Biological Psychiatry*, *54*(3), 269–282.

Spijker, S., Van Zanten, J. S., De Jong, S., Penninx, B. W., van Dyck, R., Zitman, F. G., et al. (2010). Stimulated gene expression profiles as a blood marker of major depressive disorder. *Biological Psychiatry*, *68*(2), 179–186.

Stuart, M. J., & Baune, B. T. (2012). Depression and type 2 diabetes: inflammatory mechanisms of a psychoneuroendocrine co-morbidity. *Neuroscience and Biobehavioral Reviews*, *36*(1), 658–676.

Sullivan, P. F., Neale, M. C., & Kendler, K. S. (2000). Genetic epidemiology of major depression: review and meta-analysis. *American Journal of Psychiatry*, *157*(10), 1552–1562.

Suravajhala, P., Kogelman, L. J. A., & Kadarmideen, H. N. (2016). Multi-omic data integration and analysis using systems genomics approaches: methods and applications in animal production, health and welfare. *Genetics Selection Evolution*, *48*(1), 38.

Szebeni, A., Szebeni, K., DiPeri, T., Chandley, M. J., Crawford, J. D., Stockmeier, C. A., et al. (2014). Shortened telomere length in white matter oligodendrocytes in major depression: potential role of oxidative stress. *International Journal of Neuropsychopharmacology*, *17*(10), 1579–1589.

Teeling, J. L., Cunningham, C., Newman, T. A., & Perry, V. H. (2010). The effect of non-steroidal anti-inflammatory agents on behavioural changes and cytokine production following systemic inflammation: implications for a role of COX-1. *Brain, Behavior, and Immunity*, *24*(3), 409–419.

Tochigi, M., Iwamoto, K., Bundo, M., Sasaki, T., Kato, N., & Kato, T. (2008). Gene expression profiling of major depression and suicide in the prefrontal cortex of postmortem brains. *Neuroscience Research*, *60*(2), 184–191.

Tonelli, L. H., Stiller, J., Rujescu, D., Giegling, I., Schneider, B., Maurer, K., et al. (2008). Elevated cytokine expression in the orbitofrontal cortex of victims of suicide. *Acta Psychiatrica Scandinavica*, *117*(3), 198–206.

Udina, M., Castellvi, P., Moreno-Espana, J., Navines, R., Valdes, M., Forns, X., et al. (2012). Interferon-induced depression in chronic hepatitis C: a systematic review and meta-analysis. *Journal of Clinical Psychiatry*, *73*(8), 1128–1138.

van den Akker, E. B., Passtoors, W. M., Jansen, R., van Zwet, E. W., Goeman, J. J., Hulsman, M., et al. (2014). Meta-analysis on blood transcriptomic studies identifies consistently coexpressed protein-protein interaction modules as robust markers of human aging. *Aging Cell*, *13*(2), 216–225.

Wang, T., Ji, Y. L., Yang, Y. Y., Xiong, X. Y., Wang, I. M., Sandford, A. J., et al. (2015). Transcriptomic profiling of peripheral blood CD4(+) T-cells in asthmatics with and without depression. *Gene*, *565*(2), 282–287.

Wise, T., Cleare, A. J., Herane, A., Young, A. H., & Arnone, D. (2014). Diagnostic and therapeutic utility of neuroimaging in depression: an overview. *Neuropsychiatric Disease and Treatment*, *10*, 1509–1522.

Wright, F. A., & Sullivan, P. F. (2014). Heritability and genomics of gene expression in peripheral blood. *Nature Genetics*, *46*(5), 430–437.

Zhang, B., & Horvath, S. (2005). A general framework for weighted gene co-expression network analysis. *Statistical Applications in Genetics and Molecular Biology*, *4*.

Zhou, F., Zhang, W., Wei, Y., Zhou, D., Su, Z., Meng, X., et al. (2007). The changes of oxidative stress and human 8-hydroxyguanine glycosylase1 gene expression in depressive patients with acute leukemia. *Leukemia Research*, *31*(3), 387–393.

12

Neuroimmunopharmacology at the Interface of Inflammation and Pharmacology Relevant to Depression

Joshua Holmes, Frances Corrigan, Mark R. Hutchinson

University of Adelaide, Adelaide, SA, Australia

INTRODUCTION

The involvement of the immune system in the pathogenesis of depression is supported by data identifying increased inflammatory cytokines, cytokine receptors, chemokines, and acute-phase proteins in patients suffering depression (Köhler et al., 2017). These increases in proinflammatory mediators have been observed peripherally and centrally in depressed individuals (Levine et al., 1999; Maes et al., 1997). Additional evidence has been drawn from the sickness response, where inflammation caused by infection or tissue damage leads to behavior like anhedonia, social withdrawal, irritability, and fatigue, all of which are frequently observed in depression (Dantzer, O'Connor, Freund, Johnson, & Kelley, 2008). This sickness response induced by inflammation is hypothesized to have evolved as a protective mechanism to help prevent further spreading of pathogens to our kin and divert biological resources to fighting disease (Hart, 1988). More recent theories as to the evolution of inflammatory-induced depressive behavior suggest that our ancestors' constant interaction with pathogens in the environment directed an evolutionary advantage of a more inflammatory phenotype; hence, alleles for genes associated with inflammation and depression are common in the modern gene pool (Raison & Miller, 2013). Post-mortem analysis of depressed suicide victims' brains has found that these subjects have significant increases in protein and gene expression of the innate immune receptors toll-like receptor 3 (TLR3) and toll-like receptor 4 (TLR4) supporting a role of neuroinflammatory involvement in depression (Pandey et al., 2012). The observation that a significant proportion of patients receiving interferon alpha (IFN-α) therapy, for hepatitis C, suffers depression as a direct side effect provides further evidence that cytokines can explicitly influence emotion processing and behavior (Bonaccorso et al., 2002; Renault et al., 1987). The increased inflammatory proteins in depressed patients and the depression-invoking ability of cytokine therapy suggest that cytokines may precipitate the behavior and mood associated with depression. Indeed, this has been extensively supported by animal studies where peripherally administered immune-stimulating molecules induce a state of sterile inflammation and cause both sickness-like behaviors and depressive-like behaviors (Dantzer et al., 2008; O'Connor et al., 2009).

A central tenet of inflammation-associated depression is the ability of peripheral inflammatory signals to be translated to the brain where they can affect behavior. Initially, it was thought that the brain was an immune privileged organ, owing to the protective blood-brain barrier (BBB). However, decades of research have now identified resident immune cells of the central nervous system (microglia), and in several ways, large immune molecules like cytokines can influence the brain and behavior. The three main routes of immune-to-brain communication are known as the humoral, neural, and cellular routes (Maier, 2003; Miller & Raison, 2016). The humoral route consists of cytokines crossing the blood-brain barrier through more permeable regions like the circumventricular organs and choroid plexus or through saturable transporters on the BBB. Once cytokines activate microglial cells in these regions, it has been proposed that the subsequent release of prostaglandin E2 (PE2) from these cells is what rapidly transmits the inflammatory signal throughout the brain and has implications for glial and neuronal signaling (Herkenham, Lee, & Baker, 1998; Maier, 2003). The neural route revolves around the vagal nerve afferents being directly activated by peripheral cytokines that can induce changes in behavior. A cellular route of immune-to-brain communication has been identified where stress or peripheral immune activation can cause blood monocytes to cross the BBB; this is mediated by microglial release of monocyte chemoattractant protein 1 (MCP-1) and its cognate receptor on monocyte chemokine receptor type 2.

Glial cells of the central nervous system are responsible for mediating neuroinflammatory signaling in the brain. Glia consist of three types of cells: microglia, which are the main immunocompetent cells; astrocytes, which respond to inflammatory signals and can affect neuronal signaling; and oligodendrocytes, which provide the myelin insulation around cells and assist in speed of neuronal spread of action potentials. The two main glia pertinent to inflammatory-mediated behavior are microglia and astrocytes and their interaction at the synaptic level to form the tetrapartite synapse and modulate behavior (De Leo, Tawfik, & LaCroix-Fralish, 2006). In normal conditions, microglia survey the cellular milieu of the brain responding to immune molecules like cytokines and detect danger-associated molecular patterns (DAMPs), like high-mobility group box 1 (HMGB1) and heat-shock proteins that are released by stressed and damaged cells, or pathogen-associated molecular patterns (PAMPs), such as lipopolysaccharide (LPS), a component of the cell wall of bacteria. When microglia are stimulated by these molecules, they transform into an activated state, often highly proinflammatory, and further release cytokines. In addition to this immune-surveillance role, microglia are actively involved in the neuroplasticity involved in normal learning and neuronal development and are constantly interacting with neurons detecting and responding to immune molecules in the absence of overt inflammation (Wu, Dissing-Olesen, MacVicar, & Stevens, 2015). Initially, it was thought that astrocytes' main function was to provide nutrient support to neuronal cells and balance the extracellular ion levels. It is now appreciated that astrocyte functions go far beyond this, and it has been identified that astrocytes are integral for healthy neuronal function and plasticity. Astrocytes have been shown to release gliotransmitters, adenosine triphosphate and glutamate, that can act directly on neurons to elicit a response (Takano et al., 2005; Zhang et al., 2013). The release of gliotransmitters from astrocytes appears to be mediated by astrocytic metabotropic glutamate receptors (mGluRs), which when activated by glutamate induced increase in astrocytic calcium currents that causes several downstream neuromodulatory effects (Bazargani & Attwell, 2016). These effects include further release of glutamate by astrocytes, changes in local extracellular ion balance leading to hyperpolarization of neurons, and increased expression of astrocytic gamma-aminobutyric acid (GABA) and glutamate transporters all of which can effect neuronal signaling (Shigetomi, Tong, Kwan, Corey, & Khakh, 2011).

When activated by an inflammatory stimulus, via downstream activation of the nuclear factor kappa B (NFκB) transcription factor or the NOD-like receptor protein 3 (NLRP3) inflammasome, microglia produce and release proinflammatory cytokines including interleukin-1β (IL-1β), interleukin-6 (IL-6), and tumor necrosis factor alpha (TNF-α). These proinflammatory cytokines have been shown to have several modulatory actions at the synaptic level that can influence behavior through regulating the availability of neurotransmitters. TNF-α can cause the release of the excitatory neurotransmitter glutamate from astrocytes, which had lasting effects on neuroplasticity in the hippocampus (Habbas et al., 2015). Currently, the most compelling mechanistic evidence for cytokine effects on behavior is their ability to reduce the availability of the monoamine neurotransmitters. Predominately, the proinflammatory cytokines appear to induce and inhibit enzymes and coenzymes involved in the production of the monoamines. Through their ability to induce reactive oxygen species (ROS), cytokines can inactivate tetrahydrobiopterin (BH4), an essential coenzyme in the production of all monoamines; this significantly reduces monoamine production (Swardfager, Rosenblat, Benlamri, & McIntyre, 2016). A study on patients receiving IFN-α for hepatitis C, a treatment that may induce depression as a side effect, found that the IFN-α caused a decrease in BH4 activity and this was associated with reduced cerebral spinal fluid dopamine levels (Felger et al., 2013). Proinflammatory cytokines have also been shown to induce the expression and activation of the enzyme indoleamine

2,3-dioxygenase (IDO) in microglia, which converts tryptophan, the precursor of serotonin, to kynurenine (Maes, Leonard, Myint, Kubera, & Verkerk, 2011). Kynurenine can then further be metabolized to quinolinic acid that is a potent N-methyl-D-aspartate receptor (NMDAR) agonist. This cytokine induced increase in quinolinic acid, and astrocytic release of glutamate is suggested to cause glutamate excitotoxicity that can further increase the generation of ROS and neuroinflammation (Maes et al., 1997).

In addition to inhibiting the production of monoamines, inflammatory cytokines have been shown to reduce the release of these neurotransmitters. Importantly, it has been reported that inflammatory cytokines significantly reduce the release of dopamine in reward-related centers of the brain, like the ventral striatum, which is associated with reduced motivation and anhedonia (Swardfager et al., 2016). In healthy individuals administered a low dose of LPS, which induces a rapid peripheral cytokine response, there was an increase in depressive symptoms that were associated with reduced ventral striatum activity in response to a typically rewarding activity (Eisenberger et al., 2010). Capuron et al. (2012) further identified that reduced ventral striatal activity was indeed associated with changes in dopamine release in response to rewarding stimuli. This was achieved by using neuroimaging techniques that utilized radiolabeled fluorodopa F18. Fluorodopa F18 is an amino acid precursor of dopamine, and by measuring its activity in the ventral striatum by positron emission tomography, it can be used as a surrogate indicator of dopamine activity. Capuron and colleagues identified that patients receiving IFN-α for hepatitis C had significantly reduced the release of dopamine in ventral striatum that correlated with anhedonia and depression (Capuron et al., 2012).

The full etiology of this inflammation-associated depression is still unknown and confounded by comorbid diseases and unhealthy lifestyles making it difficult to elucidate.

Nevertheless, despite the absence of the precise nuances of the molecular mechanisms involved in inflammatory-associated depression, the investigations continue into the study of novel treatments for depression that may constrain these pathways and deliver therapeutic efficacy.

CLASSICAL ANTIDEPRESSANTS

Current antidepressants have been shown to effect both the innate and adaptive immune responses, thus further implicating inflammatory signaling in depression (Eyre, Lavretsky, Kartika, Qassim, & Baune, 2016). Several studies have shown that reductions in peripheral inflammatory cytokine concentrations are associated with improvements in various depression scales. For example, serum IL-6, a proinflammatory cytokine found to be increased in depression, was found to be significantly decreased in responders to the selective serotonin reuptake inhibitors (SSRIs) paroxetine or sertraline following treatment (Yoshimura et al., 2013). Indeed, administration of all the current classes of antidepressants including SSRIs, serotonin-norepinephrine reuptake inhibitors (SNRIs), and tricyclic antidepressants has been associated with reduced plasma inflammatory cytokines in depressed patients (Gałecki, Mossakowska-Wójcik, & Talarowska, 2017).

Whether antidepressant effects on reducing inflammatory molecules are direct inhibition of immune signaling or a flow on effect from their neuronal mechanisms is still elusive, and there is evidence for both. Evidence from rodent studies show that many classical antidepressants have direct effects on production and release of proinflammatory cytokines from immune cells. Desipramine, when administered 30 min prior to an intraperitoneal LPS injection, prevented neuroinflammation and inhibited the increased IDO expression. Furthermore, when peripheral blood mononuclear cells (PBMCs) were stimulated with interferon gamma (IFN-γ) and treated with desipramine, the induction

of IDO expression was significantly blunted (Brooks et al., 2017).

Despite the reported effects of classical antidepressants reducing inflammation-associated depression, there are still a significant number of patients who do not respond to traditional antidepressants. In contrast, studies have also identified nonresponders to classical antidepressant treatment who often have higher inflammatory profile, suggesting that the antiinflammatory effects of classical antidepressants are not their primary effect in depression (Maes et al., 1997). Estimates suggest that classical antidepressants improve depressive symptoms in approximately 50% of patients suffering from unipolar depression, and the effect size of the improved depression symptoms is often minimal in responders (Ghaemi, 2008). Combine this relatively low efficacy in many patients with the 6–8 weeks before a treatment effect is observed, the common often severe side effects that contribute to morbidity, and the fact that many patients will have to sample multiple different classical antidepressants until they find the optimal balance between treatment effect and minimal side effects (Howland, 2008). These obvious shortcomings of classical antidepressants highlight the need to fully investigate the physiological mechanisms associated with depression and develop targeted treatments with improved efficacy and reduced side effects (Rush et al., 2006). The search is well underway for novel antidepressants that target neuroinflammation for the treatment of depression.

STRESS AND CORTISOL IN INFLAMMATION-ASSOCIATED DEPRESSION

Although inflammatory rodent models have been extremely useful at unraveling some specific mechanisms involved in cytokine-induced depressive-like behavior, they are limited in that the innate inflammatory response induced by these peripheral or even central injections may not adequately model the low-grade inflammatory signaling observed in depression and the neurological changes that may precede them. Therefore, using these models to screen novel antidepressants needs to be treated with caution. However, the inflammatory relationship with depression is supported in the chronic unpredictable mild stress-induced depression model; in this model, animals are exposed to random repeated stressors often over weeks; these animals show increased peripheral immune signaling and neuroimmune activation (Grippo, Francis, Beltz, & Felder, 2005; Liu, Peng, et al., 2015). This model of stress-induced depressive-like behavior has good face and construct validity, as it not only produces the increased inflammation as seen in humans but also models the chronic stress that often precipitates depression (Willner, 2017). Chronic stress, which causes chronic activation of the hypothalamic pituitary adrenal axis (HPA), is well established as a causative factor for depression (Hammen, 2005; Kendler, Karkowski, & Prescott, 1999). This chronic activation of the HPA axis, which leads to chronically elevated cortisol levels, is hypothesized to cause the glucocorticoid resistance observed in depression (Tsigos & Chrousos, 2002). This glucocorticoid insensitivity leads to a dysregulated HPA axis, as it no longer responds to cortisol as a negative feedback regulator. Interestingly, elevated levels of cortisol are simultaneously observed with increased inflammatory cytokines in a subset of patients with depression (Carvalho et al., 2008). This elevated cortisol and simultaneous increase in inflammatory cytokines observed in depression appear counterintuitive, as cortisol in high concentrations is a potent antiinflammatory. As well as its regulatory actions maintaining blood glucose, cortisol, through glucocorticoid receptor activation, inhibits NFκB and activator protein 1 (AP-1) transcription factors, hence preventing cytokine gene expression and release (Scheinman, Gualberto, Jewell, Cidlowski, & Baldwin, 1995).

Proinflammatory cytokines have been shown to be involved in both the hyperactivation of the HPA axis and the cellular mechanisms involved in glucocorticoid resistance further implicating inflammation in the pathology of depression (Miller, Maletic, & Raison, 2009). IL-1α has been shown to cause deficits in GR ability to inhibit NFκB at the cellular level and hence prevents the inhibitory effect of cortisol on cytokines (Wang, Wu, & Miller, 2004). This bidirectional relationship between the inflammatory cytokines and the neuroendocrine system and their relationship with depression is incredibly complex. Future theories need to consolidate these system biological effects to fully elucidate the molecular foundations of depression and stress-related illness. Given cortisol's profound antiinflammatory actions, particularly in reducing proinflammatory cytokine concentrations, and its current use therapeutically for inflammatory-related diseases like rheumatoid arthritis and asthma, the idea of using cortisol or hydrocortisone to treat inflammatory-related depression, initially, would not seem absurd. However, the therapeutic use of steroid antiinflammatories for inflammation-associated depression is undermined by the hypercortisolism, impaired glucocorticoid sensitivity, and HPA axis dysregulation observed in depression.

NONSTEROIDAL ANTIINFLAMMATORY DRUGS

Given the limited efficacy of using steroid antiinflammatories for the treatment of inflammation-associated depression, the next obvious candidates are the nonsteroidal antiinflammatory drugs (NSAIDs). NSAIDs conventionally refer to the cyclooxygenase-inhibiting drugs like ibuprofen, aspirin, celecoxib, and diclofenac. In humans, there are two cyclooxygenases, COX-1 and COX-2. COX enzymes are primarily responsible for the biosynthesis of prostaglandins and thromboxanes from membrane arachidonic acid. It is the downstream effects, particularly of PE2, that are responsible for their proinflammatory actions. COX-1 is constitutively expressed in most tissue, and its activation is predominantly required for maintaining tissue homeostasis. The induction of COX-2 is initiated by proinflammatory cytokines, inflammatory inducers like LPS, growth factors, and hormones. It is thought that COX-2 has more proinflammatory effects, whereas COX-1 has more homeostatic properties.

Several studies have investigated the use of specific COX inhibitors for the treatment of depression; though some are randomized double-blind placebo-controlled studies, their effect on depression score can be confounded by improvement in preexisting inflammatory conditions. For example, the COX inhibitors ibuprofen, celecoxib, and naproxen were found to reduce depressive score in osteoarthritis sufferers; however, treatment was also successful at reducing pain and patient function that can directly impact patient depressive symptoms potentially confounding these results (Iyengar et al., 2013). Similarly, positive results have been published that investigate the adjunctive administration of COX inhibitors combined with classical antidepressants. In a small double-blind placebo-controlled trial investigating the adjunctive treatment of celecoxib with fluoxetine for major depression, it was found that patients receiving both drugs had significantly improved scores on the Hamilton Rating Scale for Depression compared with placebo and fluoxetine (Akhondzadeh et al., 2009). A meta-analysis by Köhler et al. (2014), which included the abovementioned studies, among others, found that antiinflammatory treatments improve depressive symptoms over placebo. However, the authors do note that the heterogeneity of the pooled studies made the mean estimate of improvement uncertain. To further complicate the evidence for the efficacy of NSAIDs for the treatment of depression, studies have shown that common antiinflammatory

drugs can attenuate the antidepressant effect of selective serotonin reuptake inhibitors and contribute to detrimental side effects (Mort, Aparasu, & Baer, 2006; Warner-Schmidt, Vanover, Chen, Marshall, & Greengard, 2011).

Currently, the data pertaining to the efficacy of NSAIDs for the treatment of inflammation-associated depression are not robust enough to implement their widespread use. This is mainly due to the lack of double-blind placebo-controlled studies for the use of NSAIDs in unmedicated and otherwise medically healthy individuals with depression. Future studies of this nature will provide the evidence required to determine the therapeutic efficacy of the NSAIDs in inflammation-associated depression (Fig. 1).

ANTICYTOKINE ANTIBODY THERAPY

Monoclonal antibodies are large protein molecules that can be manipulated in the laboratory to recognize and bind specific epitopes on molecules of interest. By binding molecules like cytokines or cell surface proteins, the antibody inactivates the molecule, which can then be taken up by immune cells and destroyed. The advantage of antibody therapy is that they are highly specific for the molecule of interest. Anticytokine therapy exploits this principle, and monoclonal antibodies are produced that bind specific cytokines and prevent the cytokine from interacting with their receptors and hence inhibit their inflammatory effects.

Interestingly, antibodies and recombinant proteins that have been developed against inflammatory cytokines have limited bioavailability in the central nervous system (CNS), as they are large protein molecules that do not cross the blood-brain barrier in normal conditions. Although unable to cross the blood-brain barrier, animal models have shown that therapeutic proteins can modulate neuroinflammatory

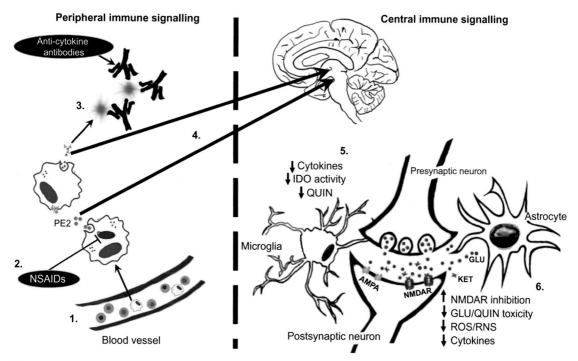

Peripheral immune signalling

Central immune signalling

Anti-cytokine antibodies

3.

4.

2.

NSAIDs

PE2

1.

Blood vessel

5.
↓ Cytokines
↓ IDO activity
↓ QUIN

Presynaptic neuron

Astrocyte

Microglia

GLU

AMPA

NMDAR

KET

6.
↑ NMDAR inhibition
↓ GLU/QUIN toxicity
↓ ROS/RNS
↓ Cytokines

Postsynaptic neuron

FIG. 1 1. Peripheral immune cells including monocytes release inflammatory cytokines into the periphery. 2. Nonsteroidal antiinflammatory drugs (NSAIDs) inhibit the production of prostaglandins and thromboxanes, preventing their further proinflammatory actions on immune cells.3. Proinflammatory cytokines released by immune cells in the periphery are inhibited when anticytokine antibodies, like infliximab, bind to the cytokine preventing their ability to signal through their cognate receptor. 4. NSAIDs, anticytokine antibodies, minocycline, and cannabidiol can reduce peripheral inflammation preventing the transmission of the proinflammatory signal from the periphery to the brain. 5. Antiinflammatory drugs capable of penetrating the blood-brain barrier, like minocycline and cannabidiol, can have direct antiinflammatory effects on microglial cells in the central nervous system. This culminates in reduced inflammatory cytokine release, which prevents the excessive activation of the enzyme indoleamine 2,3-dioxygenase and prevents increases in the neurotoxic metabolite quinolinic acid (QUIN). 6. Ketamine's (KET) exact mechanism of action responsible for its antidepressant effect is still unknown. However, evidence suggests that N-methyl-D-aspartate receptor (NMDAR) inhibition may influence glutamate (GLU)- and QUIN-induced excitotoxicity, thereby reducing reactive oxygen and nitrogen species and preventing cytokine release and neuroinflammation, having profound immediate antidepressant effects.

processes. In a rat model of hepatic encephalopathy, infliximab was successful at reducing neuroinflammation and glial morphological changes characteristic of the model (Dadsetan et al., 2016). In another study, etanercept, which acts as a TNF-α inhibitor and has similarly poor penetration into the CNS, significantly reduced the cytokine secretion and glial reactivity in a viral induced encephalitis in mice (Ye et al., 2014). Importantly, in a rodent model of depressive-like

behavior, the anti-TNF-α etanercept administered peripherally prevented the stress-induced depressive-like behavior (Krügel, Fischer, Radicke, & Sack, 2013).

A recent meta-analysis analyzing seven randomized control trials of anticytokine therapy for chronic inflammatory diseases, with depressive symptoms measured as a secondary outcome, found a significant effect of anticytokine therapy on depressive symptoms (Kappelmann,

Lewis, Dantzer, Jones, & Khandaker, 2016). Of these studies, the primary conditions being treated included psoriasis (four studies), Crohn's disease, and atopic dermatitis, and only one study investigated infliximab for treatment-resistant depression as a primary outcome (Raison et al., 2013).

A key limitation of this meta-analysis is that the antidepressant effect cannot be delineated from improvement in the underlying inflammatory condition. Currently, there are few studies that specifically investigate anticytokine treatment for major depression in otherwise medically healthy patients and no studies that first separate the subset of depressed patient with abnormal serum inflammatory profile. In one randomized, double-blind, placebo-controlled study, Raison et al. (2013) found that the monoclonal antibody anti-TNF-α, infliximab, did not significantly reduce depressive symptoms in treatment-resistant depression overall. However, their data did show a correlation with serum inflammatory markers at the beginning of treatment and treatment response to infliximab. This highlights the limitations of clinical studies of this type, as responders where those who had initially higher levels of inflammatory markers at the beginning of treatment suggesting that anticytokine therapies may be efficacious in treating this subset of depressed patients. Clinical trials are now more commonly designed to include these pre and post hoc analyses of inflammatory markers to further identify inflammatory markers of depression and response potential to antidepressants (Eyre et al., 2016). Another significant limitation of anticytokine therapy for inflammation-associated depression is that it targets only one molecule. Cytokines are at the tail end of the inflammatory cascade with many prior intermediates and cell types responsible for their release, which if targeted may prevent the production and effect of several cytokines and inflammatory molecules therefore having a wider antiinflammatory effect.

MINOCYCLINE AS AN ANTIINFLAMMATORY AGENT

Although targeting specific effector molecules like prostaglandins and cytokines is effective in reducing symptoms in other inflammatory-associated diseases, perhaps their specificity in inhibiting these downstream inflammatory molecules is a limitation when treating inflammation-associated depression. Minocycline may represent a pharmacotherapy that has antiinflammatory properties on a broad range of immune cell types and has shown promise in animal models of depressive-like behavior. Minocycline is a broad-spectrum tetracycline antibiotic; it has a long half-life, excellent CNS penetration, and safety profile (Soczynska et al., 2012). Minocycline is also a no-specific antiinflammatory drug; this combined with its lipophilicity and safety profile makes it a unique candidate for the treatment of inflammatory-associated neurological disorders like depression.

Although many studies investigate minocycline as a "specific" microglial inhibitor, its antiinflammatory effects go far beyond this. Minocycline's antiinflammatory effects include inhibition of T-cell activation, inhibition of matrix metalloproteinases, inhibition of cytokine secretion from peripheral immune cells, inhibition of ROS production, and inhibition of LPS-induced inflammation in monocytes (Cazalis, Tanabe, Gagnon, Sorsa, & Grenier, 2009; Kloppenburg et al., 1995; Kraus et al., 2005; Miyachi, Yoshioka, Imamura, & Niwa, 1986; Paemen et al., 1996; Pang, Wang, Benicky, & Saavedra, 2012). Given that the antiinflammatory properties of minocycline have been extensively studied, it is surprising that its exact mechanism of action has not been fully elucidated. Despite this, minocycline is unquestionably antiinflammatory.

In addition to and because of its antiinflammatory effects, minocycline also has effects on modulating glutamatergic systems in the brain. Minocycline has been shown to reduce neuronal glutamatergic transmission in neuronal cell cultures, which was shown to be induced by blockade of voltage-gated sodium channels and voltage-gated calcium channels, thereby reducing glutamate release (Gonzalez et al., 2007). Further, chronic unpredictable mild stress is associated with a significant decrease in glutamate receptor 1 (GluR1) phosphorylation resulting in a decrease in activity; this decrease is normalized by chronic treatment with minocycline (Banke et al., 2000; Liu, Li, et al., 2015). The effect of minocycline in the above study is suggested to be due to its antiinflammatory effects on microglial cells not its direct effects on glutamate transmission. In a rat neuropathic pain model, minocycline treatment increased glial glutamate transporters and normalized the changes in synaptic NMDAR function (Nie, Zhang, & Weng, 2010). Another study investigating the effects of chronic unpredictable mild stress in rats found that stress increased measures of oxidative stress in specific brain regions and that intracerebral ventricular injection of minocycline was able to reverse the oxidative stress in the prefrontal cortex and hippocampus but not the nucleus accumbens (Reus et al., 2015). These data taken together suggest that minocycline's neuroprotective effect may be due to its antiinflammatory mechanisms and resultant alterations in glutamate signaling.

Regardless of its exact mechanism, minocycline has been shown to reduce the depressive-like behavior in many rodent models of inflammation-associated depressive-like behavior. Although not specifically investigating the antiinflammatory properties of minocycline, Arakawa et al. (2012) found that injecting minocycline into the cerebral ventricle of rats in a learned helplessness model reduced subsequent depressive-like behavior. Recent studies have identified that the learned helplessness model of depressive-like behavior is associated with significant hippocampal microglial polarization to a more proinflammatory phenotype in rats (Iwata, Ishida, Kaneko, & Shirayama, 2016). Interestingly, Arakawa et al. (2012) found that these animals also had increased dopamine levels in the amygdala and no change in the learned helplessness induced increased serotonin turnover and decreased levels of BDNF. This study is important as it identified that minocycline has a brain-specific mechanism and its modulation of behavior goes beyond its peripheral antiinflammatory actions (Arakawa et al., 2012). Rats chronically exposed to restraint stress had increased microglial reactivity in the mPFC, which was associated with significantly impaired spatial working memory; both were rescued by chronic treatment with minocycline (Hinwood, Morandini, Day, & Walker, 2012).

Currently, there are no published double-blind placebo-controlled studies investigating minocycline specifically in depressed patients without comorbidities; however, some promising human results do exist. In a 6-week open-label study, 150 mg/day of minocycline combined with classical antidepressants had a significant effect on the 21-item Hamilton Psychiatric Rating Scale for Depression (HAMD-21) scores in patients suffering from unipolar psychotic depression. Minocycline combined with classical antidepressants achieved a statistically significant reduction in the mean HAMD-21 score at 2 weeks following treatment and was further reduced at 6 weeks of treatment (Miyaoka et al., 2012). Although this was not a double-blind placebo-controlled clinical trial, it does provide support for the safety of minocycline combined with classical antidepressants and evidence for their synergistic effect.

In the only published double-blind placebo-controlled trial of minocycline for the treatment of mild to moderate depression in human immunodeficiency virus (HIV), patients receiving minocycline 100 mg twice daily for 6 weeks reported a significant reduction in the mean Hamilton Depression Rating Scale compared with placebo (Emadi-Kouchak et al., 2016). Acute neurological effects of minocycline have also been observed in healthy humans. Following an acute dose of 200 mg of minocycline, transcranial magnetic stimulation (TMS) was used to tonically activate a muscle in the hand; the investigators observed an increase in cortical silent periods, a period of silence on the electromyography following TMS, suggesting minocycline has central inhibitory effects (Lang, Rothkegel, Terney, Antal, & Paulus, 2013). It is hypothesized that this inhibition is exclusively cortical and is an indication of activation of GABAergic systems (Paulus et al., 2008).

The broad-reaching antiinflammatory effects of minocycline and its modulation of neurotransmitter systems involved in depression and anxiety suggest that minocycline may be a potential pharmacotherapy for inflammation-associated depression. However, additional human clinical studies are required to clearly determine its efficacy for the treatment of inflammation-associated depression.

NMDA RECEPTOR ANTAGONISTS: RELEVANCE TO INFLAMMATION-ASSOCIATED DEPRESSION

As mentioned, glutamatergic neurotransmission has been heavily implicated in inflammation-associated depression (Dantzer & Walker, 2014). This has led to studies investigating NMDAR antagonists for treatment-resistant depression (Kim & Na, 2016). The most consistent and promising results have been from studies investigating low doses of the NMDAR antagonist, ketamine. Interestingly, preclinical studies have shown that low-dose ketamine may have positive effects beyond just NMDAR antagonism in the treatment of depressive-like behavior. For instance, ketamine's reversal of chronic unpredictable mild stress-induced depressive-like behavior in rats was associated with reduced hippocampal inflammatory cytokines, IL-1β, IL-6, and TNF-α (Wang et al., 2015). Of note, this study analyzed hippocampal inflammatory signaling within 4 h of ketamine administration highlighting the rapid effect ketamine can have on neuroinflammation. In contrast, Abelaira et al. found that the effects of ketamine on depressive-like behavior in rats increased serum and brain inflammatory cytokine TNF-α.

Although both studies used similar behavioral tests, the forced swim test (FST), to measure depressive-like behavior, they differed in their model of depressive-like behavior. Wang et al. (2015) used a chronic unpredictable mild stress (CUMS) to induce a depressive-like phenotype, a model previously associated with increased neuroinflammation; subsequently, they found that ketamine reduced FST immobility and prevented the increase in CUMS-induced inflammatory signaling, whereas Abelaira et al. (2016) utilized the traditional Porsolt's FST, where the acute effects of traditional antidepressants reduce immobility, and administered an acute dose of ketamine to naive animals (Porsolt, Le Pichon, & Jalfre, 1977). In this study, they observed an antidepressant effect of ketamine, but this was associated with increased inflammatory signaling and oxidative stress in several brain regions related to depression. The difference in ketamine's effects on inflammatory cytokines in these studies is likely due to the significant differences in study design; this highlights the limitation of using behavioral paradigms that are based on the activity of traditional antidepressants in naive animals, which have significantly different mechanisms of action that influences behavior and the immune system. Given the time course of ketamine's activity seen in human studies, lasting several days to 1 week, investigators need to be careful when interpreting results from classical behavioral tests like FST and tail suspension test (TST) when acute doses of ketamine are tested. Indeed, several studies have observed increased motor activity in the open-field test with low-dose ketamine, a result that may confound observations from FST or TST studies

(Hetzler & Wautlet, 1985; Imre, Fokkema, Boer, & Horst, 2006).

Further evidence for ketamine's antiinflammatory effects comes from several *in vitro* studies that have shown ketamine can suppress LPS-induced cytokine release. One of the earlier studies that investigated the cytokine suppression effect of ketamine found that when whole human blood was stimulated with LPS, ketamine at concentrations of 20–100 μM suppressed the release of proinflammatory cytokines (Kawasaki et al., 1999). In cultured astrocytes from rats, ketamine at 100 and 1000 μM significantly inhibits the LPS-induced release of proinflammatory cytokines; this was associated with a reduced expression of TLR4 and attenuated phosphorylation of the transcription factor NFκB (Wu et al., 2012).

In another rodent model of inflammation-induced depression, Walker et al. (2013) identified that ketamine, administered at various time points, does not affect LPS-induced inflammation or sickness-like behavior, but significantly reduced depressive-like behavior in mice. This study also provides evidence that LPS-induced inflammation causes an increase in NMDAR signaling evident by increased microglial production of quinolinic acid, a NMDAR agonist. Therefore, this study suggests that ketamine's direct antiinflammatory effect is not primarily responsible for its rapid antidepressant effect in this inflammatory model of depression. The exact mechanisms of ketamine's antidepressant effect and its potential to modulate the immune system are still not clear; however, its ability to exert positive effects on inflammation-associated depression is supported by the current literature.

The first report of a crossover double-blind placebo-controlled clinical trial of ketamine for the treatment of depression was by Berman et al. (2000). This small clinical trial reported a rapid antidepressant effect emerging within 3 days of an intravenous infusion of 0.5 mg/kg ketamine hydrochloride. Since then, a plethora of clinical trials have been published identifying a robust antidepressant effect of low-dose ketamine (Blier & Blier, 2016; Murrough et al., 2013; Rasmussen et al., 2013). Recent meta-analyses support low-dose ketamine as an effective rapid antidepressant with positive effect beginning within hours and lasting up to 1 week post treatment (Kishimoto et al., 2016; Romeo, Choucha, Fossati, & Rotge, 2015; Xu et al., 2016). A Cochrane report on the efficacy of 11 different NMDAR modulators for their ability to treat unipolar depression found that only ketamine was significantly more effective than placebo (Caddy et al., 2015). This is interesting as it may suggest that ketamine's lasting positive effect on mood may be due to its unique properties beyond NMDA receptor antagonism.

In an exploratory study investigating plasma cytokine levels and treatment response to subanesthetic dose of ketamine, there was no correlation found between changes in plasma cytokine levels and treatment response to ketamine (Park et al., 2017). This further supports the notion that the antiinflammatory effects of ketamine are not responsible for its rapid antidepressant activity. Future studies like this are required to determine the long-term effects of multiple-dose ketamine on depression and if long-term recovery is associated with reductions in inflammatory markers.

Although ketamine's rapid antidepressant effect may not be due to its antiinflammatory properties, its use in inflammation-associated depression may still be warranted. Studies on the long-term effects of ketamine for inflammation-associated depression have not yet been conducted. Ketamine's disruption to the glutamatergic synapse and prevention of quinolinic acid-induced neurotoxicity may, if sustained long term, allow the neuroinflammatory mechanisms discussed earlier to resolve, pushing a more permanent recovery state rather than simply blocking the downstream inflammatory cytokines. Indeed, ketamine as an adjunctive therapy for inflammatory-associated depression combined with classical antiinflammatory drugs may prove superior than either alone.

CANNABINOIDS ANTIINFLAMMATORY POTENTIAL

Cannabinoids represent a novel class of immunomodulatory drugs as suggested by the functional expression of both the receptors for cannabinoids and production of endocannabinoids by immune cells (Galiegue, Mary, & Marchand, 1995; Rom & Persidsky, 2013). Interestingly, both endocannabinoids and phytocannabinoids have been shown to modulate the inflammatory response (Boorman, Zajkowska, Ahmed, Pariante, & Zunszain, 2016; Burstein, 2015).

The schedule I status of medical marijuana in the United States (USA) and illegality in other countries have significantly inhibited intensive research into the therapeutic potential of not only cannabis but also cannabinoids in general. The illegality and the stigma behind the depressogenic effect of cannabis have perhaps serendipitously guided research toward exposing cannabinoid antidepressant potential. As the therapeutic efficacy of cannabis is obvious to many pharmacologists but the illegality inhibits clinical investigations, research has focused on the individual constituents of cannabis, their synthetic analogues and derivatives, and their potential therapeutic antiinflammatory activity (Klein, 2005).

There are two main cannabinoids from cannabis, cannabidiol (CBD) and delta-9-tetrahydrocannabinol (THC): CBD has limited direct psychoactivity and potential immunomodulatory effects, whereas THC is the key psychoactive component of cannabis and causes many undesirable side effects (Burstein, 2015; Mizrahi, Watts, & Tseng, 2017). Since the nonpsychoactive cannabinoid (CBD) seems to be the most promising of the two cannabinoids in

terms of antiinflammatory efficacy, it may be particularly useful in mood disorders over THC due to the lack of direct psychoactivity. CBD has been shown to have neuroprotective effects in experimental models of neurodegenerative diseases (Fernández-Ruiz et al., 2013). In a model of arterial ischemic stroke, a single intraperitoneal injection of 5 mg/kg cannabidiol prevented the behavioral and neuroinflammatory consequences of cerebral artery occlusion including a reduction in neuronal loss, reduced excitotoxicity, and significant modulation of glial function (Ceprián et al., 2017). Cannabidiol has also been shown to reduce neuroinflammation in a rodent model of Alzheimer's disease (Watt & Karl, 2017). In one such model, human beta-amyloid peptide is injected into the hippocampus of mice that induces similar neuroinflammatory pathology of that seen in Alzheimer's disease. When these animals were treated with systemic cannabidiol injection for 7 days following beta-amyloid injection, CBD dose-dependently reduced the production and release of IL-1β and inducible nitric oxide synthase (Esposito et al., 2007).

The classical inflammatory model involving a systemic LPS injection, often used as a depressive-like model, can be modulated by CBD in mice. For instance, the rise in TNF-α following intraperitoneal injection of 10 mg/kg LPS was prevented by simultaneous injection of CBD in mice (Malfait et al., 2000). This was later shown by Carrier et al. to be dependent on adenosine A2A receptors, as when a CBD 1 mg/kg and an adenosine A2A receptor blocker were injected together, the CBD no longer had an antiinflammatory effect in LPS exposed mice (Carrier, Auchampach, & Hillard, 2006). In BV2 microglial cells, cannabidiol dose-dependently reduced the release of IL-1β and IL-6 when stimulated with LPS. Cannabidiol inhibition of inflammatory cytokines was associated with reduced NFκB pathway activity and an increase in the activation of the transcription factor STAT3 (Kozela et al., 2010).

Interestingly, the average daily doses for prescription drugs for depression filled per physician annually are significantly lower in US states that have legalized medical marijuana (Bradford & Bradford, 2016). Interestingly, prescription for pain drugs, another inflammatory-related condition, exhibited the largest difference compared with states where marijuana is illegal. This evidence taken together with the good CNS penetrability of cannabinoids and safety profile supports a potential for therapeutic application of cannabinoids, particularly cannabidiol, for the treatment of inflammatory-associated depression. Future studies investigating the potential antidepressant effect of cannabinoids would benefit from including analysis of serum inflammatory markers in the trial. The potential of THC to treat depression is likely limited due to its significant psychoactive effects and abuse potential; however, the investigations into synthetic derivatives with diverse pharmacodynamics will undoubtedly continue.

CONCLUSIONS

Given the large proportion of depressed individuals who do not respond to classical antidepressant therapies, there is a significant need to discover novel treatments that will provide relief for these patients. The current evidence from observations in depressed patients and preclinical animal models presents the immune system and inflammatory pathways as potential new therapeutic targets. Currently, published clinical studies on antiinflammatory agents like NSAIDs and anticytokine monoclonal antibodies, due to their limitations, have not provided a clear therapeutic efficacy for these drugs to begin clinical application. However, these essential preliminary studies have provided much-needed information to aid the design of future clinical studies into the use of novel pharmacotherapies for inflammation-associated depression. The successful use of

the NMDA receptor antagonist ketamine for treatment-resistant depression is intriguing, as it highlights the essential role of aberrant glutamate signaling in depression, which has long been associated with neuroinflammation. The recent use of ketamine for depression exemplifies the value of investigating existing drugs to be used in novel ways for the treatment of depression. This is one strategy that may soon prove useful in the investigation of novel antiinflammatory drugs, like cannabidiol, for the treatment of inflammation-associated depression. The long-term outcomes of antiinflammatory therapeutics for the treatment of depression are still to be conducted, and their implementation at different stages of depression must be investigated to fully uncover their therapeutic potential.

References

Abelaira, H. M., Réus, G. Z., Ignácio, Z. M., Santos, D., Maria, A., Moura, d., et al. (2016). Ketamine exhibits different neuroanatomical profile after mammalian target of Rapamycin inhibition in the prefrontal cortex: The role of inflammation and oxidative stress. *Molecular Neurobiology.* https://doi.org/10.1007/s12035-016-0071-4.

Akhondzadeh, S., Jafari, S., Raisi, F., Nasehi, A. A., Ghoreishi, A., Salehi, B., et al. (2009). Clinical trial of adjunctive celecoxib treatment in patients with major depression: A double blind and placebo controlled trial. *Depression and Anxiety, 26*(7), 607–611. https://doi.org/10.1002/da.20589.

Arakawa, S., Shirayama, Y., Fujita, Y., Ishima, T., Horio, M., Muneoka, K., et al. (2012). Minocycline produced antidepressant-like effects on the learned helplessness rats with alterations in levels of monoamine in the amygdala and no changes in BDNF levels in the hippocampus at baseline. *Pharmacology, Biochemistry, and Behavior, 100*(3), 601–606. https://doi.org/10.1016/j.pbb.2011.09.008.

Banke, T. G., Bowie, D., Lee, H., Huganir, R. L., Schousboe, A., & Traynelis, S. F. (2000). Control of GluR1 AMPA receptor function by cAMP-dependent protein kinase. *The Journal of Neuroscience: The Official Journal of the Society for Neuroscience, 20*(1), 89–102.

Bazargani, N., & Attwell, D. (2016). Astrocyte calcium signaling: The third wave. *Nature Neuroscience, 19*(2), 182–189. https://doi.org/10.1038/nn.4201.

Berman, R. M., Cappiello, A., Anand, A., Oren, D. A., Heninger, G. R., Charney, D. S., et al. (2000). Antidepressant effects of ketamine in depressed patients. *Biological Psychiatry, 47*(4), 351–354.

Blier, P., & Blier, J. (2016). Ketamine: clinical studies in treatment-resistant depressive disorders. In *Ketamine for treatment-resistant depression* (pp. 31–42). Cham: Adis.

Bonaccorso, S., Marino, V., Biondi, M., Grimaldi, F., Ippoliti, F., & Maes, M. (2002). Depression induced by treatment with interferon-alpha in patients affected by hepatitis C virus. *Journal of Affective Disorders, 72*(3), 237–241.

Boorman, E., Zajkowska, Z., Ahmed, R., Pariante, C. M., & Zunszain, P. A. (2016). Crosstalk between endocannabinoid and immune systems: A potential dysregulation in depression? *Psychopharmacology, 233*(9), 1591–1604. https://doi.org/10.1007/s00213-015-4105-9.

Bradford, A. C., & Bradford, W. D. (2016). Medical marijuana laws reduce prescription medication use in medicare part D. *Health Affairs, 35*(7), 1230–1236. https://doi.org/10.1377/hlthaff.2015.1661.

Brooks, A. K., Janda, T. M., Lawson, M. A., Rytych, J. L., Smith, R. A., Ocampo-Solis, C., et al. (2017). Desipramine decreases expression of human and murine indoleamine-2,3-dioxygenases. *Brain, Behavior, and Immunity.* https://doi.org/10.1016/j.bbi.2017.02.010.

Burstein, S. (2015). Cannabidiol (CBD) and its analogs: A review of their effects on inflammation. *Bioorganic & Medicinal Chemistry, 23*(7), 1377–1385. https://doi.org/10.1016/j.bmc.2015.01.059.

Caddy, C., Amit, B. H., McCloud, T. L., Rendell, J. M., Furukawa, T. A., McShane, R., et al. (2015). Ketamine and other glutamate receptor modulators for depression in adults. In *The Cochrane Library*.

Capuron, L., Pagnoni, G., Drake, D. F., Woolwine, B. J., Spivey, J. R., Crowe, R. J., et al. (2012). Dopaminergic mechanisms of reduced basal ganglia responses to hedonic reward during interferon alfa administration. *Archives of General Psychiatry, 69*(10), 1044–1053. https://doi.org/10.1001/archgenpsychiatry.2011.2094.

Carrier, E. J., Auchampach, J. A., & Hillard, C. J. (2006). Inhibition of an equilibrative nucleoside transporter by cannabidiol: A mechanism of cannabinoid immunosuppression. *Proceedings of the National Academy of Sciences of the United States of America, 103*(20), 7895–7900. https://doi.org/10.1073/pnas.0511232103.

Carvalho, L. A., Juruena, M. F., Papadopoulos, A. S., Poon, L., Kerwin, R., Cleare, A. J., et al. (2008). Clomipramine in vitro reduces glucocorticoid receptor function in healthy subjects but not in patients with major depression. *Neuropsychopharmacology, 33*(13), 3182–3189. https://doi.org/10.1038/npp.2008.44.

Cazalis, J., Tanabe, S.-i., Gagnon, G., Sorsa, T., & Grenier, D. (2009). Tetracyclines and chemically modified tetracycline-3 (CMT-3) modulate cytokine secretion by lipopolysaccharide-stimulated whole blood. *Inflammation, 32*(2), 130–137. https://doi.org/10.1007/s10753-009-9111-9.

Ceprián, M., Jiménez-Sánchez, L., Vargas, C., Barata, L., Hind, W., & Martínez-Orgado, J. (2017). Cannabidiol reduces brain damage and improves functional recovery in a neonatal rat model of arterial ischemic stroke. *Neuropharmacology, 116*, 151–159. https://doi.org/10.1016/j.neuropharm.2016.12.017.

Dadsetan, S., Balzano, T., Forteza, J., Agusti, A., Cabrera-Pastor, A., Taoro-Gonzalez, L., et al. (2016). Infliximab reduces peripheral inflammation, neuroinflammation, and extracellular GABA in the cerebellum and improves learning and motor coordination in rats with hepatic encephalopathy. *Journal of Neuroinflammation, 13*(1), 245. https://doi.org/10.1186/s12974-016-0710-8.

Dantzer, R., O'Connor, J. C., Freund, G. G., Johnson, R. W., & Kelley, K. W. (2008). From inflammation to sickness and depression: When the immune system subjugates the brain. *Nature Reviews Neuroscience, 9*(1), 46–56. https://doi.org/10.1038/nrn2297.

Dantzer, R., & Walker, A. K. (2014). Is there a role for glutamate-mediated excitotoxicity in inflammation-induced depression? *Journal of Neural Transmission, 121*(8), 925–932. https://doi.org/10.1007/s00702-014-1187-1.

De Leo, J. A., Tawfik, V. L., & LaCroix-Fralish, M. L. (2006). The tetrapartite synapse: Path to CNS sensitization and chronic pain. *Pain, 122*(1), 17–21.

Eisenberger, N. I., Berkman, E. T., Inagaki, T. K., Rameson, L. T., Mashal, N. M., & Irwin, M. R. (2010). Inflammation-induced anhedonia: Endotoxin reduces ventral striatum responses to reward. *Biological Psychiatry, 68*(8), 748–754. https://doi.org/10.1016/j.biopsych.2010.06.010.

Emadi-Kouchak, H., Mohammadinejad, P., Asadollahi-Amin, A., Rasoulinejad, M., Zeinoddini, A., Yalda, A., et al. (2016). Therapeutic effects of minocycline on mild-to-moderate depression in HIV patients: A double-blind, placebo-controlled, randomized trial. *International Clinical Psychopharmacology, 31*(1), 20–26. https://doi.org/10.1097/yic.0000000000000098.

Esposito, G., Scuderi, C., Savani, C., Steardo, L., Filippis, D., Cottone, P., et al. (2007). Cannabidiol in vivo blunts amyloid induced neuroinflammation by suppressing IL 1 and iNOS expression. *British Journal of Pharmacology, 151*(8), 1272–1279. https://doi.org/10.1038/sj.bjp.0707337.

Eyre, H. A., Lavretsky, H., Kartika, J., Qassim, A., & Baune, B. T. (2016). Modulatory effects of antidepressant classes on the innate and adaptive immune system in depression. *Pharmacopsychiatry, 49*(03), 85–96.

Felger, J. C., Li, L., Marvar, P. J., Woolwine, B. J., Harrison, D. G., Raison, C. L., et al. (2013). Tyrosine metabolism during interferon-alpha administration: Association with fatigue and CSF dopamine concentrations. *Brain, Behavior, and Immunity, 31*, 153–160. https://doi.org/10.1016/j.bbi.2012.10.010.

Fernández-Ruiz, J., Sagredo, O., Pazos, M. R., García, C., Pertwee, R., Mechoulam, R., et al. (2013). Cannabidiol for neurodegenerative disorders: Important new clinical applications for this phytocannabinoid? *British Journal of Clinical Pharmacology, 75*(2), 323–333. https://doi.org/10.1111/j.1365-2125. 2012.04341.x.

Gałecki, P., Mossakowska-Wójcik, J., & Talarowska, M. (2017). The anti-inflammatory mechanism of antidepressants—SSRIs, SNRIs. *Progress in Neuro-Psychopharmacology & Biological Psychiatry.* https://doi.org/10.1016/j.pnpbp.2017.03.016.

Galiegue, S., Mary, S., & Marchand, J. (1995). Expression of central and peripheral cannabinoid receptors in human immune tissues and leukocyte subpopulations. *European Journal of Biochemistry.* https://doi.org/10.1111/j.1432-1033.1995.tb20780.x.

Ghaemi, S. N. (2008). Why antidepressants are not antidepressants: STEP-BD, STAR*D, and the return of neurotic depression. *Bipolar Disorders, 10*(8), 957–968. https://doi.org/10.1111/j.1399-5618.2008.00639.x.

Gonzalez, J. C., Egea, J., Del Carmen Godino, M., Fernandez-Gomez, F. J., Sanchez-Prieto, J., Gandia, L., et al. (2007). Neuroprotectant minocycline depresses glutamatergic neurotransmission and Ca(2+) signalling in hippocampal neurons. *The European Journal of Neuroscience, 26*(9), 2481–2495. https://doi.org/10.1111/j.1460-9568.2007.05873.x.

Grippo, A. J., Francis, J., Beltz, T. G., & Felder, R. B. (2005). Neuroendocrine and cytokine profile of chronic mild stress-induced anhedonia. *Physiology & Behavior, 84*(5), 697–706.

Habbas, S., Santello, M., Becker, D., Stubbe, H., Zappia, G., Liaudet, N., et al. (2015). Neuroinflammatory TNFα impairs memory via astrocyte signaling. *Cell, 163*(7), 1730–1741. https://doi.org/10.1016/j.cell.2015.11.023.

Hammen, C. (2005). Stress and depression. *Annual Review of Clinical Psychology, 1,* 293–319. https://doi.org/10.1146/annurev.clinpsy.1.102803.143938.

Hart, B. L. (1988). Biological basis of the behavior of sick animals. *Neuroscience & Biobehavioral Reviews, 12*(2), 123–137.

Herkenham, M., Lee, H. Y., & Baker, R. A. (1998). Temporal and spatial patterns of c-fos mRNA induced by intravenous interleukin-1: A cascade of non-neuronal cellular activation at the blood-brain barrier. *The Journal of Comparative Neurology, 400*(2), 175–196.

Hetzler, B. E., & Wautlet, B. (1985). Ketamine-induced locomotion in rats in an open-field. *Pharmacology Biochemistry and Behavior, 22*(4), 653–655.

Hinwood, M., Morandini, J., Day, T. A., & Walker, F. R. (2012). Evidence that microglia mediate the neurobiological effects of chronic psychological stress on the medial prefrontal cortex. *Cerebral Cortex, 22*(6), 1442–1454. https://doi.org/10.1093/cercor/bhr229.

Howland, R. H. (2008). Sequenced treatment alternatives to relieve depression (STAR*D). Part 2: Study outcomes.

Journal of Psychosocial Nursing and Mental Health Services, 46(10), 21–24.

Imre, G., Fokkema, D. S., Boer, D. J. A., & Horst, T. G. J. (2006). Dose–response characteristics of ketamine effect on locomotion, cognitive function and central neuronal activity. *Brain Research Bulletin, 69*(3), 338–345.

Iwata, M., Ishida, H., Kaneko, K., & Shirayama, Y. (2016). Learned helplessness activates hippocampal microglia in rats: A potential target for the antidepressant imipramine. *Pharmacology, Biochemistry, and Behavior, 150-151,* 138–146. https://doi.org/10.1016/j.pbb.2016.10.005.

Iyengar, R. L., Gandhi, S., Aneja, A., Thorpe, K., Razzouk, L., Greenberg, J., et al. (2013). NSAIDs are associated with lower depression scores in patients with osteoarthritis. *The American Journal of Medicine. 126*(11). https://doi.org/10.1016/j.amjmed.2013.02.037.

Kappelmann, N., Lewis, G., Dantzer, R., Jones, P. B., & Khandaker, G. M. (2016). Antidepressant activity of anti-cytokine treatment: A systematic review and meta-analysis of clinical trials of chronic inflammatory conditions. *Molecular Psychiatry.* https://doi.org/10.1038/mp.2016.167.

Kawasaki, T., Ogata, M., Kawasaki, C., Ogata, J., Inoue, Y., & Shigematsu, A. (1999). Ketamine suppresses proinflammatory cytokine production in human whole blood in vitro. *Anesthesia and Analgesia, 89*(3), 665–669.

Kendler, K. S., Karkowski, L. M., & Prescott, C. A. (1999). Causal relationship between stressful life events and the onset of major depression. *The American Journal of Psychiatry, 156*(6), 837–841. https://doi.org/10.1176/ajp.156.6.837.

Kim, Y.-K., & Na, K.-S. (2016). Role of glutamate receptors and glial cells in the pathophysiology of treatment-resistant depression. *Progress in Neuro-Psychopharmacology & Biological Psychiatry, 70,* 117–126. https://doi.org/10.1016/j.pnpbp.2016.03.009.

Kishimoto, T., Chawla, J. M., Hagi, K., Zarate, C. A., Kane, J. M., Bauer, M., et al. (2016). Single-dose infusion ketamine and non-ketamine N-methyl-D-aspartate receptor antagonists for unipolar and bipolar depression: A meta-analysis of efficacy, safety and time trajectories. *Psychological Medicine, 46*(7), 1459–1472. https://doi.org/10.1017/S0033291716000064.

Klein, T. W. (2005). Cannabinoid-based drugs as anti-inflammatory therapeutics. *Nature Reviews Immunology, 5*(5), 400–411. https://doi.org/10.1038/nri1602.

Kloppenburg, M., Verweij, C. L., Miltenburg, A. M., Verhoeven, A. J., Daha, M. R., Dijkmans, B. A., et al. (1995). The influence of tetracyclines on T cell activation. *Clinical and Experimental Immunology, 102*(3), 635–641.

Köhler, O., Benros, M. E., Nordentoft, M., Farkouh, M. E., Iyengar, R. L., Mors, O., et al. (2014). Effect of anti-inflammatory treatment on depression, depressive symptoms, and adverse effects: A systematic review and meta-analysis of randomized clinical trials. *JAMA*

Psychiatry, 71(12), 1381–1391. https://doi.org/10.1001/jamapsychiatry.2014.1611.

Köhler, C. A., Freitas, T. H., Maes, M., de Andrade, N. Q., Liu, C. S., Fernandes, B. S., et al. (2017). Peripheral cytokine and chemokine alterations in depression: A meta-analysis of 82 studies. *Acta Psychiatrica Scandinavica.* https://doi.org/10.1111/acps.12698.

Kozela, E., Pietr, M., Juknat, A., Rimmerman, N., Levy, R., & Vogel, Z. (2010). Cannabinoids Delta(9)-tetrahydrocannabinol and cannabidiol differentially inhibit the lipopolysaccharide-activated NF-kappaB and interferon-beta/STAT proinflammatory pathways in BV-2 microglial cells. *The Journal of Biological Chemistry*, 285(3), 1616–1626. https://doi.org/10.1074/jbc.M109.069294.

Kraus, R. L., Pasieczny, R., Willingham, K., Turner, M. S., Jiang, A., & Trauger, J. W. (2005). Antioxidant properties of minocycline: Neuroprotection in an oxidative stress assay and direct radical scavenging activity. *Journal of Neurochemistry*, 94(3), 819–827. https://doi.org/10.1111/j.1471-4159.2005.03219.x.

Krügel, U., Fischer, J., Radicke, S., & Sack, U. (2013). Antidepressant effects of TNF-α blockade in an animal model of depression. *Journal of Psychiatric Research*, 47(5), 611–616.

Lang, N., Rothkegel, H., Terney, D., Antal, A., & Paulus, W. (2013). Minocycline exerts acute inhibitory effects on cerebral cortex excitability in humans. *Epilepsy Research*, 107(3), 302–305. https://doi.org/10.1016/j.eplepsyres.2013.09.006.

Levine, J., Barak, Y., Chengappa, K. N. R., Rapoport, A., Rebey, M., & Barak, V. (1999). Cerebrospinal cytokine levels in patients with acute depression. *Neuropsychobiology*, 40(4), 171–176.

Liu, M., Li, J., Dai, P., Zhao, F., Zheng, G., Jing, J., et al. (2015). Microglia activation regulates GluR1 phosphorylation in chronic unpredictable stress-induced cognitive dysfunction. *Stress*, 18(1), 96–106. https://doi.org/10.3109/10253890.2014.995085.

Liu, Y.-N., Peng, Y.-L., Liu, L., Wu, T.-Y., Zhang, Y., Lian, Y.-J., et al. (2015). TNFα mediates stress-induced depression by upregulating indoleamine 2,3-dioxygenase in a mouse model of unpredictable chronic mild stress. *European Cytokine Network*, 26(1), 15–25. https://doi.org/10.1684/ecn.2015.0362.

Maes, M., Bosmans, E., De Jongh, R., Kenis, G., Vandoolaeghe, E., & Neels, H. (1997). Increased serum IL-6 and IL-1 receptor antagonist concentrations in major depression and treatment resistant depression. *Cytokine*, 9(11), 853–858. https://doi.org/10.1006/cyto.1997.0238.

Maes, M., Leonard, B. E., Myint, A. M., Kubera, M., & Verkerk, R. (2011). The new '5-HT' hypothesis of depression: Cell-mediated immune activation induces indoleamine 2,3-dioxygenase, which leads to lower plasma tryptophan and an increased synthesis of detrimental tryptophan catabolites (TRYCATs), both of which contribute to the onset of depression. *Progress in Neuro-Psychopharmacology & Biological Psychiatry*, 35(3), 702–721. https://doi.org/10.1016/j.pnpbp.2010.12.017.

Maier, S. F. (2003). Bi-directional immune-brain communication: Implications for understanding stress, pain, and cognition. *Brain, Behavior, and Immunity*, 17(2), 69–85.

Malfait, A. M., Gallily, R., Sumariwalla, P. F., Malik, A. S., Andreakos, E., Mechoulam, R., et al. (2000). The nonpsychoactive cannabis constituent cannabidiol is an oral anti-arthritic therapeutic in murine collagen-induced arthritis. *Proceedings of the National Academy of Sciences of the United States of America*, 97(17), 9561–9566. https://doi.org/10.1073/pnas.160105897.

Miller, A. H., Maletic, V., & Raison, C. L. (2009). Inflammation and its discontents: The role of cytokines in the pathophysiology of major depression. *Biological Psychiatry*, 65(9), 732–741. https://doi.org/10.1016/j.biopsych.2008.11.029.

Miller, A. H., & Raison, C. L. (2016). The role of inflammation in depression: From evolutionary imperative to modern treatment target. *Nature Reviews Immunology*, 16(1), 22–34. https://doi.org/10.1038/nri.2015.5.

Miyachi, Y., Yoshioka, A., Imamura, S., & Niwa, Y. (1986). Effect of antibiotics on the generation of reactive oxygen species. *The Journal of Investigative Dermatology*, 86(4), 449–453.

Miyaoka, T., Wake, R., Furuya, M., Liaury, K., Ieda, M., Kawakami, K., et al. (2012). Minocycline as adjunctive therapy for patients with unipolar psychotic depression: An open-label study. *Progress in Neuro-Psychopharmacology & Biological Psychiatry*, 37(2), 222–226. https://doi.org/10.1016/j.pnpbp.2012.02.002.

Mizrahi, R., Watts, J. J., & Tseng, K. Y. (2017). Mechanisms contributing to cognitive deficits in cannabis users. *Neuropharmacology*. https://doi.org/10.1016/j.neuropharm.2017.04.018.

Mort, J. R., Aparasu, R. R., & Baer, R. K. (2006). Interaction between selective serotonin reuptake inhibitors and nonsteroidal antiinflammatory drugs: Review of the literature. *Pharmacotherapy*, 26(9), 1307–1313. https://doi.org/10.1592/phco.26.9.1307.

Murrough, J. W., Iosifescu, D. V., Chang, L. C., Al Jurdi, R. K., Green, C. E., Perez, A. M., et al. (2013). Antidepressant efficacy of ketamine in treatment-resistant major depression: A two-site randomized controlled trial. *The American Journal of Psychiatry*, 170(10), 1134–1142. https://doi.org/10.1176/appi.ajp.2013.13030392.

Nie, H., Zhang, H., & Weng, H. R. (2010). Minocycline prevents impaired glial glutamate uptake in the spinal sensory synapses of neuropathic rats. *Neuroscience*, 170(3), 901–912. https://doi.org/10.1016/j.neuroscience.2010.07.049.

O'Connor, J. C., Lawson, M. A., André, C., Moreau, M., Lestage, J., Castanon, N., et al. (2009). Lipopolysaccharide-induced depressive-like behavior is mediated by indoleamine 2,3-dioxygenase activation in mice. *Molecular Psychiatry*, 14(5), 511–522. https://doi.org/10.1038/sj. mp.4002148.

Paemen, L., Martens, E., Norga, K., Masure, S., Roets, E., Hoogmartens, J., et al. (1996). The gelatinase inhibitory activity of tetracyclines and chemically modified tetracycline analogues as measured by a novel microtiter assay for inhibitors. *Biochemical Pharmacology*, 52(1), 105–111.

Pandey, G. N., Rizavi, H. S., Ren, X., Fareed, J., Hoppensteadt, D. A., Roberts, R. C., et al. (2012). Proinflammatory cytokines in the prefrontal cortex of teenage suicide victims. *Journal of Psychiatric Research*, 46(1), 57–63. https://doi.org/10.1016/j.jpsychires.2011.08. 006.

Pang, T., Wang, J., Benicky, J., & Saavedra, J. M. (2012). Minocycline ameliorates LPS-induced inflammation in human monocytes by novel mechanisms including LOX-1, Nur77 and LITAF inhibition. *Biochimica et Biophysica Acta*, 1820(4), 503–510. https://doi.org/10.1016/j. bbagen.2012.01.011.

Park, M., Newman, L. E., Gold, P. W., Luckenbaugh, D. A., Yuan, P., Machado-Vieira, R., et al. (2017). Change in cytokine levels is not associated with rapid antidepressant response to ketamine in treatment-resistant depression. *Journal of Psychiatric Research*, 84, 113–118. https:// doi.org/10.1016/j.jpsychires.2016.09.025.

Paulus, W., Classen, J., Cohen, L. G., Large, C. H., Di Lazzaro, V., Nitsche, M., et al. (2008). State of the art: Pharmacologic effects on cortical excitability measures tested by transcranial magnetic stimulation. *Brain Stimulation*, 1(3), 151–163. https://doi.org/10.1016/j.brs. 2008.06.002.

Porsolt, R. D., Le Pichon, M., & Jalfre, M. (1977). Depression: A new animal model sensitive to antidepressant treatments. *Nature*, 266(5604), 730–732. https://doi.org/ 10.1038/266730a0.

Raison, C. L., & Miller, A. H. (2013). The evolutionary significance of depression in pathogen host defense (PATHOS-D). *Molecular Psychiatry*, 18(1), 15–37. https://doi.org/ 10.1038/mp.2012.2.

Raison, C. L., Rutherford, R. E., Woolwine, B. J., Shuo, C., Schettler, P., Drake, D. F., et al. (2013). A randomized controlled trial of the tumor necrosis factor antagonist infliximab for treatment-resistant depression: The role of baseline inflammatory biomarkers. *JAMA Psychiatry*, 70(1), 31–41. https://doi.org/10.1001/2013. jamapsychiatry.4.

Rasmussen, K. G., Lineberry, T. W., Galardy, C. W., Kung, S., Lapid, M. I., Palmer, B. A., et al. (2013). Serial infusions of low-dose ketamine for major depression. *Journal of Psychopharmacology*, 27(5), 444–450. https://doi.org/ 10.1177/0269881113478283.

Renault, P. F., Hoofnagle, J. H., Park, Y., Mullen, K. D., Peters, M., Jones, D. B., et al. (1987). Psychiatric complications of long-term interferon alfa therapy. *Archives of Internal Medicine*, 147(9), 1577–1580.

Reus, G. Z., Abelaira, H. M., Maciel, A. L., Dos Santos, M. A., Carlessi, A. S., Steckert, A. V., et al. (2015). Minocycline protects against oxidative damage and alters energy metabolism parameters in the brain of rats subjected to chronic mild stress. *Metabolic Brain Disease*, 30(2), 545–553. https://doi.org/10.1007/s11011-014-9602-8.

Rom, S., & Persidsky, Y. (2013). Cannabinoid receptor 2: Potential role in immunomodulation and neuroinflammation. *Journal of Neuroimmune Pharmacology*, 8(3), 608–620. https://doi.org/10.1007/s11481-013-9445-9.

Romeo, B., Choucha, W., Fossati, P., & Rotge, J.-Y. (2015). Meta-analysis of short- and mid-term efficacy of ketamine in unipolar and bipolar depression. *Psychiatry Research*, 230(2), 682–688. https://doi.org/10.1016/j. psychres.2015.10.032.

Rush, A. J., Trivedi, M. H., Wisniewski, S. R., Stewart, J. W., Nierenberg, A. A., Thase, M. E., et al. (2006). Bupropion-SR, sertraline, or venlafaxine-XR after failure of SSRIs for depression. *The New England Journal of Medicine*, 354(12), 1231–1242. https://doi.org/10.1056/NEJMoa052963.

Scheinman, R. I., Gualberto, A., Jewell, C. M., Cidlowski, J. A., & Baldwin, A. S. (1995). Characterization of mechanisms involved in transrepression of NF-kappa B by activated glucocorticoid receptors. *Molecular and Cellular Biology*, 15(2), 943–953.

Shigetomi, E., Tong, X., Kwan, K. Y., Corey, D. P., & Khakh, B. S. (2011). TRPA1 channels regulate astrocyte resting calcium and inhibitory synapse efficacy through GAT-3. *Nature Neuroscience*, 15(1), 70–80. https://doi.org/ 10.1038/nn.3000.

Soczynska, J. K., Mansur, R. B., Brietzke, E., Swardfager, W., Kennedy, S. H., Woldeyohannes, H. O., et al. (2012). Novel therapeutic targets in depression: Minocycline as a candidate treatment. *Behavioural Brain Research*, 235(2), 302–317. https://doi.org/10.1016/j.bbr. 2012.07.026.

Swardfager, W., Rosenblat, J. D., Benlamri, M., & McIntyre, R. S. (2016). Mapping inflammation onto mood: Inflammatory mediators of anhedonia. *Neuroscience and Biobehavioral Reviews*, 64, 148–166. https://doi.org/10.1016/j. neubiorev.2016.02.017.

Takano, T., Kang, J., Jaiswal, J. K., Simon, S. M., Lin, J. H., Yu, Y., et al. (2005). Receptor-mediated glutamate release from volume sensitive channels in astrocytes. *Proceedings of the National Academy of Sciences of the United States of America*, 102(45), 16466–16471. https://doi.org/ 10.1073/pnas.0506382102.

Tsigos, C., & Chrousos, G. P. (2002). Hypothalamic–pituitary–adrenal axis, neuroendocrine factors and stress. *Journal of Psychosomatic Research*, 53(4), 865–871.

Walker, A. K., Budac, D. P., Bisulco, S., Lee, A. W., Smith, R. A., Beenders, B., et al. (2013). NMDA receptor blockade by ketamine abrogates lipopolysaccharide-induced depressive-like behavior in C57BL/6J mice. *Neuropsychopharmacology*, *38*(9), 1609–1616. https://doi.org/10.1038/npp.2013.71.

Wang, X., Wu, H., & Miller, A. H. (2004). Interleukin 1alpha (IL-1alpha) induced activation of p38 mitogen-activated protein kinase inhibits glucocorticoid receptor function. *Molecular Psychiatry*, *9*(1), 65–75. https://doi.org/10.1038/sj.mp.4001339.

Wang, N., Yu, H.-Y., Shen, X.-F., Gao, Z.-Q., Yang, C., Yang, J.-J., et al. (2015). The rapid antidepressant effect of ketamine in rats is associated with down-regulation of proinflammatory cytokines in the hippocampus. *Upsala Journal of Medical Sciences*, *120*(4), 241–248. https://doi.org/10.3109/03009734.2015.1060281.

Warner-Schmidt, J. L., Vanover, K. E., Chen, E. Y., Marshall, J. J., & Greengard, P. (2011). Antidepressant effects of selective serotonin reuptake inhibitors (SSRIs) are attenuated by antiinflammatory drugs in mice and humans. *Proceedings of the National Academy of Sciences of the United States of America*, *108*(22), 9262–9267. https://doi.org/10.1073/pnas.1104836108.

Watt, G., & Karl, T. (2017). In vivo evidence for therapeutic properties of cannabidiol (CBD) for Alzheimer's disease. *Frontiers in Pharmacology*. 8. https://doi.org/10.3389/fphar.2017.00020.

Willner, P. (2017). The chronic mild stress (CMS) model of depression: History, evaluation and usage. *Neurobiology of Stress*, *6*, 78–93. https://doi.org/10.1016/j.ynstr.2016.08.002.

Wu, Y., Dissing-Olesen, L., MacVicar, B. A., & Stevens, B. (2015). Microglia: Dynamic mediators of synapse development and plasticity. *Trends in Immunology*, *36*(10), 605–613. https://doi.org/10.1016/j.it.2015.08.008.

Wu, Y., Li, W., Zhou, C., Lu, F., Gao, T., Liu, Y., et al. (2012). Ketamine inhibits lipopolysaccharide-induced astrocytes activation by suppressing TLR4/NF-κB pathway. *Cellular Physiology and Biochemistry: International Journal of Experimental Cellular Physiology, Biochemistry, and Pharmacology*, *30*(3), 609–617. https://doi.org/10.1159/000341442.

Xu, Y., Hackett, M., Carter, G., Loo, C., Gálvez, V., Glozier, N., et al. (2016). Effects of low-dose and very low-dose ketamine among patients with major depression: A systematic review and meta-analysis. *The International Journal of Neuropsychopharmacology*, *19*(4). https://doi.org/10.1093/ijnp/pyv124.

Ye, J., Jiang, R., Cui, M., Zhu, B., Sun, L., Wang, Y., et al. (2014). Etanercept reduces neuroinflammation and lethality in mouse model of Japanese encephalitis. *Journal of Infectious Diseases*, *210*(6), 875–889.

Yoshimura, R., Hori, H., Ikenouchi-Sugita, A., Umene-Nakano, W., Katsuki, A., Atake, K., et al. (2013). Plasma levels of interleukin-6 and selective serotonin reuptake inhibitor response in patients with major depressive disorder. *Human Psychopharmacology*, *28*(5), 466–470. https://doi.org/10.1002/hup.2333.

Zhang, J., Wang, H., Ye, C., Ge, W., Chen, Y., Jiang, Z., et al. (2013). ATP released by astrocytes mediates glutamatergic activity-dependent heterosynaptic suppression. *Neuron*, *40*(5), 971–982.

The Gut-Brain-Microbe Interaction: Relevance in Inflammation and Depression

Natalie Parletta

University of South Australia, Adelaide, SA, Australia

Hippocrates, known as the father of medicine, proposed that all diseases begin in the gut. The intestinal tract has disproportionately high requirements for energy and protein to support its functions and maintenance (McBride & Kelly, 1990), has its own "enteric" nervous system (Grundy & Schemann, 2006), and contains two-thirds of our immune cells (Mayer, 2011; Pelaseyed et al., 2014). The gut epithelium is intricately designed to maximize its surface area with villi and microvilli, which if spread out would cover an estimated $200\,m^2$—the size of a tennis court (Sekirov, Russell, Antunes, & Finlay, 2010) and is about 100 times larger than the surface area of our skin (Mayer, 2011). It acts as a gateway with incredible, complex functions that enable us to receive life-giving nourishment while at the same time protecting us from foreign substances and incompletely digested food. These observations lend support to a pivotal role for the gastrointestinal tract in metabolism, health, and disease.

We often talk about "gut feelings"—but what does that really mean? The enteric nervous system, now dubbed the second brain, was discovered in the mid-nineteenth century. This discovery provided critical insights into the multiple, complex interactions between our brain and gut, known as the "gut-brain axis" (Mayer, 2011). The two-way communication between the gut and the brain takes place via neurological, immunologic, hormonal, and metabolic pathways and is required for maintaining homeostasis (Carabotti, Scirocco, Maselli, & Severi, 2015; Cryan & O'Mahony, 2011; Mayer, 2011). This bidirectional and interactive communication has multiple roles in physiological functions (Romijn, Corssmit, Havekes, & Pijl, 2008). Evidence suggests that the connection between the gut and the brain also affects higher-order emotional and cognitive functions and is implicated in brain disorders such as autism, anxiety, depression, chronic pain, and Parkinson's disease (Mayer, Tillisch, & Gupta, 2015).

Recent advances in microbiome research suggest that the commensal bacteria living in our gut play a crucial role in gut-brain interactions. Much of the evidence to date is preclinical, and there is a great deal more to understand about mechanisms and therapeutic applications in humans. It is also not currently clear whether the directions of interactions are "bottom-up" or "top-down," which makes it difficult to extrapolate to clinical treatments (Mayer et al., 2015; Smith, 2015). This chapter is not an exhaustive review but rather aims to give a taste of current evidence and understanding regarding the role of the gut microbiome in the gut-brain axis, its relevance to the pathophysiology of depression, and clinical applications.

GUT MICROBIOME

Increasingly sophisticated technologies have enabled identification and mapping of the estimated 100 trillion microbes that live symbiotically in our bodies, primarily in the large intestine—comprising more than 10 times our human cells and carrying over 150 times more genes than our own genome (Cryan & Dinan, 2012). The microbes include bacteria, archaea, viruses, and unicellular eukaryotes (Sekirov et al., 2010). These collectively make up the microbiome—which is now considered to be so important that it has been referred to as our forgotten organ (O'Hara & Shanahan, 2006). Human bacteria have been an intense focus of research; less is known about the other microbes. It is estimated that the commensal bacteria contain over 1000 species and 7000 strains. Bacteroidetes (gram-negative and anaerobic) and Firmicutes (gram-positive and anaerobic) bacteria are our two predominating phylotypes or enterotypes, comprising up to 90% of the human microbiome. Other less abundant phyla include Proteobacteria, Actinobacteria, Fusobacteria, and Verrucomicrobia. Although these groups appear to be consistent between people, there is considerable individual variation in their proportions and the composition of the species within these phyla (Cryan & Dinan, 2012; Foster & Neufeld, 2013; Sekirov et al., 2010).

Our individual microbiome is as distinct as our fingerprint, although it is dynamic and changeable. There is evidence that our microbes coevolved with us, showing a high level of adaptation to their environment (Sekirov et al., 2010). It was originally thought that our gut is first colonized at birth, when microbes are maternally transmitted; however, more recent evidence suggests that some microbial transfer takes place in the womb (Funkhouser & Bordenstein, 2013). Babies born via cesarean section have a markedly different microbiome profile, reflecting the bacteria present on the mother's skin rather than in her vagina (Sekirov et al., 2010). The microbial composition starts to stabilize at around 1–2 years of age. Other factors that can influence the microbiome include breast versus formula feeding, diet, antibiotics, excessive hygiene, geography, number of siblings, pet ownership, surgery, smoking, lack of exercise, infection, and disease (Conlon & Bird, 2015; Cryan & Dinan, 2012; Dash, Clarke, Berk, & Jacka, 2015; Evans, Morris, & Marchesi, 2013; Liu, 2015; Rook & Lowry, 2008; Sekirov et al., 2010).

Our gut bacteria have been associated with many important functions that are necessary for health. These include the physiology, maturation, and function of our gastrointestinal tract; maintenance of the gut epithelium; lipid metabolism; cellular and mucosal immunity; metabolism of short-chain fatty acids; nutrient synthesis and absorption; production of enzymes; metabolism; and energy balance (Conlon & Bird, 2015; Hooper, Littman, & Macpherson, 2012; Le Chatelier et al., 2013; Sekirov et al., 2010). The bacteria most often used as probiotics (defined as beneficial bacteria with positive health properties) are the *Bifidobacterium* and *Lactobacillus* families. It appears that their presence has multiple direct and indirect effects on achieving a healthy balance of microbiota in the gut (Sanders, 2011). When the composition of the commensal bacteria is compromised, for instance, through indiscriminate use of antibiotics, medications, poor diet, or stress, this can leave the host in a state of "dysbiosis," vulnerable to pathogenic or opportunistic bacteria, inflammation, and disease. Accordingly, research has started to identify gut microbiota involvement in a range of diseases including inflammatory bowel disease, irritable bowel syndrome (IBS), bowel cancer, obesity, diabetes, liver disease, allergies, atherosclerosis, and pancreatitis (Sekirov et al., 2010).

GUT-BRAIN AXIS AND MICROBIOTA

For a long time, the dominating paradigm in psychiatry focused exclusively on the brain in understanding and addressing problems with mental health and behavior. Research on the gut-brain axis, which has a long history (Bested, Logan, & Selhub, 2013a), was essentially disregarded. The discovery of the human microbiome has added a new, compelling dimension to our understanding of links between the gut and the brain, with significant paradigm-changing implications for psychiatry (Mayer, Knight, Mazmanian, Cryan, & Tillisch, 2014). Gut-brain communication takes place via multiple, complex neural, immune, endocrine, and metabolic pathways, including the hypothalamic-pituitary-adrenal (HPA) axis, the sympathoadrenal axis and descending monoaminergic pathways, and the vagus nerve (see detailed review by Mayer, Knight, et al., 2014). Bacteria produce neurotransmitters that impact brain function and mental health such as GABA, serotonin, melatonin, histamine, and acetylcholine (Iyer, Aravind, Coon, Klein, & Koonin, 2004). Understanding of the microbiome may help to explain the role of autoimmunity in psychiatry. Immune dysfunction is associated with mental illness, and there is evidence of autoimmune antibodies that target brain areas. These are associated with GI risk factors, with evidence of microbiota

involvement (reviewed in depth by Severance, Tveiten, Lindström, Yolken, & Reichelt, 2016). It is unclear whether the dysbiosis that is observed in people with brain disorders reflects the pathogenesis of the disease or results from treatment. However, inflamed intestines were identified in postmortem investigations of people with schizophrenia in the 1950s, before the widespread use of antipsychotic medications. Nearly 90% of these people had evidence of gastritis, enteritis, or colitis (Severance et al., 2016).

An insight to potential microbial influence on the brain was the remarkable discovery that a brain parasite, *Toxoplasma gondii*, could control the behavior of a mammalian host. This parasite can alter neural activity in a rat to not only switch off innate defensive behavior in response to cat odor but also increase activity in the limbic regions of the brain that create sexual attraction (House, Vyas, & Sapolsky, 2011). This observation is reminiscent of the well-known effect of the rabies virus, which causes aggression, agitation, and fear of water (Smith, 2015). Intriguingly, gut bacteria may even influence mating preferences. Sharon, Segal, Zilber-Rosenberg, and Rosenberg (2011) fed fruit flies a molasses or starch medium and observed that the flies then mated exclusively with other flies that had been fed the same medium. The flies fed with starch had notably higher levels of *Lactobacillus* species. Antibiotic administration stopped their mating preferences, which was then reestablished after they were given cultures of the *Lactobacillus* species. This suggests the bacteria were responsible for the flies' mating behavior.

DEPRESSION, INFLAMMATION AND THE GUT MICROBIOME

Of all the mental health disorders, depression is the most prevalent and now carries one of the leading global burdens of disease (Vos et al., 2015). Observations such that risk factors for IBS include depression and anxiety (Marshall et al., 2010) support observed links between the gut and the brain and further point to the possibility that psychiatric symptoms could be related to chronic inflammation or infection caused by dysbiosis (Smith, 2015). Inflammation has been increasingly indicated in the pathophysiology of depression, evidenced by elevated inflammatory cytokines in both cross-sectional and longitudinal studies (Miller & Raison, 2016; Raison & Miller, 2011; Zunszain, Hepgul, & Pariante, 2013). This could help to explain the high comorbidity between depression and inflammatory diseases like heart disease, inflammatory bowel disease, multiple sclerosis, and rheumatoid arthritis (Maes, Kubera, & Leunis, 2008). Indeed, a major depressive disorder can even be induced by exposure to inflammatory mediators (Raison & Miller, 2011). Conversely, psychological stress can induce activation of the inflammatory response system and alter the microbiome (Bailey et al., 2011).

Evidence suggests that gut microbes entering the bloodstream via compromised integrity of the gut epithelium ("leaky gut") can induce an immune response, which mediates the neuroinflammation underlying depressive symptoms. The response is associated with fatigue, sadness, gastrointestinal symptoms, muscular tension, and a subjective feeling of infection (Maes et al., 2008). The microbiota and their metabolites have been shown to influence other gastrointestinal functions such as mucosal immunity, intestinal motility, and sensitivity (Forsythe & Kunze, 2013). This suggests that gastrointestinal symptoms in people with major depressive disorder may not be attributed to increased psychological distress as commonly thought, but reflect and/or be exacerbated by compromised gastrointestinal structure and function.

In addition to gut microbial influence on immunity and inflammation, there is evidence from animal studies that pathogenic and commensal gut bacteria influence associated pathways that have been identified in gut-brain

communication (Foster & Neufeld, 2013). This includes effects on HPA programming in early life and reactivity to stress via immune and neural circuits, which in turn is associated with depression. Interestingly, pretreatment with probiotics in rats was shown to reduce intestinal permeability that normally results from stress and prevented associated HPA overreactivity to acute psychological stress (Ait-Belgnaoui et al., 2012). Bacterial infection increased anxiety-like behavior in mice that were challenged with the pathogenic bacteria *Campylobacter jejuni* or saline. The *C. jejuni*-challenged mice also showed alterations in regions of the brain including the amygdala (Goehler, Park, Opitz, Lyte, & Gaykema, 2008), which was likely mediated by the vagus nerve (Goehler et al., 2005). This supports other studies that have shown vagal mediation of gut-brain-microbe interactions using vagotomized animal models (Bested, Logan, & Selhub, 2013b).

Other evidence suggests gut bacteria can alter electrophysiological properties of the nervous system. Dysfunctions in the signaling of neurotransmitters such as GABA and serotonin have been traditionally associated with depression and anxiety. The bacteria *Lactobacillus* and *Bifidobacterium* metabolize glutamate to produce GABA *in vivo*, and preliminary studies show an association between microbiota and serotonin signaling. These studies suggest that gut microbiota may impact neurotransmission as an additional pathway to influence depression (see reviews by Forsythe & Kunze, 2013; Foster & Neufeld, 2013). An altered gut microbiome can also impact brain-derived neurotrophic factor (BDNF), which is seen in behavioral abnormalities and depression. Infection with an intestinal parasite caused intestinal inflammation in mice. This reduced expression of hippocampal BDNF messenger RNA (mRNA) and was associated with increased anxiety-like behaviors—independently of the vagus nerve. Administration of *Bifidobacterium longum* normalized the behavior and levels of

BDNF mRNA (Bercik et al., 2010). In another experiment, oral antimicrobials changed the composition of the microbiome in mice. This was associated with increased exploratory behavior and expression of BDNF, independently of the autonomic nervous system, enteric neurotransmitters, or inflammation (Bercik et al., 2011).

ANIMAL INTERVENTION STUDIES

Evidence for the impact of gut microbes on brain physiology/neurochemistry and behavior comes mostly from animal studies to date (Foster & Neufeld, 2013; Mayer et al., 2015; Smith, 2015). As indicated above, study methodologies include altering the gut microbiome with antibiotics; using fecal microbial transplants, germ-free animal models, and experimental infection; and inducing psychological distress (Bested et al., 2013b; Forsythe & Kunze, 2013; Mayer et al., 2015; Mayer, Knight, et al., 2014). Many studies have independently shown reduced anxiety-like behavior in adult germ-free mice using various outcome measures. Gut inflammation and infection with pathogenic bacteria have resulted in increased anxiety-like behaviors. In other studies, treatment with probiotics decreased anxiety- and depressive-like behaviors, including those associated with inflammation (Foster & Neufeld, 2013).

One early observation was an increased adult stress response in germ-free mice, which was reversed by early recolonization of the gut with the commensal bacteria *Bifidobacterium infantis*. Effects were not reversed successfully at a later stage, indicating the importance of microbial exposure during early development (Sudo et al., 2004). Early-life stress is a risk factor for mental illness (Lewis, Galbally, Gannon, & Symeonides, 2014) and other disorders like IBS. Preclinical evidence suggests that early-life stress alters the gut-brain axis via systemic immune response and altered fecal microbiota (O'Mahony et al., 2009). This suggests a potential role for probiotic administration to mediate the effect of early-life stress exposure on risk of mental illness.

Probiotics have ameliorated depressive symptoms in animal models that induced depression via various different environmental stressors. A maternal separation model of depression induced in rats increased inflammatory markers and decreased noradrenaline in the brain. Subsequent administration with *B. infantis* normalized the depressive behavior, decreased inflammatory cytokines, and restored noradrenaline levels (Desbonnet et al., 2010). These researchers further tested *B. infantis* in rats with induced depression from a forced swim test. Although there was no effect on swim behaviors, inflammatory cytokines significantly reduced and plasma tryptophan increased (Desbonnet, Garrett, Clarke, Bienenstock, & Dinan, 2008). Attenuation of the HPA axis stress response was shown after administration of *Lactobacillus farciminis* in female rats. This was attributed to the prevention of gut permeability and decreased translocation of lipopolysaccharides, which can induce depressive symptoms (Ait-Belgnaoui et al., 2012).

Arseneault-Bréard et al. (2012) tested the effect of *Lactobacillus helveticus* and *B. longum* on depression and impaired gut barrier integrity in mice following induced myocardial infarction. The probiotics reversed the depressive behavior and decreased intestinal permeability. In another mouse model, *B. longum* reduced behaviors related to stress, anxiety, and depression, while *Bifidobacterium breve* only reduced anxiety and induced weight loss. These effects were superior to an antidepressant, which also induced weight gain (Savignac, Kiely, Dinan, & Cryan, 2014). Similarly, supplementation with *Lactobacillus rhamnosus* reduced anxious and depressive behaviors of mice compared with controls. Furthermore, the expression of GABA was altered. These effects were not observed in vagotomized mice, supporting indications that the vagus nerve acts as a communication pathway between gut bacteria and the brain (Bravo et al., 2011). Gut bacteria may also affect personality and behavior. Fecal transplants between mice produced more exploratory behavior in normally shy mice and vice versa. As mentioned before, this was accompanied by alterations in microbial composition and levels of BDNF (Bercik et al., 2011).

Probiotics may also attenuate the high comorbidity that occurs between metabolic disorders, such as obesity, diabetes, and metabolic syndrome, and psychiatric illness. Abildgaard, Elfving, Hokland, Lund, and Wegener (2017) demonstrated that a high-fat diet increased depressive-like behavior in rats that are inherently depressed. They then showed that multispecies probiotics with a range of *Bifidobacteria* and *Lactobacilli* strains completely negated the effect of the high-fat diet on depressive behavior, while rats on the control diet were unaffected. The authors speculate that the noneffect on the control rats may have been because both groups of rats had a genetic strain of major depressive disorder. Probiotics appear to alleviate depressive symptoms that are induced by environmental influences, such as diet, myocardial infarction, stress, and maternal separation, and may not affect the genetic component. The authors highlight the clinical relevance of this finding, as obesity tends to be associated with poorer response to antidepressants.

Inconsistent results in animal studies of probiotics and behavior could be in part attributed to the fact that specific bacteria seem to have specific functions and effects that are not observed by other bacteria, as well as the complexities of the communication systems involved (Forsythe & Kunze, 2013). Outcome measures are important as well; for instance, significantly reduced inflammatory markers have been detected in the absence of significant behavior change (Gibson et al., 2010). Timing and length of probiotic administration may influence

findings. It is also important to consider that how animals are handled can affect their stress levels and in turn their microbiome (Smith, 2015). All in all, this growing evidence points to probiotics as a potential emerging treatment for depression and has highlighted a variety of mediating mechanisms. Indeed, as a result of this growing work, probiotics and prebiotics (nondigestible carbohydrates that feed and stimulate the growth of commensal bacteria in the colon) have been named "psychobiotics," referring to their influence on bacteria-brain relationships (Deans, 2016; Sarkar et al., 2016).

HUMAN INTERVENTION STUDIES

The American Gut project has collected thousands of human stool samples, providing exciting opportunities to investigate the human microbiome and its relationship with mental illness and related disorders (Mayer, Knight, et al., 2014). Some preliminary correlations in humans suggest that potentially harmful groups of bacteria are higher in people with depression and that beneficial bacterial strains are lower. A study of 37 patients with depression and nondepressed controls identified several associations between depression and fecal *Bacteroidales*, *Lachnospiraceae*, *Oscillibacter*, and *Alistipes* (Naseribafrouei et al., 2014). Jiang et al. (2015) found that *Enterobacteriaceae* and *Alistipes* were higher and *Faecalibacterium* were lower in 46 patients with major depressive disorder, compared with 30 healthy controls.

From a small number of clinical trials, there is some evidence that various probiotic strains can influence inflammatory conditions including diarrhea, IBS, ulcerative colitis, and allergies. However, findings are inconsistent, and considerably more research is required with attention to study methodologies and specific species and strains (Lomax & Calder, 2009a, 2009b). Prebiotics may also improve immune function with a beneficial influence on conditions such as inflammatory bowel disease and atopic dermatitis (Lomax & Calder, 2009a, 2009b). IBS in particular, frequently associated with depression (Lydiard, 2001), is also associated with an altered microbiome (Kassinen et al., 2007). One study randomized 77 adults suffering IBS to receive *Lactobacillus salivarius*, *B. infantis*, or placebo for 8 weeks. The group receiving *B. infantis* had a greater reduction in symptoms than placebo and a more favorable inflammatory profile after supplementation (O'Mahony et al., 2005).

A few studies have investigated the effects of psychobiotics on the brain and mental health in humans. The high comorbidity between gastrointestinal symptoms and psychiatric illness, particularly autism spectrum disorders (ASD), has long been recognized. Symptoms that can be attributed to microbial dysbiosis include poor digestion, dysmotility, inflammation, intestinal permeability, and food intolerance (Severance et al., 2016). Mayer, Knight, et al. (2014) and Mayer, Padua, and Tillisch (2014) reviewed converging and preclinical evidence for a role of the gut microbiome in ASD, which is beyond the scope of this chapter. However, a recent landmark study warrants noting in which 75 human infants were randomized to probiotics or placebo. At 13-year follow-up, the researchers reported *zero* cases of neurodevelopmental disorders in the probiotic group versus 17.1% cases of attention deficit hyperactivity disorder (ADHD) or Asperger's syndrome in the placebo group. Importantly, lower numbers of *Bifidobacterium* species bacteria were detected in the feces of children who developed the neurodevelopmental disorders than healthy children during the first 3 months of life, and lower *Lactobacillus-Enterococcus* group bacteria and *Bacteroides* were found at 6 months (Pärtty, Kalliomaki, Wacklin, Salminen, & Isolauri, 2015).

Turning toward mood and depression, a relatively early study provided 124 healthy adults with a milk drink containing *Lactobacillus casei Sharota* or placebo and found no effects on mood

after 10 and 20 days. However, the participants generally reported good mood at baseline. When results of those with baseline mood scores in the bottom third were analyzed, they reported being happier after probiotic consumption compared with placebo. Unexpectedly, the probiotic resulted in slightly poorer performance on two memory tests (Benton, Williams, & Brown, 2007). A later study tested the effects of *L. helveticus* and *B. longum* or placebo on anxiety, depression, stress, and coping strategies in 55 healthy human adults. In contrast to the previous study, they reported significantly reduced psychological distress and mood on a range of psychological questionnaires in the probiotic group after 30 days (Messaoudi, Lalonde, et al., 2011). The probiotic group also showed decreased levels of urinary cortisol, a physiological marker of reduced stress. These researchers found no evidence for detrimental effects of the probiotics on learning and memory (Messaoudi, Violle, et al., 2011). Notably, this study was longer than the previous one and combined *Lactobacillus* and *Bifidobacterium* strains. Recently, similar beneficial effects of a multispecies probiotic on mood were reported. Forty healthy volunteers were given probiotics containing an array of the *Lactobacillus* and *Bifidobacterium* strains or placebo for 4 weeks. The probiotic group showed significantly reduced cognitive reactivity to sad mood, a marker of vulnerability to depression (Steenbergen, Sellaro, van Hemert, Bosch, & Colzato, 2015).

A landmark study has extended preclinical data by showing that gut microbiota can influence brain function in humans. In this neuroimaging study, 36 healthy women were given a fermented milk product (containing *Bifidobacterium animalis* subsp. *lactis*, *Streptococcus thermophilus*, *Lactobacillus bulgaricus*, and *Lactococcus lactis* subsp. *lactis*), a nonfermented milk product, or nothing for 4 weeks. When the women were shown emotional faces, functional magnetic resonance imaging (fMRI) detected significantly decreased neural emotional reactivity following fermented milk consumption compared with increased emotional reactivity in the placebo group (Tillisch et al., 2013).

CLINICAL APPLICATIONS

Considerably more research is needed to clarify the effect of gut bacteria on depression and identify specific treatment strategies via targeted modulation of the gut microbiome. However, increased understanding of the microbiome adds a new perspective to the importance of lifestyle factors for health. In particular, recent years have yielded growing evidence that diet is important for not only physical health but also mental health (Jacka et al., 2016; Jacka, Mykletun, & Berk, 2012; Opie et al., 2017; Parletta, Milte, & Meyer, 2013; Parletta et al., 2017; Zarnowiecki et al., 2016). There is also evidence that diet—particularly a variety of dietary fibers from plant foods such as legumes, vegetables, fruit, nuts, seeds, and whole grains—can be an effective way to sustain a healthy gut microbiome (Conlon & Bird, 2015; Dash et al., 2015). In fact, Zhang et al. (2010) reported that dietary changes account for 57% of gut microbial diversity in contrast to 12% accounted for by genetic background. This may help to explain the multitude of health benefits that have been demonstrated with a traditional plant-based Mediterranean diet. As well as a variety of prebiotic dietary fibers and nutrients, the Mediterranean diet includes probiotics through fermented foods such as yogurt and cheese (Del Chierico, Vernocchi, Dallapiccola, & Putignani, 2014).

Gut bacteria ferment dietary fiber and proteins, producing short-chain fatty acids (butyrate, propionate, and acetate) that have an array of important functions in the intestines, including the provision of energy for intestinal tissues and bacteria, maintaining tissue integrity, supporting healthy immune function, and reducing inflammation. *Bifidobacterium* and other bacteria synthesize vitamins that are important for healthy brain function, such as B_{12}, biotin, folate, and thiamin. Other nutritional components of food such as polyphenols and micronutrients can support healthy gut bacteria. Conversely, obesogenic western diets that are high in saturated fat may promote pro-inflammatory microbes and increase intestinal permeability (Conlon & Bird, 2015; Kim, Gu, Lee, Joh, & Kim, 2012). A high-fat, high-sugar western diet can rapidly alter the microbiome, which was shown through fecal transplants to mediate diet-induced adiposity (Zhang et al., 2010). Western diets that are high in processed foods and low in plant foods provide little fiber or nutrients and are likely to promote poor microbial diversity. This is supported by a comparison of European children with modern western diets with African children with traditional high fiber diets similar to their ancestors. The African children had significantly higher diversity of gut microbiota and short-chain fatty acids. The authors hypothesize that the microbial diversity protected their ancestors from inflammation and noncommunicable diseases (De Filippo et al., 2010).

Beyond diet, several modifiable lifestyle factors have been associated with mental health. Emerging microbiome research provides converging evidence to support their importance including exercise (Kang et al., 2014), smoking (Biedermann et al., 2013), playing in the dirt (Lowry et al., 2007), and sleep (Everson & Toth, 2000). People with mental illness may have reduced knowledge about the impact of lifestyle factors on their health and may benefit from education around these (Parletta, Aljeesh, & Baune, 2016).

CONCLUSIONS

The alarming pandemic of noncommunicable, comorbid inflammatory diseases such as depression, allergies, obesity, diabetes, and cardiovascular disease appears at least in part to be mediated by the effects of modern lifestyles and environmental changes (poor diet, lack of exercise, stress, pharmacological intervention, pollution, global warming, and reduced biodiversity) on our microbiome (Prescott, 2013). Preclinical data provide compelling evidence that our gut microbiota have numerous pivotal functions that impact on mental health through gut-brain links, including the integrity of the gut barrier, immune regulation, inflammation, and neurotransmission. However, clinical trials in humans are sparse (Romijn & Rucklidge, 2015), so it is premature to extrapolate to treatment for depression and mental illness. In the interim, lifestyle factors that support good physical and mental health present appropriate targets for treatment. In addition, evidence suggests various therapeutic benefits of supplementing with probiotics containing *Lactobacillus* and *Bifidobacterium* strains to help restore a healthy bacterial ecology (Sanders, 2011).

References

Abildgaard, A., Elfving, B., Hokland, M., Lund, S., & Wegener, G. (2017). Probiotic treatment protects against the pro-depressant-like effect of high-fat diet in flinders sensitive line rats. *Brain, Behavior, and Immunity 65*, 33–42.

Ait-Belgnaoui, A., Durand, H., Cartier, C., Chaumaz, G., Eutamene, H., Ferrier, L., et al. (2012). Prevention of gut leakiness by a probiotic treatment leads to attenuated HPA response to an acute psychological stress in rats. *Psychoneuroendocrinology, 37*(11), 1885–1895.

Arseneault-Bréard, J., Rondeau, I., Gilbert, K., Girard, S.-A., Tompkins, T. A., Godbout, R., et al. (2012). Combination of *Lactobacillus helveticus* R0052 and *Bifidobacterium longum* R0175 reduces post-myocardial infarction depression symptoms and restores intestinal permeability in a rat model. *British Journal of Nutrition, 107*(12), 1793–1799.

Bailey, M. T., Dowd, S. E., Galley, J. D., Hufnagle, A. R., Allen, R. G., & Lyte, M. (2011). Exposure to a social stressor alters the structure of the intestinal microbiota: implications for stressor-induced immunomodulation. *Brain, Behavior, and Immunity, 25*(3), 397–407.

Benton, D., Williams, C., & Brown, A. (2007). Impact of consuming a milk drink containing a probiotic on mood and cognition. *European Journal of Clinical Nutrition, 61*(3), 355–361.

Bercik, P., Denou, E., Collins, J., Jackson, W., Lu, J., Jury, J., et al. (2011). The intestinal microbiota affect central levels of brain-derived neurotropic factor and behavior in mice. *Gastroenterology, 141*(2), 599–609. e593.

Bercik, P., Verdu, E. F., Foster, J. A., Macri, J., Potter, M., Huang, X., et al. (2010). Chronic gastrointestinal inflammation induces anxiety-like behavior and alters central nervous system biochemistry in mice. *Gastroenterology, 139*(6), 2102–2112. e2101.

Bested, A. C., Logan, A. C., & Selhub, E. M. (2013a). Intestinal microbiota, probiotics and mental health: from Metchnikoff to modern advances: Part I—autointoxication revisited. *Gut Pathogens, 5*(1), 5.

Bested, A. C., Logan, A. C., & Selhub, E. M. (2013b). Intestinal microbiota, probiotics and mental health: from Metchnikoff to modern advances: Part III—convergence toward clinical trials. *Gut Pathogens, 5*(1), 4.

Biedermann, L., Zeitz, J., Mwinyi, J., Sutter-Minder, E., Rehman, A., Ott, S. J., et al. (2013). Smoking cessation induces profound changes in the composition of the intestinal microbiota in humans. *PLoS One, 8*(3).

Bravo, J. A., Forsythe, P., Chew, M. V., Escaravage, E., Savignac, H. M., Dinan, T. G., et al. (2011). Ingestion of *Lactobacillus* strain regulates emotional behavior and central GABA receptor expression in a mouse via the vagus nerve. *Proceedings of the National Academy of Sciences, 108* (38), 16050–16055.

Carabotti, M., Scirocco, A., Maselli, M. A., & Severi, C. (2015). The gut-brain axis: interactions between enteric microbiota, central and enteric nervous systems. *Annals of Gastroenterology, 28*(2), 203.

Conlon, M. A., & Bird, A. R. (2015). The impact of diet and lifestyle on gut microbiota and human health. *Nutrients, 7*, 17–44.

Cryan, J. F., & Dinan, T. G. (2012). Mind-altering microorganisms: the impact of the gut microbiota on brain and behaviour. *Nature Reviews Neuroscience, 13*(10), 701–712.

Cryan, J. F., & O'Mahony, S. (2011). The microbiome-gut-brain axis: from bowel to behavior. *Neurogastroenterology and Motility, 23*(3), 187–192.

Dash, S., Clarke, G., Berk, M., & Jacka, F. N. (2015). The gut microbiome and diet in psychiatry: focus on depression. *Current Opinion in Psychiatry, 28*, 1–6.

De Filippo, C., Cavalieri, D., Di Paola, M., Ramazzotti, M., Poullet, J. B., Massart, S. , et al. (2010). Impact of diet in shaping gut microbiota revealed by a comparative study in children from Europe and rural Africa. *Proceedings of the National Academy of Sciences, 107*(33), 14691–14696.

Deans, E. (2016). Microbiome and mental health in the modern environment. *Journal of Physiological Anthropology, 36*(1), 1.

Del Chierico, F., Vernocchi, P., Dallapiccola, B., & Putignani, L. (2014). Mediterranean diet and health: food effects on gut microbiota and disease control. *International Journal of Molecular Sciences, 15*(7), 11678–11699.

Desbonnet, L., Garrett, L., Clarke, G., Bienenstock, J., & Dinan, T. G. (2008). The probiotic *Bifidobacteria infantis*:

an assessment of potential antidepressant properties in the rat. *Journal of Psychiatric Research, 43*(2), 164–174.

Desbonnet, L., Garrett, L., Clarke, G., Kiely, B., Cryan, J., & Dinan, T. (2010). Effects of the probiotic *Bifidobacterium infantis* in the maternal separation model of depression. *Neuroscience, 170*(4), 1179–1188.

Evans, J. M., Morris, L. S., & Marchesi, J. R. (2013). The gut microbiome: the role of a virtual organ in the endocrinology of the host. *Journal of Endocrinology, 218*(3), R37–R47.

Everson, C. A., & Toth, L. A. (2000). Systemic bacterial invasion induced by sleep deprivation. *American Journal of Physiology Regulatory, Integrative and Comparative Physiology, 278*(4), R905–R916.

Forsythe, P., & Kunze, W. A. (2013). Voices from within: gut microbes and the CNS. *Cellular and Molecular Life Sciences, 70*(1), 55–69.

Foster, J. A., & Neufeld, K.-A. M. (2013). Gut-brain axis: how the microbiome influences anxiety and depression. *Trends in Neurosciences, 36*(5), 305–312.

Funkhouser, L. J., & Bordenstein, S. R. (2013). Mom knows best: the universality of maternal microbial transmission. *PLoS Biology, 11*(8).

Gibson, G. R., Scott, K. P., Rastall, R. A., Tuohy, K. M., Hotchkiss, A., Dubert-Ferrandon, A., et al. (2010). Dietary prebiotics: current status and new definition. *Food Science & Technology Bulletin Functional Foods, 7*, 1–19.

Goehler, L. E., Gaykema, R. P., Opitz, N., Reddaway, R., Badr, N., & Lyte, M. (2005). Activation in vagal afferents and central autonomic pathways: early responses to intestinal infection with *Campylobacter jejuni*. *Brain, Behavior, and Immunity, 19*(4), 334–344.

Goehler, L. E., Park, S. M., Opitz, N., Lyte, M., & Gaykema, R. P. (2008). *Campylobacter jejuni* infection increases anxiety-like behavior in the holeboard: possible anatomical substrates for viscerosensory modulation of exploratory behavior. *Brain, Behavior, and Immunity, 22*(3), 354–366.

Grundy, D., & Schemann, M. (2006). Enteric nervous system. *Current Opinion in Gastroenterology, 22*(2), 102–110.

Hooper, L. V., Littman, D. R., & Macpherson, A. J. (2012). Interactions between the microbiota and the immune system. *Science, 336*(6086), 1268–1273.

House, P. K., Vyas, A., & Sapolsky, R. (2011). Predator cat odors activate sexual arousal pathways in brains of *Toxoplasma gondii* infected rats. *PLoS One, 6*(8).

Iyer, L. M., Aravind, L., Coon, S. L., Klein, D. C., & Koonin, E. V. (2004). Evolution of cell-cell signaling in animals: did late horizontal gene transfer from bacteria have a role? *Trends in Genetics, 20*(7), 292–299.

Jacka, F. N., Mykletun, A., & Berk, M. (2012). Moving towards a population health approach to the primary prevention of common mental disorders. *BMC Medicine, 10*, 149.

Jacka, F. N., O'Neil, A., Itsiopoulos, C., Opie, R., Cotton, S., Mohebbi, M., et al. (2016). A randomised, controlled trial of dietary improvement for adults with major depression (the "SMILES" trial). *BMC Medicine, 15*, 23. https://doi.org/10.1186/s12916-017-0791-y.

Jiang, H., Ling, Z., Zhang, Y., Mao, H., Ma, Z., Yin, Y., et al. (2015). Altered fecal microbiota composition in patients with major depressive disorder. *Brain, Behavior, and Immunity, 48*, 186–194.

Kang, S. S., Jeraldo, P. R., Kurti, A., Miller, M. E. B., Cook, M. D., Whitlock, K., et al. (2014). Diet and exercise orthogonally alter the gut microbiome and reveal independent associations with anxiety and cognition. *Molecular Neurodegeneration, 9*(1), 36.

Kassinen, A., Krogius-Kurikka, L., Mäkivuokko, H., Rinttilä, T., Paulin, L., Corander, J., et al. (2007). The fecal microbiota of irritable bowel syndrome patients differs significantly from that of healthy subjects. *Gastroenterology, 133*(1), 24–33.

Kim, K.-A., Gu, W., Lee, I.-A., Joh, E.-H., & Kim, D.-H. (2012). High fat diet-induced gut microbiota exacerbates inflammation and obesity in mice via the TLR4 signaling pathway. *PLoS One, 7*(10).

Le Chatelier, E., Nielsen, T., Qin, J., Prifti, E., Hildebrand, F., Falony, G., et al. (2013). Richness of human gut microbiome correlates with metabolic markers. *Nature, 500* (7464), 541–546.

Lewis, A. J., Galbally, M., Gannon, T., & Symeonides, C. (2014). Early life programming as a target for prevention of child and adolescent mental disorders. *BMC Medicine, 12*(1), 33.

Liu, A. H. (2015). Revisiting the hygiene hypothesis for allergy and asthma. *Journal of Allergy and Clinical Immunology, 136*(4), 860–865.

Lomax, A., & Calder, P. (2009a). Probiotics, immune function, infection and inflammation: a review of the evidence from studies conducted in humans. *Current Pharmaceutical Design, 15*(13), 1428–1518.

Lomax, A. R., & Calder, P. C. (2009b). Prebiotics, immune function, infection and inflammation: a review of the evidence. *British Journal of Nutrition, 101*(05), 633–658.

Lowry, C. A., Hollis, J. H., De Vries, A., Pan, B., Brunet, L. R., Hunt, J. R., et al. (2007). Identification of an immune-responsive mesolimbocortical serotonergic system: potential role in regulation of emotional behavior. *Neuroscience, 146*(2), 756–772.

Lydiard, R. B. (2001). Irritable bowel syndrome, anxiety, and depression: what are the links? *Journal of Clinical Psychiatry 62*(Suppl. 8), 38–45.

Maes, M., Kubera, M., & Leunis, J.-C. (2008). The gut-brain barrier in major depression: intestinal mucosal dysfunction with an increased translocation of LPS from gram negative enterobacteria (leaky gut) plays a role in the inflammatory pathophysiology of depression. *Neuroendocrinology Letters, 29*(1), 117–124.

Marshall, J. K., Thabane, M., Garg, A. X., Clark, W. F., Moayyedi, P., Collins, S. M., et al. (2010). Eight year prognosis of postinfectious irritable bowel syndrome following waterborne bacterial dysentery. *Gut, 59*(5), 605–611.

Mayer, E. A. (2011). Gut feelings: the emerging biology of gut-brain communication. *Nature Reviews Neuroscience, 12*(8), 453–466.

Mayer, E. A., Knight, R., Mazmanian, S. K., Cryan, J. F., & Tillisch, K. (2014). Gut microbes and the brain: paradigm shift in neuroscience. *Journal of Neuroscience, 34*(46), 15490–15496.

Mayer, E. A., Padua, D., & Tillisch, K. (2014). Altered brain-gut axis in autism: comorbidity or causative mechanisms? *BioEssays, 36*(10), 933–939.

Mayer, E. A., Tillisch, K., & Gupta, A. (2015). Gut/brain axis and the microbiota. *The Journal of Clinical Investigation, 125* (3), 926–938.

McBride, B., & Kelly, J. (1990). Energy cost of absorption and metabolism in the ruminant gastrointestinal tract and liver: a review. *Journal of Animal Science, 68*(9), 2997–3010.

Messaoudi, M., Lalonde, R., Violle, N., Javelot, H., Desor, D., Nejdi, A., et al. (2011). Assessment of psychotropic-like properties of a probiotic formulation (*Lactobacillus helveticus* R0052 and *Bifidobacterium longum* R0175) in rats and human subjects. *British Journal of Nutrition, 105*(05), 755–764.

Messaoudi, M., Violle, N., Bisson, J.-F., Desor, D., Javelot, H., & Rougeot, C. (2011). Beneficial psychological effects of a probiotic formulation (*Lactobacillus helveticus* R0052 and *Bifidobacterium longum* R0175) in healthy human volunteers. *Gut Microbes, 2*(4), 256–261.

Miller, A. H., & Raison, C. L. (2016). The role of inflammation in depression: from evolutionary imperative to modern treatment target. *Nature Reviews Immunology, 16*(1), 22–34.

Naseribafrouei, A., Hestad, K., Avershina, E., Sekelja, M., Linløkken, A., Wilson, R., et al. (2014). Correlation between the human fecal microbiota and depression. *Neurogastroenterology and Motility, 26*(8), 1155–1162.

O'Hara, A. M., & Shanahan, F. (2006). The gut flora as a forgotten organ. *EMBO Reports, 7*(7), 688–693.

O'Mahony, L., McCarthy, J., Kelly, P., Hurley, G., Luo, F., Chen, K., et al. (2005). *Lactobacillus* and *Bifidobacterium* in irritable bowel syndrome: symptom responses and relationship to cytokine profiles. *Gastroenterology, 128* (3), 541–551.

O'Mahony, S. M., Marchesi, J. R., Scully, P., Codling, C., Ceolho, A.-M., Quigley, E. M., et al. (2009). Early life stress alters behavior, immunity, and microbiota in rats: implications for irritable bowel syndrome and psychiatric illnesses. *Biological Psychiatry, 65*(3), 263–267.

Opie, R., Itsiopoulos, C., Parletta, N., Sanchez-Villegas, A., Akbaraly, T., Ruusunen, A., et al. (2017). Dietary recommendations for the prevention of depression. *Nutritional Neuroscience, 20*(3), 161–171.

Parletta, N., Aljeesh, Y., & Baune, B. T. (2016). Health behaviours, knowledge, life satisfaction and wellbeing in people with mental illness across four countries and comparisons with normative sample. *Frontiers in Psychiatry, 7*, 1–8. https://doi.org/10.3389/fpsyt.2016.00145.

Parletta, N., Milte, C. M., & Meyer, B. (2013). Nutritional modulation of cognitive function and mental health. *Journal of Nutritional Biochemistry, 24*(5), 725–743.

Parletta, N., Zarnowiecki, D., Cho, J., Wilson, A., Bogomolova, S., Villani, V., et al. (2017). A Mediterranean-style dietary intervention supplemented with fish oil improves diet quality and mental health in people with depression: a randomized controlled trial (HELFIMED). *Nutritional Neuroscience*. https://doi.org/10.1080/1028415X.2017.1411320.

Pärtty, A., Kalliomaki, M., Wacklin, P., Salminen, S., & Isolauri, E. (2015). A possible link between early probiotic intervention and the risk of neuropsychiatric disorders later in childhood: a randomized trial. *Pediatric Research, 77*(6), 823–828.

Pelaseyed, T., Bergström, J. H., Gustafsson, J. K., Ermund, A., Birchenough, G. M., Schütte, A., et al. (2014). The mucus and mucins of the goblet cells and enterocytes provide the first defense line of the gastrointestinal tract and interact with the immune system. *Immunological Reviews, 260* (1), 8–20.

Prescott, S. L. (2013). Early-life environmental determinants of allergic diseases and the wider pandemic of inflammatory noncommunicable diseases. *Journal of Allergy and Clinical Immunology, 131*(1), 23–30.

Raison, C. L., & Miller, A. H. (2011). Is depression an inflammatory disorder? *Current Psychiatry Reports, 13*(6), 467–475.

Romijn, A. R., & Rucklidge, J. J. (2015). Systematic review of evidence to support the theory of psychobiotics. *Nutrition Reviews, 73*(10), 675–693.

Romijn, J. A., Corssmit, E. P., Havekes, L. M., & Pijl, H. (2008). Gut-brain axis. *Current Opinion in Clinical Nutrition and Metabolic Care, 11*(4), 518–521.

Rook, G. A., & Lowry, C. A. (2008). The hygiene hypothesis and psychiatric disorders. *Trends in Immunology, 29*(4), 150–158.

Sanders, M. E. (2011). Impact of probiotics on colonizing microbiota of the gut. *Journal of Clinical Gastroenterology, 45*, S115–S119.

Sarkar, A., Lehto, S. M., Harty, S., Dinan, T. G., Cryan, J. F., & Burnet, P. W. (2016). Psychobiotics and the manipulation of bacteria-gut-brain signals. *Trends in Neurosciences, 39* (11), 763–781.

Savignac, H., Kiely, B., Dinan, T., & Cryan, J. (2014). Bifidobacteria exert strain-specific effects on stress-related behavior and physiology in BALB/c mice. *Neurogastroenterology and Motility, 26*(11), 1615–1627.

Sekirov, I., Russell, S. L., Antunes, L. C. M., & Finlay, B. B. (2010). Gut microbiota in health and disease. *Physiological Reviews, 90*(3), 859–904.

Severance, E. G., Tveiten, D., Lindström, L. H., Yolken, R. H., & Reichelt, K. L. (2016). The gut microbiota and the emergence of autoimmunity: relevance for psychiatric disorders. *Current Pharmaceutical Design, 22*, 6076–6086.

Sharon, G., Segal, D., Zilber-Rosenberg, I., & Rosenberg, E. (2011). Symbiotic bacteria are responsible for diet-induced mating preference in Drosophila melanogaster, providing support for the hologenome concept of evolution. *Gut Microbes*, *2*(3), 190–192.

Smith, P. A. (2015). The tantalizing links between gut microbes and the brain. *Nature*, *526*(7573), 312–314.

Steenbergen, L., Sellaro, R., van Hemert, S., Bosch, J. A., & Colzato, L. S. (2015). A randomized controlled trial to test the effect of multispecies probiotics on cognitive reactivity to sad mood. *Brain, Behavior, and Immunity*, *48*, 258–264.

Sudo, N., Chida, Y., Aiba, Y., Sonoda, J., Oyama, N., Yu, X. N., et al. (2004). Postnatal microbial colonization programs the hypothalamic-pituitary-adrenal system for stress response in mice. *The Journal of Physiology*, *558*(1), 263–275.

Tillisch, K., Labus, J., Kilpatrick, L., Jiang, Z., Stains, J., Ebrat, B., et al. (2013). Consumption of fermented milk product with probiotic modulates brain activity. *Gastroenterology*, *144*, 1394–1401.

Vos, T., Barber, R. M., Bell, B., Bertozzi-Villa, A., Biryukov, S., Bolliger, I., et al. (2015). Global, regional, and national incidence, prevalence, and years lived with disability for 301 acute and chronic diseases and injuries in 188 countries, 1990–2013: a systematic analysis for the Global Burden of Disease Study 2013. *The Lancet*, *386* (9995), 743.

Zarnowiecki, D., Cho, J., Wilson, A. M., Bogomolova, S., Villani, A., Itsiopoulos, C., et al. (2016). A 6-month randomised controlled trial investigating effects of Mediterranean-style diet and fish oil supplementation on dietary behaviour change, mental and cardiometabolic health and health-related quality of life in adults with depression (HELFIMED): study protocol. *BMC Nutrition*, *2*, 52. https://doi.org/10.1186/s40795-40016-40095-40791.

Zhang, C., Zhang, M., Wang, S., Han, R., Cao, Y., Hua, W., et al. (2010). Interactions between gut microbiota, host genetics and diet relevant to development of metabolic syndromes in mice. *The ISME Journal*, *4*(2), 232–241.

Zunszain, P. A., Hepgul, N., & Pariante, C. M. (2013). Inflammation and depression. *Current Topics in Behavioral Neurosciences*, *14*, 135–151.

14

Childhood Trauma and Adulthood Immune Activation

Maria A. Nettis*, *Valeria Mondelli**,†

*Institute of Psychiatry, Psychology and Neuroscience, King's College London, London, United Kingdom
†National Institute for Health Research (NIHR) Mental Health Biomedical Research Centre at South London and Maudsley NHS Foundation Trust and King's College London, London, United Kingdom

INTRODUCTION: CHILDHOOD TRAUMA AND DEPRESSION

Childhood adversities, including childhood maltreatment, socioeconomic disadvantage, and social isolation, have been increasingly found to be associated with age-related diseases and psychiatric conditions, including depression (Danese & McEwen, 2012; Khan, McCormack, et al., 2015). Unfortunately, childhood maltreatment is highly prevalent in our society, although it is often unreported because of stigma and societal acceptance (Pinheiro, 2006). The WHO Consultation on Child Abuse Prevention, held in Geneva in 1999, identified four types of child maltreatment: sexual abuse (CSA), physical abuse (CPA), emotional and psychological abuse (CEA), and neglect. In particular, neglect, the failure to provide for all aspects of the child's well-being, is considered the most frequent form with 78.5% of children exposed in the general population; the second most frequent, reported in 17.6%, is physical abuse, defined as the use of physical force that harms the child's health, survival, development, or dignity (Dubowitz, Pitts, et al.,

2004; Goldman, Salus, et al., 2003). Approximately 8% of males and 20% of females universally experienced childhood sexual abuse (the involvement in sexual activity that a child is unable to give consent to or is not developmentally prepared for), with the highest prevalence rate in Africa (34.4%), while Europe, America, and Asia had prevalence rates of 9.2%, 10.1%, and 23.9%, respectively (Verdolini, Attademo, et al., 2015). Finally, emotional abuse is the failure to provide children with a supportive environment, and it seems to have higher prevalence than sexual and physical abuse, but it is more difficult to measure and quantify (Holmes & Slap, 1998).

In the last decades, with the increased recognition of the role of the environment in susceptibility to psychiatric disorders, scientific interest has progressively focused on the study of early traumatic events as potential risk factors.

In this context, Nanni, Uher, et al. (2012) showed that maltreated children are twice more likely to develop recurrent and persistent depressive episodes than those without a history of maltreatment; this was consistent with what was reported by a recent study (Mandelli, Petrelli, et al., 2015)

showing that early adversities overall increase the risk of developing depressive symptoms with odds ratio (OR) ranging from 2.00 to 3.00. Childhood adversities not only appear to increase the lifetime risk of depression but also have been shown to affect the course of illness and treatment outcome. Depressed individuals with history of childhood maltreatment have a poorer response to pharmacological treatment compared with nonmaltreated patients (OR = 1.26, 95% CI 1.01–1.56) (Nanni, Uher, et al., 2012).

Few studies have also explored whether the type or the timing of childhood maltreatment is important in increasing susceptibility to depression. The Adverse Childhood Experiences (ACE) Study (Dube, Anda, et al., 2001; Dube, Felitti, et al., 2003) suggested that the number of types of maltreatment has a cumulative effect on risk for depression. More recent literature has proposed that susceptibility to develop depression may be better explained by a complex interaction between type and timing of maltreatment. The strongest support for this hypothesis comes from neuroimaging studies. Exposure to specific types of maltreatment, such as parental verbal abuse or witnessing domestic violence, appears to be associated with alterations in brain white matter tracts connecting frontal, temporal, and limbic regions (Choi, Jeong, et al., 2009, 2012). Furthermore, several studies looking at gray matter volume across a lifetime indicated early (3–5 years of age) and late (adolescence) windows of vulnerability (Andersen, Tomada, et al., 2008; Pechtel & Pizzagalli, 2011), when the brain may be more sensitive to exposure to childhood maltreatment.

In relation to the type of adversities, most research has focused on sexual and physical abuse as major risk factors for depression. In one community-based study of approximately 2000 women of various socioeconomic levels, those with a history of childhood physical or sexual abuse had more severe symptoms of depression and anxiety and were more likely to abuse drugs or attempt suicide compared with women without such history (McCauley, Kern, et al., 1997).

Abused subjects had also a greater susceptibility to anxiety disorders, such as panic disorder, posttraumatic stress disorder (PTSD), and generalized anxiety disorder (Nemeroff, 2004). Although sexual and physical abuse have been more frequently investigated, neglect and emotional abuse have been gradually more recognized, and nowadays, they seem to confer higher and more specific risk for depressive states in adulthood, particularly in females (Mandelli et al., 2015). In this context, a recent meta-analysis by Infurna, Reichl, et al. (2016) investigating the specific association between depression and different types of maltreatment in studies using the same instrument (Childhood Experience of Care and Abuse interview (CECA)) found that psychological abuse and neglect were most strongly associated with later depression.

In relation to the timing of maltreatment, Khan et al. (2015) not only provided evidence of sensitive exposure periods when maltreatment maximally impacts risk of depression but also suggested a gender-specific predisposition. In their study on 560 young adults, the most important predictor of lifetime history of depression was emotional abuse in males and peer emotional abuse in females at 14 years of age. Suicidal ideation was predicted by parental verbal abuse at age 5 in males and sexual abuse at age 18 in females. The findings of a sensitive time of exposure have important implication for prevention and treatment of depression.

PHYSIOLOGICAL EFFECTS OF CHILDHOOD ADVERSITIES: FROM STRESS TO IMMUNE ACTIVATION

Although the association between early-life adversities and onset of depression has been increasingly acknowledged, the biological mechanisms underlying this association remain partly unclear. A number of interrelated factors may be implicated (McCrory, De Brito, et al., 2010), such as structural and functional remodeling of limbic and cortical brain systems (Eiland & Romeo, 2013;

Fernald & Maruska, 2012; Pascual-Leone, Amedi, et al., 2005), exaggerated or prolonged response of the sympathetic nervous system (SNS) or of the hypothalamic-pituitary-adrenal (HPA) axis (Boyce, Sokolowski, et al., 2012; Gunnar & Quevedo, 2007), genetic and epigenetic processes (Slavich & Cole, 2013), DNA methylation and histone modification (Danese & McEwen, 2012; Vialou, Feng, et al., 2013), and immune system dynamics, such as increased pro-inflammatory cytokine responses to challenge and reduced immune cell sensitivity to antiinflammatory signals (Miller, Chen, et al., 2011). What seems to be clear is the key role of childhood adversities in predisposing to major depression in adulthood and that this process may be mediated by inflammation. In fact, Grosse, Ambree, et al. (2016) have recently found a linear relationship between childhood sexual abuse and increased pro-inflammatory cytokine levels in patients with major depressive disorder (MDD). Interestingly, more recent stressful life events were not related to these inflammatory markers. An explanation of the underpinning mechanism has been provided in a review by Danese and McEwen (2012). They suggested that adverse childhood experiences are associated with enduring changes in an allostatic system that persist throughout adult life. Indeed, because the nervous, the endocrine, and the immune systems are highly integrated, activation of one of these systems commonly triggers responses in the others.

The HPA axis, the main biological system involved in the stress response, seems to play a critical role in the biological embedding of childhood trauma. Chronic and repeated exposure to stress can lead to HPA axis hyperactivity, which results in increased levels of cortisol, the final hormone produced during the activation of the HPA axis. Cortisol exerts its activity by binding to glucocorticoid receptors; persistent elevated basal cortisol levels can be due to an impaired cortisol ability to exert physiological effects including the negative feedback on the HPA axis itself and antiinflammatory effects at peripheral level (Pariante, 2006; Pariante &

Miller, 2001). It has been also postulated that impaired glucocorticoid receptor (GR) sensitivity (glucocorticoid resistance) may occur as a result of chronic stress, then leading to prolonged/increased immune activation (Miller, Pariante, et al., 1999). Both HPA axis hyperactivity and increased immune activation are present in maltreated children (Danese & McEwen, 2012) and persist in adulthood possibly affecting subsequent responses to stressful situations.

Similarly, the activation of the SNS in response to stress also results in a cascade of signals that can affect the immune system. The SNS regulates pro-inflammatory cytokine production by releasing the neurotransmitter norepinephrine into peripheral tissues. By binding to β-adrenergic (Irwin & Cole, 2011; Nance & Sanders, 2007; Sandiego, Gallezot, et al., 2015) and α-adrenergic receptors (Grisanti, Perez, et al., 2011; Huang, Zhang, et al., 2012), norepinephrine regulates the transcription of pro-inflammatory cytokine genes interleukin (IL)1 and tumor necrosis factor (TNF)-α, leading to systemic immune activation (Cole, Arevalo, et al., 2010; Grebe, Takeda, et al., 2010).

Evidence of a relationship between childhood adversities and increased immune activation, which persists into adulthood, also raises questions about putative underlying mechanisms that may involve epigenetic regulation of gene expression. Early trauma has been proved to lead to greater methylation of the GR, which is associated with reduced GR function and impaired negative feedback of the HPA axis. In addition, early trauma also causes greater demethylation and thereby greater activation of FKBP5, a heat-shock protein that binds and inhibits the cytosolic GR that is also a crucial regulator of immune response. In particular, lower expression and function of GR may lead to an increased immune activation, which in turn contributes to maintain GR impairment and resistance (Slavich & Irwin, 2014).

Levine, Cole, et al. (2015) recently tested if exposure to childhood traumatic events is associated with later elevated expression of pro-inflammatory genes. They calculated a composite gene expression score of the pro-inflammatory IL1β, IL8, and cyclooxygenase 2 (PTGS2) in adulthood, and they found it was significantly associated with childhood trauma; in addition, this association may exacerbate the effect of some adverse events in adulthood, such as low socioeconomic status, on inflammatory gene expression.

Boeck, Koenig, et al. (2016) have more recently suggested another possible biological mechanism linking childhood maltreatment to adulthood inflammation involving mitochondrial function and oxidative stress. They analyzed pro-inflammatory biomarkers (levels of C-reactive protein, cytokine secretion by peripheral blood mononuclear cells (PBMCs) *in vitro*, PBMC composition, and lysophosphatidylcholine levels), serum oxidative stress levels (arginine: citrulline ratio, L-carnitine, and acetylcarnitine levels), and mitochondrial function (respiratory activity and density of mitochondria in PBMCs) in peripheral blood samples collected from 30 women with different degrees of childhood maltreatment experiences including abuse and neglect. Exposure to childhood maltreatment was found to be associated with an increased ROS production, higher levels of oxidative stress, and an increased mitochondrial activity in a dose-response relationship. Moreover, the increase in mitochondrial activity and ROS production was positively associated with the release of pro-inflammatory cytokines by PBMCs. Even by looking at a different underlying biological mechanism, these findings again emphasize how the experience of maltreatment during childhood might have life-long consequences for physical and mental health.

Evidence Linking Early Stress and Immune Activation in Early Life/Childhood

Given the link between stress and immune activation, various studies have focused on the effect of early stress on the production of acute-phase proteins that modulate inflammation, like C-reactive protein and pro-inflammatory cytokines, such as IL-1, IL-6, and TNF-α (Coelho, Viola, et al., 2014). Pro-inflammatory cytokines together coordinate a variety of cell functions that stimulate and enhance immune activation. In particular, IL-1, IL-6, and TNF-α promote the differentiation of lymphocytes called *cytotoxic T cells*, which kill pathogens that are introduced into the body during physical wounding. These cytokines also promote increased vascular permeability and cellular adhesion, which allows immune cells to leave the blood vessels and migrate to tissues where they can neutralize or eliminate pathogens (Dhabhar, Saul, et al., 2012).

Additional data on animal models showed immune responses being compromised following removal from the mother in young monkeys (Coelho et al., 2014). Consistently with these findings, Danese and McEwen (2012) reported evidence of innate immune system changes in young people exposed to adverse psychosocial experiences: they observed elevated levels of C-reactive protein (CRP) in 12-year-old children exposed to physical abuse and experiencing current depression. Also, a recent study (Schmeer & Yoon, 2016) provides new evidence that the family socioeconomic status (SES) is associated with low-grade inflammation in children (CRP levels) and that this association may be particularly strong during early and mid-childhood. According to this view, SES may be another way leading to chronic stress and impacting children's physiology with implications for later health inequalities.

Evidence Linking Early Stress and Immune Activation in Adulthood

Immune dysregulation often persists later in life: studies on adults with history of childhood adversities have been explored in a systematic literature review (Coelho et al., 2014) that highlighted a significant association between early-life stress and later production of IL-6, which is also known to be involved in the regulation of metabolic, regenerative, and neural processes. Firstly, animal models in adult rats have demonstrated that maternal separation is associated with elevated TNF-α levels in periphery and cerebrospinal fluid, as well as in prefrontal and hippocampal brain regions. Further supporting these findings, we have recently conducted a meta-analysis of all studies investigating history of childhood maltreatment and inflammatory markers in adulthood (Baumeister, Akhtar, et al., 2016). The meta-analysis included studies on both general and clinical population (with psychiatric and physical disorders) and focused on CRP, IL-6, and TNF-α as these are the most examined inflammatory markers in psychiatric research. Results showed a significant association between childhood trauma and inflammatory markers in adulthood, with greatest effect size for TNF-α ($z = 0.20$, 95% CI$= 0.10$–0.29) followed by IL-6 ($z = 0.09$, 95% CI$= 0.04$–0.15) and then CRP ($z = 0.08$, 95% CI$= 0.04$–0.11). Another important finding from this study was that different types of trauma exposure impacted differentially on the inflammatory markers: physical and sexual abuse was associated with significant increased TNF-α and IL-6, but not CRP. Conversely, CRP was primarily related to parental absence during early development. How different types of trauma may impact different aspects of immune activity remains unknown; however, it has been suggested that factors such as context, time, and duration of stress exposure may interact with individual trauma type in modulating immune response.

FROM IMMUNE ACTIVATION TO DEPRESSION

Is Immune Activation in Depression Due to Early-Life Stress?

Early-life stress has been associated with increased immune activation in the pathogenesis of depression. Danese and colleagues demonstrated not only that elevated CRP blood levels in adults are significantly associated with childhood maltreatment but also that such association is particularly strong in individuals who developed depression later in life (Danese, Moffitt, et al., 2008, 2009). Furthermore, depressed subjects with history of early-life stress show an increased inflammatory response when reexposed to an acute psychological stress in adulthood, with increased IL-6 and DNA binding (measured in PBMCs) of the key pro-inflammatory transcription factor, nuclear factor kappa-light-chain-enhancer of activated B cells (NF-κB) in PBMCs (Pace, Mletzko, et al., 2006). Depressed adults who experienced severe forms of early-life stress (e.g., maternal rejection, harsh discipline, and physical or sexual abuse) are 1.48 times more likely to have clinically high levels of CRP (>3 mg/L) than depressed adults who did not experience these severe forms of early-life stress (Danese et al., 2008).

Interestingly, although depression is also strongly associated with high CRP levels, this association is no longer significant when analyses are adjusted for the effects of early-life stress (Danese et al., 2008). In a recent study that followed up adolescent women at elevated risk for depression over 2.5 years, individuals with a history of more common forms of early-life stress, such as low SES or parental separation, had greater increases in both IL-6 and CRP when becoming depressed than their counterparts without a history of early-life adversities. Moreover, in this study, while in adolescents who did not have a history of early-life stress, CRP levels decreased as their depressive symptoms abated, in those *with* a history of early-life stress, CRP remained elevated even when recovered from depression (Miller & Cole, 2012). Of note, the association between stress, inflammation, and depression may be detectable at a relatively young age. A recent longitudinal study found that children who were exposed to physical maltreatment before age 10 and depressed at age 12 exhibited significantly higher levels of CRP relative to children with depression only, maltreatment only, and neither depression nor maltreatment (Danese, Caspi, et al., 2011).

Considered together, these studies suggest that exposure to early-life stress may highly and possibly uniquely contribute to the increased immune activation, which persists later in life and can be detected in patients with depression. One of the suggested mechanisms underlining such process is that childhood adversities lead to exaggerated biological responses to stress (Boyce & Ellis, 2005), which in turn result in the increased development of a pro-inflammatory phenotype (see Miller et al., 2011) and an increased risk to develop depression. According to Danese and McEwen (2012), maltreated children develop acute responses to face a threatening environment and to cope with it, by involving the interconnected neural, metabolic, and immune system. For instance, chronic elevation in cortisol levels could support the increased metabolic demands by mobilizing stored energy, and stress-related immune activation may be an adaptive strategy to prepare children to face possible physical injuries. Nevertheless, these strategies could be detrimental in the longer term for children development, by leading to important metabolic imbalances and nonspecific immune responses, even when the surrounding environment in adulthood is not anymore hostile, but favorable. Therefore, stress during critical periods of development may have lasting biological effects throughout late life and contribute to exacerbate maladaptive responses to future events.

Inflammation, Depression and Physical Comorbidities

Of note, depression is often associated with several somatic conditions, which have immune activation as common biological basis in their pathophysiology. For instance, in a prospective study of >1200 young adults over a 21-year period, asthma in adolescence and young adulthood was associated with a significantly increased likelihood of being diagnosed with MDD (OR=1.7, 95% confidence interval (CI) (1.3, 2.3)) (Goodwin, Fergusson, et al., 2004). Moreover, several studies have shown that individuals with rheumatoid arthritis and inflammatory bowel disease are two to three times more likely to have major depression than the general population (Graff, Walker, et al., 2009). Also, a recent meta-analysis revealed that across nine prospective cohort samples, the pooled adjusted odds of metabolic syndrome predicting future risk for depression were 1.49 (95% CI (1.19, 1.89)) (Pan, Keum, et al., 2012). Likewise, risk of developing depression has been found to be twice higher in individuals with coronary heart disease and three times as likely in persons with congestive heart failure compared with the general population. Interestingly, among patients with coronary heart disease, those with comorbid depression have been found to have elevated levels of inflammation in the context of HPA axis hypoactivity, GR resistance, and increased activation of the kynurenine pathway (Nikkheslat, Zunszain, et al., 2015). Finally, chronic pain, often highly comorbid with depression, is associated with elevated levels of inflammation, and such inflammation drives hyperalgesia or increased sensitivity to pain (Slavich & Irwin, 2014).

HOW DOES IMMUNE ACTIVATION LEAD TO DEPRESSION?

Peripheral and Central Inflammation

Supporting the role of inflammation in linking early trauma and depression in adulthood, increase in IL-6 and CRP prospectively predicts the development of depressive symptoms (Gimeno, Kivimaki, et al., 2009; Wium-Andersen, Orsted, et al., 2013). Moreover, in some studies, antidepressant treatments have been shown to decrease levels of the pro-inflammatory cytokines IL-1β, IL-2, and IL-6 along a reduction in depressive symptoms (Baumeister, Ciufolini, et al., 2016; Slavich & Irwin, 2014).

An interesting aspect of peripheral inflammation effect on the brain is the discovered ability of cytokines to communicate with the brain and to alter neural activity through different pathways: by active transport, by involving macrophage-like cells residing in circumventricular organs, or by the release of second messengers, which in turn stimulate local cytokines. These, in turn, can cross the blood-brain barrier and determine microglia activation. Microglia are myeloid cells that provide the main form of adaptive immune response in the central nervous system (CNS); they influence cell proliferation, and survival depends on the inflammatory state (Mondelli, Vernon, et al., 2017). The above described cellular and molecular mechanisms allow the brain to actively monitor the internal inflammatory state. In response to harmful stimuli, microglial cells undergo a number of changes (Walker, Beynon, et al., 2014) including the production of pro-inflammatory cytokines and the expression of several cell surface antigens.

Calcia, Bonsall, et al. (2016) identified 18 preclinical studies showing that psychosocial stress increases expression of the microglial marker Iba-1 in the hippocampus; in the prefrontal cortex; and, in many cases, in multiple brain areas. Results from these studies suggest that exposure to stress can lead to significant increases in microglial activity in humans. Consequently, changes in cortical microglia may in turn be associated with structural and functional changes in the brain that predispose individuals to mental illnesses (Couch, Anthony, et al., 2013; Reus, Fries, et al., 2015).

There are many evidences supporting the role of microglia in depression. Several preclinical studies show an association between greater microglial activity and behavioral outcomes resembling depressive symptoms, in animal models. For instance, a combined model of social defeat, restraint, and tail suspension (Couch et al., 2013) leads to a pro-inflammatory profile and heightened microglial activity associated with the development of stress-induced anhedonia in mice susceptible to depression following exposure to stress. Similarly, social isolation may be used as a stress paradigm in weaning, adolescent, or adult rats and has been associated with reduced preference for sucrose (Krishnan & Nestler, 2011), considered a mode of depressive-like behavior. Thus, the animal models used in these studies provide a useful indicator that a range of psychosocial stressors are capable of increasing microglial activity. Furthermore, the antibiotic minocycline has been found to reduce depressive-like behavioral deficits and reduce microglial activity in mice (Burke, Kerr, et al., 2014; Henry, Huang, et al., 2008; Hinwood, Tynan, et al., 2013), further supporting a mechanistic link between microglial activity and depressive symptoms.

Later, postmortem studies in humans have reported increased microglia activation in prefrontal cortex of depressed patients (Torres-Platas, Cruceanu, et al., 2014). In patients, to determine whether neuroinflammation occurs in MDD, positron-emission tomography may be applied to measure translocator protein (TSPO) binding in vivo. TSPO is an 18 kDa protein located on outer mitochondrial membranes in microglia; increased expression of TSPO has been suggested to occur when microglial cells are activated during neuroinflammation. Although the first study (Hannestad, DellaGioia, et al., 2013) conducted on 10 patients with mild to moderate MDD showed no differences with healthy controls, a second study on 20 patients with a major depressive episode in MDD (Setiawan, Wilson, et al., 2015) showed that TSPO V_T was increased, on average, by 30% in the prefrontal cortex, the anterior cingulate cortex, and the insula, in patients compared with healthy subjects. Moreover, greater TSPO V_T (so, greater microglia activation) in the anterior cingulate cortex correlated with greater depression severity. These findings suggest a key role of microglia in the development of depression and its possible role as biomarker of psychiatric disorders. Several ongoing studies are further investigating this aspect.

Mechanisms Linking Inflammation to Depression

Various mechanisms have been suggested linking inflammation to the development of depressive symptoms (Fig. 1). One of the main mechanisms appears to involve the kynurenine pathway, which is involved in the metabolism of tryptophan, a serotonin precursor. The enzyme that metabolizes tryptophan to kynurenine (KYN), indoleamine 2,3-dioxygenase (IDO, Fig. 1), is potently upregulated by proinflammatory cytokines. Furthermore, when a physical or psychological stress occurs, the increased glucocorticoid secretion stimulates the activity of the liver enzyme tryptophan 2,3-dioxygenase TDO (Knox, 1951; Salter & Pogson, 1985) that, as well as IDO, degrades tryptophan into kynurenine. This activation of TDO by stress hormones induces a further increase in the formation of KYN and other downstream metabolites. As a consequence, by breaking down tryptophan into kynurenine, levels of serotonin are dramatically reduced. Given that serotonin is intimately involved in regulating mood, motivation, and behavior, cytokine-related tryptophan depletion may well represent a key event in the pathogenesis of depression.

Moreover, KYN, once converted into quinolinic acid, may lead to neurotoxicity through the activation of the glutamatergic system (Myint & Kim, 2014). In fact, the quinolinic acid is a strong NMDAR agonist, and its accumulation could result in excitotoxicity to the neurons and disturb the glutamatergic neurotransmission (Bender & McCreanor, 1985). Interestingly, microglia are the predominant cells expressing the enzyme kynurenine 3-monooxygenase (KMO) for the generation of neurotoxic kynurenine metabolites. As already mentioned, inflammation can alter neuronal functioning through a dysregulation of microglia, which in turn affect neurogenesis. Specifically, activated microglia not only promote the release of cytokines and chemokines but also reduce levels of serotonin and generate oxidative stress molecules altering

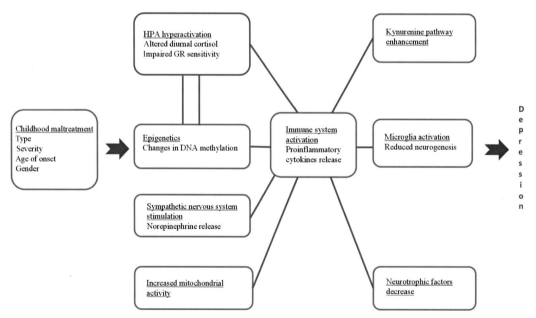

FIG. 1 Suggested mechanisms underpinning the pathway from early stress to inflammation and, then, to depression.

local metabolic processes (Madeeh Hashmi, Awais Aftab, et al., 2013).

Another mechanism through which cytokines may affect the brain is their ability in hindering neuronal synaptic plasticity by decreasing levels of neurotrophic factors such as the brain-derived neurotrophic factor (BDNF) and reducing neurogenesis in specific brain areas, such as the hippocampus (Madeeh Hashmi et al., 2013). Of note, decreased levels of BDNF have been reported to be a robust biological marker of major depression. In support of this, serotonin-specific reuptake inhibitors (SSRIs—commonly used antidepressants) have proved to increase neurogenesis in the hippocampus and potentiate the effects of BDNF. In this context, several studies have investigated the role of inflammation in affecting specifically the hippocampus, which in turn has important implication in learning and memory.

CHILDHOOD TRAUMA AND NOVEL TREATMENTS AND INTERVENTIONS TARGETING IMMUNE SYSTEM

Immune activation in depression has also important implications for treatment strategies. Not only in some cases antidepressants have proved to reduce cytokine levels in depressed patients (Basterzi, Aydemir, et al., 2005), but also patients resistant to antidepressant therapy show increased inflammatory markers than responders (Uher, Tansey, et al., 2014).

Cattaneo, Ferrari, et al. (2016) have done a first step toward translation of such evidences into clinical practice. They measured the *absolute* mRNA values of macrophage migration inhibitory factor and IL1β in 74 patients from a randomized controlled trial comparing escitalopram with nortriptyline, and then, they were able to calculate mRNA value cutoffs that best discriminated between responders and nonresponders after 12-week treatment with antidepressants.

As a consequence of these results, a number of trials are investigating the effect of antiinflammatory medications, such as cyclooxygenase inhibitors (Gamble-George, Baldi, et al., 2016) and minocycline, in treatment-resistant depressed patients. There is now also evidence that cognitive- and meditation-based interventions (such as mindfulness) that are known to alleviate depression may also lead to reduction in levels of inflammation (Walsh, Eisenlohr-Moul, et al., 2016).

Interestingly, some studies emphasize the need of screening for childhood trauma in routine clinical management of depression, to identify those patients that may not benefit from standard first line of antidepressants and may require additional therapy to better address the impact of trauma. For instance, a multisite, randomized clinical trial on 1008 patients with depression in treatment with escitalopram, sertraline, and venlafaxine (Williams, Debattista, et al., 2016) showed that patients with history of childhood abuse occurring at <7 years of age had poorer outcome after 8 weeks of antidepressants. In addition, the abuses occurring between 4 and 7 years of age predicted the poorest outcome following the treatment with sertraline.

Among possible treatment approaches, the importance of early intervention for maltreated children warrants even further attention in order to prevent the biological embedding of disease. Findings from a study of Pace, Negi, et al. (2013) suggest that engagement with cognitively based compassion training (CBCT) may buffer the detrimental effects of early-life adversities on inflammation in adolescents placed in foster care. Consistently with a previous study of the effects of CBCT on inflammatory biomarkers (Pace, Negi, et al., 2009), no difference was observed in adolescents treated with CBCT sessions in comparison with an untreated group in the same system of foster care. However, in the CBCT group, reductions in morning salivary CRP concentrations across the 6-week study period were associated with the number of CBCT practice sessions during the study ($r_s = -0.58$, $P = .002$). Thus, the degree of engagement with the practice—and perhaps the act of practicing itself—appeared to be influencing inflammation levels.

Furthermore, according to a review on clinical interventions in childhood maltreatment (Gonzalez, 2013), the dysregulation of HPA axis activity described in maltreated children may be modified by improving the sensitivity and responsiveness of caregivers or through placements in environments fostering healthy and positive relationships. Family-based interventions with infants, toddlers, and preschoolers in foster care, such as the Attachment and Biobehavioral Catch-up program and the Early Intervention Foster Care program, proved to be effective in increasing morning cortisol levels and diurnal variation in cortisol levels (Dozier, Peloso, et al., 2008; Fisher, Van Ryzin, et al., 2011). Another study found that within maltreated children, infants whose mothers were randomly assigned to a child-parent psychotherapy intervention or to a psychoeducational parenting intervention showed increased levels of morning cortisol levels over time compared with infants whose mothers were receiving standard community services (Cicchetti, Rogosch, et al., 2011). Interestingly, children's morning cortisol levels in the intervention group became indistinguishable from children from nonmaltreating families, whereas infants in the community standard care group progressively exhibited lower cortisol levels over time. These findings were also maintained at a one-year follow-up. Despite these evidences, the impact of interventions on biological systems is relatively understudied and certain types of maltreatment, such as childhood neglect.

In conclusion, there is evidence suggesting that psychosocial or cognitively based interventions have the potential to circumvent the biological embedding of childhood maltreatment. However, future research on biomarkers and outcome measures is needed to provide new strategy for interventions in reversing the negative impact on biological systems associated with early-life adversities.

Acknowledgments

This research has been supported by the National Institute for Health Research (NIHR) Mental Health Biomedical Research Centre at South London and Maudsley NHS Foundation Trust and King's College London. The views expressed are those of the authors and not necessarily those of the NHS, the NIHR, or the Department of Health. This research has also been supported by an ECNP Young Scientist Award and a Starter Grant for Clinical Lecturers from the Academy of Medical Sciences, the Wellcome Trust, and the British Heart Foundation to V. Mondelli.

References

Andersen, S. L., Tomada, A., et al. (2008). Preliminary evidence for sensitive periods in the effect of childhood sexual abuse on regional brain development. *The Journal of Neuropsychiatry and Clinical Neurosciences, 20*(3), 292–301.

Basterzi, A. D., Aydemir, C., et al. (2005). IL-6 levels decrease with SSRI treatment in patients with major depression. *Human Psychopharmacology, 20*(7), 473–476.

Baumeister, D., Akhtar, R., et al. (2016). Childhood trauma and adulthood inflammation: a meta-analysis of peripheral C-reactive protein, interleukin-6 and tumour necrosis factor-alpha. *Molecular Psychiatry, 21*(5), 642–649.

Baumeister, D., Ciufolini, S., et al. (2016). Effects of psychotropic drugs on inflammation: consequence or mediator of therapeutic effects in psychiatric treatment? *Psychopharmacology, 233*(9), 1575–1589.

Bender, D. A., & McCreanor, G. M. (1985). Kynurenine hydroxylase: a potential rate-limiting enzyme in tryptophan metabolism. *Biochemical Society Transactions, 13*(2), 441–443.

Boeck, C., Koenig, A. M., et al. (2016). Inflammation in adult women with a history of child maltreatment: the involvement of mitochondrial alterations and oxidative stress. *Mitochondrion, 30*, 197–207.

Boyce, W. T., & Ellis, B. J. (2005). Biological sensitivity to context: I. An evolutionary-developmental theory of the origins and functions of stress reactivity. *Development and Psychopathology, 17*(2), 271–301.

Boyce, W. T., Sokolowski, M. B., et al. (2012). Toward a new biology of social adversity. *Proceedings of the National Academy of Sciences of the United States of America, 109*(Suppl. 2), 17143–17148.

Burke, N. N., Kerr, D. M., et al. (2014). Minocycline modulates neuropathic pain behaviour and cortical M1-M2 microglial gene expression in a rat model of depression. *Brain, Behavior, and Immunity, 42*, 147–156.

Calcia, M. A., Bonsall, D. R., et al. (2016). Stress and neuroinflammation: a systematic review of the effects of stress on microglia and the implications for mental illness. *Psychopharmacology, 233*(9), 1637–1650.

Cattaneo, A., Ferrari, C., et al. (2016). Absolute measurements of macrophage migration inhibitory factor and Interleukin-1-beta mRNA levels accurately predict treatment response in depressed patients. *The International Journal of Neuropsychopharmacology, 19*(10), 1–10.

Choi, J., Jeong, B., et al. (2009). Preliminary evidence for white matter tract abnormalities in young adults exposed to parental verbal abuse. *Biological Psychiatry, 65*(3), 227–234.

Choi, J., Jeong, B., et al. (2012). Reduced fractional anisotropy in the visual limbic pathway of young adults witnessing domestic violence in childhood. *NeuroImage, 59*(2), 1071–1079.

Cicchetti, D., Rogosch, F. A., et al. (2011). Normalizing the development of cortisol regulation in maltreated infants through preventive interventions. *Development and Psychopathology, 23*(3), 789–800.

Coelho, R., Viola, T. W., et al. (2014). Childhood maltreatment and inflammatory markers: a systematic review. *Acta Psychiatrica Scandinavica, 129*(3), 180–192.

Cole, S. W., Arevalo, J. M., et al. (2010). Computational identification of gene-social environment interaction at the human IL6 locus. *Proceedings of the National Academy of Sciences of the United States of America, 107*(12), 5681–5686.

Couch, Y., Anthony, D. C., et al. (2013). Microglial activation, increased TNF and SERT expression in the prefrontal cortex define stress-altered behaviour in mice susceptible to anhedonia. *Brain, Behavior, and Immunity, 29*, 136–146.

Danese, A., Caspi, A., et al. (2011). Biological embedding of stress through inflammation processes in childhood. *Molecular Psychiatry, 16*(3), 244–246.

Danese, A., & McEwen, B. S. (2012). Adverse childhood experiences, allostasis, allostatic load, and age-related disease. *Physiology & Behavior, 106*(1), 29–39.

Danese, A., Moffitt, T. E., et al. (2008). Elevated inflammation levels in depressed adults with a history of childhood maltreatment. *Archives of General Psychiatry, 65*(4), 409–415.

Danese, A., Moffitt, T. E., et al. (2009). Adverse childhood experiences and adult risk factors for age-related disease: depression, inflammation, and clustering of metabolic risk markers. *Archives of Pediatrics & Adolescent Medicine, 163*(12), 1135–1143.

Dhabhar, F. S., Saul, A. N., et al. (2012). High-anxious individuals show increased chronic stress burden, decreased protective immunity, and increased cancer progression in a mouse model of squamous cell carcinoma. *PLoS One, 7*(4).

Dozier, M., Peloso, E., et al. (2008). Effects of an attachment-based intervention on the cortisol production of infants and toddlers in foster care. *Development and Psychopathology, 20*(3), 845–859.

Dube, S. R., Anda, R. F., et al. (2001). Childhood abuse, household dysfunction, and the risk of attempted suicide throughout the life span—findings from the adverse childhood experiences study. *JAMA: The Journal of the American Medical Association, 286*(24), 3089–3096.

Dube, S. R., Felitti, V. J., et al. (2003). Childhood abuse, neglect, and household dysfunction and the risk of illicit drug use: the adverse childhood experiences study. *Pediatrics, 111*(3), 564–572.

Dubowitz, H., Pitts, S. C., et al. (2004). Measurement of three major subtypes of child neglect. *Child Maltreatment, 9*(4), 344–356.

Eiland, L., & Romeo, R. D. (2013). Stress and the developing adolescent brain. *Neuroscience, 249*, 162–171.

Fernald, R. D., & Maruska, K. P. (2012). Social information changes the brain. *Proceedings of the National Academy of Sciences of the United States of America, 109*(Suppl. 2), 17194–17199.

Fisher, P. A., Van Ryzin, M. J., et al. (2011). Mitigating HPA axis dysregulation associated with placement changes in foster care. *Psychoneuroendocrinology, 36*(4), 531–539.

Gamble-George, J. C., Baldi, R., et al. (2016). Cyclooxygenase-2 inhibition reduces stress-induced affective pathology. *eLife, 5*, 1–20.

Gimeno, D., Kivimaki, M., et al. (2009). Associations of C-reactive protein and interleukin-6 with cognitive symptoms of depression: 12-year follow-up of the Whitehall II study. *Psychological Medicine, 39*(3), 413–423.

Goldman, J., Salus, M. K., et al. (2003). A coordinated response to child abuse and neglect: The foundation for practice. *Child abuse and neglect user manual series*, ERIC.

Gonzalez, A. (2013). The impact of childhood maltreatment on biological systems: implications for clinical interventions. *Paediatrics & Child Health, 18*(8), 415–418.

Goodwin, R. D., Fergusson, D. M., et al. (2004). Asthma and depressive and anxiety disorders among young persons in the community. *Psychological Medicine, 34*(8), 1465–1474.

Graff, L. A., Walker, J. R., et al. (2009). Depression and anxiety in inflammatory bowel disease: a review of comorbidity and management. *Inflammatory Bowel Diseases, 15*(7), 1105–1118.

Grebe, K. M., Takeda, K., et al. (2010). Cutting edge: sympathetic nervous system increases proinflammatory cytokines and exacerbates influenza A virus pathogenesis. *Journal of Immunology, 184*(?), 540–544

Grisanti, L. A., Perez, D. M., et al. (2011). Modulation of immune cell function by alpha(1)-adrenergic receptor activation. *Current Topics in Membranes, 67,* 113–138.

Grosse, L., Ambree, O., et al. (2016). Cytokine levels in major depression are related to childhood trauma but not to recent stressors. *Psychoneuroendocrinology, 73,* 24–31.

Gunnar, M., & Quevedo, K. (2007). The neurobiology of stress and development. *Annual Review of Psychology, 58,* 145–173.

Hannestad, J., DellaGioia, N., et al. (2013). The neuroinflammation marker translocator protein is not elevated in individuals with mild-to-moderate depression: a [(1)(1)C]PBR28 PET study. *Brain, Behavior, and Immunity, 33,* 131–138.

Henry, C. J., Huang, Y., et al. (2008). Minocycline attenuates lipopolysaccharide (LPS)-induced neuroinflammation, sickness behavior, and anhedonia. *Journal of Neuroinflammation, 5,* 15.

Hinwood, M., Tynan, R. J., et al. (2013). Chronic stress induced remodeling of the prefrontal cortex: structural re-organization of microglia and the inhibitory effect of minocycline. *Cerebral Cortex, 23*(8), 1784–1797.

Holmes, W. C., & Slap, G. B. (1998). Sexual abuse of boys—definition, prevalence, correlates, sequelae, and management. *JAMA: The Journal of the American Medical Association, 280*(21), 1855–1862.

Huang, J. L., Zhang, Y. L., et al. (2012). Enhanced phosphorylation of MAPKs by NE promotes TNF-alpha production by macrophage through alpha adrenergic receptor. *Inflammation, 35*(2), 527–534.

Infurna, M. R., Reichl, C., et al. (2016). Associations between depression and specific childhood experiences of abuse and neglect: a meta-analysis. *Journal of Affective Disorders, 190,* 47–55.

Irwin, M. R., & Cole, S. W. (2011). Reciprocal regulation of the neural and innate immune systems. *Nature Reviews. Immunology, 11*(9), 625–632.

Khan, A., McCormack, H. C., et al. (2015). Childhood maltreatment, depression, and suicidal ideation: critical importance of parental and peer emotional abuse during developmental sensitive periods in males and females. *Frontiers in Psychiatry, 6,* 42.

Knox, W. E. (1951). Two mechanisms which increase in vivo the liver tryptophan peroxidase activity: specific enzyme adaptation and stimulation of the pituitary adrenal system. *British Journal of Experimental Pathology, 32*(5), 462–469.

Krishnan, V., & Nestler, E. J. (2011). Animal models of depression: molecular perspectives. *Current Topics in Behavioral Neurosciences, 7,* 121–147.

Levine, M. E., Cole, S. W., et al. (2015). Childhood and later life stressors and increased inflammatory gene expression at older ages. *Social Science & Medicine, 130,* 16–22.

Madeeh Hashmi, A., Awais Aftab, M., et al. (2013 May-Jun). The fiery landscape of depression: a review of the inflammatory hypothesis. *The Pakistan Journal of Medical Sciences, 29*(3), 877–884.

Mandelli, L., Petrelli, C., et al. (2015). The role of specific early trauma in adult depression: a meta-analysis of published literature. Childhood trauma and adult depression. *European Psychiatry, 30*(6), 665–680.

McCauley, J., Kern, D. E., et al. (1997). Clinical characteristics of women with a history of childhood abuse: unhealed wounds. *JAMA, 277*(17), 1362–1368.

McCrory, E., De Brito, S. A., et al. (2010). Research review: the neurobiology and genetics of maltreatment and adversity. *Journal of Child Psychology and Psychiatry, 51*(10), 1079–1095.

Miller, A. H., Pariante, C. M., et al. (1999). Effects of cytokines on glucocorticoid receptor expression and function. Glucocorticoid resistance and relevance to depression. *Advances in Experimental Medicine and Biology, 461,* 107–116.

Miller, G. E., Chen, E., et al. (2011). Psychological stress in childhood and susceptibility to the chronic diseases of aging: moving toward a model of behavioral and biological mechanisms. *Psychological Bulletin, 137*(6), 959–997.

Miller, G. E., & Cole, S. W. (2012). Clustering of depression and inflammation in adolescents previously exposed to childhood adversity. *Biological Psychiatry, 72*(1), 34–40.

Mondelli, V., Vernon, A. C., et al. (2017). Brain microglia in psychiatric disorders. *Lancet Psychiatry, 4*(7), 563–572.

Myint, A. M., & Kim, Y. K. (2014). Network beyond IDO in psychiatric disorders: revisiting neurodegeneration hypothesis. *Progress in Neuro-Psychopharmacology & Biological Psychiatry, 48,* 304–313.

Nance, D. M., & Sanders, V. M. (2007). Autonomic innervation and regulation of the immune system (1987–2007). *Brain, Behavior, and Immunity, 21*(6), 736–745.

Nanni, V., Uher, R., et al. (2012). Childhood maltreatment predicts unfavorable course of illness and treatment outcome in depression: a meta-analysis. *The American Journal of Psychiatry, 169*(2), 141–151.

Nemeroff, C. B. (2004). Neurobiological consequences of childhood trauma. *The Journal of Clinical Psychiatry, 65*(Suppl. 1), 18–28.

Nikkheslat, N., Zunszain, P. A., et al. (2015). Insufficient glucocorticoid signaling and elevated inflammation in coronary heart disease patients with comorbid depression. *Brain, Behavior, and Immunity, 48,* 8–18.

Pace, T. W., Mletzko, T. C., et al. (2006). Increased stress-induced inflammatory responses in male patients with major depression and increased early life stress. *The American Journal of Psychiatry, 163*(9), 1630–1633.

Pace, T. W., Negi, L. T., et al. (2009). Effect of compassion meditation on neuroendocrine, innate immune and behavioral responses to psychosocial stress. *Psychoneuroendocrinology, 34*(1), 87–98.

Pace, T. W., Negi, L. T., et al. (2013). Engagement with cognitively-based compassion training is associated with reduced salivary C-reactive protein from before to after training in foster care program adolescents. *Psychoneuroendocrinology, 38*(2), 294–299.

Pan, A., Keum, N., et al. (2012). Bidirectional association between depression and metabolic syndrome: a systematic review and meta-analysis of epidemiological studies. *Diabetes Care, 35*(5), 1171–1180.

Pariante, C. M. (2006). The glucocorticoid receptor: part of the solution or part of the problem? *Journal of Psychopharmacology, 20*(4 Suppl), 79–84.

Pariante, C. M., & Miller, A. H. (2001). Glucocorticoid receptors in major depression: relevance to pathophysiology and treatment. *Biological Psychiatry, 49*(5), 391–404.

Pascual-Leone, A., Amedi, A., et al. (2005). The plastic human brain cortex. *Annual Review of Neuroscience, 28*, 377–401.

Pechtel, P., & Pizzagalli, D. A. (2011). Effects of early life stress on cognitive and affective function: an integrated review of human literature. *Psychopharmacology, 214*(1), 55–70.

Pinheiro, P. S. (2006). *World report on violence against children.* United Nations Publishing Services (UN Geneva).

Reus, G. Z., Fries, G. R., et al. (2015). The role of inflammation and microglial activation in the pathophysiology of psychiatric disorders. *Neuroscience, 300*, 141–154.

Salter, M., & Pogson, C. I. (1985). The role of tryptophan 2,3-dioxygenase in the hormonal control of tryptophan metabolism in isolated rat liver cells. Effects of glucocorticoids and experimental diabetes. *The Biochemical Journal, 229*(2), 499–504.

Sandiego, C. M., Gallezot, J. D., et al. (2015). Imaging robust microglial activation after lipopolysaccharide administration in humans with PET. *Proceedings of the National Academy of Sciences of the United States of America, 112*(40), 12468–12473.

Schmeer, K. K., & Yoon, A. (2016). Socioeconomic status inequalities in low-grade inflammation during childhood. *Archives of Disease in Childhood, 101*(11), 1043–1047.

Setiawan, E., Wilson, A. A., et al. (2015). Role of translocator protein density, a marker of neuroinflammation, in the brain during major depressive episodes. *JAMA Psychiatry, 72*(3), 268–275.

Slavich, G. M., & Cole, S. W. (2013). The emerging field of human social genomics. *Clinical Psychological Science: A Journal of the Association for Psychological Science, 1*(3), 331–348.

Slavich, G. M., & Irwin, M. R. (2014). From stress to inflammation and major depressive disorder: a social signal transduction theory of depression. *Psychological Bulletin, 140*(3), 774–815.

Torres-Platas, S. G., Cruceanu, C., et al. (2014). Evidence for increased microglial priming and macrophage recruitment in the dorsal anterior cingulate white matter of depressed suicides. *Brain, Behavior, and Immunity, 42*, 50–59.

Uher, R., Tansey, K. E., et al. (2014). An inflammatory biomarker as a differential predictor of outcome of depression treatment with escitalopram and nortriptyline. *The American Journal of Psychiatry, 171*(12), 1278–1286.

Verdolini, N., Attademo, L., et al. (2015). Traumatic events in childhood and their association with psychiatric illness in the adult. *Psychiatria Danubina, 27*(Suppl. 1), S60–70.

Vialou, V., Feng, J., et al. (2013). Epigenetic mechanisms of depression and antidepressant action. *Annual Review of Pharmacology and Toxicology, 53*, 59–87.

Walker, F. R., Beynon, S. B., et al. (2014). Dynamic structural remodelling of microglia in health and disease: a review of the models, the signals and the mechanisms. *Brain, Behavior, and Immunity, 37*, 1–14.

Walsh, E., Eisenlohr-Moul, T., et al. (2016). Brief mindfulness training reduces salivary IL-6 and TNF-alpha in young women with depressive symptomatology. *Journal of Consulting and Clinical Psychology, 84*(10), 887–897.

Williams, L. M., Debattista, C., et al. (2016). Childhood trauma predicts antidepressant response in adults with major depression: data from the randomized international study to predict optimized treatment for depression. *Translational Psychiatry, 6.*

Wium-Andersen, M. K., Orsted, D. D., et al. (2013). Elevated C-reactive protein levels, psychological distress, and depression in 73, 131 individuals. *JAMA Psychiatry, 70*(2), 176–184.

Stress, Maltreatment, Inflammation, and Functional Brain Changes in Depression

Kelly Doolin,†, Leonardo Tozzi*,†, Johann Steiner*,
Thomas Frodl*,†*

*Otto von Guericke University, Magdeburg, Germany
†Institute of Neuroscience, Trinity College Dublin, Dublin, Ireland

STRESS SYSTEM

Stress is the physiological response of an organism to a potentially threatening situation. It involves the activation of the hypothalamic-pituitary-adrenal (HPA) axis, leading to the secretion of cortisol (de Kloet, Joëls, & Holsboer, 2005). This hormone then acts through mineralocorticoid and glucocorticoid receptors (GRs) on several organs throughout the body. Within the brain, GRs are expressed in regions such as the hippocampus, amygdala, and prefrontal cortex (Reul & de Kloet, 1985), which are involved in cognition and mood regulation. These areas are also able to regulate the HPA axis itself through feedback mechanisms (Herman, Ostrander, Mueller, & Figueiredo, 2005).

Stress is believed to play an important role in the pathogenesis of major depressive disorder (MDD). This disorder is clinically recognized as a highly stress-sensitive illness (Kessler, 1997), and the hippocampus, which is altered in MDD, is a highly stress-sensitive brain region (Thomas, Hotsenpiller, & Peterson, 2007).

Chronic stress can increase rates of depression in susceptible individuals, but the detailed pathophysiology underlying this process remains unknown (Tsankova et al., 2006). Currently, there is much debate about the possible associations of early-life adversity with the specific course of stress-related and depressive illnesses, long-lasting emotional problems (Carboni et al., 2010; Heim, 2001; Mann & Currier, 2010), and hippocampal volumetric changes (Chen, Hamilton, & Gotlib, 2010).

HPA Axis and Homeostasis

To a large extent, how our bodies respond to stressors determines our overall health. Physiological function is altered in response to acute stressors but ultimately returns to a homeostatic or baseline control of these systems. This adaptive process is known as allostasis. If the stressor is excessive or our stress systems are faulty, the strain exerted on our physiological systems by chronic stress prevents a return to a healthy state of homeostasis. The pathophysiological

Inflammation and Immunity in Depression
https://doi.org/10.1016/B978-0-12-811073-7.00015-5

alterations in our stress-sensitive neuroendocrine, cardiovascular, immune, and neural systems brought about by this excessive strain are known as the allostatic load (Juster, McEwen, & Lupien, 2010). It is important to note that in order to keep our physiological state relatively constant, stress responses are necessary to continually adapt to an ever-changing environment.

The brain is the main regulatory organ for stress responses. The HPA axis, the major stress system in the body, is a neuroendocrine system involved in the production of the stress hormone cortisol by the adrenal glands. Cortisol is a glucocorticoid, which alters the function of numerous tissues in order to mobilize energy and change systems functionally to meet the demands of the stress challenge (de Kloet et al., 2005). Among the many processes affected by cortisol are glucose and fat metabolism, bone metabolism, cardiovascular responsiveness, and immune function. Glucocorticoids also modify brain functions via two nuclear receptors that also function as transcriptional factors: the high-affinity glucocorticoid receptor (GR) in the hippocampus and the low-affinity mineralocorticoid receptor (MR) distributed throughout the brain (de Kloet et al., 2005; Heim & Nemeroff, 2001; Oberlander et al., 2008; Weaver et al., 2004).

Most of our knowledge about the chronic effects of high levels of cortisol on the brain is inferred from human studies of HPA axis function in MDD. Cortisol is released in a pulsatile ultradian rhythm that varies in amplitude under different conditions of stress. Abnormal rhythms of cortisol secretion occur in MDD, with increased secretion of cortisol and blunting of the normal dip that should occur in the evening, leading to increased 24h production of cortisol (Pariante & Lightman, 2008).

Glucocorticoids in Depression

Clinical evidence for the role of glucocorticoid hormones in depression has long been established (Carpenter & Bunney, 1971; Gibbons & Mc, 1962). Indeed, depressive symptoms are a common side effect of long-term corticosteroid treatment (Patten, Williams, & Love, 1995), and conversely, normalization of corticosteroid levels in Cushing's syndrome improves patients' mood (Sonino, Fava, Raffi, Boscaro, & Fallo, 1998). Furthermore, signs of a hyperactivation of the HPA axis are found in many depressed patients (Nestler et al., 2002), as well as impaired response to dexamethasone suppression, adrenal gland hyperplasia (Rubin, Phillips, Sadow, & McCracken, 1995), blunted adrenocorticotropic hormone (ACTH), and increased cortisol release after corticotropin-releasing hormone (CRH) stimulation (Holsboer, 1986). An impaired dexamethasone suppression was also detected in nondepressed people with high family risk for depression, suggesting that genetic regulation of the HPA axis might be involved in determining vulnerability to depression (Holsboer, 2000).

Stress and Maltreatment

Exposure to neglect and abuse in childhood plays a crucial role in the development of MDD (Frodl & O'Keane, 2013; Nusslock & Miller, 2016; Trotta, Murray, & Fisher, 2015). The relationship between childhood adversity and depression is mediated by sex, genetic risk, parental psychopathology, stressful life events during adulthood, and social support (Pagliaccio & Barch, 2015). Childhood adversity and the abovementioned factors may contribute to treatment resistance in MDD (Pagliaccio & Barch, 2015; Tunnard et al., 2014). As such, understanding the role of early-life adversity is important to improve assessment and treatment of depression (Teicher & Samson, 2013).

As described in our recent review, childhood adversity has been implicated as a key factor associated with structural brain abnormalities in subjects who developed psychiatric disorders (Frodl & O'Keane, 2013). It has been demonstrated that childhood adversity and MDD are associated with structural brain changes, including reduction of gray-matter volumes in the hippocampus and other limbic structures (Dannlowski et al., 2012; Frodl et al., 2010; Gerritsen et al., 2015).

IMMUNE SYSTEM

The stress hormone system and the immune system are closely linked to each other. The HPA axis exerts inhibitory control over the inflammatory response system, a major component of the immune system. Alterations of inflammatory biomarkers have been repeatedly reported in patients with MDD compared with healthy controls, with indirect evidence that neuroinflammation plays a role in MDD (Eisenberger, Inagaki, Rameson, Mashal, & Irwin, 2009; Miller, Maletic, & Raison, 2009; Reichenberg et al., 2001; Wright, Strike, Brydon, & Steptoe, 2005). Interestingly, it is known that sickness behavior during the course of an infection is associated with psychological and behavioral changes that are similar to depression. They usually include lethargy, depressed mood, anxiety, malaise, the loss of appetite, sleepiness, hyperalgesia, the loss of social interests, and failure to concentrate (Dantzer, O'Connor, Freund, Johnson, & Kelley, 2008). Pro-inflammatory cytokines like interleukin-1β, interleukin-6, and tumor necrosis factor-α are thought to be responsible for these alterations by activating the HPA axis and reducing the availability of serotonin and noradrenalin in the brain (Barkhudaryan & Dunn, 1999).

Inflammation and MDD

Many research studies in the field of immunology and psychiatry have reported the association between immune system changes detected in the peripheral blood or tissue or in cerebrospinal fluid and related changes in the brain in terms of imaging findings or postmortem findings. Several findings have been reported regarding the link between immune system changes, psychological abnormalities, and impairment in other systems such as energy metabolism (Myint & Kim, 2014). Moreover, changes to immune molecules in the brain are also reported to be associated with other neurochemical changes, such as serotonergic, noradrenergic, GABAergic, and glutamatergic neurotransmissions and stress hormones. Some studies have reported metabolic pathways such as tryptophan/kynurenine metabolism and tyrosine metabolism as the link between changes in the immune system and other neurochemicals. These results have provided the missing links between the glial-neuronal network in our brains and opportunities to develop some new therapeutics and biomarkers in the field of psychiatry (Myint & Kim, 2014).

There are clinical findings that support the link between depression and inflammation. Numerous studies have reported on the antiinflammatory properties of antidepressant medication (Kenis & Maes, 2002; Sluzewska et al., 1995). Moreover, it has been shown that cytokine immunotherapy can induce depression in patients being treated for hepatitis C and certain cancers (Valentine, Meyers, Kling, Richelson, & Hauser, 1998). Also, findings from a meta-analysis of clinical trials suggest that the use of NSAIDS is associated with an improved antidepressant treatment response (Kohler et al., 2014). This bidirectional relationship between depression and inflammation provides a strong rationale for further investigation into the relationship between inflammation and other biological pathways of depression.

Cytokines are known to have central effects, either by crossing the blood-brain barrier or through the transmission of signals across the vagus nerve (Dantzer et al., 2008). Importantly, cytokines can affect dopamine and serotonin levels in the brain more generally (Miller et al., 2009) and in the striatum specifically. In this regard, it has been suggested that cytokines could influence reward processing and promote the increase of depressive symptoms through altered activity of dopaminergic and serotonergic systems (Ikemoto & Panksepp, 1999; Ressler & Nemeroff, 2000). A large epidemiological study by Benros et al. (2013) has indeed pointed out that hospitalization because of infections or autoimmune diseases are risk factors for the subsequent manifestation of mood disorders. Moreover, as revealed by recent meta-analyses, the levels of interleukin-6 and tumor necrosis factor-α appear to be slightly elevated in the peripheral blood of acutely and chronically ill patients with the clinical psychiatric diagnosis of MDD (Goldsmith, Rapaport, & Miller, 2016). Accordingly, increased concentrations of interleukin-6 have been observed in the cerebrospinal fluid of patients with MDD (Wang & Miller, 2018). Interestingly, pro-inflammatory cytokines can contribute to glutamate toxicity and also stimulate microglial cells to release IL-1 and TNF-α (reviewed in McNally, Bhagwagar, & Hannestad, 2008). Both IL-1 and TNF-α are expressed in the normal brain and play an active role in cellular events that induce structural changes at the synaptic level. However, increased levels of TNF-α have been found to inhibit long-term potentiation (LTP) and impair memory formation (reviewed in Khairova, Machado-Vieira, Du, & Manji, 2009).

Tryptophan Metabolic Pathway

In MDD, the activity of the tryptophan/kynurenine metabolic pathway has been shown to modulate inflammatory processes, with kynurenine levels being associated with depressive symptoms and/or cognitive decline (Oxenkrug, 2007). Products of the kynurenine pathway include N-methyl-D-aspartate (NMDA) agonists; quinolinic and picolinic acids (Jhamandas, Boegman, Beninger, Miranda, & Lipic, 2000); and an NMDA receptor antagonist,

kynurenic acid. It also includes free radical generators, 3-hydroxykynurenine and 3-hydroxyanthranilic acids (Forrest et al., 2004; Thomas & Stocker, 1999). An increased production of quinolinic acid has been observed in subregions of the anterior cingulate cortex (ACC) in suicidal patients with MDD in a human postmortem study (Steiner et al., 2011) and in a study of cerebrospinal fluid of suicide attempters with depression (Bay-Richter et al., 2015). Striatal injection of quinolinic acid in rats significantly stimulates the activity of cytokine-inducible nitric oxide (NO) synthase (iNOS) (Perez-Severiano, Escalante, & Rios, 1998; Ryu, Choi, & McLarnon, 2006), the enzyme that is responsible for NO synthesis from arginine. While NO is essential for maintaining normal physiological functions in blood vessels at low concentrations (Oxenkrug, 2005), it can also induce cellular death at higher concentrations following iNOS activation (mainly) in response to inflammation (Brown, 2007; Brown & Neher, 2010). As a free radical, NO damages metabolic enzymes and takes part in highly oxidative reactions (Akhtar, Sunico, Nakamura, & Lipton, 2012). Importantly, iNOS-derived NO is thought to be capable of triggering prooxidant and pro-inflammatory changes in the endothelium of brain microvessels, which MRI scans can detect as white-matter hyperintensities (WMHs) (Sloane, Hollander, Moss, Rosene, & Abraham, 1999). It has been suggested that the iNOS-derived NO might be responsible for the initial-stage microvessel inflammation in the brain, usually associated with working memory deficits and WMHs (Oxenkrug, 2007). Interestingly, research has shown that both physiological and psychological stress can induce the production of pro-inflammatory mediators, which in turn can stimulate tryptophan catabolism in the brain (Myint, Schwarz, & Muller, 2012), with consequences on neurotransmitter metabolism, neuroendocrine function, synaptic transmission, and neurocircuits that regulate mood, motor activity, motivation, anxiety, and alarm (Capuron & Miller, 2011).

BRAIN IMAGING OF DEPRESSION

Structural Changes in MDD

Recent meta-analyses of structural imaging studies have detected volumetric changes in several brain regions in MDD relative to controls. The most robust finding in the literature was reduced hippocampal volume in patients (Campbell & MacQueen, 2004; Schmaal et al., 2016), but some reports have also highlighted a decrease in the volume of dorsolateral prefrontal, dorsomedial prefrontal, orbitofrontal, and cingulate cortices; striatum; and amygdala. All of these regions are of key importance in regulating emotional responses and behavior (Drevets, Price, & Furey, 2008).

Functional Changes in MDD

As a recurrent psychiatric disorder, functional magnetic resonance imaging (fMRI) is particularly important in order to investigate neural circuits involved in depressive states and those associated with vulnerability for MDD.

Interest is growing in the use of resting-state fMRI, which does not require the use of a task and has become a popular means of complementing the results of task-based fMRI studies. Resting-state fMRI allows for the examination of large-scale neural systems that exhibit spontaneous synchronous fluctuations during goal-directed and nongoal-directed behavior (Castellanos & Proal, 2012). These low-frequency (<0.1 Hz) spontaneous fluctuations in blood-oxygen-level-dependent (BOLD) signals correlate with interactions between adjacent and nonadjacent brain areas that form spatially distributed networks of brain function (Raichle et al., 2001). Functional connectivity is the observed correlations in spontaneous neural

activity between brain areas at rest (Deco & Corbetta, 2011). In particular, four resting-state brain networks have been studies in MDD: the default mode network, the salience network, the affective network, and the task-positive network.

The *default mode network (DMN)* consists of an anterior part that covers the medial prefrontal cortex (including parts of the ACC) and a posterior part including the posterior cingulate cortex, precuneus, inferior parietal cortex, hippocampal area, and lateral temporal cortex (Buckner, Andrews-Hanna, & Schacter, 2008). Though most active during rest, activity in the DMN has been shown to be decreased during demanding cognitive tasks, and the DMN has been related to spontaneous cognition. The DMN was found to be related to emotion regulation, self-reference, and obsessive ruminations (Gusnard, Akbudak, Shulman, & Raichle, 2001). Its anterior regions are mostly related to self-referential activity and have strong connections to limbic regions such as the amygdala.

The *salience network (SN)* consists of the frontoinsular cortex, dorsal ACC, and temporal poles and activates to various salient stimuli. It's hypothesized to play a central role in emotional control due to extensive subcortical connectivity (Seeley et al., 2007).

The *affective network (AN)* contains integrated regions of the affective subdivision of the ACC, amygdala, nucleus accumbens, hypothalamus, anterior insula, hippocampus, and orbitofrontal cortex with reciprocal connections to autonomic, visceromotor, and endocrine systems (Bush, Luu, & Posner, 2000; Öngür, Ferry, & Price, 2003; Sheline, Price, Yan, & Mintun, 2010). This network is involved in emotional regulation and monitoring the salience of motivational stimuli (Bush et al., 2000; Fox, Corbetta, Snyder, Vincent, & Raichle, 2006; Sheline et al., 2010) and thus linked to the SN.

In depression, several resting-state studies have found increased resting-state functional connectivity in the DMN (Grimm et al., 2009; Sheline, 2011; Zhou et al., 2010). Furthermore, levels of metabolic activity within the DMN might be increased in MDD (Price & Drevets, 2012). A recent meta-analysis, including 32 resting-state fMRI studies, confirmed hyperactivity and hyperconnectivity in midline structures related to the DMN and in lateral frontal and parietal areas related to the CEN (Sundermann, Olde Lutke Beverborg, & Pfleiderer, 2014).

The majority of task-related fMRI studies in depression have targeted emotion-processing brain networks through the use of affect-laden stimuli. Overall, brain function in MDD seems to be associated with hyperactivity when exposed to negative affect in ventral prefrontal regions, such as the cingulate cortex, and in subcortical areas belonging to the limbic system like the insula and hippocampus. The dorsolateral prefrontal cortex shows hypoactivity in the same condition, possibly hinting at a deficit of top-down control on affect generation. Exposure to positive emotional stimuli showed opposite patterns of activation (Groenewold, Opmeer, de Jonge, Aleman, & Costafreda, 2013). Differences in regions involved in reward processing, such as the ventral striatum, have been observed as well (Leppänen, 2006).

Since many researchers assume that the depressive syndrome might arise from abnormal interactions between brain regions, functional neuroimaging studies have also examined the functional coupling between brain regions during emotional processing. A "top-down" and "bottom-up" dysregulated prefrontal-subcortical circuitry has been suggested to underlie the failure to regulate mood in patients with MDD (Ochsner et al., 2004). We have also demonstrated frontolimbic functional disconnections during emotional face matching in medication-free patients with MDD compared with controls (Preuss et al., 2010), and these changes were found to be associated with treatment response (Lisiecka et al., 2011).

Furthermore, a study involving 15 unmedicated patients with major depression and 15 healthy volunteers found decreased correlations between ACC and limbic regions, which is consistent with the hypothesis that decreased cortical regulation of limbic activation in response to negative stimuli may be present in depression (Anand et al., 2005). Again, the amygdala was not only negatively coupled with the ACC but also positively coupled bilaterally with medial temporal and ventral occipital regions in 19 unmedicated patients with major depression and 19 healthy volunteers (Chen et al., 2008). Other studies on task-related functional connectivity in patients with major depression receiving antidepressant medication, on the other hand, achieved conflicting results. Overall, coupling within an emotion-processing network consisting of the anterior cingulate region, prefrontal cortical regions, amygdala, and caudate-putamen may play a key role in MDD (Dannlowski et al., 2009; Hamilton & Gotlib, 2008; Vasic, Walter, Sambataro, & Wolf, 2008).

Structural Connectivity Changes in MDD

Some of the functional changes may be a consequence of structural changes, in particular changes in white-matter fiber systems. Magnetic resonance diffusion tensor imaging (DTI) is a novel neuroimaging technique that can evaluate both the orientation and the diffusion characteristics of white-matter tracts *in vivo* (Sexton, Mackay, & Ebmeier, 2009), and tractography methods allow extraction of white-matter fiber bundles relevant to MDD. Interestingly, a significant positive correlation has been found between the average fractional anisotropy value of the cingulum tract and the level of functional connectivity within the DMN (van den Heuvel, Mandl, Luigjes, & Hulshoff Pol, 2008). Thus, changes in structural connectivity may provide extra parameters in the attempt to form connectivity signatures.

A meta-analysis of DTI studies of patients with MDD consistently identified decreased fractional anisotropy in the white-matter fascicles connecting the prefrontal cortex within the frontal, temporal, and occipital lobes and amygdala and hippocampus (Liao et al., 2013). Recently, we reviewed the literature in DTI imaging in MDD and found, in our meta-analysis, a significant reduction in fractional anisotropy (FA) in the left superior longitudinal fasciculus in MDD (Murphy & Frodl, 2011). FA is a measure of the diffusion longitudinal, in relation to the diffusion perpendicular to white-matter tracts, often related to neural integrity. Moreover, childhood adversity measures were statistically related to the cingulum, uncinated fasciculus, and fronto-occipital fasciculus white-matter tracts (Ugwu, Amico, Carballedo, Fagan, & Frodl, 2015). Overall, however, findings from structural connectivity analyses show heterogeneous results without a clear definition of white-matter pathways. Therefore, more research is needed, in particular to identify those altered white-matter connections associated with common and more distinct psychopathology.

Brain Metabolite Changes in MDD

Some of the biological underpinnings of altered functional connectivity might also be explored by identifying their associations with metabolic (neurochemical) neuron- and glia-related changes measured using MR spectroscopy. Indeed, glutamate, glutamine, *N*-acetylaspartate (NAA), and γ-aminobutyric acid (GABA) levels seem to impact on BOLD responses (Walter et al., 2009). Most interestingly, the glutamine deficit that has been reported in melancholic/high-anhedonia MDD seems to be related to lower activation in the pregenual ACC (pgACC) (Walter et al., 2009), a key player within the DMN. Furthermore, Horn et al. (2010) reported that glutamatergic reductions in pgACC were related to depression severity and to deviant functional connectivity between DMN and salience network, namely, between pgACC and anterior insula. Importantly, these changes in functional connectivity

have not yet been related to metabolites in the insula itself. While there is converging evidence for reduced NAA levels and total concentration of glutamate and glutamine in the pgACC (Ende, Demirakca, & Tost, 2006), its GABA levels remain unchanged in MDD (Hasler et al., 2007). Recently, a meta-analysis found that ACC reduction in glutamate in patients with MDD was of sufficient effect size to discriminate them from bipolar patients (Taylor, 2014).

INTEGRATING STRESS SYSTEM, IMMUNE SYSTEM AND BRAIN CIRCUITS

HPA Axis and Structural Changes in MDD

There is evidence that specific neuronal circuits, particularly in the developing brain, are damaged by environmental stress-inducing changes in the HPA and inflammatory pathways (Krishnan & Nestler, 2008). Moreover, chronic hypercortisolism has been shown to enhance tryptophan breakdown in the brain and induce neurodegenerative changes (Capuron & Miller, 2011). Some studies, in particular, have shown that stress or cortisol administration may lead to depressive-like states and atrophy of neurons in the hippocampus (Duman, 2002) and that therapy with antidepressants reverses these changes (Santarelli et al., 2003). Importantly, chronic social stress has also been shown to induce glucocorticoid-mediated pyramidal dendrite retraction in the hippocampus and changes in dendrite arborization in the prefrontal cortex (PFC) (Kole, Czeh, & Fuchs, 2004; Magarinos, McEwen, Flugge, & Fuchs, 1996; Wellman, 2001; Woolley, Gould, Frankfurt, & McEwen, 1990), which might be associated with the behavioral manifestations of stress-related disorders like MDD (Macqueen & Frodl, 2010).

The associations between glucocorticoids, stress, and neuronal damage in the hippocampus support the theory that the potentially neurotoxic effects of glucocorticoids on this structure can be visualized in terms of its overall volume changes. Indeed, the hippocampus is vulnerable to stress, particularly during the early developmental period (Teicher et al., 2003). This damage may be species-specific and time-dependent: while the separation of monkeys from their siblings or mothers seems not to affect hippocampal volumes (Lyons, Yang, Sawyer-Glover, Moseley, & Schatzberg, 2001; Spinelli et al., 2009), chronic stress exposure in rats or tree shrews does result in reduced hippocampal volumes (Lee, Jarome, Li, Kim, & Helmstetter, 2009; Ohl, Michaelis, Vollmann-Honsdorf, Kirschbaum, & Fuchs, 2000). Transient mild stress may enhance hippocampal function (Luine, Martinez, Villegas, Magarinos, & McEwen, 1996), but chronic or severe stress disrupts hippocampus-dependent memory in experimental animals (reviewed in Sapolsky (2003)).

Effects of hypercortisolaemia on brain structure and function are difficult to separate from the effects of CRH and/or arginine vasopressin (AVP), as both are probably raised in MDD. Cushing's syndrome is characterized by hypercortisolaemia and low levels of hypothalamic ACTH secretagogues and provides a model for understanding the selective effects of high cortisol levels in the brain. Depression occurs in about 60% of patients with Cushing's syndrome and tends to be atypical in type, and the majority remit when cortisol levels are normalized (Kelly, Kelly, & Faragher, 1996). Also, premature cortical atrophy and cognitive impairments occur in Cushing's syndrome (Simmons, Do, Lipper, & Laws Jr, 2000).

Human studies demonstrate that impairments in learning and memory occur in subjects with artificially acutely elevated or reduced cortisol levels (Lupien et al., 2002). A similar impairment has been observed with artificial but chronic elevation of cortisol levels (de Quervain, Roozendaal, Nitsch, McGaugh, & Hock, 2000; Newcomer et al., 1999; Young, Sahakian, Robbins, & Cowen, 1999). Murine

models demonstrate that these effects result from glucocorticoid-induced changes in the hippocampus such as the decrease of synaptic plasticity, reduction in neurogenesis, retraction of the apical dendrites of hippocampal pyramidal cells, and in some situations neuronal atrophy and cell death (Goosens & Sapolsky, 2007). A reduction in the amount of neuropil without frank cell loss has also been observed, a finding that appears consistent with observations from postmortem studies of the hippocampus in patients with MDD (Stockmeier et al., 2004).

The hippocampal damage brought about by excessive exposure to glucocorticoids should cause reduction in the feedback inhibition mediated by cortisol, via the hippocampus, on CRH secretion, resulting in further excessive cortisol secretion and creating a cascade of hippocampal damage (Sapolsky, Krey, & McEwen, 1986). The "glucocorticoid cascade hypothesis" was formulated in 1986 and is now widely accepted as a pathophysiological pathway leading to brain changes associated with severe and enduring stress. This mechanism explains how chronic stress can lead to depression-like effects and brain changes that result in dysfunctional central control of the HPA axis and subsequent depression. It should be noted that in normal senescence, a similar process occurs leading to reduced ability of the hippocampus to control the HPA axis and a gradual reduction in cognitive tasks mediated by the hippocampus.

HPA Axis and Functional Changes in MDD

Since stressful life events and childhood maltreatment have been shown to impact on brain function using neuroimaging tests, there is the expectation that measures of the HPA axis might also be associated with brain function. Studies investigating pre- to poststressor change in cortisol showed variable results with some studies reporting increases in the prefrontal cortex and orbitofrontal cortex related to pre- to poststressor change in cortisol (Wang et al., 2008), while others reported decreases in the hypothalamus, amygdala, hippocampus, orbitofrontal cortex, and anterior cingulate cortex in participants reacting with an increase in cortisol following stressors (Pruessner et al., 2008).

Negative associations between cortisol reactivity during passively viewing of negative stimuli and ventromedial PFC activity (Root et al., 2009) and negative correlations between cortisol levels and activation of the amygdala were observed (Cunningham-Bussel et al., 2009; Pruessner et al., 2008). On the other hand, positive association between limbic activation and cortisol change has also been shown (Root et al., 2009; van Stegeren et al., 2007).

Exogenous cortisol administration, conversely, is associated with decreased activation of the amygdala and hippocampus during rest in healthy controls (Lovallo, Robinson, Glahn, & Fox, 2010). In a study of depression, 15 patients with depression in remission and 15 healthy controls were investigated. A visual stimuli task paradigm was presented to evoke a mild stress response; cortisol and ACTH were measured at baseline and following the onset of the stress paradigm. Interestingly, among remitted MDD patients, percent signal change in the right amygdala was negatively related to peak cortisol change (Holsen et al., 2013).

Therefore, some studies report that during stress, an increase of cortisol is associated with a decrease of amygdala reactivity, and others report the opposite. Thus, currently, no conclusion can be drawn about the direction of association between cortisol reactivity and neural reactivity of limbic brain circuitries.

Inflammation and Structural Changes in MDD

Preclinical and clinical MRI research suggests that neuroinflammation in MDD might be associated with structural and functional anomalies in various regions of the central nervous system.

Evidence points to an overall role for the immune system and inflammation in brain volume integrity (Glass, Saijo, Winner, Marchetto, & Gage, 2010). The presence of chronic inflammation in the subarachnoid and perivascular spaces might be responsible for an abnormal and microglial overactivation that can be harmful to neurons and subsequently lead to a dysregulation of astrocyte functions and glutamate metabolism, resulting in additional neuronal and synaptic damage (Calabrese et al., 2015). It has been shown that hippocampal volume reduction is associated with an increase in inflammatory cytokines including IL-6 and TNF-α (Kesler et al., 2013). While gray-matter loss is considered to be part of natural aging, this process is accelerated in MDD patients compared with healthy controls (Grieve, Korgaonkar, Koslow, Gordon, & Williams, 2013). Early-life stressors have been shown to diminish cell proliferation and dampen the production of new neurons during adulthood in several animal models, reinforcing the possible involvement of stress in depression (Coe et al., 2003; Lemaire, Koehl, Le Moal, & Abrous, 2000).

Recently, we reported a significant association of IL-6 and CRP with volume in the left and right hippocampus in 40 patients with MDD, independent of demographic variables (Frodl et al., 2012). Moreover, there was an association between IL-6 blood levels and hippocampal volumes in healthy controls. We found smaller hippocampal volumes in MDD patients who had lower expression of glucocorticoid-induced leucine zipper (GILZ) mRNA or SGK-1 mRNA, as markers of reduced activation of the glucocorticoid system, compared with those with higher serum SGK-1 and GILZ mRNA expression (Frodl et al., 2012). There is mounting evidence that hippocampal volume reduction could be related to immune activation (Frodl & Amico, 2014; Goshen et al., 2008; Marsland et al., 2015). For example, inflammatory cytokines such as IL-6 have been shown to be inversely correlated to hippocampal volume (Marsland, Gianaros, Abramowitch, Manuck, & Hariri, 2008). Pro-inflammatory cytokines produced by inflammatory cells such as macrophages, CD4-positive T cells, and B cells surrounding gray matter might result in the neuronal change that is at the source of gray-matter volume reduction in depression (Fig. 1).

However, these findings need replication in a larger sample size in order to confirm the relationships between inflammation, stress hormone system, and brain structure. Although possible antidepressant effects should be taken into account, these findings suggest that reduced glucocorticoid responsiveness and increased inflammation might have a role in the neuroplasticity-neurotoxicity cascade in MDD.

Inflammation and Functional Changes in MDD

Moreover, with regard to functional brain activity, an increase in TNF-α has been found to be associated with more activity in the dorsal anterior cingulate cortex and anterior insula in subjects experiencing social rejection (Slavich, Way, Eisenberger, & Taylor, 2010). In another fMRI study, it was shown that higher TNF-α plasma concentrations are correlated with increased activation of the right inferior orbitofrontal cortex in response to emotional visual stimuli (Kullmann et al., 2013). In another study, inflammation was found to alter reward-related neural activity, and this in turn led to a depressed mood (Eisenberger et al., 2010). Peripheral IL-6, in a further study, was found to modulate the association between mood states and reduced connectivity of subgenual anterior cingulate cortex to amygdala, medial prefrontal cortex, nucleus accumbens, and superior temporal sulcus (Harrison et al., 2009). Thus, initial research suggests that there is an association between inflammatory markers and brain function and structure, although more research is needed in this area.

Epigenetics and Brain Imaging

In general, recent findings are indicating that gene-by-environment interactions are of central importance in the development of depression. Although the specific mechanisms remain unknown, a number of studies suggest that DNA methylation may be an underlying mechanism mediating the impact of adverse social environments on gene function (Booij, Tremblay, Szyf, & Benkelfat, 2015; Booij, Wang, Levesque, Tremblay, & Szyf, 2013). Studies in patients with MDD have shown an association of differential DNA methylation in white blood cells with early maltreatment and depressive symptomatology (Booij et al., 2013; Nestler, 2014). The findings are not consistent, but the inconsistent effects might be due to the differences in methodology, as different methodologies are being employed in different centers (Booij et al., 2013).

Animal models tracking the trajectory from early-life stress to adult depression indicate that sustained stress during development leads to hypermethylation of the GR promoter gene, leading to reduced function of the GR and inability to shut down stress responses (McGowan et al., 2011). An impact of parental care on epigenetic regulation of hippocampal GR was demonstrated in a study observing that suicide victims with a history of early maltreatment display decreased GR mRNA expression and increased cytosine methylation of a neuron-specific GR (NR3C1) promoter in the postmortem hippocampus compared with either suicide victims with no childhood maltreatment or controls (McGowan et al., 2009).

Peripheral *SLC6A4* DNA methylation was found to be related to childhood maltreatment in a sample of pregnant women and a sample of adoptees (Beach, Brody, Todorov, Gunter, & Philibert, 2010, 2011; Devlin, Brain, Austin, & Oberlander, 2010; Kang et al., 2013). In a study on prenatal and postnatal exposure to maternal depression, it was detected that increased

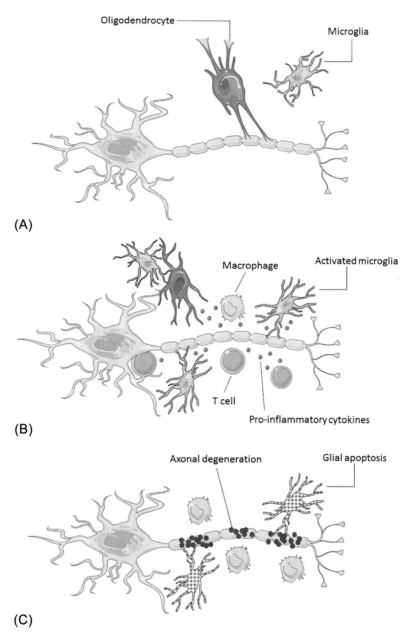

FIG. 1 Neuronal plasticity changes caused by inflammation. (A) In the healthy and undisturbed gray matter, microglia survey the environment for protrusions but are generally in a resting state. (B) Pro-inflammatory cytokines released by macrophages, CD4-positive T cells, or B cells may lead to microglia or macrophage activation and cause damage to the oligodendrocyte. (C) This process may result in neuronal cell death or morphological alterations such as pycnotic nuclei, shrinkage of dendrites, or axonal degeneration.

second trimester maternal depressed mood was associated with decreased maternal and infant *SLC6A4* promoter methylation (Devlin et al., 2010). This correlational finding seems to be in the opposite direction than research on childhood adversity, warranting further research to understand the impact of *SLC6A4* methylation and its function from a developmental view.

The functional relevance of DNA methylation in *SLC6A4* promoter regulation is that it suppresses transcriptional activity (Wang et al., 2012), although DNA methylation should be cell-type-specific so that DNA methylation changes that are relevant to brain function should be detected only in the brain. Nevertheless, we have previously reported differential methylation of a regulatory region of the *SLC6A4* gene in peripheral T cells that are associated with differences in in vivo measures of lower 5-HT synthesis, measured with positron-emission tomography (Wang et al., 2012). This T-cell methylation is also associated with hippocampal volume detected by MRI in patients with MDD and healthy controls (Booij, Szyf, et al., 2014). Moreover, an association between peripheral *SLC6A4* methylation with several gray-matter structures including the hippocampus, insula, amygdala, and nucleus caudatus has been reported (Dannlowski et al., 2014).

Previously, we reported an interactive effect between the promoter polymorphism of *SLC6A4* and childhood adversity on brain structure: a finding present in patients with depression and not in healthy controls (Frodl et al., 2010). In a recent study, a significant effect of state of *SLC6A4* methylation in whole blood DNA on fMRI BOLD responses during emotional attention processing was shown (Frodl et al., 2015). BOLD responses elicited by negative emotional stimuli in the left anterior insula/frontal operculum area were found to be positively associated with methylation of *SLC6A4* regulatory region. Patients with MDD showed increased activation in the insula and other emotional brain regions elicited by emotion-relevant stimuli and lower activity in

the hippocampal area during higher-order cognitive processing compared with controls, therefore suggesting that the state of *SLC6A4* methylation and depression influence brain function in the same direction (Frodl et al., 2015).

CONCLUSION

In summary, cumulative evidence provided by the last several decades of depression research indicates that an altered biological network composed of HPA axis dysregulation, an activated immune system, and altered synaptogenesis may be the cause of the disorder. This altered biological network is associated with the structural and functional brain changes observed in depressed patients through MRI studies, providing a link between peripheral and central changes in depression. Moreover, HPA axis dysfunction and subsequent glucocorticoid resistance are associated with early-life stressors such as childhood abuse, a risk factor for MDD, indicating the involvement of epigenetic alterations. Epigenetic alterations have been witnessed in depressed patients via methylation differences within genes hypothesized to play a role in HPA axis and immune functioning, and methylation at these sites has been found to be highly correlated to reduced hippocampal volumes in depressed patients, reiterating the likelihood of a mechanistically unified network of dysregulation in MDD.

Future depression research should focus on the development of computational models involving markers of HPA axis, immune, epigenetic, and imaging measures. For example, specific parameters of this biological network's expression may be associated with a particular depressive symptom subtype, such as atypical or melancholic depression. A greater understanding of the unique biological profiles that correspond to specific depression subtypes has the potential to assist in the development of antidepressant treatments targeted toward specific subsets of patients, such as a primarily HPA-driven type of depression.

Longitudinal data collection would be essential in determining whether specific deficits of the pathways described in this chapter or combinations of deficits may be indicative of treatment resistance or recurrence. The ability to identify resistance to a specific antidepressant drug through peripheral blood markers at the start of a patient's engagement with psychiatric services would be highly beneficial. Development of this kind of blood test would reduce the "trial-and-error" approach to depression treatment that bears the risk to leave patients at high risk of suicide and self-harm when unremitted from their depressive episode. A longitudinal research model may lead to the possibility of predicting outcomes based on specific characteristics of the altered biological network that has shown to be involved in the disorder.

Undoubtedly, the stress hormone system and the immune system play an important role in the pathophysiology of depression and are linked to altered brain structure and function. While case-control studies provide evidence of associations between stress, immune system, epigenetics, and brain function, this knowledge has not yet been used to develop a computational model that has potential to guide therapy and diagnostics. The strong evidence pointing toward dysregulation of a network encompassing hypercortisolaemia, inflammation, and reduced gray-matter volumes should continue to be investigated in the context of treatment response. Future hypotheses of the biological etiology of depression should acknowledge the involvement of deficits of each of the systems discussed as they are functionally connected, and disruptions of these systems have been shown to be correlated to one another in depressed populations.

References

Akhtar, M. W., Sunico, C. R., Nakamura, T., & Lipton, S. A. (2012). Redox regulation of protein function via cysteine S-nitrosylation and its relevance to neurodegenerative diseases. *International Journal of Cell Biology, 2012*.

Anand, A., Li, Y., Wang, Y., Wu, J., Gao, S., Bukhari, L., et al. (2005). Activity and connectivity of brain mood regulating circuit in depression: a functional magnetic resonance study. *Biological Psychiatry, 57*, 1079–1088.

Barkhudaryan, N., & Dunn, A. J. (1999). Molecular mechanisms of actions of interleukin-6 on the brain, with special reference to serotonin and the hypothalamo-pituitary-adrenocortical axis. *Neurochemical Research, 24*, 1169–1180.

Bay-Richter, C., Linderholm, K. R., Lim, C. K., Samuelsson, M., Traskman-Bendz, L., Guillemin, G. J., et al. (2015). A role for inflammatory metabolites as modulators of the glutamate N-methyl-D-aspartate receptor in depression and suicidality. *Brain, Behavior, and Immunity, 43*, 110–117.

Beach, S. R. H., Brody, G. H., Todorov, A. A., Gunter, T. D., & Philibert, R. A. (2010). Methylation at SLC6A4 is linked to family history of child abuse: an examination of the Iowa adoptee sample. *American Journal of Medical Genetics. Part B, Neuropsychiatric Genetics, 153B*, 710–713.

Beach, S. R. H., Brody, G. H., Todorov, A. A., Gunter, T. D., & Philibert, R. A. (2011). Methylation at 5HTT mediates the impact of child sex abuse on Women's antisocial behavior: an examination of the Iowa adoptee sample. *Psychosomatic Medicine, 73*, 83–87.

Benros, M. E., Waltoft, B. L., Nordentoft, M., Ostergaard, S. D., Eaton, W. W., Krogh, J., et al. (2013). Autoimmune diseases and severe infections as risk factors for mood disorders: a nationwide study. *JAMA Psychiatry, 70*, 812–820.

Booij, L., Szyf, M., Carballedo, A., Morris, D., Ly, V., Fahey, C., et al. (2014). *The role of SLC6A4 DNA methylation in stress-related changes in hippocampal volume: A study in depressed patients and healthy controls*. Vancouver: CINP.

Booij, L., Tremblay, R. E., Szyf, M., & Benkelfat, C. (2015). Genetic and early environmental influences on the serotonin system: consequences for brain development and risk for psychopathology. *Journal of Psychiatry and Neuroscience, 40*(1), 5–18.

Booij, L., Wang, D., Levesque, M. L., Tremblay, R. E., & Szyf, M. (2013). Looking beyond the DNA sequence: the relevance of DNA methylation processes for the stress-diathesis model of depression. *Philosophical Transactions of the Royal Society of London Series B, Biological Sciences, 368*.

Brown, G. C. (2007). Mechanisms of inflammatory neurodegeneration: iNOS and NADPH oxidase. *Biochemical Society Transactions, 35*, 1119–1121.

Brown, G. C., & Neher, J. J. (2010). Inflammatory neurodegeneration and mechanisms of microglial killing of neurons. *Molecular Neurobiology, 41*, 242–247.

Buckner, R. L., Andrews-Hanna, J. R., & Schacter, D. L. (2008). The brain's default network: anatomy, function, and relevance to disease. *Annals of the New York Academy of Sciences, 1124*, 1–38.

Bush, G., Luu, P., & Posner, M. I. (2000). Cognitive and emotional influences in anterior cingulate cortex. *Trends in Cognitive Sciences, 4*, 215–222.

Calabrese, M., Magliozzi, R., Ciccarelli, O., Geurts, J. J., Reynolds, R., & Martin, R. (2015). Exploring the origins of grey matter damage in multiple sclerosis. *Nature Reviews. Neuroscience, 16*, 147–158.

Campbell, S., & Macqueen, G. (2004). The role of the hippocampus in the pathophysiology of major depression. *Journal of Psychiatry & Neuroscience, 29*(6), 417–426.

Capuron, L., & Miller, A. H. (2011). Immune system to brain signaling: neuropsychopharmacological implications. *Pharmacology & Therapeutics, 130*, 226–238.

Carboni, L., Becchi, S., Piubelli, C., Mallei, A., Giambelli, R., Razzoli, M., et al. (2010). Early-life stress and antidepressants modulate peripheral biomarkers in a gene-environment rat model of depression. *Progress in Neuro-Psychopharmacology & Biological Psychiatry 34*(6), 1037–1048.

Carpenter, W. T., Jr., & Bunney, W. E., Jr. (1971). Adrenal cortical activity in depressive illness. *The American Journal of Psychiatry, 128*, 31–40.

Castellanos, F. X., & Proal, E. (2012). Large-scale brain systems in ADHD: beyond the prefrontal-striatal model. *Trends in Cognitive Sciences, 16*, 17–26.

Chen, C. H., Suckling, J., Ooi, C., Fu, C. H., Williams, S. C., Walsh, N. D., et al. (2008). Functional coupling of the amygdala in depressed patients treated with antidepressant medication. *Neuropsychopharmacology, 33*, 1909–1918.

Chen, M. C., Hamilton, J. P., & Gotlib, I. H. (2010). Decreased hippocampal volume in healthy girls at risk of depression. *Archives of General Psychiatry, 67*, 270–276.

Coe, C. L., Kramer, M., Czeh, B., Gould, E., Reeves, A. J., Kirschbaum, C., et al. (2003). Prenatal stress diminishes neurogenesis in the dentate gyrus of juvenile rhesus monkeys. *Biological Psychiatry, 54*, 1025–1034.

Cunningham-Bussel, A. C., Root, J. C., Butler, T., Tuescher, O., Pan, H., Epstein, J., et al. (2009). Diurnal cortisol amplitude and fronto-limbic activity in response to stressful stimuli. *Psychoneuroendocrinology, 34*, 694–704.

Dannlowski, U., Kugel, H., Redlich, R., Halik, A., Schneider, I., Opel, N., et al. (2014). Serotonin transporter gene methylation is associated with hippocampal gray matter volume. *Human Brain Mapping 35*(11), 5356–5367.

Dannlowski, U., Ohrmann, P., Konrad, C., Domschke, K., Bauer, J., Kugel, H., et al. (2009). Reduced amygdala-prefrontal coupling in major depression: association with MAOA genotype and illness severity. *The International Journal of Neuropsychopharmacology, 12*, 11–22.

Dannlowski, U., Stuhrmann, A., Beutelmann, V., Zwanzger, P., Lenzen, T., Grotegerd, D., et al. (2012). Limbic scars: long-term consequences of childhood maltreatment revealed by functional and structural magnetic resonance imaging. *Biological Psychiatry, 71*, 286–293.

Dantzer, R., O'Connor, J. C., Freund, G. G., Johnson, R. W., & Kelley, K. W. (2008). From inflammation to sickness and depression: when the immune system subjugates the brain. *Nature Reviews Neuroscience, 9*, 46–56.

de Kloet, E., Joëls, M., & Holsboer, F. (2005). Stress and the brain: from adaptation to disease. *Nature Reviews Neuroscience, 6*, 463–475.

de Quervain, D. J., Roozendaal, B., Nitsch, R. M., McGaugh, J. L., & Hock, C. (2000). Acute cortisone administration impairs retrieval of long-term declarative memory in humans. *Nature Neuroscience, 3*, 313–314.

Deco, G., & Corbetta, M. (2011). The dynamical balance of the brain at rest. *The Neuroscientist, 17*, 107–123.

Devlin, A. M., Brain, U., Austin, J., & Oberlander, T. F. (2010). Prenatal exposure to maternal depressed mood and the MTHFR C677T variant affect SLC6A4 methylation in infants at birth. *PLoS One, 5*.

Drevets, W. C., Price, J. L., & Furey, M. L. (2008). Brain structural and functional abnormalities in mood disorders: implications for neurocircuitry models of depression. *Brain Structure and Function, 213*(1–2), 93–118. https://doi.org/10.1007/s00429-008-0189-x.

Duman, R. S. (2002). Pathophysiology of depression: the concept of synaptic plasticity. *European Psychiatry, 17* (Suppl. 3), 306–310.

Eisenberger, N. I., Berkman, E. T., Inagaki, T. K., Rameson, L. T., Mashal, N. M., & Irwin, M. R. (2010). Inflammation-induced anhedonia: endotoxin reduces ventral striatum responses to reward. *Biological Psychiatry, 68*, 748–754.

Eisenberger, N. I., Inagaki, T. K., Rameson, L. T., Mashal, N. M., & Irwin, M. R. (2009). An fMRI study of cytokine-induced depressed mood and social pain: the role of sex differences. *NeuroImage, 47*, 881–890.

Ende, G., Demirakca, T., & Tost, H. (2006). The biochemistry of dysfunctional emotions: proton MR spectroscopic findings in major depressive disorder. *Progress in Brain Research, 156*, 481–501.

Forrest, C. M., Mackay, G. M., Stoy, N., Egerton, M., Christofides, J., Stone, T. W., et al. (2004). Tryptophan loading induces oxidative stress. *Free Radical Research, 38*, 1167–1171.

Fox, M. D., Corbetta, M., Snyder, A. Z., Vincent, J. L., & Raichle, M. E. (2006). Spontaneous neuronal activity distinguishes human dorsal and ventral attention systems. *Proceedings of the National Academy of Sciences, 103*, 10046–10051.

Frodl, T., & Amico, F. (2014). Is there an association between peripheral immune markers and structural/functional neuroimaging findings? *Progress in Neuro-Psychopharmacology & Biological Psychiatry, 48*, 295–303.

Frodl, T., Carballedo, A., Hughes, M. M., Saleh, K., Fagan, A., Skokauskas, N., et al. (2012). Reduced expression of glucocorticoid-inducible genes GILZ and SGK-1: high IL-6 levels are associated with reduced hippocampal volumes in major depressive disorder. *Translational Psychiatry, 2*.

Frodl, T., & O'Keane, V. (2013). How does the brain deal with cumulative stress? a review with focus on developmental stress, HPA axis function and hippocampal structure in humans. *Neurobiology of Disease, 52*, 24–37.

Frodl, T., Reinhold, E., Koutsouleris, N., Donohoe, G., Bondy, B., Reiser, M., et al. (2010). Childhood stress, serotonin transporter gene and brain structures in major depression. *Neuropsychopharmacology, 35*, 1383–1390.

Frodl, T., Szyf, M., Carballedo, A., Ly, V., Dymov, S., Vaisheva, F., et al. (2015). DNA methylation of the serotonin transporter gene (SLC6A4) is associated with brain function involved in processing emotional stimuli. *Journal of Psychiatry & Neuroscience, 40*, 296–305.

Gerritsen, L., van Velzen, L., Schmaal, L., van der Graaf, Y., van der Wee, N., van Tol, M. J., et al. (2015). Childhood maltreatment modifies the relationship of depression with hippocampal volume. *Psychological Medicine, 45*, 3517–3526.

Gibbons, J. L., & Mc, H. P. (1962). Plasma cortisol in depressive illness. *Journal of Psychiatric Research, 1*, 162–171.

Glass, C. K., Saijo, K., Winner, B., Marchetto, M. C., & Gage, F. H. (2010). Mechanisms underlying inflammation in neurodegeneration. *Cell, 140*, 918–934.

Goldsmith, D. R., Rapaport, M. H., & Miller, B. J. (2016). A meta-analysis of blood cytokine network alterations in psychiatric patients: comparisons between schizophrenia, bipolar disorder and depression. *Molecular Psychiatry, 21*, 1696–1709.

Goosens, K. A., & Sapolsky, R. M. (2007). Chapter 13: Stress and glucocorticoid contributions to normal and pathological aging. In D. R. Riddle (Ed.), *Brain aging: Models, methods, and mechanisms.* Boca Raton: CRC Press/Taylor & Francis.

Goshen, I., Kreisel, T., Ben-Menachem-Zidon, O., Licht, T., Weidenfeld, J., Ben-Hur, T., et al. (2008). Brain interleukin-1 mediates chronic stress-induced depression in mice via adrenocortical activation and hippocampal neurogenesis suppression. *Molecular Psychiatry, 13*, 717–728.

Grieve, S. M., Korgaonkar, M. S., Koslow, S. H., Gordon, E., & Williams, L. M. (2013). Widespread reductions in gray matter volume in depression. *NeuroImage: Clinical, 3*, 332–339.

Grimm, S., Boesiger, P., Beck, J., Schuepbach, D., Bermpohl, F., Walter, M., et al. (2009). Altered negative BOLD responses in the default-mode network during emotion processing in depressed subjects. *Neuropsychopharmacology, 34*, 932–943.

Groenewold, N. A., Opmeer, E. M., de Jonge, P., Aleman, A., & Costafreda, S. G. (2013). Emotional valence modulates brain functional abnormalities in depression: evidence from a meta-analysis of fMRI studies. *Neuroscience and Biobehavioral Reviews, 37*(2), 152–163.

Gusnard, D. A., Akbudak, E., Shulman, G. L., & Raichle, M. E. (2001). Medial prefrontal cortex and self-referential mental activity: relation to a default mode of brain function. *Proceedings of the National Academy of Sciences of the United States of America, 98*, 4259–4264.

Hamilton, J. P., & Gotlib, I. H. (2008). Neural substrates of increased memory sensitivity for negative stimuli in major depression. *Biological Psychiatry, 63*, 1155–1162.

Harrison, N. A., Brydon, L., Walker, C., Gray, M. A., Steptoe, A., & Critchley, H. D. (2009). Inflammation causes mood changes through alterations in subgenual cingulate activity and mesolimbic connectivity. *Biological Psychiatry, 66*, 407–414.

Hasler, G., van der Veen, J. W., Tumonis, T., Meyers, N., Shen, J., & Drevets, W. C. (2007). Reduced prefrontal glutamate/glutamine and gamma-aminobutyric acid levels in major depression determined using proton magnetic resonance spectroscopy. *Archives of General Psychiatry, 64*, 193–200.

Heim, C., & Nemeroff, C. (2001). The role of childhood trauma in the neurobiology of mood and anxiety disorders: preclinical and clinical studies. *Biological Psychiatry, 49*, 1023–1039.

Heim, C. N. C. (2001). The role of childhood trauma in the neurobiology of mood and anxiety disorders: preclinical and clinical studies. *Biological Psychiatry, 49*(12), 1023–1039.

Herman, J. P., Ostrander, M. M., Mueller, N. K., & Figueiredo, H. (2005). Limbic system mechanisms of stress regulation: hypothalamo-pituitary-adrenocortical axis. *Progress in Neuro-Psychopharmacology & Biological Psychiatry, 29*, 1201–1213.

Holsboer, F. (1986). Corticotropin-releasing hormone—a new tool to investigate hypothalamic-pituitary-adrenocortical physiology in psychiatric patients. *Psychopharmacology Bulletin, 22*, 907–912.

Holsboer, F. (2000). The corticosteroid receptor hypothesis of depression. *Neuropsychopharmacology, 23*, 477–501.

Holsen, L. M., Lancaster, K., Klibanski, A., Whitfield-Gabrieli, S., Cherkerzian, S., Buka, S., et al. (2013). HPA-axis hormone modulation of stress response circuitry activity in women with remitted major depression. *Neuroscience, 250*, 733–742.

Horn, D. I., Yu, C., Steiner, J., Buchmann, J., Kaufmann, J., Osoba, A., et al. (2010). Glutamatergic and resting-state functional connectivity correlates of severity in major depression—the role of pregenual anterior cingulate cortex and anterior insula. *Frontiers in Systems Neuroscience, 4*, pii: 33.

Ikemoto, S., & Panksepp, J. (1999). The role of nucleus accumbens dopamine in motivated behavior: a unifying interpretation with special reference to reward-seeking. *Brain Research Reviews, 31*, 6–41.

Jhamandas, K. H., Boegman, R. J., Beninger, R. J., Miranda, A. F., & Lipic, K. A. (2000). Excitotoxicity of quinolinic acid: modulation by endogenous antagonists. *Neurotoxicity Research, 2*, 139–155.

Juster, R. P., McEwen, B. S., & Lupien, S. J. (2010). Allostatic load biomarkers of chronic stress and impact on health and cognition. *Neuroscience and Biobehavioral Reviews, 35*, 2–16.

Kang, H. J., Kim, J. M., Stewart, R., Kim, S. Y., Bae, K. Y., Kim, S. W., et al. (2013). Association of SLC6A4 methylation with early adversity, characteristics and outcomes in depression. *Progress in Neuro-Psychopharmacology & Biological Psychiatry, 44,* 23–28.

Kelly, W. F., Kelly, M. J., & Faragher, B. (1996). A prospective study of psychiatric and psychological aspects of Cushing's syndrome. *Clinical Endocrinology, 45,* 715–720.

Kenis, G., & Maes, M. (2002). Effects of antidepressants on the production of cytokines. *The International Journal of Neuropsychopharmacology, 5,* 401–412.

Kesler, S., Janelsins, M., Koovakkattu, D., Palesh, O., Mustian, K., Morrow, G., et al. (2013). Reduced hippocampal volume and verbal memory performance associated with interleukin-6 and tumor necrosis factor-alpha levels in chemotherapy-treated breast cancer survivors. *Brain, Behavior, and Immunity, 30*(Suppl), S109–116.

Kessler, R. C. (1997). The effects of stressful life events on depression. *Annual Review of Psychology, 48,* 191–214.

Khairova, R. A., Machado-Vieira, R., Du, J., & Manji, H. K. (2009). A potential role for pro-inflammatory cytokines in regulating synaptic plasticity in major depressive disorder. *The International Journal of Neuropsychopharmacology, 12,* 561–578.

Kohler, O., Benros, M. E., Nordentoft, M., Farkouh, M. E., Iyengar, R. L., Mors, O., et al. (2014). Effect of anti-inflammatory treatment on depression, depressive symptoms, and adverse effects: a systematic review and meta-analysis of randomized clinical trials. *JAMA Psychiatry, 71,* 1381–1391.

Kole, M. H., Czeh, B., & Fuchs, E. (2004). Homeostatic maintenance in excitability of tree shrew hippocampal CA3 pyramidal neurons after chronic stress. *Hippocampus, 14,* 742–751.

Krishnan, V., & Nestler, E. J. (2008). The molecular neurobiology of depression. *Nature, 455,* 894–902.

Kullmann, J. S., Grigoleit, J. S., Lichte, P., Kobbe, P., Rosenberger, C., Banner, C., et al. (2013). Neural response to emotional stimuli during experimental human endotoxemia. *Human Brain Mapping 34*(9), 2217–2227.

Lee, T., Jarome, T., Li, S. J., Kim, J. J., & Helmstetter, F. J. (2009). Chronic stress selectively reduces hippocampal volume in rats: a longitudinal magnetic resonance imaging study. *Neuroreport, 20,* 1554–1558.

Lemaire, V., Koehl, M., Le Moal, M., & Abrous, D. N. (2000). Prenatal stress produces learning deficits associated with an inhibition of neurogenesis in the hippocampus. *Proceedings of the National Academy of Sciences of the United States of America, 97,* 11032–11037.

Leppänen, J. M. (2006). Emotional information processing in mood disorders: a review of behavioral and neuroimaging findings. *Current Opinion in Psychiatry, 19*(1), 34–39.

Liao, Y., Huang, X., Wu, Q., Yang, C., Kuang, W., Du, M., et al. (2013). Is depression a disconnection syndrome? Meta-analysis of diffusion tensor imaging studies in patients with MDD. *Journal of Psychiatry & Neuroscience, 38,* 49–56.

Lisiecka, D., Meisenzahl, E., Scheuerecker, J., Schopf, V., Whitty, P., Chaney, A., et al. (2011). Neural correlates of treatment outcome in major depression. *The International Journal of Neuropsychopharmacology, 14,* 521–534.

Lovallo, W. R., Robinson, J. L., Glahn, D. C., & Fox, P. T. (2010). Acute effects of hydrocortisone on the human brain: an fMRI study. *Psychoneuroendocrinology, 35,* 15–20.

Luine, V., Martinez, C., Villegas, M., Magarinos, A. M., & McEwen, B. S. (1996). Restraint stress reversibly enhances spatial memory performance. *Physiology & Behavior, 59,* 27–32.

Lupien, S. J., Wilkinson, C. W., Briere, S., Menard, C., Ng Ying Kin, N. M., & Nair, N. P. (2002). The modulatory effects of corticosteroids on cognition: studies in young human populations. *Psychoneuroendocrinology, 27,* 401–416.

Lyons, D. M., Yang, C., Sawyer-Glover, A. M., Moseley, M. E., & Schatzberg, A. F. (2001). Early life stress and inherited variation in monkey hippocampal volumes. *Archives of General Psychiatry, 58,* 1145–1151.

Macqueen, G., & Frodl, T. (2010). The hippocampus in major depression: evidence for the convergence of the bench and bedside in psychiatric research? *Molecular Psychiatry*.

Magarinos, A. M., McEwen, B. S., Flugge, G., & Fuchs, E. (1996). Chronic psychosocial stress causes apical dendritic atrophy of hippocampal CA3 pyramidal neurons in subordinate tree shrews. *The Journal of Neuroscience, 16,* 3534–3540.

Mann, J. J., & Currier, D. M. (2010). Stress, genetics and epigenetic effects on the neurobiology of suicidal behavior and depression. *European Psychiatry, 25,* 268–271.

Marsland, A. L., Gianaros, P. J., Abramowitch, S. M., Manuck, S. B., & Hariri, A. R. (2008). Interleukin-6 covaries inversely with hippocampal grey matter volume in middle-aged adults. *Biological Psychiatry, 64,* 484–490.

Marsland, A. L., Gianaros, P. J., Kuan, D. C., Sheu, L. K., Krajina, K., & Manuck, S. B. (2015). Brain morphology links systemic inflammation to cognitive function in midlife adults. *Brain, Behavior, and Immunity, 48,* 195–204.

McGowan, P. O., Sasaki, A., D'Alessio, A. C., Dymov, S., Labonte, B., Szyf, M., et al. (2009). Epigenetic regulation of the glucocorticoid receptor in human brain associates with childhood abuse. *Nature Neuroscience, 12,* 342–348.

McGowan, P. O., Suderman, M., Sasaki, A., Huang, T. C., Hallett, M., Meaney, M. J., et al. (2011). Broad epigenetic signature of maternal care in the brain of adult rats. *PLoS One, 6.*

McNally, L., Bhagwagar, Z., & Hannestad, J. (2008). Inflammation, glutamate, and glia in depression: a literature review. *CNS Spectrums, 13*, 501–510.

Miller, A. H., Maletic, V., & Raison, C. L. (2009). Inflammation and its discontents: the role of cytokines in the pathophysiology of major depression. *Biological Psychiatry, 65*, 732–741.

Murphy, M. L., & Frodl, T. (2011). Meta-analysis of diffusion tensor imaging studies shows altered fractional anisotropy occurring in distinct brain areas in association with depression. *Biology of Mood & Anxiety Disorders, 1*, 3.

Myint, A. M., & Kim, Y. K. (2014). Network beyond IDO in psychiatric disorders: revisiting neurodegeneration hypothesis. *Progress in Neuro-Psychopharmacology & Biological Psychiatry, 48*, 304–313.

Myint, A. M., Schwarz, M. J., & Muller, N. (2012). The role of the kynurenine metabolism in major depression. *Journal of Neural Transmission, 119*, 245–251.

Nestler, E. J. (2014). Epigenetic mechanisms of depression. *JAMA Psychiatry 71*(4), 454–456.

Nestler, E. J., Barrot, M., DiLeone, R. J., Eisch, A. J., Gold, S. J., & Monteggia, L. M. (2002). Neurobiology of depression. *Neuron, 34*, 13–25.

Newcomer, J. W., Selke, G., Melson, A. K., Hershey, T., Craft, S., Richards, K., et al. (1999). Decreased memory performance in healthy humans induced by stress-level cortisol treatment. *Archives of General Psychiatry, 56*, 527–533.

Nusslock, R., & Miller, G. E. (2016). Early-life adversity and physical and emotional health across the lifespan: a neuroimmune network hypothesis. *Biological Psychiatry 80*(1), 23–32.

Oberlander, T., Weinberg, J., Papsdorf, M., Grunau, R., Misri, S., & Devlin, A. (2008). Prenatal exposure to maternal depression, neonatal methylation of human glucocorticoid receptor gene (NR3C1) and infant cortisol stress responses. *Epigenetics, 3*, 97–106.

Ochsner, K. N., Ray, R. D., Cooper, J. C., Robertson, E. R., Chopra, S., Gabrieli, J. D., et al. (2004). For better or for worse: neural systems supporting the cognitive down- and up-regulation of negative emotion. *NeuroImage, 23*, 483–499.

Ohl, F., Michaelis, T., Vollmann-Honsdorf, G. K., Kirschbaum, C., & Fuchs, E. (2000). Effect of chronic psychosocial stress and long-term cortisol treatment on hippocampus-mediated memory and hippocampal volume: a pilot-study in tree shrews. *Psychoneuroendocrinology, 25*, 357–363.

Öngür, D., Ferry, A. T., & Price, J. L. (2003). Architectonic subdivision of the human orbital and medial prefrontal cortex. *The Journal of Comparative Neurology, 460*, 425–449.

Oxenkrug, G. (2005). Antioxidant effects of N-acetylserotonin: possible mechanisms and clinical implications. *Annals of the New York Academy of Sciences, 1053*, 334–347.

Oxenkrug, G. F. (2007). Genetic and hormonal regulation of tryptophan kynurenine metabolism: implications for vascular cognitive impairment, major depressive disorder, and aging. *Annals of the New York Academy of Sciences, 1122*, 35–49.

Pagliaccio, D., & Barch, D. M. (2015). *Early life adversity and risk for depression: Alterations in cortisol and brain structure and function as mediating mechanisms.* Amsterdam: Elsevier.

Pariante, C. M., & Lightman, S. L. (2008). The HPA axis in major depression: classical theories and new developments. *Trends in Neurosciences, 31*, 464–468.

Patten, S. B., Williams, J. V., & Love, E. J. (1995). A case-control study of corticosteroid exposure as a risk factor for clinically-diagnosed depressive disorders in a hospitalized population. *Canadian Journal of Psychiatry, 40*, 396–400.

Perez-Severiano, F., Escalante, B., & Rios, C. (1998). Nitric oxide synthase inhibition prevents acute quinolinate-induced striatal neurotoxicity. *Neurochemical Research, 23*, 1297–1302.

Preuss, U. W., Zetzsche, T., Pogarell, O., Mulert, C., Frodl, T., Muller, D., et al. (2010). Anterior cingulum volumetry, auditory P300 in schizophrenia with negative symptoms. *Psychiatry Research, 183*, 133–139.

Price, J. L., & Drevets, W. C. (2012). Neural circuits underlying the pathophysiology of mood disorders. *Trends in Cognitive Sciences, 16*, 61–71.

Pruessner, J. C., Dedovic, K., Khalili-Mahani, N., Engert, V., Pruessner, M., Buss, C., et al. (2008). Deactivation of the limbic system during acute psychosocial stress: evidence from positron emission tomography and functional magnetic resonance imaging studies. *Biological Psychiatry, 63*, 234–240.

Raichle, M. E., MacLeod, A. M., Snyder, A. Z., Powers, W. J., Gusnard, D. A., & Shulman, G. L. (2001). A default mode of brain function. *Proceedings of the National Academy of Sciences of the United States of America, 98*, 676–682.

Reichenberg, A., Yirmiya, R., Schuld, A., Kraus, T., Haack, M., Morag, A., et al. (2001). Cytokine-associated emotional and cognitive disturbances in humans. *Archives of General Psychiatry, 58*, 445–452.

Ressler, K. J., & Nemeroff, C. B. (2000). Role of serotonergic and noradrenergic systems in the pathophysiology of depression and anxiety disorders. *Depression and Anxiety, 12*(Suppl. 1), 2–19.

Reul, J. M., & de Kloet, E. R. (1985). Two receptor systems for corticosterone in rat brain: microdistribution and differential occupation. *Endocrinology, 117*, 2505–2511.

Root, J. C., Tuescher, O., Cunningham-Bussel, A., Pan, H., Epstein, J., Altemus, M., et al. (2009). Frontolimbic function and cortisol reactivity in response to emotional stimuli. *Neuroreport, 20*, 429–434.

Rubin, R. T., Phillips, J. J., Sadow, T. F., & McCracken, J. T. (1995). Adrenal gland volume in major depression. Increase during the depressive episode and decrease with successful treatment. *Archives of General Psychiatry*, *52*, 213–218.

Ryu, J. K., Choi, H. B., & McLarnon, J. G. (2006). Combined minocycline plus pyruvate treatment enhances effects of each agent to inhibit inflammation, oxidative damage, and neuronal loss in an excitotoxic animal model of Huntington's disease. *Neuroscience*, *141*, 1835–1848.

Santarelli, L., Saxe, M., Gross, C., Surget, A., Battaglia, F., Dulawa, S., et al. (2003). Requirement of hippocampal neurogenesis for the behavioral effects of antidepressants. *Science*, *301*, 805–809.

Sapolsky, R. M. (2003). Stress and plasticity in the limbic system. *Neurochemical Research*, *28*, 1735–1742.

Sapolsky, R. M., Krey, L. C., & McEwen, B. S. (1986). The neuroendocrinology of stress and aging: the glucocorticoid cascade hypothesis. *Endocrine Reviews*, *7*, 284–301.

Schmaal, L., Veltman, D. J., van Erp, T. G., Sämann, P. G., Frodl, T., Jahanshad, N., et al. (2016). Subcortical brain alterations in major depressive disorder: findings from the ENIGMA Major Depressive Disorder working group. *Molecular Psychiatry*, *21*(6), 806–812.

Seeley, W. W., Menon, V., Schatzberg, A. F., Keller, J., Glover, G. H., Kenna, H., et al. (2007). Dissociable intrinsic connectivity networks for salience processing and executive control. *The Journal of Neuroscience*, *27*, 2349–2356.

Sexton, C., Mackay, C., & Ebmeier, K. (2009). A systematic review of diffusion tensor imaging studies in affective disorders. *Biological Psychiatry*, *66*, 814–823.

Sheline, Y. I. (2011). Depression and the hippocampus: cause or effect? *Biological Psychiatry*, *70*, 308–309.

Sheline, Y. I., Price, J. L., Yan, Z., & Mintun, M. A. (2010). Resting-state functional MRI in depression unmasks increased connectivity between networks via the dorsal nexus. *Proceedings of the National Academy of Sciences*, *107*, 11020–11025.

Simmons, N. E., Do, H. M., Lipper, M. H., & Laws, E. R., Jr. (2000). Cerebral atrophy in Cushing's disease. *Surgical Neurology*, *53*, 72–76.

Slavich, G. M., Way, B. M., Eisenberger, N. I., & Taylor, S. E. (2010). Neural sensitivity to social rejection is associated with inflammatory responses to social stress. *Proceedings of the National Academy of Sciences of the United States of America*, *107*, 14817–14822.

Sloane, J. A., Hollander, W., Moss, M. B., Rosene, D. L., & Abraham, C. R. (1999). Increased microglial activation and protein nitration in white matter of the aging monkey. *Neurobiology of Aging*, *20*, 395–405.

Sluzewska, A., Rybakowski, J. K., Laciak, M., Mackiewicz, A., Sobieska, M., & Wiktorowicz, K. (1995). Interleukin-6 serum levels in depressed patients before

and after treatment with fluoxetine. *Annals of the New York Academy of Sciences*, *762*, 474–476.

Sonino, N., Fava, G. A., Raffi, A. R., Boscaro, M., & Fallo, F. (1998). Clinical correlates of major depression in Cushing's disease. *Psychopathology*, *31*, 302–306.

Spinelli, S., Chefer, S., Suomi, S. J., Higley, J. D., Barr, C. S., & Stein, E. (2009). Early-life stress induces long-term morphologic changes in primate brain. *Archives of General Psychiatry*, *66*, 658–665.

Steiner, J., Walter, M., Gos, T., Guillemin, G. J., Bernstein, H. G., Sarnyai, Z., et al. (2011). Severe depression is associated with increased microglial quinolinic acid in subregions of the anterior cingulate gyrus: evidence for an immune-modulated glutamatergic neurotransmission? *Journal of Neuroinflammation*, *8*, 94.

Stockmeier, C. A., Mahajan, G. J., Konick, L. C., Overholser, J. C., Jurjus, G. J., Meltzer, H. Y., et al. (2004). Cellular changes in the postmortem hippocampus in major depression. *Biological Psychiatry*, *56*, 640–650.

Sundermann, B., Olde Lutke Beverborg, M., & Pfleiderer, B. (2014). Toward literature-based feature selection for diagnostic classification: a meta-analysis of resting-state fMRI in depression. *Frontiers in Human Neuroscience*, *8*, 692.

Taylor, M. J. (2014). Could glutamate spectroscopy differentiate bipolar depression from unipolar? *Journal of Affective Disorders*, *167*, 80–84.

Teicher, M. H., Andersen, S. L., Polcari, A., Anderson, C. M., Navalta, C. P., & Kim, D. M. (2003). The neurobiological consequences of early stress and childhood maltreatment. *Neuroscience and Biobehavioral Reviews*, *27*, 33–44.

Teicher, M. H., & Samson, J. A. (2013). Childhood maltreatment and psychopathology: a case for ecophenotypic variants as clinically and neurobiologically distinct subtypes. *The American Journal of Psychiatry*, *170*, 1114–1133.

Thomas, R., Hotsenpiller, G., & Peterson, D. (2007). Acute psychosocial stress reduces cell survival in adult hippocampal neurogenesis without altering proliferation. *Journal of Neuroscience*, *27*, 2734.

Thomas, S. R., & Stocker, R. (1999). Redox reactions related to indoleamine 2,3-dioxygenase and tryptophan metabolism along the kynurenine pathway. *Redox Report*, *4*, 199–220.

Trotta, A., Murray, R. M., & Fisher, H. L. (2015). The impact of childhood adversity on the persistence of psychotic symptoms: a systematic review and meta-analysis. *Psychological Medicine*, *45*, 2481–2498.

Tsankova, N., Berton, O., Renthal, W., Kumar, A., Neve, R., & Nestler, E. (2006). Sustained hippocampal chromatin regulation in a mouse model of depression and antidepressant action. *Nature Neuroscience*, *9*, 519–525.

Tunnard, C., Rane, L. J., Wooderson, S. C., Markopoulou, K., Poon, L., Fekadu, A., et al. (2014). The impact of childhood adversity on suicidality and clinical course in treatment-resistant depression. *Journal of Affective Disorders, 152–154*, 122–130.

Ugwu, I. D., Amico, F., Carballedo, A., Fagan, A. J., & Frodl, T. (2015). Childhood adversity, depression, age and gender effects on white matter microstructure: a DTI study. *Brain Structure & Function 220*(4), 1997–2009.

Valentine, A. D., Meyers, C. A., Kling, M. A., Richelson, E., & Hauser, P. (1998). Mood and cognitive side effects of interferon-alpha therapy. *Seminars in Oncology, 25*, 39–47.

van den Heuvel, M., Mandl, R., Luigjes, J., & Hulshoff Pol, H. (2008). Microstructural organization of the cingulum tract and the level of default mode functional connectivity. *The Journal of Neuroscience, 28*, 10844–10851.

van Stegeren, A. H., Wolf, O. T., Everaerd, W., Scheltens, P., Barkhof, F., & Rombouts, S. A. (2007). Endogenous cortisol level interacts with noradrenergic activation in the human amygdala. *Neurobiology of Learning and Memory, 87*, 57–66.

Vasic, N., Walter, H., Sambataro, F., & Wolf, R. C. (2008). Aberrant functional connectivity of dorsolateral prefrontal and cingulate networks in patients with major depression during working memory processing. *Psychological Medicine, 39*, 1–11.

Walter, M., Henning, A., Grimm, S., Schulte, R. F., Beck, J., Dydak, U., et al. (2009). The relationship between aberrant neuronal activation in the pregenual anterior cingulate, altered glutamatergic metabolism, and anhedonia in major depression. *Archives of General Psychiatry, 66*, 478–486.

Wang, A. K., & Miller, B. J. (2018). Meta-analysis of cerebrospinal fluid cytokine and tryptophan catabolite alterations in psychiatric patients: xomparisons between schizophrenia, bipolar disorder, and depression. *Schizophrenia Bulletin 44*(1), 75–83.

Wang, D., Szyf, M., Benkelfat, C., Provencal, N., Turecki, G., Caramaschi, D., et al. (2012). Peripheral SLC6A4 DNA methylation is associated with in vivo measures of human brain serotonin synthesis and childhood physical aggression. *PLoS One, 7*.

Wang, L., Krishnan, K. R., Steffens, D. C., Potter, G. G., Dolcos, F., & McCarthy, G. (2008). Depressive state- and disease-related alterations in neural responses to affective and executive challenges in geriatric depression. *The American Journal of Psychiatry, 165*, 863–871.

Weaver, I., Cervoni, N., Champagne, F., D'Alessio, A., Sharma, S., Seckl, J., et al. (2004). Epigenetic programming by maternal behavior. *Nature Neuroscience, 7*, 847–854.

Wellman, C. L. (2001). Dendritic reorganization in pyramidal neurons in medial prefrontal cortex after chronic corticosterone administration. *Journal of Neurobiology, 49*, 245–253.

Woolley, C. S., Gould, E., Frankfurt, M., & McEwen, B. S. (1990). Naturally occurring fluctuation in dendritic spine density on adult hippocampal pyramidal neurons. *Journal of Neuroscience, 10*, 4035–4039.

Wright, C. E., Strike, P. C., Brydon, L., & Steptoe, A. (2005). Acute inflammation and negative mood: mediation by cytokine activation. *Brain, Behavior, and Immunity, 19*, 345–350.

Young, A. H., Sahakian, B. J., Robbins, T. W., & Cowen, P. J. (1999). The effects of chronic administration of hydrocortisone on cognitive function in normal male volunteers. *Psychopharmacology, 145*, 260–266.

Zhou, Y., Yu, C., Zheng, H., Liu, Y., Song, M., Qin, W., et al. (2010). Increased neural resources recruitment in the intrinsic organization in major depression. *Journal of Affective Disorders, 121*, 220–230.

16

Structural Neuroimaging of Maltreatment and Inflammation in Depression

Ronny Redlich*, Nils Opel*, Katharina Förster*, Jennifer Engelen[†], Udo Dannlowski*

*University of Münster, Münster, Germany
[†]University of Marburg, Marburg, Germany

INTRODUCTION

Major depressive disorder (MDD) is one of the most debilitating diseases worldwide (World Health Organization, 2001). Its social and economic burden is one of the major challenges in public health. Since effective treatment and prevention require knowledge of risk factors and their neurobiological mechanisms, there is a need for more detailed understanding of the neurobiological implications of this disease.

Emerging evidence suggests that MDD entails extensive alterations in the inflammatory system, including elevated levels of proinflammatory cytokines. Furthermore, neuroimaging studies repeatedly have demonstrated functional and structural aberrations in cerebral areas related to emotion and reward processing in patients with MDD (Cusi, Nazarov, Holshausen, Macqueen, & McKinnon, 2012; Redlich et al., 2015; Schmaal et al., 2017; Stuhrmann, Suslow, & Dannlowski, 2011). However, it is still unknown whether these inflammatory and neurostructural alterations

represent a consequence or a predisposition of depression.

Childhood maltreatment is a leading risk factor for depression. Although the association between maltreatment and depression has been extensively proven, there is an urgent need to understand how maltreatment increases the risk of depression. In the last decade, there has been increasing focus on the role of inflammation and its cross talk with neurocircuits as a potential mechanism to bridge this gap. There is rising evidence that early-life stress could act through both a modulation of inflammatory responses and neurostructural alterations up to adulthood or over the life span.

The aim of the present chapter is to provide an overview of our current knowledge about the aforementioned associations. First, the effects of maltreatment and early traumatic experiences on both brain structure and inflammatory processes are represented. Subsequently, we summarize the current knowledge of the relationship between childhood maltreatment and depression by linking inflammatory processes

Inflammation and Immunity in Depression
https://doi.org/10.1016/B978-0-12-811073-7.00016-7

and neurostructural alterations. Finally, the clinical importance and implications are summarized.

EARLY-LIFE STRESS, INFLAMMATION, AND BRAIN STRUCTURE

Especially in early life, the brain is sensitive to stress, most probably caused by its continuous developmental changes during childhood and adolescence. Therefore, it is hardly surprising that negative life events such as singular trauma or the experience of maltreatment have not only psychopathological (Edwards, Holden, Felitti, & Anda, 2003; Green et al., 2010) but also brain developmental consequences. It is well known that early-life stress leads to extensive structural brain alterations in both human and animal studies (Lupien, McEwen, Gunnar, & Heim, 2009; Paus, Keshavan, & Giedd, 2008; Sánchez, Ladd, & Plotsky, 2001; Teicher, Samson, Anderson, & Ohashi, 2016). Furthermore, emerging evidence suggests a role of increased inflammation or increased sensitivity of inflammatory responses in maltreated subjects (Cattaneo et al., 2015; Grosse et al., 2016).

Effects of Early-Life Stress on Brain Structure

Traumatic early-life experiences have been shown to exacerbate transformation processes of cerebral structures (Teicher et al., 2004; van Harmelen et al., 2010). In particular, hippocampal volume reductions have repeatedly been detected in subjects affected by childhood maltreatment (Dannlowski et al., 2012; Teicher & Samson, 2013) and in subjects who developed a posttraumatic stress disorder (PTSD) (Gilbertson et al., 2002; Woon, Sood, & Hedges, 2010).

In one of the first studies on brain structural alterations and early-life stress, De Bellis et al. (1999) reported a reduced intracranial and cerebral volume in maltreated children who had developed PTSD, compared with nonmaltreated healthy subjects. This volume reduction correlated positively with the age of onset of maltreatment. Further analysis showed a negative correlation of both intracranial and corpus callosum volume with duration of maltreatment, while the volume of lateral ventricles correlated positively with the duration of maltreatment (De Bellis et al., 1999).

In the last decade, emerging evidence has been pointing to the hippocampal formation as the key brain area affected by both maltreatment and PTSD. A metaanalysis including adult patients suffering from PTSD, trauma-exposed controls who did not develop PTSD, and a second trauma-unexposed healthy control group showed reduced total, left, and right hippocampal volumes in the two trauma-exposed groups compared with the trauma-unexposed control group (Woon et al., 2010). An early pilot study on monozygotic twins showed comparable hippocampal volumes in twins who were not themselves exposed to combat and their combat-exposed brothers. The authors conclude that smaller hippocampal volumes are preexisting vulnerability factors rather than products of combat-related trauma or PTSD (Gilbertson et al., 2002). In line with these results are the findings of retrospective studies on depressed patients and patients suffering from PTSD, showing reduced volume of the left hippocampus only in those patients who reported childhood physical or sexual abuse and a reduction of the left hippocampus correlating to severity of emotional neglect, duration of sexual abuse, or severity of PTSD symptoms (Bremner et al., 1997; Frodl, Reinhold, Koutsouleris, Reiser, & Meisenzahl, 2010; Stein, Koverola, Hanna, Torchia, & McClarty, 1997; Vythilingam et al., 2002).

Moreover, there is growing evidence that early childhood maltreatment is associated with brain structural changes irrespective of the history of psychiatric disorders (Chaney et al., 2014; Edmiston et al., 2011; Frodl, Reinhold,

Koutsouleris, Reiser, et al., 2010; Vythilingam et al., 2002). In a study with maltreated and non-maltreated hospitalized children, Teicher et al. (2004) reported decreased corpus callosum volumes, which were associated with maltreatment rather than disease. Furthermore, Teicher, Anderson, and Polcari (2012) showed that hippocampal alterations in maltreated subjects are not mediated by the presence or absence of life-time affective disorders. In line with these results is a study that included only healthy adults without any history of psychiatric disorders. As in comparable patient-based research, this study showed that hippocampal atrophy correlated positively with maltreatment, while physical neglect and emotional abuse were the strongest predictors for decreased left hippocampal volumes (Dannlowski et al., 2012). Even in the absence of sexual or other forms of physical abuse, emotional maltreatment in early life is associated with a significant volume reduction predominantly of the left dorsal medial prefrontal cortex, independent of gender and psychiatric status (van Harmelen et al., 2010). This notion was further corroborated by Opel et al., who demonstrated a strong impact of childhood maltreatment on hippocampal volumes, in patients and healthy subjects alike. In this study, no significant disparities in hippocampal volumes between patients and healthy controls could be discerned when the impact of childhood maltreatment was ruled out in the analysis of group differences, suggesting that frequently observed hippocampal atrophy in MDD patients could be a function of previous maltreatment experiences rather than the diagnosis itself (Opel et al., 2014).

Effects of Early-Life Stress on Hypothalamic-Pituitary-Adrenal (HPA) Axis and Inflammation

Apart from the reported brain alterations, childhood maltreatment modulates biological processes like the inflammatory or neuroendocrine system. Severe stress, like the experience of maltreatment, activates the hypothalamic-pituitary-adrenal (HPA) axis leading to a release of glucocorticoid hormones (Herman et al., 2003), which in turn affects the inflammatory system by enhancing or suppressing the immune response (Cain & Cidlowski, 2017). Studies comparing maltreated children to non-maltreated ones showed that alterations in HPA axis activity are associated with the duration of maltreatment and increased psychopathological symptom severity (De Bellis et al., 1999; Puetz et al., 2016), and they tend to persist into adulthood (Heim et al., 2002; Heim, Newport, Mletzko, Miller, & Nemeroff, 2008; Tarullo & Gunnar, 2006). The hippocampal and prefrontal cortical areas are sensitive to glucocorticoids due to their high density of glucocorticoid receptors, and therefore, these brain areas are specifically susceptible to chronic stress. Consequently, glucocorticoids have repeatedly been demonstrated to mediate the effect of stress on the hippocampal and prefrontal brain structural changes (Cerqueira et al., 2005; Conrad, 2008; Wellman, 2001). Glucocorticoid-driven effects could include impaired neurogenesis, atrophy of dendritic processes, and—at an extreme—neurotoxicity (Diorio, Viau, & Meaney, 1993; Sapolsky, 2000).

In the last decade, inflammatory processes have been investigated as potential mediators of the effects between glucocorticoids and brain structure since glucocorticoids are potent regulators of inflammation. Although our understanding of mechanisms that underlie the immune-regulatory effects of glucocorticoids is limited, there is emerging evidence to suggest that inflammatory processes in general have wide spread effects on brain structure in both individuals with neuropsychiatric disorders and healthy individuals (for a detailed summary please view Box 1).

Although studies directly investigating childhood maltreatment, inflammation, and brain structures are lacking, some studies have investigated the influence of childhood maltreatment on inflammation in general. These studies have

BOX 1

INFLAMMATORY EFFECTS ON BRAIN STRUCTURE IN HEALTHY CONTROLS AND INDIVIDUALS WITH PSYCHIATRIC DISORDERS

Only a few studies have investigated the interaction of pro-inflammatory cytokines and brain structure in healthy adults. For example, Marsland, Gianaros, Abramowitch, Manuck, and Hariri (2008) found in a healthy control group that higher periphery Interleukin 6 (IL-6) levels were associated with smaller hippocampal volume. Baune et al. (2012) investigated variations of the tumor necrosis factor-α (TNFα) gene on brain structure in a large population. It was found that the hippocampal volume was associated with two independent TNFα SNPs (rs1800629 and rs361525). Since TNFα has been found to play a neuroprotective and neurodegenerative role (Sriram & O'Callaghan, 2007), another study has investigated the effects of TNF receptors TFNR1 and TFNR2 on the hippocampal and striatal volume (Stacey et al., 2017). The authors found a brain region-specific effect—TFNR1 was associated with hippocampal morphology, while TFNR2 was associated with striatal morphology. Additionally, it has been shown in different cohorts that greater C-reactive protein (CRP) levels are associated with cortical thinning (Krishnadas et al., 2013). Also, Jefferson et al. (2007) have shown a smaller whole brain volume for higher inflammatory markers in total, especially IL-6. Further evidence for a link between inflammation and brain structure comes from animal studies. The systemic administration of lipopolysaccharide (LPS) induces the expression of pro-inflammatory cytokine mRNAs and proteins in the brain (Layé, Parnet, Goujon, & Dantzer, 1994; Quan, Stern, Whiteside, & Herkenham, 1999; van Dam, Brouns, Louisse, & Berkenbosch, 1992). Pang, Cai, and Rhodes (2003) injected LPS into neonatal

rat brains in order to induce inflammation. The inflamed rat brains were less myelinated, had a disrupted development of oligodendrocytes, and damaged white matter compared with the control group. Stone, Lehmann, Lin, and Quartermain (2006) used LPS in combination with the immediate early gene (IEG) c-fos to map the brain areas that are affected by inflammation. IEGs are activated as a response mechanism during transcription, before new proteins are synthesized. The results of the study by Stone et al. indicate a possible role of the hippocampus, extended amygdala, and hypothalamus in the neural processing of pro-inflammatory cytokines from periphery. This would be consistent with the proposed implication of these brain areas in affective disorders (Phillips, Drevets, Rauch, & Lane, 2003a, 2003b). However, the presented results do not imply that there is a full overlap between brain areas that are involved in MDDs and those that are associated with higher cytokine levels (Dantzer, O'Connor, Freund, Johnson, & Kelley, 2008).

Apart from the studies in healthy individuals, inflammatory effects on brain structure have been observed in schizophrenia, MDD, and several neurodegenerative disease such as Alzheimer's disease (AD) (Frodl & Amico, 2014). In the first episode of psychosis, in line with findings in healthy subjects, IL-6 levels can predict the hippocampal volume (Mondelli et al., 2011). Also, studies using a peripheral benzodiazepine receptor ligand that binds activated microglia with positron emission tomography have elucidated the pathogenesis of psychosis. They found a higher binding potential in the hippocampus, indicating focal neuroinflammation in this region

BOX 1 *(cont'd)*

(Doorduin et al., 2009). In AD, higher levels of pro-inflammatory cytokines in cerebrospinal fluid and plasma were associated with higher AD severity and lower medial temporal lobe volume (Matsumoto et al., 2008). Due to the common AD diagnostic procedure, research concerning the role of pro-inflammatory cytokines in brain structure has the advantage of using measures of cerebrospinal fluid, which improve prediction accuracy (Frodl & Amico, 2014). A recent review by Byrne, Whittle, and Allen (2016) found that there are only three studies investigating the structural changes and inflammation with respect to MDD. Plenty of research on the role of inflammation and brain structure within the pathogenesis of this disease has been conducted, but only a few studies have actually examined whether and how inflammation affects brain structure. Savitz et al. (2013) investigated this concern with the whole-genome expression analysis of peripheral blood mononuclear cells and morphometric scans. They found 12 inflammation and neurological disease-related protein-coding genes that differed in expression between the MDD patients and the control group. Their transcripts were significantly correlated with the thickness of the left subgenual ACC and the volume of hippocampus and caudate. All three structures have previously been associated with

MDD. The authors concluded that it is possible to map depression-associated immune dysfunction onto depression-related morphological abnormalities in the brain. Dowell et al. (2016) found that inflammation during interferon-alpha (IFN-α) treatment induced acute changes in the brain. IFN-α is commonly used to treat hepatitis C and is associated with depression-like behavioral changes such as fatigue. The authors used quantitative magnetization transfer that allows the investigation of microstructural changes in the MRI. They found acute changes in the ventral striatum—which predicted the subsequent onset of fatigue—and more constrained changes to the insula that however were not correlated with depression-like symptoms. These findings point out the sensitivity of striatal structures, explicitly the basal ganglia, to acute changes in peripheral IFN-α. Furthermore, a genetic variant of IFN-γ moderates effects of early-life stress on emotion processing, which reiterate the importance that inflammatory genes play in the interaction with early-life stress and the regulation of emotion processing (Redlich et al., 2015). Finally, Frodl et al. (2012) found that higher IL-6 concentration was associated with smaller hippocampal volume in patients with MDD, but not in healthy controls.

shown that adverse events during middle childhood are associated with increased levels of IL-6, TNFα, and CRP in adolescence (Grosse et al., 2016; Slopen, Kubzansky, McLaughlin, & Koenen, 2013). Like the neuroendocrine effects of maltreatment, these alterations in the inflammatory system seem to persist into adulthood too. Accordingly, studies using an acute stress challenge in depressed adults with childhood

maltreatment experiences showed increased levels of IL-6 and CRP after stress exposure compared with healthy controls (Danese et al., 2008; Pace et al., 2006). In line with these results, recent metaanalyses reported an association between childhood trauma and CRP and IL-6 and TNFα responses (Baumeister, Akhtar, Ciufolini, Pariante, & Mondelli, 2016; Coelho, Viola, Walss-Bass, Brietzke, & Grassi-Oliveira, 2014).

Even in healthy adults, the former experience of childhood maltreatment leads to increased basal levels of IL-6, TNFα, and IL-1β and an increased IL-6 response to an acute stress challenge. Therefore, enhanced inflammatory responsiveness not only might be a function of stress-related disorders but also is induced by stress—in this context maltreatment—itself (Carpenter et al., 2010; Hartwell et al., 2013). At the same time, these results point to the role of trauma-related inflammatory alterations for an increased risk of stress-related disorders.

Taken together, the reactivity of the HPA axis and the inflammatory response to stressful life events are widely either directly or indirectly influenced by childhood maltreatment. Furthermore, several brain structures are affected by childhood maltreatment, particularly prefrontal cortical areas and most consistent hippocampal volumes whereby different brain structures seem to have different sensitive periods depending on their developmental stages. Given the pivotal role of glucocorticoids, the abovementioned findings point to broader implications of altered immune responses in brain structural development (Frodl & Amico, 2014; Hanning, Roesler, Peters, Berger, & Baune, 2016) whereby these alterations seem to persist into adulthood and therefore could be an important factor for the development of stress-related psychiatric disorders in later life (Miller & Raison, 2016).

DEPRESSION—A FUNCTION OF EARLY-LIFE STRESS?

Similar to findings on maltreatment, neuroimaging research in affective disorders has consistently shown alterations in limbic structures, including the hippocampus, the striatum, and the cingulate cortex (Arnone, McIntosh, Ebmeier, Munafò, & Anderson, 2012; Redlich et al., 2014). Of these findings, reduction in hippocampal volume is probably the most frequently reported structural correlate in

neuroimaging studies when comparing MDD patients with a healthy control group (Cole, Costafreda, McGuffin, & Fu, 2011; MacQueen & Frodl, 2011). This observation is corroborated by recent metaanalyses; however, the debate on its underpinnings prevails (Schmaal et al., 2016). In fact, some studies indicate that reduced hippocampal volumes might be apparent before the onset of disease in subjects at risk of depression (Amico et al., 2011). Hence, hippocampal alterations have been suspected to mediate the effect of long-term stressful life events on depression risk rather than as a constituting disease marker (Frodl, Reinhold, Koutsouleris, Reiser, et al., 2010).

Stress and Brain Structure in MDD—Possible Mechanisms

The hippocampus is a structure showing the highest capacity for neuroplasticity in the human brain as the genesis of hippocampal neurons has been shown to occur throughout life that might be triggered by environmental and genetic influences (Covic, Karaca, & Lie, 2010; Eriksson et al., 1998). Thus, impaired hippocampal neurogenesis and synaptic dysfunction are susceptibility factors to play a key role in the pathophysiology of depressive disorders (Duman & Aghajanian, 2012). The prefrontal cortex—another key structure thought to be deeply involved in the etiology of MDD—has also been shown to be highly sensitive to acute and chronic exposure to stress (Arnsten, 2009; McEwen & Morrison, 2013). Moreover, it has been suggested that compared with structural impairment in the hippocampus, alterations in prefrontal architecture might be caused by much shorter periods of stress exposure (Brown, Henning, & Wellman, 2005; McEwen, 2004).

As discussed above, the pronounced neuroplastic capacity of brain structures such as the hippocampus and the prefrontal cortex is influenced by elevated levels of glucocorticoids typically involved in stress regulation. Hence, gray matter loss in MDD might be induced by stressful life events and adverse experiences (Conrad, 2008; Krishnan & Nestler, 2008; Wang, Huang, & Hsu, 2010). More precisely, the detrimental impact of chronic stress and adverse stressful life events seems to be dramatically increased in decisive periods of brain development during childhood (Heim et al., 2008).

Maltreatment as a Confounding Factor in MDD Research

Childhood maltreatment is one of the most specific risk factors for the development of affective disorders (Gilbert et al., 2009). Unsurprisingly, its prevalence is highly elevated in MDD populations (Scott, McLaughlin, Smith, & Ellis, 2012). Moreover, it has been shown in MDD patients that the extent of adverse early-life experiences aggravates the course and outcome of the disorder (Nanni, Uher, Ph, & Danese, 2012). Given the broad evidence for the detrimental impact of maltreatment and chronic stress on brain structure and the immune system as detailed above, these findings reveal several important consequences for the concept of morphometric abnormalities in MDD.

First, since childhood maltreatment appears to show a massive impact on the hippocampal volume and structure in patients and controls alike, it can be assumed that adverse early-life experiences are a powerful confounder in the direct comparisons of these groups regarding brain structure alterations. It should also be considered that patients who have suffered from maltreatment experiences might even constitute a distinct clinical subtype that shares a common set of neurobiological alterations. Corresponding to this notion, it has already been proposed that childhood maltreatment massively alters the phenotype and associated neurobiological markers (Teicher & Samson, 2013), including brain morphology (Teicher et al., 2012), neural responsiveness (Dannlowski et al., 2012), and HPA axis homeostasis (Heim et al., 2008).

As a consequence, future studies on MDD should control for the possible impact of maltreatment experiences to avoid any confounding of neurobiological findings in depression by this important factor.

Trait Characteristic of Limbic Alterations: Impact of Early Life Stress and Genetic Influence on Brain Structure in Depression

The second implication concerns the debate regarding state- or trait-related characteristics of hippocampal volume loss in depressed patients and structural changes in the brain in general. The present literature provides evidence that hippocampal atrophy is at least partly acquired through early-life experiences and might therefore constitute a trait-like risk factor for developing depression in later life. The idea of a "limbic scar" is supported by imaging studies that show similar associations of MDD, maltreatment, and hippocampal changes (Teicher et al., 2012; Vythilingam et al., 2002). Remarkably, a study by Rao et al. (2010) demonstrated that smaller hippocampal volumes indeed partially mediate the effect of early-life adversity on the development of MDD during longitudinal follow-up. These findings imply that hippocampal aberrations could act as the connecting element in the development of severe psychopathology in maltreated individuals. Given the evidence presented above on the impact of inflammatory processes on brain structure and function, it appears reasonable to assume that chronic stress might trigger inflammatory responses, which in turn affect brain structure development and might lead to increased vulnerability to develop severe psychopathology.

Further evidence for a trait characteristic of brain structural alterations in depression comes from studies showing morphometric anomalies in healthy individuals at familial risk of depression (Chen, Hamilton, & Gotlib, 2010; Opel et al., 2016). Moreover, reports on the heritability of brain structure in general and, more particularly, of hippocampal morphometry (Hibar et al., 2015; Stein et al., 2012) and several reports on genetic variations influencing the shape alterations in the hippocampus support the hypothesis of hippocampal volume reductions as a predisposing trait of MDD (Frodl et al., 2008; Frodl, Reinhold, Koutsouleris, Donohoe, et al., 2010; Stein et al., 2012). In addition, in a study on monozygotic twins, discordant for trauma exposure showed that smaller hippocampi indeed constitute a risk factor for the development of stress-related psychopathology (Gilbertson et al., 2002). Results of recent epigenetic studies supplement this hypothesis by revealing significantly altered methylation in the genes involved in pathophysiological mechanisms, such as neuronal plasticity in individuals suffering from early-life abuse (Labonté et al., 2012).

Taken together, there is evidence suggesting that altered brain structure—first and foremost hippocampal atrophy—is likely to be a preexisting condition that renders the brain more vulnerable to the development of affective disorders.

Clinical Features and Antidepressant Treatment—Evidence for State Dependent Brain Structural Changes in Depression

The concept of trait-related limbic aberrations is further supported by studies revealing that a reduction in hippocampal volumes in depressed patients is independent of the severity of depressive symptoms, age at onset, and duration of illness (Frodl et al., 2002). In contrast, other neuroimaging studies found hippocampal volume loss to be linked to clinical characteristics such as the duration of illness and number of depressive episodes (Mckinnon, Yucel, Nazarov, & Macqueen, 2009) that indicated a progressive

loss of hippocampal volume in the course of disease. Similarly, Sheline, Gado, and Kraemer (2003) demonstrated an association between reduced hippocampal volume and duration of untreated depressive periods. However, these findings are compatible with the concept of smaller hippocampal volumes as a characteristic of high-risk subjects, indicating a greater risk of a more severe course of disease. Moreover, reports on HPA axis dysregulation in acutely depressed subjects suggest that each depressive episode can be regarded as a stressful life event, which has adverse effects on brain structure comparable to the mechanisms described above, which link chronic stress to altered brain structure (Krishnan & Nestler, 2008).

Another aspect to be considered is the growing evidence that changes in hippocampal volume may depend on receiving psychiatric treatment, such as an increase in hippocampal volume following several weeks of antidepressant or electroconvulsive therapy (ECT) treatment (Arnone et al., 2013; Nordanskog et al., 2010; Redlich et al., 2016). At first sight, these state-dependent limbic alterations seem to challenge the notion of a trait-like preexisting condition. However, these different findings are not necessarily contradictory, since it could well be that preclinically existent alterations in limbic pathways are still influenced by antidepressant medication or electroconvulsive therapy. In fact, neuronal deficits caused by stressful life events have been found to be at least partly reversed by antidepressant medication (Duman & Aghajanian, 2012). Furthermore, it has been demonstrated that an increase in hippocampal volume during antidepressant treatment is not associated with depressive symptom relief, which might again be interpreted as another hint suggesting that hippocampal structure does not exert a direct or causal effect on illness symptomatology, that is, on depressive state (Redlich et al., 2016).

In summary, there is broad evidence suggesting that chronic stress in early periods of life alters the physiological brain structure development. HPA axis dysregulation and neuroinflammatory processes are the top candidates among different possible pathways that might mediate this connection. Specific brain structure alterations—particularly hippocampal volume loss—might thus be preferably characterized as a preexisting condition that is present before the first occurrence of depressive symptoms and that indicates high vulnerability to the onset and a possible unfavorable course of the disorder.

CONCLUSIONS AND CLINICAL IMPLICATIONS

This chapter discusses how emerging evidence links early-life stress and maltreatment, peripheral inflammation, alterations in neuroplastic mechanisms, and depression. In summary, there is broad evidence suggesting that chronic stress in early periods of life alters the physiological brain structural development, potentially mediated by HPA axis dysregulation and neuroinflammatory processes. In turn, these structural alterations and dysfunctional neuroendocrine response to stressful life events could be an important factor in the development of stress-related psychiatric disorders—particularly depression—in later life by increasing the vulnerability to the onset and to an unfavorable course of mental illness. However, the exact underlying biological mechanisms and mediating pathways still need to be clarified.

Clinically, this highlights the importance of prevention particularly in children and young people who have experienced childhood maltreatment. Children who have experienced childhood maltreatment and also show corresponding neurobiological and inflammatory markers might benefit from special attention in order to avert a potential onset of depression. A prevention strategy should preferably aim to reduce further loss in hippocampal volume through additional stressors, to avoid further

increase in the vulnerability to depression. One approach to accomplish this goal might be psychotherapy, pharmacotherapy, or other treatments, probably even before onset of depression, to reduce the limbic sensitivity and normalize the hippocampus volume, as has been shown by several recent studies in adults (Arnone et al., 2013; Harmer, Mackay, Reid, Cowen, & Goodwin, 2006; Redlich et al., 2016). Another approach could aim to reduce glucocorticoid levels by using antiglucocorticoid agents and antagonists. It is well known that glucocorticoids are deeply involved in immune response, which might be a potential mediator for the reduction in depression symptoms. The clinical efficacy of glucocorticoids in treating inflammatory and autoimmune disease is clear, yet the molecular mechanisms that underlie the immune-regulatory effects of glucocorticoids are still being elucidated, and there are still substantial gaps in our knowledge of glucocorticoid-mediated regulation of immunity (Cain & Cidlowski, 2017). However, the first results are promising although this therapy is still at a proof-of-concept stage (Gallagher et al., 2008). In sum, apart from family and childhood maltreatment anamnesis, additional information about the hippocampal morphometry and inflammation markers could enable us to identify high-risk subjects and develop preventive measures for those high-risk subjects even before the onset of depression. However, to date, preventive neurobiological interventions are still in a proof-of-concept stage, and further research is urgently needed in this field.

References

Amico, F., Meisenzahl, E. M., Koutsouleris, N., Reiser, M., Möller, H.-J., & Frodl, T. (2011). Structural MRI correlates for vulnerability and resilience to major depressive disorder. *Journal of Psychiatry & Neuroscience, 36*(1), 15–22. https://doi.org/10.1503/jpn.090186.

Arnone, D., McIntosh, A. M., Ebmeier, K. P., Munafò, M. R., & Anderson, I. M. (2012). Magnetic resonance imaging studies in unipolar depression: systematic review and meta-regression analyses. *European Neuropsychopharmacology*, *22*(1), 1–16. https://doi.org/10.1016/j.euroneuro.2011.05.003.

Arnone, D., McKie, S., Elliott, R., Juhasz, G., Thomas, E. J., Downey, D., et al. (2013). State-dependent changes in hippocampal grey matter in depression. *Molecular Psychiatry*, *18*(12), 1265–1272. https://doi.org/10.1038/mp.2012.150.

Arnsten, A. F. T. (2009). Stress signalling pathways that impair prefrontal cortex structure and function. Nature Reviews. Neuroscience, *10*(6), 410–422. https://doi.org/10.1038/nrn2648.

Baumeister, D., Akhtar, R., Ciufolini, S., Pariante, C. M., & Mondelli, V. (2016). Childhood trauma and adulthood inflammation: a meta-analysis of peripheral C-reactive protein, interleukin-6 and tumour necrosis factor-α. *Molecular Psychiatry*, *21*, 642–649. https://doi.org/10.1038/mp.2015.67.

Baune, B. T., Konrad, C., Grotegerd, D., Suslow, T., Ohrmann, P., Bauer, J., et al. (2012). Tumor necrosis factor gene variation predicts hippocampus volume in healthy individuals. *Biological Psychiatry*, *72*(8), 655–662. https://doi.org/10.1016/j.biopsych.2012.04.002.

Bremner, J. D., Randall, P., Verrnetten, E., Staib, L. H., Bronen, R. A., Mazure, C., et al. (1997). Magnetic resonance imaging-based measurement of hippocampal volume in posttraumatic stress disorder related to childhood physical and sexual abuse—a preliminary report. *Biological Psychiatry*, *41*, 23–32.

Brown, S. M., Henning, S., & Wellman, C. L. (2005). Mild, short-term stress alters dendritic morphology in rat medial prefrontal cortex. *Cerebral Cortex*, *15*(11), 1714–1722. https://doi.org/10.1093/cercor/bhi048.

Byrne, M. L., Whittle, S., & Allen, N. B. (2016). The role of brain structure and function in the association between inflammation and depressive symptoms: a systematic review. *Psychosomatic Medicine*, *78*(4), 389–400. https://doi.org/10.1097/PSY.0000000000000311.

Cain, D. W., & Cidlowski, J. A. (2017). Immune regulation by glucocorticoids. *Nature Reviews Immunology*, *17*, 233–247. https://doi.org/10.1038/nri.2017.1.

Carpenter, L. L., Gawuga, C. E., Tyrka, A. R., Lee, J. K., Anderson, G. M., & Price, L. H. (2010). Association between plasma IL-6 response to acute stress and early-life adversity in healthy adults. *Neuropsychopharmacology*, *35*, 2617–2623. https://doi.org/10.1038/npp.2010.159.

Cattaneo, A., Macchi, F., Plazzotta, G., Veronica, B., Bocchio-Chiavetto, L., Riva, M. A., et al. (2015). Inflammation and neuronal plasticity: a link between childhood trauma and depression pathogenesis. *Frontiers in Cellular Neuroscience*, *9*, 40. https://doi.org/10.3389/fncel.2015.00040.

Cerqueira, J. J., Pêgo, J. M., Taipa, R., Bessa, J. M., Almeida, O. F. X., & Sousa, N. (2005). Morphological correlates of corticosteroid-induced changes in prefrontal cortex-dependent behaviors. *The Journal of Neuroscience: The Official Journal of the Society for Neuroscience*, *25*(34), 7792–7800. https://doi.org/10.1523/JNEUROSCI.1598-05.2005.

Chaney, A., Carballedo, A., Amico, F., Fagan, A., Skokauskas, N., Meaney, J., et al. (2014). Effect of childhood maltreatment on brain structure in adult patients with major depressive disorder and healthy participants. *Journal of Psychiatry & Neuroscience*, *39*(1), 50–59. https://doi.org/10.1503/jpn.120208.

Chen, M. C., Hamilton, J. P., & Gotlib, I. H. (2010). Decreased hippocampal volume in healthy girls at risk of depression. *Archives of General Psychiatry*, *67*(3), 270–276. https://doi.org/10.1001/archgenpsychiatry.2009.202.

Coelho, R., Viola, T. W., Walss-Bass, C., Brietzke, E., & Grassi-Oliveira, R. (2014). Childhood maltreatment and inflammatory markers: a systematic review. *Acta Psychiatrica Scandinavica*, *129*(3), 180–192. https://doi.org/10.1111/acps.12217.

Cole, J. H., Costafreda, S. G., McGuffin, P., & Fu, C. H. Y. (2011). Hippocampal atrophy in first episode depression: a meta-analysis of magnetic resonance imaging studies. *Journal of Affective Disorders*, *134*(1–3), 483–487. https://doi.org/10.1016/j.jad.2011.05.057.

Conrad, C. D. (2008). Chronic stress-induced hippocampal vulnerability: the glucocorticoid vulnerability hypothesis. *Reviews in the Neurosciences*, *19*(6), 395–411.

Covic, M., Karaca, E., & Lie, D. C. (2010). Epigenetic regulation of neurogenesis in the adult hippocampus. *Heredity*, *105*(1), 122–134. https://doi.org/10.1038/hdy.2010.27.

Cusi, A. M., Nazarov, A., Holshausen, K., Macqueen, G. M., & McKinnon, M. C. (2012). Systematic review of the neural basis of social cognition in patients with mood disorders. *Journal of Psychiatry & Neuroscience*, *37*(3), 154–169. https://doi.org/10.1503/jpn.100179.

Danese, A., Moffitt, T. E., Pariante, C. M., Ambler, A., Poulton, R., & Caspi, A. (2008). Elevated inflammation levels in depressed adults with a history of childhood maltreatment. *Archives of General Psychiatry*, *65*(4), 409–417.

Dannlowski, U., Stuhrmann, A., Beutelmann, V., Zwanzger, P., Lenzen, T., Grotegerd, D., et al. (2012). Limbic scars: long-term consequences of childhood maltreatment revealed by functional and structural magnetic resonance imaging. *Biological Psychiatry*, *71*(4), 286–293. https://doi.org/10.1016/j.biopsych.2011.10.021.

Dantzer, R., O'Connor, J. C., Freund, G. G., Johnson, R. W., & Kelley, K. W. (2008). From inflammation to sickness and depression: when the immune system subjugates the brain. Nature Reviews. Neuroscience, *9*(1), 46–56. https://doi.org/10.1038/nrn2297.

De Bellis, M. D., Baum, A. S., Birmaher, B., Keshavan, M. S., Eccard, C. H., Boring, A. M., et al. (1999). Developmental traumatology part I: biological stress systems. *Biological Psychiatry, 45*, 1259–1270.

Diorio, D., Viau, V., & Meaney, M. J. (1993). The role of the medial prefrontal cortex (cingulate gyrus) in the regulation of hypothalamic-pituitary-adrenal responses to stress. *The Journal of Neuroscience, 13*(9), 3839–3847.

Doorduin, J., de Vries, E. F. J., Willemsen, A. T. M., de Groot, J. C., Dierckx, R. A., & Klein, H. C. (2009). Neuroinflammation in schizophrenia-related psychosis: a PET study. *Journal of Nuclear Medicine, 50*(11), 1801–1807. https://doi.org/10.2967/jnumed.109.066647.

Dowell, N. G., Cooper, E. A., Tibble, J., Voon, V., Critchley, H. D., Cercignani, M., et al. (2016). Acute changes in striatal microstructure predict the development of interferon-alpha induced fatigue. *Biological Psychiatry, 79*(4), 320–328. https://doi.org/10.1016/j.biopsych.2015.05.015.

Duman, R. S., & Aghajanian, G. K. (2012). Synaptic dysfunction in depression: potential therapeutic targets. *Science (New York, N.Y.), 338*(6103), 68–72. https://doi.org/10.1126/science.1222939.

Edmiston, E. E., Wang, F., Mazure, C. M., Guiney, J., Sinha, R., Mayes, L. C., et al. (2011). Corticostriatal-limbic gray matter morphology in adolescents with self-reported exposure to childhood maltreatment. *Archives of Pediatrics & Adolescent Medicine, 165*(12), 1069–1077. https://doi.org/10.1001/archpediatrics.2011.565.

Edwards, V. J., Holden, G. W., Felitti, V. J., & Anda, R. F. (2003). Relationship between multiple forms of childhood maltreatment and adult mental health in community respondents: results from the adverse childhood experiences study. *The American Journal of Psychiatry, 160*(8), 1453–1460. https://doi.org/10.1176/appi.ajp.160.8.1453.

Eriksson, P. S., Perfilieva, E., Björk-Eriksson, T., Alborn, A. M., Nordborg, C., Peterson, D. A., et al. (1998). Neurogenesis in the adult human hippocampus. *Nature Medicine, 4*(11), 1313–1317. https://doi.org/10.1038/3305.

Frodl, T., & Amico, F. (2014). Is there an association between peripheral immune markers and structural/functional neuroimaging findings? *Progress in Neuro-Psychopharmacology and Biological Psychiatry, 48*, 295–303. https://doi.org/10.1016/j.pnpbp.2012.12.013.

Frodl, T., Carballedo, A., Hughes, M. M., Saleh, K., Fagan, A., Skokauskas, N., et al. (2012). Reduced expression of glucocorticoid-inducible genes GILZ and SGK-1: high IL-6 levels are associated with reduced hippocampal volumes in major depressive disorder. *Translational Psychiatry 2*(November 2011)https://doi.org/10.1038/tp.2012.14.

Frodl, T., Koutsouleris, N., Bottlender, R., Born, C., Jäger, M., Mörgenthaler, M., et al. (2008). Reduced gray matter brain volumes are associated with variants of the serotonin transporter gene in major depression. *Molecular Psychiatry, 13*(12), 1093–1101. https://doi.org/10.1038/mp.2008.62.

Frodl, T., Meisenzahl, E. M., Zetzsche, T., Born, C., Groll, C., Jäger, M., et al. (2002). Hippocampal changes in patients with a first episode of major depression. *The American Journal of Psychiatry, 159*(7), 1112–1118. https://doi.org/10.1176/appi.ajp.159.7.1112.

Frodl, T., Reinhold, E., Koutsouleris, N., Donohoe, G., Bondy, B., Reiser, M., et al. (2010). Childhood stress, serotonin transporter gene and brain structures in major depression. *Neuropsychopharmacology, 35*(6), 1383–1390. https://doi.org/10.1038/npp.2010.8.

Frodl, T., Reinhold, E., Koutsouleris, N., Reiser, M., & Meisenzahl, E. M. (2010). Interaction of childhood stress with hippocampus and prefrontal cortex volume reduction in major depression. *Journal of Psychiatric Research, 44*(13), 799–807. https://doi.org/10.1016/j.jpsychires.2010.01.006.

Gallagher, P., Malik, N., Newham, J., Young, A. H., Ferrier, I. N., & Mackin, P. (2008). Antiglucocorticoid treatments for mood disorders. *The Cochrane Database of Systematic Reviews, 1*, CD005168. https://doi.org/10.1002/14651858.CD005168.pub2.

Gilbert, R., Widom, C. S., Browne, K., Fergusson, D., Webb, E., & Janson, S. (2009). Burden and consequences of child maltreatment in high-income countries. *Lancet, 373*(9657), 68–81. https://doi.org/10.1016/S0140-6736(08)61706-7.

Gilbertson, M. W., Shenton, M. E., Ciszewski, A., Kasai, K., Lasko, N. B., Orr, S. P., et al. (2002). Smaller hippocampal volume predicts pathologic vulnerability to psychological trauma. *Nature Neuroscience, 5*(11), 1242–1247. https://doi.org/10.1038/nn958.

Green, J. G., McLaughlin, K. A., Berglund, P. A., Gruber, M. J., Sampson, N. A., Zaslawsky, A. M., et al. (2010). Childhood adversities and adult psychiatric disorders in the National Comorbidity Survey Replication I: associations with first onset of DSM-IV disorders. *Archives of General Psychiatry, 67*(2), 113–123.

Grosse, L., Ambrée, O., Jörgens, S., Jawahar, M. C., Singhal, G., Stacey, D., et al. (2016). Cytokine levels in major depression are related to childhood trauma but not to recent stressors. *Psychoneuroendocrinology, 73*, 24–31. https://doi.org/10.1016/j.psyneuen.2016.07.205.

Hanning, U., Roesler, A., Peters, A., Berger, K., & Baune, B. T. (2016). Structural brain changes and all-cause mortality in the elderly population—the mediating role of inflammation. *Age (Dordrecht, Netherlands), 38*, 455–464. https://doi.org/10.1007/s11357-016-9951-9.

Harmer, C. J., Mackay, C. E., Reid, C. B., Cowen, P. J., & Goodwin, G. M. (2006). Antidepressant drug treatment modifies the neural processing of nonconscious threat cues. *Biological Psychiatry*, *59*(9), 816–820. https://doi.org/10.1016/j.biopsych.2005.10.015.

Hartwell, K. J., Moran-Santa Maria, M. M., Twal, W. O., Shaftman, S., DeSantis, S. M., McRae-Clark, A. L., et al. (2013). Association of elevated cytokines with childhood adversity in a sample of healthy adults. *Journal of Psychiatric Research*, *47*(5), 604–610.

Heim, C., Newport, D. J., Mletzko, T. C., Miller, A. H., & Nemeroff, C. B. (2008). The link between childhood trauma and depression: insights from HPA axis studies in humans. *Psychoneuroendocrinology*, *33*, 693–710. https://doi.org/10.1016/j.psyneuen.2008.03.008.

Heim, C., Newport, D. J., Wagner, D., Wilcox, M. M., Miller, A. H., & Nemeroff, C. B. (2002). The role of early adverse experience and adulthood stress in the prediction of neuroendocrine stress reactivity in women: a multiple regression analysis. *Depression and Anxiety*, *15*, 117–125. https://doi.org/10.1002/da.10015.

Herman, J. P., Figueiredo, H., Mueller, N. K., Ulrich-Lai, Y., Ostrander, M. M., Choi, D. C., et al. (2003). Central mechanisms of stress integration: hierarchical circuitry controlling hypothalamo–pituitary–adrenocortical responsiveness. *Frontiers in Neuroendocrinology*, *24*, 151–180. https://doi.org/10.1016/j.yfrne.2003.07.001.

Hibar, D. P., Stein, J. L., Renteria, M. E., Arias-Vasquez, A., Desrivières, S., Jahanshad, N., et al. (2015). Common genetic variants influence human subcortical brain structures. *Nature*, *520*(7546), 224–229. https://doi.org/10.1038/nature14101.

Jefferson, A. L., Massaro, J. M., Wolf, P. A., Seshadri, S., Vasan, R. S., Larson, M. G., et al. (2007). Inflammatory biomarkers are associated with total brain volume The Framingham Heart Study. *Neurology*, *68*, 1032–1038. https://doi.org/10.1212/01.wnl.0000257815.20548.df. Inflammatory.

Krishnadas, R., McLean, J., Batty, D. G., Burns, H., Deans, K. A., Ford, I., et al. (2013). Cardio-metabolic risk factors and cortical thickness in a neurologically healthy male population: results from the psychological, social and biological determinants of ill health (pSoBid) study. *NeuroImage: Clinical*, *2*(1), 646–657. https://doi.org/10.1016/j.nicl.2013.04.012.

Krishnan, V., & Nestler, E. J. (2008). The molecular neurobiology of depression. *Nature*, *455*(7215), 894–902. https://doi.org/10.1038/nature07455.

Labonté, B., Suderman, M., Maussion, G., Navaro, L., Yerko, V., Mahar, I., et al. (2012). Genome-wide epigenetic regulation by early-life trauma. *Archives of General Psychiatry*, *69*(7), 722–731. https://doi.org/10.1001/archgenpsychiatry.2011.2287.

Layé, S., Parnet, P., Goujon, E., & Dantzer, R. (1994). Peripheral administration of lipopolysaccharide induces the expression of cytokine transcripts in the brain and pituitary of mice. *Molecular Brain Research*, *27*(1), 157–162. https://doi.org/10.1016/0169-328X(94)90197-X.

Lupien, S. J., McEwen, B. S., Gunnar, M. R., & Heim, C. (2009). Effects of stress throughout the lifespan on the brain, behaviour and cognition. Nature Reviews. Neuroscience, *10*(June), 434–445. https://doi.org/10.1038/nrn2639.

MacQueen, G. M., & Frodl, T. (2011). The hippocampus in major depression: evidence for the convergence of the bench and bedside in psychiatric research? *Molecular Psychiatry*, *16*(3), 252–264. https://doi.org/10.1038/mp.2010.80.

Marsland, A. L., Gianaros, P. J., Abramowitch, S. M., Manuck, S. B., & Hariri, A. R. (2008). Interleukin-6 covaries inversely with hippocampal grey matter volume in middle-aged adults. *Biological Psychiatry*, *64*, 484–490.

Matsumoto, Y., Yanase, D., Noguchi-Shinohara, M., Ono, K., Yoshita, M., & Yamada, M. (2008). Cerebrospinal fluid/serum IgG index is correlated with medial temporal lobe atrophy in Alzheimer's disease. *Dementia and Geriatric Cognitive Disorders*, *25*(2), 144–147. https://doi.org/10.1159/000112555.

McEwen, B. S. (2004). Protection and damage from acute and chronic stress: allostasis and allostatic overload and relevance to the pathophysiology of psychiatric disorders. *Annals of the New York Academy of Sciences*, *1032*, 1–7. https://doi.org/10.1196/annals.1314.001.

McEwen, B. S., & Morrison, J. H. (2013). The brain on stress: vulnerability and plasticity of the prefrontal cortex over the life course. *Neuron*, *79*(1), 16–29. https://doi.org/10.1016/j.neuron.2013.06.028.

Mckinnon, M. C., Yucel, K., Nazarov, A., & Macqueen, G. M. (2009). A meta-analysis examining clinical predictors of hippocampal volume in patients with major depressive disorder. *Journal of Psychiatry & Neuroscience*, *34*(1), 41–54.

Miller, A. H., & Raison, C. L. (2016). The role of inflammation in depression: from evolutionary imperative to modern treatment target. Nature Reviews. Immunology, *16*(1), 22–34. https://doi.org/10.1038/nri.2015.5.

Mondelli, V., Cattaneo, A., Murri, M. B., Di Forti, M., Handley, R., Hepgul, N., et al. (2011). Stress and inflammation reduce brain-derived neurotrophic factor expression in first-episode psychosis: a pathway to smaller hippocampal volume. *Journal of Clinical Psychiatry*, *72*, 1677–1684.

Nanni, V., Uher, R., Ph, D., & Danese, A. (2012). Childhood maltreatment predicts unfavorable course of illness and treatment outcome in depression: a meta-analysis. *The American Journal of Psychiatry*, *169*, 141–151.

Nordanskog, P., Dahlstrand, U., Larsson, M. R., Larsson, E.-M., Knutsson, L., & Johanson, A. (2010). Increase in hippocampal volume after electroconvulsive therapy in patients with depression: a volumetric magnetic resonance imaging study. *The Journal of ECT, 26*(1), 62–67. https://doi.org/10.1097/YCT.0b013e3181a95da8.

Opel, N., Redlich, R., Zwanzger, P., Grotegerd, D., Arolt, V., Heindel, W., et al. (2014). Hippocampal atrophy in major depression: a function of childhood maltreatment rather than diagnosis? *Neuropsychopharmacology, 39*(12), 2723–2731. Retrieved from. http://www.ncbi.nlm.nih.gov/pubmed/24924799.

Opel, N., Zwanzger, P., Redlich, R., Grotegerd, D., Dohm, K., Arolt, V., et al. (2016). Differing brain structural correlates of familial and environmental risk for major depressive disorder revealed by a combined VBM/pattern recognition approach. *Psychological Medicine, 46*(2), 277–290. https://doi.org/10.1017/S0033291715001683.

Pace, T. W. W., Mletzko, T. C., Alagbe, O., Musselman, D. L., Nemeroff, C. B., Miller, A. H., et al. (2006). Increased stress-induced inflammatory responses in male patients with major depression and increased early life stress. *The American Journal of Psychiatry, 163*(9), 1630–1633. https://doi.org/10.1176/ajp.2006.163.9.1630.

Pang, Y., Cai, Z., & Rhodes, P. G. (2003). Disturbance of oligodendrocyte development, hypomyelination and white matter injury in the neonatal rat brain after intracerebral injection of lipopolysaccharide. *Developmental Brain Research, 140*(2), 205–214. https://doi.org/10.1016/S0165-3806(02)00606-5.

Paus, T., Keshavan, M., & Giedd, J. N. (2008). Why do many psychiatric disorders emerge during adolescence? Nature Reviews. Neuroscience, 9, 947–957.

Phillips, M. L., Drevets, W. C., Rauch, S. L., & Lane, R. (2003a). Neurobiology of emotion perception I: the neural basis of normal emotion perception. *Biological Psychiatry, 54*(5), 504–514. https://doi.org/10.1016/S0006-3223(03)00168-9.

Phillips, M. L., Drevets, W. C., Rauch, S. L., & Lane, R. (2003b). Neurobiology of emotion perception II: implications for major psychiatric disorders. *Biological Psychiatry, 54*(5), 515–528. https://doi.org/10.1016/S0006-3223(03)00171-9.

Puetz, V. B., Zweerings, J., Dahmen, B., Ruf, C., Herpertz-Dahlmann, W. S. B., & Konrad, K. (2016). Multidimensional assessment of neuroendocrine and psychopathological profiles in maltreated youth. *Journal of Neural Transmission, 123*(9), 1095–1106. https://doi.org/10.1007/s00702-016-1509-6.

Quan, N., Stern, E. L., Whiteside, M. B., & Herkenham, M. (1999). Induction of pro-inflammatory cytokine mRNAs in the brain after peripheral injection of subseptic doses of lipopolysaccharide in the rat. *Journal of Neuroimmunology, 93*(1–2), 72–80. https://doi.org/10.1016/S0165-5728(98)00193-3.

Rao, U., Chen, L.-A., Bidesi, A. S., Shad, M. U., Thomas, M. A., & Hammen, C. L. (2010). Hippocampal changes associated with early-life adversity and vulnerability to depression. *Biological Psychiatry, 67*(4), 357–364. https://doi.org/10.1016/j.biopsych.2009.10.017.

Redlich, R., Almeida, J. R. C., Grotegerd, D., Opel, N., Kugel, H., Heindel, W., et al. (2014). Brain morphometric biomarkers distinguishing unipolar and bipolar depression: a voxel-based morphometry-pattern classification approach. *JAMA Psychiatry, 71*(11), 1222–1230. https://doi.org/10.1001/jamapsychiatry.2014.1100.

Redlich, R., Dohm, K., Grotegerd, D., Opel, N., Zwitserlood, P., Heindel, W., et al. (2015). Reward processing in unipolar and bipolar depression: a functional MRI study. *Neuropsychopharmacology, 40*(11), 2623–2631. https://doi.org/10.1038/npp.2015.110.

Redlich, R., Opel, N., Grotegerd, D., Dohm, K., Zaremba, D., Bürger, C., et al. (2016). Prediction of individual response to electroconvulsive therapy via machine learning on structural magnetic resonance imaging data. *JAMA Psychiatry, 73*(6), 557–564. https://doi.org/10.1001/jamapsychiatry.2016.0316.

Redlich, R., Stacey, D., Opel, N., Grotegerd, D., Dohm, K., Kugel, H., et al. (2015). Evidence of an IFN-gamma by early life stress interaction in the regulation of amygdala reactivity to emotional stimuli. *Psychoneuroendocrinology, 62*, 166–173. https://doi.org/10.1016/j.psyneuen.2015.08.008.

Sánchez, M. M., Ladd, C. O., & Plotsky, P. M. (2001). Early adverse experience as a developmental risk factor for later psychopathology: evidence from rodent and primate models. *Development and Psychopathology, 13*(3), 419–449.

Sapolsky, R. M. (2000). Glucocorticoids and hippocampal atrophy in neuropsychiatric disorders. *Archives of General Psychiatry, 57*, 925–935.

Savitz, J., Frank, M. B., Victor, T., Bebak, M., Marino, J. H., Bellgowan, P. S. F., et al. (2013). Inflammation and neurological disease-related genes are differentially expressed in depressed patients with mood disorders and correlate with morphometric and functional imaging abnormalities. *Brain, Behavior, and Immunity, 31*, 161–171. https://doi.org/10.1016/j.bbi.2012.10.007.

Schmaal, L., Hibar, D. P., Sämann, P. G., Hall, G. B., Baune, B. T., Jahanshad, N., et al. (2017). Cortical abnormalities in adults and adolescents with major depression based on brain scans from 20 cohorts worldwide in the ENIGMA Major Depressive Disorder working group. *Molecular Psychiatry, 22*, 900–909.

Schmaal, L., Veltman, D. J., van Erp, T. G. M., Sämann, P. G., Frodl, T., Jahanshad, N., et al. (2016). Subcortical brain alterations in major depressive disorder: findings from the ENIGMA Major Depressive Disorder working group. *Molecular Psychiatry, 21*, 806–812.

Scott, K. M., McLaughlin, K. A., Smith, D. A. R., & Ellis, P. M. (2012). Childhood maltreatment and DSM-IV adult mental disorders: comparison of prospective and retrospective findings. *British Journal of Psychiatry, 200,* 469–475.

Sheline, Y. I., Gado, M. H., & Kraemer, H. C. (2003). Untreated depression and hippocampal volume loss. *The American Journal of Psychiatry, 160*(8), 1516–1518. Retrieved from. http://www.ncbi.nlm.nih.gov/pubmed/12900317.

Slopen, N., Kubzansky, L. D., McLaughlin, K. A., & Koenen, K. C. (2013). Childhood adversity and inflammatory processes in youth: a prospective study. *Psychoneuroendocrinology, 38,* 188–200. https://doi.org/10.1016/j.psyneuen.2012.05.013.

Sriram, K., & O'Callaghan, J. P. (2007). Divergent roles for tumor necrosis factor-α in the brain. *Journal of NeuroImmune Pharmacology, 2*(2), 140–153. https://doi.org/10.1007/s11481-007-9070-6.

Stacey, D., Redlich, R., Buschel, A., Opel, N., Grotegerd, D., Zaremba, D., et al. (2017). TNF receptors 1 and 2 exert distinct region-specific effects on striatal and hippocampal grey matter volumes (VBM) in healthy adults. *Genes, Brain, and Behavior, 16,* 352–360.

Stein, M. B., Koverola, C., Hanna, C., Torchia, M., & McClarty, B. (1997). Hippocampal volume in women victimized by childhood sexual abuse. *Psychological Medicine, 27*(4), 951–959.

Stein, J. L., Medland, S. E., Vasquez, A. A., Hibar, D. P., Senstad, R. E., Winkler, A. M., et al. (2012). Identification of common variants associated with human hippocampal and intracranial volumes. *Nature Genetics, 44*(5), 552–561. https://doi.org/10.1038/ng.2250.

Stone, E. A., Lehmann, M. L., Lin, Y., & Quartermain, D. (2006). Depressive behavior in mice due to immune stimulation is accompanied by reduced neural activity in brain regions involved in positively motivated behavior. *Biological Psychiatry, 60*(8), 803–811. https://doi.org/10.1016/j.biopsych.2006.04.020.

Stuhrmann, A., Suslow, T., & Dannlowski, U. (2011). Facial emotion processing in major depression: a systematic review of neuroimaging findings. *Biology of Mood & Anxiety Disorders, 1*(1), 1–17. https://doi.org/10.1186/2045-5380-1-10.

Tarullo, A. R., & Gunnar, M. R. (2006). Child maltreatment and the developing HPA axis. *Hormones and Behavior, 50,* 632–639. https://doi.org/10.1016/j.yhbeh.2006.06.010.

Teicher, M. H., Anderson, C. M., & Polcari, A. (2012). Childhood maltreatment is associated with reduced volume in the hippocampal subfields CA3, dentate gyrus, and subiculum. *Proceedings of the National Academy of Sciences of the United States of America, 109*(9), E563–72. https://doi.org/10.1073/pnas.1115396109.

Teicher, M. H., Dumont, N. L., Ito, Y., Vaituzis, C., Giedd, J. N., & Andersen, S. L. (2004). Childhood neglect is associated with reduced corpus callosum area. *Biological Psychiatry, 56,* 80–85. https://doi.org/10.1016/j.biopsych.2004.03.016.

Teicher, M. H., & Samson, J. A. (2013). Childhood maltreatment and psychopathology: a case for ecophenotypic variants as clinically and neurobiologically distinct subtypes. *The American Journal of Psychiatry, 170,* 1114–1133. https://doi.org/10.1176/appi.ajp.2013.12070957.

Teicher, M. H., Samson, J. A., Anderson, C. M., & Ohashi, K. (2016). The effects of childhood maltreatment on brain structure, function and connectivity. Nature Reviews. Neuroscience, *17*(10), 652–666. https://doi.org/10.1038/nrn.2016.111.

van Dam, A. M., Brouns, M., Louisse, S., & Berkenbosch, F. (1992). Appearance of interleukin-1 in macrophages and in ramified microglia in the brain of endotoxin-treated rats: a pathway for the induction of non-specific symptoms of sickness? *Brain Research, 588*(2), 291–296. https://doi.org/10.1016/0006-8993(92)91588-6.

van Harmelen, A.-L., van Tol, M.-J., van der Wee, N. J. A., Veltman, D. J., Aleman, A., Spinhoven, P., et al. (2010). Reduced medial prefrontal cortex volume in adults reporting childhood emotional maltreatment. *Biological Psychiatry, 68*(9), 832–838. https://doi.org/10.1016/j.biopsych.2010.06.011.

Vythilingam, M., Heim, C., Newport, J., Miller, A. H., Anderson, E., Bronen, R., et al. (2002). Childhood trauma associated with smaller hippocampal volume in women with major depression. *The American Journal of Psychiatry, 159*(12), 2072–2080. https://doi.org/10.1176/appi.ajp.159.12.2072.

Wang, Y.-C., Huang, C.-C., & Hsu, K.-S. (2010). The role of growth retardation in lasting effects of neonatal dexamethasone treatment on hippocampal synaptic function. *PLoS One, 5*(9), e12806. https://doi.org/10.1371/journal.pone.0012806.

Wellman, C. L. (2001). Dendritic reorganization in pyramidal neurons in medial prefrontal cortex after chronic corticosterone administration. *Journal of Neurobiology, 49*(3), 245–253.

Woon, F. L., Sood, S., & Hedges, D. W. (2010). Hippocampal volume deficits associated with exposure to psychological trauma and posttraumatic stress disorder in adults: a meta-analysis. *Progress in Neuropsychopharmacology and Biological Psychiatry, 34,* 1181–1188. https://doi.org/10.1016/j.pnpbp.2010.06.016.

World Health Organization. *Mental health: New understanding, new hope.* 2001 Geneva: World Health Organization.

Biological Embedding of Childhood Maltreatment in Adult Depression

Magdalene C. Jawahar, Bernhard T. Baune

University of Adelaide, Adelaide, SA, Australia

INTRODUCTION

Exposure to adverse childhood events is known to increase the risk for poor physical and mental health outcomes later in adulthood (Bernet & Stein, 1999; Green et al., 2010; Scott et al., 2008, 2011; Varese et al., 2012). Childhood adversities include maltreatment, parental illness/loss, family violence, physical illnesses, and socioeconomic diversities. According to the World Health Organization's World Mental Health Survey, childhood adversities attribute to 30% of all psychiatric disorders with 22.9% in mood disorders such as major depressive disorder and bipolar disorder (Kessler et al., 2010). Depression, in particular, is predicted to become the second leading global burden of disease by 2020 (Lopez & Murray, 1998). In this context, the key research question is how adverse experiences that are physical and psychological affect biological processes and increase susceptibility to psychopathology decades later?

Researchers working on the early origins of disease risks define this early experience-dependent reshaping or programming as "biological embedding" or "early-life programming" (Danese et al., 2011; Miller, Chen, & Parker, 2011; Tarry-Adkins & Ozanne, 2011). This chapter will focus on the biological embedding of childhood maltreatment (CM) only, wherein CM includes both abuse (sexual, physical, and verbal) and neglect in early childhood. The critical factors known to influence the effects of CM includes the nature of stressors (Gershon, Sudheimer, Tirouvanziam, Williams, & O'Hara, 2013), developmental period of exposure (Kaplow & Widom, 2007), severity and frequency of exposures, gender, age at assessment, and any predisposing genetic polymorphisms associated with the pathology, in this case depression (Buchmann et al., 2013; Caspi et al., 2003).

CHILDHOOD MALTREATMENT AND ADULT DEPRESSION

A range of epidemiological and clinical studies have provided evidence for a strong association between CM and depression. McCauley et al. (1997) found that childhood sexual or physical abuse was associated with increases in symptoms of depression and anxiety. Similar results have

Inflammation and Immunity in Depression
https://doi.org/10.1016/B978-0-12-811073-7.00017-9

been reported in the National Comorbidity Survey (Molnar, Buka, & Kessler, 2001), the Ontario Health Survey (Lipman, MacMillan, & Boyle, 2001), and a New Zealand community survey (Mullen, Romans-Clarkson, Walton, & Herbison, 1988). Further evidence has been found for a strong dose-response relationship between childhood adversities (sexual abuse, physical abuse, and witnessing paternal violence) and general mental health problems in adulthood (Chapman et al., 2004; Edwards, Holden, Felitti, & Anda, 2003). In addition, multiple metaanalyses has now revealed that nearly 50% of people diagnosed with depression have a history of CM and in addition are more likely to develop early-onset, chronic, and treatment-resistant depression (Infurna et al., 2016; Mandelli, Petrelli, & Serretti, 2015; Nanni, Uher, & Danese, 2012; Nelson, Klumparendt, Doebler, & Ehring, 2017). These studies suggest that exposure to CM significantly programs behavior and biology to increase vulnerability to subsequent stressors in life, ultimately leading to long-term susceptibility to depressive disorders. However, the biological processes and mechanisms that contribute to this long-lasting increased susceptibility to develop depression later in life are unclear.

CM occurs in a period where the brain and neuroendocrine and immune systems exhibit high levels of plasticity enabling environmental signals to impact their developmental trajectories. Increasing evidence points to disruptions in the neuroendocrine-immune networks following exposure to early adversities and also in depression. These include disruptions in the hypothalamus-pituitary-adrenal (HPA) axis that regulates homeostasis (Holsboer, 2000; Van Voorhees & Scarpa, 2004), alterations in immune response such as increased pro-inflammatory cytokines and other immune factors (Carpenter et al., 2010; Danese, Pariante, Caspi, Taylor, & Poulton, 2007; Zeugmann et al., 2013), and alterations in neurotransmitters and their receptors and neurotrophic factors and their receptors (Gatt et al., 2009; Harkness et al., 2015).

These changes all lead to structural alterations in important brain regions mediating psychosocial stress, such as the amygdala, hippocampus (HC), and prefrontal cortex (PFC). Epigenetic mechanisms are considered to be one of the key molecular machinery in mediating the long-lasting effects of CM in the above listed biological systems. This chapter aims to achieve a comprehensive understanding of the biological embedding of early stress in depression by reviewing current evidence and trends. Evidence from both human and animal models of maltreatment or neglect will be reviewed in the chapter. Ultimately, we aim to present a model of networked biological processes that may possibly explain CM effects in susceptibility to depression in adulthood.

CM EFFECTS ON HPA FUNCTION

Increased glucocorticoids (GCs, cortisol in humans and corticosterone in rodents) in the plasma, increased corticotrophin-releasing hormone (CRH) and arginine vasopressin (AVP)-producing neurons in the postmortem brains of depressed individuals, and reduction in the negative feedback of HPA led to the HPA dysfunction or glucocorticoid theory of depression (Meynen et al., 2006; Raadsheer, Hoogendijk, Stam, Tilders, & Swaab, 1994; Weber et al., 2000) (for detailed reviews, see Holsboer, 2000, and Swaab, Bao, & Lucassen, 2005). The early stress effect of CM on HPA function is reviewed across life span below.

CM Effects on HPA Function: Childhood and Adolescence

A study of 56 children exposed to maltreatment with or without a diagnosis of major depression found that maltreated children with major depression displayed increased cortisol levels from morning to midday as opposed to decreasing (Kaufman, 1991). Alterations in cortisol levels were also reported by other studies showing elevated basal and average daily cortisol levels in maltreated children or children institutionalized for a longer duration (Cicchetti & Rogosch, 2001; Gunnar, Morison, Chisholm, & Schuder, 2001). In addition, maltreated children also exhibited a blunted HPA response (lower cortisol/ACTH response to acute stress) to psychosocial stress (Gunnar, Frenn, Wewerka, & Van Ryzin, 2009; Ouellet-Morin et al., 2011) or pharmacological challenge (Kaufman et al., 1997). In response to psychosocial stress, where participants underwent a trier social stress test (TSST) involving two stressors (public speaking and mental arithmetic) in front of two judges, maltreated adolescents had a blunted cortisol response compared with control adolescents suggesting a dysregulated HPA function in response to acute stress (MacMillan et al., 2009). In a similar study, blunted cortisol response was found in those with moderate/severe depression and a heightened cortisol response to TSST in mild/moderate depression group (Harkness, Stewart, & Wynne-Edwards, 2011). The difference in cortisol response in the depression groups was explained as an effect of decreased glucocorticoid receptor sensitivity, suggesting a reduced negative feedback in the moderate/severe depressed and maltreated adolescents.

In another study comparing sexually abused and age-matched nonabused control girls, ovine CRH (oCRH) stimulation showed reduced basal ACTH and lower net post-oCRH ACTH levels, but no changes in the post-oCRH cortisol levels, suggesting a dysregulated HPA function particularly with pituitary hyporesponsiveness (De Bellis et al., 1994). On the contrary, maltreated children with depression displayed an increased ACTH response to pharmacological CRH challenge compared with maltreated and not depressed and healthy control children (Kaufman et al., 1997). The authors found that this contrast in response was due to two response groups in the depressed and

maltreated children, those that were ACTH responders and the nonresponders. Further analysis revealed that the children in ACTH responders group were living in an environment of continuous emotional maltreatment as opposed to a relatively stable environment in the other studies.

CM Effects on HPA Function: Adulthood

Heim et al. (2000, 2002) conducted a series of studies to determine if CM leads to a persistent alteration in HPA function in adulthood and if that contributed to increased risk of depression. They reported that women with a history of childhood sexual and physical abuse showed increased ACTH response to TSST compared with nonmaltreated controls, and those suffering from current depression showed increased responses than those without depression (Heim et al., 2000). A similar increase in cortisol reactivity was also observed in maltreated women with current depression. Increased CRH levels in the cerebrospinal fluid (CSF) of maltreated individuals also suggests increased neuroendocrine activity in response to CM (Carpenter et al., 2004). However, healthy individuals with a history of CM had reduced ACTH and cortisol reactivity to the psychosocial stressor TSST suggesting a diminished HPA responsiveness in the absence of current depression (Carpenter et al., 2007). Further analyses of the effect of later-life stressors in addition to CM revealed that interactions of early- and later-life stressors explained increased ACTH responsiveness, suggesting that CM-induced neuroendocrine response is enhanced by later-life stressors (Heim et al., 2002). Pharmacological test for HPA activity such as the dexamethasone suppression test in combination with CRH stimulation had shown similar HPA responses in maltreated individuals (Heim, Mletzko, Purselle, Musselman, & Nemeroff, 2008). Like the TSST test on females with CM history and current depression, men with a history of CM and current depression had increased ACTH and cortisol response to the dex/CRH test, suggesting hyperresponsive HPA function. These findings are in line with neuroendocrine activity in maltreated children with and without current depression. In addition, known genetic polymorphisms within the HPA axis genes confer further susceptibility to depression. A single nucleotide polymorphism (SNP) rs1360780 (C/T) in the gene FK506-binding protein 51 (FKBP5) was found to be associated with depression risk in adults with a history of maltreatment (Appel et al., 2011). FKBP5 is a cochaperone for heat shock protein hsp90 and is known to regulate glucocorticoid receptor (GR) sensitivity (Binder, 2009) and therefore plays an important role in stress-mediated GR resistance often seen in CM.

HPA Function in Animal Models of ELS

Animal models of ELS target the early maternal bonding affecting the stress hyporesponsive period, and various models exist that are parallel to maltreatment or neglect in humans. Neuroendocrine responses to early-life stress (ELS) in animal models in general report hyperresponsive HPA function similar to humans. When bonnet macaques (nonhuman primates) were raised by mothers exposed to either high, low, or variable foraging demand (VFD)-based environments, it was observed that as adults those raised by mothers in VFD had increased CRH levels in their CSF, suggesting persistent activity of the HPA response as a signature of exposure to early stress (Coplan et al., 1996). Similar evidence was also reported in rodent models of ELS such as increased CRH in the hypothalamus and other stress-responsive brain regions (Aisa, Tordera, Lasheras, Del Rio, & Ramirez, 2007; Ladd, Owens, & Nemeroff, 1996; Marais, van Rensburg, van Zyl, Stein, & Daniels, 2008; Plotsky & Meaney, 1993; van Oers, de Kloet, & Levine, 1998). Increased levels

of arginine vasopressin (AVP), involved in the activation of HPA response to stress (Murgatroyd et al., 2009; Veenema, Blume, Niederle, Buwalda, & Neumann, 2006), and persistently increased corticosterone levels along with decreased glucocorticoid receptor expression have also been reported in different animal models of ELS (Aisa et al., 2007; Meaney et al., 1996). It is evident that ELS induces alterations in the HPA response leading to an impaired state of homeostasis, whereby the system is susceptible to further stressors, likely increasing the risk of developing depressive-like behaviors in these models. Epigenetic changes in the genes regulating HPA function contribute to this long-lasting persistent effect of ELS.

Epigenetic Regulation of Genes in HPA-Axis in Response to ELS

A study on the effect of varied early maternal care in rats led to the first evidence on epigenetic mechanisms mediating the long-lasting effect of ELS on HPA function (Weaver et al., 2004). Results showed increased methylation of CpG sites within the *Nr3c1* (glucocorticoid receptor (GR) gene) exon 1_7 promoter along with decreased H3K9 acetylation that corresponded to decreased GR expression in the HC of pups raised by low licking, grooming, and arched-back nursing (LG-ABN) mothers. The later assessment revealed that this exon 17 of the GR gene was also the binding site for transcription factor *NGFI-A* (nerve growth factor-inducible protein A), therefore reducing the expression of GR gene resulting in HPA hyperactivity (Weaver et al., 2007). This finding was then translated in humans when McGowan et al. (2009) analyzed the brains of suicide victims with and without CM compared with controls and observed similar hypermethylation of *NR3C1* and reduced NGFI-A binding in the HC of those exposed to CM. Hypermethylation of the exon 17 of *NR3C1* promoter has also been observed in the peripheral blood DNA of maltreated children and adults with a history of CM (Martin-Blanco et al., 2014; Perroud et al., 2011; Romens, McDonald, Svaren, & Pollak, 2015). Differential methylation and histone modifications of other key HPA genes such as *CRH* receptor and *AVP* have also been reported in response to ELS (Chen et al., 2012; Murgatroyd et al., 2009). These epigenetic changes ultimately contribute to glucocorticoid resistance and HPA dysfunction observed in depression.

CM EFFECTS ON MONOAMINE FUNCTION

Monoamines hypothesis states that a decrease or deficiency of monoamine neurotransmitters in the synapse, for example, reduced serotonin, leads to a reduction in monoaminergic neurotransmission and ultimately depression. Clinical and preclinical studies on serotonin (also known as 5-hydroxytryptamine, 5-HT), monoamine oxidase A (MAO-A), the serotonin transporter 5-HTT, and serotonin receptors have all shown evidence for monoamines in depression (Drevets et al., 1999; Meyer et al., 2006; Neumeister et al., 2002, 2004). Genetic polymorphisms in some of these genes have been extensively studied in association with CM as seen below.

CM Effects on Monoamine Function: Focus on Serotonin in Childhood and Adolescence

Kaufman et al. (2004) were the first to report the association between the 5-HTTLPR (genetic polymorphisms in the promoter region of the serotonin transporter gene *SLC6A4*) with CM effects in children (ages 5–15, mean age 10 ± 2.3 years) showing that maltreated children who carried two copies of the short allele (s/s)

and were in poor social support system had twice the risk for depression compared with nonmaltreated children with the s/s allele. Cicchetti, Rogosch, Sturge-Apple, and Toth (2010) reported that maltreated children from lower socioeconomic (SES) backgrounds (ages 6–13, mean age 9.19 ± 1.7 years) with the s/s or s/l genotype of 5-HTTLPR had increased suicidal ideation compared with nonmaltreated children from the same SES background. However, their study failed to replicate the gene-environment interaction to predict depression in maltreated children. Similar interaction of the short allele (s/s) of the 5-HTTLPR with sexual abuse was shown to predict higher depression scores in adolescents (mean age 16.7 ± 1.31 years) from a low-SES background with or without CM (Cicchetti, Rogosch, & Sturge-Apple, 2007). Interestingly, this effect was further enhanced in children who also had low MAO-A activity, a polymorphism in the MAO-A gene resulting in the high or low activity of MAO-A, thereby showing gene-gene-environment interaction (G × G × E) effects mediating depression in adolescents.

CM Effects on Monoamine Function: Focus on Serotonin in Adulthood

Caspi et al. (2003) reported a significant interaction of the s/s genotype of the 5-HTTLPR with childhood maltreatment in predicting depressive symptoms in young adults (26 years). Their study also reported significant interaction of the s allele with stressful life events in moderating adult depression and generated large interest with close to 100 studies to date all aimed at replicating the original findings. Multiple metaanalyses were conducted on various studies with some reporting 5-HTTLPR and stressful life events or specifically CM interaction effect, while some were not. For example, the first two metaanalyses both reported a lack of interaction with stressful life events, however, did not stratify the studies based on stressors, such as CM or life stress only (Munafo, Durrant, Lewis, & Flint, 2009; Risch et al., 2009). Two different later metaanalyses analyzing 54 and 81 studies, respectively, found significant interaction effect of CM and the s allele of 5-HTTLPR (Karg, Burmeister, Shedden, & Sen, 2011; Sharpley, Palanisamy, Glyde, Dillingham, & Agnew, 2014). More recently, a metaanalysis through international collaborations had reported no interaction effects of 5-HTTLPR and CM or stressful life events (Culverhouse et al., 2018). Variations in the methods for assessment of stress and depression and statistical models compared with the original study could all be factors contributing to the inconsistent results in this gene-environment study. It is likely that the 5-HTTLPR along with other risk genes/polymorphisms moderate depression risk that could explain the negative results in some studies and the lower significance for $G \times E$ effect in others.

Serotonin Function in Animal Models of ELS

In rats, MS reduced serotonin in the HC and increased 5HIAA (5-hydroxyindoleacetic acid) a metabolite of serotonin in the frontal cortex and HC (Daniels, Pietersen, Carstens, & Stein, 2004; Matthews, Dalley, Matthews, Tsai, & Robbins, 2001). In contrast, studies also report an increase in serotonin in the raphe nucleus in response to MS (Arborelius & Eklund, 2007; Ruedi-Bettschen et al., 2006). MS alone lead to depression-like behavior and reduced expression of serotonin and the serotonin reuptake transporter 5-HTT in HC and raphe nucleus (Lee et al., 2007); however, in the presence of an adulthood, stressor showed increased 5-HTT in the dorsal raphe nucleus (Gardner, Hale, Lightman, Plotsky, & Lowry, 2009). Maternally separated dams were also reported to exhibit depressive-like behaviors and lower levels of serotonin in the dorsal raphe along with a reduction in cell proliferation in the HC

(Sung et al., 2010). Gene-environmental interaction of the rh5-HTTLPR polymorphism association to ELS was assessed in rhesus macaque where nursery or peer-raised infant monkeys with the l/s genotype demonstrated poor affective and orientation scores and increased activation of the HPA compared with l/l genotype (Barr et al., 2004; Champoux et al., 2002). Contrary to the above finding, infant rhesus macaques also showed an absence of $G \times E$ interaction between rh5-HTTLPR and rhMAO-A-LPR with rearing condition providing a parallel inconsistent report as in human studies (Kinnally, Karere, et al., 2010; Kinnally, Lyons, Abel, Mendoza, & Capitanio, 2008). These studies, therefore, provide clear evidence for ELS effects on serotonergic neurotransmission along with genetic risks in mediating affective functions via region-specific alterations and atrophy, for example, the HC (Sung et al., 2010).

Epigenetic Regulation of Genes Within Serotonergic Neurotransmission in Response to ELS

Initial evidence for epigenetic mechanisms in serotonin transporter comes from the Iowa Adoptee Study that reported a significant effect of CM on overall methylation levels of the promoter CpG Island (800 bp region) in 5-HTT gene *SLC6A4* promoter region ($P = .004$). Analysis of gender, CM, and interaction of the two revealed that abused males had hypermethylation across all CpG sites in the promoter region; however, in abused females, only two CpG sites, CpG1 (25586514) and CpG3 (25586527), were significantly hypermethylated (Beach, Brody, Todorov, Gunter, & Philibert, 2010). A follow-up study by the same group on childhood sexual abuse in women found that 5-HTT promoter hypermethylation significantly associated with antisocial personality disorder (ASPD) (Beach, Brody, Todorov, Gunter, & Philibert, 2011). Both of these studies did not associate the CM-induced hypermethylation of the *SLC6A4* gene to depression.

Kang et al. (2013) were first to investigate this association to depression and treatment response in depression reporting hypermethylation of the promoter region significantly associated with depression scores but not with treatment response. Kinnaly and colleagues analyzed the 5-HTT promoter methylation in rhesus macaque (mother vs nursery raised) and bonnet macaque (variable foraging demand) model. Although they did not find an effect of early stress on overall methylation, they did find that carriers of the s allele of the rh5-HTTLPR had higher mean methylation compared with the L allele carriers (Kinnally, Capitanio, et al., 2010; Kinnally et al., 2011). Cumulatively, these studies suggest that promoter methylation of *SLC6A4* could be a potential epigenetic biomarker in predicting CM-induced health trajectories in adulthood.

CM EFFECTS ON INFLAMMATION

The cytokine theory (also known as macrophage or monocyte theory) states that depression is caused by increased secretion of pro-inflammatory cytokines such as interleukin 1 (IL-1), tumor necrosis factor (TNF), and interferon alpha (IFNα) by activated macrophages (Maes et al., 1990; Smith, 1991). Both stress and depression have been associated with elevated levels of pro-inflammatory cytokines in the periphery and brain and in behavioral dysfunctions (Bierhaus et al., 2003; Dantzer, O'Connor, Freund, Johnson, & Kelley, 2008; Dowlati et al., 2010; McAfoose & Baune, 2009; Steptoe, Hamer, & Chida, 2007).

CM Effects on Inflammation: Childhood and Adolescence

A key study in 2011 assessed 12-year-old children from a longitudinal cohort to identify CM effects on inflammation and depression status (Danese et al., 2011). In analyzing four different groups of children (heathy controls, depressed only, maltreated only, and

maltreated and depressed), it was found that CRP, an acute-phase inflammatory protein, was higher in the maltreated and depressed children than healthy controls. This difference was significant even after accounting for other confounding variables and replicated prior findings in adults, therefore suggesting that a process of biological embedding of CM effects was evident even at adolescence. A similar increase in CRP and IL-6 levels as a consequence of exposure to CM between birth and 8 years of age was reported in adolescents (Slopen, Kubzansky, McLaughlin, & Koenen, 2013). Another recent study analyzed polymorphism in the IL-1β gene (rs1143633) in mediating CM effects on internalizing behavior, depression, and PTSD in maltreated preschoolers (3–5 years) (Ridout et al., 2014). The study reported an association of maltreatment to all behavioral outcomes and IL-1β predicted depression; however, no GxE effect was observed suggesting that maltreatment had a strong effect on depression irrespective of the IL-1β genotype.

CM Effects on Inflammation: Adulthood

The first report on the effects of CM on inflammatory markers was on male patients with depression, where in response to an acute psychosocial stress test (TSST), increased IL-6 and NF-κB DNA binding was observed only in those with a history of CM (Pace et al., 2006). A subsequent study analyzed high-sensitivity CRP (hsCRP) in the peripheral blood of adults with or without a history of CM [large cohort ($n = 1015$, age $= 32$)] (Danese et al., 2007). Those with a history of CM had higher hsCRP compared with those without CM history, and this was significant even after correcting for multiple co-occurring childhood and adulthood risk factors. Further, within the depressed adults, those with a history of CM had higher hsCRP, and this association was significant after corrections for other risk factors (Danese et al., 2008). Elevated peripheral cytokines have also been observed in adults with a history of CM and current depression (Grosse et al., 2016; Lu et al., 2013). Maltreated adults also show more rapid T-cell responses as observed by a greater delayed-type hypersensitivity (Altemus, Cloitre, & Dhabhar, 2003) and a higher percentage of effector T cells to memory T cells compared with controls (Lemieux, Coe, & Carnes, 2008).

Inflammation in Animal Models of ELS

Increased inflammation of the inner lining of the colon (colitis) was reported in mice when chemically stimulated, and this increase was only observed in those mice that had gone through both MS (3 h/day, PND1-14) and an adult chronic mild stress (Veenema, Reber, Selch, Obermeier, & Neumann, 2008). Increased TNFα and IFNγ and reduced colon length were also observed along with increased anxiety-like behavior. MS was also shown to modulate cytokine responses in sickness behavior when MS and control animals were treated with IL1β, TNFα, or LPS (lipopolysaccharides, endotoxin) (Avitsur, Maayan, & Weizman, 2013). It was shown that MS mice treated with IL-1β and TNFα showed long-lasting behavioral effects compared with controls and TNFα and IL1β were increased in response to LPS in a dosage- and sex-dependent manner. Increased activation of microglia within the PFC and HC of mice exposed to MS was reported along with imbalances in the tryptophan-kynurenine pathway (Gracia-Rubio et al., 2016). Proteomic analyses on Flinders sensitive line (FSL) rats exposed to MS and treated with two different antidepressants revealed increased inflammatory cytokines and acute-phase proteins such as CRP, albumin, and complement component in the rats that underwent MS, and this was reversed upon treatment (Carboni et al., 2010). Similarly, decreased parvalbumin and increased cyclooxygenase-2 (COX-2) in PFC were found in response to MS. A COX-2 inhibitor

(antiinflammatory treatment) was shown to reduce the parvalbumin loss and improve working memory deficits (Brenhouse & Andersen, 2011). While a majority of the studies with different ELS models showed similar increased pro-inflammatory milieu in the developing brain post ELS (Hennesssy, Paik, Caraway, Schiml, & Deak, 2011; Reus et al., 2017; Wieck, Andersen, & Brenhouse, 2013), some other have reported a reduction in cytokine gene expression as well (Dimatelis et al., 2012).

Epigenetic Regulation of Inflammatory Genes in Response to CM

Despite increased evidence for CM-induced inflammation in adult depression, there is no published study to date that analyzed the epigenetic mechanism mediating early stress effects in immune genes in relation to depression. However, one recent study has investigated the epigenetic profile of the IL-6 gene in late-life depression (Ryan et al., 2017). This study reports that hypomethylation of the promoter region of the IL-6 gene is associated with depression in late life and antidepressant treatment leads to hypermethylation of the gene promoter. This finding explains the presence of increased circulating levels of IL-6 in depression, and given that other studies (reviewed earlier) have shown an effect of CM on IL-6, it could be hypothesized that similar methylation changes in IL-6 also possibly occur in CM-induced depression.

CM EFFECTS IN NEUROTROPHIC FUNCTION

The neurotrophic theory states that a reduction in the levels of neurotrophic factors in the brain leads to decreased hippocampal neurogenesis, neuronal plasticity and atrophy, and the loss of glial cells, all shown to be reversed by chronic antidepressant treatment (Duman, Heninger, & Nestler, 1997; Duman & Monteggia,

2006). BDNF is the most studied neurotrophin in stress and stress-related mood disorders and carries a functional polymorphism (amino acid substitution at codon 66 changing valine to methionine (val66met)) often studied in the context of gene-environment interaction (Brunoni, Lopes, & Fregni, 2008; Egan et al., 2003; Joffe et al., 2009; Lee & Kim, 2009; Shimizu et al., 2003).

CM Effects on BDNF: Childhood and Adolescence

Kaufman and colleagues were the first to analyze the effect of the BDNF val66met polymorphism in a G x G x E model in depressed children (age 5–15, mean age 9.3 years). The study reported that maltreated children carrying the met/met allele of BDNF and s/s allele for the 5-HTTLPR showed highest depression scores (Kaufman et al., 2006). Replication of this initial finding in three separate larger cohorts of children and adolescents exposed to CM either found association with CM and depression [this study had association to a different SNP in BDNF, (Cicchetti & Rogosch, 2014)] or failed to find an association (Nederhof, Bouma, Oldehinkel, & Ormel, 2010) ($n = 1096$) or found only association among abused girls and with the opposite alleles, that is, carriers of val/val of BDNF and s/s or s/l of 5-HTTLPR had higher depression scores (Comasco, Aslund, Oreland, & Nilsson, 2013) ($n = 1393$).

CM Effects on BDNF: Adulthood

The epistatic and gene-environment interaction effect of val66met BDNF and 5-HTTLPR reported by Kaufman was replicated in adults by Wichers et al. (2008). They found a positive three-way interaction effect similar to Kaufman et al. and also found a significant BDNF-CM effect with met carriers at increased risk of depression. Gatt et al. (2009) examine the effect of BDNF val66met polymorphism and CM in predicting brain arousal and in depression

in healthy adults. Carriers of the met allele and maltreated as children had smaller hippocampus and amygdala volume, with greater gray matter loss and associated impairment in cognition. The BDNF met allele carriers have also been shown to have reduced serum BDNF, and this depends on exposure to CM in a dose-dependent manner with those without a history of CM having higher serum BDNF than the val carriers (Elzinga et al., 2011). Together these studies show a strong evidence for a GxE effect of BDNF and CM in depression. However, studies have also reported no association of BDNF and CM to depression; in fact, one of these was a metaanalysis that included around 22 studies that assessed CM, BDNF polymorphism, and depression across age groups and found only a weak interaction with the met allele carriers (Brown et al., 2014; Hosang, Shiles, Tansey, McGuffin, & Uher, 2014).

Animal Models of CM and BDNF

A single maternal deprivation (MD, 24 h at PND 9) led to a reduction in the expression of BDNF in the adult rat HC (Roceri, Hendriks, Racagni, Ellenbroek, & Riva, 2002) and was found to cause neuronal and glial cell death in several white matter tracts in the infant rat brains (Zhang et al., 2002). A repeated MD or MS stimulus showed age-dependent effects in selected brain regions with some younger rats showing an increase in BDNF and adult rats a selective reduction in the PFC (Roceri et al., 2004). A reduction in mature BDNF in HC and increase in pro-BDNF in ventral tegmental area (VTA) has also been shown in rats that underwent repeated MS (Lippmann, Bress, Nemeroff, Plotsky, & Monteggia, 2007). In contrary, increased expression of BDNF in adult rats exposed to MS as infants have been shown to act as a compensatory mechanism to protect the brain from further insults (Faure, Uys, Marais, Stein, & Daniels, 2007; Greisen, Altar, Bolwig, Whitehead, & Wortwein, 2005). The GxE effect of BDNF and 5-HTT polymorphism found in humans was analyzed with 5-HTT mutant rats subject to ELS, results showing similar observations with mutant mice exposed to ELS having an overall reduction of BDNF in the HC and PFC (Calabrese et al., 2015). Taken together, the stress effects of CM leads to reduced BDNF in adulthood and could be either maladaptive or adaptive depending on further insults in combination with other predisposing risk genes.

Epigenetic Regulation of Neurotrophic Genes in Response to CM

Increased DNA methylation and reduction in expression of *Bdnf* were observed in the PFC of rat pups that were exposed to abusive mothers (Roth, Lubin, Funk, & Sweatt, 2009). This altered methylation of BDNF was also observed in the female offspring of stressed rats showing a transgenerational effect of stress. The microRNA miR-16 was significantly increased in the HC of MD rats resulting in decreased Bdnf mRNA and protein compared with rats that underwent chronic stress in adulthood or controls with no stress (Bai et al., 2012). Suri and colleagues report increased Bdnf levels and decreased histone methylation of the Bdnf promoter in younger MS rats and the opposite effect in middle-aged rats suggesting an age-dependent effect in HC BDNF expression (Suri et al., 2013). Increased methylation has also been observed in BDNF gene promoter regions in response to increased CM; however, the association was with borderline personality rather than depression (Perroud et al., 2013).

Structural Brain Differences in Response to CM

As a direct consequence of CM and due to changes in neuroendocrine-immune systems as described above, significant changes in the brain structures underlying stress responsivity, cognition, and emotion process are observed. In maltreated children, these include smaller PFC volume (Hanson et al., 2010) and greater amygdala volume (Mehta et al., 2009; Tottenham et al., 2010). However, no significant volume change is reported for HC (Mehta et al., 2009; Woon & Hedges, 2008). These morphometric changes are reflected in the behavioral functions including deficits in cognitive functions such as attention, problem solving, executive functioning that are dependent on the PFC (Beers & De Bellis, 2002; Carrion et al., 2009; Cohen, Lojkasek, Zadeh, Pugliese, & Kiefer, 2008; De Bellis, Hooper, Spratt, & Woolley, 2009; Loman, Wiik, Frenn, Pollak, & Gunnar, 2009), and impaired emotional regulation mediated by amygdala (Tottenham et al., 2010).

Morphometric changes in brains of adults with CM are reduced PFC volume (Tomoda et al., 2009; van Harmelen et al., 2010) and more commonly, reduced hippocampal volume (Frodl et al., 2010; Karl et al., 2006; Vythilingam et al., 2002; Woon, Sood, & Hedges, 2010). There are mixed reports on volume changes in amygdala with some reporting a reduction in volume (Driessen et al., 2000; Vermetten, Schmahl, Lindner, Loewenstein, & Bremner, 2006) while some report no difference (Cohen et al., 2006). Consistent with these structural changes, significant behavioral impairments have been found across multiple studies of adults maltreated as children. These include deficits in cognitive functions (Bos, Fox, Zeanah, & Nelson Iii, 2009; Bremner et al., 2003; Navalta, Polcari, Webster, Boghossian, & Teicher, 2006), depression, and anxiety-related disorders (Graham, Heim, Goodman, Miller, & Nemeroff, 1999; Heim & Nemeroff, 2001; Mello, Mello, Carpenter, & Price, 2003).

DISCUSSION

It is evident from the current review that CM affects various biological mechanisms associated with depression and mediates long-lasting effects in the etiology of adult depression, thereby also explaining part of the disease heterogeneity. Further, the effects of these ELS (first hit of stress) are evident even in childhood and appears to be more primed into the underlying neurobiology by adolescence (see Table 1 for summary of CM effects), suggesting that exposures to further stressors at this stage and young adulthood (second hit of stress) can significantly increase the risk for psychopathology later in life. There is a considerable amount of interaction or cross talk that occurs between each of these major mechanisms that might present a networked model of the biological embedding of CM stress in depression (Fig. 1).

HPA FUNCTION AND INFLAMMATION—MODULATION BY EARLY STRESS

HPA response and increased inflammation are both adaptive processes to survive in varied environments. However, under chronic stress as in CM, these two systems are overworked leading to a heightened endocrine and inflammatory state as seen in children, adolescents, and adults with a history of CM (Table 1). It is known that glucocorticoids (GC) released from HPA activation is generally immunosuppressive, and GC receptors (GRs) are found on almost all immune cells. Upon exposure to CM, there is a reduction in the expression of GRs through stress-induced hypermethylation of the GR gene (NR3C1) leading to reduced negative feedback of HPA at various levels (McGowan et al., 2009; Perroud et al., 2011; Romens et al., 2015; Weaver et al., 2004), including peripheral immune cells, which have also been shown to have hypermethylated NR3C1 (Tyrka, Price, Marsit, Walters, & Carpenter, 2012). Therefore, CM mediated reduction in GR levels along with glucocorticoid resistance (desensitization of GRs in the presence of excessive GCs (Holsboer, 2000)) may lead to poor inhibition of activated immune responses resulting in elevated levels of pro-inflammatory cytokines in both periphery and the CNS as reviewed earlier.

Chronic stress can initiate immune cell signaling via activation of noradrenergic receptors and the pro-inflammatory transcription factor nuclear factor kappa B (NF-κB) to increase levels of pro-inflammatory cytokines in response to stress (Bierhaus et al., 2003). In addition, a recent preclinical study demonstrated that psychological stress results in excess release of adenine triphosphate (ATP) from glial cells in the hippocampus that then signals via the purinergic receptor type 2X7 (P2X7) and activates the nucleotide-binding, leucine-rich repeat, pyrin domain-containing 3 (NLRP3) inflammasome (Iwata et al., 2016). This results in increased production of pro-inflammatory cytokines IL-1β and TNFα with the latter expressed significantly higher for a longer duration. There is a considerable amount of evidence showing that pro-inflammatory cytokines such as IL-1β, IL-6, and TNFα can activate HPA axis and the release of GCs, therefore rendering the HPA more hyperresponsive (Pace, Hu, & Miller, 2007; Turnbull & Rivier, 1995). This bidirectional communication between the HPA axis and inflammatory cytokines, therefore, could be considered the primary effector systems in mediating CM effects.

SEROTONIN AND BDNF— MODULATION BY GCs AND CYTOKINES

HPA activation and inflammation both can affect the levels of serotonin and neurotrophic factor BDNF. GCs can induce the transcription of serotonin transporter gene SLC6A4, thereby increasing the transporter and decreasing serotonin levels in the brain synapses (Tafet, Toister-Achituv, & Shinitzky, 2001). In addition, GCs also reduce the levels of tryptophan hydroxylase 2 (TPH2), the rate-limiting enzyme for serotonin in the brain, therefore reducing serotonin levels (Clark et al., 2008). It is shown that increased GCs as a result of stress can inhibit BDNF-dependent synaptic proteins by suppressing the activation of mitogen-activated

TABLE 1 Summary of CM Effects in Major Biological Hypotheses of Depression

Major Biological Theories	Childhood/Adolescence	Adulthood
HPA dysfunction	Hyperactivation of HPA response - ↑ CRH in hypothalamus and basal cortisol - Blunted cortisol and ACTH response to TSST - ↓ Negative feedback inhibition of HPA Hypermethylation of the exon 1_7 promoter of NR3C1 - Reduced NR3C1 mRNA	Hyperactive HPA response - ↑ Or blunted ACTH and cortisol (TSST and DEX test) - ↓ Negative feedback of HPA Hypermethylation of the exon 1_7 promoter of NR3C1 - Reduced NR3C1 mRNA
Monoamine theory	G × E effect - s/s & s/l of 5-HTTLPR × CM = ↑ depression risk, ↑ suicide ideation - s/s genotype of 5-HTTLPR and ↓ MAO-A activity Brain levels - ↓ Serotonin and 5-HTT levels in HC[a]	GxE effect - s/s genotype with CM predicted depression (inconsistent replications) Methylation - Hypermethylation of the SLC6A4 gene (5-HTT) promoter region - s allele of 5-HTTLPR associated with ↑ methylation[a]
Cytokine theory	Serum levels - ↑ CRP, IL-6 Genetic polymorphism effect IL 1β predicted depression but no CM effect	Serum levels - ↑ CRP, IL-6, TNFα in serum and ↑ NF-κB binding Cellular responses - ↑ Percentage of effector memory T cells and delayed-type hypersensitivity - ↑ Microglial activation in PFC and HC ↓ parvalbumin and ↑ COX-2 in PFC[a]
Neurotrophin theory	G × E effect - met/met of BDNF x s/s of 5-HTTLPR = ↑ depression Brain levels - ↑ BDNF in PFC[a]	G × E effect - met/met of BDNF × s/s of 5-HTTLPR = ↑ depression - ↓ Serum BDNF, ↓ HC and amygdala volume (CM dose-dependent) in met allele carriers Brain levels - ↓ BDNF in HC, PFC, and VTA[a]
Structural/ behavioral changes	Structural - ↓ PFC, amygdala and ↔ HC Behavioral - ↓ Cognition, attention, problem solving, executive functions, and emotion regulation	Structural - ↓ PFC, HC, mixed reports on amygdala Behavioral - ↓ Cognition, depression, and anxiety

Legend: ↑, increased; ↓, decreased; ↔, no change; *COX-2*, cyclooxygenase-2; *CRP*, C-reactive protein; *HC*, hippocampus; *IL-6*, interleukin 6; *IL-1β*, interleukin 1 beta; *NF-κB*, nuclear factor kappa B; *NR3C1*, nuclear receptor subfamily 3 group C; *PFC*, prefrontal cortex; *SLC6A4*, solute carrier family 6 member 4; *TNFα*, tumor necrosis factor alpha; *VTA*, ventral tegmental area.
[a] *Results from animal models only.*

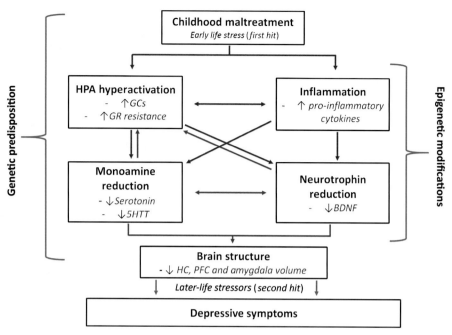

FIG. 1 Biological embedding of CM in adult depression risk. The figure depicts the cascade of events that might confer increased risk for depression in adulthood after early stress. The *red arrows* depict the primary effector systems mediating early stress, and the *blue arrows* are downstream events after activation of primary processes within HPA and inflammation. Predisposing genetic factors and epigenetic modifications induced by early stress increase the risk substantially. *5-HTT*, 5-hydroxytryptamine transporter; *BDNF*, brain-derived neurotrophic factor; *GC*, glucocorticoids; *GR*, glucocorticoid receptors; *HC*, hippocampus; *HPA*, hypothalamic-pituitary-adrenal axis; *PFC*, prefrontal cortex.

protein kinases (MAPK) in the hippocampal neurons (Kumamaru et al., 2008). Further, GC-mediated control of BDNF gene expression and signaling via its receptors tropomyosin receptor kinase B (Trkb) is also extensively studied and reviewed (Suri & Vaidya, 2013). Increased pro-inflammatory cytokines have been shown to affect serotonin levels by affecting serotonin metabolism. Specifically, interferon gamma (IFNγ) has been shown to deplete tryptophan, the amino acid required for serotonin biosynthesis by activation of the alternate tryptophan metabolism pathway via indoleamine-2,3-dioxygenase (IDO) that converts tryptophan into kynurenine and then quinolinic acid (Capuron et al., 2003). Increased

pro-inflammatory cytokines specifically IL-1 have also been shown to downregulate the expression of BDNF mRNA in the hippocampus of rat brains along with memory impairments that were reversed by an IL-1 receptor antagonist (Barrientos et al., 2003). In light of these known mechanisms linking GCs and pro-inflammatory cytokines to serotonin and BDNF levels, it is highly likely that similar interacting mechanisms are activated by CM albeit at an early stage. As seen from the summary Table 1, genetic polymorphisms in both serotonin transporter and BDNF genes could enhance this GC and cytokine-mediated suppression, thereby building a molecular foundation that is primed to malfunction.

CONCLUSION

We, therefore, conclude that upon exposure to CM both GCs and pro-inflammatory cytokines are activated in parallel and are therefore considered primary effector molecules in mediating CM stress. These primary effectors then inhibit the neurotransmitter and neurotrophic molecules leading to poor neurotransmission, lowered neurogenesis, reduced synaptic plasticity and increased neurodegeneration, resulting in atrophy of key brain regions such as HC, PFC, and amygdala mediating emotional and cognitive dysfunctions, often associated with depression. Epigenetic mechanisms in key genes within these processes contribute to the long-lasting effects of early exposure and later development of a clinical phenotype. Future research could focus on these primary effectors and downstream biological processes through clinical and preclinical studies at the transcriptomic, epigenomic, and genomic levels to understand the complex interplay of molecular mechanisms in the biological embedding of CM. This research will provide us subsequently with valuable cumulative risk prediction models for depression and inform early intervention, specifically in those with a history of CM.

References

Aisa, B., Tordera, R., Lasheras, B., Del Rio, J., & Ramirez, M. J. (2007). Cognitive impairment associated to HPA axis hyperactivity after maternal separation in rats. *Psychoneuroendocrinology, 32*(3), 256–266. https://doi.org/10.1016/j.psyneuen.2006.12.013.

Altemus, M., Cloitre, M., & Dhabhar, F. S. (2003). Enhanced cellular immune response in women with PTSD related to childhood abuse. *The American Journal of Psychiatry, 160*(9), 1705–1707.

Appel, K., Schwahn, C., Mahler, J., Schulz, A., Spitzer, C., Fenske, K., et al. (2011). Moderation of adult depression by a polymorphism in the FKBP5 gene and childhood physical abuse in the general population. *Neuropsychopharmacology, 36*(10), 1982–1991. https://doi.org/10.1038/npp.2011.81.

Arborelius, L., & Eklund, M. B. (2007). Both long and brief maternal separation produces persistent changes in tissue levels of brain monoamines in middle-aged female rats. *Neuroscience, 145*(2), 738–750. https://doi.org/10.1016/j.neuroscience.2006.12.007.

Avitsur, R., Maayan, R., & Weizman, A. (2013). Neonatal stress modulates sickness behavior: role for proinflammatory cytokines. *Journal of Neuroimmunology, 257*(1–2), 59–66. https://doi.org/10.1016/j.jneuroim.2013.02.009.

Bai, M., Zhu, X., Zhang, Y., Zhang, S., Zhang, L., Xue, L., et al. (2012). Abnormal hippocampal BDNF and miR-16 expression is associated with depression-like behaviors induced by stress during early life. *PLoS One, 7*(10), e46921. https://doi.org/10.1371/journal.pone.0046921.

Barr, C. S., Newman, T. K., Shannon, C., Parker, C., Dvoskin, R. L., Becker, M. L., et al. (2004). Rearing condition and rh5-HTTLPR interact to influence limbic-hypothalamic-pituitary-adrenal axis response to stress in infant macaques. *Biological Psychiatry, 55*(7), 733–738. https://doi.org/10.1016/j.biopsych.2003.12.008.

Barrientos, R. M., Sprunger, D. B., Campeau, S., Higgins, E. A., Watkins, L. R., Rudy, J. W., et al. (2003). Brain-derived neurotrophic factor mRNA downregulation produced by social isolation is blocked by intrahippocampal interleukin-1 receptor antagonist. *Neuroscience, 121*(4), 847–853.

Beach, S. R. H., Brody, G. H., Todorov, A. A., Gunter, T. D., & Philibert, R. A. (2010). Methylation at SLC6A4 is linked to family history of child abuse: an examination of the Iowa Adoptee sample. *American Journal of Medical Genetics. Part B, Neuropsychiatric Genetics, 153B*(2), 710–713. https://doi.org/10.1002/ajmg.b.31028.

Beach, S. R., Brody, G. H., Todorov, A. A., Gunter, T. D., & Philibert, R. A. (2011). Methylation at 5HTT mediates the impact of child sex abuse on women's antisocial behavior: an examination of the Iowa adoptee sample. *Psychosomatic Medicine, 73*(1), 83–87. https://doi.org/10.1097/PSY.0b013e3181fdd074.

Beers, S. R., & De Bellis, M. D. (2002). Neuropsychological function in children with maltreatment-related posttraumatic stress disorder. *The American Journal of Psychiatry, 159*(3), 483–486.

Bernet, C. Z., & Stein, M. B. (1999). Relationship of childhood maltreatment to the onset and course of major depression in adulthood. *Depression and Anxiety, 9*(4), 169–174.

Bierhaus, A., Wolf, J., Andrassy, M., Rohleder, N., Humpert, P. M., Petrov, D., et al. (2003). A mechanism converting psychosocial stress into mononuclear cell activation. *Proceedings of the National Academy of Sciences of the United States of America, 100*(4), 1920–1925. https://doi.org/10.1073/pnas.0438019100.

Binder, E. B. (2009). The role of FKBP5, a co-chaperone of the glucocorticoid receptor in the pathogenesis and therapy of affective and anxiety disorders. *Psychoneuroendocrinology, 34*(Suppl. 1), S186–195. https://doi.org/10.1016/j.psyneuen.2009.05.021.

Bos, K. J., Fox, N., Zeanah, C. H., & Nelson Iii, C. A. (2009). Effects of early psychosocial deprivation on the development of memory and executive function. *Frontiers in Behavioral Neuroscience*, 3, 16. https://doi.org/10.3389/neuro.08.016.2009.

Bremner, J. D., Vythilingam, M., Vermetten, E., Southwick, S. M., McGlashan, T., Nazeer, A., et al. (2003). MRI and PET study of deficits in hippocampal structure and function in women with childhood sexual abuse and posttraumatic stress disorder. *The American Journal of Psychiatry*, 160(5), 924–932.

Brenhouse, H. C., & Andersen, S. L. (2011). Nonsteroidal anti-inflammatory treatment prevents delayed effects of early life stress in rats. *Biological Psychiatry*, 70(5), 434–440. https://doi.org/10.1016/j.biopsych.2011.05.006.

Brown, G. W., Craig, T. K., Harris, T. O., Herbert, J., Hodgson, K., Tansey, K. E., et al. (2014). Functional polymorphism in the brain-derived neurotrophic factor gene interacts with stressful life events but not childhood maltreatment in the etiology of depression. *Depression and Anxiety*, 31(4), 326–334. https://doi.org/10.1002/da.22221.

Brunoni, A. R., Lopes, M., & Fregni, F. (2008). A systematic review and meta-analysis of clinical studies on major depression and BDNF levels: implications for the role of neuroplasticity in depression. *The International Journal of Neuropsychopharmacology*, 11(8), 1169–1180. https://doi.org/10.1017/S1461145708009309.

Buchmann, A. F., Hellweg, R., Rietschel, M., Treutlein, J., Witt, S. H., Zimmermann, U. S., et al. (2013). BDNF Val 66 Met and 5-HTTLPR genotype moderate the impact of early psychosocial adversity on plasma brain-derived neurotrophic factor and depressive symptoms: a prospective study. *European Neuropsychopharmacology*, 23(8), 902–909. https://doi.org/10.1016/j.euroneuro.2012.09.003.

Calabrese, F., van der Doelen, R. H., Guidotti, G., Racagni, G., Kozicz, T., Homberg, J. R., et al. (2015). Exposure to early life stress regulates Bdnf expression in SERT mutant rats in an anatomically selective fashion. *Journal of Neurochemistry*, 132(1), 146–154. https://doi.org/10.1111/jnc.12846.

Capuron, L., Neurauter, G., Musselman, D. L., Lawson, D. H., Nemeroff, C. B., Fuchs, D., et al. (2003). Interferon-alpha-induced changes in tryptophan metabolism. relationship to depression and paroxetine treatment. *Biological Psychiatry*, 54(9), 906–914.

Carboni, L., Becchi, S., Piubelli, C., Mallei, A., Giambelli, R., Razzoli, M., et al. (2010). Early-life stress and antidepressants modulate peripheral biomarkers in a gene-environment rat model of depression. *Progress in Neuro-Psychopharmacology & Biological Psychiatry*, 34(6), 1037–1048. https://doi.org/10.1016/j.pnpbp.2010.05.019.

Carpenter, L. L., Carvalho, J. P., Tyrka, A. R., Wier, L. M., Mello, A. F., Mello, M. F., et al. (2007). Decreased adrenocorticotropic hormone and cortisol responses to stress in healthy adults reporting significant childhood maltreatment. *Biological Psychiatry*, 62(10), 1080–1087. https://doi.org/10.1016/j.biopsych.2007.05.002.

Carpenter, L. L., Gawuga, C. E., Tyrka, A. R., Lee, J. K., Anderson, G. M., & Price, L. H. (2010). Association between plasma IL-6 response to acute stress and early-life adversity in healthy adults. *Neuropsychopharmacology*, 35(13), 2617–2623. https://doi.org/10.1038/npp.2010.159.

Carpenter, L. L., Tyrka, A. R., McDougle, C. J., Malison, R. T., Owens, M. J., Nemeroff, C. B., et al. (2004). Cerebrospinal fluid corticotropin-releasing factor and perceived early-life stress in depressed patients and healthy control subjects. *Neuropsychopharmacology*, 29(4), 777–784. https://doi.org/10.1038/sj.npp.1300375.

Carrion, V. G., Weems, C. F., Watson, C., Eliez, S., Menon, V., & Reiss, A. L. (2009). Converging evidence for abnormalities of the prefrontal cortex and evaluation of midsagittal structures in pediatric posttraumatic stress disorder: an MRI study. *Psychiatry Research*, 172(3), 226–234. pii: S0925-4927(08)00099-1(2009). https://doi.org/10.1016/j.pscychresns.2008.07.008.

Caspi, A., Sugden, K., Moffitt, T. E., Taylor, A., Craig, I. W., Harrington, H., et al. (2003). Influence of life stress on depression: moderation by a polymorphism in the 5-HTT gene. *Science*, 301(5631), 386–389. https://doi.org/10.1126/science.1083968.

Champoux, M., Bennett, A., Shannon, C., Higley, J. D., Lesch, K. P., & Suomi, S. J. (2002). Serotonin transporter gene polymorphism, differential early rearing, and behavior in rhesus monkey neonates. *Molecular Psychiatry*, 7(10), 1058–1063. https://doi.org/10.1038/sj.mp.4001157.

Chapman, D. P., Whitfield, C. L., Felitti, V. J., Dube, S. R., Edwards, V. J., & Anda, R. F. (2004). Adverse childhood experiences and the risk of depressive disorders in adulthood. *Journal of Affective Disorders*, 82(2), 217–225. pii: S016503270400028X(2004). https://doi.org/10.1016/j.jad.2003.12.013.

Chen, J., Evans, A., Liu, Y., Honda, M., Saavedra, J., & Aguilera, G. (2012). Maternal deprivation in rats is associated with corticotrophin-releasing hormone (CRH) promoter hypomethylation and enhances CRH transcriptional responses to stress in adulthood. *Journal of Neuroendocrinology*, 24(7), 1055–1064. https://doi.org/10.1111/j.1365-2826.2012.02306.x.

Cicchetti, D., & Rogosch, F. A. (2001). The impact of child maltreatment and psychopathology on neuroendocrine functioning. *Development and Psychopathology*, 13(4), 783–804.

Cicchetti, D., & Rogosch, F. A. (2014). Genetic moderation of child maltreatment effects on depression and internalizing symptoms by serotonin transporter linked polymorphic region (5-HTTLPR), brain-derived neurotrophic factor (BDNF), norepinephrine transporter (NET), and corticotropin releasing hormone receptor 1 (CRHR1) genes in African American children. *Development and*

Psychopathology, 26(4 Pt 2), 1219–1239. https://doi.org/10.1017/S0954579414000984.

Cicchetti, D., Rogosch, F. A., & Sturge-Apple, M. L. (2007). Interactions of child maltreatment and serotonin transporter and monoamine oxidase A polymorphisms: depressive symptomatology among adolescents from low socioeconomic status backgrounds. *Development and Psychopathology, 19*(4), 1161–1180. https://doi.org/10.1017/S0954579407000600.

Cicchetti, D., Rogosch, F. A., Sturge-Apple, M., & Toth, S. L. (2010). Interaction of child maltreatment and 5-HTT polymorphisms: suicidal ideation among children from low-SES backgrounds. *Journal of Pediatric Psychology, 35* (5), 536–546. https://doi.org/10.1093/jpepsy/jsp078.

Clark, J. A., Flick, R. B., Pai, L. Y., Szalayova, I., Key, S., Conley, R. K., et al. (2008). Glucocorticoid modulation of tryptophan hydroxylase-2 protein in raphe nuclei and 5-hydroxytryptophan concentrations in frontal cortex of C57/Bl6 mice. *Molecular Psychiatry, 13*(5), 498–506. https://doi.org/10.1038/sj.mp.4002041.

Cohen, R. A., Grieve, S., Hoth, K. F., Paul, R. H., Sweet, L., Tate, D., et al. (2006). Early life stress and morphometry of the adult anterior cingulate cortex and caudate nuclei. *Biological Psychiatry, 59*(10), 975–982. pii: S0006-3223(06)00140-5 (2006). https://doi.org/10.1016/j.biopsych.2005.12.016.

Cohen, N. J., Lojkasek, M., Zadeh, Z. Y., Pugliese, M., & Kiefer, H. (2008). Children adopted from China: a prospective study of their growth and development. *Journal of Child Psychology and Psychiatry, 49*(4), 458–468. pii: JCPP1853(2008). https://doi.org/10.1111/j.1469-7610.2007.01853.x.

Comasco, E., Aslund, C., Oreland, L., & Nilsson, K. W. (2013). Three-way interaction effect of 5-HTTLPR, BDNF Val66-Met, and childhood adversity on depression: a replication study. *European Neuropsychopharmacology, 23*(10), 1300–1306. https://doi.org/10.1016/j.euroneuro.2013.01.010.

Coplan, J. D., Andrews, M. W., Rosenblum, L. A., Owens, M. J., Friedman, S., Gorman, J. M., et al. (1996). Persistent elevations of cerebrospinal fluid concentrations of corticotropin-releasing factor in adult nonhuman primates exposed to early-life stressors: implications for the pathophysiology of mood and anxiety disorders. *Proceedings of the National Academy of Sciences of the United States of America, 93*(4), 1619–1623.

Culverhouse, R. C., Saccone, N. L., Horton, A. C., Ma, Y., Anstey, K. J., Banaschewski, T., et al. (2018). Collaborative meta-analysis finds no evidence of a strong interaction between stress and 5-HTTLPR genotype contributing to the development of depression. *Molecular Psychiatry, 23*(1), 133–142. https://doi.org/10.1038/mp.2017.44.

Danese, A., Caspi, A., Williams, B., Ambler, A., Sugden, K., Mika, J., et al. (2011). Biological embedding of stress through inflammation processes in childhood. *Molecular Psychiatry, 16*(3), 244–246. https://doi.org/10.1038/mp.2010.5.

Danese, A., Moffitt, T. E., Pariante, C. M., Ambler, A., Poulton, R., & Caspi, A. (2008). Elevated inflammation levels in depressed adults with a history of childhood maltreatment. *Archives of General Psychiatry, 65*(4), 409–415. https://doi.org/10.1001/archpsyc.65.4.409.

Danese, A., Pariante, C. M., Caspi, A., Taylor, A., & Poulton, R. (2007). Childhood maltreatment predicts adult inflammation in a life-course study. *Proceedings of the National Academy of Sciences of the United States of America, 104*(4), 1319–1324. https://doi.org/10.1073/pnas.0610362104.

Daniels, W. M., Pietersen, C. Y., Carstens, M. E., & Stein, D. J. (2004). Maternal separation in rats leads to anxiety-like behavior and a blunted ACTH response and altered neurotransmitter levels in response to a subsequent stressor. *Metabolic Brain Disease, 19*(1–2), 3–14.

Dantzer, R., O'Connor, J. C., Freund, G. G., Johnson, R. W., & Kelley, K. W. (2008). From inflammation to sickness and depression: when the immune system subjugates the brain. *Nature Reviews Neuroscience, 9*(1), 46–56. https://doi.org/10.1038/nrn2297.

De Bellis, M. D., Chrousos, G. P., Dorn, L. D., Burke, L., Helmers, K., Kling, M. A., et al. (1994). Hypothalamic-pituitary-adrenal axis dysregulation in sexually abused girls. *The Journal of Clinical Endocrinology and Metabolism, 78*(2), 249–255.

De Bellis, M. D., Hooper, S. R., Spratt, E. G., & Woolley, D. P. (2009). Neuropsychological findings in childhood neglect and their relationships to pediatric PTSD. *Journal of the International Neuropsychological Society, 15*(6), 868–878. pii: S1355617709990464(2009). https://doi.org/10.1017/S1355617709990464.

Dimatelis, J. J., Pillay, N. S., Mutyaba, A. K., Russell, V. A., Daniels, W. M., & Stein, D. J. (2012). Early maternal separation leads to down-regulation of cytokine gene expression. *Metabolic Brain Disease, 27*(3), 393–397. https://doi.org/10.1007/s11011-012-9304-z.

Dowlati, Y., Herrmann, N., Swardfager, W., Liu, H., Sham, L., Reim, E. K., et al. (2010). A meta-analysis of cytokines in major depression. *Biological Psychiatry, 67*(5), 446–457. https://doi.org/10.1016/j.biopsych.2009.09.033.

Drevets, W. C., Frank, E., Price, J. C., Kupfer, D. J., Holt, D., Greer, P. J., et al. (1999). PET imaging of serotonin 1A receptor binding in depression. *Biological Psychiatry, 46* (10), 1375–1387.

Driessen, M., Herrmann, J., Stahl, K., Zwaan, M., Meier, S., Hill, A., et al. (2000). Magnetic resonance imaging volumes of the hippocampus and the amygdala in women with borderline personality disorder and early traumatization. *Archives of General Psychiatry, 57*(12), 1115–1122 pii: yoa9416.

Duman, R. S., Heninger, G. R., & Nestler, E. J. (1997). A molecular and cellular theory of depression. *Archives of General Psychiatry, 54*(7), 597–606.

Duman, R. S., & Monteggia, L. M. (2006). A neurotrophic model for stress-related mood disorders. *Biological Psychiatry, 59*(12), 1116–1127. https://doi.org/10.1016/j.biopsych.2006.02.013.

Edwards, V. J., Holden, G. W., Felitti, V. J., & Anda, R. F. (2003). Relationship between multiple forms of childhood maltreatment and adult mental health in community respondents: results from the adverse childhood experiences study. *The American Journal of Psychiatry, 160*(8), 1453–1460.

Egan, M. F., Kojima, M., Callicott, J. H., Goldberg, T. E., Kolachana, B. S., Bertolino, A., et al. (2003). The BDNF val66met polymorphism affects activity-dependent secretion of BDNF and human memory and hippocampal function. *Cell, 112*(2), 257–269.

Elzinga, B. M., Molendijk, M. L., Oude Voshaar, R. C., Bus, B. A., Prickaerts, J., Spinhoven, P., et al. (2011). The impact of childhood abuse and recent stress on serum brain-derived neurotrophic factor and the moderating role of BDNF Val66Met. *Psychopharmacology, 214*(1), 319–328. https://doi.org/10.1007/s00213-010-1961-1.

Faure, J., Uys, J. D., Marais, L., Stein, D. J., & Daniels, W. M. (2007). Early maternal separation alters the response to traumatization: resulting in increased levels of hippocampal neurotrophic factors. *Metabolic Brain Disease, 22*(2), 183–195. https://doi.org/10.1007/s11011-007-9048-3.

Frodl, T., Reinhold, E., Koutsouleris, N., Donohoe, G., Bondy, B., Reiser, M., et al. (2010). Childhood stress, serotonin transporter gene and brain structures in major depression. *Neuropsychopharmacology, 35*(6), 1383–1390. pii: npp20108 (2010). https://doi.org/10.1038/npp.2010.8.

Gardner, K. L., Hale, M. W., Lightman, S. L., Plotsky, P. M., & Lowry, C. A. (2009). Adverse early life experience and social stress during adulthood interact to increase serotonin transporter mRNA expression. *Brain Research, 1305*, 47–63. https://doi.org/10.1016/j.brainres.2009.09.065.

Gatt, J. M., Nemeroff, C. B., Dobson-Stone, C., Paul, R. H., Bryant, R. A., Schofield, P. R., et al. (2009). Interactions between BDNF Val66Met polymorphism and early life stress predict brain and arousal pathways to syndromal depression and anxiety. *Molecular Psychiatry, 14*(7), 681–695. https://doi.org/10.1038/mp.2008.143.

Gershon, A., Sudheimer, K., Tirouvanziam, R., Williams, L. M., & O'Hara, R. (2013). The long-term impact of early adversity on late-life psychiatric disorders. *Current Psychiatry Reports, 15*(4), 013–0352.

Gracia-Rubio, I., Moscoso-Castro, M., Pozo, O. J., Marcos, J., Nadal, R., & Valverde, O. (2016). Maternal separation induces neuroinflammation and long-lasting emotional alterations in mice. *Progress in Neuro-Psychopharmacology & Biological Psychiatry, 65*, 104–117. https://doi.org/10.1016/j.pnpbp.2015.09.003.

Graham, Y. P., Heim, C., Goodman, S. H., Miller, A. H., & Nemeroff, C. B. (1999). The effects of neonatal stress on brain development: implications for psychopathology. *Development and Psychopathology, 11*(3), 545–565.

Green, J. G., McLaughlin, K. A., Berglund, P. A., Gruber, M. J., Sampson, N. A., Zaslavsky, A. M., et al. (2010). Childhood adversities and adult psychiatric disorders in the national comorbidity survey replication I: associations with first onset of DSM-IV disorders. *Archives of General Psychiatry, 67*(2), 113–123. https://doi.org/10.1001/archgenpsychiatry.2009.186.

Greisen, M. H., Altar, C. A., Bolwig, T. G., Whitehead, R., & Wortwein, G. (2005). Increased adult hippocampal brain-derived neurotrophic factor and normal levels of neurogenesis in maternal separation rats. *Journal of Neuroscience Research, 79*(6), 772–778. https://doi.org/10.1002/jnr.20418.

Grosse, L., Ambree, O., Jorgens, S., Jawahar, M. C., Singhal, G., Stacey, D., et al. (2016). Cytokine levels in major depression are related to childhood trauma but not to recent stressors. *Psychoneuroendocrinology, 73*, 24–31. https://doi.org/10.1016/j.psyneuen.2016.07.205.

Gunnar, M. R., Frenn, K., Wewerka, S. S., & Van Ryzin, M. J. (2009). Moderate versus severe early life stress: associations with stress reactivity and regulation in 10-12-year-old children. *Psychoneuroendocrinology, 34*(1), 62–75. https://doi.org/10.1016/j.psyneuen.2008.08.013.

Gunnar, M. R., Morison, S. J., Chisholm, K., & Schuder, M. (2001). Salivary cortisol levels in children adopted from romanian orphanages. *Development and Psychopathology, 13*(3), 611–628.

Hanson, J. L., Chung, M. K., Avants, B. B., Shirtcliff, E. A., Gee, J. C., Davidson, R. J., et al. (2010). Early stress is associated with alterations in the orbitofrontal cortex: a tensor-based morphometry investigation of brain structure and behavioral risk. *The Journal of Neuroscience, 30*(22), 7466–7472. pii: 30/22/7466(2010). https://doi.org/10.1523/JNEUROSCI.0859-10.2010.

Harkness, K. L., Stewart, J. G., & Wynne-Edwards, K. E. (2011). Cortisol reactivity to social stress in adolescents: role of depression severity and child maltreatment. *Psychoneuroendocrinology, 36*(2), 173–181. https://doi.org/10.1016/j.psyneuen.2010.07.006.

Harkness, K. L., Strauss, J., Michael Bagby, R., Stewart, J. G., Larocque, C., Mazurka, R., et al. (2015). Interactions between childhood maltreatment and brain-derived neurotrophic factor and serotonin transporter polymorphisms on depression symptoms. *Psychiatry Research, 229*(1–2), 609–612. https://doi.org/10.1016/j.psychres.2015.04.040.

Heim, C., Mletzko, T., Purselle, D., Musselman, D. L., & Nemeroff, C. B. (2008). The dexamethasone/corticotropin-releasing factor test in men with major depression: role of childhood trauma. *Biological Psychiatry, 63*(4), 398–405. pii: S0006-3223(07)00640-3(2008). https://doi.org/10.1016/j.biopsych.2007.07.002.

Heim, C., & Nemeroff, C. B. (2001). The role of childhood trauma in the neurobiology of mood and anxiety disorders: preclinical and clinical studies. *Biological Psychiatry*, 49(12), 1023–1039. pii: S000632230101157X.

Heim, C., Newport, D. J., Heit, S., Graham, Y. P., Wilcox, M., Bonsall, R., et al. (2000). Pituitary-adrenal and autonomic responses to stress in women after sexual and physical abuse in childhood. *JAMA*, 284(5), 592–597.

Heim, C., Newport, D. J., Wagner, D., Wilcox, M. M., Miller, A. H., & Nemeroff, C. B. (2002). The role of early adverse experience and adulthood stress in the prediction of neuroendocrine stress reactivity in women: a multiple regression analysis. *Depression and Anxiety*, 15(3), 117–125. https://doi.org/10.1002/da.10015.

Hennesssy, M., Paik, K., Caraway, J., Schiml, P., & Deak, T. (2011). Proinflammatory activity and the sensitization of depressive-like behavior during maternal separation. *Behavioral Neuroscience*, 125(3), 426–433.

Holsboer, F. (2000). The corticosteroid receptor hypothesis of depression. *Neuropsychopharmacology*, 23(5), 477–501. https://doi.org/10.1016/S0893-133X(00)00159-7.

Hosang, G. M., Shiles, C., Tansey, K. E., McGuffin, P., & Uher, R. (2014). Interaction between stress and the BDNF Val66Met polymorphism in depression: a systematic review and meta-analysis. *BMC Medicine*, 12, 7. https://doi.org/10.1186/1741-7015-12-7.

Infurna, M. R., Reichl, C., Parzer, P., Schimmenti, A., Bifulco, A., & Kaess, M. (2016). Associations between depression and specific childhood experiences of abuse and neglect: a meta-analysis. *Journal of Affective Disorders*, 190, 47–55. https://doi.org/10.1016/j.jad.2015.09.006.

Iwata, M., Ota, K. T., Li, X. Y., Sakaue, F., Li, N., Dutheil, S., et al. (2016). Psychological stress activates the inflammasome via release of adenosine triphosphate and stimulation of the purinergic type 2X7 receptor. *Biological Psychiatry*, 80(1), 12–22. https://doi.org/10.1016/j.biopsych.2015.11.026.

Joffe, R. T., Gatt, J. M., Kemp, A. H., Grieve, S., Dobson-Stone, C., Kuan, S. A., et al. (2009). Brain derived neurotrophic factor Val66Met polymorphism, the five factor model of personality and hippocampal volume: implications for depressive illness. *Human Brain Mapping*, 30(4), 1246–1256. https://doi.org/10.1002/hbm.20592.

Kang, H. J., Kim, J. M., Stewart, R., Kim, S. Y., Bae, K. Y., Kim, S. W., et al. (2013). Association of SLC6A4 methylation with early adversity, characteristics and outcomes in depression. *Progress in Neuro-Psychopharmacology & Biological Psychiatry*, 44, 23–28. https://doi.org/10.1016/j.pnpbp.2013.01.006.

Kaplow, J. B., & Widom, C. S. (2007). Age of onset of child maltreatment predicts long-term mental health outcomes. *Journal of Abnormal Psychology*, 116(1), 176–187. https://doi.org/10.1037/0021-843X.116.1.176.

Karg, K., Burmeister, M., Shedden, K., & Sen, S. (2011). The serotonin transporter promoter variant (5-HTTLPR), stress, and depression meta-analysis revisited: evidence of genetic moderation. *Archives of General Psychiatry*, 68(5), 444–454. https://doi.org/10.1001/archgenpsychiatry.2010.189.

Karl, A., Schaefer, M., Malta, L. S., Dorfel, D., Rohleder, N., & Werner, A. (2006). A meta-analysis of structural brain abnormalities in PTSD. *Neuroscience and Biobehavioral Reviews*, 30(7), 1004–1031. pii: S0149-7634(06)00028-5 (2006). https://doi.org/10.1016/j.neubiorev.2006.03.004.

Kaufman, J. (1991). Depressive disorders in maltreated children. *Journal of the American Academy of Child and Adolescent Psychiatry*, 30(2), 257–265. https://doi.org/10.1097/00004583-199103000-00014.

Kaufman, J., Birmaher, B., Perel, J., Dahl, R. E., Moreci, P., Nelson, B., et al. (1997). The corticotropin-releasing hormone challenge in depressed abused, depressed nonabused, and normal control children. *Biological Psychiatry*, 42(8), 669–679.

Kaufman, J., Yang, B. Z., Douglas-Palumberi, H., Grasso, D., Lipschitz, D., Houshyar, S., et al. (2006). Brain-derived neurotrophic factor-5-HTTLPR gene interactions and environmental modifiers of depression in children. *Biological Psychiatry*, 59(8), 673–680. https://doi.org/10.1016/j.biopsych.2005.10.026.

Kaufman, J., Yang, B. Z., Douglas-Palumberi, H., Houshyar, S., Lipschitz, D., Krystal, J. H., et al. (2004). Social supports and serotonin transporter gene moderate depression in maltreated children. *Proceedings of the National Academy of Sciences of the United States of America*, 101 (49), 17316–17321. https://doi.org/10.1073/pnas.0404376101.

Kessler, R. C., McLaughlin, K. A., Green, J. G., Gruber, M. J., Sampson, N. A., Zaslavsky, A. M., et al. (2010). Childhood adversities and adult psychopathology in the WHO World Mental Health Surveys. *The British Journal of Psychiatry*, 197(5), 378–385. pii: 197/5/378(2010). https://doi.org/10.1192/bjp.bp.110.080499.

Kinnally, E. L., Capitanio, J. P., Leibel, R., Deng, L., LeDuc, C., Haghighi, F., et al. (2010). Epigenetic regulation of serotonin transporter expression and behavior in infant rhesus macaques. *Genes, Brain, and Behavior*, 9(6), 575–582. https://doi.org/10.1111/j.1601-183X.2010.00588.x.

Kinnally, E. L., Feinberg, C., Kim, D., Ferguson, K., Leibel, R., Coplan, J. D., et al. (2011). DNA methylation as a risk factor in the effects of early life stress. *Brain, Behavior, and Immunity*, 25(8), 1548–1553.

Kinnally, E. L., Karere, G. M., Lyons, L. A., Mendoza, S. P., Mason, W. A., & Capitanio, J. P. (2010). Serotonin pathway gene-gene and gene-environment interactions influence behavioral stress response in infant rhesus macaques. *Development and Psychopathology*, 22(1), 35–44. https://doi.org/10.1017/S0954579409990241.

Kinnally, E. L., Lyons, L. A., Abel, K., Mendoza, S., & Capitanio, J. P. (2008). Effects of early experience and genotype on serotonin transporter regulation in infant rhesus macaques. *Genes, Brain, and Behavior*, 7(4), 481–486. https://doi.org/10.1111/j.1601-183X.2007.00383.x.

Kumamaru, E., Numakawa, T., Adachi, N., Yagasaki, Y., Izumi, A., Niyaz, M., et al. (2008). Glucocorticoid prevents brain-derived neurotrophic factor-mediated maturation of synaptic function in developing hippocampal neurons through reduction in the activity of mitogen-activated protein kinase. *Molecular Endocrinology*, 22(3), 546–558. https://doi.org/10.1210/me.2007-0264.

Ladd, C. O., Owens, M. J., & Nemeroff, C. B. (1996). Persistent changes in corticotropin-releasing factor neuronal systems induced by maternal deprivation. *Endocrinology*, 137(4), 1212–1218. https://doi.org/10.1210/endo.137.4.8625891.

Lee, B. H., & Kim, Y. K. (2009). Reduced platelet BDNF level in patients with major depression. *Progress in Neuro-Psychopharmacology & Biological Psychiatry*, 33(5), 849–853. https://doi.org/10.1016/j.pnpbp.2009.04.002.

Lee, J. H., Kim, H. J., Kim, J. G., Ryu, V., Kim, B. T., Kang, D. W., et al. (2007). Depressive behaviors and decreased expression of serotonin reuptake transporter in rats that experienced neonatal maternal separation. *Neuroscience Research*, 58(1), 32–39. https://doi.org/10.1016/j.neures.2007.01.008.

Lemieux, A., Coe, C. L., & Carnes, M. (2008). Symptom severity predicts degree of T cell activation in adult women following childhood maltreatment. *Brain, Behavior, and Immunity*, 22(6), 994–1003. https://doi.org/10.1016/j.bbi.2008.02.005.

Lipman, E. L., MacMillan, H. L., & Boyle, M. H. (2001). Childhood abuse and psychiatric disorders among single and married mothers. *The American Journal of Psychiatry*, 158 (1), 73–77.

Lippmann, M., Bress, A., Nemeroff, C. B., Plotsky, P. M., & Monteggia, L. M. (2007). Long-term behavioural and molecular alterations associated with maternal separation in rats. *The European Journal of Neuroscience*, 25(10), 3091–3098. https://doi.org/10.1111/j.1460-9568.2007.05522.x.

Loman, M. M., Wiik, K. L., Frenn, K. A., Pollak, S. D., & Gunnar, M. R. (2009). Postinstitutionalized children's development: growth, cognitive, and language outcomes. *Journal of Developmental and Behavioral Pediatrics*, 30(5), 426–434. https://doi.org/10.1097/DBP.0b013e3181b1fd08.

Lopez, A. D., & Murray, C. C. (1998). The global burden of disease, 1990-2020. *Nature Medicine*, 4(11), 1241–1243. https://doi.org/10.1038/3218.

Lu, S., Peng, H., Wang, L., Vasish, S., Zhang, Y., Gao, W., et al. (2013). Elevated specific peripheral cytokines found in major depressive disorder patients with childhood trauma exposure: a cytokine antibody array analysis. *Comprehensive Psychiatry*, 54(7), 953–961. https://doi.org/10.1016/j.comppsych.2013.03.026.

MacMillan, H. L., Georgiades, K., Duku, E. K., Shea, A., Steiner, M., Niec, A., et al. (2009). Cortisol response to stress in female youths exposed to childhood maltreatment: results of the youth mood project. *Biological Psychiatry*, 66(1), 62–68. https://doi.org/10.1016/j.biopsych.2008.12.014.

Maes, M., Bosmans, E., Suy, E., Vandervorst, C., De Jonckheere, C., & Raus, J. (1990). Immune disturbances during major depression: upregulated expression of interleukin-2 receptors. *Neuropsychobiology*, 24(3), 115–120.

Mandelli, L., Petrelli, C., & Serretti, A. (2015). The role of specific early trauma in adult depression: a meta-analysis of published literature. Childhood trauma and adult depression. *European Psychiatry*, 30(6), 665–680. https://doi.org/10.1016/j.eurpsy.2015.04.007.

Marais, L., van Rensburg, S. J., van Zyl, J. M., Stein, D. J., & Daniels, W. M. (2008). Maternal separation of rat pups increases the risk of developing depressive-like behavior after subsequent chronic stress by altering corticosterone and neurotrophin levels in the hippocampus. *Neuroscience Research*, 61(1), 106–112. https://doi.org/10.1016/j.neures.2008.01.011.

Martin-Blanco, A., Ferrer, M., Soler, J., Salazar, J., Vega, D., Andion, O., et al. (2014). Association between methylation of the glucocorticoid receptor gene, childhood maltreatment, and clinical severity in borderline personality disorder. *Journal of Psychiatric Research*, 57, 34–40. https://doi.org/10.1016/j.jpsychires.2014.06.011.

Matthews, K., Dalley, J. W., Matthews, C., Tsai, T. H., & Robbins, T. W. (2001). Periodic maternal separation of neonatal rats produces region- and gender-specific effects on biogenic amine content in postmortem adult brain. *Synapse*, 40(1), 1–10. https://doi.org/10.1002/1098-2396(200104)40:1<1::AID-SYN1020>3.0.CO;2-E.

McAfoose, J., & Baune, B. T. (2009). Evidence for a cytokine model of cognitive function. *Neuroscience and Biobehavioral Reviews*, 33(3), 355–366. https://doi.org/10.1016/j.neubiorev.2008.10.005.

McCauley, J., Kern, D. E., Kolodner, K., Dill, L., Schroeder, A. F., DeChant, H. K., et al. (1997). Clinical characteristics of women with a history of childhood abuse: unhealed wounds. *JAMA*, 277(17), 1362–1368.

McGowan, P., Sasaki, A., D'Alessio, A. C., Dymov, S., Labonte, B., Szyf, M., et al. (2009). Epigenetic regulation of the glucocorticoid receptor in human brain associates with childhood abuse. *Nature Neuroscience*, 12(3), 342–348. https://doi.org/10.1038/nn.2270.

Meaney, M. J., Diorio, J., Francis, D., Widdowson, J., LaPlante, P., Caldji, C., et al. (1996). Early environmental regulation of forebrain glucocorticoid receptor gene expression: implications for adrenocortical responses to stress. *Developmental Neuroscience*, 18(1–2), 49–72.

Mehta, M. A., Golembo, N. I., Nosarti, C., Colvert, E., Mota, A., Williams, S. C., et al. (2009). Amygdala, hippocampal and corpus callosum size following severe early institutional deprivation: the English and Romanian Adoptees study pilot. *Journal of Child Psychology and Psychiatry, 50* (8), 943–951. pii: JCPP2084(2009). https://doi.org/10.1111/j.1469-7610.2009.02084.x.

Mello, A. F., Mello, M. F., Carpenter, L. L., & Price, L. H. (2003). Update on stress and depression: the role of the hypothalamic-pituitary-adrenal (HPA) axis. *Revista Brasileira de Psiquiatria, 25*(4), 231–238. https://doi.org/10.1590/S1516-44462003000400010.

Meyer, J. H., Ginovart, N., Boovariwala, A., Sagrati, S., Hussey, D., Garcia, A., et al. (2006). Elevated monoamine oxidase a levels in the brain: an explanation for the monoamine imbalance of major depression. *Archives of General Psychiatry, 63*(11), 1209–1216. https://doi.org/10.1001/archpsyc.63.11.1209.

Meynen, G., Unmehopa, U. A., van Heerikhuize, J. J., Hofman, M. A., Swaab, D. F., & Hoogendijk, W. J. (2006). Increased arginine vasopressin mRNA expression in the human hypothalamus in depression: a preliminary report. *Biological Psychiatry, 60*(8), 892–895. https://doi.org/10.1016/j.biopsych.2005.12.010.

Miller, G. E., Chen, E., & Parker, K. J. (2011). Psychological stress in childhood and susceptibility to the chronic diseases of aging: moving toward a model of behavioral and biological mechanisms. *Psychological Bulletin, 137* (6), 959–997. https://doi.org/10.1037/a0024768.

Molnar, B. E., Buka, S. L., & Kessler, R. C. (2001). Child sexual abuse and subsequent psychopathology: results from the National Comorbidity Survey. *American Journal of Public Health, 91*(5), 753–760.

Mullen, P. E., Romans-Clarkson, S. E., Walton, V. A., & Herbison, G. P. (1988). Impact of sexual and physical abuse on women's mental health. *Lancet, 1*(8590), 841–845.

Munafo, M. R., Durrant, C., Lewis, G., & Flint, J. (2009). Gene X environment interactions at the serotonin transporter locus. *Biological Psychiatry, 65*(3), 211–219. https://doi.org/10.1016/j.biopsych.2008.06.009.

Murgatroyd, C., Patchev, A., Wu, Y., Micale, V., Bockmuhl, Y., Fischer, D., et al. (2009). Dynamic DNA methylation programs persistent adverse effects of early-life stress. *Nature Neuroscience, 12*(12), 1559–1566. https://doi.org/10.1038/nn.2436 Epub 2009 Nov 1558.

Nanni, V., Uher, R., & Danese, A. (2012). Childhood maltreatment predicts unfavorable course of illness and treatment outcome in depression: a meta-analysis. *The American Journal of Psychiatry, 169*(2), 141–151. https://doi.org/10.1176/appi.ajp.2011.11020335.

Navalta, C. P., Polcari, A., Webster, D. M., Boghossian, A., & Teicher, M. H. (2006). Effects of childhood sexual abuse on neuropsychological and cognitive function in college women. *The Journal of Neuropsychiatry and Clinical Neurosciences, 18*(1), 45–53. pii: 18/1/45(2006). https://doi.org/10.1176/appi.neuropsych.18.1.45.

Nederhof, E., Bouma, E. M., Oldehinkel, A. J., & Ormel, J. (2010). Interaction between childhood adversity, brain-derived neurotrophic factor val/met and serotonin transporter promoter polymorphism on depression: the TRAILS study. *Biological Psychiatry, 68*(2), 209–212. https://doi.org/10.1016/j.biopsych.2010.04.006.

Nelson, J., Klumparendt, A., Doebler, P., & Ehring, T. (2017). Childhood maltreatment and characteristics of adult depression: meta-analysis. *The British Journal of Psychiatry, 210*(2), 96–104. https://doi.org/10.1192/bjp.bp.115.180752.

Neumeister, A., Konstantinidis, A., Stastny, J., Schwarz, M. J., Vitouch, O., Willeit, M., et al. (2002). Association between serotonin transporter gene promoter polymorphism (5HTTLPR) and behavioral responses to tryptophan depletion in healthy women with and without family history of depression. *Archives of General Psychiatry, 59*(7), 613–620.

Neumeister, A., Nugent, A. C., Waldeck, T., Geraci, M., Schwarz, M., Bonne, O., et al. (2004). Neural and behavioral responses to tryptophan depletion in unmedicated patients with remitted major depressive disorder and controls. *Archives of General Psychiatry, 61*(8), 765–773. https://doi.org/10.1001/archpsyc.61.8.765.

Ouellet-Morin, I., Odgers, C. L., Danese, A., Bowes, L., Shakoor, S., Papadopoulos, A. S., et al. (2011). Blunted cortisol responses to stress signal social and behavioral problems among maltreated/bullied 12-year-old children. *Biological Psychiatry, 70*(11), 1016–1023. https://doi.org/10.1016/j.biopsych.2011.06.017.

Pace, T. W., Hu, F., & Miller, A. H. (2007). Cytokine-effects on glucocorticoid receptor function: relevance to glucocorticoid resistance and the pathophysiology and treatment of major depression. *Brain, Behavior, and Immunity, 21*(1), 9–19. https://doi.org/10.1016/j.bbi.2006.08.009.

Pace, T. W., Mletzko, T. C., Alagbe, O., Musselman, D. L., Nemeroff, C. B., Miller, A. H., et al. (2006). Increased stress-induced inflammatory responses in male patients with major depression and increased early life stress. *The American Journal of Psychiatry, 163*(9), 1630–1633. https://doi.org/10.1176/ajp.2006.163.9.1630.

Perroud, N., Paoloni-Giacobino, A., Prada, P., Olie, E., Salzmann, A., Nicastro, R., et al. (2011). Increased methylation of glucocorticoid receptor gene (NR3C1) in adults with a history of childhood maltreatment: a link with the severity and type of trauma. *Translational Psychiatry, 1*, e59. https://doi.org/10.1038/tp.2011.60.

Perroud, N., Salzmann, A., Prada, P., Nicastro, R., Hoeppli, M. E., Furrer, S., et al. (2013). Response to psychotherapy in borderline personality disorder and methylation status of the BDNF gene. *Translational Psychiatry, 3*, e207. https://doi.org/10.1038/tp.2012.140.

Plotsky, P., & Meaney, M. (1993). Early, postnatal experience alters hypothalamic corticotropin-releasing factor (CRF) mRNA, median eminence CRF content and stress-induced release in adult rats. *Brain Research Molecular Brain Research, 18*(3), 195–200.

Raadsheer, F. C., Hoogendijk, W. J., Stam, F. C., Tilders, F. J., & Swaab, D. F. (1994). Increased numbers of corticotropin-releasing hormone expressing neurons in the hypothalamic paraventricular nucleus of depressed patients. *Neuroendocrinology, 60*(4), 436–444.

Reus, G. Z., Fernandes, G. C., de Moura, A. B., Silva, R. H., Darabas, A. C., de Souza, T. G., et al. (2017). Early life experience contributes to the developmental programming of depressive-like behaviour, neuroinflammation and oxidative stress. *Journal of Psychiatric Research, 95,* 196–207. https://doi.org/10.1016/j.jpsychires.2017.08.020.

Ridout, K. K., Parade, S. H., Seifer, R., Price, L. H., Gelernter, J., Feliz, P., et al. (2014). Interleukin 1B gene (IL1B) variation and internalizing symptoms in maltreated preschoolers. *Development and Psychopathology, 26*(4 Pt 2), 1277–1287. https://doi.org/10.1017/S0954579414001023.

Risch, N., Herrell, R., Lehner, T., Liang, K. Y., Eaves, L., Hoh, J., et al. (2009). Interaction between the serotonin transporter gene (5-HTTLPR), stressful life events, and risk of depression: a meta-analysis. *JAMA, 301*(23), 2462–2471. https://doi.org/10.1001/jama.2009.878.

Roceri, M., Cirulli, F., Pressina, C., Peretto, P., Racagni, G., & Riva, M. A. (2004). Postnatal repeated maternal deprivation produces age-dependent changes of brain-derived neurotrophic factor expression in selected rat brain regions. *Biological Psychiatry, 55*(7), 708–714.

Roceri, M., Hendriks, W., Racagni, G., Ellenbroek, B. A., & Riva, M. A. (2002). Early maternal deprivation reduces the expression of BDNF and NMDA receptor subunits in rat hippocampus. *Molecular Psychiatry, 7*(6), 609–616. https://doi.org/10.1038/sj.mp.4001036.

Romens, S. E., McDonald, J., Svaren, J., & Pollak, S. D. (2015). Associations between early life stress and gene methylation in children. *Child Development, 86*(1), 303–309. https://doi.org/10.1111/cdev.12270.

Roth, T. L., Lubin, F. D., Funk, A., & Sweatt, J. D. (2009). Lasting epigenetic influence of early-life adversity on the BDNF gene. *Biological Psychiatry, 65*(9), 760–769. https://doi.org/10.1016/j.biopsych.2008.1011.1028 Epub 2009 Jan 1015.

Ruedi-Bettschen, D., Zhang, W., Russig, H., Ferger, B., Weston, A., Pedersen, E. M., et al. (2006). Early deprivation leads to altered behavioural, autonomic and endocrine responses to environmental challenge in adult Fischer rats. *The European Journal of Neuroscience, 24*(10), 2879–2893. https://doi.org/10.1111/j.1460-9568.2006.05158.x.

Ryan, J., Pilkington, L., Neuhaus, K., Ritchie, K., Ancelin, M. L., & Saffery, R. (2017). Investigating the epigenetic profile of the inflammatory gene IL-6 in late-life depression. *BMC Psychiatry, 17*(1), 354. https://doi.org/10.1186/s12888-017-1515-8.

Scott, K. M., Von Korff, M., Alonso, J., Angermeyer, M. C., Benjet, C., Bruffaerts, R., et al. (2008). Childhood adversity, early-onset depressive/anxiety disorders, and adult-onset asthma. *Psychosomatic Medicine, 70*(9), 1035–1043. https://doi.org/10.1097/PSY.0b013e318187a2fb.

Scott, K. M., Von Korff, M., Angermeyer, M. C., Benjet, C., Bruffaerts, R., de Girolamo, G., et al. (2011). Association of childhood adversities and early-onset mental disorders with adult-onset chronic physical conditions. *Archives of General Psychiatry, 68*(8), 838–844. https://doi.org/10.1001/archgenpsychiatry.2011.77.

Sharpley, C. F., Palanisamy, S. K., Glyde, N. S., Dillingham, P. W., & Agnew, L. L. (2014). An update on the interaction between the serotonin transporter promoter variant (5-HTTLPR), stress and depression, plus an exploration of non-confirming findings. *Behavioural Brain Research, 273,* 89–105. https://doi.org/10.1016/j.bbr.2014.07.030.

Shimizu, E., Hashimoto, K., Okamura, N., Koike, K., Komatsu, N., Kumakiri, C., et al. (2003). Alterations of serum levels of brain-derived neurotrophic factor (BDNF) in depressed patients with or without antidepressants. *Biological Psychiatry, 54*(1), 70–75.

Slopen, N., Kubzansky, L. D., McLaughlin, K. A., & Koenen, K. C. (2013). Childhood adversity and inflammatory processes in youth: a prospective study. *Psychoneuroendocrinology, 38*(2), 188–200. https://doi.org/10.1016/j.psyneuen.2012.05.013.

Smith, R. S. (1991). The macrophage theory of depression. *Medical Hypotheses, 35*(4), 298–306.

Steptoe, A., Hamer, M., & Chida, Y. (2007). The effects of acute psychological stress on circulating inflammatory factors in humans: a review and meta-analysis. *Brain, Behavior, and Immunity, 21*(7), 901–912. https://doi.org/10.1016/j.bbi.2007.03.011.

Sung, Y. H., Shin, M. S., Cho, S., Baik, H. H., Jin, B. K., Chang, H. K., et al. (2010). Depression-like state in maternal rats induced by repeated separation of pups is accompanied by a decrease of cell proliferation and an increase of apoptosis in the hippocampus. *Neuroscience Letters, 470*(1), 86–90. https://doi.org/10.1016/j.neulet.2009.12.063.

Suri, D., & Vaidya, V. A. (2013). Glucocorticoid regulation of brain-derived neurotrophic factor: relevance to hippocampal structural and functional plasticity. *Neuroscience, 239,* 196–213. https://doi.org/10.1016/j.neuroscience.2012.08.065.

Suri, D., Veenit, V., Sarkar, A., Thiagarajan, D., Kumar, A., Nestler, E. J., et al. (2013). Early stress evokes age-dependent biphasic changes in hippocampal neurogenesis, BDNF expression, and cognition. *Biological Psychiatry, 73*(7), 658–666. https://doi.org/10.1016/j.biopsych.2012.10.023.

Swaab, D. F., Bao, A. M., & Lucassen, P. J. (2005). The stress system in the human brain in depression and neurode-generation. *Ageing Research Reviews*, 4(2), 141–194. https://doi.org/10.1016/j.arr.2005.03.003.

Tafet, G. E., Toister-Achituv, M., & Shinitzky, M. (2001). Enhancement of serotonin uptake by cortisol: a possible link between stress and depression. *Cognitive, Affective, & Behavioral Neuroscience*, 1(1), 96–104.

Tarry-Adkins, J. L., & Ozanne, S. E. (2011). Mechanisms of early life programming: current knowledge and future directions. *The American Journal of Clinical Nutrition*, 94(6 Suppl), 1765S–1771S. https://doi.org/10.3945/ajcn.110.000620.

Tomoda, A., Suzuki, H., Rabi, K., Sheu, Y. S., Polcari, A., & Teicher, M. H. (2009). Reduced prefrontal cortical gray matter volume in young adults exposed to harsh corporal punishment. *NeuroImage*, 47(Suppl. 2), T66–71. pii: S1053-8119(09)00228-6(2009). https://doi.org/10.1016/j.neuroimage.2009.03.005.

Tottenham, N., Hare, T. A., Quinn, B. T., McCarry, T. W., Nurse, M., Gilhooly, T., et al. (2010). Prolonged institu-tional rearing is associated with atypically large amyg-dala volume and difficulties in emotion regulation. *Developmental Science*, 13(1), 46–61. pii: DESC852(2010). https://doi.org/10.1111/j.1467-7687.2009.00852.x.

Turnbull, A. V., & Rivier, C. (1995). Regulation of the HPA axis by cytokines. *Brain, Behavior, and Immunity*, 9(4), 253–275. https://doi.org/10.1006/brbi.1995.1026.

Tyrka, A. R., Price, L. H., Marsit, C., Walters, O. C., & Carpenter, L. L. (2012). Childhood adversity and epige-netic modulation of the leukocyte glucocorticoid recep-tor: preliminary findings in healthy adults. *PLoS One*, 7 (1), e30148. https://doi.org/10.1371/journal.pone.0030148.

van Harmelen, A. L., van Tol, M. J., van der Wee, N. J., Veltman, D. J., Aleman, A., Spinhoven, P., et al. (2010). Reduced medial prefrontal cortex volume in adults reporting childhood emotional maltreatment. *Biological Psychiatry*, 68(9), 832–838. https://doi.org/10.1016/j.biopsych.2010.06.011.

van Oers, H. J., de Kloet, E. R., & Levine, S. (1998). Early vs. late maternal deprivation differentially alters the endo-crine and hypothalamic responses to stress. *Brain Research. Developmental Brain Research*, 111(2), 245–252.

Van Voorhees, E., & Scarpa, A. (2004). The effects of child maltreatment on the hypothalamic-pituitary-adrenal axis. *Trauma Violence Abuse*, 5(4), 333–352. https://doi.org/10.1177/1524838004269486.

Varese, F., Smeets, F., Drukker, M., Lieverse, R., Lataster, T., Viechtbauer, W., et al. (2012). Childhood adversities increase the risk of psychosis: a meta-analysis of patient-control, prospective- and cross-sectional cohort studies. *Schizophrenia Bulletin*, 38(4), 661–671. https://doi.org/10.1093/schbul/sbs050.

Veenema, A. H., Blume, A., Niederle, D., Buwalda, B., & Neumann, I. D. (2006). Effects of early life stress on adult male aggression and hypothalamic vasopressin and sero-tonin. *The European Journal of Neuroscience*, 24(6), 1711–1720. https://doi.org/10.1111/j.1460-9568.2006.05045.x.

Veenema, A. H., Reber, S. O., Selch, S., Obermeier, F., & Neumann, I. D. (2008). Early life stress enhances the vul-nerability to chronic psychosocial stress and experimen-tal colitis in adult mice. *Endocrinology*, 149(6), 2727–2736. https://doi.org/10.1210/en.2007-1469.

Vermetten, E., Schmahl, C., Lindner, S., Loewenstein, R. J., & Bremner, J. D. (2006). Hippocampal and amygdalar vol-umes in dissociative identity disorder. *The American Jour-nal of Psychiatry*, 163(4), 630–636. pii: 163/4/630(2006). https://doi.org/10.1176/appi.ajp.163.4.630.

Vythilingam, M., Heim, C., Newport, J., Miller, A. H., Anderson, E., Bronen, R., et al. (2002). Childhood trauma associated with smaller hippocampal volume in women with major depression. *The American Journal of Psychiatry*, 159(12), 2072–2080.

Weaver, I. C., Cervoni, N., Champagne, F. A., D'Alessio, A. C., Sharma, S., Seckl, J. R., et al. (2004). Epigenetic pro-gramming by maternal behavior. *Nature Neuroscience*, 7 (8), 847–854. https://doi.org/10.1038/nn1276.

Weaver, I. C. G., D'Alessio, A. C., Brown, S. E., Hellstrom, I. C., Dymov, S., Sharma, S., et al. (2007). The transcription factor nerve growth factor-inducible protein a mediates epigenetic programming: altering epigenetic marks by immediate-early genes. *Journal of Neuroscience*, 27(7), 1756–1768. https://doi.org/10.1523/jneurosci.4164-06.2007.

Weber, B., Lewicka, S., Deuschle, M., Colla, M., Vecsei, P., & Heuser, I. (2000). Increased diurnal plasma concentra-tions of cortisone in depressed patients. *The Journal of Clinical Endocrinology and Metabolism*, 85(3), 1133–1136. https://doi.org/10.1210/jcem.85.3.6469.

Wichers, M., Kenis, G., Jacobs, N., Mengelers, R., Derom, C., Vlietinck, R., et al. (2008). The BDNF Val(66)Met x 5-HTTLPR x child adversity interaction and depressive symptoms: an attempt at replication. *American Journal of Medical Genetics. Part B, Neuropsychiatric Genetics*, 147B(1), 120–123. https://doi.org/10.1002/ajmg.b.30576.

Wieck, A., Andersen, S. L., & Brenhouse, H. C. (2013). Evi-dence for a neuroinflammatory mechanism in delayed effects of early life adversity in rats: relationship to corti-cal NMDA receptor expression. *Brain, Behavior, and Immunity*, 28, 218–226. https://doi.org/10.1016/j.bbi.2012.11.012.

Woon, F. L., & Hedges, D. W. (2008). Hippocampal and amygdala volumes in children and adults with childhood maltreatment-related posttraumatic stress disorder: a meta-analysis. *Hippocampus*, *18*(8), 729–736. https://doi.org/10.1002/hipo.20437.

Woon, F. L., Sood, S., & Hedges, D. W. (2010). Hippocampal volume deficits associated with exposure to psychological trauma and posttraumatic stress disorder in adults: a meta-analysis. *Progress in Neuro-Psychopharmacology & Biological Psychiatry*, *34*(7), 1181–1188. pii: S0278-5846 (10)00233-2(2010). https://doi.org/10.1016/j.pnpbp.2010.06.016.

Zeugmann, S., Buehrsch, N., Bajbouj, M., Heuser, I., Anghelescu, I., & Quante, A. (2013). Childhood maltreatment and adult proinflammatory status in patients with major depression. *Psychiatria Danubina*, *25*(3), 227–235.

Zhang, L. X., Levine, S., Dent, G., Zhan, Y., Xing, G., Okimoto, D., et al. (2002). Maternal deprivation increases cell death in the infant rat brain. *Brain Research. Developmental Brain Research*, *133*(1), 1–11.

Epigenetic Changes in the Immune Systems Following Early-Life Stress

Chris Murgatroyd

Manchester Metropolitan University, Manchester, United Kingdom

INTRODUCTION

The close relationship between the quality of early life and mental health in later life is well described. Many human and rodent studies illustrate that aspects of the early environment can lead to dramatic changes in physical and mental development compromising altered cognition, mood, and behavior. In particular, stress and adverse conditions during early-life periods of development can shape individual differences in vulnerability to stress-related disorders throughout life (*for review, see* Murgatroyd & Spengler, 2011b). This raises the question of how environmental experiences become incorporated at the cellular and molecular level leading to long-term alterations in various neuronal functions, including stress responses and risk to mental disease.

In this review, we highlight important animal and human studies addressing epigenetic regulation of gene expression in sustaining the effects of early-life experiences. We focus on studies that have investigated how adversity is able to shape biological stress systems, particularly the hypothalamic-pituitary-adrenal (HPA) axis and immune systems, leading to lasting alterations in stress responsivity. We discuss what we know about the function of epigenetic systems and their roles in programming immune and brain development and disease in response to prenatal and postnatal environmental stressors.

EARLY LIFE ADVERSITY SHAPES LATER LIFE STRESS RESPONSIVITY

Upon exposure to a stressor, the autonomic nervous system initiates a rapid and relatively short-lived "fight-or-flight" response, while the HPA axis is slower, instigating a more protracted response. Tight regulation of the HPA axis is core to the long-term control of systems governing stress responsivity. Following a stressor, the neuropeptides corticotropin-releasing hormone (CRH) and arginine vasopressin (AVP) are released from the paraventricular nucleus (PVN) of the hypothalamus that stimulates the release of adrenocorticotropic hormone (ACTH) from the anterior pituitary that in turn acts on the adrenal cortex to release glucocorticoid hormones, that is, cortisol (humans) and corticosterone (animals).

A negative feedback loop, through glucocorticoid receptors (GR), ensures efficient return to a homeostatic balance when it is no longer challenged. It has been shown that this negative feedback can become dysregulated, particularly following periods of chronic stress. This may critically impact the development of affective disorders; altered HPA activity is one of the most commonly observed neuroendocrine symptoms in major depressive disorder (*for review, see* Holsboer, 2000). Childhood stress is a strong predictor of impaired inhibitory feedback regulation of the HPA axis, with dysregulation occurring during the period of childhood in which the adversity occurs and lasting into adulthood, long after the early adverse experience (*for review, see* Gonzalez, 2013). Studies in rodent models further support the concept that exposure to a chronic stressor can lead to long-term changes in HPA regulation and behavior (*for review, see* Murgatroyd & Spengler, 2011a).

Such chronic dysregulation of the HPA axis has serious consequences for later health and development being implicated in the pathogenesis of and increasing risk for numerous chronic physical and psychiatric diseases.

BIDIRECTIONAL REGULATION OF THE HPA AXIS AND IMMUNE SYSTEM

The inflammatory response is critical for infections and injuries but must be carefully regulated so that it does not lead to tissue damage and disease. An imbalance in this system, such as from sustained activity, can lead to chronic low-grade inflammation; this has been related to many diseases of aging, such as diabetes, obesity-related problems, cardiovascular disease, autoimmune disease, and cancer, as well as psychiatric diseases (Furtado & Katzman, 2015).

A bidirectional communication exists between the HPA axis and the immune system,

and HPA dysregulation further results in a generalized increase in inflammation (*for review, see* Bellavance & Rivest, 2014). This may link to some of the above long-term health conditions, especially when HPA axis alterations are chronic and long lasting. The HPA axis can be activated, leading to the release of glucocorticoids, by immune cells through the production of cytokines such as tumor necrosis factor-alpha (TNF-α), interleukins (e.g., IL-1 and IL-6), and interferons (e.g., IFN-γ) (*for review, see* Dumbell, Matveeva, & Oster, 2016). The traditional view of glucocorticoids is as immunosuppressive hormones that ultimately function to repress the inflammatory response, playing an important role in reducing acute inflammatory responses. However, at the same time, glucocorticoids serve an important complementary role in regulating pro-inflammatory functions of immune cells to coordinate inflammatory responses (Busillo & Cidlowski, 2013). As such, glucocorticoids can be regarded as immune modulators, and stress acutely enhances and chronically suppresses the peripheral immune response (Dhabhar, 2002).

DEPRESSION AND INFLAMMATION

There is strong association between depression and activation of the innate immune system. This is supported by numerous studies showing alterations in the functional activity of the immune system in the blood and in the brain of patients with depression (Herkenham & Kigar, 2016). A significant number of depressed individuals exhibit activated inflammatory responses as measured by increased circulating levels of inflammatory cytokines such as interleukins IL-1β and IL-6, IFN-γ, and TNF-α (Dowlati et al., 2010), for example, increased TNF-α plasma levels and increased IL-6 in plasma in subjects with depressive disorders compared with healthy controls (Pike & Irwin, 2006). Several studies support that these serum changes relate to regulatory changes of the cytokine genes within the peripheral blood cells. For example, higher mRNA levels of TNF-α, IL-1β, IL-6, and INF-α have been found in the leukocytes of patients suffering from major depression (Cattaneo et al., 2013; Tsao, Lin, Chen, Bai, & Wu, 2006). A genome-wide longitudinal study on 1846 individuals found that the strongest expression differences between the depressed and control groups were in genes enriched for IL-6 signaling and natural killer (NK) cell pathways (Jansen et al., 2016). Furthermore, the above cytokines have been significantly correlated with several clinical depressive "traits." In particular, higher cytokine levels in serum have been associated with higher depression severity (Bunck et al., 2009) and with poor antidepressant response (Cattaneo et al., 2013). Immune abnormalities and increased inflammatory activities are also found in postmortem brains of depressed and suicide patients. For example, a gene expression array on prefrontal cortex revealed increases in pro-inflammatory cytokines (IL-1α, IL-2, IL-3, IL-5, IL-8, IL-9, IL-10, IL-12A, IL-13, IL-15, IL-18, IFN-γ, and TNF-α) in depressed suicide patients compared with controls (Shelton et al., 2011). Finally, further solidifying the relationship between cytokines and immune activation in depression, several functional allelic variants and single-nucleotide polymorphisms (SNPs) of genes encoding immune and inflammatory molecules have been identified in association with depression (*for review, see* Bufalino, Hepgul, Aguglia, & Pariante, 2013).

EARLY ADVERSITY AND INFLAMMATION

Mounting evidence supports the idea that early-life adversity is linked to long-lasting alterations in inflammatory processes. Genome-wide transcriptional profiling in healthy adults who were either low or high in socioeconomic status in early life revealed that in subjects with low early-life socioeconomic status, there was significant upregulation of genes bearing response elements for the CREB/ATF family of transcription factors, increased expression of genes bearing response elements for NF-κB, downregulation of genes with response elements for the glucocorticoid receptor, increased output of cortisol, and greater stimulated production of IL-6. This suggests that low early-life socioeconomic status programs a defensive phenotype characterized by resistance to glucocorticoid signaling, which in turn facilitates exaggerated adrenocortical and inflammatory responses (Miller et al., 2009). A further study demonstrated that an increased IL-6 production in adolescents with histories of childhood adversity preceded the subsequent development of depression 6 months later (Miller & Cole, 2012). Together, this suggests that childhood adversity programs immune cells to have a pro-inflammatory phenotype, which manifests in relatively aggressive inflammatory responses to stimuli and lower insensitivity to signals that dampen this response. Fitting with the biological embedding model, though these response patterns could serve adaptive functions, over the long term is thought to contribute to the low-grade chronic inflammation implicated in the development and progression of age-related illnesses. This early-life stress would also stimulate inflammatory signaling between the brain and periphery, coupling depression and inflammation, increasing the risk for psychiatric disorders (Fig. 1).

This leads to the question of how early-life environment could program changes in the immune system.

EPIGENETIC MECHANISMS

"Epigenetics" describes the study of stable alterations in gene expression potential that arise during development and differentiation

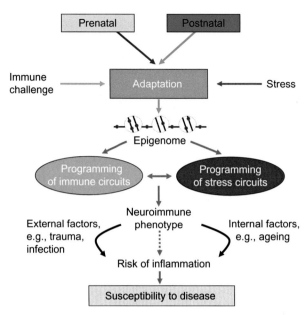

FIG. 1   Epigenetic programming by early-life stress. Early-life experience can program behavior, stress responsivity, and immune systems throughout the later life persistently through alteration of the expression levels of key genes involved in stress regulation and inflammation through epigenetic marking. The nature of the environment throughout the later life may exacerbate the effects of programming established during early life resulting in increased vulnerability to inflammation and psychiatric disease.

and is under the influence of the environment (*for review*, *see* Doherty & Roth, 2016). Genomes can therefore be considered to contain two layers of information: Firstly, there is the DNA sequence that is conserved throughout life and mostly identical in all cells of the body. Secondly, there are the epigenetic marks that are cell-specific and can be dynamic.

Epigenetic mechanisms enable the regulation of a gene, independent of DNA sequence. DNA methylation and modification of core histones that package the DNA into chromatin represent the best-understood epigenetic marks. These marks are able to govern accessibility of the DNA to the machinery driving gene expression; inaccessible genes become silenced, whereas accessible genes are actively transcribed.

These epigenetic processes enable a cell and organism to integrate intrinsic and environmental signals into the genome resulting in

regulatory control of gene expression and thus facilitating adaptation. In this way, epigenetic mechanisms could be thought of as conferring plasticity to the hard-coded genome. In the context of the early-life environment, epigenetic changes offer a plausible mechanism by which early experiences could be integrated into the genome as a kind of memory to program adult hormonal and behavioral responses.

DNA methylation describes the addition of a methyl group to DNA at the cytosine side chain in cytosine-guanine (CpG) dinucleotides, known as a CpG. The conventional view is that DNA methylation is a silencing mark and that, in general, the end point of DNA methylation is either long-term silencing or fine-tuning of gene expression potential. Hence, promoters and enhancer elements of transcriptioally active genes are usually methylated but may become silenced once targeted by DNA methylation. The insertion of methyl groups changes the appearance and structure of DNA, which may either (1) directly block DNA binding of transcription factors or (2) attract factors that preferentially bind to methylated or unmethylated DNA to interfere with transcription factor accessibility.

The vast majority (70%–80%) of all CpGs are methylated. Around 85% of these are located in repetitive sequences such as transposons that constitute around half of the human genome, while the other 15% typically cluster within GC-rich regions known as "CpG islands." CpG islands are defined as regions of greater than 500 bp that have cytosine/guanine content of >55%. Up to 60% of CpG islands are in the 5′ regulatory (promoter) regions of genes. Although CpG islands, promoters, and enhancer regions appear to attract most attention in regard to DNA methylation and gene transcription, it is becoming increasingly clear that methylation at other gene elements plays important regulatory roles. Indeed, genome-wide sequencing of DNA methylation patterns among different tissues is revealing the presence and locations of tissue-specific differentially methylated regions. Strikingly, many of these do not seem to occur in CpG islands themselves but tend to locate in regions next to CpG islands around 2 kb away within sequences of intermediate CpG density. These are now referred to as "CpG shores" (Gomez et al., 2016). Further regions can be divided up as "CpG shelves" as 2 kb regions extending from the shores, while CpGs located yet further away from CpG islands are defined as being in "open sea." In contrast to repression-associated DNA methylation in promoters and enhancers, higher CpG methylation within gene bodies is being frequently observed in transcriptionally active genes perhaps suggesting a further functional role of DNA methylation in transcriptional regulation.

The process of DNA methylation depends on a family of enzymes known as DNA methyltransferases (DNMTs). DNMT1 recognizes hemimethylated DNA and methylates appropriate cytosines in newly synthesized daughter strands formed during replication, acting as the maintenance DNMT. DNMT3A and DNMT3B can methylate unmethylated DNA supportive of their roles as de novo methylases. DNA methylation can also be reversed. For example, during development, there are global and gene-specific increases and decreases in the levels of 5mC. The TET family of enzymes can modify 5mC to 5-hydroxymethylcytosine (5hmC) as part of the processes of active demethylation. This modification appears to be particularly abundant in brain (Wen & Tang, 2014) and immune cells (Ichiyama et al., 2015) and might allow dynamic regulation of DNA methylation (Fig. 2).

In general, patterns of DNA methylation tend to correlate with chromatin structure. For example, regions with hypomethylated DNA associate with active mark chromatin, while hypermethylated DNA is found with inactive chromatin. The major proteins of chromatin, histones, can play both positive and negative roles in gene expression, through various modifications forming the basis of a "histone code" for gene regulation (Jenuwein, 2001). For example, histone acetylation is known to be a predominant signal for active chromatin configurations, while some specific histone methylation reactions are associated with either gene silencing or activation. These are regulated through an array of histone-modifying enzymes that catalyze the addition or removal of an array of covalent modifications.

This leads to the concept that epigenetic changes at sensitive gene promoters for immune regulators could explain the persistence of early-life effects into adulthood of programmed inflammatory and HPA axis dysregulation, increasing the risk of psychiatric disorders.

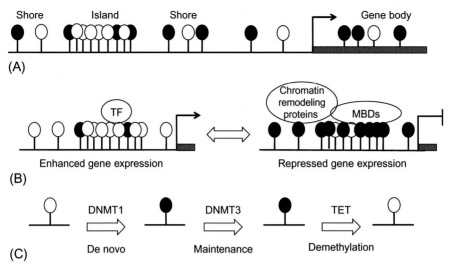

FIG. 2   DNA methylation. (A) GC-rich regions are known as CpG islands, and areas around 2 kb away, within sequences of intermediate CpG density, are referred to as CpG shores (*black circle*, methylated, and *empty circle*, unmethylated). (B) DNA methylation can lead to gene repression through inhibiting transcription factor (TF) binding and recruitment of methylated DNA-binding proteins (MBDs) and chromatin-remodeling factors. (C) DNMT1 is considered the primary maintenance methyltransferase, while DNMT3A and DNMT3B are considered de novo methyltransferases important in establishing methylation.

IMPORTANCE OF EPIGENETIC MECHANISMS IN IMMUNE AND NEURONAL DEVELOPMENT

All cells within the brain have the same DNA sequence yet are differentiated for their diverse functions through epigenetic programming during pre and postnatal development and possibly throughout life. Alterations in the molecular machinery regulating these processes can lead to either silencing or inappropriate expression of specific sets of genes, important for development, that can lead to neurological disease. There are a large number of genes encoding epigenetic regulators that, when mutated, can give rise to neurodevelopmental disease (*for review, see* Murgatroyd & Spengler, 2012) and immune dysregulation. Though one might assume that different epigenetic factors would orchestrate the expression of a large number of potentially unrelated genes, disruptions in distinct epigenetic regulators seemingly lead to

symptomatically similar neurodevelopmental syndromes. Ergo, it is conceivable that many neurodevelopmental diseases do not take root in the changes of specific target gene(s) but by the inability of concerned neurons to respond adequately to environmental signals under conditions of greatly distorted transcriptional homeostasis. One of the most common causes of mental retardation in females is Rett syndrome, a progressive neurodevelopmental disorder resulting from mutations in the methyl-CpG-binding protein *MeCP2* located on the X chromosome. Interestingly, it has further been shown that Rett syndrome is characterized by a deregulated cytokine/chemokine profile together with morphologically altered immune cells (Pecorelli et al., 2016) supporting MECP2's dual role in T-lymphocyte growth (Balmer, Arredondo, Samaco, & LaSalle, 2002). Further components of DNA methylation machinery also link immune and neuronal regulation. For example, mutations in DNMT1 and DNMT3b lead to disorders characterized by neurological syndromes together with immune disturbances (e.g., Fox, Ealing, Murphy, Gow, & Gosal, 2016), while loss of various MBD proteins also associates with various alterations in immune and neurological development (*for review, see* Wood & Zhou, 2016). Together, this supports that important shared epigenetic pathways underlie immune and neurological development.

EPIGENETIC PROGRAMMING BY PRENATAL STRESS

Aside from controlling constitutive gene expression, epigenetic mechanisms can also serve to fine-tune gene expression potential in response to environmental cues. Hence, it has been proposed that conditions of prenatal and postnatal environment can evoke changes in DNA methylation facilitating epigenetic programming of critical genes involved in regulating stress responsivity and immune function that may in turn manifest with neuroendocrine and behavioral symptoms in adulthood (Fig. 1).

Animal and clinical studies support that prenatal stress associates with increased risks in the offspring of behavioral and neurobiological problems including anxiety, depression, and alterations in HPA axis in later life (*for review, see* Babenko, Kovalchuk, & Metz, 2015; Waters, Hay, Simmonds, & van Goozen, 2014) and can be an important programming factor of postnatal immunity (*for review, see* Howerton & Bale, 2012). It has been further demonstrated that these long-term neurological and immune effects of early-life stress are mediated through alterations in the maternal and fetal HPA axes leading to exposure of the fetus to excess maternal glucocorticoids that may pass the placental barrier and disrupt fetal brain and immune development. An important factor in regulating the supply of maternal glucocorticoids to the developing fetus is the placental enzyme, 11β-hydroxysteroid dehydrogenase 2 (11β-HSD2),

that deactivates cortisol (Stewart, Rogerson, & Mason, 1995). It is important to control this placental diffusion as maternal glucocorticoid levels are up to 10-fold higher than in the fetus (Seckl, 2001). Regulation of placental 11β-HSD2 is altered in response to a number of maternal components including maternal undernutrition and protein deficiency (Bertram, Trowern, Copin, Jackson, & Whorwood, 2001). Rodent studies have further shown that it is upregulated following acute stress in pregnancy (Welberg, 2005) and that chronic stress and prenatal maternal anxiety cause a reduction of *11β-HSD2* placental expression that is associated with increased CpG methylation of the *11β-HSD2* gene in the placenta (Jensen Peña, Monk, & Champagne, 2012).

Activity of the human *11β-HSD2* gene has also been shown to be regulated by methylation [*for review, see* Togher et al. (2014). In agreement with findings from the aforementioned animal models, the study by O'Donnell et al. (2012)] revealed that prenatal trait anxiety is negatively correlated with placental *11β-HSD2* mRNA expression. When testing infants, a study by Conradt, Lester, Appleton, Armstrong, and Marsit (2013) revealed that intrauterine exposure to maternal depression negatively affected infant behavior and this is associated with increased methylation of the placental *11β-HSD2* and *NR3C1* (glucocorticoid receptor) genes. Appleton, Lester, Armstrong, Lesseur, and Marsit (2015) also found that infants whose mothers experienced the greatest levels of socioeconomic adversity during pregnancy had the lowest extent of methylation of the placental *11β-HSD2* gene. This suggests that epigenetic regulation of *11β-HSD2* links environmental cues transmitted from the mother to the fetus during pregnancy that could program the response to adverse postnatal environment via less exposure to glucocorticoids during development. Further studies have also identified differential epigenetic regulation of glucocorticoids. For example, a study by Oberlander et al. (2008) revealed that prenatal exposure to maternal depressed/anxious mood during the third trimester of gestational development resulted in increased methylation of a CpG-rich region in the promoter and exon1F of the *NR3C1* gene in the cord blood of newborns, which correlated with salivary cortisol levels in infants at 3 months of age. This supports that prenatal stress can alter epigenetic regulation in gene expression affecting glucocorticoid regulation that could program immune and stress res-ponsivity.

An alternate pathway through which prenatal stress might affect fetal development is via epigenetic programming of cytokine genes. A study examining prenatal response to maternal stress during the 1998 Quebec ice storm showed that the objective degree of maternal exposure to the ice storm resulted in significant alterations in the offspring's cytokine levels (including TNF-α, IL-1β, IL-6, IL-4, and IL-13) together with DNA methylation changes in T lymphocyte at 13½ years of age, in genes predominantly associated with immune function. In particular, CpGs in genes involved in NF-κB signaling pathways were significant mediators of the relationship between objective prenatal maternal stress and IFN-γ levels after correcting for multiple testing (Cao-Lei et al., 2016).

EPIGENETIC PROGRAMMING BY POSTNATAL STRESS

Human studies reveal compelling evidence that adversity during early childhood may cause a range of physical and mental problems in adulthood including depression-related disorders and alterations in HPA axis activity (*for review, see* Murgatroyd & Spengler, 2011c). Although the association between early-life stressful events and depression may occur via several biological processes, a number of studies have also suggested a role for increased inflammation or increased sensitivity of inflammatory responses (*for review, see* Cattaneo et al., 2015).

Numerous rodent studies have shown that postnatal stress (such as maternal stress, separation, or deprivation) can program behaviors such as anxiety and depression together with neuroendocrine dysregulation of the HPA axis and glucocorticoids (Murgatroyd & Bradburn, 2016). Studies show that this can also program premature activation of the immune system that can significantly shift long-term patterns of immune activation. Vice versa, further rodent studies show that early-life immune exposure can program HPA axis activity. For example, neonatal lipopolysaccharide exposure in rats induces robust increases in HPA axis activity together with increased expression of *Crh* and *Nr3c1* genes in various brain regions (Sominsky et al., 2013).

One seminal animal study investigated the epigenetic effects of early-life environment on stress programming through variations in the quality of early postnatal maternal care, as measured by levels of licking and grooming. Rats who received high levels of maternal care during early life developed sustained elevations in *Nr3c1* expression within the hippocampus and reduced HPA axis responses to stress. Meaney and colleagues further found an important role for epigenetic regulation revealing that the enhanced *Nr3c1* expression is associated with a persistent DNA hypomethylation at specific CpG dinucleotides within the hippocampal *Nr3c1* exons 1–7 promoter and increased histone acetylation. The lower CpG methylation facilitated the binding of the transcriptional activator nerve growth factor-inducible protein A (NGF1a) to this region (Weaver et al., 2004).

A number of key studies demonstrate that the epigenetic programming seen in the early-life stress animal models may extrapolate to human studies. For example, McGowan and colleagues (McGowan et al., 2008) found hypermethylation of *NR3C1* promoter among suicide victims with a history of abuse in childhood, but not among controls or suicide victims who did not suffer such early-life stress. These data appear consistent with the previously described study at the same homologous region of the rat *Nr3c1* promoter (Weaver et al., 2004). Numerous further studies support *NR3C1* to be susceptible to alterations in DNA methylation, in response to early environmental stress (*for review, see* Smart, Strathdee, Watson, Murgatroyd, & McAllister-Williams, 2015). Studies have shown that childhood adversity-induced epigenetic alterations are not limited to brain tissues, but may occur in peripheral tissues as well perhaps suggesting alterations in glucocorticoid receptors in peripheral blood cells that may link to possible changes in glucocorticoid receptor resistance, which have yet to be tested. For example, a significant negative association between the mothers' reports on the parenting quality provided to their children and the offspring's methylation levels of the GR gene and other candidate gene (*MIF*, a macrophage migration inhibitory factor gene that is functionally involved in GR expression and immune responses) in blood samples 5–10 years after assessing the caregiving quality was observed in the research by Bick et al. (2012). Other studies using peripheral DNA, from blood or saliva of infants, adolescents, or adults, have shown increased levels of *NR3C1* methylation in response to perinatal stress (Mulligan, D'Errico, Stees, & Hughes, 2012; Murgatroyd, Quinn, Sharp, Pickles, & Hill, 2015; Oberlander et al., 2008; Radtke et al., 2011) and abuse or neglect during childhood (Perroud et al., 2011; Tyrka, Price, Marsit, Walters, & Carpenter, 2012).

While the aforementioned candidate-gene-targeted epigenetic studies mainly focused on genes involved in HPA axis regulation, epigenome-wide studies have allowed identification of broad methylation changes that introduce new potential biological pathways related to childhood adversity. Labonté et al. (2012) in a genome-wide study of promoter methylation identified 362 differentially methylated promoters in hippocampal neurons from postmortem brain tissues of individuals with a history of severe childhood abuse compared with control subjects. Genes involved in cellular/neuronal plasticity were among the most significantly differentially methylated. Suderman et al. (2014) demonstrated that 997 gene promoters in whole blood DNA of adult subjects were differentially methylated in association with childhood abuse. Most of these genes are known to be involved in key cell signaling pathways linked to the development and regulation of transcription. Provençal et al. (2014) in a genome-wide promoter DNA methylation profiles in T cells from adult men found 448 differentially methylated gene promoters in individuals with a history of parental physical aggression from 6 to 15 years of age compared with a control group. Most of these genes have been previously demonstrated to play a substantial role in aggression and were enriched in biological pathways that are affected by behavior.

Some of the above studies investigating blood cells raise the question of whether epigenetic changes in blood provide a clinically valuable surrogate to what is happening in the brain. Studies in rhesus macaques by Provençal et al. (2012) showed that variations in mothering (surrogate vs mother reared) led to differential methylation rates including the *A2D681* gene that is the homologue of *NR3C1* in humans. Importantly, a weak but significant correlation was seen in differential methylation between prefrontal cortex samples and T-lymphocyte cells with differential rearing inducing methylation marks common between both tissues. However, results as a whole tended to suggest that the T cells were not direct surrogate markers of the brain tissue, but probably reflected the response of the immune system to early-life stress, again supporting the important relation between blood and the brain. Therefore, the prospect that the epigenomic response to early-life stress is not limited to the brain and can be studied in peripheral lymphocytes provides a major opportunity to study epigenetic marks in children and perform longitudinal studies.

Numerous studies have demonstrated that childhood trauma increases the risk for developing depression in adulthood, particularly in response to additional stress (*for review, see* Heim & Binder, 2012). The hypothesis is that neuroendocrine, particularly HPA activity, resultant from early-life stress could increase risk to develop depression in response to stress later in life. A study by Klengel et al. (2012) showed that an increased risk of developing adult stress-related psychiatric disorders is associated with allele-specific, childhood trauma-dependent DNA demethylation at glucocorticoid response DNA elements of the FK506-binding protein 51 (*FKBP5*) gene, which encodes a cochaperone that regulates GR sensitivity. The demethylation of *FKBP5* related to an increased stress-dependent gene transcription followed by a long-term dysregulation of the stress hormone system and a global effect on immune functions and brain areas is linked to stress regulation.

CONCLUSIONS

In this chapter, we have reviewed studies linking early-life adversity with risk for mood disorders, alterations in glucocorticoid and HPA axis regulation, and peripheral inflammation. We have covered the role of epigenetics, particularly DNA methylation, as one mechanism in which gene expression can be programmed. We discuss findings showing that prenatal and postnatal environments can activate epigenetic mechanisms at global levels and at the promoter regions of key target genes, producing long-lasting and stable changes in gene expression, which can last into adulthood and may be responsible of an increased vulnerability to develop inflammation and psychiatric disorders. It is important to understand the mechanisms underlying epigenetic programming in response to early-life adversity to enable to pinpoint alterations in the regulation of specific genes that, in turn, contribute to the pathogenesis of these disorders.

One major idea is that epigenetic tuning in response to early-life environment may prepare genes for responses to future environment, that is, enabling the organism to adapt to changing environmental conditions. These adaptive processes can, however, enhance the risk of pathology in the later life especially if there is mismatch between early- and later-life environments. Thereby, epigenetic research promises to improve the understanding of the genes and pathways by which organisms show widely diverse responses to stressful events in early life. Several specific genes linked to GR regulation such as NR3C1 are proving to be important markers, for both immune activation and mood disorders as we have discussed.

As there are associations between immune activation and mood disorders, it will be important to investigate whether increased inflammation identifies specific gene pathways underlying psychiatric disease, which together with inflammatory biomarkers may be used as a strategy to screen patients with mood disorders who may benefit from drugs that target inflammatory mechanisms. Through a better understanding of how epigenetic mechanisms underlie psychiatric disorders, we could also better characterize how these modifications can have an impact on specific genes that, in turn, contribute to the pathogenesis of these disorders. Furthermore, as increased inflammation is observed in depressed patients and, in particular, in those who do not respond to antidepressant therapies (*for review, see* Cattaneo et al., 2015), future research could clarify whether increased inflammation actually identifies a single group of depressed patients that has experienced childhood maltreatment and is also resistant to conventional antidepressants. Finally, future studies should also provide new insights on the reversibility of the damage associated with childhood stress experiences, including studies testing whether pharmacological and nonpharmacological interventions could reverse the abnormalities induced by childhood adversities on the functionality of the immune and stress response systems and thus also minimize the risk for mood disorders, both in the individuals affected and in the next generations.

In sum, understanding how early-life experiences can give rise to lasting epigenetic memories conferring increased risk for mental disorders is emerging at the epicenter of modern psychiatry. Whether suitable social or pharmacological interventions could reverse deleterious epigenetic programming triggered by adverse conditions during early life should receive highest priority on future research agendas. Progress in this field will further garner public interest, a general understanding and appreciation of the consequences of childhood abuse and neglect for victims in later life.

References

Appleton, A. A., Lester, B. M., Armstrong, D. A., Lesseur, C., & Marsit, C. J. (2015). Examining the joint contribution of placental NR3C1 and HSD11B2 methylation for infant neurobehavior. *Psychoneuroendocrinology*, 52, 32–42. https://doi.org/10.1016/j.psyneuen.2014.11.004.

Babenko, O., Kovalchuk, I., & Metz, G. A. S. (2015). Stress-induced perinatal and transgenerational epigenetic programming of brain development and mental health. *Neuroscience & Biobehavioral Reviews*, 48, 70–91. https://doi.org/10.1016/j.neubiorev.2014.11.013.

Balmer, D., Arredondo, J., Samaco, R. C., & LaSalle, J. M. (2002). MECP2 mutations in Rett syndrome adversely affect lymphocyte growth, but do not affect imprinted gene expression in blood or brain. *Human Genetics*, 110(6), 545–552. https://doi.org/10.1007/s00439-002-0724-4.

Bellavance, M.-A., & Rivest, S. (2014). The HPA—immune axis and the immunomodulatory actions of glucocorticoids in the brain. *Frontiers in Immunology*, 5, 136. https://doi.org/10.3389/fimmu.2014.00136.

Bertram, C., Trowern, A. R., Copin, N., Jackson, A. A., & Whorwood, C. B. (2001). The maternal diet during pregnancy programs altered expression of the glucocorticoid receptor and type 2 11β-hydroxysteroid dehydrogenase: potential molecular mechanisms underlying the programming of hypertension *in utero*. *Endocrinology*, 142(7), 2841–2853. https://doi.org/10.1210/endo.142.7.8238.

Bick, J., Naumova, O., Hunter, S., Barbot, B., Lee, M., Luthar, S. S., et al. (2012). Childhood adversity and DNA methylation of genes involved in the hypothalamus-pituitary-adrenal axis and immune system: whole-genome and candidate-gene associations. *Developmental Psychopatholgy*, 24(4), 1417–1425. https://doi.org/10.1017/S0954579412000806.

Bufalino, C., Hepgul, N., Aguglia, E., & Pariante, C. M. (2013). The role of immune genes in the association between depression and inflammation: a review of recent clinical studies. *Brain, Behavior, and Immunity*, 31, 31–47. https://doi.org/10.1016/j.bbi.2012.04.009.

Bunck, M., Czibere, L., Horvath, C., Graf, C., Frank, E., Kessler, M. S., et al. (2009). A hypomorphic vasopressin allele prevents anxiety-related behavior. *PLoS One*, 4(4), e5129. https://doi.org/10.1371/journal.pone.0005129.

Busillo, J. M., & Cidlowski, J. A. (2013). The five Rs of glucocorticoid action during inflammation: ready, reinforce, repress, resolve, and restore. *Trends in Endocrinology and Metabolism: TEM*, 24(3), 109–119. https://doi.org/10.1016/j.tem.2012.11.005.

Cao-Lei, L., Veru, F., Elgbeili, G., Szyf, M., Laplante, D. P., & King, S. (2016). DNA methylation mediates the effect of exposure to prenatal maternal stress on cytokine production in children at age 13½ years: Project Ice Storm. *Clinical Epigenetics*, 8(1), 54. https://doi.org/10.1186/s13148-016-0219-0.

Cattaneo, A., Gennarelli, M., Uher, R., Breen, G., Farmer, A., Aitchison, K. J., et al. (2013). Candidate genes expression profile associated with antidepressants response in the GENDEP study: differentiating between baseline "predictors" and longitudinal "targets". *Neuropsychopharmacology*, 38(3), 377–385. https://doi.org/10.1038/npp.2012.191.

Cattaneo, A., Macchi, F., Plazzotta, G., Veronica, B., Bocchio-Chiavetto, L., Riva, M. A., et al. (2015). Inflammation and neuronal plasticity: a link between childhood trauma and depression pathogenesis. *Frontiers in Cellular Neuroscience*, 9, 40. https://doi.org/10.3389/fncel.2015.00040.

Conradt, E., Lester, B. M., Appleton, A. A., Armstrong, D. A., & Marsit, C. J. (2013). The roles of DNA methylation of NR3C1 and 11β-HSD2 and exposure to maternal mood disorder in utero on newborn neurobehavior. *Epigenetics*, 8(12), 1321–1329. https://doi.org/10.4161/epi.26634.

Dhabhar, F. S. (2002). Stress-induced augmentation of immune function—the role of stress hormones, leukocyte trafficking, and cytokines. *Brain, Behavior, and Immunity*, 16(6), 785–798. Retrieved from http://www.ncbi.nlm.nih.gov/pubmed/12480507.

Doherty, T. S., & Roth, T. L. (2016). Insight from animal models of environmentally driven epigenetic changes in the developing and adult brain. *Development and Psychopathology*, 28(4pt2), 1229–1243. https://doi.org/10.1017/S095457941600081X.

Dowlati, Y., Herrmann, N., Swardfager, W., Liu, H., Sham, L., Reim, E. K., et al. (2010). A meta-analysis of cytokines in major depression. *Biological Psychiatry*, 67(5), 446–457. https://doi.org/10.1016/j.biopsych.2009.09.033.

Dumbell, R., Matveeva, O., & Oster, H. (2016). Circadian clocks, stress, and immunity. *Frontiers in Endocrinology*, 7, 37. https://doi.org/10.3389/fendo.2016.00037.

Fox, R., Ealing, J., Murphy, H., Gow, D. P., & Gosal, D. (2016). A novel *DNMT1* mutation associated with early onset hereditary sensory and autonomic neuropathy, cataplexy, cerebellar atrophy, scleroderma, endocrinopathy, and common variable immune deficiency. *Journal of the Peripheral Nervous System*, 21(3), 150–153. https://doi.org/10.1111/jns.12178.

Furtado, M., & Katzman, M. A. (2015). Neuroinflammatory pathways in anxiety, posttraumatic stress, and obsessive compulsive disorders. *Psychiatry Research*, 229(1–2), 37–48. https://doi.org/10.1016/j.psychres.2015.05.036.

Gomez, S., Diawara, A., Gbeha, E., Awadalla, P., Sanni, A., Idaghdour, Y., et al. (2016). Comparative analysis of iron homeostasis in sub-Saharan African children with sickle cell disease and their unaffected siblings. *Frontiers in Pediatrics*, 4, 8. https://doi.org/10.3389/fped.2016.00008.

Gonzalez, A. (2013). The impact of childhood maltreatment on biological systems: implications for clinical interventions. *Paediatrics & Child Health*, 18(8), 415–418. Retrieved from http://www.ncbi.nlm.nih.gov/pubmed/24426793.

Heim, C., & Binder, E. B. (2012). Current research trends in early life stress and depression: review of human studies on sensitive periods, gene–environment interactions, and epigenetics. *Experimental Neurology*, 233(1), 102–111. https://doi.org/10.1016/j.expneurol.2011.10.032.

Herkenham, M., & Kigar, S. L. (2016). Contributions of the adaptive immune system to mood regulation: mechanisms and pathways of neuroimmune interactions. *Progress in Neuro-Psychopharmacology and Biological Psychiatry*, 79(Part A), 49–57. https://doi.org/10.1016/j.pnpbp.2016.09.003.

Holsboer, F. (2000). The corticosteroid receptor hypothesis of depression. *Neuropsychopharmacology*, 23(5), 477–501. https://doi.org/10.1016/S0893-133X(00)00159-7.

Howerton, C. L., & Bale, T. L. (2012). Prenatal programing: at the intersection of maternal stress and immune activation. *Hormones and Behavior*, 62(3), 237–242. https://doi.org/10.1016/j.yhbeh.2012.03.007.

Ichiyama, K., Chen, T., Wang, X., Yan, X., Kim, B.-S., Tanaka, S., et al. (2015). The methylcytosine dioxygenase Tet2 promotes DNA demethylation and activation of cytokine gene expression in T cells. *Immunity*, 42(4), 613–626. https://doi.org/10.1016/j.immuni.2015.03.005.

Jansen, R., Penninx, B. W. J. H., Madar, V., Xia, K., Milaneschi, Y., Hottenga, J. J., et al. (2016). Gene expression in major depressive disorder. *Molecular Psychiatry*, 21(3), 339–347. https://doi.org/10.1038/mp.2015.57.

Jensen Peña, C., Monk, C., & Champagne, F. A. (2012). Epigenetic effects of prenatal stress on 11β-hydroxysteroid dehydrogenase-2 in the placenta and fetal brain. *PLoS One*, 7(6), e39791. https://doi.org/10.1371/journal.pone.0039791.

Jenuwein, T. (2001). Translating the histone code. *Science*, 293(5532), 1074–1080. https://doi.org/10.1126/science.1063127.

Klengel, T., Mehta, D., Anacker, C., Rex-Haffner, M., Pruessner, J. C., Pariante, C. M., et al. (2012). Allele-specific FKBP5 DNA demethylation mediates gene–childhood trauma interactions. *Nature Neuroscience*, 16(1), 33–41. https://doi.org/10.1038/nn.3275.

Labonté, B., Suderman, M., Maussion, G., Navaro, L., Yerko, V., Mahar, I., et al. (2012). Genome-wide epigenetic regulation by early-life trauma. *Archives of General Psychiatry*, 69(7), 722–731. https://doi.org/10.1001/archgenpsychiatry.2011.2287.

McGowan, P. O., Sasaki, A., Huang, T. C. T., Unterberger, A., Suderman, M., Ernst, C., et al. (2008). Promoter-wide hypermethylation of the ribosomal RNA gene promoter in the suicide brain. *PLoS One*, 3(5), e2085. https://doi.org/10.1371/journal.pone.0002085.

Miller, G. E., Chen, E., Fok, A. K., Walker, H., Lim, A., Nicholls, E. F., et al. (2009). Low early-life social class leaves a biological residue manifested by decreased glucocorticoid and increased proinflammatory signaling. *Proceedings of the National Academy of Sciences*, 106(34), 14716–14721. https://doi.org/10.1073/pnas.0902971106.

Miller, G. E., & Cole, S. W. (2012). Clustering of depression and inflammation in adolescents previously exposed to childhood adversity. *Biological Psychiatry*, 72(1), 34–40. https://doi.org/10.1016/j.biopsych.2012.02.034.

Mulligan, C. J., D'Errico, N. C., Stees, J., & Hughes, D. A. (2012). Methylation changes at NR3C1 in newborns associate with maternal prenatal stress exposure and newborn birth weight. *Epigenetics*, 7, 853–857. https://doi.org/10.4161/epi.21180.

Murgatroyd, C., & Bradburn, S. (2016). Translational animal models for the study of epigenetics and the environment. In *Epigenetics, the environment, and children's health across lifespans* (pp. 207–229). Cham: Springer International Publishing. https://doi.org/10.1007/978-3-319-25325-1_8.

Murgatroyd, C., Quinn, J. P., Sharp, H. M., Pickles, A., & Hill, J. (2015). Effects of prenatal and postnatal depression, and maternal stroking, at the glucocorticoid receptor gene. *Translational Psychiatry*, 5, e560. https://doi.org/10.1038/tp.2014.140.

Murgatroyd, C., & Spengler, D. (2011a). Epigenetic programming of the HPA axis: early life decides. *Stress (Amsterdam, Netherlands)*, 14(6), 581–589. https://doi.org/10.3109/10253890.2011.602146.

Murgatroyd, C., & Spengler, D. (2011b). Epigenetics of early child development. *Frontiers in Psychiatry*, 2, 15. Retrieved from http://www.pubmedcentral.nih.gov/articlerender.fcgi?artid=3102328&tool=pmcentrez&rendertype=abstract.

Murgatroyd, C., & Spengler, D. (2011c). Epigenetics of early child development. *Frontiers in Psychiatry, 2,* 16. https://doi.org/10.3389/fpsyt.2011.00016.

Murgatroyd, C., & Spengler, D. (2012). Genetic variation in the epigenetic machinery and mental health. *Current Psychiatry Reports, 14*(2), 138–149. https://doi.org/10.1007/s11920-012-0255-1.

Oberlander, T. F., Weinberg, J., Papsdorf, M., Grunau, R., Misri, S., & Devlin, A. M. (2008). Prenatal exposure to maternal depression, neonatal methylation of human glucocorticoid receptor gene (NR3C1) and infant cortisol stress responses. *Epigenetics, 3*(2), 97–106. Retrieved from http://www.ncbi.nlm.nih.gov/pubmed/18536531.

O'Donnell, K. J., Bugge Jensen, A., Freeman, L., Khalife, N., O'Connor, T. G., & Glover, V. (2012). Maternal prenatal anxiety and downregulation of placental 11β-HSD2. *Psychoneuroendocrinology, 37*(6), 818–826. https://doi.org/10.1016/j.psyneuen.2011.09.014.

Pecorelli, A., Cervellati, F., Belmonte, G., Montagner, G., Waldon, P., Hayek, J., et al. (2016). Cytokines profile and peripheral blood mononuclear cells morphology in Rett and autistic patients. *Cytokine, 77,* 180–188. https://doi.org/10.1016/j.cyto.2015.10.002.

Perroud, N., Paoloni-Giacobino, A., Prada, P., Olie, E., Salzmann, A., & Nicastro, R. (2011). Increased methylation of glucocorticoid receptor gene (NR3C1) in adults with a history of childhood maltreatment: a link with the severity and type of trauma. *Translational Psychiatry, 1,* e59. Retrieved from. https://doi.org/10.1038/tp.2011.60.

Pike, J. L., & Irwin, M. R. (2006). Dissociation of inflammatory markers and natural killer cell activity in major depressive disorder. *Brain, Behavior, and Immunity, 20*(2), 169–174. https://doi.org/10.1016/j.bbi.2005.05.004.

Provençal, N., Suderman, M. J., Guillemin, C., Massart, R., Ruggiero, A., Wang, D., et al. (2012). The signature of maternal rearing in the methylome in rhesus macaque prefrontal cortex and T cells. *Journal of Neuroscience, 32*(44), 15626–15642. https://doi.org/10.1523/JNEUROSCI.1470-12.2012.

Provençal, N., Suderman, M. J., Guillemin, C., Vitaro, F., Côté, S. M., Hallett, M., et al. (2014). Association of childhood chronic physical aggression with a DNA methylation signature in adult human T cells. *PLoS One, 9*(4), e89839. https://doi.org/10.1371/journal.pone.0089839.

Radtke, K. M., Ruf, M., Gunter, H. M., Dohrmann, K., Schauer, M., & Meyer, A. (2011). Transgenerational impact of intimate partner violence on methylation in the promoter of the glucocorticoid receptor. *Translational Psychiatry, 1,* e21. https://doi.org/10.1038/tp.2011.21.

Seckl, J. R. (2001). Glucocorticoid programming of the fetus; adult phenotypes and molecular mechanisms. *Molecular and Cellular Endocrinology, 185*(1–2), 61–71. Retrieved from http://www.ncbi.nlm.nih.gov/pubmed/11738795.

Shelton, R. C., Claiborne, J., Sidoryk-Wegrzynowicz, M., Reddy, R., Aschner, M., Lewis, D. A., et al. (2011). Altered expression of genes involved in inflammation and apoptosis in frontal cortex in major depression. *Molecular Psychiatry, 16*(7), 751–762. https://doi.org/10.1038/mp.2010.52.

Smart, C., Strathdee, G., Watson, S., Murgatroyd, C., & McAllister-Williams, R. H. (2015). Early life trauma, depression and the glucocorticoid receptor gene—an epigenetic perspective. *Psychological Medicine, 45*(16), 3393–3410. https://doi.org/10.1017/S0033291715001555.

Sominsky, L., Fuller, E. A., Bondarenko, E., Ong, L. K., Averell, L., Nalivaiko, E., et al. (2013). Functional programming of the autonomic nervous system by early life immune exposure: implications for anxiety. *PLoS One, 8*(3), e57700. https://doi.org/10.1371/journal.pone.0057700.

Stewart, P. M., Rogerson, F. M., & Mason, J. I. (1995). Type 2 11 beta-hydroxysteroid dehydrogenase messenger ribonucleic acid and activity in human placenta and fetal membranes: its relationship to birth weight and putative role in fetal adrenal steroidogenesis. *The Journal of Clinical Endocrinology & Metabolism, 80*(3), 885–890. https://doi.org/10.1210/jcem.80.3.7883847.

Suderman, M., Borghol, N., Pappas, J. J., Pinto Pereira, S. M., Pembrey, M., Hertzman, C., et al. (2014). Childhood abuse is associated with methylation of multiple loci in adult DNA. *BMC Medical Genomics, 7*(1), 13. https://doi.org/10.1186/1755-8794-7-13.

Togher, K. L., Togher, K. L., O'Keeffe, M. M., O'Keeffe, M. M., Khashan, A. S., Khashan, A. S., et al. (2014). Epigenetic regulation of the placental HSD11B2 barrier and its role as a critical regulator of fetal development. *Epigenetics, 9*(6), 816–822. https://doi.org/10.4161/epi.28703.

Tsao, C.-W., Lin, Y.-S., Chen, C.-C., Bai, C.-H., & Wu, S.-R. (2006). Cytokines and serotonin transporter in patients with major depression. *Progress in Neuro-Psychopharmacology and Biological Psychiatry, 30*(5), 899–905. https://doi.org/10.1016/j.pnpbp.2006.01.029.

Tyrka, A. R., Price, L. H., Marsit, C., Walters, O. C., & Carpenter, L. L. (2012). Childhood adversity and epigenetic modulation of the leukocyte glucocorticoid receptor: preliminary findings in healthy adults. *PLoS One, 7,* e30148. https://doi.org/10.1371/journal.pone.0030148.

Waters, C. S., Hay, D. F., Simmonds, J. R., & van Goozen, S. H. M. (2014). Antenatal depression and children's developmental outcomes: potential mechanisms and treatment options. *European Child & Adolescent Psychiatry, 23*(10), 957–971. https://doi.org/10.1007/s00787-014-0582-3.

Weaver, I. C. G., Cervoni, N., Champagne, F. A., D'Alessio, A. C., Sharma, S., Seckl, J. R., et al. (2004). Epigenetic programming by maternal behavior. *Nature Neuroscience, 7* (8), 847–854. https://doi.org/10.1038/nn1276.

Welberg, L. A. M. (2005). Chronic maternal stress inhibits the capacity to up-regulate placental 11 -hydroxysteroid dehydrogenase type 2 activity. *Journal of Endocrinology, 186*(3), R7–R12. https://doi.org/10.1677/joe.1.06374.

Wen, L., & Tang, F. (2014). Genomic distribution and possible functions of DNA hydroxymethylation in the brain. *Genomics, 104*(5), 341–346. https://doi.org/10.1016/j.ygeno.2014.08.020.

Wood, K. H., & Zhou, Z. (2016). Emerging molecular and biological functions of MBD2, a reader of DNA methylation. *Frontiers in Genetics, 7*, 93. https://doi.org/10.3389/fgene.2016.00093.

Further Reading

Hammen, C., Shih, J. H., & Brennan, P. A. (2004). Intergenerational transmission of depression: test of an interpersonal stress model in a community sample. *Journal of Consulting and Clinical Psychology, 72*(3), 511–522. https://doi.org/10.1037/0022-006X.72.3.511.

Lyons-Ruth, K., Easterbrooks, M. A., & Cibelli, C. D. (1997). Infant attachment strategies, infant mental lag, and maternal depressive symptoms: predictors of internalizing and externalizing problems at age 7. *Developmental Psychology, 33*(4), 681–692. Retrieved from(1997). http://www.ncbi.nlm.nih.gov/pubmed/9232383.

Murgatroyd, C. A., & Nephew, B. C. (2013). Effects of early life social stress on maternal behavior and neuroendocrinology. *Psychoneuroendocrinology, 38*(2), 219–228. https://doi.org/10.1016/j.psyneuen.2012.05.020.

National Research Council (US) and Institute of Medicine (US) Committee on Depression, Parenting Practices, and the Healthy Development of Children (2009). In M. J. England & L. J. Sim (Eds.), *Depression in parents, parenting, and children: Opportunities to improve identification, treatment, and prevention.* Washington, DC: National Academies Press Retrieved from. *http://www.ncbi.nlm.nih.gov/pubmed/25009931.*

Olino, T. M., Pettit, J. W., Klein, D. N., Allen, N. B., Seeley, J. R., & Lewinsohn, P. M. (2008). Influence of parental and grandparental major depressive disorder on behavior problems in early childhood: a three-generation study. *Journal of the American Academy of Child and Adolescent Psychiatry, 47*(1), 53–60. https://doi.org/10.1097/chi.0b013e31815a6ae6.

Pettit, J. W., Olino, T. M., Roberts, R. E., Seeley, J. R., & Lewinsohn, P. M. (2008). Intergenerational transmission of internalizing problems: effects of parental and grandparental major depressive disorder on child behavior. *Journal of Clinical Child and Adolescent Psychology: The Official Journal for the Society of Clinical Child and Adolescent Psychology, American Psychological Association, Division 53, 37*(3), 640–650. https://doi.org/10.1080/15374410802148129.

Weissman, M. M., Berry, O. O., Warner, V., Gameroff, M. J., Skipper, J., Talati, A., et al. (2016). A 30-year study of 3 generations at high risk and low risk for depression. *JAMA Psychiatry, 73*(9), 970. https://doi.org/10.1001/jamapsychiatry.2016.1586.

Mechanisms Linking Depression, Immune System and Epigenetics During Aging

Steven Bradburn

Manchester Metropolitan University, Manchester, United Kingdom

INTRODUCTION

It is becoming increasingly apparent that inflammation plays a central role in the etiology of depression. Interestingly, this inflammatory imbalance shares many similarities with the insidious inflammaging processes that manifest during later biological aging. Despite many reports of immune-related polymorphic contributors to depression and inflammaging, genetic influences do not explain the majority of cases. This can be seen by studying monozygotic (MZ) twins, that is, those that are genetically identical, whereby the majority of inflammatory parameters are largely driven by nonheritable influences, which are more evident in later life (Brodin et al., 2015). With regard to depression, additive genetic determinants have been shown to contribute to approximately 37% of the cases of major depression disorder (MDD), with the rest accounted for by other factors including environmental (Sullivan, Neale, & Kendler, 2000). Together, these observations suggest that environmental and gene-environment (GxE)

interactions may be the predominant causes of inflammaging and depression.

This chapter will explore the recent developments of epigenetic, specifically DNA methylation, histone acetylation, and microRNA (miRNA), involvements in immune-related genes during aging and depression.

INFLAMMAGING

Franceschi et al. (2000) postulated the inflammaging process at the start of the millennium to describe the chronic, pro-inflammatory state that accumulates in later life. The researchers describe a two-hit hypothesis of inflammaging: (first hit) the gradual pro-inflammatory favorability following a lifetime of recurrent exposures to inflammatory stressors including chemical, physical, and antigenic and (second hit) the absence of robust and/or the presence of frail gene variants (Franceschi et al., 2000). Over time, inflammaging results in a vicious cycle of immune overactivation and a failure to control inflammatory homeostasis ultimately leading to the development of age-related diseases that have an inflammatory pathogenesis (Baylis, Bartlett, Patel, & Roberts, 2013).

Systemic inflammation in aging humans is best observed when analyzing cytokines, cellular messenger peptides, many of which contain diverse immunologic properties in aging cohorts. Interleukin-6 (IL-6) and tumor necrosis factor alpha (TNF-α) are established examples of proinflammatory cytokines that have been shown to significantly increase within the circulation during aging (Bruunsgaard et al., 1999; Ferrucci et al., 2005; Mariani et al., 2006; Miles et al., 2008; Paolisso et al., 1998). Further, these age effects are independent of inflammatory confounders such as cardiovascular risk factors (Ferrucci et al., 2005; Miles et al., 2008).

The genetic influence for the second hit of inflammaging can also be observed when analyzing the frequency of functionally significant cytokine polymorphisms in longevity cohorts. For example, in Italian men, a higher frequency of the IL-6-174 GG genotype, whereby the G allele has been associated with increased IL-6 expression and plasma protein levels compared with the C allele (Fishman et al., 1998), has been associated with a twofold reduction in reaching extreme old age compared with those with CC or CG genotypes (Di Bona et al., 2009).

Additional inflammatory parameters dysregulated with age that reflects and further contributes to the inflammaging process, including the exacerbated release of cytokines from peripheral blood mononuclear cells (PBMCs), changes in lymphocyte subsets, and inflammatory-related polymorphic variations, have been reviewed in detail elsewhere (Minciullo et al., 2016).

EPIGENETICS

Epigenetics, in simple terms, refers to the study of modifications to chromatin that affects gene expression without altering the underlying DNA sequence itself (Berger, Kouzarides, Shiekhattar, & Shilatifard, 2009). Waddington (2012) initially coined the term to describe the complex connection between a stable genotype with differing phenotypes. Indeed, it is now appreciated that GxE actions are controlled by multiple epigenetic mechanisms, such as DNA methylation, histone modifications (e.g., acetylation), and miRNAs (Fig. 1), which interact to fine-tune gene expression in response to environmental stimuli. Each aforementioned mechanism will be discussed in further detail, including their significance to immune-related gene control in aging and depression.

GENETIC CONTROL THROUGH DNA METHYLATION

DNA methylation is the process whereby a methyl ($-CH_3$) group is added to specific areas of the DNA. The majority of DNA methylation is associated to the fifth carbon atom of the cytosine ring to form 5-methylcytosine (5-mC) that predominately occurs when a cytosine is adjacent to a guanine base, referred to as a CpG site (Fig. 1). The "p" denotes the phosphodiester bond connecting the two bases. DNA methyltransferase (DNMT) enzymes are responsible for the transfer of methyl groups from S-adenosylmethionine (SAM), the universal donor substrate in mammals, to CpG sites. DNMT1 is the primary enzyme responsible for somatic cell DNA methylation maintenance, by utilizing hemimethylated CpG sites as a template, whereas DNMT3A and DNMT3B are mainly responsible for de novo methylation marks (Smith & Meissner, 2013).

Functionally, methylation at CpG sites is an established mark of gene repression via direct and indirect actions. The former involves the inhibition of transcription factor binding when methylated CpG sites are within binding motifs (Watt & Molloy, 1988). The latter involves methyl-CpG-binding proteins (MBPs) that are

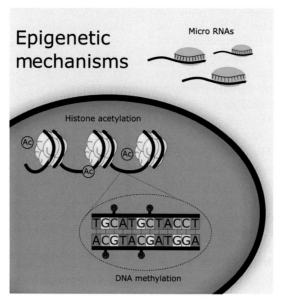

FIG. 1 A representation of the DNA methylation, histone acetylation, and microRNA epigenetic mechanisms discussed within this chapter. DNA methylation and histone acetylation take effect within the cell nucleus, whereas microRNAs mediate their post-transcriptional control in the cytoplasm. Histone acetylation (Ac) occurs along the histone tails to promote gene induction. DNA methylation (M) occurs at CpG sites along the DNA sequence to form 5-methylcytosine. MicroRNAs, within the RNA-induced silencing complex, target the regulation of complementary mRNA transcripts.

able to bind to 5-mC to form a protein complex capable of inhibiting transcription (Boyes & Bird, 1991).

ALTERATIONS TO THE DNA METHYLATION MACHINERY IN AGING AND DEPRESSION

By examining older adults and those in depressive states, striking alterations to the DNA methylation system can be observed.

A prominent example can be seen when assessing folate status. The vitamin folate is a dietary factor involved in one-carbon metabolism and ultimately methyl group formation (Crider, Yang, Berry, & Bailey, 2012). Since methyl groups are required for DNA methylation synthesis, disruptions in folate availability can have pronounced influences on DNA methylation

regulation. The prevalence of metabolically significant folate deficiency increases with age, with approximately 10% of those aged 75 years and older being affected (Clarke et al., 2004). Causes of such deficiency include intestinal malabsorption, reduced dietary intake, and drug interactions (Araújo, Martel, Borges, Araújo, & Keating, 2015). Low folate status has also been linked to enhanced inflammatory responses, as seen in microglial (Cianciulli, Salvatore, Porro, Trotta, & Panaro, 2016) and macrophage (Kolb & Petrie, 2013) models, and has been implicated in the development of various neuropsychiatric disorders including depression (Araújo et al., 2015). The association between low folate and depression risk is also apparent even when considering younger adults (Gilbody, Lightfoot, & Sheldon, 2007).

Another alteration to the DNA methylation machinery can be seen when assessing DNMT activity in aging and depression. For example, in a large-scale ($n = 2453$) cohort of the European general population, both *DNMT1* and *DNMT3B* gene expression from PBMCs are reduced from the ages of 35–64 years old (Ciccarone et al., 2016). Further, *DNMT1* and *DNMT3B* expressions are significantly reduced and increased, respectively, in the blood from those with MDD in a depressive, but not remissive, state compared with controls (Higuchi et al., 2011). *DNMT1* and *DNMT3B* were also dysregulated in various brain regions, including the prefrontal cortex, amygdala, and paraventricular nucleus, of suicide victims diagnosed with major depression (Poulter et al., 2008; Smalheiser et al., 2012). Reduced actions of the DNMT enzymes may therefore contribute to the loss and maintenance of methylation marks in aging and depression. In addition, both DNMT1 and DNMT3A have critical roles in immune function (Lee et al., 2001) and inflammation (Yu et al., 2012).

Taken together, the reduced availability of methyl groups as reflected by low folate and the reduced activity in the DNMT enzymes in immune function, aging, and depression suggest disruptions to the regulation of the DNA methylation network.

EFFECTS OF AGING AND DEPRESSION ON GENOME-WIDE DNA METHYLATION

The best example for the effect of aging and depression on the methylome comes from studies on MZ twins. MZ twins are powerful subjects for epigenetic studies because their underlying DNA sequence is virtually identical between pairs. Furthermore, they usually share common environments, both pre- and postnatally, which limits stochastic methylation influences.

There is strong evidence that younger MZ pairs have comparable levels of genome-wide 5-mC marks, as shown in a variety of tissue types, whereas older MZ pairs present with a higher divergence in their overall 5-mC content

(Fraga et al., 2005; Kaminsky et al., 2009). This same age-related dysregulation can also be seen in observational studies using leukocyte DNA on methylation arrays, suggesting an increased intra-individual variability and, in most cases, a global demethylation with advancing age (Heyn et al., 2012; Johansson, Enroth, & Gyllensten, 2013).

The aforementioned methylome-wide divergence is similarly presented in MZ pairs when relating to their psychopathological status. Studies have shown substantial blood and buccal cell DNA methylation variability between the MDD-affected and unaffected MZ pairs (Byrne et al., 2013; Córdova-Palomera et al., 2015; Davies et al., 2014; Dempster et al., 2014).

Collectively, these studies suggest that DNA methylation is significantly altered in aging and depressive states throughout the genome. In most cases, a global demethylation is observed in those affected. Depending upon the genetic location of such alterations, DNA methylation variability can impact the regulation of genes, including those encoding cytokines, some of which will be explored in more detail.

EFFECTS OF AGING AND DEPRESSION ON INTERFERON-GAMMA DNA METHYLATION

Interferon-gamma (IFN-γ), a type II IFN, is essential for immune control against intracellular pathogens, especially viruses (Schroder, Hertzog, Ravasi, & Hume, 2004). It is primarily secreted by Th1 lymphocytes and natural killer cells in the periphery and microglia in the central nervous system (CNS). Binding of IFN-γ to the IFN-γ receptor 1/2 (IFNGR1/2) complex results in the activation of Janus kinase (JAK) 1 and 2 tyrosine kinases. This leads to the recruitment and phosphorylation of signal transducer and activator of transcription 1 that translocate to the nucleus to regulate IFN-stimulated genes. IFN-γ also results in microglia and macrophage priming, which exacerbates their inflammatory potential upon stimulation.

With regard to aging, Tserel et al. (2015) noted lower DNA methylation levels in and around the *IFN-γ* gene in CD8[+] T cells from older adults compared with younger controls. The researchers further discovered a strong inverse correlation between methylation at these CpG sites and *IFN-γ* gene expression, suggesting methylation at this locus is controlling gene expression. Previous investigations have also found a positive correlation between the subject age and the number of IFN-γ producing CD8[+] T cells in a population of healthy controls aged 17–62 years old (Bandrés et al., 2000), which could be a reflection of the aforementioned hypomethylation of the *IFN-γ* gene with age.

Nemoda et al. (2015) highlighted remarkable immune-related DNA methylation alterations in neonates and adult offspring exposed to maternal depression. The researchers found at least 145 differentially methylated CpG sites, the majority of which were hypomethylated, in cord blood T lymphocytes of neonates exposed to maternal depression compared with the control group. Furthermore, up to 294 differentially methylated sites were detected from an independent selection of male hippocampal samples who had a history of maternal depression. Remarkably, there was a significant overlap of 33 genes, including many IFN-stimulated genes, differentially methylated between the maternally depressed neonatal cord blood and adult brains.

The significance of *IFN-γ* DNA methylation on genetic control has been best characterized in T cells. It is known that the *IFN-γ* gene promoter rapidly demethylates in memory CD8[+] T cells following antigenic stimulation, resulting in a surge in *IFN-γ* expression (Kersh et al., 2006). Excessive IFN-γ secretion from T cells is best observed following cytomegalovirus (CMV) stimulation, a beta-herpesvirus that infects approximately 80%–90% of older adults (Pawelec, Larbi, & Derhovanessian, 2010), whereby the production of IFN-γ is up to eightfold higher in very old compared with younger controls (Vescovini et al., 2007). Aberrant IFN-γ secretion is also observed through the use of ex vivo stimulation on PBMCs from patients with and without MDD, whereby the MDD-affected individuals produce higher IFN-γ than the controls (Maes et al., 1994). Interestingly, in a community-based sample of older adults, those with high CMV-specific antibody titers were 1.18 times more likely to report depressive symptoms, as measured with the General Health Questionnaire, compared with those with lower antibody titers (Phillips, Carroll, Khan, & Moss, 2008). Scores on the Beck Depression Inventory have also been significantly related to IFN-γ production in ex vivo stimulated blood mononuclear cells from those diagnosed with relapsing-remitting multiple sclerosis, which were both reduced following a 16-week intervention therapy strategy (Mohr, Goodkin, Islar, Hauser, & Genain, 2001).

DNA methylation investigations within the *IFN-γ* gene in depressive states remain to be seen. However, there is evidence for an upregulation of *IFN-γ* expression within the prefrontal cortex of patients with a history of MDD (Shelton et al., 2011). There are also some reports of higher peripheral IFN-γ concentrations in adolescents (Mills, Scott, Wray, Cohen-Woods, & Baune, 2013), but not adults (Dowlati et al., 2010), with MDD compared with controls. Future studies will elucidate the potential involvement of DNA methylation in the regulation of the *IFN-γ* gene in MDD.

EFFECTS OF AGING AND DEPRESSION ON TOLL-LIKE RECEPTORS DNA METHYLATION

Toll-like receptors (TLRs) are a family of 10 (TLR1–TLR10) receptor proteins found on the cells of the innate immune system that are vital for the recognition of microbial components known as pathogen-associated molecular patterns (PAMPs). Upon PAMP recognition, TLRs recruit cytoplasmic toll/IL-1 receptor adapter proteins (TIR), such as myeloid differentiation primary response gene 88 (MyD88) and TIR-domain-containing adapter-inducinginterferon-β (TRIF), which trigger the activation of transcription factors including nuclear factor-kappa B (NF-κB) to further regulate the expression of inflammatory cytokines and chemokines (Kawasaki & Kawai, 2014). TLRs, therefore, are vital regulators of the immune response to foreign material.

There is very little known about the effects of aging on *TLR* DNA methylation. One study however did find a reduction of blood methylation at five CpG sites within the *TLR2* promoter, equating to 0.03% loss per year over a 10-year period, thus suggesting an increase in gene activity with age (Madrigano et al., 2012).

Dysregulations of the *TLR* gene family are more evident in neuropsychiatric conditions. For example, Uddin et al. (2010) analyzed DNA methylation signatures in the blood of those with and without posttraumatic stress disorder (PTSD) from the Detroit Neighborhood Health Study. Through functional annotation clustering analysis, the researchers discovered a unique set of unmethylated genes among the PTSD-affected individuals that included *TLR1* and *TLR3*. The observation of reduced methylation in the psychological distressed is further supported by the significant upregulation of *TLR3* and *TLR4* in the prefrontal cortex of depressed suicide victims (Pandey, Rizavi, Ren, Bhaumik, & Dwivedi, 2014).

Functionally, previous studies have shown that hypomethylating the *TLR2* and *TLR4* genes, through the use of DNMT inhibitors, results in an enhanced gene expression in numerous human cell line models (Furuta et al., 2008; Shuto et al., 2006; Takahashi, Sugi, Hosono, & Kaminogawa, 2009). Within the *TLR2* promoter region, a specificity protein 1 (SP1) binding site, maintained within a CpG-rich region, has been shown to be influential for basal promoter activity (Furuta et al., 2008). SP1-dependent transcriptional activity has been further shown

to be indirectly abolished by DNA methylation, most likely through the interaction of MBPs over the SP1 binding region (Furuta et al., 2008), ultimately reducing *TLR2* promoter activity. DNA methylation therefore can directly block SP1 binding to the *TLR2* promoter to inhibit gene expression.

EFFECTS OF AGING AND DEPRESSION ON TNF-α DNA METHYLATION

TNF-α is a potent immunoregulatory cytokine notably produced during the acute-phase response to induce peripheral inflammation in the defense against pathogens. Dominantly produced, but not exclusively, by activated myeloid-lineage cells, TNF-α is able to bind to two receptors: TNF receptors 1 (TNFR1) and 2 (TNFR2). The majority of focus regarding inflammation is associated with TNFR1 activation by either soluble or transmembrane TNF-α. Simply, TNFR1 activation recruits TNFR1-associated death domain protein (TRADD) and the complex I that ultimately activates NF-κB and mitogen-activated protein kinases to induce inflammatory genes (Kalliolias & Ivashkiv, 2016).

Evidence suggests that DNA methylation at the *TNF-α* promoter region is reduced in the periphery during aging. For example, one report described an age-related demethylation of 0.8% and 1.4% per decade from blood- and monocyte-derived macrophage DNA, respectively (Gowers et al., 2011). Methylation at the *TNF-α* promoter has previously been shown to reduce promoter activity by up to 78% (Gowers et al., 2011) or more (Pieper et al., 2008). It is known that the binding affinity of both the transcription factors activator protein 2 and SP1, integral activators of *TNF-α*, is reduced against methylated *TNF-α* promoter constructs compared with those that are unmethylated (Pieper et al., 2008). Further, treatment of cell models with SAM is able to attenuate lipopolysaccharide (LPS)-induced *TNF-α* expression (Veal et al., 2004; Watson, Zhao, & Chawla, 1999), and demethylating the *TNF-α* gene, through the use of the DNMT inhibitor 5-azacytidine, augments the ability of cells to respond to an inflammatory stimulant (Sullivan et al., 2007). Collectively, age-related hypomethylation of the *TNF-α* promoter contributes to an exacerbation of gene activity.

Whether there is a similar loss of methylation within the regulatory regions of *TNF-α* in depression remains to be explored. Current evidence does suggest an increased action of TNF-α in depression, as reflected by higher blood TNF-α concentrations in those suffering from depression compared with controls (Dowlati et al., 2010), and serum TNF-α has been independently associated with depressive symptoms in systemic lupus erythematosus patients (Postal et al., 2016). Therefore, it would be interesting to see if DNA methylation is also dysregulated in depression.

HISTONE MODIFICATIONS AS CHROMATIN MODELERS

Histones are a family of core (H2A, H2B, H3, and H4) and linker (H1 and H5) proteins that package DNA into the chromatin structure (Bannister & Kouzarides, 2011). Specifically, two of each core histone subunits collate to form an octamer that wraps approximately 150 base pairs of DNA to form a nucleosome. This is achieved through the positively charged properties of the histone proteins, enabling the attraction of the negative phosphate backbone of the DNA molecule. With the aid of linker histones, nucleosomes connect to form and maintain a higher order chromatin structure.

Each histone unit is composed of a head and protruding amino (N)-terminal tail structure. The latter is the main site for post-translational modifications (PTM) that include methylation,

acetylation, phosphorylation, and ubiquitylation (Bannister & Kouzarides, 2011). The properties of histone-DNA interactions are determined by the type and location of the PTM.

Acetylation is one of the most established histone PTM known. Specifically, acetylation occurs at the lysine residues present in the histone tails (Fig. 1) catalyzed by histone acetyl transferase (HAT) enzymes through the transfer of an acetyl group from the cofactor acetyl coenzyme A. Acetylation removes the lysine's positive charge, therefore reducing histone affinity toward DNA that, in turn, relaxes the surrounding chromatin to permit gene expression. Conversely, the removal of acetyl marks is induced by the histone deacetylase (HDAC) family that condenses chromatin and impedes transcriptional access (Bannister & Kouzarides, 2011). In theory, since the majority of histone acetylation marks are associated with gene promoter and enhancer regions (Wang et al., 2008), any alterations could have profound implications in gene regulation.

GENOME-WIDE ALTERATIONS TO HISTONE ACETYLATION IN AGING AND DEPRESSION

The influence of aging on histone acetylation has been evident since the discovery of the PTM, whereby a reduced rate of histone acetylation was demonstrated in aging human diploid cells (Ryan & Cristofalo, 1972). More recently, MZ twin studies have elaborated on this aging effect. Briefly, Fraga et al. (2005) observed comparable levels of both blood histone H3 and H4 acetylation between MZ pairs at the age of 3 years old; however, within 50-year-old twins, there was a significant divergence of histone acetylation between pairs.

The majority of evidence relating genome-wide histone acetylation alterations in depression has so far been performed in the brain tissue of rodent models (Sun, Kennedy, & Nestler, 2013).

For example, Hollis, Duclot, Gunjan, and Kabbaj (2011) observed significant reductions of H4, H3, and H2B acetylation from the hippocampi of rats following a social defeat. In humans, clinical investigations have highlighted the potential use of HDAC inhibitors for the treatment of depression (Fuchikami et al., 2016), which further supports the significance genome-wide histone acetylation in depressive cases.

Overall, there are clear alterations to the patterns of histone acetylation across the genome in both aging and depression. Accumulating evidence has thus far suggested a reduction in the histone acetylation process in the affected.

SIRTUIN 1 (SIRT1): AN INFLAMMATORY-RELATED HDAC DYSREGULATED IN AGING AND DEPRESSION

Knowledge is lacking regarding gene-specific histone acetylation changes with age and depression. Contrary to this, a dysregulation of the HDAC and HAT enzymes is evident with relation to immune-related gene control, for example, the HDAC sirtuin 1 (SIRT1), which will be explored in more detail.

Sirtuins are a family of seven (SIRT1–7) highly conserved nicotinamide adenine dinucleotide (NAD)-dependent deacetylases (Bonkowski & Sinclair, 2016). Most of the research so far has concentrated on SIRT1, the closest mammalian ortholog of the silent information regulator 2 (Sir2) enzyme originally discovered to extend life span through calorie restriction in the *Saccharomyces cerevisiae* budding yeast (Lin, Defossez, & Guarente, 2000). Functionally, SIRT1 is an HDAC predominately localized in the nucleus that can deacetylate a variety of histones, such as H3 and H4, and over 50 transcription factors and DNA repair proteins (Nakagawa & Guarente, 2014), including NF-κB (Nakagawa & Guarente, 2014; Yeung et al., 2004).

Aging effects on SIRT1 activity have been extensively studied in animal models, with very little evidence in humans. In the senescence-accelerated mouse model, it is known that there are age-dependent reductions in *Sirt1* expression patterns in a variety of tissues, including the brain, liver, and skeletal muscle, which is most significant in the brain (Gong et al., 2014). Tissue-specific Sirt1 alterations are also evident between young and old wild-type mice (Kwon, Kim, Lee, & Kim, 2015). Further, reductions of Sirt1 are observed in the rat hippocampus with aging (Quintas, de Solís, Díez-Guerra, Carrascosa, & Bogónez, 2012). SIRT1 plasma protein levels in humans, however, are positively associated with age (Kilic et al., 2015), which may be a compensatory reaction to account for the reduction of NAD$^+$ with age (Chaleckis, Murakami, Takada, Kondoh, & Yanagida, 2016). Genetic links between the *SIRT1* gene and longevity are also apparent. For example, a *SIRT1* promoter polymorphism was predominantly represented in the oldest people who also had the highest level of SIRT1 plasma protein concentrations (Kilic et al., 2015).

Alterations of SIRT1 in depression are also in agreement with the reduction of activity during aging. Those with MDD or bipolar disorder in a depressive, but not remissive, state have reduced *SIRT1* mRNA levels in PBMC compared with healthy controls (Abe et al., 2011). Hippocampal Sirt1 activation has been recently shown to alleviate chronic stress-elicited depressive symptoms in a mouse model, whereas reducing the Sirt1 pathway exacerbates prodepressive behaviors (Abe-Higuchi et al., 2016). Genetic analysis within and around the *SIRT1* gene has also identified numerous polymorphisms associated with MDD (Converge Consortium, 2015; Kishi et al., 2010).

The functional loss of SIRT1 activity can favor the pro-inflammatory state. For example, within SIRT1-depleted human fibroblasts, mRNAs and proteins of IL-8 and IL-6 rapidly accumulate

(Hayakawa et al., 2015). This effect is also seen during cellular senescence coinciding with acetylation of histones adjacent to the promoter regions of the inflammatory genes (Hayakawa et al., 2015). Also, SIRT1 knockdown in mouse macrophages leads to the activation of the c-Jun N-terminal kinase (JNK) and IκB kinase (IKK) pathways, which promotes the upregulation of various cytokines such as *TNF-α*, *IL-1β*, and *IL-6* (Schug et al., 2010; Yoshizaki et al., 2010). Conversely, promoting SIRT1 activity through the use of activators, for example, resveratrol, attenuates LPS-stimulated JNK and IKK activation (Yoshizaki et al., 2010). SIRT1 has also been shown to deacetylate the RelA/p65 subunit of NF-κB at lysine 310 (Schug et al., 2010; Yeung et al., 2004) and the nucleosome histone H4 at lysine 16 (Liu, Yoza, Gazzar, Vachharajani, & McCall, 2011). Further, SIRT1 promotes RelB, a sequester of the RelA/p65 subunit (Marienfeld et al., 2003; Yoza, Hu, Cousart, Forrest, & McCall, 2006), to form a repressible complex at promoter regions such as *TNF-α* (Liu et al., 2011). Collectively, these investigations suggest that SIRT1 represses the pro-inflammatory response associated with senescence and aging.

Interestingly, there is evidence for an involvement of SIRT1 in regulating DNMT activity. Specifically, SIRT1 has been shown to physically deacetylate DNMT1 *in vitro* and *in vivo* that can enhance or reduce DNMT1 activity depending upon the position of lysine deacetylation on the enzyme (Peng et al., 2011). Also, SIRT1 deficiency in mice reduces DNMT activity in myeloid cells, which reduces methylation of the proximal promoter of *IL-1β*, leading to transcriptional upregulation (Cho et al., 2015). Cho and colleagues further describe an age-related hypomethylation of the *IL-1β* promoter in humans that coincided with the aforementioned reduction of SIRT1 with age. These studies highlight the cross talk between multiple epigenetic mechanisms.

miRNAs AS POST-TRANSCRIPTIONAL REGULATORS OF mRNA

miRNAs are an extensive family of small noncoding RNA transcripts that, conventionally speaking, induce post-transcriptional silencing of complementary mRNA molecules (Ha & Kim, 2014) but in certain states actually promote gene expression (Vasudevan, Tong, & Steitz, 2007). According to the miRBase database, there are 2588 known mature miRNAs annotated in the human genome (http://www.mirbase.org/index.shtml; Release 21; GRCh38), a number that is set to keep on rising. Interestingly, over 60% of human genes are believed to contain miRNA binding sites (Friedman, Farh, Burge, & Bartel, 2009).

To modulate post-transcriptional effects on mRNA, miRNAs are first biosynthesized in the nucleus before maturation in the cytoplasm (Ha & Kim, 2014). Briefly, miRNAs are transcribed by RNA polymerase II to form large precursor transcripts known as pre-miRNAs. Pre-miRNA is further processed in the nucleus by the Drosha and DGCR8/Pasha enzymes before being exported to the cytoplasm via the exportin 5 and Ran-GTP complex. Here, the RNase enzyme Dicer finalizes the conversion of double-stranded pre-miRNA to single-stranded miRNA approximately 22 nucleotides long. Some of the mature miRNAs, along with the Argonaute proteins, are incorporated into an RNA-induced silencing complex (RISC). RISC is able to bind to the 3′ untranslated regions of their complementary mRNA targets either to directly block translation or to initiate mRNA degradation (Fig. 1). Since different genes contain the same predicted miRNA recognition sites, each RISC can theoretically target and regulate multiple genes simultaneously.

ALTERATIONS TO miRNA BIOSYNTHESIS IN AGING AND DEPRESSION

Dysregulations for some of the key enzymes, such as Dicer, involved in miRNA biosynthesis have been implicated in inflammation (Hartmann et al., 2016), aging, and depression.

An age-related downregulation of *DICER1* expression, which encodes the Dicer enzyme, has been observed in human preadipocytes (Mori et al., 2012), PBMCs (Hooten et al., 2016) and rat small cerebral vessels (Ungvari et al., 2013). A similar decline of *Dicer* expression is also evident in the aged *Caenorhabditis elegans* worm (Mori et al., 2012). Conversely, restoring *DICER1* levels through life-extending interventions, such as metformin treatment or caloric restriction, is able to upregulate several miRNAs associated with senescence and aging (Hooten et al., 2016).

A connection between Dicer function has also been found in those with posttraumatic stress disorder with comorbid depression (PTSD&Dep). Specifically, blood *DICER1* expression was significantly lower in those affected, compared with controls, which was replicable in two further independent cohorts (Wingo et al., 2015). Further, a genetic variant of the *DICER1* gene (rs10144436), located within the 3′ untranslated region, was significantly associated with *DICR1* expression and PTSD&Dep (Wingo et al., 2015).

The reduction of Dicer activity suggests that the enzymes responsible for miRNA biogenesis are dysregulated by aging and depression. Specifically, in the context of Dicer, this may reflect on RISC generation and alterations to the miRNome.

EFFECTS OF AGING AND DEPRESSION ON THE miRNome

Insights into the aging miRNA network have only recently been explored. Hackl et al. (2010) shown that between 10% and 20% of the miRNAs

in their investigation were significantly regulated, many of which were downregulated, in a variety of replicative and organismal aging models. Extensive insights into the relationship between PBMC miRNA profiles and aging have been demonstrated by Hooten et al. (2010), including an analysis of over 800 miRNAs. In a separate analysis involving only females, 34 differentially expressed miRNAs were found between an age group gap of just 10 years (Sredni, Gadd, Jafari, & Huang, 2011). Further investigations have also highlighted significant age-associated changes to miRNA expression profiles even in noncellular plasma and serum samples (Ameling et al., 2015; Hooten et al., 2013).

Peripheral alterations of the miRNome are also evident in those with MDD. For example, through the analysis of 1733 mature miRNAs in whole blood, 5 miRNAs were specifically altered in MDD, compared with healthy controls (Maffioletti et al., 2016). Interestingly, a separate study also identified dysregulations of numerous miRNAs, many of which were significantly reduced, in the prefrontal cortex of those with depression, compared with nonpsychiatric controls (Smalheiser et al., 2012). These studies highlight the dynamic alteration of miRNAs in both blood and the brain of those with depression.

There are clear genome-wide alterations in aged and depressed subjects, compared with young and nondepressed controls, respectively. The consequence of such changes is only just becoming apparent. Even in its infancy, miRNA explorations are providing compelling evidence for the control over inflammatory genes, such as cytokines. Specifically, miR-155 and miR-181a will be discussed in more detail.

EFFECTS OF AGING AND DEPRESSION ON miR-155 REGULATION

MiR-155 is a pleiotropic miRNA possessing both activating and inhibiting properties (Olivieri, Rippo, Procopio, & Fazioli, 2013;

Vigorito, Kohlhaas, Lu, & Leyland, 2013). Due to the strong association with immune-related genes, including predicted mRNA targets of TNF-α, NF-κB, and TLRs, it is also regarded as so-called inflamma-miR (Olivieri et al., 2013; Vigorito et al., 2013). *MiR-155* is substantially upregulated by a variety of TLR ligands through the MyD88- and TRIF-dependent pathways (O'Connell, Taganov, Boldin, Cheng, & Baltimore, 2007; Taganov, Boldin, Chang, & Baltimore, 2006) and by IFN-γ via TNF-α autocrine/paracrine signaling (O'Connell et al., 2007). Additionally, *miR-155* is upregulated in the synovial membrane and leads to the exacerbated release of pro-inflammatory cytokines from peripheral blood monocytes in clinical and experimental rheumatoid arthritis (Kurowska-Stolarska et al., 2011). Collectively, in the context of inflammation, available evidence suggests that miR-155 is a potent activator, rather than repressor, of the inflammatory network.

Insights into the effects of aging on miR-155 regulation are relatively unknown. However, one study based on peripheral blood analysis did identify *miR-155* as one of the most upregulated miRNAs between young and older adult women (Sredni et al., 2011). This initial result suggests a link between miR-155 and inflammaging.

Consistent with this evidence, miR-155 dysregulation has also been implicated in depressive states. For example, transgenic mice with elevated expressions of miR-155, via neural and hematopoietic stem cells, have increased numbers of activated microglia and neurogenic deficits (Woodbury et al., 2015). Further, miR-155 knockout mice present reduced anxiety- and depressive-like behaviors and also have reduced expression of inflammatory genes (*Tnf-α* and *Il-6*) in their hippocampi, when compared with wild types (Fonken, Gaudet, Gaier, Nelson, & Popovich, 2016). In humans, *miR-155* gene expression is upregulated in the cerebrospinal fluid of those with MDD compared with controls (Wan et al., 2015). However, *miR-155* expression was low in the prefrontal cortex of depressed suicide victims compared with nonpsychiatric controls (Smalheiser et al., 2012).

EFFECTS OF AGING AND DEPRESSION ON miR-181a REGULATION

MiR-181a has numerous predicted inflammatory targets including IL-1α, TLR4, and TNF-α (Xie et al., 2013), thus suggesting some control over the pro-inflammatory phenotype. In monocyte and macrophage cell models, miR-181a inhibited IL-1α, IL-1β, IL-6, and TNF-α levels (Xie et al., 2013). Furthermore, miR-181a has been shown to reduce NF-κB signaling activity in B-cell lymphoma pathogenesis (Kozloski et al., 2016). Therefore, unlike miR-155, evidence suggests that miR-181a acts as a repressor of certain inflammatory pathways.

Reports of miR-181a in aging have so far suggested a reduction in activity. Serum *miR-181a* expression is significantly reduced in both older adult humans and aged rhesus monkeys, compared with their younger counterparts (Hooten et al., 2013). Complementing this observation, a separate investigation found a threefold reduction of *miR-181a* expression in CD4$^+$ T cells in the older adult group (90–85 years old) compared with the younger controls (20–35 years old) (Li et al., 2012). Consequently, the age-associated loss of miR-181a contributes to an increased *DUSP6* expression, a cytoplasmic phosphatase responsible for ERK inactivation and ultimately T-cell receptor activation, whereas restoring miR-181a has been shown to improve CD4$^+$ T-cell responses (Li et al., 2012).

In the context of depression, one study did find a 50% reduction in *miR-181a* expression within the prefrontal cortex of those with either bipolar, depression, or schizophrenia, compared with those that died from other factors (Smalheiser et al., 2014). Disease-specific alterations to *miR-181a* activity have also been observed in fibroblasts from patients with MDD following metabolic stressors (Garbett et al., 2015).

CONCLUSIONS

This chapter has introduced the topic of inflammaging that denotes the chronic inflammatory state frequently observed in later life. Indeed, inflammaging shares many mechanistic commonalities with the inflammatory pathophysiology of depression. Given this, epigenetic mechanisms, specifically DNA methylation, histone acetylation, and miRNA mechanisms, have been discussed in greater detail with relation to genome-wide and immune-specific alterations during aging and depression. These include the loss of DNA methylation at functionally significant regions associated with pro-inflammatory cytokine genes, such as *IFN-γ* and *TNF-α*; the reduced capacity of HDAC function, as seen in SIRT1, to control histone acetylation around immune-related genes; and the disruptions to miRNAs that have inflammatory repressive and promoting properties. These are just some of the many underlying mechanisms that are forming a complex network dysregulated in aging and depression.

The majority of research on depression to date has so far concentrated on younger adults, mainly due to the higher prevalence among the young (Kessler et al., 2005), with little applied to the older population. Based on the aforementioned molecular disruptions, it could be speculative that older adults are more prone to the inflammatory insults associated with depression. Therefore, future work should elaborate on these epigenetic components regulating the immune system in older adults with and without depression. Given the rapid advances in economical genome-wide technologies and bioinformatic capabilities, such investigations will elucidate novel therapeutic targets. Additionally, the use of epigenetic modifying agents, both pharmacological and nonpharmacological, as therapeutic agents for specifically reducing age- and depression-associated inflammation should be explored. For example, the use of HDAC inhibitors for the treatment of depression has produced promising benefits to the behavioral aspects of the disorder (Fuchikami et al., 2016); however, very little is known about their benefits to the immune system in depressive states. Increasing evidence suggests that certain HDAC inhibitors have potent anti-inflammatory properties (Hull, Montgomery, & Leyva, 2016). It would be interesting to apply these as a means of preventing or even reversing the underlying inflammation.

In sum, the epigenome corroborates with the underlying genetic code to infer GxE interactions that shape the immune function. Disruptions to these regulatory mechanisms as seen in aging and depression can ultimately augment inflammation both systemically and within the CNS.

References

Abe, N., Uchida, S., Otsuki, K., Hobara, T., Yamagata, H., Higuchi, F., et al. (2011). Altered sirtuin deacetylase gene expression in patients with a mood disorder. *Journal of Psychiatric Research*, 45(8), 1106–1112. https://doi.org/10.1016/j.jpsychires.2011.01.016.

Abe-Higuchi, N., Uchida, S., Yamagata, H., Higuchi, F., Hobara, T., Hara, K., et al. (2016). Hippocampal sirtuin 1 signaling mediates depression-like behavior. *Biological Psychiatry*. https://doi.org/10.1016/j.biopsych.2016.01.009.

Ameling, S., Kacprowski, T., Chilukoti, R. K., Malsch, C., Liebscher, V., Suhre, K., et al. (2015). Associations of circulating plasma microRNAs with age, body mass index and sex in a population-based study. *BMC Medical Genomics*, 8, 61. https://doi.org/10.1186/s12920-015-0136-7.

Araújo, J. R., Martel, F., Borges, N., Araújo, J. M., & Keating, E. (2015). Folates and aging: role in mild cognitive impairment, dementia and depression. *Ageing Research Reviews*, 22, 9–19. https://doi.org/10.1016/j.arr.2015.04.005.

Bandrés, E., Merino, J., Vázquez, B., Inogés, S., Moreno, C., Subirá, M. L., et al. (2000). The increase of IFN-γ production through aging correlates with the expanded

CD8+highCD28−CD57+ subpopulation. *Clinical Immunology, 96*(3), 230–235. https://doi.org/10.1006/clim.2000.4894.

Bannister, A. J., & Kouzarides, T. (2011). Regulation of chromatin by histone modifications. *Cell Research, 21*(3), 381–395. https://doi.org/10.1038/cr.2011.22.

Baylis, D., Bartlett, D. B., Patel, H. P., & Roberts, H. C. (2013). Understanding how we age: insights into inflammaging. *Longevity & Healthspan, 2*(1), 8. https://doi.org/10.1186/2046-2395-2-8.

Berger, S. L., Kouzarides, T., Shiekhattar, R., & Shilatifard, A. (2009). An operational definition of epigenetics. *Genes & Development, 23*(7), 781–783. https://doi.org/10.1101/gad.1787609.

Bonkowski, M. S., & Sinclair, D. A. (2016). Slowing ageing by design: the rise of NAD+ and sirtuin-activating compounds. *Nature Reviews Molecular Cell Biology, 17*(11), 679–690. https://doi.org/10.1038/nrm.2016.93.

Boyes, J., & Bird, A. (1991). DNA methylation inhibits transcription indirectly via a methyl-CpG binding protein. *Cell, 64*(6), 1123–1134.

Brodin, P., Jojic, V., Gao, T., Bhattacharya, S., Angel, C. J. L., Furman, D., et al. (2015). Variation in the human immune system is largely driven by non-heritable influences. *Cell, 160*, 37–47. https://doi.org/10.1016/j.cell.2014.12.020.

Bruunsgaard, H., Andersen-Ranberg, K., Jeune, B., Pedersen, A. N., Skinhøj, P., & Pedersen, B. K. (1999). A high plasma concentration of TNF-alpha is associated with dementia in centenarians. *Journals of Gerontology. Series A, Biological Sciences and Medical Sciences, 54*(7), M357–364.

Byrne, E. M., Carrillo-Roa, T., Henders, A. K., Bowdler, L., McRae, A. F., Heath, A. C., et al. (2013). Monozygotic twins affected with major depressive disorder have greater variance in methylation than their unaffected co-twin. *Translational Psychiatry, 3*. https://doi.org/10.1038/tp.2013.45.

Chaleckis, R., Murakami, I., Takada, J., Kondoh, H., & Yanagida, M. (2016). Individual variability in human blood metabolites identifies age-related differences. *Proceedings of the National Academy of Sciences, 113*(16), 4252–4259. https://doi.org/10.1073/pnas.1603023113.

Cho, S.-H., Chen, J. A., Sayed, F., Ward, M. E., Gao, F., Nguyen, T. A., et al. (2015). SIRT1 deficiency in microglia contributes to cognitive decline in aging and neurodegeneration via epigenetic regulation of IL-1β. *Journal of Neuroscience: The Official Journal of the Society for Neuroscience, 35*(2), 807–818. https://doi.org/10.1523/JNEUROSCI.2939-14.2015.

Cianciulli, A., Salvatore, R., Porro, C., Trotta, T., & Panaro, M. A. (2016). Folic acid is able to polarize the inflammatory response in LPS activated microglia by regulating multiple signaling pathways. *Mediators of Inflammation, 2016*. https://doi.org/10.1155/2016/5240127.

Ciccarone, F., Malavolta, M., Calabrese, R., Guastafierro, T., Bacalini, M. G., Reale, A., et al. (2016). Age-dependent expression of DNMT1 and DNMT3B in PBMCs from a large European population enrolled in the MARK-AGE study. *Aging Cell, 15*(4), 755–765. https://doi.org/10.1111/acel.12485.

Clarke, R., Grimley Evans, J., Schneede, J., Nexo, E., Bates, C., Fletcher, A., et al. (2004). Vitamin B12 and folate deficiency in later life. *Age and Ageing, 33*(1), 34–41. https://doi.org/10.1093/ageing/afg109.

Converge Consortium. (2015). Sparse whole-genome sequencing identifies two loci for major depressive disorder. *Nature, 523*(7562), 588–591. https://doi.org/10.1038/nature14659.

Córdova-Palomera, A., Fatjó-Vilas, M., Gastó, C., Navarro, V., Krebs, M.-O., & Fañanás, L. (2015). Genome-wide methylation study on depression: differential methylation and variable methylation in monozygotic twins. *Translational Psychiatry, 5*(4). https://doi.org/10.1038/tp.2015.49.

Crider, K. S., Yang, T. P., Berry, R. J., & Bailey, L. B. (2012). Folate and DNA methylation: a review of molecular mechanisms and the evidence for folate's role. *Advances in nutrition (Bethesda, Md.), 3*(1), 21–38. https://doi.org/10.3945/an.111.000992.

Davies, M. N., Krause, L., Bell, J. T., Gao, F., Ward, K. J., Wu, H., et al. (2014). Hypermethylation in the ZBTB20 gene is associated with major depressive disorder. *Genome Biology, 15*(4), R56. https://doi.org/10.1186/gb-2014-15-4-r56.

Dempster, E. L., Wong, C. C. Y., Lester, K. J., Burrage, J., Gregory, A. M., Mill, J., et al. (2014). Genome-wide methylomic analysis of monozygotic twins discordant for adolescent depression. *Biological Psychiatry, 76*(12), 977–983. https://doi.org/10.1016/j.biopsych.2014.04.013.

Di Bona, D., Vasto, S., Capurso, C., Christiansen, L., Deiana, L., Franceschi, C., et al. (2009). Effect of interleukin-6 polymorphisms on human longevity: a systematic review and meta-analysis. *Ageing Research Reviews, 8*(1), 36–42. https://doi.org/10.1016/j.arr.2008.09.001.

Dowlati, Y., Herrmann, N., Swardfager, W., Liu, H., Sham, L., Reim, E. K., et al. (2010). A meta-analysis of cytokines in major depression. *Biological Psychiatry, 67*(5), 446–457. https://doi.org/10.1016/j.biopsych.2009.09.033.

Ferrucci, L., Corsi, A., Lauretani, F., Bandinelli, S., Bartali, B., Taub, D. D., et al. (2005). The origins of age-related proinflammatory state. *Blood, 105*(6), 2294–2299. https://doi.org/10.1182/blood-2004-07-2599.

Fishman, D., Faulds, G., Jeffery, R., Mohamed-Ali, V., Yudkin, J. S., Humphries, S., et al. (1998). The effect of novel polymorphisms in the interleukin-6 (IL-6) gene on IL-6 transcription and plasma IL-6 levels, and an association with systemic-onset juvenile chronic arthritis. *Journal of Clinical Investigation*, 102(7), 1369–1376.

Fonken, L. K., Gaudet, A. D., Gaier, K. R., Nelson, R. J., & Popovich, P. G. (2016). MicroRNA-155 deletion reduces anxiety- and depressive-like behaviors in mice. *Psychoneuroendocrinology*, 63, 362–369. https://doi.org/10.1016/j.psyneuen.2015.10.019.

Fraga, M. F., Ballestar, E., Paz, M. F., Ropero, S., Setien, F., Ballestar, M. L., et al. (2005). Epigenetic differences arise during the lifetime of monozygotic twins. *Proceedings of the National Academy of Sciences of the United States of America*, 102(30), 10604–10609. https://doi.org/10.1073/pnas.0500398102.

Franceschi, C., Bonafè, M., Valensin, S., Olivieri, F., De Luca, M., Ottaviani, E., et al. (2000). Inflamm-aging. An evolutionary perspective on immunosenescence. *Annals of the New York Academy of Sciences*, 908, 244–254.

Friedman, R. C., Farh, K. K.-H., Burge, C. B., & Bartel, D. P. (2009). Most mammalian mRNAs are conserved targets of microRNAs. *Genome Research*, 19(1), 92–105. https://doi.org/10.1101/gr.082701.108.

Fuchikami, M., Yamamoto, S., Morinobu, S., Okada, S., Yamawaki, Y., & Yamawaki, S. (2016). The potential use of histone deacetylase inhibitors in the treatment of depression. *Progress in Neuro-Psychopharmacology & Biological Psychiatry*, 64, 320–324. https://doi.org/10.1016/j.pnpbp.2015.03.010.

Furuta, T., Shuto, T., Shimasaki, S., Ohira, Y., Suico, M. A., Gruenert, D. C., et al. (2008). DNA demethylation-dependent enhancement of toll-like receptor-2 gene expression in cystic fibrosis epithelial cells involves SP1-activated transcription. *BMC Molecular Biology*, 9, 39. https://doi.org/10.1186/1471-2199-9-39.

Garbett, K. A., Vereczkei, A., Kálmán, S., Wang, L., Korade, Ž., Shelton, R. C., et al. (2015). Fibroblasts from patients with major depressive disorder show distinct transcriptional response to metabolic stressors. *Translational Psychiatry*, 5(3). https://doi.org/10.1038/tp.2015.14.

Gilbody, S., Lightfoot, T., & Sheldon, T. (2007). Is low folate a risk factor for depression? A meta-analysis and exploration of heterogeneity. *Journal of Epidemiology and Community Health*, 61(7), 631–637. https://doi.org/10.1136/jech.2006.050385.

Gong, H., Pang, J., Han, Y., Dai, Y., Dai, D., Cai, J., et al. (2014). Age-dependent tissue expression patterns of Sirt1 in senescence-accelerated mice. *Molecular Medicine Reports*, 10(6), 3296–3302. https://doi.org/10.3892/mmr.2014.2648.

Gowers, I. R., Walters, K., Kiss-Toth, E., Read, R. C., Duff, G. W., & Wilson, A. G. (2011). Age-related loss of CpG methylation in the tumour necrosis factor promoter. *Cytokine*, 56(3), 792–797. https://doi.org/10.1016/j.cyto.2011.09.009.

Ha, M., & Kim, V. N. (2014). Regulation of microRNA biogenesis. *Nature Reviews Molecular Cell Biology*, 15(8), 509–524. https://doi.org/10.1038/nrm3838.

Hackl, M., Brunner, S., Fortschegger, K., Schreiner, C., Micutkova, L., Mück, C., et al. (2010). miR-17, miR-19b, miR-20a, and miR-106a are down-regulated in human aging. *Aging Cell*, 9(2), 291–296. https://doi.org/10.1111/j.1474-9726.2010.00549.x.

Hartmann, P., Zhou, Z., Natarelli, L., Wei, Y., Nazari-Jahantigh, M., Zhu, M., et al. (2016). Endothelial Dicer promotes atherosclerosis and vascular inflammation by miRNA-103-mediated suppression of KLF4. *Nature Communications*, 7, 10521. https://doi.org/10.1038/ncomms10521.

Hayakawa, T., Iwai, M., Aoki, S., Takimoto, K., Maruyama, M., Maruyama, W., et al. (2015). SIRT1 suppresses the senescence-associated secretory phenotype through epigenetic gene regulation. *PLoS One*, 10(1). https://doi.org/10.1371/journal.pone.0116480.

Heyn, H., Li, N., Ferreira, H. J., Moran, S., Pisano, D. G., Gomez, A., et al. (2012). Distinct DNA methylomes of newborns and centenarians. *Proceedings of the National Academy of Sciences*, 109(26), 10522–10527. https://doi.org/10.1073/pnas.1120658109.

Higuchi, F., Uchida, S., Yamagata, H., Otsuki, K., Hobara, T., Abe, N., et al. (2011). State-dependent changes in the expression of DNA methyltransferases in mood disorder patients. *Journal of Psychiatric Research*, 45(10), 1295–1300. https://doi.org/10.1016/j.jpsychires.2011.04.008.

Hollis, F., Duclot, F., Gunjan, A., & Kabbaj, M. (2011). Individual differences in the effect of social defeat on anhedonia and histone acetylation in the rat hippocampus. *Hormones and Behavior*, 59(3), 331–337. https://doi.org/10.1016/j.yhbeh.2010.09.005.

Hooten, N. N., Abdelmohsen, K., Gorospe, M., Ejiogu, N., Zonderman, A. B., & Evans, M. K. (2010). microRNA expression patterns reveal differential expression of target genes with age. *PLoS One*, 5(5). https://doi.org/10.1371/journal.pone.0010724.

Hooten, N. N., Fitzpatrick, M., Wood, W. H., De, S., Ejiogu, N., Zhang, Y., et al. (2013). Age-related changes in microRNA levels in serum. *Aging (Albany, NY)*, 5(10), 725–740.

Hooten, N., Martin-Montalvo, A., Dluzen, D. F., Zhang, Y., Bernier, M., Zonderman, A. B., et al. (2016). Metformin-mediated increase in DICER1 regulates microRNA expression and cellular senescence. *Aging Cell*, 15(3), 572–581. https://doi.org/10.1111/acel.12469.

Hull, E. E., Montgomery, M. R., & Leyva, K. J. (2016). HDAC inhibitors as epigenetic regulators of the immune system: impacts on cancer therapy and inflammatory diseases. *BioMed Research International*, 2016. https://doi.org/10.1155/2016/8797206.

Johansson, A., Enroth, S., & Gyllensten, U. (2013). Continuous aging of the human DNA methylome throughout the human lifespan. *PLoS One, 8*(6). https://doi.org/10.1371/journal.pone.0067378.

Kalliolias, G. D., & Ivashkiv, L. B. (2016). TNF biology, pathogenic mechanisms and emerging therapeutic strategies. *Nature Reviews Rheumatology, 12*(1), 49–62. https://doi.org/10.1038/nrrheum.2015.169.

Kaminsky, Z. A., Tang, T., Wang, S.-C., Ptak, C., Oh, G. H. T., Wong, A. H. C., et al. (2009). DNA methylation profiles in monozygotic and dizygotic twins. *Nature Genetics, 41*(2), 240–245. https://doi.org/10.1038/ng.286.

Kawasaki, T., & Kawai, T. (2014). Toll-like receptor signaling pathways. *Frontiers in Immunology, 5*. https://doi.org/10.3389/fimmu.2014.00461.

Kersh, E. N., Fitzpatrick, D. R., Murali-Krishna, K., Shires, J., Speck, S. H., Boss, J. M., et al. (2006). Rapid demethylation of the IFN-γ gene occurs in memory but not naive CD8 T cells. *Journal of Immunology, 176*(7), 4083–4093. https://doi.org/10.4049/jimmunol.176.7.4083.

Kessler, R. C., Berglund, P., Demler, O., Jin, R., Merikangas, K. R., & Walters, E. E. (2005). Lifetime prevalence and age-of-onset distributions of DSM-IV disorders in the National Comorbidity Survey Replication. *Archives of General Psychiatry, 62*(6), 593–602. https://doi.org/10.1001/archpsyc.62.6.593.

Kilic, U., Gok, O., Erenberk, U., Dundaroz, M. R., Torun, E., Kucukardali, Y., et al. (2015). A remarkable age-related increase in SIRT1 protein expression against oxidative stress in elderly: SIRT1 gene variants and longevity in human. *PLoS One, 10*(3). https://doi.org/10.1371/journal.pone.0117954.

Kishi, T., Yoshimura, R., Kitajima, T., Okochi, T., Okumura, T., Tsunoka, T., et al. (2010). SIRT1 gene is associated with major depressive disorder in the Japanese population. *Journal of Affective Disorders, 126*(1–2), 167–173. https://doi.org/10.1016/j.jad.2010.04.003.

Kolb, A. F., & Petrie, L. (2013). Folate deficiency enhances the inflammatory response of macrophages. *Molecular Immunology, 54*(2), 164–172. https://doi.org/10.1016/j.molimm.2012.11.012.

Kozloski, G. A., Jiang, X., Bhatt, S., Ruiz, J., Vega, F., Shaknovich, R., et al. (2016). miR-181a negatively regulates NF-κB signaling and affects activated B-cell-like diffuse large B-cell lymphoma pathogenesis. *Blood, 127*(23), 2856–2866. https://doi.org/10.1182/blood-2015-11-680462.

Kurowska-Stolarska, M., Alivernini, S., Ballantine, L. E., Asquith, D. L., Millar, N. L., Gilchrist, D. S., et al. (2011). MicroRNA-155 as a proinflammatory regulator in clinical and experimental arthritis. *Proceedings of the National Academy of Sciences of the United States of America, 108*(27), 11193–11198. https://doi.org/10.1073/pnas.1019536108.

Kwon, Y., Kim, J., Lee, C.-Y., & Kim, H. (2015). Expression of SIRT1 and SIRT3 varies according to age in mice. *Anatomy and Cell Biology, 48*(1), 54–61. https://doi.org/10.5115/acb.2015.48.1.54.

Lee, P. P., Fitzpatrick, D. R., Beard, C., Jessup, H. K., Lehar, S., Makar, K. W., et al. (2001). A critical role for Dnmt1 and DNA methylation in T cell development, function, and survival. *Immunity, 15*(5), 763–774.

Li, G., Yu, M., Lee, W.-W., Tsang, M., Krishnan, E., Weyand, C. M., et al. (2012). Decline in miR-181a expression with age impairs T cell receptor sensitivity by increasing DUSP6 activity. *Nature Medicine, 18*(10), 1518–1524. https://doi.org/10.1038/nm.2963.

Lin, S.-J., Defossez, P.-A., & Guarente, L. (2000). Requirement of NAD and SIR2 for life-span extension by calorie restriction in Saccharomyces cerevisiae. *Science, 289*(5487), 2126–2128. https://doi.org/10.1126/science.289.5487.2126.

Liu, T. F., Yoza, B. K., Gazzar, M. E., Vachharajani, V. T., & McCall, C. E. (2011). NAD+-dependent SIRT1 deacetylase participates in epigenetic reprogramming during endotoxin tolerance. *Journal of Biological Chemistry, 286*(11), 9856–9864. https://doi.org/10.1074/jbc.M110.196790.

Madrigano, J., Baccarelli, A., Mittleman, M. A., Sparrow, D., Vokonas, P. S., Tarantini, L., et al. (2012). Aging and epigenetics: longitudinal changes in gene-specific DNA methylation. *Epigenetics, 7*(1), 63–70. https://doi.org/10.4161/epi.7.1.18749.

Maes, M., Scharpé, S., Meltzer, H. Y., Okayli, G., Bosmans, E., D'Hondt, P., et al. (1994). Increased neopterin and interferon-gamma secretion and lower availability of L-tryptophan in major depression: further evidence for an immune response. *Psychiatry Research, 54*(2), 143–160.

Maffioletti, E., Cattaneo, A., Rosso, G., Maina, G., Maj, C., Gennarelli, M., et al. (2016). Peripheral whole blood microRNA alterations in major depression and bipolar disorder. *Journal of Affective Disorders, 200*, 250–258. https://doi.org/10.1016/j.jad.2016.04.021.

Mariani, E., Cattini, L., Neri, S., Malavolta, M., Mocchegiani, E., Ravaglia, G., et al. (2006). Simultaneous evaluation of circulating chemokine and cytokine profiles in elderly subjects by multiplex technology: relationship with zinc status. *Biogerontology, 7*(5–6), 449–459. https://doi.org/10.1007/s10522-006-9060-8.

Marienfeld, R., May, M. J., Berberich, I., Serfling, E., Ghosh, S., & Neumann, M. (2003). RelB forms transcriptionally inactive complexes with RelA/p65. *Journal of Biological Chemistry, 278*(22), 19852–19860. https://doi.org/10.1074/jbc.M301945200.

Miles, E. A., Rees, D., Banerjee, T., Cazzola, R., Lewis, S., Wood, R., et al. (2008). Age-related increases in circulating inflammatory markers in men are independent of BMI, blood pressure and blood lipid concentrations. *Atherosclerosis*, 196(1), 298–305. https://doi.org/10.1016/j.atherosclerosis.2006.11.002.

Mills, N. T., Scott, J. G., Wray, N. R., Cohen-Woods, S., & Baune, B. T. (2013). Research review: the role of cytokines in depression in adolescents: a systematic review. *Journal of Child Psychology and Psychiatry*, 54(8), 816–835. https://doi.org/10.1111/jcpp.12080.

Minciullo, P. L., Catalano, A., Mandraffino, G., Casciaro, M., Crucitti, A., Maltese, G., et al. (2016). Inflammaging and anti-inflammaging: the role of cytokines in extreme longevity. *Archivum Immunologiae et Therapiae Experimentalis*, 64(2), 111–126. https://doi.org/10.1007/s00005-015-0377-3.

Mohr, D. C., Goodkin, D. E., Islar, J., Hauser, S. L., & Genain, C. P. (2001). Treatment of depression is associated with suppression of nonspecific and antigen-specific TH1 responses in multiple sclerosis. *Archives of Neurology*, 58(7), 1081–1086. https://doi.org/10.1001/archneur.58.7.1081.

Mori, M. A., Raghavan, P., Thomou, T., Boucher, J., Robida-Stubbs, S., Macotela, Y., et al. (2012). Role of microRNA processing in adipose tissue in stress defense and longevity. *Cell Metabolism*, 16(3), 336–347. https://doi.org/10.1016/j.cmet.2012.07.017.

Nakagawa, T., & Guarente, L. (2014). SnapShot: sirtuins, NAD, and aging. *Cell Metabolism*, 20(1), 192–192.e1. https://doi.org/10.1016/j.cmet.2014.06.001.

Nemoda, Z., Massart, R., Suderman, M., Hallett, M., Li, T., Coote, M., et al. (2015). Maternal depression is associated with DNA methylation changes in cord blood T lymphocytes and adult hippocampi. *Translational Psychiatry*, 5(4) https://doi.org/10.1038/tp.2015.32.

O'Connell, R. M., Taganov, K. D., Boldin, M. P., Cheng, G., & Baltimore, D. (2007). MicroRNA-155 is induced during the macrophage inflammatory response. *Proceedings of the National Academy of Sciences of the United States of America*, 104(5), 1604–1609. https://doi.org/10.1073/pnas.0610731104.

Olivieri, F., Rippo, M. R., Procopio, A. D., & Fazioli, F. (2013). Circulating inflamma-miRs in aging and age-related diseases. *Frontiers in Genetics*, 4. https://doi.org/10.3389/fgene.2013.00121.

Pandey, G. N., Rizavi, H. S., Ren, X., Bhaumik, R., & Dwivedi, Y. (2014). Toll-like receptors in the depressed and suicide brain. *Journal of Psychiatric Research*, 53, 62–68. https://doi.org/10.1016/j.jpsychires.2014.01.021.

Paolisso, G., Rizzo, M. R., Mazziotti, G., Tagliamonte, M. R., Gambardella, A., Rotondi, M., et al. (1998). Advancing age and insulin resistance: role of plasma tumor necrosis factor-alpha. *American Journal of Physiology*, 275(2 Pt 1), E294–299.

Pawelec, G., Larbi, A., & Derhovanessian, E. (2010). Senescence of the human immune system. *Journal of Comparative Pathology*, 142(Supplement 1), S39–S44. https://doi.org/10.1016/j.jcpa.2009.09.005.

Peng, L., Yuan, Z., Ling, H., Fukasawa, K., Robertson, K., Olashaw, N., et al. (2011). SIRT1 deacetylates the DNA methyltransferase 1 (DNMT1) protein and alters its activities. *Molecular and Cellular Biology*, 31(23), 4720–4734. https://doi.org/10.1128/MCB.06147-11.

Phillips, A. C., Carroll, D., Khan, N., & Moss, P. (2008). Cytomegalovirus is associated with depression and anxiety in older adults. *Brain, Behavior, and Immunity*, 22(1), 52–55. https://doi.org/10.1016/j.bbi.2007.06.012.

Pieper, H. C., Evert, B. O., Kaut, O., Riederer, P. F., Waha, A., & Wüllner, U. (2008). Different methylation of the TNF-alpha promoter in cortex and substantia nigra: implications for selective neuronal vulnerability. *Neurobiology of Disease*, 32(3), 521–527. https://doi.org/10.1016/j.nbd.2008.09.010.

Postal, M., Lapa, A. T., Sinicato, N. A., de Oliveira Peliçari, K., Peres, F. A., Costallat, L. T. L., et al. (2016). Depressive symptoms are associated with tumor necrosis factor alpha in systemic lupus erythematosus. *Journal of Neuroinflammation*, 13, 5. https://doi.org/10.1186/s12974-015-0471-9.

Poulter, M. O., Du, L., Weaver, I. C. G., Palkovits, M., Faludi, G., Merali, Z., et al. (2008). GABAA receptor promoter hypermethylation in suicide brain: implications for the involvement of epigenetic processes. *Biological Psychiatry*, 64(8), 645–652. https://doi.org/10.1016/j.biopsych.2008.05.028.

Quintas, A., de Solís, A. J., Díez-Guerra, F. J., Carrascosa, J. M., & Bogónez, E. (2012). Age-associated decrease of SIRT1 expression in rat hippocampus: prevention by late onset caloric restriction. *Experimental Gerontology*, 47(2), 198–201. https://doi.org/10.1016/j.exger.2011.11.010.

Ryan, J. M., & Cristofalo, V. J. (1972). Histone acetylation during aging of human cells in culture. *Biochemical and Biophysical Research Communications*, 48(4), 735–742. https://doi.org/10.1016/0006-291X(72)90668-7.

Schroder, K., Hertzog, P. J., Ravasi, T., & Hume, D. A. (2004). Interferon-γ: an overview of signals, mechanisms and functions. *Journal of Leukocyte Biology*, 75(2), 163–189. https://doi.org/10.1189/jlb.0603252.

Schug, T. T., Xu, Q., Gao, H., Peres-da-Silva, A., Draper, D. W., Fessler, M. B., et al. (2010). Myeloid deletion of SIRT1 induces inflammatory signaling in response to environmental stress. *Molecular and Cellular Biology*, 30(19), 4712–4721. https://doi.org/10.1128/MCB.00657-10.

Shelton, R. C., Claiborne, J., Sidoryk-Wegrzynowicz, M., Reddy, R., Aschner, M., Lewis, D. A., et al. (2011). Altered expression of genes involved in inflammation and apoptosis in frontal cortex in major depression. *Molecular Psychiatry*, 16(7), 751–762. https://doi.org/10.1038/mp.2010.52.

Shuto, T., Furuta, T., Oba, M., Xu, H., Li, J.-D., Cheung, J., et al. (2006). Promoter hypomethylation of Toll-like receptor-2 gene is associated with increased proinflammatory response toward bacterial peptidoglycan in cystic fibrosis bronchial epithelial cells. *FASEB Journal: Official Publication of the Federation of American Societies for Experimental Biology*, 20(6), 782–784. https://doi.org/10.1096/fj.05-4934fje.

Smalheiser, N. R., Lugli, G., Rizavi, H. S., Torvik, V. I., Turecki, G., & Dwivedi, Y. (2012). MicroRNA expression is down-regulated and reorganized in prefrontal cortex of depressed suicide subjects. *PLoS One*, 7(3). https://doi.org/10.1371/journal.pone.0033201.

Smalheiser, N. R., Lugli, G., Zhang, H., Rizavi, H., Cook, E. H., & Dwivedi, Y. (2014). Expression of microRNAs and other small RNAs in prefrontal cortex in schizophrenia, bipolar disorder and depressed subjects. *PLoS One*, 9(1). https://doi.org/10.1371/journal.pone.0086469.

Smith, Z. D., & Meissner, A. (2013). DNA methylation: roles in mammalian development. *Nature Reviews Genetics*, 14(3), 204–220. https://doi.org/10.1038/nrg3354.

Sredni, S. T., Gadd, S., Jafari, N., & Huang, C.-C. (2011). A parallel study of mRNA and microRNA profiling of peripheral blood in young adult women. *Frontiers in Genetics*, 2. https://doi.org/10.3389/fgene.2011.00049.

Sullivan, P. F., Neale, M. C., & Kendler, K. S. (2000). Genetic epidemiology of major depression: review and meta-analysis. *American Journal of Psychiatry*, 157(10), 1552–1562. https://doi.org/10.1176/appi.ajp.157.10.1552.

Sullivan, K. E., Reddy, A. B. M., Dietzmann, K., Suriano, A. R., Kocieda, V. P., Stewart, M., et al. (2007). Epigenetic regulation of tumor necrosis factor alpha. *Molecular and Cellular Biology*, 27(14), 5147–5160. https://doi.org/10.1128/MCB.02429-06.

Sun, H., Kennedy, P. J., & Nestler, E. J. (2013). Epigenetics of the depressed brain: role of histone acetylation and methylation. *Neuropsychopharmacology*, 38(1), 124–137. https://doi.org/10.1038/npp.2012.73.

Taganov, K. D., Boldin, M. P., Chang, K.-J., & Baltimore, D. (2006). NF-kappaB-dependent induction of microRNA miR-146, an inhibitor targeted to signaling proteins of innate immune responses. *Proceedings of the National Academy of Sciences of the United States of America*, 103(33), 12481–12486. https://doi.org/10.1073/pnas.0605298103.

Takahashi, K., Sugi, Y., Hosono, A., & Kaminogawa, S. (2009). Epigenetic regulation of TLR4 gene expression in intestinal epithelial cells for the maintenance of

intestinal homeostasis. *Journal of Immunology*, 183(10), 6522–6529. https://doi.org/10.4049/jimmunol.0901271.

Tserel, L., Kolde, R., Limbach, M., Tretyakov, K., Kasela, S., Kisand, K., et al. (2015). Age-related profiling of DNA methylation in CD8+ T cells reveals changes in immune response and transcriptional regulator genes. *Scientific Reports*, 5. https://doi.org/10.1038/srep13107.

Uddin, M., Aiello, A. E., Wildman, D. E., Koenen, K. C., Pawelec, G., de los Santos, R., et al. (2010). Epigenetic and immune function profiles associated with posttraumatic stress disorder. *Proceedings of the National Academy of Sciences of the United States of America*, 107(20), 9470–9475. https://doi.org/10.1073/pnas.0910794107.

Ungvari, Z., Tucsek, Z., Sosnowska, D., Toth, P., Gautam, T., Podlutsky, A., et al. (2013). Aging-induced dysregulation of dicer1-dependent microRNA expression impairs angiogenic capacity of rat cerebromicrovascular endothelial cells. *Journals of Gerontology Series A: Biological Sciences and Medical Sciences*, 68(8), 877–891. https://doi.org/10.1093/gerona/gls242.

Vasudevan, S., Tong, Y., & Steitz, J. A. (2007). Switching from repression to activation: microRNAs can up-regulate translation. *Science*, 318(5858), 1931–1934.

Veal, N., Hsieh, C.-L., Xiong, S., Mato, J. M., Lu, S., & Tsukamoto, H. (2004). Inhibition of lipopolysaccharide-stimulated TNF-alpha promoter activity by S-adenosylmethionine and 5′-methylthioadenosine. *American Journal of Physiology. Gastrointestinal and Liver Physiology*, 287(2), G352–362. https://doi.org/10.1152/ajpgi.00316.2003.

Vescovini, R., Biasini, C., Fagnoni, F. F., Telera, A. R., Zanlari, L., Pedrazzoni, M., et al. (2007). Massive load of functional effector CD4+ and CD8+ T cells against cytomegalovirus in very old subjects. *Journal of Immunology*, 179(6), 4283–4291.

Vigorito, E., Kohlhaas, S., Lu, D., & Leyland, R. (2013). miR-155: an ancient regulator of the immune system. *Immunological Reviews*, 253(1), 146–157. https://doi.org/10.1111/imr.12057.

Waddington, C. H. (2012). The epigenotype. 1942. *International Journal of Epidemiology*, 41(1), 10–13. https://doi.org/10.1093/ije/dyr184.

Wan, Y., Liu, Y., Wang, X., Wu, J., Liu, K., Zhou, J., et al. (2015). Identification of differential microRNAs in cerebrospinal fluid and serum of patients with major depressive disorder. *PLoS One*, 10(3). https://doi.org/10.1371/journal.pone.0121975.

Wang, Z., Zang, C., Rosenfeld, J. A., Schones, D. E., Barski, A., Cuddapah, S., et al. (2008). Combinatorial patterns of histone acetylations and methylations in the human genome. *Nature Genetics*, 40(7), 897–903. https://doi.org/10.1038/ng.154.

Watson, W. H., Zhao, Y., & Chawla, R. K. (1999). S-adenosylmethionine attenuates the lipopolysaccharide-induced expression of the gene for tumour necrosis factor alpha. *Biochemical Journal*, *342*(Pt 1), 21–25.

Watt, F., & Molloy, P. L. (1988). Cytosine methylation prevents binding to DNA of a HeLa cell transcription factor required for optimal expression of the adenovirus major late promoter. *Genes & Development*, *2*(9), 1136–1143.

Wingo, A. P., Almli, L. M., Stevens, J. S., Klengel, T., Uddin, M., Li, Y., et al. (2015). DICER1 and microRNA regulation in post-traumatic stress disorder with comorbid depression. *Nature Communications*, *6*. https://doi.org/10.1038/ncomms10106.

Woodbury, M. E., Freilich, R. W., Cheng, C. J., Asai, H., Ikezu, S., Boucher, J. D., et al. (2015). miR-155 is essential for inflammation-induced hippocampal neurogenic dysfunction. *Journal of Neuroscience*, *35*(26), 9764–9781. https://doi.org/10.1523/JNEUROSCI.4790-14.2015.

Xie, W., Li, M., Xu, N., Lv, Q., Huang, N., He, J., et al. (2013). miR-181a regulates inflammation responses in monocytes and macrophages. *PLoS One*, *8*(3). https://doi.org/10.1371/journal.pone.0058639.

Yeung, F., Hoberg, J. E., Ramsey, C. S., Keller, M. D., Jones, D. R., Frye, R. A., et al. (2004). Modulation of NF-kappaB-dependent transcription and cell survival by the SIRT1 deacetylase. *EMBO Journal*, *23*(12), 2369–2380. https://doi.org/10.1038/sj.emboj.7600244.

Yoshizaki, T., Schenk, S., Imamura, T., Babendure, J. L., Sonoda, N., Bae, E. J., et al. (2010). SIRT1 inhibits inflammatory pathways in macrophages and modulates insulin sensitivity. *American Journal of Physiology. Endocrinology and Metabolism*, *298*(3), E419–428. https://doi.org/10.1152/ajpendo.00417.2009.

Yoza, B. K., Hu, J. Y.-Q., Cousart, S. L., Forrest, L. M., & McCall, C. E. (2006). Induction of RelB participates in endotoxin tolerance. *Journal of Immunology (Baltimore, Md: 1950)*, *177*(6), 4080–4085.

Yu, Q., Zhou, B., Zhang, Y., Nguyen, E. T., Du, J., Glosson, N. L., et al. (2012). DNA methyltransferase 3a limits the expression of interleukin-13 in T helper 2 cells and allergic airway inflammation. *Proceedings of the National Academy of Sciences of the United States of America*, *109*(2), 541–546. https://doi.org/10.1073/pnas.1103803109.

Role of Inflammation in Neuropsychiatric Comorbidity of Obesity: Experimental and Clinical Evidence

Célia Fourrier, Lucile Capuron, Nathalie Castanon

INRA, Bordeaux, France
University of Bordeaux, Bordeaux, France

INTRODUCTION

Mood disorders currently represent a global public health concern, and they are strongly associated with a high risk of morbidity and death (Ferrari et al., 2013). Their prevalence is continuously rising, concurrently with the increasing prevalence of other severe chronic medical conditions, in particular obesity and metabolic syndrome, which reciprocally present a high vulnerability for neuropsychiatric comorbidities (Francis & Stevenson, 2013; Luppino et al., 2010). Mounting evidence points to a tricky bidirectional relationship between those conditions and mood disorders. Depressed patients are more likely to gain weight and to develop obesity than healthy individuals (Luppino et al., 2010), including later in life (Carpenter, Hasin, Allison, & Faith, 2000), whereas neuropsychiatric symptoms associated with obesity have been shown to compromise weight control, treatment compliance, and the management of obesity-related metabolic complications (Brunault et al., 2012; Dixon,

Dixon, & O'Brien, 2003; Kinzl et al., 2006). To worsen the picture, both obesity and mood disorders share the ability of promoting the development of other severe chronic diseases, such as type 2 diabetes and cardiovascular diseases (Amare, Schubert, Klingler-Hoffmann, Cohen-Woods, & Baune, 2017; Despres, 2012; Foguet-Boreu et al., 2016; Scherer & Hill, 2016). All together, these findings highlight the necessity of better understanding the pathophysiological mechanisms underlying the comorbid association between obesity and mood disorders and their combined deleterious impact on well-being and health. Beyond psychosocial or personality factors that likely participate to the development of mood disorders, biological factors able to modulate brain function and shown to be altered in these medical conditions are also good candidates. In the present chapter, we aim to provide and discuss experimental and clinical evidence supporting the relevance of inflammation, a common feature of both obesity and depression, in neuropsychiatric comorbidity of obesity.

INFLAMMATION AND NEUROPSYCHIATRIC SYMPTOMS

From the Immune System to the Brain

Abundant evidence documents the strong and intricate relationships between the immune system and the brain. A large communication network, including humoral, nervous, and chemical pathways, has been described by which inflammatory cytokines (e.g., interleukin (IL)-1β, IL-6, and tumor necrosis factor (TNF)-α) released peripherally by activated immune cells during the host response to pathogen invasion reach and act within the central nervous system (Capuron & Miller, 2011). These neuroimmune interactions are well-known for decades to be involved in the induction of sickness symptoms that accompany infections and tissue lesions (Dantzer, O'Connor, Freund, Johnson, & Kelley, 2008). Interestingly, they have later been shown to also play a key role in the development of neuropsychiatric symptoms (Allison & Ditor, 2014; Capuron & Castanon, 2017; Dantzer et al., 2008; Dantzer, O'Connor, Lawson, & Kelley, 2011; Schedlowski, Engler, & Grigoleit, 2014). Within the brain, cytokines can modulate the function of several systems involved in behavioral regulation, such as neuroendocrine and neurotransmitter systems, and neural plasticity. During an infection, brain cytokines coordinate a set of behavioral alterations, such as fatigue, apathy, feeding and sleep dysregulations, and anhedonia, collectively referred to as sickness behavior (Dantzer et al., 2008). This behavioral response is adaptive and allows the organism to appropriately fight against the infection. It is time-limited and is supposed to disappear after the elimination of the infectious agent. However, in the context of chronic inflammation, the sustained production of peripheral and brain cytokines can lead to the development of neuropsychiatric symptoms, in particular when brain structures involved in the regulation of mood and emotions, such as the hippocampus, are affected (Castanon, Luheshi, & Laye, 2015). Indeed, the administration of cytokines or induction of their synthesis triggers mood alterations in humans, independently of sickness behavior. Systemic administration of the cytokine inducer lipopolysaccharide (LPS) in healthy men induces depressive symptoms that correlate with increased cerebrospinal fluid levels of IL-6 (Engler et al., 2017). In addition, clinical studies revealed that 21%–58% of patients suffering from cancer or hepatitis C and receiving cytokine therapy develop depressive symptoms (Capuron, Gumnick, et al., 2002; Kawase et al., 2016; Raison, Demetrashvili, Capuron, & Miller, 2005). Consistent with clinical findings, preclinical studies in rodents report that an immune challenge similarly induces depressive-like and anxiety-like behaviors and more importantly that this induction depends on IFN-γ and TNF-α (Goshen et al., 2008; Klaus et al., 2016; Lacosta, Merali, & Anisman, 1999; Merali, Brennan, Brau, & Anisman, 2003; O'Connor, Andre, Wang, Lawson, et al., 2009). As in humans, these behavioral alterations are independent from sickness behavior, since they persist much longer (Frenois et al., 2007; Moreau et al., 2008). Further reinforcing the role of inflammation in the development of mood alterations, using anti-inflammatory compounds such as ibuprofen or aspirin, decreases neuropsychiatric symptoms in animal models of inflammatory diseases such as cancer (Llorens-Martin et al., 2014; Norden et al., 2015), Alzheimer's disease (Llorens-Martin et al., 2014), and Parkinson's disease (Zaminelli et al., 2014). Similarly, directly targeting inflammatory cytokines decreases mood alterations in both humans and rodents. For instance, patients receiving TNF-α antagonist as a treatment in the context of psoriasis or rheumatoid polyarthritis display decreased depressive symptoms and a general improvement of their well-being (Fleming et al., 2015; Kekow et al., 2011). Similarly, inhibiting TNF-α in rodent decreases anxiety-like and depressive-like behaviors, both in control animals (Bayramgurler, Karson, Ozer, & Utkan, 2013) and in a mouse model of multiple sclerosis (Haji et al., 2012).

Neurobiological Bases of Inflammation-Related Neuropsychiatric Symptoms

We and others have shown that among the mechanisms potentially involved in the effects of cytokines on mood, alterations of enzymatic activities, particularly affecting the enzymes indoleamine 2,3-dioxygenase (IDO) and GTP-cyclohydrolase 1 (GTP-CH1), play a major role (Capuron, Poitou, et al., 2011; Capuron & Castanon, 2017; Capuron & Miller, 2011; Castanon et al., 2015; O'Connor, Andre, Wang, Lawson, et al., 2009; O'Connor, Lawson, Andre, Briley, et al., 2009; O'Connor, Lawson, Andre, Moreau, et al., 2009). These alterations lead to substantial changes in the biosynthesis of monoamines known to participate in mood regulation, particularly serotonin and dopamine. IDO is the enzyme catalyzing the first and rate-limiting step of tryptophan degradation, an essential precursor of serotonin, through the kynurenine pathway. Whereas acute IDO activation is generally beneficial, it can become deleterious upon conditions of chronic activation because it then alters serotoninergic neurotransmission, by decreasing serotonin synthesis, and concomitantly leads to the production of kynurenine derivatives with high neurotoxic properties (Reus et al., 2015). Interestingly, neuropsychiatric symptoms displayed by Alzheimer's disease patients are associated with increased circulating levels of kynurenine (Gulaj, Pawlak, Bien, & Pawlak, 2010). Increased kynurenine levels have been reported in other inflammatory conditions also displaying neuropsychiatric comorbidities, such as aging (Capuron, Schroecksnadel, et al., 2011) and stroke (Gold et al., 2011; Stone, Forrest, Stoy, & Darlington, 2012), or in patients undergoing immunotherapy (Raison et al.,

2010). Additional clinical studies also report that cytokine-induced decrease in circulating tryptophan levels correlates with the severity of depression (Capuron, Neurauter, et al., 2003; Capuron, Ravaud, et al., 2002). In rodents, we deeply described the causal role of IDO activation in the induction of depressive-like behaviors following an immune challenge (Godbout et al., 2008; Moreau et al., 2008; O'Connor, Andre, Wang, Lawson, et al., 2009; O'Connor, Lawson, Andre, Briley, et al., 2009; O'Connor, Lawson, Andre, Moreau, et al., 2009). Similarly, IDO activation has been associated with the induction of both anxiety-like behavior and cognitive alterations following immune activation (Barichello et al., 2013; Corona et al., 2013; Gibney, McGuinness, Prendergast, Harkin, & Connor, 2013; Lawson, Kelley, & Dantzer, 2011; Salazar, Gonzalez-Rivera, Redus, Parrott, & O'Connor, 2012; Xie et al., 2014). In addition, chronic stress-induced depressive-like behaviors are reduced by TNF-α antagonism through decreased IDO activation (Fu et al., 2016). Interestingly, IDO inhibition in rodents submitted to an immune challenge impedes the induction of emotional alterations, without impacting the development of sickness behavior (Barichello et al., 2013; Henry, Huang, Wynne, & Godbout, 2009; O'Connor, Andre, Wang, Lawson, et al., 2009; O'Connor, Lawson, Andre, Briley, et al., 2009; O'Connor, Lawson, Andre, Moreau, et al., 2009; Salazar et al., 2012; Xie et al., 2014). Importantly, direct brain immune stimulation and subsequent activation of the kynurenine pathway are sufficient to alter emotional behaviors in rodents (Dobos et al., 2012; Fu et al., 2010; Lawson et al., 2013; Park, Lawson, Dantzer, Kelley, & McCusker, 2011), especially when it occurs within the hippocampus (Amare et al., 2017; Andre et al., 2008; Frenois et al., 2007; Fu et al., 2010; Henry et al., 2009). Concomitantly, cytokines alter the activity of the enzyme GTP-CH1, which induces the production of neopterin and decreases the synthesis of tetrahydrobiopterin (BH4), an essential cofactor of the rate-limiting enzyme of dopamine biosynthesis, tyrosine hydroxylase (Capuron & Miller, 2011; Murr, Widner, Wirleitner, & Fuchs, 2002). Hence, alteration in BH4 pathway has been reported to interfere with the production of dopamine (Murr et al., 2002). In depressed patients, increased blood neopterin levels, which classically reflect inflammation-induced GTP-CH1 mobilization, have been associated with the incidence of depressive episodes (Celik et al., 2010). In addition, BH4 was found to be decreased in the plasma of patients suffering from depression (Hashimoto et al., 1990), and aging was found to be associated with increased neopterin levels and concomitant increase in depressive symptomatology (Capuron, Schroecksnadel, et al., 2011). Interestingly, at the preclinical level, congenital BH4 deficiency has been recently reported to increase anxiety-like and depressive-like behaviors, as compared with wild-type mice (Nasser, Moller, Olesen, Konradsen Refsgaard, & Andreasen, 2014).

Beyond alterations in enzymatic function and monoamine transmission, recent preclinical studies also point to the impairment of neuronal plasticity as additional potential mediator of inflammation in the development of neuropsychiatric symptoms. In several inflammatory diseases, mood alterations are associated with changes in miniature excitatory postsynaptic currents (mEPSC) in brain areas known to participate in the regulation of mood. For instance, anxiety-like behaviors displayed by a rodent model of chronic pain are associated with increased mEPSC frequency and amplitude within the amygdala (Wang, Tan, Yu, & Tan, 2015). Similar changes are reported in a model of inflammatory bowel disease (Liu et al., 2015), also characterized by emotional alterations (Zhang et al., 2014). In a mouse model of multiple sclerosis, TNF-α has been shown to contribute to the development of anxiety-like behavior through modulation of striatal glutamatergic transmission (Haji et al., 2012). Lastly,

it is important to mention that other biological systems known to often be altered in depressed patients and to interact with cytokines likely participate to the development of inflammation-related neuropsychiatric symptoms. This is in particular the case for the hypothalamic-pituitary-adrenal (HPA) axis, whose hyperreactivity has been shown to constitute an important risk factor for the development of those symptoms (Capuron, Raison, et al., 2003; Capuron & Miller, 2004, 2011; Pariante & Miller, 2001).

RELEVANCE OF NEUROIMMUNE INTERACTIONS TO DEPRESSIVE COMORBIDITY IN OBESITY

Inflammation as a Fundamental Feature of Obesity

Beyond being a metabolic disorder, obesity is nowadays also considered as an inflammatory condition, characterized by a chronic low-grade inflammatory state, occurring both at the periphery and within the brain (Gregor & Hotamisligil, 2011). Increased levels of inflammatory cytokines and overactivation of associated signaling pathways have been reported in the plasma and adipose tissue of both obese subjects (Hansen et al., 2010; Lasselin et al., 2014) and rodent models of obesity (Cani et al., 2008; de Cossio et al., 2017; Dinel et al., 2011, 2014). Different mechanisms contribute to the instauration of this chronic low-grade inflammatory state in obesity (Capuron, Lasselin, & Castanon, 2017), one of the main player being the white adipose tissue (Cancello & Clement, 2006; Lasselin et al., 2014). Consistent with this, associations have been reported between circulating markers of inflammation, including C-reactive protein (CRP), TNF-α, and IL-6, and measures of central adiposity or waist-to-hip ratio (reflecting central adiposity) (Park, Park, & Yu, 2005; Visser, Bouter, McQuillan, Wener, & Harris, 1999). Conversely, weight loss in obese patients and animals is associated with decreased peripheral inflammation (Erion et al., 2014; Forsythe, Wallace, & Livingstone, 2008; Rao, 2012). Obesity-induced changes in the morphology and secretory function of adipocytes, especially hypertrophy and hyperplasia, stimulate the secretion of adipokines, which in turn locally recruit and activate macrophages. Both adipocytes and activated macrophages then produce inflammatory mediators contributing to reinforce macrophage recruitment and cytokine production (Heilbronn & Campbell, 2008). In addition, infiltration of activated immune cells in other organs, such as the liver and muscles, concomitantly supports the chronic inflammatory state associated with obesity (McNelis & Olefsky, 2014; Pedersen & Febbraio, 2012). More recently, converging studies have also highlighted the involvement of the gut microbiota in the development of obesity-related low-grade inflammation (Cani et al., 2009; Cani, Osto, Geurts, & Everard, 2012; Nakamura & Omaye, 2012). Gut microbiota diversity is decreased in obese subjects in comparison with nonobese ones. This decreased diversity is associated with increased adiposity and peripheral inflammation (Le Chatelier et al., 2013). Moreover, increased gut permeability has been reported in diet-induced obesity (DIO) animal models, leading to the instauration of a metabolic endotoxemia (i.e., the translocation of bacteria compounds and endotoxins within the blood) (Cani, 2016). Recent studies reporting that modulating the composition of the gut microbiota with specific compounds (prebiotics or probiotics) improves metabolic endotoxemia in obese patients and rodents (Cani, 2016; Cani & Delzenne, 2007, 2009) nicely support its role in obesity-related low-grade inflammation.

Over the last decades, the abundant literature dealing with the host response to infection has clearly established that the systemic inflammatory response locally triggered by the infectious agent extends to the brain, where peripheral cytokines induce the production of brain cytokines, which are ultimately responsible for infection-related behavioral changes (Dantzer et al., 2008). In agreement with this, mounting evidence indicates that systemic low-grade inflammation observed in obese subjects and in animal models of obesity is similarly associated with increased central inflammation. DIO is associated with increased immune cell entry into the brain (Buckman et al., 2014; Rummel, Inoue, Poole, & Luheshi, 2010). Other studies also highlight increased inflammatory processes in the hypothalamus of obese rodents and humans, a brain area playing a key role in the control of energy homeostasis (Dorfman & Thaler, 2015). For example, increased hypothalamic expression of inflammatory cytokines has been repeatedly reported in several rodent models of obesity (Cai & Liu, 2012; de Cossio et al., 2017; De Souza et al., 2005; Dinel et al., 2011, 2014; Pascoal et al., 2017; Thaler et al., 2012). Moreover, increased brain inflammation has been related to adiposity (Erion et al., 2014). Interestingly, sensitivity and resistance to DIO in these models correlate respectively with the presence or absence of hypothalamic inflammation and reactive gliosis (Dorfman & Thaler, 2015). Similarly, hypothalamic gliosis has been positively correlated with body mass index in obese individuals (Thaler et al., 2012). Of note, obesity is associated not only with basal low-grade brain inflammation but also with exacerbated inflammatory reactivity to an immune challenge, as revealed by the overproduction of hypothalamic inflammatory cytokines in response to systemic administration of LPS (Andre, Dinel, Ferreira, Laye, & Castanon, 2014; Pohl, Woodside, & Luheshi, 2009). Interestingly, this enhanced inflammatory response also occurs in other brain areas, including those rather involved in the regulation of mood and emotions such as the cortex, the amygdala, and the hippocampus (Almeida-Suhett, Graham, Chen, & Deuster, 2017; Andre et al., 2014; Boitard et al., 2014; Dinel et al., 2011,

2014). These findings extend those reporting increased unstimulated production of inflammatory cytokines in the hippocampus or the cortex of animal obesity models (Almeida-Suhett et al., 2017; de Sousa Rodrigues et al., 2017; Dinel et al., 2011, 2014; Erion et al., 2014; Kang et al., 2016; Pistell et al., 2010). Consistent with this, recent studies show that obesity is also associated with increased activation of hippocampal microglial cells (Hao, Dey, Yu, & Stranahan, 2016; Kang et al., 2016).

Depressive Symptoms in Obesity

Mounting epidemiological and clinical studies converge to show that obese subjects are at increased risk of depression in comparison with nonobese individuals (Atlantis & Baker, 2008; Capuron et al., 2017; Castanon et al., 2015; Castanon, Lasselin, & Capuron, 2014; Faulconbridge & Bechtel, 2014; Hryhorczuk, Sharma, & Fulton, 2013; Kaidanovich-Beilin, Cha, & McIntyre, 2012; Soczynska et al., 2011). Data from adult cohorts report a prevalence of depressive symptoms in up to 23%–30% of obese people, whereas this rate is only about 11% in the general population (Carey et al., 2014; Dawes et al., 2016; Lasselin & Capuron, 2014). Moreover, longitudinal studies conducted in young and adult subjects also show that obesity represents a potent risk factor for the development of depressive and anxiety disorders at follow-up, especially in women (Anderson, Cohen, Naumova, Jacques, & Must, 2007; Luppino et al., 2010). Interestingly, weight loss in obese patients, notably after bariatric surgery, was found to be associated with a reduced frequency and/or severity of depressive symptoms (Andersen et al., 2010; Dawes et al., 2016; Dixon et al., 2003; Lasselin & Capuron, 2014).

Similar to what has been reported in clinical studies, preclinical investigations in animal models of obesity support the link between obesity and depressive-like behaviors (de Cossio et al., 2017; Dinel et al., 2011; Finger, Dinan, &

Cryan, 2010). Interestingly, these investigations provide a unique opportunity to further study the physiopathologic mechanisms underlying this association (Castanon et al., 2015). Since overconsumption of food rich in fat and/or sugar is one of the main contributors of obesity in humans, a lot of rodent models of obesity have been based on diet modifications. Interestingly, these DIO models not only lead to the progressive development of overweight/obesity and related metabolic dysfunctions but also reproduce some of the behavioral abnormalities displayed by obese patients. Accordingly, we and others have reported increased anxiety-like and depressive-like behaviors, as well as altered emotional memory in DIO models (Andre et al., 2014; Boitard et al., 2015; da Costa Estrela et al., 2015; Krishna et al., 2016; Kurhe & Mahesh, 2015a; Park, Lee, Cho, Park, & Kim, 2017; Sharma & Fulton, 2013). Life-span DIO exposure from weaning induces a progressive development of behavioral alterations, going from early onset of spatial memory alterations to later development of anxiety-like behaviors (Andre et al., 2014). Interestingly, depressive-like behaviors are not altered in such moderate DIO models, but they are increased when more severe models of obesity, such as those using very-high-fat diets, are used (Sharma & Fulton, 2013; Yamada et al., 2011). Similarly, postnatal ablation of POMC neurons induces an obese phenotype, which is associated with increased anxiety-like behavior (Greenman et al., 2013). Moreover, exercise-induced weight loss decreases depressive-like and anxiety-like behaviors in obese mice (Park et al., 2017), similarly to what has been previously reported in obese subjects. Severe obesity is also modeled through the use of strains of genetically obese mice, such as the *ob/ob* and *db/db* mice. *Ob/ob* mice display increased anxiety-like behaviors (Finger et al., 2010), which can be reduced following chronic leptin administration (Asakawa et al., 2003). We also reported such increased anxiety-like behavior in *db/db* mice (Dinel et al., 2011, 2014). Interestingly, while *db/db* mice display similar depressive-like behaviors than their lean controls in basal conditions (Dinel et al., 2011), those behaviors are exacerbated in stressful conditions (Collin, Hakansson-Ovesjo, Misane, Ogren, & Meister, 2000; Yamada et al., 2011).

Inflammation as a Link Between Obesity and Depressive Comorbidity: Evidence and Mechanisms

Based on the abundant literature linking inflammation and neuropsychiatric symptoms, it has been hypothesized that inflammation may similarly be involved in obesity-related depressive comorbidity (Capuron et al., 2017; Castanon et al., 2014; Lasselin & Capuron, 2014). Consistent with this hypothesis, we found that inflammation represents one major determinant of neuropsychiatric symptoms in patients with metabolic syndrome (Capuron et al., 2008), and plasma levels of inflammatory factors (e.g., CRP and IL-6 levels) were found to be significantly associated with depressive symptoms in obese patients (Daly, 2013; Dixon et al., 2008; Martinac et al., 2017). Moreover, caloric restriction, which is known to display anti-inflammatory properties at the periphery and within the brain (Higami et al., 2006; Kim, Kim, No, Chung, & Fernandes, 2006; Radler, Wright, Walker, Hale, & Kent, 2015), has been shown to improve emotional reactivity in both humans and rodents (Incollingo Belsky, Epel, & Tomiyama, 2014; Redman & Ravussin, 2011). Reinforcing the link between adiposity, inflammation, and mood symptoms, bariatric surgery-induced weight loss is associated with a reduction of peripheral inflammatory markers (especially CRP), concomitantly with decreased depressive symptoms (Capuron, Poitou, et al., 2011; Emery et al., 2007).

Experiments performed in animal models of obesity are particularly useful to deeply study the complex relationship between obesity, inflammation, and neuropsychiatric comorbidities. The hippocampus, a brain area well-known to be crucial for mood control (Bannerman et al., 2014), is particularly targeted by inflammation in the context of obesity. In the obese *db/db* mice, we reported that anxiety-like behaviors are associated with increased expression of IL-6, IL-1β, and TNF-α in the hippocampus (de Cossio et al., 2017; Dinel et al., 2011, 2014). Similarly, IL-1β expression in this brain area is increased after prolonged consumption of a very-high-fat diet, along with an increase of both anxiety-like and depressive-like behaviors (Almeida-Suhett et al., 2017). In addition, previous studies report a role of hippocampal IL-1β in cognitive deficits that are associated with emotional alterations in *db/db* mice (Erion et al., 2014). In rodent models rather mimicking moderate obesity, hippocampal inflammation is not significantly increased in nonstimulated conditions, but it is markedly exacerbated after an immune challenge with LPS (Andre et al., 2014; Boitard et al., 2014). More importantly, exacerbated hippocampal inflammation is associated with increased depressive-like behavior, as compared with LPS-treated lean controls (Andre et al., 2014).

There are several mechanisms by which chronic inflammation may promote the development of depressive comorbidity in obese subjects. Based on the pathways that have been described to support neuroimmune interactions, these mechanisms may include effects on neurotransmitter metabolism via enzymatic alterations, impaired insulin function, reduced synaptic plasticity, and neuroendocrine alterations (Capuron et al., 2017; Castanon et al., 2014, 2015; Lasselin & Capuron, 2014; Miller & Spencer, 2014). Supporting the role of enzymatic effects, studies have shown that obese subjects with neuropsychiatric symptoms display elevated activation of IDO and hippocampal atrophy, as compared with nonobese individuals (Brandacher et al., 2006; Brandacher, Hoeller, Fuchs, & Weiss, 2007). In DIO mice,

LPS-induced IDO activation increases brain kynurenine concentrations and exacerbates depressive-like behaviors (Andre et al., 2014). In addition, several studies show dysregulations of serotoninergic activity in obese animals. Circulating serotonin levels are decreased in DIO mice when compared with lean mice (Kim, Bae, & Lim, 2013), whereas pharmacological facilitation of serotoninergic neurotransmission attenuates comorbid depression and anxiety associated with obesity (Kurhe & Mahesh, 2015b; Kurhe, Mahesh, & Devadoss, 2017). Similarly, increased anxiety- and depressive-like behaviors in DIO mice have been reported to be associated with the impairment of the serotoninergic system in the dorsal raphe, and treadmill exercise improves both serotoninergic neurotransmission and associated behavioral alterations (Park et al., 2017). Concomitantly, other studies also report increased neopterin levels in both obese humans and rodents (Brandacher et al., 2006; Mangge et al., 2014; Oxenkrug, Tucker, Requintina, & Summergrad, 2011), pointing to a likely role for GTP-CH1-induced impairment of dopamine neurotransmission in obesity-related depressive comorbidity. Accordingly, emotional alterations displayed by DIO mice are associated with decreased dopamine turnover in the ventral hippocampus (Krishna et al., 2015). In addition, DIO induces plasticity-related alterations in dopaminergic system, namely, higher D2R and reduced D1R expression in the nucleus accumbens, which are associated with depressive-like behavior (Sharma & Fulton, 2013).

Along with enzymatic alterations, recent data also suggest that brain insulin resistance may act as a potential additional player to link obesity-related inflammation and mood alterations. Inflammatory cytokines, in particular IL-1β and TNF-α, can interfere with the activation of the intracellular signaling pathway associated with the insulin receptor, both at the periphery and within the brain (De Souza et al., 2005; Hotamisligil, 2006; Lann & LeRoith, 2007; Xu

et al., 2003). TNF-α treatment has been shown to decrease, in different cell types, the catalytic activity of the insulin receptor and consequently the inhibition of insulin receptor substrate-1 (IRS-1) phosphorylation (Hotamisligil, Murray, Choy, & Spiegelman, 1994; Liu, Spelleken, Rohrig, Hauner, & Eckel, 1998). Alternatively, TNF-α is also able to directly phosphorylate IRS-1 (Hotamisligil et al., 1996; Kanety, Feinstein, Papa, Hemi, & Karasik, 1995), further reinforcing its effect on downstream insulin signaling pathway. Interestingly, both the high density of insulin receptors in the hippocampus and the alterations of insulin signaling pathway reported in several neuropsychiatric disorders (Blazquez, Velazquez, Hurtado-Carneiro, & Ruiz-Albusac, 2014; Ghasemi, Haeri, Dargahi, Mohamed, & Ahmadiani, 2013; Yates, Sweat, Yau, Turchiano, & Convit, 2012) are consistent with a role for insulin resistance in the development of obesity-related neuropsychiatric comorbidity. Supporting this assumption, insulin resistance has been reported in the hippocampus of *db/db* mice (Dey, Hao, Erion, Wosiski-Kuhn, & Stranahan, 2014; Kim, Sullivan, Backus, & Feldman, 2011), which also display emotional alterations (de Cossio et al., 2017; Dinel et al., 2011, 2014). Conversely, antidiabetic compounds improve depressive symptoms in humans and depressive-like behaviors in rodents (Cline et al., 2012; Gupta et al., 2014; Pomytkin et al., 2015). In rats, chronic central administration of the GLP-1 agonist exendin-4 significantly reduces depressive-like behaviors, independently of body weight loss and reduced food intake (Anderberg et al., 2016). In *db/db* mice, dietary administration of the PPAR-γ agonist rosiglitazone reverses depressive-like behavior (Sharma, Elased, & Lucot, 2012) and improves both memory and hippocampal synaptic plasticity (Kariharan et al., 2015). DIO-induced anxiety-like and depressive-like behaviors are also associated with decreased synaptic plasticity, as evidenced by reduced long-term potentiation in the ventral hippocampus (Krishna et al., 2015). Previous studies have also reported altered long-term potentiation in the hippocampus of *db/db* mice (Erion et al., 2014; Li et al., 2002; Stranahan, 2015; Wosiski-Kuhn, Erion, Gomez-Sanchez, Gomez-Sanchez, & Stranahan, 2014). Interestingly, IL-1β contributes to the development of these synaptic dysregulations and associated cognitive impairments (Erion et al., 2014), consistent with the notion that inflammation participates in the development of behavioral alterations in obesity by affecting synaptic transmission, in particular in the hippocampus.

Beyond synaptic transmission, brain cytokines are also likely to affect the HPA axis (Capuron & Miller, 2011; Raison & Miller, 2003), which is highly impaired in obesity (Dey et al., 2014; Dinel et al., 2011, 2014). Both DIO and genetic models of obesity display elevated plasma glucocorticoids levels (Andre et al., 2014; Dinel et al., 2011; Wosiski-Kuhn et al., 2014), which are able in that context to impact the brain immune system by sensitizing microglia and participating in the increased levels of hippocampal TNF-α and IL-1β reported in obese *db/db* mice (Dey et al., 2014). Moreover, glucocorticoids have been previously shown to contribute to the development of both cognitive (Stranahan et al., 2008) and emotional alterations (Boitard et al., 2015) in obese animals. Further supporting the interactions between the different pathways potentially involved in neuropsychiatric comorbidity of obesity, facilitating serotoninergic neurotransmission in DIO mice improves both emotional alterations and HPA axis hyperactivity (Kurhe & Mahesh, 2015a). Lastly, recent studies have described the role of gut microbiota in the regulation of mood disorders, including in inflammatory disorders (Dinan & Cryan, 2017; Evrensel & Ceylan, 2015; Fung, Olson, & Hsiao, 2017). Given the role of gut microbiota in the induction of peripheral and central low-grade inflammation in obese animals, this suggests it can represent an additional mechanism in the development of mood

disorders in obese patients. In support of this, transplantation of gut microbiota from DIO mice to lean mice increases both anxiety-like behavior and microglial activation (Bruce-Keller et al., 2015).

Therapeutic Implications

Identifying therapeutic strategies able to improve mood alterations associated with obesity is a major socioeconomic and public health challenge. Among those strategies, pharmacological approaches aiming at reducing or alleviating chronic inflammation in patients with obesity appear particularly relevant. Nevertheless, the potential side effects associated with the prolonged use of anti-inflammatory agents clearly limit enthusiasm. In that context, nonpharmacological therapeutic strategies, including weight management programs and nutritional interventions, may represent a promising alternative associated with a lower risk of complications and reduced economic cost. Supporting this notion, bariatric surgery-induced weight loss or low-fat diet consumption were found to be effective at decreasing neuropsychiatric symptoms in obese patients, in addition to their ability to decrease inflammation (Magkos et al., 2016; Moller et al., 2016; Schmatz et al., 2017). Similarly, chronic consumption of natural anti-inflammatory agents such as *n*-3 polyunsaturated fatty acids (*n*-3 PUFAs), antioxidants, and compounds able to modulate microbiota composition is likely to improve not only metabolic alterations but also mood symptoms in obese subjects, as shown in other medical conditions (Bazinet & Laye, 2014; Dinan & Cryan, 2017; Su, Matsuoka, & Pae, 2015; Vauzour, Martinsen, & Laye, 2015). Epidemiological studies have contributed to show the beneficial effect of n-3 PUFAs' supplementation on depressive symptoms (Appleton, Rogers, & Ness, 2010; Bozzatello, Brignolo, De Grandi, & Bellino, 2016; Grosso et al., 2016). Moreover, patients suffering from hepatitis C and treated with interferon-alpha are at higher risk of developing interferon-induced depression when they display low blood n-3 PUFAs' contents (Su, Huang, et al., 2010; Su, Lai, et al. 2014). In rodents, n-3 PUFAs' supplementation through chronic fish oil consumption prevents LPS-induced depressive-like behaviors (Shi et al., 2017). This behavioral improvement is associated with decreased production of inflammatory cytokines in the brain, especially in the hippocampus. Although the effects of n-3 PUFAs' supplementation remain understudied in the context of obesity, recent data reveal a negative association between depressive symptoms and plasma n-3 PUFAs' content in patients suffering from severe obesity (Chalut-Carpentier, Pataky, Golay, & Bobbioni-Harsch, 2015).

Similarly, chronic consumption of antioxidants, including blueberry polyphenols, curcumin, and vitamins C and E, has been shown to decrease plasma cytokine concentrations in healthy individuals (Karlsen et al., 2007) and in obese or type 2 diabetic patients (Ganjali et al., 2014; Jamalan, Rezazadeh, Zeinali, & Ghaffari, 2015). Moreover, coenzyme Q10 supplementation was found to decrease brain inflammation in a mouse model of multiple sclerosis (Soleimani, Jameie, Barati, Mehdizadeh, & Kerdari, 2014). Relevant to neuropsychiatric symptoms, α-tocopherol decreases depressive-like behaviors induced by TNF-α administration in mice (Manosso et al., 2013), and supplement with curcumin improves anxiety in obese patients (Esmaily et al., 2015).

As previously mentioned, modifications of the gut microbiota are an important player in the induction of peripheral and central low-grade inflammation associated with obesity. Hence, modulating gut microbiota composition through the use of pre- or probiotics could decrease inflammation and therefore improve mood in obese patients. Interestingly, such interventions have previously been reported to decrease peripheral inflammation (Cani et al., 2007; Cani & Delzenne, 2007, 2009). However, their effect on central inflammation associated with inflammatory diseases remains understudied. It is known that gut microbiota is able to impact brain functions (Gareau, 2016; Moloney, Desbonnet, Clarke, Dinan, & Cryan, 2014; Neufeld, Kang, Bienenstock, & Foster, 2011), although the underlying mechanisms are not yet completely understood (Cryan & Dinan, 2012; Moloney et al., 2014). Of note, it has recently been described that gut microbiota transplantation from obese mice to lean mice induces microglial activation along with anxiety-like behavior (Bruce-Keller et al., 2015). This supports the assumption that modulating gut microbiota composition could decrease neuropsychiatric symptoms in obese patients by decreasing peripheral and/or central inflammation.

CONCLUSION

Nowadays, obesity is considered as a real pandemic, and its associated neuropsychiatric comorbidities, which markedly interfere with well-being and quality of life, worsen the picture even more. The clinical and experimental data presented in this chapter support a critical role for inflammatory processes in the development of depressive comorbidity in obese individuals. Moreover, they describe several mechanisms and pathways likely to promote inflammation-induced depressive symptoms in obesity (Fig. 1). Further investigations are needed to progress in the elucidation and better understanding of these mechanisms in order to identify, in particular, new therapeutic strategies. In addition to immunomodulatory pharmacological approaches, nonpharmacological strategies, including weight-loss programs and nutritional interventions with anti-inflammatory properties, appear currently very promising in the management of depressive comorbidity in obesity.

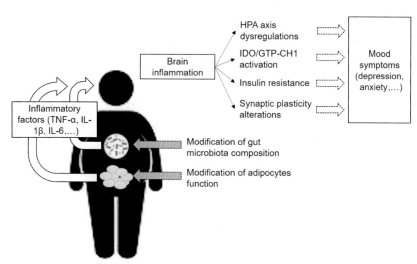

FIG. 1 Proposed role of inflammation in the development of neuropsychiatric symptoms in obesity. Obesity is associated with chronic low-grade inflammation originating mainly from the white adipose tissue and altered gut microbiota composition. This peripheral inflammation contributes to the instauration of brain inflammation through different immune-to-brain communication pathways. Neuroinflammation in brain areas participating to mood regulation (e.g., the hippocampus) leads to alterations of neurobiological systems including HPA axis; synaptic plasticity; and IDO, GTP-CH1, and insulin pathways. Altogether, these cytokine-induced biological dysregulations participate to the development of neuropsychiatric symptoms in obese patients. *TNF-α*, tumor necrosis factor-α; *IL-1β*, interleukin-1β; *IL-6*, interleukin-6; *HPA*, hypothalamic-pituitary-adrenal; *IDO*, indoleamine 2,3-dioxygenase; *GTP-CH1*, GTP-cyclohydrolase 1.

References

Allison, D. J., & Ditor, D. S. (2014). The common inflammatory etiology of depression and cognitive impairment: a therapeutic target. *Journal of Neuroinflammation*, 11, 151.

Almeida-Suhett, C. P., Graham, A., Chen, Y., & Deuster, P. (2017). Behavioral changes in male mice fed a high-fat diet are associated with IL-1beta expression in specific brain regions. *Physiology & Behavior*, 169, 130–140.

Amare, A. T., Schubert, K. O., Klingler-Hoffmann, M., Cohen-Woods, S., & Baune, B. T. (2017). The genetic overlap between mood disorders and cardiometabolic diseases: a systematic review of genome wide and candidate gene studies. *Translational Psychiatry*, 7(1).

Anderberg, R. H., Richard, J. E., Hansson, C., Nissbrandt, H., Bergquist, F., & Skibicka, K. P. (2016). GLP-1 is both anxiogenic and antidepressant; divergent effects of acute and chronic GLP-1 on emotionality. *Psychoneuroendocrinology*, 65, 54–66.

Andersen, J. R., Aasprang, A., Bergsholm, P., Sletteskog, N., Vage, V., & Natvig, G. K. (2010). Anxiety and depression in association with morbid obesity: changes with improved physical health after duodenal switch. *Health and Quality of Life Outcomes*, 8, 52.

Anderson, S. E., Cohen, P., Naumova, E. N., Jacques, P. F., & Must, A. (2007). Adolescent obesity and risk for subsequent major depressive disorder and anxiety disorder: prospective evidence. *Psychosomatic Medicine*, 69(8), 740–747.

Andre, C., Dinel, A. L., Ferreira, G., Laye, S., & Castanon, N. (2014). Diet-induced obesity progressively alters cognition, anxiety-like behavior and lipopolysaccharide-induced depressive-like behavior: focus on brain indoleamine 2,3-dioxygenase activation. *Brain, Behavior, and Immunity*, 41, 10–21.

Andre, C., O'Connor, J. C., Kelley, K. W., Lestage, J., Dantzer, R., & Castanon, N. (2008). Spatio-temporal differences in the profile of murine brain expression of proinflammatory cytokines and indoleamine 2,3-dioxygenase in response to peripheral lipopolysaccharide administration. *Journal of Neuroimmunology*, 200(1–2), 90–99.

Appleton, K. M., Rogers, P. J., & Ness, A. R. (2010). Updated systematic review and meta-analysis of the effects of n-3 long-chain polyunsaturated fatty acids on depressed mood. *American Journal of Clinical Nutrition*, 91(3), 757–770.

Asakawa, A., Inui, A., Inui, T., Katsuura, G., Fujino, M. A., & Kasuga, M. (2003). Leptin treatment ameliorates anxiety in ob/ob obese mice. *Journal of Diabetes and Its Complications*, 17(2), 105–107.

Atlantis, E., & Baker, M. (2008). Obesity effects on depression: systematic review of epidemiological studies. *International Journal of Obesity, 32*(6), 881–891.

Bannerman, D. M., Sprengel, R., Sanderson, D. J., McHugh, S. B., Rawlins, J. N., Monyer, H., et al. (2014). Hippocampal synaptic plasticity, spatial memory and anxiety. *Nature Reviews. Neuroscience, 15*(3), 181–192.

Barichello, T., Generoso, J. S., Simoes, L. R., Elias, S. G., Tashiro, M. H., Dominguini, D., et al. (2013). Inhibition of indoleamine 2,3-dioxygenase prevented cognitive impairment in adult Wistar rats subjected to pneumococcal meningitis. *Translational Research, 162*(6), 390–397.

Bayramgurler, D., Karson, A., Ozer, C., & Utkan, T. (2013). Effects of long-term etanercept treatment on anxiety- and depression-like neurobehaviors in rats. *Physiology & Behavior, 119*, 145–148.

Bazinet, R. P., & Laye, S. (2014). Polyunsaturated fatty acids and their metabolites in brain function and disease. *Nature Reviews. Neuroscience, 15*(12), 771–785.

Blazquez, E., Velazquez, E., Hurtado-Carneiro, V., & Ruiz-Albusac, J. M. (2014). Insulin in the brain: its pathophysiological implications for States related with central insulin resistance, type 2 diabetes and Alzheimer's disease. *Frontiers in Endocrinology, 5*, 161.

Boitard, C., Cavaroc, A., Sauvant, J., Aubert, A., Castanon, N., Laye, S., et al. (2014). Impairment of hippocampal-dependent memory induced by juvenile high-fat diet intake is associated with enhanced hippocampal inflammation in rats. *Brain, Behavior, and Immunity, 40*, 9–17.

Boitard, C., Maroun, M., Tantot, F., Cavaroc, A., Sauvant, J., Marchand, A., et al. (2015). Juvenile obesity enhances emotional memory and amygdala plasticity through glucocorticoids. *Journal of Neuroscience, 35*(9), 4092–4103.

Bozzatello, P., Brignolo, E., De Grandi, E., & Bellino, S. (2016). Supplementation with omega-3 fatty acids in psychiatric disorders: a review of literature data. *Journal of Clinical Medicine, 5*(8), 1–26.

Brandacher, G., Hoeller, E., Fuchs, D., & Weiss, H. G. (2007). Chronic immune activation underlies morbid obesity: is IDO a key player? *Current Drug Metabolism, 8*(3), 289–295.

Brandacher, G., Winkler, C., Aigner, F., Schwelberger, H., Schroecksnadel, K., Margreiter, R., et al. (2006). Bariatric surgery cannot prevent tryptophan depletion due to chronic immune activation in morbidly obese patients. *Obesity Surgery, 16*(5), 541–548.

Bruce-Keller, A. J., Salbaum, J. M., Luo, M., Blanchard, E. t., Taylor, C. M., Welsh, D. A., et al. (2015). Obese-type gut microbiota induce neurobehavioral changes in the absence of obesity. *Biological Psychiatry, 77*(7), 607–615.

Brunault, P., Jacobi, D., Miknius, V., Bourbao-Tournois, C., Huten, N., Gaillard, P., et al. (2012). High preoperative depression, phobic anxiety, and binge eating scores and low medium-term weight loss in sleeve gastrectomy obese patients: a preliminary cohort study. *Psychosomatics, 53*(4), 363–370.

Buckman, L. B., Hasty, A. H., Flaherty, D. K., Buckman, C. T., Thompson, M. M., Matlock, B. K., et al. (2014). Obesity induced by a high-fat diet is associated with increased immune cell entry into the central nervous system. *Brain, Behavior, and Immunity, 35*, 33–42.

Cai, D., & Liu, T. (2012). Inflammatory cause of metabolic syndrome via brain stress and NF-kappaB. *Aging (Albany, NY), 4*(2), 98–115.

Cancello, R., & Clement, K. (2006). Is obesity an inflammatory illness? Role of low-grade inflammation and macrophage infiltration in human white adipose tissue. *BJOG, 113*(10), 1141–1147.

Cani, P. D. (2016). Gut microbiota: changes in gut microbes and host metabolism: squaring the circle? *Nature Reviews. Gastroenterology & Hepatology, 13*(10), 563–564.

Cani, P. D., Bibiloni, R., Knauf, C., Waget, A., Neyrinck, A. M., Delzenne, N. M., et al. (2008). Changes in gut microbiota control metabolic endotoxemia-induced inflammation in high-fat diet-induced obesity and diabetes in mice. *Diabetes, 57*(6), 1470–1481.

Cani, P. D., & Delzenne, N. M. (2007). Gut microflora as a target for energy and metabolic homeostasis. *Current Opinion in Clinical Nutrition and Metabolic Care, 10*(6), 729–734.

Cani, P. D., & Delzenne, N. M. (2009). Interplay between obesity and associated metabolic disorders: new insights into the gut microbiota. *Current Opinion in Pharmacology, 9*(6), 737–743.

Cani, P. D., Neyrinck, A. M., Fava, F., Knauf, C., Burcelin, R. G., Tuohy, K. M., et al. (2007). Selective increases of bifidobacteria in gut microflora improve high-fat-diet-induced diabetes in mice through a mechanism associated with endotoxaemia. *Diabetologia, 50*(11), 2374–2383.

Cani, P. D., Osto, M., Geurts, L., & Everard, A. (2012). Involvement of gut microbiota in the development of low-grade inflammation and type 2 diabetes associated with obesity. *Gut Microbes, 3*(4), 279–288.

Cani, P. D., Possemiers, S., Van de Wiele, T., Guiot, Y., Everard, A., Rottier, O., et al. (2009). Changes in gut microbiota control inflammation in obese mice through a mechanism involving GLP-2-driven improvement of gut permeability. *Gut, 58*(8), 1091–1103.

Capuron, L., & Castanon, N. (2017). Role of inflammation in the development of neuropsychiatric symptom domains: evidence and mechanisms. *Current Topics in Behavioral Neurosciences, 31*, 31–44.

Capuron, L., Gumnick, J. F., Musselman, D. L., Lawson, D. H., Reemsnyder, A., Nemeroff, C. B., et al. (2002). Neurobehavioral effects of interferon-alpha in cancer patients: phenomenology and paroxetine responsiveness of symptom dimensions. *Neuropsychopharmacology, 26*(5), 643–652.

Capuron, L., Lasselin, J., & Castanon, N. (2017). Role of adiposity-driven inflammation in depressive morbidity. *Neuropsychopharmacology, 42*(1), 115–128.

Capuron, L., & Miller, A. H. (2004). Cytokines and psychopathology: lessons from interferon-alpha. *Biological Psychiatry*, 56(11), 819–824.

Capuron, L., & Miller, A. H. (2011). Immune system to brain signaling: neuropsychopharmacological implications. *Pharmacology & Therapeutics*, 130(2), 226–238.

Capuron, L., Neurauter, G., Musselman, D. L., Lawson, D. H., Nemeroff, C. B., Fuchs, D., et al. (2003). Interferon-alpha-induced changes in tryptophan metabolism. Relationship to depression and paroxetine treatment. *Biological Psychiatry*, 54(9), 906–914.

Capuron, L., Poitou, C., Machaux-Tholliez, D., Frochot, V., Bouillot, J. L., Basdevant, A., et al. (2011). Relationship between adiposity, emotional status and eating behaviour in obese women: role of inflammation. *Psychological Medicine*, 41(7), 1517–1528.

Capuron, L., Raison, C. L., Musselman, D. L., Lawson, D. H., Nemeroff, C. B., & Miller, A. H. (2003). Association of exaggerated HPA axis response to the initial injection of interferon-alpha with development of depression during interferon-alpha therapy. *American Journal of Psychiatry*, 160(7), 1342–1345.

Capuron, L., Ravaud, A., Neveu, P. J., Miller, A. H., Maes, M., & Dantzer, R. (2002). Association between decreased serum tryptophan concentrations and depressive symptoms in cancer patients undergoing cytokine therapy. *Molecular Psychiatry*, 7(5), 468–473.

Capuron, L., Schroecksnadel, S., Feart, C., Aubert, A., Higueret, D., Barberger-Gateau, P., et al. (2011). Chronic low-grade inflammation in elderly persons is associated with altered tryptophan and tyrosine metabolism: role in neuropsychiatric symptoms. *Biological Psychiatry*, 70(2), 175–182.

Capuron, L., Su, S., Miller, A. H., Bremner, J. D., Goldberg, J., Vogt, G. J., et al. (2008). Depressive symptoms and metabolic syndrome: is inflammation the underlying link? *Biological Psychiatry*, 64(10), 896–900.

Carey, M., Small, H., Yoong, S. L., Boyes, A., Bisquera, A., & Sanson-Fisher, R. (2014). Prevalence of comorbid depression and obesity in general practice: a cross-sectional survey. *British Journal of General Practice*, 64(620), e122–127.

Carpenter, K. M., Hasin, D. S., Allison, D. B., & Faith, M. S. (2000). Relationships between obesity and DSM-IV major depressive disorder, suicide ideation, and suicide attempts: results from a general population study. *American Journal of Public Health*, 90(2), 251–257.

Castanon, N., Lasselin, J., & Capuron, L. (2014). Neuropsychiatric comorbidity in obesity: role of inflammatory processes. *Frontiers in Endocrinology*, 5, 74.

Castanon, N., Luheshi, G., & Laye, S. (2015). Role of neuroinflammation in the emotional and cognitive alterations displayed by animal models of obesity. *Frontiers in Neuroscience*, 9, 229.

Celik, C., Erdem, M., Cayci, T., Ozdemir, B., Ozgur Akgul, E., Kurt, Y. G., et al. (2010). The association between serum levels of neopterin and number of depressive episodes of major depression. *Progress in Neuro-Psychopharmacology & Biological Psychiatry*, 34(2), 372–375.

Chalut-Carpentier, A., Pataky, Z., Golay, A., & Bobbioni-Harsch, E. (2015). Involvement of dietary fatty acids in multiple biological and psychological functions, in morbidly obese subjects. *Obesity Surgery*, 25(6), 1031–1038.

Cline, B. H., Steinbusch, H. W., Malin, D., Revishchin, A. V., Pavlova, G. V., Cespuglio, R., et al. (2012). The neuronal insulin sensitizer dicholine succinate reduces stress-induced depressive traits and memory deficit: possible role of insulin-like growth factor 2. *BMC Neuroscience*, 13, 110.

Collin, M., Hakansson-Ovesjo, M. L., Misane, I., Ogren, S. O., & Meister, B. (2000). Decreased 5-HT transporter mRNA in neurons of the dorsal raphe nucleus and behavioral depression in the obese leptin-deficient ob/ob mouse. *Brain Research. Molecular Brain Research*, 81(1–2), 51–61.

Corona, A. W., Norden, D. M., Skendelas, J. P., Huang, Y., O'Connor, J. C., Lawson, M., et al. (2013). Indoleamine 2,3-dioxygenase inhibition attenuates lipopolysaccharide induced persistent microglial activation and depressive-like complications in fractalkine receptor (CX(3)CR1)-deficient mice. *Brain, Behavior, and Immunity*, 31, 134–142.

Cryan, J. F., & Dinan, T. G. (2012). Mind-altering microorganisms: the impact of the gut microbiota on brain and behaviour. *Nature Reviews. Neuroscience*, 13(10), 701–712.

da Costa Estrela, D., da Silva, W. A., Guimaraes, A. T., de Oliveira Mendes, B., da Silva Castro, A. L., da Silva Torres, I. L., et al. (2015). Predictive behaviors for anxiety and depression in female wistar rats subjected to cafeteria diet and stress. *Physiology & Behavior*, 151, 252–263.

Daly, M. (2013). The relationship of C-reactive protein to obesity-related depressive symptoms: a longitudinal study. *Obesity (Silver Spring)*, 21(2), 248–250.

Dantzer, R., O'Connor, J. C., Freund, G. G., Johnson, R. W., & Kelley, K. W. (2008). From inflammation to sickness and depression: when the immune system subjugates the brain. *Nature Reviews. Neuroscience*, 9(1), 46–56.

Dantzer, R., O'Connor, J. C., Lawson, M. A., & Kelley, K. W. (2011). Inflammation-associated depression: from serotonin to kynurenine. *Psychoneuroendocrinology*, 36(3), 426–436.

Dawes, A. J., Maggard-Gibbons, M., Maher, A. R., Booth, M. J., Miake-Lye, I., Beroes, J. M., et al. (2016). Mental health conditions among patients seeking and undergoing bariatric surgery: a meta-analysis. *JAMA*, 315(2), 150–163.

de Cossio, L. F., Fourrier, C., Sauvant, J., Everard, A., Capuron, L., Cani, P. D., et al. (2017). Impact of prebiotics on metabolic and behavioral alterations in a mouse model of metabolic syndrome. *Brain, Behavior, and Immunity*, 64, 33–49.

de Sousa Rodrigues, M. E., Bekhbat, M., Houser, M. C., Chang, J., Walker, D. I., Jones, D. P., et al. (2017). Chronic psychological stress and high-fat high-fructose diet disrupt metabolic and inflammatory gene networks in the brain, liver, and gut and promote behavioral deficits in mice. *Brain, Behavior, and Immunity, 59*, 158–172.

De Souza, C. T., Araujo, E. P., Bordin, S., Ashimine, R., Zollner, R. L., Boschero, A. C., et al. (2005). Consumption of a fat-rich diet activates a proinflammatory response and induces insulin resistance in the hypothalamus. *Endocrinology, 146*(10), 4192–4199.

Despres, J. P. (2012). Body fat distribution and risk of cardiovascular disease: an update. *Circulation, 126*(10), 1301–1313.

Dey, A., Hao, S., Erion, J. R., Wosiski-Kuhn, M., & Stranahan, A. M. (2014). Glucocorticoid sensitization of microglia in a genetic mouse model of obesity and diabetes. *Journal of Neuroimmunology, 269*(1–2), 20–27.

Dinan, T. G., & Cryan, J. F. (2017). The microbiome-gut-brain axis in health and disease. *Gastroenterology Clinics of North America, 46*(1), 77–89.

Dinel, A. L., Andre, C., Aubert, A., Ferreira, G., Laye, S., & Castanon, N. (2011). Cognitive and emotional alterations are related to hippocampal inflammation in a mouse model of metabolic syndrome. *PLoS One, 6*(9).

Dinel, A. L., Andre, C., Aubert, A., Ferreira, G., Laye, S., & Castanon, N. (2014). Lipopolysaccharide-induced brain activation of the indoleamine 2,3-dioxygenase and depressive-like behavior are impaired in a mouse model of metabolic syndrome. *Psychoneuroendocrinology, 40*, 48–59.

Dixon, J. B., Dixon, M. E., & O'Brien, P. E. (2003). Depression in association with severe obesity: changes with weight loss. *Archives of Internal Medicine, 163*(17), 2058–2065.

Dixon, J. B., Hayden, M. J., Lambert, G. W., Dawood, T., Anderson, M. L., Dixon, M. E., et al. (2008). Raised CRP levels in obese patients: symptoms of depression have an independent positive association. *Obesity (Silver Spring), 16*(9), 2010–2015.

Doboo, N., de Vries, E. F., Kema, I. P., Patas, K., Prins, M., Nijholt, I. M., et al. (2012). The role of indoleamine 2,3-dioxygenase in a mouse model of neuroinflammation-induced depression. *Journal of Alzheimer's Disease, 28*(4), 905–915.

Dorfman, M. D., & Thaler, J. P. (2015). Hypothalamic inflammation and gliosis in obesity. *Current Opinion in Endocrinology, Diabetes, and Obesity, 22*(5), 325–330.

Emery, C. F., Fondow, M. D., Schneider, C. M., Christofi, F. L., Hunt, C., Busby, A. K., et al. (2007). Gastric bypass surgery is associated with reduced inflammation and less depression: a preliminary investigation. *Obesity Surgery, 17*(6), 759–763.

Engler, H., Brendt, P., Wischermann, J., Wegner, A., Rohling, R., Schoemberg, T., et al. (2017). Selective increase of cerebrospinal fluid IL-6 during experimental systemic inflammation in humans: association with depressive symptoms. *Molecular Psychiatry, 22*(10), 1448–1454.

Erion, J. R., Wosiski-Kuhn, M., Dey, A., Hao, S., Davis, C. L., Pollock, N. K., et al. (2014). Obesity elicits interleukin 1-mediated deficits in hippocampal synaptic plasticity. *Journal of Neuroscience, 34*(7), 2618–2631.

Esmaily, H., Sahebkar, A., Iranshahi, M., Ganjali, S., Mohammadi, A., Ferns, G., et al. (2015). An investigation of the effects of curcumin on anxiety and depression in obese individuals: a randomized controlled trial. *Chinese Journal of Integrative Medicine, 21*(5), 332–338.

Evrensel, A., & Ceylan, M. E. (2015). The gut-brain axis: the missing link in depression. *Clinical Psychopharmacology and Neuroscience: The Official Scientific Journal of the Korean College of Neuropsychopharmacology, 13*(3), 239–244.

Faulconbridge, L. F., & Bechtel, C. F. (2014). Depression and disordered eating in the obese person. *Current Obesity Reports, 3*(1), 127–136.

Ferrari, A. J., Charlson, F. J., Norman, R. E., Patten, S. B., Freedman, G., Murray, C. J., et al. (2013). Burden of depressive disorders by country, sex, age, and year: findings from the global burden of disease study 2010. *PLoS Medicine, 10*(11).

Finger, B. C., Dinan, T. G., & Cryan, J. F. (2010). Leptin-deficient mice retain normal appetitive spatial learning yet exhibit marked increases in anxiety-related behaviours. *Psychopharmacology, 210*(4), 559–568.

Fleming, P., Roubille, C., Richer, V., Starnino, T., McCourt, C., McFarlane, A., et al. (2015). Effect of biologics on depressive symptoms in patients with psoriasis: a systematic review. *Journal of the European Academy of Dermatology and Venereology, 29*(6), 1063–1070.

Foguet-Boreu, Q., Fernandez San Martin, M. I., Flores Mateo, G., Zabaleta Del Olmo, E., Ayerbe Garcia-Morzon, L., Perez-Pinar Lopez, M., et al. (2016). Cardiovascular risk assessment in patients with a severe mental illness: a systematic review and meta-analysis. *BMC Psychiatry, 16*, 141.

Forsythe, L. K., Wallace, J. M., & Livingstone, M. B. (2008). Obesity and inflammation: the effects of weight loss. *Nutrition Research Reviews, 21*(2), 117–133.

Francis, H., & Stevenson, R. (2013). The longer-term impacts of Western diet on human cognition and the brain. *Appetite, 63*, 119–128.

Frenois, F., Moreau, M., O'Connor, J., Lawson, M., Micon, C., Lestage, J., et al. (2007). Lipopolysaccharide induces delayed FosB/DeltaFosB immunostaining within the mouse extended amygdala, hippocampus and hypothalamus, that parallel the expression of depressive-like behavior. *Psychoneuroendocrinology, 32*(5), 516–531.

Fu, X. Y., Li, H. Y., Jiang, Q. S., Cui, T., Jiang, X. H., Zhou, Q. X., et al. (2016). Infliximab ameliorating depression-like behavior through inhibiting the activation of the IDO-HAAO pathway mediated by tumor necrosis factor-alpha in a rat model. *Neuroreport, 27*(13), 953–959.

Fu, X., Zunich, S. M., O'Connor, J. C., Kavelaars, A., Dantzer, R., & Kelley, K. W. (2010). Central administration of lipopolysaccharide induces depressive-like behavior in vivo and activates brain indoleamine 2,3 dioxygenase in murine organotypic hippocampal slice cultures. *Journal of Neuroinflammation, 7,* 43.

Fung, T. C., Olson, C. A., & Hsiao, E. Y. (2017). Interactions between the microbiota, immune and nervous systems in health and disease. *Nature Neuroscience, 20*(2), 145–155.

Ganjali, S., Sahebkar, A., Mahdipour, E., Jamialahmadi, K., Torabi, S., Akhlaghi, S., et al. (2014). Investigation of the effects of curcumin on serum cytokines in obese individuals: a randomized controlled trial. *ScientificWorldJournal, 2014,* 898361.

Gareau, M. G. (2016). Cognitive function and the microbiome. *International Review of Neurobiology, 131,* 227–246.

Ghasemi, R., Haeri, A., Dargahi, L., Mohamed, Z., & Ahmadiani, A. (2013). Insulin in the brain: sources, localization and functions. *Molecular Neurobiology, 47*(1), 145–171.

Gibney, S. M., McGuinness, B., Prendergast, C., Harkin, A., & Connor, T. J. (2013). Poly I:C-induced activation of the immune response is accompanied by depression and anxiety-like behaviours, kynurenine pathway activation and reduced BDNF expression. *Brain, Behavior, and Immunity, 28,* 170–181.

Godbout, J. P., Moreau, M., Lestage, J., Chen, J., Sparkman, N. L., O'Connor, J., et al. (2008). Aging exacerbates depressive-like behavior in mice in response to activation of the peripheral innate immune system. *Neuropsychopharmacology, 33*(10), 2341–2351.

Gold, A. B., Herrmann, N., Swardfager, W., Black, S. E., Aviv, R. I., Tennen, G., et al. (2011). The relationship between indoleamine 2,3-dioxygenase activity and post-stroke cognitive impairment. *Journal of Neuroinflammation, 8,* 17.

Goshen, I., Kreisel, T., Ben-Menachem-Zidon, O., Licht, T., Weidenfeld, J., Ben-Hur, T., et al. (2008). Brain interleukin-1 mediates chronic stress-induced depression in mice via adrenocortical activation and hippocampal neurogenesis suppression. *Molecular Psychiatry, 13*(7), 717–728.

Greenman, Y., Kuperman, Y., Drori, Y., Asa, S. L., Navon, I., Forkosh, O., et al. (2013). Postnatal ablation of POMC neurons induces an obese phenotype characterized by decreased food intake and enhanced anxiety-like behavior. *Molecular Endocrinology, 27*(7), 1091–1102.

Gregor, M. F., & Hotamisligil, G. S. (2011). Inflammatory mechanisms in obesity. *Annual Review of Immunology, 29,* 415–445.

Grosso, G., Micek, A., Marventano, S., Castellano, S., Mistretta, A., Pajak, A., et al. (2016). Dietary n-3 PUFA, fish consumption and depression: a systematic review and meta-analysis of observational studies. *Journal of Affective Disorders, 205,* 269–281.

Gulaj, E., Pawlak, K., Bien, B., & Pawlak, D. (2010). Kynurenine and its metabolites in Alzheimer's disease patients. *Advances in Medical Sciences, 55*(2), 204–211.

Gupta, S. C., Tyagi, A. K., Deshmukh-Taskar, P., Hinojosa, M., Prasad, S., & Aggarwal, B. B. (2014). Downregulation of tumor necrosis factor and other proinflammatory biomarkers by polyphenols. *Archives of Biochemistry and Biophysics, 559,* 91–99.

Haji, N., Mandolesi, G., Gentile, A., Sacchetti, L., Fresegna, D., Rossi, S., et al. (2012). TNF-alpha-mediated anxiety in a mouse model of multiple sclerosis. *Experimental Neurology, 237*(2), 296–303.

Hansen, D., Dendale, P., Beelen, M., Jonkers, R. A., Mullens, A., Corluy, L., et al. (2010). Plasma adipokine and inflammatory marker concentrations are altered in obese, as opposed to non-obese, type 2 diabetes patients. *European Journal of Applied Physiology, 109*(3), 397–404.

Hao, S., Dey, A., Yu, X., & Stranahan, A. M. (2016). Dietary obesity reversibly induces synaptic stripping by microglia and impairs hippocampal plasticity. *Brain, Behavior, and Immunity, 51,* 230–239.

Hashimoto, R., Ozaki, N., Ohta, T., Kasahara, Y., Kaneda, N., & Nagatsu, T. (1990). Plasma tetrahydrobiopterin levels in patients with psychiatric disorders. *Neuropsychobiology, 23*(3), 140–143.

Heilbronn, L. K., & Campbell, L. V. (2008). Adipose tissue macrophages, low grade inflammation and insulin resistance in human obesity. *Current Pharmaceutical Design, 14*(12), 1225–1230.

Henry, C. J., Huang, Y., Wynne, A. M., & Godbout, J. P. (2009). Peripheral lipopolysaccharide (LPS) challenge promotes microglial hyperactivity in aged mice that is associated with exaggerated induction of both pro-inflammatory IL-1beta and anti-inflammatory IL-10 cytokines. *Brain, Behavior, and Immunity, 23*(3), 309–317.

Higami, Y., Barger, J. L., Page, G. P., Allison, D. B., Smith, S. R., Prolla, T. A., et al. (2006). Energy restriction lowers the expression of genes linked to inflammation, the cytoskeleton, the extracellular matrix, and angiogenesis in mouse adipose tissue. *Journal of Nutrition, 136*(2), 343–352.

Hotamisligil, G. S. (2006). Inflammation and metabolic disorders. *Nature, 444*(7121), 860–867.

Hotamisligil, G. S., Murray, D. L., Choy, L. N., & Spiegelman, B. M. (1994). Tumor necrosis factor alpha inhibits signaling from the insulin receptor. *Proceedings of the National Academy of Sciences of the United States of America, 91*(11), 4854–4858.

Hotamisligil, G. S., Peraldi, P., Budavari, A., Ellis, R., White, M. F., & Spiegelman, B. M. (1996). IRS-1-mediated inhibition of insulin receptor tyrosine kinase activity in TNF-alpha- and obesity-induced insulin resistance. *Science, 271*(5249), 665–668.

Hryhorczuk, C., Sharma, S., & Fulton, S. E. (2013). Metabolic disturbances connecting obesity and depression. *Frontiers in Neuroscience, 7*, 177.

Incollingo Belsky, A. C., Epel, E. S., & Tomiyama, A. J. (2014). Clues to maintaining calorie restriction? Psychosocial profiles of successful long-term restrictors. *Appetite, 79*, 106–112.

Jamalan, M., Rezazadeh, M., Zeinali, M., & Ghaffari, M. A. (2015). Effect of ascorbic acid and alpha-tocopherol supplementations on serum leptin, tumor necrosis factor alpha, and serum amyloid A levels in individuals with type 2 diabetes mellitus. *Avicenna Journal of Phytomedicine, 5*(6), 531–539.

Kaidanovich-Beilin, O., Cha, D. S., & McIntyre, R. S. (2012). Crosstalk between metabolic and neuropsychiatric disorders. *F1000 Biology Reports, 4*, 14.

Kanety, H., Feinstein, R., Papa, M. Z., Hemi, R., & Karasik, A. (1995). Tumor necrosis factor alpha-induced phosphorylation of insulin receptor substrate-1 (IRS-1). Possible mechanism for suppression of insulin-stimulated tyrosine phosphorylation of IRS-1. *Journal of Biological Chemistry, 270*(40), 23780–23784.

Kang, E. B., Koo, J. H., Jang, Y. C., Yang, C. H., Lee, Y., Cosio-Lima, L. M., et al. (2016). Neuroprotective effects of endurance exercise against high-fat diet-induced hippocampal neuroinflammation. *Journal of Neuroendocrinology, 28*(5), 1–10.

Kariharan, T., Nanayakkara, G., Parameshwaran, K., Bagasrawala, I., Ahuja, M., Abdel-Rahman, E., et al. (2015). Central activation of PPAR-gamma ameliorates diabetes induced cognitive dysfunction and improves BDNF expression. *Neurobiology of Aging, 36* (3), 1451–1461.

Karlsen, A., Retterstol, L., Laake, P., Paur, I., Bohn, S. K., Sandvik, L., et al. (2007). Anthocyanins inhibit nuclear factor-kappaB activation in monocytes and reduce plasma concentrations of pro-inflammatory mediators in healthy adults. *Journal of Nutrition, 137*(8), 1951–1954.

Kawase, K., Kondo, K., Saito, T., Shimasaki, A., Takahashi, A., Kamatani, Y., et al. (2016). Risk factors and clinical characteristics of the depressive state induced by pegylated interferon therapy in patients with hepatitis C virus infection: a prospective study. *Psychiatry and Clinical Neurosciences, 70*(11), 489–497.

Kekow, J., Moots, R., Khandker, R., Melin, J., Freundlich, B., & Singh, A. (2011). Improvements in patient-reported outcomes, symptoms of depression and anxiety, and their association with clinical remission among patients with moderate-to-severe active early rheumatoid arthritis. *Rheumatology (Oxford), 50*(2), 401–409.

Kim, M., Bae, S., & Lim, K. M. (2013). Impact of high fat diet-induced obesity on the plasma levels of monoamine neurotransmitters in C57BL/6 mice. *Biomolecules & Therapeutics, 21*(6), 476–480.

Kim, Y. J., Kim, H. J., No, J. K., Chung, H. Y., & Fernandes, G. (2006). Anti-inflammatory action of dietary fish oil and calorie restriction. *Life Sciences, 78*(21), 2523–2532.

Kim, B., Sullivan, K. A., Backus, C., & Feldman, E. L. (2011). Cortical neurons develop insulin resistance and blunted Akt signaling: a potential mechanism contributing to enhanced ischemic injury in diabetes. *Antioxidants & Redox Signaling, 14*(10), 1829–1839.

Kinzl, J. F., Schrattenecker, M., Traweger, C., Mattesich, M., Fiala, M., & Biebl, W. (2006). Psychosocial predictors of weight loss after bariatric surgery. *Obesity Surgery, 16*(12), 1609–1614.

Klaus, F., Paterna, J. C., Marzorati, E., Sigrist, H., Gotze, L., Schwendener, S., et al. (2016). Differential effects of peripheral and brain tumor necrosis factor on inflammation, sickness, emotional behavior and memory in mice. *Brain, Behavior, and Immunity, 58*, 310–326.

Krishna, S., Keralapurath, M. M., Lin, Z., Wagner, J. J., de La Serre, C. B., Harn, D. A., et al. (2015). Neurochemical and electrophysiological deficits in the ventral hippocampus and selective behavioral alterations caused by high-fat diet in female C57BL/6 mice. *Neuroscience, 297*, 170–181.

Krishna, S., Lin, Z., de La Serre, C. B., Wagner, J. J., Harn, D. H., Pepples, L. M., et al. (2016). Time-dependent behavioral, neurochemical, and metabolic dysregulation in female C57BL/6 mice caused by chronic high-fat diet intake. *Physiology & Behavior, 157*, 196–208.

Kurhe, Y., & Mahesh, R. (2015a). Mechanisms linking depression co-morbid with obesity: an approach for serotonergic type 3 receptor antagonist as novel therapeutic intervention. *Asian Journal of Psychiatry, 17*, 3–9.

Kurhe, Y., & Mahesh, R. (2015b). Ondansetron attenuates co-morbid depression and anxiety associated with obesity by inhibiting the biochemical alterations and improving serotonergic neurotransmission. *Pharmacology, Biochemistry, and Behavior, 136*, 107–116.

Kurhe, Y., Mahesh, R., & Devadoss, T. (2017). Novel 5-HT3 receptor antagonist QCM-4 attenuates depressive-like phenotype associated with obesity in high-fat-diet-fed mice. *Psychopharmacology, 234*(7), 1165–1179.

Lacosta, S., Merali, Z., & Anisman, H. (1999). Behavioral and neurochemical consequences of lipopolysaccharide in mice: anxiogenic-like effects. *Brain Research, 818*(2), 291–303.

Lann, D., & LeRoith, D. (2007). Insulin resistance as the underlying cause for the metabolic syndrome. *Medical Clinics of North America, 91*(6), 1063–1077. viii.

Lasselin, J., & Capuron, L. (2014). Chronic low-grade inflammation in metabolic disorders: relevance for behavioral symptoms. *Neuroimmunomodulation, 21*(2–3), 95–101.

Lasselin, J., Magne, E., Beau, C., Ledaguenel, P., Dexpert, S., Aubert, A., et al. (2014). Adipose inflammation in obesity: relationship with circulating levels of inflammatory markers and association with surgery-induced weight loss.

Journal of Clinical Endocrinology and Metabolism, 99(1), E53–61.

Lawson, M. A., Kelley, K. W., & Dantzer, R. (2011). Intracerebroventricular administration of HIV-1 Tat induces brain cytokine and indoleamine 2,3-dioxygenase expression: a possible mechanism for AIDS comorbid depression. *Brain, Behavior, and Immunity, 25*(8), 1569–1575.

Lawson, M. A., Parrott, J. M., McCusker, R. H., Dantzer, R., Kelley, K. W., & JC, O. C. (2013). Intracerebroventricular administration of lipopolysaccharide induces indoleamine-2,3-dioxygenase-dependent depression-like behaviors. *Journal of Neuroinflammation, 10*(1), 87.

Le Chatelier, E., Nielsen, T., Qin, J., Prifti, E., Hildebrand, F., Falony, G., et al. (2013). Richness of human gut microbiome correlates with metabolic markers. *Nature, 500* (7464), 541–546.

Li, X. L., Aou, S., Oomura, Y., Hori, N., Fukunaga, K., & Hori, T. (2002). Impairment of long-term potentiation and spatial memory in leptin receptor-deficient rodents. *Neuroscience, 113*(3), 607–615.

Liu, L. S., Spelleken, M., Rohrig, K., Hauner, H., & Eckel, J. (1998). Tumor necrosis factor-alpha acutely inhibits insulin signaling in human adipocytes: implication of the p80 tumor necrosis factor receptor. *Diabetes, 47*(4), 515–522.

Liu, Y., Wu, X. M., Luo, Q. Q., Huang, S., Yang, Q. W., Wang, F. X., et al. (2015). CX3CL1/CX3CR1-mediated microglia activation plays a detrimental role in ischemic mice brain via p38MAPK/PKC pathway. *Journal of Cerebral Blood Flow and Metabolism, 35*(10), 1623–1631.

Llorens-Martin, M., Jurado-Arjona, J., Fuster-Matanzo, A., Hernandez, F., Rabano, A., & Avila, J. (2014). Peripherally triggered and GSK-3beta-driven brain inflammation differentially skew adult hippocampal neurogenesis, behavioral pattern separation and microglial activation in response to ibuprofen. *Translational Psychiatry, 4.*

Luppino, F. S., de Wit, L. M., Bouvy, P. F., Stijnen, T., Cuijpers, P., Penninx, B. W., et al. (2010). Overweight, obesity, and depression: a systematic review and meta-analysis of longitudinal studies. *Archives of General Psychiatry, 67*(3), 220–229.

Magkos, F., Fraterrigo, G., Yoshino, J., Luecking, C., Kirbach, K., Kelly, S. C., et al. (2016). Effects of moderate and subsequent progressive weight loss on metabolic function and adipose tissue biology in humans with obesity. *Cell Metabolism, 23*(4), 591–601.

Mangge, H., Summers, K. L., Meinitzer, A., Zelzer, S., Almer, G., Prassl, R., et al. (2014). Obesity-related dysregulation of the tryptophan-kynurenine metabolism: role of age and parameters of the metabolic syndrome. *Obesity (Silver Spring), 22*(1), 195–201.

Manosso, L. M., Neis, V. B., Moretti, M., Daufenbach, J. F., Freitas, A. E., Colla, A. R., et al. (2013). Antidepressant-like effect of alpha-tocopherol in a mouse model of depressive-like behavior induced by TNF-alpha. *Progress in Neuro-Psychopharmacology & Biological Psychiatry, 46,* 48–57.

Martinac, M., Babic, D., Bevanda, M., Vasilj, I., Glibo, D. B., Karlovic, D., et al. (2017). Activity of the hypothalamic-pituitary-adrenal axis and inflammatory mediators in major depressive disorder with or without metabolic syndrome. *Psychiatria Danubina, 29*(1), 39–50.

McNelis, J. C., & Olefsky, J. M. (2014). Macrophages, immunity, and metabolic disease. *Immunity, 41*(1), 36–48.

Merali, Z., Brennan, K., Brau, P., & Anisman, H. (2003). Dissociating anorexia and anhedonia elicited by interleukin-1beta: antidepressant and gender effects on responding for "free chow" and "earned" sucrose intake. *Psychopharmacology, 165*(4), 413–418.

Miller, A. A., & Spencer, S. J. (2014). Obesity and neuroinflammation: a pathway to cognitive impairment. *Brain, Behavior, and Immunity, 42,* 10–21.

Moller, K., Ostermann, A. I., Rund, K., Thoms, S., Blume, C., Stahl, F., et al. (2016). Influence of weight reduction on blood levels of C-reactive protein, tumor necrosis factor-alpha, interleukin-6, and oxylipins in obese subjects. *Prostaglandins, Leukotrienes, and Essential Fatty Acids, 106,* 39–49.

Moloney, R. D., Desbonnet, L., Clarke, G., Dinan, T. G., & Cryan, J. F. (2014). The microbiome: stress, health and disease. *Mammalian Genome, 25*(1–2), 49–74.

Moreau, M., Andre, C., O'Connor, J. C., Dumich, S. A., Woods, J. A., Kelley, K. W., et al. (2008). Inoculation of Bacillus Calmette-Guerin to mice induces an acute episode of sickness behavior followed by chronic depressive-like behavior. *Brain, Behavior, and Immunity, 22*(7), 1087–1095.

Murr, C., Widner, B., Wirleitner, B., & Fuchs, D. (2002). Neopterin as a marker for immune system activation. *Current Drug Metabolism, 3*(2), 175–187.

Nakamura, Y. K., & Omaye, S. T. (2012). Metabolic diseases and pro- and prebiotics: mechanistic insights. *Nutrition & Metabolism (London), 9*(1), 60.

Nasser, A., Moller, L. B., Olesen, J. H., Konradsen Refsgaard, L., & Andreasen, J. T. (2014). Anxiety- and depression-like phenotype of hph-1 mice deficient in tetrahydrobiopterin. *Neuroscience Research, 89,* 44–53.

Neufeld, K. A., Kang, N., Bienenstock, J., & Foster, J. A. (2011). Effects of intestinal microbiota on anxiety-like behavior. *Communicative & Integrative Biology, 4*(4), 492–494.

Norden, D. M., McCarthy, D. O., Bicer, S., Devine, R. D., Reiser, P. J., Godbout, J. P., et al. (2015). Ibuprofen ameliorates fatigue- and depressive-like behavior in tumor-bearing mice. *Life Sciences, 143,* 65–70.

O'Connor, J. C., Andre, C., Wang, Y., Lawson, M. A., Szegedi, S. S., Lestage, J., et al. (2009). Interferon-gamma and tumor necrosis factor-alpha mediate the upregulation of indoleamine 2,3-dioxygenase and the induction of depressive-like behavior in mice in response to bacillus Calmette-Guerin. *Journal of Neuroscience*, 29(13), 4200–4209.

O'Connor, J. C., Lawson, M. A., Andre, C., Briley, E. M., Szegedi, S. S., Lestage, J., et al. (2009). Induction of IDO by bacille Calmette-Guerin is responsible for development of murine depressive-like behavior. *Journal of Immunology*, 182(5), 3202–3212.

O'Connor, J. C., Lawson, M. A., Andre, C., Moreau, M., Lestage, J., Castanon, N., et al. (2009). Lipopolysaccharide-induced depressive-like behavior is mediated by indoleamine 2,3-dioxygenase activation in mice. *Molecular Psychiatry*, 14(5), 511–522.

Oxenkrug, G., Tucker, K. L., Requintina, P., & Summergrad, P. (2011). Neopterin, a marker of interferon-gamma-inducible inflammation, correlates with pyridoxal-5'-phosphate, waist circumference, HDL-cholesterol, insulin resistance and mortality risk in adult Boston community dwellers of Puerto Rican Origin. *American Journal of Neuroprotection and Neuroregeneration*, 3(1), 48–52.

Pariante, C. M., & Miller, A. H. (2001). Glucocorticoid receptors in major depression: relevance to pathophysiology and treatment. *Biological Psychiatry*, 49(5), 391–404.

Park, S. E., Lawson, M., Dantzer, R., Kelley, K. W., & McCusker, R. H. (2011). Insulin-like growth factor-I peptides act centrally to decrease depression-like behavior of mice treated intraperitoneally with lipopolysaccharide. *Journal of Neuroinflammation*, 8, 179.

Park, H. S., Lee, J. M., Cho, H. S., Park, S. S., & Kim, T. W. (2017). Physical exercise ameliorates mood disorder-like behavior on high fat diet-induced obesity in mice. *Psychiatry Research*, 250, 71–77.

Park, H. S., Park, J. Y., & Yu, R. (2005). Relationship of obesity and visceral adiposity with serum concentrations of CRP, TNF-alpha and IL-6. *Diabetes Research and Clinical Practice*, 69(1), 29–35.

Pascoal, L. B., Bombassaro, B., Ramalho, A. F., Coope, A., Moura, R. F., Correa-da-Silva, F., et al. (2017). Resolvin RvD2 reduces hypothalamic inflammation and rescues mice from diet-induced obesity. *Journal of Neuroinflammation*, 14(1), 5.

Pedersen, B. K., & Febbraio, M. A. (2012). Muscles, exercise and obesity: skeletal muscle as a secretory organ. *Nature Reviews. Endocrinology*, 8(8), 457–465.

Pistell, P. J., Morrison, C. D., Gupta, S., Knight, A. G., Keller, J. N., Ingram, D. K., et al. (2010). Cognitive impairment following high fat diet consumption is associated with brain inflammation. *Journal of Neuroimmunology*, 219 (1–2), 25–32.

Pohl, J., Woodside, B., & Luheshi, G. N. (2009). Changes in hypothalamically mediated acute-phase inflammatory responses to lipopolysaccharide in diet-induced obese rats. *Endocrinology*, 150(11), 4901–4910.

Pomytkin, I. A., Cline, B. H., Anthony, D. C., Steinbusch, H. W., Lesch, K. P., & Strekalova, T. (2015). Endotoxaemia resulting from decreased serotonin tranporter (5-HTT) function: a reciprocal risk factor for depression and insulin resistance? *Behavioural Brain Research*, 276, 111–117.

Radler, M. E., Wright, B. J., Walker, F. R., Hale, M. W., & Kent, S. (2015). Calorie restriction increases lipopolysaccharide-induced neuropeptide Y immunolabeling and reduces microglial cell area in the arcuate hypothalamic nucleus. *Neuroscience*, 285, 236–247.

Raison, C. L., Dantzer, R., Kelley, K. W., Lawson, M. A., Woolwine, B. J., Vogt, G., et al. (2010). CSF concentrations of brain tryptophan and kynurenines during immune stimulation with IFN-alpha: relationship to CNS immune responses and depression. *Molecular Psychiatry*, 15(4), 393–403.

Raison, C. L., Demetrashvili, M., Capuron, L., & Miller, A. H. (2005). Neuropsychiatric adverse effects of interferon-alpha: recognition and management. *CNS Drugs*, 19(2), 105–123.

Raison, C. L., & Miller, A. H. (2003). When not enough is too much: the role of insufficient glucocorticoid signaling in the pathophysiology of stress-related disorders. *American Journal of Psychiatry*, 160(9), 1554–1565.

Rao, S. R. (2012). Inflammatory markers and bariatric surgery: a meta-analysis. *Inflammation Research*, 61(8), 789–807.

Redman, L. M., & Ravussin, E. (2011). Caloric restriction in humans: impact on physiological, psychological, and behavioral outcomes. *Antioxidants & Redox Signaling*, 14(2), 275–287.

Reus, G. Z., Jansen, K., Titus, S., Carvalho, A. F., Gabbay, V., & Quevedo, J. (2015). Kynurenine pathway dysfunction in the pathophysiology and treatment of depression: evidences from animal and human studies. *Journal of Psychiatric Research*, 68, 316–328.

Rummel, C., Inoue, W., Poole, S., & Luheshi, G. N. (2010). Leptin regulates leukocyte recruitment into the brain following systemic LPS-induced inflammation. *Molecular Psychiatry*, 15(5), 523–534.

Salazar, A., Gonzalez-Rivera, B. L., Redus, L., Parrott, J. M., & O'Connor, J. C. (2012). Indoleamine 2,3-dioxygenase mediates anhedonia and anxiety-like behaviors caused by peripheral lipopolysaccharide immune challenge. *Hormones and Behavior*, 62(3), 202–209.

Schedlowski, M., Engler, H., & Grigoleit, J. S. (2014). Endotoxin-induced experimental systemic inflammation in humans: a model to disentangle immune-to-brain communication. *Brain, Behavior, and Immunity*, 35, 1–8.

Scherer, P. E., & Hill, J. A. (2016). Obesity, diabetes, and cardiovascular diseases: a compendium. *Circulation Research*, 118(11), 1703–1705.

Schmatz, R., Bitencourt, M. R., Patias, L. D., Beck, M., da, C. A. G., Zanini, D., et al. (2017). Evaluation of the biochemical, inflammatory and oxidative profile of obese patients given clinical treatment and bariatric surgery. *Clinica Chimica Acta, 465,* 72–79.

Sharma, A. N., Elased, K. M., & Lucot, J. B. (2012). Rosiglitazone treatment reversed depression- but not psychosis-like behavior of db/db diabetic mice. *Journal of Psychopharmacology, 26*(5), 724–732.

Sharma, S., & Fulton, S. (2013). Diet-induced obesity promotes depressive-like behaviour that is associated with neural adaptations in brain reward circuitry. *International Journal of Obesity, 37*(3), 382–389.

Shi, Z., Ren, H., Huang, Z., Peng, Y., He, B., Yao, X., et al. (2017). Fish oil prevents lipopolysaccharide-induced depressive-like behavior by inhibiting neuroinflammation. *Molecular Neurobiology, 54*(9), 7327–7334.

Soczynska, J. K., Kennedy, S. H., Woldeyohannes, H. O., Liauw, S. S., Alsuwaidan, M., Yim, C. Y., et al. (2011). Mood disorders and obesity: understanding inflammation as a pathophysiological nexus. *Neuromolecular Medicine, 13*(2), 93–116.

Soleimani, M., Jameie, S. B., Barati, M., Mehdizadeh, M., & Kerdari, M. (2014). Effects of coenzyme Q10 on the ratio of TH1/TH2 in experimental autoimmune encephalomyelitis model of multiple sclerosis in C57BL/6. *Iranian Biomedical Journal, 18*(4), 203–211.

Stone, T. W., Forrest, C. M., Stoy, N., & Darlington, L. G. (2012). Involvement of kynurenines in Huntington's disease and stroke-induced brain damage. *Journal of Neural Transmission (Vienna, Austria: 1996), 119*(2), 261–274.

Stranahan, A. M. (2015). Models and mechanisms for hippocampal dysfunction in obesity and diabetes. *Neuroscience, 309,* 125–139.

Stranahan, A. M., Arumugam, T. V., Cutler, R. G., Lee, K., Egan, J. M., & Mattson, M. P. (2008). Diabetes impairs hippocampal function through glucocorticoid-mediated effects on new and mature neurons. *Nature Neuroscience, 11*(3), 309–317.

Su, K. P., Huang, S. Y., Peng, C. Y., Lai, H. C., Huang, C. L., Chen, Y. C., et al. (2010). Phospholipase A2 and cyclooxygenase 2 genes influence the risk of interferon-alpha-induced depression by regulating polyunsaturated fatty acids levels. *Biological Psychiatry, 67*(6), 550–557.

Su, K. P., Lai, H. C., Yang, H. T., Su, W. P., Peng, C. Y., Chang, J. P., et al. (2014). Omega-3 fatty acids in the prevention of interferon-alpha-induced depression: results from a randomized, controlled trial. *Biological Psychiatry, 76*(7), 559–566.

Su, K. P., Matsuoka, Y., & Pae, C. U. (2015). Omega-3 polyunsaturated fatty acids in prevention of mood and anxiety disorders. *Clinical Psychopharmacology and Neuroscience: The Official Scientific Journal of the Korean College of Neuropsychopharmacology, 13*(2), 129–137.

Thaler, J. P., Yi, C. X., Schur, E. A., Guyenet, S. J., Hwang, B. H., Dietrich, M. O., et al. (2012). Obesity is associated with hypothalamic injury in rodents and humans. *Journal of Clinical Investigation, 122*(1), 153–162.

Vauzour, D., Martinsen, A., & Laye, S. (2015). Neuroinflammatory processes in cognitive disorders: is there a role for flavonoids and n-3 polyunsaturated fatty acids in counteracting their detrimental effects? *Neurochemistry International 89,* 63–74.

Visser, M., Bouter, L. M., McQuillan, G. M., Wener, M. H., & Harris, T. B. (1999). Elevated C-reactive protein levels in overweight and obese adults. *JAMA, 282*(22), 2131–2135.

Wang, W. Y., Tan, M. S., Yu, J. T., & Tan, L. (2015). Role of pro-inflammatory cytokines released from microglia in Alzheimer's disease. *Annals of Translational Medicine, 3* (10), 136.

Wosiski-Kuhn, M., Erion, J. R., Gomez-Sanchez, E. P., Gomez-Sanchez, C. E., & Stranahan, A. M. (2014). Glucocorticoid receptor activation impairs hippocampal plasticity by suppressing BDNF expression in obese mice. *Psychoneuroendocrinology, 42,* 165–177.

Xie, W., Cai, L., Yu, Y., Gao, L., Xiao, L., He, Q., et al. (2014). Activation of brain indoleamine 2,3-dioxygenase contributes to epilepsy-associated depressive-like behavior in rats with chronic temporal lobe epilepsy. *Journal of Neuroinflammation, 11,* 41.

Xu, H., Barnes, G. T., Yang, Q., Tan, G., Yang, D., Chou, C. J., et al. (2003). Chronic inflammation in fat plays a crucial role in the development of obesity-related insulin resistance. *Journal of Clinical Investigation, 112*(12), 1821–1830.

Yamada, N., Katsuura, G., Ochi, Y., Ebihara, K., Kusakabe, T., Hosoda, K., et al. (2011). Impaired CNS leptin action is implicated in depression associated with obesity. *Endocrinology, 152*(7), 2634–2643.

Yates, K. F., Sweat, V., Yau, P. L., Turchiano, M. M., & Convit, A. (2012). Impact of metabolic syndrome on cognition and brain: a selected review of the literature. *Arteriosclerosis, Thrombosis, and Vascular Biology, 32*(9), 2060–2067.

Zaminelli, T., Gradowski, R. W., Bassani, T. B., Barbiero, J. K., Santiago, R. M., Maria-Ferreira, D., et al. (2014). Antidepressant and antioxidative effect of Ibuprofen in the rotenone model of Parkinson's disease. *Neurotoxicity Research, 26*(4), 351–362.

Zhang, M. M., Liu, S. B., Chen, T., Koga, K., Zhang, T., Li, Y. Q., et al. (2014). Effects of NB001 and gabapentin on irritable bowel syndrome-induced behavioral anxiety and spontaneous pain. *Molecular Brain, 7,* 47.

Inflammation and Depression in Patients With Autoimmune Disease, Diabetes, and Obesity

Jonathan M. Gregory*, Michael Mak*, Roger S. McIntyre[†,‡]

*Western University, London, ON, Canada
[†]University of Toronto, Toronto, ON, Canada
[‡]University Health Network, Toronto, ON, Canada

INTRODUCTION

Depression is the leading cause of disability worldwide. More than 300 million people suffer from depression globally, and the disease burden of depression is on the rise (World Health Organization, 2017). Compounding this burden is the high likelihood of comorbid medical illnesses in those with depression. Multimorbidity (the presence of two or more chronic medical conditions) is the norm rather than the exception for this population (Smith et al., 2014). Moreover, the relationship between physical illness and depression is bidirectional. Not only are those with depression at increased risk of developing physical illness, but those with physical illness are also at increased risk of developing depression (Patten, 2001).

There is compelling evidence that inflammation is an important factor underlying the complex relationship between depression and physical comorbidities. In brief, the factors that support the role of inflammation in depression include the following: (1) Major depressive disorder is associated with modest increases in levels of pro-inflammatory biomarkers when compared with the general population (Young, Bruno, & Pomara, 2014); (2) indicators of systemic inflammation can predict the incidence of depression in a depression-free population (Khandaker, Pearson, Zammit, Lewis, & Jones, 2014; Milaneschi et al., 2009); (3) induction of a pro-inflammatory state via pathogens or immunomodulating treatments can produce depressive symptoms ("sickness behavior") or major depressive episodes (e.g., following treatment with interferon) (Rosenblat, Cha, Mansur, & McIntyre, 2014); (4) anti-inflammatory treatments (including both pharmacological and lifestyle-based interventions) can be effective for depression (Köhler et al., 2014); (5) plausible pathophysiological mechanisms have been proposed involving cytokine-induced monoamine alterations, dysregulation of the hypothalamic-pituitary axis, pathological microglial activation, and alterations in neurotrophins and neuroplasticity (Berk et al., 2013; Kim, Na, Myint, & Leonard, 2016; Rosenblat et al., 2014); and (6) illnesses characterized by systemic inflammation (e.g., autoimmune disease and metabolic disease) are highly comorbid with depression.

The latter factor—depression in those with inflammatory illnesses—is the subject of this chapter. Examining the epidemiology and clinical correlates of depression in inflammatory illnesses can elucidate the complex relationship between depression and physical illness, allowing for an exploration of shared pathoetiologic mechanisms and informing future treatment strategies. Rather than creating an exhaustive review of the clinical correlates of depression in all inflammation-related diseases, the discussion will focus on the two broad areas of inflammatory disease: well-studied autoimmune disorders that have a known association with depression (rheumatoid arthritis, psoriasis, inflammatory bowel disease, systemic lupus erythematosus, and multiple sclerosis) and common metabolic disorders that are characterized by pro-inflammatory states (diabetes and obesity).

EPIDEMIOLOGY AND CLINICAL CORRELATES

In a population-based longitudinal study of 5437 patients, Caneo, Marston, Bellón, and King (2016) found that the odds of incident depression are higher in almost all inflammatory illnesses than in noninflammatory illnesses. After the adjustment for confounders, those with autoimmune disease had the highest odds ratio (OR) of new-onset depression at 1-year follow-up compared with healthy controls (OR = 2.27; 95% confidence interval (CI), 1.06–4.86). For those with metabolic syndrome, the OR was 1.83 (CI, 1.26–2.63). Further adjustments for illness load revealed that autoimmune illness contributed the most to the risk of

depression, followed by cardiometabolic diseases, whereas noninflammatory medical illnesses did not contribute to the risk. Autoimmune disease is generally associated with a higher grade of systemic inflammation than cardiometabolic disease, although severe inflammation often occurs in a relapsing-remitting fashion rather than the consistent low-grade inflammation seen in diabetes or obesity. This finding, utilizing a categorical division of inflammatory versus noninflammatory disease, is consistent with previous longitudinal studies that have found associations between the dimensional measures of systemic inflammation (levels of pro-inflammatory cytokines or C-reactive protein) and the incidence of depression (Khandaker et al., 2014; Milaneschi et al., 2009). Based on the Danish registry (which includes 91,637 people who have had hospital contact for mood disorders), prior hospital contact because of an autoimmune disease increased the risk of a subsequent mood disorder diagnosis by 45% (incident rate ratio, 1.45; CI, 1.39–1.52), and 5% of those with mood disorders had a previous hospital contact for autoimmune disease (Benros et al., 2013). Increasingly, depression and autoimmune diseases are being shown to have a bidirectional relationship. Based on the data from the British birth cohort study (including 315 cases with autoimmune diseases and 1499 cases with depression), autoimmune disorder onset was associated with subsequent increased hazard of depression onset (HR = 1.39; CI, 1.11–1.74), and depression onset was associated with increased subsequent hazard of autoimmune disorder onset (HR = 1.40; CI, 1.09–1.80) (Euesden, Danese, Lewis, & Maughan, 2017).

The following subsections review the epidemiology of depression when it is comorbid with specific inflammatory diseases. Reviewing the phenotypes of patients with combined depression and inflammatory disorders can provide insight into a possible "inflamed depression" phenotype. Each section will also review the correlation (or the lack thereof) between depressive symptoms and inflammatory markers and the impact of disease modification on depressive symptoms.

Inflammatory Bowel Disease

Inflammatory bowel diseases (IBD), encompassing Crohn's disease (CD) and ulcerative colitis (UC), are chronic illnesses characterized by inflammation of the small and large intestines. Common symptoms include bloody diarrhea, vomiting, intense abdominal pain, and weight loss; diagnosis is made by colonoscopy and biopsies of lesions. Those with IBD often suffer from depression as a comorbidity. A recent metaanalysis of 13 cross-sectional studies found that 21.2% of patients with IBD had depression versus 13.4% of controls (Mikocka-Walus, Knowles, Keefer, & Graff, 2016). Another 2016 systematic review noted that the pooled prevalence of diagnosed depressive disorders (15.2%) was lower than the prevalence of depressive symptoms (21.6%) in IBD (Neuendorf, Harding, Stello, Hanes, & Wahbeh, 2016). The high rate of depression is present in both UC and CD, although the risk seems to be higher in CD (Mikocka-Walus et al., 2016; Neuendorf et al., 2016). A population-based study using the gold standard for diagnosis of depression (structured clinical interviews) found that the lifetime prevalence of major depression was significantly higher in the IBD group (27.2%) as compared with age, gender, and region-matched controls (12.3%; OR 2.20; CI, 1.64–2.95) with 12-month prevalence at 9.1% in IBD as compared with 5.5% (OR 1.53; CI, 0.96–2.45) in controls (Walker et al., 2008). Compounding this high coprevalence is the finding that comorbid psychiatric disorders are largely untreated in this patient population. A Dutch study showed that over 60% of adult IBD patients attending a tertiary care center with comorbid depression or anxiety did not receive treatment (Evertsz'

et al., 2012). Similarly, American data demonstrated that only 36% of individuals with IBD and depressive symptoms had visited mental health professionals in the prior year (Bhandari, Larson, Kumar, & Stein, 2017).

The data supporting IBD as a risk factor for depression are more robust than the data supporting depression as a risk factor for IBD, although there is some evidence of a bidirectional relationship. Adults with CD and UC are more likely to develop depression before IBD onset, but a significant proportion also develops depression after IBD onset (Mikocka-Walus et al., 2016). Prospective data show that IBD is associated with a significantly higher risk of future depression; relative risk for depression was 1.21 (CI, 0.95–1.54) in patients with ulcerative colitis and 1.67 (CI, 1.31–2.09) in patients with Crohn's disease (Kurina, Goldacre, Yeates, & Gill, 2001). The same study found that the risk for IBD in those with prior depression was also significantly elevated for UC [relative risk 1.4 (CI, 1.13–1.72)] but not for CD [relative risk 1.14 (CI, 0.85–1.5)].

For those with IBD, there is an increased prevalence of depression with active disease versus remission. Pooled prevalence data from a recent systematic review found that patients with active disease had a significantly higher prevalence of depressive symptoms (40.7%; CI, 31.1%–50.3%) compared with those in remission (16.5%) (Neuendorf et al., 2016). Similarly, increased severity of depression [as measured by the beck depression inventory (BDI)] has been independently associated with disease activity in CD (Mittermaier et al., 2004). A number of relapses and time to recurrence of disease were shown to be significantly correlated with higher BDI scores at baseline in a prospective study of 60 clinically inactive IBD patients over 18 months. Interestingly, immunosuppressive therapy has been shown to improve depression in patients with IBD (Horst et al., 2015). In a recent study of 160 patients with IBD, treatment with either vedolizumab or standard anti-TNF treatments (infliximab or adalimumab) significantly improved both sleep and depression within 6 weeks, and these reductions in mood and sleep symptoms persisted to the end of the study at 54 weeks (Stevens et al., 2016).

Notably, depressive symptoms tend to manifest after IBD diagnosis in pediatric IBD (Mikocka-Walus et al., 2016). In an attempt to characterize the phenotype of depression in patients with IBD, Szigethy et al. performed a multivariate modeling analysis of depressive symptoms in a pediatric population with IBD. These subtypes, based on items in the children's depression rating scale-revised, included (1) mild depression (fatigue, irritability, and depressed feelings), (2) somatic depression (anhedonia, change in appetite, fatigue, physical complaints, irritability, depressed feelings, depressed facial affect, listless speech, and hypoactivity), and (3) cognitive despair (morbid ideation, suicidal ideation, and weeping) (Szigethy et al., 2014).

Rheumatoid Arthritis

Rheumatoid arthritis (RA) is an autoimmune disease that primarily affects joints, causing them to become swollen, painful, and stiff due to inflammation; the skin, lungs, kidneys, and vascular system are also affected. In a comprehensive metaanalysis of 72 studies, there was a significantly increased prevalence of depression in those with RA compared with the general population, but the degree of this difference varied by the methods used to define depression (Matcham, Rayner, Steer, & Hotopf, 2013). Using DSM-defined criteria through clinical interviews, the prevalence was 16.8% (CI, 10%–24%); using the Patient Health Questionnaire-9, depression prevalence was 38.8% (CI, 34%–43%); and using the Hospital Anxiety and Depression Scale (HADS) with a cutoff of 8 or 11, depression prevalence was 34.2% (CI, 25%–44%) and 14.8% (CI, 12%–18%), respectively.

In a large nationwide cohort study from Finland focusing on older community-dwelling patients (> 50 years old), 10% of rheumatoid arthritis sufferers had used antidepressants in the year prior to diagnosis (Jyrkkä et al., 2014). Post-RA diagnosis, increasing number of medical comorbidities showed a linear association with new antidepressant use. Cumulative incidence of antidepressant initiations during follow-up (mean of 4.4 years) was 11.4% in men and 16.2% in women in RA patients who had been antidepressant naive (Jyrkkä et al., 2014). In a recent German longitudinal study (involving 7301 cases of late-onset RA and 7301 controls), RA was found to be a strong risk factor for the development of depression [hazard ratio (HR) 1.55; $P < .001$] (Drosselmeyer, Jacob, Rathmann, Rapp, & Kostev, 2017). In the 5-year follow-up, 22% of those with RA were newly diagnosed with depression, whereas only 14.3% of the control group received this diagnosis. In a similar finding, a recent British longitudinal study noted that ∼30% of RA patients develop depression within 5 years of RA diagnosis, with a significantly higher risk in women than in men (36.5% of women vs 23.7% of men had incident depression; $P < .001$) (Jacob, Rockel, & Kostev, 2017).

There is recent evidence of a bidirectional relationship between depression and RA (i.e., depression increases the risk of RA, and RA increases the risk of depression) based on a large Taiwanese cohort study (Lu et al., 2016). The incidence of RA was higher in those with depression compared with those without depression, even after adjusting for confounders [2.07 vs 1.21/1000 person-years (PY); adjusted HR = 1.65; CI, 1.41–1.77]. The incidence of depression in those with RA was higher than those without RA (15.69 vs 8.95/1000 PY; adjusted HR = 1.69; CI, 1.51–1.87) (Lu et al., 2016).

There is growing evidence that patients with depression or a history of depression tend to achieve worse long-term outcomes and increased mortality, possibly mediated by the bidirectional relationship between pain and depression (Ang, Choi, Kroenke, & Wolfe, 2005; Irwin,

Davis, & Zautra, 2008). Depressive symptoms are more common in RA patients with higher pain levels, and those with moderate-to-severe pain and concurrent depressive symptoms have substantially higher opioid use (Jobski, Luque Ramos, Albrecht, & Hoffmann, 2017). In a 1-year prospective study involving outpatient rheumatoid arthritis patients, tender joint count and patient global assessment (which are the subjective components of the DAS28 and the gold standard measure for RA disease activity) were significantly associated with higher depressive scores (HADS), although there was no significant association between depression at study baseline and odds of remission at study end (Matcham, Ali, Irving, Hotopf, & Chalder, 2016). Conversely, a recent study showed lower RA remission rate symptoms after 6 months of treatment with biological disease-modifying antirheumatic drugs in those with higher baseline depressive symptoms (Miwa et al., 2017).

Fatigue and sleep disruption not only are particularly common in RA but also represent somatic symptoms of depression. In a recent prospective study of RA patients, the incidence of fatigue, sleep disorders, and depression at baseline was 89%, 95%, and 67%, respectively. Following 3 months of biological therapy, improvement in these three categories was observed in 58.6%, 26.3%, and 34.3% of cases, respectively (Genty et al., 2017). Notably, post hoc analysis of the data from the trials of sirukumab (an IL-6 antibody) for RA has shown that patients with prevalent depressed mood and anhedonia but not receiving antidepressants (26% of patients at study entry) achieved significant improvements in these symptoms ($P = .0006$) (Hsu et al., 2015). Interestingly, this improvement was independent of clinical response in RA symptoms.

Psoriasis

Psoriasis is an autoimmune disease characterized by the development of scaly patches and plaques on the skin due to premature skin cell regeneration; arthritis is a common complication. The underlying pathophysiology is a disruption of the inflammatory cascade leading to secretion of pro-inflammatory cytokines [tumor necrosis factor (TNF)-α, interleukin (IL)-1β, IL-6, IL-22, and IL-36] in the epidermis. In a robust metaanalysis of 98 cross-sectional studies, the prevalence of depression in patients with psoriasis was greater than would be expected in the general population (Dowlatshahi, Wakkee, Arends, & Nijsten, 2014). The pooled prevalence of depression in psoriasis varied depending on classification criteria; the prevalence was 12% for studies using ICD codes and 19% when clinical depression was diagnosed with DSM-IV criteria (Dowlatshahi et al., 2014). When using validated questionnaires, the pooled prevalence of depressive symptoms ranged from 23% (using the HADS with a cutoff of 8) to 36% (using the BDI with a cutoff of 10). In studies with a control population, psoriasis patients were much more likely to use antidepressant medications (OR 4.24 and CI, 1.53–11.76) and have an ICD code for depression (OR 1.57; CI, 1.40–1.76) (Dowlatshahi et al., 2014).

A population-based cohort study including 150,000 cases of psoriasis found that those with psoriasis had an increased risk of depression (adjusted HR 1.39; CI, 1.37–1.41) (Kurd, Troxel, Crits-Christoph, & Gelfand, 2010). The adjusted HR was higher in severe psoriasis than in mild psoriasis (1.72 vs 1.38). Similarly, a recent Danish study found that severe psoriasis appeared to be a risk factor for developing depression (incident rate ratio 1.36 and CI, 1.27–1.46); however, after adjustment for comorbidity, the increased risk was only significant for those younger than 50 years old (Jensen et al., 2015). Of particular note is the significant increase in suicidality in those with psoriasis (Kurd et al., 2010). The increased risk of depression for women with psoriasis appears higher for those with concomitant psoriatic arthritis (adjusted RR 1.52; CI, 1.06–2.19) than those without arthritis (adjusted RR 1.29; CI, 1.10–1.52) (Dommasch et al., 2015).

In a prospective study of 86,880 American women, those who reported high depressive symptomatology or who were using antidepressants had a multivariate relative risk of 1.59 (CI, 1.21–2.08) for the development of subsequent psoriasis, suggesting there may be a bidirectional relationship (Dominguez, Han, Li, Ascherio, & Qureshi, 2013).

There is growing evidence that the use of biologics to treat psoriasis can reduce depressive symptoms. A cohort study of 980 patients found a 40% reduction in the use of antidepressant medication after 2 years for those taking biological therapy; furthermore, there was a significantly faster and more sustained reduction in depression in the group that had continuous biological therapy compared with the group that had interrupted biological therapy (Wu et al., 2016). In three randomized, double-blind, placebo-controlled clinical trials, biological therapies have been associated with improvement in depressive symptoms: adalimumab significantly reduced depressive symptoms as measured by the Zung self-rating depression scale, an improvement that was correlated with improvement in psoriatic symptoms (Menter et al., 2010); ustekinumab significantly reduced symptoms as measured by the HADS (Langley et al., 2010); and etanercept significantly reduced depressive symptoms as measured by the HAMD and BDI (Krishnan et al., 2007; Tyring et al., 2006).

Systemic Lupus Erythematosus

Systemic lupus erythematosus (SLE) is a multisystem autoimmune disease with a relapsing-remitting course caused by autoantibodies that attack healthy connective tissue, including that of the central nervous system. Common symptoms include malar rash, Raynaud's phenomenon, arthritis, and vasculitis. Neuropsychiatric manifestations also occur frequently, with two recognized pathoetiologic pathways: ischemic-thrombotic-vascular and inflammatory-neurotoxic (Govoni

et al., 2016). The prevalence of depression is higher in SLE populations when compared with the general population. A recent metaanalysis including 10,828 adult SLE patients found the prevalence of depression to be 24% (CI, 16%–31%) according to clinical interviews, 30% (CI, 22%–38%) based on a HADS threshold of 8, and 39% (CI 29%–49%) based on a BDI threshold of 14 (Zhang, Fu, Yin, Zhang, & Shen, 2017). Depressive symptoms also appear to be undertreated in the SLE population, given relatively low rates of antidepressant use (Karol, Criscione-Schreiber, Lin, & Clowse, 2013; van Exel et al., 2013).

Based on a prospective study of 1827 SLE patients, the estimated cumulative incidence of any mood disorder was 17.8% (CI, 15.1%–20%); lower risk was associated with Asian ethnicity ($P = .01$) and treatment with immunosuppressive drugs ($P = .003$) (Hanly et al., 2015). Similarly, data from the Hopkins lupus cohort revealed that the incidence of new-onset depression in those with SLE was 29.7 episodes/1000 PY (Huang, Magder, & Petri, 2014). Independent risk factors for depression include recent SLE diagnosis, cutaneous activity, longitudinal myelitis, non-Asian ethnicity, and disability.

It is not clear whether depression increases the risk of developing SLE. In addition, there is contradictory data on the relationship between SLE disease activity and depressive symptoms. Although several studies to date have examined this relationship, some have demonstrated higher depressive symptoms during active SLE, whereas others have found no association (Nery et al., 2007; Palagini et al., 2013). Better prospective studies are needed in order to elucidate the temporal relationship between depression and SLE.

Certain depressive symptoms appear to be more common in SLE-related depression. Cognitive dysfunction is present in up to 80% of SLE patients, although a substantial portion of this dysfunction is not attributable to depression (Meszaros, Perl, & Faraone, 2012). Fatigue, weakness, and sleep disruption are also prominent symptoms in those with SLE and depression (Palagini et al., 2013). Of note, suicidal ideation is particularly high in the population with SLE and depressive symptoms (Mok, Chan, Cheung, & Yip, 2014; Xie et al., 2012).

Multiple Sclerosis

Multiple sclerosis (MS) is the most common autoimmune disease of the central nervous system. Characterized by demyelination, MS can take a relapsing-remitting or progressive course with common symptoms including ataxia, fatigue, muscle weakness, paraesthesia/hypoesthesia, and impaired vision. Fatigue is the most common depressive symptom in MS, occurring in up to 90% of MS patients. Metaanalysis of data from 58 studies including 87,756 MS patients demonstrated a pooled mean prevalence of 30.5% for depression (CI, 26.3%–35.1%). The results varied based on the means of classification, as the prevalence of clinical significant depressive symptoms (35%) exceeded the prevalence of clinician-diagnosed depression (21%) (Boeschoten et al., 2017). There is substantial heterogeneity in the MS-depression data, with prevalence rates of depression ranging from 4.98% to 70.1% depending on the population studies and method of classification (Marrie et al., 2015). Nevertheless, there are consistent data that the depression is more prevalent in those with MS than the general population. It has also been well-established that the many individuals with MS reporting depressive symptoms do not receive treatment for depression (Feinstein, Magalhaes, Richard, Audet, & Moore, 2014).

The temporal relationship between MS and depression is less clear. A 2015 systematic review only found two studies that analyzed the incidence of depression in MS patients; one study found an incidence of 4.0% over 1 year, whereas the other found it to be 34.7% over a 5-year period (Marrie et al., 2015). A recent Canadian study including 9624 cases of MS showed that the risk of incident depression was higher in the MS population than matched controls (HR 1.92; 1.82–2.04), even after the adjustment for other physical comorbidities (Marrie et al., 2016). Recent longitudinal studies suggest that depressive symptoms remain stable over 4 years of follow-up even as disability scores tended to increase, suggesting a chronic (rather than episodic) course for depressive symptoms in MS (Koch et al., 2015). However, there are also data suggesting that there is a higher rate of depressive symptoms acutely during an MS relapse (Moore et al., 2012). Moreover, it is well documented that clinical depression is associated with increased morbidity and mortality in MS patients (Feinstein, 2011).

The impact of MS treatments on depressive symptoms is controversial. Although interferon beta has been suggested to increase the risk of depressive symptoms in MS patients, this has not been substantiated by the data from MS clinical trials (Schippling et al., 2016). More longitudinal data are needed to clarify the relationship between depression and MS treatments.

Diabetes

Pro-inflammatory cytokines are implicated in both type 1 diabetes mellitus (T1DM) and type 2 diabetes mellitus (T2DM). T1DM is well defined as an autoimmune disease in which the adaptive immune system attacks pancreatic beta cells. The chronic low-grade inflammation in T2DM chiefly involves derangements of the innate immune system; inflammatory processes underlie the development of T2DM, its relationship with obesity, and the development of vascular complications in both T1DM and T2DM (Nunemaker, 2016).

Based on metaanalysis of cross-sectional data, depression has a prevalence of 17.6% among those with T2DM versus 9.8% in those without T2DM (OR = 1.6; CI = 1.2–2.0) (Roy & Lloyd, 2012). A 2012 metaanalysis of 16 longitudinal studies reported an adjusted relative risk of 1.25 (CI, 1.10–1.44) of incident depression in the population with T2DM (Rotella & Mannucci, 2013).

For adults with T1DM, the prevalence of depression was 12% compared with 3.2% among controls, although heterogeneity limited the ability to perform metaanalysis (Barnard, Skinner, & Peveler, 2006). Subsequent metaanalytic data suggest that the prevalence of depressive symptoms [as measured by the Children's Depression Inventory (CDI)] may be as high as 30% in youth with T1DM (Buchberger et al., 2016).

It is important to note that depression itself is also a strong risk factor for the development of T2DM in longitudinal studies (relative risk, 1.60, and CI = 1.37–1.88) and that inflammation seems to mediate this increased risk (Gregory, Rosenblat, & McIntyre, 2016; Knol et al., 2006; Moulton, Pickup, & Ismail, 2015). The precise role of chronic inflammation in the development of new-onset depression for those with T2DM is not clearly elucidated. In primary-care patients with new T2DM diagnoses, patients with comorbid depression tended to be more overweight and have higher concentrations of white blood cells, CRP, interleukin-1 receptor antagonist (IL-1Ra), and IL-1β (Laake et al., 2014). In a large cohort study of patients with T2DM, there was a significant association between elevated CRP and depression, although with multivariate analysis this association remained only for patients with high body mass index (BMI) (Hayashino et al., 2014). Interleukin-6, in particular, has been found to be significantly higher in patients with both diabetes and depression than either illness alone (Doyle et al., 2013). More broadly, there has been some evidence associating poorer glycemic control with depressive symptoms; however, it is difficult to determine whether hyperglycemia worsens depressive symptoms, depression leads to worse glycemic control, or both depression and poor glycemic control relate to shared risk factors (such as lifestyle or obesity) (Moulton et al., 2015).

A recent metaanalysis of four small RCTs shows that pioglitazone (a selective agonist of the nuclear transcription factor peroxisome proliferator-activated receptor-gamma (PPAR-γ)

used in T2DM treatment) can induce remission of a major depressive episode (OR = 3.3; CI, 1.4–7.8; $P = .008$) (Colle et al., 2017). Recent data also suggest that liraglutide (a glucagon-like peptide-1 agonist used in T2DM treatment) can have beneficial effects on cognition in a nondiabetic population with mood disorders (Mansur et al., 2017). Unfortunately, a high-quality RCT examining the use of intranasal insulin in nondiabetic depression did not demonstrate improvements in mood or cognition (Cha et al., 2017).

Obesity

Lipid accumulation in the body, particularly in abdominal adipose tissue, can cause a proinflammatory state. Adipocytes secrete IL-6, TNF-α, and monocyte chemoattractant protein-1 (MCP-1), which lead to an accumulation of leukocytes and chronic systemic inflammation. Inflammation mediates the associations between obesity and conditions such as cardiovascular disease or diabetes, and thus, adiposity-related inflammation might similarly contribute to depression risk (Shelton & Miller, 2010). Obese individuals tend to develop atypical features of depression, including increased appetite, mood reactivity, hypersomnia, and fatigue, and some have proposed that atypical depression might represent a metabolic subtype of depression related to adiposity (Capuron, Lasselin, & Castanon, 2017).

The prevalence of depression in obese patients varies depending on methods of assessment; however, metaanalysis of cross-sectional studies has found a significant association between depression and obesity, with an odds ratio of 1.26 (CI, 1.17–1.36) (de Wit et al., 2010). Among patients seeking bariatric surgery, a recent metaanalysis found a prevalence of depression of 19% (CI, 14%–25%) (Dawes et al., 2016). A recent metaanalysis of longitudinal studies determined that those who were obese had an 18% increased risk of depression (RR, 1.18; CI = 1.04–1.35) while those with depression had a 37% increased risk of being obese (RR, 1.37;

CI = 1.17–1.48) (Mannan, Mamun, Doi, & Clavarino, 2016). There is also metaanalytic evidence (based on both cross-sectional studies and longitudinal data) of an increased risk of depression in those with metabolic syndrome (MetS) (Pan et al., 2012). When the diagnostic components of MetS are analyzed individually, central obesity, hypertriglyceridemia, and low high-density lipoprotein (HDL) concentrations are the factors that associate significantly with depression (Pan et al., 2012).

Elevated CRP and IL-6 portend the increased risk of depressive symptoms in patients with obesity or metabolic syndrome (Capuron et al., 2017). CRP levels can explain 20% of the increase in depression scores over time in obese patients (Daly, 2013). Furthermore, metaanalytic data have found a significant improvement in depressive symptoms following bariatric surgery-induced weight loss; this improvement in emotional status or depressive symptom postbariatric surgery seems to correlate with reductions in levels of inflammatory markers (Capuron et al., 2011; Dawes et al., 2016; Emery et al., 2007).

FACTORS UNDERPINNING THE RELATIONSHIP BETWEEN DEPRESSION AND INFLAMMATORY ILLNESS

Traditionally, the development of depression in a patient with a medical illness has been attributed to a psychological or cognitive mechanism; a severe or chronic illness can threaten an individual's purpose, meaning, and sense of self, thus providing the life event that triggers a depressive episode in a psychologically vulnerable individual (Goodwin, 2006). It is easy to formulate psychosocial factors that might precipitate depression in the inflammatory diseases described in the preceding sections: distress due to the diagnosis (e.g., difficulty accepting the reality of a chronic medical illness), development of disability (e.g., impaired mobility due to joint

destruction in RA, weakness in MS, or amputation in T2DM), direct and indirect costs of treatment (e.g., medications and hospitalizations), or perceived stigma (e.g., weight discrimination in obesity and concerns about physical appearance in psoriasis). However, the development of depression in medically ill people is also increasing being related to underlying neurobiological disturbances (i.e., systemic inflammation) inherent to the medical illness. In fact, because psychological stressors provoke an acute inflammatory response, even so-called "psychological" explanations of increased depression in medical illness may be subserved by biological mechanisms (Steptoe, Hamer, & Chida, 2007).

A pro-inflammatory state can cause "sickness behavior," a behavioral phenotype resembling depression that might confer evolutionary benefit by allowing for energy conservation and social isolation in response to infections (Dantzer, O'Connor, Freund, Johnson, & Kelley, 2008). However, sickness behavior in response to inflammation for individuals with autoimmune disease would be inherently maladaptive because the immune system is attacking the individual's tissues rather than a foreign pathogen. For autoimmune diseases such as RA, MS, SLE, psoriasis, and IBD, the chronic relapsing and remitting inflammation can sensitize the immune-inflammatory pathways and activate oxidative and nitrosative pathways, thereby causing progressive damage and underlying the transition from "sickness behavior" to depression (Maes et al., 2012).

The proposed pathophysiology underlying the relationship between inflammation and depression has been well described in the literature (Dantzer, O'Connor, Lawson, & Kelley, 2011; Rosenblat et al., 2014; Stuart & Baune, 2012). Pro-inflammatory cytokines (IL-2, IL-6, TNF-α, and IFN) can decrease serotonin levels by (1) increasing the activity of indoleamine 2,3-dioxygenase (IDO) on tryptophan (thus increasing kynurenine and decreasing serotonin production) and (2) increasing the breakdown of serotonin to

5-hydroxyindoleacetic acid. Pro-inflammatory cytokines also activate the HPA axis and, with chronic overactivation, can impair the negative feedback mechanisms by inducing glucocorticoid receptor resistance. In addition, pro-inflammatory cytokines activate microglial cells, which then amplify the innate immune response by further secretion of inflammatory cytokines. Microglial activation contributes to decreased neurogenesis and increased synaptic pruning and apoptosis. The end result of this neuroinflammatory response is oxidative and nitrosative stress, decreased neuroplasticity, and neurodegeneration, which underlies the mood disorder risk.

Some aspects of these pathophysiological pathways have been demonstrated in specific autoimmune illnesses. High levels of pro-inflammatory cytokines are characteristic of the active phases in psoriasis, IBD, SLE, and RA. Overexpression of IDO and high levels of tryptophan metabolites have been found peripherally in patients with IBD (Hisamatsu et al., 2012). Pathological microglial activation has been implicated specifically in multiple sclerosis. Certain inflammatory cytokines are associated with pain, a particularly common symptom in RA and IBD, and may also underlie the transition from acute to chronic pain (Zhang & An, 2007). Sleep disruption, prominent in autoimmune disease, is strongly linked to acute and chronic inflammation (Opp & Krueger, 2015). In RA models, for example, there are concurrent bidirectional impacts of sleep disruption on psychological distress, pain on psychological distress, and pain on sleep disruption—in which one symptom promotes the other and vice versa due to a synergistic inflammatory response—leading to progressive deterioration in clinical outcomes (Irwin et al., 2008).

Although inflammatory illness increases the incidence of depression, depression also appears to increase the incidence of inflammatory illnesses. Furthermore, shared risk factors for both categories of illness may be another explanation for the high coprevalence. As reviewed earlier, the relative risk of T2DM in depression (1.60) is larger than the relative risk of depression in T2DM (1.25), so the depression-T2DM provides a good example to explore these alternative explanations for the association between depression and inflammatory illness. Proposed factors underlying the increased risk of T2DM in depression include the following: (1) Depression is associated with behavioral factors that increase the risk of T2DM, including poor diet, decreased physical activity, high body mass index, and smoking; (2) depression is associated with biochemical factors (e.g., HPA axis dysregulation and increased levels of pro-inflammatory cytokines) that are linked to the development of T2DM; and (3) treatments for depression may increase T2DM risk, as some antidepressants are associated with significant weight gain and atypical antipsychotics (directly implicated in the development of T2DM) are first-line agents for augmentation (Gregory et al., 2016; Moulton et al., 2015; Stuart & Baune, 2012; Tabák, Akbaraly, Batty, & Kivimäki, 2014).

T2DM and depression share several risk factors that all contribute to overactivation of the innate immune system. Indicators of early-life adversity, from low birthweight to childhood abuse and neglect, are associated with both T2DM and depression (Moulton et al., 2015). Children who are disadvantaged or mistreated develop a pro-inflammatory phenotype and have measurable changes in their HPA responses and autonomic nervous system (Miller, Chen, & Parker, 2011; Nusslock & Miller, 2016). The accumulation of wear and tear from successive inflammatory insults (e.g., psychological stressors, environmental exposures, and lifestyle choices), comprising part of one's "allostatic load," may predispose individuals to both depression and inflammatory illnesses. At the population level, the allostatic load model helps to understand the strong association between socioeconomic deprivation and earlier onset of physical and mental health disorders (e.g., a population-based study showed that the onset of multimorbidity occurred 10–15 years earlier in those living in the most deprived areas compared with the most affluent) (Barnett et al., 2012). In a search for sources of inflammation in depression, a range of factors have been identified: psychosocial stressors, poor diet, physical inactivity, obesity, smoking, altered gut permeability, atopy, dental caries, sleep disruption, and vitamin D deficiency (Berk et al., 2013). In many cases, these risk factors echo the risk factors for the inflammatory illnesses described previously, supporting a shared cause conceptualization for inflammatory and depressive illness.

CONCLUSIONS

Depression is frequently comorbid with autoimmune disease, diabetes, and obesity. In many cases, there is evidence of a bidirectional relationship. Inflammatory processes may mediate the bidirectionally increased risk yet also may underlie common risk factors for both depression and inflammatory comorbidity. The "inflamed depression" phenotype may be the behavioral result of sustained inflammation, a maladaptive derivative of "sickness behavior." Comorbid depression and inflammatory illness are associated with a worse prognosis for both conditions. Nevertheless, disease-specific anti-inflammatory treatments, particularly biological agents, can specifically target depressive symptoms in autoimmune illnesses. Inflammatory pathways might represent a therapeutic target during acute episodes of depression and also could provide a target for the prevention of depression.

References

Ang, D. C., Choi, H., Kroenke, K., & Wolfe, F. (2005). Comorbid depression is an independent risk factor for mortality in patients with rheumatoid arthritis. *The Journal of Rheumatology*, 32(6), 1013–1019.

Barnard, K. D., Skinner, T. C., & Peveler, R. (2006). The prevalence of co-morbid depression in adults with Type 1 diabetes: systematic literature review. *Diabetic Medicine*, 23(4), 445–448. https://doi.org/10.1111/j.1464-5491.2006.01814.x.

Barnett, K., Mercer, S. W., Norbury, M., Watt, G., Wyke, S., & Guthrie, B. (2012). Epidemiology of multimorbidity and implications for health care, research, and medical education: a cross-sectional study. *Lancet (London, England)*, 380(9836), 37–43. https://doi.org/10.1016/S0140-6736(12)60240-2.

Benros, M. E., Waltoft, B. L., Nordentoft, M., Østergaard, S. D., Eaton, W. W., Krogh, J., et al. (2013). Autoimmune diseases and severe infections as risk factors for mood disorders. *JAMA Psychiatry*, 70(8), 812. https://doi.org/10.1001/jamapsychiatry.2013.1111.

Berk, M., Williams, L. J., Jacka, F. N., O'Neil, A., Pasco, J. A., Moylan, S., et al. (2013). So depression is an inflammatory disease, but where does the inflammation come from? *BMC Medicine*, 11(1), 200. https://doi.org/10.1186/1741-7015-11-200.

Bhandari, S., Larson, M. E., Kumar, N., & Stein, D. (2017). Association of inflammatory bowel disease (IBD) with depressive symptoms in the United States population and independent predictors of depressive symptoms in an IBD population: a NHANES study. *Gut and Liver*, 11(4), 512–519. https://doi.org/10.5009/gnl16347.

Boeschoten, R. E., Braamse, A. M. J., Beekman, A. T. F., Cuijpers, P., van Oppen, P., Dekker, J., et al. (2017). Prevalence of depression and anxiety in multiple sclerosis: a systematic review and meta-analysis. *Journal of the Neurological Sciences*, 372, 331–341. https://doi.org/10.1016/j.jns.2016.11.067.

Buchberger, B., Huppertz, H., Krabbe, L., Lux, B., Mattivi, J. T., & Siafarikas, A. (2016). Symptoms of depression and anxiety in youth with type 1 diabetes: a systematic review and meta-analysis. *Psychoneuroendocrinology*, 70, 70–84. https://doi.org/10.1016/j.psyneuen.2016.04.019.

Caneo, C., Marston, L., Bellón, J. Á., & King, M. (2016). Examining the relationship between physical illness and depression: is there a difference between inflammatory and non inflammatory diseases? A cohort study. *General Hospital Psychiatry*, 43, 71–77. https://doi.org/10.1016/j.genhosppsych.2016.09.007.

Capuron, L., Lasselin, J., & Castanon, N. (2017). Role of adiposity-driven inflammation in depressive morbidity. *Neuropsychopharmacology*, 42(1), 115–128. https://doi.org/10.1038/npp.2016.123.

Capuron, L., Poitou, C., Machaux-Tholliez, D., Frochot, V., Bouillot, J.-L., Basdevant, A., et al. (2011). Relationship between adiposity, emotional status and eating behaviour in obese women: role of inflammation. *Psychological Medicine*, 41(7), 1517–1528. https://doi.org/10.1017/S0033291710001984.

Cha, D. S., Best, M. W., Bowie, C. R., Gallaugher, L. A., Woldeyohannes, H. O., Soczynska, J. K., et al. (2017). A randomized, double-blind, placebo-controlled, crossover trial evaluating the effect of intranasal insulin on cognition and mood in individuals with treatment-resistant major depressive disorder. *Journal of Affective Disorders*, 210, 57–65. https://doi.org/10.1016/j.jad.2016.12.006.

Colle, R., de Larminat, D., Rotenberg, S., Hozer, F., Hardy, P., Verstuyft, C., et al. (2017). Pioglitazone could induce remission in major depression: a meta-analysis. *Neuropsychiatric Disease and Treatment*, 13, 9–16. https://doi.org/10.2147/NDT.S121149.

Daly, M. (2013). The relationship of C-reactive protein to obesity-related depressive symptoms: a longitudinal study. *Obesity*, 21(2), 248–250. https://doi.org/10.1002/oby.20051.

Dantzer, R., O'Connor, J. C., Freund, G. G., Johnson, R. W., & Kelley, K. W. (2008). From inflammation to sickness and depression: when the immune system subjugates the brain. *Nature Reviews. Neuroscience*, 9(1), 46–56. https://doi.org/10.1038/nrn2297.

Dantzer, R., O'Connor, J. C., Lawson, M. A., & Kelley, K. W. (2011). Inflammation-associated depression: from serotonin to kynurenine. *Psychoneuroendocrinology*, 36(3), 426–436. https://doi.org/10.1016/j.psyneuen.2010.09.012.

Dawes, A. J., Maggard-Gibbons, M., Maher, A. R., Booth, M. J., Miake-Lye, I., Beroes, J. M., et al. (2016). Mental health conditions among patients seeking and undergoing bariatric surgery. *JAMA*, 315(2), 150. https://doi.org/10.1001/jama.2015.18118.

de Wit, L., Luppino, F., van Straten, A., Penninx, B., Zitman, F., & Cuijpers, P. (2010). Depression and obesity: a meta-analysis of community-based studies. *Psychiatry Research*, 178(2), 230–235. https://doi.org/10.1016/j.psychres.2009.04.015.

Dominguez, P. L., Han, J., Li, T., Ascherio, A., & Qureshi, A. A. (2013). Depression and the risk of psoriasis in US women. *Journal of the European Academy of Dermatology and Venereology*, 27(9), 1163–1167. https://doi.org/10.1111/j.1468-3083.2012.04703.x.

Dommasch, E. D., Li, T., Okereke, O. I., Li, Y., Qureshi, A. A., & Cho, E. (2015). Risk of depression in women with psoriasis: a cohort study. *British Journal of Dermatology*, 173(4), 975–980. https://doi.org/10.1111/bjd.14032.

Dowlatshahi, E. A., Wakkee, M., Arends, L. R., & Nijsten, T. (2014). The prevalence and odds of depressive symptoms and clinical depression in psoriasis patients: a systematic review and meta-analysis. *The Journal of Investigative Dermatology*, 134(6), 1542–1551. https://doi.org/10.1038/jid.2013.508.

Doyle, T. A., de Groot, M., Harris, T., Schwartz, F., Strotmeyer, E. S., Johnson, K. C., et al. (2013). Diabetes, depressive symptoms, and inflammation in older adults: results from the health, aging, and body

composition study. *Journal of Psychosomatic Research*, 75(5), 419–424. https://doi.org/10.1016/j.jpsychores.2013.08.006.

Drosselmeyer, J., Jacob, L., Rathmann, W., Rapp, M. A., & Kostev, K. (2017). Depression risk in patients with late-onset rheumatoid arthritis in Germany. *Quality of Life Research*, 26(2), 437–443. https://doi.org/10.1007/s11136-016-1387-2.

Emery, C. F., Fondow, M. D. M., Schneider, C. M., Christofi, F. L., Hunt, C., Busby, A. K., et al. (2007). Gastric bypass surgery is associated with reduced inflammation and less depression: a preliminary investigation. *Obesity Surgery*, 17(6), 759–763. https://doi.org/10.1007/s11695-007-9140-0.

Euesden, J., Danese, A., Lewis, C. M., & Maughan, B. (2017). A bidirectional relationship between depression and the autoimmune disorders—new perspectives from the national child development study. *PLoS One*, 12(3), e0173015. https://doi.org/10.1371/journal.pone.0173015.

Evertsz', F. B., Bockting, C. L., Stokkers, P. C., Hinnen, C., Sanderman, R., & Sprangers, M. A. (2012). The effectiveness of cognitive behavioral therapy on the quality of life of patients with inflammatory bowel disease: multicenter design and study protocol (KL!C-study). *BMC Psychiatry*, 12(1), 227. https://doi.org/10.1186/1471-244X-12-227.

Feinstein, A. (2011). Multiple sclerosis and depression. *Multiple Sclerosis Journal*, 17(11), 1276–1281. https://doi.org/10.1177/1352458511417835.

Feinstein, A., Magalhaes, S., Richard, J.-F., Audet, B., & Moore, C. (2014). The link between multiple sclerosis and depression. *Nature Reviews Neurology*, 10(9), 507–517. https://doi.org/10.1038/nrneurol.2014.139.

Genty, M., Combe, B., Kostine, M., Ardouin, E., Morel, J., & Lukas, C. (2017). Improvement of fatigue in patients with rheumatoid arthritis treated with biologics: relationship with sleep disorders, depression and clinical efficacy. A prospective, multicentre study. *Clinical and Experimental Rheumatology*, 35(1), 85–92. https://doi.org/10.1136/annrheumdis-2014-eular.4124.

Goodwin, G. M. (2006). Depression and associated physical diseases and symptoms. *Dialogues in Clinical Neuroscience*, 8(2), 259–265.

Govoni, M., Bortoluzzi, A., Padovan, M., Silvagni, E., Borrelli, M., Donelli, F., et al. (2016). The diagnosis and clinical management of the neuropsychiatric manifestations of lupus. *Journal of Autoimmunity*, 74, 41–72. https://doi.org/10.1016/j.jaut.2016.06.013.

Gregory, J. M., Rosenblat, J. D., & McIntyre, R. S. (2016). Deconstructing diabetes and depression: clinical context, treatment strategies, and new directions. *Focus*, 14(2), 184–193. https://doi.org/10.1176/appi.focus.20150040.

Hanly, J. G., Su, L., Urowitz, M. B., Romero-Diaz, J., Gordon, C., Bae, S.-C., et al. (2015). Mood disorders in systemic lupus erythematosus: results from an international inception cohort study. *Arthritis & Rheumatology*, 67(7), 1837–1847. https://doi.org/10.1002/art.39111.

Hayashino, Y., Mashitani, T., Tsujii, S., Ishii, H., & Diabetes Distress and Care Registry at Tenri Study Group. (2014). Elevated levels of hs-CRP are associated with high prevalence of depression in Japanese patients with type 2 diabetes: the diabetes distress and care registry at Tenri (DDCRT 6). *Diabetes Care*, 37(9), 2459–2465. https://doi.org/10.2337/dc13-2312.

Hisamatsu, T., Okamoto, S., Hashimoto, M., Muramatsu, T., Andou, A., Uo, M., et al. (2012). Novel, objective, multivariate biomarkers composed of plasma amino acid profiles for the diagnosis and assessment of inflammatory bowel disease. *PLoS One*, 7(1), e31131. https://doi.org/10.1371/journal.pone.0031131.

Horst, S., Chao, A., Rosen, M., Nohl, A., Duley, C., Wagnon, J. H., et al. (2015). Treatment with immunosuppressive therapy may improve depressive symptoms in patients with inflammatory bowel disease. *Digestive Diseases and Sciences*, 60(2), 465–470. https://doi.org/10.1007/s10620-014-3375-0.

Hsu, B., Wang, D., Sun, Y., Salvadore, G., Singh, J., Curran, M., et al. (2015). SAT0182 improvement in measures of depressed mood and anhedonia, and fatigue, in a randomized, placebo-controlled, phase 2 study of Sirukumab, a human anti-Interleukin-6 antibody, in patients with rheumatoid arthritis. *Annals of the Rheumatic Diseases*, 74(Suppl. 2), 720.3–721. https://doi.org/10.1136/annrheumdis-2015-eular.4081.

Huang, X., Magder, L. S., & Petri, M. (2014). Predictors of incident depression in systemic lupus erythematosus. *The Journal of Rheumatology*, 41(9), 1823–1833.

Irwin, M. R., Davis, M., & Zautra, A. (2008). Behavioral comorbidities in rheumatoid arthritis: a psychoneuroimmunological perspective. *The Psychiatric Times*, 25(9), 1.

Jacob, L., Rockel, T., & Kostev, K. (2017). Depression risk in patients with rheumatoid arthritis in the United Kingdom. *Rheumatology and Therapy*, 4(1), 195–200. https://doi.org/10.1007/s40744-017-0058-2.

Jensen, P., Ahlehoff, O., Egeberg, A., Gislason, G., Hansen, P. R., & Skov, L. (2015). Psoriasis and new-onset depression: a Danish nationwide cohort study. *Acta Dermato-Venereologica*, 96(1), 39–42. https://doi.org/10.2340/00015555-2183.

Jobski, K., Luque Ramos, A., Albrecht, K., & Hoffmann, F. (2017). Pain, depressive symptoms and medication in German patients with rheumatoid arthritis-results from the linking patient-reported outcomes with claims data for health services research in rheumatology (PROCLAIR) study. *Pharmacoepidemiology and Drug Safety*, 26(7), 766–774. https://doi.org/10.1002/pds.4202.

Jyrkkä, J., Kautiainen, H., Koponen, H., Puolakka, K., Virta, L., Pohjolainen, T., et al. (2014). Antidepressant use among persons with recent-onset rheumatoid arthritis: a nationwide register-based study in Finland. *Scandinavian Journal of Rheumatology*, 43(5), 364–370. https://doi.org/10.3109/03009742.2013.878386.

Karol, D. E., Criscione-Schreiber, L. G., Lin, M., & Clowse, M. E. B. (2013). Depressive symptoms and associated factors in systemic lupus erythematosus. *Psychosomatics*, 54(5), 443–450. https://doi.org/10.1016/j.psym.2012.09.004.

Khandaker, G. M., Pearson, R. M., Zammit, S., Lewis, G., & Jones, P. B. (2014). Association of serum interleukin 6 and C-reactive protein in childhood with depression and psychosis in young adult life. *JAMA Psychiatry*, 71(10), 1121. https://doi.org/10.1001/jamapsychiatry.2014.1332.

Kim, Y. K., Na, K. S., Myint, A. M., & Leonard, B. E. (2016). The role of pro-inflammatory cytokines in neuroinflammation, neurogenesis and the neuroendocrine system in major depression. *Progress in Neuro-Psychopharmacology and Biological Psychiatry*, 64, 277–284. https://doi.org/10.1016/j.pnpbp.2015.06.008.

Knol, M. J., Twisk, J. W. R., Beekman, A. T. F., Heine, R. J., Snoek, F. J., & Pouwer, F. (2006). Depression as a risk factor for the onset of type 2 diabetes mellitus. A meta-analysis. *Diabetologia*, 49(5), 837–845. https://doi.org/10.1007/s00125-006-0159-x.

Koch, M. W., Patten, S., Berzins, S., Zhornitsky, S., Greenfield, J., Wall, W., et al. (2015). Depression in multiple sclerosis: a long-term longitudinal study. *Multiple Sclerosis Journal*, 21(1), 76–82. https://doi.org/10.1177/1352458514536086.

Köhler, O., Benros, M. E., Nordentoft, M., Farkouh, M. E., Iyengar, R. L., Mors, O., et al. (2014). Effect of anti-inflammatory treatment on depression, depressive symptoms, and adverse effects. *JAMA Psychiatry*, 71(12), 1381. https://doi.org/10.1001/jamapsychiatry.2014.1611.

Krishnan, R., Cella, D., Leonardi, C., Papp, K., Gottlieb, A. B., Dunn, M., et al. (2007). Effects of etanercept therapy on fatigue and symptoms of depression in subjects treated for moderate to severe plaque psoriasis for up to 96 weeks. *British Journal of Dermatology*, 157(6), 1275–1277. https://doi.org/10.1111/j.1365-2133.2007.08205.x.

Kurd, S. K., Troxel, A. B., Crits-Christoph, P., & Gelfand, J. M. (2010). The risk of depression, anxiety, and suicidality in patients with psoriasis: a population-based cohort study. *Archives of Dermatology*, 146(8), 891–895. https://doi.org/10.1001/archdermatol.2010.186.

Kurina, L. M., Goldacre, M. J., Yeates, D., & Gill, L. E. (2001). Depression and anxiety in people with inflammatory bowel disease. *Journal of Epidemiology and Community Health*, 55(10), 716–720. https://doi.org/10.1136/JECH.55.10.716.

Laake, J.-P. S., Stahl, D., Amiel, S. A., Petrak, F., Sherwood, R. A., Pickup, J. C., et al. (2014). The association between depressive symptoms and systemic inflammation in people with type 2 diabetes: findings from the South London diabetes study. *Diabetes Care*, 37(8), 2186–2192.

Langley, R. G., Feldman, S. R., Han, C., Schenkel, B., Szapary, P., Hsu, M.-C., et al. (2010). Ustekinumab significantly improves symptoms of anxiety, depression, and skin-related quality of life in patients with moderate-to-severe psoriasis: results from a randomized, double-blind, placebo-controlled phase III trial. *Journal of the American Academy of Dermatology*, 63(3), 457–465. https://doi.org/10.1016/j.jaad.2009.09.014.

Lu, M.-C., Guo, H.-R., Lin, M.-C., Livneh, H., Lai, N.-S., & Tsai, T.-Y. (2016). Bidirectional associations between rheumatoid arthritis and depression: a nationwide longitudinal study. *Scientific Reports*, 6, 20647. https://doi.org/10.1038/srep20647.

Maes, M., Berk, M., Goehler, L., Song, C., Anderson, G., Gałecki, P., et al. (2012). Depression and sickness behavior are Janus-faced responses to shared inflammatory pathways. *BMC Medicine*, 10(1), 66. https://doi.org/10.1186/1741-7015-10-66.

Mannan, M., Mamun, A., Doi, S., & Clavarino, A. (2016). Is there a bi-directional relationship between depression and obesity among adult men and women? Systematic review and bias-adjusted meta analysis. *Asian Journal of Psychiatry*, 21, 51–66. https://doi.org/10.1016/j.ajp.2015.12.008.

Mansur, R. B., Ahmed, J., Cha, D. S., Woldeyohannes, H. O., Subramaniapillai, M., Lovshin, J., et al. (2017). Liraglutide promotes improvements in objective measures of cognitive dysfunction in individuals with mood disorders: a pilot, open-label study. *Journal of Affective Disorders*, 207, 114–120. https://doi.org/10.1016/j.jad.2016.09.056.

Marrie, R. A., Patten, S. B., Greenfield, J., Svenson, L. W., Jette, N., Tremlett, H., et al. (2016). Physical comorbidities increase the risk of psychiatric comorbidity in multiple sclerosis. *Brain and Behavior*, 6(9), e00493. https://doi.org/10.1002/brb3.493.

Marrie, R. A., Reingold, S., Cohen, J., Stuve, O., Trojano, M., Sorensen, P. S., et al. (2015). The incidence and prevalence of psychiatric disorders in multiple sclerosis: a systematic review. *Multiple Sclerosis (Houndmills, Basingstoke, England)*, 21(3), 305–317. https://doi.org/10.1177/1352458514564487.

Matcham, F., Ali, S., Irving, K., Hotopf, M., & Chalder, T. (2016). Are depression and anxiety associated with disease activity in rheumatoid arthritis? A prospective study. *BMC Musculoskeletal Disorders*, 17(1), 155. https://doi.org/10.1186/s12891-016-1011-1.

Matcham, F., Rayner, L., Steer, S., & Hotopf, M. (2013). The prevalence of depression in rheumatoid arthritis: a systematic review and meta-analysis. *Rheumatology*

(Oxford, England), 52(12), 2136–2148. https://doi.org/10.1093/rheumatology/ket169.

Menter, A., Augustin, M., Signorovitch, J., Yu, A. P., Wu, E. Q., Gupta, S. R., et al. (2010). The effect of adalimumab on reducing depression symptoms in patients with moderate to severe psoriasis: a randomized clinical trial. *Journal of the American Academy of Dermatology*, 62(5), 812–818. https://doi.org/10.1016/j.jaad.2009.07.022.

Meszaros, Z. S., Perl, A., & Faraone, S. V. (2012). Psychiatric symptoms in systemic lupus erythematosus: a systematic review. *The Journal of Clinical Psychiatry*, 73(7), 993–1001. https://doi.org/10.4088/JCP.11R07425.

Mikocka-Walus, A., Knowles, S. R., Keefer, L., & Graff, L. (2016). Controversies revisited. *Inflammatory Bowel Diseases*, 22(3), 752–762. https://doi.org/10.1097/MIB.0000000000000620.

Milaneschi, Y., Corsi, A. M., Penninx, B. W., Bandinelli, S., Guralnik, J. M., & Ferrucci, L. (2009). Interleukin-1 receptor antagonist and incident depressive symptoms over 6 years in older persons: the InCHIANTI study. *Biological Psychiatry*, 65(11), 973–978. https://doi.org/10.1016/j.biopsych.2008.11.011.

Miller, G. E., Chen, E., & Parker, K. J. (2011). Psychological stress in childhood and susceptibility to the chronic diseases of aging: moving toward a model of behavioral and biological mechanisms. *Psychological Bulletin*, 137(6), 959–997. https://doi.org/10.1037/a0024768.

Mittermaier, C., Dejaco, C., Waldhoer, T., Oefferlbauer-Ernst, A., Miehsler, W., Beier, M., et al. (2004). Impact of depressive mood on relapse in patients with inflammatory bowel disease: a prospective 18-month follow-up study. *Psychosomatic Medicine*, 66(1), 79–84. https://doi.org/10.1097/01.PSY.0000106907.24881.F2.

Miwa, Y., Takahashi, R., Ikari, Y., Maeoka, A., Nishimi, S., Oguro, N., et al. (2017). Clinical characteristics of rheumatoid arthritis patients achieving functional remission with six months of biological DMARDs treatment. *Internal Medicine*, 56(8), 903–906. https://doi.org/10.2169/internalmedicine.56.8039.

Mok, C. C., Chan, K. L., Cheung, E. F. C., & Yip, P. S. F. (2014). Suicidal ideation in patients with systemic lupus erythematosus: incidence and risk factors. *Rheumatology*, 53(4), 714–721. https://doi.org/10.1093/rheumatology/ket404.

Moore, P., Hirst, C., Harding, K. E., Clarkson, H., Pickersgill, T. P., & Robertson, N. P. (2012). Multiple sclerosis relapses and depression. *Journal of Psychosomatic Research*, 73(4), 272–276. https://doi.org/10.1016/j.jpsychores.2012.08.004.

Moulton, C. D., Pickup, J. C., & Ismail, K. (2015). The link between depression and diabetes: the search for shared mechanisms. *The Lancet Diabetes and Endocrinology*, 3(6), 461–471. https://doi.org/10.1016/S2213-8587(15)00134-5.

Nery, F. G., Borba, E. F., Hatch, J. P., Soares, J. C., Bonfá, E., & Neto, F. L. (2007). Major depressive disorder and disease activity in systemic lupus erythematosus. *Comprehensive Psychiatry*, 48(1), 14–19. https://doi.org/10.1016/j.comppsych.2006.04.002.

Neuendorf, R., Harding, A., Stello, N., Hanes, D., & Wahbeh, H. (2016). Depression and anxiety in patients with inflammatory bowel disease: a systematic review. *Journal of Psychosomatic Research*, 87, 70–80. https://doi.org/10.1016/j.jpsychores.2016.06.001.

Nunemaker, C. S. (2016). Considerations for defining cytokine dose, duration, and milieu that are appropriate for modeling chronic low-grade inflammation in type 2 diabetes. *Journal of Diabetes Research*, 2016, 2846570. https://doi.org/10.1155/2016/2846570.

Nusslock, R., & Miller, G. E. (2016). Early-life adversity and physical and emotional health across the lifespan: a neuroimmune network hypothesis. *Biological Psychiatry*, 80(1), 23–32. https://doi.org/10.1016/j.biopsych.2015.05.017.

Opp, M. R., & Krueger, J. M. (2015). Sleep and immunity: a growing field with clinical impact. *Brain, Behavior, and Immunity*, 47, 1–3. https://doi.org/10.1016/j.bbi.2015.03.011.

Palagini, L., Mosca, M., Tani, C., Gemignani, A., Mauri, M., & Bombardieri, S. (2013). Depression and systemic lupus erythematosus: a systematic review. *Lupus*, 22(5), 409–416. https://doi.org/10.1177/0961203313477227.

Pan, A., Keum, N., Okereke, O. I., Sun, Q., Kivimaki, M., Rubin, R. R., et al. (2012). Bidirectional association between depression and metabolic syndrome: a systematic review and meta-analysis of epidemiological studies. *Diabetes Care*, 35(5), 1171–1180. https://doi.org/10.2337/dc11-2055.

Patten, S. B. (2001). Long-term medical conditions and major depression in a Canadian population study at waves 1 and 2. *Journal of Affective Disorders*, 63(1–3), 35–41. https://doi.org/10.1016/S0165-0327(00)00186-5.

Rosenblat, J. D., Cha, D. S., Mansur, R. B., & McIntyre, R. S. (2014). Inflamed moods: a review of the interactions between inflammation and mood disorders. *Progress in Neuro-Psychopharmacology and Biological Psychiatry*, 53, 23–34. https://doi.org/10.1016/j.pnpbp.2014.01.013.

Rotella, F., & Mannucci, E. (2013). Diabetes mellitus as a risk factor for depression. A meta-analysis of longitudinal studies. *Diabetes Research and Clinical Practice*, 99(2), 98–104. https://doi.org/10.1016/j.diabres.2012.11.022.

Roy, T., & Lloyd, C. E. (2012). Epidemiology of depression and diabetes: a systematic review. *Journal of Affective Disorders*, 142(Suppl), S8–S21. https://doi.org/10.1016/S0165-0327(12)70004-6.

Schippling, S., O'Connor, P., Knappertz, V., Pohl, C., Bogumil, T., Suarez, G., et al. (2016). Incidence and course of depression in multiple sclerosis in the multinational BEYOND trial. *Journal of Neurology*, 263(7), 1418–1426. https://doi.org/10.1007/s00415-016-8146-8.

Shelton, R. C., & Miller, A. H. (2010). Eating ourselves to death (and despair): the contribution of adiposity and inflammation to depression. *Progress in Neurobiology*, 91(4), 275–299. https://doi.org/10.1016/j.pneurobio.2010.04.004.

Smith, D. J., Court, H., McLean, G., Martin, D., Martin, J. L., Guthrie, B., et al. (2014). Depression and multimorbidity. *The Journal of Clinical Psychiatry*, 75(11), 1202–1208. https://doi.org/10.4088/JCP.14m09147.

Steptoe, A., Hamer, M., & Chida, Y. (2007). The effects of acute psychological stress on circulating inflammatory factors in humans: a review and meta-analysis. *Brain, Behavior, and Immunity*, 21(7), 901–912. https://doi.org/10.1016/j.bbi.2007.03.011.

Stevens, B., Cleland, T., Conway, G., Velonias, G. M., Garber, J., Giallourakis, C., et al. (2016). Vedolizumab therapy is associated with improvement in sleep quality in inflammatory bowel diseases: a prospective study. *Gastroenterology*, 1, S88.

Stuart, M. J., & Baune, B. T. (2012). Depression and type 2 diabetes: inflammatory mechanisms of a psychoneuroendocrine co-morbidity. *Neuroscience and Biobehavioral Reviews*, 36(1), 658–676. https://doi.org/10.1016/j.neubiorev.2011.10.001.

Szigethy, E. M., Youk, A. O., Benhayon, D., Fairclough, D. L., Newara, M. C., Kirshner, M. A., et al. (2014). Depression subtypes in pediatric inflammatory bowel disease. *Journal of Pediatric Gastroenterology and Nutrition*, 58(5), 574–581. https://doi.org/10.1097/MPG.0000000000000262.

Tabák, A. G., Akbaraly, T. N., Batty, G. D., & Kivimäki, M. (2014). Depression and type 2 diabetes: a causal association? *The Lancet Diabetes and Endocrinology*, 2(3), 236–2452. https://doi.org/10.1016/S2213-8587(13)70139-6.

Tyring, S., Gottlieb, A., Papp, K., Gordon, K., Leonardi, C., Wang, A., et al. (2006). Etanercept and clinical outcomes, fatigue, and depression in psoriasis: double-blind placebo-controlled randomised phase III trial. *Lancet (London, England)*, 367(9504), 29–35. https://doi.org/10.1016/S0140-6736(05)67763-X.

van Exel, E., Jacobs, J., Korswagen, L.-A., Voskuyl, A., Stek, M., Dekker, J., et al. (2013). Depression in systemic lupus erythematosus, dependent on or independent of severity of disease. *Lupus*, 22(14), 1462–1469. https://doi.org/10.1177/0961203313508443.

Walker, J. R., Ediger, J. P., Graff, L. A., Greenfeld, J. M., Clara, I., Lix, L., et al. (2008). The Manitoba IBD cohort study: a population-based study of the prevalence of lifetime and 12-month anxiety and mood disorders. *The American Journal of Gastroenterology*, 103(8), 1989–1997. https://doi.org/10.1111/j.1572-0241.2008.01980.x.

World Health Organization. (2017). *Depression fact sheet.* April 22, Retrieved from http://www.who.int/mediacentre/factsheets/ fs369/en/.

Wu, C.-Y., Chang, Y.-T., Juan, C.-K., Shen, J.-L., Lin, Y.-P., Shieh, J.-J., et al. (2016). Depression and insomnia in patients with psoriasis and psoriatic arthritis taking tumor necrosis factor antagonists. *Medicine, 95*(22), e3816. https://doi.org/10.1097/MD.0000000000003816.

Xie, L.-F., Chen, P.-L., Pan, H.-F., Tao, J.-H., Li, X.-P., Zhang, Y.-J., et al. (2012). Prevalence and correlates of suicidal ideation in SLE inpatients: Chinese experience. *Rheumatology International, 32*(9), 2707–2714. https://doi.org/10.1007/s00296-011-2043-3.

Young, J. J., Bruno, D., & Pomara, N. (2014). A review of the relationship between proinflammatory cytokines and major depressive disorder. *Journal of Affective Disorders, 169*, 15–20. https://doi.org/10.1016/j.jad.2014.07.032.

Zhang, J.-M., & An, J. (2007). Cytokines, inflammation, and pain. *International Anesthesiology Clinics, 45*(2), 27–37. https://doi.org/10.1097/AIA.0b013e318034194e.

Zhang, L., Fu, T., Yin, R., Zhang, Q., & Shen, B. (2017). Prevalence of depression and anxiety in systemic lupus erythematosus: a systematic review and meta-analysis. *BMC Psychiatry, 17*(1), 70. https://doi.org/10.1186/s12888-017-1234-1.

22

Does Inflammation Link Clinical Depression and Coronary Artery Disease?

Silke Jörgens, Volker Arolt

University Hospital Muenster, Münster, Germany

INTRODUCTION

Both coronary artery disease (CAD) and depression are common disorders associated with significant burdens of disease and, in many cases, with substantial disabilities. According to the data of the World Health Organization (WHO, 2017), depression affects >300 million people worldwide, and 800,000 people died of suicide every year. For cardiovascular disorders, there were estimated 422 million cases and 17.92 million deaths in 2015 underlining the importance of both disorders in permanent disabilities and death worldwide (Roth et al., 2017; Vos et al., 2015; Whiteford et al., 2013). There is now cumulating evidence that both disorders may be associated with each other etiologically, clinically, and biologically and can influence each other in a bidirectional way (Baune et al., 2012; Whooley & Wong, 2013). Several studies have shown that patients with depression have a higher risk to develop coronary artery disease (Rugulies, 2002), myocardial infarction (MI) (Pereira, Cerqueira, Palha, & Sousa, 2013), and heart failure (Gustad, Laugsand, Janszky, Dalen, & Bjerkeset, 2014). A recent review supports the validity of this effect by summarizing

moderate effect sizes from 1.5 to 2.7 (Baune et al., 2012). A meta-analysis integrating 21 longitudinal etiologic studies shows a pooled relative risk of 1.81 (95% CI 1.53–2.15) for depressive patients to develop CAD, with a higher risk for those who fulfill the diagnostic criteria of depression than for those who only report depressive symptoms (Nicholson, Kuper, & Hemingway, 2006).

There is also substantial evidence for an increased risk to develop depression subsequent to the onset of CAD. Nearly 65% of patients with acute myocardial infarction exhibit depressive symptoms, while 15%–25% fulfill the criteria for major depressive disorder (MDD) (Frasure-Smith, Lesperance, & Talajic, 1993; Thombs, Grace, & Ziegelstein, 2006). This relationship has been shown to be independent of typical cardiovascular risk factors such as hypertension, high cholesterol, disease severity, and history of cardiac events (Barefoot & Schroll, 1996; Grippo & Johnson, 2009; Penninx et al., 2001). Furthermore, there is increasing evidence that depression contributes to not only the onset but also the progression of CAD. For example, depression predicts the risk of a subsequent cardiovascular event in preexisting cardiovascular diseases and is associated with poorer outcome and higher rates of mortality independent of other risk factors such as arrhythmias, severity of cardiovascular disease, or smoking (Barefoot & Schroll, 1996; Carney et al., 2008; Doyle et al., 2015; Frasure-Smith et al., 1993; Meyer et al., 2011). Studies suggest that depressive patients are at 1.6–2.7-fold higher risk to develop further cardiovascular events in the next 2 years (Baune et al., 2012; Nicholson et al., 2006). A recent meta-analysis indicates that depression is a risk factor for mortality after bypass surgery (Stenman, Holzmann, & Sartipy, 2016). Also, the severity of depressive symptoms seems to be related to cardiovascular risk as shown in a study by Barefoot et al. Authors followed CAD patients over a period up to 19.4 years and found that moderate-to-severe depression scores were associated with higher odds ratios for subsequent cardiac death and for total mortality in contrast to nondepressed patients. This increased risk was not confined to the initial month after the event but stable over the follow-up period, whereas patients with mild depression show only intermediate levels of risk (Barefoot et al., 1996). On the contrary, the severity of cardiovascular event, for example, the size of affected tissue in myocardial infarction or the volume of ejection fraction, appears not to be related to the occurrence of depressive symptoms (Shimizu, Suzuki, Okumura, & Yamada, 2014).

Several factors from behavioral and biological dimensions have been discussed as putative mechanisms that link depression and cardiovascular disease. An unhealthy lifestyle (smoking, alcohol use, lack of physical activity, unhealthy nutrition, overweight, and noncompliance during pharmacotherapy) is one of these behavioral factors (Penninx, 2017). Because the association between cardiovascular disease and depression is still significant after controlling for these confounders, unhealthy lifestyle factors appear not to represent a sufficient explanation for this association (Carnevali, Montano, Statello, &

Sgoifo, 2017). In contrast, another environmental factor, namely, chronic stress exposure seems to be a relevant factor for developing both depression (Anisman & Zacharko, 1992; Kendler, Karkowski, & Prescott, 1999) and CAD (Grippo & Johnson, 2009).

Many studies illustrate the impact of physiological disturbances such as autonomic dysfunction (Dao et al., 2010), endothelial dysfunction (Adibfar, Saleem, Lanctot, & Herrmann, 2016), coagulation factors (Savoy, Van Lieshout, & Steiner, 2017), HPA dysregulation (Jokinen & Nordstrom, 2009), and immunologic changes with increased concentrations of proinflammatory cytokines (Dantzer, O'Connor, Freund, Johnson, & Kelley, 2008; Johnson & Grippo, 2006) on the relationship between depression and cardiovascular events (Penninx, 2017). A meta-analysis of genome-wide association studies and also candidate-gene studies have identified genetic variants associated with both depression and cardiovascular disease (Amare, Schubert, Klingler-Hoffmann, Cohen-Woods, & Baune, 2017) indicating that shared genetic factors in the cooccurrence of depression and cardiovascular disease may play an important role. But still, the underlying biological interactions are poorly understood. It was hypothesized that the cooccurrence of both diseases depends on how the underlying biological mechanisms are affected by inflammation (Wright, Simpson, Van Lieshout, & Steiner, 2014). The following book will explore how factors of inflammation present as a possible mechanism for the relationship between depression and coronary artery disease (CAD).

INFLAMMATION IN DEPRESSION

Since the putative role of the immune system has been brought into focus of biomarker studies on depression, several investigations have reported evidence for a low-grade proinflammatory activity, mainly by reporting mild to moderately elevated levels of C-reactive protein (CRP) and an increase of proinflammatory cytokines in the sera of depressed patients (Baumeister, Russell, Pariante, & Mondelli, 2014; Dantzer et al., 2008; Dowlati et al., 2010).

Recent meta-analyses show a dysregulation of not only proinflammatory cytokines with elevated levels of interleukins (IL-6, IL-10, IL-12, IL-13, IL-18, and TNF-α) but also C-reactive protein (CRP) in depression, but the heterogeneity between the respective studies was large (Dowlati et al., 2010; Howren, Lamkin, & Suls, 2009; Kohler et al., 2017; Liu, Ho, & Mak, 2012). In a meta-analysis using follow-up data of depressed patients, raised IL-6 and CRP have the strongest association with the development of depressive symptoms (Valkanova, Ebmeier, & Allan, 2013). In line with this, elevated levels of IL-6 were reported in the pathophysiology of suicidal behavior (Gananca et al., 2016). Recently, it could be shown that the level of childhood maltreatment is correlated with an increase of TNF-α and IL-6 (Grosse et al., 2016). Early investigations indicate also an influence of depression subtype with a decrease of proinflammatory cytokines in melancholic depression that is linked to HPA axis activity (Kaestner et al., 2005; Rothermundt et al., 2001).

Brain imaging studies in healthy subjects show that proinflammatory cytokines influence brain areas involved in mood regulation (Kim & Won, 2017). For example, a study by Harrison reveals that disturbed mood after cytokine-induced inflammation is correlated with activity in the subgenual anterior cingulate gyrus and reduced connectivity to areas of emotion network like amygdala, nucleus accumbens, and medial prefrontal cortex, whereas levels of circulation IL-6 influence the connectivity of anterior cingulate cortex to these regions (Harrison et al., 2009). Also, in patients with depression, increased inflammatory markers as CRP and cytokines are associated with decreased functional connectivity in corticostriatal reward

system (Felger et al., 2016). These observations underline an interaction of peripheral immune response and brain function.

With regard to the symptomatology of depression, it may be the case that proinflammatory and inflammatory cytokines seem to be inversely correlated to depression symptoms and severity as shown by a recent study (Schmidt et al., 2016). At baseline and after 4 weeks of antidepressant, treatment levels of TNF-α, INF-γ, IL-2, IL-4, IL-5, IL-10, IL-12, IL-13, and CRP were measured in 30 patients and 30 healthy controls. At baseline, cytokines but not CRP were negatively correlated with severity of symptoms measured with BDI-II in patients, whereas no effect can be observed during follow-up and in healthy controls. In treatment responders, levels of pro- and antiinflammatory cytokines were higher at both time points, but after controlling for BDI score, only IL-2 at baseline remains significant.

There are also lines of evidence that proinflammatory activity may trigger the onset of depression. It has been demonstrated in several studies that immune therapy with proinflammatory agents like interferon-alpha (INF-α) can induce depressive symptoms in approximately one-third of the patients and increase pituitary–adrenal activity after first INF-α application (Beratis et al., 2005; Capuron, Ravaud, Miller, & Dantzer, 2004; Friebe et al., 2010). In contrast to this, parallel treatment with an SSRI (paroxetine) in cancer patients during interferon therapy shows a fourfold reduction of risk to develop a major depression (Musselman et al., 2001). Furthermore, there are also evidences from a number of studies that mood disorders can be triggered by inflammation after infection or vaccination (Gunaratne, Lloyd, & Vollmer-Conna, 2013). One study using *Salmonella abortus* endotoxin found in healthy subjects a significant transient increase in mood disturbances in contrast to placebo. This was accompanied with elevated levels of IL-6, IL-1 receptor antagonist, and TNF-α (Reichenberg et al., 2001). Interestingly, inflammation induced by typhoid vaccination impairs endothelial-dependent vascular dilatation in humans, implicating transient endothelial dysfunction (Hingorani et al., 2000). Also, numerous animal studies support findings in humans by showing that depression-equivalent behavior can be induced by application of inflammatory stimuli (Dantzer et al., 2008).

As mentioned above, treatment with antidepressants has shown regulatory effects to cytokines both *in vivo* and *in vitro*, while clomipramine and fluoxetine tend to consistently decrease proinflammatory cytokines, whereas mirtazapine and venlafaxine tend to increase their levels (Baumeister, Ciufolini, & Mondelli, 2016; Kraus et al., 2002). Remission from depression is, in a number of patients, accompanied with normalization of HPA axis and a decrease of TNF-α (Himmerich et al., 2006). Recently, it could be shown that IL-2 at baseline could be a predictor for treatment responders after 4 weeks of pharmacotherapy as described above (Schmidt et al., 2016). However, it remains unclear whether the normalization of immune dysregulations is caused by antidepressant compounds themselves or only represents a side effect of remission of the depressive state.

Nevertheless, it is in accordance with the evidence that points to a putative influence of proinflammatory mechanisms, that antiinflammatory treatment, especially celecoxib as an add-on medication, reduces depressive symptoms, as shown by a first study of Muller et al. (2006) and a recent meta-analysis by Kohler and colleagues (Kohler et al., 2014). Additionally, there are several hints that augmentation with acetylsalicylic acid accelerates the effect of SSRIs in animal model and in humans (Brunello et al., 2006; Kohler, Petersen, Mors, & Gasse, 2015; Mendlewicz et al., 2006; Yang et al., 2014). Interestingly, also phytopharmacotherapy with *Artemisia absinthium*, a TNF-α blocker, is efficient in reducing depressive symptoms in Crohn's disease (Krebs, Omer, & Omer, 2010).

All investigations mentioned above contribute to a putative role of proinflammatory immunologic functions in depression and underline the hypothesis of disturbed immune function. Despite often replicated and seemingly robust findings, it must be pointed out that all of these observations are based on group comparisons between patients with (major) depression and healthy controls and that the interindividual heterogeneity in the depressive cohort is high (Kohler et al., 2017; Raison & Miller, 2011). On one hand, not every patient suffering from major depression shows signs of proinflammatory activation, and on the other hand, the majority of people with clinical inflammation do not develop depression. Hence, it is compelling to speculate (although this could, as yet, never be clearly shown) that, in specific subgroups of patients, depression is associated with a pronounced, however mild, proinflammatory activity that is mainly provided by the innate immune system.

INFLAMMATION IN CORONARY ARTERY DISEASE

The term "coronary artery disease" (CAD) also known as "coronary heart disease" or "ischemic heart disease" labels a group of disorders that affect the blood vessels of the heart and includes stable and unstable angina as well as myocardial infarction and sudden cardiac death. The primary underlying pathology of coronary artery disease is atherosclerosis, a chronic inflammatory disease of large- to medium-sized artery walls. Development of atherosclerosis starts early in life. Since progression is slow, relevant clinical diseases like angina pectoris or myocardial infarction (MI) usually occur many years later (Libby, Ridker, & Hansson, 2011). Because inflammation is one key factor in the development of arteriosclerosis, the role of inflammation in the pathogenesis of CAD is widely accepted. Several markers such as acute-phase proteins like CRP, fibrinogen, immunoglobulins, adhesion molecules, and cytokines (e.g., TNF-α, IL-1, and IL-6) have been identified as markers for inflammation and atherosclerosis (Montecucco et al., 2017). Accordingly, patients with coronary artery disease (CAD) show elevated levels of proinflammatory cytokines and CRP as markers for clinical inflammation and higher expression of genes that code for proinflammatory proteins. Low-grade inflammation seems to be predictive of MI, and a few hours after acute MI, CRP increased dramatically (Kushner, Broder, & Karp, 1978). Furthermore, the magnitude of the acute inflammatory response seems to be predictive for cardiac outcome. In a meta-analysis investigating the influence of CRP levels, a relationship between CRP levels and outcome after MI was found, indicating that enhanced CRP levels measured within 3 days after myocardial infarction were associated with a relative risk of 2.18 (95%, 1.77–2.68) for recurrent cardiovascular events or even death (He, Tang, Ling, Chen, & Chen, 2010). Also, increased levels of other proinflammatory cytokines like TNF-α, IL-1 beta, IL-2, IL-6, and IL-10 have been found in plasma of patients with acute coronary infarction (Blum, Schneider, Sobel, & Dauerman, 2004;

Gouweleeuw et al., 2015), and also, patients with congestive heart failure showed elevated cytokines like TNF-α and IL-6 (Levine, Kalman, Mayer, Fillit, & Packer, 1990). Because the local inflammatory responses are extremely different between individuals with acute coronary syndrome, some authors hypothesize that there may be individual cytokine expression phenotypes, which could explain different risks of adverse cardiac events (Cirillo et al., 2014; Granville Smith, Parker, Cvejic, & Vollmer-Conna, 2015).

MECHANISM POSSIBLY LINKING CAD AND DEPRESSION

Several mechanisms are involved in the complex interplay between depression and CAD. Here, we will focus on mechanisms possibly influenced by proinflammatory cytokines. See also Fig. 1 for an illustration.

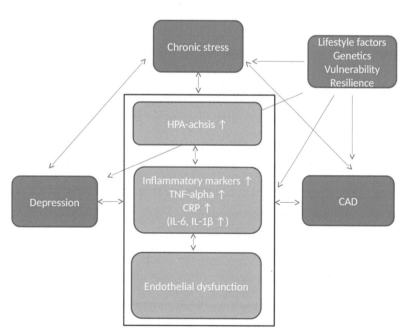

FIG. 1 Inflammatory markers and the interplay of CAD and depression.

THE ROLE OF PROINFLAMMATORY CYTOKINES IN CAD AND DEPRESSION

As described above, alterations of immune functions are observed in both cardiovascular disease and depression. Due to this fact and because proinflammatory cytokines globally interact in the body mediating numerous pathways that are involved in both disorders, several authors suggest that disturbed immune function is the key marker in comorbidity of depression and CAD (Baune et al., 2012; Mondelli & Pariante, 2015; Wright et al., 2014). A wide range of studies have investigated proinflammatory cytokines in the cooccurrence of depression and cardiovascular disease, however with heterogeneous results. When patients with CAD and cooccurring depression were compared with CAD patients without depression, increased levels of CRP in the first group were described (Bankier, Barajas, Martinez-Rumayor, & Januzzi, 2009). Also after acute MI, higher TNF-α levels were observed in patients with depression in contrast to those without depression (Shang et al., 2014). In patients with heart failure due to CAD, elevated TNF-α levels were reported in association with depressive symptoms (Ferketich, Ferguson, & Binkley, 2005). However, in the "Heart and Soul" study, higher IL-6 and hsCRP levels related to depressive symptoms in patients with CAD were found, but after correction for health behavior, these differences were no longer significant. Interestingly, elevated levels were not predictive for depressive symptoms in this study. Another important finding was presented by Steptoe and colleagues showing that white blood cell counts but not CRP predict the onset of first-time depression within the next 6 months, whereas the severity of this proinflammatory reaction seems not to be related to the onset of depression (Steptoe, Wikman, Molloy, Messerli-Burgy, & Kaski, 2013). In a cohort prospective epidemiological study of myocardial infarction (PRIME), it was found that depressed mood is related to elevated

CRP and IL-6 levels in patients with MI and, furthermore, that each inflammatory marker contributed significantly to CAD event risk (Empana et al., 2005). In contrast to this, others found no significant difference in CRP levels between patients with and without depression during hospital stay (Shimbo, Rieckmann, Paulino, & Davidson, 2006) or failed to detect elevated levels of CRP, IL-6, and TNF-α in relationship to depression after MI, while in contrast to healthy subjects, the levels of sIL-6R and sTNF-RII were elevated in both groups (Schins et al., 2005). In the MONICA-KORA Augsburg cohort study, by which initially healthy men were investigated, a combination of depressed mood and higher levels of high-sensitive (hs) CRP was able to predict a subsequent MI, whereas hsCRP alone failed (Ladwig, Marten-Mittag, Lowel, Doring, & Koenig, 2005).

Nevertheless, a recent review that addresses relevant biomarkers for the comorbidity between depression and coronary artery disease exhibits the most consistent associations for TNF-α (Adibfar et al., 2016). Especially investigating the TNF-α serum levels after MI, a positive correlation with BDI score was detected, while other cytokines (CRP, IL1-β, and IL-6) seem to be unrelated (Shang et al., 2014). Depending on their results, the authors concluded that "evidence supporting the use of inflammatory cytokines as biomarkers is inconsistent" (Adibfar et al., 2016).

ENDOTHELIAL DYSFUNCTION IN COMORBIDITY OF DEPRESSION AND CAD

In contrast to these findings, this review explicates that flow-mediated dilation, endothelin-1, endothelial progenitor cells, brain-derived neurotrophic factor, and docosahexaenoic acid were associated with depression and coronary artery disease. These findings indicate a key role of endothelial dysfunction in linking depression and coronary artery disease. Interestingly, some researchers hypothesized that endothelial dysfunction is triggered by TNF-α, because in patients with MI, therapy with TNF-α antagonists (Etanercept) has reduced TNF-α levels and improved endothelial vasodilator capacity indicating a relationship between both (Fichtlscherer et al., 2001; Liu et al., 2012). TNF-α is also increased in depression, so it could be a mediating factor between both disorders, because proinflammatory cytokines have been postulated to be an intermediating factor for inducing endothelial damage and it has been proposed that this influences the development of atherosclerosis during complex signaling cascades resulting in both depression and CAD (Halaris, 2013). Several studies have shown that also elevated CRP levels are associated to a blunted endothelial vasodilator response indicative for a systemic immune response. Moreover, some authors argued that "the identification of elevated CRP levels as a transient independent risk factor for endothelial dysfunction might provide an important clue to link a systemic marker of inflammation to atherosclerotic disease progression" (Fichtlscherer & Zeiher, 2000). In line with this, some authors found impairments in endothelial functions in depression (Rybakowski, Wykretowicz, Heymann-Szlachcinska, & Wysocki, 2006), and a recent review described an association between atherosclerosis and depression, even when the impact of this relation remains unclear (Saleptsis, Labropoulos, Halaris, Angelopoulos, & Giannoukas, 2011). Furthermore, depression in people suffering from acute coronary syndrome is associated with endothelial dysfunction, as indicated by higher levels of cellular adhesion molecules (Lesperance, Frasure-Smith, Theroux, & Irwin, 2004). The Maastricht Study, a population-based cohort study, shows that sum scores of inflammatory and endothelial markers are both associated with an increased risk for depression (MDD), even after correction

for somatic and for behavioral and lifestyle factors (van Dooren et al., 2016). These findings indicate that endothelial dysfunction could be a mediating factor in the relationship between chronic inflammation, depression, and CAD.

Another focus of research lies on the possible dysfunction of the blood-brain barrier (BBB) that might represent a mechanism that links inflammation and depression. After MI, the occurrence of systemic inflammation may induce endothelial leakage and loss of integrity of BBB. An increased permeability of BBB results in the penetration of the brain by inflammatory cytokines, a process that seems to increase the risk for depression (Shalev, Serlin, & Friedman, 2009). An increased permeability of the BBB is also associated with elevated CRP levels, and it can be argued that higher CRP levels observed in CAD patients could therefore be associated with higher risk to develop depression (Nikkheslat et al., 2015).

INFLUENCE OF STRESS AND HPA-AXIS-DYSFUNCTION IN DEPRESSION AND CAD

On a behavioral level, different lines of evidence suggest that chronic stress represents an important factor in the interplay between depression and CAD and maybe inflammation is also involved in this interplay. The HPA axis is one crucial and probably the best investigated pathway that is involved in the regulation of stress response. In response to acute stress, cortisol is secreted and than, by eliciting a negative feedback in the HPA axis, facilitates recovery from stressors. As a result of chronic stress, however, the downregulation of peripheral glucocorticoid receptors and glucocorticoid resistance impair this recovery function. Furthermore, while these mechanisms are appearing, also elevated inflammatory markers have been observed. In patients with depression and in patients with CAD, an upregulation of the HPA axis was described. In a substantial proportion of patients, depression is associated with an increased activation of HPA axis that results in hypercortisolemia and increased levels of corticotropin-releasing hormone (CRH) (Holsboer & Ising, 2008). This was interpreted as a reduced ability of cortisol, probably due to the downregulation of the peripheral glucocorticoid receptors, to inhibit the activity of HPA axis and to end the inflammatory response (Pariante, 2006). Regarding cardiovascular functions that respond to acute stress, the HPA axis is involved in the regulation of blood pressure and blood vessel dilatation (Steptoe & Kivimaki, 2013). In response to chronic stress, an upregulation of the HPA axis and increased cortisol levels in CAD patients are correlated with the calcification of arteries. This was discussed as a factor by which psychosocial stress influences the risk of CAD (Hamer, Batty, & Kivimaki, 2012). In a recent study by Nikkheslat and colleagues, the proinflammatory activity in CAD patients with and without depression is measured, also addressing the function of the HPA axis and the glucocorticoid receptor (Nikkheslat et al., 2015). Authors described that CAD patients with depression have higher levels of CRP, IL-6 gene expression, and plasma vascular endothelial growth factor (VEGF) and reduced plasma and saliva cortisol levels in contrast to a nondepressed group with CAD. Also, in patients with depression, a reduction in expression of the glucocorticoid receptor, lower tryptophan levels, and an increased kynurenine-tryptophan ratio has been observed. Regarding the kynurenine pathway, it is now well known (particularly from animal studies) that proinflammatory cytokines increase the activity of the indoleamine-2,3-dioxygenase (IDO) and, by this mechanism, facilitate the degradation of tryptophan to kynurenine instead of serotonin. This mechanism leads to a relative serotonin deficiency (like a tryptophan depletion diet) that is a causal factor for depression in vulnerable

individuals (Capuron et al., 2004; Dantzer, O'Connor, Lawson, & Kelley, 2011; Myint & Kim, 2003). The authors conclude that CAD patients with depression have increased levels of inflammation in the context of HPA axis hypoactivity, GR resistance, and an increased activation of the kynurenine pathway connected to insufficient glucocorticoid signaling (Nikkheslat et al., 2015). Increased proinflammatory activity is hence suggested as a relevant factor for developing depressive symptoms.

ANIMAL STRESS PARADIGMS FOR MIMICKING DEPRESSION-CARDIOVASCULAR COMORBIDITY

Moreover, rodent models have demonstrated that stress paradigms can modulate biological aspects of the depression-cardiovascular comorbidity. A recent review (Carnevali et al., 2017) focuses on rodent models of chronic stress exposure using social defeat (SD), social isolation (SI), and chronic mild stress (CMS) mimicking stressful human life for inducing depression in combination with cardiovascular disease. All models of stress exposure can induce depression-like symptoms that were measured, for example, lower locomotor activity, reduced sucrose solution, circadian and sleep disturbances, and body weight alterations. Similar as in depressive humans, also changes in neural, physiological, and neuroendocrine parameters like reduced hippocampal neurogenesis, decreased BDNF levels, autonomic and hypothalamic-pituitary-adrenocortical (HPA) axis dysfunctions, and increased proinflammatory cytokines were described (Sgoifo, Carnevali, Alfonso Mde, & Amore, 2015). The behavioral and biological changes in chronically stressed rodents with depression-like symptoms have prompted the investigation of cardiovascular changes. Others reported elevated resting heart rate, reduced heart rate variability, and larger heart rate responses to ß-adrenergic receptor blockade

(Johnson & Grippo, 2006). Additional CMS-exposed rats showed physiological alterations, which results in a disturbance of endothelial control and hormonal changes in catecholamines, aldosterone, and corticosterone (Grippo, Francis, Beltz, Felder, & Johnson, 2005). Looking for proinflammatory cytokines, increased plasma levels of TNF-α and IL-1ß can be observed (Carnevali et al., 2017; Wann, Audet, Gibb, & Anisman, 2010). Interestingly, after the end of the stress period, the cardiovascular changes occur over a longer period, while the behavioral changes recover (Grippo, Francis, Weiss, Felder, & Johnson, 2003). This was similar in humans, where cardiovascular symptoms after remission of acute depression persist. In contrast, the authors described that in rodents, chronic stressors applied in perinatal or postnatal period increase vulnerability for depression in later life but have only moderate impact on cardiovascular function in adulthood (Wulsin, Wick-Carlson, Packard, Morano, & Herman, 2016). On the basis of the reviewed rodent stress models, the authors conclude that increased cardiovascular risk after stress and depression could be mediated by automatic dysfunction, HPA axis hyperactivity, increased levels of proinflammatory cytokines, endothelial dysfunction, and remodeling of electrical and cellular properties of the heart (Carnevali et al., 2017). Additionally, several rodent studies investigated the link between depression and CVD from the cardiovascular perspective. In an animal model conducted by Grippo and colleagues (Grippo et al., 2003), rodents develop depression-like symptoms after experimental induced heart failures. The observed anhedonia can be reversed by application of the TNF-alpha blocker Etanercept. Interestingly, TNF-α infusion in mice can also induce a depression-like behavior (Kaster, Gadotti, Calixto, Santos, & Rodrigues, 2012), indicating a relationship between TNF-α and depression.

Inducing of myocardial infarction (MI) in rats results in depression-like behavior, which was associated with elevated levels of TNF-α and apoptosis in the limbic system (Wann et al., 2007, 2009) and in the ventral hippocampus, which was associated with decreased gray matter concentration, neurite outgrow, and neurogenesis (Suzuki et al., 2015). This finding underlines the hypothesis that the interplay between both diseases was linked by an interaction of immune function and the brain, probably involving proinflammatory cytokines (Carnevali et al., 2017).

IS THE SICKNESS-BEHAVIOR-MODEL A MODEL FOR DEVELOPMENT OF DEPRESSION AFTER ACI?

Similar to this, some authors argued according to a sickness behavior model that human depression after acute cardiac ischemia (ACI) is different from common mental disorder (de Miranda Azevedo, Roest, Hoen, & de Jonge, 2014; Granville Smith et al., 2015; Poole et al., 2014; Poole, Dickens, & Steptoe, 2011; Vollmer-Conna, Cvejic, Granville Smith, Hadzi-Pavlovic, & Parker, 2015). This understanding is supported by three observations. First, depression after ACI is time-correlated and an obvious risk factor for later morbidity and mortality, independent of cardiac severity and other prognostic factors (Nicholson et al., 2006). A strong association was described with new-onset depression after acute coronary incident (Carney, Freedland, & Jaffe, 2009), but others describe an association with a depression onset during 1 month after the event independent of being a first or recurrent episode (Parker et al., 2008). Secondly, patients with ACI seem to have a different symptom profile with more somatic than cognitive symptoms. One study looking at depression symptoms found that particularly somatic-affective symptoms of depression (like anhedonia and fatigue) in contrast to cognitive-affective symptoms (like social withdrawal) are present in patients with CAD and are associated with a poorer prognosis (de Miranda Azevedo et al., 2014). And thirdly, depression in CAD seems to be relatively treatment resistant: patients with depression after CAD showed a lower treatment response to conventional antidepressants as found in large-scale studies (Poole et al., 2011). Interestingly, a subgroup analysis of the "Canadian Cardiac Randomized Evaluation of Antidepressants and Psychotherapy Efficacy Trial" (Habra et al., 2010) found a better effect of citalopram in recurrent than in new-onset depression indicating different underlying biological processes in both diseases. A recent review by Granville Smith and colleagues addresses this question of a distinct subtype of ACS-associated depression and evaluates a sickness behavior analogy of infection as an explanatory model for this (Granville Smith et al., 2015). The acute sickness response is a short period during an acute illness that is triggered by increased proinflammatory cytokines (interferon-gamma, IL-1β, IL-6, and TNF-α) (Dantzer & Kelley, 2006; Dantzer et al., 2008). This is accompanied with behavioral and motivational changes including disturbed mood, anorexia, anhedonia, impaired concentration, fatigue, and altered sleep. The biological function seems to be a survival strategy shifting the individual priorities to recovery from illness. Although this response is reversible under normal circumstances, the sickness response model also implicates that, in vulnerable individuals, a major depressive episode may be elicited, and prospective cohort studies and experimental studies that were reviewed by Gunaratne et al. (2013) indicate a relationship between immune activation and depression. Similar to a stress-responsive activation of the HPA axis and the sympathetic nervous system, a systemic immune activation also may function as a trigger. Given that ACS is associated with a strong systemic immune activation, the hypothesis that depression after

ACS may have an immunologic basis seems to be conclusive at least in a proportion of cases. It has, however, to be taken into account that patients with ACS differ significantly in activation of inflammatory pathways and only about one-half of them show an increase of cytokines (Cirillo et al., 2014).

Furthermore, only one-third of patients develop depression although the immunogenic stressor was the same (Raison & Miller, 2011). A recent study investigated a wide range of biomarkers in ACS patients during their initial hospital stay including several immunologic markers like IL-6, IL-10, IL-1 beta, and TNF-α. Disappointingly, there was no support for distinguishing patients that had developed depression after ACI from those who had not only on the basis of their biological profiles (Vollmer-Conna et al., 2015). But looking from the biological marker side, a subgroup of individuals with an enhanced CRP level, low heart rate variability, and poor nutritional status could be described, which includes most of the depressive patients in this cohort, independent of recurrent or new-onset depression. Although it seems to be plausible to explain ACS-associated depression as a sickness response analogy, many authors point out that, at the moment, the evidence for depression after ACS being a distinct high-risk subtype of depression with an associated inflammatory profile is still not completely convincing, due to a lack of fully conclusive findings. Given the available evidence, it seems plausible that, in a subgroup of patients, additional factors like genetic or acquired vulnerabilities occur, which leads to a disturbed systemic immune stress response after ACI resulting in depression. So, it might be helpful for the understanding of underlying immunologic mechanisms to separately study (1) depressive patients, who develop cardiovascular disease, and (2) patients with cardiovascular disease, especially acute coronary artery disease, who subsequently develop depression.

CONCLUSION

There is now some evidence indicating that disturbances in immune function may play a central role in the interplay of CAD and depression. There are also findings pointing to a complex interaction between immune function and biological systems that are involved in both depression and CAD, such as HPA axis, endothelial dysfunctions, and brain function. Especially, TNF-α seems to play an important role in this interplay, as shown in a recent review (Adibfar et al., 2016).TNF-α is increased in depression and in CAD. Additionally, it was hypothesized that TNF-α triggers endothelial dysfunction (Liu et al., 2012). Hence, TNF-α blocker could be an interesting therapeutic option. Studies using TNF-α blocker in patients with rheumatoid arthritis (Ursini et al., 2017) and in patients with heart failure show positive effects on endothelial dysfunction (Fichtlscherer et al., 2001). Also, a reduction of depressive symptoms in Crohn's disease during therapeutic intervention with a TNF-a blocker was reported by one group (Krebs et al., 2010). Interestingly in treatment-resistant depression, baseline hsCRP (>5mg/L) was predictive for significant reduction in HAM-D scores during therapy with infliximab, a TNF-α blocker, in contrast to placebo (Raison et al., 2013). These findings possibly indicate a novel therapeutic option for a subgroup of depressive patients with high inflammatory markers.

However, the anticipated complexity of the interaction between depression and CAD and the fact that most studies analyze only single biomarkers instead of complex biomarker sets frustrated the identification of the underlying mechanism. Another problem is the large inhomogeneity between studies in classification of depression and CAD. While some studies investigated heart disease in general, others concentrated on acute or chronic CAD. The same can be observed in classification in depression using the criteria for "major depression" on the one side and the "Public Health Questionnaire" (PHQ-9) to detect depressive symptoms on the other side. Also, different outcome parameters and periods as well as different physiological markers were used. This contributes in difficulties to integrate the existing results, and so, it is not surprising that no simple physiological model that can explain the relationship in detail exists.

Further prospective longitudinal cohort studies with exactly defined clinical categories and a large set of biomarkers (Teismann et al., 2014) and studies focusing on small and homogeneous patient groups with a controlled study design are required to investigate the interaction of these mechanisms and create a complex model of the relationship between CVD and depression. Extending the current knowledge would be helpful to develop new therapeutic strategies in order to modulate disturbed biological pathways, especially in the field of endothelial dysfunction and immune system. This could also be helpful for identifying individual cytokine expression phenotypes, which on the one hand would explain different risks of adverse cardiac events and on the other hand could be helpful in developing new therapeutic agents depending on these profiles. Driving research further along this avenue seems to yield promising options to lessen the disease burden of this complex comorbidity in the future.

References

Adibfar, A., Saleem, M., Lanctot, K. L., & Herrmann, N. (2016). Potential biomarkers for depression associated with coronary artery disease: A critical review. *Current Molecular Medicine, 16*(2), 137–164.

Amare, A. T., Schubert, K. O., Klingler-Hoffmann, M., Cohen-Woods, S., & Baune, B. T. (2017). The genetic overlap between mood disorders and cardiometabolic diseases: A systematic review of genome wide and candidate gene studies. *Translational Psychiatry, 7*(1). https://doi.org/10.1038/tp.2016.261.

Anisman, H., & Zacharko, R. M. (1992). Depression as a consequence of inadequate neurochemical adaptation in response to stressors. *The British Journal of Psychiatry, Suppl*(15), 36–43.

Bankier, B., Barajas, J., Martinez-Rumayor, A., & Januzzi, J. L. (2009). Association between major depressive disorder and C-reactive protein levels in stable coronary heart disease patients. *Journal of Psychosomatic Research, 66*(3), 189–194. https://doi.org/10.1016/j.jpsychores.2008.09.010.

Barefoot, J. C., Helms, M. J., Mark, D. B., Blumenthal, J. A., Califf, R. M., Haney, T. L., et al. (1996). Depression and long-term mortality risk in patients with coronary artery disease. *The American Journal of Cardiology, 78*(6), 613–617.

Barefoot, J. C., & Schroll, M. (1996). Symptoms of depression, acute myocardial infarction, and total mortality in a community sample. *Circulation, 93*(11), 1976–1980.

Baumeister, D., Ciufolini, S., & Mondelli, V. (2016). Effects of psychotropic drugs on inflammation: Consequence or mediator of therapeutic effects in psychiatric treatment? *Psychopharmacology, 233*(9), 1575–1589. https://doi.org/10.1007/s00213-015-4044-5.

Baumeister, D., Russell, A., Pariante, C. M., & Mondelli, V. (2014). Inflammatory biomarker profiles of mental disorders and their relation to clinical, social and lifestyle factors. *Social Psychiatry and Psychiatric Epidemiology, 49*(6), 841–849. https://doi.org/10.1007/s00127-014-0887-z.

Baune, B. T., Stuart, M., Gilmour, A., Wersching, H., Heindel, W., Arolt, V., et al. (2012). The relationship between subtypes of depression and cardiovascular disease: A systematic review of biological models. *Translational Psychiatry, 2*. https://doi.org/10.1038/tp.2012.18.

Beratis, S., Katrivanou, A., Georgiou, S., Monastirli, A., Pasmatzi, E., Gourzis, P., et al. (2005). Major depression and risk of depressive symptomatology associated with short-term and low-dose interferon-alpha treatment. *Journal of Psychosomatic Research, 58*(1), 15–18. https://doi.org/10.1016/j.jpsychores.2004.03.010.

Blum, A., Schneider, D. J., Sobel, B. E., & Dauerman, H. L. (2004). Endothelial dysfunction and inflammation after percutaneous coronary intervention. *The American Journal of Cardiology, 94*(11), 1420–1423. https://doi.org/10.1016/j.amjcard.2004.07.146.

Brunello, N., Alboni, S., Capone, G., Benatti, C., Blom, J. M., Tascedda, F., et al. (2006). Acetylsalicylic acid accelerates the antidepressant effect of fluoxetine in the chronic escape deficit model of depression. *International Clinical Psychopharmacology, 21*(4), 219–225.

Capuron, L., Ravaud, A., Miller, A. H., & Dantzer, R. (2004). Baseline mood and psychosocial characteristics of patients developing depressive symptoms during interleukin-2 and/or interferon-alpha cancer therapy. *Brain, Behavior, and Immunity, 18*(3), 205–213. https://doi.org/10.1016/j.bbi.2003.11.004.

Carnevali, L., Montano, N., Statello, R., & Sgoifo, A. (2017). Rodent models of depression-cardiovascular comorbidity:

Bridging the known to the new. *Neuroscience & Biobehavioral Reviews, 76*(Pt A), 144–153. https://doi.org/10.1016/j.neubiorev.2016.11.006.

Carney, R. M., Freedland, K. E., & Jaffe, A. S. (2009). Depression screening in patients with heart disease. *JAMA, 301*(13), 1337. author reply 1338. . https://doi.org/10.1001/jama.2009.408.

Carney, R. M., Freedland, K. E., Steinmeyer, B., Blumenthal, J. A., Berkman, L. F., Watkins, L. L., et al. (2008). Depression and five year survival following acute myocardial infarction: A prospective study. *Journal of Affective Disorders, 109*(1–2), 133–138. https://doi.org/10.1016/j.jad.2007.12.005.

Cirillo, P., Cimmino, G., D'Aiuto, E., Di Palma, V., Abbate, G., Piscione, F., et al. (2014). Local cytokine production in patients with acute coronary syndromes: A look into the eye of the perfect (cytokine) storm. *International Journal of Cardiology, 176*(1), 227–229. https://doi.org/10.1016/j.ijcard.2014.05.035.

Dantzer, R., & Kelley, K. W. (2006). Twenty years of research on cytokine-induced sickness behavior. *Brain, Behavior, and Immunity, 21*(2), 153–160.

Dantzer, R., O'Connor, J. C., Freund, G. G., Johnson, R. W., & Kelley, K. W. (2008). From inflammation to sickness and depression: When the immune system subjugates the brain. Nature Reviews. Neuroscience, 9(1), 46–56. https://doi.org/10.1038/nrn2297.

Dantzer, R., O'Connor, J. C., Lawson, M. A., & Kelley, K. W. (2011). Inflammation-associated depression: From serotonin to kynurenine. *Psychoneuroendocrinology, 36*(3), 426–436. https://doi.org/10.1016/j.psyneuen.2010.09.012.

Dao, T. K., Youssef, N. A., Gopaldas, R. R., Chu, D., Bakaeen, F., Wear, E., et al. (2010). Autonomic cardiovascular dysregulation as a potential mechanism underlying depression and coronary artery bypass grafting surgery outcomes. *Journal of Cardiothoracic Surgery, 5*. https://doi.org/10.1186/1749-8090-5-36.

de Miranda Azevedo, R., Roest, A. M., Hoen, P. W., & de Jonge, P. (2014). Cognitive/affective and somatic/affective symptoms of depression in patients with heart disease and their association with cardiovascular prognosis: A meta-analysis. *Psychological Medicine, 44*(13), 2689–2703. https://doi.org/10.1017/s0033291714000063.

Dowlati, Y., Herrmann, N., Swardfager, W., Liu, H., Sham, L., Reim, E. K., et al. (2010). A meta-analysis of cytokines in major depression. *Biological Psychiatry, 67*(5), 446–457. https://doi.org/10.1016/j.biopsych.2009.09.033.

Doyle, F., McGee, H., Conroy, R., Conradi, H. J., Meijer, A., Steeds, R., et al. (2015). Systematic review and individual patient data meta-analysis of sex differences in depression and prognosis in persons with myocardial infarction: A MINDMAPS study. *Psychosomatic Medicine, 77*(4), 419–428. https://doi.org/10.1097/psy. 0000000000000174.

Empana, J. P., Sykes, D. H., Luc, G., Juhan-Vague, I., Arveiler, D., Ferrieres, J., et al. (2005). Contributions of depressive mood and circulating inflammatory markers to coronary heart disease in healthy European men: The Prospective Epidemiological Study of Myocardial Infarction (PRIME). *Circulation, 111*(18), 2299–2305. https://doi.org/10.1161/01.cir.0000164203.54111.ae.

Felger, J. C., Li, Z., Haroon, E., Woolwine, B. J., Jung, M. Y., Hu, X., et al. (2016). Inflammation is associated with decreased functional connectivity within corticostriatal reward circuitry in depression. *Molecular Psychiatry, 21* (10), 1358–1365. https://doi.org/10.1038/mp.2015.168.

Ferketich, A. K., Ferguson, J. P., & Binkley, P. F. (2005). Depressive symptoms and inflammation among heart failure patients. *American Heart Journal, 150*(1), 132–136. https://doi.org/10.1016/j.ahj.2004.08.029.

Fichtlscherer, S., Rossig, L., Breuer, S., Vasa, M., Dimmeler, S., & Zeiher, A. M. (2001). Tumor necrosis factor antagonism with etanercept improves systemic endothelial vasoreactivity in patients with advanced heart failure. *Circulation, 104*(25), 3023–3025.

Fichtlscherer, S., & Zeiher, A. M. (2000). Endothelial dysfunction in acute coronary syndromes: Association with elevated C-reactive protein levels. *Annals of Medicine, 32*(8), 515–518. https://doi.org/10.3109/07853890008998830.

Frasure-Smith, N., Lesperance, F., & Talajic, M. (1993). Depression following myocardial infarction. Impact on 6-month survival. *JAMA, 270*(15), 1819–1825.

Friebe, A., Horn, M., Schmidt, F., Janssen, G., Schmid-Wendtner, M. H., Volkenandt, M., et al. (2010). Dose-dependent development of depressive symptoms during adjuvant interferon-{alpha} treatment of patients with malignant melanoma. *Psychosomatics, 51*(6), 466–473. https://doi.org/10.1176/appi.psy.51.6.466.

Gananca, L., Oquendo, M. A., Tyrka, A. R., Cisneros-Trujillo, S., Mann, J. J., & Sublette, M. E. (2016). The role of cytokines in the pathophysiology of suicidal behavior. *Psychoneuroendocrinology, 63*, 296–310. https://doi.org/10.1016/j.psyneuen.2015.10.008.

Gouweleeuw, L., Naude, P. J., Rots, M., DeJongste, M. J., Eisel, U. L., & Schoemaker, R. G. (2015). The role of neutrophil gelatinase associated lipocalin (NGAL) as biological constituent linking depression and cardiovascular disease. *Brain, Behavior, and Immunity, 46*, 23–32. https://doi.org/10.1016/j.bbi.2014.12.026.

Granville Smith, I., Parker, G., Cvejic, E., & Vollmer-Conna, U. (2015). Acute coronary syndrome-associated depression: The salience of a sickness response analogy? *Brain, Behavior, and Immunity, 49*, 18–24. https://doi.org/10.1016/j.bbi.2015.02.025.

Grippo, A. J., Francis, J., Beltz, T. G., Felder, R. B., & Johnson, A. K. (2005). Neuroendocrine and cytokine profile of chronic mild stress-induced anhedonia. *Physiology & Behavior, 84*(5), 697–706. https://doi.org/10.1016/j.physbeh.2005.02.011.

Grippo, A. J., Francis, J., Weiss, R. M., Felder, R. B., & Johnson, A. K. (2003). Cytokine mediation of experimental heart failure-induced anhedonia. *American Journal of Physiology. Regulatory, Integrative and Comparative Physiology, 284*(3), R666–673.

Grippo, A. J., & Johnson, A. K. (2009). Stress, depression and cardiovascular dysregulation: A review of neurobiological mechanisms and the integration of research from preclinical disease models. *Stress, 12*(1), 1–21. https://doi.org/10.1080/10253890802046281.

Grosse, L., Ambree, O., Jorgens, S., Jawahar, M. C., Singhal, G., Stacey, D., et al. (2016). Cytokine levels in major depression are related to childhood trauma but not to recent stressors. *Psychoneuroendocrinology, 73*, 24–31. https://doi.org/10.1016/j.psyneuen.2016.07.205.

Gunaratne, P., Lloyd, A. R., & Vollmer-Conna, U. (2013). Mood disturbance after infection. *The Australian and New Zealand Journal of Psychiatry, 47*(12), 1152–1164. https://doi.org/10.1177/0004867413503718.

Gustad, L. T., Laugsand, L. E., Janszky, I., Dalen, H., & Bjerkeset, O. (2014). Symptoms of anxiety and depression and risk of heart failure: the HUNT study. *European Journal of Heart Failure, 16*(8), 861–870. https://doi.org/10.1002/ejhf.133.

Habra, M. E., Baker, B., Frasure-Smith, N., Swenson, J. R., Koszycki, D., Butler, G., et al. (2010). First episode of major depressive disorder and vascular factors in coronary artery disease patients: Baseline characteristics and response to antidepressant treatment in the CREATE trial. *Journal of Psychosomatic Research, 69*(2), 133–141. https://doi.org/10.1016/j.jpsychores.2010.02.010.

Halaris, A. (2013). Inflammation, heart disease, and depression. *Current Psychiatry Reports, 15*(10), 400. https://doi.org/10.1007/s11920-013-0400-5.

Hamer, M., Batty, G. D., & Kivimaki, M. (2012). Sleep loss due to worry and future risk of cardiovascular disease and all-cause mortality: The Scottish Health Survey. *European Journal of Preventive Cardiology, 19*(6), 1437–1443. https://doi.org/10.1177/1741826711426092.

Harrison, N. A., Brydon, L., Walker, C., Gray, M. A., Steptoe, A., & Critchley, H. D. (2009). Inflammation causes mood changes through alterations in subgenual cingulate activity and mesolimbic connectivity. *Biological Psychiatry, 66*(5), 407–414. https://doi.org/10.1016/j.biopsych.2009.03.015.

He, L. P., Tang, X. Y., Ling, W. H., Chen, W. Q., & Chen, Y. M. (2010). Early C-reactive protein in the prediction of long-term outcomes after acute coronary syndromes: A meta-analysis of longitudinal studies. *Heart, 96*(5), 339–346. https://doi.org/10.1136/hrt.2009.174912.

Himmerich, H., Binder, E. B., Kunzel, H. E., Schuld, A., Lucae, S., Uhr, M., et al. (2006). Successful antidepressant therapy restores the disturbed interplay between TNF-alpha system and HPA axis. *Biological Psychiatry, 60*(8), 882–888. https://doi.org/10.1016/j.biopsych.2006.03.075.

Hingorani, A. D., Cross, J., Kharbanda, R. K., Mullen, M. J., Bhagat, K., Taylor, M., et al. (2000). Acute systemic inflammation impairs endothelium-dependent dilatation in humans. *Circulation, 102*(9), 994–999.

Holsboer, F., & Ising, M. (2008). Central CRH system in depression and anxiety—Evidence from clinical studies with CRH1 receptor antagonists. *European Journal of Pharmacology, 583*(2–3), 350–357. https://doi.org/10.1016/j.ejphar.2007.12.032.

Howren, M. B., Lamkin, D. M., & Suls, J. (2009). Associations of depression with C-reactive protein, IL-1, and IL-6: A meta-analysis. *Psychosomatic Medicine, 71*(2), 171–186. https://doi.org/10.1097/PSY.0b013e3181907c1b.

Johnson, A. K., & Grippo, A. J. (2006). Sadness and broken hearts: Neurohumoral mechanisms and co-morbidity of ischemic heart disease and psychological depression. *Journal of Physiology and Pharmacology, 57* (Suppl 11), 5–29.

Jokinen, J., & Nordstrom, P. (2009). HPA axis hyperactivity and cardiovascular mortality in mood disorder inpatients. *Journal of Affective Disorders, 116*(1–2), 88–92. https://doi.org/10.1016/j.jad.2008.10.025.

Kaestner, F., Hettich, M., Peters, M., Sibrowski, W., Hetzel, G., Ponath, G., et al. (2005). Different activation patterns of proinflammatory cytokines in melancholic and nonmelancholic major depression are associated with HPA axis activity. *Journal of Affective Disorders, 87*(2–3), 305–311. https://doi.org/10.1016/j.jad.2005.03.012.

Kaster, M. P., Gadotti, V. M., Calixto, J. B., Santos, A. R., & Rodrigues, A. L. (2012). Depressive-like behavior induced by tumor necrosis factor-alpha in mice. *Neuropharmacology, 62*(1), 419–426. https://doi.org/10.1016/j.neuropharm.2011.08.018.

Kendler, K. S., Karkowski, L. M., & Prescott, C. A. (1999). Causal relationship between stressful life events and the onset of major depression. *The American Journal of Psychiatry, 156*(6), 837–841. https://doi.org/10.1176/ajp.156.6.837.

Kim, Y. K., & Won, E. (2017). The influence of stress on neuroinflammation and alterations in brain structure and function in major depressive disorder. *Behavioural Brain Research, 329,* 6–11. https://doi.org/10.1016/j.bbr.2017.04.020.

Kohler, O., Benros, M. E., Nordentoft, M., Farkouh, M. E., Iyengar, R. L., Mors, O., et al. (2014). Effect of anti-inflammatory treatment on depression, depressive symptoms, and adverse effects: A systematic review and meta-analysis of randomized clinical trials. *JAMA Psychiatry, 71*(12), 1381–1391. https://doi.org/10.1001/jamapsychiatry.2014.1611.

Kohler, C. A., Freitas, T. H., Maes, M., de Andrade, N. Q., Liu, C. S., Fernandes, B. S., et al. (2017). Peripheral cytokine and chemokine alterations in depression: A meta-analysis of 82 studies. *Acta Psychiatrica Scandinavica, 135* (5), 373–387. https://doi.org/10.1111/acps.12698.

Kohler, O., Petersen, L., Mors, O., & Gasse, C. (2015). Inflammation and depression: Combined use of selective serotonin reuptake inhibitors and NSAIDs or paracetamol

and psychiatric outcomes. *Brain and Behavior: A Cognitive Neuroscience Perspective. 5*(8)https://doi.org/10.1002/brb3.338.

Kraus, T., Haack, M., Schuld, A., Hinze-Selch, D., Koethe, D., & Pollmacher, T. (2002). Body weight, the tumor necrosis factor system, and leptin production during treatment with mirtazapine or venlafaxine. *Pharmacopsychiatry, 35*(6), 220–225. https://doi.org/10.1055/s-2002-36390.

Krebs, S., Omer, T. N., & Omer, B. (2010). Wormwood (*Artemisia absinthium*) suppresses tumour necrosis factor alpha and accelerates healing in patients with Crohn's disease—A controlled clinical trial. *Phytomedicine, 17*(5), 305–309. https://doi.org/10.1016/j.phymed.2009.10.013.

Kushner, I., Broder, M. L., & Karp, D. (1978). Control of the acute phase response. Serum C-reactive protein kinetics after acute myocardial infarction. *Journal of Clinical Investigation, 61*(2), 235–242. https://doi.org/10.1172/jci108932.

Ladwig, K. H., Marten-Mittag, B., Lowel, H., Doring, A., & Koenig, W. (2005). C-reactive protein, depressed mood, and the prediction of coronary heart disease in initially healthy men: Results from the MONICA-KORA Augsburg Cohort Study 1984-1998. *European Heart Journal, 26*(23), 2537–2542. https://doi.org/10.1093/eurheartj/ehi456.

Lesperance, F., Frasure-Smith, N., Theroux, P., & Irwin, M. (2004). The association between major depression and levels of soluble intercellular adhesion molecule 1, interleukin-6, and C-reactive protein in patients with recent acute coronary syndromes. *The American Journal of Psychiatry, 161*(2), 271–277. https://doi.org/10.1176/appi.ajp.161.2.271.

Levine, B., Kalman, J., Mayer, L., Fillit, H. M., & Packer, M. (1990). Elevated circulating levels of tumor necrosis factor in severe chronic heart failure. *The New England Journal of Medicine, 323*(4), 236–241. https://doi.org/10.1056/nejm199007263230405.

Libby, P., Ridker, P. M., & Hansson, G. K. (2011). Progress and challenges in translating the biology of atherosclerosis. *Nature, 473*(7347), 317–325. https://doi.org/10.1038/nature10146.

Liu, Y., Ho, R. C., & Mak, A. (2012). Interleukin (IL)-6, tumour necrosis factor alpha (TNF-alpha) and soluble interleukin-2 receptors (sIL-2R) are elevated in patients with major depressive disorder: A meta-analysis and meta-regression. *Journal of Affective Disorders, 139*(3), 230–239. https://doi.org/10.1016/j.jad.2011.08.003.

Mendlewicz, J., Kriwin, P., Oswald, P., Souery, D., Alboni, S., & Brunello, N. (2006). Shortened onset of action of antidepressants in major depression using acetylsalicylic acid augmentation: A pilot open-label study. *International Clinical Psychopharmacology, 21*(4), 227–231.

Meyer, T., Stanske, B., Kochen, M. M., Cordes, A., Yuksel, I., Wachter, R., et al. (2011). Serum levels of interleukin-6 and interleukin-10 in relation to depression scores in patients with cardiovascular risk factors. *Behavioral Medicine*, *37*(3), 105–112. https://doi.org/10.1080/08964289.2011.609192.

Mondelli, V., & Pariante, C. M. (2015). On the heart, the mind, and how inflammation killed the Cartesian dualism. Commentary on the 2015 named series: Psychological risk factors and immune system involvement in cardiovascular disease. *Brain, Behavior, and Immunity*, *50*, 14–17. https://doi.org/10.1016/j.bbi.2015.09.010.

Montecucco, F., Liberale, L., Bonaventura, A., Vecchie, A., Dallegri, F., & Carbone, F. (2017). The role of inflammation in cardiovascular outcome. *Current Atherosclerosis Reports*, *19*(3), 11. https://doi.org/10.1007/s11883-017-0646-1.

Muller, N., Schwarz, M. J., Dehning, S., Douhe, A., Cerovecki, A., Goldstein-Muller, B., et al. (2006). The cyclooxygenase-2 inhibitor celecoxib has therapeutic effects in major depression: Results of a double-blind, randomized, placebo controlled, add-on pilot study to reboxetine. *Molecular Psychiatry*, *11*(7), 680–684. https://doi.org/10.1038/sj.mp.4001805.

Musselman, D. L., Lawson, D. H., Gumnick, J. F., Manatunga, A. K., Penna, S., Goodkin, R. S., et al. (2001). Paroxetine for the prevention of depression induced by high-dose interferon alfa. *The New England Journal of Medicine*, *344*(13), 961–966. https://doi.org/10.1056/nejm200103293441303.

Myint, A. M., & Kim, Y. K. (2003). Cytokine-serotonin interaction through IDO: A neurodegeneration hypothesis of depression. *Medical Hypotheses*, *61*(5–6), 519–525.

Nicholson, A., Kuper, H., & Hemingway, H. (2006). Depression as an aetiologic and prognostic factor in coronary heart disease: A meta-analysis of 6362 events among 146 538 participants in 54 observational studies. *European Heart Journal*, *27*(23), 2763–2774. https://doi.org/10.1093/eurheartj/ehl338.

Nikkheslat, N., Zunszain, P. A., Horowitz, M. A., Barbosa, I. G., Parker, J. A., Myint, A. M., et al. (2015). Insufficient glucocorticoid signaling and elevated inflammation in coronary heart disease patients with comorbid depression. *Brain, Behavior, and Immunity*, *48*, 8–18. https://doi.org/10.1016/j.bbi.2015.02.002.

Pariante, C. M. (2006). The glucocorticoid receptor: Part of the solution or part of the problem? *Journal of Psychopharmacology*, *20*(4 Suppl), 79–84. https://doi.org/10.1177/1359786806066063.

Parker, G. B., Hilton, T. M., Walsh, W. F., Owen, C. A., Heruc, G. A., Olley, A., et al. (2008). Timing is everything: The onset of depression and acute coronary syndrome outcome. *Biological Psychiatry*, *64*(8), 660–666. https://doi.org/10.1016/j.biopsych.2008.05.021.

Penninx, B. W. (2017). Depression and cardiovascular disease: Epidemiological evidence on their linking mechanisms. *Neuroscience & Biobehavioral Reviews*, *74*(Pt B), 277–286. https://doi.org/10.1016/j.neubiorev.2016.07.003.

Penninx, B. W., Beekman, A. T., Honig, A., Deeg, D. J., Schoevers, R. A., van Eijk, J. T., et al. (2001). Depression and cardiac mortality: Results from a community-based longitudinal study. *Archives of General Psychiatry*, *58*(3), 221–227.

Pereira, V. H., Cerqueira, J. J., Palha, J. A., & Sousa, N. (2013). Stressed brain, diseased heart: A review on the pathophysiologic mechanisms of neurocardiology. *International Journal of Cardiology*, *166*(1), 30–37. https://doi.org/10.1016/j.ijcard.2012.03.165.

Poole, L., Dickens, C., & Steptoe, A. (2011). The puzzle of depression and acute coronary syndrome: Reviewing the role of acute inflammation. *Journal of Psychosomatic Research*, *71*(2), 61–68. https://doi.org/10.1016/j.jpsychores.2010.12.009.

Poole, L., Kidd, T., Leigh, E., Ronaldson, A., Jahangiri, M., & Steptoe, A. (2014). Depression, C-reactive protein and length of post-operative hospital stay in coronary artery bypass graft surgery patients. *Brain, Behavior, and Immunity*, *37*, 115–121. https://doi.org/10.1016/j.bbi.2013.11.008.

Raison, C. L., & Miller, A. H. (2011). Is depression an inflammatory disorder? *Current Psychiatry Reports*, *13*(6), 467–475. https://doi.org/10.1007/s11920-011-0232-0.

Raison, C. L., Rutherford, R. E., Woolwine, B. J., Shuo, C., Schettler, P., Drake, D. F., et al. (2013). A randomized controlled trial of the tumor necrosis factor antagonist infliximab for treatment-resistant depression: The role of baseline inflammatory biomarkers. *JAMA Psychiatry*, *70*(1), 31–41. https://doi.org/10.1001/2013.jamapsychiatry.4.

Reichenberg, A., Yirmiya, R., Schuld, A., Kraus, T., Haack, M., Morag, A., et al. (2001). Cytokine-associated emotional and cognitive disturbances in humans. *Archives of General Psychiatry*, *58*(5), 445–452.

Roth, G. A., Johnson, C., Abajobir, A., Abd-Allah, F., Abera, S. F., Abyu, G., et al. (2017). Global, regional, and national burden of cardiovascular diseases for 10 causes, 1990 to 2015. *Journal of the American College of Cardiology*. https://doi.org/10.1016/j.jacc.2017.04.052.

Rothermundt, M., Arolt, V., Peters, M., Gutbrodt, H., Fenker, J., Kersting, A., et al. (2001). Inflammatory markers in major depression and melancholia. *Journal of Affective Disorders*, *63*(1–3), 93–102.

Rugulies, R. (2002). Depression as a predictor for coronary heart disease. A review and meta-analysis. *American Journal of Preventive Medicine*, *23*(1), 51–61.

Rybakowski, J. K., Wykretowicz, A., Heymann-Szlachcinska, A., & Wysocki, H. (2006). Impairment of endothelial function in unipolar and bipolar depression. *Biological Psychiatry*, *60*(8), 889–891. https://doi.org/10.1016/j.biopsych.2006.03.025.

Saleptsis, V. G., Labropoulos, N., Halaris, A., Angelopoulos, N. V., & Giannoukas, A. D. (2011). Depression and atherosclerosis. *International Angiology*, *30*(2), 97–104.

Savoy, C., Van Lieshout, R. J., & Steiner, M. (2017). Is plasminogen activator inhibitor-1 a physiological bottleneck bridging major depressive disorder and cardiovascular disease? *Acta Physiologica (Oxford, England)*, 219(4), 715–727. https://doi.org/10.1111/apha.12726.

Schins, A., Tulner, D., Lousberg, R., Kenis, G., Delanghe, J., Crijns, H. J., et al. (2005). Inflammatory markers in depressed post-myocardial infarction patients. *Journal of Psychiatric Research*, 39(2), 137–144. https://doi.org/10.1016/j.jpsychires.2004.05.009.

Schmidt, F. M., Schroder, T., Kirkby, K. C., Sander, C., Suslow, T., Holdt, L. M., et al. (2016). Pro- and anti-inflammatory cytokines, but not CRP, are inversely correlated with severity and symptoms of major depression. *Psychiatry Research*, 239, 85–91. https://doi.org/10.1016/j.psychres.2016.02.052.

Sgoifo, A., Carnevali, L., Alfonso Mde, L., & Amore, M. (2015). Autonomic dysfunction and heart rate variability in depression. *Stress*, 18(3), 343–352. https://doi.org/10.3109/10253890.2015.1045868.

Shalev, H., Serlin, Y., & Friedman, A. (2009). Breaching the blood-brain barrier as a gate to psychiatric disorder. *Cardiovascular Psychiatry and Neurology*, 2009. https://doi.org/10.1155/2009/278531.

Shang, Y. X., Ding, W. Q., Qiu, H. Y., Zhu, F. P., Yan, S. Z., & Wang, X. L. (2014). Association of depression with inflammation in hospitalized patients of myocardial infarction. *Pakistan Journal of Medical Sciences*, 30(4), 692–697.

Shimbo, D., Rieckmann, N., Paulino, R., & Davidson, K. W. (2006). Relation between C reactive protein and depression remission status in patients presenting with acute coronary syndrome. *Heart*, 92(9), 1316–1318. https://doi.org/10.1136/hrt.2005.075861.

Shimizu, Y., Suzuki, M., Okumura, H., & Yamada, S. (2014). Risk factors for onset of depression after heart failure hospitalization. *Journal of Cardiology*, 64(1), 37–42. https://doi.org/10.1016/j.jjcc.2013.11.003.

Stenman, M., Holzmann, M. J., & Sartipy, U. (2016). Association between preoperative depression and long-term survival following coronary artery bypass surgery—A systematic review and meta-analysis. *International Journal of Cardiology*, 222, 462–466. https://doi.org/10.1016/j.ijcard.2016.07.216.

Steptoe, A., & Kivimaki, M. (2013). Stress and cardiovascular disease: An update on current knowledge. *Annual Review of Public Health*, 34, 337–354. https://doi.org/10.1146/annurev-publhealth-031912-114452.

Steptoe, A., Wikman, A., Molloy, G. J., Messerli-Burgy, N., & Kaski, J. C. (2013). Inflammation and symptoms of depression and anxiety in patients with acute coronary heart disease. *Brain, Behavior, and Immunity*, 31, 183–188. https://doi.org/10.1016/j.bbi.2012.09.002.

Suzuki, H., Sumiyoshi, A., Matsumoto, Y., Duffy, B. A., Yoshikawa, T., Lythgoe, M. F., et al. (2015). Structural abnormality of the hippocampus associated with depressive symptoms in heart failure rats. *NeuroImage*, 105, 84–92. https://doi.org/10.1016/j.neuroimage.2014.10.040.

Teismann, H., Wersching, H., Nagel, M., Arolt, V., Heindel, W., Baune, B. T., et al. (2014). Establishing the bidirectional relationship between depression and subclinical arteriosclerosis—Rationale, design, and characteristics of the BiDirect study. *BMC Psychiatry*. 14, https://doi.org/10.1186/1471-244x-14-174.

Thombs, B. D., Grace, S. L., & Ziegelstein, R. C. (2006). Do symptom dimensions of depression following myocardial infarction relate differently to physical health indicators and cardiac prognosis? *The American Journal of Psychiatry*, 163(7), 1295–1296. [Author reply 1296]. https://doi.org/10.1176/appi.ajp.163.7.1295-a.

Ursini, F., Leporini, C., Bene, F., D'Angelo, S., Mauro, D., Russo, E., et al. (2017). Anti-TNF-alpha agents and endothelial function in rheumatoid arthritis: A systematic review and meta-analysis. *Scientific Reports*, 7(1), 5346. https://doi.org/10.1038/s41598-017-05759-2.

Valkanova, V., Ebmeier, K. P., & Allan, C. L. (2013). CRP, IL-6 and depression: A systematic review and meta-analysis of longitudinal studies. *Journal of Affective Disorders*, 150(3), 736–744. https://doi.org/10.1016/j.jad.2013.06.004.

van Dooren, F. E., Schram, M. T., Schalkwijk, C. G., Stehouwer, C. D., Henry, R. M., Dagnelie, P. C., et al. (2016). Associations of low grade inflammation and endothelial dysfunction with depression—The Maastricht study. *Brain, Behavior, and Immunity*, 56, 390–396. https://doi.org/10.1016/j.bbi.2016.03.004.

Vollmer-Conna, U., Cvejic, E., Granville Smith, I., Hadzi-Pavlovic, D., & Parker, G. (2015). Characterising acute coronary syndrome-associated depression: Let the data speak. *Brain, Behavior, and Immunity*, 48, 19–28. https://doi.org/10.1016/j.bbi.2015.03.001.

Vos, T., Barber, R. M., Bell, B., Bertozzi-Villa, A., Biryukov, S., Bolliger, I., et al. (2015). Global, regional, and national incidence, prevalence, and years lived with disability for 301 acute and chronic diseases and injuries in 188 countries, 1990-2013: A systematic analysis for the Global Burden of Disease Study 2013. *Lancet*, 386 (9995), 743–800. https://doi.org/10.1016/s0140-6736 (15)60692-4.

Wann, B. P., Audet, M. C., Gibb, J., & Anisman, H. (2010). Anhedonia and altered cardiac atrial natriuretic peptide following chronic stressor and endotoxin treatment in mice. *Psychoneuroendocrinology*, 35(2), 233–240. https://doi.org/10.1016/j.psyneuen.2009.06.010.

Wann, B. P., Bah, T. M., Boucher, M., Courtemanche, J., Le Marec, N., Rousseau, G., et al. (2007). Vulnerability for apoptosis in the limbic system after myocardial infarction in rats: A possible model for human postinfarct major depression. *Journal of Psychiatry & Neuroscience*, 32(1), 11–16.

Wann, B. P., Bah, T. M., Kaloustian, S., Boucher, M., Dufort, A. M., Le Marec, N., et al. (2009). Behavioural signs of depression and apoptosis in the limbic system following myocardial infarction: Effects of sertraline. *Journal of Psychopharmacology*, 23(4), 451–459. https://doi.org/10.1177/0269881108089820.

Whiteford, H. A., Degenhardt, L., Rehm, J., Baxter, A. J., Ferrari, A. J., Erskine, H. E., et al. (2013). Global burden of disease attributable to mental and substance use disorders: Findings from the Global Burden of Disease Study 2010. *Lancet*, 382(9904), 1575–1586. https://doi.org/10.1016/s0140-6736(13)61611-6.

WHO (2017). *Depression fact sheet*. http://www.who.int/mediacentre/factsheets/fs369/en.2017.

Whooley, M. A., & Wong, J. M. (2013). Depression and cardiovascular disorders. *Annual Review of Clinical Psychology*, 9, 327–354. https://doi.org/10.1146/annurev-clinpsy-050212-185526.

Wright, L., Simpson, W., Van Lieshout, R. J., & Steiner, M. (2014). Depression and cardiovascular disease in women: Is there a common immunological basis? A theoretical synthesis. *Therapeutic Advances in Cardiovascular Disease*, 8(2), 56–69. https://doi.org/10.1177/1753944714521671.

Wulsin, A. C., Wick-Carlson, D., Packard, B. A., Morano, R., & Herman, J. P. (2016). Adolescent chronic stress causes hypothalamo-pituitary-adrenocortical hypo-responsiveness and depression-like behavior in adult female rats. *Psychoneuroendocrinology*, 65, 109–117. https://doi.org/10.1016/j.psyneuen.2015.12.004.

Yang, J. M., Rui, B. B., Chen, C., Chen, H., Xu, T. J., Xu, W. P., et al. (2014). Acetylsalicylic acid enhances the anti-inflammatory effect of fluoxetine through inhibition of NF-kappaB, p38-MAPK and ERK1/2 activation in lipopolysaccharide-induced BV-2 microglia cells. *Neuroscience*, 275, 296–304. https://doi.org/10.1016/j.neuroscience.2014.06.016.

23

Inflammation Genetics of Depression

Michael Musker*,†, Julio Licinio*,†,‡, Ma-Li Wong*,†

*South Australian Health and Medical Research Institute, Adelaide, SA, Australia
†Flinders University, Bedford Park, SA, Australia
‡South Ural State University Biomedical School, Chelyabinsk, Russian Federation

INTRODUCTION

Major depressive disorder (MDD) is considered to be a multigenic disorder, in that no single gene has been identified so far as a definitive causal factor but the disease is known to affect 10%–15% of the population. There is evidence to demonstrate the hereditary risk of depression using both adoption and twin studies. A gene is a segment of DNA that contributes to phenotype or function, and the biological effects of inflammation are one such function. We know that MDD is a genetic disorder because genetic studies have shown a strong association between probands (the person reporting depression) and a first-degree relative, particularly in atypical depression (Lamers et al., 2016). In a meta-analysis of hereditary risk, the point estimate of heritability of liability = 37% (95% CI = 31%–42%) with a Mantel-Haenszel odds ratio of 2.84 (95% CI = 2.31–3.49) that there is a familial risk of depression (Sullivan, Neale, & Kendler, 2000). Research focused on proinflammatory markers linked to specific single-nucleotide polymorphisms (SNPs, often pronounced as Snips) has tried to discover the biological reason

for this genetic risk to depression (Schmidt et al., 2016). A polymorphism exists when two (or more) nucleotides can be found in one location creating two different alleles, at a frequency ≥1% of the population (Jobling, Hollox, Hurles, Kivisild, & Tyler-Smith, 2014). Sometimes, it is useful to link a combination of SNPs in an area of DNA to form "haplotypes" (haploid genotype) that is a set of polymorphisms on a single chromosome that has been inherited together from a single parent. These groups of SNPs that are inherited together enable us to identify ancestral patterns of inheritance within a population, which we call a "haplogroup." Murine models allow us to investigate variant SNPs on candidate genes, and this approach provides the opportunity to examine the pathophysiology of depressive symptoms in a controlled way. A candidate gene is a gene that is thought to cause a particular outcome or phenotype/disease, a known example being the gene for cystic fibrosis (cystic fibrosis transmembrane conductance regulator gene—*CFTR*). We are able to provide a cytogenetic location for this gene (e.g., 7q31.2) that is the chromosome location but even more specifically the genomic location (7:117,478,366-117,668,664). The cystic fibrosis is known to be a Mendelian disorder, in which the *CFTR* gene is inherited in an autosomal recessive inheritance pattern, requiring a copy of the mutated *CFTR* gene from each parent for the disorder to eventuate (Sosnay et al., 2017). Researchers can further assess the effects of candidate genes as they are able to manipulate the genetic makeup of various species to create transgenic strains that may have parts of a gene added or deleted, and then comparing these new strains to unaltered wild-type strains. Additionally, we can manipulate their environment to mimic human conditions such as chronic mild stress or provide a predetermined consistent diet and sleep pattern. Providing similar structured and controlled conditions in human environments is not so easy, if not impossible. Creating comparative stress reactions in the laboratory enables researchers to evaluate effects of the environment and expression of particular genotypes and phenotypic outcomes. A combination of both human and animal models over the last two decades has enabled the identification of some key genetic pathways that demonstrate strong links between inflammatory mechanisms and MDD.

HUMAN GENETICS

Owing to the advances in genetic reading technology that use array chip technology machines, such as the Illumina NextSeq 500 series, and the rapidly reducing cost of genotyping chips, we can take advantage of examining much larger areas of the genome or even the whole genome (i.e., genome wide). Genome-wide association studies (GWAS) use a case control method, comparing the differences in genetic variation of a case group and a control group. Such studies do not require a hypothetical or a priori rationale to research a specific gene loci, allowing more extensive/inclusive

research across the whole genome (Narimatsu, 2017). A key difference of the GWAS approach is that it requires extremely large sample sizes to find significant outcomes, but this is the new accelerated battlefront in our search for the link between the inflammatory system and MDD.

Over recent decades, both candidate gene approaches and linkage studies have helped demonstrate the connection of various polymorphisms to depression, the most convincing evidence being provided by the work with candidate genes. Due to the complexity and multifactorial aspects between genotype (genetic code), phenotype (traits, symptoms, and behavior), and the environment (stressors/habitat), these results have provided only limited evidence and have shown poor replication (Bosker et al., 2011). Inflammatory mediators that are commonly cited to have made genetic contributions to MDD include C-reactive protein (CRP), cytosolic phospholipase A2 (cPLA2), monocyte chemoattractant protein (MCP1), tumor necrosis factor alpha (TNF), interleukin (IL) 1 beta (IL1ß), IL6, and IL10 (Barnes, Mondelli, & Pariante, 2017). This chapter will present the role of just a few of these inflammatory mediators and their genetic variants, exploring the part they play in the cross talk between the immune system and MDD or antidepressant response.

CANDIDATE GENES, LINKAGE, AND GENOME WIDE ASSOCIATION STUDIES

Many early studies on depression have focused on functional polymorphisms (alleles that alter the function a gene) of the monoaminergic neurotransmitters, for example, the serotonin transport (*SLC6A4*) region 5-HTTLPR—a 44-base pair polymorphic region (Levinson, 2006). While polymorphisms in the 5-HTTLPR region have been linked to depression, anxiety, and stress, as a lone genetic marker, it does not predict depression. The role of the transporter gene *SLC6A4* is to activate the return of the neurotransmitter serotonin (5-hydroxytryptamine; 5-HT) from the synaptic cleft to the presynaptic pool (Ramamoorthy et al., 1993). Most antidepressants play a role in inhibiting this transport process and reuptake, increasing the availability of serotonin. MDD is believed to be of polygenic origin that has differing phenotypic outcomes and may be dependent on exposure to particular types of stress at different life stages. There is growing evidence that the immune system, mediated by the inflammatory network, has a two-way pathophysiological relationship with neurological networks and responses to stress, inhibiting the production of neurotransmitters by the production of inflammatory metabolites (Drachmann Bukh et al., 2009). The mechanistic relationship between inflammatory mediators and stress can be seen in the inflammasome response whereby our body uses "pattern recognition receptors" to identify danger. An example of this is the NOD-like receptor (NLR) family, pyrin domain containing 3 (NLRP3) inflammasome that provides a bidirectional pathway between depression and comorbid systemic illnesses (Iwata, Ota, & Duman, 2013).

These danger signals set off a cascade of bodily responses (host defense systems) including T cells, which in turn raise the plasma levels and expression of proinflammatory cytokines, for example, interleukins (IL1ß and IL6), and have been associated with the biofeedback mechanism of the hypothalamic-pituitary-adrenal (HPA) axis. Stressful experiences can hinder hippocampal neurogenesis, which in turn can contribute to hippocampal atrophy and anhedonic behavior (Koo & Duman, 2008). Research studies have demonstrated raised peripheral plasma levels and cerebrospinal fluid levels of various inflammatory mediators in participants that have been identified to have MDD, with a similar opposing fall in proinflammatory mediators such as IL10 (Wong, Dong, Maestre-Mesa, & Licinio, 2008). This interaction

of an immune related response has been used to investigate the effectiveness of some treatment modalities such as antidepressant responsiveness and speed of recovery to remission. Inflammatory mediators may also be responsible for physical changes to particular brain regions such as the amygdala, hippocampus, basal ganglia, and anterior cingulate cortex, which are associated with depressive symptomatology when compared with control groups (Baune et al., 2010).

MDD is seen as a biologically distinct condition when compared with other mood disorders, for example, bipolar disorder. A diagnosis of MDD is based on a group of descriptors agreed upon by experts assigned by the American Psychiatric Association who have developed the Diagnostic and Statistical Manual of Mental Disorders, Fifth Edition (DSM-5) or the equivalent International Classification of Diseases (ICD-10) produced by the World Health Organization. The lack of precision or consistency in diagnosis can prove problematic when making comparisons across research studies that often use different measurement tools or techniques. Data can be elicited from much larger studies using GWAS, which are not as dependent on the accuracy of the diagnosis or cut-off scores due to the massive cohorts and the equally large amount of SNPs that are used when compared with candidate gene studies (van der Sluis, Posthuma, Nivard, Verhage, & Dolan, 2013). One example that used 33,332 cases and 27,888 controls of European ancestry showed that there are overlap between five of the major mental health disorders and a relationship of aggregate molecular genetic risk factors or trait SNPs that are statistically significant (Zhao & Nyholt, 2017).

CANDIDATE GENES

The most frequently studied genes and biological markers have been developed from candidate gene studies in the inflammatory pathway of MDD, which include the following genes: *IL1A*, *IL1B*, *IL6*, interferon gamma (*IFNG*), tumor necrosis factor (*TNF*), and C-reactive protein (*CRP*). Candidate gene studies have been used to investigate the relationship of inflammation genetic variants and depression; however, the difficulty in replicating the results has caused doubt in the field concerning methodologies and the evidence provided so far (Arango, 2017). To identify an immune-depression candidate gene or variant, one might hypothesize that a particular SNP, for example, rs16944 (rs = reference SNP found in database of SNP (dbSNP)) on the *IL1B* gene and physically based in band 14, subband 1 on the long arm of chromosome 2 (2q14.1), has a polymorphic allele of guanine/adenine (G/A) that is likely to be more prevalent or less prevalent in the case group (composed of individuals who have that condition/disease) in comparison with the control group (composed of individuals who do not have the condition/disease but are otherwise similar). The actual polymorphism of interest, as there may be many on a gene, will also have a specific location, that is, chr2:112837290-112837290, and this can be found by searching the dbSNP database https://www.ncbi.nlm.nih.gov/SNP/ using the reference SNP ID, that is, rs16944. Part of the case group may have a different genotype such as being homozygous adenine/adenine (A/A) that produces a clear difference in phenotype. Positions are described as polymorphic as they can exist in any combination of the base pairs, such as homozygous A/A, heterozygous A/G, heterozygous G/A, or homozygous G/G. If these SNPs are found in the protein coding region (or exonic region), they can be synonymous in that they have no effect on the amino acid sequence or nonsynonymous, which means they alter the amino acid sequence (Hartwell, Hood, Goldberg, Reynolds, & Silver, 2011). Armed with the genotype using a particular SNP reference, it is necessary to provide a statistical frequency that demonstrates the relationship and possible risk to specific depressive symptoms

(e.g., neuroticism is associated with rs25531) or a diagnosis of MDD (Chang, Chang, Fang, Chang, & Huang, 2017). This SNP rs25531 is based on the *SLC6A4* solute carrier family 6 member 4 gene and has a minor allele frequency (MAF) of C = 0.1376/689, which means that C is the minor allele at a frequency of 13% in 1000 genomes, and the SNP was observed 689 times within the population. You can read more details about this gene and its attributes at the dbSNP database referenced earlier.

Linkage studies are another method used to discover important genetic effects, by examining two genetic loci that have a close proximity on the same chromosome, and do not always separate during meiosis. The closer they are on the chromosome, the lower the probability that they will separate during recombination. These are inherited within family groups or clusters at a higher rate than would be expected in families who do not have any history of the disease or trait (Lohoff, 2010). As technology has advanced, out of all the techniques described, it is the genome-wide associate studies that are likely to provide us with clearer direction in our search for relevant genetic markers in the links between proinflammatory mechanisms and MDD. Below, we review some significant findings in the genetics field, reporting on commonly examined inflammatory mediators, metabolic markers of stress, gene expression of SNPs, and their link to MDD.

THE INFLAMMATORY BIOMARKERS ASSOCIATED WITH MDD

The Interleukin Family

The importance of the *IL1B* polymorphism and its role in the immune system was described by Pociot, MØLvig, Wogensen, Worsaae, and Nerup (1992) who found that the polymorphism correlated with the secretion of IL1ß in the peripheral system. The authors suggested that it played a role not only in depression but also in multiple inflammatory-associated disorders including arthritis, thyroid disease, and septic shock. *IL1B* SNPs—rs16944 (-511C/T variation)—located on chromosome 2q14 were investigated by Yu, Chen, Hong, Chen, and Tsai (2003) to understand whether this biallelic polymorphism in the promoter region (-511) had any relationship with the treatment outcomes of fluoxetine ($n = 157$ cases and 112 controls). While they did not find any relationship in the response to antidepressants, they did note that those who were homozygous for the -511T allele of *IL1B* had less severe symptoms than those who were carriers of the *IL1B*-511CT or -511CC genotype (Yu et al., 2003). Tadic et al. (2008) completed a randomized double-blind controlled clinical trial with a group of clinically diagnosed depressed patients ($n = 101$; females/males $= 74:27$) and tested their biological response to paroxetine or mirtazapine. They found that carriers of the homozygous T allele at position -511 were more responsive to paroxetine. There was no difference in the response to mirtazapine (Tadic et al., 2008). Building on these findings, Baune et al. (2010) looked at three polymorphism of *IL1B* (rs16944, rs1143643, and rs1143634) in 256 Caucasian patients with MDD ($n = 145$ women and 111 men) and their response to antidepressant treatment after 6 weeks. They found associations between two variants rs16944 and rs1143643 and nonremission after antidepressant treatment, showing that carriers of the GG genotype were less likely to have a response to treatment than those with the AA genotype.

Inflammatory mediators may differ in their effects across the life span. A study of 407 elderly Chinese patients (125 cases and 282 controls) found that those homozygous for the *IL1B* gene (rs16944) -511T allele were more likely to have a later onset of depression of approximately 7 years but found no differences in the severity of depression (Hwang et al., 2009). At the other

end of the age spectrum, a group of 444 Australian youths aged 22–25 years (92% Caucasian, 49% female) participated in a study comparing plasma levels of IL1ß to test whether variants that enhance immune reactivity may promote a negative reaction to stress, creating vulnerability to depression. The -511C allele carriers of *IL1B* were associated with higher IL1ß expression and were noted to have more severe depressive symptoms following chronic interpersonal stress (Tartter, Hammen, Bower, Brennan, & Cole, 2015). In a recent study of 1053 Hungarian subjects completed as part of the "NewMood study," the authors identified that the rs16944 minor (A) allele interacted with childhood adversity and increased the likelihood of depressive and anxiety symptoms, while the obverse reaction for the SNP rs1143643 minor (A) allele showed protective effects against depression following recent life stress (Kovacs et al., 2016).

IL1ß has been identified as a marker for life stress, and *IL1B* variants have frequently been used to analyze stress and depression across the life span as part of gene environment studies. Preschoolers aged 3–5 years ($n = 198$; 108 females and 90 males) who had experienced maltreatment were examined for polymorphisms in the rs1143633, an intronic variant in the *IL1B* gene. The researchers found that homozygotes for the A allele had greater MDD symptoms than carriers of the G allele (Ridout et al., 2014). The importance of such childhood studies is the effects these inflammatory mediators may have on the growing brain and hence their potential irreversible neuropsychological outcomes. There are many stressors in early life that may bring about such a biological response including different types of psychological stressors from child abuse, early trauma, bullying, or physical diseases like asthma, psoriasis, and allergies. The *IL1B* gene has the ability to mobilize other cytokines as part of an autoimmune regulatory network and may be responsible for destruction of cells such as beta cells in the pancreas causing diabetic disorders and may have significant effects in other autoimmune diseases (Pociot et al., 1992). Caution needs to be taken when comparing results across studies with attributes such as haplotype, age range, gender, ethnicity, and life history, as these factors may have varying differences when investigating genetic profiles. These potential confounding issues may explain some of the difficulties with the diversity of results and differing experimental outcomes. The consistent research evidence for *IL1B* genetic variants rs16944 and rs1143634 provides a strong foundation for future genomic studies.

There have been further associations with other members of the interleukin family, particularly IL6, a proinflammatory cytokine and an antiinflammatory myokine, in the C-174G polymorphism where G carriers (GG and CG genotypes) have been found to increase depression risk (Frydecka, Pawlowski, Pawlak, & Malyszczak, 2017). This research used a population being treated for chronic hepatitis C (CHC), whose treatment can induce depressive-like disorders. CHC treatment response has provided a major opportunity to explore the effects of inflammatory mediators and their effect on mood in humans. In a genetic rat model of depression, elevation of *Il6* gene expression was found in the prefrontal cortex (PFC) that has been linked with the expression of the lethal-7 (let-7) miRNA family suspected to be part of the biological pathway in the inflammatory process (Wei et al., 2016). In this animal model, exercise such as running was seen as a protective factor inhibiting proinflammatory mediators. Interestingly, increased levels of IL6 have been found in patients 4h following electroconvulsive therapy (ECT). This effect reduces over time but may take weeks to alter, as patients recover the reduction in IL6 levels that correlate with gradual remission (Jarventausta et al., 2017). This suggests that downregulation of inflammatory mediators may be a way of measuring

recovery from MDD and the effectiveness of treatment.

IL10 is an antiinflammatory cytokine and inhibits indoleamine 2,3-dioxygenase (IDO) reducing the production of neurotoxic metabolites. Other members of the interleukin family that have been implicated in a biological depressogenic response include IL18, IL19, IL20, and IL24 linking them with the following nominated SNPs (rs2243193, rs2981572, rs1150253, rs1800872, rs2243188, rs1518108, and rs1150258) as they are all implicated in the inflammatory process. Furthermore, these SNPs act on the Th1 (type 1 T helper cell)/Th2 cytokine balance and T cell functioning (Traks et al., 2008). Some of these are located in the *IL10* gene cluster in a 200 kb region of chromosome 1, within the locus q31-32, and were used to produce four common haplotypes that showed association with depression in the *IL20* and *IL24* haplotypes (which are recently discovered paralogs of *IL10*) (Traks et al., 2008). Almost every interleukin has been described as having some relationship with the development of depressive symptoms and such a large publication body precludes that we include of all of them here.

T-Cell Function

In 1977, one of the first studies to show that severe psychological stress had a measurable effect on immune function was completed by Bartrop and colleagues (Bartrop, Luckhurst, Lazarus, Kiloh, & Penny, 1977). It involved the analysis of blood samples of 26 bereaved spouses using the lymphocyte transformation test at 2 weeks and 6 weeks following bereavement measuring responses to phytohemagglutinin (PHA) and concanavalin A (Con A). The results indicated that T-cell function was significantly reduced, particularly at 6 weeks following bereavement, linking this type of severe psychological stress to an immune response.

Wong et al. (2008) described the association of two polymorphisms in genes critical for T-cell function *TBX21* (rs17244587—risk genotypes AA=2, AG=1, and GG=0) and *PBSM4* (rs2296840—risk genotypes TT=2, TC=1, and CC=0) with depression in a Mexican American cohort. In this study, it was reported that 47.8% of the population risk was attributable to these two risk genotypes. The risk alleles showed a significant dose effect in which having a greater number of the risk alleles, listed in brackets, would indicate a greater cumulative risk of being in the MDD group, one risk allele (2.3 times), two (3.2 times), and three (9.8 times).

The effects of depression and stress on CD4$^+$ T cells are considered to cause accelerated spontaneous apoptosis, and this may relate to a reduced tryptophan environment (Miller, 2010). Tryptophan is the precursor to the neurotransmitter serotonin and melatonin. The potential immune imbalance and causative relation-ship to depression were further supported by a study that examined circulating T lymphocyte subsets, which showed that the proportion of cytotoxic T cells (CD3$^+$ and CD8$^+$) was significantly reduced in the MDD group when compared with controls. Immune checkpoint inhibitors involved in this interplay between the immune system and affective disorders are the expression of T-cell immunoglobulin and mucin-domain containing-3 (TIM-3), programmed cell death protein 1 (PDCD1) and its ligands, PDL1 and PDL2, and their effects on T lymphocytes and monocytes (Wu et al., 2017).

Tumor Necrosis Factor Alpha (*TNF*)

TNF may be the underlying mechanism for depression in initiating activation of the hypothalamic-pituitary-adrenal (HPA) axis and can initiate cell death or stimulate IL1 secretion (Postal & Appenzeller, 2015). Administration of lipopolysaccharide (LPS) a component of the gram-negative bacteria wall causes neuroinflammation and microglial activation, characterized by increased levels of TNF. These levels can remain raised for months and induce "sickness behavior" associated with depression, an adaptive response that enhances recovery by conserving energy (Maes et al., 2012). In a study of 24 teenage suicide victims and an equal number of matched controls, the mRNA and protein expression levels of IL1ß, IL6, and TNF were significantly increased in Brodmann area 10 (BA-10) demonstrating a significant increase of inflammatory mediators in the prefrontal cortex of teenage suicide victims (Pandey et al., 2012). A study that examined the role of *TNF* included 294 people who had attempted suicide; the GG genotype of the *TNF* −308 (G/A) polymorphism significantly increased the risk for suicide attempt, when compared with 97 controls (Kim et al., 2013). In a group of 50 genotyped elderly patients with depression and free from dementia, when compared with 240 healthy matched controls, it was found that there was a raised risk of developing depression in those with the GG genotype of the *TNF* −308 (G/A) polymorphism—with an odds ratio of 2.43 (Cerri et al., 2009). Conversely, in an analysis of cognitive functioning in 369 elderly subjects, the GA/AA genotype of *TNF* −308 (G/A) polymorphism showed a significant association with better processing speed when compared with homozygous GG carriers suggesting a possible neuroprotective effect (Baune et al., 2008).

Interferon Gamma Gene (*IFNG*) (+874) T/A Genotypes

Many papers describe the role of interferon alpha (IFNα)-induced depression following treatment in hepatitis C patients with as many as 30%–70% experiencing psychiatric symptoms (Belvederi Murri et al., 2017). The causal pathway, however, remains unclear as many immune factors may be at play, but IFNα stimulates IFNγ that in turn induces indoleamine

2,3-dioxygenase (IDO) resulting in a rise in the depressogenic metabolites such as kynurenine and quinolinic acid (Young et al., 2016). One systematic review confirmed that there are many mechanisms involved in the reaction to IFNα including IL6, salivary cortisol, and arachidonic acid/eicosapentaenoic acid plus docosahexaenoic acid ratio and that genetic polymorphisms may present variations that are linked to predisposition to depression (Machado et al., 2017).

Oxenkrug et al. (2011) retrospectively studied 170 Caucasian hepatitis C virus (HCV) patients treated by IFNα and examined their genotype for distribution of *IFNG* (+874). They found that the T/A allele was more common among depressed patients as opposed to homozygous A/A allele that was more common in the nondepressed patients. This suggests that being a carrier of the T allele is a possible risk factor for IFNα treatment-induced depression. The *IFNG* gene has a variable length of CA repeat in its first intron, and in a study of 218 participants ($n = 125$ cases and 93 controls), it was found that those homozygous for the CA repeat allele 2 showed higher levels of kynurenine and increased tryptophan breakdown (Myint et al., 2013). The buildup of kynurenine can cause neurotoxic effects, inhibit production of the mood-related monoamine neurotransmitter serotonin, and inhibit effective response to antidepressant treatment (Dantzer, 2017).

C-Reactive Protein (CRP) Related Polymorphisms

CRP is used as an inflammatory sensitive marker and is upregulated as part of the positive acute-phase protein response to inflammation. Higher basal serum concentrations of CRP have been associated with symptoms of MDD, but this is moderated by weight and gender (Y. Liu et al., 2014). In a large cohort of 14,276

from the National Health and Nutrition Examination Survey (NHANES), it was found that those with depression had CRP levels that were 31% higher than those without depression (Cepeda, Stang, & Makadia, 2016). CRP levels >3 mg/L are considered to place individuals at high risk for cardiac disease, another inflammatory disorder. It was observed in a group of female patients, who even though they had remitted from depression continued to have higher levels of CRP when compared with the control group (Kling et al., 2007). Carlson et al. (2005) identified several SNPs of interest in the *CRP* promoter region that are associated with changes in CRP levels (rs3093058 (790)), (rs3091244 (1440)), (rs1417938 (1919)), (rs1800947 (2667)), (rs3093066 (3006)), (rs1205 (3872)), and (rs2808630 (5237)). The authors examined two large biracial cohorts (African Americans and European Americans) with CRP levels taken across their life span, comparing common *CRP* SNPs between these two ethnic groups. Four associated haplotype groups in relation to *CRP* polymorphisms were formed, produced from a total combination of eight clades of haplotypes. When combined as groups, they showed an association with marked differences in circulating CRP levels, the highest being in the H6 grouped haplotype (tagSNP TTAGCGA). A similar haplotype study ($n = 868$ healthy volunteers) that focused on just three polymorphisms (rs1417938 A/T, rs1800947 C/G, and rs1205 C/T) to generate three-locus haplotypes discovered that there was an association between haplotypic variation (A-G-T) and higher depression scores. The AGT haplotype showed that depression scores correlated with higher circulating CRP, but this was mediated by body mass index (Halder et al., 2010).

A study of a community sample of 3700 men aged ≥70 years investigated the relationship of two SNPs rs1130864 and rs1205 of the *CRP* gene and serum levels of CRP. They found that levels of CRP are not associated with the onset of depression in old age, but there is an association with the SNP rs1205 AA variant that indicated a reduction of up to 24% less circulating CRP in these homozygous carriers compared with men with other variants of the gene. An inability to respond effectively to psychological or physical threat by an increase in CRP levels may be problematic in older age groups, as this mechanism is an adaptive protective response to danger (Almeida et al., 2009). In another longitudinal study that examined the modifying effects of variants in the rs1205 gene and rs3093068 ($n = 3035$), a group aged 53 years who had been studied as a cohort on 21 occasions since birth failed to show an association with depression and the two nominated *CRP* SNPs. However, those homozygous for the major C allele of rs1205 were more probable to have expressed adolescent emotional problems when compared with carriers of the T allele (Gaysina et al., 2011). This demonstrates the importance of collecting information about life experiences, environmental factors, and investigating their interaction with depressive symptoms. In a systematic review of longitudinal studies ($n = 14,832$ participants) examining the relationship of raised CRP levels and depression, it was found that there is evidence of a causal pathway between CRP levels and the development of depression, particularly in older age where there tends to be an exaggerated inflammatory response (Valkanova, Ebmeier, & Allan, 2013). The literature is not consistent in ascribing serum CRP levels to a relationship with MDD, with many authors stating that there is only a link when we include another phenotypic dependent variable such as body mass index or emotional stress such as child maltreatment. CRP measures continue to be used in studies as a significant biological marker to investigate the association between stress, mental health, childhood trauma, and inflammatory disorders (Baumeister, Akhtar, Ciufolini, Pariante, & Mondelli, 2015; Zahn et al., 2016).

GWAS STUDIES

The search for SNPs, genes, or even polygenic links using GWAS remains elusive with only limited findings compared with other psychiatric disorders such as schizophrenia and bipolar disorder (Zhao & Nyholt, 2017). Bosker et al. (2011) report that only one inflammatory-related gene *TNF* (rs76917; OR T = 1.35, 95% CI 1.13–1.63; $P = 0.0034$) was identified as having any association with MDD using GWAS studies. In a significant GWAS study of 34,549, no loci were found to be of genome-wide significance, but seven SNPS were considered in a replication set (16,709) finding only a single SNP that displayed suggestive association to MDD (rs161645, 5q21, $p = 9.19 \times 10^{-3}$) (Hek et al., 2013). Samples of at least 50,000 are expected to be required for any future genome-wide association studies in MDD for them to be sufficiently powered to detect genes involved in depression, its traits or symptoms.

One significant breakthrough using GWAS was in a study using low-coverage whole-genome sequencing in a cohort of 5303 Chinese women with recurrent depression. Two loci exceeded genome-wide significance: one 5′ to the sirtuin-1 (*SIRT1*) gene on chromosome 10 (SNP5 rs12415800, chromosome 10:69624180) and the second in an intron of the phospholysine phosphohistidine inorganic pyrophosphate phosphatase (*LHPP*) gene SNP rs35936514, chromosome 10:126244970 (CONVERGE consortium, 2015). The authors were able to replicate these findings in a cohort of Han Chinese consisting of 3231 cases, and they place some of their success down to using a relatively homogenous group of severe cases of depression.

Another significant finding was in a study of rare functional variants associated with MDD using GWAS with Mexican Americans (203 with mild to moderate MDD cases; 196 controls) compared with a group of European Ancestry (499 MDD cases and 473 controls). The *PHF21B* gene (PHD finger protein 21B in chromosome 22q13.31) was identified near a genome-wide significant locus. Three areas overrepresented in the gene ontology (GO) processes were found in the innate immune response, glutamate receptor signaling, and detection of chemical stimulus in smell sensory perception. They were able to replicate this association of the rare genetic variant *PHF21B* gene in the ethnically unrelated European ancestry cohort (Wong et al., 2016). PHF21B is believed to be involved in tumor suppression. Using the UK Brain Expression Consortium, variants of this gene were found to significantly change brain expression quantitative trait loci in 10 brain regions including the hippocampus, frontal cortex, and substantia nigra. The hippocampal expression of *Phf21b* was modulated by chronic restraint stress; rodents resilient to chronic stress had significantly lower levels of *Phf21b* mRNA in the hippocampus than nonstressed animals. In this study, the pathway and network analysis of genes associated to MDD in the Mexican American cohort revealed that the top gene ontology GO) process was natural immune response, followed by G-protein coupled glutamate receptor signaling pathway and detection of smell sensory perception.

In one of the largest GWAS studies to date, there was a comparison between subjective well-being ($N = 298{,}420$), depressive symptoms ($N = 161{,}460$), and neuroticism trait ($N = 170{,}910$) that found 15 variants of significance related to these areas: three variants associated with subjective well-being, two with depressive symptoms, and 11 with neuroticism including two inversion polymorphisms. The SNPs that were associated with depression were rs4346787 (Chr:5:165056788) and rs4481363 (Chr:5:165047713), and the authors were able to replicate these findings in an independent depression sample (Okbay et al., 2016). Researchers are collaborating with companies such as "23andMe" or "Ancestry" to assess the predictive power of their findings using polygenic prediction methods. Using large convenient cohorts provides mass participants and data that may successfully identify genetic associations with highly polygenic phenotypes. Another way to investigate MDD is to examine gene environment interaction. In a first Japanese study examining the link between genes and environment using GWAS, it analyzed 534,848 SNPs and looked at the environmental responses using the questionnaire for Social Readjustment Rating Scale (SRRS) and correlated this social data with depression using the Center for Epidemiologic Studies Depression (CES-D) scale. In this study, one SNP was found to be marginally associated with depression rs10510057 (located on chromosome 10q26 near the regulators of G-protein signaling 10 gene—*RGS10*), and this is understood to be involved in the stress response (Otowa et al., 2016). Another research methodology is to use age stratification, and to this end, 22,158 cases with MDD and 133,749 case controls were examined. Using this method as part of a GWAS study, the team found only one replicated genome-wide significant locus associated with SNP related to late onset >27 years of depression, rs7647854 (odds ratio, 1.16) (Power et al., 2016). At present, there are no pronounced findings from GWAS or other research methods that would lead an investigator to feel confident in diagnosing immune related MDD that would provide effective treatment outcomes. As sampling sizes increase with companies similar to 23andMe or the development of biobanks that are now being developed across the world, researchers will be able to use evolving technology and global cooperation to test the most recently identified MDD-related SNPs. There is a thirst from the general public to know more about their genetic makeup and health status that coincides with an increase in the literature on transcriptomics, proteomics, and pharmacogenetics or indeed the "omics." GWAS technology has helped to develop advances in genomics across almost every field of medicine.

CONCLUSION

Earlier literature indicated that there may be a relationship of SNPs associated with MDD and inflammatory mediators, which has been gradually consolidated over the last two decades. Many articles indicate an evidence-based relationship to depression, anxiety, and stress, but there are a reasonable number of researchers concluding that there is no clear genetic link suggesting that we should accept some of these results with caution (Bosker et al., 2011). For example, in an investigation of genetic variants of *TNF, IL1A, IL1B, IL6, IL1RN* (IL1 receptor antagonist gene), and *IL10*, the predictive outcomes for depression were not supported by either individual polymorphisms nor haplotypes (Misener et al., 2008). In contrast however, there is a huge body of work and more recent systematic reviews that have found positive links to proinflammatory mediators and MDD (Kohler et al., 2017; Levinson, 2006; Lohoff, 2010; Machado et al., 2017; Maes et al., 2012; Miller & Raison, 2016). One recent systematic review and meta-analysis study (of candidate gene approaches and one GWAS study) looked at a distinct life phase examining "late-life" depression (Tsang, Mather, Sachdev, & Reppermund, 2017). It identified 46 potential candidate genes but only one GWAS location was identified and it significantly associated one locus in chromosome 5q21 (in a gene desert location) for depressive symptoms in older adults (Hek et al., 2013). MDD is thought to be prevalent in 10% of the population (about 300 million people worldwide), and if there is a biological key to this disorder, then finding it would help many people across the world and possibly prevent people from committing suicide, which accounts for 800,000 deaths worldwide every year (WHO, 2017).

The ultimate goal is finding combinations of risk alleles across multiple genes that form a polygenic risk profile that can be reliably tested across populations. The risk profile can then be used to assess cohorts against the interaction of endophenotypic behavior such as diet, exercise, and body mass index that may promote better clinical management or even prevention (Peterson et al., 2014). Bioinformatics is a mathematical science that analyzes such mass data and is assisting in matching SNP's and phenotypes. Proinflammatory genes are also providing clues to the comorbidity of disorders such as heart disease and MDD resulting in discoveries of shared pathways and possible preventive interventions, promoting further research in these connections (van der Wall, 2016). The apparent interaction of environment, personal characteristics, and genetic predisposition in determining the risk outcome of MDD is complex, but by examining these health factors together results in humans being examined from a holistic perspective. It may also reduce the stigma of depression. Genotypic risk does not predetermine an inevitable phenotypic outcome, because protective behaviors such as exercise can be changed to remediate against these identified risks (Barnes et al., 2017). One of the major challenges in the treatment of MDD is finding a medication regimen that will match the individual's symptoms and depression subtype. Polymorphic profiles of risk factors will be extremely useful in the area of pharmacogenomics, utilizing genetic data to assist in tailoring medication to the individual. Targeting treatments that consider biological determinants will ensure greater efficacy and an increased chance of remission in those with inflammatory-associated MDD (Liu, Adibfar, Herrmann, Gallagher, & Lanctot, 2016). Translating results into practice will be a challenge, but it is clear that depression is not merely a social and psychological disorder, but one that is embedded in systemic biological phenomena.

References

Almeida, O. P., Norman, P. E., Allcock, R., van Bockxmeer, F., Hankey, G. J., & Jamrozik, K. (2009). Polymorphisms of the CRP gene inhibit inflammatory response and increase susceptibility to depression: The Health in Men Study. *International Journal of Epidemiology*, *38*, 1049–1059.

Arango, C. (2017). Candidate gene associations studies in psychiatry: Time to move forward. *European Archives of Psychiatry and Clinical Neuroscience*, *267*(1), 1–2. https://doi.org/10.1007/s00406-016-0765-7.

Barnes, J., Mondelli, V., & Pariante, C. M. (2017). Genetic contributions of inflammation to depression. *Neuropsychopharmacology*, *42*(1), 81–98. https://doi.org/10.1038/npp.2016.169.

Bartrop, R. W., Luckhurst, E., Lazarus, L., Kiloh, L. G., & Penny, R. (1977). Depressed lymphocyte function after bereavement. *Lancet*, *1*(8016), 834–836.

Baumeister, D., Akhtar, R., Ciufolini, S., Pariante, C., & Mondelli, V. (2015). Childhood trauma and adulthood inflammation: A meta-analysis of peripheral C-reactive protein, interleukin-6 and tumour necrosis factor-[alpha]. *Molecular Psychiatry*, *21*, 642–649.

Baune, B. T., Dannlowski, U., Domschke, K., Janssen, D. G., Jordan, M. A., Ohrmann, P., et al. (2010). The interleukin 1 beta (IL1B) gene is associated with failure to achieve remission and impaired emotion processing in major depression. *Biological Psychiatry*, *67*(6), 543–549. https://doi.org/10.1016/j.biopsych.2009.11.004.

Baune, B. T., Ponath, G., Rothermundt, M., Riess, O., Funke, H., & Berger, K. (2008). Association between genetic variants of IL-1beta, IL-6 and TNF-alpha cytokines and cognitive performance in the elderly general population of the MEMO-study. *Psychoneuroendocrinology*, *33*. https://doi.org/10.1016/j.psyneuen.2007.10.002.

Belvederi Murri, M., Cecere, A. C., Masotti, M., Sammito, G., la Marca, A., Torres, G. V., et al. (2017). Biopsychosocial predictors of interferon-related depression in patients with Hepatitis C. *Asian Journal of Psychiatry*, *26*, 24–28. https://doi.org/10.1016/j.ajp.2017.01.001.

Bosker, F. J., Hartman, C. A., Nolte, I. M., Prins, B. P., Terpstra, P., & Posthuma, D. (2011). Poor replication of candidate genes for major depressive disorder using genome-wide association data. *Molecular Psychiatry*, *16*, 516–532.

Carlson, C. S., Aldred, S. F., Lee, P. K., Tracy, R. P., Schwartz, S. M., Rieder, M., et al. (2005). Polymorphisms within the C-reactive protein (CRP) promoter region are associated with plasma CRP levels. *The American Journal of Human Genetics*, *77*(1), 64–77. https://doi.org/10.1086/431366.

Cepeda, M. S., Stang, P., & Makadia, R. (2016). Depression is associated with high levels of C-reactive protein and low levels of fractional exhaled nitric oxide: Results from the 2007–2012 National Health and Nutrition Examination Surveys. *The Journal of Clinical Psychiatry*. https://doi.org/10.4088/JCP.15m10267.

Cerri, A. P., Arosio, B., Viazzoli, C., Confalonieri, R., Teruzzi, F., & Annoni, G. (2009). 308(G/A) TNF-alpha gene polymorphism and risk of depression late in the life. *Archives of Gerontology and Geriatrics*, *49*, 29–34.

Chang, C. C., Chang, H. A., Fang, W. H., Chang, T. C., & Huang, S. Y. (2017). Gender-specific association between serotonin transporter polymorphisms (5-HTTLPR and rs25531) and neuroticism, anxiety and depression in well-defined healthy Han Chinese. *Journal of Affective Disorders*, *207*, 422–428. https://doi.org/10.1016/j.jad.2016.08.055.

CONVERGE consortium. (2015). Sparse whole-genome sequencing identifies two loci for major depressive disorder. *Nature*, *523*(7562), 588–591. https://doi.org/10.1038/nature14659. (2015). http://www.nature.com/nature/journal/v523/n7562/abs/nature14659.html#supplementary-information.

Dantzer, R. (2017). Role of the kynurenine metabolism pathway in inflammation-induced depression: Preclinical approaches. *Current Topics in Behavioral Neurosciences*, *31*, 117–138. https://doi.org/10.1007/7854_2016_6.

Drachmann Bukh, J., Bock, C., Vinberg, M., Werge, T., Gether, U., & Vedel Kessing, L. (2009). Interaction between genetic polymorphisms and stressful life events in first episode depression. *Journal of Affective Disorders*, *119*(1–3), 107–115. https://doi.org/10.1016/j.jad.2009.02.023.

Frydecka, D., Pawlowski, T., Pawlak, D., & Malyszczak, K. (2017). Functional polymorphism in the interleukin 6 (IL6) gene with respect to depression induced in the course of interferon-alpha and ribavirin treatment in chronic hepatitis patients. *Archivum Immunologiae et Therapiae Experimentalis (Warsz)*. https://doi.org/10.1007/s00005-016-0441-7.

Gaysina, D., Pierce, M., Richards, M., Hotopf, M., Kuh, D., & Hardy, R. (2011). Association between adolescent emotional problems and metabolic syndrome: The modifying effect of C-reactive protein gene (CRP) polymorphisms. *Brain, Behavior, and Immunity*, *25*(4), 750–758. https://doi.org/10.1016/j.bbi.2011.01.019.

Halder, I., Marsland, A. L., Cheong, J., Muldoon, M. F., Ferrell, R. E., & Manuck, S. B. (2010). Polymorphisms in the CRP gene moderate an association between depressive symptoms and circulating levels of C-reactive protein. *Brain, Behavior, and Immunity*, *24*, 160–167.

Hartwell, L. H., Hood, L., Goldberg, M. L., Reynolds, A. E., & Silver, L. M. (2011). *Genetics: From genes to genomes* (4th ed.). New York: McGraw Hill.

Hek, K., Demirkan, A., Lahti, J., Terracciano, A., Teumer, A., & Cornelis, M. C. (2013). A genome-wide association study of depressive symptoms. *Biological Psychiatry, 73*, 667–678.

Hwang, J. P., Tsai, S. J., Hong, C. J., Yang, C. H., Hsu, C. D., & Liou, Y. J. (2009). Interleukin-1 beta -511C/T genetic polymorphism is associated with age of onset of geriatric depression. *Neuromolecular Medicine, 11*, 322–327.

Iwata, M., Ota, K., & Duman, R. (2013). The inflammasome: Pathways linking psychological stress, depression, and systemic illnesses. *Brain, Behavior, and Immunity, 31*, 105–114.

Jarventausta, K., Sorri, A., Kampman, O., Bjorkqvist, M., Tuohimaa, K., Hamalainen, M., et al. (2017). Changes in interleukin-6 levels during electroconvulsive therapy may reflect the therapeutic response in major depression. *Acta Psychiatrica Scandinavica, 135*(1), 87–92. https://doi.org/10.1111/acps.12665.

Jobling, M., Hollox, E., Hurles, M., Kivisild, T., & Tyler-Smith, C. (2014). *Human evolutionairy genetics* (2nd ed.). New York: Garland Science.

Kim, Y. K., Hong, J. P., Hwang, J. A., Lee, H. J., Yoon, H. K., Lee, B. H., et al. (2013). TNF-alpha -308G>A polymorphism is associated with suicide attempts in major depressive disorder. *Journal of Affective Disorders, 150*(2), 668–672. https://doi.org/10.1016/j.jad.2013.03.019.

Kling, M. A., Alesci, S., Csako, G., Costello, R., Luckenbaugh, D. A., Bonne, O., et al. (2007). Sustained low-grade pro-inflammatory state in unmedicated, remitted women with major depressive disorder as evidenced by elevated serum levels of the acute phase proteins C-reactive protein and serum amyloid A. *Biological Psychiatry, 62*(4), 309–313. https://doi.org/10.1016/j.biopsych.2006.09.033.

Kohler, C. A., Freitas, T. H., Maes, M., de Andrade, N. Q., Liu, C. S., Fernandes, B. S., et al. (2017). Peripheral cytokine and chemokine alterations in depression: A meta-analysis of 82 studies. *Acta Psychiatrica Scandinavica.* https://doi.org/10.1111/acps.12698.

Koo, J. W., & Duman, R. S. (2008). IL-1β is an essential mediator of the antineurogenic and anhedonic effects of stress. *Proceedings of the National Academy of Sciences of the United States of America, 105*(2), 751–756. https://doi.org/10.1073/pnas.0708092105.

Kovacs, D., Eszlari, N., Petschner, P., Pap, D., Vas, S., & Kovacs, P. (2016). Effects of IL1B single nucleotide polymorphisms on depressive and anxiety symptoms are determined by severity and type of life stress. *Brain, Behavior, and Immunity, 56*, 96–104.

Lamers, F., Cui, L., Hickie, I. B., Roca, C., Machado-Vieira, R., Zarate, C. A., Jr., et al. (2016). Familial aggregation and heritability of the melancholic and atypical subtypes of depression. *Journal of Affective Disorders, 204*, 241–246. https://doi.org/10.1016/j.jad.2016.06.040.

Levinson, D. F. (2006). The genetics of depression: A review. *Biological Psychiatry, 60*, 84–92.

Liu, C., Adibfar, A., Herrmann, N., Gallagher, D., & Lanctot, K. (2016). Evidence for inflammation-associated depression. *Current Topics in Behavioral Neurosciences.* https://doi.org/10.1007/7854_2016_2.

Liu, Y., Al-Sayegh, H., Jabrah, R., Wang, W., Yan, F., & Zhang, J. (2014). Association between C-reactive protein and depression: Modulated by gender and mediated by body weight. *Psychiatry Research, 219*(1), 103–108. https://doi.org/10.1016/j.psychres.2014.05.025.

Lohoff, F. W. (2010). Overview of the genetics of major depressive disorder. *Current Psychiatry Reports, 12*(6), 539–546. https://doi.org/10.1007/s11920-010-0150-6.

Machado, M. O., Oriolo, G., Bortolato, B., Köhler, C. A., Maes, M., Solmi, M., et al. (2017). Biological mechanisms of depression following treatment with interferon for chronic hepatitis C: A critical systematic review. *Journal of Affective Disorders, 209*, 235–245. https://doi.org/10.1016/j.jad.2016.11.039.

Maes, M., Berk, M., Goehler, L., Song, C., Anderson, G., Gałecki, P., et al. (2012). Depression and sickness behavior are Janus-faced responses to shared inflammatory pathways. *BMC Medicine, 10*(1), 66. https://doi.org/10.1186/1741-7015-10-66.

Miller, A. (2010). Depression and immunity: A role for T cells? *Brain, Behavior, and Immunity, 24*(1), 1–8. https://doi.org/10.1016/j.bbi.2009.09.009.

Miller, A., & Raison, C. (2016). The role of inflammation in depression: From evolutionary imperative to modern treatment target. *Nature Reviews. Immunology, 16*(1), 22–34. https://doi.org/10.1038/nri.2015.5.

Misener, V. L., Gomez, L., Wigg, K. G., Luca, P., King, N., & Kiss, E. (2008). Cytokine genes TNF, IL1A, IL1B, IL6, IL1RN and IL10, and childhood-onset mood disorders. *Neuropsychobiology, 58*, 71–80.

Myint, A., Bondy, B., Baghai, T., Eser, D., Nothdurfter, C., & Schule, C. (2013). Tryptophan metabolism and immunogenetics in major depression: A role for interferon-[gamma] gene. *Brain, Behavior, and Immunity, 31*, 128–133.

Narimatsu, H. (2017). Gene-environment interactions in preventive medicine: Current status and expectations for the future. *International Journal of Molecular Sciences, 18*(2) https://doi.org/10.3390/ijms18020302.

Okbay, A., Baselmans, B., De Neve, J., Turley, P., Nivard, M., & Fontana, M. (2016). Genetic variants associated with subjective well-being, depressive symptoms, and neuroticism identified through genome-wide analyses. *Nature Genetics, 48*, 624–633.

Otowa, T., Kawamura, Y., Tsutsumi, A., Kawakami, N., Kan, C., Shimada, T., et al. (2016). The first pilot genome-wide gene-environment study of depression in the Japanese population. *PLoS One, 11*(8). https://doi.org/10.1371/journal.pone.0160823.

Oxenkrug, G., Perianayagam, M., Mikolich, D., Requintina, P., Shick, L., & Ruthazer, R. (2011). Interferon-gamma (+874) T/A genotypes and risk of IFN-alpha-induced depression. *Journal of Neural Transmission, 118*, 271–274.

Pandey, G. N., Rizavi, H. S., Ren, X., Fareed, J., Hoppensteadt, D. A., Roberts, R. C., et al. (2012). Proinflammatory cytokines in the prefrontal cortex of teenage suicide victims. *Journal of Psychiatric Research, 46*(1), 57–63. https://doi.org/10.1016/j.jpsychires.2011.08.006.

Peterson, B. S., Wang, Z., Horga, G., Warner, V., Rutherford, B., Klahr, K. W., et al. (2014). Discriminating risk and resilience endophenotypes from lifetime illness effects in familial major depressive disorder. *JAMA Psychiatry, 71*(2), 136–148. https://doi.org/10.1001/jamapsychiatry.2013.4048.

Pociot, F., MØLvig, J., Wogensen, L., Worsaae, H., & Nerup, J. (1992). A TaqI polymorphism in the human interleukin-1β (IL-1β) gene correlates with IL-1β secretion in vitro. *European Journal of Clinical Investigation, 22*(6), 396–402. https://doi.org/10.1111/j.1365-2362.1992.tb01480.x.

Postal, M., & Appenzeller, S. (2015). The importance of cytokines and autoantibodies in depression. *Autoimmunity Reviews, 14*(1), 30–35. https://doi.org/10.1016/j.autrev.2014.09.001.

Power, R. A., Tansey, K. E., Buttenschon, H. N., Cohen-Woods, S., Bigdeli, T., Hall, L. S., et al. (2016). Genome-wide association for major depression through age at onset stratification: Major depressive disorder working group of the psychiatric genomics consortium. *Biological Psychiatry, 81*(4), 325–335. https://doi.org/10.1016/j.biopsych.2016.05.010.

Ramamoorthy, S., Bauman, A. L., Moore, K. R., Han, H., Yang-Feng, T., Chang, A. S., et al. (1993). Antidepressant- and cocaine-sensitive human serotonin transporter: Molecular cloning, expression, and chromosomal localization. *Proceedings of the National Academy of Sciences of the United States of America, 90*(6), 2542–2546.

Ridout, K., Parade, S., Seifer, R., Price, L., Gelernter, J., & Feliz, P. (2014). Interleukin 1B gene (IL1B) variation and internalizing symptoms in maltreated preschoolers. *Development and Psychopathology, 26*, 1277–1287.

Schmidt, F. M., Schröder, T., Kirkby, K. C., Sander, C., Suslow, T., Holdt, L. M., et al. (2016). Pro- and anti-inflammatory cytokines, but not CRP, are inversely correlated with severity and symptoms of major depression. *Psychiatry Research, 239*, 85–91. https://doi.org/10.1016/j.psychres.2016.02.052.

Sosnay, P. R., Salinas, D. B., White, T. B., Ren, C. L., Farrell, P. M., Raraigh, K. S., et al. (2017). Applying cystic fibrosis transmembrane conductance regulator genetics and CFTR2 data to facilitate diagnoses. *The Journal of Pediatrics, 181s*, S27–S32.e21. https://doi.org/10.1016/j.jpeds.2016.09.063.

Sullivan, P., Neale, M., & Kendler, K. (2000). Genetic epidemiology of major depression: Review and meta-analysis. *The American Journal of Psychiatry, 157*, 1552–1562.

Tadic, A., Rujescu, D., Muller, M. J., Kohnen, R., Stassen, H. H., & Szegedi, A. (2008). Association analysis between variants of the interleukin-1beta and the interleukin-1 receptor antagonist gene and antidepressant treatment response in major depression. *Neuropsychiatric Disease and Treatment, 4*, 269–276.

Tartter, M., Hammen, C., Bower, J., Brennan, P., & Cole, S. (2015). Effects of chronic interpersonal stress exposure on depressive symptoms are moderated by genetic variation at IL6 and IL1[beta] in youth. *Brain, Behavior, and Immunity, 46*, 104–111.

Traks, T., Koido, K., Eller, T., Maron, E., Kingo, K., Vasar, V., et al. (2008). Polymorphisms in the interleukin-10 gene cluster are possibly involved in the increased risk for major depressive disorder. *BMC Medical Genetics, 9*. https://doi.org/10.1186/1471-2350-9-111.

Tsang, R. S., Mather, K. A., Sachdev, P. S., & Reppermund, S. (2017). Systematic review and meta-analysis of genetic studies of late-life depression. *Neuroscience and Biobehavioral Reviews, 75*, 129–139. https://doi.org/10.1016/j.neubiorev.2017.01.028.

Valkanova, V., Ebmeier, K., & Allan, C. (2013). CRP, IL-6 and depression: A systematic review and meta-analysis of longitudinal studies. *Journal of Affective Disorders, 150*, 736–744.

van der Sluis, S., Posthuma, D., Nivard, M. G., Verhage, M., & Dolan, C. V. (2013). Power in GWAS: Lifting the curse of the clinical cut-off. *Molecular Psychiatry, 18*, 2–3.

van der Wall, E. E. (2016). Cardiac disease and depression; a direct association? *Netherlands Heart Journal, 24*(9), 495–497. https://doi.org/10.1007/s12471-016-0868-9.

Wei, Y. B., Liu, J. J., Villaescusa, J. C., Aberg, E., Brene, S., Wegener, G., et al. (2016). Elevation of Il6 is associated with disturbed let-7 biogenesis in a genetic model of depression. *Translational Psychiatry. 6*, https://doi.org/10.1038/tp.2016.136.

WHO (2017). *Depression fact sheet: Updated February 2017.* Geneva: World Health Organization.

Wong, M. L., Arcos-Burgos, M., Liu, S., Velez, J. I., Yu, C., Baune, B. T., et al. (2016). The PHF21B gene is associated with major depression and modulates the stress response. *Molecular Psychiatry.* https://doi.org/10.1038/mp.2016.174.

Wong, M. L., Dong, C., Maestre-Mesa, J., & Licinio, J. (2008). Polymorphisms in inflammation-related genes are associated with susceptibility to major depression and antidepressant response. *Molecular Psychiatry, 13*(8), 800–812. https://doi.org/10.1038/mp.2008.59.

Wu, W., Zheng, Y. L., Tian, L. P., Lai, J. B., Hu, C. C., Zhang, P., et al. (2017). Circulating T lymphocyte subsets, cytokines, and immune checkpoint inhibitors in patients with bipolar II or major depression: A preliminary study. *Scientific Reports. 7*, https://doi.org/10.1038/srep40530.

Young, K. D., Drevets, W. C., Dantzer, R., Teague, T. K., Bodurka, J., & Savitz, J. (2016). Kynurenine pathway metabolites are associated with hippocampal activity during autobiographical memory recall in patients with depression. *Brain, Behavior, and Immunity, 56*, 335–342. https://doi.org/10.1016/j.bbi.2016.04.007.

Yu, Y. W., Chen, T. J., Hong, C. J., Chen, H. M., & Tsai, S. J. (2003). Association study of the interleukin-1 beta (C-511T) genetic polymorphism with major depressive disorder, associated symptomatology, and antidepressant response. *Neuropsychopharmacology, 28*, 1182–1185.

Zahn, D., Herpertz, S., Albus, C., Hermanns, N., Hiemke, C., Hiller, W., et al. (2016). hs-CRP predicts improvement in depression in patients with type 1 diabetes and major depression undergoing depression treatment: Results from the diabetes and depression (DAD) study. *Diabetes Care, 39*(10), e171–e173. https://doi.org/10.2337/dc16-0710.

Zhao, H., & Nyholt, D. R. (2017). Gene-based analyses reveal novel genetic overlap and allelic heterogeneity across five major psychiatric disorders. *Human Genetics, 136*(2), 263–274. https://doi.org/10.1007/s00439-016-1755-6.

Adolescent-Onset Depressive Disorders and Inflammation

Ian B. Hickie*, Joanne S. Carpenter*, Elizabeth M. Scott*,†

*University of Sydney, Camperdown, NSW, Australia
†Notre Dame Medical School, Sydney, NSW, Australia

There is international recognition of the premature death and disability costs attributable to common mental disorders (Bloom et al., 2012; Erskine et al., 2015; Gustavsson et al., 2011). The largest proportion of this excessive morbidity is attributable to depressive disorders—reflecting their early age of onset, high population prevalence, chronicity, and comorbidity with physical illness (Gore et al., 2011; Lopez, Mathers, Ezzati, Jamison, & Murray, 2006). There is also recognition that the evidence-base for choosing new or existing treatments for young persons with mood disorders or providing cost-effective treatments at scale is sparse (Allen, Hetrick, Simmons, & Hickie, 2007; Catania, Hetrick, Newman, & Purcell, 2014; Das et al., 2016; Hickie, 2011; Insel, 2007). More specialized depression research programs typically lack a specific focus on the adolescent-onset period or an early intervention or secondary prevention perspective. Functional outcomes from current pharmacological or psychological treatments remain limited, with accumulating evidence of poor long-term education and employment outcomes even among those receiving clinical care (Gitlin & Miklowitz, 2017; Lee et al., 2013; O'Dea et al., 2016).

A FOCUS ON THE ADOLESCENT PERIOD

Epidemiological evidence confirms that while 75% of major mental disorders occur before 25 years of age, over 50% have their onset before 15 years (Kessler et al., 2005). Consequently, from both a research and clinical perspective, adolescent-onset and early intervention perspectives are critical to uncovering the multiple developmental and pathophysiological paths that lead to persistent and disabling adult mood disorders (Allen et al., 2007; Catania et al., 2014; Cross et al., 2014; Hickie, Scott, et al., 2013; Merikangas & Kalaydjian, 2007). These life-trajectory perspectives (Fairweather, Anstey, Rodgers, Jorm, & Christensen, 2007; Insel, 2007; Mastronardi et al., 2011; Narayan & Manji, 2016) are also essential to maximize the opportunities for

Inflammation and Immunity in Depression
https://doi.org/10.1016/B978-0-12-811073-7.00024-6

secondary prevention of other psychological disorders (e.g., alcohol or other substance misuse; Merikangas & Kalaydjian, 2007), self-harm or suicidal behavior (Fairweather et al., 2007), or medical ill health (e.g., obesity, diabetes, and premature death from ischemic heart disease; Bunker et al., 2003; Mastronardi et al., 2011; Stuart & Baune, 2012). New early intervention services focusing on the adolescent period and with emphasis on the major mood and psychotic disorders have been developing internationally (Cross et al., 2014). These new health system developments have been a priority in Australia and have created a platform for more integrated clinical, neurobiological, and longitudinal investigation of early clinical stages of major mood and psychotic disorders (Iorfino et al., 2018). Consequently, we now have a new capacity to study the ways in which neurobiological, immune, metabolic, and circadian processes develop in individuals with mood disorders and are impacted by prior (e.g., childhood trauma and childhood central nervous system (CNS) infection or autoimmune disorder) or concurrent (adolescent infection-inflammation and substance misuse) environmental exposures. Additionally, these platforms provide a unique opportunity for identifying novel pathways to early illness onset and testing novel (notably immune regulatory, glutamatergic, and circadian) treatments.

DEPRESSION AND IMMUNE DYSFUNCTION

There has been a long history of epidemiological and clinical interest in the relationships between depressive disorders and immune dysfunction. Historically, this was based on observations of associations between depressive disorders and other medical conditions, notably infective, autoimmune, and malignant disorders (Hickie, Banati, Stewart, & Lloyd, 2009; Miller & Raison, 2016). In more recent times, increased emphasis has been placed on associations with cardiovascular disease, due both to increased evidence of a causal relationship between major depression and onset of coronary artery disease (Bunker et al., 2003; Winning, Glymour, McCormick, Gilsanz, & Kubzansky, 2015) and associations between chronic inflammation and adverse cardiac events (Libby, 2012; Pfeiler, Winkels, Kelm, & Gerdes, 2017). Previously, the dominant working assumption was that perturbed immune function (notably impaired cell-mediated immunity assessed both *in vitro* and *in vivo*, see Hickie, Silove, Hickie, Wakefield, & Lloyd, 1990) was a consequence of more severe or persistent depressive disorders and this impairment had the capacity to put the affected individual at risk of serious medical conditions, notably infective or malignant conditions (Dantzer, 2012; Hickie et al., 1990). In this conceptualization, objective *in vitro* and *in vivo* evidence of immune dysfunction was expected to be in the direction of downregulation or impaired responsiveness to external challenges. It was assumed that this adverse effect of depressive disorders was imparted via prolonged activation of the hypothalamic-pituitary-adrenal (HPA) axis (due to impaired feedback and persistent hypercortisolemia) and/or persistent sympathetic nervous system (SNS) activation. For both autoimmune and cardiovascular disorders, however, persistent immune dysregulation (e.g., as evidenced by raised pro-inflammatory cytokines or C-reactive protein) has emerged as another plausible explanation, and the direction of causality has become a more open question. That is, while the presence of increased inflammatory markers may still be a consequence of persistent immune impairment, low-grade inflammation may itself give rise to adverse medical and psychological consequences (via negative feedback loops on both the CNS and other peripheral tissues) (Halaris, 2017; Rosenblat & McIntyre, 2017; Saltiel & Olefsky, 2017; Soczynska et al., 2011).

DEPRESSION AND INFLAMMATION: CAUSE OR EFFECT?

Over the last two decades, there has been rapidly growing interest in the associations not simply between depression and immune impairment, but more specifically depression and evidence of persisting or low-grade inflammation (Khandaker, Dantzer, & Jones, 2017; Rosenblat & McIntyre, 2017; Zalli, Jovanova, Hoogendijk, Tiemeier, & Carvalho, 2016). Here, the concept of "low-grade" inflammation is based largely on differentiation from the inflammatory responses that characterize acute infection, chronic infection, or very active autoimmune or inflammatory disorders. Previously, it was assumed that such "low-grade" activity (as detected outside the CNS by standard laboratory measures) was unlikely to have any pathophysiological consequences and, most importantly, no significant impact on CNS signaling or CNS-regulated homeostatic mechanisms. Consequently, it was thought that there was no plausible mechanism by which such "low-grade" changes could impact directly on behavior or cognition. Perhaps the most important insight over the last three decades is that previous assumptions about the signaling roles and the potential impacts of inflammatory factors detected in the peripheral circulation on the CNS need to be radically rethought (see Dantzer, 2017; D'Mello & Swain, 2017; Haapakoski, Ebmeier, Alenius, & Kivimaki, 2016). While several very novel and intriguing lines of evidence in both human and other animal studies have underpinned this movement, substantial academic interest has focused on whether persistent inflammation may actually constitute a novel pathway to the onset (or persistence or recurrence) of depressive disorders (D'Mello & Swain, 2017; Herkenham & Kigar, 2017; Rosenblat & McIntyre, 2017; Zalli et al., 2016). If so, then the active use of novel antiinflammatory or other immune modulatory strategies for prevention, early intervention, or ongoing management of depression would have a sound pathophysiological basis (Bhattacharya & Drevets, 2017; Jeon & Kim, 2017).

While there have been very many cross-sectional studies of adults with major depression that demonstrate at least some evidence of low-grade or persistent inflammatory markers, there are also many inconsistencies. These are most likely to reflect the considerable heterogeneity of major depressive and other related mood disorders (see Fernandes et al., 2016; Haapakoski, Mathieu, Ebmeier, Alenius, & Kivimaki, 2015; Liu, Adibfar, Herrmann, Gallagher, & Lanctot, 2017; Modabbernia, Taslimi, Brietzke, & Ashrafi, 2013; Rosenblat & McIntyre, 2017). Depressive subtyping studies (e.g., focusing on melancholia or atypical depression) and retrospective selective risk factor investigations (e.g., the role of childhood trauma within adult cohorts of persons with current depression) have sought to resolve the extent to which inflammation may have greater relevance for subgroups of those experiencing depression (Baumeister, Akhtar, Ciufolini, Pariante, & Mondelli, 2016; Capuron & Castanon, 2017; Euteneuer et al., 2017; Grosse et al., 2016; Lamers, Vogelzangs, et al., 2013; Lamers, Milaneschi, de Jonge, Giltay, & Penninx, 2017). Importantly, there are now some prospective or longitudinal studies that suggest that low-level inflammation may indeed have a causal relationship with later onset or recurrence of depressive disorders (Hayes et al., 2017; Khandaker, Stochl, et al., 2017; Valkanova, Ebmeier, & Allan, 2013; Zalli et al., 2016). Certainly, it appears that persistent or recurrent depression, typically commencing in the adolescent period, is likely to be associated with low-grade inflammation and comorbid metabolic disturbances (Mills, Scott, Wray, Cohen-Woods, & Baune, 2013). From an epidemiological and clinical perspective, however, this remains an important area for further research—notably in childhood, adolescent, and young adult populations. In these groups, prospective or longitudinal associations are less likely to be explained by other confounding factors related to chronicity of illness, treatments received, or other secondary medical morbidity (Kim, Szigethy, Melham, Saghafi, & Brent, 2014; Milkowska, Popko, Demkow, & Wolanczyk, 2017; Mills et al., 2013).

POSTINFECTIVE AND POSTINFLAMMATORY MODELS OF DEPRESSION ONSET

While epidemiological questions related to low-grade inflammation and the onset or course of depressive disorders have been slow to resolve, this intriguing area of inflammatory correlates of major depressive disorders has been much more ably assisted by direct examination of humans experiencing postinfective and postinflammatory syndromes (Dantzer, O'Connor, Lawson, & Kelley, 2011; de Barros et al., 2017; Jeon & Kim, 2017). Concurrently, experimentation with infective or inflammatory challenges in relevant animal models has helped to highlight plausible bidirectional pathways between changes in CNS signaling pathways affecting mood, cognition, and sleep and peripheral activation of inflammatory-immune factors (Liu et al., 2017; Ma et al., 2017; Ramirez, Fornaguera-Trias, & Sheridan, 2017; Straley et al., 2017). Both humans and other mammals experiencing acute infective insults (typically associated with fever) display characteristic and stereotypical behavioral responses (e.g., reduced motor activity, reduced feeding, and increased slow-wave sleep), now characterized as acute "sickness behavior" (see Dantzer et al., 2011; Dantzer, O'Connor, Freund, Johnson, & Kelley, 2008; D'Mello & Swain, 2017). For humans, this phenomenon is typically associated with reports of profound fatigue, anhedonia, muscle aches and pains, loss of appetite, poor concentration, and—to varying extents—irritable or depressed mood. In animal models, along with the relevant reduction in motor behavior, other "depressive-like" behaviors (e.g., loss of sucrose preference in rodents) are typically observed.

The observational studies of illness course in human postinfective cohorts and active immune treatment studies in humans with other concurrent infective or malignant illness (Dantzer et al., 2008; Felger, Haroon, Woolwine, Raison, & Miller, 2016; Hickie et al., 2006; Hickie & Lloyd, 1995; Miller & Raison, 2016; Vollmer-Conna et al., 2004) have provided highly plausible evidence that acute inflammatory responses (in these cases in response to significant external stimuli and associated with laboratory evidence of immune activation or dysregulation) result in not only acute "sickness behavior" but also persistent "depressive" disorders—at least in a subgroup of those individuals exposed. While these "depressive" disorders share common core elements with major depression that is not associated with concurrent medical morbidity (e.g., depressed mood and anhedonia), these "postinfective/inflammatory" syndromes are classically more "somatic." That is, patients emphasize typically profound and persisting fatigue, musculoskeletal pain, and sleep-wake cycle dysregulation. These syndromes and most notably the altered mood, anhedonia, profound fatigue, and impaired cognition last well beyond the acute sickness phase in vulnerable individuals (e.g., Hickie et al., 2006). In de novo postinfective cases, these postinfective disorders may last from 3 to 12 months. These ongoing and stereotypical phenomena suggest secondary and ongoing disruption of other CNS-based homeostatic systems (e.g., monoamine neurosignaling, circadian, or HPA) that drive the disorders beyond the acute inflammatory insults. The extent to which these "depressive" syndromes (or at least the mood component) can be prevented by administration of antidepressant therapy prior to active immune therapy strongly supports this assumption (Miller, Haroon, & Felger, 2017; Miller & Raison, 2016).

While these studies provide a strong empirical basis for the assumption that some major depressive disorders can result from acute immune activation or persistent inflammation and provide key insights into potential mechanistic factors, the implications for more common forms of major depression are often not appreciated. However, there is now a resurgence of interest in the neurobiological and transgenerational characteristics of certain depressive subtypes (i.e., atypical depression and depressed phase of bipolar disorder) and of certain critical phenomena (motor activation and 24 h sleep-wake cycle patterns) and the wider recognition of both immune and linked metabolic disturbance in individuals with major depression without other medical morbidity or precipitating environmental challenges (see Fig. 1) (Bhattacharya & Drevets, 2017; Haapakoski et al., 2016; Halaris, 2017; Jeon & Kim, 2017). In this context, it is very timely to reexamine the time course; phenomenological features; and immune, metabolic, and circadian correlates of child and adolescent-onset depressive disorders.

A FOCUS ON DEVELOPMENTAL PERSPECTIVES

Concurrently, our conceptual approach of childhood and adolescent pathways to adult depressive disorders needs to integrate the knowledge of age of onset, the key hormonal and brain developmental events of adolescence, the biochemical pathways of neuroprogression, and the likely childhood precursors of adolescent-onset disorders. Importantly, our conceptualization of the potential role of immune-inflammatory factors needs to incorporate how the age of exposure to these challenges (at different points along the developmental path) intersects with other key neurobiological and CNS-dependent hormonal, circadian, and metabolic processes (see Fig. 2) (Brenhouse &

Schwarz, 2016; Du Preez, Leveson, Zunszain, & Pariante, 2016; Mills et al., 2013; Walther, Penz, Ijacic, & Rice, 2017). The extent to which observed changes in immune and metabolic parameters are indicative of causative processes and are not simply nonspecific markers of other systemic changes requires ongoing and careful study of relevant longitudinal population-based and clinical cohorts. As we explore these developmental continuities, it is also important to consider the mechanisms by which earlier developmental experiences may have their impact on adolescent-onset mood disorders. While some may be via direct impacts on CNS or other key homeostatic mechanisms at the time of exposure, others may act indirectly (e.g., via immune memory or sensitized secondary microglial activation) and only have pathophysiological relevance following later additional exposures in the adolescent period.

POTENTIAL RELEVANCE OF DEPRESSIVE SUBTYPES—BY DEVELOPMENTAL PATH

With regard to specific pathophysiological paths to adolescent-onset depression, we propose three key developmental dimensions, each of which has potential implications for concurrent immune-metabolic disturbances. These three pathways are as follows:

(a) *The anxious depression dimension* (approx. 50% of adolescent-onset cases; Hickie, Hermens, et al., 2013): Longitudinal studies emphasize childhood anxiety in those who are at greatest risk of developing depression in early adolescence (Kim-Cohen et al., 2003; Lewinsohn, Rohde, & Seeley, 1998; Merikangas et al., 2003). Genetic analyses and twin and family studies demonstrate that such disorders are inherited together—that is, they are age-dependent expressions of the same basic traits

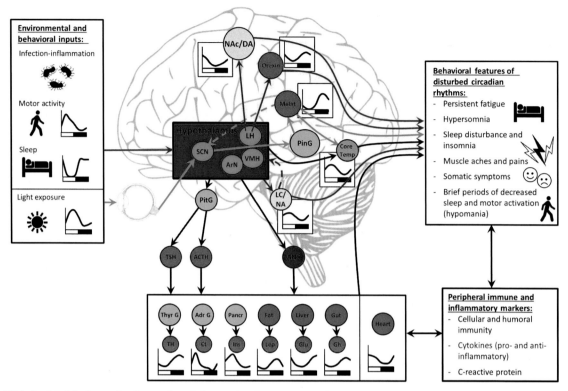

FIG. 1 Model of neural pathways from behavioral and environmental inputs to circadian rhythm disturbance. *SCN*, supra-chiasmatic nucleus; *LH*, lateral hypothalamus; *VMH*, ventromedial hypothalamus; *ArN*, arcuate nucleus; *NAc*, nucleus accumbens; *PinG*, pineal gland; *PitG*, pituitary gland; *LC*, locus coeruleus; *ANS*, autonomic nervous system; *Thyr G*, thyroid gland; *Adr G*, adrenal gland; *Pancr*, pancreas; *DA*, dopamine; *Or*, orexin; *Mel*, melatonin; *NA*, noradrenaline; *ACTH*, adreno-corticotropic hormone; *TSH*, thyroid-stimulating hormone; *TH*, thyroid hormone; *Ct*, cortisol; *Ins*, insulin; *Lep*, leptin; *Glu*, glucose; *Gh*, ghrelin; *Temp*, temperature. *Blue circles* indicate hypothalamic nuclei. *Yellow circles* indicate monoamine-rich nuclei. *Orange circles* indicate hormonal outputs. *Green circles* indicate glands. *Purple circles* indicate organs. Each of the hypothalamic nuclei represented here has its own complex pattern of bidirectional cortical, subcortical, and spinothalamic pathways, and only major interactions between systems are shown here. In this figure, we have emphasized those environ-mental inputs that impact on the 24 h regulation of sleep-wake cycles in humans. Humans are able to remain entrained with the total 24 h light-dark period (and the seasonal variations in the proportions of light and darkness) via two major brain-behavior circuits (one operating largely via the so-called master clock of the suprachiasmatic nuclei (SCN) and other largely independent of the SCN). Pathways using *red arrows* form the direct (or fast changing) pathways for regulation of those behaviors that vary continuously across the course of 24 h and are largely independent of the SCN. *Blue arrows* indicate the indirect (slow changing) pathways for the regulation of those behaviors that are typically closely tied to 24 h light-dark periods and resultant sleep-wake cycles. *Black arrows* indicate output pathways from hypothalamic-pituitary structures to peripheral organs and interactions with immune and inflammatory markers. The loss of coordination between all of these internal timing systems can result in a state of internal desynchrony—such that key parameters such as core body temperature, cortisol release, and metabolic and immune regulation are not optimally aligned with each other or relevant external behaviors (e.g., feeding and sleeping).

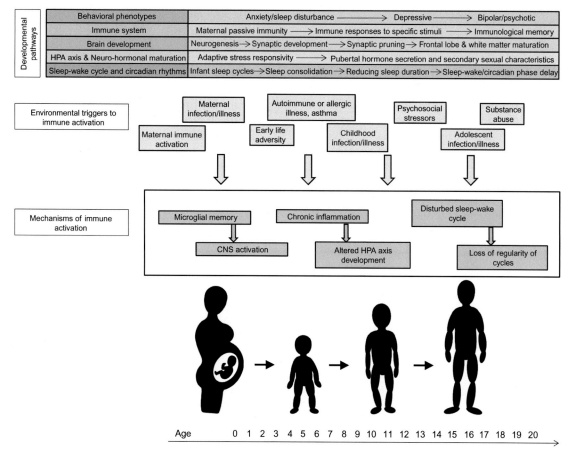

FIG. 2 Pathways influencing immune activation from prenatal to adolescence. *HPA,* hypothalamus pituitary adrenal; *CNS,* central nervous system. Various environmental influences occur at different ages to trigger immune activation. Multiple mechanisms may contribute to abnormalities following immune activation including microglial memory and secondary CNS activation, chronic inflammation and associated alterations in HPA axis development, and immune-mediated disruptions in the sleep-wake cycle leading to a loss of regularity of cycles. These processes occur in the context of developmental pathways across multiple brain, body, and behavioral systems, with interactions between all of these systems contributing to the expression of behavioral mood disorder phenotypes.

(Middeldorp, Cath, Van Dyck, & Boomsma, 2005; Mosing et al., 2009). The key transition to depression occurs in the adolescent period (Merikangas et al., 2010; Paus, Keshavan, & Giedd, 2008), with fundamental concepts of high arousal, excessive SNS reactivity, and development of emotional circuitry reinforcing catastrophic responses to adverse life events (Gold, 2015; Kircanski, LeMoult, Ordaz, & Gotlib, 2017). The consequences of prolonged childhood and adolescent arousal-anxiety on both HPA axis and SNS development and ongoing function and their downstream and continuing effects on immune and metabolic responses require continued investigation (Camacho, 2013; Gaspersz et al., 2017; Gold, 2015; Miller & Raison, 2016). There are unresolved issues with regard to this subtype of depression (and its course) and its relationship to immune disturbance. Historically, excessive activation (or nonresponse of the central aspects) of the HPA system in severe depression (i.e., melancholia) was typically associated with raised and persisting peripheral cortisol and concurrent suppression of key immune factors, including key aspects of cell-mediated immunity *in vivo* (Hickie et al., 1990; Miller & Raison, 2016). Consequently, one might also expect a lack of association with other cytokine-based measures suggesting immune activation in studies focusing on this subtype (see Lamers et al., 2017). However, it is likely that states of high arousal persist between overt depressive episodes and that ongoing HPA axis and SNS dysfunction are associated with varying patterns of immune activation at different stages of depressive illness (Miller & Raison, 2016; Mills et al., 2013).

(b) *The mania/hypomania/fatigue/atypical depression (circadian depression) dimension* (approx. 25% of adolescent-onset cases; Hickie, Hermens, et al., 2013): This dimension challenges existing models of the genetic architecture of both unipolar and bipolar disorders, favoring a pathophysiological model based on phenotypes related to high and low energy/activation states. Specifically, these typically include persistent fatigue states, "atypical" depression, and the depressed phase of bipolar disorders (Angst et al., 2006; Benazzi, 2000; Hickie, Bennett, Lloyd, Heath, & Martin, 1999; Mitchell, Goodwin, Johnson, & Hirschfeld, 2008; Scott et al., 2017). Key supporting data are derived from epidemiological studies in the United States and Europe (Addington, Gallo, Ford, & Eaton, 2001; Angst, Gamma, Sellaro, Zhang, & Merikangas, 2002; Lamers et al., 2012) and NIMH and Swiss family studies

(Merikangas et al., 2008, 2012; Merikangas & Lamers, 2012). It has a strong developmental perspective (see Fig. 3), whereby the interaction of genetic architecture and environmental exposures at different time points may result in different age-dependent mood and fatigue phenotypes (Hansell et al., 2012; Hickie et al., 1999; Hickie, Hermens, et al., 2013; Lamers, Hickie, & Merikangas, 2013; Pryce & Fontana, 2017). Most importantly, these phenotypes share common elements (e.g., 24 h sleep-wake cycle disruption, fatigue, nonspecific somatic symptoms, and unstable mood) and continued disruption of fundamental circadian-dependent neurobiological and peripheral hormonal and metabolic mechanisms (see Fig. 1) (Felger & Treadway, 2017; Pryce & Fontana, 2017; Zadka, Dziegiel, Kulus, & Olajossy, 2017). From a circadian perspective, we have shown that those with early stages of these depressive disorders have a loss of the normal pattern of melatonin onset and that (based on actigraphy measurement) young people with bipolar and unipolar disorders differ markedly from controls, with a pattern of significant circadian phase delay (Carpenter et al., 2017; Naismith et al., 2012; Robillard et al., 2013). Disturbed circadian function, notably that associated with phase delay, predicts a poorer course of depressive disorder and persistent hypomanic features (Gruber et al., 2011; Robillard et al., 2016; Staton, 2008).

(c) *The impaired development dimension* (approx. 25% of adolescent-onset cases; Hickie, Hermens, et al., 2013): Longitudinal studies indicate those with childhood evidence of impaired brain development (indicated phenotypically, through educational difficulties or through formal testing of specific neuropsychological abilities) are at increased risk of a wide range of adolescent-onset neuropsychological disorders

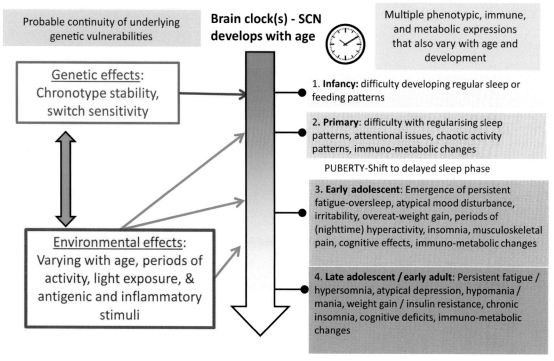

FIG. 3 Model of emergence of age-dependent mood disorder and immunometabolic phenotypes. SCN, suprachiasmatic nucleus. Genetic predisposition for variations in the 24 h sleep-wake cycle (e.g., morning vs. evenings and period length of the circadian cycle) interacts with differing environmental exposures across the life span from infancy to early adulthood. This continuous interaction results in various age-dependent clinical phenotypes with concurrent immune and metabolic expressions.

including depressive disorders, psychotic disorders, and alcohol or other substance misuse (Colman, Ploubidis, Wadsworth, Jones, & Croudace, 2007; Hermens et al., 2013; Khandaker, Stochl, et al., 2017; Koenen et al., 2009; Zammit et al., 2004). These children are particularly likely to have been exposed to CNS insults in utero or early childhood including CNS infection, noninfective "encephalitis," and other central and peripherally focused autoimmune events affecting the CNS (Capuron & Castanon, 2017; Du Preez et al., 2016; Hickie et al., 2009; Khandaker, Zimbron, Lewis, & Jones, 2013; Mills et al.,

2013). In adolescence and particularly in association with more severe psychotic symptoms, evidence emerges of active CNS inflammation or immune activation during the most severe episodes of illness (Estes & McAllister, 2016; Khandaker, Dantzer, et al., 2017; Khandaker, Zammit, Lewis, & Jones, 2014; Mills et al., 2013). At times, these are characterized by other CNS-dependent phenomena including disturbed sleep-wake cycles (e.g., narcolepsy-like), motor tics, or other extrapyramidal phenomena and obsessive or compulsive behavior features (Mohammad & Dale, 2017; Moldofsky & Dickstein, 1999; Perez-Vigil et al., 2016).

A FOCUS ON THE CIRCADIAN SYSTEM AND LINKS TO IMMUNE AND METABOLIC DYSFUNCTION

In recent times, there has been an increasing focus on the extent to which 24 h rhythms are intrinsic to most, if not all, homeostatic systems and that disruption of these systems is associated with major changes in neurobiological, metabolic, immune, and hormonal function at both cellular signaling and higher-order system levels (see Fig. 1) (Turek, 2016). Chronic disruption of circadian systems, independent of the precipitating cause(s), is particularly likely to be associated with a wide range of inflammatory and metabolic conditions. The sensitivity of the circadian system to disruption by immune-inflammatory factors (precipitated by events such as external infection or internal autoimmune processes) is particularly notable (Musiek & Holtzman, 2016; Turek, 2016). It is likely that this is one of the mechanisms by which inflammation-precipitated acute "sickness behavior" is induced in both humans and other animals. The phenotypic overlap between atypical mood disorders, "sickness behavior," and those disorders known to result directly from disruption of circadian systems (e.g., shift work and jet lag) strongly suggests that they share at least some core pathophysiological mechanisms. Additionally, there is increasing epidemiological evidence that "atypical" forms of depression (not just depression severity) are particularly associated with strong evidence of immune activation and metabolic disruption (Lamers et al., 2017; Lamers, Vogelzangs, et al., 2013).

POTENTIAL FOR NEW THERAPEUTIC APPROACHES

There is an ongoing gap between the community demand for better treatments—particularly early in the course of major mood disorders—and the evidence for providing safe, effective, and personally optimized forms of care. This gap is, in part, due to the lack of knowledge of how specific pathophysiological elements operate at the level of individuals who present for care. Currently, there are a range of new therapeutic opportunities deserving for testing larger clinical trials (e.g., cannabidiol, ketamine, and estrogen supplementation) either alone or in combination with standard antidepressant therapies.

Among those of greatest interest, at least to those who are interested in the potential pathophysiological roles of inflammatory factors, are new or existing treatments that

(i) directly target inflammatory processes, via either nonspecific (e.g., corticosteroids, mycophenolate, minocycline, and NSAIDs) or specific (e.g., monoclonal antibodies targeting TNF-alpha) immune modulatory therapies;

(ii) directly target autoimmune processes (e.g., immune suppression or plasmapheresis);

(iii) focus from a phenotypic perspective on individuals who present with "atypical" features (e.g., sleep-wake cycle disturbance, fatigue, muscle aches and pains, and other nonspecific somatic symptoms) that most closely mimic the stereotypical "sickness behavior" observed in both humans and other animals directly exposed to inflammatory or infective stimuli;

(iv) target the "circadian" system (e.g., behavioral or light regulation of circadian systems, orexin antagonists, or melatonin agonists), which when disrupted by infective inflammatory or other environmental stimuli results in not only "atypical" features but also a wider network of hormonal and metabolic disturbances.

SELECTING SPECIFIC PHENOTYPES FOR MORE TARGETED INTERVENTIONS

With regard to acute episodes of adolescent depression, as yet, there is insufficient evidence to make specific recommendations about the optimal use of any antiinflammatory or immune regulatory therapies as primary treatments (Baune, 2017; Husain, Strawbridge, Stokes, & Young, 2017). While there has been much enthusiasm for the use of nonspecific and low-risk strategies, such as omega-3 fatty acids, there is a lack of clear evidence for the use of these agents as stand-alone therapeutic agents in other youth-onset neuropsychiatric disorders (Bozzatello, Brignolo, De Grandi, & Bellino, 2016; Markulev et al., 2017). The potential role of such agents as adjunctive antidepressant agents is yet to be clearly established. In addition, however, a major consideration for adolescent-onset depression is the prevention of the longer-term adverse effects of recurrent or persistent disorders on both CNS structures and linked peripheral systems, such as immune and metabolic pathways (Goldstein et al., 2015; Penninx, Milaneschi, Lamers, & Vogelzangs, 2013; Suglia, Demmer, Wahi, Keyes, & Koenen, 2016). This consideration, of secondary prevention of other adverse neurobiological, immune, and metabolic outcomes, is one where the deployment of antiinflammatory or immune regulatory therapies is worthy of more serious longitudinal research.

Recognition of potential differential developmental pathways—as outlined above—may assist to inform individual therapeutic choices. For example, in those with a clear "circadian"-type pattern of illness, atypical depressive features, major disruptions of sleep or 24 h sleep-wake cycle, or concurrent (clinical or laboratory) evidence of metabolic and immune dysfunction, an earlier focus on circadian-targeted behavioral or pharmacological therapies may well be justified (see Hickie, Scott, et al., 2013). If, along with other standard psychological or pharmacological approaches, this is unsuccessful, then this may constitute a group in whom more antiinflammatory or immune modulatory therapies are justified. For those with a "neurodevelopmental"-type pathway, particularly

where there is clinical evidence of other relevant factors (e.g., motor tics, rapid emergence of extrapyramidal, or other neurological side effects with standard pharmacotherapies), we are also more likely to consider progression to antiinflammatory or immune modulatory therapies at an earlier phase of illness.

ROLE OF MORE DETAILED CLINICAL INVESTIGATION

There are four clinical groups where we do specifically consider whether it is reasonable to conduct more extensive and potentially more invasive clinical investigations. When there is either evidence of active autoimmune disease (i.e., specific autoantibodies) or evidence of nonspecific inflammation affecting the CNS, acute use of immunomodulatory therapies may be justified. Typically, these groups respond poorly to conventional psychological and pharmacological treatments for adolescent depression and also often develop significant side effects early in the course of conventional therapies.

These four groups are adolescents and young adults who present with the following:

(i) Clear concurrent medical comorbidity suggesting autoimmune disease. Most typically, these are young women with persisting or treatment-resistant depressive disorders and evidence of metabolic (e.g., obesity and insulin resistance) or hormonal (e.g., hirsutism and disturbed menstrual function, often diagnosed with polycystic ovarian syndrome) disturbances. This group also typically has a strong family history of autoimmune or other chronic inflammatory disorders.
(ii) Profound fatigue, musculoskeletal pain, and other nonspecific somatic symptoms (often previously diagnosed with chronic fatigue syndrome).
(iii) Neurodevelopmental and atypical psychotic syndromes with strong affective features.

(iv) Significant concurrent neurological (e.g., motor tics, narcolepsy-like, and epilepsy-like) features or who develop such features soon after exposure to pharmacological therapies.

CONCLUSION

Consistent with advances in epidemiology, clinical and pathophysiological research in young people, and concurrent investigation of relevant postinfective and inflammatory illness models, we are now in a position to incorporate a new perspective on the role of inflammatory and metabolic perturbations in the onset and course of adolescent depression. Strongly aligned with these developments is the increasing recognition of the potential role of perturbed circadian systems in both primary pathophysiology and ongoing perpetuation of risk and recurrence. These developments recognize the importance of more comprehensive assessment, consideration of new therapeutic approaches, and incorporation of new perspectives on secondary prevention of adverse medical and psychological outcomes. Currently, trials of new antiinflammatory approaches as both acute and secondary preventive therapies are underway and the development of new therapies targeting circadian dysfunction. Additionally, prior stratification of adolescent cohorts on the basis of relevant clinical phenotypes (e.g., "atypical" features and relevant neurological or medical comorbidity) or laboratory markers (e.g., evidence of low-grade inflammation or specific autoimmune processes) may well lead to much greater personalization of treatment choices.

References

Addington, A. M., Gallo, J. J., Ford, D. E., & Eaton, W. W. (2001). Epidemiology of unexplained fatigue and major depression in the community: the Baltimore ECA follow-up, 1981–1994. *Psychological Medicine, 31*(6), 1037–1044.

Allen, N. B., Hetrick, S. E., Simmons, J. G., & Hickie, I. B. (2007). Early intervention for depressive disorders in young people: the opportunity and the (lack of) evidence. *The Medical Journal of Australia, 187*(7), S15.

Angst, J., Gamma, A., Benazzi, F., Silverstein, B., Ajdacic-Gross, V., Eich, D., et al. (2006). Atypical depressive syndromes in varying definitions. *European Archives of Psychiatry and Clinical Neuroscience, 256*(1), 44–54.

Angst, J., Gamma, A., Sellaro, R., Zhang, H., & Merikangas, K. (2002). Toward validation of atypical depression in the community: results of the Zurich cohort study. *Journal of Affective Disorders, 72*(2), 125–138.

Baumeister, D., Akhtar, R., Ciufolini, S., Pariante, C. M., & Mondelli, V. (2016). Childhood trauma and adulthood inflammation: a meta-analysis of peripheral C-reactive protein, interleukin-6 and tumour necrosis factor-alpha. *Molecular Psychiatry, 21*(5), 642–649.

Baune, B. T. (2017). Are non-steroidal anti-inflammatory drugs clinically suitable for the treatment of symptoms in depression-associated inflammation? in R. Dantzer, & L. Capuron (Eds.), *Vol. 31. Inflammation-associated depression: Evidence, mechanisms and implications current topics in behavioral neurosciences* (pp. 303–319). Cham: Springer.

Benazzi, F. (2000). Depression with DSM-IV atypical features: a marker for bipolar II disorder. *European Archives of Psychiatry and Clinical Neuroscience, 250*(1), 53–55.

Bhattacharya, A., & Drevets, W. C. (2017). Role of neuroimmunological factors in the pathophysiology of mood disorders: implications for novel therapeutics for treatment resistant depression. in R. Dantzer, & L. Capuron (Eds.), *Vol. 31. Inflammation-associated depression: Evidence, mechanisms and implications current topics in behavioral neurosciences* (pp. 339–356). Cham: Springer.

Bloom, D., Cafiero, E., Jane-Llopis, E., Abrahams-Gessel, S., Bloom, L., Fathima, S., et al. (2012). *The global economic burden of non-communicable diseases*. Program on the Global Demography of Aging.

Bozzatello, P., Brignolo, E., De Grandi, E., & Bellino, S. (2016). Supplementation with omega-3 fatty acids in psychiatric disorders: a review of literature data. *Journal of Clinical Medicine, 5*(8), E67.

Brenhouse, H. C., & Schwarz, J. M. (2016). Immunoadolescence: neuroimmune development and adolescent behavior. *Neuroscience and Biobehavioral Reviews, 70,* 288–299.

Bunker, S. J., Colquhoun, D. M., Esler, M. D., Hickie, I. B., Hunt, D., Jelinek, M., et al. (2003). "Stress" and coronary heart disease: psychosocial risk factors. *The Medical Journal of Australia, 178*(6), 272–276.

Camacho, A. (2013). Is anxious-depression an inflammatory state? *Medical Hypotheses, 81*(4), 577–581.

Capuron, L., & Castanon, N. (2017). Role of inflammation in the development of neuropsychiatric symptom domains:

evidence and mechanisms. R. Dantzer, & L. Capuron (Eds.), *Vol. 31. Inflammation-associated depression: Evidence, mechanisms and implications current topics in behavioral neurosciences* (pp. 31–44). Cham: Springer.

Carpenter, J. S., Robillard, R., Hermens, D. F., Naismith, S. L., Gordon, C., Scott, E. M., et al. (2017). Sleep-wake profiles and circadian rhythms of core temperature and melatonin in young people with affective disorders. *Journal of Psychiatric Research, 94*, 131–138.

Catania, L. S., Hetrick, S. E., Newman, L. K., & Purcell, R. (2014). Prevention and early intervention for mental health problems in 0–25 year olds: are there evidence-based models of care? *Advances in Mental Health, 10*(1), 6–19.

Colman, I., Ploubidis, G. B., Wadsworth, M. E., Jones, P. B., & Croudace, T. J. (2007). A longitudinal typology of symptoms of depression and anxiety over the life course. *Biological Psychiatry, 62*(11), 1265–1271.

Cross, S. P. M., Hermens, D. F., Scott, E. M., Ottavio, A., McGorry, P. D., & Hickie, I. B. (2014). A clinical staging model for early intervention youth mental health services. *Psychiatric Services, 65*(7), 939–943.

Dantzer, R. (2012). Depression and inflammation: an intricate relationship. *Biological Psychiatry, 71*(1), 4–5.

Dantzer, R. (2017). Neuroimmune interactions: from the brain to the immune system and vice versa. *Physiological Reviews, 98*(1), 477–504.

Dantzer, R., O'Connor, J. C., Freund, G. G., Johnson, R. W., & Kelley, K. W. (2008). From inflammation to sickness and depression: when the immune system subjugates the brain. *Nature Reviews. Neuroscience, 9*(1), 46–56.

Dantzer, R., O'Connor, J. C., Lawson, M. A., & Kelley, K. W. (2011). Inflammation-associated depression: from serotonin to kynurenine. *Psychoneuroendocrinology, 36* (3), 426–436.

Das, J. K., Salam, R. A., Lassi, Z. S., Khan, M. N., Mahmood, W., Patel, V., et al. (2016). Interventions for adolescent mental health: an overview of systematic reviews. *The Journal of Adolescent Health, 59*(4S), S49–S60.

de Barros, J. L., Barbosa, I. G., Salem, H., Rocha, N. P., Kummer, A., Okusaga, O. O., et al. (2017). Is there any association between toxoplasma gondii infection and bipolar disorder? A systematic review and meta-analysis. *Journal of Affective Disorders, 209*, 59–65.

D'Mello, C., & Swain, M. G. (2017). Immune-to-brain communication pathways in inflammation-associated sickness and depression. R. Dantzer, & L. Capuron (Eds.), *Vol. 31. Inflammation-associated depression: Evidence, mechanisms and implications current topics in behavioral neurosciences* (pp. 73–94). Cham: Springer.

Du Preez, A., Leveson, J., Zunszain, P. A., & Pariante, C. M. (2016). Inflammatory insults and mental health consequences: does timing matter when it comes to depression? *Psychological Medicine, 46*(10), 2041–2057.

Erskine, H. E., Moffitt, T. E., Copeland, W. E., Costello, E. J., Ferrari, A. J., Patton, G., et al. (2015). A heavy burden on young minds: the global burden of mental and substance use disorders in children and youth. *Psychological Medicine, 45*(7), 1511–1563.

Estes, M. L., & McAllister, K. (2016). Maternal immune activation: implications for neuropsychiatric disorders. *Science, 353*(6301), 772–777.

Euteneuer, F., Dannehl, K., Del Rey, A., Engler, H., Schedlowski, M., & Rief, W. (2017). Peripheral immune alterations in major depression: the role of subtypes and pathogenetic characteristics. *Frontiers in Psychiatry. 8*, https://doi.org/10.3389/fpsyt.2017.00250.

Fairweather, A. K., Anstey, K. J., Rodgers, B., Jorm, A. F., & Christensen, H. (2007). Age and gender differences among Australian suicide ideators: prevalence and correlates. *The Journal of Nervous and Mental Disease, 195*(2), 130–136.

Felger, J. C., Haroon, E., Woolwine, B. J., Raison, C. L., & Miller, A. H. (2016). Interferon-alpha-induced inflammation is associated with reduced glucocorticoid negative feedback sensitivity and depression in patients with hepatitis C virus. *Physiology & Behavior, 166*, 14–21.

Felger, J. C., & Treadway, M. T. (2017). Inflammation effects on motivation and motor activity: role of dopamine. *Neuropsychopharmacology, 42*(1), 216–241.

Fernandes, B. S., Steiner, J., Molendijk, M. L., Dodd, S., Nardin, P., Gonçalves, C.-A., et al. (2016). C-reactive protein concentrations across the mood spectrum in bipolar disorder: a systematic review and meta-analysis. *Lancet Psychiatry, 3*(12), 1147–1156.

Gaspersz, R., Lamers, F., Wittenberg, G., Beekman, A. T. F., van Hemert, A. M., Schoevers, R. A., et al. (2017). The role of anxious distress in immune dysregulation in patients with major depressive disorder. *Translational Psychiatry, 7*(12), 1268.

Gitlin, M. J., & Miklowitz, D. J. (2017). The difficult lives of individuals with bipolar disorder: a review of functional outcomes and their implications for treatment. *Journal of Affective Disorders, 209*, 147–154.

Gold, P. W. (2015). The organization of the stress system and its dysregulation in depressive illness. *Molecular Psychiatry, 20*(1), 32–47.

Goldstein, B. I., Carnethon, M. R., Matthews, K. A., McIntyre, R. S., Miller, G. E., Raghuveer, G., et al. (2015). Major depressive disorder and bipolar disorder predispose youth to accelerated atherosclerosis and early cardiovascular disease: a scientific statement from the American Heart Association. *Circulation, 132*(10), 965–986.

Gore, F. M., Bloem, P. J. N., Patton, G. C., Ferguson, J., Joseph, V., Coffey, C., et al. (2011). Global burden of disease in young people aged 10–24 years: a systematic analysis. *Lancet, 377*(9783), 2093–2102.

Grosse, L., Ambree, O., Jorgens, S., Jawahar, M. C., Singhal, G., Stacey, D., et al. (2016). Cytokine levels in major depression are related to childhood trauma but not to recent stressors. *Psychoneuroendocrinology, 73*, 24–31.

Gruber, J., Miklowitz, D. J., Harvey, A. G., Frank, E., Kupfer, D., Thase, M. E., et al. (2011). Sleep matters: sleep functioning and course of illness in bipolar disorder. *Journal of Affective Disorders, 134*(1–3), 416–420.

Gustavsson, A., Svensson, M., Jacobi, F., Allgulander, C., Alonso, J., Beghi, E., et al. (2011). Cost of disorders of the brain in Europe 2010. *European Neuropsychopharmacology, 21*(10), 718–779.

Haapakoski, R., Ebmeier, K. P., Alenius, H., & Kivimaki, M. (2016). Innate and adaptive immunity in the development of depression: an update on current knowledge and technological advances. *Progress in Neuro-Psychopharmacology & Biological Psychiatry, 66*, 63–72.

Haapakoski, R., Mathieu, J., Ebmeier, K. P., Alenius, H., & Kivimaki, M. (2015). Cumulative meta-analysis of interleukins 6 and 1beta, tumour necrosis factor alpha and C-reactive protein in patients with major depressive disorder. *Brain, Behavior, and Immunity, 49*, 206–215.

Halaris, A. (2017). Inflammation-associated co-morbidity between depression and cardiovascular disease. in R. Dantzer, & L. Capuron (Eds.), *Vol. 31. Inflammation-associated depression: Evidence, mechanisms and implications current topics in behavioral neurosciences* (pp. 45–70). Cham: Springer.

Hansell, N. K., Wright, M. J., Medland, S. E., Davenport, T. A., Wray, N. R., Martin, N. G., et al. (2012). Genetic co-morbidity between neuroticism, anxiety/depression and somatic distress in a population sample of adolescent and young adult twins. *Psychological Medicine, 42*(6), 1249–1260.

Hayes, J. F., Khandaker, G. M., Anderson, J., Mackay, D., Zammit, S., Lewis, G., et al. (2017). Childhood interleukin-6, C-reactive protein and atopic disorders as risk factors for hypomanic symptoms in young adulthood: a longitudinal birth cohort study. *Psychological Medicine, 47*(1), 23–33.

Herkenham, M., & Kigar, S. L. (2017). Contributions of the adaptive immune system to mood regulation: mechanisms and pathways of neuroimmune interactions. *Progress in Neuro-Psychopharmacology & Biological Psychiatry, 79*, 49–57.

Hermens, D. F., Lagopoulos, J., Tobias-Webb, J., De Regt, T., Dore, G., Juckes, L., et al. (2013). Pathways to alcohol-induced brain impairment in young people: a review. *Cortex, 49*(1), 3–17.

Hickie, I., Bennett, B., Lloyd, A., Heath, A., & Martin, N. (1999). Complex genetic and environmental relationships between psychological distress, fatigue and immune functioning: a twin study. *Psychological Medicine, 29*(2), 269–277.

Hickie, I., Davenport, T., Wakefield, D., Vollmer-Conna, U., Cameron, B., Vernon, S. D., et al. (2006). Post-infective and chronic fatigue syndromes precipitated by viral and non-viral pathogens: prospective cohort study. *BMJ, 333*(7568), 575.

Hickie, I., & Lloyd, A. (1995). Are cytokines associated with neuropsychiatric syndromes in humans. *International Journal of Immunopharmacology, 17*(8), 677–683.

Hickie, I., Silove, D., Hickie, C., Wakefield, D., & Lloyd, A. (1990). Is there immune dysfunction in depressive disorders? *Psychological Medicine, 20*(4), 755–761.

Hickie, I. B. (2011). Youth mental health: we know where we are and we can now say where we need to go next. *Early Intervention in Psychiatry, 5*(Suppl. 1), 63–69.

Hickie, I. B., Banati, R., Stewart, C. H., & Lloyd, A. R. (2009). Are common childhood or adolescent infections risk factors for schizophrenia and other psychotic disorders? *The Medical Journal of Australia, 190*(4), S17–S21.

Hickie, I. B., Hermens, D. F., Naismith, S. L., Guastella, A. J., Glozier, N., Scott, J., et al. (2013). Evaluating differential developmental trajectories to adolescent-onset mood and psychotic disorders. *BMC Psychiatry, 13*(303).

Hickie, I. B., Scott, J., Hermens, D. F., Scott, E. M., Naismith, S. L., Guastella, A. J., et al. (2013). Clinical classification in mental health at the cross-roads: which direction next? *BMC Medicine, 11*(125).

Husain, M. I., Strawbridge, R., Stokes, P. R., & Young, A. H. (2017). Anti-inflammatory treatments for mood disorders: systematic review and meta-analysis. *Journal of Psychopharmacology (Thousand Oaks, CA), 31*(9), 1137–1148.

Insel, T. R. (2007). The arrival of preemptive psychiatry. *Early Intervention in Psychiatry, 1*(1), 5–6.

Iorfino, F., Hermens, D. H., Cross, S. P. M., Zmicerevska, N., Nichles, A., Badcock, C., et al. (2018). Delineating the trajectories of social and occupational functioning of young people attending early intervention mental health services in Australia: a longitudinal study. *BMJ, e020678.*

Jeon, S. W., & Kim, Y. K. (2017). Inflammation-induced depression: its pathophysiology and therapeutic implications. *Journal of Neuroimmunology, 313*, 92–98.

Kessler, R. C., Berglund, P., Demler, O., Jin, R., Merikangas, K. R., & Walters, E. E. (2005). Lifetime prevalence and age-of-onset distributions of DSM-IV disorders in the National Comorbidity Survey Replication. *Archives of General Psychiatry, 62*(6), 593–602.

Khandaker, G. M., Dantzer, R., & Jones, P. B. (2017). Immunopsychiatry: important facts. *Psychological Medicine, 47*(13), 2229–2237.

Khandaker, G. M., Stochl, J., Zammit, S., Goodyer, I., Lewis, G., & Jones, P. B. (2017). Childhood inflammatory markers and intelligence as predictors of subsequent persistent depressive symptoms: a longitudinal cohort study. *Psychological Medicine.*

Khandaker, G. M., Zammit, S., Lewis, G., & Jones, P. B. (2014). A population-based study of atopic disorders and inflammatory markers in childhood before psychotic experiences in adolescence. *Schizophrenia Research, 152*(1), 139–145.

Khandaker, G. M., Zimbron, J., Lewis, G., & Jones, P. B. (2013). Prenatal maternal infection, neurodevelopment and adult schizophrenia: a systematic review of population-based studies. *Psychological Medicine, 43*(2), 239–257.

Kim, J.-W., Szigethy, E. M., Melham, N. M., Saghafi, E. M., & Brent, D. A. (2014). Inflammatory markers and the pathogenesis of pediatric depression and suicide: a systematic review of the literature. *The Journal of Clinical Psychiatry, 75*(11), 1242–1253.

Kim-Cohen, J., Caspi, A., Moffitt, T. E., Harrington, H., Milne, B. J., & Poulton, R. (2003). Prior juvenile diagnoses in adults with mental disorder: developmental followback of a prospective-longitudinal cohort. *Archives of General Psychiatry, 60*(7), 709–717.

Kircanski, K., LeMoult, J., Ordaz, S., & Gotlib, I. H. (2017). Investigating the nature of co-occurring depression and anxiety: comparing diagnostic and dimensional research approaches. *Journal of Affective Disorders, 216*, 123–135.

Koenen, K. C., Moffitt, T. E., Roberts, A. L., Martin, L. T., Kubzansky, L., Harrington, H., et al. (2009). Childhood IQ and adult mental disorders: a test of the cognitive reserve hypothesis. *The American Journal of Psychiatry, 166*(1), 50–57.

Lamers, F., Burstein, M., He, J. P., Avenevoli, S., Angst, J., & Merikangas, K. R. (2012). Structure of major depressive disorder in adolescents and adults in the US general population. *The British Journal of Psychiatry, 201*(2), 143–150.

Lamers, F., Hickie, I., & Merikangas, K. R. (2013). Prevalence and correlates of prolonged fatigue in a U.S. sample of adolescents. *The American Journal of Psychiatry, 170*(5), 502–510.

Lamers, F., Milaneschi, Y., de Jonge, P., Giltay, E. J., & Penninx, B. (2017). Metabolic and inflammatory markers: associations with individual depressive symptoms. *Psychological Medicine*, 1–11.

Lamers, F., Vogelzangs, N., Merikangas, K. R., de Jonge, P., Beekman, A. T., & Penninx, B. W. (2013). Evidence for a differential role of HPA-axis function, inflammation and metabolic syndrome in melancholic versus atypical depression. *Molecular Psychiatry, 18*(6), 692–699.

Lee, R. S., Hermens, D. F., Redoblado-Hodge, M. A., Naismith, S. L., Porter, M. A., Kaur, M., et al. (2013). Neuropsychological and socio-occupational functioning in young psychiatric outpatients: a longitudinal investigation. *PLoS One, 8*(3).

Lewinsohn, P. M., Rohde, P., & Seeley, J. R. (1998). Major depressive disorder in older adolescents: prevalence, risk factors, and clinical implications. *Clinical Psychology Review, 18*(7), 765–794.

Libby, P. (2012). Inflammation in atherosclerosis. *Arteriosclerosis, Thrombosis, and Vascular Biology, 32*(9), 2045–2051.

Liu, C. S., Adibfar, A., Herrmann, N., Gallagher, D., & Lanctot, K. L. (2017). Evidence for inflammation-associated depression. in R. Dantzer, & L. Capuron (Eds.), *Vol. 31. Inflammation-associated depression: Evidence, mechanisms and implications current topics in behavioral neurosciences* (pp. 3–30). Cham: Springer.

Lopez, A. D., Mathers, C. D., Ezzati, M., Jamison, D. T., & Murray, C. J. L. (2006). Global and regional burden of disease and risk factors, 2001: systematic analysis of population health data. *Lancet, 367*(9524), 1747–1757.

Ma, L., Demin, K. A., Kolesnikova, T. O., Kharsko, S. L., Zhu, X., Yuan, X., et al. (2017). Animal inflammation-based models of depression and their application to drug discovery. *Expert Opinion on Drug Discovery, 12*(10), 995–1009.

Markulev, C., McGorry, P. D., Nelson, B., Yuen, H. P., Schaefer, M., Yung, A. R., et al. (2017). NEURAPRO-E study protocol: a multicentre randomized controlled trial of omega-3 fatty acids and cognitive-behavioural case management for patients at ultra high risk of schizophrenia and other psychotic disorders. *Early Intervention in Psychiatry, 11*(4), 418–428.

Mastronardi, C., Paz-Filho, G. J., Valdez, E., Maestre-Mesa, J., Licinio, J., & Wong, M. L. (2011). Long-term body weight outcomes of antidepressant-environment interactions. *Molecular Psychiatry, 16*(3), 265–272.

Merikangas, K. R., Cui, L., Kattan, G., Carlson, G. A., Youngstrom, E. A., & Angst, J. (2012). Mania with and without depression in a community sample of US adolescents. *Archives of General Psychiatry, 69*(9), 943–951.

Merikangas, K. R., He, J.-P., Burstein, M., Swanson, S. A., Avenevoli, S., Cui, L., et al. (2010). Lifetime prevalence of mental disorders in U.S. adolescents: results from the National Comorbidity Survey Replication—Adolescent Supplement (NCS-A). *Journal of the American Academy of Child and Adolescent Psychiatry, 49*(10), 980–989.

Merikangas, K. R., Herrell, R., Swendsen, J., Rossler, W., Ajdacic-Gross, V., & Angst, J. (2008). Specificity of bipolar spectrum conditions in the comorbidity of mood and substance use disorders: results from the Zurich Cohort Study. *Archives of General Psychiatry, 65*(1), 47–52.

Merikangas, K. R., & Kalaydjian, A. (2007). Magnitude and impact of comorbidity of mental disorders from epidemiologic surveys. *Current Opinion in Psychiatry, 20*(4), 353–358.

Merikangas, K. R., & Lamers, F. (2012). The "true" prevalence of bipolar II disorder. *Current Opinion in Psychiatry, 25*(1), 19–23.

Merikangas, K. R., Zhang, H., Avenevoli, S., Acharyya, S., Neuenschwander, M., & Angst, J. (2003). Longitudinal trajectories of depression and anxiety in a prospective community study. *Archives of General Psychiatry*, 60(10), 993–1000.

Middeldorp, C. M., Cath, D. C., Van Dyck, R., & Boomsma, D. I. (2005). The co-morbidity of anxiety and depression in the perspective of genetic epidemiology. A review of twin and family studies. *Psychological Medicine*, 35(5), 611–624.

Milkowska, P., Popko, K., Demkow, U., & Wolanczyk, T. (2017). ss. in M. Pokorski (Ed.), *Vol. 1021. Pulmonary care and clinical medicine advances in experimental medicine and biology* (pp. 73–80). Cham: Springer.

Miller, A. H., Haroon, E., & Felger, J. C. (2017). Therapeutic implications of brain-immune interactions: treatment in translation. *Neuropsychopharmacology*, 42(1), 334–359.

Miller, A. H., & Raison, C. L. (2016). The role of inflammation in depression: from evolutionary imperative to modern treatment target. *Nature Reviews. Immunology*, 16(1), 22–34.

Mills, N. T., Scott, J. G., Wray, N. R., Cohen-Woods, S., & Baune, B. T. (2013). Research review: the role of cytokines in depression in adolescents: a systematic review. *Journal of Child Psychology and Psychiatry*, 54(8), 816–835.

Mitchell, P. B., Goodwin, G. M., Johnson, G. F., & Hirschfeld, R. M. A. (2008). Diagnostic guidelines for bipolar depression: a probabilistic approach. *Bipolar Disorders*, 10(1p2), 144–152.

Modabbernia, A., Taslimi, S., Brietzke, E., & Ashrafi, M. (2013). Cytokine alterations in bipolar disorder: a meta-analysis of 30 studies. *Biological Psychiatry*, 74(1), 15–25.

Mohammad, S. S., & Dale, R. C. (2017). Principles and approaches to the treatment of immune-mediated movement disorders. *European Journal of Paediatric Neurology*, 22(2), 292–300.

Moldofsky, H., & Dickstein, J. B. (1999). Sleep and cytokine—immune functions in medical, psychiatric and primary sleep disorders. *Sleep Medicine Reviews*, 3(4), 325–337.

Mosing, M. A., Gordon, S. D., Medland, S. E., Statham, D. J., Nelson, E. C., Heath, A. C., et al. (2009). Genetic and environmental influences on the co-morbidity between depression, panic disorder, agoraphobia, and social phobia: a twin study. *Depression and Anxiety*, 26(11), 1004–1011.

Musiek, E. S., & Holtzman, D. M. (2016). Mechanisms linking circadian clocks, sleep, and neurodegeneration. *Science*, 354(6315), 1004–1008.

Naismith, S. L., Hermens, D. F., Ip, T. K., Bolitho, S., Scott, E., Rogers, N. L., et al. (2012). Circadian profiles in young people during the early stages of affective disorder. *Translational Psychiatry*, 2(5).

Narayan, V. A., & Manji, H. K. (2016). Moving from "diagnose and treat" to "predict and pre-empt" in neuropsychiatric disorders. *Nature Reviews. Drug Discovery*, 15 (2), 71–72.

O'Dea, B., Lee, R. S., McGorry, P. D., Hickie, I. B., Scott, J., Hermens, D. F., et al. (2016). A prospective cohort study of depression course, functional disability, and NEET status in help-seeking young adults. *Social Psychiatry and Psychiatric Epidemiology*, 51(10), 1395–1404.

Paus, T., Keshavan, M., & Giedd, J. N. (2008). Why do many psychiatric disorders emerge during adolescence? *Nature Reviews Neuroscience*, 9(12), 947–957.

Penninx, B. W., Milaneschi, Y., Lamers, F., & Vogelzangs, N. (2013). Understanding the somatic consequences of depression: biological mechanisms and the role of depression symptom profile. *BMC Medicine*, 11(129).

Perez-Vigil, A., Fernandez de la Cruz, L., Brander, G., Isomura, K., Gromark, C., & Mataix-Cols, D. (2016). The link between autoimmune diseases and obsessive-compulsive and tic disorders: a systematic review. *Neuroscience and Biobehavioral Reviews*, 71, 542–562.

Pfeiler, S., Winkels, H., Kelm, M., & Gerdes, N. (2017). IL-1 family cytokines in cardiovascular disease. *Cytokine*.

Pryce, C. R., & Fontana, A. (2017). Depression in autoimmune diseases. in R. Dantzer, & L. Capuron (Eds.), *Vol. 31. Inflammation-associated depression: Evidence, mechanisms and implications current topics in behavioral neurosciences* (pp. 139–154). Cham: Springer.

Ramirez, K., Fornaguera-Trias, J., & Sheridan, J. F. (2017). Stress-induced microglia activation and monocyte trafficking to the brain underlie the development of anxiety and depression. R. Dantzer, & L. Capuron (Eds.), *Vol. 31. Inflammation-associated depression: Evidence, mechanisms and implications current topics in behavioral neurosciences* (pp. 155–172). Cham: Springer.

Robillard, R., Hermens, D. F., Lee, R. S., Jones, A., Carpenter, J. S., White, D., et al. (2016). Sleep-wake profiles predict longitudinal changes in manic symptoms and memory in young people with mood disorders. *Journal of Sleep Research*, 25(5), 549–555.

Robillard, R., Naismith, S. L., Rogers, N. L., Ip, T. K., Hermens, D. F., Scott, E. M., et al. (2013). Delayed sleep phase in young people with unipolar or bipolar affective disorders. *Journal of Affective Disorders*, 145(2), 260–263.

Rosenblat, J. D., & McIntyre, R. S. (2017). Bipolar disorder and immune dysfunction: epidemiological findings, proposed pathophysiology and clinical implications. *Brain Sciences*, 7(11), 144.

Saltiel, A. R., & Olefsky, J. M. (2017). Inflammatory mechanisms linking obesity and metabolic disease. *The Journal of Clinical Investigation*, 127(1), 1–4.

Scott, J., Murray, G., Henry, C., Morken, G., Scott, E., Angst, J., et al. (2017). Activation in bipolar disorders: a systematic review. *JAMA Psychiatry*, 74(2), 189–196.

Soczynska, J. K., Kennedy, S. H., Woldeyohannes, H. O., Liauw, S. S., Alsuwaidan, M., Yim, C. Y., et al. (2011). Mood disorders and obesity: understanding inflammation as a pathophysiological nexus. *Neuromolecular Medicine, 13* (2), 93–116.

Staton, D. (2008). The impairment of pediatric bipolar sleep: hypotheses regarding a core defect and phenotype-specific sleep disturbances. *Journal of Affective Disorders, 108*(3), 199–206.

Straley, M. E., Van Oeffelen, W., Theze, S., Sullivan, A. M., O'Mahony, S. M., Cryan, J. F., et al. (2017). Distinct alterations in motor & reward seeking behavior are dependent on the gestational age of exposure to LPS-induced maternal immune activation. *Brain, Behavior, and Immunity, 63*, 21–34.

Stuart, M. J., & Baune, B. T. (2012). Depression and type 2 diabetes: inflammatory mechanisms of a psychoneuroendocrine co-morbidity. *Neuroscience and Biobehavioral Reviews, 36*(1), 658–676.

Suglia, S. F., Demmer, R. T., Wahi, R., Keyes, K. M., & Koenen, K. C. (2016). Depressive symptoms during adolescence and young adulthood and the development of type 2 diabetes mellitus. *American Journal of Epidemiology, 183*(4), 269–276.

Turek, F. W. (2016). Circadian clocks: not your grandfather's clock. *Science, 364*(3615), 992–993.

Valkanova, V., Ebmeier, K. P., & Allan, C. L. (2013). CRP, IL-6 and depression: a systematic review and meta-analysis of longitudinal studies. *Journal of Affective Disorders, 150*(3), 736–744.

Vollmer-Conna, U., Fazou, C., Cameron, B., Li, H., Brennan, C., Luck, L., et al. (2004). Production of pro-inflammatory cytokines correlates with the symptoms of acute sickness behaviour in humans. *Psychological Medicine, 34*(7), 1289–1297.

Walther, A., Penz, M., Ijacic, D., & Rice, T. R. (2017). Bipolar spectrum disorders in male youth: the interplay between symptom severity, inflammation, steroid secretion, and body composition. *Frontiers in Psychiatry, 8*, https://doi.org/10.3389/fpsyt.2017.00207.

Winning, A., Glymour, M. M., McCormick, M. C., Gilsanz, P., & Kubzansky, L. D. (2015). Psychological distress across the life course and cardiometabolic risk: findings from the 1958 British Birth Cohort Study. *Journal of the American College of Cardiology, 66*(14), 1577–1586.

Zadka, L., Dziegiel, P., Kulus, M., & Olajossy, M. (2017). Clinical phenotype of depression affects interleukin-6 synthesis. *Journal of Interferon & Cytokine Research, 37*(6), 231–245.

Zalli, A., Jovanova, O., Hoogendijk, W. J., Tiemeier, H., & Carvalho, L. A. (2016). Low-grade inflammation predicts persistence of depressive symptoms. *Psychopharmacology, 233*(9), 1669–1678.

Zammit, S., Allebeck, P., David, A. S., Dalman, C., Hemmingsson, T., Lundberg, I., et al. (2004). A longitudinal study of premorbid IQ score and risk of developing schizophrenia, bipolar disorder, severe depression, and other nonaffective psychoses. *Archives of General Psychiatry, 61*(4), 354–360.

Inflammation in Bipolar Disorder

Joshua D. Rosenblat*, Jonathan M. Gregory[†],
Sophie Flor-Henry*, Roger S. McIntyre*,[‡]

*University of Toronto, Toronto, ON, Canada
[†]Western University, London, ON, Canada
[‡]University Health Network, Toronto, ON, Canada

INTRODUCTION

Bipolar disorder (BD) is a severe and persistent mental illness associated with significant morbidity and mortality. Current treatments are often ineffective with high rates of relapse, treatment resistance, and persistence of symptoms during periods of euthymia. Additionally, pharmacological treatments are often poorly tolerated due to significant adverse effects. Historically, treatments for BD, such as lithium and valproic acid, were discovered somewhat serendipitously, rather than through knowledge of specific biological targets. The development of new treatments with improved clinical outcomes continues to be limited by the poor understanding of the etiology of BD. Decades of research have yielded numerous hypotheses regarding the potential biological causes of BD; however, the pathophysiology of BD remains largely unknown.

Increasingly, immune dysfunction has been implicated in the pathophysiology of BD (Rosenblat & McIntyre, 2016). The hypothesis that immune dysfunction may contribute to BD pathophysiology initially emerged through observations of the immunomodulating effects of lithium. Horrobin and Lieb (1981) hypothesized that immune modulation may be a key mechanism of action of lithium's mood-stabilizing effects. They further hypothesized that the relapsing-remitting nature of BD may be driven by the immune system, as seen in other relapsing-remitting inflammatory disorders, such as multiple sclerosis (MS) (Horrobin & Lieb, 1981). Epidemiological observations of high rates of inflammatory medical comorbidities in BD added further support for the potential link between immune dysfunction and BD (Rosenblat & McIntyre, 2015).

Accordingly, numerous studies have been conducted to understand the immune profile of patients with BD (Modabbernia, Taslimi, Brietzke, & Ashrafi, 2013). Preclinical and clinical investigations have also sought to understand potential mechanisms whereby immune changes may lead to mood episodes (Rosenblat, Cha, Mansur, & McIntyre, 2014). More recently, these findings have been translated into clinical trials evaluating the effects

Inflammation and Immunity in Depression
https://doi.org/10.1016/B978-0-12-811073-7.00025-8

of immunomodulating agents in the treatment of BD (Ayorech, Tracy, Baumeister, & Giaroli, 2015; Rosenblat et al., 2016). The objective of the current chapter is to summarize and synthesize studies examining the interactions between BD and immune dysfunction through exploring the following: (1) the association between BD and inflammatory medical comorbidities, (2) cytokine changes associated with BD, (3) proposed mechanisms linking BD with immune dysfunction, and (4) studies assessing the effects of immune modulating agents in the treatment of BD.

BD AND INFLAMMATORY MEDICAL COMORBIDITIES

Epidemiological studies have consistently shown increased rates of several medical and psychiatric comorbidities in BD compared with healthy controls (Kupfer, 2005). Notably, the increased prevalence of medical comorbidities, particularly cardiovascular disease, is primarily responsible for the 10- to 20-year decrease in life expectancy in BD compared with the rest of the population (Kessing, Vradi, McIntyre, & Andersen, 2015). Understanding the factors that mediate or moderate the bidirectional interaction between BD and medical comorbidities may potentially serve to improve both medical and psychiatric outcomes. As most of the medical disorders associated with BD are inflammatory in nature, immune dysfunction has been identified as one key factor that may mediate this complex interaction between BD and medical comorbidity (Goldstein, Kemp, Soczynska, & McIntyre, 2009).

Autoimmune disorders are the prototypical inflammatory conditions with immune dysfunction as the key etiologic factor driving the onset and progression of the disorder. In autoimmune diseases, the immune system falsely recognizes normal host tissue as pathogenic and launches an immune response to destroy this tissue. Locally, immune cells will attempt to induce cell death and clearance of the tissue (Abbas, Lichtman, & Pillai, 2012). This local response also increases inflammatory cytokine levels systemically leading to an increased level of inflammatory processes throughout the body, including in the brain (Dantzer, O'Connor, Freund, Johnson, & Kelley, 2008). While tissues elsewhere will not be destroyed to the same extent as the site of the local inflammatory response, the subtler effects of low-grade systemic inflammation on off-target areas (e.g., the brain) have been increasingly recognized as important.

For example, in psoriasis, while the psoriatic plaques are present on the skin due to a local inappropriate inflammatory response, the pro-inflammatory cytokines are circulated systemically through the lymphatic and circulatory system and may have significant off-target effects on the central nervous system (CNS), potentially leading to mood symptomatology (Walker, Graff, Dutz, & Bernstein, 2011). This conceptual understanding of the potential interplay between autoimmune disorders and BD has been further supported by epidemiological studies showing increased rates of inflammatory bowel disease, systemic lupus erythematosis, autoimmune thyroiditis, psoriasis, Guillain-Barré syndrome, autoimmune hepatitis, MS, and rheumatoid arthritis (RA) in patients with BD (Bachen, Chesney, & Criswell, 2009; Eaton, Pedersen, Nielsen, & Mortensen, 2010; Edwards & Constantinescu, 2004; Han, Lofland, Zhao, & Schenkel, 2011; Hsu et al., 2014; Kupka et al., 2002; Perugi et al., 2014).

While autoimmune disorders are the prototypical inflammatory conditions, many other medical comorbidities associated with BD have immune dysfunction as a key factor in their pathophysiology. Specifically, BD has been strongly associated with increased rates of cardiovascular disease, diabetes, and obesity (Calkin et al., 2009; McIntyre, Konarski, Misener, & Kennedy, 2005; Perugi et al., 2014). In cardiovascular disease, inflammations play a key role in the propagation of atherosclerotic plaques (Libby, Ridker, & Maseri, 2002). Type 2 diabetes mellitus has also been recognized as an inflammatory condition associated with chronic low-grade inflammation that is directly related to the progression of disease (Boutzios & Kaltsas, 2000). Central obesity has also been shown to be inflammatory in nature as excess adipose tissue produces adipokines and cytokines that chronically increase systemic levels of inflammation (Matsuzawa, 2006). The increased coprevalence of these disorders with BD provides further support of an association between BD and immune dysfunction.

These epidemiological studies provide evidence of an association between BD, immune dysfunction, and inflammatory disorders; however, causation has yet to be definitively established. The temporal relationship of BD and inflammatory conditions suggests that the interaction between BD, inflammation, and medical comorbidities is likely bidirectional (Perugi et al., 2014). If inflammation is a pathophysiological nexus between BD and medical comorbidity, targeting inflammation may potentially provide disease-modifying effects for both disease processes simultaneously (Goldstein et al., 2009).

CYTOKINE CHANGES ASSOCIATED WITH BD

Cytokines are signaling molecules of the immune system that may increase or decrease local and systemic inflammation. Measuring cytokine levels peripherally (i.e., serum levels) and centrally (i.e., cerebral spinal fluid levels) provides insight into immune system activity. Cytokine levels can identify current levels of inflammation and specifically identify which specific part of the immune system is over- or underactive leading to the observed immune dysfunction in BD. Moreover, as signaling molecules, specific cytokines may be directly implicated in the pathophysiology of BD and may therefore present as potential novel targets of treatment.

Cytokines have significant fluctuations and variability; however, some trends have emerged through over 40 cytokine studies of BD patients compared with healthy controls (Modabbernia et al., 2013; Munkholm, Vinberg, & Vedel Kessing, 2013). Based on these studies, BD appears to be associated with chronic low-grade inflammation whereby pro-inflammatory cytokines are in excess of anti-inflammatory cytokines. Serum levels of pro-inflammatory molecules interleukin-4 (IL-4), tumor necrosis factor-alpha (TNF-α), soluble IL-2 receptor, IL-1 beta (IL-1β), IL-6, soluble receptor of TNF-alpha type 1 (STNFR1), and C-reactive protein (CRP) are elevated in BD patients compared with healthy controls (Barbosa, Bauer, Machado-Vieira, & Teixeira, 2014; Brietzke, Kauer-Sant'Anna, Teixeira, & Kapczinski, 2009; Brietzke, Stertz, et al., 2009; Modabbernia et al., 2013). This cytokine profile indicates dysfunction of the *innate* immune system.

Another key observation has been variability in cytokine profiles depending on current mood state. Different cytokine profiles appear to be associated with different mood states (i.e., differing profiles during periods of depression, mania, hypomania, and euthymia). Replicated evidence has demonstrated that BD is associated with chronic low-grade inflammation, even during euthymic periods, with periods of increased inflammation that are sometimes associated with mood episodes (i.e., mania, hypomania, or depression) (Modabbernia et al., 2013; Munkholm et al., 2013). This variability in cytokine profiles may be suggestive of a differential involvement of immune dysfunction in depression versus mania versus euthymia. Significant heterogeneity in BD cytokine studies has been problematic, and as such, there has been no definitive inflammatory "fingerprint" that is reproducibly associated with each mood state. This significant heterogeneity also suggests that inflammation is likely a pertinent pathogenic factor for only a *subset* of BD; this subset of BD may potentially represent an "inflammatory BD" that may be pathophysiologically and phenomenologically different from other BD patients. This subtyping of BD is currently being investigated with important treatment implications. Within the context of this substantial heterogeneity, the following BD mood-dependent cytokine profiles have been identified.

The most robust evidence exists for an association between pro-inflammatory cytokines and depressive episodes, in both bipolar and unipolar depression. During depressive episodes, serum levels of CRP, TNF-α, IL-6, IL-1β, sTNFR1, and CXCL10 are elevated (Barbosa, Bauer, et al., 2014; Barbosa, Machado-Vieira, Soares, & Teixeira, 2014; Rosenblat et al., 2014). Increased depression severity is associated with greater elevations of pro-inflammatory cytokine (Siwek et al., 2016). During manic episodes, serum levels of IL-6, TNF-α, sTNFR1, IL-RA, CXCL10, CXCL11, and IL-4 have been shown to be elevated (Barbosa, Bauer, et al., 2014; Barbosa, Machado-Vieira, et al., 2014). During euthymic periods, sTNFR1 is the only consistently elevated inflammatory marker (Barbosa, Bauer, et al., 2014; Brietzke, Kauer-Sant'Anna, et al., 2009). One significant limitation of these cytokine studies is their cross-sectional nature (i.e., serum levels are usually only taken at one point in time). Longitudinal studies are needed to measure cytokine levels within the same group of BD subjects to determine how they alter during mood episodes. Understanding this chronological relationship (e.g., if cytokines are elevated prior to vs after mood episode onset) would also provide further insight into the cross talk between BD and immune dysfunction.

PROPOSED MECHANISMS: THE INFLAMMATORY-MOOD PATHWAY

Several biologically plausible mechanisms have been identified to understand the bidirectional interaction between immune dysfunction and BD. The immune system directly interacts with the CNS, as peripheral cytokines may traverse the blood-brain barrier (BBB) leading to downstream effects in the brain. Systemically, circulating cytokines may traverse the BBB via active transport channels and through leaky regions of the BBB. Recent findings in animal models have also suggested the presence of lymphatic vessels in the brain that could provide another direct entrance to the CNS for cytokines and other signaling molecules (Louveau et al., 2015). Cytokines may then signal several downstream effects that alter the structure and function of key brain regions subserving mood and cognitive function. Cytokines can directly alter monoamine levels, cause overactivation of microglial cells, and lead to increased oxidative stress in the brain (McNamara & Lotrich, 2012). The net effect of these changes is neurodegeneration and decreased neuroplasticity in key brain regions that may lead to the phenotypic changes observed in BD. These pathways will each be briefly discussed in turn.

Pro-inflammatory cytokines TNF-α, IL-2, and IL-6 have been shown to directly alter monoamine levels (Capuron et al., 2003). IL-2 and interferon (IFN) increase the enzymatic activity of indoleamine 2,3-dioxygenase, thereby increasing the breakdown of tryptophan to depressogenic tryptophan catabolites (TRYCATs). Serotonin (5-HT) levels may be further modulated through the IL-6- and TNF-α-dependent breakdown of 5-HT to 5-hydroxyindoleacetic acid (Wang & Dunn, 1998). Depletion of tryptophan and decreased levels of 5-HT can directly impair affective and cognitive function (Arango, Underwood, & Mann, 2002).

Inflammation in BD has also been associated with overactivation of microglia, the macrophages of the CNS (Stertz, Magalhaes, & Kapczinski, 2013). Microglia, under physiological conditions, perform an important role in neuroplasticity, facilitating neural network pruning via inducing apoptosis of unneeded neurons and underutilized neural pathways. Pruning of these pathways is essential to allow for the growth and maintenance of more important neural pathways (Harry & Kraft, 2012). With chronic low-grade inflammation, as is often seen in BD, microglia may be overactive, aberrantly destroying important neural pathways. Microglia may be overactivated in key brain regions subserving mood and cognition (e.g., the prefrontal cortex, amygdala, hippocampus, insula, and anterior cingulate cortex) (Frick, Williams, & Pittenger, 2013; Haarman et al., 2014). This process results in a positive feedforward loop whereby activated microglia release cytokines, which further increases inflammation and further microglia recruitment and activation. The release of cytokines from activated microglia may also further perpetuate the previously discussed monoamine changes. Lastly, the overactivation of microglia increases the production of reactive oxygen species leading to local oxidative stress, further damaging neural circuitry in key brain regions subserving mood and cognition (Stertz et al., 2013).

Another key mechanism by which inflammation perpetuates mood dysfunction in BD is hypothalamic-pituitary-adrenal (HPA) axis dysregulation. Increased levels of pro-inflammatory cytokines IFN, TNF-α, and IL-6 significantly upregulate HPA activity, thereby increasing systemic cortisol levels (Beishuizen & Thijs, 2003). Increased cortisol levels can potently alter mood and induce either mania or depression, a well-documented phenomenon with increased levels of endogenous and/or exogenous corticosteroids (Murphy, 1991). Elevated levels of cortisol also increase the activity of hepatic tryptophan 2,3-dioxygenase activity, thereby increasing the breakdown of tryptophan to TRYCATs (Maes, Leonard, Myint, Kubera, & Verkerk, 2011). Elevated levels of inflammatory cytokines also decrease glucocorticoid receptor synthesis, transport, and sensitivity in the hypothalamus and pituitary (Pace & Miller, 2009). Therefore, the negative feedback loop, which usually downregulates cortisol production, is disabled, thus leading to chronic cortisol elevation. Further, impaired cortisol suppression itself has long been recognized a strong predictor of mood disorders (Cowen, 2010).

TREATMENT IMPLICATIONS

Current psychopharmacological treatments of BD are associated with high rates of treatment resistance, relapse, and poor tolerability. Given the replicated and convergent evidence implicating inflammation in the pathophysiology of BD, the immune system presents as a novel treatment target, providing hope for improved outcomes and tolerability. Several studies have sought to determine whether adjunctive anti-inflammatory agents may have antidepressant effects in BD (Ayorech et al., 2015). Indeed, several randomized controlled trials (RCTs) have been conducted to assess the antidepressant effects of N-acetylcysteine (NAC), nonsteroidal anti-inflammatory drugs (NSAIDs), omega-3 polyunsaturated fatty acids (omega-3s), pioglitazone, and minocycline in BD (Rosenblat et al., 2016). Several studies have also assessed the antimanic effects of adjunctive anti-inflammatory agents, specifically NSAIDs and NAC. Notably, lithium, one of the oldest and most effective treatments for BD, has potent immunomodulating effects (Horrobin & Lieb, 1981).

In a recent systematic review and meta-analysis conducted by our group, we identified 10 RCTs assessing the acute antidepressant effects of adjunctive anti-inflammatory agents in the treatment of bipolar depression (Rosenblat et al., 2016). Eight RCTs ($n = 312$) assessing adjunctive NSAIDs ($n = 53$), omega-3s ($n = 140$), NAC ($n = 76$), and pioglitazone ($n = 44$) were included in the quantitative analysis. The overall pooled effect size of adjunctive anti-inflammatories on depressive symptoms was -0.40 (95% confidence interval from -0.14 to -0.65 ($P = 0.002$)), indicative of a moderate and statistically significant antidepressant effect compared with conventional therapy alone. No manic/hypomanic induction or significant treatment-emergent adverse events were reported. The small number of studies, diversity of agents, and small sample sizes were significant limitations of this analysis, and at this point, it would be premature to recommend routine clinical use of anti-inflammatory agents in this population. Nevertheless, this provides a proof of concept that targeting inflammation may improve potentially outcomes in the treatment of bipolar depression.

Among all anti-inflammatory agents, NAC has the strongest evidence in support of both efficacy and tolerability as an adjunctive treatment of BD (Berk et al., 2008, 2012). In an RCT with 75 BD subjects, adjunctive NAC was shown to lower depression severity scores throughout the trial with a statistically and clinically significant difference compared with conventional therapy alone at the primary end point of 24 weeks (Berk et al., 2008). Additionally, post hoc analysis of 17 participants from this sample who met the criteria for a current MDE at baseline revealed that 8 out of 10 participants in the NAC group had a clinical response (i.e., >50% reduction in depression severity) compared with only 1 out of 7 participants in the placebo group (Magalhaes et al., 2011). An 8-week open-label trial of NAC also showed a significant reduction in depressive symptoms in BD subjects (Berk et al., 2011). The effect of adjunctive NAC in mania/hypomania was also explored in one study; a small post hoc analysis of 15 BD subjects experiencing an acute manic/hypomanic episode compared subjects receiving adjunctive NAC ($n=8$) with subjects receiving adjunctive placebo ($n=7$). This analysis revealed a greater improvement in symptoms of mania in the NAC group compared with placebo (Magalhaes et al., 2013). Overall, NAC shows promise as an adjunctive treatment for BD during all phases of illness; however, evidence is strongest for use during depressive episodes, and further studies are still required.

Several RCTs have also been conducted evaluating the effects of adjunctive omega-3s, a naturally occurring and well-tolerated anti-inflammatory agent (Bloch & Hannestad, 2012). Results have been mixed with some trials showing an antidepressant effect in BD

(Frangou, Lewis, & McCrone, 2006; Stoll et al., 1999) and others reporting no antidepressant effect compared with conventional therapy alone (Frangou, Lewis, Wollard, & Simmons, 2007; Hirashima et al., 2004; Keck et al., 2006). When pooling these results together in a meta-analysis, a moderate and statistically significant antidepressant effect of adjunctive omega-3s in BD was found compared with conventional therapy alone (Bloch & Hannestad, 2012).

The antidepressant effect of adjunctive NSAIDs has also been evaluated in BD. Nery et al. (2008) assessed adjunctive celecoxib in BD subjects ($n=28$) during an acute depressive or mixed episode. Adjunctive celecoxib lowered depression severity by week 1; however, the primary outcome was negative as change in depression severity converged with the placebo group by the end of week 6. Saroukhani et al. (2013) assessed the effect of adjunctive aspirin in male BD subjects ($n=32$) and found no significant difference between treatment groups by the end of the 6-week RCT.

Three studies have also evaluated the effect of NSAIDs during acute manic/hypomanic episodes. In a small, proof-of-concept RCT, Arabzadeh et al. (2015) compared adjunctive celecoxib with treatment at usual for acute mania in BD inpatients ($n=46$). They found that a significantly higher remission rate was observed in the celecoxib group (87.0%) compared with the placebo group (43.5%) by the week 6 primary end point ($P=0.005$). The same investigators also evaluated adjunctive celecoxib in a separate RCT of adolescent inpatients ($n=42$) during an acute manic episode (Mousavi et al., 2017). There was no significant difference in remission rates by the primary end point of 8 weeks; however, significantly greater improvement was observed in Young Mania Rating Scale (YMRS) scores in the celecoxib group compared with the placebo group from baseline YMRS score at week 8 ($P=0.04$). In another RCT including BD inpatients ($n=35$) with mania receiving electroconvulsive therapy (ECT), participants received either celecoxib or placebo from 1 day before the first ECT session throughout the sixth session. Brain-derived neurotrophic factor (BDNF) levels were also measured before and during the trial. Adding celecoxib was not associated with a significant rise in BDNF levels following ECT. No difference was noted between groups in terms of treatment response (Kargar et al., 2015).

TNF-α inhibitors have also been of interest. One key RCT assessed infliximab in treatment-resistant depression, including both bipolar and unipolar depressed subjects in their sample. Although the overall antidepressant effect was negative for this study, a significant antidepressant effect was observed for a subgroup of subjects, namely, those with elevated levels of serum CRP and TNF-α (Raison et al., 2013). The results of this trial suggested that stratification using inflammatory biomarkers might help determine which patients may benefit from anti-inflammatory treatments. A 12-week RCT evaluating the efficacy, safety, and tolerability of adjunctive infliximab for the treatment of BD subjects with an elevated serum CRP is currently underway (NCT02363738).

Taken together, several anti-inflammatory agents have shown promising results as adjunctive treatments for BD, adding merit to the concept of targeting the immune system in BD treatment. There is greater evidence for adjunctive anti-inflammatories for bipolar depression, while evidence is more limited and mixed for mania and hypomania. Additional research is needed to further establish safety and efficacy profiles to determine which specific agents may be clinically useful.

CONCLUSION

Inflammation is likely a key etiologic factor of BD, especially for a subset of patients. For this subset of inflammatory BD patients, targeting inflammation may be a more effective strategy compared with current treatments because this approach targets the underlying cause of the disorder, rather than only targeting downstream effects (such as monoamine changes). Several recent proof-of-concept clinical trials have assessed the antidepressant and antimanic effects of anti-inflammatory agents showing very promising results. Further research is still needed to determine the role of immune dysfunction in BD and to translate these findings into novel treatments and improved clinical outcomes.

References

Abbas, A. K., Lichtman, A. H., & Pillai, S. (2012). *Cellular and molecular immunology*. Vol. 7. Philadelphia, PA: Elsevier Saunders.

Arabzadeh, S., Ameli, N., Zeinoddini, A., Rezaei, F., Farokhnia, M., Mohammadinejad, P., et al. (2015). Celecoxib adjunctive therapy for acute bipolar mania: a randomized, double-blind, placebo-controlled trial. *Bipolar Disorders*, 17(6), 606–614. http://dx.doi.org/10.1111/bdi.12324.

Arango, V., Underwood, M. D., & Mann, J. J. (2002). Serotonin brain circuits involved in major depression and suicide. *Progress in Brain Research*, 136, 443–453.

Ayorech, Z., Tracy, D. K., Baumeister, D., & Giaroli, G. (2015). Taking the fuel out of the fire: evidence for the use of anti-inflammatory agents in the treatment of bipolar disorders. *Journal of Affective Disorders*, 174, 467–478. http://dx.doi.org/10.1016/j.jad.2014.12.015.

Bachen, E. A., Chesney, M. A., & Criswell, L. A. (2009). Prevalence of mood and anxiety disorders in women with systemic lupus erythematosus. *Arthritis and Rheumatism*, 61(6), 822–829. http://dx.doi.org/10.1002/art.24519.

Barbosa, I. G., Bauer, M. E., Machado-Vieira, R., & Teixeira, A. L. (2014). Cytokines in bipolar disorder: paving the way for neuroprogression. *Neural Plasticity*, 2014, 360481. http://dx.doi.org/10.1155/2014/360481.

Barbosa, I. G., Machado-Vieira, R., Soares, J. C., & Teixeira, A. L. (2014). The immunology of bipolar disorder. *Neuroimmunomodulation*, 21(2–3), 117–122. http://dx.doi.org/10.1159/000356539.

Beishuizen, A., & Thijs, L. G. (2003). Endotoxin and the hypothalamo-pituitary-adrenal (HPA) axis. *Journal of Endotoxin Research*, 9(1), 3–24. http://dx.doi.org/10.1179/096805103125001298.

Berk, M., Copolov, D. L., Dean, O., Lu, K., Jeavons, S., Schapkaitz, I., et al. (2008). N-acetyl cysteine for depressive symptoms in bipolar disorder—a double-blind randomized placebo-controlled trial. *Biological Psychiatry*, *64*(6), 468–475. http://dx.doi.org/10.1016/j.biopsych.2008.04.022.

Berk, M., Dean, O., Cotton, S. M., Gama, C. S., Kapczinski, F., Fernandes, B. S., et al. (2011). The efficacy of N-acetylcysteine as an adjunctive treatment in bipolar depression: an open label trial. *Journal of Affective Disorders*, *135*(1–3), 389–394. http://dx.doi.org/10.1016/j.jad.2011.06.005.

Berk, M., Dean, O. M., Cotton, S. M., Gama, C. S., Kapczinski, F., Fernandes, B., et al. (2012). Maintenance N-acetyl cysteine treatment for bipolar disorder: a double-blind randomized placebo controlled trial. *BMC Medicine*, *10*, 91. http://dx.doi.org/10.1186/1741-7015-10-91.

Bloch, M. H., & Hannestad, J. (2012). Omega-3 fatty acids for the treatment of depression: systematic review and meta-analysis. *Molecular Psychiatry*, *17*(12), 1272–1282. http://dx.doi.org/10.1038/mp.2011.100.

Boutzios, G., & Kaltsas, G. (2000). Immune system effects on the endocrine system. In L. J. De Groot, G. Chrousos, K. Dungan, et al. (Eds.), *Endotext*. South Dartmouth (MA): MDText.com, Inc. https://www.ncbi.nlm.nih.gov/books/NBK279139/

Brietzke, E., Kauer-Sant'Anna, M., Teixeira, A. L., & Kapczinski, F. (2009). Abnormalities in serum chemokine levels in euthymic patients with bipolar disorder. *Brain, Behavior, and Immunity*, *23*(8), 1079–1082. http://dx.doi.org/10.1016/j.bbi.2009.04.008.

Brietzke, E., Stertz, L., Fernandes, B. S., Kauer-Sant'anna, M., Mascarenhas, M., Escosteguy Vargas, A., et al. (2009). Comparison of cytokine levels in depressed, manic and euthymic patients with bipolar disorder. *Journal of Affective Disorders*, *116*(3), 214–217. http://dx.doi.org/10.1016/j.jad.2008.12.001.

Calkin, C., van de Velde, C., Ruzickova, M., Slaney, C., Garnham, J., Hajek, T., et al. (2009). Can body mass index help predict outcome in patients with bipolar disorder? *Bipolar Disorders*, *11*(6), 650–656. http://dx.doi.org/10.1111/j.1399-5618.2009.00730.x.

Capuron, L., Neurauter, G., Musselman, D. L., Lawson, D. H., Nemeroff, C. B., Fuchs, D., et al. (2003). Interferon-alpha-induced changes in tryptophan metabolism: relationship to depression and paroxetine treatment. *Biological Psychiatry*, *54*(9), 906–914.

Cowen, P. J. (2010). Not fade away: the HPA axis and depression. *Psychological Medicine*, *40*(1), 1–4. http://dx.doi.org/10.1017/S0033291709005558.

Dantzer, R., O'Connor, J. C., Freund, G. G., Johnson, R. W., & Kelley, K. W. (2008). From inflammation to sickness and depression: when the immune system subjugates the brain. *Nature Reviews. Neuroscience*, *9*(1), 46–56. http://dx.doi.org/10.1038/nrn2297.

Eaton, W. W., Pedersen, M. G., Nielsen, P. R., & Mortensen, P. B. (2010). Autoimmune diseases, bipolar disorder, and non-affective psychosis. *Bipolar Disorders*, *12*(6), 638–646. http://dx.doi.org/10.1111/j.1399-5618.2010.00853.x.

Edwards, L. J., & Constantinescu, C. S. (2004). A prospective study of conditions associated with multiple sclerosis in a cohort of 658 consecutive outpatients attending a multiple sclerosis clinic. *Multiple Sclerosis*, *10*(5), 575–581.

Frangou, S., Lewis, M., & McCrone, P. (2006). Efficacy of ethyl-eicosapentaenoic acid in bipolar depression: randomised double-blind placebo-controlled study. *British Journal of Psychiatry*, *188*, 46–50. http://dx.doi.org/10.1192/bjp.188.1.46.

Frangou, S., Lewis, M., Wollard, J., & Simmons, A. (2007). Preliminary in vivo evidence of increased N-acetyl-aspartate following eicosapentanoic acid treatment in patients with bipolar disorder. *Journal of Psychopharmacology*, *21*(4), 435–439. http://dx.doi.org/10.1177/0269881106067787.

Frick, L. R., Williams, K., & Pittenger, C. (2013). Microglial dysregulation in psychiatric disease. *Clinical & Developmental Immunology*, *2013*, 608654. http://dx.doi.org/10.1155/2013/608654.

Goldstein, B. I., Kemp, D. E., Soczynska, J. K., & McIntyre, R. S. (2009). Inflammation and the phenomenology, pathophysiology, comorbidity, and treatment of bipolar disorder: a systematic review of the literature. *Journal of Clinical Psychiatry*, *70*(8), 1078–1090. http://dx.doi.org/10.4088/JCP.08r04505.

Haarman, B. C., Riemersma-Van der Lek, R. F., de Groot, J. C., Ruhe, H. G., Klein, H. C., Zandstra, T. E., et al. (2014). Neuroinflammation in bipolar disorder—a [(11)C]-(R)-PK11195 positron emission tomography study. *Brain, Behavior, and Immunity*, *40*, 219–225. http://dx.doi.org/10.1016/j.bbi.2014.03.016.

Han, C., Lofland, J. H., Zhao, N., & Schenkel, B. (2011). Increased prevalence of psychiatric disorders and health care-associated costs among patients with moderate-to-severe psoriasis. *Journal of Drugs in Dermatology*, *10*(8), 843–850.

Harry, G. J., & Kraft, A. D. (2012). Microglia in the developing brain: a potential target with lifetime effects. *Neurotoxicology*, *33*(2), 191–206. http://dx.doi.org/10.1016/j.neuro.2012.01.012.

Hirashima, F., Parow, A. M., Stoll, A. L., Demopulos, C. M., Damico, K. E., Rohan, M. L., et al. (2004). Omega-3 fatty acid treatment and T(2) whole brain relaxation times in bipolar disorder. *American Journal of Psychiatry*, *161*(10), 1922–1924. http://dx.doi.org/10.1176/appi.ajp.161.10.1922.

Horrobin, D. F., & Lieb, J. (1981). A biochemical basis for the actions of lithium on behaviour and on immunity: relapsing and remitting disorders of inflammation and immunity such as multiple sclerosis or recurrent herpes as manic-depression of the immune system. *Medical Hypotheses, 7*(7), 891–905.

Hsu, C. C., Chen, S. C., Liu, C. J., Lu, T., Shen, C. C., Hu, Y. W., et al. (2014). Rheumatoid arthritis and the risk of bipolar disorder: a nationwide population-based study. *PLoS One, 9*(9). http://dx.doi.org/10.1371/journal.pone.0107512.

Kargar, M., Yoosefi, A., Akhondzadeh, S., Artonian, V., Ashouri, A., & Ghaeli, P. (2015). Effect of adjunctive celecoxib on BDNF in manic patients undergoing electroconvulsive therapy: a randomized double blind controlled trial. *Pharmacopsychiatry, 48*(7), 268–273. http://dx.doi.org/10.1055/s-0035-1559667.

Keck, P. E., Jr., Mintz, J., McElroy, S. L., Freeman, M. P., Suppes, T., Frye, M. A., et al. (2006). Double-blind, randomized, placebo-controlled trials of ethyl-eicosapentanoate in the treatment of bipolar depression and rapid cycling bipolar disorder. *Biological Psychiatry, 60*(9), 1020–1022. http://dx.doi.org/10.1016/j.biopsych.2006.03.056.

Kessing, L. V., Vradi, E., McIntyre, R. S., & Andersen, P. K. (2015). Causes of decreased life expectancy over the life span in bipolar disorder. *Journal of Affective Disorders, 180*, 142–147. http://dx.doi.org/10.1016/j.jad.2015.03.027.

Kupfer, D. J. (2005). The increasing medical burden in bipolar disorder. *JAMA, 293*(20), 2528–2530. http://dx.doi.org/10.1001/jama.293.20.2528.

Kupka, R. W., Nolen, W. A., Post, R. M., McElroy, S. L., Altshuler, L. L., Denicoff, K. D., et al. (2002). High rate of autoimmune thyroiditis in bipolar disorder: lack of association with lithium exposure. *Biological Psychiatry, 51*(4), 305–311.

Libby, P., Ridker, P. M., & Maseri, A. (2002). Inflammation and atherosclerosis. *Circulation, 105*(9), 1135–1143.

Louveau, A., Smirnov, I., Keyes, T. J., Eccles, J. D., Rouhani, S. J., Peske, J. D., et al. (2015). Structural and functional features of central nervous system lymphatic vessels. *Nature, 523*(7560), 337–341. http://dx.doi.org/10.1038/nature14432.

Maes, M., Leonard, B. E., Myint, A. M., Kubera, M., & Verkerk, R. (2011). The new '5-HT' hypothesis of depression: cell-mediated immune activation induces indoleamine 2,3-dioxygenase, which leads to lower plasma tryptophan and an increased synthesis of detrimental tryptophan catabolites (TRYCATs), both of which contribute to the onset of depression. *Progress in Neuro-Psychopharmacology & Biological Psychiatry, 35*(3), 702–721. http://dx.doi.org/10.1016/j.pnpbp.2010.12.017.

Magalhaes, P. V., Dean, O. M., Bush, A. I., Copolov, D. L., Malhi, G. S., Kohlmann, K., et al. (2011). N-acetylcysteine for major depressive episodes in bipolar disorder. *Revista Brasileira de Psiquiatria, 33*(4), 374–378.

Magalhaes, P. V., Dean, O. M., Bush, A. I., Copolov, D. L., Malhi, G. S., Kohlmann, K., et al. (2013). A preliminary investigation on the efficacy of N-acetyl cysteine for

mania or hypomania. *Australian and New Zealand Journal of Psychiatry, 47*(6), 564–568. http://dx.doi.org/10.1177/0004867413481631.

Matsuzawa, Y. (2006). The metabolic syndrome and adipocytokines. *FEBS Letters, 580*(12), 2917–2921. http://dx.doi.org/10.1016/j.febslet.2006.04.028.

McIntyre, R. S., Konarski, J. Z., Misener, V. L., & Kennedy, S. H. (2005). Bipolar disorder and diabetes mellitus: epidemiology, etiology, and treatment implications. *Annals of Clinical Psychiatry, 17*(2), 83–93.

McNamara, R. K., & Lotrich, F. E. (2012). Elevated immune-inflammatory signaling in mood disorders: a new therapeutic target? *Expert Review of Neurotherapeutics, 12*(9), 1143–1161. http://dx.doi.org/10.1586/ern.12.98.

Modabbernia, A., Taslimi, S., Brietzke, E., & Ashrafi, M. (2013). Cytokine alterations in bipolar disorder: a meta-analysis of 30 studies. *Biological Psychiatry, 74*(1), 15–25. http://dx.doi.org/10.1016/j.biopsych.2013.01.007.

Mousavi, S. Y., Khezri, R., Karkhaneh-Yousefi, M. A., Mohammadinejad, P., Gholamian, F., Mohammadi, M. R., et al. (2017). A randomized, double-blind placebo-controlled trial on effectiveness and safety of celecoxib adjunctive therapy in adolescents with acute bipolar mania. *Journal of Child and Adolescent Psychopharmacology*. http://dx.doi.org/10.1089/cap.2016.0207.

Munkholm, K., Vinberg, M., & Vedel Kessing, L. (2013). Cytokines in bipolar disorder: a systematic review and meta-analysis. *Journal of Affective Disorders, 144*(1-2), 16–27. http://dx.doi.org/10.1016/j.jad.2012.06.010.

Murphy, B. E. (1991). Steroids and depression. *Journal of Steroid Biochemistry and Molecular Biology, 38*(5), 537–559.

Nery, F. G., Monkul, E. S., Hatch, J. P., Fonseca, M., Zunta-Soares, G. B., Frey, B. N., et al. (2008). Celecoxib as an adjunct in the treatment of depressive or mixed episodes of bipolar disorder: a double-blind, randomized, placebo-controlled study. *Human Psychopharmacology, 23*(2), 87–94. http://dx.doi.org/10.1002/hup.912.

Pace, T. W., & Miller, A. H. (2009). Cytokines and glucocorticoid receptor signaling. Relevance to major depression. *Annals of the New York Academy of Sciences, 1179*, 86–105. http://dx.doi.org/10.1111/j.1749-6632.2009.04984.x.

Perugi, G., Quaranta, G., Belletti, S., Casalini, F., Mosti, N., Toni, C., et al. (2014). General medical conditions in 347 bipolar disorder patients: clinical correlates of metabolic and autoimmune-allergic diseases. *Journal of Affective Disorders, 170C*, 95–103. http://dx.doi.org/10.1016/j.jad.2014.08.052.

Raison, C. L., Rutherford, R. E., Woolwine, B. J., Shuo, C., Schettler, P., Drake, D. F., et al. (2013). A randomized controlled trial of the tumor necrosis factor antagonist infliximab for treatment-resistant depression: the role of baseline inflammatory biomarkers. *JAMA Psychiatry, 70*(1), 31–41. http://dx.doi.org/10.1001/2013.jamapsychiatry.4.

Rosenblat, J. D., Cha, D. S., Mansur, R. B., & McIntyre, R. S. (2014). Inflamed moods: a review of the interactions between inflammation and mood disorders. *Progress in Neuro-Psychopharmacology & Biological Psychiatry*, *53*, 23–34. http://dx.doi.org/10.1016/j.pnpbp.2014.01.013.

Rosenblat, J. D., Kakar, R., Berk, M., Kessing, L. V., Vinberg, M., Baune, B. T., et al. (2016). Anti-inflammatory agents in the treatment of bipolar depression: a systematic review and meta-analysis. *Bipolar Disorders*, *18*(2), 89–101. http://dx.doi.org/10.1111/bdi.12373.

Rosenblat, J. D., & McIntyre, R. S. (2015). Are medical comorbid conditions of bipolar disorder due to immune dysfunction? *Acta Psychiatrica Scandinavica*, http://dx.doi.org/10.1111/acps.12414.

Rosenblat, J. D., & McIntyre, R. S. (2016). Bipolar disorder and inflammation. *Psychiatric Clinics of North America*, *39*(1), 125–137. http://dx.doi.org/10.1016/j.psc.2015.09.006.

Saroukhani, S., Emami-Parsa, M., Modabbernia, A., Ashrafi, M., Farokhnia, M., Hajiaghaee, R., et al. (2013). Aspirin for treatment of lithium-associated sexual dysfunction in men: randomized double-blind placebo-controlled study. *Bipolar Disorders*, *15*(6), 650–656. http://dx.doi.org/10.1111/bdi.12108.

Siwek, M., Sowa-Kucma, M., Styczen, K., Misztak, P., Nowak, R. J., Szewczyk, B., et al. (2016). Associations of serum cytokine receptor levels with melancholia, staging of illness, depressive and manic phases, and severity of depression in bipolar disorder. *Molecular Neurobiology*, http://dx.doi.org/10.1007/s12035-016-0124-8.

Stertz, L., Magalhaes, P. V., & Kapczinski, F. (2013). Is bipolar disorder an inflammatory condition? The relevance of microglial activation. *Current Opinion in Psychiatry*, *26*(1), 19–26. http://dx.doi.org/10.1097/YCO.0b013e32835aa4b4.

Stoll, A. L., Severus, W. E., Freeman, M. P., Rueter, S., Zboyan, H. A., Diamond, E., et al. (1999). Omega 3 fatty acids in bipolar disorder: a preliminary double-blind, placebo-controlled trial. *Archives of General Psychiatry*, *56*(5), 407–412.

Walker, J. R., Graff, L. A., Dutz, J. P., & Bernstein, C. N. (2011). Psychiatric disorders in patients with immune-mediated inflammatory diseases: prevalence, association with disease activity, and overall patient well-being. *Journal of Rheumatology. Supplement*, *88*, 31–35. http://dx.doi.org/10.3899/jrheum.110900.

Wang, J., & Dunn, A. J. (1998). Mouse interleukin-6 stimulates the HPA axis and increases brain tryptophan and serotonin metabolism. *Neurochemistry International*, *33*(2), 143–154.

26

Depression Subtypes and Inflammation: Atypical Rather Than Melancholic Depression Is Linked With Immunometabolic Dysregulations

Femke Lamers, Yuri Milaneschi, Brenda W.J.H. Penninx

VU University Medical Center & GGZ inGeest, Amsterdam Public Health Research Institute, Amsterdam, The Netherlands

INFLAMMATION AND DEPRESSION

The link between inflammation and major depressive disorder (MDD) has been widely established. Since the first papers on sickness behavior and the introduction of the cytokine hypothesis, postulating a causal role for pro-inflammatory mediators in depression pathophysiology, the number of studies evaluating the association of MDD and inflammation has rapidly increased (Dantzer & Kelley, 2007). Several meta-analyses describing this body of literature have been published since (Dowlati et al., 2010; Hiles, Baker, de Malmanche, & Attia, 2012a; Howren, Lamkin, & Suls, 2009; Köhler et al., 2017; Liu, Ho, & Mak, 2012), reporting on significantly higher levels of C-reactive protein (CRP), and pro-inflammatory and anti-inflammatory cytokines, including interleukin (IL)-6, IL-1, IL-10, IL-12, IL-13, tumor necrosis factor-alpha (TNF-

α), soluble IL-2 receptor, C-C chemokine ligand 1, IL-18, IL-1 receptor antagonist (IL1-RA), and soluble TNF receptor 2, and significantly lower levels of interferon gamma, in depressed persons versus controls. The link between inflammation and depression has also been confirmed in gene-expression studies (Jansen et al., 2016; Mostafavi et al., 2014). In newer hypotheses on the origin of depression, such as the Pathogen Host Defense hypothesis (Raison & Miller, 2013), inflammation and associated sickness behavior also play a prominent role. In this hypothesis, depression is viewed as an evolutionary adaptation, and an integral part of immune-mediated host defense against pathogens in the ancestral environment. The hypothesis poses that depression enhances chances of survival in the context of acute infections, and genes for depression are predicted to be the same genes that are associated with successful host immune responses.

455

Besides cross-sectional associations, longitudinal associations also exist. In a meta-analyses it was found that high baseline CRP levels had a small effect on depressive symptoms at follow-up, while no effects were found of baseline IL-6 (Valkanova, Ebmeier, & Allan, 2013). More recent studies have also found a longitudinal association between CRP and later depressive symptoms (Au, Smith, Gariépy, & Schmitz, 2015; Gallagher, Kiss, Lanctot, & Herrmann, 2016; Jones et al., 2015; Matsushima et al., 2015; Tully et al., 2015; Zalli, Jovanova, Hoogendijk, Tiemeier, & Carvalho, 2015), although one study did not find this association (Eurelings, Richard, Eikelenboom, van Gool, & Moll van Charante, 2015). Two studies found IL-6 to predict subsequent depressive symptoms (Khandaker, Pearson, Zammit, Lewis, & Jones, 2014; Zalli et al., 2015) and depression diagnoses (Khandaker et al., 2014), and another study found only a marginal effect of IL-6 (Hughes et al., 2014). Limitations of these longitudinal studies are that they have mostly focused on depressive symptoms and not on clinical diagnoses of depression. Also, most studies were conducted in elderly samples, which makes it unclear if associations hold in younger adults who experience less somatic comorbidity than elderly do. Of the few longitudinal studies using clinical diagnoses of depression as outcome, two were conducted in children. These studies found mixed results: CRP predicted onset in one (Pasco et al., 2010) out of three studies (Copeland, Shanahan, Worthman, Angold, & Costello, 2012; Khandaker et al., 2014), and the only study evaluating IL-6 found it to be associated with onset of depression (Khandaker et al., 2014).

HETEROGENEITY OF DEPRESSION

While the links between MDD and inflammation thus seem firm, they may not exist in all cases of depression. The diagnosis of MDD is highly heterogeneous, and different subtypes are being distinguished in the literature. It is increasingly recognized that the heterogeneity of depression is hindering the identification of biomarkers of depression, as different subtypes of depression may be the result of different pathophysiological mechanisms (Antonijevic, 2006). Studying more homogenous groups of MDD patients therefore may help us to unravel pathophysiology that is specific to certain subtypes of MDD.

Classifications of subtypes of depression can be made from different starting points. In Baumeister and Parker's (2012) review of depression subtyping systems, the three most important starting points to make subclassifications are time of onset-based, etiology-based, and symptom-based subtypes. Time of onset-based types includes not only early and late onset types but also seasonal affective disorder; etiology-based subtypes include drug-induced depression, reproductive depression, and perinatal depression. Symptom-based subtypes include psychotic, melancholic, atypical, and anxious depression. An important aspect to note is that these starting points are not mutually exclusive; subtypes from different starting points may overlap. For this chapter, we will focus on symptom-based subtypes.

The DSM-5 diagnostic criteria for MDD comprise nine symptom criteria, with 227 combinations of symptoms that can result in an MDD diagnosis (American Psychiatric Association, 2013). But within those criteria, more than nine symptoms can be distinguished as there are two directions (increase and decrease) to changes in (1) sleep, (2) appetite/weight, and (3) psychomotor activity, resulting in 945 possible symptom combinations. Of the main four symptom-based subtypes (e.g., melancholic, atypical psychotic, and anxious depression), melancholic and atypical depression has been considered a higher "class" within depression, while psychotic and anxious depression has been identified mostly based on dominant core symptoms of psychotic features and comorbid anxiety (Baumeister & Parker, 2012). In this chapter, we will focus on melancholic and atypical depressive subtypes; in the next paragraph, we describe the diagnostic criteria for each subtype and the reasons why these two subtypes in particular are relevant in the context of inflammation.

ATYPICAL AND MELANCHOLIC DEPRESSION SUBTYPES

The DSM-5 defines melancholic and atypical depression as follows (American Psychiatric Association, 2013). Melancholic depression is characterized by loss of pleasure or the lack of mood reactivity plus three or more of the following symptoms: distinct quality of the depressed mood, mood that is worse in the morning, early morning awakening, psychomotor changes, weight loss or decreased appetite, and excessive guilt. Atypical depression on the other hand is characterized by mood reactivity plus two or more of the following symptoms: weight gain or increased appetite, hypersomnia, leaden paralysis, and interpersonal rejection sensitivity. It should be noted though that the definition of atypical depression has been debated (Angst, Gamma, Sellaro, Zhang, & Merikangas, 2002; Antonijevic, 2006; Parker et al., 2002), as some studies do not find that mood reactivity is correlated to the other atypical symptoms. Also, the criterion of interpersonal rejection sensitivity has been critiqued for being a personality trait rather than a symptom. In epidemiological research, the definition of atypical depression is often simplified as having increased appetite/weight, hypersomnia and leaden paralysis (Angst et al., 2006), or even based on data-driven derived subgroups that strongly resemble DSM atypical depression (Lamers et al., 2010).

While DSM-5 distinguishes more subtypes besides the atypical and melancholic subtype, these two are of particular interest as they seem

opposites not only in terms of symptom presentation (i.e., appetite/weight and sleep criteria) but also with respect to certain neurobiological correlates (Antonijevic, 2006; Gold & Chrousos, 2002). Gold and Chrousos (2002) postulated that the stress system is differently dysregulated in the two subtypes, with melancholic depression showing a hyperactive stress response and atypical depression a hypoactive stress response. Meta-analyses of cortisol levels in depression confirm that effect sizes are larger when more melancholic cases are included and lower with inclusion of more atypical cases (Stetler & Miller, 2011). As the stress response system and immune system are closely linked, the supposed differences in stress response dysregulation between subtypes can potentially also be observed in the immune system.

INFLAMMATION AND DEPRESSIVE SUBTYPES

In the past decade, some evidence has indeed been found for differential associations between depression subtypes and inflammation (Baune et al., 2012; Penninx, Milaneschi, Lamers, & Vogelzangs, 2013). Such findings are highly relevant as they can contribute to an increased understanding of etiological pathways, identification of potential treatment targets, and thus contribute to personalized treatment. Since our last overview of the literature, more studies have examined melancholic and atypical subtypes and their associations with inflammation, and in this chapter, we will provide an updated overview on these studies, integrate the findings, and discuss their relevance for research and clinic. To achieve this, we evaluated relevant literature on atypical and melancholic depression and markers of inflammation to date. We considered papers reporting on CRP, TNF-α, and ILs.

Table 1 provides an overview of included studies, sample source and size, used subtypes

TABLE 1 Overview of Study Characteristics of Included Studies

Author	Year	Sample Source	Definitions Used	Mitogen-Stimulated	Markers	Control	Atypical	Nonatypical	Unspecified[a]	Nonmelancholic	Melancholic
									MDD		
Anisman	1999	Outpatients (cases), blue- and white-collar hospital employees (controls)	Atypical: Columbia criteria	PHA	IL-1β, IL-2	27	31	14			
Bai	2015	Outpatients	Atypical: adapted DSM-IV criteria; without interpersonal rejection sensitivity Melancholic: adapted DSM-IV criteria; without distinct quality of mood and without diurnal variation	—	IL-2R, IL-6R, TNF-R1, CRP	130	17		106		26
Dunjic-Kostic	2013	Inpatients (cases); clinic employees (controls)	Atypical and melancholic: DSM-IV criteria	—	IL-6, TNF-α	39	18				29
Glaus[a]	2014	Population-based (PsyCoLaus study)	Atypical and melancholic: DSM-IV criteria	—	IL-1β, IL-6, TNF-α, CRP	3059 1466	98 235		258 721		198 499
Hickman et al.[b]	2014	Population-based (NHANES (1999–2004))	Atypical: reversed neurovegetative symptoms (both hypersomnia and increased appetite/weight)	—	CRP	1682	16	93			
Huang	2007	Inpatients (cases); medical staff and students (controls)	Melancholic: DSM-IV criteria	—	IL-1β, IL-10, TNF-α	31				17	25

Kaestner	2005	Inpatients (cases); healthy blood donors (controls)	Melancholic: DSM-IV criteria and Newcastle endogenicity scale ≥ 6	PHA	IL-1β, IL1-RA	37		16	21
Karlovic	2012	Inpatients (cases); hospital workers (controls)	Atypical and melancholic: DSM-IV criteria	—	IL-6, CRP, TNF-α	18	23		32
Lamers	2012	Community, primary care, and mental health clinics (NESDA study)	Atypical and melancholic: latent class derived atypical and melancholic depression (Lamers et al, 2010)	—	IL-6, CRP, TNF-α	543	122		111
Maes	1993	Inpatients (cases); local volunteers (controls)	Melancholic: DSM-III criteria	—	IL-6	8		7	10
Maes	1995	Inpatients (cases); staff members and their relatives (controls)	Melancholic: DSM-III-R criteria	—	IL-1RA	22		25	22
Maes	2012	Outpatients (cases); lab staff and their relatives (controls)	Melancholic: DSM-IV-R criteria	—	IL-1β, TNF-α	26		54	29
Marques-Deak	2007	Outpatients (cases); source controls not mentioned	Melancholic: based on SCID	LPS	IL-1β, IL-6	41		18	28

Continued

TABLE 1 Overview of Study Characteristics of Included Studies—cont'd

Author	Year	Sample Source	Definitions Used	Mitogen-Stimulated	Markers	Control	MDD				
							Atypical	Nonatypical	Unspecified[a]	Nonmelancholic	Melancholic
Rothermundt, Arolt, Fenker, et al.	2001	Inpatients (cases); healthy blood donors (controls)	Melancholic: DSM-IV criteria	PHA	IL-2, IL-10	43				21	22
Rothermundt, Arolt, Peters, et al.	2001	Inpatients (cases); healthy blood donors (controls)	Melancholic: DSM-IV criteria	PHA (IL-1β)	IL-1β, CRP	43				21	22
Rudolf[c]	2014	Inpatients (cases); volunteers recruited through university bulletin boards (controls)	Melancholic (typical): insomnia and weight loss Atypical: hypersomnia and weight gain	—	IL-6	24	8				24[d]
Schlatter	2014	Inpatients (cases); source controls not mentioned	Melancholic: DSM-IV criteria	—	IL-2	20				28	14
Yoon	2012	Inpatients	Atypical: reversed neurovegetative symptoms (both hypersomnia and increased appetite/weight)	PHA and LPS	IL-2, IL-4, IL-6, TNF-α		35	70			

PHA, phytohemagglutinin; LPA, lipopolysaccharide.
[a] Unspecified depression refers to absence of both the atypical and melancholic specifier.
[b] Unweighted numbers are listed; sampling weights were applied in analyses.
[c] Diagnosis of typical depression rather than melancholic depression.
[d] First row of numbers are current diagnoses and the second row lifetime diagnoses.

definitions, and the investigated markers. Of the 18 studies, 3 only reported on the atypical subtype, 9 only on the melancholic subtype, and 6 on both the atypical and melancholic subtypes. Sixteen studies also included a healthy control group, but not all studies always compared each depressive subtype with healthy controls. Rather than studying basal inflammation levels, six studies applied ex vivo mitogen stimulation to their samples. Such stimulation induces an inflammatory reaction reflecting the innate production capacity of markers (van der Linden, Huizinga, Stoeken, Sturk, & Westendorp, 1998). Sample sizes varied largely across studies, with the number of atypically depressed patients ranging from as little as 8 to 235 and the number of melancholic patients from 14 to 499. Most studies were clinical samples ($n=4$ outpatient and $n=11$ inpatient), with the exception of the population-based PsyCoLaus cohort study, the US population representative NHANES survey (Hickman, Khambaty, & Stewart, 2014), and the longitudinal NESDA cohort study (Lamers et al., 2012). These latter three studies also had the largest total sample sizes of included studies. In the following section, we will describe results per inflammatory marker.

IL-1β/α & IL1-RA

Mitogen Stimulated Inflammation Studies

Four studies evaluated mitogen-stimulated IL-1β (Anisman, Ravindran, Griffiths, & Merali, 1999; Kaestner et al., 2005; Marques-Deak et al., 2007; Rothermundt, Arolt, Peters, et al., 2001), but all had included <35 cases per subtype. The first found no difference in stimulated IL-1β between atypically depressed persons and controls (Anisman et al., 1999). Kaestner et al. (2005) found stimulated IL-1β to be higher in nonmelancholic depression versus melancholic depression and controls. Others did not observe differences in mitogen-stimulated IL-1β between melancholic cases and controls (Marques-Deak

et al., 2007; Rothermundt, Arolt, Peters, et al., 2001) and no difference between melancholic and nonmelancholic cases (Marques-Deak et al., 2007).

In a small study, mitogen-stimulated IL-1RA levels were higher in nonmelancholic depression versus controls and melancholic depression (Kaestner et al., 2005).

Basal Inflammation Studies

Three studies evaluated basal IL-1β. The largest study, with 198 current/499 lifetime melancholic cases and 98 current/235 lifetime atypical cases, found no differences in IL-1β among atypical, melancholic, and unspecified depressed cases and controls (Glaus et al., 2014). Each of the other two studies studies <30 cases of the melancholic subtype. One study found IL-1β and IL-1α to be higher in depressed persons than in controls, but did not observe differences between melancholic ($n=29$) and nonmelancholic depression ($n=59$) (Maes, Mihaylova, Kubera, & Ringel, 2012), in contrast to another study that found IL-1β levels to be higher in melancholic ($n=25$) versus nonmelancholic depression ($n=17$) (Huang & Lee, 2007).

One study evaluating IL-1RA among 47 cases and 22 controls found these levels to be higher in cases versus controls but no differences between melancholic and unspecified depression (Maes et al., 1995).

IL-2 AND IL-2R

Mitogen Stimulated Inflammation Studies

Three studies evaluated mitogen-stimulated IL-2 in depressive subtypes; however, none of these studies directly compared atypical with melancholic depression, and all included 35 of fewer cases of a given subtype. One study found lower IL-2 levels after PHA stimulation in those with atypical depression versus controls

(Anisman et al., 1999); another study using lipopolysaccharide (LPS) stimulation observed higher IL-2 levels in atypical versus nonatypical depression (Yoon, Kim, Lee, Kwon, & Kim, 2012). Lastly, Rothermundt et al. reported lower stimulated IL-2 in melancholic depression versus controls and those with nonmelancholic depression (Rothermundt, Arolt, Fenker, et al., 2001).

Basal Inflammation Studies

The only study evaluating basal IL-2 levels compared 28 nonmelancholic cases, 14 melancholic cases, and 20 controls. IL-2 levels were higher in melancholic and nonmelancholic depression versus controls, but no differences between melancholic and nonmelancholic depression were observed (Schlatter, Ortuno, & Cervera-Enguix, 2004). No differences in IL-2R were found between the subtypes atypical depression, melancholic depression, and unspecified depression (Bai et al., 2015). This study too had only a limited number of atypical ($n = 17$) and melancholic ($n = 26$) depression under study.

IL-6 & IL-6R

Mitogen Stimulated Inflammation Studies

Two studies evaluated the association between mitogen-stimulated IL-6 and depression subtypes. No differences in LPS-stimulated IL-6 were observed between atypical ($n = 35$) and nonatypical patients ($n = 70$) (Yoon et al., 2012) or among melancholic cases ($n = 28$), nonmelancholic cases ($n = 18$), and controls ($n = 41$) (Marques-Deak et al., 2007).

Basal Inflammation Studies

Of the six studies evaluating basal IL-6 levels, the largest (198 current/499 lifetime melancholic cases and 98 current/235 lifetime atypical cases) found no differences across atypical and melancholic subtypes in IL-6 (Glaus et al., 2014). In the

NESDA sample of 122 atypical cases and 111 melancholic cases, we found IL-6 levels to be higher in atypical cases versus controls and also higher versus melancholic cases of depression (Lamers et al., 2012). Of the four other, much smaller studies, one study showed IL-6 levels to be higher in atypical cases versus controls and also higher versus melancholic cases of depression (Rudolf, Greggersen, Kahl, Huppe, & Schweiger, 2014). In the studies evaluating melancholic depression, IL-6 was higher in melancholic depression compared with nonmelancholic depression and minor depression (Maes et al., 1993) and compared with control subjects (Maes et al., 1993) (Dunjic-Kostic et al., 2013). Two studies observed no IL-6 difference between melancholic and atypical cases (Dunjic-Kostic et al., 2013; Karlovic, Serretti, Vrkic, Martinac, & Marcinko, 2012). One study evaluated associations with the IL-6 receptor and reported no differences between atypical ($n = 17$), melancholic ($n = 26$), and unspecified subtypes of depression ($n = 106$) (Bai et al., 2015).

OTHER ILS

A few small studies (all with 35 of fewer cases of a given subtype) also analyzed other ILs. LPS-stimulated IL-4 levels were found to be higher in atypical depression versus nonatypical depression (Yoon et al., 2012). Lower mitogen-stimulated IL-10 were observed in melancholic cases versus nonmelancholic cases and versus controls (Rothermundt, Arolt, Fenker, et al., 2001). Basal IL-10 levels on the other hand did not differ between depressed cases and controls nor between melancholic and nonmelancholic depression (Huang & Lee, 2007).

C-REACTIVE PROTEIN

Six studies measured basal CRP levels and evaluated its association to different depression subtypes. The three largest and/or population-based studies all found CRP to be higher in atypical depression versus controls (Glaus et al., 2014; Hickman et al., 2014; Lamers et al., 2012), versus melancholic depression (Glaus et al., 2014; Lamers et al., 2012), and versus nonatypical and unspecific depression (Glaus et al., 2014; Hickman et al., 2014). Effects in the PsyCoLaus studies were attenuated when controlling for additional health and lifestyle factors, including body mass index (BMI) (Glaus et al., 2014).

In the smaller studies, mixed results were found. One reported higher CRP levels in atypical depression versus controls, while CRP levels did not differ between atypical and melancholic depression (Karlovic et al., 2012), whereas no differences in the level of CRP between atypical, melancholic, and unspecified depression were observed in another study (Bai et al., 2015). And the third smaller study reported that those with nonmelancholic depression had lower CRP than controls, but no difference was observed compared with melancholic cases (Rothermundt, Arolt, Peters, et al., 2001).

TUMOR NECROSIS FACTOR-ALPHA

Mitogen Stimulated Inflammation Studies

The only small study looking at LPS-stimulated TNF-α found marginally lower levels in atypical versus nonatypical depression ($P = 0.06$) (Yoon et al., 2012).

Basal Inflammation Studies

Six studies evaluated basal TNF-α among depression subtypes. The largest study of these found no differences in TNF-α across subtypes and controls (Glaus et al., 2014), but the second largest did observe higher TNF-α levels in atypical depression versus melancholic depression and controls (Lamers et al., 2012). Of the remaining smaller studies, two studies found

that while TNF-α was higher in depressed cases than controls (Huang & Lee, 2007; Maes et al., 2012) but where one observed no difference in levels between melancholic and nonmelancholic cases (Huang & Lee, 2007), the other reported higher TNF-α levels in melancholic versus non-melancholic cases (Maes et al., 2012). No differences in TNF-α across subtypes and controls were observed by one study (Karlovic et al., 2012), while another found TNF-α to be lower in atypical depression versus controls, but did not observe statistically different levels in atypical versus melancholic depression (Dunjic-Kostic et al., 2013).

Higher levels of the soluble TNF receptor type 1 were reported in depressed cases versus controls, but no differences between atypical, melancholic, and unspecified depression cases were detected (Bai et al., 2015).

SUMMARY OF RESULTS

Atypical depression seems to be consistently linked to higher levels of CRP, with most studies' results pointing in the same direction and with consistent findings also across the studies with largest sample sizes (providing the most robust evidence). The three small studies of mitogen-stimulated IL-2 also show some consistency, with atypical depression possibly having higher and melancholic depression having lower stimulated IL-2 level than controls. However, this should be confirmed in larger samples. For other markers, either there are inconsistent results (IL-1β, IL-6, and TNF-α), or they are only investigated in one or two studies with small sample sizes (IL-2R, IL1-α, IL-1RA, IL-6R, IL-4, IL-10, and TNF-R1). Table 2 summarizes the findings of each study, with studies sorted alphabetically per marker, and stratified by mitogen-stimulated and basal inflammation studies. The fact that for CRP, there is more consistency than for other markers could have to do with the fact that measurement of CRP—as a general inflammatory marker—is easier to

TABLE 2 Summary of Results

Marker	N Studies	Atypical	Melancholic	References
MITOGEN-STIMULATED MARKERS				
IL-1β	4	= C		Anisman et al. (1999)
			↓ Versus nonmel	Kaestner et al. (2005)
			= Nonmel=C	Marques-Deak et al. (2007)
			= C	Rothermundt, Arolt, Peters, et al. (2001)
IL-1RA	1		↓ Versus nonmel	Kaestner et al. (2005)
IL-2	3	↓ Versus C		Anisman et al. (1999)
		↑ Versus nonatyp		Yoon et al. (2012)
			↓ Versus C, nonmel	Rothermundt, Arolt, Fenker, et al. (2001)
IL-6	2		= Nonmel=C	Marques-Deak et al. (2007)
		= Nonatyp		Yoon et al. (2012)
IL-4	1	↑ Versus nonatyp		Yoon et al. (2012)
IL-10	1		↓ Versus nonmel, C	Rothermundt, Arolt, Fenker, et al. (2001)
TNF-α	1	= Nonatyp		Yoon et al. (2012)
BASAL INFLAMMATION MARKERS				
IL-1β	3	= Mel=unspec=C		Glaus et al. (2014)
			↑ Versus nonmel	Huang and Lee (2007)
			= Nonmel	Maes et al. (2012)
IL-1α	1		= Nonmel	Maes et al. (2012)
IL-1RA	1		= Unspec	Maes et al. (1995)
IL-2	1		= Nonmel	Schlatter et al. (2004)
IL-2R	1	= Mel=unspec		Bai et al. (2015)
IL-6	6		= Atyp, ↑ versus C	Dunjic-Kostic et al. (2013)
		= Mel		Glaus et al. (2014)
			↑ Versus nonmel, C	Maes et al. (1993)
		= Mel		Karlovic et al. (2012)
		↑ Versus C, mel		Lamers et al. (2012)
		↑ Versus C, mel		Rudolf et al. (2014)
IL-6R	1	= Mel=unspec		Bai et al. (2015)
IL-10	1		= Nonmel	Huang and Lee (2007)

TABLE 2 Summary of Results—cont'd

Marker	N Studies	Atypical	Melancholic	References
CRP	6	= Mel = unspec		Bai et al. (2015)
		↑ Versus C, mel, unspec		Glaus et al. (2014)[a]
		↑ Versus C		Hickman et al. (2014)
		↑ Versus C; = mel		Karlovic et al. (2012)
		↑ Versus C, mel		Lamers et al. (2012)
			= Nonmel; nonmel ↓ versus C	Rothermundt, Arolt, Peters, et al. (2001)
TNF-α	6	↓ Versus C; = mel		Dunjic-Kostic et al. (2013)
		= Mel = unspec = C		Glaus et al. (2014)
			= Nonmel	Huang and Lee (2007)
		= Mel = unspec = C		Karlovic et al. (2012)
		↑ Versus C, mel		Lamers et al. (2012)
			↑ Versus nonmel	Maes et al. (2012)
TNF-R1	1	= Mel = unspec		Bai et al. (2015)

C, control; *nonatyp*, nonatypical; *nonmel*, nonmelancholic; *mel*, melancholic; *atyp*, atypical; *unspec*, unspecified depression.
[a] *Corrected for sociodemographics only.*

implement in studies and measured with a higher reliability than that of other markers. If this overview makes one thing clear, it is that we need larger studies to draw firm conclusions about cytokines and depression subtypes.

METHODOLOGICAL LIMITATIONS

As said, small sample size makes it harder to draw conclusions from the papers discussed here. Also, because more than half of the studies did not include both atypical and melancholic subtypes, drawing conclusions on how atypical and melancholic subtypes differ in inflammatory marker levels is further complicated. Furthermore, in studies only evaluating one subtype, it is sometimes unclear what the comparison MDD group comprises; often labeled as nonatypical or nonmelancholic, this group is not necessarily the same as MDD without any other DSM specifier. It is therefore recommendable for future studies to study multiple subtypes, rather than a single subtype, and to clearly define the comparison depression group. On a more general level, depression researchers should be encouraged to include enough detailed questions in their clinical diagnostic interviews and questionnaires to be able to distinguish atypical and melancholic subtypes in all of their biomarker studies. The lack of such information hinders replication; in a recent study on serum proteomic profiles in atypical and melancholic subtypes, there was a lack of independent proteomic samples to replicate our findings of significantly more metabolic and immune dysregulation in the atypical subtype versus controls and melancholic cases (Lamers et al., 2016). Also, as the definition of atypical depression has been debated, and specific atypical symptoms seem to be driving associations with metabolic and immune markers (Lamers, Milaneschi, de Jonge, Giltay, & Penninx, 2017), more research should focus on symptom or dimension specificity of associations with biological correlates, as this can contribute to a better definition of depression subtypes. Such definitions could then also be further investigated in genetic studies.

The covariates included in statistical models also widely differ across studies, further complicating comparison of results. Especially when it comes to lifestyle factors or BMI, it could be argued that these are mediators of the association between depression and inflammation, thus potentially leading to an overcorrection. Another recommendation would therefore be to add lifestyle factors and BMI only in a second step of the statistical modeling, so that their effect on the estimates can be evaluated.

INFLAMMATION: CLINICAL IMPLICATIONS AND RELEVANCE TO TREATMENT

Personalized medicine and personalized psychiatry are given much attention in research in recent years. Inflammation can be considered a key component in personalized psychiatry. Previously, it was demonstrated that TCA users had more inflammatory dysregulation than medication- naïve depressed persons (Hamer, Batty, Marmot, Singh-Manoux, & Kivimäki, 2011; van Reedt Dortland, Giltay, van Veen, Zitman, & Penninx, 2010; Vogelzangs et al., 2012), while those on SSRIs had somewhat lower inflammation levels than medication-naïve cases (Vogelzangs et al., 2012). The anti-inflammatory effects of SSRIs—but not of other antidepressants—have also been demonstrated in two meta-analyses (Hannestad, DellaGioia, & Bloch, 2011; Hiles, Baker, de Malmanche, & Attia, 2012b). With respect to differential effects of antidepressants in depression subtypes, a recent meta-analysis did not provide evidence for differential effects across subtypes (Cuijpers et al., 2017). Another study not included in the meta-analysis also found that melancholic and atypical groups responded similarly to SSRIs and SNRIs (Arnow et al., 2015).

Besides potential anti-inflammatory effects of antidepressant themselves, an increasing number of studies are evaluating the effect of adjunctive anti-inflammatory drugs. A meta-analysis including 14 placebo-controlled trials with in total over 6000 participants concluded that anti-inflammatory treatment decreased depressive symptoms more than did placebo, and it improved antidepressant response (Kohler et al., 2014). While this meta-analysis found encouraging effects, there was a high level of heterogeneity in the meta-analysis, and it is not unthinkable that effects are driven by depression cases exhibiting elevated levels of inflammatory markers. Indeed, the authors noted the need to identify subgroups that may benefit more from anti-inflammatory medication. Raison et al. (2013) demonstrated previously in their RCT that effects of augmentation with an TNF-α antagonist seemed effective only in those with high CRP levels at baseline.

Another treatment for which differential effects in those with high versus low inflammation can be expected is exercise therapy, as exercise is known to normalize immune dysregulation (You & Nicklas, 2008). Rethorst et al. (2013) showed larger effects of an exercise intervention for depression in those with high baseline inflammation levels versus low baseline inflammation and in subsequent work showed that effect sizes were also increased in those with atypical depression (Rethorst, Tu, Carmody, Greer, & Trivedi, 2016). These results point to increased inflammation in atypical depression.

MECHANISMS AND OTHER METABOLIC DYSREGULATIONS IN (ATYPICAL) DEPRESSION

One major factor known to contribute to peripheral inflammation is BMI. Adipose tissue is a large endocrine organ (Galic, Oakhill, & Steinberg, 2010) that secretes various factors including cytokines, leptin and adiponectin, and obesity can therefore lead to a state of low-grade inflammation. BMI and depression often co-occur, and meta-analyses have shown that BMI predicts depression and vice versa (Luppino et al., 2010).

There is evidence to suggest that these associations too are specific to atypical depression. Higher levels of BMI have been observed in atypical versus nonatypical depressed persons (Kaestner et al., 2005; Lamers et al., 2012; Rudolf et al., 2014; Seppala et al., 2012; Sullivan, Prescott, & Kendler, 2002; Yoon et al., 2012), and longitudinal studies have found atypical depression to predict weight gain (Hasler et al., 2004), obesity, and metabolic syndrome (Lasserre et al., 2014, 2016).

Besides BMI, obesity, and metabolic syndrome, specific metabolic and immune markers were previously found by our group to differ in atypical depression from both controls and melancholic depression in a proteomics approach (Lamers et al., 2016). Compared with controls and persons with melancholic depression, persons with atypical depression had higher levels of leptin, fatty-acid-binding protein adipocyte, complement C3, insulin, and beta-2-microglobulin while having lower levels of insulin-like growth factor-binding protein 1, insulin-like growth factor-binding protein 2, and mesothelin (MSLN). Leptin—an appetite-decreasing hormone with antidepressant-like effects excreted by white fat cells—has been previously found to be increased in atypical depression only (Gecici et al., 2005; Milaneschi, Lamers, Bot, Drent, & Penninx, 2015). High levels of leptin could suggest leptin resistance, which is

thought to contribute to depressive symptomatology (Lu, 2007). These findings could suggest that the atypical subtype is a more metabolic and inflammatory disorder on a more fundamental level, and perhaps the term "immunometabolic depression" is a better suited term than is atypical depression. This hypothesis is supported by genetic findings. We recently found independent associations of an atypical depressive subtype with the rs9939609 SNP of the FTO gene (i.e., fat mass and obesity related gene) (Milaneschi et al., 2014) and with genomic profile risk scores for BMI and triglycerides (Milaneschi, Lamers, Peyrot, et al., 2015).

As both leptin and the FTO gene are involved in the regulation of energy intake, these findings allude to a potential importance of the atypical symptom of increased appetite. Indeed, analyses on the individual symptoms of atypical depression show that it is mostly the symptom of increased appetite that is associated dysregulations in metabolic and immune measures (Lamers et al., 2017). What could trigger increased appetite is unclear, as associations between food intake and mood state are highly complex (Singh, 2014). But it has been suggested that increased appetite in atypical depression could be resulting from disturbances on reward perceived from food that is regulated by the mesolimbic reward centers (Korte et al., 2015; Simmons et al., 2016).

CONCLUSION

While the existing literature may not provide a definite answer to the question whether atypical depression indeed captures the MDD cases with high inflammation, there is some evidence that CRP is indeed increased in atypical cases compared with controls and melancholic depression. Larger studies should be conducted to fully elucidate inflammation in melancholic and atypical depression. The observed CRP-atypical depression link, combined with evidence that the same atypical subtype is linked with metabolic dysregulation, could imply that the atypical depressive subtype may be better described as an immunometabolic form of depression. While anti-inflammatory treatments and exercise therapy may sound promising, in particular for atypical depression, more replication in RCTs that can distinguish atypical or immunometabolic depressive subtypes is needed.

References

American Psychiatric Association (2013). *Diagnostic and statistical manual of mental disorders* (5th ed.). American Psychiatric Association. Washington, DC.

Angst, J., Gamma, A., Benazzi, F., Silverstein, B., Jdacic-Gross, V., Eich, D., et al. (2006). Atypical depressive syndromes in varying definitions. *European Archives of Psychiatry and Clinical Neuroscience, 256*(1), 44–54.

Angst, J., Gamma, A., Sellaro, R., Zhang, H., & Merikangas, K. (2002). Toward validation of atypical depression in the community: results of the Zurich cohort study. *Journal of Affective Disorders, 72*(2), 125–138.

Anisman, H., Ravindran, A. V., Griffiths, J., & Merali, Z. (1999). Endocrine and cytokine correlates of major depression and dysthymia with typical or atypical features. *Molecular Psychiatry, 4*(2), 182–188.

Antonijevic, I. A. (2006). Depressive disorders—is it time to endorse different pathophysiologies? *Psychoneuroendocrinology, 31*(1), 1–15.

Arnow, B. A., Blasey, C., Williams, L. M., Palmer, D. M., Rekshan, W., Schatzberg, A. F., et al. (2015). Depression subtypes in predicting antidepressant response: a report from the iSPOT-D trial. *American Journal of Psychiatry, 172*(8), 743–750. http://dx.doi.org/10.1176/appi.ajp.2015.14020181.

Au, B., Smith, K. J., Gariépy, G., & Schmitz, N. (2015). The longitudinal associations between C-reactive protein and depressive symptoms: evidence from the English Longitudinal Study of Ageing (ELSA). *International Journal of Geriatric Psychiatry, 30*(9), 976–984. http://dx.doi.org/10.1002/gps.4250.

Bai, Y.-M., Su, T.-P., Li, C.-T., Tsai, S.-J., Chen, M.-H., Tu, P.-C., et al. (2015). Comparison of pro-inflammatory cytokines among patients with bipolar disorder and unipolar depression and normal controls. *Bipolar Disorders, 17*(3), 269–277. http://dx.doi.org/10.1111/bdi.12259.

Baumeister, H., & Parker, G. (2012). Meta-review of depressive subtyping models. *Journal of Affective Disorders, 139*(2), 126–140.

Baune, B. T., Stuart, M., Gilmour, A., Wersching, H., Heindel, W., Arolt, V., et al. (2012). The relationship between subtypes of depression and cardiovascular disease: a systematic review of biological models. *Translational Psychiatry, 2*(3). http://dx.doi.org/10.1038/tp.2012.18.

Copeland, W. E., Shanahan, L., Worthman, C., Angold, A., & Costello, E. J. (2012). Cumulative depression episodes predict later C-reactive protein levels: a prospective analysis. *Biological Psychiatry, 71*(1), 15–21. http://dx.doi.org/10.1016/j.biopsych.2011.09.023.

Cuijpers, P., Weitz, E., Lamers, F., Penninx, B. W., Twisk, J., DeRubeis, R. J., et al. (2017). Melancholic and atypical depression as predictor and moderator of outcome in cognitive behavior therapy and pharmacotherapy for adult depression. *Depression and Anxiety, 34*(3), 246–256. http://dx.doi.org/10.1002/da.22580.

Dantzer, R., & Kelley, K. W. (2007). Twenty years of research on cytokine-induced sickness behavior. *Brain, Behavior, and Immunity, 21*(2), 153–160. http://dx.doi.org/10.1016/j.bbi.2006.09.006.

Dowlati, Y., Herrmann, N., Swardfager, W., Liu, H., Sham, L., Reim, E. K., et al. (2010). A meta-analysis of cytokines in major depression. *Biological Psychiatry, 67*(5), 446–457.

Dunjic-Kostic, B., Ivkovic, M., Radonjic, N. V., Petronijevic, N. D., Pantovic, M., Damjanovic, A., et al. (2013). Melancholic and atypical major depression—connection between cytokines, psychopathology and treatment. *Progress in Neuropsychopharmacology and Biological Psychiatry, 43*(1878–4216 (Electronic)), 1–6.

Eurelings, L. S. M., Richard, E., Eikelenboom, P., van Gool, W. A., & Moll van Charante, E. P. (2015). Low-grade inflammation differentiates between symptoms of apathy and depression in community-dwelling older individuals. *International Psychogeriatrics, 27*(4), 639–647. http://dx.doi.org/10.1017/S1041610214002683.

Galic, S., Oakhill, J. S., & Steinberg, G. R. (2010). Adipose tissue as an endocrine organ. *Molecular and Cellular Endocrinology, 316*(2), 129–139. http://dx.doi.org/10.1016/j.mce.2009.08.018.

Gallagher, D., Kiss, A., Lanctot, K., & Herrmann, N. (2016). Depression with inflammation: longitudinal analysis of a proposed depressive subtype in community dwelling older adults. *International Journal of Geriatric Psychiatry.* http://dx.doi.org/10.1002/gps.4645.

Gecici, O., Kuloglu, M., Atmaca, M., Tezcan, A. E., Tunckol, H., Emül, H. M., et al. (2005). High serum leptin levels in depressive disorders with atypical features. *Psychiatry and Clinical Neurosciences, 59*(6), 736–738. http://dx.doi.org/10.1111/j.1440-1819.2005.01445.x.

Glaus, J., Vandeleur, C. L., Von, K. R., Lasserre, A. M., Strippoli, M. P., Gholam-Rezaee, M., et al. (2014). Associations between mood, anxiety or substance use disorders and inflammatory markers after adjustment for multiple covariates in a population-based study. *Journal of Psychiatric Research*, 58(1879–1379 (Electronic)), 36–45.

Gold, P. W., & Chrousos, G. P. (2002). Organization of the stress system and its dysregulation in melancholic and atypical depression: high vs low CRH/NE states. *Molecular Psychiatry*, 7(3), 254–275.

Hamer, M., Batty, G. D., Marmot, M. G., Singh-Manoux, A., & Kivimäki, M. (2011). Anti-depressant medication use and C-reactive protein: results from two population-based studies. *Brain, Behavior, and Immunity*, 25(1), 168–173. http://dx.doi.org/10.1016/j.bbi.2010.09.013.

Hannestad, J., DellaGioia, N., & Bloch, M. (2011). The effect of antidepressant medication treatment on serum levels of inflammatory cytokines: a meta-analysis. *Neuropsychopharmacology: Official Publication of the American College of Neuropsychopharmacology*, 36(12), 2452–2459. http://dx.doi.org/10.1038/npp.2011.132.

Hasler, G., Pine, D. S., Gamma, A., Milos, G., Ajdacic, V., Eich, D., et al. (2004). The associations between psychopathology and being overweight: a 20-year prospective study. *Psychological Medicine*, 34(6), 1047–1057.

Hickman, R. J., Khambaty, T., & Stewart, J. C. (2014). C-reactive protein is elevated in atypical but not nonatypical depression: data from the National Health and Nutrition Examination Survey (NHANES) 1999–2004. *Journal of Behavioral Medicine*, 37(1573–3521 (Electronic)), 621–629.

Hiles, S. A., Baker, A. L., de Malmanche, T., & Attia, J. (2012a). A meta-analysis of differences in IL-6 and IL-10 between people with and without depression: exploring the causes of heterogeneity. *Brain, Behavior, and Immunity*, 26(7), 1180–1188. http://dx.doi.org/10.1016/j.bbi.2012.06.001.

Hiles, S. A., Baker, A. L., de Malmanche, T., & Attia, J. (2012b). Interleukin-6, C-reactive protein and interleukin-10 after antidepressant treatment in people with depression: a meta-analysis. *Psychological Medicine*, 42(10), 2015–2026. http://dx.doi.org/10.1017/S0033291712000128.

Howren, M. B., Lamkin, D. M., & Suls, J. (2009). Associations of depression with C-reactive protein, IL-1, and IL-6: a meta-analysis. *Psychosomatic Medicine*, 71(2), 171–186.

Huang, T. L., & Lee, C. T. (2007). T-helper 1/T-helper 2 cytokine imbalance and clinical phenotypes of acute-phase major depression. *Psychiatry and Clinical Neurosciences*, 61(4), 415–420.

Hughes, S., Jaremka, L. M., Alfano, C. M., Glaser, R., Povoski, S. P., Lipari, A. M., et al. (2014). Social support predicts inflammation, pain, and depressive symptoms: longitudinal relationships among breast cancer survivors. *Psychoneuroendocrinology*, 42, 38–44. http://dx.doi.org/10.1016/j.psyneuen.2013.12.016.

Jansen, R., Penninx, B. W. J. H., Madar, V., Xia, K., Milaneschi, Y., Hottenga, J. J., et al. (2016). Gene expression in major depressive disorder. *Molecular Psychiatry*, 21(3), 339–347. http://dx.doi.org/10.1038/mp.2015.57.

Jones, S. M., Weitlauf, J., Danhauer, S. C., Qi, L., Zaslavsky, O., Wassertheil-Smoller, S., et al. (2015). Prospective data from the Women's Health Initiative on depressive symptoms, stress, and inflammation. *Journal of Health Psychology*. http://dx.doi.org/10.1177/1359105315603701.

Kaestner, F., Hettich, M., Peters, M., Sibrowski, W., Hetzel, G., Ponath, G., et al. (2005). Different activation patterns of proinflammatory cytokines in melancholic and non-melancholic major depression are associated with HPA axis activity. *Journal of Affective Disorders*, 87(2–3), 305–311.

Karlovic, D., Serretti, A., Vrkic, N., Martinac, M., & Marcinko, D. (2012). Serum concentrations of CRP, IL-6, TNF-alpha and cortisol in major depressive disorder with melancholic or atypical features. *Psychiatry Research*, 198(1), 74–80.

Khandaker, G. M., Pearson, R. M., Zammit, S., Lewis, G., & Jones, P. B. (2014). Association of serum interleukin 6 and C-reactive protein in childhood with depression and psychosis in young adult life: a population-based longitudinal study. *JAMA Psychiatry*, 71(10), 1121–1128. http://dx.doi.org/10.1001/jamapsychiatry.2014.1332.

Kohler, O., Benros, M. E., Nordentoft, M., Farkouh, M. E., Iyengar, R. L., Mors, O., et al. (2014). Effect of anti-inflammatory treatment on depression, depressive symptoms, and adverse effects: a systematic review and meta-analysis of randomized clinical trials. *JAMA Psychiatry*, 17(2168–6238 (Electronic)), 1381–1391.

Köhler, C. A., Freitas, T. H., Maes, M., de Andrade, N. Q., Liu, C. S., Fernandes, B. S., et al. (2017). Peripheral cytokine and chemokine alterations in depression: a meta-analysis of 82 studies. *Acta Psychiatrica Scandinavica* http://dx.doi.org/10.1111/acps.12698.

Korte, S. M., Prins, J., Krajnc, A. M., Hendriksen, H., Oosting, R. S., Westphal, K. G., et al. (2015). The many different faces of major depression: it is time for personalized medicine. *European Journal of Pharmacology*, 753, 88–104.

Lamers, F., Bot, M., Jansen, R., Chan, M. K., Cooper, J. D., Bahn, S., et al. (2016). Serum proteomic profiles of depressive subtypes. *Translational Psychiatry*. 6(7)http://dx.doi.org/10.1038/tp.2016.115.

Lamers, F., de Jonge, P., Nolen, W. A., Smit, J. H., Zitman, F. G., Beekman, A. T., et al. (2010). Identifying depressive subtypes in a large cohort study: results from the Netherlands Study of Depression and Anxiety (NESDA). *Journal of Clinical Psychiatry*, 71(12), 1582–1589.

Lamers, F., Milaneschi, Y., de Jonge, P., Giltay, E., & Penninx, B. W. J. H. (2017). Metabolic and inflammatory markers: associations with individual depressive symptoms. *Psychological Medicine*, 1–11. http://dx.doi.org/10.1017/S0033291717002483.

Lamers, F., Vogelzangs, N., Merikangas, K. R., de Jonge, P., Beekman, A. T. F., & Penninx, B. W. J. H. (2012). Evidence for a differential role of HPA-axis function, inflammation and metabolic syndrome in melancholic versus atypical depression. *Molecular Psychiatry*, *18*(6), 692–699. http://dx.doi.org/10.1038/mp.2012.144.

Lasserre, A. M., Glaus, J., Vandeleur, C. L., Marques-Vidal, P., Vaucher, J., Bastardot, F., et al. (2014). Depression with atypical features and increase in obesity, body mass index, waist circumference, and fat mass: a prospective, population-based study. *JAMA Psychiatry*, *71*(2168–6238 (Electronic)), 880–888.

Lasserre, A. M., Strippoli, M.-P. F., Glaus, J., Gholam-Rezaee, M., Vandeleur, C. L., Castelao, E., et al. (2016). Prospective associations of depression subtypes with cardio-metabolic risk factors in the general population. *Molecular Psychiatry*. http://dx.doi.org/10.1038/mp.2016.178.

Liu, Y., Ho, R. C.-M., & Mak, A. (2012). Interleukin (IL)-6, tumour necrosis factor alpha (TNF-α) and soluble interleukin-2 receptors (sIL-2R) are elevated in patients with major depressive disorder: a meta-analysis and meta-regression. *Journal of Affective Disorders*, *139*(3), 230–239. http://dx.doi.org/10.1016/j.jad.2011.08.003.

Lu, X. Y. (2007). The leptin hypothesis of depression: a potential link between mood disorders and obesity? *Current Opinion in Pharmacology*, *7*(6), 648–652.

Luppino, F. S., de Wit, L. M., Bouvy, P. F., Stijnen, T., Cuijpers, P., Penninx, B. W., et al. (2010). Overweight, obesity, and depression: a systematic review and meta-analysis of longitudinal studies. *Archives of General Psychiatry*, *67*(3), 220–229.

Maes, M., Mihaylova, I., Kubera, M., & Ringel, K. (2012). Activation of cell-mediated immunity in depression: association with inflammation, melancholia, clinical staging and the fatigue and somatic symptom cluster of depression. *Progress in Neuropsychopharmacology and Biological Psychiatry*, *36*(1), 169–175. http://dx.doi.org/10.1016/j.pnpbp.2011.09.006.

Maes, M., Scharpé, S., Meltzer, H. Y., Bosmans, E., Suy, E., Calabrese, J., et al. (1993). Relationships between interleukin-6 activity, acute phase proteins, and function of the hypothalamic-pituitary-adrenal axis in severe depression. *Psychiatry Research*, *49*(1), 11–27.

Maes, M., Vandoolaeghe, E., Ranjan, R., Bosmans, E., Bergmans, R., & Desnyder, R. (1995). Increased serum interleukin-1-receptor-antagonist concentrations in major depression. *Journal of Affective Disorders*, *36*(1–2), 29–36.

Marques-Deak, A. H., Neto, F. L., Dominguez, W. V., Solis, A. C., Kurcgant, D., Sato, F., et al. (2007). Cytokine profiles in women with different subtypes of major depressive disorder. *Journal of Psychiatric Research*, *41*(1–2), 152–159.

Matsushima, J., Kawashima, T., Nabeta, H., Imamura, Y., Watanabe, I., Mizoguchi, Y., et al. (2015). Association of inflammatory biomarkers with depressive symptoms and cognitive decline in a community-dwelling healthy

older sample: a 3-year follow-up study. *Journal of Affective Disorders*, *173*, 9–14. http://dx.doi.org/10.1016/j.jad.2014.10.030.

Milaneschi, Y., Lamers, F., Bot, M., Drent, M. L., & Penninx, B. W. J. H. J. H. (2015). Leptin dysregulation is specifically associated with major depression with atypical features: evidence for a mechanism connecting obesity and depression. *Biological Psychiatry*. http://dx.doi.org/10.1016/j.biopsych.2015.10.023.

Milaneschi, Y., Lamers, F., Mbarek, H., Hottenga, J.-J., Boomsma, D. l., & Penninx, B. W. J. H. (2014). The effect of FTO rs9939609 on major depression differs across MDD subtypes. *Molecular Psychiatry*, *19*(9), 960–962. http://dx.doi.org/10.1038/mp.2014.4.

Milaneschi, Y., Lamers, F., Peyrot, W. J., Abdellaoui, A., Willemsen, G., Hottenga, J.-J., et al. (2015). Polygenic dissection of major depression clinical heterogeneity. *Molecular Psychiatry*. http://dx.doi.org/10.1038/mp.2015.86.

Mostafavi, S., Battle, A., Zhu, X., Potash, J. B., Weissman, M. M., Shi, J., et al. (2014). Type I interferon signaling genes in recurrent major depression: increased expression detected by whole-blood RNA sequencing. *Molecular Psychiatry*, *19*(1476–5578 (Electronic)), 1267–1274.

Parker, G., Roy, K., Mitchell, P., Wilhelm, K., Malhi, G., & Hadzi-Pavlovic, D. (2002). Atypical depression: a reappraisal. *American Journal of Psychiatry*, *159*(9), 1470–1479.

Pasco, J. A., Nicholson, G. C., Williams, L. J., Jacka, F. N., Henry, M. J., Kotowicz, M. A., et al. (2010). Association of high-sensitivity C-reactive protein with de novo major depression. *British Journal of Psychiatry: The Journal of Mental Science*, *197*(5), 372–377. http://dx.doi.org/10.1192/bjp.bp.109.076430.

Penninx, B. W. J. H., Milaneschi, Y., Lamers, F., & Vogelzangs, N. (2013). Understanding the somatic consequences of depression: biological mechanisms and the role of depression symptom profile. *BMC Medicine*, *11*(1), 129. http://dx.doi.org/10.1186/1741-7015-11-129.

Raison, C. L., & Miller, A. H. (2013). The evolutionary significance of depression in Pathogen Host Defense (PATHOS-D). *Molecular Psychiatry*, *18*(1), 15–37. http://dx.doi.org/10.1038/mp.2012.2.

Raison, C. L., Rutherford, R. E., Woolwine, B. J., Shuo, C., Schettler, P., Drake, D. F., et al. (2013). A randomized controlled trial of the tumor necrosis factor antagonist infliximab for treatment-resistant depression: the role of baseline inflammatory biomarkers. *JAMA Psychiatry*, *70*(1), 31–41.

Rethorst, C. D., Toups, M. S., Greer, T. L., Nakonezny, P. A., Carmody, T. J., Grannemann, B. D., et al. (2013). Proinflammatory cytokines as predictors of antidepressant effects of exercise in major depressive disorder. *Molecular Psychiatry*, *18*(10), 1119–1124. http://dx.doi.org/10.1038/mp.2012.125.

Rethorst, C. D., Tu, J., Carmody, T. J., Greer, T. L., & Trivedi, M. H. (2016). Atypical depressive symptoms as a predictor of treatment response to exercise in major depressive disorder. *Journal of Affective Disorders, 200*, 156–158. http://dx.doi.org/10.1016/j.jad.2016.01.052.

Rothermundt, M., Arolt, V., Fenker, J., Gutbrodt, H., Peters, M., & Kirchner, H. (2001). Different immune patterns in melancholic and non-melancholic major depression. *European Archoves of Psychiatry and Clinical Neuroscience, 251*(2), 90–97.

Rothermundt, M., Arolt, V., Peters, M., Gutbrodt, H., Fenker, J., Kersting, A., et al. (2001). Inflammatory markers in major depression and melancholia. *Journal of Affective Disorders, 63*(1–3), 93–102.

Rudolf, S., Greggersen, W., Kahl, K. G., Huppe, M., & Schweiger, U. (2014). Elevated IL-6 levels in patients with atypical depression but not in patients with typical depression. *Psychiatry Research, 217*(1872–7123 (Electronic)), 34–38.

Schlatter, J., Ortuno, F., & Cervera-Enguix, S. (2004). Lymphocyte subsets and lymphokine production in patients with melancholic versus nonmelancholic depression. *Psychiatry Research, 128*(3), 259–265. http://dx.doi.org/10.1016/j.psychres.2004.06.004.

Seppala, J., Vanhala, M., Kautiainen, H., Eriksson, J., Kampman, O., Mantyselka, P., et al. (2012). Prevalence of metabolic syndrome in subjects with melancholic and non-melancholic depressive symptoms. A Finnish population-based study. *Journal of Affective Disorders, 136*(3), 543–549.

Simmons, W. K., Burrows, K., Avery, J. A., Kerr, K. L., Bodurka, J., Savage, C. R., et al. (2016). Depression-related increases and decreases in appetite: dissociable patterns of aberrant activity in reward and interoceptive neurocircuitry. *American Journal of Psychiatry, 173*(4), 418–428. http://dx.doi.org/10.1176/appi.ajp.2015.15020162.

Singh, M. (2014). Mood, food, and obesity. *Frontiers in Psychology, 5*(1664–1078 (Electronic)), 925.

Stetler, C., & Miller, G. E. (2011). Depression and hypothalamic-pituitary-adrenal activation: a quantitative summary of four decades of research. *Psychosomatic Medicine, 73*(2), 114–126.

Sullivan, P. F., Prescott, C. A., & Kendler, K. S. (2002). The subtypes of major depression in a twin registry. *Journal of Affective Disorders, 68*(2–3), 273–284.

Tully, P. J., Baumeister, H., Bengel, J., Jenkins, A., Januszewski, A., Martin, S., et al. (2015). The longitudinal association between inflammation and incident depressive symptoms in men: the effects of hs-CRP are independent of abdominal obesity and metabolic disturbances. *Physiology & Behavior, 139*, 328–335. http://dx.doi.org/10.1016/j.physbeh.2014.11.058.

Valkanova, V., Ebmeier, K. P., & Allan, C. L. (2013). CRP, IL-6 and depression: a systematic review and meta-analysis of longitudinal studies. *Journal of Affective Disorders, 150*(3), 736–744. http://dx.doi.org/10.1016/j.jad.2013.06.004.

van der Linden, M. W., Huizinga, T. W., Stoeken, D. J., Sturk, A., & Westendorp, R. G. (1998). Determination of tumour necrosis factor-alpha and interleukin-10 production in a whole blood stimulation system: assessment of laboratory error and individual variation. *Journal of Immunological Methods, 218*(1–2), 63–71.

van Reedt Dortland, A. K. B., Giltay, E. J., van Veen, T., Zitman, F. G., & Penninx, B. W. J. H. (2010). Metabolic syndrome abnormalities are associated with severity of anxiety and depression and with tricyclic antidepressant use. *Acta Psychiatrica Scandinavica, 122*(1), 30–39. http://dx.doi.org/10.1111/j.1600-0447.2010.01565.x.

Vogelzangs, N., Duivis, H. E., Beekman, A. T. F., Kluft, C., Neuteboom, J., Hoogendijk, W., et al. (2012). Association of depressive disorders, depression characteristics and antidepressant medication with inflammation. *Translational Psychiatry. 2*(2)http://dx.doi.org/10.1038/tp.2012.8.

Yoon, H. K., Kim, Y. K., Lee, H. J., Kwon, D. Y., & Kim, L. (2012). Role of cytokines in atypical depression. *Nordic Journal of Psychiatry, 66*(3), 183–188.

You, T., & Nicklas, B. J. (2008). Effects of exercise on adipokines and the metabolic syndrome. *Current Diabetes Reports, 8*(1), 7–11.

Zalli, A., Jovanova, O., Hoogendijk, W. J. G., Tiemeier, H., & Carvalho, L. A. (2015). Low-grade inflammation predicts persistence of depressive symptoms. *Psychopharmacology.* http://dx.doi.org/10.1007/s00213-015-3919-9.

Inflammation as a Marker of Clinical Response to Treatment: A Focus on Treatment-Resistant Depression

Rebecca Strawbridge*, *Allan H. Young**,[†], *Anthony J. Cleare**,[†]

*Institute of Psychiatry, Psychology & Neuroscience, King's College London, London, United Kingdom
[†]South London and Maudsley NHS Foundation Trust, London, United Kingdom

THE BURDEN OF TREATMENT-RESISTANCE IN DEPRESSION

Treatment resistance is a serious problem in affective disorders (Keller, 2005). While only about 30% of patients achieve full remission to their first treatment for depression, the likelihood of response gradually reduces with subsequent treatments (Gaynes et al., 2009). The pioneering STAR*D study also demonstrated that after a total of four treatment trials, approximately two-thirds of patients achieve remission but concluded that after two unsuccessful treatments, the likelihood of remitting is much reduced. There is not a universally agreed definition of treatment-resistant depression (TRD), despite attempts (e.g., Fava, 2003). Many researchers use that of two antidepressant trials failing to achieve response, but it has been argued that an accurate measurement of TRD requires a multifactorial staging model, incorporating duration, severity, and treatment count (Fekadu, Wooderson, Markopoulou, & Cleare, 2009; Fekadu, Wooderson, Donaldson, et al., 2009).

Those who go on to develop TRD frequently do not respond to numerous treatments and experience chronic and disabling affective illness (Fekadu, Wooderson, Donaldson, et al., 2009). Patients with TRD present with more long-term physical and mental comorbidities with an increased mortality risk and experienced greater difficulties with global and cognitive functioning (Fekadu, Wooderson, Donaldson, et al., 2009, 2012). However, even the most treatment-resistant patients can achieve remission with specialized, multidisciplinary, intensive, and careful intervention programs (Wooderson et al., 2011). Particularly in conjunction with a strong social support network, clinical remission can also be sustained in the longer term (Fekadu et al., 2012; Wooderson et al., 2014). Other general predictors of nonresponse might include incomplete remission of previous episodes, a longer history of psychiatric disorder, a number of comorbidities, and particular personality characteristics, as well as specific genetic variants (Bennabi et al., 2015; Kornstein & Schneider, 2001).

Identifying a set of factors that reliably associates with treatment response for different interventions will be extremely useful for improving and individualizing treatment choices in both bipolar and unipolar depression and for identifying potential new targets for novel treatments. Due to the burden of TRD, an improvement to outcomes in this population has particular scope to reduce the wider costs associated with depressive disorders (Fineberg et al., 2013).

Biomarkers might be valuable candidates for identification as predictors of response to various interventions (Thase, 2014); the evidence to date suggests that circulating markers in blood (reflecting activity of neurotransmitter, neurotrophic, neuroendocrine, metabolic, and inflammatory systems) can predict mental and physical health outcomes, but these reports remain far from consistent at present (Jani et al., 2015). In addition, thorough and direct comparisons of biomarkers between TRD and non-TRD groups are scarce (Fagiolini & Kupfer, 2003).

INFLAMMATION AS A PUTATIVE PREDICTOR OF CLINICAL RESPONSE TO TREATMENT FOR DEPRESSION

Jani et al. (2015) identified 14 studies examining blood-based markers as potential predictors of response to treatments for major depressive disorder (MDD). Pertaining to the inflammatory response, low interleukin-12p70 (IL-12p70) and high interleukin-8 (IL-8) predicted nonremission of depression symptoms 2 years later (Baune et al., 2012), while C-reactive protein (CRP) elevations predicted poorer physical health outcomes in the coming years (Ladwig, Marten-Mittag, Löwel, Döring, & Koenig, 2005). In addition, a high IL-6 and a high lymphocyte/monocyte ratio were found in subsequent nonresponders to amitriptyline (Lanquillon, Krieg, Bening-Abu-Shach, & Vedder, 2000). Only the latter study specifically

assessed response to an antidepressant treatment. In contrast, in our own meta-analysis assessing peripheral IL-6, CRP, and tumor necrosis factor-alpha (TNFα) in longitudinal treatment studies for depression, high IL-6 did not predict treatment response when results from five studies were combined. However, findings suggested that higher pretreatment levels of a composite measure of pro-inflammatory proteins are found in subsequent antidepressant nonresponders (Strawbridge et al., 2015). This effect was stronger in studies of outpatient than inpatient treatment interventions and in investigations with a higher quality rating. This meta-analysis also found that TNFα is persistently elevated following antidepressant treatment in nonresponders but reduces with treatment in patients whose depression improves. It is possible that an inability to suppress TNFα production could be a physiological barrier to clinical improvement. In line with this, Eyre, Stuart, and Baune (2014) propose a staging model of inflammation that normalizes only in patients achieving remission; see Fig. 1.

The existence of nonbiological interventions for depression can facilitate an observation of inflammatory changes during treatment that are not purely attributable to biological pharmacological effects. However, this opportunity has rarely been exploited in assessing response to treatments. Harley, Luty, Carter, Mulder, and Joyce (2010) pooled two studies whose methodology was almost identical: an antidepressant randomized controlled trial (RCT) and a psychological intervention RCT for MDD. Elevated CRP predicted greater nonresponse to interpersonal psychotherapy or cognitive behavioral therapy (CBT) but a good response to nortriptyline or fluoxetine. Uher et al. (2014) also identified this effect for nortriptyline but found the opposite for escitalopram response, and Chang et al. (2012) reported both fluoxetine and venlafaxine as more efficacious for those with high CRP. This set of findings might indicate psychological or psychotropic treatment categorization as a potential candidate on which to stratify interventions using inflammatory markers if sufficiently replicated.

Studies assessing the inflammatory effects of antidepressants that have not accounted for depression severity or clinical improvement have often concluded that these medications tend to have some antiinflammatory effects but these are inconsistent and incompletely understood (Eyre, Lavretsky, Kartika, Qassim, & Baune, 2016; Janssen, Caniato, Verster, & Baune, 2010; Maes, Bosmans, et al., 1997). Different classes of antidepressants could have distinct effects: Selective serotonin reuptake inhibitors (SSRIs) may direct a shift toward Th2 activity, while serotonin and noradrenaline reuptake inhibitors (SNRIs) could yield a Th1 shift in inflammation (Hashimoto, 2015); overall, a down-regulation in Th1 activity has been proposed across antidepressant classes (Eyre, Lavretsky, et al., 2016). Individual medications have difficult-to-model, distinct neurobiological effects that will be mediated by other factors. Inconsistencies between research findings to date have arisen from a multitude of factors, including clinical and methodological heterogeneity, concomitant medication use, and treatment length and dosage. Most studies investigating inflammation longitudinally have been short-term trials (less than 12 weeks) and assessed baseline inflammation in unmedicated participants, to reduce heterogeneity (Strawbridge, Young, & Cleare, 2017). Thus, the effects described above may be initial, transient influences that may also have been subject to influences of antidepressant use prior to a washout period. Hernandez et al. (2008) suggest that there are different patterns

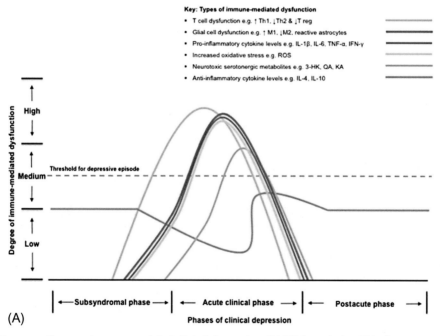

FIG. 1 A phase-specific neuroimmune model of clinical depression. (A) With remission: This figure represents an acute clinical depressive episode with full remission in the context of the three phases of the phase-specific neuroimmune model of clinical depression—subsyndromal, acute clinical, and postacute phases. The x-axis shows the relevant phases; the y-axis shows the level of immune-mediated dysfunction that can occur. The colored lines represent the various types of immune-mediated dysfunction. The gray dashed line shows the immune dysfunction threshold line whereby a clinically significant depressive episode is diagnosable.

(Continued)

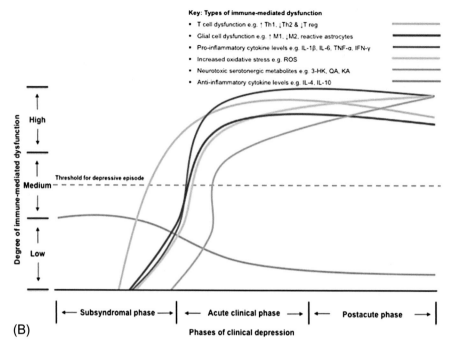

Key: Types of immune-mediated dysfunction

- T cell dysfunction e.g. ↑Th1, ↓Th2 & ↓T reg
- Glial cell dysfunction e.g. ↑M1, ↓M2, reactive astrocytes
- Pro-inflammatory cytokine levels e.g. IL-1β, IL-6, TNF-α, IFN-γ
- Increased oxidative stress e.g. ROS
- Neurotoxic serotonergic metabolites e.g. 3-HK, QA, KA
- Anti-inflammatory cytokine levels e.g. IL-4, IL-10

FIG. 1, CONT'D (B) Chronic major depressive episode with progressive depressive features and cognitive dysfunction. This figure represents a chronic major depressive episode with progressive depressive features and cognitive dysfunction. The x-axis shows the relevant phases (subsyndromal, acute clinical, and postacute); the y-axis shows the level of immune-mediated dysfunction that can occur. The colored lines represent the various types of immune-mediated dysfunction. The gray dashed line shows the immune dysfunction threshold line whereby a clinically significant depressive episode is diagnosable. *Abbreviations: IL*, interleukin; *TNF*, tumor necrosis factor; *IFN*, interferon; *BDNF*, brain-derived neurotrophic factor; *LTP*, long-term potentiation; *ROS*, reactive oxygen species; *3-HK*, 3-hydroxykynurenine; *QA*, quinolinic acid; *KA*, kynurenic acid; *Th*, T helper; *T reg*, T regulatory cell. *Reprinted with permission from Eyre, H., Stuart, M., & Baune, B. T. (2014). A phase-specific neuroimmune model of clinical depression. Progress in Neuro-Psychopharmacology and Biological Psychiatry, 54, (A) p. 270, (B) p. 271. Copyright 2014 by Elsevier.*

seen after 1 year of taking antidepressants compared with those occurring at either 5- or 20-week time points: IL-2 and interferon-γ (IFNγ) increased over the 12 months after an initial decrease in levels; antiinflammatory cytokines IL-10 and IL-13 decreased gradually between each assessment; IL-1β rose after an initial lack of change; IL-4 fluctuated markedly between measurements. Like many other studies, this investigation recruited a small sample size (particularly at follow-up), where all participants were unmedicated at baseline and in clinical remission at end point, having received nonstandardized treatment. Some studies excluded only

psychotropic medication use, meaning that other medications could be influencing inflammatory activity: use of oral contraceptives is often not excluded in studies or adjusted for in analyses, despite evidence that it can increase inflammatory activity (Divani, Luo, Datta, Flaherty, & Panoskaltsis-Mortari, 2015).

Thus, the insights on the relationship between inflammation, depressive state, and treatment to date are limited in their ability to explain the effects of specific treatments over a variety of time frames, their attribution to treatment mechanisms or clinical states, and how this varies across individuals. The inflammatory effects of medications are especially problematic for patients with TRD because treatment for TRD typically comprises multiple antidepressant trials (including dose escalation and switching and/or combining medications) in conjunction with occupational and psychological therapies. Ethical concerns about titrating TRD patients off medication may contribute to the relative scarcity of studies testing this phenomenon in specifically treatment-resistant populations. However, this may present an opportunity to undertake naturalistic studies where participants continue on stable medication programs in order to examine inflammatory changes with treatment that are more enduring and ecologically valid and that distinguish between participants based on clinical improvements to permit an indication of the confluence between treatment outcomes and inflammation longitudinally.

INFLAMMATION AS A PREDICTOR OF TREATMENT OUTCOMES IN TREATMENT-RESISTANT DEPRESSION

It has been estimated that almost 900 articles are published per year on the topic of biomarkers for depression but less than 50 for specifically TRD (Smith, 2013).

Early studies assessing markers of inflammation in conjunction with response to treatment identified no associations between responders and nonresponders or changes alongside treatment, but did report abnormalities in TRD compared with control participants: for dipeptidyl peptidase-IV (DPP-IV) (Maes, De Meester, et al., 1997), zinc and T-cell number (Maes, Delange, et al., 1997), and the more traditional inflammatory markers IL-6 and IL-1RA (Maes, Vandoolaeghe, et al., 1997). The latter study indicated similar immune alterations in non-TRD and TRD depression compared with controls, while most studies of TRD populations do not include an MDD comparison group of

patients. These studies all defined TRD as having two unsuccessful antidepressant trials of different classes but contained small numbers of non-TRD participants. Another inpatient TRD investigation (where patients had experienced a minimum of nonresponse to one previous treatment) more recently identified low MCP1 as a predictor of nonresponse to treatment (Carvalho et al., 2013) despite no overall difference between patients and controls and contrary to an overall elevation of this chemokine seen in MDD versus control participants (Eyre, Air, et al., 2016). In a similar, highly TRD sample, we recently reported that attenuated pretreatment IL-2 was linked to short-term nonresponse to multidisciplinary inpatient intervention (comprising pharmacological, occupational, and psychological therapies), but elevated posttreatment CRP, IL-6, and monocyte chemoattractant protein-4 (MCP4) predicted a poorer long-term outcome over the 3–12 months following discharge from hospital (Strawbridge, Powell, Breen, Young, & Cleare, 2016). This occurred despite an increase in many cytokines during the inpatient intervention. The participants in this study had TRD as assessed by the multifactorial Maudsley Staging Method (Fekadu, Wooderson, Markopoulou, et al., 2009), with scores above a cutoff of 7.5 as has been proposed to define TRD (Trevino, 2012).

While possessing high inflammation is posited to predict a poorer overall response, it may be that these patients would derive more pronounced benefit from a specifically targeted antiinflammatory treatment program. Those who were resistant to standard antidepressant regimes and had high levels of CRP were more likely to respond to the TNFα antagonist infliximab than those with less than 5 mg/L CRP (Raison, Felger, & Miller, 2013). Responders and nonresponders in this trial were later found to possess a network of different gene expression markers of glucose and lipid (including cholesterol) metabolism (Mehta et al., 2013). As well as infliximab, etanercept is another

TNFα antagonist that may even be useful as a monotherapy for depression (Schmidt, Kirkby, & Himmerich, 2014). It is possible that reducing the levels of TNFα improves well-being in a subset of patients through a positive influence on sleep quality (Weinberger et al., 2015). Minocycline has more generalized effects on inflammation and in one trial significantly reduces depression severity in a TRD sample (Husain et al., 2017).

Other novel treatments for depression, such as ketamine or erythropoietin (EPO), may indirectly affect the inflammatory response. In a similar fashion to electroconvulsive therapy (ECT) (Guloksuz, Rutten, Arts, van Os, & Kenis, 2014), ketamine may have acute stimulatory effects on pro-inflammatory markers (e.g., IL-6) that could reflect a nonspecific physiological stress response (Park et al., 2017) but decrease inflammatory activity over a more prolonged period. However, despite preclinical indications that high inflammation (particularly TNFα) might predict a good response to ketamine (Walker et al., 2015), this was not verified by Park et al. (2017) in a TRD population. Similarly, while EPO is likely to have nonspecific antiinflammatory effects, this was only observed over a 9-week trial in TRD patients who had elevated CRP levels prior to treatment (Vinberg, Weikop, Olsen, Kessing, & Miskowiak, 2016). Pramipexole or other D3 dopamine agonists have been heralded as putative treatments for TRD that may act through inflammatory pathways, but evidence is needed to establish this (Escalona & Fawcett, 2017).

AN "INFLAMMATORY SUBTYPE" OF DEPRESSION

Although much evidence has previously linked abnormal levels of inflammatory proteins with the presence of depression, there is inconsistency in the literature, and it may be that inflammation is only dysregulated in some subpopulations of mood disorders: Depression is a highly heterogeneous condition, and inflammation is a highly complex system with an extensive array of biomarkers representing different aspects of the inflammatory response that has been reported as overactive, normal, or suppressed in MDD (Blume, Douglas, & Evans, 2011).

The quantity of research and findings signifying the importance of this biological system has led to many articles positing that a subgroup of depression exists for which inflammation is important (Krishnadas & Cavanagh, 2012; Vogelzangs et al., 2012), but few strategic attempts have been made to characterize this group.

In general, attempts to identify homogenous subsets of patients with depression have not assisted to date with improving estimations of which patients are likely to respond to any specific treatments (Arnow et al., 2015). Most hypothesized subtypes have explored symptom-based experiences, but biomarker-based subtypes could be more reliably distinguished between patients. Kunugi, Hori, and Ogawa (2015) proposed that different neurobiological systems might display clinically relevant subtypes in depression, including groups displaying hypercortisolism or hypocortisolism (reflecting melancholic and atypical subtypes, respectively), a dopamine-related subset (prominently experiencing anhedonia, for whom dopamine agonists may be beneficial), and a further group characterized by elevated inflammation.

The most frequent subtype delineation in depression has been between atypical and melancholic depression. Melancholia has been associated with higher cortisol but lower inflammation than atypical depression: in the largest study comparing subtypes to date, only people with atypical depression had significantly elevated inflammation compared with controls. These individuals also reported more extensive metabolic symptoms; controlling for metabolic syndrome rendered the differences in CRP and

IL-6 nonsignificant, though TNFα remained significantly higher (Lamers et al., 2013). Other clinical factors differing between these groups are of note: atypical depression has been more frequently associated with childhood trauma (Withers, Tarasoff, & Stewart, 2013), chronic life stressors, and an early age of onset (Gold & Chrousos, 2002), which have all been associated with an overactive inflammatory response (Vogelzangs et al., 2012). Sleep dysregulations also appear to cause elevated cytokine levels; both insomnia (Irwin, 2015) and hypersomnia (Szymusiak & Gvilia, 2012) are often experienced by patients with melancholic and atypical depression, respectively.

It is, however, apparent that the often used melancholic and atypical subtypes do not represent the entire depressed population and the differing symptoms are not mutually exclusive. The existence of inflammatory abnormalities in melancholic or atypical depression is not consistently reported: Dahl et al. (2014) did not find associations with inflammation for either, while Karlović, Serretti, Vrkić, Martinac, and Marčinko (2012) report aberrant inflammation in both subtypes. Both melancholic and atypical depression appear more prevalent in treatment-resistant than non-resistant depression. Given the overlap between these subtypes, the physical symptoms of sickness behavior and somatic depression, there is scope for more than one symptomatic subgroup within depressed patients to be considered as potential inflammatory subtypes.

Sickness behavior is an adaptive response to immune challenge that developed early in human evolution (Miller & Raison, 2016) and phenomenologically resembles depression in some aspects, involving dysregulated sleep and appetite, fatigue and lassitude, aches or pain, anhedonia, and cognitive difficulties. Maes et al. (2012) outline the substantial overlap between the characteristics of depression and sickness behavior alongside the release of cytokines, also theorizing that as the number of depressive episodes increases, rises in inflammation become more sensitized, rendering individuals more susceptible to developing recurrent and treatment-resistant depression. Similarly, there have been indications that an inflammation-related subgroup of depression might be confined to those with prominently somatic symptoms (Penninx, Milaneschi, Lamers, & Vogelzangs, 2013) or explicitly as TRD patients (Raison, Rutherford, et al., 2013).

CORRELATES BETWEEN TRD AND INFLAMMATORY STATES

Many parallels are apparent between the characteristics of TRD and factors that stimulate inflammatory activity. Table 1 outlines some of these. Prominent correlates include adiposity and metabolic syndrome, as well as a wide variety of medical illnesses, including autoimmune disorders. It has been reported that depression diagnosed as occurring secondary to a physical disorder is more likely to enter a chronic illness course (in spite of comparable treatment) than for patients with primary MDD (Keller, Lavori, Rice, Coryell, & Hirschfeld, 1986). People with TRD similarly take more medications, both psychotropic and nonpsychotropic: many compounds indirectly influence inflammatory activity via pleiotropic mechanisms. Even within mood disorder diagnoses, depression that is psychotic is more likely treated with atypical antipsychotics, atypical depression may be better managed by monoamine inhibitors (MAOIs), bipolar depression with mood stabilizers, and menstrual-related depression with SSRIs (Kornstein & Schneider, 2001). These classes of medication all likely have different influences on inflammation, and these subpopulations may relate differently to inflammation than one another.

The clinical and biological effects of stress (both in the short and long term) are likely to be important; both acute and chronic

TABLE 1   Features Associated With Both Treatment Resistance and Inflammation

Parallel Between Inflammation and Treatment Resistance	Summary of Relationship/Key Findings	Key Reference(s)
Physical illness (e.g., diabetes, multiple sclerosis, rheumatoid arthritis, irritable bowel syndrome, cardiovascular disease, Parkinson's disease, hypothyroidism, Cushing's disease, Addison's disease, coronary artery disease, cancer, fibromyalgia, chronic fatigue syndrome, and HIV)	Many of the most common physical comorbidities with depression have an inflammatory link. Recurrent depressive symptoms are associated with future inflammation in stable coronary heart disease patients	Maes, Kubera, Obuchowiczwa, Goehler, and Brzeszcz (2011) Duivis et al. (2011)
Psychiatric comorbidity (e.g., anxiety, substance use, personality, and eating disorders)	Aberrant inflammatory activity has been identified across many diagnoses in psychiatry	Mondelli et al. (2015) Najjar, Pearlman, Alper, Najjar, and Devinsky (2013)
Severity of depressive symptoms	Correlations between proteomic inflammatory markers and severity of depression have often been identified	Anisman, Ravindran, Griffiths, and Merali (1999)
Chronicity of depression	Indications of chronicity have been associated with higher inflammatory markers	Vogelzangs et al. (2014)
Recurrence of episodes	Recurrence of depression has been linked with heightened inflammation, although this appears partially mediated by health and lifestyle factors	Duivis et al. (2011)
Early-life stress (ELS)	ELS increases inflammatory sensitivity, especially in people with concurrent depression, but depression may be secondary in this relationship	Danese and McEwen (2012)
Cognitive deficits	Cognitive deficits appear cumulative alongside treatment resistance. Elevated inflammation has been linked with cognitive decline and likely affects cognitive functioning in mood disorders even in euthymic states	Allison and Ditor (2014) Bauer, Pascoe, Wollenhaupt-Aguiar, Kapczinski, and Soares (2014)
Metabolic factors	The lack of adjustment for BMI confounds large differences reported between depressed and control groups. Adipose tissue stimulates macrophages and cytokines. Obesity is associated with TRD or atypical depression (may also relate to psychotropic use)	Lamers et al. (2013) Howren, Lamkin, and Suls (2009)
Psychotic symptoms	Both affective and nonaffective psychoses are associated with elevated inflammatory states	Bergink, Gibney, and Drexhage (2014)
Bipolar disorder	Patients classified as TRD can be found to have undetected symptoms of bipolar disorders. Bipolar disorder is also associated with aberrant inflammation, which may be more pronounced in manic phases or bipolar compared with unipolar depression, although this has not been confirmed	Hirschfeld, Lewis, and Vornik (2003) Bai et al. (2014)

Additional factors such as gender, family history, and age of onset have more inconsistently been found to affect likelihood of treatment resistance and/or inflammatory activity.
Abbreviations: HIV, human immunodeficiency virus; *ELS*, early-life stress; *BMI*, body mass index; *TRD*, treatment-resistant depression.

psychological stressors augment the inflammatory response, in the shorter and longer term (Miller, Rohleder, & Cole, 2009; Steptoe, Hamer, & Chida, 2007). People who experience depression and ongoing stress may consequently undergo an increasingly sensitized inflammatory response that could magnify the likelihood of treatment-resistant mood disorders in a progressive pattern. Cognitive dysfunction in refractory depression may also develop progressively alongside more long-lasting or severe episodes (Gallagher, Robinson, Gray, Young, & Porter, 2007; Kemp, Gordon, Rush, & Williams, 2008). In turn, peripheral cytokine increases (whether transient or chronic) can stimulate neuroinflammatory states that are associated with reduced brain-derived neurotrophic factor (BDNF) and ultimately impact upon cognitive function (Bauer et al., 2014).

CHALLENGES IN UNDERSTANDING THE ROLE OF INFLAMMATION IN TRD

Many of the factors often present in TRD complicate the interpretation of the inflammation literature. Taking multiple medications for different amounts of time will have a difficult-to-model effect on proteins or gene expression. For example, most short-term antidepressant trials have identified decreases in cytokines (Strawbridge et al., 2015), but one of the few long-term studies reported an increase in pro-inflammatory proteins over 1 year (Hernandez et al., 2008).

In addition, the underrepresentation of TRD research is accentuated in this area as many studies have utilized a definition of TRD as failure to respond to one antidepressant (e.g., Carvalho et al., 2013), or are not comparable between studies due to variation in the definition used. Related to this challenge is the importance in being able to differentiate between authentic TRD and pseudoresistant depression, associated, for example, with inadequately delivered treatment, poor adherence, or treatment intolerance. Many investigations dichotomize patients into responders and nonresponders based on a 50% symptom-score reduction on a commonly used measure of depression; while this provides a potentially clinically relevant set of individuals, dichotomization also succumbs to data loss as depression, and treatment response/resistance is a complex phenomenon that occurs across a spectrum (Altman & Royston, 2006). Remission is arguably a more reflective measure of the ultimate clinical aim for treatments (Keller, 2005) but does not account for the amount of improvement over pretreatment severity. Almost all studies only assess treatment response at a single time point at a short duration from baseline, and thus do not detect sustained well-being or occurrences of relapse; residual symptoms and a fluctuating pattern of symptoms over time may be especially common in TRD and be better assessed using longitudinal evaluations (Fekadu et al., 2011). Thus, alternative definitions of treatment success could include an average depression score over time with treatment or multiple follow-up assessments, the degree of fluctuation in depression scores, or the discrimination between early response and nonresponse during treatment to yield more valid outcome data. Assessing a longer duration of follow-up may be a practical solution that reflects enduring treatment success.

It is important to note that neurobiological differences in TRD are not confined to the inflammatory response, and likely interact with neuroendocrine (e.g., cortisol; Fischer, Strawbridge, Herane Vives, & Cleare, 2017), neurotransmitter (Perez et al., 1998), and neurotrophic (Wolkowitz et al., 2011) systems, which also may play a role in the success of treatments in individuals with depression.

TREATMENT IMPLICATIONS

There are a number of directions and opportunities that emerge from the research discussed above. Below, we discuss the questions that need to be answered and the potential benefits of these for clinical practice.

Could Inflammation Prospectively Predict Risk for Treatment-Resistance?

It has been proposed many times that markers of inflammation could predict a poor outcome in response to antidepressant treatments in depression (Cattaneo et al., 2016; Strawbridge et al., 2015), although the opposite has also been found and replications have been insufficient (Chang et al., 2012; Uher et al., 2014). One putative explanation for this is that high inflammation predicts a risk for generalized treatment-resistant or recurrent depression. It is uncertain whether this is the case as few trials examine the long-term outcome of treatments or the course of depressive illness after interventions. However, the implications of this possibility are substantial; enhanced care and monitoring could be provided for at-risk individuals to increase the probability of achieving remission (at any clinical stage) that could have profound repercussions for reducing the burden of TRD. Specifically, measures of inflammation could help determine the risk of future relapse and need for/length of continuation therapies either utilizing single biomarker values, a network or composite of markers, or biomarkers in conjunction with psychosocial or clinical information.

Does Inflammation Reflect Retrospective Accumulation of TRD?

If it becomes apparent that inflammation increases alongside longer, more severe, or recurrent depressive episodes, then it is plausible that TRD represents an "inflammatory subtype." Cytokine measurements may even provide an additional measure of TRD burden. There is not yet empirical evidence to support this theory: Maes, Vandoolaeghe, et al. (1997); Maes, Delange, et al. (1997) did not find statistically relevant differences between those with different extents of TRD, but these findings were based on very small samples of which many had low levels of treatment resistance. To our knowledge, TRD and non-TRD groups have not been comprehensively compared in terms of their inflammatory profiles, although we have identified significant positive correlations between the extent of treatment resistance and pro-inflammatory markers (including IL-6, CRP, TNFα, and TNFβ; *unpublished data*) in a sample of TRD inpatients.

Regardless of whether it comprises those with TRD or another subpopulation with MDD, if an inflammatory subtype of depression can be determined and established in the future, then this has considerable implications for targeting and optimizing interventions. It also has the potential for more accurate diagnosis and individualizing treatment selection; this group may benefit from specific treatments that target elevation inflammation, including existing or repurposed medications, or novel interventions.

Might Inflammation Provide Targets for Treating a Subset of Depressed Patients?

The potential for existing antiinflammatory treatments, such as TNFα antagonists (Raison, Felger, et al., 2013; Tyring et al., 2006) or IL-6 antagonists (Wolfe, 2016), is already showing promise for mood disorders. Multiple systematic reviews and meta-analysis now suggest that overall, short-term treatment with direct antiinflammatory medication significantly reduces the severity of depressive symptoms (Köhler et al., 2014; Rosenblat et al., 2016). The vast majority of trials have investigated non-TRD depressive disorders: while Raison, Felger, et al. (2013) and Husain et al. (2017) suggest that these are efficacious for TRD, the effectiveness of antiinflammatory medications has not been compared between potential subgroups. Longer-term efficacy and effects on relapse are also as yet untested.

Could Inflammatory Markers be Used to Personalize Treatment for Depression?

Each of the above questions could have the potential to indicate stratification to the most appropriate treatments (for instance a short-term course of an antiinflammatory medication taken concomitantly with a "stepped-up" psychotherapy program). Previously, high inflammation has predicted a poor response to psychological therapies (Harley et al., 2010), a good response to antiinflammatory medication (Raison, Felger, et al., 2013), and generally a poorer response to antidepressants, albeit with mixed findings (Strawbridge et al., 2015). Thus, patients with high inflammation could be stratified to receive antiinflammatory treatment as opposed to stand-alone psychological therapy. A more personalized treatment-selection strategy may inspect a range of biomarkers in individuals and, for example, prescribe a TNFα antagonist to a patient with particularly high TNFα levels.

CONCLUDING REMARKS

At present, it is challenging to detangle the effects of pharmacology and clinical improvements in understanding longitudinal inflammatory alterations in depression, notwithstanding other characteristics such as physical illness, cognitive difficulties, or the specific symptoms experienced. A number of studies indicate a range of negative consequences of elevated inflammation in depressed populations when measured retrospectively, cross-sectionally, or prospectively. A concerning feature of this

literature to date is the lack of reproducibility between studies; another is that very few studies have investigated a range of inflammatory proteins in severely treatment-resistant depressed populations, either in comparison with controls or longitudinally alongside treatment. The scarcity of well-defined TRD studies is partly a result of nonstandardized definitions of TRD, severity of illness as a barrier to research participation, and influence of medication on study outcomes. We recommend that future investigations assess a wide array of inflammatory markers in large samples in conjunction with salient clinical and sociodemographic factors, before and after specific (including psychological) treatments with an optimized measure of treatment outcome. Due to the burden and progressive nature of TRD, the ability to improve the rate of treatment response would be especially beneficial in this population and we highlight that the evidence supports the potential for this in a subpopulation of people with depression.

References

Allison, D. J., & Ditor, D. S. (2014). The common inflammatory etiology of depression and cognitive impairment: a therapeutic target. *Journal of Neuroinflammation, 11*(1), 151.

Altman, D. G., & Royston, P. (2006). The cost of dichotomising continuous variables. *BMJ, 332,* 1080.

Anisman, H., Ravindran, A., Griffiths, J., & Merali, Z. (1999). Endocrine and cytokine correlates of major depression and dysthymia with typical or atypical. *Molecular Psychiatry, 4,* 182–188.

Arnow, B. A., Blasey, C., Williams, L. M., Palmer, D. M., Rekshan, W., Schatzberg, A. F., et al. (2015). Depression subtypes in predicting antidepressant response: a report from the iSPOT-D trial. *American Journal of Psychiatry, 172,* 743–750.

Bai, Y. M., Su, T. P., Tsai, S. J., Wen-Fei, C., Li, C. T., Pei-Chi, T., et al. (2014). Comparison of inflammatory cytokine levels among type I/type II and manic/hypomanic/euthymic/depressive states of bipolar disorder. *Journal of Affective Disorders, 166,* 187–192.

Bauer, I. E., Pascoe, M. C., Wollenhaupt-Aguiar, B., Kapczinski, F., & Soares, J. C. (2014). Inflammatory mediators of cognitive impairment in bipolar disorder. *Journal of Psychiatric Research, 56,* 18–27.

Baune, B. T., Smith, E., Reppermund, S., Air, T., Samaras, K., Lux, O., et al. (2012). Inflammatory biomarkers predict depressive, but not anxiety symptoms during aging: the prospective Sydney Memory and Aging Study. *Psychoneuroendocrinology, 37*, 1521–1530.

Bennabi, D., Aouizerate, B., El-Hage, W., Doumy, O., Moliere, F., Courtet, P., et al. (2015). Risk factors for treatment resistance in unipolar depression: a systematic review. *Journal of Affective Disorders, 171*, 137–141.

Bergink, V., Gibney, S. M., & Drexhage, H. A. (2014). Autoimmunity, inflammation, and psychosis: a search for peripheral markers. *Biological Psychiatry, 75*(4), 324–331.

Blume, J., Douglas, S. D., & Evans, D. L. (2011). Immune suppression and immune activation in depression. *Brain, Behavior, and Immunity, 25*, 221–229.

Carvalho, L., Torre, J., Papadopoulos, A., Poon, L., Juruena, M., Markopoulou, K., et al. (2013). Lack of clinical therapeutic benefit of antidepressants is associated overall activation of the inflammatory system. *Journal of Affective Disorders, 148*, 136–140.

Cattaneo, A., Ferrari, C., Uher, R., Bocchio-Chiavetto, L., Riva, M. A., Pariante, C. M., et al. (2016). Absolute measurements of macrophage migration inhibitory factor and interleukin-1-β mRNA levels accurately predict treatment response in depressed patients. *International Journal of Neuropsychopharmacology*.

Chang, H. H., Lee, I. H., Gean, P. W., Lee, S.-Y., Chi, M. H., Yang, Y. K., et al. (2012). Treatment response and cognitive impairment in major depression: association with C-reactive protein. *Brain, Behavior, and Immunity, 26*, 90–95.

Dahl, J., Ormstad, H., Aass, H. C., Malt, U. F., Bendz, L. T., Sandvik, L., et al. (2014). The plasma levels of various cytokines are increased during ongoing depression and are reduced to normal levels after recovery. *Psychoneuroendocrinology, 45*, 77–86.

Danese, A., & McEwen, B. S. (2012). Adverse childhood experiences, allostasis, allostatic load, and age-related disease. *Physiology and Behavior, 106*(1), 29–39.

Divani, A. A., Luo, X., Datta, Y. H., Flaherty, J. D., & Panoskaltsis-Mortari, A. (2015). Effect of oral and vaginal hormonal contraceptives on inflammatory blood biomarkers. *Mediators of Inflammation, 2015*, 8. https://doi.org/10.1155/2015/379501.

Duivis, H. E., de Jonge, P., Penninx, B. W., Na, B. Y., Cohen, B. E., & Whooley, M. A. (2011). Depressive symptoms, health behaviors, and subsequent inflammation in patients with coronary heart disease: prospective findings from the heart and soul study. *American Journal of Psychiatry, 168*(9), 913–920.

Escalona, R., & Fawcett, J. (2017). Pramipexole in treatment resistant-depression, possible role of inflammatory cytokines. *Neuropsychopharmacology, 42*, 363.

Eyre, H., Lavretsky, H., Kartika, J., Qassim, A., & Baune, B. T. (2016). Modulatory effects of antidepressant classes on the innate and adaptive immune system in depression. *Pharmacopsychiatry, 49*, 85–96.

Eyre, H., Stuart, M., & Baune, B. T. (2014). A phase-specific neuroimmune model of clinical depression. *Progress in Neuro-Psychopharmacology and Biological Psychiatry, 54*, 265–274.

Eyre, H. A., Air, T., Pradhan, A., Johnston, J., Lavretsky, H., Stuart, M. J., et al. (2016). A meta-analysis of chemokines in major depression. *Progress in Neuro-Psychopharmacology and Biological Psychiatry, 68*, 1–8.

Fagiolini, A., & Kupfer, D. J. (2003). Is treatment-resistant depression a unique subtype of depression? *Biological Psychiatry, 53*, 640–648.

Fava, M. (2003). Diagnosis and definition of treatment-resistant depression. *Biological Psychiatry, 53*(8), 649–659.

Fekadu, A., Rane, L. J., Wooderson, S. C., Markopoulou, K., Poon, L., & Cleare, A. J. (2012). Prediction of longer-term outcome of treatment-resistant depression in tertiary care. *British Journal of Psychiatry, 201*(5), 369–375.

Fekadu, A., Wooderson, S. C., Donaldson, C., Markopoulou, K., Masterson, B., Poon, L., et al. (2009). A multidimensional tool to quantify treatment resistance in depression: the Maudsley staging method. *Journal of Clinical Psychiatry, 70*, 177–184.

Fekadu, A., Wooderson, S. C., Markopoulou, K., & Cleare, A. J. (2009). The Maudsley Staging Method for treatment-resistant depression: prediction of longer-term outcome and persistence of symptoms. *Journal of Clinical Psychiatry, 70*, 952–957.

Fekadu, A., Wooderson, S. C., Rane, L. J., Markopoulou, K., Poon, L., & Cleare, A. J. (2011). Long-term impact of residual symptoms in treatment-resistant depression. *The Canadian Journal of Psychiatry, 56*, 549–557.

Fineberg, N. A., Haddad, P. M., Carpenter, L., Gannon, B., Sharpe, R., Young, A. H., et al. (2013). The size, burden and cost of disorders of the brain in the UK. *Journal of Psychopharmacology, 27*(9), 761–770.

Fischer, S., Strawbridge, R., Herane Vives, A., & Cleare, A. J. (2017). Cortisol as a predictor of psychological therapy response in depressive disorders—a systematic review and meta-analysis. *British Journal of Psychiatry, 210*(2), 105–109.

Gallagher, P., Robinson, L., Gray, J., Young, A. H., & Porter, R. (2007). Neurocognitive function following remission in major depressive disorder: potential objective marker of response? *Australian and New Zealand Journal of Psychiatry, 41*, 54–61.

Gaynes, B. N., Warden, D., Trivedi, M. H., Wisniewski, S. R., Fava, M., & Rush, A. J. (2009). What did STAR*D teach us? Results from a large-scale, practical, clinical trial for patients with depression. *Psychiatric Services, 60*, 1439–1445.

Gold, P., & Chrousos, G. (2002). Organization of the stress system and its dysregulation in melancholic and atypical depression: high vs low CRH/NE states. *Molecular Psychiatry, 7*, 254–275.

Guloksuz, S., Rutten, B. P., Arts, B., van Os, J., & Kenis, G. (2014). The immune system and electroconvulsive therapy for depression. *The Journal of ECT, 30*, 132–137.

Harley, J., Luty, S., Carter, J., Mulder, R., & Joyce, P. (2010). Elevated C-reactive protein in depression: a predictor of good long-term outcome with antidepressants and poor outcome with psychotherapy. *Journal of Psychopharmacology, 24*, 625–626.

Hashimoto, K. (2015). Inflammatory biomarkers as differential predictors of antidepressant response. *International Journal of Molecular Sciences, 16*, 7796–7801.

Hernandez, M. E., Mendieta, D., Martinez-Fong, D., Loria, F., Moreno, J., Estrada, I., et al. (2008). Variations in circulating cytokine levels during 52 week course of treatment with SSRI for major depressive disorder. *European Neuropsychopharmacology, 18*, 917–924.

Hirschfeld, R. M., Lewis, L., & Vornik, L. A. (2003). Perceptions and impact of bipolar disorder: how far have we really come? Results of the national depressive and manic-depressive association 2000 survey of individuals with bipolar disorder. *The Journal of Clinical Psychiatry, 64* (2), 161–174.

Howren, M. B., Lamkin, D. M., & Suls, J. (2009). Associations of depression with C-reactive protein, IL-1, and IL-6: a meta-analysis. *Psychosomatic Medicine, 71*(2), 171–186.

Husain, M. I., Chaudhry, I. B., Husain, N., Khoso, A. B., Rahman, R. R., Hamirani, M. M., et al. (2017). Minocycline as an adjunct for treatment-resistant depressive symptoms: a pilot randomised placebo-controlled trial. *Journal of Psychopharmacology, 31*(9), 1166–1175.

Irwin, M. R. (2015). Why sleep is important for health: a psychoneuroimmunology perspective. *Annual Review of Psychology, 66*, 143–172.

Jani, B. D., McLean, G., Nicholl, B. I., Barry, S. J., Sattar, N., Mair, F. S., et al. (2015). Risk assessment and predicting outcomes in patients with depressive symptoms: a review of potential role of peripheral blood based biomarkers. *Frontiers in Human Neuroscience, 9*, 18.

Janssen, D. G., Caniato, R. N., Verster, J. C., & Baune, B. T. (2010). A psychoneuroimmunological review on cytokines involved in antidepressant treatment response. *Human Psychopharmacology, 25*, 201–215.

Karlović, D., Serretti, A., Vrkić, N., Martinac, M., & Marčinko, D. (2012). Serum concentrations of CRP, IL-6, TNF-α and cortisol in major depressive disorder with melancholic or atypical features. *Psychiatry Research, 198*, 74–80.

Keller, M. B. (2005). Issues in treatment-resistant depression. *Journal of Clinical Psychiatry, 66*(Suppl 8), 5–12.

Keller, M. B., Lavori, P. W., Rice, J., Coryell, W., & Hirschfeld, R. M. (1986). The persistent risk of chronicity in recurrent episodes of nonbipolar major depressive disorder: a prospective follow-up. *American Journal of Psychiatry, 143*, 24–28.

Kemp, A. H., Gordon, E., Rush, A. J., & Williams, L. M. (2008). Improving the prediction of treatment response in depression: integration of clinical, cognitive, psychophysiological, neuroimaging, and genetic measures. *CNS Spectrums, 13*, 1066–1086.

Köhler, O., Benros, M. E., Nordentoft, M., Farkouh, M. E., Iyengar, R. L., Mors, O., et al. (2014). Effect of anti-inflammatory treatment on depression, depressive symptoms, and adverse effects: a systematic review and meta-analysis of randomized clinical trials. *JAMA Psychiatry, 71*, 1381–1391.

Kornstein, S. G., & Schneider, R. K. (2001). Clinical features of treatment-resistant depression. *Journal of Clinical Psychiatry, 62*, 18–25.

Krishnadas, R., & Cavanagh, J. (2012). Depression: an inflammatory illness? *Journal of Neurology, Neurosurgery & Psychiatry, 83*, 495–502.

Kunugi, H., Hori, H., & Ogawa, S. (2015). Biochemical markers subtyping major depressive disorder. *Psychiatry and Clinical Neurosciences, 69*, 597–608.

Ladwig, K.-H., Marten-Mittag, B., Löwel, H., Döring, A., & Koenig, W. (2005). C-reactive protein, depressed mood, and the prediction of coronary heart disease in initially healthy men: results from the MONICA-KORA Augsburg Cohort Study 1984–1998. *European Heart Journal, 26*, 2537–2542.

Lamers, F., Vogelzangs, N., Merikangas, K., De Jonge, P., Beekman, A., & Penninx, B. (2013). Evidence for a differential role of HPA-axis function, inflammation and metabolic syndrome in melancholic versus atypical depression. *Molecular Psychiatry, 18*, 692–699.

Lanquillon, S., Krieg, J. C., Bening-Abu-Shach, U., & Vedder, H. (2000). Cytokine production and treatment response in major depressive disorder. *Neuropsychopharmacology, 22*, 370–379.

Maes, M., Berk, M., Goehler, L., Song, C., Anderson, G., Galecki, P., et al. (2012). Depression and sickness behavior are Janus-faced responses to shared inflammatory pathways. *BMC Medicine, 10*, 66.

Maes, M., Bosmans, E., De Jongh, R., Kenis, G., Vandoolaeghe, E., & Neels, H. (1997). Increased serum IL-6 and IL-1 receptor antagonist concentrations in major depression and treatment resistant depression. *Cytokine, 9*, 853–858.

Maes, M., De Meester, I., Verkerk, R., De Medts, P., Wauters, A., Vanhoof, G., et al. (1997). Lower serum dipeptidyl peptidase IV activity in treatment resistant major depression: relationships with immune-inflammatory markers. *Psychoneuroendocrinology, 22*, 65–78.

Maes, M., Delange, J., Ranjan, R., Meltzer, H. Y., Desnyder, R., Cooremans, W., et al. (1997). Acute phase proteins in schizophrenia, mania and major depression: modulation by psychotropic drugs. *Psychiatry Research, 66*, 1–11.

Maes, M., Kubera, M., Obuchowiczwa, E., Goehler, L., & Brzeszcz, J. (2011). Depression's multiple comorbidities explained by (neuro)inflammatory and oxidative & nitrosative stress pathways. *Neuroendocrinology Letters, 32*(1), 7–24.

Maes, M., Vandoolaeghe, E., Neels, H., Demedts, P., Wauters, A., Meltzer, H. Y., et al. (1997). Lower serum zinc in major depression is a sensitive marker of treatment resistance and of the immune/inflammatory response in that illness. *Biological Psychiatry*, 42, 349–358.

Mehta, D., Raison, C. L., Woolwine, B. J., Haroon, E., Binder, E. B., Miller, A. H., et al. (2013). Transcriptional signatures related to glucose and lipid metabolism predict treatment response to the tumor necrosis factor antagonist infliximab in patients with treatment-resistant depression. *Brain, Behavior, and Immunity*, 31, 205–215.

Miller, A. H., & Raison, C. L. (2016). The role of inflammation in depression: from evolutionary imperative to modern treatment target. *Nature Reviews Immunology*, 16(1), 22–34.

Miller, G., Rohleder, N., & Cole, S. W. (2009). Chronic interpersonal stress predicts activation of pro-and anti-inflammatory signaling pathways six months later. *Psychosomatic Medicine*, 71, 57.

Mondelli, V., Ciufolini, S., Murri, M. B., Bonaccorso, S., Di Forti, M., Giordano, A., et al. (2015). Cortisol and inflammatory biomarkers predict poor treatment response in first episode psychosis. *Schizophrenia Bulletin*, 028.

Najjar, S., Pearlman, D. M., Alper, K., Najjar, A., & Devinsky, O. (2013). Neuroinflammation and psychiatric illness. *Journal of Neuroinflammation*, 10(1), 43.

Park, M., Newman, L. E., Gold, P. W., Luckenbaugh, D. A., Yuan, P., Machado-Vieira, R., et al. (2017). Change in cytokine levels is not associated with rapid antidepressant response to ketamine in treatment-resistant depression. *Journal of Psychiatric Research*, 84, 113–118.

Penninx, B. W., Milaneschi, Y., Lamers, F., & Vogelzangs, N. (2013). Understanding the somatic consequences of depression: biological mechanisms and the role of depression symptom profile. *BMC Medicine*, 11(1), 129.

Perez, V., Bel, N., Celada, P., Ortiz, J., Alvarez, E., & Artigas, F. (1998). Relationship between blood serotonergic variables, melancholic traits, and response to antidepressant treatments. *Journal of Clinical Psychopharmacology*, 18, 222–230.

Raison, C. L., Felger, J. C., & Miller, A. H. (2013). Inflammation and treatment resistance in major depression: the perfect storm. *The Psychiatric Times*, 30(9).

Raison, C. L., Rutherford, R. E., Woolwine, B. J., Shuo, C., Schettler, P., Drake, D. F., et al. (2013). A randomized controlled trial of the tumor necrosis factor antagonist infliximab for treatment-resistant depression: the role of baseline inflammatory biomarkers. *JAMA Psychiatry*, 70, 31–41.

Rosenblat, J. D., Kakar, R., Berk, M., Kessing, L. V., Vinberg, M., Baune, B. T., et al. (2016). Anti-inflammatory agents in the treatment of bipolar depression: a systematic review and meta-analysis. *Bipolar Disorders*, 18, 89–101.

Schmidt, F. M., Kirkby, K. C., & Himmerich, H. (2014). The TNF-alpha inhibitor etanercept as monotherapy in treatment-resistant depression-report of two cases. *Psychiatria Danubina*, 26, 288–290.

Smith, D. F. (2013). Quest for biomarkers of treatment-resistant depression: shifting the paradigm toward risk. *Frontiers in Psychiatry*, 4, 57.

Steptoe, A., Hamer, M., & Chida, Y. (2007). The effects of acute psychological stress on circulating inflammatory factors in humans: a review and meta-analysis. *Brain, Behavior, and Immunity*, 21, 901–912.

Strawbridge, R., Arnone, D., Danese, A., Papadopoulos, A., Herane Vives, A., & Cleare, A. J. (2015). Inflammation and clinical response to treatment in depression: a meta-analysis. *European Neuropsychopharmacology*, 25, 1532–1543.

Strawbridge, R., Powell, T. R., Breen, G., Young, A. H., & Cleare, A. J. (2016). *Inflammatory predictors of treatment-response in treatment-resistant bipolar and unipolar depressed inpatients*. Amsterdam, NL: International Society of Bipolar Disorders & International Society of Affective Disorders.

Strawbridge, R., Young, A. H., & Cleare, A. J. (2017). Biomarkers for depression: recent insights, current challenges and future prospects. *Neuropsychiatric Disease and Treatment*, 13, 1245.

Szymusiak, R., & Gvilia, I. (2012). Neurophysiology and neurochemistry of hypersomnia. *Sleep Medicine Clinics*, 7, 179–190.

Thase, M. E. (2014). Using biomarkers to predict treatment response in major depressive disorder: evidence from past and present studies. *Dialogues in Clinical Neuroscience*, 16, 539–544.

Trevino, K. (2012). *Defining and differentiating treatment-resistant depression*. [Ph.D. thesis]. The University of Texas.

Tyring, S., Gottlieb, A., Papp, K., Gordon, K., Leonardi, C., Wang, A., et al. (2006). Etanercept and clinical outcomes, fatigue, and depression in psoriasis: double-blind placebo-controlled randomised phase III trial. *The Lancet*, 367, 29–35.

Uher, R., Tansey, K. E., Dew, T., Maier, W., Mors, O., Hauser, J., et al. (2014). An inflammatory biomarker as a differential predictor of outcome of depression treatment with escitalopram and nortriptyline. *The American Journal of Psychiatry*, 171(12), 278–286.

Vinberg, M., Weikop, P., Olsen, N. V., Kessing, L. V., & Miskowiak, K. (2016). Effect of recombinant erythropoietin on inflammatory markers in patients with affective disorders: a randomised controlled study. *Brain, Behavior, and Immunity*, 57, 53–57.

Vogelzangs, N., Beekman, A. T., van Reedt Dortland, A. K., Schoevers, R. A., Giltay, E. J., De Jonge, P., et al. (2014). Inflammatory and metabolic dysregulation and the 2-year course of depressive disorders in antidepressant users. *Neuropsychopharmacology*, 39(7), 1624–1634.

Vogelzangs, N., Duivis, H. E., Beekman, A. T., Kluft, C., Neuteboom, J., Hoogendijk, W., et al. (2012). Association of depressive disorders, depression characteristics and antidepressant medication with inflammation. *Translational Psychiatry*, 2.

Walker, A. J., Foley, B. M., Sutor, S. L., McGillivray, J. A., Frye, M. A., & Tye, S. J. (2015). Peripheral proinflammatory markers associated with ketamine response in a preclinical model of antidepressant-resistance. *Behavioural Brain Research, 293,* 198–202.

Weinberger, J. F., Raison, C. L., Rye, D. B., Montague, A. R., Woolwine, B. J., Felger, J. C., et al. (2015). Inhibition of tumor necrosis factor improves sleep continuity in patients with treatment resistant depression and high inflammation. *Brain, Behavior, and Immunity, 47,* 193–200.

Withers, A. C., Tarasoff, J. M., & Stewart, J. W. (2013). Is depression with atypical features associated with trauma history? *Journal of Clinical Psychiatry, 74,* 500–506.

Wolfe, D. J. (2016). *Tocilizumab augmentation in treatment-refractory major depressive disorder. Vol. 2016,* Available from: (2016). www.clinicaltrials.gov/show/NCT02660528(2016). ClinicalTrials.gov.

Wolkowitz, O. M., Wolf, J., Shelly, W., Rosser, R., Burke, H. M., Lerner, G. K., et al. (2011). Serum BDNF levels before treatment predict SSRI response in depression. *Progress in Neuro-Psychopharmacology and Biological Psychiatry, 35,* 1623–1630.

Wooderson, S. C., Fekadu, A., Markopoulou, K., Rane, L. J., Poon, L., Juruena, M. F., et al. (2014). Long-term symptomatic and functional outcome following an intensive inpatient multidisciplinary intervention for treatment-resistant affective disorders. *Journal of Affective Disorders, 166,* 334–342.

Wooderson, S. C., Juruena, M. F., Fekadu, A., Commane, C., Donaldson, C., Cowan, M., et al. (2011). Prospective evaluation of specialist inpatient treatment for refractory affective disorders. *Journal of Affective Disorders, 131,* 92–103.

Clinical Trials of Anti-Inflammatory Treatments of Major Depression

Norbert Müller

Ludwig-Maximilian University of Munich, Munich, Germany
Marion von Tessin Memory Center, Munich, Germany

INTRODUCTION

Although modern antidepressants, such as selective serotonin reuptake inhibitors (SSRIs), are effective and well tolerated in the majority of patients with major depression (MD), research on the role of monoaminergic neurotransmission in depression seems to have reached its limitations over the last few years. One important argument for the necessity of widening the view on the role of inflammation and the immune system in depression is that no pathogenetic cause has been found for the dysfunction of monoaminergic neurotransmission, despite intense genetic and biochemical research over the last 30 years. Inflammation and immune dysfunction, however, have direct and indirect influences on monoaminergic neurotransmission and thus may explain the monoaminergic dysfunction in MD. Activation of the inflammatory response system in MD is well documented (Maes, 1994; Maes, Stevens, DeClerck, et al., 1992; Muller, Hofschuster, Ackenheil, et al., 1993; Myint, Leonard, Steinbusch, et al., 2005; Rothermundt, Arolt,

Peters, et al., 2001), and two meta-analyses on the role of cytokines clearly showed elevated interleukin-6 (IL-6) levels in patients with MD (Dowlati, Herrmann, Swardfager, et al., 2010; Howren, Lamkin, & Suls, 2009). However, the findings of these two meta-analyses differed regarding levels of the inflammatory markers C-reactive protein (CRP), IL-1, IL-1RA, and TNF-α. In general, the inflammatory response system appears to be activated, but the levels of the different markers vary across studies.

Microglial cells and astrocytes are the immune cells of the brain, because they produce and release pro-inflammatory molecules there. Interestingly, there are reports that the higher the levels of pro-inflammatory cytokines, the more treatment-resistant depressed patients are to established antidepressants (Lanquillon, Krieg, Bening-Abu-Shach, et al., 2000).

MD is a disorder often triggered by stress. Early-life or separation stress, often coupled with a genetic disposition, has been shown to be associated with an increase of pro-inflammatory cytokines, leading to an activation of the immune system and pro-inflammatory

Inflammation and Immunity in Depression
https://doi.org/10.1016/B978-0-12-811073-7.00028-3

prostaglandins (Avitsur & Sheridan, 2009; Hennessy, Schiml-Webb, Miller, et al., 2007). Prostaglandin E2 (PGE2) is an important mediator of inflammation (Song, Lin, Bonaccorso, et al., 1998), and increased PGE2 has been described previously in the saliva, serum, and cerebrospinal fluid (CSF) of depressed patients (Calabrese, Skwerer, Barna, et al., 1986; Linnoila, Whorton, Rubinow, et al., 1983; Nishino, Ueno, Ohishi, et al., 1989; Ohishi, Ueno, Nishino, et al., 1988). The interactions between the immune system and neurotransmitters, the tryptophan-kynurenine system, and the glutamatergic neurotransmission represent links between stress, depression, and the immune system. Accordingly, anti-inflammatory treatment approaches, on the one hand with nonspecific anti-inflammatory agents such as the cyclooxygenase-2 (COX-2) inhibitor celecoxib and on the other with specific antibodies against pro-inflammatory cytokines such as TNF-α or IL-6, have been studied in patients with MD.

THE PRO-INFLAMMATORY IMMUNE STATE IN MD

MD can represent the first symptoms of an inflammatory or autoimmune disorder of a peripheral organ (Muller, Gizycki-Nienhaus, Gunther, et al., 1992). A high blood level of CRP is a common marker for an inflammatory process. Higher-than-normal CRP levels have been repeatedly observed in depression, for example, in severely depressed inpatients (Lanquillon et al., 2000), and high CRP levels have been found to be associated with the severity of depression (Hafner, Baghai, Eser, et al., 2008). Higher CRP levels were observed also in patients remitted from a depressive state, in both men (Danner, Kasl, Abramson, et al., 2003; Ford & Erlinger, 2004) and women (Cizza, Eskandari, Coyle, et al., 2009; Kling, Alesci, Csako, et al., 2007). In a sample of older healthy persons, CRP levels (and IL-6 levels) were predictive of cognitive symptoms of depression 12 years later (Gimeno, Marmot, & Singh-Manoux, 2008).

Characteristics of immune activation in MD include increased numbers of circulating lymphocytes and phagocytic cells; upregulated serum levels of markers of immune activation (neopterin and soluble IL-2 receptors); higher serum concentrations of positive acute-phase proteins (APPs), coupled with reduced levels of negative APPs; and increased release of pro-inflammatory cytokines, such as IL-1β, IL-2, TNF-α, and IL-6 through activated macrophages and interferon (IFN)-γ through activated T-cells (Irwin, 1999; Maes, Meltzer, Bosmans, et al., 1995; Maes, Meltzer, Buckley, et al., 1995; Mikova, Yakimova, Bosmans, et al., 2001; Muller et al., 1993; Müller & Schwarz, 2002; Nunes, Reiche, Morimoto, et al., 2002) (see Table 1).

TABLE 1   Candidates for Immune Markers related to Major Depression

Disease-Related Markers	Markers for Antidepressant Response	Markers for Response to Immune-Related Therapy
Il-6	IL-6	IL-6
TNF-α	Quinolinic acid	CRP
CRP	TNF-α	TNF-α
Neopterin		TNFR1
		TNFR2
		Kynurenine/tryptophan

Various research groups have described increased numbers of peripheral mononuclear cells in MD (Herbert & Cohen, 1993; Rothermundt, Arolt, Fenker, et al., 2001; Seidel, Arolt, Hunstiger, et al., 1996). In line with the findings of increased monocytes and macrophages, an increased level of neopterin also has been described (Bonaccorso, Lin, Verkerk, et al., 1998; Duch, Woolf, Nichol, et al., 1984; Dunbar, Hill, Neale, et al., 1992; Maes, Scharpe, Meltzer, et al., 1994). Reviews describe the role of cellular immunity, cytokines, and innate and adaptive immune systems in depression (Maes, 2011; Muller, Myint, & Schwarz, 2011), and it has been well known for many years that efficient antidepressant treatment is associated with a decrease of pro-inflammatory cytokines, in particular IL-1 and IL-6 and partly TNF-α (Dowlati et al., 2010; Hannestad, DellaGioia, & Bloch, 2011; Liu, Ho, & Mak, 2012).

PRO-INFLAMMATORY CYTOKINES IN THE CSF IN MD

With regard to the possible inflammatory pathology of MD, the localization of the inflammatory process has been a matter of discussion for many years. The central nervous system (CNS) has long been considered as an immune-privileged organ. Today, we know that there are strong and multiple communications between the peripheral immune system and the brain. Peripheral inflammatory processes may involve the CNS and vice versa. The results of an interesting population-based Danish register study support the view that an infection or autoimmune disease significantly increases the risk to later develop a depressive disorder: hospitalization for an infection significantly increased the risk for a later mood disorder by 62% (incidence rate ratio (IRR) 1.62), while hospitalization for an autoimmune disease significantly increased it by 45% (IRR 1.45) (Benros, Waltoft, Nordentoft, et al., 2013). Both risk

factors interacted and increased the risk to IRR 2.35. Interestingly, the findings do not support the idea that primarily, CNS infections lead to later symptoms of mood disorders because the risks were higher for infections in the periphery, for example, for hepatitis (IRR 2.82), than for sepsis (IRR 2.06) or CNS infections (IRR 1.65). Intriguingly, the risk for a mood disorder increased with the proximity of time to the infection and was highest within the first year (IRR 2.70) (Benros et al., 2013). Because the study recorded only infections and autoimmune disorders that led to a hospital contact, the risk for a mood disorder might further increase if all infections are considered.

This study supports the view that a peripheral inflammatory process involves the CNS but does not answer the question whether markers in the peripheral blood or CSF better represent the inflammatory process in the CNS. Nevertheless, cytokines in the CSF are "closer" to the pathological process in the CNS in depression than cytokines in the peripheral blood. Accordingly, increased levels of pro-inflammatory cytokines in the CSF of patients with MD have been described repeatedly, for example, for IL-6 (Kern, Skoog, Borjesson-Hanson, et al., 2014; Sasayama, Hattori, Wakabayashi, et al., 2013) and IL-8 (Kern et al., 2014).

THE INFLUENCE OF STRESS ON THE IMMUNE SYSTEM

The influence of stress on the immune system should be seen to be as varied as the way stress is experienced and processed. Stress research differentiates between eustress and distress, acute and chronic stress, and somatic and mental stress. One and the same stressor can have quite different effects on the immune system, depending, for example, on personality factors, coping mechanisms, and current mental and physical conditions. Acute stress causes a rapid upregulation of immune parameters, for example, in the paradigm of the first parachute jump: the number and activity of natural killer cells in the blood—parts of the innate immune response that react very quickly as the first "barrier" of the immune system—increase sharply right before the jump but then fall again quickly, namely, within minutes after the jump.

The effect of chronic stress on the peripheral immune system and its relevance for MD have been discussed extensively (O'Brien, Scott, & Dinan, 2004). *In vivo* evidence suggests that stress-induced elevation of glucocorticoids also enhances immune function within the CNS through microglia activation and proliferation. Animal studies show that stress induces an enhanced expression of pro-inflammatory factors, such as IL-1β (Nguyen, Deak, Owens, et al., 1998; Pugh, Nguyen, Gonyea, et al., 1999), macrophage migration inhibitory factor (MIF) (Bacher, Meinhardt, Lan, et al., 1998; Niino, Ogata, Kikuchi, et al., 2000; Suzuki, Ogata, Tashiro, et al., 2000), and COX-2 (Madrigal, Garcia-Bueno, Moro, et al., 2003), in the brain. Pro-inflammatory cytokines, such as IL-1 and IL-6, are known to stimulate the hypothalamic–pituitary–adrenal (HPA) axis via hypothalamic neurons. For example, IL-1 stimulates the release of corticotropin-releasing hormone (CRH) and growth hormone-releasing hormone (Berkenbosch, van Oers, del Rey, et al., 1987; Besedovsky, del Rey, Sorkin, et al., 1986), and in turn, central IL-1 upregulation leads to stimulation of the CRH, the HPA axis, and the sympathetic nervous system (Sundar, Cierpial, Kilts, et al., 1990; Weiss, Quan, & Sundar, 1994). Therefore, a vicious circle may be induced if the stress response is not limited, as is discussed in MD.

Elevation of these pro-inflammatory factors is accompanied by dendritic atrophy and neuronal death within the hippocampus (Sapolsky, 1985; Woolley, Gould, & McEwen, 1990), both of which are found also in the brains of people with MD (Campbell & Macqueen, 2004). These detrimental effects of glucocorticoids in the CNS are mediated by a rise in extracellular glutamate (Moghaddam, Bolinao, Stein-Behrens, et al., 1994; Stein-Behrens, Lin, & Sapolsky, 1994) and subsequent overstimulation of the N-methyl-D-aspartate (NMDA) receptor, which in turn results in excitotoxic neuronal damage (Takahashi, Kimoto, Tanabe, et al., 2002). Nair and Bonneau (2006) could demonstrate that restraint-induced psychological stress stimulates the proliferation of microglia and that this process is prevented by a blockade of the corticosterone synthesis, the glucocorticoid receptor, or the NMDA receptor. These data show that stress-induced proliferation of microglia is mediated by corticosterone-induced, NMDA receptor-mediated activation within the CNS. Moreover, NMDA receptor activation during stress leads to increased expression of COX-2 and PGE2, and both COX-2 and PGE2 are able per se to stimulate microglia activation.

The association between chronic, mostly aversively experienced stress and downregulation of the immune system has been known for a long time. It often seems plausible from one's own experience and was systematically investigated as early as the 1980s. One finding was that both the exacerbation and the course of viral infections are affected by stress (Glaser & Kiecolt-Glaser, 1986; Laudenslager, Fleshner, Hofstadter, et al., 1988). When stress levels are high, exacerbations are more common, and the course is less favorable. The antibody titers of neurotropic viruses also are affected by stress (Glaser & Kiecolt-Glaser, 1986). Stress has been known for a long time to have (unfavorable) effects on the exacerbation and course of autoimmune diseases. Stress "weakens" the immune defense, whereby speaking of a "strong" or "weak" immune system does not do justice to the complex regulation mechanisms of the immune system, with its upregulation of pro-inflammatory cytokines and downregulation of anti-inflammatory cytokines. The key element is the dysbalance of the immune response, which is related to inadequate immune activation.

THE VULNERABILITY-STRESS-INFLAMMATION MODEL OF DEPRESSION

The vulnerability-stress model of mental disorders, which was first postulated over 30 years ago for schizophrenic disorders (Zubin & Spring, 1977), focuses on the role of physical and mental stress as the trigger for a psychotic episode. It has also been known for a long time, however, that different types of stress can trigger a depressive episode and may even play a role in the pathogenesis of depression. The concept states that in people with an increased vulnerability for depression, stress can trigger an exacerbation of depressive symptoms. This vulnerability could represent the result of a genetic disposition or could be acquired during early childhood development.

The inflammation theory of depression fitted the vulnerability-stress model well, as experiments in animal models and also data from studies in patients showed and gave rise to the vulnerability-stress-inflammation model of depression (see Fig. 1). A study in men with a depressive illness who had been exposed to higher levels of stress in early childhood showed increased reactivity of inflammatory parameters to psychosocial stress, that is, after stress, the inflammatory parameters increased more in this group than in healthy controls (Pace, Mletzko, Alagbe, et al., 2006). A similar result was obtained in a birth cohort of 1000 individuals with a history of maltreatments in childhood: people with a history of maltreatments more often had depression and increased inflammatory parameters (Danese, Moffitt, Pariante, et al., 2008). These data show that stress in early childhood is a vulnerability factor for the later occurrence of a depressive syndrome associated with inflammation.

The mechanisms underlying the joint occurrence of stress and inflammation were studied in animal experiments, and stress was repeatedly shown to be associated with an increase in pro-inflammatory cytokines (Sparkman & Johnson, 2008).

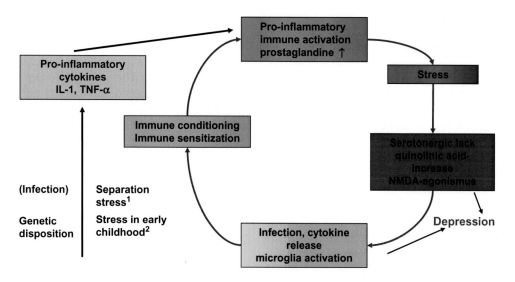

[1]Hennessy et al. (2007), [2]Avitsur and Sheridan (2009)

FIG. 1 Vulnerability-stress-inflammation hypothesis of depression.

INFLAMMATION INFLUENCES THE METABOLISM OF SEROTONIN AND NORADRENALIN IN DEPRESSION

Overwhelming evidence collected over the last 40 years suggests that disturbances in serotonergic and noradrenergic neurotransmission are crucial factors in MD (Coppen & Swade, 1988; Matussek, 1966). Although the pathogenesis of the disturbed serotonergic and noradrenergic mechanisms is still unclear, the involvement of the pro-inflammatory immune state might be crucial. The pro-inflammatory cytokine IL-1β increases the metabolism of serotonin and noradrenalin within the hypothalamus, prefrontal cortex, hippocampus, and amygdala (Anisman & Merali, 1999; Day, Curran, Watson Jr, et al., 1999; Linthorst, Flachskamm, Muller-Preuss, et al., 1995; Merali, Lacosta, & Anisman, 1997; Shintani, Nakaki, Kanba, et al., 1995; Song, Merali, & Anisman, 1999). Similar but less pronounced effects on central monoamine activity after stimulation with lipopolysaccharide or poly I:C have been observed for several mediators, such as IL-6 and TNF-α (Song et al., 1999; Zalcman, Green-Johnson, Murray, et al., 1994). Similarly, administration of the pro-inflammatory cytokine IFN-α is associated with reduced levels of serotonin in the prefrontal cortex (Asnis, De La Garza 2nd, Kohn, et al., 2003). Accordingly, symptoms of depression were observed in many patients treated with IFN-α (Raison, Capuron, & Miller, 2006; Schaefer, Horn, Schmidt, et al., 2004). Moreover, two indirect pathways of tryptophan/kynurenine metabolism contribute to the induction of symptoms of depression by pro-inflammatory cytokines: (1) the increased metabolism of serotonin and (2) the increased production of NMDA agonists, that is, glutamatergic products of kynurenine metabolism, after the activation of the enzyme indoleamine 2,3-dioxygenase (IDO) by pro-inflammatory cytokines (Muller & Schwarz, 2007; Müller, Schwarz, & Riedel, 2005). Pro-inflammatory molecules such as PGE2 or TNF-α, however, induce—synergistically with IFN—the increase of IDO activity (Braun, Longman, & Albert, 2005; Kwidzinski, Bunse, Aktas, et al., 2005; Robinson, Hale, & Carlin, 2005).

Increased activity of the glutamatergic system in the peripheral blood of depressive patients has been shown repeatedly (Altamura, Mauri, Ferrara, et al., 1993; Kim, Schmid-Burgk, Claus, et al., 1982; Mauri, Ferrara, Boscati, et al., 1998), although this result could not be replicated by all groups (Maes, Verkerk, Vandoolaeghe, et al., 1998). The inconsistency of the findings, however, might be due to methodological problems (Kugaya & Sanacora, 2005). Support for increased glutamatergic activity in depression comes from magnetic resonance spectroscopy: elevated glutamate levels were found in the occipital cortex of unmedicated patients with MD (Sanacora, Gueorguieva, Epperson, et al., 2004). Furthermore, NMDA antagonists such as MK-801 (Maj, Rogoz, Skuza, et al., 1992; Trullas & Skolnick, 1990), ketamine (Yilmaz, Schulz, Aksoy, et al., 2002), memantine (Ossowska, Klenk-Majewska, & Szymczyk, 1997), amantadine (Huber, Dietrich, & Emrich, 1999; Stryjer, Strous, Shaked, et al., 2003), and others (Kugaya & Sanacora, 2005) have exhibited antidepressant effects in humans. The partial NMDA receptor agonist D-cycloserine demonstrated antidepressant effects at high doses (Crane, 1959). Several mechanisms can cause depressive states: (1) a direct influence of pro-inflammatory cytokines on serotonin and noradrenaline metabolism (Besedovsky, del Rey, Sorkin, et al., 1983; Song & Leonard, 2000; Zalcman et al., 1994); (2) an imbalance of the type 1 and type 2 immune responses, leading to increased tryptophan and serotonin metabolism by the activation of IDO in the CNS (Myint & Kim, 2003; Schwarz, Chiang, Muller, et al., 2001); (3) a decreased availability of tryptophan and serotonin (Maes et al., 1994); and (4) a disturbance of kynurenine metabolism, with an imbalance in favor of the production of the NMDA receptor agonist quinolinic acid (Myint & Kim, 2003; Myint, Kim, Verkerk, et al., 2007).

EFFECTS OF THE PRO-INFLAMMATORY IMMUNE ACTIVATION ON KYNURENINE METABOLISM IN DEPRESSION

The enzyme IDO metabolizes tryptophan to kynurenine, which is then converted to quinolinic acid via the intermediate 3-HK by the enzyme kynurenine hydroxylase. Both IDO and kynurenine hydroxylase are induced by the type 1 cytokine IFN-γ. IDO is an important regulatory component in the control of lymphocyte proliferation, the activation of the type 1 immune response, and the regulation of tryptophan metabolism (Mellor & Munn, 1999). It halts the lymphocyte cell cycle by catabolizing tryptophan (Munn, Shafizadeh, Attwood, et al., 1999). In contrast to type 1 cytokines, the type 2 cytokines IL-4 and IL-10 inhibit IFN-γ-induced, IDO-mediated tryptophan catabolism (Weiss, Murr, Zoller, et al., 1999) (see Fig. 2). IDO is located in several

cell types, including monocytes and microglial cells (Alberati-Giani, Ricciardi-Castagnoli, Kohler, et al., 1996). An IFN-γ-induced, IDO-mediated decrease in CNS tryptophan availability may lead to a serotonergic deficiency in the CNS, because tryptophan availability is one of the limitations of serotonin synthesis. Other pro-inflammatory molecules such as PGE2 or TNF-α, however, induce synergistically with IFN-γ the increase of IDO activity (Braun et al., 2005; Kwidzinski et al., 2005; Robinson et al., 2005). Accordingly, increased levels of PGE2 and TNF-α have been described in MD (e.g., Linnoila et al., 1983; Mikova et al., 2001).

Low levels of 5-hydroxyindoleacetic acid—the metabolite of serotonin—in the CSF of suicidal people have been observed repeatedly (Lidberg, Belfrage, Bertilsson, et al., 2000; Mann & Malone, 1997; Nordstrom, Samuelsson, Asberg, et al., 1994). This gives additional evidence for a possible link between the type 1 cytokine IFN-γ and the IDO-related reduction of serotonin availability in the CNS of suicidal patients.

An interesting study showed that immunotherapy with IFN-α was followed by an increase

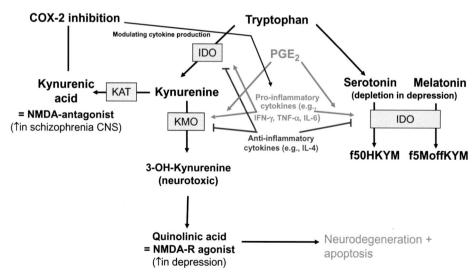

FIG. 2 Tryptophan/kynurenine metabolism and possible implications for psychiatric disorders.

in depressive symptoms and serum kynurenine concentrations on the one hand and a decrease in concentrations of tryptophan and serotonin on the other (Bonaccorso, Marino, Puzella, et al., 2002). The kynurenine/tryptophan ratio, which reflects IDO activity, increased significantly. Changes in depressive symptoms were significantly positively correlated with kynurenine concentrations and significantly negatively correlated with serotonin concentrations (Bonaccorso et al., 2002). This study and others (Capuron, Neurauter, Musselman, et al., 2003) clearly show that IDO activity is increased by IFN, leading to increased kynurenine production and tryptophan and serotonin depletion. Further metabolism of kynurenine, however, seems to play an additional crucial role in psychopathological states, in particular the NMDA receptor agonist quinolinic acid in depression (Leonard & Myint, 2006; Steiner, Walter, Gos, et al., 2011) and the NMDA receptor antagonist kynurenic acid in schizophrenia (Olsson, Andersson, Linderholm, et al., 2009).

In addition to the effects of the proinflammatory immune response on serotonin metabolism, other neurotransmitter systems, in particular the catecholaminergic system, also are involved in depression (Matussek, 1988). Although the relationship of immune activation and the lack of catecholaminergic neurotransmission has not been well studied, the increase in monoamine oxidase activity, which leads to decreased noradrenergic neurotransmission, might be an indirect effect of the increased production of kynurenine and quinolinic acid (Schiepers, Wichers, & Maes, 2005).

CNS VOLUME LOSS IN NEUROIMAGING STUDIES—A CONSEQUENCE OF AN INFLAMMATORY PROCESS?

A loss of brain volume has been observed in MD. Male patients with a first episode of MD had significantly smaller hippocampal total and gray matter volumes than healthy comparison males (Frodl, Meisenzahl, Zetzsche, et al., 2002), and a long-term study found a significantly higher decline of volume in several CNS regions compared with healthy controls (Frodl, Koutsouleris, Bottlender, et al., 2008). The pathophysiology of this volume loss is unclear. Glial reductions have been consistently found in brain circuits known to be involved in mood disorders, such as the limbic and prefrontal cortex (Cotter, Pariante, & Rajkowska, 2002; Ongur, Drevets, & Price, 1998; Rajkowska, 2003; Rajkowska, Halaris, & Selemon, 2001; Rajkowska, Miguel-Hidalgo, Wei, et al., 1999). Recent studies have shown that the number of astrocytes is reduced in patients with MD (Johnston-Wilson, Sims, Hofmann, et al., 2000; Miguel-Hidalgo, Baucom, Dilley, et al., 2000; Si, Miguel-Hidalgo, O'Dwyer, et al., 2004), although the data are not fully consistent (Davis, Thomas, Perry, et al., 2002).

CNS INFLAMMATION IN MD: FINDINGS FROM POSITRON-EMISSION TOMOGRAPHY

A recently published *in vivo* study used positron-emission tomography (PET) to measure translocator protein density as a marker for the activation of microglia, which was interpreted as a measure of neuroinflammation in MD patients. The study had a very interesting outcome: the authors observed an increase of activated microglia in 20 patients during an episode of MD compared with 20 healthy controls. The patients had been medication-free for at least 6 weeks, and all participants were otherwise healthy and nonsmokers. The significant elevation of microglia activation was primarily found in the prefrontal cortex (elevated by 26%), nucleus accumbens (ACC; 32%), and insula (33%), all structures that are known to be involved in MD. Moreover, a statistically significant correlation between microglia activation in the ACC and the severity of depression ($r = 0.63$; $P < 0.001$) was described, that is, the higher the translocator protein density was, the more pronounced was the severity of MD (Setiawan, Wilson, Mizrahi, et al., 2015).

CYCLO-OXYGENASE-2 (COX-2) INHIBITION AS AN EXAMPLE FOR AN ANTI-INFLAMMATORY THERAPEUTIC APPROACH IN MD

COX-2 inhibitors influence the CNS serotonergic system, either directly or via CNS immune mechanisms. In a rat model, treatment with rofecoxib was followed by an increase of serotonin in the frontal and temporoparietal cortex (Sandrini, Vitale, & Pini, 2002). Hence, COX-2 inhibitors would be expected to show a clinical antidepressant effect. In the depression animal model of the bulbectomized rat, a decrease in hypothalamic cytokine levels and a change in behavior have been observed after chronic celecoxib treatment (Myint, Steinbusch, Goeghegan, et al., 2007). In another animal model of depression, however, the mixed COX-1/COX-2 inhibitor acetylsalicylic acid showed an additional antidepressant effect by accelerating the antidepressant effect of fluoxetine (Brunello, Alboni, Capone, et al., 2006). A significant therapeutic effect of the COX-2 inhibitor celecoxib in MD was also found in a randomized, double-blind pilot add-on study of celecoxib and reboxetine versus placebo (Muller, Schwarz, Dehning, et al., 2006). Interestingly, the ratio of kynurenine to tryptophan, which represents the activity of the pro-inflammatory cytokine-driven enzyme IDO, predicted the antidepressant response to celecoxib. Patients with a high activity of IDO, that is, a high pro-inflammatory activity, responded better to celecoxib (Krause, Myint, Schuett, et al., 2017). Another randomized, double-blind study in 50 patients with MD also showed a significantly

better outcome with fluoxetine plus the COX-2 inhibitor celecoxib than with fluoxetine alone (Akhondzadeh, Jafari, Raisi, et al., 2009). This finding was recently replicated with the combination of sertraline and celecoxib in 40 depressed patients (Abbasi, Hosseini, Modabbernia, et al., 2012). Interestingly, the blood levels of IL-6 predicted the antidepressant response in both the sertraline (plus placebo) and the celecoxib (plus sertraline) groups.

A meta-analysis on the efficiency of adjunctive celecoxib treatment for patients with MD included 150 patients. The analysis concluded that adjunctive treatment with nonsteroidal anti-inflammatory drugs (NSAIDs), particularly celecoxib, can be a promising strategy for patients with depressive disorder. However, future studies with a larger sample size and longer study duration are needed to confirm the efficacy and tolerability of NSAIDs for depression (Na, Lee, Lee, et al., 2014).

A double-blind, randomized, placebo-controlled international multicenter study of the selective COX-2 inhibitor cimicoxib as an add-on to sertraline in patients with MD showed also interesting results: while there was no benefit of add-on cimicoxib over sertraline alone or placebo in the whole group of depressed patients, the combined treatment showed a statistically significantly better outcome in the subgroup of severely depressed patients (Hamilton depression score <25 at baseline) (Müller et al., manuscript in preparation). As discussed below, because in general the placebo response is very high in depression studies and higher in mild to moderately depressed patients than in severely depressed patients, a therapeutic benefit can be shown more easily in severely depressed patients.

As part of the European-wide study on the role of inflammation in mood disorders (MOODINFLAME), a double-blind, randomized, placebo-controlled multicenter study compared the COX-2 inhibitor celecoxib as an add-on to sertraline with sertraline alone and placebo in a subgroup of patients with MD. This study had a small sample size (53 patients) and could show only a limited benefit of the COX-2 treatment: celecoxib showed significantly better efficacy in the completer analysis, that is, in the group of patients who completed the study, but not in the last observation carried forward analysis (Leitner B., Müller N., et al., manuscript in preparation).

Despite the limitations of these studies, they represent further small pieces in the puzzle of the effects of anti-inflammatory treatment and in particular of COX-2 inhibition in MD. Future studies should consider more closely the sample size; the high rate of placebo response, particularly in studies of an add-on to an effective antidepressant; and the severity of depression.

METHODOLOGICAL ISSUES REGARDING ANTI-INFLAMMATORY TREATMENT IN MD

Several methodological issues have to be raised regarding anti-inflammatory treatment in MD. From a scientific point of view, a head-to-head comparison of an anti-inflammatory compound, such as a COX-2 inhibitor or an anticytokine antibody, with an appropriate placebo in a randomized controlled trial (RCT) would be the gold standard of clinical trials. However, for ethical reasons, it would not be justifiable to treat a depressed patient only with an anti-inflammatory substance (or placebo) without any antidepressant medication. Therefore, in nearly all trials, the anti-inflammatory substance is given as an add-on to an antidepressant. Consequently, the anti-inflammatory agent has to have a large effect in order to prove an additional benefit over the antidepressant alone and over the antidepressant combined with placebo.

One has to keep in mind that over the last few years, the number of "failed studies" in MD, that is, studies of antidepressants that could not show a benefit compared with placebo, has

increased dramatically. This is partly not only due to methodological reasons such as sample size and the lack of statistical power but also due to the high placebo response rate of up to 40%. This high placebo response may also contribute to difficulties in proving an advantage of anti-inflammatory substances (added to an antidepressant) over placebo.

The benefit of an antidepressant compound over placebo is well known to increase with the severity of depression. Therefore, clinical trials should include more severely depressed patients, if possible. The antidepressant effect of anti-inflammatory compounds also might increase with depression severity, as we could show in a study with the COX-2 inhibitor cimicoxib (Müller et al., manuscript in preparation).

A very important point discussed in most of the studies of anti-inflammatories in depression is the fact that different pathologies might play a role in depression, and an inflammatory process is unlikely to be involved in all cases—although that view cannot yet be ruled out. Treatment resistance to monoaminergic antidepressants, such as SSRIs, might indicate an inflammatory origin of the depressive syndrome because increased pro-inflammatory cytokines have been found in treatment-resistant patients with MD (Lanquillon et al., 2000). The more patients with an inflammatory pathology who are involved in clinical trials of anti-inflammatory compounds, the higher the probability is to see an effect of anti-inflammatory treatment. So far, unfortunately, no valid, reliable marker is available to identify—on a biological or psychopathological basis—the subgroup of patients to whom inflammation is involved in the pathological process.

ANTI-INFLAMMATORY COMPOUNDS OTHER THAN NSAIDs

The anti-TNF-α antibody infliximab, which blocks the interaction of TNF-α with cell-surface receptors and was developed for the treatment of inflammatory joint disorders and psoriasis, showed in a first study a highly significant effect on symptoms of depression in psoriasis patients (Tyring, Gottlieb, Papp, et al., 2006). Further studies of anti-TNF antibodies also provided antidepressant effects (overview in Kappelmann, Lewis, Dantzer, et al., 2018).

A random-effect meta-analysis of seven RCTs involving 2370 participants showed a significant antidepressant effect of anticytokine treatment compared with placebo (standardized mean difference (SMD) = 0.40; 95% confidence interval (CI), 0.22–0.59) (Kappelmann et al., 2018). Anti-TNF-α drugs (adalimumab, etanercept, and infliximab) were most commonly studied (five RCTs); SMD = 0.33 (95% CI, 0.06–0.60). Separate meta-analyses of two RCTs of adjunctive treatment with anticytokine therapy and eight nonrandomized and/or nonplacebo studies yielded similar small-to-medium effect estimates favoring anticytokine therapy; SMD = 0.19 (95% CI, 0.00–0.37), and SMD = 0.51 (95% CI, 0.34–0.67), respectively. The anti-TNF antibodies adalimumab, etanercept, and infliximab and the anti-IL-6 antibody tocilizumab all showed statistically significant improvements in depressive symptoms. Metaregression exploring predictors of response found that the antidepressant effect was associated with baseline symptom severity ($P = 0.018$) but not with improvement in primary physical illness, sex, age, or study duration. An important limitation of these studies is that these patient samples exhibited concomitant symptoms of anxiety or depression or both in addition to their (primary) diagnosis of an inflammatory disease such as rheumatoid arthritis, Crohn's disease, or psoriasis. Moreover, overall, the symptoms of depression were mild to modest, and depression was not formally diagnosed.

So far, only one study has used an anti-TNF-α antibody in patients diagnosed with MD. In this 12-week, placebo-controlled study ($n = 60$), partly medication-free ($n = 23$) nonresponders to antidepressant therapy received three infusions of infliximab or placebo. No overall better outcome of infliximab could be shown compared with placebo. There was, however, a significant interaction between treatment, time, and baseline CRP ($\leq 5 \, mg/L$): patients with higher baseline CRP had a better response rate to infliximab (62%) compared with placebo (33%). Moreover, the baseline concentrations of TNF-α, sTNFR1, and sTNFR2 were significantly higher in infliximab responders ($P \leq 0.01$). Additionally, infliximab responders exhibited a significantly higher decrease in CRP ($P \leq 0.01$) than nonresponders (Raison, Rutherford, Woolwine, et al., 2013). This result is promising, particularly in light of the fact that treatment-resistant depressed patients mirror a negative selection for treatment outcome. The CRP level might be a possible biological marker for the outcome of treatment with anti-TNF-α antibodies.

ANTI-IL-6 COMPLEX AS A THERAPEUTIC TARGET IN MD

As mentioned earlier, elevated levels of IL-6 have been reported in the peripheral blood and CSF of depressed patients. Therefore, in addition to TNF-α, the IL-6 complex is an interesting target for anticytokine treatment. Two open-label studies of the anti-IL-6 antibody tocilizumab (Gossec, Steinberg, Rouanet, et al., 2015; Traki, Rostom, Tahiri, et al., 2014) were included in the meta-analysis mentioned earlier (Kappelmann et al., 2018). The concomitant symptoms of anxiety and depression improved in both studies. Nevertheless, valid and reliable data on anti-IL-6 therapy in patients with MD are still missing. Furthermore, the right target for anti-IL-6 therapy remains a matter of discussion: IL-6 levels are well known to be higher in the CSF than in the periphery, and depressed patients are well known to have higher IL-6 CSF levels than controls; therefore, CSF IL-6 was postulated to be the most promising therapeutic target. Moreover, the preclinical model of chronically stressed rats shows an overexpression of IL-6 in the cortex. On the other hand, peripheral CSF IL-6 levels cannot be seen as completely distinct from IL-6 concentration in the CSF or IL-6 expression in the brain because there are many interconnections between central and peripheral IL-6 concentrations. Therefore, peripheral IL-6 might be a promising target to treat depressive symptoms (Yang & Hashimoto, 2015). This view is supported by the findings that peripheral IL-6 promotes resilience versus susceptibility to chronic inescapable electric stress in an animal model of depression (Yang, Shirayama, Zhang, et al., 2015) and serum IL-6 predicts the antidepressant response to ketamine (Yang, Wang, Yang, et al., 2015) in depressed patients. Maes, Anderson, Kubera, et al. (2014) propose, however, that a more promising strategy to inhibit IL-6 might be to increase the inhibition of IL-6 trans-signaling by soluble glycoprotein 130 (sGP 130), a part of the IL-6 complex, while allowing the maintenance of IL-6 receptor (IL-6R) signaling (Maes et al., 2014). Moreover, sirukumab, a monoclonal antibody against IL-6 that is distinct from tocilizumab, also has been proposed for use in depression (Zhou, Lee, Salvadore, et al., 2017). Sirukumab targets the IL-6 signaling pathway and inhibits both the pro- and anti-inflammatory effects of the pleiotropic cytokine IL-6 (Zhou et al., 2017). Sirukumab has been shown to have beneficial effects in other inflammatory diseases, such as lupus erythematosus and rheumatoid arthritis.

OTHER IMMUNE-RELATED SUBSTANCES IN THE THERAPY OF MD

Interestingly, there are also preliminary findings that angiotensin II AT1 receptor blockade has anti-inflammatory effects in the CNS and ameliorates stress, anxiety, and CNS inflammation (Benicky, Sanchez-Lemus, Honda, et al.,

2011; Saavedra, Sanchez-Lemus, & Benicky, 2011).

A recently published, broader meta-analysis of inflammation-related therapeutic approaches included 10 publications reporting on 14 trials (6262 participants). The analysis showed very interesting results of anti-inflammatory treatment in MD: 10 trials evaluated the use of NSAIDs ($n = 4258$), and four investigated cytokine inhibitors ($n = 2004$). The pooled effect estimate suggested that anti-inflammatory treatment reduced depressive symptoms (SMD, −0.34; 95% CI, from −0.57 to −0.11; I2 = 90%) compared with placebo. This effect was observed in studies that included patients with depression (SMD, −0.54; 95% CI, from −1.08 to −0.01; I2 = 68%) and patients with depressive symptoms (SMD, −0.27; 95% CI, from −0.53 to −0.01; I2 = 68%). The heterogeneity of the studies was not explained by differences in inclusion of clinical depression versus depressive symptoms or the use of NSAIDs versus cytokine inhibitors. Subanalyses emphasized the antidepressant properties of the selective COX-2 inhibitor celecoxib (SMD, −0.29; 95% CI, from −0.49 to −0.08; I2 = 73%) on remission (OR, 7.89; 95% CI, 2.94–21.17; I2 = 0%) and response (OR, 6.59; 95% CI, 2.24–19.42; I2 = 0%). Among the six studies reporting on adverse effects, no evidence of an increased number of gastrointestinal or cardiovascular events after 6 weeks or infections after 12 weeks of anti-inflammatory treatment was found compared with placebo. All trials were associated with a high risk of bias owing to potentially compromised internal validity. The analysis suggests that anti-inflammatory treatment, in particular with celecoxib, decreases depressive symptoms without increased risks of adverse effects. This study supports a proof of concept concerning the use of anti-inflammatory treatment in depression (Kohler, Benros, Nordentoft, et al., 2014).

Despite being used primarily for their lipid-lowering properties, statins have direct anti-inflammatory effects that are not mediated by their hypocholesterolemic activity (Kohler, Gasse, Petersen, et al., 2016; Weitz-Schmidt, 2002). Therefore, a large population-based study was performed in nearly 900,000 SSRI users, more than 110,000 of whom used a statin concomitantly, in order to evaluate a possible antidepressant effect of statins under the aspect of the inflammatory hypothesis of depression. The concomitant treatment of SSRI and statins resulted in a robust advantage regarding the risk for (a relapse to) depression compared with an SSRI alone (Kohler et al., 2016).

CONCLUSION

The data discussed earlier show clearly that immunomodulatory therapeutic substances are beneficial in MD. For ethical and methodological reasons, most studies evaluated immunomodulatory substances as an add-on to antidepressants. The data also show that not all patients benefit equally from immunomodulatory or anti-inflammatory treatment. Various conclusions can be drawn. First, an important goal for further research is to establish reliable and valid criteria, for example, based on clinical characteristics or biomarkers (possible examples are shown in Table 1), for the identification of those patients with MD who will benefit from immunomodulatory therapy; studies indicate that this subgroup probably comprises 30%–40% of patients with MD. Second, studies of immunomodulatory treatment are often performed in patients who are resistant to established antidepressants. It is well known, however, that this group of patients represents a negative selection for treatment studies. We know from schizophrenia studies that the anti-inflammatory drug celecoxib shows therapeutic effects primarily in the early stage of the disease, but not in chronic patients. Whether the same is true for patients with MD has yet to be evaluated. Third, from a methodological perspective, studies of monotherapy with an anti-inflammatory or immunomodulatory substance have a much higher scientific explorative power than studies with an add-on design. Recent depression studies report a placebo response of 30%–40%. Thus, to show a statistically significant difference over a potent antidepressant plus placebo (e.g., in a study of sertraline plus placebo versus sertraline plus TNF), a new drug needs to be evaluated in studies with large sample sizes and to have a powerful add-on effect. Fourth, the exact immune mechanism leading to pro-inflammatory activation in MD and, possibly in a second step, to chronicity of the immune activation is still unclear, but no therapy acting on the pathological immune mechanism can be developed before this mechanism is further elucidated. Anti-inflammatory treatment, for example, with celecoxib, is a "broad" treatment approach, and more targeted interventions might have a better outcome or fewer side effects. Fifth, no studies have examined whether different immunomodulatory approaches, for example, celecoxib or anti-TNF antibodies, may show a benefit in different kinds of patients with MD. To tailor anti-inflammatory treatment for an individual patient, we may need to compare different immunomodulatory approaches. Although immunotherapy is a promising and in the meantime well-justified experimental therapeutic approach, many research questions have to be answered before it can become a routine therapy for MD.

ACKNOWLEDGMENT

Parts of the manuscript have been published before. The work was supported by the foundation Immunität und Seele.

STATEMENT OF CONFLICT OF INTEREST

None to declare.

References

Abbasi, S. H., Hosseini, F., Modabbernia, A., et al. (2012). Effect of celecoxib add-on treatment on symptoms and serum IL-6 concentrations in patients with major depressive disorder: randomized double-blind placebo-controlled study. *Journal of Affective Disorders, 141*, 308–314.

Akhondzadeh, S., Jafari, S., Raisi, F., et al. (2009). Clinical trial of adjunctive celecoxib treatment in patients with major depression: a double blind and placebo controlled trial. *Depression and Anxiety, 26*, 607–611.

Alberati-Giani, D., Ricciardi-Castagnoli, P., Kohler, C., et al. (1996). Regulation of the kynurenine metabolic pathway by interferon-gamma in murine cloned macrophages and microglial cells. *Journal of Neurochemistry, 66*, 996–1004.

Altamura, C. A., Mauri, M. C., Ferrara, A., et al. (1993). Plasma and platelet excitatory amino acids in psychiatric disorders. *American Journal of Psychiatry, 150*, 1731–1733.

Anisman, H., & Merali, Z. (1999). Anhedonic and anxiogenic effects of cytokine exposure. *Advances in Experimental Medicine and Biology, 461*, 199–233.

Asnis, G. M., De La Garza, R., 2nd, Kohn, S. R., et al. (2003). IFN-induced depression: a role for NSAIDs. *Psychopharmacology Bulletin, 37*, 29–50.

Avitsur, R., & Sheridan, J. F. (2009). Neonatal stress modulates sickness behavior. *Brain, Behavior, and Immunity, 23*, 977–985.

Bacher, M., Meinhardt, A., Lan, H. Y., et al. (1998). MIF expression in the rat brain: implications for neuronal function. *Molecular Medicine, 4*, 217–230.

Benicky, J., Sanchez-Lemus, E., Honda, M., et al. (2011). Angiotensin II AT1 receptor blockade ameliorates brain inflammation. *Neuropsychopharmacology, 36*, 857–870.

Benros, M. E., Waltoft, B. L., Nordentoft, M., et al. (2013). Autoimmune diseases and severe infections as risk factors for mood disorders: a nationwide study. *JAMA Psychiatry, 70*, 812–820.

Berkenbosch, F., van Oers, J., del Rey, A., et al. (1987). Corticotropin-releasing factor-producing neurons in the rat activated by interleukin-1. *Science, 238*, 524–526.

Besedovsky, H., del Rey, A., Sorkin, E., et al. (1983). The immune response evokes changes in brain noradrenergic neurons. *Science, 221*, 564–566.

Besedovsky, H., del Rey, A., Sorkin, E., et al. (1986). Immunoregulatory feedback between interleukin-1 and glucocorticoid hormones. *Science, 233*, 652–654.

Bonaccorso, S., Lin, A. H., Verkerk, R., et al. (1998). Immune markers in fibromyalgia: comparison with major depressed patients and normal volunteers. *Journal of Affective Disorders, 48*, 75–82.

Bonaccorso, S., Marino, V., Puzella, A., et al. (2002). Increased depressive ratings in patients with hepatitis C receiving interferon-alpha-based immunotherapy are related to interferon-alpha-induced changes in the serotonergic system. *Journal of Clinical Psychopharmacology, 22*, 86–90.

Braun, D., Longman, R. S., & Albert, M. L. (2005). A two-step induction of indoleamine 2,3 dioxygenase (IDO) activity during dendritic-cell maturation. *Blood, 106*, 2375–2381.

Brunello, N., Alboni, S., Capone, G., et al. (2006). Acetylsalicylic acid accelerates the antidepressant effect of fluoxetine in the chronic escape deficit model of depression. *International Clinical Psychopharmacology, 21*, 219–225.

Calabrese, J. R., Skwerer, R. G., Barna, B., et al. (1986). Depression, immunocompetence, and prostaglandins of the E series. *Psychiatry Research, 17*, 41–47.

Campbell, S., & Macqueen, G. (2004). The role of the hippocampus in the pathophysiology of major depression. *Journal of Psychiatry & Neuroscience, 29*, 417–426.

Capuron, L., Neurauter, G., Musselman, D. L., et al. (2003). Interferon-alpha-induced changes in tryptophan metabolism. Relationship to depression and paroxetine treatment. *Biological Psychiatry, 54*, 906–914.

Cizza, G., Eskandari, F., Coyle, M., et al. (2009). Plasma CRP levels in premenopausal women with major depression: a 12-month controlled study. *Hormone and Metabolic Research, 41*, 641–648.

Coppen, A., & Swade, C. (1988). 5-HT and depression: the present position. In M. Briley & G. Fillion (Eds.), *New concepts in depression. Pierre Fabre monograph series*. London: MacMillan Press.

Cotter, D., Pariante, C., & Rajkowska, G. (2002). Glial pathology in major psychiatric disorders. In G. Agam, R. H. Belmaker, & I. Everall (Eds.), *The post-mortem brain in psychiatric research* (pp. 291–324). Boston: Kluwer Academic Publishers.

Crane, G. E. (1959). Cyloserine as an antidepressant agent. *American Journal of Psychiatry, 115*, 1025–1026.

Danese, A., Moffitt, T. E., Pariante, C. M., et al. (2008). Elevated inflammation levels in depressed adults with a history of childhood maltreatment. *Archives of General Psychiatry, 65*, 409–415.

Danner, M., Kasl, S. V., Abramson, J. L., et al. (2003). Association between depression and elevated C-reactive protein. *Psychosomatic Medicine, 65*, 347–356.

Davis, S., Thomas, A., Perry, R., et al. (2002). Glial fibrillary acidic protein in late life major depressive disorder: an immunocytochemical study. *Journal of Neurology, Neurosurgery, and Psychiatry, 73*, 556–560.

Day, H. E., Curran, E. J., Watson, S. J., Jr., et al. (1999). Distinct neurochemical populations in the rat central nucleus of the amygdala and bed nucleus of the stria terminalis: evidence for their selective activation by interleukin-1beta. *Journal of Comparative Neurology, 413*, 113–128.

Dowlati, Y., Herrmann, N., Swardfager, W., et al. (2010). A meta-analysis of cytokines in major depression. *Biological Psychiatry, 67*, 446–457.

Duch, D. S., Woolf, J. H., Nichol, C. A., et al. (1984). Urinary excretion of biopterin and neopterin in psychiatric disorders. *Psychiatry Research, 11,* 83–89.

Dunbar, P. R., Hill, J., Neale, T. J., et al. (1992). Neopterin measurement provides evidence of altered cell-mediated immunity in patients with depression, but not with schizophrenia. *Psychological Medicine, 22,* 1051–1057.

Ford, D. E., & Erlinger, T. P. (2004). Depression and C-reactive protein in US adults: data from the Third National Health and Nutrition Examination Survey. *Archives of Internal Medicine, 164,* 1010–1014.

Frodl, T. S., Koutsouleris, N., Bottlender, R., et al. (2008). Depression-related variation in brain morphology over 3 years: effects of stress? *Archives of General Psychiatry, 65,* 1156–1165.

Frodl, T., Meisenzahl, E. M., Zetzsche, T., et al. (2002). Hippocampal changes in patients with a first episode of major depression. *American Journal of Psychiatry, 159,* 1112–1118.

Gimeno, D., Marmot, M. G., & Singh-Manoux, A. (2008). Inflammatory markers and cognitive function in middle-aged adults: the Whitehall II study. *Psychoneuroendocrinology, 33,* 1322–1334.

Glaser, R., & Kiecolt-Glaser, J. K. (1986). Stress and immune function. *Clinical Neuropharmacology, 9*(Suppl. 4), 485–487.

Gossec, L., Steinberg, G., Rouanet, S., et al. (2015). Fatigue in rheumatoid arthritis: quantitative findings on the efficacy of tocilizumab and on factors associated with fatigue. The French multicentre prospective PEPS study. *Clinical and Experimental Rheumatology, 33,* 664–670.

Hafner, S., Baghai, T. C., Eser, D., et al. (2008). C-reactive protein is associated with polymorphisms of the angiotensin-converting enzyme gene in major depressed patients. *Journal of Psychiatric Research, 42,* 163–165.

Hannestad, J., DellaGioia, N., & Bloch, M. (2011). The effect of antidepressant medication treatment on serum levels of inflammatory cytokines: a meta-analysis. *Neuropsychopharmacology, 36,* 2452–2459.

Hennessy, M. B., Schiml-Webb, P. A., Miller, E. E., et al. (2007). Anti-inflammatory agents attenuate the passive responses of guinea pig pups: evidence for stress-induced sickness behavior during maternal separation. *Psychoneuroendocrinology, 32,* 508–515.

Herbert, T. B., & Cohen, S. (1993). Depression and immunity: a meta-analytic review. *Psychological Bulletin, 113,* 472–486.

Howren, M. B., Lamkin, D. M., & Suls, J. (2009). Associations of depression with C-reactive protein, IL-1, and IL-6: a meta-analysis. *Psychosomatic Medicine, 71,* 171–186.

Huber, T. J., Dietrich, D. E., & Emrich, H. M. (1999). Possible use of amantadine in depression. *Pharmacopsychiatry, 32,* 47–55.

Irwin, M. (1999). Immune correlates of depression. *Advances in Experimental Medicine and Biology, 461,* 1–24.

Johnston-Wilson, N. L., Sims, C. D., Hofmann, J. P., et al. (2000). Disease-specific alterations in frontal cortex brain proteins in schizophrenia, bipolar disorder, and major depressive disorder. The Stanley Neuropathology Consortium. *Molecular Psychiatry, 5,* 142–149.

Kappelmann, N., Lewis, G., Dantzer, R., et al. (2018). Antidepressant activity of anti-cytokine treatment: a systematic review and meta-analysis of clinical trials of chronic inflammatory conditions. *Molecular Psychiatry, 23*(2), 335–343.

Kern, S., Skoog, I., Borjesson-Hanson, A., et al. (2014). Higher CSF interleukin-6 and CSF interleukin-8 in current depression in older women. Results from a population-based sample. *Brain, Behavior, and Immunity, 41,* 55–58.

Kim, J. S., Schmid-Burgk, W., Claus, D., et al. (1982). Increased serum glutamate in depressed patients. *Archiv für Psychiatrie und Nervenkrankheiten, 232,* 299–304.

Kling, M. A., Alesci, S., Csako, G., et al. (2007). Sustained low-grade pro-inflammatory state in unmedicated, remitted women with major depressive disorder as evidenced by elevated serum levels of the acute phase proteins C-reactive protein and serum amyloid A. *Biological Psychiatry, 62,* 309–313.

Kohler, O., Benros, M. E., Nordentoft, M., et al. (2014). Effect of anti-inflammatory treatment on depression, depressive symptoms, and adverse effects: a systematic review and meta-analysis of randomized clinical trials. *JAMA Psychiatry, 71,* 1381–1391.

Kohler, O., Gasse, C., Petersen, L., et al. (2016). The effect of concomitant treatment with SSRIs and statins: a population-based study. *American Journal of Psychiatry, 173,* 807–815.

Krause, D., Myint, A. M., Schuett, C., et al. (2017). High kynurenine (a tryptophan metabolite) predicts remission in patients with major depression to add-on treatment with celecoxib. *Front Psychiatry, 8,* 16.

Kugaya, A., & Sanacora, G. (2005). Beyond monoamines: glutamatergic function in mood disorders. *CNS Spectrums, 10,* 808–819.

Kwidzinski, E., Bunse, J., Aktas, O., et al. (2005). Indolamine 2,3-dioxygenase is expressed in the CNS and down-regulates autoimmune inflammation. *FASEB Journal, 19,* 1347–1349.

Lanquillon, S., Krieg, J. C., Bening-Abu-Shach, U., et al. (2000). Cytokine production and treatment response in major depressive disorder. *Neuropsychopharmacology, 22,* 370–379.

Laudenslager, M. L., Fleshner, M., Hofstadter, P., et al. (1988). Suppression of specific antibody production by inescapable shock: stability under varying conditions. *Brain, Behavior, and Immunity, 2,* 92–101.

Leonard, B. E., & Myint, A. (2006). Inflammation and depression: is there a causal connection with dementia? *Neurotoxicity Research, 10,* 149–160.

Lidberg, L., Belfrage, H., Bertilsson, L., et al. (2000). Suicide attempts and impulse control disorder are related to low cerebrospinal fluid 5-HIAA in mentally disordered violent offenders. *Acta Psychiatrica Scandinavica, 101,* 395–402.

Linnoila, M., Whorton, A. R., Rubinow, D. R., et al. (1983). CSF prostaglandin levels in depressed and schizophrenic patients. *Archives of General Psychiatry, 40,* 405–406.

Linthorst, A. C., Flachskamm, C., Muller-Preuss, P., et al. (1995). Effect of bacterial endotoxin and interleukin-1 beta on hippocampal serotonergic neurotransmission, behavioral activity, and free corticosterone levels: an in vivo microdialysis study. *Journal of Neuroscience, 15,* 2920–2934.

Liu, Y., Ho, R. C., & Mak, A. (2012). Interleukin (IL)-6, tumour necrosis factor alpha (TNF-alpha) and soluble interleukin-2 receptors (sIL-2R) are elevated in patients with major depressive disorder: a meta-analysis and meta-regression. *Journal of Affective Disorders, 139,* 230–239.

Madrigal, J. L., Garcia-Bueno, B., Moro, M. A., et al. (2003). Relationship between cyclooxygenase-2 and nitric oxide synthase-2 in rat cortex after stress. *European Journal of Neuroscience, 18,* 1701–1705.

Maes, M. (1994). Cytokines in major depression. *Biological Psychiatry, 36,* 498–499.

Maes, M. (2011). Depression is an inflammatory disease, but cell-mediated immune activation is the key component of depression. *Progress in Neuro-Psychopharmacology & Biological Psychiatry, 35,* 664–675.

Maes, M., Anderson, G., Kubera, M., et al. (2014). Targeting classical IL-6 signalling or IL-6 trans-signalling in depression? *Expert Opinion on Therapeutic Targets, 18,* 495–512.

Maes, M., Meltzer, H. Y., Bosmans, E., et al. (1995). Increased plasma concentrations of interleukin-6, soluble interleukin-6, soluble interleukin-2 and transferrin receptor in major depression. *Journal of Affective Disorders, 34,* 301–309.

Maes, M., Meltzer, H. Y., Buckley, P., et al. (1995). Plasma-soluble interleukin-2 and transferrin receptor in schizophrenia and major depression. *European Archives of Psychiatry and Clinical Neuroscience, 244,* 325–329.

Maes, M., Scharpe, S., Meltzer, H. Y., et al. (1994). Increased neopterin and interferon-gamma secretion and lower availability of L-tryptophan in major depression: further evidence for an immune response. *Psychiatry Research, 54,* 143–160.

Maes, M., Stevens, W., DeClerck, L., et al. (1992). Immune disorders in depression: higher T helper/T suppressor-cytotoxic cell ratio. *Acta Psychiatrica Scandinavica, 86,* 423–431.

Maes, M., Verkerk, R., Vandoolaeghe, E., et al. (1998). Serum levels of excitatory amino acids, serine, glycine, histidine, threonine, taurine, alanine and arginine in treatment-resistant depression: modulation by treatment with antidepressants and prediction of clinical responsivity. *Acta Psychiatrica Scandinavica, 97,* 302–308.

Maj, J., Rogoz, Z., Skuza, G., et al. (1992). Effects of MK-801 and antidepressant drugs in the forced swimming test in rats. *European Neuropsychopharmacology, 2,* 37–41.

Mann, J. J., & Malone, K. M. (1997). Cerebrospinal fluid amines and higher-lethality suicide attempts in depressed inpatients. *Biological Psychiatry, 41,* 162–171.

Matussek, N. (1966). Neurobiologie und depression. *Medizinische Monatsschrift, 3,* 109–112.

Matussek, N. (1988). Catecholamines and mood: neuroendocrine aspects. In D. Ganten, D. Pfaff, & K. Fuxe (Eds.), *Neuroendocrinology of mood* (pp. 141–182). Heidelberg, New York: Springer.

Mauri, M. C., Ferrara, A., Boscati, L., et al. (1998). Plasma and platelet amino acid concentrations in patients affected by major depression and under fluvoxamine treatment. *Neuropsychobiology, 37,* 124–129.

Mellor, A. L., & Munn, D. H. (1999). Tryptophan catabolism and T-cell tolerance: immunosuppression by starvation? *Immunology Today, 20,* 469–473.

Merali, Z., Lacosta, S., & Anisman, H. (1997). Effects of interleukin-1beta and mild stress on alterations of norepinephrine, dopamine and serotonin neurotransmission: a regional microdialysis study. *Brain Research, 761,* 225–235.

Miguel-Hidalgo, J. J., Baucom, C., Dilley, G., et al. (2000). Glial fibrillary acidic protein immunoreactivity in the prefrontal cortex distinguishes younger from older adults in major depressive disorder. *Biological Psychiatry, 48,* 861–873.

Mikova, O., Yakimova, R., Bosmans, E., et al. (2001). Increased serum tumor necrosis factor alpha concentrations in major depression and multiple sclerosis. *European Neuropsychopharmacology, 11,* 203–208.

Moghaddam, B., Bolinao, M. L., Stein-Behrens, B., et al. (1994). Glucocorticoids mediate the stress-induced extracellular accumulation of glutamate. *Brain Research, 655,* 251–254.

Muller, N., Gizycki-Nienhaus, B., Gunther, W., et al. (1992). Depression as a cerebral manifestation of scleroderma: immunological findings in serum and cerebrospinal fluid. *Biological Psychiatry, 31,* 1151–1156.

Muller, N., Hofschuster, E., Ackenheil, M., et al. (1993). Investigations of the cellular immunity during depression and the free interval: evidence for an immune activation in affective psychosis. *Progress in Neuro-Psychopharmacology & Biological Psychiatry, 17,* 713–730.

Muller, N., Myint, A. M., & Schwarz, M. J. (2011). Inflammatory biomarkers and depression. *Neurotoxicity Research, 19,* 308–318.

Müller, N., & Schwarz, M. J. (2002). Immunology in anxiety and depression. In S. Kasper, J. A. den Boer, & J. M. A.

Sitsen (Eds.), *Handbook of depression and anxiety* (pp. 267–288). New York: Marcel Dekker.

Muller, N., & Schwarz, M. J. (2007). The immune-mediated alteration of serotonin and glutamate: towards an integrated view of depression. *Molecular Psychiatry, 12,* 988–1000.

Muller, N., Schwarz, M. J., Dehning, S., et al. (2006). The cyclooxygenase-2 inhibitor celecoxib has therapeutic effects in major depression: results of a double-blind, randomized, placebo controlled, add-on pilot study to reboxetine. *Molecular Psychiatry, 11,* 680–684.

Müller, N., Schwarz, M. J., & Riedel, M. (2005). COX-2 inhibition in schizophrenia: focus on clinical effects of celecoxib therapy and the role of TNF-alpha. In W. W. Eaton (Ed.), *Medical and psychiatric comorbidity over the course of life* (pp. 265–276). Washington, DC: American Psychiatric Publishing.

Munn, D. H., Shafizadeh, E., Attwood, J. T., et al. (1999). Inhibition of T cell proliferation by macrophage tryptophan catabolism. *Journal of Experimental Medicine, 189,* 1363–1372.

Myint, A. M., & Kim, Y. K. (2003). Cytokine-serotonin interaction through IDO: a neurodegeneration hypothesis of depression. *Medical Hypotheses, 61,* 519–525.

Myint, A. M., Kim, Y. K., Verkerk, R., et al. (2007). Kynurenine pathway in major depression: evidence of impaired neuroprotection. *Journal of Affective Disorders, 98,* 143–151.

Myint, A. M., Leonard, B. E., Steinbusch, H. W., et al. (2005). Th1, Th2, and Th3 cytokine alterations in major depression. *Journal of Affective Disorders, 88,* 167–173.

Myint, A. M., Steinbusch, H. W., Goeghegan, L., et al. (2007). Effect of the COX-2 inhibitor celecoxib on behavioural and immune changes in an olfactory bulbectomised rat model of depression. *Neuroimmunomodulation, 14,* 65–71.

Na, K. S., Lee, K. J., Lee, J. S., et al. (2014). Efficacy of adjunctive celecoxib treatment for patients with major depressive disorder: a meta-analysis. *Progress in Neuro-Psychopharmacology & Biological Psychiatry, 48,* 79–85.

Nair, A., & Bonneau, R. H. (2006). Stress-induced elevation of glucocorticoids increases microglia proliferation through NMDA receptor activation. *Journal of Neuroimmunology, 171,* 72–85.

Nguyen, K. T., Deak, T., Owens, S. M., et al. (1998). Exposure to acute stress induces brain interleukin-1beta protein in the rat. *Journal of Neuroscience, 18,* 2239–2246.

Niino, M., Ogata, A., Kikuchi, S., et al. (2000). Macrophage migration inhibitory factor in the cerebrospinal fluid of patients with conventional and optic-spinal forms of multiple sclerosis and neuro-Behcet's disease. *Journal of the Neurological Sciences, 179,* 127–131.

Nishino, S., Ueno, R., Ohishi, K., et al. (1989). Salivary prostaglandin concentrations: possible state indicators for major depression. *American Journal of Psychiatry, 146,* 365–368.

Nordstrom, P., Samuelsson, M., Asberg, M., et al. (1994). CSF 5-HIAA predicts suicide risk after attempted suicide. *Suicide & Life-Threatening Behavior, 24,* 1–9.

Nunes, S. O., Reiche, E. M., Morimoto, H. K., et al. (2002). Immune and hormonal activity in adults suffering from depression. *Brazilian Journal of Medical and Biological Research, 35,* 581–587.

O'Brien, S. M., Scott, L. V., & Dinan, T. G. (2004). Cytokines: abnormalities in major depression and implications for pharmacological treatment. *Human Psychopharmacology, 19,* 397–403.

Ohishi, K., Ueno, R., Nishino, S., et al. (1988). Increased level of salivary prostaglandins in patients with major depression. *Biological Psychiatry, 23,* 326–334.

Olsson, S. K., Andersson, A. S., Linderholm, K. R., et al. (2009). Elevated levels of kynurenic acid change the dopaminergic response to amphetamine: implications for schizophrenia. *International Journal of Neuropsychopharmacology, 12,* 501–512.

Ongur, D., Drevets, W. C., & Price, J. L. (1998). Glial reduction in the subgenual prefrontal cortex in mood disorders. *Proceedings of the National Academy of Sciences of the United States of America, 95,* 13290–13295.

Ossowska, G., Klenk-Majewska, B., & Szymczyk, G. (1997). The effect of NMDA antagonists on footshock-induced fighting behavior in chronically stressed rats. *Journal of Physiology and Pharmacology, 48,* 127–135.

Pace, T. W., Mletzko, T. C., Alagbe, O., et al. (2006). Increased stress-induced inflammatory responses in male patients with major depression and increased early life stress. *American Journal of Psychiatry, 163,* 1630–1633.

Pugh, C. R., Nguyen, K. T., Gonyea, J. L., et al. (1999). Role of interleukin-1 beta in impairment of contextual fear conditioning caused by social isolation. *Behavioural Brain Research, 106,* 109–118.

Raison, C. L., Capuron, L., & Miller, A. H. (2006). Cytokines sing the blues: inflammation and the pathogenesis of depression. *Trends in Immunology, 27,* 24–31.

Raison, C. L., Rutherford, R. E., Woolwine, B. J., et al. (2013). A randomized controlled trial of the tumor necrosis factor antagonist infliximab for treatment-resistant depression: the role of baseline inflammatory biomarkers. *JAMA Psychiatry, 70,* 31–41.

Rajkowska, G. (2003). Depression: what we can learn from postmortem studies. *Neuroscientist, 9,* 273–284.

Rajkowska, G., Halaris, A., & Selemon, L. D. (2001). Reductions in neuronal and glial density characterize the dorsolateral prefrontal cortex in bipolar disorder. *Biological Psychiatry, 49,* 741–752.

Rajkowska, G., Miguel-Hidalgo, J. J., Wei, J., et al. (1999). Morphometric evidence for neuronal and glial prefrontal cell pathology in major depression. *Biological Psychiatry, 45,* 1085–1098.

Robinson, C. M., Hale, P. T., & Carlin, J. M. (2005). The role of IFN-gamma and TNF-alpha-responsive regulatory elements in the synergistic induction of indoleamine dioxygenase. *Journal of Interferon & Cytokine Research, 25*, 20–30.

Rothermundt, M., Arolt, V., Fenker, J., et al. (2001). Different immune patterns in melancholic and non-melancholic major depression. *European Archives of Psychiatry and Clinical Neuroscience, 251*, 90–97.

Rothermundt, M., Arolt, V., Peters, M., et al. (2001). Inflammatory markers in major depression and melancholia. *Journal of Affective Disorders, 63*, 93–102.

Saavedra, J. M., Sanchez-Lemus, E., & Benicky, J. (2011). Blockade of brain angiotensin II AT1 receptors ameliorates stress, anxiety, brain inflammation and ischemia: therapeutic implications. *Psychoneuroendocrinology, 36*, 1–18.

Sanacora, G., Gueorguieva, R., Epperson, C. N., et al. (2004). Subtype-specific alterations of gamma-aminobutyric acid and glutamate in patients with major depression. *Archives of General Psychiatry, 61*, 705–713.

Sandrini, M., Vitale, G., & Pini, L. A. (2002). Effect of rofecoxib on nociception and the serotonin system in the rat brain. *Inflammation Research, 51*, 154–159.

Sapolsky, R. M. (1985). A mechanism for glucocorticoid toxicity in the hippocampus: increased neuronal vulnerability to metabolic insults. *Journal of Neuroscience, 5*, 1228–1232.

Sasayama, D., Hattori, K., Wakabayashi, C., et al. (2013). Increased cerebrospinal fluid interleukin-6 levels in patients with schizophrenia and those with major depressive disorder. *Journal of Psychiatric Research, 47*, 401–406.

Schaefer, M., Horn, M., Schmidt, F., et al. (2004). Correlation between sICAM-1 and depressive symptoms during adjuvant treatment of melanoma with interferon-alpha. *Brain, Behavior, and Immunity, 18*, 555–562.

Schiepers, O. J., Wichers, M. C., & Maes, M. (2005). Cytokines and major depression. *Progress in Neuro-Psychopharmacology & Biological Psychiatry, 29*, 201–217.

Schwarz, M. J., Chiang, S., Muller, N., et al. (2001). T-helper-1 and T-helper-2 responses in psychiatric disorders. *Brain, Behavior, and Immunity, 15*, 340–370.

Seidel, A., Arolt, V., Hunstiger, M., et al. (1996). Major depressive disorder is associated with elevated monocyte counts. *Acta Psychiatrica Scandinavica, 94*, 198–204.

Setiawan, E., Wilson, A. A., Mizrahi, R., et al. (2015). Role of translocator protein density, a marker of neuroinflammation, in the brain during major depressive episodes. *JAMA Psychiatry, 72*, 268–275.

Shintani, F., Nakaki, T., Kanba, S., et al. (1995). Involvement of interleukin-1 in immobilization stress-induced increase in plasma adrenocorticotropic hormone and in release of hypothalamic monoamines in the rat. *Journal of Neuroscience, 15*, 1961–1970.

Si, X., Miguel-Hidalgo, J. J., O'Dwyer, G., et al. (2004). Age-dependent reductions in the level of glial fibrillary acidic protein in the prefrontal cortex in major depression. *Neuropsychopharmacology, 29*, 2088–2096.

Song, C., & Leonard, B. E. (2000). *Fundamentals of psychoneuroimmunology.* J Wiley and Sons: Chichester, New York.

Song, C., Lin, A., Bonaccorso, S., et al. (1998). The inflammatory response system and the availability of plasma tryptophan in patients with primary sleep disorders and major depression. *Journal of Affective Disorders, 49*, 211–219.

Song, C., Merali, Z., & Anisman, H. (1999). Variations of nucleus accumbens dopamine and serotonin following systemic interleukin-1, interleukin-2 or interleukin-6 treatment. *Neuroscience, 88*, 823–836.

Sparkman, N. L., & Johnson, R. W. (2008). Neuroinflammation associated with aging sensitizes the brain to the effects of infection or stress. *Neuroimmunomodulation, 15*, 323–330.

Stein-Behrens, B. A., Lin, W. J., & Sapolsky, R. M. (1994). Physiological elevations of glucocorticoids potentiate glutamate accumulation in the hippocampus. *Journal of Neurochemistry, 63*, 596–602.

Steiner, J., Walter, M., Gos, T., et al. (2011). Severe depression is associated with increased microglial quinolinic acid in subregions of the anterior cingulate gyrus: evidence for an immune-modulated glutamatergic neurotransmission? *Journal of Neuroinflammation, 8*, 94.

Stryjer, R., Strous, R. D., Shaked, G., et al. (2003). Amantadine as augmentation therapy in the management of treatment-resistant depression. *International Clinical Psychopharmacology, 18*, 93–96.

Sundar, S. K., Cierpial, M. A., Kilts, C., et al. (1990). Brain IL-1-induced immunosuppression occurs through activation of both pituitary-adrenal axis and sympathetic nervous system by corticotropin-releasing factor. *Journal of Neuroscience, 10*, 3701–3706.

Suzuki, T., Ogata, A., Tashiro, K., et al. (2000). Japanese encephalitis virus up-regulates expression of macrophage migration inhibitory factor (MIF) mRNA in the mouse brain. *Biochimica et Biophysica Acta, 1517*, 100–106.

Takahashi, T., Kimoto, T., Tanabe, N., et al. (2002). Corticosterone acutely prolonged N-methyl-d-aspartate receptor-mediated Ca^{2+} elevation in cultured rat hippocampal neurons. *Journal of Neurochemistry, 83*, 1441–1451.

Traki, L., Rostom, S., Tahiri, L., et al. (2014). Responsiveness of the EuroQol EQ-5D and Hospital Anxiety and Depression Scale (HADS) in rheumatoid arthritis patients receiving tocilizumab. *Clinical Rheumatology, 33*, 1055–1060.

Trullas, R., & Skolnick, P. (1990). Functional antagonists at the NMDA receptor complex exhibit antidepressant actions. *European Journal of Pharmacology, 185*, 1–10.

Tyring, S., Gottlieb, A., Papp, K., et al. (2006). Etanercept and clinical outcomes, fatigue, and depression in psoriasis: double-blind placebo-controlled randomised phase III trial. *Lancet*, *367*, 29–35.

Weiss, G., Murr, C., Zoller, H., et al. (1999). Modulation of neopterin formation and tryptophan degradation by Th1- and Th2-derived cytokines in human monocytic cells. *Clinical and Experimental Immunology*, *116*, 435–440.

Weiss, J. M., Quan, N., & Sundar, S. K. (1994). Immunological consequences of Interleukin-1 in the brain. *Neuropsychopharmacology*, *10*, 833.

Weitz-Schmidt, G. (2002). Statins as anti-inflammatory agents. *Trends in Pharmacological Sciences*, *23*, 482–486.

Woolley, C. S., Gould, E., & McEwen, B. S. (1990). Exposure to excess glucocorticoids alters dendritic morphology of adult hippocampal pyramidal neurons. *Brain Research*, *531*, 225–231.

Yang, C., & Hashimoto, K. (2015). Peripheral IL-6 signaling: a promising therapeutic target for depression? *Expert Opinion on Investigational Drugs*, *24*, 989–990.

Yang, C., Shirayama, Y., Zhang, J. C., et al. (2015). Peripheral interleukin-6 promotes resilience versus susceptibility to inescapable electric stress. *Acta Neuropsychiatrica*, *27*, 312–316.

Yang, J. J., Wang, N., Yang, C., et al. (2015). Serum interleukin-6 is a predictive biomarker for ketamine's antidepressant effect in treatment-resistant patients with major depression. *Biological Psychiatry*, *77*, e19–20.

Yilmaz, A., Schulz, D., Aksoy, A., et al. (2002). Prolonged effect of an anesthetic dose of ketamine on behavioral despair. *Pharmacology, Biochemistry, and Behavior*, *71*, 341–344.

Zalcman, S., Green-Johnson, J. M., Murray, L., et al. (1994). Cytokine-specific central monoamine alterations induced by interleukin-1, -2 and -6. *Brain Research*, *643*, 40–49.

Zhou, A. J., Lee, Y., Salvadore, G., et al. (2017). Sirukumab: a potential treatment for mood disorders? *Advances in Therapy*, *34*, 78–90.

Zubin, J., & Spring, B. (1977). Vulnerability—a new view of schizophrenia. *Journal of Abnormal Psychology*, *86*, 103–126.

29

Alcohol, Inflammation, and Depression: The Gut-Brain Axis

Abigail R. Cannon, Adam M. Hammer, Mashkoor A. Choudhry

Loyola University Chicago Health Sciences Division, Maywood, IL, United States

INTRODUCTION

Alcohol misuse remains one of the largest clinical complications worldwide (Control CfD, 2015; WHO, 2011, 2015). The problems associated with alcohol misuse are wide-ranging, but the most shocking is that alcohol-related deaths are one of the largest premature losses of life (WHO, 2015). However, even when intoxication does not lead to premature mortality, the negative effects of intoxication are wide-ranging from physiological to social and behavioral (WHO, 2015). The goal of this chapter is to review studies in the area of alcohol misuse and their effects on human physiology, specifically in the brain and gut, and how these changes contribute to depression.

ALCOHOL MISUSE: CURRENT DEFINITIONS AND EPIDEMIOLOGY

Misuse of alcohol can take several different forms, as defined by the National Institute on Alcohol Abuse and Alcoholism (NIAAA). Binge drinking, the most common form of alcohol misuse, is defined by any pattern of drinking that raises one's blood alcohol content to 0.08 g/dL or higher (NIAAA, 2017b). This is more generally described as four drinks for women or five drinks for men in a 2 h period (NIAAA, 2017b). Chronic alcohol misuse, or heavy drinking, is defined as binge drinking 5 or more days per month (NIAAA, 2017b). Definitions of alcohol use disorders are fluid as researchers and clinicians reevaluate what constitutes healthy versus unhealthy drinking behavior. Currently, diagnosis of an alcohol use disorder is made by answering "yes" to two or more questions of a questionnaire generated by the American Psychiatric Association in May of 2013 (NIAAA, 2013). Some of the following questions help determine whether the patient has an existing alcohol use disorder (NIAAA, 2013).

In the past year, have you
(1) had times when you ended up drinking more, or longer, than you intended?
(2) continued to drink even though it was making you feel depressed or anxious or adding to another health problem or after having had a memory blackout?

(3) had to drink much more than you once did to get the effect you want or found that your usual number of drinks had much less effect than before?

It is with this questionnaire and other tools that the NIAAA was able to conduct a survey in 2015 that found that 26.9% of people ages 18 and older reported engaging in binge drinking within the past month and 7.0% of people reported heavy drinking behavior (NIAAA, 2017a). The statistics of heavy drinkers are roughly in line with the 6.2% of adults diagnosed with an alcohol use disorder (NIAAA, 2017a). In adults, nearly twice as many men (9.8 million) have an alcohol use disorder compared with women (5.3 million) (NIAAA, 2017a). These statistics will be important later as we discuss the relationship between alcohol, physiology, and consequences for depression.

ALCOHOL AND DEPRESSION (ALCOHOL MISUSE: COMORBIDITY OF DEPRESSION)

A major challenge when assessing the relationship between alcohol misuse and its possible effects on depression is determining whether alcohol misuse causes depression or depression causes alcohol misuse behavior. Not surprisingly, there are data to support the latter, as many individuals diagnosed with depression have an alcohol use disorder as a way to self-medicate. Therefore, another important question is whether alcohol can induce or perpetuate a depressive episode?

One of the first studies to assess the relationship between alcohol misuse and depression was performed by Kessler et al. (1996). Among other psychological disorders, the study found that individuals with alcohol dependence (alcoholism) were 3.9 times more likely to have major depressive disorder (MDD) compared with those without alcohol dependence (Kessler et al., 1996). Another large epidemiological study in 2005 by Hasin et al. found that the incidence of alcohol use disorders in individuals with a MDD is staggering in comparison with individuals without depression. Of individuals with MDD within the previous 12 months, 14.1% had an alcohol use disorder (twice the incidence compared with the general population) (Hasin, Goodwin, Stinson, & Grant, 2005). Even more shocking, of those with lifetime MDD, 40.3% had an alcohol use disorder (6.5 times the incidence compared with the general population) (Hasin et al., 2005). These figures alone suggest a strong correlation between alcohol misuse and depression.

Depression is a mood disorder characterized by melancholic symptoms, anxiety, and fatigue and somatic symptoms (Eyre, Papps, & Baune, 2013). Until the early 2000s, depression was believed to be a result of immunosuppression, as depressed patients are presented with decreased percentages of B cells, T cells, and T-helper cells; decreased ratio of T suppressor/T cytotoxic; decreased lymphocyte proliferation in response to antigen; and decreased natural killer cell number and activity (Raison & Miller, 2001). However, the thinking behind the etiology of depression has taken a dramatic 180 degree shift away from linking depression to a state of immunosuppression to now linking depression to a hyperactivation of innate immune inflammatory responses (Maes, 2011).

The activation of this inflammatory cascade can be partly attributed to consumption of alcohol leading to increases in gut permeability. As the gut becomes leaky, this creates a favorable environment for bacteria and bacterial endotoxins to translocate through the compromised intestinal barrier and enter the blood stream. Specifically, LPS utilizes two routes to enter the systemic circulation: the first through the portal vein and the second through the lymphatic vessels in the GI tract (Moore et al., 1991; Mowat, 2003; Wang, Zakhari, & Jung, 2010). However, the lymphatic route predominates in determining the amount of LPS in circulation. This is due to the fact that as LPS travels through the lymphatics, it does not encounter a detoxifying organ (i.e., the liver) while in transit. Thus, most of the bioactive LPS is released via the lymphatics making it available to other organs and subsequently inducing inflammatory injury in both proximal and remote organs.

LPS is one of the most common types of pathogen-associated molecular patterns or PAMPs (Mogensen, 2009). Recognition of PAMPs in both invertebrates and vertebrates is accomplished via a family of endogenous receptors named toll-like receptors (TLRs) (Mogensen, 2009). Interestingly, TLRs have a dichotomous role in the brain in that there is evidence supporting their activation in both neurogenesis and also neuropathogenesis (Okun, Griffioen, & Mattson, 2011). Of the 13 known TLRs, TLR4 is the most widely studied and is responsible for detecting components of Gram-negative bacteria, specifically LPS (Kawai & Akira, 2008; Mogensen, 2009). Microglial cells are the "resident macrophages" of the brain, and they express high levels of TLR4 and are known to rapidly respond to LPS by releasing pro-inflammatory mediators (Dheen, Kaur, & Ling, 2007; Lehnardt, 2010). Activation of microglial cells via LPS detection by TLR4 initiates a pro-inflammatory cascade as this is required for the astrocyte pro-inflammatory response (Holm, Draeby, & Owens, 2012). Furthermore, TLR4 has been found to be expressed on neurons, themselves, which in the presence of LPS perpetuates the inflammatory response while demonstrating cross communication between microglia, astrocytes, and neurons (Hua et al., 2007; Pandey, 2012; Tang et al., 2007). The fact that alcohol leads to increases in systemic LPS as a result of gut permeability is of critical importance as increases in pro-inflammatory cytokines within the brain have been heavily linked to symptoms of depression (Dantzer, O'Connor, Freund, Johnson, & Kelley, 2008).

One pivotal role of the brain is coordinating defense against systemic infection or injury by restoring homeostasis, the sole intention of which is restoring homeostasis to the organism. Circulating pro-inflammatory cytokines, as a result of alcohol-induced LPS leakage from the gut, are one such way systemic infection or injury is communicated to the central nervous system (CNS) that a homeostatic imbalance exists (Quan & Banks, 2007). The CNS then activates the hypothalamic-pituitary-adrenal (HPA) axis and sympathetic nervous system (SNS), both of which are largely responsible for initiating anti-inflammatory mechanisms to counteract increases in inflammation. Once activated, the HPA axis initiates the synthesis of corticotropin-releasing hormone (CRH) by the hypothalamus, which leads to adrenocorticotropic hormone release from the pituitary gland, thus inducing synthesis and release of glucocorticoid cortisol from the adrenal cortex (Beishuizen & Thijs, 2003; Kunugi et al., 2006). Glucocorticoids are of critical importance to the anti-inflammatory response as they promote the release of the anti-inflammatory cytokine, IL-10 (Miyamasu et al., 1998; Ogawa et al., 2005; Richards, Fernandez, Caulfield, & Hawrylowicz, 2000; Scheinman, Gualberto, Jewell, Cidlowski, & Baldwin, 1995; Sewell, Scurr, Orphanides, Kinder, & Ludowyke, 1998).

Likewise, the SNS becomes activated due to circulating levels of pro-inflammatory cytokines. SNS activation leads to the peripheral release of the catecholamines, epinephrine and norepinephrine, which have systemic anti-inflammatory effects (Wang et al., 2003). Furthermore, binding of either epinephrine or norepinephrine to specific receptors on innate immune cells promotes the release of anti-inflammatory cytokines (Wang et al., 2003).

The concerted effort of the HPA axis and the SNS to contain increases in inflammation by the initiation of anti-inflammatory signaling pathways is not only the body's natural response but also vital to the integrity of healthy tissue (Wang et al., 2010). The consumption of alcohol alone is perceived as a stressful event by the

HPA axis resulting in elevated cortisol levels (Spencer & Hutchison, 1999). At first glance, the neuroimmune sensing of alcohol as a stressor and subsequent activation of the HPA axis anti-inflammatory signaling transduction pathway mentioned above is a good thing because alcohol alone leads to increased systemic and neuronal inflammation. Unfortunately, chronic consumption of alcohol desensitizes the HPA axis response to alcohol, such that compared with nonalcohol controls, serum cortisol is lower after alcohol consumption (Spencer & Hutchison, 1999). This blunting of the HPA axis response would, therefore, eliminate one of checkpoints and control mechanism to increased inflammation that gives evidence to the findings of persistent low-grade inflammation in alcoholics (Richardson, Lee, O'Dell, Koob, & Rivier, 2008).

However, low-grade chronic inflammation is associated with chronic alcohol consumption even in the absence of any obvious infection as defined by the mechanisms outlined above (Wang et al., 2010). TNF-α and IL-1β are the two main pro-inflammatory cytokines associated with alcohol-induced intestinal and neuroinflammation. Surprisingly, these two cytokines are directly connected to a whole host of increased sickness and depressive behaviors in mouse models in both a dose- and time-dependent manner (Dantzer, 2001). Systemic or central administration of either TNF-α or IL-1β caused animals to retreat to corners of their cages in a hunched posture, shows little or no interest in the physical or social environment unless stimulated, decreased motor activity and social withdrawal, reduced food and water intake, and altered cognition (Pugh et al., 1998; Vereker, O'Donnell, & Lynch, 2000). IL-1 in the brain is known to play a role in the occurrence of fatigue as assessed by decreased resistance to forced exercise on a treadmill (Carmichael et al., 2006). Furthermore, the inflammatory cytokine, interferon, is regularly used in treatment regimens for many types of cancer such as kidney cancer and malignant melanomas (Cancer Research UK, n.d.; Goldstein & Laszlo, 1988). Interestingly, one of the deleterious side effects of interferon treatment is depressive episodes in cancer patients, thus linking systemic inflammation to feelings of depression (Fig. 1; Illman et al., 2005). The fact that administration of exogenous TNF-α, IL-1, or interferon, all pro-inflammatory cytokines, results in severe

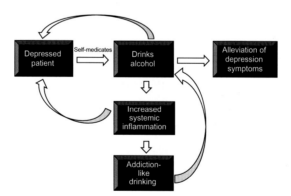

FIG. 1   Depression and alcohol abuse feed into one another. Firstly, depressed patients will self-medicate with alcohol to alleviate symptoms of depression (Hasin et al., 2005; Kessler et al., 1996). However, this activates the inflammatory response. Activation of both neurointestinal and intestinal inflammation following alcohol use favors addiction-like drinking (Crews et al., 2011; Mulligan et al., 2006). Addiction-like drinking, thus, creates a positive feedback loop by increasing alcohol consumption and, therefore, systemic inflammation, which then loops back to increases in depressive-like symptoms in patients (Carmichael et al., 2006; Illman et al., 2005; Pugh et al., 1998; Vereker et al., 2000).

symptoms of depression highlights the trifold connection between depression, inflammation, and alcohol use.

The relationship between alcohol and depression is of interest because alcohol itself inhibits CNS responses leading to symptoms of depression such as sadness, hopelessness, and lethargy. Yet, depressed patients will often self-medicate with alcohol in attempts to lift their mood or numb depressive thoughts. As a result, depression and alcohol abuse feed into one another with one condition often perpetuating the other. The indefinite cycle of alcohol being used to alleviate feelings of depression but actually propagating symptoms of depression can also link to activation of the innate inflammatory response feeding into the cycle of alcohol addiction (Crews, Zou, & Qin, 2011; Krishnan, Sakharkar, Teppen, Berkel, & Pandey, 2014; Fig. 1). Increased activation of nuclear factor-kB (NF-kB), its regulatory proteins, and innate immune genes was found via a meta-analysis of both rats and mice with addiction-like drinking behavior (Mulligan et al., 2006). Remarkably, a single injection of LPS produced long-lasting increases in alcohol consumption (Blednov et al., 2011). Thus, serendipitously, bringing us full circle to that of alcohol consumption leading to increased gut permeability allowing increased leakage of LPS into the systemic circulation consequently resulting in neuroinflammation that has been shown to have direct effects on symptoms of depression.

ALCOHOL AND NEUROINFLAMMATION

A number of enzymes are responsible for breaking down alcohol after its consumption, the most well-known being alcohol dehydrogenase (ADH) and aldehyde dehydrogenase (ALDH) in the liver (Cederbaum, 2012). As alcohol is absorbed by different cells in the mouth, stomach, and intestines, a small amount is metabolized in the cells that absorb it. However, a majority of alcohol that is consumed passes directly into the blood stream due to its hydrophobicity. Once in the blood stream, alcohol travels to the liver, the organ responsible for the bulk of alcohol metabolism (Cederbaum, 2012). The breakdown of alcohol in the liver occurs in two sequential steps. First, ADH metabolizes alcohol into acetaldehyde, a highly toxic molecule that is a known carcinogen. Acetaldehyde has been shown to promote alcohol dependence and is found in all regions of the brain following alcohol consumption (Radcliffe et al., 2009). Additionally, acetaldehyde protein adducts are highly pro-inflammatory through activating complement cascade, recruiting neutrophils, and eliciting the production of reactive oxygen species (ROS) (Setshedi, Wands, & Monte, 2010). In this regard, acetaldehyde can cause or perpetuate an inflammatory state following alcohol consumption. While acetaldehyde has high toxicity, it is rather short lived, as it is quickly metabolized by ALDH to convert it into acetate. Acetate is finally broken down into carbon dioxide and water, which are both easily removed (Cederbaum, 2012).

While this mechanism is how a majority of the alcohol that is consumed is metabolized, not all organs contain ADH. In the brain, alcohol is metabolized by two different enzymes, cytochrome P450 and catalase (Zakhari, 2006). Cytochrome P450 is located within vesicles and the endoplasmic reticulum of cells and has three different isoforms. Upon oxidation, ethanol produces a number of ROS that can contribute to tissue damage (Zakhari, 2006). Catalase is another enzyme that is able to oxidize ethanol but is a minor player.

The interplay between alcohol/alcohol metabolites and inflammation is an ever-expanding area of research with vast amounts of evidence linking the use of alcohol or alcohol use disorders to that of chronic inflammatory conditions. Patients who suffer from alcohol use disorders consistently have high circulating levels of pro-inflammatory cytokines, which greatly contribute to disease initiation and progression (McClain, Barve, Deaciuc, Kugelmas, & Hill, 1999). Two main sources of alcohol-mediated activation of inflammation are that of alcohol damaged cells and changes to the gut microflora following alcohol use. Alcohol and alcohol's metabolites can lead to damaged cells in many tissue types with particular focus on cells of the brain and those of the intestinal tract (Agarwal, 2001). This alcohol-induced damage can be a result of the production of ROS following the metabolism of alcohol (Collins & Neafsey, 2012). NAPDH oxidase (NOX) is a major oxidant-generating enzyme found in activated phagocytes. Therefore, activation of NOX increases levels of ROS (Babior, 1999). Researchers found dramatic increases in the catalytic subunit of NOX, which is responsible for activation of pro-inflammatory signals, in the brains of mice following treatment with ethanol. Interestingly, these increases in the NOX catalytic subunit persisted long after the cessation of drinking alcohol (Qin & Crews, 2012). Furthermore, increases in the ROS, O_2^- and O_2^--derived oxidants, were found in the mouse brain following ethanol, thus indicating that ethanol increases the expression of NOX that leads to the formation of ROS (Qin et al., 2008; Qin & Crews, 2012). ROS can activate key pro-inflammatory signaling pathways, specifically NF-kB resulting in the production of both pro-inflammatory cytokines and chemokines (Qin & Crews, 2012; Ward et al., 1996).

Elevations of pro-inflammatory cytokines, such as TNF-α, are closely linked to neurodegeneration (Block & Hong, 2007; Qin et al., 2008; Tajuddin et al., 2014). The neurodegeneration following increases in TNF-α was found to be a result of an inhibition of both proliferation and differentiation of hippocampal neuroprogenetors (Nixon & Crews, 2002). Long-term exposure to alcohol has been shown to increase other inflammatory cytokines and mediators, such as IL-1β, caspase-3, MCP-1, and HMGB1,

which are associated with disease symptoms in animal models. Specific increases in IL-1β were shown to be a result of alcohol- induced activation of the NLRP3/ASC inflammasome, which further amplified neuroinflammation by increasing the levels of TNF-α and MCP-1 (Lippai et al., 2013).

ALCOHOL AND INTESTINAL INFLAMMATION

Ethanol-induced increases in inflammation are not limited to the tissue of the brain, as elevated levels of both pro-inflammatory cytokines and chemokines have been found in both the small and large intestinal tract following exposure to alcohol (Bode & Bode, 2003). The intestinal epithelium and gut-associated lymphoid tissue (GALT) create both a physical and immunologic barrier restricting the passage of potential harmful toxins, such as those from intestinal bacteria, from the luminal space to extraintestinal sites (Sonnenberg & Artis, 2015; Swanson, Sedghi, Farhadi, & Keshavarzian, 2010). T and B cells, macrophages, dendritic cells of Peyer's patches, mesenteric lymph nodes, and lamina propria (LP) of the GALT make up the intestinal immune barrier. The key components of the physical intestinal barrier include tight junctional complexes, adherens junctions, and desmosomes between intestinal epithelial cells (IECs). In particular, tight junctional complexes of the small and large intestine are made up of the proteins: claudin, occludin, and zonal occludin. These proteins are imperative to the maintenance of the physical intestinal barrier prohibiting translocation of bacteria out of the lumen while allowing the selective absorption of critical nutrients required by the host. Any perturbation to this tightly regulated intestinal barrier not only could lead to a so-called leaky gut, with deleterious effects at the level of the gastrointestinal tract but also allow bacterial endotoxin to penetrate the mucosa and enter systemic circulation. Alcohol consumption alone is known to disrupt the functional and structural integrity of IECs contributing to increased gut leakiness by a variety of mechanisms (Banan, Choudhary, Zhang, Fields, & Keshavarzian, 1999; Keshavarzian, Fields, Vaeth, & Holmes, 1994; Tang et al., 2008). Firstly, as increases in ROS following ethanol exposure contributed to disruptions in hypothalamic homeostasis, ethanol-induced increases in ROS also led to disruptions in intestinal homeostasis. Researchers have attributed increases in gut leakiness following ethanol treatment to increases in oxidative stress, specifically by nitric oxide (NO). At basal levels, NO is involved in maintaining normal intestinal barrier function (Alican & Kubes, 1996). However, when NO is in excess, as is found after chronic exposure to alcohol, it results in barrier disruption culminating in increased gut leakiness (Banan, Fields, Decker, Zhang, & Keshavarzian, 2000). Secondly, alcohol use can result in intestinal barrier structure defects via damage to the mucosa observed as the loss of epithelium at the apexes of villi, hemorrhagic erosions, and hemorrhage in the LP (Beck & Dinda, 1981). Thirdly, alcohol is known to alter the intestinal microbiome, by changing the intestinal microenvironment, which could potentially favor the growth of pathogenic bacteria over that of nonpathogenic, commensal bacteria. This dysbiosis compounded with the destructive nature of alcohol on the integrity of the intestinal epithelium could allow for bacterial translocation out of gut and into the circulation resulting in systemic inflammation. Finally, alcohol has a stimulatory effect on neuroendocrine hormones, such as CRH, which can directly lead to increased gut permeability (Soderholm & Perdue, 2001). Increases in CRH within the gut microenvironment can lead to degranulation of mast cells (Jacob et al., 2005; Wang et al., 2010). This degranulation triggers the synthesis and paracrine-like secretion of mediators to gut epithelial cells resulting in epithelial cell F-actin rearrangements increasing gut permeability (Jacob et al., 2005).

These increases in gut permeability following alcohol consumption allow both bacteria and bacterial products including LPS to translocate outside the lumen and into the circulation (Bode & Bode, 2003; Bode, Kugler, & Bode, 1987). LPS that has breached the gastrointestinal barrier induces the synthesis and release of IL-1β from mononuclear myeloid cells into the circulation (Malarkey & Mills, 2007). Furthermore, circulating LPS has been found to increase the pro-inflammatory cytokine, TNF-α, in the brain, liver, and serum after 1 h (Qin et al., 2008). As discussed earlier, both IL-1β and TNF-α are known to induce neurodegeneration (Simi, Tsakiri, Wang, & Rothwell, 2007). As a compensatory mechanism to increases in pro-inflammatory cytokines, there exist anti-inflammatory cytokines such as IL-10, which is known to downregulate the expression of Th1 pro-inflammatory cytokines. However, when LPS leaks into the circulation as a result of increased gut permeability following exposure to alcohol, IL-10 levels are decreased both in brain and intestinal tissues (Qin et al., 2008).

Therefore, consuming alcohol has direct effects on the brain and the intestine by inducing both neuroinflammation and intestinal inflammation, respectively. However, the real pathological implication lies in the fact that alcohol-induced intestinal inflammation can initiate a destructive signaling pathway that results in gut permeability allowing leakage of LPS or potentially pathogenic bacteria into the circulation, inducing increases in circulating pro-inflammatory cytokines, and thus feeding forward the neuroinflammation and neurodegeneration already induced by alcohol consumption. Moreover, it is this gut-brain axis that becomes of critical importance when discussing the connections between alcohol consumption, intestinal inflammation, and depression.

ALCOHOL AND THE INTESTINAL MICROBIOME

Due to the close relationships of alcohol metabolism to the gastrointestinal tract, liver, and brain, the connections between these organs may contribute to depression. Recently, a significant amount of work has been performed examining the effects of the intestinal microbiome on various conditions pertaining to human health. Not surprisingly, a number of these studies have examined the possible roles for intestinal bacteria and their influence on depression. The intestines are tied much more closely to the CNS, through a complex network collectively termed the gut-brain axis.

The human intestines are home to an estimated 100 trillion bacteria (Hattori & Taylor, 2009). These microbes are involved in functions including digestion, metabolism, and development of host-immune defense. While the intestinal microbiome of an individual varies depending on a multitude of factors (diet, age, race, geographic location, etc.), the general balance of the major phyla that make up the majority of the intestinal microbiome is fairly similar. Not surprisingly, new technologies such as deep 16S ribosomal sequencing have allowed scientists to gain a much better understanding of the makeup of the intestinal microbiome, and how it varies between a healthy individual and during a diseased state (Hattori & Taylor, 2009). Alcohol use has been demonstrated by a number of groups to drastically change the makeup of the intestinal microbiome (Engen, Green, Voigt, Forsyth, & Keshavarzian, 2015). Specifically, data show a general reduction in the ratio of the most prevalent phyla, Firmicutes and *Bacteroides* groups, which generally make up about 85%–90% of the bacteria within the gut (Chen et al., 2011; Engen et al., 2015; Mutlu et al., 2012). Additionally, a relative increase in the ratio of Gram-negative Proteobacteria occurs following both acute and chronic alcohol consumption, which may lead to increased inflammation in the intestines and other sites following intoxication (Chen et al., 2011; Engen et al., 2015; Mutlu et al., 2012). It is important to note that not all alcohol consumption is detrimental to microbial populations within the intestine. A study performed by Queipo-Ortuno et al. showed that people that consumed a large glass of red wine (272 mL) every day for 20 days had significantly increased Proteobacteria, Fusobacteria, Firmicutes, and Bacteroidetes. Conversely, those who consumed gin (100 mL) per day had a significant decrease in the same phyla (Queipo-Ortuno et al., 2012). These data suggest that both the amount and type of alcohol consumed can influence the intestinal microbial communities, and this may have a significant impact on the stress and inflammatory response and directly on neurochemical signaling itself.

ALCOHOL AND THE GUT-BRAIN AXIS

In 2004, the first study was published linking variations in the intestinal microbiome to changes in the HPA axis (Sudo et al., 2004). This study demonstrated that stress resulted in an elevated corticosterone and adrenocorticotropin response in germ-free mice (mice that are devoid of any intestinal bacteria). The HPA axis is extremely important in generating a proper stress response and is slowly developed through adolescence. Interestingly, it appears that germ-free mice do not properly develop HPA axis in a normal fashion, and colonizing germ-free mice with probiotics helps to attenuate depressive-like behavior in response to a stressor (Bravo et al., 2011; Desbonnet et al., 2010). While groups have not looked directly at how alcohol may induce depressive-like behavior in mice with an altered HPA axis, alcohol has been shown by several groups to activate the HPA axis leading to an increase in corticosterone in the circulation of mice and rats following a single binge (Li, Rana, Schwacha, Chaudry, & Choudhry, 2006).

Interestingly, newer data have shown that the intestinal microbiota are directly responsible for controlling a portion of the inflammatory response (Allen et al., 2012) and that this phenomenon is shaped immediately following birth when an infant is first exposed to microbes. These data suggest that not only the environment that an individual is raised in but also how early in life and how frequent their alcohol exposure is may potentially shape how the intestinal microbiome contributes to depressive episodes. Many groups have shown that adolescent binge drinking has profound effects on the brain, and potentiating depression is one possible consequence of heavy drinking from a young age (Allen, Rivier, & Lee, 2011; Allen, Lee, Koob, & Rivier, 2011; Evans, Greaves-Lord, Euser, Franken, & Huizink, 2012). However, many more studies will be required to elucidate if this is indeed the case.

Intestinal microbes are able to directly interact with the CNS through vagal nerve stimulation and the enteric nervous system (Forsythe, Bienenstock, & Kunze, 2014). Studies have shown that exposure to certain Gram-negative species of bacteria, which can induce inflammation, directly leads to changes in neuronal activity in the hypothalamus and central brain regions (Goehler, Park, Opitz, Lyte, & Gaykema, 2008; Lyte, Li, Opitz, Gaykema, & Goehler, 2006). In addition to bacteria themselves possessing the ability to modulate the gut-brain axis, bacterial metabolites have also been shown to influence inflammation and neuronal signaling in the CNS. The potential signaling mechanism by which this neuromodulation could occur in the presence and absence of alcohol is outlined in Fig. 2. In 2014, Stilling et al. published a study demonstrating that bacterial metabolites are capable of epigenetic regulation in the brain (Stilling, Dinan, & Cryan, 2014).

However, the most direct evidence for the intestinal microbiome playing a role in behavioral modulation in the brain is how these bacteria are able to alter neurochemical signaling. The most well-studied alterations in neurochemical signaling are brain-derived neurotrophic factor (BDNF), gamma-aminobutyric acid (GABA) ergic signaling, and serotonergic signaling. BDNF is extremely important for the proliferation and plasticity of neurons in the brain. Studies in infection models have shown a significant reduction in BDNF protein expression in the brain, which was associated with anxiety-like behavior in mice (Bercik et al., 2011). GABA signaling, a main inhibitory neurotransmitter signaling process in the CNS, is dysregulated in anxiety and depression (Cryan & Kaupmann, 2005). Studies have shown that treatment with probiotics is able to induce GABA production, leading to significantly improved behavioral phenotypes (Higuchi, Hayashi, & Abe, 1997). Finally, serotonin is one major neuromodulator that can affect mood disorders. Many studies have established the gut as one of the largest producers of serotonin in the body, but few studies have examined how altered serotonin expression due to dysbiosis in the intestine can affect neuronal signaling in the brain (Diaz Heijtz et al., 2011; Uribe, Alam, Johansson, Midtvedt, & Theodorsson, 1994; Wikoff et al., 2009).

While altered bacterial populations in the gut appear to have numerous negative effects on the CNS, there is some clinical evidence that probiotic treatment has some efficacy (Benton, Williams, & Brown, 2007; Messaoudi et al., 2011; Rao et al., 2009). For example, one study used a treatment of *Lactobacillus helveticus* and *Bifidobacterium longum* or placebo for 1 month to healthy subjects in order to test for psychological stress. The group receiving probiotic treatment exhibited far less psychological stress than the placebo control group (Messaoudi et al., 2011).

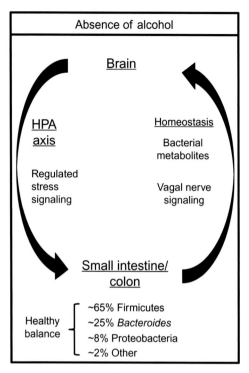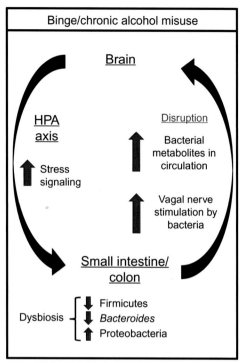

FIG. 2 Possible feedback mechanism between the HPA axis and gut-brain axis in the context of alcohol misuse. Under homeostatic conditions, the HPA axis and gut-brain axis are able to maintain balance through various negative feedback mechanisms (Wang et al., 2010) (left panel). Following alcohol misuse, elevated HPA axis signaling leads to an increased systemic stress response (Spencer & Hutchison, 1999). One downstream consequence is the disruption of the intestinal microbiome that results in decreased healthy Firmicutes and *Bacteroides* and increased Gram-negative Proteobacteria that contribute to inflammation (Chen et al., 2011; Engen et al., 2015; Mutlu et al., 2012; Queipo-Ortuno et al., 2012). The result of intestinal dysbiosis is increased vagal nerve stimulation by these bacteria (Goehler et al., 2008; Lyte et al., 2006) and bacterial metabolites entering systemic circulation. These stimuli result in further HPA axis stress signaling, perpetuating the cycle (Stilling et al., 2014) (right panel).

The take-home message from this section is twofold: (1) Alcohol intoxication can significantly alter the intestinal microbiome, and (2) dysbiosis in the intestine can mediate changes to the CNS through many different mechanisms including direct signaling via the vagal nerve, CNS epigenetic alterations by bacterial metabolites, or changing the expression of neurochemical signaling (Cryan & Kaupmann, 2005; Foster & McVey Neufeld, 2013). We have attempted to summarize the possible feedback relationships between HPA axis signaling and the gut-brain axis following alcohol misuse in Fig. 2.

CONCLUSION

Understanding the link between alcohol, inflammation, and depression remains of critical importance not only to elucidate the cellular mechanisms involved but also to allow for novel treatment regimens for patients suffering from alcohol use disorders and depression. Serotonin-norepinephrine reuptake inhibitors and tricyclic antidepressants are two common antidepressant medications prescribed to patients with depression (Anderson, 2000). Interestingly, a study done by Vogelzangs et al. (2012) found that although patients taking these antidepressants experience alleviation of depressive symptoms, these patients also had increases in the systemic inflammatory markers CRP and IL-6. As discussed above, patients with MDDs in conjunction with an AUD experience compounded increases in inflammation stemming from both alcohol-induced and depression-induced inflammatory responses (Maes, 2011). Therefore, the "chicken or the egg" question, "do alcohol use disorders lead to symptoms of depression or does depression lead to alcohol use disorders?", becomes significantly more important when determining treatment plans for patients with both an MDD and an AUD. Kessler et al. (1996), Hasin et al. (2005), and others have accumulated evidence suggesting that the latter is true in that over 40% of patients with a MDD also had an alcohol use disorder, which suggests a level of self-medication. Now that depression is understood as an overactivation of the inflammatory response, the use of alcohol in conjunction with depression only adds to the inflammatory response (Maes, 2011). The inflammatory response is twofold as alcohol increases gut leakiness to bacteria and bacterial endotoxins (i.e., LPS), thus increasing inflammation locally and extraintestinally as far reaching as the brain via the pro-inflammatory TLR4 and increases in ROS (Dheen et al., 2007; Lehnardt, 2010; Qin et al., 2008; Qin & Crews, 2012; Ward et al., 1996). Increases in inflammation directly correlated with increases in depressive symptoms such as little interest in physical or social environment, decreased motor activity, social withdrawal, and reduced food and water intake (Carmichael et al., 2006; Pugh et al., 1998; Vereker et al., 2000). Therefore, the gut-brain axis connection to alcohol use and depression allows us to suggest a potential positive feedback loop connecting symptoms of depression to alcohol use to inflammation, which then leads back to increases in depression starting the cycle over again (Fig. 1).

Future research is critical to uncover the best treatment plans for patients stuck in this cycle of depression, alcohol use, and inflammation as outlined in Fig. 1. Treatment of depression with antidepressants alone in MDD and AUD patients may alleviate depressive symptoms temporarily, but the consequential rise in systemic inflammation and its connection to alcohol use underlies the need for multitarget or the combination of antidepressant and anti-inflammatory treatment regimens. Determining the mechanistic interplay between all three arms—depression, alcohol use, and inflammation—should be the greatest focus going forward in order to employ optimal treatment plans for patients suffering from both MDDs and AUDs.

ACKNOWLEDGMENT

Supported by NIH R01AA015731, R21AA022324, T32AA013527, F31AA024367, and F31AA025536.

CONFLICTS OF INTEREST AND FUNDING SOURCES

The authors have no conflicts of interest to declare.

References

Agarwal, D. P. (2001). Genetic polymorphisms of alcohol metabolizing enzymes. *Pathologie-Biologie*, 49(9), 703–709. pii: S0369-8114(01)00242-5.

Alican, I,& Kubes, P. (1996). A critical role for nitric oxide in intestinal barrier function and dysfunction. *American Journal of Physiology*, 270(2 Pt 1), G225–37.

Allen, R. G., Lafuse, W. P., Galley, J. D., Ali, M. M., Ahmer, B. M., & Bailey, M. T. (2012). The intestinal microbiota are necessary for stressor-induced enhancement of splenic macrophage microbicidal activity. *Brain, Behavior, and Immunity*, 26(3), 371–382. https://doi.org/10.1016/j.bbi.2011.11.002.

Allen, C. D., Lee, S., Koob, G. F., & Rivier, C. (2011). Immediate and prolonged effects of alcohol exposure on the activity of the hypothalamic-pituitary-adrenal axis in adult and adolescent rats. *Brain, Behavior, and Immunity*, 25(*Suppl. 1*), S50–60. https://doi.org/10.1016/j.bbi. 2011.01.016.

Allen, C. D., Rivier, C. L., & Lee, S. Y. (2011). Adolescent alcohol exposure alters the central brain circuits known to regulate the stress response. *Neuroscience*, 182, 162–168. https://doi.org/10.1016/j.neuroscience.2011.03.003.

Anderson, I. M. (2000). Selective serotonin reuptake inhibitors versus tricyclic antidepressants: a meta-analysis of efficacy and tolerability. *Journal of Affective Disorders*, 58(1), 19–36. pii:S0165-0327(99)00092-0.

Babior, B. M. (1999). NADPH oxidase: an update. *Blood, 93* (5), 1464–1476.

Banan, A., Choudhary, S., Zhang, Y., Fields, J. Z., & Keshavarzian, A. (1999). Ethanol-induced barrier dysfunction and its prevention by growth factors in human intestinal monolayers: evidence for oxidative and cytoskeletal mechanisms. *Journal of Pharmacology and Experimental Therapeutics*, 291(3), 1075–1085.

Banan, A., Fields, J. Z., Decker, H., Zhang, Y., & Keshavarzian, A. (2000). Nitric oxide and its metabolites mediate ethanol-induced microtubule disruption and intestinal barrier dysfunction. *Journal of Pharmacology and Experimental Therapeutics*, 294(3), 997–1008.

Beck, I. T., & Dinda, P. K. (1981). Acute exposure of small intestine to ethanol: effects on morphology and function. *Digestive Diseases and Sciences*, 26(9), 817–838.

Beishuizen, A., & Thijs, L. G. (2003). Endotoxin and the hypothalamo-pituitary-adrenal (HPA) axis. *Journal of Endotoxin Research*, 9(1), 3–24. https://doi.org/10.1179/096805103125001298.

Benton, D., Williams, C., & Brown, A. (2007). Impact of consuming a milk drink containing a probiotic on mood and cognition. *European Journal of Clinical Nutrition*, 61(3), 355–361. pii:1602546.

Bercik, P., Denou, E., Collins, J., Jackson, W., Lu, J., Jury, J., et al. (2011). The intestinal microbiota affect central levels of brain-derived neurotropic factor and behavior in mice. *Gastroenterology*, 141(2), 599–609. 609.e1-3. (2011). https://doi.org/10.1053/j.gastro.2011.04.052.

Blednov, Y. A., Benavidez, J. M., Geil, C., Perra, S., Morikawa, H., & Harris, R. A. (2011). Activation of inflammatory signaling by lipopolysaccharide produces a prolonged increase of voluntary alcohol intake in mice. *Brain, Behavior, and Immunity*, 25(*Suppl. 1*), S92–S105. https://doi.org/10.1016/j.bbi.2011.01.008.

Block, M. L., & Hong, J. S. (2007). Chronic microglial activation and progressive dopaminergic neurotoxicity. *Biochemical Society Transactions*, 35(Pt 5), 1127–1132. pii: BST0351127.

Bode, C., & Bode, J. C. (2003). Effect of alcohol consumption on the gut. *Best Practice & Research. Clinical Gastroenterology*, 17(4), 575–592. pii:S1521691803000349.

Bode, C., Kugler, V., & Bode, J. C. (1987). Endotoxemia in patients with alcoholic and non-alcoholic cirrhosis and in subjects with no evidence of chronic liver disease following acute alcohol excess. *Journal of Hepatology*, 4(1), 8–14. pii:S0168-8278(87)80003-X.

Bravo, J. A., Forsythe, P., Chew, M. V., Escaravage, E., Savignac, H. M., Dinan, T. G., et al. (2011). Ingestion of lactobacillus strain regulates emotional behavior and central GABA receptor expression in a mouse via the vagus nerve. *Proceedings of the National Academy of Sciences of the United States of America*, 108(38), 16050–16055. https://doi.org/10.1073/pnas.1102999108.

Cancer Research UK (). *Interferon (intron A). Retrieved from http://www.cancerresearchuk.org/about-cancer/cancer-in-general/treatment/cancer-drugs/drugs/interferon.*

Carmichael, M. D., Davis, J. M., Murphy, E. A., Brown, A. S., Carson, J. A., Mayer, E. P., et al. (2006). Role of brain IL-1beta on fatigue after exercise-induced muscle damage. *American Journal of Physiology. Regulatory, Integrative and Comparative Physiology*, 291(5), R1344–8. pii:00141.2006.

Cederbaum, A. I. (2012). Alcohol metabolism. *Clinics in Liver Disease*, 16(4), 667–685. https://doi.org/10.1016/j.cld.2012.08.002.

Chen, Y., Yang, F., Lu, H., Wang, B., Chen, Y., Lei, D., et al. (2011). Characterization of fecal microbial communities in patients with liver cirrhosis. *Hepatology (Baltimore, Md.)*, 54(2), 562–572. https://doi.org/10.1002/hep.24423.

Collins, M. A., & Neafsey, E. J. (2012). Neuroinflammatory pathways in binge alcohol-induced neuronal degeneration: oxidative stress cascade involving aquaporin, brain edema, and phospholipase A2 activation. *Neurotoxicity Research*, 21(1), 70–78. https://doi.org/10.1007/s12640-011-9276-5.

Control CfD. (2015). *Fact sheets—binge drinking*

Crews, F. T., Zou, J., & Qin, L. (2011). Induction of innate immune genes in brain create the neurobiology of addiction. *Brain, Behavior, and Immunity*, 25(*Suppl. 1*), S4–S12. https://doi.org/10.1016/j.bbi.2011.03.003.

Cryan, J. F., & Kaupmann, K. (2005). Don't worry 'B' happy!: a role for GABA(B) receptors in anxiety and depression. *Trends in Pharmacological Sciences*, 26(1), 36–43. pii:S0165-6147(04)00310-4.

Dantzer, R. (2001). Cytokine-induced sickness behavior: where do we stand? *Brain, Behavior, and Immunity*, 15(1), 7–24. https://doi.org/10.1006/brbi.2000.0613.

Dantzer, R., O'Connor, J. C., Freund, G. G., Johnson, R. W., & Kelley, K. W. (2008). From inflammation to sickness and depression: when the immune system subjugates the brain. Nature Reviews. *Neuroscience*, 9(1), 46–56. pii: nrn2297.

Desbonnet, L., Garrett, L., Clarke, G., Kiely, B., Cryan, J. F., & Dinan, T. G. (2010). Effects of the probiotic bifidobacterium infantis in the maternal separation model of depression. *Neuroscience, 170*(4), 1179–1188. https://doi.org/10.1016/j.neuroscience.2010.08.005.

Dheen, S. T., Kaur, C., & Ling, E. A. (2007). Microglial activation and its implications in the brain diseases. *Current Medicinal Chemistry, 14*(11), 1189–1197.

Diaz Heijtz, R., Wang, S., Anuar, F., Qian, Y., Bjorkholm, B., Samuelsson, A., et al. (2011). Normal gut microbiota modulates brain development and behavior. *Proceedings of the National Academy of Sciences of the United States of America, 108*(7), 3047–3052. https://doi.org/10.1073/pnas.1010529108.

Engen, P. A., Green, S. J., Voigt, R. M., Forsyth, C. B., & Keshavarzian, A. (2015). The gastrointestinal microbiome: alcohol effects on the composition of intestinal microbiota. *Alcohol Research: Current Reviews, 37*(2), 223–236.

Evans, B. E., Greaves-Lord, K., Euser, A. S., Franken, I. H., & Huizink, A. C. (2012). The relation between hypothalamic-pituitary-adrenal (HPA) axis activity and age of onset of alcohol use. *Addiction (Abingdon, England), 107*(2), 312–322. https://doi.org/10.1111/j.1360-0443.2011. 03568.x.

Eyre, H. A., Papps, E., & Baune, B. T. (2013). Treating depression and depression-like behavior with physical activity: an immune perspective. *Frontiers in Psychiatry 4*, 3.

Forsythe, P., Bienenstock, J., & Kunze, W. A. (2014). Vagal pathways for microbiome-brain-gut axis communication. *Advances in Experimental Medicine and Biology, 817*, 115–133. https://doi.org/10.1007/978-1-4939-0897-4_5.

Foster, J. A., & McVey Neufeld, K. A. (2013). Gut-brain axis: how the microbiome influences anxiety and depression. *Trends in Neurosciences, 36*(5), 305–312. https://doi.org/10.1016/j.tins.2013.01.005.

Goehler, L. E., Park, S. M., Opitz, N., Lyte, M., & Gaykema, R. P. (2008). Campylobacter jejuni infection increases anxiety-like behavior in the holeboard: possible anatomical substrates for viscerosensory modulation of exploratory behavior. *Brain, Behavior, and Immunity, 22*(3), 354–366. pii:S0889-1591(07)00215-2.

Goldstein, D., & Laszlo, J. (1988). The role of interferon in cancer therapy: a current perspective. *CA: A Cancer Journal for Clinicians, 38*(5), 258–277.

Hasin, D. S., Goodwin, R. D., Stinson, F. S., & Grant, B. F. (2005). Epidemiology of major depressive disorder: results from the national epidemiologic survey on alcoholism and related conditions. *Archives of General Psychiatry, 62*(10), 1097–1106. pii:62/10/1097.

Hattori, M., & Taylor, T. D. (2009). The human intestinal microbiome: a new frontier of human biology. *DNA Research: An International Journal for Rapid Publication of Reports on Genes and Genomes, 16*(1), 1–12. https://doi.org/10.1093/dnares/dsn033.

Higuchi, T., Hayashi, H., & Abe, K. (1997). Exchange of glutamate and gamma-aminobutyrate in a lactobacillus strain. *Journal of Bacteriology, 179*(10), 3362–3364.

Holm, T. H., Draeby, D., & Owens, T. (2012). Microglia are required for astroglial toll-like receptor 4 response and for optimal TLR2 and TLR3 response. *Glia, 60*(4), 630–638. https://doi.org/10.1002/glia.22296.

Hua, F., Ma, J., Ha, T., Xia, Y., Kelley, J., Williams, D. L., et al. (2007). Activation of toll-like receptor 4 signaling contributes to hippocampal neuronal death following global cerebral ischemia/reperfusion. *Journal of Neuroimmunology, 190*(1–2), 101–111. pii:S0165-5728(07)00301-3.

Illman, J., Corringham, R., Robinson, D., Jr., Davis, H. M., Rossi, J. F., Cella, D., et al. (2005). Are inflammatory cytokines the common link between cancer-associated cachexia and depression? *Journal of Supportive Oncology, 3*(1), 37–50.

Jacob, C., Yang, P. C., Darmoul, D., Amadesi, S., Saito, T., Cottrell, G. S., et al. (2005). Mast cell tryptase controls paracellular permeability of the intestine. Role of protease-activated receptor 2 and beta-arrestins. *Journal of Biological Chemistry, 280*(36), 31936–31948. pii:M506338200.

Kawai, T., & Akira, S. (2008). Toll-like receptor and RIG-I-like receptor signaling. *Annals of the New York Academy of Sciences, 1143*, 1–20. https://doi.org/10.1196/annals.1443.020.

Keshavarzian, A., Fields, J. Z., Vaeth, J., & Holmes, E. W. (1994). The differing effects of acute and chronic alcohol on gastric and intestinal permeability. *American Journal of Gastroenterology, 89*(12), 2205–2211.

Kessler, R. C., Nelson, C. B., McGonagle, K. A., Edlund, M. J., Frank, R. G., & Leaf, P. J. (1996). The epidemiology of co-occurring addictive and mental disorders: implications for prevention and service utilization. *American Journal of Orthopsychiatry, 66*(1), 17–31.

Krishnan, H. R., Sakharkar, A. J., Teppen, T. L., Berkel, T. D., & Pandey, S. C. (2014). The epigenetic landscape of alcoholism. *International Review of Neurobiology, 115*, 75–116. https://doi.org/10.1016/B978-0-12-801311-3.00003-2.

Kunugi, H., Ida, I., Owashi, T., Kimura, M., Inoue, Y., Nakagawa, S., et al. (2006). Assessment of the dexamethasone/CRH test as a state-dependent marker for hypothalamic-pituitary-adrenal (HPA) axis abnormalities in major depressive episode: a multicenter study. *Neuropsychopharmacology: Official Publication of the American College of Neuropsychopharmacology, 31*(1), 212–220. pii:1300868.

Lehnardt, S. (2010). Innate immunity and neuroinflammation in the CNS: the role of microglia in toll-like receptor-mediated neuronal injury. *Glia, 58*(3), 253–263. https://doi.org/10.1002/glia.20928.

Li, X., Rana, S. N., Schwacha, M. G., Chaudry, I. H., & Choudhry, M. A. (2006). A novel role for IL-18 in corticosterone-mediated intestinal damage in a two-hit rodent model of alcohol intoxication and injury. *Journal of Leukocyte Biology, 80*(2), 367–375. pii:jlb.1205745.

Lippai, D., Bala, S., Petrasek, J., Csak, T., Levin, I., Kurt-Jones, E. A., et al. (2013). Alcohol-induced IL-1beta in the brain is mediated by NLRP3/ASC inflammasome activation that amplifies neuroinflammation. *Journal of Leukocyte Biology, 94*(1), 171–182. https://doi.org/10.1189/jlb.1212659.

Lyte, M., Li, W., Opitz, N., Gaykema, R. P., & Goehler, L. E. (2006). Induction of anxiety-like behavior in mice during the initial stages of infection with the agent of murine colonic hyperplasia citrobacter rodentium. *Physiology & Behavior, 89*(3), 350–357. pii:S0031-9384(06)00284-8.

Maes, M. (2011). Depression is an inflammatory disease, but cell-mediated immune activation is the key component of depression. *Progress in Neuro-Psychopharmacology & Biological Psychiatry, 35*(3), 664–675. https://doi.org/10.1016/j.pnpbp.2010.06.014.

Malarkey, W. B., & Mills, P. J. (2007). Endocrinology: the active partner in PNI research. *Brain, Behavior, and Immunity, 21*(2), 161–168. pii:S0889-1591(06)00345-X.

McClain, C. J., Barve, S., Deaciuc, I., Kugelmas, M., & Hill, D. (1999). Cytokines in alcoholic liver disease. *Seminars in Liver Disease, 19*(2), 205–219. https://doi.org/10.1055/s-2007-1007110.

Messaoudi, M., Violle, N., Bisson, J. F., Desor, D., Javelot, H., & Rougeot, C. (2011). Beneficial psychological effects of a probiotic formulation (lactobacillus helveticus R0052 and bifidobacterium longum R0175) in healthy human volunteers. *Gut Microbes, 2*(4), 256–261. https://doi.org/10.4161/gmic.2.4.16108.

Miyamasu, M., Misaki, Y., Izumi, S., Takaishi, T., Morita, Y., Nakamura, H., et al. (1998). Glucocorticoids inhibit chemokine generation by human eosinophils. *Journal of Allergy and Clinical Immunology, 101*(1 Pt 1), 75–83. pii:S0091-6749(98)70196-4.

Mogensen, T. H. (2009). Pathogen recognition and inflammatory signaling in innate immune defenses. *Clinical Microbiology Reviews, 22*(2), 240–273. Table of Contents. https://doi.org/10.1128/CMR.00046-08.

Moore, F. A., Moore, E. E., Poggetti, R., McAnena, O. J., Peterson, V. M., Abernathy, C. M., et al. (1991). Gut bacterial translocation via the portal vein: a clinical perspective with major torso trauma. *Journal of Trauma, 31*(5), 629–636 [discussion 636–638].

Mowat, A. M. (2003). Anatomical basis of tolerance and immunity to intestinal antigens. *Nature Reviews. Immunology, 3*(4), 331–341. https://doi.org/10.1038/nri1057.

Mulligan, M. K., Ponomarev, I., Hitzemann, R. J., Belknap, J. K., Tabakoff, B., Harris, R. A., et al. (2006). Toward understanding the genetics of alcohol drinking through transcriptome meta-analysis. *Proceedings of the National Academy of Sciences of the United States of America, 103* (16), 6368–6373. pii:0510188103.

Mutlu, E. A., Gillevet, P. M., Rangwala, H., Sikaroodi, M., Naqvi, A., Engen, P. A., et al. (2012). Colonic microbiome is altered in alcoholism. *American Journal of Physiology. Gastrointestinal and Liver Physiology, 302*(9), G966–78. https://doi.org/10.1152/ajpgi.00380.2011.

National Institute on Alcohol Abuse and Alcoholism (NIAAA). (2013). *Alcohol use disorder: A comparison between DSM–IV and DSM–5.* http://pubs.niaaa.nih.gov/publications/dsmfactsheet/dsmfact.pdf. (Accessed 2017).

National Institute on Alcohol Abuse and Alcoholism (NIAAA). (2017a). *Alcohol facts and statistics.* https://www.niaaa.nih.gov/alcohol-health/overview-alcohol-consumption/alcohol-facts-and-statistics. (Accessed 2017).

National Institute on Alcohol Abuse and Alcoholism. (2017b). *Drinking levels defined.* https://www.niaaa.nih.gov/alcohol-health/overview-alcohol-consumption/moderate-binge-drinking. (Accessed 2017).

Nixon, K., & Crews, F. T. (2002). Binge ethanol exposure decreases neurogenesis in adult rat hippocampus. *Journal of Neurochemistry, 83*(5), 1087–1093. pii:1214.

Ogawa, S., Lozach, J., Benner, C., Pascual, G., Tangirala, R. K., Westin, S., et al. (2005). Molecular determinants of crosstalk between nuclear receptors and toll-like receptors. *Cell, 122*(5), 707–721. pii:S0092-8674(05)00648-3.

Okun, E., Griffioen, K. J., & Mattson, M. P. (2011). Toll-like receptor signaling in neural plasticity and disease. *Trends in Neurosciences, 34*(5), 269–281. https://doi.org/10.1016/j.tins.2011.02.005.

Pandey, S. C. (2012). TLR4-MyD88 signalling: a molecular target for alcohol actions. *British Journal of Pharmacology, 165*(5), 1316–1318. https://doi.org/10.1111/j.1476-5381.2011.01695.x.

Pugh, C. R., Kumagawa, K., Fleshner, M., Watkins, L. R., Maier, S. F., & Rudy, J. W. (1998). Selective effects of peripheral lipopolysaccharide administration on contextual and auditory-cue fear conditioning. *Brain, Behavior, and Immunity, 12*(3), 212–229 pii:S0889-1591(98)90524-4.

Qin, L., & Crews, F. T. (2012). NADPH oxidase and reactive oxygen species contribute to alcohol-induced microglial activation and neurodegeneration. *Journal of Neuroinflammation, 9*, 5. https://doi.org/10.1186/1742-2094-9-5.

Qin, L., He, J., Hanes, R. N., Pluzarev, O., Hong, J. S., & Crews, F. T. (2008). Increased systemic and brain cytokine production and neuroinflammation by endotoxin following ethanol treatment. *Journal of Neuroinflammation, 5*, 10. https://doi.org/10.1186/1742-2094-5-10.

Quan, N., & Banks, W. A. (2007). Brain-immune communication pathways. *Brain, Behavior, and Immunity, 21*(6), 727–735. pii:S0889-1591(07)00115-8.

Queipo-Ortuno, M. I., Boto-Ordonez, M., Murri, M., Gomez-Zumaquero, J. M., Clemente-Postigo, M., Estruch, R., et al. (2012). Influence of red wine polyphenols and ethanol on the gut microbiota ecology and biochemical biomarkers. *American Journal of Clinical Nutrition, 95*(6), 1323–1334. https://doi.org/10.3945/ajcn.111.027847.

Radcliffe, R. A., Erwin, V. G., Bludeau, P., Deng, X., Fay, T., Floyd, K. L., et al. (2009). A major QTL for acute ethanol sensitivity in the alcohol tolerant and non-tolerant selected rat lines. *Genes, Brain, and Behavior*, 8(6), 611–625. https://doi.org/10.1111/j.1601-183X.2009.00496.x.

Raison, C. L., & Miller, A. H. (2001). The neuroimmunology of stress and depression. *Seminars in Clinical Neuropsychiatry*, 6(4), 277–294. pii:ascnp0060277.

Rao, A. V., Bested, A. C., Beaulne, T. M., Katzman, M. A., Iorio, C., Berardi, J. M., et al. (2009). A randomized, double-blind, placebo-controlled pilot study of a probiotic in emotional symptoms of chronic fatigue syndrome. *Gut Pathogens*, 1(1), 6. https://doi.org/10.1186/1757-4749-1-6.

Richards, D. F., Fernandez, M., Caulfield, J., & Hawrylowicz, C. M. (2000). Glucocorticoids drive human CD8(+) T cell differentiation towards a phenotype with high IL-10 and reduced IL-4, IL-5 and IL-13 production. *European Journal of Immunology*, 30(8), 2344–2354. https://doi.org/10.1002/1521-4141(2000)30:8.3.0.CO;2-7.

Richardson, H. N., Lee, S. Y., O'Dell, L. E., Koob, G. F., & Rivier, C. L. (2008). Alcohol self-administration acutely stimulates the hypothalamic-pituitary-adrenal axis, but alcohol dependence leads to a dampened neuroendocrine state. *European Journal of Neuroscience*, 28(8), 1641–1653.

Scheinman, R. I., Gualberto, A., Jewell, C. M., Cidlowski, J. A., & Baldwin, A. S., Jr. (1995). Characterization of mechanisms involved in transrepression of NF-kappa B by activated glucocorticoid receptors. *Molecular and Cellular Biology*, 15(2), 943–953.

Setshedi, M., Wands, J. R., & Monte, S. M. (2010). Acetaldehyde adducts in alcoholic liver disease. *Oxidative Medicine and Cellular Longevity*, 3(3), 178–185. https://doi.org/10.4161/oxim.3.3.12288.

Sewell, W. A., Scurr, L. L., Orphanides, H., Kinder, S., & Ludowyke, R. I. (1998). Induction of interleukin-4 and interleukin-5 expression in mast cells is inhibited by glucocorticoids. *Clinical and Diagnostic Laboratory Immunology*, 5(1), 18–23.

Simi, A., Tsakiri, N., Wang, P., & Rothwell, N. J. (2007). Interleukin-1 and inflammatory neurodegeneration. *Biochemical Society Transactions*, 35(Pt 5), 1122–1126. pii:BST0351122.

Soderholm, J. D., & Perdue, M. H. (2001). Stress and gastrointestinal tract. II. stress and intestinal barrier function. *American Journal of Physiology. Gastrointestinal and Liver Physiology*, 280(1), G7–G13.

Sonnenberg, G. F., & Artis, D. (2015). Innate lymphoid cells in the initiation, regulation and resolution of inflammation. *Nature Medicine*, 21(7), 698–708. https://doi.org/10.1038/nm.3892.

Spencer, R. L., & Hutchison, K. E. (1999). Alcohol, aging, and the stress response. *Alcohol Research & Health: The Journal of the National Institute on Alcohol Abuse and Alcoholism*, 23(4), 272–283.

Stilling, R. M., Dinan, T. G., & Cryan, J. F. (2014). Microbial genes, brain & behaviour - epigenetic regulation of the gut-brain axis. *Genes, Brain, and Behavior*, 13(1), 69–86. https://doi.org/10.1111/gbb.12109.

Sudo, N., Chida, Y., Aiba, Y., Sonoda, J., Oyama, N., Yu, X. N., et al. (2004). Postnatal microbial colonization programs the hypothalamic-pituitary-adrenal system for stress response in mice. *Journal of Physiology*, 558(Pt 1), 263–275. https://doi.org/10.1113/jphysiol.2004.063388.

Swanson, G. R., Sedghi, S., Farhadi, A., & Keshavarzian, A. (2010). Pattern of alcohol consumption and its effect on gastrointestinal symptoms in inflammatory bowel disease. *Alcohol (Fayetteville, NY)*, 44(3), 223–228. https://doi.org/10.1016/j.alcohol.2009.10.019.

Tajuddin, N., Moon, K. H., Marshall, S. A., Nixon, K., Neafsey, E. J., Kim, H. Y., et al. (2014). Neuroinflammation and neurodegeneration in adult rat brain from binge ethanol exposure: abrogation by docosahexaenoic acid. *PLoS One*, 9(7). https://doi.org/10.1371/journal.pone.0101223.

Tang, S. C., Arumugam, T. V., Xu, X., Cheng, A., Mughal, M. R., Jo, D. G., et al. (2007). Pivotal role for neuronal toll-like receptors in ischemic brain injury and functional deficits. *Proceedings of the National Academy of Sciences of the United States of America*, 104(34), 13798–13803. pii:0702553104.

Tang, Y., Banan, A., Forsyth, C. B., Fields, J. Z., Lau, C. K., Zhang, L. J., et al. (2008). Effect of alcohol on miR-212 expression in intestinal epithelial cells and its potential role in alcoholic liver disease. *Alcoholism, Clinical and Experimental Research*, 32(2), 355–364. pii:ACER584.

Uribe, A., Alam, M., Johansson, O., Midtvedt, T., & Theodorsson, E. (1994). Microflora modulates endocrine cells in the gastrointestinal mucosa of the rat. *Gastroenterology*, 107(5), 1259–1269. pii: S0016508594003215.

Vereker, E., O'Donnell, E., & Lynch, M. A. (2000). The inhibitory effect of interleukin-1beta on long-term potentiation is coupled with increased activity of stress-activated protein kinases. *Journal of Neuroscience: The Official Journal of the Society for Neuroscience*, 20(18), 6811–6819. pii:20/18/6811.

Vogelzangs, N., Duivis, H. E., Beekman, A. T., Kluft, C., Neuteboom, J., Hoogendijk, W., et al. (2012). Association of depressive disorders, depression characteristics and antidepressant medication with inflammation. *Translational Psychiatry*, 2, e79. https://doi.org/10.1038/tp.2012.8.

Wang, H., Yu, M., Ochani, M., Amella, C. A., Tanovic, M., Susarla, S., et al. (2003). Nicotinic acetylcholine receptor alpha7 subunit is an essential regulator of inflammation. *Nature*, 421(6921), 384–388. https://doi.org/10.1038/nature01339.

Wang, H. J., Zakhari, S., & Jung, M. K. (2010). Alcohol, inflammation, and gut-liver-brain interactions in tissue damage and disease development. *World Journal of Gastroenterology*, 16(11), 1304–1313.

Ward, R. J., Zhang, Y., Crichton, R. R., Piret, B., Piette, J., & de Witte, P. (1996). Identification of the nuclear transcription factor NFkappaB in rat after in vivo ethanol administration. *FEBS Letters, 389*(2), 119–122. pii:0014579396005455.

World Health Organization (WHO) – Management of Substance Abuse Team. (2011). *Global status report on alcohol and health.* http://www.who.int/substance_abuse/publications/global_alcohol_report/msbgsruprofiles.pdf. (Accessed 2017).

World Health Organization (WHO). (2015). *Alcohol fact sheet.* http://www.who.int/mediacentre/factsheets/fs349/en/. (Accessed 2017).

Wikoff, W. R., Anfora, A. T., Liu, J., Schultz, P. G., Lesley, S. A., Peters, E. C., et al. (2009). Metabolomics analysis reveals large effects of gut microflora on mammalian blood metabolites. *Proceedings of the National Academy of Sciences of the United States of America, 106*(10), 3698–3703. https://doi.org/10.1073/pnas.0812874106.

Zakhari, S. (2006). Overview: how is alcohol metabolized by the body? *Alcohol Research & Health: The Journal of the National Institute on Alcohol Abuse and Alcoholism, 29*(4), 245–254.

30

Efficacy of Anti-Inflammatory Treatment in Depression

Ole Köhler-Forsberg[*,†,‡], *Michael Eriksen Benros*[‡]

[*]Psychosis Research Unit, Aarhus University Hospital, Risskov, Denmark
[†]Department of Clinical Medicine, Aarhus University, Denmark
[‡]Mental Health Centre, Copenhagen University Hospital, Copenhagen, Denmark

INTRODUCTION

Increasing evidence has accumulated during the recent decades associating depression with inflammatory processes (Howren, Lamkin, & Suls, 2009; Smith, 1991), which has led researchers to hypothesize that anti-inflammatory treatment may yield antidepressant properties (Muller et al., 2006; Tyring et al., 2006). Indeed, recent meta-analyses on clinical trial data have provided proof-of-concept results indicating that particularly anti-inflammatory add-on treatment to antidepressants yields additional beneficial effect on depression symptoms (Kappelmann, Lewis, Dantzer, Jones, & Khandaker, 2016; Kohler et al., 2014; Na, Lee, Lee, Cho, & Jung, 2013). These are intriguing findings in view of the frequently emphasized need for new and improved antidepressant treatment strategies with the possibility of developing personalized treatments in individuals with depression, such as immune modulating treatment in subgroups with inflammatory pathophysiological mechanisms contributing to the depression. However, the potential bidirectional associations between the inflammatory

cascade and depression are highly complex with the clinical findings still being limited and controversial due to small studies and methodological heterogeneity (Eyre, Stuart, & Baune, 2014; Kohler et al., 2014). The purpose of this chapter is therefore to provide an overview of this field and to address clinically important questions, including timing and duration of anti-inflammatory treatment in potential subgroups of patients with depression and the application of biological immune markers to identify subgroups benefitting from immune modulating treatment potentially improving the antidepressant treatment effects.

Inflammation—A Short Overview

The term inflammation covers immune-related processes within the body. Inflammation is a protective immunovascular response involving many different cell types and may be activated through external (e.g., an infection) and internal causes (e.g., atherosclerosis). Furthermore, also psychological stress has been shown to activate inflammatory processes (Iwata et al.,

Inflammation and Immunity in Depression
https://doi.org/10.1016/B978-0-12-811073-7.00030-1

2016). The primary purpose is to maintain homeostasis, and under normal circumstances, the immune system produces both pro- and anti-inflammatory mediators. However, when these self-regulatory actions are not able to inhibit the pro-inflammatory response, a chronic inflammatory reaction may develop. Hence, inflammation is divided into an acute response and chronic inflammation, which may be present as low-grade inflammation for several years, not necessarily causing clinical symptoms (Anthony, Couch, Losey, & Evans, 2012; Sokol & Luster, 2015).

Potential Interplay Between the Peripheral and the Central Immune System

The inflammatory processes are further divided into peripheral and central, the latter covering immune-related reactions within the central nervous system (CNS), that is, neuroinflammation. During a peripheral inflammatory response, macrophages play an important role by stimulating other cells of the immune system to produce a variety of pro-inflammatory (e.g., tumor necrosis factor alpha (TNF-α) or interleukin-6 (IL-6)) and anti-inflammatory (e.g., IL-10) prostaglandins and cytokines (Jiang, Jiang, & Zhang, 2014; Sokol & Luster, 2015). IL-6 in turn stimulates the liver to produce C-reactive protein (CRP). These molecular mediators communicate with other parts of the immune system, such as the adaptive immune system (Sokol & Luster, 2015).

The CNS has for long been considered an immune privileged organ, but several studies have suggested bidirectional communication with the peripheral immune system (Anthony et al., 2012). This has very recently been indicated by the identification of lymphatic vessels within the CNS (Louveau et al., 2015). Furthermore, the blood-brain barrier may become more permeable during inflammation (Engelhardt & Sorokin, 2009), and the endothelial cells of the blood-brain barrier may transmit signals from the periphery into the CNS (Banks, Kastin, & Broadwell, 1995; Quan, Whiteside, & Herkenham, 1998). Indeed,

TNF-α, IL-6, and other pro-inflammatory cytokines have been shown to cross the blood-brain barrier by an active transport system (Banks et al., 1995). In addition, fever represents a well-known example of the direct effect of peripheral cytokines affecting the hypothalamus, which is an important part of the CNS. Within the CNS, microglia and astrocytes represent the main immunocompetent cells (Bentivoglio, Mariotti, & Bertini, 2011) regulating both the initiation and limitation of neuroinflammatory processes (Bentivoglio et al., 2011; Jiang et al., 2014). Microglia represent the resting macrophages of the CNS (Block, Zecca, & Hong, 2007), whereas astrocytes have several functions, including maintenance of the extracellular ion balance and support of the blood-brain barrier. Hence, the potential ways of communication between the peripheral and central immune system are numerous, and the research field and pharma are working to further understand this important and complex interplay.

The Inflammatory Hypothesis in Depression

The main findings associating inflammatory processes with depression can be described as the three "cornerstones" of the inflammatory hypothesis in depression:

(1) Numerous studies have consistently associated inflammation and somatic diseases comprising inflammatory processes, such as infections and autoimmune diseases, with an increased risk of depression (Benros et al., 2013; Dickens, McGowan, Clark-Carter, & Creed, 2002; Korczak, Pereira, Koulajian, Matejcek, & Giacca, 2011; Wium-Andersen, Orsted, Nielsen, & Nordestgaard, 2013).

(2) Studies have consistently shown increased levels of pro-inflammatory markers among individuals with depression, particularly during acute phases (Dowlati et al., 2010; Goldsmith, Rapaport, & Miller, 2016; Howren et al., 2009). Meta-analyses have

gathered the large evidence, with particularly CRP (Dahl et al., 2014; Howren et al., 2009), IL-6 (Dahl et al., 2014; Dowlati et al., 2010; Goldsmith et al., 2016; Howren et al., 2009; Liu, Ho, & Mak, 2012), TNF-α (Dowlati et al., 2010; Goldsmith et al., 2016; Liu et al., 2012), and IL-1 receptor antagonist (Dahl et al., 2014; Goldsmith et al., 2016; Howren et al., 2009) being elevated among patients with depression. Interestingly, levels of, for example, IL-6 decreased after treatment of the acute depressive episode (Goldsmith et al., 2016). However, these trials have only investigated peripheral markers of inflammation, and it is still unclear whether waist circumference or BMI may confound these findings (Krogh et al., 2014). Therefore, it is noteworthy that a recent brain imaging study found increased microglial activation, indicating neuroinflammation, among 20 individuals suffering of an active depressive episode as compared with 20 healthy matched controls (Setiawan et al., 2015).

(3) Pro-inflammatory agents can induce depressive symptoms, with up to 80% of IFN-α-treated patients suffering from mild to moderate depressive symptoms (Eggermont et al., 2008; Friebe et al., 2010; Reichenberg, Gorman, & Dieterich, 2005). Interestingly, these symptoms can be treated with antidepressants (Friebe et al., 2010; Nery et al., 2008).

The Sickness Behavior Theory: Does Inflammation Affect Specific Depressive Symptoms?

Since depression is a very heterogeneous disorder, it is noteworthy that recent studies have suggested that inflammatory components may be used to characterize a specific subgroup of patients with depression. Increased levels of pro-inflammatory markers such as CRP have been linked to greater symptom severity in general (Hope et al., 2013; Jokela, Virtanen, Batty, & Kivimaki, 2015; Kohler et al., 2017; Kohler-Forsberg et al., 2017; Krogh et al., 2014) and also to greater severity on specific symptoms. Interestingly, the neurovegetative symptoms (i.e., sleep and appetite) were associated with increased inflammatory markers (Jokela et al., 2015; Kohler-Forsberg et al., 2017). These findings may help to identify those individuals who may benefit from other treatment approaches, for example, anti-inflammatory treatment, potentially leading to more "personalized treatment." Studies that have investigated this anti-inflammatory treatment approach will be discussed in the next paragraph.

ANTI-INFLAMMATORY TREATMENT IN DEPRESSION— EFFICACY VERSUS SIDE EFFECTS

The following section will review the evidence from randomized clinical trials (RCTs) that have investigated potential antidepressant treatment effects of anti-inflammatory agents, describing both antidepressant treatment effects and side effects.

Non-Steroidal Anti-Inflammatory Drugs

Nonsteroidal anti-inflammatory drugs (NSAIDs) are the most frequently investigated anti-inflammatory drugs in clinical trials regarding potential antidepressant treatment effects. Since the selective cyclooxygenase-2 (COX-2) inhibitors have been suggested to have a more pronounced anti-inflammatory and thus a better antidepressant effect compared with the other NSAIDs, this section is divided into "selective COX-2 inhibitors" and "nonselective COX inhibitors."

Selective COX-2 Inhibitors

Add-on therapy to antidepressants with celecoxib has been investigated by four trials ($N=160$), three trials exploring 6-week treatment with 400 mg/day (Abbasi, Hosseini, Modabbernia, Ashrafi, & Akhondzadeh, 2012;

Akhondzadeh et al., 2009; Muller et al., 2006), and one trial of 8-week treatment with 200 mg/day (Hashemian et al., 2011). All four trials suggested improved antidepressant treatment effects for celecoxib add-on treatment compared with antidepressants and placebo. Recent meta-analyses associated celecoxib with a large improved antidepressant effect by a standard mean difference (SMD) of 0.82 (95% CI, 0.46–1.17) and without heterogeneity between the studies ($I^2 = 0\%$) (Kohler et al., 2014; Na et al., 2013). Furthermore, remission of depression was improved by an odds ratio (OR) of 7.89 (95% CI, 2.94–21.17) and treatment response by an OR of 6.59 (95% CI, 2.24–19.42) by this 6–8-week add-on treatment. Finally, one trial included peripheral blood tests and found that higher levels of the pro-inflammatory marker IL-6 predicted better antidepressant response to celecoxib add-on (Abbasi et al., 2012).

Celecoxib monotherapy has been studied by two trials ($N = 3846$) (Fields, Drye, Vaidya, & Lyketsos, 2012; Iyengar et al., 2013). One trial including patients with active osteoarthritis ($N = 1497$) found that celecoxib 200 mg/day during 6 weeks significantly reduced the depressive symptoms, a finding that was independent of the pain-relieving effects (Iyengar et al., 2013). However, another trial ($N = 2528$) including healthy individuals aged ≥70 years found no effect of 12 months treatment with celecoxib 400 mg/day (Fields et al., 2012).

Non-Selective COX-Inhibitors

Monotherapy with the nonselective COX inhibitors naproxen and ibuprofen showed antidepressant effects after 6 weeks among 890 patients with active osteoarthritis (Iyengar et al., 2013). On the other hand, 12-month treatment with naproxen among 1757 healthy users had no impact on depressive symptoms (Fields et al., 2012). Hence, the effect of the NSAIDs seems to be more pronounced among individuals with somatic comorbidity.

No RCTs have investigated the effect of add-on treatment with other NSAIDs than celecoxib.

One open-label trial among 24 patients, who were nonresponders to their first selective serotonin reuptake inhibitor (SSRI) treatment, found that 4-week add-on treatment to the SSRIs with acetylsalicylic acid (ASA) in doses of 160 mg/day was associated with high rates of treatment response and remission of depression (Mendlewicz et al., 2006). However, another trial had to be stopped since patients randomized to citalopram and ASA experienced anxiety and akathisia (Ghanizadeh & Hedayati, 2014).

Side Effects of NSAIDs

The traditional NSAIDs (nonselective COX inhibitors) have repeatedly been associated with an increased risk for several side effects. The most important include cardiovascular events (Schjerning Olsen et al., 2011) and gastrointestinal (GI) bleeding (de Abajo & Garcia-Rodriguez, 2008). The selective COX-2 inhibitors were marketed in 1999 and were considered to have a targeted anti-inflammatory effect and a decreased risk for GI adverse events compared with traditional NSAIDs. However, the selective COX-2 inhibitors were subsequently associated with an increased risk for severe cardiovascular events (Bresalier et al., 2005; Solomon et al., 2005), leading to the withdrawal of rofecoxib in 2004. Since then, clinical use of the selective COX-2 inhibitors declined. Nevertheless, studies have also found that celecoxib did not increase the risk for cardiovascular events, contrasting the rofecoxib findings (Solomon et al., 2006). Furthermore, it was emphasized that the risk for cardiovascular events seemed to depend on dosage; age; treatment length; and, in particular, baseline cardiovascular risk factors (Solomon et al., 2008). Therefore, it is noteworthy that the antidepressant add-on effects of celecoxib were present after few weeks and among relatively young individuals. This may possibly indicate a subgroup with a low a priori risk for cardiovascular events that may benefit from short-term add-on treatment with celecoxib.

ANTI-INFLAMMATORY TREATMENT IN DEPRESSION—EFFICACY VERSUS SIDE EFFECTS

Cytokine Inhibitors

Cytokine inhibitors are interesting due to their direct anti-inflammatory effects, and several trials have studied their potential antidepressant effects (Kappelmann et al., 2016). A recent meta-analysis pooled the findings from all trials investigating cytokine inhibitors and indicated that cytokine modulators may be novel drugs for depression in chronically inflamed subjects (Kappelmann et al., 2016). The most frequently studied drugs were TNF inhibitors.

Three trials including patients with psoriasis ($N=1944$) found that monotherapy during 12–24 weeks significantly reduced depressive symptoms (Langley et al., 2010; Menter et al., 2010; Tyring et al., 2006). One trial ($N=60$) included individuals suffering from treatment-resistant depression and found no overall antidepressant effect during 12-week monotherapy with infliximab (TNF-α inhibitor) compared with placebo (Raison et al., 2013). However, in the secondary analyses of the study, infliximab showed better antidepressant effects among patients with CRP > 5 mg/L, and the authors argued that elevated CRP may be used as a biomarker for treatment response (Raison et al., 2013).

The most important side effects of cytokine-inhibitor treatment are infections (Toussi, Pan, Walters, & Walsh, 2013); however, a recent meta-analysis could not identify an increased risk of adverse events among the abovementioned studies (Kohler et al., 2014).

Statins

Three RCTs on the antidepressant effects of statins have been published (Ghanizadeh & Hedayati, 2013; Gougol et al., 2015; Haghighi et al., 2014). In 2013, Ghanizadeh et al. showed that 6-week treatment ($N=68$) with fluoxetine and lovastatin had better antidepressant treatment effects compared with fluoxetine and placebo (Ghanizadeh & Hedayati, 2013). Subsequently, Haghighi and colleagues showed that citalopram and atorvastatin ($N=60$) were

associated with greater response compared with citalopram and placebo after 12 weeks (Haghighi et al., 2014). In 2015, Gougol and colleagues found that individuals randomized to fluoxetine and simvastatin ($N=44$) improved significantly more compared with individuals on fluoxetine and placebo over 6 weeks (Gougol et al., 2015). The abovementioned trials were pooled in a recent meta-analysis suggesting that the combination of an SSRI and a statin was associated with significantly higher reductions in Hamilton Depression Rating Scale (HAM-D) scores (SMD of -0.73; 95% CI, from -1.04 to -0.42; $P < 0.001$) without heterogeneity between studies ($I^2 = 0\%$; $P = 0.99$) (Salagre, Fernandes, Dodd, Brownstein, & Berk, 2016). This effect size indicates a clinically relevant effect. However, the trials were small, and all were conducted in the same country, limiting generalizability (Salagre et al., 2016).

In terms of head-to-head comparisons, simvastatin showed greater improvement in depression scores after 6 weeks compared with atorvastatin among individuals who had undergone coronary artery bypass graft and had mild to moderate depression (Abbasi et al., 2015).

Regarding side effects, the good safety profile of statins has been repeatedly emphasized (Beckett, Schepers, & Gordon, 2015; Collins et al., 2016). A recent meta-analysis pooled the risks associated with statin treatment and estimated that treatment of 10,000 patients for 5 years with a standard statin regimen would be expected to cause five cases of myopathy, 50–100 new cases of diabetes, and 5–10 hemorrhagic strokes (Collins et al., 2016). Furthermore, approximately two to three new cases of rhabdomyolysis would occur per 100,000 treated individuals (Collins et al., 2016).

Omega-3 Fatty Acids

Polyunsaturated fatty acids (PUFAs) have been shown to yield anti-inflammatory properties, and recent meta-analyses have indicated a minor antidepressant effect of PUFAs based on 13 RCTs including 731 participants (Bloch & Hannestad, 2012; Martins, Bentsen, & Puri, 2012). This small effect has been suggested to depend on the content of eicosapentaenoic acid (EPA). An analysis of studies using $\geq 60\%$ EPA resulted in a significant pooled SMD estimate of 0.37 (95% CI = 0.33–0.41), whereas studies using $\leq 60\%$ EPA found no significant antidepressant effects (Martins et al., 2012).

Pioglitazone

Pioglitazone is a second-line antidiabetic drug with potential anti-inflammatory effects. The antidepressant treatment effects of pioglitazone have been investigated in two trials (Kashani et al., 2013; Sepanjnia, Modabbernia, Ashrafi, Modabbernia, & Akhondzadeh, 2012). One trial ($N = 40$) included individuals suffering of major depression and found that 6 weeks of pioglitazone 30 mg/day add-on to SSRI treatment improved the antidepressant effects compared with SSRI and placebo (Sepanjnia et al., 2012). Another trial ($N = 40$) aimed to test whether pioglitazone may be superior to other antidiabetic drugs and found that 6-week monotherapy with 30 mg/day pioglitazone showed better effects on depressive symptoms compared with 1500 mg/day metformin among women with polycystic ovarian syndrome (Kashani et al., 2013).

Pioglitazone has been associated with several side effects, including an increased risk for fractures, weight increase, and cardiovascular events (Della-Morte et al., 2014); hence, cautiousness regarding the possible use of pioglitazone among individuals suffering of depression is needed.

Minocycline

Minocycline is a second generation tetracyclic antibiotic and has gained interest during the recent years since it is able to cross the blood-brain barrier more easily than the other tetracycline antibiotics (Tomas-Camardiel et al., 2004)

and may exert potential antidepressant effects through its robust neuroprotective activities. These include increased neurogenesis and antioxidation, antiglutamate excitotoxicity, and downregulation of pro-inflammatory markers (Miyaoka et al., 2012). A small nonrandomized open-label trial including 25 adult inpatients with major depression with psychotic features taking minocycline 150 mg/day in combination with antidepressants (fluvoxamine, paroxetine, or sertraline) found significant improvement in depression (Miyaoka et al., 2012). More clinical trials are currently under way (Dean et al., 2014; Soczynska et al., 2012).

Corticosteroids

Two early trials have studied whether corticosteroids may exert antidepressant treatment effects (Arana et al., 1995; DeBattista, Posener, Kalehzan, & Schatzberg, 2000). One trial among 37 outpatients with MDD studied 4 days of dexamethasone 4 mg/day and found that 37% of the dexamethasone patients responded ($\geq 50\%$ HAM-D reduction) after 14 days compared with 6% of the placebo patients ($P = 0.03$) (Arana et al., 1995). Another trial among 22 patients with MDD could associate one infusion of 15 mg hydrocortisone with a mean reduction in HAM-D$_{21}$ of 8.4 points compared with a 1.3 point reduction in the placebo group (DeBattista et al., 2000).

Several well-known side effects are associated with corticosteroid treatment, with the risk and severity increasing with higher dose and longer treatment duration. These include endocrine disturbances, diabetes, osteoporosis, and increased blood pressure.

Modafinil

Finally, another emerging drug is modafinil, which is primarily used against narcolepsy. Modafinil has also been associated with anti-inflammatory properties (Jung et al., 2012), and furthermore, it has shown effective antidepressant effects in augmentation strategies for acute depressive episodes, including symptoms of fatigue, in both unipolar and bipolar disorders (Abolfazli et al., 2011; Goss, Kaser, Costafreda, Sahakian, & Fu, 2013). However, due to several side effects, cautious use of modafinil is recommended (Kumar, 2008).

DISCUSSION

As outlined above, RCTs have suggested antidepressant properties for several anti-inflammatory agents, both as add-on and monotherapy. Most evidence has been gathered for NSAIDs, cytokine inhibitors, and statins, which have shown promising results, whereas the evidence for the other agents is more limited. Noteworthily, the antidepressant effects may be independent of their primary pharmacological effects. However, the risk for side effects associated with these agents emphasizes cautious use. Therefore, we will discuss possibilities for identifying the patients who may benefit from anti-inflammatory treatment for depression, which is of utmost importance, in particular with regard to timing and duration of such treatment.

Personalized Treatment With Anti-Inflammatory Agents—Which Patient at What Disease Status and for How Long?

Timing of Anti-Inflammatory Treatment—Acute Treatment of Depressive Episodes

Studies have shown that acutely developed depressive episodes may be associated with a more pronounced inflammatory response (Howren et al., 2009; Setiawan et al., 2015). Hence, it is noteworthy that RCTs have found adjunctive antidepressant effects of celecoxib (Abbasi et al., 2012; Akhondzadeh et al., 2009; Hashemian et al., 2011; Muller et al., 2006) and statins (Ghanizadeh & Hedayati, 2013; Gougol

et al., 2015) among acutely depressed patients already after 4–6 weeks. Furthermore, the pro-inflammatory marker IL-6 decreased after treatment of the acute depressive episode (Goldsmith et al., 2016). Interestingly, regarding the benefit-risk assessment, celecoxib has not been found to increase the risk for acute cardiovascular events within the first 60 days of treatment in contrary to treatment with rofecoxib, which was withdrawn from the market (Solomon et al., 2006). Moreover, side effects associated with statins after such short treatment episodes are very limited (Collins et al., 2016). Also monotherapy with monoclonal antibodies has shown better antidepressant treatment effects compared with placebo after 12 (Menter et al., 2010; Raison et al., 2013; Tyring et al., 2006) and 24 weeks (Langley et al., 2010) without increased risks for infections (Kohler et al., 2014). These findings may indicate that intervention only lasting few weeks or few months may be beneficial in the acute treatment of depressive episodes, while also minimizing the risk for adverse events (Eyre et al., 2014; Krogh et al., 2014). However, as with antidepressants, a detailed assessment of the individual patient regarding baseline risk factors for the potential side effects of the specific anti-inflammatory agents would be of utmost importance.

Can Anti-Inflammatory Drugs Treat Patients With Specific Depressive Symptoms Concordant With the Sickness Behavior Theory?

Several recent studies have shown that an increased inflammatory response seems to be associated with a greater severity of the depression symptoms (Hope et al., 2013; Jokela et al., 2015; Kohler et al., 2017; Kohler-Forsberg et al., 2017). Interestingly, patients with increased pro-inflammatory markers, for example, CRP, have been found to have greater severity of the neurovegetative symptoms, comprising sleep and appetite, which are also specific parts of the depression symptoms (Jokela et al., 2015;

Kohler-Forsberg et al., 2017). However, evidence whether anti-inflammatory agents may decrease specific depressive symptoms is still missing. In schizophrenia though, celecoxib has been found to have beneficial effects on cognition—a core feature of depression (Muller, Riedel, Schwarz, & Engel, 2005). In a rat model of depression, infliximab prevented cognitive decline, that is, spatial and emotional memory impairments, which was accompanied by prevention of reduction of hippocampal brain-derived neurotropic factor (Sahin et al., 2015). Another cardinal symptom in depression is fatigue. In multiple sclerosis (MS) patients, NSAID treatment lowered fatigue (Wingerchuk et al., 2005), and elevated body temperature may be linked to worse fatigue (Sumowski & Leavitt, 2014). These findings suggest that the antipyretic effects of NSAIDs might result in reduced fatigue among MS patients and that higher body temperature may indicate treatment with NSAIDs. However, effects on specific symptoms, such as fatigue or cognition, and body temperature as potential biomarkers, still need to be explored in depressed individuals.

Somatic Comorbidities and the Antidepressant Effects of Anti-Inflammatory Agents

Specific somatic comorbidities, possibly indicating that an active inflammatory process is part of the depression etiology, may predict better treatment response. Monotherapy with the NSAIDs celecoxib, naproxen, and ibuprofen showed better antidepressant effects compared with placebo among patients with active osteoarthritis (Iyengar et al., 2013), but not among healthy individuals aged 70 years or above (Fields et al., 2012). Similarly, monotherapy with the monoclonal antibodies etanercept (Tyring et al., 2006), ustekinumab (Langley et al., 2010), and adalimumab (Menter et al., 2010) showed better antidepressant treatment effects on depressive symptoms among psoriasis patients as compared with placebo, whereas in individuals with depression but no known somatic

comorbidity, monoclonal antibodies were only effective on depression in individuals with CRP > 5. Also, antidepressant effects have been found for pioglitazone among obese women (BMI ≥27) with polycystic ovarian syndrome (Kashani et al., 2013) and for omega-3 fatty acids among patients undergoing maintenance dialysis (Gharekhani et al., 2014).

Despite not including specific pro-inflammatory markers, the presence of these comorbid somatic states indicates the presence of an active inflammatory state and hence supports the notion that anti-inflammatory intervention may have effects among individuals with depressive symptoms and an inflammatory response.

The Potential Value of Pro-inflammatory Biomarkers in Predicting Antidepressant Treatment Response

Few studies have tested whether peripheral levels of pro-inflammatory markers among individuals with depression can be used to predict the antidepressant effects of anti-inflammatory drugs. It has been shown that increased IL-6 levels (Abbasi et al., 2012) and MIF levels (Musil et al., 2011) are associated with higher remission rates and better response among individuals with depression treated with celecoxib add-on to antidepressants. Similarly, depressed patients with CRP > 5 mg/L responded better to infliximab treatment compared with patients with CRP < 5 mg/L (Raison et al., 2013).

In addition, the level of inflammatory markers has also been suggested to predict the effect of standard antidepressant drugs. A recent trial including 241 adults with depression indicated that patients with CRP > 1 mg/L responded better to the tricyclic antidepressant nortriptyline compared with the SSRI escitalopram, whereas patients with CRP < 1 mg/L responded better to escitalopram (Uher et al., 2014). However, the abovementioned studies included markers from the peripheral blood only, and future studies need to include cerebrospinal fluid and/or brain scans for more precise measures of CNS inflammation.

Safety Issues—Which Agents to Be Used in Patients With Specific Somatic Comorbidites?

A high bidirectional comorbidity between depression and cardiovascular diseases is well known. This aspect complicates, for example, the use of NSAIDs, which have been associated with cardiovascular side effects (Schjerning Olsen et al., 2011). Hence, in depressed individuals with cardiovascular comorbidity or risk factors, the use of cardioprotective agents with anti-inflammatory effects would be preferable, such as statins (Ghanizadeh & Hedayati, 2013; Kohler et al., 2016), low-dose ASA (Kohler, Petersen, Mors, & Gasse, 2015; Mendlewicz et al., 2006), or PUFAs (Gharekhani et al., 2014). In addition, NSAIDs increase the risk of GI bleeding, in particular when used concomitantly with SSRIs (de Abajo & Garcia-Rodriguez, 2008). Hence, additional caution must be present among patients with prior GI bleeding or risk factors for GI bleeding and anti-inflammatory agents not affecting the GI tract should be preferred.

SUMMARY

In conclusion, the abovementioned results suggest an increased inflammatory response during acute depressive episodes (Howren et al., 2009; Setiawan et al., 2015) and an effect on specific depressive symptoms (i.e., neurovegetative symptoms) (Muller et al., 2005; Wingerchuk et al., 2005), and an assessment of markers of systemic inflammation to predict better response seems highly relevant (Abbasi et al., 2012; Raison et al., 2013; Sumowski & Leavitt, 2014). Interestingly, anti-inflammatory treatment lasting only few weeks may improve antidepressant treatment effects among acutely depressed

individuals (Abbasi et al., 2012; Akhondzadeh et al., 2009; Ghanizadeh & Hedayati, 2013; Gougol et al., 2015; Hashemian et al., 2011; Muller et al., 2006) and individuals with somatic diseases suffering of single depressive symptoms (Iyengar et al., 2013; Langley et al., 2010; Menter et al., 2010; Tyring et al., 2006). These findings should encourage future clinical trials to further investigate potential subgroups with immune markers possibly predicting better treatment response to anti-inflammatory agents together with the timing and needed duration of the intervention.

PERSPECTIVES

Based on the potential role of inflammatory processes in the etiology of depression, anti-inflammatory intervention may represent a possibility for more personalized and improved treatment regimens among specific subgroups of patients with depression. In particular, studies have indicated that add-on treatment with celecoxib or statins for 6–8 weeks and monotherapy with cytokine inhibitors for 12–24 weeks may represent safe treatments among subgroups of depressed patients. The potential subgroups may be patients with elevated pro-inflammatory markers, such as CRP or IL-6, and/or patients with somatic comorbidity including an inflammatory etiology, such as osteoarthritis. Interestingly, treatment regimens from neurology regarding autoimmune NMDA receptor encephalitis may encourage and inspire this new approach in psychiatry (Kayser & Dalmau, 2011). Among patients with NMDA receptor encephalitis, 70% have (and often present with) psychiatric symptoms, and current treatment flowcharts include short-term treatment with high potential immunosuppressants, including steroids, intravenous immunoglobulin, cytokine inhibitors, and plasmapheresis, which also improves the psychiatric symptoms. However, despite these initial encouraging findings, more research is needed on the identification of markers or specific symptoms predicting response, dosages, and timing of immune modulating intervention for subgroups of individuals with depression. Furthermore, the risk for side effects always needs to be included in benefit-risk assessments, and widespread use of anti-inflammatory agents for longer periods is not recommendable.

Finally, anti-inflammatory intervention only represents one approach for personalized treatment regimens. Other promising targets include nitrosative and oxidative stress pathways (Anderson, Berk, Dean, Moylan, & Maes, 2013) and increased glutathione levels (Morris et al., 2014). Thus, in light of the low remission and response rates among patients with depression, a better understanding of personalized antidepressant treatment regimens is needed. Anti-inflammatory intervention may represent one new line of treatment options that can be used in personalized treatment of individuals with depression and an inflammatory component. Furthermore, more research into the specific underlying mechanisms between inflammation and depression may lead to the development of more targeted antidepressive medicine with an anti-inflammatory component with a greater effect on the possible subgroup of patients with immune-related depression. This may furthermore include a more detailed assessment of the nature of the inflammatory response observed in some patients with depression.

References

Abbasi, S. H., Hosseini, F., Modabbernia, A., Ashrafi, M., & Akhondzadeh, S. (2012). Effect of celecoxib add-on treatment on symptoms and serum IL-6 concentrations in patients with major depressive disorder: randomized double-blind placebo-controlled study. *Journal of Affective Disorders*, 141(2–3), 308–314. https://doi.org/10.1016/j.jad.2012.03.033.

Abbasi, S. H., Mohammadinejad, P., Shahmansouri, N., Salehiomran, A., Beglar, A. A., Zeinoddini, A.,

et al. (2015). Simvastatin versus atorvastatin for improving mild to moderate depression in post-coronary artery bypass graft patients: a double-blind, placebo-controlled, randomized trial. *Journal of Affective Disorders*, 183, 149–155. https://doi.org/10.1016/j.jad.2015.04.049.

Abolfazli, R., Hosseini, M., Ghanizadeh, A., Ghaleiha, A., Tabrizi, M., Raznahan, M., et al. (2011). Double-blind randomized parallel-group clinical trial of efficacy of the combination fluoxetine plus modafinil versus fluoxetine plus placebo in the treatment of major depression. *Depression and Anxiety*, 28(4), 297–302. https://doi.org/10.1002/da.20801.

Akhondzadeh, S., Jafari, S., Raisi, F., Nasehi, A. A., Ghoreishi, A., Salehi, B., et al. (2009). Clinical trial of adjunctive celecoxib treatment in patients with major depression: a double blind and placebo controlled trial. *Depression and Anxiety*, 26(7), 607–611.

Anderson, G., Berk, M., Dean, O., Moylan, S., & Maes, M. (2013). Role of immune-inflammatory and oxidative and nitrosative stress pathways in the etiology of depression: therapeutic implications. *CNS Drugs*. https://doi.org/10.1007/s40263-013-0119-1.

Anthony, D. C., Couch, Y., Losey, P., & Evans, M. C. (2012). The systemic response to brain injury and disease. *Brain, Behavior, and Immunity*, 26(4), 534–540. https://doi.org/10.1016/j.bbi.2011.10.011.

Arana, G. W., Santos, A. B., Laraia, M. T., McLeod-Bryant, S., Beale, M. D., Rames, L. J., et al. (1995). Dexamethasone for the treatment of depression: a randomized, placebo-controlled, double-blind trial. *American Journal of Psychiatry*, 152(2), 265–267.

Banks, W. A., Kastin, A. J., & Broadwell, R. D. (1995). Passage of cytokines across the blood-brain barrier. *Neuroimmunomodulation*, 2(4), 241–248.

Beckett, R. D., Schepers, S. M., & Gordon, S. K. (2015). Risk of new-onset diabetes associated with statin use. *SAGE Open Medicine*. 3, https://doi.org/10.1177/2050312115605518.

Benros, M. E., Waltoft, B. L., Nordentoft, M., Ostergaard, S. D., Eaton, W. W., Krogh, J., et al. (2013). Autoimmune diseases and severe infections as risk factors for mood disorders: a nationwide study. *JAMA Psychiatry*, 70(8), 812–820. https://doi.org/10.1001/jamapsychiatry.2013.1111.

Bentivoglio, M., Mariotti, R., & Bertini, G. (2011). Neuroinflammation and brain infections: historical context and current perspectives. *Brain Research Reviews*, 66(1–2), 152–173. https://doi.org/10.1016/j.brainresrev.2010.09.008.

Bloch, M. H., & Hannestad, J. (2012). Omega-3 fatty acids for the treatment of depression: systematic review and meta-analysis. *Molecular Psychiatry*, 17(12), 1272–1282. https://doi.org/10.1038/mp.2011.100.

Block, M. L., Zecca, L., & Hong, J. S. (2007). Microglia-mediated neurotoxicity: uncovering the molecular mechanisms. *Nature Reviews Neuroscience*, 8(1), 57–69 pii:nrn2038.

Bresalier, R. S., Sandler, R. S., Quan, H., Bolognese, J. A., Oxenius, B., Horgan, K., et al. (2005). Cardiovascular events associated with rofecoxib in a colorectal adenoma chemoprevention trial. *New England Journal of Medicine*, 352(11), 1092–1102 pii: NEJMoa050493.

Collins, R., Reith, C., Emberson, J., Armitage, J., Baigent, C., Blackwell, L., et al. (2016). Interpretation of the evidence for the efficacy and safety of statin therapy. *Lancet (London, England)*, 388(10059), 2532–2561. https://doi.org/10.1016/S0140-6736(16)31357-5.

Dahl, J., Ormstad, H., Aass, H. C., Malt, U. F., Bendz, L. T., Sandvik, L., et al. (2014). The plasma levels of various cytokines are increased during ongoing depression and are reduced to normal levels after recovery. *Psychoneuroendocrinology*, 45, 77–86. https://doi.org/10.1016/j.psyneuen.2014.03.019.

de Abajo, F. J., & Garcia-Rodriguez, L. A. (2008). Risk of upper gastrointestinal tract bleeding associated with selective serotonin reuptake inhibitors and venlafaxine therapy: interaction with nonsteroidal anti-inflammatory drugs and effect of acid-suppressing agents. *Archives of General Psychiatry*, 65(7), 795–803. https://doi.org/10.1001/archpsyc.65.7.795.

Dean, O. M., Maes, M., Ashton, M., Berk, L., Kanchanatawan, B., Sughondhabirom, A., et al. (2014). Protocol and rationale-the efficacy of minocycline as an adjunctive treatment for major depressive disorder: a double blind, randomised, placebo controlled trial. *Clinical Psychopharmacology and Neuroscience: The Official Scientific Journal of the Korean College of Neuropsychopharmacology*, 12 (3), 180–188. https://doi.org/10.9758/cpn.2014.12.3.180.

DeBattista, C., Posener, J. A., Kalehzan, B. M., & Schatzberg, A. F. (2000). Acute antidepressant effects of intravenous hydrocortisone and CRH in depressed patients: a double-blind, placebo-controlled study. *American Journal of Psychiatry*, 157(8), 1334–1337.

Della-Morte, D., Palmirotta, R., Rehni, A. K., Pastore, D., Capuani, B., Pacifici, F., et al. (2014). Pharmacogenomics and pharmacogenetics of thiazolidinediones: role in diabetes and cardiovascular risk factors. *Pharmacogenomics*, 15(16), 2063–2082. https://doi.org/10.2217/pgs.14.162.

Dickens, C., McGowan, L., Clark-Carter, D., & Creed, F. (2002). Depression in rheumatoid arthritis: a systematic review of the literature with meta-analysis. *Psychosomatic Medicine*, 64(1), 52–60.

Dowlati, Y., Herrmann, N., Swardfager, W., Liu, H., Sham, L., Reim, E. K., et al. (2010). A meta-analysis of cytokines in major depression. *Biological Psychiatry*, 67(5), 446–457.

Eggermont, A. M., Suciu, S., Santinami, M., Testori, A., Kruit, W. H., Marsden, J., et al. (2008). Adjuvant therapy with pegylated interferon alfa-2b versus observation alone in resected stage III melanoma: final results of EORTC 18991, a randomised phase III trial. *Lancet*, 372(9633), 117–126. https://doi.org/10.1016/S0140-6736(08)61033-8.

Engelhardt, B., & Sorokin, L. (2009). The blood-brain and the blood-cerebrospinal fluid barriers: function and dysfunction. *Seminars in Immunopathology*, 31(4), 497–511. https://doi.org/10.1007/s00281-009-0177-0.

Eyre, H. A., Stuart, M. J., & Baune, B. T. (2014). A phase-specific neuroimmune model of clinical depression. *Progress in Neuro-Psychopharmacology & Biological Psychiatry*, 54, 265–274. https://doi.org/10.1016/j.pnpbp.2014.06.011.

Fields, C., Drye, L., Vaidya, V., & Lyketsos, C. (2012). Celecoxib or naproxen treatment does not benefit depressive symptoms in persons age 70 and older: findings from a randomized controlled trial. *American Journal of Geriatric Psychiatry*, 20(6), 505–513.

Friebe, A., Horn, M., Schmidt, F., Janssen, G., Schmid-Wendtner, M. H., Volkenandt, M., et al. (2010). Dose-dependent development of depressive symptoms during adjuvant interferon-{alpha} treatment of patients with malignant melanoma. *Psychosomatics*, 51(6), 466–473. https://doi.org/10.1176/appi.psy.51.6.466.

Ghanizadeh, A., & Hedayati, A. (2013). Augmentation of fluoxetine with lovastatin for treating major depressive disorder, a randomized double-blind placebo controlled-clinical trial. *Depression and Anxiety*, 30(11), 1084–1088. https://doi.org/10.1002/da.22195.

Ghanizadeh, A., & Hedayati, A. (2014). Augmentation of citalopram with aspirin for treating major depressive disorder, a double blind randomized placebo controlled clinical trial. *Anti-Inflammatory & Anti-Allergy Agents in Medicinal Chemistry*, 13(2), 108–111 pii: AIAAMC-EPUB-61601.

Gharekhani, A., Khatami, M. R., Dashti-Khavidaki, S., Razeghi, E., Noorbala, A. A., Hashemi-Nazari, S. S., et al. (2014). The effect of omega-3 fatty acids on depressive symptoms and inflammatory markers in maintenance hemodialysis patients: a randomized, placebo-controlled clinical trial. *European Journal of Clinical Pharmacology*, 70(6), 655–665. https://doi.org/10.1007/s00228-014-1666-1.

Goldsmith, D. R., Rapaport, M. H., & Miller, B. J. (2016). A meta-analysis of blood cytokine network alterations in psychiatric patients: comparisons between schizophrenia, bipolar disorder and depression. *Molecular Psychiatry*, 21 (12), 1696–1709. https://doi.org/10.1038/mp.2016.3.

Goss, A. J., Kaser, M., Costafreda, S. G., Sahakian, B. J., & Fu, C. H. (2013). Modafinil augmentation therapy in unipolar and bipolar depression: a systematic review and meta-analysis of randomized controlled trials. *Journal of Clinical Psychiatry*, 74(11), 1101–1107. https://doi.org/10.4088/JCP.13r08560.

Gougol, A., Zareh-Mohammadi, N., Raheb, S., Farokhnia, M., Salimi, S., Iranpour, N., et al. (2015). Simvastatin as an adjuvant therapy to fluoxetine in patients with moderate to severe major depression: a double-blind placebo-controlled trial. *Journal of Psychopharmacology (Oxford, England)*, 29(5), 575–581. https://doi.org/10.1177/0269881115578160.

Haghighi, M., Khodakarami, S., Jahangard, L., Ahmadpanah, M., Bajoghli, H., Holsboer-Trachsler, E., et al. (2014). In a randomized, double-blind clinical trial, adjuvant atorvastatin improved symptoms of depression and blood lipid values in patients suffering from severe major depressive disorder. *Journal of Psychiatric Research*, 58, 109–114. https://doi.org/10.1016/j.jpsychires.2014.07.018.

Hashemian, F., Majd, M., Hosseini, S. M., Sharifi, A., Panahi, M. V. S., & Bigdeli, O. (2011). A randomized, double-blind, placebo-controlled trial of celecoxib augmentation of sertraline in the treatment of a drug-naive women with major depression. *Klinik Psikofarmakoloji Bulteni*, 21, S183–S184.

Hope, S., Ueland, T., Steen, N. E., Dieset, I., Lorentzen, S., Berg, A. O., et al. (2013). Interleukin 1 receptor antagonist and soluble tumor necrosis factor receptor 1 are associated with general severity and psychotic symptoms in schizophrenia and bipolar disorder. *Schizophrenia Research*, 145(1–3), 36–42. https://doi.org/10.1016/j.schres.2012.12.023.

Howren, M. B., Lamkin, D. M., & Suls, J. (2009). Associations of depression with C-reactive protein, IL-1, and IL-6: a meta-analysis. *Psychosomatic Medicine*, 71(2), 171–186. https://doi.org/10.1097/PSY.0b013e3181907c1b.

Iwata, M., Ota, K. T., Li, X. Y., Sakaue, F., Li, N., Dutheil, S., et al. (2016). Psychological stress activates the inflammasome via release of adenosine triphosphate and stimulation of the purinergic type 2X7 receptor. *Biological Psychiatry*, 80(1), 12–22. https://doi.org/10.1016/j.biopsych.2015.11.026.

Iyengar, R. L., Gandhi, S., Aneja, A., Thorpe, K., Razzouk, L., Greenberg, J., et al. (2013). NSAIDs are associated with lower depression scores in patients with osteoarthritis. *American Journal of Medicine*. https://doi.org/10.1016/j.amjmed.2013.02.037.

Jiang, Z., Jiang, J. X., & Zhang, G. X. (2014). Macrophages: a double-edged sword in experimental autoimmune encephalomyelitis. *Immunology Letters*, 160(1), 17–22. https://doi.org/10.1016/j.imlet.2014.03.006.

Jokela, M., Virtanen, M., Batty, G. D., & Kivimaki, M. (2015). Inflammation and specific symptoms of depression. *JAMA Psychiatry*, 1–2. https://doi.org/10.1001/jamapsychiatry.2015.1977.

Jung, J. C., Lee, Y., Son, J. Y., Lim, E., Jung, M., & Oh, S. (2012). Simple synthesis of modafinil derivatives and their anti-inflammatory activity. *Molecules (Basel, Switzerland)*, 17(9), 10446–10458. https://doi.org/10.3390/molecules170910446.

Kappelmann, N., Lewis, G., Dantzer, R., Jones, P. B., & Khandaker, G. M. (2016). Antidepressant activity of anti-cytokine treatment: a systematic review and meta-analysis of clinical trials of chronic inflammatory conditions. *Molecular Psychiatry*. https://doi.org/10.1038/mp.2016.167.

Kashani, L., Omidvar, T., Farazmand, B., Modabbernia, A., Ramzanzadeh, F., Tehraninejad, E. S., et al. (2013). Does pioglitazone improve depression through insulin-sensitization? results of a randomized double-blind metformin-controlled trial in patients with polycystic ovarian syndrome and comorbid depression. *Psychoneuroendocrinology*, 38(6), 767–776.

Kayser, M. S., & Dalmau, J. (2011). Anti-NMDA receptor encephalitis in psychiatry. *Current Psychiatry Reviews*, 7(3), 189–193. https://doi.org/10.2174/157340011797183184.

Kohler, O., Benros, M. E., Nordentoft, M., Farkouh, M. E., Iyengar, R. L., Mors, O., et al. (2014). Effect of anti-inflammatory treatment on depression, depressive symptoms, and adverse effects: a systematic review and meta-analysis of randomized clinical trials. *JAMA Psychiatry*, 71(12), 1381–1391. https://doi.org/10.1001/jamapsychiatry.2014.1611.

Kohler, O., Gasse, C., Petersen, L., Ingstrup, K. G., Nierenberg, A. A., Mors, O., et al. (2016). The effect of concomitant treatment with SSRIs and statins: a population-based study. *American Journal of Psychiatry*, 173(8), 807–815. https://doi.org/10.1176/appi.ajp.2016.15040463.

Kohler, O., Petersen, L., Mors, O., & Gasse, C. (2015). Inflammation and depression: combined use of selective serotonin reuptake inhibitors and NSAIDs or paracetamol and psychiatric outcomes. *Brain and Behavior*, 5(8). https://doi.org/10.1002/brb3.338.

Kohler, O., Sylvia, L. G., Bowden, C. L., Calabrese, J. R., Thase, M., Shelton, R. C., et al. (2017). White blood cell count correlates with mood symptom severity and specific mood symptoms in bipolar disorder. *Australian and New Zealand Journal of Psychiatry*, 51(4), 355–365. https://doi.org/10.1177/0004867416644508.

Kohler-Forsberg, O., Buttenschon, H. N., Tansey, K. E., Maier, W., Hauser, J., Dernovsek, M. Z., et al. (2017). Association between C-reactive protein (CRP) with depression symptom severity and specific depressive symptoms in major depression. *Brain, Behavior, and Immunity*, 62, 344–350. pii: S0889-1591(17)30062-4.

Korczak, D. J., Pereira, S., Koulajian, K., Matejcek, A., & Giacca, A. (2011). Type 1 diabetes mellitus and major depressive disorder: evidence for a biological link. *Diabetologia*, 54(10), 2483–2493. https://doi.org/10.1007/s00125-011-2240-3.

Krogh, J., Benros, M. E., Jorgensen, M. B., Vesterager, L., Elfving, B., & Nordentoft, M. (2014). The association between depressive symptoms, cognitive function, and inflammation in major depression. *Brain, Behavior, and Immunity*, 35, 70–76. https://doi.org/10.1016/j.bbi.2013.08.014.

Kumar, R. (2008). Approved and investigational uses of modafinil: an evidence-based review. *Drugs*, 68(13), 1803–1839. pii: 68133.

Langley, R. G., Feldman, S. R., Han, C., Schenkel, B., Szapary, P., Hsu, M. C., et al. (2010). Ustekinumab significantly improves symptoms of anxiety, depression, and skin-related quality of life in patients with moderate-to-severe psoriasis: results from a randomized, double-blind, placebo-controlled phase III trial. *Journal of the American Academy of Dermatology*, 63(3), 457–465.

Liu, Y., Ho, R. C., & Mak, A. (2012). Interleukin (IL)-6, tumour necrosis factor alpha (TNF-alpha) and soluble interleukin-2 receptors (sIL-2R) are elevated in patients with major depressive disorder: a meta-analysis and meta-regression. *Journal of Affective Disorders*, 139(3), 230–239. https://doi.org/10.1016/j.jad.2011.08.003.

Louveau, A., Smirnov, I., Keyes, T. J., Eccles, J. D., Rouhani, S. J., Peske, J. D., et al. (2015). Structural and functional features of central nervous system lymphatic vessels. *Nature*, 523(7560), 337–341. https://doi.org/10.1038/nature14432.

Martins, J. G., Bentsen, H., & Puri, B. K. (2012). Eicosapentaenoic acid appears to be the key omega-3 fatty acid component associated with efficacy in major depressive disorder: a critique of bloch and hannestad and updated meta-analysis. *Molecular Psychiatry*, 17(12), 1144–1149. discussion 1163–7(2012). https://doi.org/10.1038/mp.2012.25.

Mendlewicz, J., Kriwin, P., Oswald, P., Souery, D., Alboni, S., & Brunello, N. (2006). Shortened onset of action of antidepressants in major depression using acetylsalicylic acid augmentation: a pilot open-label study. *International Clinical Psychopharmacology*, 21(4), 227–231.

Menter, A., Augustin, M., Signorovitch, J., Yu, A. P., Wu, E. Q., Gupta, S. R., et al. (2010). The effect of adalimumab on reducing depression symptoms in patients with moderate to severe psoriasis: a randomized clinical trial. *Journal of the American Academy of Dermatology*, 62(5), 812–818.

Miyaoka, T., Wake, R., Furuya, M., Liaury, K., Ieda, M., Kawakami, K., et al. (2012). Minocycline as adjunctive therapy for patients with unipolar psychotic depression: an open-label study. *Progress in Neuro-Psychopharmacology and Biological Psychiatry*, 37(2), 222–226.

Morris, G., Anderson, G., Dean, O., Berk, M., Galecki, P., Martin-Subero, M., et al. (2014). The glutathione system: a new drug target in neuroimmune disorders. *Molecular Neurobiology*, 50(3), 1059–1084. https://doi.org/10.1007/s12035-014-8705-x.

Muller, N., Riedel, M., Schwarz, M. J., & Engel, R. R. (2005). Clinical effects of COX-2 inhibitors on cognition in schizophrenia. *European Archives of Psychiatry and Clinical Neuroscience*, 255(2), 149–151. https://doi.org/10.1007/s00406-004-0548-4.

Muller, N., Schwarz, M. J., Dehning, S., Douhe, A., Cerovecki, A., Goldstein-Muller, B., et al. (2006). The cyclooxygenase-2 inhibitor celecoxib has therapeutic effects in major depression: results of a double-blind, randomized, placebo controlled, add-on pilot study to reboxetine. *Molecular Psychiatry*, 11(7), 680–684. https://doi.org/10.1038/sj.mp.4001805.

Musil, R., Schwarz, M. J., Riedel, M., Dehning, S., Cerovecki, A., Spellmann, I., et al. (2011). Elevated macrophage migration inhibitory factor and decreased transforming growth factor-beta levels in major depression—no influence of celecoxib treatment. *Journal of Affective Disorders*, 134(1–3), 217–225. https://doi.org/10.1016/j.jad.2011.05.047.

Na, K. S., Lee, K. J., Lee, J. S., Cho, Y. S., & Jung, H. Y. (2013). Efficacy of adjunctive celecoxib treatment for patients with major depressive disorder: a meta-analysis. *Progress in Neuro-Psychopharmacology & Biological Psychiatry*, 48C, 79–85. https://doi.org/10.1016/j.pnpbp.2013.09.006.

Nery, F. G., Monkul, E. S., Hatch, J. P., Fonseca, M., Zunta-Soares, G. B., Frey, B. N., et al. (2008). Celecoxib as an adjunct in the treatment of depressive or mixed episodes of bipolar disorder: a double-blind, randomized, placebo-controlled study. *Human Psychopharmacology*, 23(2), 87–94. https://doi.org/10.1002/hup.912; 10.1002/hup.912.

Quan, N., Whiteside, M., & Herkenham, M. (1998). Time course and localization patterns of interleukin-1beta messenger RNA expression in brain and pituitary after peripheral administration of lipopolysaccharide. *Neuroscience*, 83(1), 281–293. pii:S0306452297003503.

Raison, C. L., Rutherford, R. E., Woolwine, B. J., Shuo, C., Schettler, P., Drake, D. F., et al. (2013). A randomized controlled trial of the tumor necrosis factor antagonist infliximab for treatment-resistant depression: the role of baseline inflammatory biomarkers. *JAMA Psychiatry*, 70(1), 31–41.

Reichenberg, A., Gorman, J. M., & Dieterich, D. T. (2005). Interferon-induced depression and cognitive impairment in hepatitis C virus patients: a 72 week prospective study. *AIDS (London, England)*, 19(*Suppl. 3*), S174–8. pii: 00002030-200510003-00026.

Sahin, T. D., Karson, A., Balci, F., Yazir, Y., Bayramgurler, D., & Utkan, T. (2015). TNF-alpha inhibition prevents cognitive decline and maintains hippocampal BDNF levels in the unpredictable chronic mild stress rat model of depression. *Behavioural Brain Research*, 292, 233–240 pii: S0166-4328(15)00372-1.

Salagre, E., Fernandes, B. S., Dodd, S., Brownstein, D. J., & Berk, M. (2016). Statins for the treatment of depression: a meta-analysis of randomized, double-blind, placebo-controlled trials. *Journal of Affective Disorders*, 200, 235–242. https://doi.org/10.1016/j.jad.2016.04.047.

Schjerning Olsen, A. M., Fosbol, E. L., Lindhardsen, J., Folke, F., Charlot, M., Selmer, C., et al. (2011). Duration of treatment with nonsteroidal anti-inflammatory drugs and impact on risk of death and recurrent myocardial infarction in patients with prior myocardial infarction: a nationwide cohort study. *Circulation*, 123(20), 2226–2235. https://doi.org/10.1161/CIRCULATIONAHA.110.004671.

Sepanjnia, K., Modabbernia, A., Ashrafi, M., Modabbernia, M., & Akhondzadeh, S. (2012). Pioglitazone adjunctive therapy for moderate-to-severe major depressive disorder: randomized double-blind placebo-controlled trial. *Neuropsychopharmacology*, 37(9), 2093–2100.

Setiawan, E., Wilson, A. A., Mizrahi, R., Rusjan, P. M., Miler, L., Rajkowska, G., et al. (2015). Role of translocator protein density, a marker of neuroinflammation, in the brain during major depressive episodes. *JAMA Psychiatry*, 72(3), 268–275. https://doi.org/10.1001/jamapsychiatry.2014.2427.

Smith, R. S. (1991). The macrophage theory of depression. *Medical Hypotheses*, 35(4), 298–306.

Soczynska, J. K., Mansur, R. B., Brietzke, E., Swardfager, W., Kennedy, S. H., Woldeyohannes, H. O., et al. (2012). Novel therapeutic targets in depression: minocycline as a candidate treatment. *Behavioural Brain Research*, 235 (2), 302–317. https://doi.org/10.1016/j.bbr.2012.07.026.

Sokol, C. L., & Luster, A. D. (2015). The chemokine system in innate immunity. *Cold Spring Harbor Perspectives in Biology* 7(5). https://doi.org/10.1101/cshperspect.a016303 pii:a0 16303.

Solomon, D. H., Avorn, J., Sturmer, T., Glynn, R. J., Mogun, H., & Schneeweiss, S. (2006). Cardiovascular outcomes in new users of coxibs and nonsteroidal antiinflammatory drugs: high-risk subgroups and time course of risk. *Arthritis and Rheumatism*, 54(5), 1378–1389. https://doi.org/10.1002/art.21887.

Solomon, S. D., McMurray, J. J., Pfeffer, M. A., Wittes, J., Fowler, R., Finn, P., et al. (2005). Cardiovascular risk associated with celecoxib in a clinical trial for colorectal adenoma prevention. *The New England Journal of Medicine*, 352(11), 1071–1080. pii:NEJMoa050405.

Solomon, S. D., Wittes, J., Finn, P. V., Fowler, R., Viner, J., Bertagnolli, M. M., et al. (2008). Cardiovascular risk of celecoxib in 6 randomized placebo-controlled trials: the cross trial safety analysis. *Circulation*, 117(16), 2104–2113. https://doi.org/10.1161/CIRCULATIONAHA.108.764530.

Sumowski, J. F., & Leavitt, V. M. (2014). Body temperature is elevated and linked to fatigue in relapsing-remitting multiple sclerosis, even without heat exposure. *Archives of Physical Medicine and Rehabilitation*, 95(7), 1298–1302. https://doi.org/10.1016/j.apmr.2014.02.004.

Tomas-Camardiel, M., Rite, I., Herrera, A. J., de Pablos, R. M., Cano, J., Machado, A., et al. (2004). Minocycline reduces the lipopolysaccharide-induced inflammatory reaction, peroxynitrite-mediated nitration of proteins, disruption of the blood-brain barrier, and damage in the nigral dopaminergic system. *Neurobiology of Disease*, 16(1), 190–201. https://doi.org/10.1016/j.nbd.2004.01.010.

Toussi, S. S., Pan, N., Walters, H. M., & Walsh, T. J. (2013). Infections in children and adolescents with juvenile idiopathic arthritis and inflammatory bowel disease treated with tumor necrosis factor-alpha inhibitors: systematic review of the literature. *Clinical Infectious Diseases: An Official Publication of the Infectious Diseases Society of America*, 57(9), 1318–1330. https://doi.org/10.1093/cid/cit489.

Tyring, S., Gottlieb, A., Papp, K., Gordon, K., Leonardi, C., Wang, A., et al. (2006). Etanercept and clinical outcomes, fatigue, and depression in psoriasis: double-blind placebo-controlled randomised phase III trial. *Lancet*, 367(9504), 29–35. https://doi.org/10.1016/S0140-6736(05)67763-X.

Uher, R., Tansey, K. E., Dew, T., Maier, W., Mors, O., Hauser, J., et al. (2014). An inflammatory biomarker as a differential predictor of outcome of depression treatment with escitalopram and nortriptyline. *American Journal of Psychiatry*, 171(12), 1278–1286. https://doi.org/10.1176/appi.ajp.2014.14010094.

Wingerchuk, D. M., Benarroch, E. E., O'Brien, P. C., Keegan, B. M., Lucchinetti, C. F., Noseworthy, J. H., et al. (2005). A randomized controlled crossover trial of aspirin for fatigue in multiple sclerosis. *Neurology*, 64(7), 1267–1269. pii:64/7/1267.

Wium-Andersen, M. K., Orsted, D. D., Nielsen, S. F., & Nordestgaard, B. G. (2013). Elevated C-reactive protein levels, psychological distress, and depression in 73, 131 individuals. *JAMA Psychiatry*, 70(2), 176–184. https://doi.org/10.1001/2013.jamapsychiatry.102.

Modulation of Inflammation by Antidepressants

Bernhard T. Baune

University of Adelaide, Adelaide, SA, Australia

INTRODUCTION

Novel strategies for the treatment of depression are urgently needed. Global data suggest that unipolar depression currently ranks the 11th for disability-adjusted life years, a 37% increase since 1990 (Murray et al., 2012). The burden is expected to continue to grow into the 21st century (Holtzheimer et al., 2008; Murray et al., 2012). This unprecedented burden of depressive illness requires increased efforts to find novel therapeutic agents for treatment (Licinio, 2011). Additionally, more than 50% of patients on antidepressants will not achieve remission following initial treatment (Rush, Trivedi, Wisniewski, Stewart, et al., 2006), and nearly one-third will not achieve remission even following several treatment steps (Rush, Trivedi, Wisniewski, Nierenberg, et al., 2006; Rush, Trivedi, Wisniewski, Stewart, et al., 2006).

A possible role of the immune system in depression has received much attention from research that focused on the innate immune response and on inflammation in particular. Cytokines such as tumor necrosis factor (TNF)-α, interleukin (IL)-1β, IL-6, and interferon (IFN)-γ have been repeatedly shown to exert effects on key processes such as neuroplasticity, neurotransmission, oxidative stress, and neuroendocrinological functions that are considered to be central to the development of depression (Dantzer, O'Connor, Freund, Johnson, & Kelley, 2008; Eyre & Baune, 2012; Haroon, Raison, & Miller, 2012; McAfoose & Baune, 2009; Miller, Maletic, & Raison, 2009). These pro-inflammatory cytokines have been shown to be elevated in the brain and the periphery (Dantzer et al., 2008; Eyre & Baune, 2012; Leonard & Maes, 2012; Miller et al., 2009; Moylan, Maes, Wray, & Berk, 2013). Importantly, pro-inflammatory cytokines may exert a number of neurobiological effects relevant to clinical depression including impairment of hippocampal (HC) neuroplasticity (e.g., neurogenesis, synaptic plasticity, and long-term potentiation), induction of glucocorticoid insensitivity of the hypothalamic-pituitary-adrenal axis, increase of oxidative stress in the HC, reduction of serotonin (5-HT) levels, and induction of neurotoxic serotonergic metabolites (i.e., 3-hydroxykynurenine and quinolinic acid (QA)) (Dantzer et al., 2008; Eyre & Baune, 2012; Leonard & Maes, 2012; Miller et al., 2009;

Inflammation and Immunity in Depression
https://doi.org/10.1016/B978-0-12-811073-7.00031-3

Moylan et al., 2013). Additional immune-relevant cells such as astrocytes and microglia have been found to produce pro-inflammatory cytokines in depression, and these cells show regulatory effects on HC neuroplasticity, oxidative stress, and QA (Eyre & Baune, 2012; Eyre, Stuart, & Baune, 2014). Moreover, astrocytes have shown a role in releasing neurotropic and antioxidant factors in depression (Eyre et al., 2014).

Given the evidence for a role of the immune system in depression published since the early 1990s, reports from the past few years suggest that antidepressants may exert some of their neurobiological effects via the immune system (Janssen, Caniato, Verster, & Baune, 2010; Martino, Rocchi, Escelsior, & Fornaro, 2012). Besides a global effect of antidepressants on immune function, it is scientifically and clinically important as to whether there is a distinction of antidepressant classes that exert class-specific effects on immune functions. For example, a meta-analysis of 22 studies by Hannestad, Dellagioia, and Bloch (2011) indicated that selective serotonin reuptake inhibitors (SSRIs) may reduce the levels of IL-6 and TNFα, whereas other types of antidepressants— while efficacious in overall antidepressant effects—did not appear to reduce cytokine levels. Beyond this meta-analysis, it has been suggested that while serotonin-noradrenaline reuptake inhibitor (SNRI) antidepressants suppress Th1-type cytokines (e.g., IFN-γ, IL-2, and TNF-α) and shift the balance toward humoral immunity, SSRIs reduce the production of Th2-type cytokines (e.g., IL-4, IL-5, IL-6, IL-9, IL-10, and IL-13) and shift the balance toward cellular immune response (Martino et al., 2012; Uher et al., 2014). These findings suggest the possibility of differential mechanistic effects of different antidepressant classes on immune function.

Several studies have explored the effects of SSRIs on adaptive immune cells (Frank, Hendricks, Johnson, Wieseler, & Burke, 1999; Hernandez et al., 2010; Kook, Mizruchin, Odnopozov, Gershon, & Segev, 1995; Sluzewska et al., 1995). Early studies in this area explored the effect of SSRIs for 4–12 weeks in major

depressive disorder (MDD) patients and noted as a main finding that SSRIs decrease the number of circulating natural killer (NK) cells without affecting other lymphocyte subsets (Frank, Hendricks, Burke, & Johnson, 2004; Gladkevich, Kauffman, & Korf, 2004; Ravindran, Griffiths, Merali, & Anisman, 1995; Rothermundt et al., 2001). Increased counts of NK cells might occur due to the stimulation of their serotonergic receptors as the result of increased levels of 5-HT caused by long-term SSRI treatment (Hernandez et al., 2010). A paper by Hernandez et al. (2010) explored the effect of a 52-week-long treatment with SSRIs (various types) on lymphocyte subsets. This study included 31 adult MDD subjects and 22 healthy controls. The patients showed remission of depressive episodes after 20 weeks of treatment along with an increase in NK-cell and B-cell populations, which remained increased until the end of the study. At the 52nd week of treatment, patients showed an increase in the counts of NK cells (396 ± 101 cells/mL) and B cells (268 ± 64 cells/mL) compared with healthy volunteers (NK, 159 ± 30 cells/mL, and B cells, 179 ± 37 cells/mL). Activated mature B lymphocytes proliferated in a time-dose-dependent manner with regard to either 5-HT concentration or 5-HT$_{1A}$ receptor agonist concentration (Iken, Chheng, Fargin, Goulet, & Kouassi, 1995). Therefore, the authors suggest that increased extracellular levels of 5-HT caused by long-term SSRI treatment may have stimulated the proliferation of B lymphocytes in patients with MDD. The increase in T-cell subsets at W36 might have resulted from the proliferative effect of IL-2, whose serum levels increased at W20 during SSRI treatment (Hernandez et al., 2008; Figs. 1 and 2).

EFFECTS OF CLASSES OF ANTIDEPRESSANT ON THE IMMUNE SYSTEM

In addition to the abovementioned meta-analysis (Hannestad et al., 2011), results of several studies exploring the effects of SRIs on

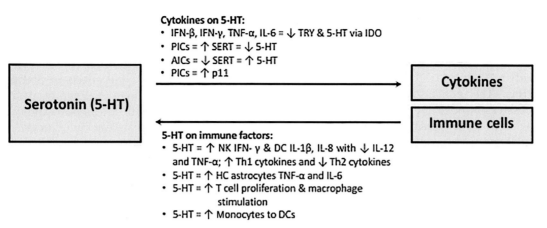

FIG. 1   Bidirectional relationships between serotonin and the immune system. *TNF*, tumor necrosis factor; *IL*, interleukin; *IFN*, interferon; *PICs*, pro-inflammatory cytokines; *AICs*, anti-inflammatory cytokines; *NK*, natural killer; *Th*, T helper.

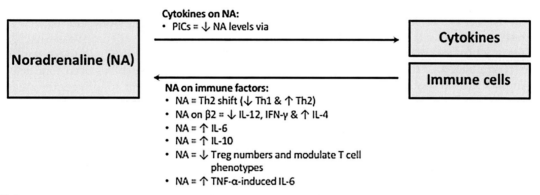

FIG. 2   Bidirectional relationships between noradrenaline and the immune system. *NA*, noradrenaline; *TNF*, tumor necrosis factor; *IL*, interleukin; *IFN*, interferon; *PICs*, pro-inflammatory cytokines; *Th*, T helper; *Treg*, regulatory T cells.

the innate immune system indicate an overall anti-inflammatory effect of SSRIs (Basterzi et al., 2005; Eller, Vasar, Shlik, & Maron, 2008; Hernandez et al., 2008; Leo et al., 2006; Mamdani et al., 2011; Mamdani, Berlim, Beaulieu, & Turecki, 2014; O'Brien, Scott, & Dinan, 2006; Sluzewska et al., 1995; Sutcigil et al., 2007; Tsao, Lin, Chen, Bai, & Wu, 2006; Tuglu, Kara, Caliyurt, Vardar, & Abay, 2003). Moreover, SSRIs have shown to upregulate *interferon regulatory factor 7 (IRF7)* gene activity, a transcription regulator of IFN-α (Mamdani et al., 2011).

Less explored are the potential effects of SSRIs on the adaptive immune system (Frank et al., 1999; Hernandez et al., 2010; Kook et al., 1995; Sluzewska et al., 1995). Early studies in this area explored the effect of SSRIs for 4–12 weeks in MDD patients and noted a main finding—SSRIs decrease the number of circulating NK cells without affecting other lymphocyte subsets (Frank et al., 2004; Gladkevich et al., 2004; Ravindran et al., 1995; Rothermundt et al., 2001). Increased counts of NK cells might occur due to the stimulation of their serotonergic receptors as the result of increased levels of 5-HT caused by long-term SSRI treatment (Hernandez et al., 2010). A study by Hernandez et al. (2010) explored the effect of a 52-week-long treatment with SSRIs (various types) on lymphocyte subsets. This study included 31 adult MDD subjects and 22 healthy controls. The patients showed remission of depressive episodes after 20 weeks of treatment along with an increase in NK-cell and B-cell populations, which remained increased until the end of the study. At the 52nd week of treatment, patients showed an increase in the counts of NK cells (396±101 cells/mL) and B cells (268±64 cells/mL) compared with healthy volunteers (NK, 159±30 cells/mL, and B cells, 179±37 cells/mL). Activated mature B lymphocytes proliferate in a time-dose-dependent manner with regard to either 5-HT concentration or $5-HT_{1A}$ receptor agonist concentration (Iken et al., 1995). Therefore, the authors suggest that increased extracellular levels of 5-HT caused by long-term SSRI treatment may have stimulated the proliferation of B lymphocytes in patients with MDD. The increase in T-cell subsets at W36 might result from the proliferative effect of IL-2, whose serum levels increase at W20 during SSRI treatment (Hernandez et al., 2008). While these studies are of interest for further exploration, their findings need to be regarded as preliminary, however, worthwhile for further scientific investigation.

In contrast to studies involving SSRIs, a relatively limited number of studies explored the effect of SNRI antidepressants on innate and adaptive immune factors with only one study exploring the leukocyte gene expression profile changes due to SNRIs. Neither a meta-analysis nor a systematic review exists for SNRI effects on the immune system. Some research suggests that SNRI antidepressants may suppress Th1-type cytokines (e.g., IFN-γ, IL-2, and TNF-α) and shift the balance toward humoral immunity (Martino et al., 2012; Uher et al., 2014). However, the possible anti-inflammatory effects of SNRIs may depend on the dose range such as of venlafaxine (Li et al., 2013; Piletz et al., 2009) or

depend on whether patients are early responders or early nonresponders as shown in atrial with duloxetine (Fornaro, Rocchi, Escelsior, Contini, & Martino, 2013). Very few studies have explored the effects of SNRIs on cellular immune markers (Kalman et al., 2005; Piletz et al., 2009), with first reports indicating the effects of venlafaxine on lymphocyte gene expression, affecting genes responsible for cell migration, lymphocyte cytoskeletal remodeling, and lymphocyte activation (Kalman et al., 2005). Clearly, follow-up studies are required that look at these immune effects in larger cohorts and in varying age ranges.

Similarly, tricyclic antidepressants (TCAs) appear to exert anti-inflammatory effects, and if confirmed in independent trials, TNF-α and IL-6 may be interesting biomarkers predicting treatment response (Hannestad et al., 2011; Lanquillon, Krieg, Bening-Abu-Shach, & Vedder, 2000; Uher et al., 2014). In contrast, only little systematic research has been conducted as to whether TCAs affect adaptive immune function: in those early and preliminary studies, it was shown that treatment with TCAs was associated with decreased T, CD4+, CD29+, and CD45RA+ lymphocytes and T-cell mitogen responses (Lanquillon et al., 2000; Schleifer, Keller, & Bartlett, 1999).

COMPARATIVE STUDIES EXPLORING DIFFERENTIAL EFFECTS OF ANTIDEPRESSANT ON INNATE IMMUNE FACTORS

In total, five studies have antidepressant classes for effects on humoral immune factors (Cattaneo et al., 2013; Chang et al., 2012; Narita et al., 2006; Uher et al., 2014; Yoshimura et al., 2009). While some studies find no difference between antidepressant classes, others find a significant difference of antidepressant (AD) classes on humoral immune factors. A multicenter open-label randomized clinical trial by Uher et al. (2014) aimed to determine if C-reactive protein (CRP) predicted differential response to escitalopram (SSRI) and nortriptyline. These drugs were administered between 5 and 30 mg/day and 50–200 mg/day, respectively. This study involved 241 adult subjects with MDD and followed them for 12 weeks. The study found that the CRP level at baseline differentially predicted treatment outcome with the two antidepressants. In cases with low levels of CRP ($<$1 mg/L) at baseline, the improvement on the Montgomery–Åsberg Depression Rating Scale (MADRS) after 12 weeks of treatment score was 3 points greater in patients treated with escitalopram compared with nortriptyline. On the contrary, in patients showing high CRP levels at baseline, an improvement on the MADRS score was 3 points higher in nortriptyline compared with escitalopram-treated patients. Unfortunately, there was no follow-up measurement of CRP.

Another study by Cattaneo et al. (2013) employing the same study design as above compared the effects of escitalopram and nortriptyline on the immune system; however, they explored leukocyte gene expression profiles: *IL-6* levels decreased significantly in responders ($P < 0.0001$), and this was present for both responders to escitalopram (-12%, $P = 0.001$) and to nortriptyline (-6%, $P = 0.02$). While overall in nonresponders, no effect of the ADs was observed on *IL-6* ($+1\%$, $P = 0.5$), a differential effect was found when the two drugs were analyzed separately: *IL-6* failed to change in the nonresponders to escitalopram (-2%, $P = 0.5$), and this cytokine increased in the nonresponders to nortriptyline ($+7\%$, $P = 0.037$). Antidepressant treatment (irrespective of type) significantly reduced the expression levels of *IL-1β* (-6%, $P = 0.006$) and *MIF* (-24%, $P < 0.0001$) and increased *GR* mRNA levels ($+5\%$, $P = 0.009$) and *p11* levels ($+8\%$, $P = 0.005$) during the treatment trial. Therefore, except for IL-6, all examined genes were modulated in the same way by both drugs.

In addition to the above comparison, two other studies compared immune effects during treatments with SSRIs and SNRIs. A direct comparison of immune effects of SSRIs (various types) and SNRIs was conducted in a study by Yoshimura et al. (2009). In the study by Yoshimura et al. (2009) among 51 adult MDD patients and 30 age- and sex-matched controls, plasma IL-6 level, but not plasma TNF-α level, was higher in SSRI and SNRI refractory but not in SSRI- and SNRI-responsive patients. These findings suggest that patients with higher plasma IL-6 levels might develop SSRI or SNRI treatment-resistant depression. In a 6-week study in MDD patients treated with either venlafaxine (75 mg/day) or fluoxetine (80 mg/day), the group by Chang et al. (2012) showed that patients with higher CRP at baseline had poorer treatment response regardless of antidepressant; however, both patient groups showed reduced CRP levels at the end of the trial.

A COMPARATIVE ANTIDEPRESSANT STUDY EXPLORES DIFFERENTIAL EFFECTS ON *ADAPTIVE* IMMUNE FACTORS

Only a single study has been published that compares the effects of antidepressant classes on cellular immune factors. The study by Basterzi et al. (2010) explored whether venlafaxine (75–150 mg/day) and fluoxetine (20–40 mg/day) exerted differential effects on lymphocyte subsets in 69 MDD patients and 36 healthy controls over 6 weeks. At baseline, patients with MDD had a significantly lower CD16/CD56 ratio and higher CD45 ratio compared with controls. Although numerically higher in the venlafaxine-treated patients, treatment response rates between the fluoxetine (53%) and the venlafaxine (75%) groups were not statistically different. CD45 values were decreased significantly in the venlafaxine group at the end of the 6-week treatment period, whereas no difference was observed in the fluoxetine

```
Escitalopram (SSRI) versus nortriptyline (NRI; TCA)
                  Humoral factors:
Low CRP at baseline = antidepressant effect—SSRI > TCA
High CRP at baseline = antidepressant effect—TCA > SSRI
         Both SSRI & TCA responders = ↓ IL-6
   NR for SSRI = no Δ IL-6; NR for TCA = ↑ IL-6
    Both SSRI & TCA = ↓ IL-1β, MIF; ↑ GR, p11
```

```
                 SSRI versus SNRI
                  Humoral factors:
High IL-6, not TNF-α = SSRI and SNRI resistance
High CRP = SSRI (fluoxetine) and SNRI resistance

                  Cellular factors:
Venlafaxine (SNRI) = ↓ CD45; fluoxetine had no effect
```

FIG. 3 Comparison of immune-modulatory effects of antidepressant classes. *TNF*, tumor necrosis factor; *IL*, interleukin; *SSRI*, selective serotonin reuptake inhibitor; *SNRI*, serotonin-noradrenaline reuptake inhibitor; *TCA*, tricyclic antidepressant; *CRP*, C-reactive protein.

group. By the sixth week of study, treatment responders showed a significantly higher CD16/CD56 ratio compared with nonresponders in both treatment groups. Interestingly, no changes in the absolute number of circulating B or T cells or helper/inducer (CD4) or suppressor/cytotoxic (CD8) subsets were observed (Fig. 3).

DISCUSSION

Given the relatively high rate of treatment resistance in antidepressant treatment trials and the lack of a full understanding of how antidepressants exert their biological effects, it is imperative to explore other mechanistic avenues such as the innate and adaptive immune system that not only are involved in the pathophysiology of depression but also may represent important antidepressant treatment targets.

The currently published evidence suggests that antidepressants do appear to have the ability to reduce pro-inflammatory cytokine levels, particularly TNF-α, IL-1β, and IL-6, with a suggestion that SSRIs exert the greatest anti-inflammatory effects (Hannestad et al., 2011). We caution the conclusion at this stage which class of antidepressant possesses greater anti-inflammatory effect. In adult depressed populations, pretreatment levels of CRP informed the efficacy of escitalopram and nortriptyline in that high CRP levels at baseline were associated with greater antidepressant effect of nortriptyline than escitalopram over time (Uher et al., 2014). This is clinically potentially important in that enriched patient groups stratified by CRP may represent subgroups with better response to treatment. Both clinical and preclinical (Tynan et al., 2012) studies are urgently required to further the understanding of the underlying pharmacological mechanisms involved. It should be noted that in this stage, only a very limited number of studies have explored the possibility of the effect of antidepressants on chemokines.

The effects of antidepressant classes on adaptive immune factors are even more complex and not well understood at this stage, which is in part contributed by the incomplete understanding of the role of adaptive immune factors in the pathophysiology of depression (Toben & Baune, 2015). Hence, a single study published so far showing that CD45 values decreased in patients treated with venlafaxine as opposed to those treated with fluoxetine (Basterzi et al., 2010) implies the need for further clinical- and basic science-oriented studies.

Since noncomparative studies show much methodological heterogeneity making meaningful conclusions across studies difficult to reach, studies that investigate differential antidepressant effects may be more informative clinically and they would be more useful for systematic reviews and meta-analyses. Heterogeneity that should be reduced affects variations in length of treatment, cohort characteristics (including medical comorbidities), dosage used, and immune factors analyzed. Since the dosage of antidepressant medications differs largely across studies or within studies by using flexible dosing, dose-dependent immune-modulatory effects are difficult to estimate. In addition, age effects on "background" inflammation during aging need consideration during sampling as well. A meta-analysis of antidepressant clinical trials in subjects 65 years and older compared with adult populations, which found antidepressant treatment to be less efficacious in the older age group (Tedeschini et al., 2011), leaves the question whether increased low-grade and age-related inflammation or inflammation associated with medical comorbidity during aging (which has a higher prevalence in older age groups) may affect antidepressant treatment response. Depression pathophysiology and hence response to antidepressant treatment may differ in late-life depression due to the higher burden of vascular-related pathology (e.g., white matter hyperintensities) (Taylor, Aizenstein, & Alexopoulos, 2013). The level of pro-inflammatory general medical condition comorbidity may also affect the comparative assessment between the antidepressant classes; this differed between the abovementioned studies. Higher burden of general medical conditions is associated with lower antidepressant treatment efficacy (Otte, 2008). The presence of heart disease, obesity, or diabetes likely creates a pro-inflammatory state (Baune et al., 2012; Stuart & Baune, 2012), hence possibly minimizing the treatment response of antidepressants. The variation in immune analyses and a wide array of cytokine, genetic, and cellular immune markers assessed also makes firm conclusions regarding the effects of antidepressant classes difficult.

While the field is premature for firm conclusions, the presented evidence suggests (A) that antidepressants may exert effects on both the innate and adaptive immune system that (B) seems to be associated with response to antidepressant treatment and with the prediction of

antidepressant response. Since both factors of the innate and the adaptive immune system seem to have a mechanistic role in the pathophysiology of clinical depression, it seems reasonable to take on the endeavor of further exploration of this field. It is recommended to develop clinical trials with comparative, head-to-head analyses between antidepressant classes to be conducted in discrete age groups to ensure a higher degree of homogeneity among trial participants. Immune analyses should be conducted at baseline and at the end of the trial not only to enable prediction modeling of treatment response but also for stratifying patient groups based on their immune profile prior to treatment selection. A variety of immune factors should be analyzed going beyond the typical two to three cytokines. These may include peripheral cytokines and chemokines, leukocytes for gene expression analyses, cellular/adaptive immune markers, and possibly future immune-based genetic testing. Finally, basic research should stimulate the exploration of the neuroimmune interaction that should be extended into both the innate and adaptive parts of the immune system. Working at the interface between the fields of psychiatry, neuroscience, and immunology may help the scientific endeavor and patients to optimize treatment outcomes in depression.

References

Basterzi, A. D., Aydemir, C., Kisa, C., Aksaray, S., Tuzer, V., Yazici, K., et al. (2005). IL-6 levels decrease with SSRI treatment in patients with major depression. *Human Psychopharmacology, 20*(7), 473–476.

Basterzi, A. D., Yazici, K., Buturak, V., Cimen, B., Yazici, A., Eskandari, G., et al. (2010). Effects of venlafaxine and fluoxetine on lymphocyte subsets in patients with major depressive disorder: a flow cytometric analysis. *Progress in Neuro-Psychopharmacology & Biological Psychiatry, 34*(1), 70–75.

Baune, B. T., Stuart, M., Gilmour, A., Wersching, H., Heindel, W., Arolt, V., et al. (2012). The relationship between subtypes of depression and cardiovascular disease: a systematic review of biological models. *Translational Psychiatry, 2*.

Cattaneo, A., Gennarelli, M., Uher, R., Breen, G., Farmer, A., Aitchison, K. J., et al. (2013). Candidate genes expression profile associated with antidepressants response in the GENDEP study: differentiating between baseline 'predictors' and longitudinal 'targets'. *Neuropsychopharmacology*, *38*(3), 377–385.

Chang, H. H., Lee, I. H., Gean, P. W., Lee, S. Y., Chi, M. H., Yang, Y. K., et al. (2012). Treatment response and cognitive impairment in major depression: association with C-reactive protein. *Brain, Behavior, and Immunity*, *26*(1), 90–95.

Dantzer, R., O'Connor, J. C., Freund, G. G., Johnson, R. W., & Kelley, K. W. (2008). From inflammation to sickness and depression: when the immune system subjugates the brain. *Nature Reviews. Neuroscience*, *9*(1), 46–56.

Eller, T., Vasar, V., Shlik, J., & Maron, E. (2008). Proinflammatory cytokines and treatment response to escitalopram in major depressive disorder. *Progress in Neuro-Psychopharmacology & Biological Psychiatry*, *32*(2), 445–450.

Eyre, H., & Baune, B. T. (2012). Neuroplastic changes in depression: a role for the immune system. *Psychoneuroendocrinology*, *37*(9), 1397–1416.

Eyre, H. A., Stuart, M. J., & Baune, B. T. (2014). A phase-specific neuroimmune model of clinical depression. *Progress in Neuro-Psychopharmacology & Biological Psychiatry*, *54*, 265–274.

Fornaro, M., Rocchi, G., Escelsior, A., Contini, P., & Martino, M. (2013). Might different cytokine trends in depressed patients receiving duloxetine indicate differential biological backgrounds. *Journal of Affective Disorders*, *145*(3), 300–307.

Frank, M. G., Hendricks, S. E., Burke, W. J., & Johnson, D. R. (2004). Clinical response augments NK cell activity independent of treatment modality: a randomized double-blind placebo controlled antidepressant trial. *Psychological Medicine*, *34*(3), 491–498.

Frank, M. G., Hendricks, S. E., Johnson, D. R., Wieseler, J. L., & Burke, W. J. (1999). Antidepressants augment natural killer cell activity: in vivo and in vitro. *Neuropsychobiology*, *39*(1), 18–24.

Gladkevich, A., Kauffman, H. F., & Korf, J. (2004). Lymphocytes as a neural probe: potential for studying psychiatric disorders. *Progress in Neuro-Psychopharmacology & Biological Psychiatry*, *28*(3), 559–576.

Hannestad, J., Dellagioia, N., & Bloch, M. (2011). The effect of antidepressant medication treatment on serum levels of inflammatory cytokines: a meta-analysis. *Neuropsychopharmacology*, *36*(12), 2452–2459.

Haroon, E., Raison, C. L., & Miller, A. H. (2012). Psychoneuroimmunology meets neuropsychopharmacology: translational implications of the impact of inflammation on behavior. [Research Support, N.I.H., Extramural Review]. *Neuropsychopharmacology*, *37*(1), 137–162.

Hernandez, M. E., Martinez-Fong, D., Perez-Tapia, M., Estrada-Garcia, I., Estrada-Parra, S., & Pavón, L. (2010). Evaluation of the effect of selective serotonin-reuptake inhibitors on lymphocyte subsets in patients with a major depressive disorder. *European Neuropsychopharmacology*, *20*(2), 88–95.

Hernandez, M. E., Mendieta, D., Martinez-Fong, D., Loria, F., Moreno, J., Estrada, I., et al. (2008). Variations in circulating cytokine levels during 52 week course of treatment with SSRI for major depressive disorder. *European Neuropsychopharmacology*, *18*(12), 917–924.

Holtzheimer, P. E., Meeks, T. W., Kelley, M. E., Mufti, M., Young, R., McWhorter, K., et al. (2008). A double blind, placebo-controlled pilot study of galantamine augmentation of antidepressant treatment in older adults with major depression. *International Journal of Geriatric Psychiatry*, *23*(6), 625–631.

Iken, K., Chheng, S., Fargin, A., Goulet, A. C., & Kouassi, E. (1995). Serotonin upregulates mitogen-stimulated B lymphocyte proliferation through 5-HT1A receptors. *Cellular Immunology*, *163*(1), 1–9.

Janssen, D. G., Caniato, R. N., Verster, J. C., & Baune, B. T. (2010). A psychoneuroimmunological review on cytokines involved in antidepressant treatment response. *Human Psychopharmacology*, *25*(3), 201–215.

Kalman, J., Palotas, A., Juhasz, A., Rimanoczy, A., Hugyecz, M., Kovacs, Z., et al. (2005). Impact of venlafaxine on gene expression profile in lymphocytes of the elderly with major depression—evolution of antidepressants and the role of the "neuro-immune" system. *Neurochemical Research*, *30*(11), 1429–1438.

Kook, A. I., Mizruchin, A., Odnopozov, N., Gershon, H., & Segev, Y. (1995). Depression and immunity: the biochemical interrelationship between the central nervous system and the immune system. *Biological Psychiatry*, *37*(11), 817–819.

Lanquillon, S., Krieg, J. C., Bening-Abu-Shach, U., & Vedder, H. (2000). Cytokine production and treatment response in major depressive disorder. *Neuropsychopharmacology*, *22*(4), 370–379.

Leo, R., Di Lorenzo, G., Tesauro, M., Razzini, C., Forleo, G. B., Chiricolo, G., et al. (2006). Association between enhanced soluble CD40 ligand and proinflammatory and prothrombotic states in major depressive disorder: pilot observations on the effects of selective serotonin reuptake inhibitor therapy. *Journal of Clinical Psychiatry*, *67*(11), 1760–1766.

Leonard, B., & Maes, M. (2012). Mechanistic explanations how cell-mediated immune activation, inflammation and oxidative and nitrosative stress pathways and their sequels and concomitants play a role in the pathophysiology of unipolar depression [Review]. *Neuroscience and Biobehavioral Reviews*, *36*(2), 764–785.

Li, Z., Qi, D., Chen, J., Zhang, C., Yi, Z., Yuan, C., et al. (2013). Venlafaxine inhibits the upregulation of plasma tumor necrosis factor-alpha (TNF-alpha) in the Chinese patients with major depressive disorder: a prospective longitudinal study. *Psychoneuroendocrinology, 38*(1), 107–114.

Licinio, J. (2011). Translational psychiatry: leading the transition from the cesspool of devastation to a place where the grass is really greener. *Translational Psychiatry, 1.*

Mamdani, F., Berlim, M. T., Beaulieu, M. M., Labbe, A., Merette, C., & Turecki, G. (2011). Gene expression biomarkers of response to citalopram treatment in major depressive disorder. *Translational Psychiatry, 1.*

Mamdani, F., Berlim, M. T., Beaulieu, M. M., & Turecki, G. (2014). Pharmacogenomic predictors of citalopram treatment outcome in major depressive disorder. *World Journal of Biological Psychiatry, 15*(2), 135–144.

Martino, M., Rocchi, G., Escelsior, A., & Fornaro, M. (2012). Immunomodulation mechanism of antidepressants: interactions between serotonin/norepinephrine balance and Th1/Th2 balance. *Current Neuropharmacology, 10*(2), 97–123.

McAfoose, J., & Baune, B. T. (2009). Evidence for a cytokine model of cognitive function. *Neuroscience and Biobehavioral Reviews, 33*(3), 355–366.

Miller, A. H., Maletic, V., & Raison, C. L. (2009). Inflammation and its discontents: the role of cytokines in the pathophysiology of major depression. *Biological Psychiatry, 65*(9), 732–741.

Moylan, S., Maes, M., Wray, N. R., & Berk, M. (2013). The neuroprogressive nature of major depressive disorder: pathways to disease evolution and resistance, and therapeutic implications. *Molecular Psychiatry, 18*(5), 595–606.

Murray, C. J., Vos, T., Lozano, R., Naghavi, M., Flaxman A. D., Michaud, C., et al. (2012). Disability-adjusted life years (DALYs) for 291 diseases and injuries in 21 regions, 1990–2010: a systematic analysis for the Global Burden of Disease Study 2010. *Lancet, 380*(9859), 2197–2223.

Narita, K., Murata, T., Takahashi, T., Kosaka, H., Omata, N., & Wada, Y. (2006). Plasma levels of adiponectin and tumor necrosis factor-alpha in patients with remitted major depression receiving long-term maintenance antidepressant therapy. *Progress in Neuro-Psychopharmacology & Biological Psychiatry, 30*(6), 1159–1162.

O'Brien, S. M., Scott, L. V., & Dinan, T. G. (2006). Antidepressant therapy and C-reactive protein levels. *British Journal of Psychiatry, 188,* 449–452.

Otte, C. (2008). Incomplete remission in depression: role of psychiatric and somatic comorbidity. *Dialogues in Clinical Neuroscience, 10*(4), 453–460.

Piletz, J. E., Halaris, A., Iqbal, O., Hoppensteadt, D., Fareed, J., Zhu, H., et al. (2009). Pro-inflammatory biomakers in depression: treatment with venlafaxine. *World Journal of Biological Psychiatry, 10*(4), 313–323.

Ravindran, A. V., Griffiths, J., Merali, Z., & Anisman, H. (1995). Lymphocyte subsets associated with major depression and dysthymia: modification by antidepressant treatment. *Psychosomatic Medicine, 57*(6), 555–563.

Rothermundt, M., Arolt, V., Fenker, J., Gutbrodt, H., Peters, M., & Kirchner, H. (2001). Different immune patterns in melancholic and non-melancholic major depression. *European Archives of Psychiatry and Clinical Neuroscience, 251*(2), 90–97.

Rush, A. J., Trivedi, M. H., Wisniewski, S. R., Nierenberg, A. A., Stewart, J. W., Warden, D., et al. (2006). Acute and longer-term outcomes in depressed outpatients requiring one or several treatment steps: a STAR*D report. *American Journal of Psychiatry, 163*(11), 1905–1917.

Rush, A. J., Trivedi, M. H., Wisniewski, S. R., Stewart, J. W., Nierenberg, A. A., Thase, M. E., et al. (2006). Bupropion-SR, sertraline, or venlafaxine-XR after failure of SSRIs for depression. [Multicenter Study Randomized Controlled Trial Research Support, N.I.H., Extramural Research Support, Non-U.S. Gov't] *The New England Journal of Medicine, 354*(12), 1231–1242.

Schleifer, S. J., Keller, S. E., & Bartlett, J. A. (1999). Depression and immunity: clinical factors and therapeutic course. *Psychiatry Research, 85*(1), 63–69.

Sluzewska, A., Rybakowski, J. K., Laciak, M., Mackiewicz, A., Sobieska, M., & Wiktorowicz, K. (1995). Interleukin-6 serum levels in depressed patients before and after treatment with fluoxetine. *Annals of the New York Academy of Sciences, 762,* 474–476.

Stuart, M. J., & Baune, B. T. (2012). Depression and type 2 diabetes: inflammatory mechanisms of a psychoneuroendocrine co-morbidity. *Neuroscience and Biobehavioral Reviews, 36*(1), 658–676.

Sutcigil, L., Oktenli, C., Musabak, U., Bozkurt, A., Cansever, A., Uzun, O., et al. (2007). Pro- and anti-inflammatory cytokine balance in major depression: effect of sertraline therapy. *Clinical & Developmental Immunology, 2007,* 76396.

Taylor, W. D., Aizenstein, H. J., & Alexopoulos, G. S. (2013). The vascular depression hypothesis: mechanisms linking vascular disease with depression. *Molecular Psychiatry, 18*(9), 963–974.

Tedeschini, E., Levkovitz, Y., Iovieno, N., Ameral, V. E., Nelson, J. C., & Papakostas, G. I. (2011). Efficacy of antidepressants for late-life depression: a meta-analysis and meta-regression of placebo-controlled randomized trials. *Journal of Clinical Psychiatry, 72*(12), 1660–1668.

Toben, C., & Baune, B. T. (2015). An act of balance between adaptive and maladaptive immunity in depression: a role for T lymphocytes. *Journal of Neuroimmune Pharmacology, 10*(4), 595–609.

Tsao, C. W., Lin, Y. S., Chen, C. C., Bai, C. H., & Wu, S. R. (2006). Cytokines and serotonin transporter in patients with major depression. *Progress in Neuro-Psychopharmacology & Biological Psychiatry, 30*(5), 899–905.

Tuglu, C., Kara, S. H., Caliyurt, O., Vardar, E., & Abay, E. (2003). Increased serum tumor necrosis factor-alpha levels and treatment response in major depressive disorder. *Psychopharmacology, 170*(4), 429–433.

Tynan, R. J., Weidenhofer, J., Hinwood, M., Cairns, M. J., Day, T. A., & Walker, F. R. (2012). A comparative examination of the anti-inflammatory effects of SSRI and SNRI antidepressants on LPS stimulated microglia. *Brain, Behavior, and Immunity, 26*(3), 469–479.

Uher, R., Tansey, K. E., Dew, T., Maier, W., Mors, O., Hauser, J., et al. (2014). An inflammatory biomarker as a differential predictor of outcome of depression treatment with escitalopram and nortriptyline. *American Journal of Psychiatry, 171*(12), 1278–1286.

Yoshimura, R., Hori, H., Ikenouchi-Sugita, A., Umene-Nakano, W., Ueda, N., & Nakamura, J. (2009). Higher plasma interleukin-6 (IL-6) level is associated with SSRI- or SNRI-refractory depression. *Progress in Neuro-Psychopharmacology & Biological Psychiatry, 33*(4), 722–726.

The Biological Underpinnings of Mood Disorders Interact With Early Trauma, Sexual Abuse and Neuroticism: Implications for Psychiatric Classification and Treatment

George Anderson*, *Michael Maes*[†]

*CRC Scotland & London, London, United Kingdom
[†]Deakin University, Geelong, VIC, Australia

Abbreviations

a7nAChR	alpha 7 nicotinic acetylcholine receptor;
AANAT	arylalkylamine *N*-acetyl-transferase
AD	Alzheimer's disease
BBB	blood-brain barrier
BDNF	brain-derived neurotropic factor
BMI	body mass index
CoQ10	coenzyme Q10
COX-2	cyclooxygenase-2
CRP	C-reactive protein
EGCG	epigallocatechin gallate
G-CSF	granulocyte colony-stimulating factor
GPX	glutathione peroxidase
GSH	glutathione
HIOMT	hydroxyindole *O*-methyltransferase
HPA	hypothalamic-pituitary-adrenal
IDO	indoleamine 2,3-dioxygenase
IFN-γ	interferon gamma
IL	interleukin
IL-1RA	IL-1 receptor antagonist
KAT	kynurenine aminotransferase
kyn	kynurenine
KYNA	kynurenic acid
MDD	major depressive disorder
MS	multiple sclerosis
mtDNA	mitochondrial DNA
NLRP3	NOD-like receptor pyrin domain-containing
NMDA	*N*-methyl D-aspartate
NO	nitric oxide
O&NS	oxidative and nitrosative stress
RNS	reactive nitrogen species
ROS	reactive oxygen species
SNPs	single-nucleotide polymorphisms
SOD	superoxide dismutase
TDO	tryptophan 2,3-dioxygenase
TNF-α	tumor necrosis factor alpha
TRYCATs	tryptophan catabolites

INTRODUCTION

Conceptualizations of depression and classification systems thereby derived have been many and various, including psychoanalytic, cognitive, and biological. Most work on the biological underpinnings and consequent treatment approaches have emphasized a decrease in monoamines, especially serotonin. Such classical conceptualizations still considerably influence clinical practice. Recent research on the biological underpinnings of depression, both in major depressive disorder (MDD) and in bipolar disorder (BD), has highlighted alterations in oxidative and nitrosative stress (O&NS) and immune-inflammatory pathways, which act to drive tryptophan away from serotonin synthesis and to the production of tryptophan catabolites (TRYCATs) (Anderson & Maes, 2014a; Maes, 1993; Maes et al., 1991, 2000; Maes, Galecki, Chang, & Berk, 2011). As well as neuroregulatory TRYCATs, such as kynurenine (kyn) and kynurenic acid (KYNA), increased O&NS and immune inflammation have consequences for a wide array of biochemical processes, including DNA repair, sirtuin levels, and mitochondria functioning (Anderson & Maes, 2014a). Such data provide a simple model, whereby changes in such neuroregulatory TRYCATS, with their differential expression in different brain regions that alters interarea neuronal activity patterning, may be seen as effectors of pro-inflammatory cytokines and O&NS (Anderson & Maes, 2014a). A key aspect of such model is the pro-inflammatory cytokine-induced indoleamine 2,3-dioxygenase (IDO) and stress−/cortisol-induced tryptophan 2,3-dioxygenase (TDO), which divert tryptophan to kynurenine pathway products. Consequences arise for a number of other cell processes, including the levels of activity in the melatonergic pathway, given that serotonin is a necessary precursor for this pathway. This may have significant implications for the biological differentiation of mood disorders into MDD and BD, with the poles of BD thought to be partly dependent on the ratio of melatonin to its immediate precursor, N-acetylserotonin (NAS) (Anderson, Jacob, Bellivier, & Geoffroy, 2016). As such, genetic susceptibilities in the melatonin system in BD, although once removed from the direct effects of pro-inflammatory cytokines and O&NS, may be relevant to the differentiation of MDD and BD (Anderson et al., 2016). The TRYCATs and melatonergic pathways are summarized in Fig. 1, with the influence of leaky gut-associated stress and inflammatory processes given as an example of a peripheral site for the induction of immune inflammation and O&NS induction, with consequences, both peripherally and centrally, for coordinated alterations in the TRYCATs and melatonergic pathways.

It should be noted that recent data on immune-inflammatory and O&NS processes in mood disorders considerably complicate their conceptualization and treatment. For example, data suggest that the biological pathways indicated above may be more associated with somatization aspects of depression (Anderson, Maes, & Berk, 2012) and/or with early abusive experiences (Grosse et al., 2016; Moraes et al., 2017, 2018). Given that these aspects are infrequently investigated in studies of MDD and BD, it is likely that significant biological subgroups have not been distinguished in most previous publications in this area. Consequently, by not taking into account relevant biologically defined subgroups, data collected are likely to be confounded. For example, an increase in the kyn/KYNA ratio in somatization, but not in MDD without high levels of somatization, will have differential effects on KYNA inhibition of the glutamate N-methyl-D-aspartate receptor (NMDAR) and the alpha 7 nicotinic acetylcholine receptor (a7nAChR) (Anderson & Maes, 2014a). As such, the wide array of data looking at changes in the NMDAR and a7nAChR in mood disorders will have been collected without taking into account relevant biologically defined subgroups that are likely to show differential changes in the levels and activity of these two receptors. As such,

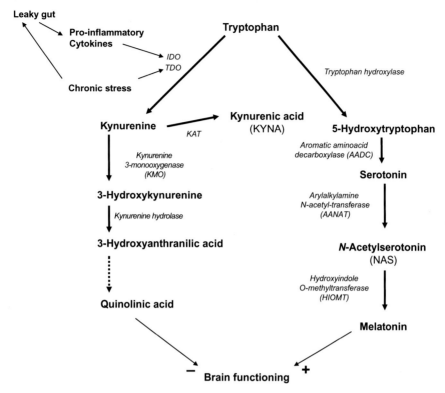

FIG. 1 The major components of the kynurenine and serotonergic/melatonergic pathways and how these are coordinated are indicated. Tryptophan conversion by tryptophan hydroxylase leads to the synthesis of serotonin and ultimately melatonin. Stress-driven changes, here exemplified by a leaky gut, can increase pro-inflammatory cytokine levels. Stress and pro-inflammatory cytokines induce TDO and IDO, respectively, thereby driving tryptophan down the kynurenine pathway to the synthesis of tryptophan catabolites (TRYCATs). Gut permeability is given as an example, with stress/cortisol increasing gut permeability and increasing TDO, while increased gut permeability can enhance pro-inflammatory cytokines, which induce IDO.

the collation of people presenting with mood disorders into DSM-defined categories is likely to have obscured significant correlations across collected biological data.

This chapter looks at these interconnected biological changes in people presenting with depression, both MDD and BD, and the implications that this may have for the etiology, course, and treatment of mood disorders. Firstly, we review O&NS before looking at their role in MDD and BD.

OXIDATIVE AND NITROSATIVE STRESS (O&NS)

Lowered levels of endogenous antioxidant enzymes and antioxidants are often evident in MDD patients and animal models of depression (Anderson & Maes, 2014a; Che et al., 2015; Maes et al., 2000, 2011; Mansur et al., 2016). Pro-inflammatory cytokines and other metabolic challenges in mitochondria generate free radicals, including reactive oxygen species

(ROS) and reactive nitrogen species (RNS), such as superoxide, peroxynitrite, hydrogen peroxide, and nitric oxide (NO). ROS and RNS have evolved for use in normal plasticity processes. However, in excess or when endogenous antioxidant defenses are low, ROS and RNS are toxic. Such toxicity includes effects in membranes, DNA, and mitochondria. Antioxidants include vitamin C, vitamin E, glutathione (GSH), and coenzyme Q10 (CoQ10), while antioxidant enzymes include superoxide dismutase (SOD), catalase, and glutathione peroxidase (GPX), reviewed in Anderson and Maes (2014a).

O&NS arises when the oxidant/antioxidant ratio is increased, which drives a number of damaging processes, including peroxynitrite driven lipid peroxidation. Lipid peroxidation involves damage to membrane lipids and DNA (Anderson & Maes, 2014a). Dysfunction of such key cellular processes can contribute to a wide array of altered responses, including an increased likelihood of cellular apoptosis. Some of these effects are in mitochondria, where ROS are highly produced. As such, inadequate antioxidant defenses, in the face of a raised oxidant challenge, results in the suboptimal functioning of mitochondria. As mitochondria are key sites for cellular energy and survival, changes in mitochondria can have significant impacts on wider cellular functions and patterned gene responses.

It should be noted that increased levels of ROS, RNS, and O&NS are evident in many medical conditions, often with relevance to the etiology, course, and management of these conditions, including inflammatory bowel disease (Martin-Subero, Anderson, Kanchanatawan, Berk, & Maes, 2016), multiple sclerosis (MS) (Anderson & Rodriguez, 2011), migraine (Anderson & Maes, 2015; Geyik, Altunısık, Neyal, & Taysi, 2016), endometriosis (Anderson & Maes, 2015; Donnez, Binda, Donnez, & Dolmans, 2016), Parkinson's disease (Anderson & Maes, 2014b), and Alzheimer's disease (AD) (Maes & Anderson, 2016), as well as BD (Mansur et al., 2016) and MDD (Moylan et al., 2014). All of these medical conditions are associated with increased levels

of depression, which is proposed to arise from the overlapping changes in O&NS, immune inflammation, TRYCATs, and melatonergic pathways. As such, the biological underpinnings of depression are intimately linked to the pathophysiological changes occurring in other medical conditions, suggesting that depression is not a psychological reaction to these other conditions, but may be biologically intertwined. This is important to note as it suggests that the biological underpinnings, including O&NS, of a wide array of diverse medical presentations may be intimately linked to the etiology and course of depression.

It is also important to note that the effects of O&NS, especially via its close association with immune-inflammatory processes, may be mediated in organs and tissues distant from the brain. O&NS and immune inflammation are long recognized contributors to increased gut permeability (Yu et al., 2016), with increased gut permeability and alterations in the gut-brain axis at the cutting edge of research on the pathophysiological changes occurring in depression. Many other seemingly distant factors may trigger increases in oxidative stress, such as small-particle air pollutants that contribute not only to cancer susceptibility (Øvrevik et al., 2017) but also to increased gut permeability (Mutlu et al., 2011) and depression (Kim et al., 2016), possibly via increased gut permeability. Overall, O&NS may be intimately linked to depression, including as to how depression integrates into wider medical conditions and their pathophysiological processes as well as wider body systems, such as the gut-brain axis.

It is also important to note that the biological underpinnings of depression may change over time, with raised levels of O&NS and immune-inflammatory processes in one episode driving changes in the presenting pathophysiological processes in subsequent episodes. This is a process referred to as neuroprogression and is evident in a host of psychiatric conditions, including MDD and BD (Kapczinski et al., 2017; Moylan, Maes, Wray, & Berk, 2013). O&NS is proposed to play an important role in neuropro-gression, including via damage to membranes that leads to the exposure of neoepitopes, thereby increasing autoimmune responses (Maes et al., 2013). Some of these effects of O&NS may then be mediated by the induction of autoimmune-associated T helper (Th) 17 cells, which are proposed to contribute to more prolonged immune-inflammatory processes (Slyepchenko et al., 2016). Neuroprogression therefore compli-cates the utility of previous studies of patients presenting with mood dysregulation, when look-ing for reliable pathophysiological changes (Anderson & Maes, 2018).

We now look in more detail at O&NS changes in depression, bearing in mind the challenges in the interpretation of this, as indicated above.

O&NS and MDD

The oxidant/antioxidant imbalance has been proposed as an inherent feature of MDD (Maes et al., 2000, 2011; Talarowska, Szemraj, Berk, Maes, & Gałecki, 2015), with an elevated oxi-dant/antioxidant ratio correlating with decre-ments in cognitive functioning in MDD patients, including when the MDD is associated with other medical conditions (Talarowska, Gałecki, Maes, Bobińska, & Kowalczyk, 2012). Studies show MDD patients to have raised levels of myeloper-oxidase, both mRNA and protein (Talarowska, Szemraj, & Gałecki, 2014); the pro-inflammation associated cyclooxygenase-2 (COX-2) (Gałecki, Talarowska, Bobińska, & Szemraj, 2015); induc-ible nitric oxide synthase (Kowalczyk, Talarowska, Zajączkowska, Szemraj, & Gałecki, 2013); and malondialdehyde (Bajpai, Verma, Srivastava, & Srivastava, 2014), which are more evident in MDD patients with recurrent episodes. Although not all data are in agreement (Black, Bot, Scheffer, & Penninx, 2017), the raised levels of ROS and O&NS in MDD patients can associate with the presence of more DNA breaks and alkali-labile sites, as well as oxidative DNA dam-age (Czarny et al., 2015), in part due to a genetic susceptibility that leads to suboptimal DNA

repair systems in MDD patients (Czarny et al., 2016). Future research will have to determine the extent to which the changes underpinning neuroprogression over the course of recurrent depressive episodes are mediated by DNA dam-age and alterations in repair response, including in comparison with first-episode presentation.

It is also of note that most antidepressant medications can lower levels of ROS, RNS, and O&NS (Behr, Moreira, & Frey, 2012), even when there is no evidence of increased oxidative stress at baseline in MDD patients (Black et al., 2017). The role of antidepressants in neuroprogressive processes in MDD patients is given support by their increasing utilization and utility in neuro-degenerative conditions (Riederer & Müller, 2017), where the role of O&NS has been long rec-ognized (Anderson & Maes, 2016). In a variety of preclinical models of depression, data also show increases in the oxidant/antioxidant ratio, in association with raised levels of O&NS, reviewed in Anderson and Maes (2014b). Again, taking into account the heterogeneity in the pathophysiological changes occurring in MDD, the majority of data indicate a role for ROS, RNS, and O&NS in MDD. As suggested by the association of recurrent MDD with processes classically associated with neurodegenerative conditions, such as Alzheimer's disease, it is of note that ROS-driven shortening of telomere length has been shown in MDD oligodendro-cytes (Szebeni et al., 2014), with some suggestion that the hippocampus may show a differential shortening of telomere length in MDD patients (Mamdani et al., 2015). The latter study suggests that ROS and O&NS effects are likely to differen-tially induce damage, including accelerated aging, in different brain regions in MDD patients (Mamdani et al., 2015).

O&NS in BD Depression

Enhanced O&NS levels are also evident in BD (Tunçel et al., 2015). A meta-analysis of serum O&NS in BD patients showed O&NS to be raised in all BD phases, including raised levels of lipid peroxidation and NO and DNA/RNA damage (Aydemir et al., 2014). An increase in the oxidant/antioxidant ratio is also evident in BD, as indicated by raised levels of myeloperoxidase coupled to a relative lowering of the antioxidant, catalase, often in association with raised levels of immune inflammation (Selek, Altindag, Saracoglu, & Aksoy, 2015). Single-nucleotide polymorphisms (SNPs) of the *NO synthase* (*NOS*) *3* gene modulate the course and severity of BD, as indicated by their association with violent suicidal behaviors (Oliveira et al., 2015).

As well as raised O&NS levels in the serum, O&NS is also raised centrally, with the products of lipid peroxidation, namely, 4-hydroxy-2-nonenal and 8-isoprostane, being elevated in the prefrontal cortex of BD patients, coupled to indicants of mitochondrial dysfunction (Andreazza, Wang, Salmasi, Shao, & Young, 2013). As with MDD, white matter changes are evident in BD, which correlate with serum lipid peroxidation levels (Versace et al., 2014). In addition, telomere shortening is evident in BD (Barbé-Tuana et al., 2016), as in MDD, which may indicate accelerated aging. Decreased levels of the longevity-associated sirtuins occur in the depressed pole of BD and in MDD, further supporting an impact of O&NS on aging-associated processes in depression per se (Abe et al., 2011). This is likely to be at least partly mediated by oxidative DNA damage, given that its repair by poly(ADP-ribose) polymerase (PARP) requires NAD^+, which will lower NAD^+-induced sirtuins. Increased O&NS in BD will also shorten life span by enhancing the risk of cardiovascular disease (CVD) in BD patients (Hatch et al., 2015), where O&NS is associated with a proxy measure of atherosclerosis.

Notably, O&NS also decreases brain-derived neurotropic factor (BDNF) levels, lowering neurotropic support in BD, as in MDD (Iuvone, Boatright, Tosini, & Ye, 2014). Serotonin is enzymatically converted to NAS, which is then converted to melatonin. NAS is a BDNF mimic, given its activation of the BDNF receptor, TrkB. As such, the driving of tryptophan down the kynurenine pathway and away from serotonin, NAS and melatonin synthesis by O&NS and immune inflammation, will further act to decrease neurotropic support in depression (Anderson & Maes, 2013). The NAS/melatonin ratio has also been proposed to play a role in the regulation of the switch between the two poles of BD (Anderson et al., 2016). As with antidepressants in MDD, the classical treatment of BD, namely, lithium, attenuates O&NS (Banerjee, Dasgupta, Rout, & Singh, 2012).

Overall, it is clear that O&NS play a significant role in depression, including in people classed as having MDD and BD. O&NS is intimately linked to alterations in the immune-inflammatory response, which is looked at next.

IMMUNE-INFLAMMATORY PROCESSES IN MDD

Heightened levels of immune-inflammatory processes, including pro-inflammatory cytokines, are evident in MDD and may be intimately intertwined with O&NS (Gardner & Boles, 2011; Maes et al., 1991, 2000, 2011). A number of pro-inflammatory cytokine levels are increased in MDD, including interleukin (IL)-1β, IL-8, IL-18, and tumor necrosis factor alpha (TNF-α), with ROS contributing to this by its positive modulation of the NOD-like receptor pyrin domain-containing (NLRP) 3 inflammasome (Alcocer-Gómez et al., 2014; Maes, 1993). Increases in pro-inflammatory markers have been shown to occur both centrally and peripherally, in positive association with depression severity and duration and indicants of neuroprogression (Dahl et al., 2014; Rotter et al., 2013). Dahl and colleagues found significantly increased IL-1β, IL-1 receptor antagonist (IL-1RA), IL-5, IL-6, IL-7, IL-8, IL-10, granulocyte colony-stimulating factor (G-CSF), and interferon gamma (IFN-γ) plasma levels in MDD patients prior to treatment versus healthy controls (Dahl et al., 2014). Following treatment for 3 months, seven out of nine cytokines were significantly decreased versus baseline and no longer differing from levels in healthy controls. This reduction in cytokines correlated with levels of mood improvement. However, recent data analyzing cytokine changes over the course of MDD treatment indicate greater individual complexity, likely reflective of the great biological heterogeneity of most samples of MDD patients (Myung et al., 2016).

Two meta-analyses of cytokines levels in MDD patients indicate increased levels of TNF-α, IL-6, and C-reactive protein (CRP) (Dowlati et al., 2010; Valkanova, Ebmeier, & Allan, 2013). CRP is a classical indicator of increased immune pro-inflammatory processes across a host of medical conditions, with interesting effects, including increasing blood-brain barrier (BBB) permeability (Hsuchou, Kastin, Mishra, & Pan, 2012). Recent work indicates that changes in such pro-inflammatory factors significantly correlate with a wider array of biochemical factors, including catecholamines and neurotransmitters such as peripheral dopamine, epinephrine, histamine, kynurenic acid, norepinephrine, and serotonin (Peacock, Scheiderer, & Kellermann, 2016). Such data highlight the interconnectivity of immune-inflammatory processes with wider regulation of neuronal activity and neuronal patterning. SNPs in TNF-α (Cerri et al., 2009) and IL-18 (Haastrup et al., 2012) increase the risk of MDD, giving further support to their role in depression susceptibility, which includes interactions of IL-18 SNPs with life stressors (Haastrup et al., 2012).

However, as may be expected given the heterogeneity of this disorder, not all previous research indicates enhanced levels of pro-inflammatory cytokines in MDD (Steptoe, Kunz-Ebrecht, & Owen, 2003).

IMMUNE-INFLAMMATORY PROCESSES IN BD

Increased levels of immune inflammation are widely reported in studies of BD patients, even during periods of remission, at least in part via reciprocated interactions with O&NS (do Prado et al., 2013). Many of the investigations of immune-inflammatory activity in BD have looked to determine changes that could differentiate the two poles of BD and comparisons to the euthymic phase. As with MDD, there can be considerable differences between individual studies, with two major meta-analyses providing further evidence of such diversity. A meta-analysis of immune-inflammatory changes by Munkholm and colleagues showed enhanced levels of soluble IL-2R, TNF-α, soluble TNF receptor 1 (TNFR1), and soluble IL-6R in BD patients versus controls (Munkholm, Braüner, Kessing, & Vinberg, 2013). However, these authors found no reliable significant changes in IL-1, IL-2, IL-5, IL-6, IL-8, IL-10, IL-12, IL-1RA, IFN-γ, transforming growth factor-beta1 (TGF-β1), and sTNFR2. In a meta-analysis of 30 studies of cytokine alterations in BD, Modabbernia and colleagues found IL-4, sIL-2R, sIL-6R, and TNF-α to be significantly increased, with IL-1RA levels only increased during manic states, while there was only a trend increase in IL-1β and IL-6 (Modabbernia, Taslimi, Brietzke, & Ashrafi, 2013). Overall, the results of these two meta-analyses suggest that immune-inflammatory processes, as determined by cytokine changes, are a highly regular occurrence in BD, although also indicative of wide differences across different studies (Modabbernia et al., 2013; Munkholm et al., 2013). A number of factors are likely to contribute to such variations, including neuroprogression in BD (Kapczinski et al., 2017).

In a recent meta-analysis of 68 studies in acutely ill patients and 46 of chronically ill patients, Goldsmith and colleagues compared serum cytokine changes in patients with schizophrenia, BD, and MDD (Goldsmith, Rapaport, & Miller, 2016). IL-6, TNF-α, sIL-2R, and IL-1RA were all significantly increased in acutely ill patients with schizophrenia, BD mania, and MDD versus controls ($P < .01$). In chronically ill patients, IL-6 levels were significantly raised in schizophrenia, euthymic BD, and MDD versus controls. IL-1β and sIL-2R levels were significantly raised in both chronic schizophrenia and euthymic BD (Goldsmith et al., 2016). Overall, these authors concluded that the pattern of cytokine alterations in schizophrenia, BD, and MDD during acute and chronic phases of illness was very similar and suggested common underlying pathways for immune dysfunction across these diagnostic groups.

OTHER IMMUNE PROCESSES IN DEPRESSION

It is important to note that other aspects of immune activity and inflammation are also evident in depression, both in MDD and BD, including raised levels of chemokines and chemoattractant cytokines, which determine circulating leukocytes attraction to sites of inflammation (Barbosa et al., 2013; Grassi-Oliveira et al., 2012); osteoprotegerin and von Willebrand factor, as well as other indications of endothelial cell activation (Barbosa et al., 2013; Bot et al., 2015); resistin and leptin, as well as other adipose tissue-derived adipokines (Barbosa et al., 2012; Milaneschi, Lamers, Bot, Drent, & Penninx, 2017), which may be driven by mood medication-induced metabolic syndrome (Hiles, Révész, Lamers, Giltay, & Penninx, 2016; Soeiro-de-Souza et al., 2014); and differential effects on CRP and other acute-phase proteins (Chang et al., 2017; Dickerson et al., 2013). Recent work has also indicated significant mucosal immune changes in psychiatric conditions (Maes, Kubera, Leunis, & Berk, 2012; Roomruangwong et al., 2017), indicating a significant role for alterations in gut regulation across a number of psychiatric conditions, including mood disorders.

WIDER IMMUNE AND O&NS MODULATORS

It is important to emphasize that a plethora of factors may contribute to the variability in specific immune and O&NS responses across

different mood disorder studies, including first or early presentations versus those with recurrent or long-standing depression; the severity of depression; the presence of psychotic features; variations in wider comorbidities, including addiction, cigarette smoking, and obesity, which are immune and O&NS modulators per se (Aguilar-Valles, Inoue, Rummel, & Luheshi, 2015; Bortolasci et al., 2014); age of depression onset; history of early traumatic events; whether somatization is a significant aspect of the depressed presentation (Anderson et al., 2012); the differential effects of diverse array of medications in the management of depression; dietary factors, such as increased inflammation derived from a saturated fat diet driving an increase in gut permeability (Rahman et al., 2016); whether medication is used alone or in combination; and whether neuroprogression is evident (Slyepchenko et al., 2016). Also of relevance in BD is the specificity of immune and O&NS changes while in the different poles of BD and how changes in the manic phase may alter the changes occurring in subsequent depressed poles.

Recent work indicates that the increases in O&NS and pro-inflammatory cytokines in MDD and BD may be only, or predominantly, in patients who have had early traumatic experiences, especially sexual abuse (Agnew-Blais & Danese, 2016; Grosse et al., 2016; Lu et al., 2013; Moraes et al., 2017, 2018). BD patients with a history of childhood maltreatment present with more severe depressive and manic episodes and a host of negative factors, including greater psychosis severity, higher risk of comorbid posttraumatic stress disorder, increased anxiety disorders, substance misuse disorders, alcohol misuse disorder, earlier age of BD onset, higher risk of rapid cycling, and higher risk of suicide attempts versus BD patients presenting without childhood maltreatment (Agnew-Blais & Danese, 2016). Given that this is an array of factors that are known to modulate the levels of O&NS and immune-inflammatory responses, it is clear that extensive further research on the role of early abusive events in the regulation of O&NS, immune activity, and immune patterning in mood disorders is required. This may be of particular importance in regard to the early biochemical changes that contribute to such later changes in responses.

Such data on the role of early abusive events in the patterning of immune and O&NS changes that underpin many depressive presentations highlight the need for more extensive investigation of the role of developmental processes in the etiology of psychiatric conditions. For example, recent work suggests that many of the benefit mediated by breastfeeding over formula feed are mediated by alterations in the gut microbiota, gut permeability, and subsequent impacts on the foundations of the immune systems (Anderson, Vaillancourt, Maes, & Reiter, 2017). Breastfeeding is associated with a decrease in offspring psychopathology (Hayatbakhsh, O'Callaghan, Bor, Williams, & Najman, 2012). There are also likely to be significant impacts at an even earlier developmental time point, with data indicating that maternal prenatal infection may increase the risk of offspring depression (Simanek & Meier, 2015), perhaps in a sexually dimorphic manner (Gilman et al., 2016). This has some links to a long-standing preclinical literature, which indicates that prenatal stress increases the risk of mood disorder, often coupled to compromised cognition (Lin et al., 2017), which recent data suggest may be mediated via alterations in the offspring gut microbiome (Gur et al., 2017). As to whether prenatal stress, maternal infection, variations in breastfeeding, and early traumatic events interact to modulate depression susceptibility and/or the patterning of immune and pro-inflammatory cytokine responses requires investigation. In this context, it may be important to note that the stress hormone cortisol increases gut permeability (Vanuytsel et al., 2014), suggesting that such stress effects in the gut may be relevant to these early developmental regulators of not only depression susceptibility but also the nature of the specific pathophysiological changes occurring in adult mood presentations. Overall, the above data indicate that a complex array of developmental processes may act to drive alterations not only in depression susceptibility but also in the specific pathophysiological changes occurring. This could be suggestive of distinct mood disorders.

A further complication relates to the nature of early abusive event reporting, which is widely thought to be significantly underreported. Data by Della Femina and colleagues indicate that the incidence of sexual abuse is likely to be highly underreported (Della Femina, Yeager, & Lewis, 1990). As to whether the levels of reporting, versus occluding the truth, are substantially higher in patients presenting with depression is unknown. It is assumed by many that a more trusting therapeutic alliance with patients will increase the reporting of such early traumatic events. However, this may be more to do with therapist/researcher narcissism and self-esteem rather than science. As it stands, it would seem that the reporting of early trauma, especially sexual abuse, is associated with increased levels of pro-inflammatory cytokines and O&NS, which may partially reflect other factors, such as the levels of personality disintegration. Alongside such data on early trauma associations with immune and O&NS responses, another body of data indicates that it is the comorbidity of personality disorders that determines increased levels of pro-inflammatory cytokines in depression (Ogłodek, Szota, Just, Szromek, & Araszkiewicz, 2016). Much to many psychiatrists' disdain, personality disorders have been classed as distinct from the Axis I disorders of classical DSM classification. The data of Ogłodek and colleagues clearly suggest that personality disorders are intimately associated with biological changes that have long been linked to depression (Ogłodek et al., 2016). Clearly, substantial research is required to clarify the interactions of such factors in driving alterations in immune inflammation and O&NS in mood and wider psychiatric presentations.

Overall, heterogeneity in factors, such as early trauma and personality, across studies may be relevant to the mixed results that have been found regarding O&NS and immune inflammation. Such data are also suggestive of biologically determined, unrecognized subgroups in depression investigations, resulting in data that are lost in meta-analysis designs. However, it is clear that increased pro-inflammatory cytokines and O&NS are a robust finding in depression, both in MDD and BD, as currently defined.

TRYCATs, MELATONERGIC PATHWAYS, a7nAChR, ARYL HYDROCARBON RECEPTOR & MITOCHONDRIA

Before looking at the research implications of the above data, it is important to note that O&NS and pro-inflammatory cytokines, including IL-1β, IL-6, IL-18, and TNF-α but especially IFN-γ, are widely reported to mediate their effects in mood disorders and in a host of other medical conditions, by increasing IDO, thereby driving tryptophan to the production of neuroregulatory TRYCATs. By decreasing the availability of tryptophan for serotonin synthesis, TRYCATs pathway activation decreases the levels of NAS and melatonin. This may be of some importance, given that melatonin is a significant antioxidant, antiinflammatory, and antinociceptive, as well as an optimizer of mitochondria functioning (Anderson & Maes, 2014c) that may be produced within mitochondria (He et al., 2016; Tan et al., 2013). Melatonin may also act in an autocrine manner to switch off inflammatory processes in immune cells (Muxel et al., 2012). KYNA can inhibit the a7nAChR, while melatonin positively regulates the a7nAChR (Markus, Silva, Franco, & Barbosa, 2010). As well as having importance in cognition and immune regulation, a7nAChR activation also decreases gut

permeability (Sommansson, Nylander, & Sjöblom, 2013) and in animal models decreases levels of pro-inflammatory cytokines and lowers depression susceptibility (Zhang et al., 2016). As well as inhibiting the a7nAChR, KYNA, like kyn, can also activate the aryl hydrocarbon receptor (AhR), leading to the induction of IDO and a possible positive feedback loop on TRYCATs production. The AhR is expressed in mitochondria, as is melatonin, the a7nAChR and the AhR, while two neurotoxic TRYCATs, 3-hydroxykynurenine (3-OHK) and 3-hydroxyanthranilic acid (3-ANA), have also been shown to be present in mitochondria and to regulate aspects of mitochondria functioning (Baran et al., 2016). This suggests potential interactions of these pathways in mitochondria, making mitochondria sites for the induction of ROS and in determining the consequences of increased levels of O&NS and pro-inflammatory cytokines, as indicated by the putative interactions of these two pathways and receptors in mitochondria (Anderson & Maes, 2017). As such, the regulation of cytokines and O&NS in depression and depression-associated factors is likely to have significant impacts on an array of fundamental cellular and intercellular processes. These wider processes are shown in Fig. 2.

FUTURE DIRECTIONS

The above raises a number of questions as to the etiology of people presenting with mood disorders, especially the role of increased O&NS and immune-inflammatory processes. Are the abusive events the drivers of later increases in pro-inflammatory cytokines that drive many changes, including in gut permeability and the driving of tryptophan to the production of neuroregulatory kynurenines, at the expense of serotonin, NAS, and melatonin? This would suggest subtypes of mood dysregulation, in both MDD and BD as currently conceived, with distinct biological underpinnings that may represent subtypes or even different disorders.

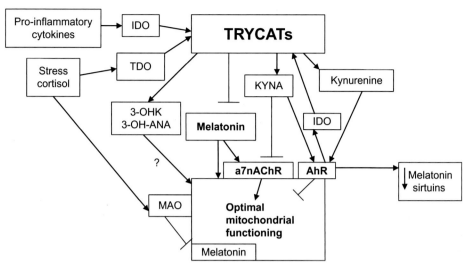

FIG. 2 The interactions of the AhR and a7nAChR with the TRYCATs and melatonergic pathways are shown. Pro-inflammatory cytokines and chronic stress increase IDO and TDO, respectively, leading to TRYCATs induction, with differential effects on mitochondria functioning. TRYCATs activation and stress-induced MAO, by decreasing serotonin availability, decrease melatonergic pathway activation. Increased KYNA may also compete with melatonin via the opposing regulation of the a7nAChR. AhR activation, via IDO induction, may positively feedback on TRYCATs synthesis and decreases the availability of melatonin by increasing its metabolism. Some data indicate that the AhR decreases sirtuins, which may also be mediated by decreased melatonin. Both 3-OHK and 3-OH-ANA can modulate mitochondria functioning. *Abbreviations*: *3-OH-ANA*, 3-hydroxyanthranilic acid; *3-OH*, 3-hydroxykynurenine; *a7nAChR*, alpha 7 nicotinic receptor; *AhR*, aryl hydrocarbon receptor; *IDO*, indoleamine 2,3-dioxygenase; *MAO*, monoamine oxidase; *TDO*, tryptophan 2,3-dioxygenase; *TRYCATs*, tryptophan catabolites.

Given that it is generally accepted that such abusive events are underreported (Della Femina et al., 1990), is the reporting of these events relevant per se, perhaps indicative of greater levels of emotional disarray and fragmentation, such that abusive events cannot be contained within, as data suggest that many people would seem to do (Della Femina et al., 1990)? The most widely accepted causal model would indicate that abusive events in childhood and adolescence are the drivers of such increased levels of pro-inflammatory cytokines in adults presenting with mood disorders, with variations in O&NS and immune-inflammatory processes indicative of wider genetic and epigenetic processes but still containable within a DSM-type formulation. However, all of these studies are cross sectional and require more thorough longitudinal studies

before such conclusions can be drawn. For example, it may be that such dysregulation in pro-inflammatory cytokine production predates early abusive events, perhaps primed by earlier events, including perinatal, leading to a state that increases the susceptibility to personality disorder and interpersonal disarray, increasing the likelihood of reporting abusive events. An example of such putative priming perinatal events is prenatal maternal depression, which is associated with an increase in hs-CRP in the offspring at aged 25 years (Plant, Pawlby, Sharp, Zunszain, & Pariante, 2016). As to whether prenatal depression would then act to prime the reporting of early abusive events or indeed as to whether it is most evident in those with a history of early abusive events has still to be determined.

This could overlap to the large literature on stress resilience, which indicates that resilience, especially the biological processes that improve emotion regulation, is a significant inhibitor of the impact of early abuse on later psychopathology (Kim et al., 2015). As such, some of the effects of early abuse and the susceptibility genes and epigenetic factors linked to mood and personality disorders may be acting on processes linked to stress resilience. Given that early abuse is linked to increased levels of pro-inflammatory factors, this could suggest that stress resilience is modulated by levels of immune-inflammatory processes. This would suggest that an increase in pro-inflammatory processes that are linked to mood disorders may be mediating their effects by acting on processes of stress resilience, which contribute to the reporting of early abusive events.

It is also of note that data in nonhuman primates indicate that early exposure to severe stress usually results in deficient coping abilities, with some forms of milder stressors able to promote subsequent resilience in these animals (Meyer & Hamel, 2014). Plausibly, this could be seen as having parallels in something like cognitive-behavioral therapy, with a focus on coping with different challenges in patient's lives, thereby enhancing stress resilience and the biological processes that underpin this. Such a perspective does not transport the causation of early abusive events forward decades to the later reporting of early abuse but suggests that the trauma of these early events is more linked to the ability and resilience to cope with more later anxieties and often day-to-day stressors.

Such a pattern may be crystallized in what has long been described as "borderline" personality disorder (BPD) presentations, which include high levels of early abuse reporting (Ferrer et al., 2017), which correlates with a fivefold increase in suicide attempts (Kaplan et al., 2016). BPD has most recently been operationalized to include frequent, often superficial, suicide attempts or deliberate self-harm, coupled to emotional, including interpersonal, instability, rejection sensitivity, and abandonment fears. BPD patients can also show significant increases in immune-inflammatory markers, including increased levels of IL-1β reactivity (Westling, Ahrén, Träskman-Bendz, & Brundin, 2011). In contrast to stress resilience, BPD is clearly associated with stress susceptibility.

BPD is highly correlated with levels of depression and is the personality disorder that is most associated with BD, with some studies looking at the difficulties in differentiating BD and BPD (Bayes et al., 2016). Most BPD patients also satisfy criteria for an MDD diagnosis (Friborg et al., 2014). Different SNPs in the hypothalamic-pituitary-adrenal (HPA) axis have been shown to differentially associate with a history of physical and sexual abuse in BPD patients (Martín-Blanco et al., 2016). Alterations in the HPA axis response are evident in BD offspring with a history of early trauma versus the offspring of healthy controls that also experienced early trauma (Schreuder et al., 2016). This could suggest that early trauma interacts with the biological underpinnings of BD to modulate changes in the stress response. A wide array of data indicate that prenatal stress, in clinical and preclinical models, increases depression risk in the offspring (Lin et al., 2017). As such, changes in the HPA axis following early abuse in BD, MDD, or BPD may be linked to a host of downstream consequences, including in the regulation of the gut microbiome and gut permeability (Gur et al., 2017). This also suggests that the dysregulation of the HPA axis in BD and MDD (Belvederi Murri et al., 2016; Keller et al., 2017) may significantly interact with the early abuse events. HPA axis response is associated with variations in stress resilience, in correlations with altered CNS responses, as indicated in an fMRI study (Henckens et al., 2016).

Personality disorders, including BPD, are generally associated with increased levels of pro-inflammatory cytokines, which is suggested to be maintained by anxiety attacks, thereby acting to modulate stress resilience (Ogłodek et al.,

2016). Given the association of personality disorders generally with increased levels of pro-inflammatory processes (Ogłodek et al., 2016), it is of note that personality researchers have argued for decades that these classical personality disorder classifications should be replaced by the widely accepted five dimensions of personality, namely, openness to experience, conscientiousness, extraversion, agreeableness, and neuroticism (McCrae & Costa, 1989). Increased levels of neuroticism are associated with all personality and mood disorders (Baryshnikov et al., 2016) and being highly associated with people presenting with trauma-related conditions, such as early abuse (Hovens, Giltay, van Hemert, & Penninx, 2016) and post-traumatic stress disorder (PTSD) (Ogle, Siegler, Beckham, & Rubin, 2017). Almost by definition, neuroticism is associated with increased levels of emotional reactivity and lower levels of stress resilience (Carter et al., 2016). Neuroticism has been shown to moderate the association of early abuse and later adult onset mood disorders (Hovens et al., 2016), while in MDD and BD patients, there are moderate to high correlations among neuroticism, self-reported early abuse, attachment style, and BPD characteristics (Baryshnikov et al., 2016).

However, it should be noted that neuroticism has classically been conceived as a personality structure, which, following convention in the early decades of the last century, has been viewed as distinct from the biological changes occurring in mood disorders and other classical psychiatric DSM Axis I conditions. Consequently, the biological underpinnings of neuroticism and the genetic and epigenetic factors that may underpin it have been relatively little explored. A difficulty in maintaining emotional stability, the opposite pole of the neuroticism dimension, is fundamental to a host of current psychiatric classifications, including mood disorders, PTSD, and BPD. The biological underpinnings of the various dimensions that underlie neuroticism (viz., anxiety, depression, susceptibility to stress, self-consciousness, hostility, and impulsiveness) more closely align to concepts such as emotional and stress resilience and may form a substrate that allows for a biologically determined set of classifications to replace the outdated and inadequate DSM classification system that is most widely used.

It is important to note that childhood abuse is directly associated not only with mood disorders but only via neuroticism (Jones, Oudenhove, Koloski, Tack, & Talley, 2013). Experimental pain increases IL-1β in correlation with measures of neuroticism, with increased levels of IL-1β correlating with decreased levels of mu-opioid receptors in the amygdala (Prossin et al., 2015). This suggests an association of alterations in peripheral pro-inflammatory cytokines with changes in emotional regulation, via the impacts of the mu-opioid receptor in the amygdala (Anderson et al., 2012). As indicated above, it is not unlikely that other developmental factors, such as maternal prenatal depression or breastfeeding regulation of infant gut permeability (Anderson et al., 2017), will also act to regulate the levels of immune responsivity that contribute to the increased levels of pro-inflammatory processes in mood disorders.

The lack of biologically determined classification for any psychiatric condition in DSM may be argued to highlight the inadequacy of DSM classification in driving meaningful research on the biological underpinnings of people with psychiatric presentations. Using dimensions such as neuroticism, early trauma, and stress resilience more clearly aligns presenting problems, such as mood disorders, with coping management of cognitive-behavioral therapy and wider psychotherapy styles while also providing a frame of reference that encourages what would seem to be more fruitful avenues of research on the pathophysiological underpinnings of presenting problems in a psychiatric setting. The above data on O&NS and immune-inflammatory processes in mood disorders highlight the relevance of early developmental processes that impact on everyday interactions that lead to traditional symptomatic presentations. The criticisms of DSM-5 have been manifold, including from the US National

Institute of Health and British Psychological Society. It is clear that DSM classification in now acting to inhibit progress in the identification of meaningful biological processes that will improve the psychological and pharmaceutical treatment of psychiatric patients.

ACKNOWLEDGMENT

None declared.

CONFLICTS OF INTEREST

The authors declare that this chapter contents have no conflicts of interest.

References

Abe, N., Uchida, S., Otsuki, K., Hobara, T., Yamagata, H., Higuchi, F., et al. (2011). Altered sirtuin deacetylase gene expression in patients with a mood disorder. *Journal of Psychiatric Research, 45*(8), 1106–1112.

Agnew-Blais, J., & Danese, A. (2016). Childhood maltreatment and unfavourable clinical outcomes in bipolar disorder: a systematic review and meta-analysis. *The Lancet Psychiatry, 3*(4), 342–349.

Aguilar-Valles, A., Inoue, W., Rummel, C., & Luheshi, G. N. (2015). Obesity, adipokines and neuroinflammation. *Neuropharmacology, 96*(Pt A), 124–134.

Alcocer-Gómez, E., de Miguel, M., Casas-Barquero, N., Núñez-Vasco, J., Sánchez-Alcazar, J. A., Fernández Rodríguez, A., et al. (2014). NLRP3 inflammasome is activated in mononuclear blood cells from patients with major depressive disorder. *Brain, Behavior, and Immunity, 36*, 111–117.

Anderson, G. (2018). Linking the biological underpinnings of depression: role of mitochondria interactions with melatonin, inflammation, sirtuins, tryptophan catabolites, DNA repair and oxidative and nitrosative stress, with consequences for classification and cognition. *Progress in Neuro-Psychopharmacology & Biological Psychiatry, 80*(Part C), 255–266.

Anderson, G., Jacob, A., Bellivier, F., & Geoffroy, P. A. (2016). Bipolar disorder: the role of the Kynurenine and Melatoninergic pathways. *Current Pharmaceutical Design, 22*(8), 987–1012.

Anderson, G., & Maes, M. (2013). Metabolic syndrome, Alzheimer's, schizophrenia and depression: role for leptin, melatonin, kynurenine pathways and neuropeptides. In T. Farooqui, & A. A. Farooqui (Eds.), *Metabolic*

syndrome and neurological disorders. Chichester: John Wiley & Sons Ltd [chap. 13].

Anderson, G., & Maes, M. (2014a). Neurodegeneration in Parkinson's disease: interactions of oxidative stress, tryptophan catabolites and depression with mitochondria and sirtuins. *Molecular Neurobiology, 49*(2), 771–783.

Anderson, G., & Maes, M. (2014b). Oxidative/nitrosative stress and immuno-inflammatory pathways in depression: treatment implications. *Current Pharmaceutical Design, 20*(23), 3812–3847.

Anderson, G., & Maes, M. (2014c). Local melatonin regulates inflammation resolution: a common factor in neurodegenerative, psychiatric and systemic inflammatory disorders. *CNS & Neurological Disorders Drug Targets, 13*(5), 817–827.

Anderson, G., & Maes, M. (2015). Melatonin: a natural homeostatic regulator—interactions with immune inflammation and trytophan catabolite pathways in the modulation of migraine and endometriosis. *Journal of Natural Products Research Updates, 1*, 7–17.

Anderson, G., & Maes, M. (2016). A role for the regulation of the melatoninergic pathways in Alzheimer's disease and other neurodegenerative and psychiatric conditions. In G. A. Ravishankar, & A. Ramakrishna (Eds.), *Serotonin and melatonin: Their functional role in plants and implications in human health* (pp. 421–444). India: Taylor & Francis [chap 29].

Anderson, G., & Maes, M. (2017). The interactions of tryptophan and its catabolites with melatonin and the alpha 7 nicotinic receptor in CNS and psychiatric disorders: role for the aryl hydrocarbon receptor and direct mitochondria regulation. *International Journal of Tryptophan Research: IJTR, 10.*

Anderson, G., Maes, M., & Berk, M. (2012). Inflammation-related disorders in the tryptophan catabolite pathway in depression and somatization. *Advances in Protein Chemistry and Structural Biology, 88*, 27–48.

Anderson, G., & Rodriguez, M. (2011). Multiple sclerosis, seizures, and antiepileptics: role of IL-18, IDO, and melatonin. *European Journal of Neurology, 18*(5), 680–685.

Anderson, G., Vaillancourt, C., Maes, M., & Reiter, R. R. (2017). Breastfeeding and the gut-brain axis: is there a role for melatonin. *Biomolecular Concepts, 8*(3–4), 185–195.

Andreazza, A. C., Wang, J. F., Salmasi, F., Shao, L., & Young, L. T. (2013). Specific subcellular changes in oxidative stress in prefrontal cortex from patients with bipolar disorder. *Journal of Neurochemistry, 127*(4), 552–561.

Aydemir, O., Çubukçuoğlu, Z., Erdin, S., Taş, C., Onur, E., & Berk, M. (2014). Oxidative stress markers, cognitive functions, and psychosocial functioning in bipolar disorder: an empirical cross-sectional study. *Revista Brasileira de Psiquiatria, 36*(4), 293–297.

Bajpai, A., Verma, A. K., Srivastava, M., & Srivastava, R. (2014). Oxidative stress and major depression. *Journal of Clinical and Diagnostic Research, 8*(12), CC04–7.

Banerjee, U., Dasgupta, A., Rout, J. K., & Singh, O. P. (2012). Effects of lithium therapy on Na+-K+-ATPase activity and lipid peroxidation in bipolar disorder. *Progress in Neuro-Psychopharmacology & Biological Psychiatry, 37*(1), 56–61.

Baran, H., Staniek, K., Bertignol-Spörr, M., Attam, M., Kronsteiner, C., & Kepplinger, B. (2016). Effects of various kynurenine metabolites on respiratory parameters of rat brain, liver and heart mitochondria. *International Journal of Tryptophan Research, 9,* 17–29.

Barbé-Tuana, F. M., Parisi, M. M., Panizzutti, B. S., Fries, G. R., Grun, L. K., Guma, F. T., et al. (2016). Shortened telomere length in bipolar disorder: a comparison of the early and late stages of disease. *Revista Brasileira de Psiquiatria, 38*(4), 281–286.

Barbosa, I. G., Rocha, N. P., Bauer, M. E., de Miranda, A. S., Huguet, R. B., Reis, H. J., et al. (2013). Chemokines in bipolar disorder: trait or state? *European Archives of Psychiatry and Clinical Neuroscience, 263,* 159–165.

Barbosa, I. G., Rocha, N. P., de Miranda, A. S., Magalhães, P. V., Huguet, R. B., de Souza, L. P., et al. (2012). Increased levels of adipokines in bipolar disorder. *Journal of Psychiatric Research, 46,* 389–393.

Baryshnikov, I., Joffe, G., Koivisto, M., Melartin, T., Aaltonen, K., Suominen, K., et al. (2016). Relationships between self-reported childhood traumatic experiences, attachment style, neuroticism and features of borderline personality disorders in patients with mood disorders. *Journal of Affective Disorders, 210,* 82–89.

Bayes, A. J., McClure, G., Fletcher, K., Román Ruiz Del Moral, Y. E., Hadzi-Pavlovic, D., Stevenson, J. L., et al. (2016). Differentiating the bipolar disorders from borderline personality disorder. *Acta Psychiatrica Scandinavica, 133* (3), 187–195.

Behr, G. A., Moreira, J. C., & Frey, B. N. (2012). Preclinical and clinical evidence of antioxidant effects of antidepressant agents: implications for the pathophysiology of major depressive disorder. *Oxidative Medicine and Cellular Longevity, 2012,* 609421.

Belvederi Murri, M., Prestia, D., Mondelli, V., Pariante, C., Patti, S., Olivieri, B., et al. (2016). The HPA axis in bipolar disorder: systematic review and meta-analysis. *Psychoneuroendocrinology, 63,* 327–342.

Black, C. N., Bot, M., Scheffer, P. G., & Penninx, B. W. (2017). Oxidative stress in major depressive and anxiety disorders, and the association with antidepressant use; results from a large adult cohort. *Psychological Medicine, 47*(5), 936–948.

Bortolasci, C. C., Vargas, H. O., Souza-Nogueira, A., Gastaldello Moreira, E., Vargas Nunes, S. O., Berk, M., et al. (2014). Paraoxonase (PON)1 Q192R functional genotypes and PON1 Q192R genotype by smoking interactions

are risk factors for the metabolic syndrome, but not overweight or obesity. *Redox Report, 19*(6), 232–241.

Bot, M., Chan, M. K., Jansen, R., Lamers, F., Vogelzangs, N., Steiner, J., et al. (2015). Serum proteomic profiling of major depressive disorder. *Translational Psychiatry, 5.*

Carter, F., Bell, C., Ali, A., McKenzie, J., Boden, J. M., Wilkinson, T., et al. (2016). Predictors of psychological resilience amongst medical students following major earthquakes. *The New Zealand Medical Journal, 129*(1434), 17–22.

Cerri, A. P., Arosio, B., Viazzoli, C., Confalonieri, R., Teruzzi, F., & Annoni, G. (2009). -308(G/A) TNF-alpha gene polymorphism and risk of depression late in the life. *Archives of Gerontology and Geriatrics, 49*(Suppl. 1), 29–34.

Chang, H. H., Wang, T. Y., Lee, I. H., Lee, S. Y., Chen, K. C., Huang, S. Y., et al. (2017). C-reactive protein: a differential biomarker for major depressive disorder and bipolar II disorder. *The World Journal of Biological Psychiatry, 18*(1), 63–70.

Che, Y., Zhou, Z., Shu, Y., Zhai, C., Zhu, Y., Gong, S., et al. (2015). Chronic unpredictable stress impairs endogenous antioxidant defense in rat brain. *Neuroscience Letters, 584,* 208–213.

Czarny, P., Kwiatkowski, D., Kacperska, D., Kawczyńska, D., Talarowska, M., Orzechowska, A., et al. (2015). Elevated level of DNA damage and impaired repair of oxidative DNA damage in patients with recurrent depressive disorder. *Medical Science Monitor, 21,* 412–418.

Czarny, P., Kwiatkowski, D., Toma, M., Galecki, P., Orzechowska, A., Bobinska, K., et al. (2016). Single-nucleotide polymorphisms of genes involved in repair of oxidative DNA damage and the risk of recurrent depressive disorder. *Medical Science Monitor: International Medical Journal of Experimental and Clinical Research, 22,* 4455–4474.

Dahl, J., Ormstad, H., Aass, H. C., Malt, U. F., Bendz, L. T., Sandvik, L., et al. (2014). The plasma levels of various cytokines are increased during ongoing depression and are reduced to normal levels after recovery. *Psychoneuroendocrinology, 45,* 77–86.

Della Femina, D., Yeager, C. A., & Lewis, D. O. (1990). Child abuse: adolescent records vs. adult recall. *Child Abuse & Neglect, 14*(2), 227–231.

Dickerson, F., Stallings, C., Origoni, A., Vaughan, C., Khushalani, S., & Yolken, R. (2013). Elevated C-reactive protein and cognitive deficits in individuals with bipolar disorder. *Journal of Affective Disorders, 150*(2), 456–459.

do Prado, C. H., Rizzo, L. B., Wieck, A., Lopes, R. P., Teixeira, A. L., Grassi-Oliveira, R., et al. (2013). Reduced regulatory T cells are associated with higher levels of Th1/TH17 cytokines and activated MAPK in type 1 bipolar disorder. *Psychoneuroendocrinology, 38* (5), 667–676.

Donnez, J., Binda, M. M., Donnez, O., & Dolmans, M. M. (2016). Oxidative stress in the pelvic cavity and its role in the pathogenesis of endometriosis. *Fertility and Sterility*, *106*(5), 1011–1017.

Dowlati, Y., Herrmann, N., Swardfager, W., Liu, H., Sham, L., Reim, E. K., et al. (2010). A meta-analysis of cytokines in major depression. *Biological Psychiatry*, *67*(5), 446–457.

Ferrer, M., Andión, Ó., Calvo, N., Ramos-Quiroga, J. A., Prat, M., Corrales, M., et al. (2017). Differences in the association between childhood trauma history and borderline personality disorder or attention deficit/hyperactivity disorder diagnoses in adulthood. *European Archives of Psychiatry and Clinical Neuroscience*, *267*(6), 541–549.

Friborg, O., Martinsen, E. W., Martinussen, M., Kaiser, S., Overgård, K. T., & Rosenvinge, J. H. (2014). Comorbidity of personality disorders in mood disorders: a meta-analytic review of 122 studies from 1988 to 2010. *Journal of Affective Disorders*, *152–154*, 1–11.

Gałecki, P., Talarowska, M., Bobińska, K., & Szemraj, J. (2015). COX-2 gene expression is correlated with cognitive function in recurrent depressive disorder. *Psychiatry Research*, *215*(2), 488–490.

Gardner, A., & Boles, R. G. (2011). Beyond the serotonin hypothesis: mitochondria, inflammation and neurodegeneration in major depression and affective spectrum disorders. *Progress in Neuro-Psychopharmacology & Biological Psychiatry*, *35*(3), 730–743.

Geyik, S., Altunısık, E., Neyal, A. M., & Taysi, S. (2016). Oxidative stress and DNA damage in patients with migraine. *The Journal of Headache and Pain*, *17*, 10.

Gilman, S. E., Cherkerzian, S., Buka, S. L., Hahn, J., Hornig, M., & Goldstein, J. M. (2016). Prenatal immune programming of the sex-dependent risk for major depression. *Translational Psychiatry*, *6*(5).

Goldsmith, D. R., Rapaport, M. H., & Miller, B. J. (2016). A meta-analysis of blood cytokine network alterations in psychiatric patients: comparisons between schizophrenia, bipolar disorder and depression. *Molecular Psychiatry*, *21*(12), 1696–1709.

Grassi-Oliveira, R., Brieztke, E., Teixeira, A., Pezzi, J. C., Zanini, M., Lopes, R. P., et al. (2012). Peripheral chemokine levels in women with recurrent major depression with suicidal ideation. *Revista Brasileira de Psiquiatria*, *34*(1), 71–75.

Grosse, L., Ambrée, O., Jörgens, S., Jawahar, M. C., Singhal, G., Stacey, D., et al. (2016). Cytokine levels in major depression are related to childhood trauma but not to recent stressors. *Psychoneuroendocrinology*, *73*, 24–31.

Gur, T. L., Shay, L., Vadodkar Palkar, A., Fisher, S., Varaljay, V. A., Dowd, S., et al. (2017). Prenatal stress affects placental cytokines and neurotrophins, commensal microbes, and anxiety-like behavior in adult female offspring. *Brain, Behavior, and Immunity*, *64*, 50–58.

Haastrup, E., Bukh, J. D., Bock, C., Vinberg, M., Thørner, L. W., Hansen, T., et al. (2012). Promoter variants in IL18 are associated with onset of depression in patients previously exposed to stressful-life events. *Journal of Affective Disorders*, *136*(1–2), 134–138.

Hatch, J., Andreazza, A., Olowoyeye, O., Rezin, G. T., Moody, A., & Goldstein, B. I. (2015). Cardiovascular and psychiatric characteristics associated with oxidative stress markers among adolescents with bipolar disorder. *Journal of Psychosomatic Research*, *79*(3), 222–227.

Hayatbakhsh, M. R., O'Callaghan, M. J., Bor, W., Williams, G. M., & Najman, J. M. (2012). Association of breastfeeding and adolescents' psychopathology: a large prospective study. *Breastfeeding Medicine*, *7*(6), 480–486.

He, C., Wang, J., Zhang, Z., Yang, M., Li, Y., Tian, X., et al. (2016). Mitochondria synthesize melatonin to ameliorate its function and improve mice Oocyte's quality under in vitro conditions. *International Journal of Molecular Sciences*, *17*(6). pii: E939.

Henckens, M. J., Klumpers, F., Everaerd, D., Kooijman, S. C., van Wingen, G. A., & Fernández, G. (2016). Interindividual differences in stress sensitivity: basal and stress-induced cortisol levels differentially predict neural vigilance processing under stress. *Social Cognitive and Affective Neuroscience*, *11*(4), 663–673.

Hiles, S. A., Révész, D., Lamers, F., Giltay, E., & Penninx, B. W. (2016). Bidirectional prospective associations of metabolic syndrome components with depression, anxiety, and antidepressant use. *Depression and Anxiety*, *33*(8), 754–764.

Hovens, J. G., Giltay, E. J., van Hemert, A. M., & Penninx, B. W. (2016). Childhood maltreatment and the course of depressive and anxiety disorders: the contribution of personality characteristics. *Depression and Anxiety*, *33*(1), 27–34.

Hsuchou, H., Kastin, A. J., Mishra, P. K., & Pan, W. (2012). C-reactive protein increases BBB permeability: implications for obesity and neuroinflammation. *Cellular Physiology and Biochemistry*, *30*(5), 1109–1119.

Iuvone, P. M., Boatright, J. H., Tosini, G., & Ye, K. (2014). N-acetylserotonin: circadian activation of the BDNF receptor and neuroprotection in the retina and brain. *Advances in Experimental Medicine and Biology*, *801*, 765–771.

Jones, M. P., Oudenhove, L. V., Koloski, N., Tack, J., & Talley, N. J. (2013). Early life factors initiate a 'vicious circle' of affective and gastrointestinal symptoms: a longitudinal study. *United European Gastroenterology Journal*, *1*(5), 394–402.

Kapczinski, N. S., Mwangi, B., Cassidy, R. M., Librenza-Garcia, D., Bermudez, M. B., Kauer-Sant'anna, M., et al. (2017). Neuroprogression and illness trajectories in bipolar disorder. *Expert Review of Neurotherapeutics*, *17*(3), 277–285.

Kaplan, C., Tarlow, N., Stewart, J. G., Aguirre, B., Galen, G., & Auerbach, R. P. (2016). Borderline personality disorder in youth: the prospective impact of child abuse on non-suicidal self-injury and suicidality. *Comprehensive Psychiatry*, *71*, 86–94.

Keller, J., Gomez, R., Williams, G., Lembke, A., Lazzeroni, L., Murphy, G. M., Jr., et al. (2017). HPA axis in major depression: cortisol, clinical symptomatology and genetic variation predict cognition. *Molecular Psychiatry*, *22*(4), 527–536.

Kim, J., Seok, J. H., Choi, K., Jon, D. I., Hong, H. J., Hong, N., et al. (2015). The protective role of resilience in attenuating emotional distress and aggression associated with early-life stress in young enlisted military service candidates. *Journal of Korean Medical Science*, *30*(11), 1667–1674.

Kim, K. N., Lim, Y. H., Bae, H. J., Kim, M., Jung, K., & Hong, Y. C. (2016). Long-term fine particulate matter exposure and major depressive disorder in a community-based urban cohort. *Environmental Health Perspectives*, *124*(10), 1547–1553.

Kowalczyk, M., Talarowska, M., Zajączkowska, M., Szemraj, J., & Gałecki, P. (2013). iNOS gene expression correlates with cognitive impairment. *Medical Science and Technology*, *54*, 16–21.

Lin, Y., Xu, J., Huang, J., Jia, Y., Zhang, J., Yan, C., et al. (2017). Effects of prenatal and postnatal maternal emotional stress on toddlers' cognitive and temperamental development. *Journal of Affective Disorders*, *207*, 9–17.

Lu, S., Peng, H., Wang, L., Vasish, S., Zhang, Y., Gao, W., et al. (2013). Elevated specific peripheral cytokines found in major depressive disorder patients with childhood trauma exposure: a cytokine antibody array analysis. *Comprehensive Psychiatry*, *54*(7), 953–961.

Maes, M. (1993). A review on the acute phase response in major depression. *Reviews in the Neurosciences*, *4*(4), 407–416.

Maes, M., & Anderson, G. (2016). Overlapping the tryptophan catabolite (TRYCAT) and melatoninergic pathways in Alzheimer's disease. *Current Pharmaceutical Design*, *22*(8), 1074–1085.

Maes, M., Bosmans, E., Suy, E., Vandervorst, C., DeJonckheere, C., & Raus, J. (1991). Depression-related disturbances in mitogen-induced lymphocyte responses and interleukin-1 beta and soluble interleukin-2 receptor production. *Acta Psychiatrica Scandinavica*, *84*(4), 379–386.

Maes, M., De Vos, N., Pioli, R., Demedts, P., Wauters, A., Neels, H., et al. (2000). Lower serum vitamin E concentrations in major depression. Another marker of lowered antioxidant defenses in that illness. *Journal of Affective Disorders*, *58*(3), 241–246.

Maes, M., Galecki, P., Chang, Y. S., & Berk, M. (2011). A review on the oxidative and nitrosative stress (O&NS) pathways in major depression and their possible contribution to the (neuro)degenerative processes in that illness. *Progress in Neuro-psychopharmacology & Biological Psychiatry*, *35*(3), 676–692.

Maes, M., Kubera, M., Leunis, J. C., & Berk, M. (2012). Increased IgA and IgM responses against gut commensals in chronic depression: further evidence for increased bacterial translocation or leaky gut. *Journal of Affective Disorders*, *141*(1), 55–62.

Maes, M., Kubera, M., Mihaylova, I., Geffard, M., Galecki, P., Leunis, J. C., et al. (2013). Increased autoimmune responses against auto-epitopes modified by oxidative and nitrosative damage in depression: implications for the pathways to chronic depression and neuroprogression. *Journal of Affective Disorders*, *149* (1–3), 23–29.

Mamdani, F., Rollins, B., Morgan, L., Myers, R. M., Barchas, J. D., Schatzberg, A. F., et al. (2015). Variable telomere length across post-mortem human brain regions and specific reduction in the hippocampus of major depressive disorder. *Translational Psychiatry*, *5*.

Mansur, R. B., Santos, C. M., Rizzo, L. B., Cunha, G. R., Asevedo, E., Noto, M. N., et al. (2016). Inter-relation between brain-derived neurotrophic factor and antioxidant enzymes in bipolar disorder. *Bipolar Disorders*, *18* (5), 433–439.

Markus, R. P., Silva, C. L., Franco, D. G., Barbosa, E. M., Jr., & Ferreira, Z. S. (2010). Is modulation of nicotinic acetylcholine receptors by melatonin relevant for therapy with cholinergic drugs? *Pharmacology & Therapeutics*, *126*(3), 251–262.

Martín-Blanco, A., Ferrer, M., Soler, J., Arranz, M. J., Vega, D., Calvo, N., et al. (2016). The role of hypothalamus-pituitary-adrenal genes and childhood trauma in borderline personality disorder. *European Archives of Psychiatry and Clinical Neuroscience*, *266*(4), 307–316.

Martin-Subero, M., Anderson, G., Kanchanatawan, B., Berk, M., & Maes, M. (2016). Comorbidity between depression and inflammatory bowel disease explained by immune-inflammatory, oxidative, and nitrosative stress; tryptophan catabolite; and gut-brain pathways. *CNS Spectrums*, *21*(2), 184–198.

McCrae, R. R., & Costa, P. T., Jr. (1989). Reinterpreting the Myers-Briggs Type Indicator from the perspective of the five-factor model of personality. *Journal of Personality*, *57*(1), 17–40.

Meyer, J. S., & Hamel, A. F. (2014). Models of stress in nonhuman primates and their relevance for human psychopathology and endocrine dysfunction. *ILAR Journal*, *55*(2), 347–360.

Milaneschi, Y., Lamers, F., Bot, M., Drent, M. L., & Penninx, B. W. (2017). Leptin dysregulation is specifically associated with major depression with atypical features: evidence for a mechanism connecting obesity and depression. *Biological Psychiatry*, *81*(9), 807–814.

Modabbernia, A., Taslimi, S., Brietzke, E., & Ashrafi, M. (2013). Cytokine alterations in bipolar disorder: a meta-analysis of 30 studies. *Biological Psychiatry, 74*(1), 15–25.

Moraes, J. B., Maes, M., Barbosa, D. S., Ferrari, T. Z., Uehara, M. K. S., Carvalho, A. F., et al. (2017). Elevated C-reactive protein levels in women with bipolar disorder may be explained by a history of childhood trauma, especially sexual abuse, body mass index and age. *CNS & Neurological Disorders Drug Targets, 16*(4), 514–521.

Moraes, J. B., Maes, M., Roomruangwong, C., Bonifacio, K. L., Barbosa, D. S., Vargas, H. O., et al. (2018). In major affective disorders, early life trauma predict increased nitrooxidative stress, lipid peroxidation and protein oxidation and recurrence of major affective disorders, suicidal behaviors and a lowered quality of life. *Metabolic Brain Disease*. https://doi.org/10.1007/s11011-018-0209-3.

Moylan, S., Berk, M., Dean, O. M., Samuni, Y., Williams, L. J., O'Neil, A., et al. (2014). Oxidative & nitrosative stress in depression: why so much stress? *Neuroscience and Biobehavioral Reviews, 45*, 46–62.

Moylan, S., Maes, M., Wray, N. R., & Berk, M. (2013). The neuroprogressive nature of major depressive disorder: pathways to disease evolution and resistance, and therapeutic implications. *Molecular Psychiatry, 18*(5), 595–606.

Munkholm, K., Bräuner, J. V., Kessing, L. V., & Vinberg, M. (2013). Cytokines in bipolar disorder vs. healthy control subjects: a systematic review and meta-analysis. *Journal of Psychiatric Research, 47*(9), 1119–1133.

Mutlu, E. A., Engen, P. A., Soberanes, S., Urich, D., Forsyth, C. B., Nigdelioglu, R., et al. (2011). Particulate matter air pollution causes oxidant-mediated increase in gut permeability in mice. *Particle and Fibre Toxicology, 8*, 19.

Muxel, S. M., Pires-Lapa, M. A., Monteiro, A. W., Cecon, E., Tamura, E. K., Floeter-Winter, L. M., et al. (2012). NF-kB drives the synthesis of melatonin in RAW 264.7 macrophages by inducing the transcription of the arylalkylamine-N-acetyltransferase (AA-NAT) gene. *PLoS ONE, 7*(12).

Myung, W., Lim, S. W., Woo, H. I., Park, J. H., Shim, S., Lee, S. Y., et al. (2016). Serum cytokine levels in major depressive disorder and its role in antidepressant response. *Psychiatry Investigation, 13*(6), 644–651.

Ogle, C. M., Siegler, I. C., Beckham, J. C., & Rubin, D. C. (2017). Neuroticism increases PTSD symptom severity by amplifying the emotionality, rehearsal, and centrality of trauma memories. *Journal of Personality, 85*(5), 702–715. https://doi.org/10.1111/jopy.12278.

Ogłodek, E. A., Szota, A. M., Just, M. J., Szromek, A. R., & Araszkiewicz, A. (2016). A study of chemokines, chemokine receptors and interleukin-6 in patients with panic disorder, personality disorders and their co-morbidity. *Pharmacological Reports, 68*(4), 756–763.

Oliveira, J., Debnath, M., Etain, B., Bennabi, M., Hamdani, N., Lajnef, M., et al. (2015). Violent suicidal behaviour in bipolar disorder is associated with nitric oxide synthase 3 gene polymorphism. *Acta Psychiatrica Scandinavica, 132*(3), 218–225.

Øvrevik, J., Refsnes, M., Låg, M., Brinchmann, B. C., Schwarze, P. E., & Holme, J. A. (2017). Triggering mechanisms and inflammatory effects of combustion exhaust particles with implication for carcinogenesis. *Basic & Clinical Pharmacology & Toxicology, 121*(Suppl. 3), 55–62.

Peacock, B. N., Scheiderer, D. J., & Kellermann, G. H. (2016). Biomolecular aspects of depression: a retrospective analysis. *Comprehensive Psychiatry, 73*, 168–180.

Plant, D. T., Pawlby, S., Sharp, D., Zunszain, P. A., & Pariante, C. M. (2016). Prenatal maternal depression is associated with offspring inflammation at 25 years: a prospective longitudinal cohort study. *Translational Psychiatry, 6*(11).

Prossin, A. R., Zalcman, S. S., Heitzeg, M. M., Koch, A. E., Campbell, P. L., Phan, K. L., et al. (2015). Dynamic interactions between plasma IL-1 family cytokines and central endogenous opioid neurotransmitter function in humans. *Neuropsychopharmacology, 40*(3), 554–565.

Rahman, K., Desai, C., Iyer, S. S., Thorn, N. E., Kumar, P., Liu, Y., et al. (2016). Loss of junctional adhesion molecule a promotes severe steatohepatitis in mice on a diet high in saturated fat, fructose, and cholesterol. *Gastroenterology, 151*(4), 733–746.e12.

Riederer, P., & Müller, T. (2017). Use of monoamine oxidase inhibitors in chronic neurodegeneration. *Expert Opinion on Drug Metabolism & Toxicology, 13*(2), 233–240.

Roomruangwong, C., Kanchanatawan, B., Sirivichayakul, S., Anderson, G., Carvalho, A. F., Duleu, S., et al. (2017). IgA/IgM responses to tryptophan and tryptophan catabolites (TRYCATs) are differently associated with prenatal depression, physio-somatic symptoms at the end of term and premenstrual syndrome. *Molecular Neurobiology, 54*(4), 3038–3049.

Rotter, A., Biermann, T., Stark, C., Decker, A., Demling, J., Zimmermann, R., et al. (2013). Changes of cytokine profiles during electroconvulsive therapy in patients with major depression. *The Journal of ECT, 29*(3), 162–169.

Schreuder, M. M., Vinkers, C. H., Mesman, E., Claes, S., Nolen, W. A., & Hillegers, M. H. (2016). Childhood trauma and HPA axis functionality in offspring of bipolar parents. *Psychoneuroendocrinology, 74*, 316–323.

Selek, S., Altindag, A., Saracoglu, G., & Aksoy, N. (2015). Oxidative markers of myeloperoxidase and catalase and their diagnostic performance in bipolar disorder. *Journal of Affective Disorders, 181*, 92–95.

Simanek, A. M., & Meier, H. C. (2015). Association between prenatal exposure to maternal infection and offspring mood disorders: a review of the literature. *Current Problems in Pediatric and Adolescent Health Care*, *45*(11), 325–364.

Slyepchenko, A., Köhler, C., Luciano de Quevedo, J., Maes, M., Alves, G., Fernandes, B., et al. (2016). T helper 17 cells may drive neuroprogression in major depressive disorder: proposal of an integrative model. *Neuroscience & Biobehavioral Reviews*, *64*, 83–100.

Soeiro-de-Souza, M. G., Gold, P. W., Brunoni, A. R., de Sousa, R. T., Zanetti, M. V., Carvalho, A. F., et al. (2014). Lithium decreases plasma adiponectin levels in bipolar depression. *Neuroscience Letters*, *564*, 111–114.

Sommansson, A., Nylander, O., & Sjöblom, M. (2013). Melatonin decreases duodenal epithelial paracellular permeability via a nicotinic receptor-dependent pathway in rats in vivo. *Journal of Pineal Research*, *54*(3), 282–291.

Steptoe, A., Kunz-Ebrecht, S. R., & Owen, N. (2003). Lack of association between depressive symptoms and markers of immune and vascular inflammation in middle-aged men and women. *Psychological Medicine*, *33*(4), 667–674.

Szebeni, A., Szebeni, K., DiPeri, T., Chandley, M. J., Crawford, J. D., Stockmeier, C. A., et al. (2014). Shortened telomere length in white matter oligodendrocytes in major depression: potential role of oxidative stress. *The International Journal of Neuropsychopharmacology*, *17*(10), 1579–1589.

Talarowska, M., Gałecki, P., Maes, M., Bobińska, K., & Kowalczyk, E. (2012). Total antioxidant status correlates with cognitive impairment in patients with recurrent depressive disorder. *Neurochemical Research*, *37* (8), 1761–1767.

Talarowska, M., Szemraj, J., Berk, M., Maes, M., & Gałecki, P. (2015). Oxidant/antioxidant imbalance is an inherent feature of depression. *BMC Psychiatry*, *15*, 71.

Talarowska, M., Szemraj, J., & Gałecki, P. (2014). Myeloperoxidase gene expression and cognitive functions in depression. *Advances in Medical Sciences*, *60*(1), 1–5.

Tan, D. X., Manchester, L. C., Liu, X., Rosales-Corral, S. A., Acuna-Castroviejo, D., & Reiter, R. J. (2013). Mitochondria and chloroplasts as the original sites of melatonin synthesis: a hypothesis related to melatonin's primary function and evolution in eukaryotes. *Journal of Pineal Research*, *54*(2), 127–138.

Tunçel, Ö. K., Sarısoy, G., Bilgici, B., Pazvantoglu, O., Çetin, E., Ünverdi, E., et al. (2015). Oxidative stress in bipolar and schizophrenia patients. *Psychiatry Research*, *228*(3), 688–694.

Valkanova, V., Ebmeier, K. P., & Allan, C. L. (2013). CRP, IL-6 and depression: a systematic review and meta-analysis of longitudinal studies. *Journal of Affective Disorders*, *150*(3), 736–744.

Vanuytsel, T., van Wanrooy, S., Vanheel, H., Vanormelingen, C., Verschueren, S., Houben, E., et al. (2014). Psychological stress and corticotropin-releasing hormone increase intestinal permeability in humans by a mast cell-dependent mechanism. *Gut*, *63*(8), 1293–1299.

Versace, A., Andreazza, A. C., Young, L. T., Fournier, J. C., Almeida, J. R., Stiffler, R. S., et al. (2014). Elevated serum measures of lipid peroxidation and abnormal prefrontal white matter in euthymic bipolar adults: toward peripheral biomarkers of bipolar disorder. *Molecular Psychiatry*, *19*(2), 200–208.

Westling, S., Ahrén, B., Träskman-Bendz, L., & Brundin, L. (2011). Increased IL-1β reactivity upon a glucose challenge in patients with deliberate self-harm. *Acta Psychiatrica Scandinavica*, *124*(4), 301–306.

Yu, L., Zhai, Q., Tian, F., Liu, X., Wang, G., Zhao, J., et al. (2016). Potential of lactobacillus plantarum CCFM639 in protecting against aluminum toxicity mediated by intestinal barrier function and oxidative stress. *Nutrients*, *8*(12). pii: E783.

Zhang, J. C., Yao, W., Ren, Q., Yang, C., Dong, C., Ma, M., et al. (2016). Depression-like phenotype by deletion of α7 nicotinic acetylcholine receptor: role of BDNF-TrkB in nucleus accumbens. *Scientific Reports*, *6*.

Is There Still Hope for Treating Depression With Antiinflammatories?

Bernhard T. Baune

University of Adelaide, Adelaide, SA, Australia

SYMPTOMS AND MECHANISMS OF DEPRESSION-ASSOCIATED INFLAMMATION

While the pathophysiology of depression still requires understanding of key mechanisms that are translatable into treatment approaches, the role of inflammation and immune activation more generally in depressive symptoms has long been discussed and recently gained momentum as a potential treatment target. Growing evidence suggests that pro-inflammatory cytokines play a major role in the pathophysiology of depression. A number of studies, both experimental and meta-analytic analyses, suggest an increased expression of pro-inflammatory cytokines, such as TNF-α, IL-1β, and IL-6, in patients with major depressive disorder (MDD) (Dowlati et al., 2010; Hannestad, DellaGioia, & Bloch, 2011; Howren, Lamkin, & Suls, 2009; Maes et al., 1997). A seminal meta-analysis (Dowlati et al., 2010) of 24 studies found significantly higher concentrations of the pro-inflammatory cytokines, tumor necrosis factor (TNF)-α and interleukin (IL)-6, in depressed subjects than control subjects. An updated meta-analysis (Haapakoski, Mathieu,

Ebmeier, Alenius, & Kivimaki, 2015) of IL-6, C-reactive protein (CRP), and TNF-α found higher levels of IL-6 and CRP in depressed patients versus controls (29 studies for IL-6 and 20 for CRP). These studies strengthen the clinical evidence that symptoms of depression can be accompanied by an activation of the inflammatory response system (Dowlati et al., 2010).

The biological pathways by which cytokines may mediate symptoms of depression are poorly understood, though several mechanisms have been proposed. An important mechanism is that pro-inflammatory cytokines affect serotonin (5-HT) metabolism by reducing tryptophan (TRP) levels. Cytokines appear to activate indoleamine-2,3-dioxygenase (IDO), an enzyme that metabolizes TRP, thereby reducing serotonin levels. Furthermore, inflammatory cytokines, such as IL-1β, may reduce extracellular 5-HT levels, via the activation of the serotonin transporter mechanisms. Moreover, pro-inflammatory cytokines have a potent direct effect on the hypothalamic-pituitary-adrenal (HPA) axis. Cytokines, including IL-1, IL-6, TNF-α, and IFN-α, have been shown to increase inflammatory responses by disrupting the function of glucocorticoid receptors (GRs). Importantly, a link between peripheral cytokines and neurobiological effects in the brain has been established. Peripheral cytokines may communicate with the cerebrum through various pathways. These include communication with the CNS through the afferent sensory fibers of the vagus nerve. Other pathways of communication can occur through passive diffusion at areas where the blood-brain barrier is deficient or by active transport of cytokines stimulated by the central noradrenergic system. Peripheral cytokines may also activate neural afferents, leading to synthesis of IL-6 within the brain by microglia and endothelial cells (O'Brien, Scott, & Dinan, 2004; Pollmacher, Haack, Schuld, Reichenberg, & Yirmiya, 2002; Szelenyi & Vizi, 2007). In the CNS, cytokines may exert their effects by activating the hypothalamic-pituitary-adrenal (HPA) axis. Pro-inflammatory cytokines induce gene expression and synthesis of corticotropin-releasing factor (CRF), which stimulates adrenocorticotropic hormone (ACTH) release and causes glucocorticoid secretion (Song, 2002). An activated HPA axis may lead to a further rise in pro-inflammatory cytokines, through a complex positive feedback loop. Stress can lead to increased cytokine levels and an induction of catecholamines via an activation of the HPA axis, which may further increase pro-inflammatory cytokines (Szelenyi & Vizi, 2007). Cytokines may also directly affect higher cognitive and emotional functions (McAfoose & Baune, 2009), possibly leading to depression and associated cognitive dys-function.

Inflammation and cytokines have been shown to exert a variety of neurobiological effects relevant to depression. Specifically, cytokines such as TNF-α, IL-1β, IL-6, and interferon (IFN)-γ impact key neurobiological processes such as neuroplasticity, neurotransmission, oxidative stress, and neuroendocrinological functions that are considered to be central to the development of depression (Dantzer, O'Connor, Freund, Johnson, & Kelley, 2008; Eyre & Baune, 2012; Haroon, Raison, & Miller, 2012; McAfoose & Baune, 2009; Miller, Maletic, & Raison, 2009). Pro-inflammatory cytokines impair hippocampal (HC) neuroplasticity (e.g., neurogenesis, synaptic plasticity, and long-term potentiation (LTP)), induce glucocorticoid insensitivity of the HPA axis, increase oxidative stress in the HC, reduce serotonin (5-HT) levels, and create neurotoxic serotoninergic metabolites (i.e., 3-hydroxykynurenine (3-HK) and quinolinic acid (QA)) (Dantzer et al., 2008; Eyre & Baune, 2012; Leonard & Maes, 2012; Miller et al., 2009; Moylan, Maes, Wray, & Berk, 2013). Astrocytes and microglia, also key components of the innate immune system, also play a role in the pathophysiology of depression (Eyre & Baune, 2012). Microglia that are a major producer of pro-inflammatory cytokines and are important regulators of HC neuroplasticity, oxidative stress, and QA have increased activity in depression (Setiawan et al., 2015). Microglia can cause detrimental processes when activated, while producing beneficial processes when quiescent. A recent study (Setiawan et al., 2015) used positron-emission tomography to better understand neuroinflammation in depression. The binding activity of the TPSO in microglia is indicative of microglial activity and related inflammatory processes in the brain. Investigators found increased binding in all brain regions, especially the prefrontal cortex, anterior cingulate cortex, and insula in depressed subjects. Moreover, astrocytes that have shown a prominent role in releasing neurotrophic and antioxidant factors have also been found to be activated in depression (Eyre, Stuart, & Baune, 2014). A range of clinical and preclinical studies indicate astrocytic abnormality in depression, leading to detrimental processes in the brain (Peng, Verkhratsky, Gu, & Li, 2015).

PHARMACOLOGICAL APPROACHES TO DEPRESSION-ASSOCIATED INFLAMMATION

Nonsteroidal Antiinflammatory Treatments in Depression

Given the above summarized clinical and mechanistic relationship between inflammation and depression, both selective cyclooxygenase (COX)-2 and nonselective COX inhibitor nonsteroidal antiinflammatory drugs (NSAIDs) have been investigated as possible adjuncts in the treatment of depression with antidepressants (Akhondzadeh et al., 2009; Almeida et al., 2012; Almeida, Alfonso, Jamrozik, Hankey, & Flicker, 2010; Fields, Drye, Vaidya, & Lyketsos, 2012; Fond et al., 2014; Gallagher et al., 2012; Muller, 2013; Muller et al., 2006; Musil et al., 2011; Nery et al., 2008; Pasco et al., 2010; Shelton, 2012; Uher et al., 2012; Warner-Schmidt, Vanover, Chen, Marshall, & Greengard, 2011). The results are mixed based on various study designs ranging from retrospective cohort studies (Almeida et al., 2010, 2012; Fields et al., 2012; Gallagher et al., 2012; Pasco et al., 2010) and randomized controlled trials (RCTs) (Fields et al., 2012; Muller et al., 2006; Musil et al., 2011; Nery et al., 2008) to nested case-control studies (Pasco et al., 2010). While some studies have found positive antidepressant effects (Akhondzadeh et al., 2009; Muller et al., 2006; Nery et al., 2008; Pasco et al., 2010), others have found no effect (Almeida et al., 2010; Fields et al., 2012; Uher et al., 2012), and yet, others have found detrimental effects suggesting NSAIDs may reduce the antidepressant effect of selective serotonin reuptake inhibitors (SSRIs) (Gallagher et al., 2012; Warner-Schmidt et al., 2011). These mixed results may be due to various reasons such as differing antidepressant classes and doses, the use of varying selective COX-2 and/or nonselective COX inhibitor NSAIDs, study design, age of study population, as well as the consideration of populations with varying degrees of depressive symptomatology and presence of comorbid medication conditions. It is important to differentiate the types of NSAIDs by key characteristics. COX-1 is a major player in mediating pro-inflammatory microglial activation (Maes, 2012); hence, blocking this enzyme should be beneficial. Moreover, another evidence indicates that depression is accompanied by increased COX-2 activity (Galecki et al., 2012; Galecki, Florkowski, Bienkiewicz, & Szemraj, 2010), hence postulating a role of COX-2 in depression pathophysiology. In support of this, Galecki et al. (2012) found that the mRNA expression of genes encoding for COX-2 was significantly increased in the peripheral blood cells of recurrent depressive disorder patients ($n = 181$) versus controls ($n = 149$). Given the observed differing efficacy of selective COX-2 versus nonselective COX inhibitor NSAIDs, this raises important questions about their pharmacological actions. Our understanding of the physiological and pathophysiological effects of COX-1 and 2 is still limited. For example, COX-1 is predominantly pro-inflammatory, mediating pro-inflammatory microglial activation (Maes, 2012), and is considered detrimental to the brain (Aid & Bosetti, 2011), whereas COX-2 can exert both beneficial and detrimental effects on the brain (see for review Berk et al., 2013). A recent meta-analysis of RCTs suggests that celecoxib, a selective COX-2 inhibitor NSAID, has a therapeutic effect when used adjunctively with antidepressants (Na, Lee, Lee, Cho, & Jung, 2013). However, mechanistic evidence suggests that selective COX-2 inhibitor NSAIDs may actually increase inflammation, Th1 immune responses, and glial cell activation (Aid & Bosetti, 2011; Aid, Langenbach, & Bosetti, 2008; Maes, 2012). Various mechanistic explanations for the inconsistent effects of NSAIDs in the clinical treatment of depression need to be considered. Some NSAIDs seem to have more efficacy than others based on COX enzyme effects; however, this is still subject to debate (Aid & Bosetti, 2011; Aid et al., 2008;

Maes, 2012). Evidence from rodent studies indicates that by-products of COX-2 metabolism (e.g., prostaglandin D2 and docosahexaenoic acid) exert antiinflammatory effects (Aid & Bosetti, 2011). Given this, aspirin is thought to be a more desirable compound as it preferentially targets COX-1 over COX-2 (Berk et al., 2013). COX-1 may play a more significant role in inflammation relevant to depression-like states; however, this is poorly understood (Aid et al., 2008; Aid & Bosetti, 2011; Berk et al., 2013). Some evidence indicates that NSAIDs may have a minimal and potentially detrimental effect on depressive symptoms by attenuating the effects of SSRIs (Warner-Schmidt et al., 2011). A rodent study by Warner-Schmidt et al. (2011) shows that SSRIs increase TNF-α, IFN-γ, and p11 levels in the frontal cortex. The NSAID, ibuprofen, reduced these levels and attenuated the antidepressant-like actions of SSRIs, but not of tricyclic or monoamine oxidase inhibitors. In a rodent study, it was attempted to differentiate the effects of COX-1 and COX-2 enzymes on systemic immune cell migration (Choi, Aid, Choi, & Bosetti, 2010). This study found that the inhibition of COX-1 activity reduces leukocyte recruitment into the inflamed CNS (via intracerebroventricular lipopolysaccharide), whereas selective COX-2 inhibition increases this recruitment (Choi et al., 2010), suggesting differential chemotactic effects of these enzymes on the brain. These partly contradictory mechanistic findings may assist with the explanation of the inconsistent findings in clinical studies found in human trials on the therapeutic effects of NSAIDs in depression. These inconsistent results by the way also suggest that the design of clinical trials requires the consideration of different pharmacological properties when using NSAIDs. In addition, effects of NSAIDs on depressive symptoms may depend upon the presence of an inflammatory state that has been shown repeatedly in patients with comorbidities such as osteoarthritis or psoriasis and depressive symptoms (Na et al., 2013).

Unfortunately, clinical trials of NSAIDs in depression have not stratified patients according to the presence of inflammation yet.

Aspirin in the Treatment of Depression-Associated Inflammation

Aspirin is a NSAID and an irreversible inhibitor of both COX-1 and COX-2. It is more potent in its inhibition of COX-1 than COX-2, and targeting COX-2 alone may be a less viable therapeutic approach in neuropsychiatric disorders such as depression. While some clinical evidence indicates beneficial effects for aspirin in mood disorders through a shortened onset of action of antidepressants (Mendlewicz et al., 2006), negative results come from epidemiological analyses of 5556 older men, which showed no association between current aspirin use and depression (Almeida et al., 2010). In an intervention study of 70 patients with depression, administration of aspirin together with fluoxetine conferred a greater reduction of oxidative stress markers compared with fluoxetine monotherapy (Galecki, Szemraj, Bienkiewicz, Zboralski, & Galecka, 2009). More generally, it is currently investigated if low-dose aspirin has the potential to extend a healthy process of aging and disability-free life among the elderly in the *ASPirin in Reducing Events in the Elderly (ASPREE)* study (ASPREE Investigator Group, 2013). As ASPREE will examine whether the potential primary prevention benefits of low-dose aspirin outweigh the risks in older healthy individuals, its secondary end points include all-cause and cause-specific mortality, fatal and nonfatal cardiovascular events, fatal and nonfatal cancer (excluding nonmelanoma skin cancer), dementia, mild cognitive impairment, depression, physical disability, and clinically significant bleeding. This study will be providing data and insight into the preventive effects of low-dose aspirin on depression development with and without comorbidities characterized by inflammation (e.g., osteoarthritis, cardiovascular diseases, and cancer). At this stage of the literature, it is too early to conclude any replicated clinical effects of aspirin in depression whether with or without concomitant inflammatory state; hence, the current state of knowledge for aspirin in depression is mainly experimental and hypothetical.

Antiinflammatory Interventions Using Targeted Antagonists

A range of biological antagonists is available such as infliximab, a chimeric bivalent IgG1 monoclonal antibody composed of a human constant region and murine variable regions; adalimumab, a humanized bivalent mouse IgG1 monoclonal antibody; and etanercept, a fusion protein composed of human IgG fused to a dimer of the extracellular regions of TNF-α-R-2. Additionally, anti-TNF therapy is being considered as an option in improving postoperative cognitive dysfunction (Terrando et al., 2010). Research into the effects of centrally and peripherally administered TNF-α blockade has shown clinical efficacy on cognition and depressive symptoms and has improved our understanding of cytokine actions in the CNS (Couch et al., 2013; Miller & Cole, 2012; Tobinick & Gross, 2008). In addition, a recently published trial in an animal model under peripheral LPS stimulation showed that centrally administered etanercept reduced anxiety-like behaviors, but not spatial memory, and was associated with a decrease in hippocampal microglia numbers being suggestive that etanercept recovers anxiety-like behavior possibly mediated by a reduction of TNF-α-related central inflammation (Camara et al., 2015).

However, there are two problems with this approach. First, the BBB prevents a deep penetration of the monoclonal antibody into the CNS tissue, and second, while basal levels of TNF-α are still required for normal functioning,

animal models have shown that the complete lack of TNF-α due to genetic modification results in cognitive impairment (Baune et al., 2008). This is possibly due to the influence that TNF-α exerts on NGF and BDNF. An imbalance in TNF-α causes subsequent deregulation in these neurotrophins ultimately leading to morphological changes in the hippocampus such as decreased arborization of pyramidal neurons (Golan, Levav, Mendelsohn, & Huleihel, 2004). Additionally, the role TNF-α plays in long-term potentiation (LTP) and depression (LTD) formation by upregulation of AMPA receptors (Beattie et al., 2002) and endocytosis of GABA receptors (Stellwagen, Beattie, Seo, & Malenka, 2005) is crucial in the development of synaptic neuroplasticity-related memory and learning (McAfoose & Baune, 2009). Therefore, though complete blockage may prove to be initially beneficial, however, long-term negative effects may manifest and make further research into these effects necessary. The complex signaling pathways of TNF-α and its receptors and the duality of its function in being both neuroprotective and neurodegenerative make for a compelling argument against the validity and long-term benefits of anti-TNF-α therapies. TNF-α may both exacerbate and attenuate cognitive dysfunction depending on the physiological context (Longhi et al., 2008; Peschon et al., 1998). Clearer perspectives into TNF-α signaling and pharmacological interventions that can target specific apoptotic factors in the TNF-α receptor-associated pathways, rather than complete blockage of TNF-α or of its receptors, would make for more effective therapeutics in the treatment of neurological disorders in which TNF-α is an active participant.

Investigations of alternative pathways to inhibit inflammation show promising results. In a study examining the effect of ubiquitin-specific processing protease 8 (USP8) in luteolin-treated microglia, it was found that luteolin inhibits microglial inflammation by enhancing USP8 protein production and USP8 might represent a novel mechanism for the treatment of inflammation and neurodegeneration (Zhu et al., 2014). In addition, it has been shown that blocking the kinase activity of RIP1, a key druggable target in the necroptosis pathway, by necrostatins inhibits the activation of necroptosis and allows cell survival and proliferation in the presence of death receptor ligands (Zhou & Yuan, 2014). Hence, targeting RIP1 kinase may provide therapeutic benefits for the treatment of human diseases characterized by necrosis and inflammation. Although still experimental and hypothetical, a robust RIP3-dependent necroptosis signaling pathway in TLR-activated microglia upon caspase blockade has been reported that suggests that TLR signaling and programmed cell death pathways are closely linked in microglia, which could contribute to neuropathology and inflammation (Kim & Li, 2013). A role of RIP3-dependent necroptosis has also been suggested for TNF-alpha-induced toxicity of hippocampal neurons. Specifically, it has been shown that TNF-alpha promotes CYLD-RIP1-RIP3-MLKL-mediated necroptosis of hippocampal neurons largely bypassing ROS accumulation and calcium influx (Liu et al., 2014). Taken together, these RIP1/RIP3 dependent pathways in TLR-activated microglia and TNF-alpha-induced toxicity may present future targets for intervention provided future research extends these experimental findings.

Caution Warranted Using Antiinflammatory Agents in Depression

It is important to note that the short- and particularly the long-term use of NSAIDs for either acute or preventive treatment of depression-associated inflammation carry a variety of caveats that need careful consideration. This is not to undermine the potential clinical utility of antiinflammatory agents in depression, but to be cautious that the apparent simplicity of the idea that antiinflammatory agents improve depressive symptoms in depression-associated inflammation does not lead to an uncritical practice. These factors include pharmacological properties of antiinflammatory agents and dynamic clinical and immunologic changes that can occur during various phases of clinical depression.

Complex Pharmacology of Antiinflammatory Agents During a Dynamic Course of Depression

Key issues based on the pharmacology of antiinflammatory agents and on the dynamic course of immune changes in various phases of depression are important to consider. These include the (1) immunophysiology of COX enzymes, (2) nuance of cytokine signaling and (3) reported phase-specific involvement of the immune system in depression. Firstly, COX enzyme immunophysiology is incompletely understood (Eyre, Air, Proctor, Rositano, & Baune, 2015). Evidence from rodent studies suggests that by-products of COX-2 metabolism (e.g., prostaglandin D2 and docosahexaenoic acid) exert antiinflammatory effects; therefore, it can be argued that selective COX-2 inhibitor NSAIDs may increase neuroinflammation, Th1 immune responses, and glial cell activation (Maes, 2012). COX-1 is predominantly pro-inflammatory (Maes, 2012). Taken together, nonselective COX inhibitor NSAIDs would theoretically be more effective than COX-2 inhibitors (Eyre et al., 2015); however, there are no quality trial exploring nonselective COX inhibitors.

Secondly, blanket blockade of pro-inflammatory cytokine signaling may not be advisable (Eyre & Baune, 2012). For example, TNF-α receptors have mostly opposing functions, that is, receptor 1 primarily mediates pro-inflammatory/neurodegenerative activities, whereas receptor 2 is primarily involved in neuroprotective processes (Eyre & Baune, 2012). The Janus-faced signaling pathways of TNF-α make an argument against anti-TNF-α therapies, and indeed, TNF-α inhibitors have small-to-negligible antidepressant effects as per the Kohler et al. publication (Kohler et al., 2014). A deeper understanding of cytokine biology and pathway-specific therapeutic manipulation is needed.

Thirdly, the phase of depression during which antiinflammatory drugs are applied prompts further thoughts. While the onset of an episode and certain symptoms of depression appear well explained by this inflammatory model, the underpinnings of the episodic and progressive nature, as well as relapse and remission status, in depression require attention (Eyre et al., 2014). There is emerging clinical and basic science evidence to suggest a phase-specific profile of immune-mediated dysfunction (Eyre et al., 2014). Antiinflammatory pharmacotherapy may need to be used according to the phase of illness and may need to be tailored based on the immune profile of the patient based on the stage of depression (subsyndromal, acute, and remitted) (see Fig. 1) (Eyre et al., 2014). As part of such a phase-specific model of inflammatory changes in depression, it is still to be established whether antiinflammatory agents only work if peripheral inflammation is established in patients. Most of the studies in the field have not measured the state of inflammation before the commencement of the antiinflammatory trial with the aim to stratify patients for the presence and absence or degree of inflammation as a decision-making biological tool for assignment of antiinflammatory treatment in depression. Further research is required to address this important gap in clinical research in depression-associated inflammation.

Potential Side-Effects of Antiinflammatory Agents

In addition to the above reasons for caution of unreflected and premature clinical use of antiinflammatory drugs, possible side effects of antiinflammatory drugs need further consideration. The current state of the literature suggests that the occurrence of potential side effects on the use of NSAIDs need further consideration, especially when considered as preventive treatments on large population scales. NSAIDs are among the most commonly used agents in clinical practice. They are employed as antiinflammatory, analgesic, and antipyretic agents for a wide spectrum of clinical conditions. Their antiinflammatory properties are primarily due to inhibition of prostaglandin synthesis. Acute CNS toxicity related to NSAID use is pervasive and varied. A prospective study looking at ibuprofen overdose noted that 30% of patients experience CNS effects ranging from drowsiness to coma. Case reports have identified numerous neurological sequelae including ataxia, vertigo, dizziness, recurrent falls, nystagmus, headache, encephalopathy, and disorientation. Seizures have also been reported, mostly after overdose ingestions, but even therapeutic doses have occasionally been associated with seizures. One of the important neurological side effects attributed to the use of NSAIDs is aseptic meningitis. The clinical signs of drug-induced meningitis are similar to those of infectious meningitis and include fever, headache, photophobia, and stiff neck. The laboratory findings are also similar, including cerebrospinal fluid (CSF) pleocytosis of several hundred or thousand cells, mainly neutrophils; elevated levels of protein; normal or low glucose levels; and negative cultures. Drug-induced meningitis is a transient disorder with an excellent prognosis.

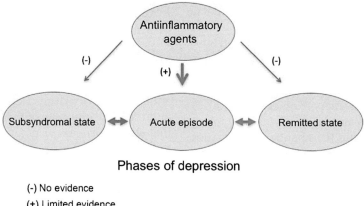

FIG. 1   Level of evidence for treatment of phases of depression with antiinflammatory agents.

Most or all drugs used for the treatment of headache, including NSAIDs, may cause a condition known as medication overuse headache—a refractory chronic daily headache that tends to resolve following discontinuation of the analgesics. Reye's syndrome is a rare severe illness strongly associated with aspirin use. Aspirin, the classic and most commonly used NSAID, has a well-documented effect in inhibiting intravascular clotting, thus reducing the occurrence of ischemic strokes and other vascular events. NSAIDs, however, have a double impact on coagulation. On the one hand, most agents inhibit the synthesis of thromboxane in the platelets, thereby inhibiting coagulation. On the other hand, they also inhibit the production of prostacyclin by endothelial cells, resulting in a prothrombotic state. Selective inhibition of COX-2 by drugs such as rofecoxib (Vioxx) and valdecoxib (Bextra) results in specific inhibition of synthesis of prostaglandins participating in inflammation and can lead to vascular complications including an increased risk for stroke (Auriel, Regev, & Korczyn, 2014).

CONCLUSIONS AND FUTURE DIRECTIONS

Although clinical depression has long been associated with inflammation, limited treatment options have resulted from this immune-associated pathophysiology of depression. The literature suggests that an increased inflammatory response may define a subgroup of individuals in ultra-high-risk states, of those in acute disease episodes, and of those with severe mental illness. Among antiinflammatory agents and NSAIDs specifically, the selective COX-2 inhibitor celecoxib shows the most evidence for antidepressant effects; however, the body of accumulated evidence on the role of selective and nonselective COX inhibitors is small and not suited to give clinical guidance at this stage. One could even argue that clinical proof-of-concept studies are still required before entering a phase of larger clinical trials. An improvement of future study methodologies would include the measure of the inflammation status of patients entering the trial. Future study results may point then to an indicated antiinflammatory treatment based on the inflammation status of patients with clinical depression at commencement of the intervention. This might particularly apply to depressed patients with—but not limited to—concomitant inflammatory conditions such as rheumatoid arthritis, psoriasis, and other chronic inflammatory conditions. Moreover, if antiinflammatory treatment proves to be useful in acute depression, it is clinically worthwhile exploring whether this treatment has potential for treating also more severe forms of depression in the acute stage and whether it is useful for relapse prevention. While the consideration of the phase of the illness is important for developing a potential phase-specific treatment approach using NSAIDs, the literature clearly points to the dynamic nature of the immune response in depression (Eyre et al., 2014) and shows an involvement of both the innate and adaptive immune system in depression (Toben & Baune, 2015). A focus on inflammation only might be shortsighted. While it has been shown that inflammation can be detrimental to the neurobiology of the brain, it has also been suggested that inflammation may lead to an increase in impaired neuroprotective mechanisms; hence, a concept of enhancing neuroprotection plus a reduction in inflammation may be an extended avenue for future interventions. While progress has been made in the understanding of the role of inflammation in depression and first antiinflammatory treatments show some potential clinical value, the field is far away from an indicated and evidence-based approach for the use of antiinflammatory agents in depression-associated inflammation. Clearly, an in-depth clinical study program including proof-of-concept studies and translational clinical studies are required to enhance the knowledge before NSAIDs can clinically be recommended as suitable agents for the treatment of symptoms in depression-associated inflammation.

References

Aid, S., & Bosetti, F. (2011). Targeting cyclooxygenases-1 and -2 in neuroinflammation: therapeutic implications. *Biochimie, 93*(1), 46–51.

Aid, S., Langenbach, R., & Bosetti, F. (2008). Neuroinflammatory response to lipopolysaccharide is exacerbated in mice genetically deficient in cyclooxygenase-2. *Journal of Neuroinflammation, 5*, 17.

Akhondzadeh, S., Jafari, S., Raisi, F., Nasehi, A. A., Ghoreishi, A., Salehi, B., et al. (2009). Clinical trial of adjunctive celecoxib treatment in patients with major depression: a double blind and placebo controlled trial. *Depression and Anxiety, 26*(7), 607–611.

Almeida, O. P., Alfonso, H., Jamrozik, K., Hankey, G. J., & Flicker, L. (2010). Aspirin use, depression, and cognitive impairment in later life: the health in men study. *Journal of the American Geriatrics Society, 58*(5), 990–992.

Almeida, O. P., Flicker, L., Yeap, B. B., Alfonso, H., McCaul, K., & Hankey, G. J. (2012). Aspirin decreases the risk of depression in older men with high plasma homocysteine. *Translational Psychiatry, 2*.

ASPREE Investigator Group. (2013). Study design of ASPirin in reducing events in the elderly (ASPREE): a randomized, controlled trial. *Contemporary Clinical Trials, 36*(2), 555–564.

Auriel, E., Regev, K., & Korczyn, A. D. (2014). Nonsteroidal anti-inflammatory drugs exposure and the central nervous system. *Handbook of Clinical Neurology, 119*, 577–584.

Baune, B. T., Wiede, F., Braun, A., Golledge, J., Arolt, V., & Koerner, H. (2008). Cognitive dysfunction in mice deficient for TNF- and its receptors. *American Journal of Medical Genetics. Part B, Neuropsychiatric Genetics, 147B*(7), 1056–1064.

Beattie, E. C., Stellwagen, D., Morishita, W., Bresnahan, J. C., Ha, B. K., Von Zastrow, M., et al. (2002). Control of synaptic strength by glial TNFalpha. *Science, 295*(5563), 2282–2285.

Berk, M., Dean, O., Drexhage, H., McNeil, J. J., Moylan, S., O'Neil, A., et al. (2013). Aspirin: a review of its neurobiological properties and therapeutic potential for mental illness. *BMC Medicine, 11*, 74.

Camara, M. L., Corrigan, F., Jaehne, E. J., Jawahar, M. C., Anscomb, H., & Baune, B. T. (2015). Effects of centrally administered etanercept on behavior, microglia, and astrocytes in mice following a peripheral immune challenge. *Neuropsychopharmacology, 40*(2), 502–512.

Choi, S. H., Aid, S., Choi, U., & Bosetti, F. (2010). Cyclooxygenases-1 and -2 differentially modulate leukocyte recruitment into the inflamed brain. *The Pharmacogenomics Journal, 10*(5), 448–457.

Couch, Y., Anthony, D. C., Dolgov, O., Revischin, A., Festoff, B., Santos, A. I., et al. (2013). Microglial activation, increased TNF and SERT expression in the prefrontal cortex define stress-altered behaviour in mice susceptible to anhedonia. *Brain, Behavior, and Immunity, 29*, 136–146.

Dantzer, R., O'Connor, J. C., Freund, G. G., Johnson, R. W., & Kelley, K. W. (2008). From inflammation to sickness and depression: when the immune system subjugates the brain. *Nature Reviews. Neuroscience, 9*(1), 46–56.

Dowlati, Y., Herrmann, N., Swardfager, W., Liu, H., Sham, L., Reim, E. K., et al. (2010). A meta-analysis of cytokines in major depression. *Biological Psychiatry, 67*(5), 446–457.

Eyre, H., Air, T., Proctor, S., Rositano, S., & Baune, B. T. (2015). A critical review of the efficacy of non-steroidal antiinflammatory drugs in depression. *Progress in Neuro-Psychopharmacology and Biological Psychiatry, 57*, 11–16.

Eyre, H., & Baune, B. T. (2012). Neuroplastic changes in depression: a role for the immune system. *Psychoneuroendocrinology, 37*(9), 1397–1416.

Eyre, H. A., Stuart, M. J., & Baune, B. T. (2014). A phase-specific neuroimmune model of clinical depression. *Progress in Neuro-Psychopharmacology & Biological Psychiatry, 54*, 265–274.

Fields, C., Drye, L., Vaidya, V., & Lyketsos, C. (2012). Celecoxib or naproxen treatment does not benefit depressive symptoms in persons age 70 and older: findings from a randomized controlled trial. *The American Journal of Geriatric Psychiatry, 20*(6), 505–513.

Fond, G., Hamdani, N., Kapczinski, F., Boukouaci, W., Drancourt, N., Dargel, A., et al. (2014). Effectiveness and tolerance of anti-inflammatory drugs' add-on therapy in major mental disorders: a systematic qualitative review. *Acta Psychiatrica Scandinavica, 129*(3), 163–179.

Galecki, P., Florkowski, A., Bienkiewicz, M., & Szemraj, J. (2010). Functional polymorphism of cyclooxygenase-2 gene (G-765C) in depressive patients. *Neuropsychobiology, 62*(2), 116–120.

Galecki, P., Galecka, E., Maes, M., Chamielec, M., Orzechowska, A., Bobinska, K., et al. (2012). The expression of genes encoding for COX-2, MPO, iNOS, and sPLA2-IIA in patients with recurrent depressive disorder. *Journal of Affective Disorders, 138*(3), 360–366.

Galecki, P., Szemraj, J., Bienkiewicz, M., Zboralski, K., & Galecka, E. (2009). Oxidative stress parameters after combined fluoxetine and acetylsalicylic acid therapy in depressive patients. *Human Psychopharmacology, 24*(4), 277–286.

Gallagher, P. J., Castro, V., Fava, M., Weilburg, J. B., Murphy, S. N., Gainer, V. S., et al. (2012). Antidepressant response in patients with major depression exposed to NSAIDs: a pharmacovigilance study. *The American Journal of Psychiatry, 169*(10), 1065–1072.

Golan, H., Levav, T., Mendelsohn, A., & Huleihel, M. (2004). Involvement of tumor necrosis factor alpha in hippocampal development and function. *Cerebral Cortex, 14*(1), 97–105.

Haapakoski, R., Mathieu, J., Ebmeier, K. P., Alenius, H., & Kivimaki, M. (2015). Cumulative meta-analysis of interleukins 6 and 1beta, tumour necrosis factor alpha and C-reactive protein in patients with major depressive disorder. *Brain, Behavior, and Immunity, 49*, 206–215.

Hannestad, J., DellaGioia, N., & Bloch, M. (2011). The effect of antidepressant medication treatment on serum levels of inflammatory cytokines: a meta-analysis. *Neuropsychopharmacology, 36*(12), 2452–2459.

Haroon, E., Raison, C. L., & Miller, A. H. (2012). Psychoneuroimmunology meets neuropsychopharmacology: translational implications of the impact of inflammation on behavior [Research Support, N.I.H., Extramural Review]. *Neuropsychopharmacology, 37*(1), 137–162.

Howren, M. B., Lamkin, D. M., & Suls, J. (2009). Associations of depression with C-reactive protein, IL-1, and IL-6: a meta-analysis. *Psychosomatic Medicine, 71*(2), 171–186.

Kim, S. J., & Li, J. (2013). Caspase blockade induces RIP3-mediated programmed necrosis in toll-like receptor-activated microglia. *Cell Death & Disease, 4*.

Kohler, O., Benros, M. E., Nordentoft, M., Farkouh, M. E., Iyengar, R. L., Mors, O., et al. (2014). Effect of anti-inflammatory treatment on depression, depressive symptoms, and adverse effects: a systematic review and meta-analysis of randomized clinical trials. *JAMA Psychiatry, 71*(12), 1381–1391.

Leonard, B., & Maes, M. (2012). Mechanistic explanations how cell-mediated immune activation, inflammation and oxidative and nitrosative stress pathways and their sequels and concomitants play a role in the pathophysiology of unipolar depression [Review]. *Neuroscience and Biobehavioral Reviews, 36*(2), 764–785.

Liu, S., Wang, X., Li, Y., Xu, L., Yu, X., Ge, L., et al. (2014). Necroptosis mediates TNF-induced toxicity of hippocampal neurons. *BioMed Research International, 2014*.

Longhi, L., Ortolano, F., Zanier, E. R., Perego, C., Stocchetti, N., & De Simoni, M. G. (2008). Effect of traumatic brain injury on cognitive function in mice lacking p55 and p75 tumor necrosis factor receptors. *Acta Neurochirurgica. Supplement, 102*, 409–413.

Maes, M. (2012). Targeting cyclooxygenase-2 in depression is not a viable therapeutic approach and may even aggravate the pathophysiology underpinning depression. *Metabolic Brain Disease, 27*(4), 405–413.

Maes, M., Bosmans, E., De Jongh, R., Kenis, G., Vandoolaeghe, E., & Neels, H. (1997). Increased serum IL-6 and IL-1 receptor antagonist concentrations in major depression and treatment resistant depression. *Cytokine, 9*(11), 853–858.

McAfoose, J., & Baune, B. T. (2009). Evidence for a cytokine model of cognitive function. *Neuroscience and Biobehavioral Reviews, 33*(3), 355–366.

Mendlewicz, J., Kriwin, P., Oswald, P., Souery, D., Alboni, S., & Brunello, N. (2006). Shortened onset of action of antidepressants in major depression using acetylsalicylic acid augmentation: a pilot open-label study. *International Clinical Psychopharmacology, 21*(4), 227–231.

Miller, A. H., Maletic, V., & Raison, C. L. (2009). Inflammation and its discontents: the role of cytokines in the pathophysiology of major depression. *Biological Psychiatry, 65* (9), 732–741.

Miller, G. E., & Cole, S. W. (2012). Clustering of depression and inflammation in adolescents previously exposed to childhood adversity. *Biological Psychiatry, 72*(1), 34–40.

Moylan, S., Maes, M., Wray, N. R., & Berk, M. (2013). The neuroprogressive nature of major depressive disorder: pathways to disease evolution and resistance, and therapeutic implications. *Molecular Psychiatry, 18*(5), 595–606.

Muller, N. (2013). The role of anti-inflammatory treatment in psychiatric disorders. *Psychiatria Danubina, 25*(3), 292–298.

Muller, N., Schwarz, M. J., Dehning, S., Douhe, A., Cerovecki, A., Goldstein-Muller, B., et al. (2006). The cyclooxygenase-2 inhibitor celecoxib has therapeutic effects in major depression: results of a double-blind, randomized, placebo controlled, add-on pilot study to reboxetine. *Molecular Psychiatry, 11*(7), 680–684.

Musil, R., Schwarz, M. J., Riedel, M., Dehning, S., Cerovecki, A., Spellmann, I., et al. (2011). Elevated macrophage migration inhibitory factor and decreased transforming growth factor-beta levels in major depression–no influence of celecoxib treatment [Randomized Controlled Trial Research Support, Non-U.S. Gov't]. *Journal of Affective Disorders, 134*(1–3), 217–225.

Na, K. S., Lee, K. J., Lee, J. S., Cho, Y. S., & Jung, H. Y. (2013). Efficacy of adjunctive celecoxib treatment for patients with major depressive disorder: a meta-analysis. *Progress in Neuro-Psychopharmacology & Biological Psychiatry, 48C*, 79–85.

Nery, F. G., Monkul, E. S., Hatch, J. P., Fonseca, M., Zunta-Soares, G. B., Frey, B. N., et al. (2008). Celecoxib as an adjunct in the treatment of depressive or mixed episodes of bipolar disorder: a double-blind, randomized, placebo-controlled study. *Human Psychopharmacology, 23*(2), 87–94.

O'Brien, S. M., Scott, L. V., & Dinan, T. G. (2004). Cytokines: abnormalities in major depression and implications for pharmacological treatment. *Human Psychopharmacology, 19*(6), 397–403.

Pasco, J. A., Jacka, F. N., Williams, L. J., Henry, M. J., Nicholson, G. C., Kotowicz, M. A., et al. (2010). Clinical implications of the cytokine hypothesis of depression: the association between use of statins and aspirin and the risk of major depression. *Psychotherapy and Psychosomatics, 79*(5), 323–325.

Peng, L., Verkhratsky, A., Gu, L., & Li, B. (2015). Targeting astrocytes in major depression. *Expert Review of Neurotherapeutics, 15*(11), 1299–1306.

Peschon, J. J., Torrance, D. S., Stocking, K. L., Glaccum, M. B., Otten, C., Willis, C. R., et al. (1998). TNF receptor-deficient mice reveal divergent roles for p55 and p75 in several models of inflammation. *Journal of Immunology, 160*(2), 943–952.

Pollmacher, T., Haack, M., Schuld, A., Reichenberg, A., & Yirmiya, R. (2002). Low levels of circulating inflammatory cytokines—do they affect human brain functions? *Brain, Behavior, and Immunity, 16*(5), 525–532.

Setiawan, E., Wilson, A. A., Mizrahi, R., Rusjan, P. M., Miler, L., Rajkowska, G., et al. (2015). Role of translocator protein density, a marker of neuroinflammation, in the brain during major depressive episodes. *JAMA Psychiatry, 72* (3), 268–275.

Shelton, R. C. (2012). Does concomitant use of NSAIDs reduce the effectiveness of antidepressants? *The American Journal of Psychiatry, 169*(10), 1012–1015.

Song, C. (2002). The effect of thymectomy and IL-1 on memory: implications for the relationship between immunity and depression. *Brain, Behavior, and Immunity, 16*(5), 557–568.

Stellwagen, D., Beattie, E. C., Seo, J. Y., & Malenka, R. C. (2005). Differential regulation of AMPA receptor and GABA receptor trafficking by tumor necrosis factor-alpha. *The Journal of Neuroscience, 25*(12), 3219–3228.

Szelenyi, J., & Vizi, E. S. (2007). The catecholamine cytokine balance: interaction between the brain and the immune system. *Annals of the New York Academy of Sciences, 1113*, 311–324.

Terrando, N., Monaco, C., Ma, D., Foxwell, B. M., Feldmann, M., & Maze, M. (2010). Tumor necrosis factor-alpha triggers a cytokine cascade yielding postoperative cognitive decline. *Proceedings of the National Academy of Sciences of the United States of America, 107*(47), 20518–20522.

Toben, C., & Baune, B. T. (2015). An act of balance between adaptive and maladaptive immunity in depression: a role for T lymphocytes. *Journal of Neuroimmune Pharmacology, 10*(4), 595–609.

Tobinick, E. L., & Gross, H. (2008). Rapid cognitive improvement in Alzheimer's disease following perispinal etanercept administration. *Journal of Neuroinflammation, 5*, 2.

Uher, R., Carver, S., Power, R. A., Mors, O., Maier, W., Rietschel, M., et al. (2012). Non-steroidal anti-inflammatory drugs and efficacy of antidepressants in major depressive disorder. *Psychological Medicine, 42*(10), 2027–2035.

Warner-Schmidt, J. L., Vanover, K. E., Chen, E. Y., Marshall, J. J., & Greengard, P. (2011). Antidepressant effects of selective serotonin reuptake inhibitors (SSRIs) are attenuated by antiinflammatory drugs in mice and humans [Research Support, N.I.H., Extramural Research Support, U.S. Gov't, Non-P.H.S.]. *Proceedings of the National Academy of Sciences of the United States of America, 108*(22), 9262–9267.

Zhou, W., & Yuan, J. (2014). Necroptosis in health and diseases. *Seminars in Cell & Developmental Biology, 35,* 14–23.

Zhu, L., Bi, W., Lu, D., Zhang, C., Shu, X., Wang, H., et al. (2014). Regulation of ubiquitin-specific processing protease 8 suppresses neuroinflammation. *Molecular and Cellular Neurosciences, 64,* 74–83.

Effects of Physical Exercise on Inflammation in Depression

Harris A. Eyre[*,†,‡,§], *Katarina Arandjelovic*[§], *David A. Merrill*[†], *Ajeet B. Singh*[§], *Helen Lavretsky*[†]

*University of Adelaide, Adelaide, SA, Australia
†UCLA, Los Angeles, CA, United States
‡University of Melbourne, Melbourne, VIC, Australia
§Deakin University, Geelong, VIC, Australia

INTRODUCTION

With the major burden of major depressive disorder (MDD), the modest treatment outcomes, novel approaches to depression treatment are greatly needed (Dalvie et al., 2016; Trivedi, 2016). Physical activity (PA) is emerging as an effective tool in the prevention and treatment of major depression (Eyre, Papps, & Baune, 2013; Schuch & de Almeida Fleck, 2013). PA is defined as "any bodily movement produced by skeletal muscles that requires energy expenditure," while physical exercise is "a subcategory of physical activity that is planned, structured, repetitive, and purposeful in the sense that the improvement or maintenance of one or more components of physical fitness is the objective" (WHO, 2010). PA subtypes include aerobic, resistance, flexibility, neuromotor (involving balance, agility, and coordination), mind-body (e.g., tai chi, qigong, and yoga, combining neuromotor, flexibility, and meditation), and mixed.

There are many reasons why PA is an attractive therapeutic option in depression. A recent Cochrane meta-analysis of 28 trials (1101 participants) by Rimer et al. (2012) that compared exercise with no treatment (control) found a moderate clinical effect in MDD. Studies have found that while PA has an initial treatment effect equal to that of antidepressants (Rimer et al., 2012), its effects are slower (Blumenthal et al., 1999) but with greater relapse prevention (Babyak et al., 2000). PA interventions have been shown to be efficacious as a stand-alone (Rethorst, Wipfli, & Landers, 2009) and as an augmentation treatment for MDD (Trivedi et al., 2011). PA has a low side-effect profile and can be adapted according to a patient's medical comorbidities and functional status (Garber et al., 2011; Knochel et al., 2012; Rimer et al., 2012). PA also enhances self-esteem

Inflammation and Immunity in Depression
https://doi.org/10.1016/B978-0-12-811073-7.00034-9

(Salmon, 2001), has less stigmatization than psychotherapy, may reduce the use of pharmacotherapies in MDD (Deslandes et al., 2010), and has a positive effect on cardiometabolic risk factors relevant to many psychiatric diseases (e.g., chronic inflammation, visceral fat mass, glucocorticoid sensitivity, glucose control, and insulin sensitivity) (Eyre & Baune, 2014).

The role of inflammation and immune activation more generally in depressive symptoms has long been discussed and recently gained momentum as a potential treatment target (Eyre, Stuart, & Baune, 2014). A number of studies, experimental, clinical, and meta-analytic, suggest an increased expression of pro-inflammatory cytokines (PICs), tumor necrosis factor (TNF)-α, interleukin (IL)-1β, and IL-6, in the brain of patients with MDD (Eyre et al., 2014). There is likewise literature surrounding the anti-inflammatory effects of antidepressant therapies (Eyre, Lavretsky, Kartika, Qassim, & Baune, 2016) and the antidepressant effects of anti-inflammatory medications (Eyre, Air, Proctor, Rositano, & Baune, 2015).

Exercise immunology in psychiatry is a rapidly evolving field given advances in immunology and greater understanding of the clinical effects of PA (Simpson & Bosch, 2014). Understanding the immunology of PA is important for determining the unique immune characteristics of various types of PA and ultimately in determining if certain patients with immune alterations would benefit from PA.

The chapter aims to explore the immune-modulatory effects of PA and to explore opportunities for targeted treatment of depression using PA based on the immune profile of patients.

THE INVOLVEMENT OF THE IMMUNE SYSTEM IN DEPRESSION

Prior to investigating clinical and immunologic effects of PA, we will first frame the most up-to-date understanding of immune-related pathophysiology in depression. These fields have been extensively reviewed in recent times and will therefore be summarized later.

Depression and the Immune System

The most established immune-based model of depression is the inflammatory or cytokine model of depression (Dantzer, O'Connor, Freund, Johnson, & Kelley, 2008; McAfoose & Baune, 2009; Miller, Maletic, & Raison, 2009). This model postulates that a pro-inflammatory state, characterized by elevations in PICs and reductions in anti-inflammatory cytokines, is involved in the development of depression-like behavior in animals and clinical depression in humans. In addition, the resulting net pro-inflammatory state, mediated via elevations in TNF-α, IL-6, interferon (IFN)-γ, and IL-1β, is found to impair hippocampal (HC) neuroplasticity (e.g., neurogenesis, synaptic plasticity, and long-term potentiation (LTP)), induce glucocorticoid insensitivity of the hypothalamic-pituitary-adrenal axis, increase oxidative stress in the HC, reduce serotonin levels, and create neurotoxic serotonergic metabolites (i.e., 3-hydroxykynurenine and quinolinic acid) (Dantzer et al., 2008; Eyre & Baune, 2012; Leonard & Maes, 2012; Miller et al., 2009; Moylan, Maes, Wray, & Berk, 2012). From a clinical perspective, a meta-analysis by Dowlati et al. (2010) concludes that the pro-inflammatory state is associated with clinical depression. This study pooled 24 studies involving unstimulated measurements of cytokines in patients meeting DSM criteria for major depression and found significantly higher concentrations of TNF-α and IL-6 in depressed subjects compared with control subjects. These findings were recently replicated for IL-6 (Hiles, Baker, de Malmanche, & Attia, 2012). Prospective studies have found associations between PIC levels (specifically IL-6, IL-8, and CRP) and incident depressive symptoms (Baune et al., 2012; Rohleder & Miller, 2008; Vogelzangs et al., 2012).

IMMUNE-MODULATORY EFFECTS OF PHYSICAL ACTIVITY IN DEPRESSION

Aerobic Physical Activity Studies

There are a small number of studies examining the effect of aerobic PA (Rethorst et al., 2012; Rethorst, Moynihan, Lyness, Heffner, & Chapman, 2011). A recent study by Rethorst et al. (2012) investigated the extent to which inflammatory markers can be used to predict response to exercise treatment after an incomplete response to a selective serotonin reuptake inhibitor (SSRI). This neuroimmune investigation was conducted with the cohort from the Treatment with Exercise Augmentation for Depression (TREAD) study, a randomized, parallel dose comparison trial (Trivedi et al., 2011). Incomplete response was qualified as having at least moderate residual depressive symptomatology, quantified by a 17-item Hamilton Depression Rating Scale (HDRS) score ≥ 14. This study randomized 73 participants aged 18–70 years to 12 weeks of either 4 or 16 kcal/kg of body weight per week (KKW), via aerobic PA. The study examined serum IFN-γ, IL-1β, IL-6, and TNF-α and reported several relevant findings. From a clinical perspective, both doses of exercise showed significant improvements over time ($F_{1,121} = 39.9$; $P < 0.0001$). Adjusted remission rates at week 12 were 28.3% versus 15.5% for the 16 and 4 KKW groups, respectively. There was a trend for higher remission rates in the higher-dose exercise group ($P < 0.06$). High-baseline TNF-α (>5.493 pg/mL) was associated with a greater reduction in depressive symptoms (measured by Inventory for Depressive Symptomatology Clinical (IDS-C)). There was also a significant correlation between reductions of IL-1β and depressive symptoms in the 16 KWW group, but not the 4 KWW group. Otherwise, there was no significant change in cytokine levels following the 12-week PA intervention and a nonsignificant association between PA dose and change in cytokine levels.

A cross-sectional study by the same authors, Rethorst et al. (2011), examined the relationship between IL-6 and depressive symptoms by participation in moderate-intensity PA in a sample of 97 primary care patients aged >40 years. The patients had a score on the Center for Epidemiological Studies Depression (CES-D) scale >15, and the type and level of PA were determined by a questionnaire. In this study, there were no correlation between IL-6 and depressive symptoms and no effect for PA on IL-6 levels ($r = 0.086$; $P = 0.40$); however, the association between IL-6 and depressive symptoms was moderated by PA ($P = 0.02$). In physically inactive subjects, higher depressive symptoms were associated with higher IL-6 levels ($r = 0.28$; $P = 0.05$).

Mind-Body PA

Several studies have investigated the effects of mind-body therapies on the immune system and depressive symptoms. A recent RCT study by Black et al. (2013) examined if yogic meditation, kirtan kriya meditation (KKM), might alter the activity of inflammatory and antiviral transcription control pathways that shape immune cell gene expression, as compared to relaxing music (RM). This study was conducted for a total of 8 weeks of active intervention, with 8-week follow-up, and involved 39 family dementia caregivers, mean age 60.5 years ± 28.2 years. Genome-wide transcription profiles from peripheral blood mononuclear cells were examined at 0 and 8 weeks; depression was measured by the HDRS. In the KKM group, 65.2% of the participants showed 50% improvement on the HDRS. In RM group, 31.2% of the participants showed 50% improvement. After adjusting for sex, illness burden, and BMI, 68 genes were differentially expressed (19 upregulated and 49 downregulated). Upregulated genes included immunoglobulin-related transcripts (e.g., IGJ and IGLL3). Downregulated transcripts included PICs (e.g., IL-8) and activation-related immediate-early genes (e.g., JUN and FOSB). IL-1β and IL-6 showed 0.94-fold reductions in KKM versus RM, whereas TNF showed a 1.07-fold increase. Transcript origin analyses identified plasmacytoid dendritic cells and B lymphocytes as the primary cellular context of these transcriptional alterations. Promoter-based bioinformatics analysis implicated reduced NF-κB signaling and increased IRF1 in structuring these effects. It was not possible to pinpoint the specific effects of KKM that are responsible for these biological effects (e.g., chanting, breathing, focused meditation, or mudras). A study by Lavretsky et al. (2011) explored whether T'ai Chi Chih (TCC), added to escitalopram, would augment the treatment of geriatric depression. In this study, 112 older adults with major depression (aged 60 years and over) were recruited and treated with escitalopram for 4 weeks. Seventy-three partial responders to escitalopram were randomly assigned to 10 weeks of TCC or HE. Subjects in the escitalopram and TCC condition were more likely to show greater reduction of depressive symptoms and greater reductions in CRP as compared with the escitalopram and HE group.

Comparative Studies

There is only one study directly comparing the immune-modulatory and antidepressive effects of PA subtypes in depression. This clinical trial by Kohut et al. (2006) randomized 87 healthy older adults (64–87 years) to either aerobic (cardio) or strength and flexibility (flex) training 3 days/week, 45 min/day for 10 months. A subgroup of subjects treated with nonselective $\beta_1\beta_2$ adrenergic antagonists were included to evaluate the potential role of β-adrenergic receptor adaptations as mediators of PA-induced change in inflammation. The study found cardio treatment resulted in significant reductions in serum CRP, IL-6, and IL-18 compared with flex treatment, whereas TNF-α

declined in both groups. Both groups had similar improvements with depressive symptoms (measured by the Geriatric Depression Scale) (F $(1, 86) = 5.943$; $P = 0.017$). Depressive symptoms in the cardio group changed from 2.5 ± 0.3 at baseline to 1.8 ± 0.7 at end of trial; depressive symptoms in the flex group changed from 3.2 ± 0.4 at baseline to 2.8 ± 0.4 at end of trial. Several psychosocial measures (i.e., depression, optimism, and sense of coherence) improved in both groups suggesting that the reduction of CRP, IL-6, and IL-18 in the cardio group was not mediated by improvements in psychosocial scores.

Additional Considerations

Research is suggestive of unique immune-modulatory profiles according to PA subtypes (i.e., resistance, aerobic, and mind-body), which are an important consideration for precision treatment in clinical care. For example, the RCT by Kohut et al. (2006) suggests aerobic treatment resulted in significant reductions in serum CRP, IL-6, and IL-18 compared with strength and flexibility treatment, whereas TNF-α declined in both groups. Another study compared resistance and aerobic exercise in sedentary, nondepressed adults and found resistance PA produced a greater reduction in CRP than aerobic PA, 32.8% versus 16.1%, respectively (Donges, Duffield, & Drinkwater, 2010). These studies hence show differences in which PA type reduces CRP the most, and the reason for this is likely due to the study cohort characteristics. Levels of various cytokines at baseline may predict unique treatment response to subtypes of PA. For example, aerobic PA is shown to be more efficacious with subjects (partial responders to SSRIs) who have a higher baseline TNF-α (>5.493 pg/mL) (Rethorst et al., 2012). Given Eller, Vasar, Shlik, and Maron (2008) found high-baseline serum TNF-α associated with nonresponse to an SSRI, and the finding by Rethorst et al. (2012) suggests TNF-α is a moderator between SSRI and exercise treatment; hence, TNF-α levels could be used to recommend exercise rather than medication as part of a personalized treatment algorithm.

BIOLOGICAL FACTORS UNDERLYING POSSIBLE IMMUNE-MODULATORY PROFILES OF PHYSICAL ACTIVITY

Biological factors that may underlie immune-modulatory profiles of PA subtypes include specific effects on adipose tissue, muscle, blood vessels, vagal tone, and the brain as discussed below. PA subtypes appear to have differing effects on adipose tissue deposits, and this may in turn affect immune-modulatory profiles. A study comparing 10 weeks of aerobic and resistance PA with control in 103 sedentary adults has found resistance training reduces CRP to a greater extent than aerobic training (Donges et al., 2010). The resistance group significantly improved total body fat mass as compared with the aerobic group, measured by dual-energy X-ray absorptiometry scanning. The aerobic group exhibited greater reductions in intra-abdominal fat mass and total body mass as compared with the resistance group. A rodent study (Cohen, Shea, Heffron, Schmelz, & Roberts, 2013) has found differing leukocyte and macrophage phenotype compositions between white subcutaneous fat and various types of white visceral fat (i.e., peritoneal serous fluid and parametrial, retroperitoneal, and omental fat). Serous visceral fluid was composed almost entirely of CD45(+) leukocytes (>99%), while omental fat contained less but still almost twofold more leukocytes than parametrial and retroperitoneal (75%, 38%, and 38% respectively; $P < 0.01$). Parametrial fat was composed primarily of macrophages, whereas retroperitoneal fat more closely resembled omental fat, denoted by high levels of B1 B-cell and monocyte populations. Further, omental fat

harbored significantly higher proportions of T cells than the other tissues, consistent with its role as a secondary lymphoid organ.

Changes in muscle physiology may belie the unique immune-modulatory profiles of PA subtypes. In 14 young untrained health male subjects, skeletal muscle biopsies were analyzed prior to, immediately after, and in the recovery period following resistance PA, aerobic PA, and control intervention (Moller et al., 2013). Resistance exercise, but not aerobic exercise, increased inhibitory κB protein kinase complex (Iκκ)β phosphorylation (a key regulator of NF-κB), possibly suggesting Iκκβ can influence the activation of mammalian target of rapamycin complex 1 (mTORC1). mTORC1 is considered a principal mediator of muscular adaptations to exercise and integrates signals from mitogenic growth factors, cellular stressors, and/or nutrients to regulate protein synthesis (Ma & Blenis, 2009).

A recent review by Roque, Hernanz, Salaices, and Briones (2013) outlines evidence to suggest PA minimizes vascular damage produced by obesity, hypertension, diabetes, dyslipidemia, and metabolic syndrome. PA training reduces endothelial dysfunction, altered vascular structure, and/or increased vascular stiffness. Mechanisms underlying these changes include reduced inflammation and pro-inflammatory adipokines, improved nitric oxide availability, and increased antioxidant production. Further research is required to understand unique vascular modifying profiles of PA subtypes.

METHODOLOGICAL LIMITATIONS

Several methodological limitations should be noted. From a clinical trial perspective, as stated in the most recent Cochrane review assessing the clinical effects of exercise in depression by Cooney et al. (2013), measures need to be taken to enhance the methodological robust nature of this field. Cooney et al. (2013) suggest clinical evidence is of moderate quality. They point out significant study bias, heterogeneity, and outcome reporting bias. They also suggest the requirement of additional large-scale high-quality studies where all participants at the time of recruitment were diagnosed through clinical interview as having depression, adhered closely to an exercise regimen as a sole intervention, and were further assessed through diagnostic clinical interview post intervention. Further work is required to improve the understanding of the biology of such PA-related immune-modulatory profiles.

FUTURE DIRECTIONS

Important future directions should be considered to advance this field of research. These include the need to develop comparative studies examining aerobic, resistance, neuromotor, and mind-body exercise from clinical and immune perspectives in depression. Creating such comparative studies will also assist in minimizing methodological heterogeneity by standardizing PA interventions, duration, intensities, and study populations. Moreover, rodent models can be utilized to assist in understanding the immune-modulatory differences between aerobic and resistance training—the study by Cassilhas et al. (2012) provides a useful model for such a study. The age of subjects is important to consider in PA interventions examining immune-modulatory and clinical effects in depression, given the age-dependent immune factors that may affect the immune-modulatory mechanisms of PA. For example, human aging is characterized by chronic, low-grade inflammation, referred to as inflammaging, which may make late-life depressed patients more prone to benefits from anti-inflammatory types of PA (Franceschi & Campisi, 2014). Clinical trials examining neuropsychological effects of PA should include concurrent immune biomarker analysis. Other immune markers and

investigative methodologies should be utilized in future comparative studies. Further, *in vivo* microglial imaging techniques utilizing positron-emission tomography are available and have been utilized in depression (see primary studies (Hannestad et al., 2013; Kreisl et al., 2013)). Finally, the immune effects of mind-body PA may be better understood through the role of this PA subtype in influencing neuroplasticity metrics in neuroimaging.

CONCLUSION

The burden of depression is a pressing public health issue. PA is increasingly recognized as an important treatment modality in depression, and there is an emerging field of exercise immunology that aims to better understand the significant immune effects of PA. Early evidence suggests immune biomarkers can be used to improve outcome in some populations of clinical depression. More research is required to replicate and extend these findings and to understand the immune signatures of various subtypes of PA.

CONFLICT OF INTEREST

All authors declare that there are no conflicts of interest.

References

Babyak, M., Blumenthal, J. A., Herman, S., Khatri, P., Doraiswamy, M., Moore, K., et al. (2000). Exercise treatment for major depression: maintenance of therapeutic benefit at 10 months. *Psychosomatic Medicine*, 62(5), 633–638.

Baune, B. T., Smith, E., Reppermund, S., Air, T., Samaras, K., Lux, O., et al. (2012). Inflammatory biomarkers predict depressive, but not anxiety symptoms during aging: the prospective Sydney Memory and Aging Study. *Psychoneuroendocrinology*. https://doi.org/10.1016/j.psyneuen.2012.02.006.

Black, D. S., Cole, S. W., Irwin, M. R., Breen, E., St Cyr, N. M., Nazarian, N., et al. (2013). Yogic meditation reverses NF-kappaB and IRF-related transcriptome dynamics in leukocytes of family dementia caregivers in a randomized controlled trial. *Psychoneuroendocrinology*, 38(3), 348–355. https://doi.org/10.1016/j.psyneuen.2012.06.011.

Blumenthal, J. A., Babyak, M. A., Moore, K. A., Craighead, W. E., Herman, S., Khatri, P., et al. (1999). Effects of exercise training on older patients with major depression. *Archives of Internal Medicine*, 159(19), 2349–2356.

Cassilhas, R. C., Lee, K. S., Fernandes, J., Oliveira, M. G., Tufik, S., Meeusen, R., et al. (2012). Spatial memory is improved by aerobic and resistance exercise through divergent molecular mechanisms. *Neuroscience*, 202, 309–317. https://doi.org/10.1016/j.neuroscience.2011.11.029.

Cohen, C. A., Shea, A. A., Heffron, C. L., Schmelz, E. M., & Roberts, P. C. (2013). Intra-abdominal fat depots represent distinct immunomodulatory microenvironments: a murine model. *PLoS One*, 8(6). https://doi.org/10.1371/journal.pone.0066477.

Cooney, G. M., Dwan, K., Greig, C. A., Lawlor, D. A., Rimer, J., Waugh, F. R., et al. (2013). Exercise for depression. *Cochrane Database of Systematic Reviews*, 9. https://doi.org/10.1002/14651858.CD004366.pub6.

Dalvie, S., Koen, N., McGregor, N., O'Connell, K., Warnich, L., Ramesar, R., et al. (2016). Toward a global roadmap for precision medicine in psychiatry: challenges and opportunities. *OMICS*, 20(10), 557–564. https://doi.org/10.1089/omi.2016.0110.

Dantzer, R., O'Connor, J. C., Freund, G. G., Johnson, R. W., & Kelley, K. W. (2008). From inflammation to sickness and depression: when the immune system subjugates the brain. *Nature Reviews. Neuroscience*, 9(1), 46–56. https://doi.org/10.1038/nrn2297. pii: nrn2297.

Deslandes, A. C., Moraes, H., Alves, H., Pompeu, F. A., Silveira, H., Mouta, R., et al. (2010). Effect of aerobic training on EEG alpha asymmetry and depressive symptoms in the elderly: a 1-year follow-up study. *Brazilian Journal of Medical and Biological Research*, 43(6), 585–592.

Donges, C. E., Duffield, R., & Drinkwater, E. J. (2010). Effects of resistance or aerobic exercise training on interleukin-6, C-reactive protein, and body composition. *Medicine and Science in Sports and Exercise*, 42(2), 304–313. https://doi.org/10.1249/MSS.0b013e3181b117ca.

Dowlati, Y., Herrmann, N., Swardfager, W., Liu, H., Sham, L., Reim, E. K., et al. (2010). A meta-analysis of cytokines in major depression. *Biological Psychiatry*, 67(5), 446–457. https://doi.org/10.1016/j.biopsych.2009.09.033. pii: S0006-3223(09)01229-3.

Eller, T., Vasar, V., Shlik, J., & Maron, E. (2008). Pro-inflammatory cytokines and treatment response to escitalopram in major depressive disorder. *Progress in Neuro-Psychopharmacology & Biological Psychiatry*, 32(2), 445–450. https://doi.org/10.1016/j.pnpbp.2007.09.015.

Eyre, H. A., Air, T., Proctor, S., Rositano, S., & Baune, B. T. (2015). A critical review of the efficacy of non-steroidal anti-inflammatory drugs in depression. *Progress in Neuro-Psychopharmacology & Biological Psychiatry*, 57, 11–16. https://doi.org/10.1016/j.pnpbp.2014.10.003.

Eyre, H., & Baune, B. T. (2012). Neuroplastic changes in depression: a role for the immune system. *Psychoneuroendocrinology*, 37(9), 1397–1416. https://doi.org/10.1016/j.psyneuen.2012.03.019.

Eyre, H. A., & Baune, B. T. (2014). Assessing for unique immunomodulatory and neuroplastic profiles of physical activity subtypes: a focus on psychiatric disorders. *Brain, Behavior, and Immunity*, 39, 42–55. https://doi.org/10.1016/j.bbi.2013.10.026.

Eyre, H. A., Lavretsky, H., Kartika, J., Qassim, A., & Baune, B. T. (2016). Modulatory effects of antidepressant classes on the innate and adaptive immune system in depression. *Pharmacopsychiatry*, 49(3), 85–96. https://doi.org/10.1055/s-0042-103159.

Eyre, H. A., Papps, E., & Baune, B. T. (2013). Treating depression and depression-like behavior with physical activity: an immune perspective. *Frontiers in Psychiatry*, 4, 3. https://doi.org/10.3389/fpsyt.2013.00003.

Eyre, H. A., Stuart, M. J., & Baune, B. T. (2014). A phase-specific neuroimmune model of clinical depression. *Progress in Neuro-Psychopharmacology & Biological Psychiatry*, 54, 265–274. https://doi.org/10.1016/j.pnpbp.2014.06.011.

Franceschi, C., & Campisi, J. (2014). Chronic inflammation (inflammaging) and its potential contribution to age-associated diseases. *Journals of Gerontology Series A, Biological Sciences and Medical Sciences*, 69(Suppl. 1), S4–9. https://doi.org/10.1093/gerona/glu057.

Garber, C. E., Blissmer, B., Deschenes, M. R., Franklin, B. A., Lamonte, M. J., Lee, I. M., et al. (2011). American College of Sports Medicine position stand. Quantity and quality of exercise for developing and maintaining cardiorespiratory, musculoskeletal, and neuromotor fitness in apparently healthy adults: guidance for prescribing exercise. *Medicine and Science in Sports and Exercise*, 43(7), 1334–1359. https://doi.org/10.1249/MSS.0b013e318213fefb.

Hannestad, J., Dellagioia, N., Gallezot, J. D., Lim, K., Nabulsi, N., Esterlis, I., et al. (2013). The neuroinflammation marker translocator protein is not elevated in individuals with mild-to-moderate depression: a [C]PBR28 PET study. *Brain, Behavior, and Immunity*. https://doi.org/10.1016/j.bbi.2013.06.010.

Hiles, S. A., Baker, A. L., de Malmanche, T., & Attia, J. (2012). A meta-analysis of differences in IL-6 and IL-10 between people with and without depression: exploring the causes of heterogeneity. *Brain, Behavior, and Immunity*, 26(7), 1180–1188. https://doi.org/10.1016/j.bbi.2012.06.001.

Knochel, C., Oertel-Knochel, V., O'Dwyer, L., Prvulovic, D., Alves, G., Kollmann, B., et al. (2012). Cognitive and behavioural effects of physical exercise in psychiatric patients. *Progress in Neurobiology*, 96(1), 46–68. https://doi.org/10.1016/j.pneurobio.2011.11.007.

Kohut, M. L., McCann, D. A., Russell, D. W., Konopka, D. N., Cunnick, J. E., Franke, W. D., et al. (2006). Aerobic exercise, but not flexibility/resistance exercise, reduces serum IL-18, CRP, and IL-6 independent of beta-blockers, BMI, and psychosocial factors in older adults. *Brain, Behavior, and Immunity*, 20(3), 201–209. https://doi.org/10.1016/j.bbi.2005.12.002.

Kreisl, W. C., Lyoo, C. H., McGwier, M., Snow, J., Jenko, K. J., Kimura, N., et al. (2013). In vivo radioligand binding to translocator protein correlates with severity of Alzheimer's disease. *Brain*, 136(Pt 7), 2228–2238. https://doi.org/10.1093/brain/awt145.

Lavretsky, H., Alstein, L. L., Olmstead, R. E., Ercoli, L. M., Riparetti-Brown, M., Cyr, N. S., et al. (2011). Complementary use of tai chi chih augments escitalopram treatment of geriatric depression: a randomized controlled trial. *American Journal of Geriatric Psychiatry*, 19(10), 839–850. https://doi.org/10.1097/JGP.0b013e31820ee9ef.

Leonard, B., & Maes, M. (2012). Mechanistic explanations how cell-mediated immune activation, inflammation and oxidative and nitrosative stress pathways and their sequels and concomitants play a role in the pathophysiology of unipolar depression. *Neuroscience and Biobehavioral Reviews*, 36(2), 764–785. https://doi.org/10.1016/j.neubiorev.2011.12.005.

Ma, X. M., & Blenis, J. (2009). Molecular mechanisms of mTOR-mediated translational control. *Nature Reviews. Molecular Cell Biology*, 10(5), 307–318. https://doi.org/10.1038/nrm2672.

McAfoose, J., & Baune, B. T. (2009). Evidence for a cytokine model of cognitive function. *Neuroscience and Biobehavioral Reviews*, 33(3), 355–366. https://doi.org/10.1016/j.neubiorev.2008.10.005.

Miller, A. H., Maletic, V., & Raison, C. L. (2009). Inflammation and its discontents: the role of cytokines in the pathophysiology of major depression. *Biological Psychiatry*, 65(9), 732–741. https://doi.org/10.1016/j.biopsych.2008.11.029. pii: S0006-3223(08)01532-1.

Moller, A. B., Vendelbo, M. H., Rahbek, S. K., Clasen, B. F., Schjerling, P., Vissing, K., et al. (2013). Resistance exercise, but not endurance exercise, induces IKKbeta phosphorylation in human skeletal muscle of training-accustomed individuals. *Pflügers Archiv*. https://doi.org/10.1007/s00424-013-1318-9.

Moylan, S., Maes, M., Wray, N. R., & Berk, M. (2012). The neuroprogressive nature of major depressive disorder: pathways to disease evolution and resistance, and therapeutic implications. *Molecular Psychiatry*. https://doi.org/10.1038/mp.2012.33.

Rethorst, C. D., Moynihan, J., Lyness, J. M., Heffner, K. L., & Chapman, B. P. (2011). Moderating effects of moderate-intensity physical activity in the relationship between depressive symptoms and interleukin-6 in primary care patients. *Psychosomatic Medicine*, 73(3), 265–269. https://doi.org/10.1097/PSY.0b013e3182108412.

Rethorst, C. D., Toups, M. S., Greer, T. L., Nakonezny, P. A., Carmody, T. J., Grannemann, B. D., et al. (2012). Pro-inflammatory cytokines as predictors of antidepressant effects of exercise in major depressive disorder. *Molecular Psychiatry*. https://doi.org/10.1038/mp.2012.125.

Rethorst, C. D., Wipfli, B. M., & Landers, D. M. (2009). The antidepressive effects of exercise: a meta-analysis of randomized trials. *Sports Medicine*, 39(6), 491–511. https://doi.org/10.2165/00007256-200939060-00004.

Rimer, J., Dwan, K., Lawlor, D. A., Greig, C. A., McMurdo, M., Morley, W., et al. (2012). Exercise for depression. *Cochrane Database of Systematic Reviews*, 7. https://doi.org/10.1002/14651858.CD004366.pub5.

Rohleder, N., & Miller, G. E. (2008). Acute deviations from long-term trait depressive symptoms predict systemic inflammatory activity. *Brain, Behavior, and Immunity*, 22(5), 709–716. https://doi.org/10.1016/j.bbi.2007.10.012.

Roque, F. R., Hernanz, R., Salaices, M., & Briones, A. M. (2013). Exercise training and cardiometabolic diseases: focus on the vascular system. *Current Hypertension Reports*, 15(3), 204–214. https://doi.org/10.1007/s11906-013-0336-5.

Salmon, P. (2001). Effects of physical exercise on anxiety, depression, and sensitivity to stress: a unifying theory. *Clinical Psychology Review*, 21(1), 33–61.

Schuch, F. B., & de Almeida Fleck, M. P. (2013). Is exercise an efficacious treatment for depression? A comment upon recent negative findings. *Frontiers in Psychiatry*, 4. https://doi.org/10.3389/fpsyt.2013.00020.

Simpson, R. J., & Bosch, J. A. (2014). Special issue on exercise immunology: current perspectives on aging, health and extreme performance. *Brain, Behavior, and Immunity*, 39, 1–7. https://doi.org/10.1016/j.bbi.2014.03.006.

Trivedi, M. H. (2016). Right patient, right treatment, right time: biosignatures and precision medicine in depression. *World Psychiatry: official journal of the World Psychiatric Association (WPA)*, 15(3), 237–238. https://doi.org/10.1002/wps.20371.

Trivedi, M. H., Greer, T. L., Church, T. S., Carmody, T. J., Grannemann, B. D., Galper, D. I., et al. (2011). Exercise as an augmentation treatment for nonremitted major depressive disorder: a randomized, parallel dose comparison. *Journal of Clinical Psychiatry*, 72(5), 677–684. https://doi.org/10.4088/JCP.10m06743.

Vogelzangs, N., Duivis, H. E., Beekman, A. T., Kluft, C., Neuteboom, J., Hoogendijk, W., et al. (2012). Association of depressive disorders, depression characteristics and antidepressant medication with inflammation. *Translational Psychiatry*, 2. https://doi.org/10.1038/tp.2012.8.

WHO (2010). In W. Press (Ed.), *Global recommendations on physical activity for health*. Geneva: WHO.

Future Perspectives on Immune-Related Treatments

Bernhard T. Baune

University of Adelaide, Adelaide, SA, Australia

Among biological theories of depression, factors of the immune system exerting effects on key processes such as neuroplasticity, neurotransmission, oxidative stress, and neuroendocrinological functions are considered to be central to the development and chronic course of depression (Dantzer, O'Connor, Freund, Johnson, & Kelley, 2008; Eyre & Baune, 2012; Haroon, Raison, & Miller, 2012; McAfoose & Baune, 2009; Miller, Maletic, & Raison, 2009). Inflammation has been associated with the clinical presentation of depression and with anatomical, functional, and biochemical disturbances relevant to depression (Baune, 2009; Mahar, Bambico, Mechawar, & Nobrega, 2014; Moylan et al., 2014; Slavich & Irwin, 2014). The most established immune-based model of depression is the inflammatory or cytokine model of depression (Dantzer et al., 2008; McAfoose & Baune, 2009; Miller et al., 2009). This model postulates that a pro-inflammatory state characterized by elevations in pro-inflammatory cytokines (PICs) and reductions in anti-inflammatory cytokines is involved in the development of depression-like behavior in animals and clinical depression in humans. The net pro-inflammatory state is found

to impair hippocampal (HC) neuroplasticity (e.g., neurogenesis, synaptic plasticity, and long-term potentiation), induce glucocorticoid insensitivity of the hypothalamic-pituitary-adrenal (HPA) axis, increase oxidative stress in the HC, reduce serotonin levels, and create neurotoxic serotonergic metabolites (i.e., 3-hydroxykynurenine and quinolinic acid) (Dantzer et al., 2008; Eyre & Baune, 2012; Leonard & Maes, 2012; Miller et al., 2009; Moylan, Maes, Wray, & Berk, 2013).

In line with a pro-inflammatory state in the brain of depressed patients is the neuroprogression model of depression (Moylan et al., 2013). This model suggests that a neuroprogressive process occurs in some patients with major depressive disorder (MDD) leading to poorer symptomatic, treatment, and functional outcomes. In such patients, longer and more frequent episodes appear to increase vulnerability to develop further episodes (Kendler, Thornton, & Gardner, 2001). The neuroprogressive nature of such a course of disease is believed to be associated with structural brain changes and changes in inflammatory conditions, neurotransmission, oxidative and nitrosative stress pathways, neuroplasticity, HPA axis modulation, and mitochondrial dysfunction

Inflammation and Immunity in Depression
https://doi.org/10.1016/B978-0-12-811073-7.00035-0

(Maes, Mihaylova, Kubera, & Ringel, 2012; Maes, Ringel, Kubera, Berk, & Rybakowski, 2012; Moylan et al., 2013; Sheline, Gado, & Kraemer, 2003; Videbech & Ravnkilde, 2004). While this model of neuroprogression may apply to a subgroup of patients diagnosed with depression, a model on phase-dynamic immune alteration may help stratify patients into subgroups resulting in different immune signatures. Hence, it has been suggested that immune alterations in depression may vary according to the different stages of clinical depression, which constitutes another important layer of immune association with clinical depression.

INFLAMMATION DURING PHASES OF DEPRESSION

Different phases of depression can be distinguished. A preclinical phase of depression is characterized by subthreshold symptoms and risk factors for the development of depression. In this phase, a proportion of people may not go on to develop a depressive episode, whereas some patients may develop a first or recurrent major depressive episode. When symptomatology is not self-limited and endures or reaches higher severity, an onset of a clinical phase of depression occurs during which the depressive symptoms may be mild, moderate, or severe and may include melancholic, atypical, psychotic, and nonmelancholic symptomatology. This clinical phase can be a first episode for an individual, and interestingly, studies suggest that 60%–90% of adolescents with depressive episode show remission within 1 year (Dunner et al., 2006; March et al., 2004). However, 50%–70% of remitted patients suffer ≥1 subsequent depressive episodes within 5 years (Lewinsohn, Rohde, Seeley, Klein, & Gotlib, 2000). Importantly, only a few adults experience a full symptomatic and functional remission between depressive episodes with a high risk of long-term loss of function (Conradi, Ormel, & de Jonge,

2011; Fava, Ruini, & Belaise, 2007). These data suggest that it is important to identify markers according to the phase or stage of the disease to be able to predict the course of depression and to determine which patients may benefit from phase-specific interventions. The activity of innate and adaptive immune factors according to the phase of depression is presented in Figs. 1 and 2.

IMMUNE-RELATED BIOMARKERS FOR DEPRESSION

An immune-based biomarker of depression might be helpful for stratifying patient groups into those patients showing association between a biomarker and a specific disease state or response to treatment. Despite intensive research and significant advances made toward the discovery of genomic biomarkers for depression, no reliable biomarker either blood-based or non-blood-based has made it to clinical practice yet. Given the often recognized clinical and biological heterogeneity of depression, recent studies have suggested that certain inflammatory components may be suitable to characterize a specific subgroup of patients with depression. Specifically, increased levels of pro-inflammatory markers such as C-reactive protein (CRP) at baseline have been linked to greater general depressive symptom severity (Hope et al., 2013; Howren, Lamkin, & Suls, 2009; Kohler et al., 2017; Kohler-Forsberg et al., 2017; Krogh et al., 2014; Valkanova, Ebmeier, & Allan, 2013) and to greater severity of specific symptoms affecting mood, interest, activity, suicidality, and cognitive function (Kohler-Forsberg et al., 2017). Interestingly, neurovegetative symptoms (i.e., sleep and appetite) were affected by increased inflammatory markers, with a trend toward a greater increase in neurovegetative symptoms specifically observed in men (Kohler-Forsberg et al., 2017). In addition, retrospective analyses have suggested that selecting antidepressant treatment

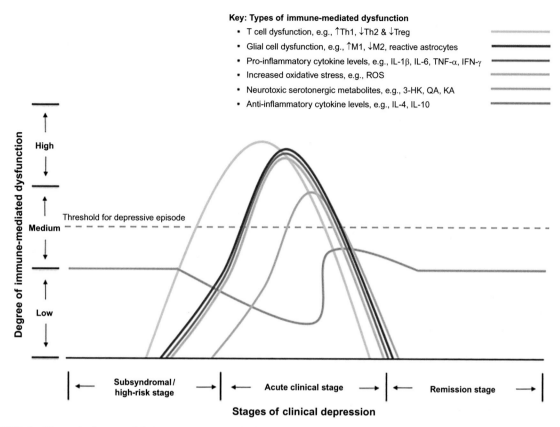

Key: Types of immune-mediated dysfunction
- T cell dysfunction, e.g., ↑Th1, ↓Th2 & ↓Treg
- Glial cell dysfunction, e.g., ↑M1, ↓M2, reactive astrocytes
- Pro-inflammatory cytokine levels, e.g., IL-1β, IL-6, TNF-α, IFN-γ
- Increased oxidative stress, e.g., ROS
- Neurotoxic serotonergic metabolites, e.g., 3-HK, QA, KA
- Anti-inflammatory cytokine levels, e.g., IL-4, IL-10

FIG. 1 Dynamic changes of the immune response during stages of clinical depression until recovery. This figure represents an acute clinical depressive episode with full remission in the context of the three phases of the phase-specific neuroimmune model of clinical depression—subsyndromal, acute clinical, and remission stages as shown on the x-axis show the relevant phases; the y-axis shows the degree of immune-mediated dysfunction. The colored lines represent the various types of immune-mediated dysfunction. The *gray dashed line* indicates an immune dysfunction threshold line whereby a clinically significant depressive episode is diagnosable. *IL*, interleukin; *TNF*, tumor necrosis factor; *IFN*, interferon; *ROS*, reactive oxygen species; *3-HK*, 3-hydroxykynurenine; *QA*, quinolinic acid; *KA*, kynurenic acid; *Th*, T helper; *Treg*, regulatory T cell.

based on CRP could potentially be advantageous to patients, where improvement in depressive symptoms was observed in the group of patients who had raised CRP levels at baseline (CRP >3 mg/L and CRP >5 mg/L) (Raison et al., 2013). Interestingly, recent work tested whether CRP levels predicted response to two different antidepressant medications (Uher et al., 2014). The study found that the antidepressant escitalopram was less effective than the antidepressant nortriptyline for individuals with high levels of CRP (CRP >3 mg/L) (Uher et al., 2014). Furthermore, in individuals treated with escitalopram, an increase in baseline CRP levels was associated with a worsening of depression scores (Uher et al., 2014). In this study, CRP levels were determined retrospectively after completing the trial and not prior to the trial as proposed in our study. The authors concluded that by providing more information on treatment choice, CRP levels may contribute to better outcomes of MDD (Uher et al., 2014). This study was reported to

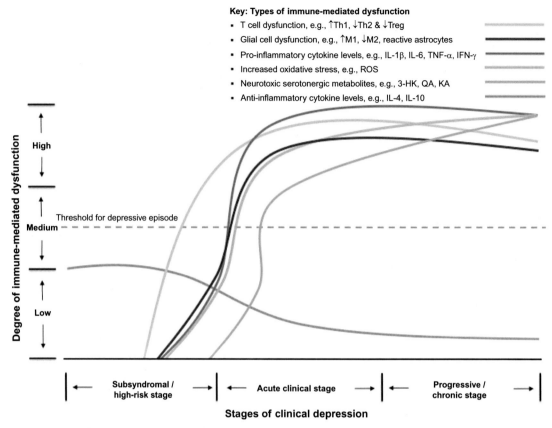

FIG. 2 Progressive immune changes during a progressive course of clinical depression. This figure represents a chronic and progressive major depressive episode. The x-axis shows the relevant phases (subsyndromal/high-risk stage, acute clinical stage, and progressive/chronic stage); the y-axis shows the degree of immune-mediated dysfunction. The coloured lines represent the various types of immune-mediated dysfunction. The *gray dashed line* indicates an immune dysfunction threshold line whereby a clinically significant depressive episode is diagnosable. *IL,* interleukin; *TNF,* tumor necrosis factor; *IFN,* interferon; *ROS,* reactive oxygen species; *3-HK,* 3-hydroxykynurenine; *QA,* quinolinic acid; *KA,* kynurenic acid; *Th,* T helper; *Treg,* regulatory T cell.

be the first to examine markers of inflammation as predictors of response to different antidepressants (Uher et al., 2014); however, it did not include the addition of anti-inflammatory medication in either of the study groups as proposed in our study. Although such findings may help in identifying individuals who could benefit from targeted treatment approaches, potentially leading to more tailored and more effective treatment in groups of MDD patients, proof-of-concept studies have not been studied prospectively yet. To take the promising results further and make use of a potential biomarker for a clinical trial, our group conducts a randomized controlled trial based on CRP-informed stratification into those with higher and those with lower CRP values in peripheral blood. These advances in clinical trials are in line with recommendations based on a recent systematic review and meta-analysis of clinical trials that the field needs randomized clinical trials, using depression as the primary outcome, in individuals with high inflammation levels (Kappelmann, Lewis, Dantzer, Jones, & Khandaker, 2018).

Although it is too early to tell whether an inflammatory state marker such as CRP is a clinically suitable biomarker for treatment indication and selection, the above is an example on the potential of how an immune-based biomarker could be derived from basic science and clinical research findings and be applied to potentially help stratify patient groups and inform treatment selection strategies. Importantly, following clinical proof-of-concept studies, greater efforts and advances need to be made in clinical experimental studies to test potential biomarker to aid patient stratification and treatment selection.

IMMUNE-RELATED TREATMENT APPROACHES

Novel treatment strategies for depression are urgently needed. Global data suggest that unipolar depression currently ranks 11th for disability-adjusted life years, a 37% increase since 1990 (Murray et al., 2012). The burden is expected to continue to grow into the 21st century (Holtzheimer et al., 2008; Murray et al., 2012). Hence, this is an unprecedented burden of depressive illness requiring increased effort to find novel therapeutic agents for treatment (Licinio, 2011). Additionally, more than 50% of patients on antidepressants will not achieve remission following initial treatment (Trivedi et al., 2006), and nearly one-third will not achieve remission following several treatment steps (Rush, Trivedi, Wisniewski, Nierenberg, et al., 2006; Rush, Trivedi, Wisniewski, Stewart, et al., 2006).

While the pathophysiology of depression still requires understanding of key mechanisms that are translatable into treatment approaches, the role of inflammation and immune activation more generally in depressive symptoms has long been discussed and recently gained momentum as a potential treatment target. Growing evidence suggests that PICs play a major role in the pathophysiology of depression. A large number of studies, both experimental and meta-analytic analyses, suggest an increased expression of PICs, tumor necrosis factor (TNF)-α, interleukin (IL)-1β, and IL-6, in the brain of patients with MDD (Dowlati et al., 2010; Hannestad, DellaGioia, & Bloch, 2011; Howren et al., 2009; Maes et al., 1997). A seminal meta-analysis (Dowlati et al., 2010) of 24 studies found significantly higher concentrations of the PICs TNF-α and IL-6 in depressed subjects compared with control subjects. An updated meta-analysis (Haapakoski, Mathieu, Ebmeier, Alenius, & Kivimaki, 2015) of IL-6, CRP, and TNF-α found higher levels of IL-6 and CRP in depressed patients versus controls (29 studies for IL-6 and 20 for CRP). These studies strengthen the clinical evidence that symptoms of depression can be accompanied by an activation of the inflammatory response system (Dowlati et al., 2010). However, has this accumulated knowledge fostered the development of biomarkers of inflammation-associated depression? The research and clinical demands of having biomarkers of depression available for both diagnostic and treatment prediction purposes are complicated by the clinical and biological heterogeneity of depression and its resulting lack of specificity of biomarkers and often recognized by the lack of replication of biomarker findings.

NEUROPROTECTION AND NEUROINFLAMMATION DURING STAGES OF DEPRESSION

It is proposed that immune-system-related pathophysiological changes in psychiatric disorders such as clinical depression are due to impaired neuroprotective immune processes and due to enhanced neuroinflammatory and possibly neurodegenerative (in subgroups of patients) immune processes.

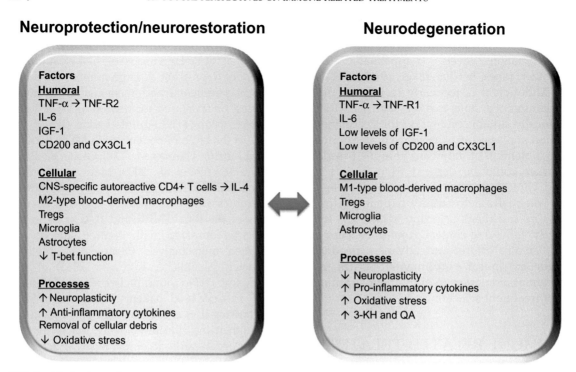

FIG. 3 Mechanisms of neuroprotection and neurodegeneration. The figure presents that both humoral and cellular immune processes affect neurobiological, inflammatory, and oxidative stress processes in specific ways during either neuroprotection or neurodegeneration. *TNF-α*, tumor necrosis factor alpha; *TNF-R2*, tumor necrosis factor receptor 2; *IL-6*, interleukin 6; *IL-4*, interleukin 4; *IGF-1*, insulin-like growth factor 1; *CD200*, OX-2 membrane glycoprotein also named CD200 (cluster of differentiation 200); *CX3CL1*, CXC3 chemokine receptor 1 also known as the fractalkine receptor or G-protein-coupled receptor 13 (GPR13); *Tregs*, regulatory T cells; *CD4⁺ T cells*, T-helper cells; *3-KH*, 3-hydroxykynurenine; *QA*, quinolinic acid.

In a multi-immune factor model involving cytokines, chemokines, glial cells, and peripheral immune cells, it is suggested that physiological and healthy conditions in adults are associated with a positive balance of neuroprotective immune effects (see Fig. 3). Physiological and healthy conditions are associated with increased IL-4, IL-10, neuroprotective functions of TNF-α (signaling via R2), and IL-6; increased levels of CNS- and Aβ-specific autoreactive CD4$^+$ T cells, M2 microglia, and quiescent astrocytes; and increased CX3CL1, CD200, and insulin-like growth factor (IGF)-1. In addition, less neurodegenerative activity of TNF-α (signaling via R1), IL-1β, M1 microglia, and reactive astrocytes is observed under physiological conditions.

On the other hand, during pathological conditions such as clinical depression, the disease course appears to be associated with predominantly neurodegenerative effects of immune factors. A long-term chronic course of depression that is contributed by aging processes is associated with increased CRP, TNF-α (via R1), IL-6, IL-1β, M1 microglia, reactive astrocytes, CCR2-deficient monocytes, and Th1 CD4$^+$ T cells, while simultaneously, neuroprotective factors decline during long-term depression and related aging processes. Based on the strong relationship between the neuroimmune system and other neurobiological systems (i.e., neuroplasticity, neurotransmission, mitochondrial physiology, and oxidative stress) (Krstic & Knuesel, 2013; Liu et al., 2013; Monsonego, Nemirovsky, & Harpaz, 2013; Naert & Rivest, 2013; Wyss-Coray & Rogers, 2012), it is suggested that a long-term course of depression may exert immune-mediated detrimental effects on these neurobiologically active systems.

The suggested model of balance/imbalance of neuroinflammation and neuroprotection during depression has several clinical and preventive implications. It is important to consider treatment and preventive strategies that may promote neuroprotective processes and hence counterbalance neurodegenerative imbalance. Physical activity (PA) (Erickson, Gildengers, & Butters, 2013; Hamer et al., 2012), the Mediterranean diet (Lourida et al., 2013), and Ω-3 polyunsaturated fatty acid (Ω-3 PUFA) supplementation (Loef & Walach, 2013) are treatment and preventive strategies that exert positive effects on the immune system. The Mediterranean diet (Barbaresko, Koch, Schulze, & Nothlings, 2013) and Ω-3 PUFA supplementation (Calviello et al., 2013) have anti-inflammatory effects as shown by a number of observational clinical studies. From an immunologic perspective, PA has been shown to increase IL-10, IL-6 (acutely), MIF, CNS-specific autoreactive $CD4^+$ T cells, M2 microglia, quiescent astrocytes, CX3CL1, and IGF-1 (Eyre, Papps, & Baune, 2013). It is also shown to reduce Th1/Th2 balance, PICs, CRP, M1 microglia, and reactive astrocytes (Eyre et al., 2013). A variety of other immune factors require further investigation. These include micro-ribonucleic acid (Ponomarev, Veremeyko, & Weiner, 2013), chemokines (Torres et al., 2012), CD200 (Blank & Prinz, 2013), IGF-1 (Park, Dantzer, Kelley, & McCusker, 2011), glial cells (e.g., oligodendrocytes and astrocytes (Nagelhus et al., 2013)), and peripheral immune cells (B cells and

neutrophils) (Aguilar-Valles, Kim, Jung, Woodside, & Luheshi, 2014; Eyre & Baune, 2012) (Fig. 4).

ANTI-INFLAMMATORY TREATMENTS OF CLINICAL DEPRESSION

Following on from the increasing and consistent evidence in support of an upregulated pro-inflammatory environment in clinical depression, the possible use of anti-inflammatory drugs has been gaining momentum in the scientific discussion and in clinical exploration as a treatment modality. However, anti-inflammatory treatments have so far been mainly explored in acute disease stages rather than throughout different disease phases.

Ω-3 POLYUNSATURATED FATTY ACIDS

Ω-3 PUFAs are a potential prevention and treatment strategy; however, their mechanism of action is poorly understood (Freund-Levi et al., 2006, 2009; Orr, Trepanier, & Bazinet, 2013). Ω-3 PUFAs exert beneficial effects on inflammation (Hjorth et al., 2013), through decreased expression of TNF-α, IL-6, and IL-1β; reduced microglial activation; and reduced astrocyte and monocyte PIC production (Gupta, Knight, Gupta, Keller, & Bruce-Keller, 2012; Labrousse et al., 2012; Mizwicki et al., 2013). The effects on other humoral and cellular immune factors are poorly understood; however, these include improved Aβ phagocytosis, prevention of astrocyte dysfunction and senescence, and possibly microglial phagocytosis of Aβ (Hjorth et al., 2013; Latour et al., 2013; Mizwicki et al., 2013). Additionally, PA (Erickson et al., 2013; Hamer et al., 2012), the Mediterranean diet (Lourida et al., 2013), and Ω-3 PUFA supplementation (Loef &

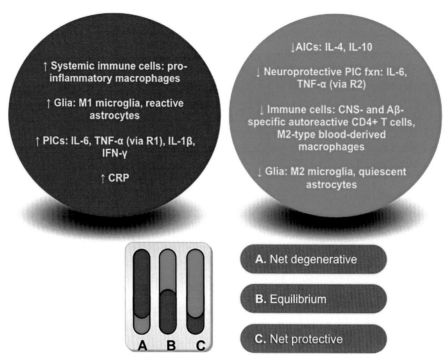

FIG. 4   Balance and imbalance of neuroprotective and neurodegenerative mechanisms. The figure indicates that depending on the degree and balance of neuroprotection and neurodegeneration, three stages can be distinguished: (A) net degenerative, (B) equilibrium, and (C) net protective.

Walach, 2013) are treatment and preventive strategies that exert positive effects on the immune system. It can therefore be speculated that these anti-inflammatory and neuroprotective effects may translate into clinical benefits in depression. Support of such a translational notion relates to the finding that long-chain Ω-3 PUFAs reduce the risk of progression to psychotic disorder in high-risk individuals and may offer a safe and efficacious strategy for indicated prevention in young people with subthreshold psychotic states (Amminger et al., 2010). It can be assumed that such a strategy would also be worthwhile exploring in individuals with high-risk depression prompting future design and conduct of appropriate studies.

NON-STEROIDAL ANTI-INFLAMMATORY DRUGS TREATMENT

Abundant literature has highlighted that inflammation is closely associated to depressive symptoms in patients with MDD (Capuron & Miller, 2004; Dantzer, 2009; Dantzer et al., 2008; Dantzer, O'Connor, Lawson, & Kelley, 2011; Eyre et al., 2016; Goldsmith, Rapaport, & Miller, 2016; Udina et al., 2014). Hence, targeting inflammation has become a promising strategy in improving depressive symptoms. Interestingly, numerous studies demonstrate a beneficial effect of anti-inflammatory medications (e.g., nonsteroidal anti-inflammatory medications or anticytokines) on depressive symptoms in patients with

MDD (Eyre, Air, Proctor, Rositano, & Baune, 2015; Eyre & Baune, 2015; Raison et al., 2013). Specifically, administering cyclooxygenase (COX)-2 inhibitors such as celecoxib in addition to antidepressant treatment for 6 or 8 weeks has shown to improve treatment response and remission in comparison with antidepressant medication alone (Abbasi, Hosseini, Modabbernia, Ashrafi, & Akhondzadeh, 2012; Akhondzadeh et al., 2009; Majd, Hashemian, Hosseini, Vahdat Shariatpanahi, & Sharifi, 2015; Muller et al., 2006). The best evidence to date has been accumulated for celecoxib. Celecoxib, a COX-2 inhibitor, has been used as an adjunct to antidepressant medication in the treatment of MDD in four trials, with three of these trials exploring 6 weeks of treatment at 400 mg daily (Abbasi et al., 2012; Akhondzadeh et al., 2009; Muller et al., 2006) and one trial of 8 weeks at 200 mg daily (Majd et al., 2015). All four of these trials suggested improved antidepressant treatment effects for celecoxib add-on treatment, compared with antidepressants plus placebo. Recent meta-analyses associated celecoxib with a large improved antidepressant effect by a standard mean difference of 0.82 (95% CI, 0.46–1.17) and without heterogeneity between the studies (I2 = 0%) (Kohler et al., 2014; Na, Lee, Lee, Cho, & Jung, 2014). Furthermore, remission of depression was improved by an odds ratio (OR) of 7.89 (95% CI, 2.94–21.17) and treatment response by an OR of 6.59 (95% CI, 2.24–19.42) by the 6–8-week add-on treatment, hence the choice of celecoxib for our study. Finally, one trial included peripheral blood tests and found that higher levels of the pro-inflammatory marker IL-6 predicted better antidepressant response to celecoxib add-on (Abbasi et al., 2012).

Despite this encouraging evidence, it is important to consider with caution other aspects of anti-inflammatory treatments in depression. Some of the previous studies have yielded inconclusive results following the investigation of both selective COX-2 and nonselective COX inhibitor nonsteroidal anti-inflammatory drugs (NSAIDs) as possible adjuncts in antidepressant treatment of depression (Akhondzadeh et al., 2009; Almeida et al., 2012; Almeida, Alfonso, Jamrozik, Hankey, & Flicker, 2010; Fields, Drye, Vaidya, Lyketsos, and Group, 2012; Fond et al., 2014; Gallagher et al., 2012; Muller, 2013; Muller et al., 2006; Musil et al., 2011; Nery et al., 2008; Pasco et al., 2010; Shelton, 2012; Uher et al., 2012; Warner-Schmidt, Vanover, Chen, Marshall, & Greengard, 2011), and some other studies reported detrimental effects suggesting NSAIDs may reduce the antidepressant effect of some selective serotonin reuptake inhibitors (Gallagher et al., 2012; Warner-Schmidt et al., 2011). These mixed results may be due to various reasons such as differing antidepressant utilization and doses, use of varying selective COX-2 and/or nonselective COX inhibitor NSAIDs, study design, age of study population, and study populations with varying degrees of depressive symptomatology and the presence of general medication conditions. Hence, it is important to note for clinical translation that only suitable anti-inflammatory treatments should be selected either as monotherapy or as an adjunct with the *right* antidepressant for the *right* duration using the *right* dosage during the *right* phase of depression. These are important further considerations for clinical translation and for stimulating the urgently needed clinical trials in this area that may help bolster clinical proof-of-concept evidence and stimulate careful clinical translation.

ANTI-INFLAMMATORY INTERVENTIONS WITH TARGETED ANTAGONISTS

A range of biological antagonists is available such as infliximab and etanercept. Additionally, anti-TNF therapy is being considered as an option in improving postoperative cognitive dysfunction (Terrando et al., 2010). Research into the effects of centrally and peripherally administered TNF-α blockade has shown clinical efficacy on cognition and depressive symptoms (Couch et al., 2013; Miller & Cole, 2012; Tobinick & Gross, 2008). In addition, a recently published trial in an animal model under peripheral lipopolysaccharide (LPS) stimulation showed that centrally administered etanercept reduced anxiety-like behaviors, but not spatial memory, and was associated with a decrease in HC microglia numbers being suggestive that etanercept recovers anxiety-like behavior possibly mediated by a reduction of TNF-α-related central inflammation (Camara et al., 2015). However, there are two important problems limiting the clinical usefulness of these techniques. First, the BBB prevents a deep penetration into the CNS tissue, and second, while basal levels of TNF-α are still required for normal functioning, animal models have shown that the complete lack of TNF-α due to genetic modification results in cognitive impairment (Baune et al., 2008). This is possibly due to the influence that TNF-α exerts on NGF and BDNF. Overall, this approach appears to be of limited use in early stages and as a long-term strategy for the treatment of clinical depression but may have a role in acute and progressive stages of the disease pending further studies. Clearer understanding of TNF α signaling and pharmacological interventions that have the property to target specific apoptotic factors in the TNF-α receptor-associated pathways, rather than complete blockage of TNF-α or of its receptors, would make for more effective therapeutics in the treatment of neuropsychiatric disorders such as depression in which TNF-α is an active participant (Fig. 5).

ALTERNATIVE APPROACHES

Investigations of alternative pathways to inhibit neuroinflammation show promising results. In a study examining the effect of ubiquitin-specific processing protease 8 (USP8) in luteolin-treated microglia, it was found that luteolin inhibits microglial inflammation by enhancing USP8 protein production and USP8 might represent a novel mechanism for the treatment of neuroinflammation and neurodegeneration (Zhu et al., 2015). In addition, it has been shown that blocking the kinase activity of RIP1, a key druggable target in the necroptosis pathway, by necrostatins inhibits the activation of necroptosis and allows cell survival and proliferation in the presence of death receptor ligands (Zhou & Yuan, 2014). Hence, targeting RIP1 kinase may provide therapeutic benefits for the treatment of human diseases characterized by necrosis and inflammation.

In addition to these examples, other types of investigations such as intermittent fasting (IF) show experimental evidence in rats to exert anti-inflammatory effects. It was found that in rats being on IF diet, IF prevented LPS-induced elevation of IL-1α, IL-1β, and TNF-α levels and prevented the LPS-induced reduction of BDNF levels in the hippocampus. IF also significantly attenuated LPS-induced elevations of serum IL-1β, IFN-γ, RANTES, TNF-α, and IL-6 levels and ameliorated cognitive impairment in rat behavior induced by LPS (Vasconcelos et al., 2014).

CONCLUSIONS

Although clinical depression has long been associated with inflammation, immune markers have only recently been discussed as potential biomarkers of depression that may stratify patient groups and that may inform treatment selection. The literature suggests that an increased inflammatory response may define a subgroup of individuals among the heterogeneous clinical and biological forms of depression. Among anti-inflammatory agents, the selective COX-2 inhibitor celecoxib shows the best evidence for antidepressant effects;

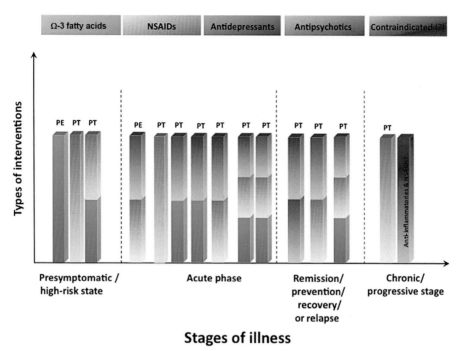

Stages of illness

FIG. 5   Anti-inflammatory interventions according to the phases of depression. The figure shows various types of treatments with evidence-based or potential anti-inflammatory efficacy at various phases of mental illness, exemplified for depression. While evidence for Ω-3 fatty acids suggests the prevention of transition from high-risk states to disease onset, its use during acute and postacute stages still lacks evidence. Likewise, NSAIDs that are widely used during acute illness could possibly also be used in combination with Ω-3 fatty acids during presymptomatic/high-risk stages to prevent transition to full acute disease stages becoming acute and during remission and chronic stages for the prevention of relapse of illness and faster recovery. However, continuous use of NSAIDs for a longer period of time could potentially negatively affect the immune system and may enhance blood clotting and may induce other relevant side effects; hence, NSAIDs may be contraindicated for chronic use. Additionally, antidepressants can be combined with NSAIDs and/or Ω-3 fatty acids to control acute illness and during remission/recovery and relapse prevention. An evidence-based treatment algorithm using the above interventions alone or in combination is still lacking and requires extensive future investigations and clinical trials. *PE*, published evidence; *PT*, potential treatment, but no/conflicting evidence.

however, the body of accumulated evidence on the role of selective and nonselective COX inhibitors is small and not entirely suited for providing conclusive clinical guidance at this stage. Further clinical proof-of-concept studies and clinical trials utilizing inflammation for prospective patient stratification are needed to guide clinical practice in the future. Experimental evidence has been produced to suggest that such subgroups of depressed patients with inflammation might benefit from targeted

anti-inflammatory drugs such as etanercept for TNF blockade. An improvement of future study methodologies would include the measure of the inflammation status of patients before entering the trial. Future study results may point to an indicated anti-inflammatory treatment based on the inflammation status of patients with clinical depression at commencement of the intervention. This might extend to depressed patients with—but not limited to—concomitant inflammatory conditions such as rheumatoid arthritis, psoriasis, and other chronic inflammatory conditions. Moreover, if anti-inflammatory treatment proves to be useful in acute depression, it is clinically worthwhile exploring whether this treatment has potential for treating also more severe forms of depression in the acute stage and whether it is useful for relapse prevention. While the consideration of the phase of the illness is important for developing a potential phase-specific treatment and prevention approach, the literature clearly points to the dynamic nature of the immune response in depression that requires consideration for a rationale of anti-inflammatory treatments beyond the acute disease stage (Eyre, Stuart, & Baune, 2014). These considerations should also extend into the observation that both the innate and adaptive immune systems are involved in depression (Toben & Baune, 2015). A focus on inflammation as a sole target for anti-inflammatory treatments might be shortsighted given the low to moderate long-term increase in inflammation during depression. While it has been shown that inflammation can be detrimental to the neurobiology of the brain, it has also been suggested that inflammation may lead to an increase in impaired neuroprotective mechanisms; hence a concept of enhancing neuroprotection plus a reduction in inflammation may be an extended avenue for future interventions.

While progress has been made in the understanding of the role of inflammation in depression and first anti-inflammatory treatments show potential clinical value, the field still has to reach out to arrive at an indicated and evidence-based approach for the use of anti-inflammatory agents in inflammation-associated depression. Clearly, an in-depth clinical study program including proof-of-concept studies and extensive clinical trials are required to enhance the knowledge before anti-inflammatory drugs can clinically be recommended as suitable agents for the treatment of symptoms in inflammation-associated depression.

References

Abbasi, S. H., Hosseini, F., Modabbernia, A., Ashrafi, M., & Akhondzadeh, S. (2012). Effect of celecoxib add-on treatment on symptoms and serum IL-6 concentrations in patients with major depressive disorder: randomized double-blind placebo-controlled study. *Journal of Affective Disorders, 141*(2–3), 308–314.

Aguilar-Valles, A., Kim, J., Jung, S., Woodside, B., & Luheshi, G. N. (2014). Role of brain transmigrating neutrophils in depression-like behavior during systemic infection. *Molecular Psychiatry, 19*(5), 599–606.

Akhondzadeh, S., Jafari, S., Raisi, F., Nasehi, A. A., Ghoreishi, A., Salehi, B., et al. (2009). Clinical trial of adjunctive celecoxib treatment in patients with major depression: a double blind and placebo controlled trial. *Depression and Anxiety, 26*(7), 607–611.

Almeida, O. P., Alfonso, H., Jamrozik, K., Hankey, G. J., & Flicker, L. (2010). Aspirin use, depression, and cognitive impairment in later life: the health in men study. *Journal of the American Geriatrics Society, 58*(5), 990–992.

Almeida, O. P., Flicker, L., Yeap, B. B., Alfonso, H., McCaul, K., & Hankey, G. J. (2012). Aspirin decreases the risk of depression in older men with high plasma homocysteine. *Translational Psychiatry, 2.*

Amminger, G. P., Schafer, M. R., Papageorgiou, K., Klier, C. M., Cotton, S. M., Harrigan, S. M., et al. (2010). Long-chain omega-3 fatty acids for indicated prevention of psychotic disorders: a randomized, placebo-controlled trial. *Archives of General Psychiatry, 67*(2), 146–154.

Barbaresko, J., Koch, M., Schulze, M. B., & Nothlings, U. (2013). Dietary pattern analysis and biomarkers of low-grade inflammation: a systematic literature review. *Nutrition Reviews, 71*(8), 511–527.

Baune, B. (2009). Conceptual challenges of a tentative model of stress-induced depression. *PLoS One, 4*(1).

Baune, B. T., Wiede, F., Braun, A., Golledge, J., Arolt, V., & Koerner, H. (2008). Cognitive dysfunction in mice deficient for TNF- and its receptors. *American Journal of Medical Genetics. Part B, Neuropsychiatric Genetics, 147B*(7), 1056–1064.

Blank, T., & Prinz, M. (2013). Microglia as modulators of cognition and neuropsychiatric disorders[Research Support, Non-U.S. Gov't]. *Glia, 61*(1), 62–70.

Calviello, G., Su, H. M., Weylandt, K. H., Fasano, E., Serini, S., & Cittadini, A. (2013). Experimental evidence of omega-3 polyunsaturated fatty acid modulation of inflammatory cytokines and bioactive lipid mediators: their potential role in inflammatory, neurodegenerative, and neoplastic diseases. *BioMed Research International, 2013.*

Camara, M. L., Corrigan, F., Jaehne, E. J., Jawahar, M. C., Anscomb, H., & Baune, B. T. (2015). Effects of centrally administered etanercept on behavior, microglia, and astrocytes in mice following a peripheral immune challenge. *Neuropsychopharmacology, 40*(2), 502–512.

Capuron, L., & Miller, A. H. (2004). Cytokines and psychopathology: lessons from interferon-alpha. *Biological Psychiatry, 56*(11), 819–824.

Conradi, H. J., Ormel, J., & de Jonge, P. (2011). Presence of individual (residual) symptoms during depressive episodes and periods of remission: a 3-year prospective study. *Psychological Medicine, 41*(6), 1165–1174.

Couch, Y., Anthony, D. C., Dolgov, O., Revischin, A., Festoff, B., Santos, A. I., et al. (2013). Microglial activation, increased TNF and SERT expression in the prefrontal cortex define stress-altered behaviour in mice susceptible to anhedonia. *Brain, Behavior, and Immunity, 29*, 136–146.

Dantzer, R. (2009). Cytokine, sickness behavior, and depression. *Immunology and Allergy Clinics of North America, 29*(2), 247–264.

Dantzer, R., O'Connor, J. C., Freund, G. G., Johnson, R. W., & Kelley, K. W. (2008). From inflammation to sickness and depression: when the immune system subjugates the brain. *Nature Reviews. Neuroscience, 9*(1), 46–56.

Dantzer, R., O'Connor, J. C., Lawson, M. A., & Kelley, K. W. (2011). Inflammation-associated depression: from serotonin to kynurenine. *Psychoneuroendocrinology, 36*(3), 426–436.

Dowlati, Y., Herrmann, N., Swardfager, W., Liu, H., Sham, L., Reim, E. K., et al. (2010). A meta-analysis of cytokines in major depression. *Biological Psychiatry, 67*(5), 446–457.

Dunner, D. L., Rush, A. J., Russell, J. M., Burke, M., Woodard, S., Wingard, P., et al. (2006). Prospective, long-term, multicenter study of the naturalistic outcomes of patients with treatment-resistant depression. *Journal of Clinical Psychiatry, 67*(5), 688–695.

Erickson, K. I., Gildengers, A. G., & Butters, M. A. (2013). Physical activity and brain plasticity in late adulthood. *Dialogues in Clinical Neuroscience, 15*(1), 99–108.

Eyre, H. A., Air, T., Pradhan, A., Johnston, J., Lavretsky, H., Stuart, M. J., et al. (2016). A meta-analysis of chemokines in major depression. *Progress in Neuro-Psychopharmacology & Biological Psychiatry, 68*, 1–8.

Eyre, H. A., Air, T., Proctor, S., Rositano, S., & Baune, B. T. (2015). A critical review of the efficacy of non-steroidal anti-inflammatory drugs in depression. *Progress in Neuro-Psychopharmacology & Biological Psychiatry, 57*, 11–16.

Eyre, H., & Baune, B. T. (2012). Neuroplastic changes in depression: a role for the immune system. *Psychoneuroendocrinology, 37*(9), 1397–1416.

Eyre, H. A., & Baune, B. T. (2015). Anti-inflammatory intervention in depression. *JAMA Psychiatry, 72*(5), 511.

Eyre, H., Papps, E., & Baune, B. T. (2013). Treating depression and depression-like behaviour with physical activity: an immune perspective. *Frontiers in Psychiatry, 4*(4), 3.

Eyre, H. A., Stuart, M. J., & Baune, B. T. (2014). A phase-specific neuroimmune model of clinical depression. *Progress in Neuro-Psychopharmacology & Biological Psychiatry, 54*, 265–274.

Fava, G. A., Ruini, C., & Belaise, C. (2007). The concept of recovery in major depression. *Psychological Medicine, 37*(3), 307–317.

Fields, C., Drye, L., Vaidya, V., Lyketsos, C., & ADAPT Research Group. (2012). Celecoxib or naproxen treatment does not benefit depressive symptoms in persons age 70 and older: findings from a randomized controlled trial. *American Journal of Geriatric Psychiatry, 20*(6), 505–513.

Fond, G., Hamdani, N., Kapczinski, F., Boukouaci, W., Drancourt, N., Dargel, A., et al. (2014). Effectiveness and tolerance of anti-inflammatory drugs' add-on therapy in major mental disorders: a systematic qualitative review. *Acta Psychiatrica Scandinavica, 129*(3), 163–179.

Freund-Levi, Y., Eriksdotter-Jonhagen, M., Cederholm, T., Basun, H., Faxen-Irving, G., Garlind, A., et al. (2006). Omega-3 fatty acid treatment in 174 patients with mild to moderate Alzheimer disease: OmegAD study: a randomized double-blind trial. *Archives of Neurology, 63*(10), 1402–1408.

Freund-Levi, Y., Hjorth, E., Lindberg, C., Cederholm, T., Faxen-Irving, G., Vedin, I., et al. (2009). Effects of omega-3 fatty acids on inflammatory markers in cerebrospinal fluid and plasma in Alzheimer's disease: the OmegAD study. *Dementia and Geriatric Cognitive Disorders, 27*(5), 481–490.

Gallagher, P. J., Castro, V., Fava, M., Weilburg, J. B., Murphy, S. N., Gainer, V. S., et al. (2012). Antidepressant response in patients with major depression exposed to NSAIDs: a pharmacovigilance study. *American Journal of Psychiatry, 169*(10), 1065–1072.

Goldsmith, D. R., Rapaport, M. H., & Miller, B. J. (2016). A meta-analysis of blood cytokine network alterations in psychiatric patients: comparisons between schizophrenia, bipolar disorder and depression. *Molecular Psychiatry, 21*(12), 1696–1709.

Gupta, S., Knight, A. G., Gupta, S., Keller, J. N., & Bruce-Keller, A. J. (2012). Saturated long-chain fatty acids activate inflammatory signaling in astrocytes[Research Support, N.I.H., Extramural]. *Journal of Neurochemistry, 120*(6), 1060–1071.

Haapakoski, R., Mathieu, J., Ebmeier, K. P., Alenius, H., & Kivimaki, M. (2015). Cumulative meta-analysis of interleukins 6 and 1beta, tumour necrosis factor alpha and C-reactive protein in patients with major depressive disorder. *Brain, Behavior, and Immunity, 49*, 206–215.

Hamer, M., Sabia, S., Batty, G. D., Shipley, M. J., Tabak, A. G., Singh-Manoux, A., et al. (2012). Physical activity and inflammatory markers over 10 years: follow-up in men and women from the Whitehall II cohort study. *Circulation*, 126(8), 928–933.

Hannestad, J., DellaGioia, N., & Bloch, M. (2011). The effect of antidepressant medication treatment on serum levels of inflammatory cytokines: a meta-analysis. *Neuropsychopharmacology*, 36(12), 2452–2459.

Haroon, E., Raison, C. L., & Miller, A. H. (2012). Psychoneuroimmunology meets neuropsychopharmacology: translational implications of the impact of inflammation on behavior. *Neuropsychopharmacology*, 37(1), 137–162.

Hjorth, E., Zhu, M., Toro, V. C., Vedin, I., Palmblad, J., Cederholm, T., et al. (2013). Omega-3 fatty acids enhance phagocytosis of Alzheimer's disease-related amyloid-beta42 by human microglia and decrease inflammatory markers. *Journal of Alzheimer's Disease*, 35(4), 697–713.

Holtzheimer, P. E., 3rd, Meeks, T. W., Kelley, M. E., Mufti, M., Young, R., McWhorter, K., et al. (2008). A double blind, placebo-controlled pilot study of galantamine augmentation of antidepressant treatment in older adults with major depression. *International Journal of Geriatric Psychiatry*, 23(6), 625–631.

Hope, S., Ueland, T., Steen, N. E., Dieset, I., Lorentzen, S., Berg, A. O., et al. (2013). Interleukin 1 receptor antagonist and soluble tumor necrosis factor receptor 1 are associated with general severity and psychotic symptoms in schizophrenia and bipolar disorder. *Schizophrenia Research*, 145(1–3), 36–42.

Howren, M. B., Lamkin, D. M., & Suls, J. (2009). Associations of depression with C-reactive protein, IL-1, and IL-6: a meta-analysis. *Psychosomatic Medicine*, 71(2), 171–186.

Kappelmann, N., Lewis, G., Dantzer, R., Jones, P. B., & Khandaker, G. M. (2018). Antidepressant activity of anti-cytokine treatment: a systematic review and meta-analysis of clinical trials of chronic inflammatory conditions. *Molecular Psychiatry* 23(2), 335–343.

Kendler, K. S., Thornton, L. M., & Gardner, C. O. (2001). Genetic risk, number of previous depressive episodes, and stressful life events in predicting onset of major depression. *American Journal of Psychiatry*, 158(4), 582–586.

Kohler, O., Benros, M. E., Nordentoft, M., Farkouh, M. E., Iyengar, R. L., Mors, O., et al. (2014). Effect of anti-inflammatory treatment on depression, depressive symptoms, and adverse effects a systematic review and meta-analysis of randomized clinical trials. *JAMA Psychiatry*, 71(12), 1381–1391.

Kohler, O., Petersen, L., Mors, O., Mortensen, P. B., Yolken, R. H., Gasse, C., et al. (2017). Infections and exposure to anti-infective agents and the risk of severe mental disorders: a nationwide study. *Acta Psychiatrica Scandinavica*, 135(2), 97–105.

Kohler-Forsberg, O., Buttenschon, H. N., Tansey, K. E., Maier, W., Hauser, J., Dernovsek, M. Z., et al. (2017). Association between C-reactive protein (CRP) with depression symptom severity and specific depressive symptoms in major depression. *Brain, Behavior, and Immunity*, 62, 344–350.

Krogh, J., Benros, M. E., Jorgensen, M. B., Vesterager, L., Elfving, B., & Nordentoft, M. (2014). The association between depressive symptoms, cognitive function, and inflammation in major depression. *Brain, Behavior, and Immunity*, 35, 70–76.

Krstic, D., & Knuesel, I. (2013). Deciphering the mechanism underlying late-onset Alzheimer disease. *Nature Reviews. Neurology*, 9(1), 25–34.

Labrousse, V. F., Nadjar, A., Joffre, C., Costes, L., Aubert, A., Gregoire, S., et al. (2012). Short-term long chain omega3 diet protects from neuroinflammatory processes and memory impairment in aged mice. *PLoS One*, 7(5).

Latour, A., Grintal, B., Champeil-Potokar, G., Hennebelle, M., Lavialle, M., Dutar, P., et al. (2013). Omega-3 fatty acids deficiency aggravates glutamatergic synapse and astroglial aging in the rat hippocampal CA1[Research Support, Non-U.S. Gov't]. *Aging Cell*, 12(1), 76–84.

Leonard, B., & Maes, M. (2012). Mechanistic explanations how cell-mediated immune activation, inflammation and oxidative and nitrosative stress pathways and their sequels and concomitants play a role in the pathophysiology of unipolar depression[Review]. *Neuroscience and Biobehavioral Reviews*, 36(2), 764–785.

Lewinsohn, P. M., Rohde, P., Seeley, J. R., Klein, D. N., & Gotlib, I. H. (2000). Natural course of adolescent major depressive disorder in a community sample: predictors of recurrence in young adults. *American Journal of Psychiatry*, 157(10), 1584–1591.

Licinio, J. (2011). Translational psychiatry: leading the transition from the cesspool of devastation to a place where the grass is really greener. *Translational Psychiatry*, 1.

Liu, Y. H., Zeng, F., Wang, Y. R., Zhou, H. D., Giunta, B., Tan, J., et al. (2013). Immunity and Alzheimer's disease: immunological perspectives on the development of novel therapies. *Drug Discovery Today*, 18(23–24), 1212–1220.

Loef, M., & Walach, H. (2013). The omega-6/omega-3 ratio and dementia or cognitive decline: a systematic review on human studies and biological evidence. *Journal of Nutrition in Gerontology and Geriatrics*, 32(1), 1–23.

Lourida, I., Soni, M., Thompson-Coon, J., Purandare, N., Lang, I. A., Ukoumunne, O. C., et al. (2013). Mediterranean diet, cognitive function, and dementia: a systematic review. *Epidemiology*, 24(4), 479–489.

Maes, M., Bosmans, E., De Jongh, R., Kenis, G., Vandoolaeghe, E., & Neels, H. (1997). Increased serum IL-6 and IL-1 receptor antagonist concentrations in major depression and treatment resistant depression. *Cytokine*, 9(11), 853–858.

Maes, M., Mihaylova, I., Kubera, M., & Ringel, K. (2012). Activation of cell-mediated immunity in depression: association with inflammation, melancholia, clinical staging and the fatigue and somatic symptom cluster of depression. *Progress in Neuro-Psychopharmacology & Biological Psychiatry, 36*(1), 169–175.

Maes, M., Ringel, K., Kubera, M., Berk, M., & Rybakowski, J. (2012). Increased autoimmune activity against 5-HT: a key component of depression that is associated with inflammation and activation of cell-mediated immunity, and with severity and staging of depression. *Journal of Affective Disorders, 136*(3), 386–392.

Mahar, I., Bambico, F. R., Mechawar, N., & Nobrega, J. N. (2014). Stress, serotonin, and hippocampal neurogenesis in relation to depression and antidepressant effects. *Neuroscience and Biobehavioral Reviews, 38*, 173–192.

Majd, M., Hashemian, F., Hosseini, S. M., Vahdat Shariatpanahi, M., & Sharifi, A. (2015). A randomized, double-blind, placebo-controlled trial of celecoxib augmentation of sertraline in treatment of drug-naive depressed women: a pilot study. *Iranian Journal of Pharmaceutical Research, 14*(3), 891–899.

March, J., Silva, S., Petrycki, S., Curry, J., Wells, K., Fairbank, J., et al. (2004). Fluoxetine, cognitive-behavioral therapy, and their combination for adolescents with depression: Treatment for Adolescents With Depression Study (TADS) randomized controlled trial. *JAMA, 292*(7), 807–820.

McAfoose, J., & Baune, B. T. (2009). Evidence for a cytokine model of cognitive function. *Neuroscience and Biobehavioral Reviews, 33*(3), 355–366.

Miller, G. E., & Cole, S. W. (2012). Clustering of depression and inflammation in adolescents previously exposed to childhood adversity. *Biological Psychiatry, 72*(1), 34–40.

Miller, A. H., Maletic, V., & Raison, C. L. (2009). Inflammation and its discontents: the role of cytokines in the pathophysiology of major depression. *Biological Psychiatry, 65*(9), 732–741.

Mizwicki, M. T., Liu, G., Fiala, M., Magpantay, L., Sayre, J., Siani, A., et al. (2013). 1alpha,25-dihydroxyvitamin D3 and resolvin D1 retune the balance between amyloid-beta phagocytosis and inflammation in Alzheimer's disease patients. *Journal of Alzheimer's Disease, 34*(1), 155–170.

Monsonego, A., Nemirovsky, A., & Harpaz, I. (2013). CD4 T cells in immunity and immunotherapy of Alzheimer's disease. *Immunology, 139*(4), 438–446.

Moylan, S., Berk, M., Dean, O. M., Samuni, Y., Williams, L. J., O'Neil, A., et al. (2014). Oxidative & nitrosative stress in depression: why so much stress? *Neuroscience and Biobehavioral Reviews, 45*, 46–62.

Moylan, S., Maes, M., Wray, N. R., & Berk, M. (2013). The neuroprogressive nature of major depressive disorder: pathways to disease evolution and resistance, and therapeutic implications. *Molecular Psychiatry, 18*(5), 595–606.

Muller, N. (2013). The role of anti-inflammatory treatment in psychiatric disorders. *Psychiatria Danubina, 25*(3), 292–298.

Muller, N., Schwarz, M. J., Dehning, S., Douhe, A., Cerovecki, A., Goldstein-Muller, B., et al. (2006). The cyclooxygenase-2 inhibitor celecoxib has therapeutic effects in major depression: results of a double-blind, randomized, placebo controlled, add-on pilot study to reboxetine. *Molecular Psychiatry, 11*(7), 680–684.

Murray, C. J., Vos, T., Lozano, R., Naghavi, M., Flaxman A. D., Michaud, C., et al. (2012). Disability-adjusted life years (DALYs) for 291 diseases and injuries in 21 regions, 1990–2010: a systematic analysis for the Global Burden of Disease Study 2010. *Lancet, 380*(9859), 2197–2223.

Musil, R., Schwarz, M. J., Riedel, M., Dehning, S., Cerovecki, A., Spellmann, I., et al. (2011). Elevated macrophage migration inhibitory factor and decreased transforming growth factor-beta levels in major depression—no influence of celecoxib treatment. *Journal of Affective Disorders, 134*(1–3), 217–225.

Na, K. S., Lee, K. J., Lee, J. S., Cho, Y. S., & Jung, H. Y. (2014). Efficacy of adjunctive celecoxib treatment for patients with major depressive disorder: a meta-analysis. *Progress in Neuro-Psychopharmacology & Biological Psychiatry, 48*, 79–85.

Naert, G., & Rivest, S. (2013). A deficiency in CCR2 + monocytes: the hidden side of Alzheimer's disease. *Journal of Molecular Cell Biology, 5*(5), 284–293.

Nagelhus, E. A., Amiry-Moghaddam, M., Bergersen, L. H., Bjaalie, J. G., Eriksson, J., Gundersen, V., et al. (2013). The glia doctrine: addressing the role of glial cells in healthy brain ageing. *Mechanisms of Ageing and Development, 134*(10), 449–459.

Nery, F. G., Monkul, E. S., Hatch, J. P., Fonseca, M., Zunta-Soares, G. B., Frey, B. N., et al. (2008). Celecoxib as an adjunct in the treatment of depressive or mixed episodes of bipolar disorder: a double-blind, randomized, placebo-controlled study. *Human Psychopharmacology, 23*(2), 87–94.

Orr, S. K., Trepanier, M. O., & Bazinet, R. P. (2013). n-3 Polyunsaturated fatty acids in animal models with neuroinflammation. *Prostaglandins, Leukotrienes, and Essential Fatty Acids, 88*(1), 97–103.

Park, S. E., Dantzer, R., Kelley, K. W., & McCusker, R. H. (2011). Central administration of insulin-like growth factor-I decreases depressive-like behavior and brain cytokine expression in mice. *Journal of Neuroinflammation, 8*, 12.

Pasco, J. A., Jacka, F. N., Williams, L. J., Henry, M. J., Nicholson, G. C., Kotowicz, M. A., et al. (2010). Clinical implications of the cytokine hypothesis of depression:

the association between use of statins and aspirin and the risk of major depression. *Psychotherapy and Psychosomatics, 79*(5), 323–325.

Ponomarev, E. D., Veremeyko, T., & Weiner, H. L. (2013). MicroRNAs are universal regulators of differentiation, activation, and polarization of microglia and macrophages in normal and diseased CNS. *Glia, 61*(1), 91–103.

Raison, C. L., Rutherford, R. E., Woolwine, B. J., Shuo, C., Schettler, P., Drake, D. F., et al. (2013). randomized controlled trial of the tumor necrosis factor antagonist infliximab for treatment-resistant depression. *JAMA Psychiatry, 70*(1), 31–41.

Rush, A. J., Trivedi, M. H., Wisniewski, S. R., Nierenberg, A. A., Stewart, J. W., Warden, D., et al. (2006). Acute and longer-term outcomes in depressed outpatients requiring one or several treatment steps: a STAR*D report. *American Journal of Psychiatry, 163*(11), 1905–1917.

Rush, A. J., Trivedi, M. H., Wisniewski, S. R., Stewart, J. W., Nierenberg, A. A., Thase, M. E., et al. (2006). Bupropion-SR, sertraline, or venlafaxine-XR after failure of SSRIs for depression. *New England Journal of Medicine, 354*(12), 1231–1242.

Sheline, Y. I., Gado, M. H., & Kraemer, H. C. (2003). Untreated depression and hippocampal volume loss. *American Journal of Psychiatry, 160*(8), 1516–1518.

Shelton, R. C. (2012). Does concomitant use of NSAIDs reduce the effectiveness of antidepressants? *American Journal of Psychiatry, 169*(10), 1012–1015.

Slavich, G. M., & Irwin, M. R. (2014). From stress to inflammation and major depressive disorder: a social signal transduction theory of depression. *Psychological Bulletin, 140*(3), 774–815.

Terrando, N., Monaco, C., Ma, D., Foxwell, B. M., Feldmann, M., & Maze, M. (2010). Tumor necrosis factor-alpha triggers a cytokine cascade yielding postoperative cognitive decline. *Proceedings of the National Academy of Sciences of the United States of America, 107*(47), 20518–20522.

Toben, C., & Baune, B. T. (2015). An act of balance between adaptive and maladaptive immunity in depression: a role for T lymphocytes. *Journal of Neuroimmune Pharmacology, 10*(4), 595–609.

Tobinick, E. L., & Gross, H. (2008). Rapid cognitive improvement in Alzheimer's disease following perispinal etanercept administration. *Journal of Neuroinflammation, 5*, 2.

Torres, K. C., Santos, R. R., de Lima, G. S., Ferreira, R. O., Mapa, F. C., Pereira, P. A., et al. (2012). Decreased expression of CCL3 in monocytes and CCR5 in lymphocytes from frontotemporal dementia as compared with Alzheimer's disease patients. *Journal of Neuropsychiatry and Clinical Neurosciences, 24*(3), E11–12.

Trivedi, M. H., Rush, A. J., Wisniewski, S. R., Nierenberg A. A., Warden, D., Ritz, L., et al. (2006). Evaluation of outcomes with citalopram for depression using measurement-based care in STAR*D: implications for clinical practice. *American Journal of Psychiatry, 163*(1), 28–40.

Udina, M., Moreno-Espana, J., Capuron, L., Navines, R., Farre, M., Vieta, E., et al. (2014). Cytokine-induced depression: current status and novel targets for depression therapy. *CNS & Neurological Disorders Drug Targets, 13*(6), 1066–1074.

Uher, R., Carver, S., Power, R. A., Mors, O., Maier, W., Rietschel, M., et al. (2012). Non-steroidal anti-inflammatory drugs and efficacy of antidepressants in major depressive disorder. *Psychological Medicine, 42*(10), 2027–2035.

Uher, R., Tansey, K. E., Dew, T., Maier, W., Mors, O., Hauser, J., et al. (2014). An inflammatory biomarker as a differential predictor of outcome of depression treatment with escitalopram and nortriptyline. *American Journal of Psychiatry, 171*(12), 1278–1286.

Valkanova, V., Ebmeier, K. P., & Allan, C. L. (2013). CRP, IL-6 and depression: a systematic review and meta-analysis of longitudinal studies. *Journal of Affective Disorders, 150*(3), 736–744.

Vasconcelos, A. R., Yshii, L. M., Viel, T. A., Buck, H. S., Mattson, M. P., Scavone, C., et al. (2014). Intermittent fasting attenuates lipopolysaccharide-induced neuroinflammation and memory impairment. *Journal of Neuroinflammation, 11*, 85.

Videbech, P., & Ravnkilde, B. (2004). Hippocampal volume and depression: a meta-analysis of MRI studies. *American Journal of Psychiatry, 161*(11), 1957–1966.

Warner-Schmidt, J. L., Vanover, K. E., Chen, E. Y., Marshall, J. J., & Greengard, P. (2011). Antidepressant effects of selective serotonin reuptake inhibitors (SSRIs) are attenuated by antiinflammatory drugs in mice and humans. *Proceedings of the National Academy of Sciences of the United States of America, 108*(22), 9262–9267.

Wyss-Coray, T., & Rogers, J. (2012). Inflammation in Alzheimer disease-a brief review of the basic science and clinical literature. *Cold Spring Harbor Perspectives in Medicine, 2*(1).

Zhou, W., & Yuan, J. (2014). Necroptosis in health and diseases. *Seminars in Cell & Developmental Biology, 35*, 14–23.

Zhu, L., Bi, W., Lu, D., Zhang, C., Shu, X., Wang, H., et al. (2015). Regulation of ubiquitin-specific processing protease 8 suppresses neuroinflammation. *Molecular and Cellular Neurosciences, 64*, 74–83.

Index

Note: Page numbers followed by *f* indicate figures, and *t* indicate tables.

Printed in the United States
By Bookmasters